Scattering of Electromagnetic Waves by Obstacles

Mario Boella Series on Electromagnetism in Information and Communication

Piergiorgio L.E. Uslenghi, PhD – Series Editor

The Mario Boella series offers textbooks and monographs in all areas of radio science, with a special emphasis on the applications of electromagnetism to information and communication technologies. The series is scientifically and financially sponsored by the Istituto Superiore Mario Boella affiliated with the Politecnico di Torino, Italy, and is scientifically cosponsored by the International Union of Radio Science (URSI). It is named to honor the memory of Professor Mario Boella of the Politecnico di Torino, who was a pioneer in the development of electronics and telecommunications in Italy for half a century, and a vice president of URSI from 1966 to 1969.

Published Titles in the Series

Fundamentals of Wave Phenomena, 2nd Edition
by Akira Hirose and Karl Lonngren (2010)

Scattering of Waves by Wedges and Cones with Impedance Boundary Conditions
by Mikhail Lyalinov and Ning Yan Zhu (2012)

Complex Space Source Theory of Spatially Localized Electromagnetic Waves
by S.R. Seshadri (2013)

The Wiener-Hopf Method in Electromagnetics
by Vito Daniele and Rodolfo Zich (2014)

Higher-order Techniques in Computational Electromagnetics
by Roberto Graglia and Andrew Peterson (2015)

Forthcoming Titles

Slotted Waveguide Array Antennas
by Sembiam Rengarajan and Lars Josefsson

Scattering of Electromagnetic Waves by Obstacles

Gerhard Kristensson

Lund University

SciTech Publishing

an imprint of the IET

scitechpub.com

SciTech Publishing
an imprint of the IET

Published by SciTech Publishing, an imprint of the IET
www.scitechpub.com
www.theiet.org

First published 2016

ISBN 978-1-61353-221-8 (hardback)
ISBN 978-1-61353-222-5 (PDF)

Typeset in India by MPS Limited

This book is dedicated to my love, Mona-Lisa
and our wonderful children, Ester and Elias

Contents

Preface

The primary method of detecting and characterizing objects surrounding us is mainly through interpretation of acoustic and electromagnetic wave interaction. Most important is perhaps perception formed by reception of light that has been scattered or reflected by "obstacles" around us. This sensing can also be accomplished artificially by means of electromagnetic waves of lower frequencies, e.g., radio or microwaves, like in radar, and the interpretation of scattering by object has been one of the cornerstones in our interpretation of the world around us. To a great extent, our knowledge of the atmosphere of the Earth and of space is obtained by interpreting the propagation of electromagnetic waves through unknown materials. In order to master the interpretation of these scattering problems, understanding and solving electromagnetic scattering by obstacles are instrumental.

This textbook is an introduction to some of the most important properties of electromagnetic waves and their interaction with passive materials and scatterers. The main purpose of the book is to give a theoretical treatment of these scattering phenomena, and to illustrate numerical computations of some canonical scattering problems for different geometries and materials. The scattering theory is also important in the theory of passive antennas, and in this book we give several examples on this topic, e.g., reflector antennas.

The book contains a lot of mathematics, but is still not a book that treats the scattering problems in full mathematical detail. This is done on purpose. The intention is to cover the scattering concepts in a mathematical framework without introducing all the mathematical details. The presentation of the full mathematical details is simply another book. The mathematical theory of electromagnetics has seen a strong development during the last 4–5 decades. A number of textbooks cover the mathematical aspect of scattering by electromagnetic waves. The following books illustrate these aspects of the subject: [22, 30, 38, 48, 49, 93, 157, 158, 195, 199, 223, 230, 289, 295]. Instead, we assume that all mathematical operations done in this textbook are legitimate. In particular, we assume:

- Fields are smooth enough to be differentiable
- Fourier transforms of all fields are admissible whenever needed
- Existence of solutions to the various scattering problems in the book
- Regularity of surfaces to ensure that the operations are adequate

The emphasis in this textbook is instead on the computational and theoretical aspects of the scattering problem, and the final goal is to take mathematics to a point where creation of computational algorithms is possible. To accomplish these endeavors, the notation in the text has been paid great attention. There exist many books with similar ambitions in the literature, e.g., References 18, 53, 57, 99, 185, just to mention a few.

The book requires good knowledge of basic electromagnetic field theory, for instance a basic course in electromagnetic field theory at an undergraduate level. We expect the Maxwell field equations to be known, as well as basic vector analysis, and calculations with the nabla operator ∇. General knowledge in differential and integral calculus and algebra is also required.

Some parts of this book are harder to digest than others. To help and guide the reader, the contents of this textbook are divided in two tracks—Track 1 and Track 2—each with a special mark in upper left or right corners, i.e., a light gray or a dark gray symbol, respectively. The Track 1 material is appropriate for a first reading of the textbook, while Track 2 contains more advanced material suited for the second reading and for reference. The notation in the book is uniform, and explained in Appendix I.

Some of the features of the book are:

- Most results in this textbook are based upon the integral representations of the solution to the Maxwell equations
- Coordinate independent formulation of the scattering problem is strived after
- The material modeling starts in a time-domain setting, and the more traditional frequency-domain modeling results follow
- Focus on causality conditions
- Two tracks to follow—one introductory track and one more advanced—make the book complete
- Unified notation throughout the book
- All chapters end with a collection of problems for the student to solve
- A collection of appendices makes the book comprehensive

In the first chapter, we review the general equations for modeling electromagnetic fields—in particular the special case of time-harmonic fields. We define a number of important concepts, such as power transport, active, passive, and lossless materials, as well as the polarization state of the electromagnetic field. This chapter serves as an introduction to modeling of electromagnetic interaction with matter. The basic mathematical tools—the Green functions and dyadics—are introduced in Chapter 2. Representations of the electromagnetic fields in volume- and surface integrals are found in Chapter 3, and they constitute the basis of the analysis of the scattering problems that we introduce in Chapter 4. This chapter contains the formulation of the scattering problem, the definitions of different fundamental scattering quantities, and an overview of the optical theorem. Some results in time-domain scattering are presented in Chapter 5. The approximation of the scattering quantities in the low-frequency limit and different high-frequency approximations are presented in Chapter 6. We also analyze scattering from several scatterers in the case where multiple scattering effects can be neglected. We develop the scattering theory further in Chapter 7 by using the spherical vector waves, and in Chapter 8 by solving scattering from spherical objects explicitly. Scattering by more general geometries is developed by the use of the Null-field method in Chapter 9. A final chapter, Chapter 10, is devoted to wave propagation in planar structures.

For the convenience of the reader, the textbook contains several appendices that contain mathematical background material that is used in the analysis.

Exercises or problems are gathered at the end of each chapter. Advanced exercises are marked with a star (*). Answers to the exercises are found at the end of the book.

Lund and Fort Collins, December 2015.

Foreword

The Mario Boella series contains textbooks and research monographs in all areas of Radio Science, with a special emphasis on the applications of electromagnetism to information and communications technologies. The series is scientifically and financially sponsored by the Istituto Superiore Mario Boella affiliated with the Politecnico di Torino, Italy, and is scientifically co-sponsored by the International Union of Radio Science (URSI). It is named to honor the memory of Professor Mario Boella of the Politecnico di Torino, who was a pioneer in the development of electronics and telecommunications in Italy for half a century and was a vice president of URSI from 1966 to 1969.

This advanced textbook by Professor Gerhard Kristensson deals with the scattering of electromagnetic waves by obstacles. The treatment is at the advanced level; it is focused on an integral formulation of the problems, and is directed to graduate students and researchers. The analytical derivations are not encumbered by excessive details and are carried out to the point where numerical implementation can be profitably inserted. The organization along dual tracks is particularly attractive, in that it allows for a preliminary understanding of the subject matter which can be followed later by an in-depth second reading. Several problems are listed at the end of each chapter. As examples, propagation in layered media and scattering by spherical objects are treated in detail. This work is a valuable addition to the literature on electromagnetic scattering and diffraction.

Piergiorgio L. E. Uslenghi
ISMB Series Editor
Chicago, March 2016

Acknowledgment

My first encounter with scattering theory was in the 1970s through the wonderful book *Scattering Theory: The Quantum Theory of Nonrelativistic Collisions* by John R. Taylor[1] [258]. This book opened my eyes to the fantastic world of scattering theory—a fascination that has given me pleasure for over 40 years. Taylor adopts a time-domain description in the modeling of the scattering process. The basis for the presentation in the present book is also a time-domain approach, particularly in the material modeling.

The first ideas to write this book were born in 2000, when I had just completed two books in Swedish for undergraduates—one on electromagnetic wave propagation and one on scattering of electromagnetic waves by obstacles, respectively. The books were well received by the Swedish audience, and it stuck me that there might exist an interest for the subject to a broader audience. At the start, writing the book was a part time project, but during recent years, it has occupied most of my time.

Several persons have contributed to the contents of this book. The first person to show me the power of the integral representations in the early 1970s was my supervisor, Professor Staffan Ström. Professor Ström made several important research contributions to the field of scattering theory, and this book has the ambition to pass on this knowledge and legacy.

I have received input and encouraging comments by many skillful scientists. Foremost, I would like to thank my colleagues and friends Samel Arslanagić and Paul Martin, who have contributed immensely with their expertise by giving valuable comments and corrections. Their remarks have contributed to a significant improvement of the text in the book. Torleif Martin has been most helpful and provided me with two-sphere scattering data and many constructive comments on the multiple scattering text.

I am also most grateful to Michael Andersson, Martin Byröd, Margaret Cheney, Casimir Ehrenborg, Andreas Ericsson, Mats Gustafsson, Jakob Helander, Magnus Herberthson, Anders Karlsson, Daniel Sjöberg, Christian Sohl, Ioannis Stratis, Doruk Tayli, Marina Wallin, and Niklas Wellander for comments, finding misprints and slips, and many inspiring discussions on scattering problems. Needless to say, all remaining errors are due to my neglect.

The entire text is typeset in LaTeX, and moreover, all figures and drawings in the book are produced by Ti*k*Z. The author is indebted to Christian Feuersänger for valuable help with the axes and the arrow heads in the log figures. Ola Hedbäck has been very helpful in finding the

[1] John Robert Taylor (1939–), American physicist.

missing details in some biographical notes in the book. I am very grateful for all his efforts. Many thanks to Elias Kristensson for the loan of the beam splitter.

This book project was completed while the author was a visiting professor at Colorado State University. It is my sincere pleasure to thank Margaret Cheney for wonderful hospitality and for providing excellent conditions for the completion of the book project.

My family has played an important role in the formation of this book, and this is one of the reasons I dedicate this book to my family.

CHAPTER 1

Basic Equations

The foundation of electromagnetics stands on the shoulders of the scientific giants of the nineteenth century. Stars like André Marie Ampère,[1] Michael Faraday,[2] and James Clerk Maxwell[3] shine brightly.

The physics of electromagnetic phenomena takes place in space and time. Therefore, a time-dependent description is a natural starting point of modeling the electromagnetic interaction with matter. In fact, this approach is the guiding principle throughout the first part of this chapter, which is devoted to modeling of electromagnetic interaction with matter. By taking this viewpoint, we avoid some of the pitfalls that might occur if you start with a frequency-domain formulation. In particular, causality is naturally included in the modeling.

In this chapter, we review the basic equations that model electromagnetic wave propagation—the Maxwell equations—and we set the notation used in this book. The chapter is rather long, and examines not only the basic equations, but also a detailed modeling of the interaction of electromagnetic fields with materials. The material modeling is general with several explicit examples, and it provides the appropriate background setting for e.g., scattering by general linear stratified materials in Chapter 10. Some of this modeling material can be ignored at first reading.

The most important consequences of the Maxwell equations, e.g., the boundary conditions between two materials and conservation of power, are presented in Section 1.1. The models of materials for time-dependent fields are presented in Section 1.2. Several important examples, such as the Debye and the Lorentz dispersion models, together with the underlying physical assumptions are presented in detail. The special case of time-harmonic fields is introduced in Section 1.3. Concepts like active and passive media and reciprocity are defined. The notion of coherence and the degree of polarization, including the Stokes parameters, is presented in Section 1.4.

1.1 The Maxwell Equations

The Maxwell equations are the fundamental mathematical model for all theoretical analysis of macroscopic electromagnetic phenomena. Maxwell presented his famous equations to the Royal Society on December 8, 1864. The formal publication was made in 1865 in a paper

[1] André-Marie Ampère (1775–1836), French physicist.

[2] Michael Faraday (1791–1867), English chemist and physicist.

[3] James Clerk Maxwell (1831–1879), Scottish physicist and mathematician.

entitled "A dynamical theory of the electromagnetic field" [177]. A decade later, in 1873, he collected the theoretical results in his textbook: *A Treatise on Electricity and Magnetism* [178, 179].

All experimental tests performed since then have verified this model, and through the years an impressive amount of evidences for the validity of these equations has been gathered in different fields of applications. However, microscopic phenomena—like quantum phenomena—require a more refined model including quantum effects also, but these effects are out of the scope of this book.

The Maxwell equations are the cornerstone in the analysis of macroscopic electromagnetic wave propagation phenomena.[4] The Maxwell equations in SI units (MKSA) are:

$$\nabla \times \boldsymbol{E}(\boldsymbol{r}, t) = -\frac{\partial \boldsymbol{B}(\boldsymbol{r}, t)}{\partial t} \tag{1.1}$$

$$\nabla \times \boldsymbol{H}(\boldsymbol{r}, t) = \boldsymbol{J}(\boldsymbol{r}, t) + \frac{\partial \boldsymbol{D}(\boldsymbol{r}, t)}{\partial t} \tag{1.2}$$

Equation (1.1) (or the corresponding integral formulation) is called the Faraday law of induction, and (1.2) is often called Ampère–Maxwell law or simply the Ampère law. The different vector fields in the Maxwell equations are:[5]

$\boldsymbol{E}(\boldsymbol{r}, t)$	Electric field intensity (V/m)
$\boldsymbol{H}(\boldsymbol{r}, t)$	Magnetic field intensity (A/m)
$\boldsymbol{D}(\boldsymbol{r}, t)$	Electric flux density (A s/m^2)
$\boldsymbol{B}(\boldsymbol{r}, t)$	Magnetic flux density or magnetic induction (V s/m^2)
$\boldsymbol{J}(\boldsymbol{r}, t)$	Current density (A/m^2)

All these fields are functions of space and time, i.e., space coordinates $\boldsymbol{r}$ and time t. Often these arguments are suppressed. However, if there is a possibility to misinterpret any equation or expression, we make sure the arguments are explicitly written out.

The electric field $\boldsymbol{E}(\boldsymbol{r}, t)$ and the magnetic flux density $\boldsymbol{B}(\boldsymbol{r}, t)$ are defined by the force, $\boldsymbol{F}(t)$, on a charged particle by Lorentz force equation[6]

$$\boldsymbol{F}(t) = q\{\boldsymbol{E}(\boldsymbol{r}, t) + \boldsymbol{v}(t) \times \boldsymbol{B}(\boldsymbol{r}, t)\} \tag{1.3}$$

where q is the electric charge of the particle located at $\boldsymbol{r}(t)$, and $\boldsymbol{v}(t)$ is its velocity. The unit of the charge q is C = A s.

The free charges in the material, e.g., the conduction electrons, are described by the current density $\boldsymbol{J}(\boldsymbol{r}, t)$. The field contributions from bound charges, e.g., the electrons bound to the kernel of the atom, are traditionally included in the electric flux density $\boldsymbol{D}(\boldsymbol{r}, t)$ and the magnetic field $\boldsymbol{H}(\boldsymbol{r}, t)$. In Section 1.2, we address the differences between the electric flux density $\boldsymbol{D}(\boldsymbol{r}, t)$ and the electric field $\boldsymbol{E}(\boldsymbol{r}, t)$, as well as the differences between the magnetic field $\boldsymbol{H}(\boldsymbol{r}, t)$ and the magnetic flux density $\boldsymbol{B}(\boldsymbol{r}, t)$, and models for the interrelations between these fields are presented.

---

[4]It is out of the scope of this textbook to present a derivation of these equations. Several excellent derivations of these macroscopic equations from a microscopic formulation are found in the literature, see, e.g., References 61, 101, 238.

[5]Sometimes, for simplicity, we use the names $\boldsymbol{E}$-field, $\boldsymbol{D}$-field, $\boldsymbol{B}$-field, and $\boldsymbol{H}$-field.

[6]Hendrik Antoon Lorentz (1853–1928), Dutch physicist.

One of the fundamental assumptions in physics is that electric charges are indestructible, i.e., the sum of all charges in a closed system (no interaction with the surroundings) is always constant. This physical law has been very carefully tested. The conservation of charges is expressed in mathematical terms as the continuity law of charges

$$\nabla \cdot \boldsymbol{J}(\boldsymbol{r},t) + \frac{\partial \rho(\boldsymbol{r},t)}{\partial t} = 0 \tag{1.4}$$

Here, $\rho(\boldsymbol{r},t)$ is the volume density of free charges (unit $\mathrm{C/m^3}$) that is associated with the current density $\boldsymbol{J}(\boldsymbol{r},t)$. As alluded to above, the contributions from bound charges are included in the electric flux density $\boldsymbol{D}(\boldsymbol{r},t)$ and the magnetic field $\boldsymbol{H}(\boldsymbol{r},t)$.

Two additional equations are usually associated with the Maxwell equations

$$\nabla \cdot \boldsymbol{B}(\boldsymbol{r},t) = 0 \tag{1.5}$$

$$\nabla \cdot \boldsymbol{D}(\boldsymbol{r},t) = \rho(\boldsymbol{r},t) \tag{1.6}$$

Equation (1.5) tells us that no magnetic charges exist, and it implies that the magnetic flux is conserved. Equation (1.6) is usually called the Gauss law.[7] Under suitable assumptions, both these equations can be derived from (1.1), (1.2), and (1.4). To see this, take the divergence of (1.1) and (1.2). This implies

$$\begin{cases} \nabla \cdot \dfrac{\partial \boldsymbol{B}(\boldsymbol{r},t)}{\partial t} = 0 \\ \nabla \cdot \boldsymbol{J}(\boldsymbol{r},t) + \nabla \cdot \dfrac{\partial \boldsymbol{D}(\boldsymbol{r},t)}{\partial t} = 0 \end{cases}$$

since $\nabla \cdot (\nabla \times \boldsymbol{A}) = 0$ for an arbitrary vector field $\boldsymbol{A}$. Interchanging the order of differentiation and using (1.4) gives

$$\begin{cases} \dfrac{\partial(\nabla \cdot \boldsymbol{B}(\boldsymbol{r},t))}{\partial t} = 0 \\ \dfrac{\partial(\nabla \cdot \boldsymbol{D}(\boldsymbol{r},t) - \rho(\boldsymbol{r},t))}{\partial t} = 0 \end{cases}$$

These equations imply

$$\begin{cases} \nabla \cdot \boldsymbol{B}(\boldsymbol{r},t) = f_1(\boldsymbol{r}) \\ \nabla \cdot \boldsymbol{D}(\boldsymbol{r},t) - \rho(\boldsymbol{r},t) = f_2(\boldsymbol{r}) \end{cases}$$

where $f_1(\boldsymbol{r})$ and $f_2(\boldsymbol{r})$ are two functions that do not depend on time t, but can depend on the spatial coordinates $\boldsymbol{r}$. If the fields $\boldsymbol{B}(\boldsymbol{r},t)$, $\boldsymbol{D}(\boldsymbol{r},t)$, and $\rho(\boldsymbol{r},t)$ are identically zero before a fixed time, τ, i.e.,

$$\begin{cases} \boldsymbol{B}(\boldsymbol{r},t) = \boldsymbol{0} \\ \boldsymbol{D}(\boldsymbol{r},t) = \boldsymbol{0} \qquad t < \tau \\ \rho(\boldsymbol{r},t) = 0 \end{cases}$$

then (1.5) and (1.6) follow. Of course, static or time-harmonic fields do not satisfy this assumption, since there is no time, τ, before which all fields are zero.[8] To conclude, under the

[7] Johann Carl Friedrich Gauss (1777–1855), German mathematician.

[8] We will return to the derivation of (1.5) and (1.6) for time-harmonic fields in Section 1.3 on page 38.

assumption that fields and charges did not exist for ever, it is sufficient to use (1.1), (1.2), and (1.4).

Equations (1.1) and (1.2) contain six different equations—one for each vector component. Provided the current density $\boldsymbol{J}(\boldsymbol{r}, t)$ is given, the Maxwell equations contain 12 unknowns—the four vector fields $\boldsymbol{E}(\boldsymbol{r}, t)$, $\boldsymbol{B}(\boldsymbol{r}, t)$, $\boldsymbol{D}(\boldsymbol{r}, t)$, and $\boldsymbol{H}(\boldsymbol{r}, t)$. The result is that we lack six equations in order to have as many equations as unknowns. The lacking six equations are called the constitutive relations and they are addressed in Section 1.2.

In vacuum, the electric field $\boldsymbol{E}(\boldsymbol{r}, t)$ and the electric flux density $\boldsymbol{D}(\boldsymbol{r}, t)$ are parallel—the difference is in the units they are measured. The same holds for the magnetic flux density $\boldsymbol{B}(\boldsymbol{r}, t)$ and the magnetic field $\boldsymbol{H}(\boldsymbol{r}, t)$. We have

$$\begin{cases} \boldsymbol{D}(\boldsymbol{r}, t) = \epsilon_0 \boldsymbol{E}(\boldsymbol{r}, t) \\ \boldsymbol{B}(\boldsymbol{r}, t) = \mu_0 \boldsymbol{H}(\boldsymbol{r}, t) \end{cases} \tag{1.7}$$

where ϵ_0 and μ_0 are the permittivity and the permeability of vacuum, respectively. Numerical values of these constants are found in Appendix J. Approximate values of these constants are: $\epsilon_0 \approx 8.854 \times 10^{-12}$ A s/V m and $\mu_0 = 4\pi \times 10^{-7}$ V s/A m $\approx 1.257 \times 10^{-6}$ V s/A m.

Inside the material, the differences between the electric field $\boldsymbol{E}(\boldsymbol{r}, t)$ and the electric flux density $\boldsymbol{D}(\boldsymbol{r}, t)$, and between the magnetic flux density $\boldsymbol{B}(\boldsymbol{r}, t)$ and the magnetic field $\boldsymbol{H}(\boldsymbol{r}, t)$ are usually more complex. These differences are a measure of the interaction between the charges in the material and the fields. Often, two new vector fields, the polarization $\boldsymbol{P}(\boldsymbol{r}, t)$ and the magnetization $\boldsymbol{M}(\boldsymbol{r}, t)$, of the material are introduced to describe the differences between these fields. The definitions of these fields are

$$\begin{cases} \boldsymbol{P}(\boldsymbol{r}, t) = \boldsymbol{D}(\boldsymbol{r}, t) - \epsilon_0 \boldsymbol{E}(\boldsymbol{r}, t) \\ \boldsymbol{M}(\boldsymbol{r}, t) = \dfrac{1}{\mu_0} \boldsymbol{B}(\boldsymbol{r}, t) - \boldsymbol{H}(\boldsymbol{r}, t) \end{cases} \tag{1.8}$$

The vector field $\boldsymbol{P}(\boldsymbol{r}, t)$ quantifies how much the bound charges in the material are displaced relative to their equilibrium positions. The polarization includes both permanent and induced polarization. In an analogous manner, the magnetization $\boldsymbol{M}(\boldsymbol{r}, t)$ is a measure of the bound currents in the material. The origin of this field can also be both permanent or induced.

The modeling of the polarization and the magnetization effects of the material is equivalent to a specification of the constitutive relations of the material, and it implies that six additional equations characterizing the material are given. In Section 1.2, we analyze several different models for the polarization and the magnetization of the material in detail. In the next two subsections, we investigate additional consequences of the Maxwell equations, namely, the boundary conditions and energy conservation.

1.1.1 Boundary Conditions at Interfaces

At an interface between two different materials some components of the electromagnetic fields are discontinuous. The way they vary is a consequence of the Maxwell equations, and in this section we give a simple derivation of the boundary conditions the fields must satisfy. Only

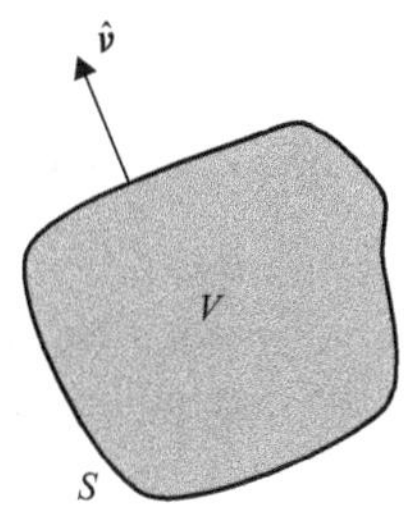

Figure 1.1 Geometry of integration.

surfaces that are fixed in time (no moving surfaces) are treated here. The boundary conditions at a moving boundary are presented in Appendix E.3.

The Maxwell equations, as they were presented in (1.1) and (1.2), assume that all field quantities are differentiable functions of space and time. At an interface between two media, the fields, as already mentioned above, are discontinuous functions of the spatial variables. Therefore, we need to reformulate the Maxwell equations in such a way that the equations can be interpreted in a weaker, more general, sense. The aim is to obtain equations that are also valid for fields that are not differentiable at all points in space.

Let V be an arbitrary (simply connected) volume with bounding surface S and outward unit normal vector $\hat{\boldsymbol{\nu}}$, see Figure 1.1. We start by assuming the fields $\boldsymbol{E}, \boldsymbol{B}, \boldsymbol{D}$, and $\boldsymbol{H}$ are continuously differentiable in V, and then we see how the differentiability property can be relaxed. Integrate the Maxwell equations, (1.1), (1.2), (1.5), and (1.6), over the volume V. We get

$$\begin{aligned} \iiint_V \nabla \times \boldsymbol{E} \, \mathrm{d}v &= -\iiint_V \frac{\partial \boldsymbol{B}}{\partial t} \, \mathrm{d}v \\ \iiint_V \nabla \times \boldsymbol{H} \, \mathrm{d}v &= \iiint_V \boldsymbol{J} \, \mathrm{d}v + \iiint_V \frac{\partial \boldsymbol{D}}{\partial t} \, \mathrm{d}v \end{aligned} \tag{1.9}$$

and

$$\begin{aligned} \iiint_V \nabla \cdot \boldsymbol{B} \, \mathrm{d}v &= 0 \\ \iiint_V \nabla \cdot \boldsymbol{D} \, \mathrm{d}v &= \iiint_V \rho \, \mathrm{d}v \end{aligned} \tag{1.10}$$

where $\mathrm{d}v$ is the differential volume element ($\mathrm{d}v = \mathrm{d}x\,\mathrm{d}y\,\mathrm{d}z$).

The following two integration theorems for vector fields are now useful:

$$\iiint_V \nabla \cdot \boldsymbol{A} \, \mathrm{d}v = \iint_S \boldsymbol{A} \cdot \hat{\boldsymbol{\nu}} \, \mathrm{d}S, \qquad \iiint_V \nabla \times \boldsymbol{A} \, \mathrm{d}v = \iint_S \hat{\boldsymbol{\nu}} \times \boldsymbol{A} \, \mathrm{d}S$$

where $\boldsymbol{A}$ is a continuously differentiable vector field in V, and $\mathrm{d}S$ is the surface element of S. The first theorem is usually called the divergence theorem or the Gauss' theorem[9] and the other Gauss' analogous theorem, see Problem 1.1.

The result after interchanging the differentiation w.r.t.[10] time t and integration over the space variables (the volume V is fixed in time and we assume all fields to be sufficiently regular) is[11]

$$\begin{cases} \displaystyle\iint_S \hat{\boldsymbol{\nu}} \times \boldsymbol{E} \, \mathrm{d}S = -\frac{\mathrm{d}}{\mathrm{d}t} \iiint_V \boldsymbol{B} \, \mathrm{d}v \\ \displaystyle\iint_S \hat{\boldsymbol{\nu}} \times \boldsymbol{H} \, \mathrm{d}S = \iiint_V \boldsymbol{J} \, \mathrm{d}v + \frac{\mathrm{d}}{\mathrm{d}t} \iiint_V \boldsymbol{D} \, \mathrm{d}v \\ \displaystyle\iint_S \boldsymbol{B} \cdot \hat{\boldsymbol{\nu}} \, \mathrm{d}S = 0 \\ \displaystyle\iint_S \boldsymbol{D} \cdot \hat{\boldsymbol{\nu}} \, \mathrm{d}S = \iiint_V \rho \, \mathrm{d}v \end{cases} \tag{1.11}$$

In a domain V, where the fields $\boldsymbol{E}$, $\boldsymbol{B}$, $\boldsymbol{D}$, and $\boldsymbol{H}$ are continuously differentiable, these integral expressions are equivalent to the differential equations in (1.1), (1.2), (1.5), and (1.6). We have proved this equivalence one way and in the other direction we do the analysis in a reversed direction and use the fact that the volume V is simply connected and arbitrary (let the volume V shrink to a point).

The integral formulation, (1.11), has, however, the advantage that the fields do not have to be differentiable in the spatial variables to make sense. In this respect, the integral formulation is more general than the differential formulation in (1.1), (1.2), (1.5), and (1.6). The fields $\boldsymbol{E}$, $\boldsymbol{B}$, $\boldsymbol{D}$, and $\boldsymbol{H}$ that satisfy the equations in (1.11) are called weak solutions to the Maxwell equations, in the case the fields are not continuously differentiable, and (1.1), (1.2), (1.5), and (1.6) lack meaning.

The integral expressions (1.11) are now employed to a special volume, V_h, that intersects the interface, S, between two different media, see Figure 1.2. The unit normal, $\hat{\boldsymbol{\nu}}$, of the interface, S, is directed from medium 2 into medium 1. We assume all electromagnetic fields $\boldsymbol{E}$, $\boldsymbol{B}$, $\boldsymbol{D}$, and $\boldsymbol{H}$, and their time derivatives, to have finite limit values from both sides of the interface. For the electric field, these limit values in media 1 and 2 are denoted $\boldsymbol{E}_1$ and $\boldsymbol{E}_2$, respectively, and a similar notation, with indices 1 and 2, is adopted for the other fields. The current density

---

[9] Distinguish between the Gauss law, (1.6), and the Gauss' theorem.

[10] w.r.t. stands for "with respect to."

[11] The boundary conditions found by the use of the first two equations are traditionally obtained by a Stokesian loop integration argument [101]. The volume integral employed here gives a more uniform approach. The two approaches give the same result.

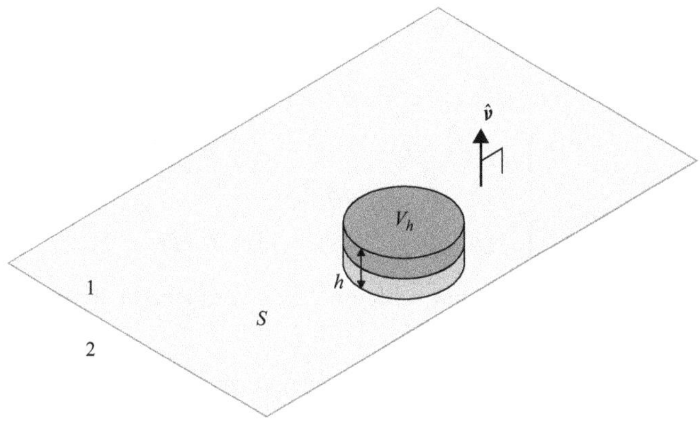

Figure 1.2 The interface S between two different media 1 and 2, and the special volume, V_h, that intersects the interface. The volume V_h is intersected by the surface S, and the height of the volume is h. The unit normal, $\hat{\boldsymbol{\nu}}$, of the interface is directed from medium 2 into medium 1.

$\boldsymbol{J}$ and the charge density ρ, however, can assume infinite values at the interface for perfectly conducting surfaces.[12] It is convenient to introduce a surface current density $\boldsymbol{J}_S$ and surface charge density ρ_S as a limit process

$$\begin{cases} \boldsymbol{J}_S = h\boldsymbol{J} \\ \rho_S = h\rho \end{cases}$$

where h is the thickness of the layer that contains the charges close to the surface. We assume that this thickness approaches zero, and that $\boldsymbol{J}$ and ρ approach infinity in such a way that $\boldsymbol{J}_S$ and ρ_S have well-defined limits as the thickness approaches zero. The surface current density $\boldsymbol{J}_S$ is assumed to be a tangential field to the surface S. We denote the height of the volume V_h by h, and the area on the upper and lower part of the bounding surface of V_h by A, which is assumed small compared to the variations of the fields and the curvature of the surface S.

The two terms containing time derivatives in (1.11), i.e.,

$$\frac{\mathrm{d}}{\mathrm{d}t}\iiint_{V_h} \boldsymbol{B}\,\mathrm{d}v, \qquad \frac{\mathrm{d}}{\mathrm{d}t}\iiint_{V_h} \boldsymbol{D}\,\mathrm{d}v$$

both approach zero as $h \to 0$, since the fields $\boldsymbol{B}$ and $\boldsymbol{D}$ and their time derivatives are assumed to be finite at the interface. Moreover, the contributions from all lateral sides (areas $\sim h$) of the surface integrals on the left-hand side in (1.11) approach zero as $h \to 0$. The contribution from the upper part (unit normal $\hat{\boldsymbol{\nu}}$) and lower part (unit normal $-\hat{\boldsymbol{\nu}}$) are proportional to the area A, provided the area is chosen sufficiently small, and the mean value theorem for integrals

[12]This is of course an idealization of a situation where the density assumes very high values in a macroscopically thin layer.

is used. The contributions from the upper and the lower parts of the surface integrals in the limit $h \to 0$ are

$$\begin{cases} A\left[\hat{\boldsymbol{\nu}} \times (\boldsymbol{E}_1 - \boldsymbol{E}_2)\right] = \boldsymbol{0} \\ A\left[\hat{\boldsymbol{\nu}} \times (\boldsymbol{H}_1 - \boldsymbol{H}_2)\right] = Ah\boldsymbol{J} = A\boldsymbol{J}_S \\ A\left[\hat{\boldsymbol{\nu}} \cdot (\boldsymbol{B}_1 - \boldsymbol{B}_2)\right] = 0 \\ A\left[\hat{\boldsymbol{\nu}} \cdot (\boldsymbol{D}_1 - \boldsymbol{D}_2)\right] = Ah\rho = A\rho_S \end{cases}$$

Simplify these expressions by dividing with the area A (and take the limit as A limits a point $\boldsymbol{r}$ on the surface S). The result is

$$\begin{cases} \hat{\boldsymbol{\nu}} \times (\boldsymbol{E}_1 - \boldsymbol{E}_2) = \boldsymbol{0} \\ \hat{\boldsymbol{\nu}} \times (\boldsymbol{H}_1 - \boldsymbol{H}_2) = \boldsymbol{J}_S \\ \hat{\boldsymbol{\nu}} \cdot (\boldsymbol{B}_1 - \boldsymbol{B}_2) = 0 \\ \hat{\boldsymbol{\nu}} \cdot (\boldsymbol{D}_1 - \boldsymbol{D}_2) = \rho_S \end{cases} \tag{1.12}$$

These boundary conditions hold at every point $\boldsymbol{r}$ on the surface S, and describe how the electromagnetic fields are related to each other on both sides of the interface (the unit normal $\hat{\boldsymbol{\nu}}$ is directed from medium 2 into medium 1). We formulate these boundary conditions in words.

- The tangential components of the electric field $\boldsymbol{E}$ are continuous across the interface S.
- The tangential components of the magnetic field $\boldsymbol{H}$ are discontinuous over the interface S. The size of the discontinuity is $\boldsymbol{J}_S$. If the surface current density is zero, e.g., if the material has finite conductivity,[13] the tangential components of the magnetic field are continuous across the interface S. In fact, the electric field $\boldsymbol{E}$ is assumed finite close to the interface. Then $\boldsymbol{J}_S = h\boldsymbol{J} = h\sigma\boldsymbol{E} \to \boldsymbol{0}$, as $h \to 0$. Thus, the discontinuity only occurs with a perfectly conducting surfaces or superconductors.
- The normal component of the magnetic flux density $\boldsymbol{B}$ is continuous across the interface S.
- The normal component of the electric flux density $\boldsymbol{D}$ is discontinuous across the interface S. The size of the discontinuity is ρ_S. If the surface charge density is zero, the normal component of the electric flux density is continuous across the interface.

A special case of (1.12) is the case where medium 2 is a perfectly conducting material (perfect electric conductor, PEC), which is a model of a material with freely moving charges, e.g., many metals. In material 2, the fields are zero, and we get from (1.12)

$$\begin{cases} \hat{\boldsymbol{\nu}} \times \boldsymbol{E}_1 = \boldsymbol{0} \\ \hat{\boldsymbol{\nu}} \times \boldsymbol{H}_1 = \boldsymbol{J}_S \\ \hat{\boldsymbol{\nu}} \cdot \boldsymbol{B}_1 = 0 \\ \hat{\boldsymbol{\nu}} \cdot \boldsymbol{D}_1 = \rho_S \end{cases} \tag{1.13}$$

where $\boldsymbol{J}_S$ and ρ_S are the surface current density and surface charge density on the surface S, respectively.

[13]The conductivity of a material is introduced below, and the concept of surface resistance is excluded.

1.1.2 Energy Conservation and Poynting's Theorem

Energy conservation is shown from the Maxwell equations (1.1) and (1.2).

$$\begin{cases} \nabla \times \boldsymbol{E} = -\dfrac{\partial \boldsymbol{B}}{\partial t} \\ \nabla \times \boldsymbol{H} = \boldsymbol{J} + \dfrac{\partial \boldsymbol{D}}{\partial t} \end{cases}$$

Scalar multiplication (dot product) of the first equation with $\boldsymbol{H}$ and the second equation with $\boldsymbol{E}$ followed by a subtraction imply

$$\boldsymbol{H} \cdot (\nabla \times \boldsymbol{E}) - \boldsymbol{E} \cdot (\nabla \times \boldsymbol{H}) + \boldsymbol{H} \cdot \frac{\partial \boldsymbol{B}}{\partial t} + \boldsymbol{E} \cdot \frac{\partial \boldsymbol{D}}{\partial t} + \boldsymbol{E} \cdot \boldsymbol{J} = 0$$

We rewrite this expression with the use of the differential rule of the divergence of a vector product, i.e., $\nabla \cdot (\boldsymbol{a} \times \boldsymbol{b}) = \boldsymbol{b} \cdot (\nabla \times \boldsymbol{a}) - \boldsymbol{a} \cdot (\nabla \times \boldsymbol{b})$. We have

$$\nabla \cdot (\boldsymbol{E} \times \boldsymbol{H}) + \boldsymbol{H} \cdot \frac{\partial \boldsymbol{B}}{\partial t} + \boldsymbol{E} \cdot \frac{\partial \boldsymbol{D}}{\partial t} + \boldsymbol{E} \cdot \boldsymbol{J} = 0$$

The vector product of the electric and the magnetic fields plays a special role, and we introduce the Poynting vector,[14] $\boldsymbol{S} = \boldsymbol{E} \times \boldsymbol{H}$. We get Poynting's theorem

$$\nabla \cdot \boldsymbol{S} + \boldsymbol{H} \cdot \frac{\partial \boldsymbol{B}}{\partial t} + \boldsymbol{E} \cdot \frac{\partial \boldsymbol{D}}{\partial t} + \boldsymbol{E} \cdot \boldsymbol{J} = 0 \tag{1.14}$$

The Poynting vector $\boldsymbol{S}$ gives the power per unit area of the electromagnetic fields[15] or the power flux density in the direction of the vector $\boldsymbol{S}$. This becomes clearer if we integrate (1.14) over a simply connected volume V, bounded by the surface S and with outward unit normal vector $\hat{\boldsymbol{\nu}}$, see Figure 1.1, and use the divergence theorem. We get

$$\begin{aligned} \iint_S \boldsymbol{S} \cdot \hat{\boldsymbol{\nu}} \, \mathrm{d}S &= \iiint_V \nabla \cdot \boldsymbol{S} \, \mathrm{d}v \\ &= -\iiint_V \left[\boldsymbol{H} \cdot \frac{\partial \boldsymbol{B}}{\partial t} + \boldsymbol{E} \cdot \frac{\partial \boldsymbol{D}}{\partial t} \right] \mathrm{d}v - \iiint_V \boldsymbol{E} \cdot \boldsymbol{J} \, \mathrm{d}v \end{aligned} \tag{1.15}$$

We recognize the two terms on the right-hand side from introductory treatments in electromagnetics. The first integral is the power stored in the electromagnetic fields in the volume V

$$\iiint_V \left[\boldsymbol{H} \cdot \frac{\partial \boldsymbol{B}}{\partial t} + \boldsymbol{E} \cdot \frac{\partial \boldsymbol{D}}{\partial t} \right] \mathrm{d}v$$

---

[14] John Henry Poynting (1852–1914), English physicist.

[15] Further details on this interpretation are given in Reference 184 and references therein.

Included in this contribution is the power needed to polarize and magnetize the material in V. The second volume integral in (1.15)

$$\iiint_V \boldsymbol{E} \cdot \boldsymbol{J} \, \mathrm{d}v$$

quantifies the work per unit time, i.e., the power, that the electric field does on the charges in V.

This interpretation implies that (1.15) expresses power balance in the volume V, i.e.,

$$\begin{aligned} &\text{Power radiated through } S + \text{Power consumption in } V \\ &\qquad = -\text{ Power bounded to the electromagnetic fields in } V \end{aligned}$$

and the integral on the left-hand side in (1.15) then is the power flux through the surface S, i.e.,

$$\iint_S \boldsymbol{S} \cdot \hat{\boldsymbol{\nu}} \, \mathrm{d}S$$

quantifies the total power radiated out of the bounding surface S, i.e., the energy per time unit, carried by the electromagnetic fields.

In the derivation above, we assumed that the volume V did not cut any surface where the fields vary discontinuously, e.g., an interface between two different media. We now prove that this assumption can easily be relaxed. If the surface S is an interface between two media, see Figure 1.2, the Poynting vector in medium 1 close to the interface is

$$\boldsymbol{S}_1 = \boldsymbol{E}_1 \times \boldsymbol{H}_1$$

and the Poynting vector close to the interface in medium 2 is

$$\boldsymbol{S}_2 = \boldsymbol{E}_2 \times \boldsymbol{H}_2$$

The boundary conditions at the interface given by (1.12) read

$$\begin{cases} \hat{\boldsymbol{\nu}} \times \boldsymbol{E}_1 = \hat{\boldsymbol{\nu}} \times \boldsymbol{E}_2 \\ \hat{\boldsymbol{\nu}} \times \boldsymbol{H}_1 = \hat{\boldsymbol{\nu}} \times \boldsymbol{H}_2 + \boldsymbol{J}_S \end{cases}$$

A cyclic permutation of the vectors and the use of the boundary conditions imply

$$\begin{aligned} \hat{\boldsymbol{\nu}} \cdot \boldsymbol{S}_1 &= \hat{\boldsymbol{\nu}} \cdot (\boldsymbol{E}_1 \times \boldsymbol{H}_1) = \boldsymbol{H}_1 \cdot (\hat{\boldsymbol{\nu}} \times \boldsymbol{E}_1) = \boldsymbol{H}_1 \cdot (\hat{\boldsymbol{\nu}} \times \boldsymbol{E}_2) \\ &= -\boldsymbol{E}_2 \cdot (\hat{\boldsymbol{\nu}} \times \boldsymbol{H}_1) = -\boldsymbol{E}_2 \cdot (\hat{\boldsymbol{\nu}} \times \boldsymbol{H}_2 + \boldsymbol{J}_S) \\ &= \hat{\boldsymbol{\nu}} \cdot (\boldsymbol{E}_2 \times \boldsymbol{H}_2) - \boldsymbol{E}_2 \cdot \boldsymbol{J}_S = \hat{\boldsymbol{\nu}} \cdot \boldsymbol{S}_2 - \boldsymbol{E}_2 \cdot \boldsymbol{J}_S \end{aligned}$$

By integration of this expression over the interface S, we obtain power conservation over the surface S as expressed as

$$\iint_S \boldsymbol{S}_1 \cdot \hat{\boldsymbol{\nu}} \, \mathrm{d}S = \iint_S \boldsymbol{S}_2 \cdot \hat{\boldsymbol{\nu}} \, \mathrm{d}S - \iint_S \boldsymbol{E}_2 \cdot \boldsymbol{J}_S \, \mathrm{d}S \tag{1.16}$$

where the surface S is an arbitrary part of the interface. Note that the unit normal $\hat{\boldsymbol{\nu}}$ is directed from medium 2 into medium 1.

The last surface integral in (1.16) gives the work per unit time the electric field does on the charges at the interface. If there are no surface currents at the interface, the normal component of the Poynting vector is continuous across the interface. It is irrelevant which electric field we use in the last surface integral in (1.16) since the surface current density $\boldsymbol{J}_S$ is parallel to the interface S and the tangential components of the electric field are continuous across the interface, i.e.,

$$\iint_S \boldsymbol{E}_1 \cdot \boldsymbol{J}_S \, \mathrm{d}S = \iint_S \boldsymbol{E}_2 \cdot \boldsymbol{J}_S \, \mathrm{d}S$$

1.2 Constitutive Relations

As already mentioned in Section 1.1, the Maxwell equations (1.1) and (1.2) are not complete. The equations contain 12 field quantities ($\boldsymbol{E}$, $\boldsymbol{B}$, $\boldsymbol{D}$, and $\boldsymbol{H}$), but there are only six equations. The remaining six equations, the so-called constitutive relations will be treated in this section. This section deals with general time-dependent fields in the same way as in Section 1.1. The special conditions that hold for time-harmonic fields are addressed in Section 1.3.

The Maxwell equations contain the fields $\boldsymbol{E}, \boldsymbol{B}, \boldsymbol{D}$, and $\boldsymbol{H}$ and their sources. These equations model the dynamics of the fields, but how the fields are related to each other is independent information. This information is, crudely speaking, the dynamics of the charges in the material. In general, the constitutive relations relate two pairs of fields, e.g., $\{\boldsymbol{D}, \boldsymbol{B}\}$ and $\{\boldsymbol{E}, \boldsymbol{H}\}$, to each other. Formally, we write this relation as

$$\{\boldsymbol{D}, \boldsymbol{B}\} = \boldsymbol{F}\left(\{\boldsymbol{E}, \boldsymbol{H}\}\right) \tag{1.17}$$

This particular form of the constitutive relations emphasizes the fields $\boldsymbol{E}$ and $\boldsymbol{H}$ as the primary pair, and the transformation (1.17) can be used to eliminate the flux densities, $\boldsymbol{D}$ and $\boldsymbol{B}$, from the Maxwell equations. The electric and the magnetic fields, $\boldsymbol{E}$ and $\boldsymbol{H}$, which define the Poynting vector, are the primary fields in this relation. In wave propagation problems, this particular form of the constitutive relations is appropriate, since power flux is an important quantity in these problems.

Other combinations of the fields in the constitutive relations often occur in the literature. Frequently, a relation between the pairs $\{\boldsymbol{D}, \boldsymbol{H}\}$ and $\{\boldsymbol{E}, \boldsymbol{B}\}$ is used. This relation emphasizes the fields $\boldsymbol{E}$ and $\boldsymbol{B}$ as the primary pair, and these fields enter in the Lorentz force in (1.3). There are also other reasons, based upon the theory of special relativity, that make this transformation between $\{\boldsymbol{D}, \boldsymbol{H}\}$ and $\{\boldsymbol{E}, \boldsymbol{B}\}$ preferable. However, in this book, we use the constitutive relations given in (1.17), due to the fact that the fields $\boldsymbol{E}$ and $\boldsymbol{H}$ define the Poynting vector.

The mapping $\boldsymbol{F}$ in (1.17) has to satisfy certain general requirements or physical assumptions to be a candidate for a realistic model. These physical assumptions or requirements are discussed in this section. The special simplifications that occur for time-harmonic fields are analyzed in Section 1.3. The first subsection deals with isotropic media with dispersion. In a subsequent subsection, these results are generalized and models for general linear media are stated.

1.2.1 Isotropic Media with Dispersion

As an introduction to the more general constitutive relations that are treated in Section 1.2.3, we first analyze the simpler case with isotropic media.

An isotropic, dispersive material is the simplest example of a constitutive relation between the fields. The isotropy implies that the material has identical properties in all directions and as a consequence of this there is no coupling between different components of the fields.[16] Moreover, we assume that there is no coupling between the electric fields, $\boldsymbol{E}$ and $\boldsymbol{D}$, and the magnetic fields, $\boldsymbol{H}$ and $\boldsymbol{B}$. The mapping in (1.17) then becomes two separate mappings that do not couple. For simplicity, we also assume that the material is non-magnetic, i.e., the magnetic fields satisfy the vacuum relations, (1.7). Equation (1.17) then simplifies to

$$\begin{cases} \boldsymbol{D} = \boldsymbol{F}(\boldsymbol{E}) \\ \boldsymbol{B} = \mu_0 \boldsymbol{H} \end{cases}$$

We now state a series of requirements that the mapping $\boldsymbol{F}$ has to satisfy as a function of time t at every point, $\boldsymbol{r}$, in the medium. All other dependence of the macroscopic variables, such as temperature or pressure, are not explicitly indicated to avoid cumbersome notation. Specifically, no functional dependence, except time t, is written out, unless it is required for the understanding. To this end, fields and other quantities do depend on the space variables $\boldsymbol{r}$, even if this is not made explicit.

The following assumptions on the mapping $\boldsymbol{F}$ are required in this textbook [118, 203]:

1. The mapping $\boldsymbol{F}$ is linear in the field $\boldsymbol{E}$, i.e., for all real constants α and β and for all fields $\boldsymbol{E}$ and $\boldsymbol{E}'$ we have

$$\boldsymbol{F}(\alpha\boldsymbol{E} + \beta\boldsymbol{E}') = \alpha\boldsymbol{F}(\boldsymbol{E}) + \beta\boldsymbol{F}(\boldsymbol{E}')$$

2. The mapping $\boldsymbol{F}$ is causal, i.e., for all τ, and for every field $\boldsymbol{E}$ such that $\boldsymbol{E}(t) = 0$, for $t < \tau$ we have

$$\boldsymbol{F}(\boldsymbol{E})(t) = \boldsymbol{0} \quad \text{for } t < \tau$$

3. The mapping $\boldsymbol{F}$ is invariant under time translations, i.e., for every pair of fields $\boldsymbol{E}$ and $\boldsymbol{D}$, where the fields are related by $\boldsymbol{D}(t) = \boldsymbol{F}(\boldsymbol{E})(t)$, and every τ and pair of fields $\boldsymbol{E}'$ and $\boldsymbol{D}'$, defined by $\boldsymbol{E}'(t) = \boldsymbol{E}(t-\tau)$ and $\boldsymbol{D}'(t) = \boldsymbol{F}(\boldsymbol{E}')(t)$, we have

$$\boldsymbol{D}'(t) = \boldsymbol{D}(t-\tau)$$

Property 1, of course, excludes all nonlinear phenomena that occur in many electromagnetic applications, but since most materials show linear behavior at a sufficiently weak field strength, this is not a serious limitation. Causality, as it is stated in item 2, implies that there is no reaction (the electric flux density $\boldsymbol{D}$ or polarization $\boldsymbol{P}$) before its cause (the electric field $\boldsymbol{E}$). Property 3 implies that the material does not change (age), and the same result is obtained

[16] Note that an isotropic material does not imply that the material is homogeneous, i.e., the material properties are the same at all points in the material. The isotropic properties of the material are a microscopic property, but the inhomogeneous variation of the material is a variation on a macroscopic length scale.

in an experiment that is repeated at a later time. These assumptions are the foundations to the constitutive relations used in this book and all linear constitutive relations must satisfy these requirements to be physically sound.[17]

The starting point of a realization of the mapping $\boldsymbol{F}$ is the following Ansatz between the polarization of the material $\boldsymbol{P}$ and the electric field $\boldsymbol{E}$:

$$\boldsymbol{P}(t) = \epsilon_0 \int_{-\infty}^{\infty} \chi(t,t')\boldsymbol{E}(t')\,\mathrm{d}t'$$

The function $\chi(t,t')$ has the unit Hz (s^{-1}). The electric flux density $\boldsymbol{D}$ then is, see (1.8),

$$\boldsymbol{D}(t) = \epsilon_0 \left\{ \boldsymbol{E}(t) + \int_{-\infty}^{\infty} \chi(t,t')\boldsymbol{E}(t')\,\mathrm{d}t' \right\}$$

This integral is, of course, linear, and, furthermore, it is in agreement with the assumption that the material is isotropic (no coupling between different vector components, and the same relation for all three Cartesian[18] components).

We now address the requirements stated in items 2 and 3, and assume the electric field to be zero before a given fixed time τ, i.e., $\boldsymbol{E}(t) = \boldsymbol{0}$, $t < \tau$. The constitutive relation then is

$$\boldsymbol{D}(t) = \epsilon_0 \left\{ \boldsymbol{E}(t) + \int_{\tau}^{\infty} \chi(t,t')\boldsymbol{E}(t')\,\mathrm{d}t' \right\}$$

Causality, item 2, immediately implies that the function $\chi(t,t')$ has to be zero when $t < t'$, i.e.,

$$\chi(t,t') = 0, \quad t < t'$$

since the causality requirement is $\boldsymbol{D}(t) = \boldsymbol{0}$, $t < \tau$ independently of the electric field values for $t > \tau$. In general, the range of integration for a general electric field therefore is limited to the interval $(-\infty, t]$. We have

$$\boldsymbol{D}(t) = \epsilon_0 \left\{ \boldsymbol{E}(t) + \int_{-\infty}^{t} \chi(t,t')\boldsymbol{E}(t')\,\mathrm{d}t' \right\}$$

This relation between the electric flux density $\boldsymbol{D}$ and the electric field $\boldsymbol{E}$ implies that $\boldsymbol{D}$ depends on the entire time history of the electric field—in other words, the material has a memory or shows dispersion. This memory is characterized or modeled by the function $\chi(t,t')$.

[17] In fact, an additional assumption is also made, but we prefer not to explicitly list this assumption in order to avoid some technical mathematical machinery that is not used elsewhere in this textbook. The cause, $\boldsymbol{D}$ (or $\boldsymbol{P}$), must depend continuously on the electric field $\boldsymbol{E}$, i.e., a small electric field implies small polarization effects in the material. For more details, see Reference 118.

[18] René Descartes (Renatus Cartesius) (1596–1650), French philosopher, mathematician, physicist, and writer.

In order to investigate what conditions the time invariance imply, we create two field pairs $\{\boldsymbol{D},\boldsymbol{E}\}$ and $\{\boldsymbol{D}',\boldsymbol{E}'\}$, defined by

$$\begin{cases} \boldsymbol{D}(t) = \epsilon_0 \left\{\boldsymbol{E}(t) + \displaystyle\int_{-\infty}^{t} \chi(t,t')\boldsymbol{E}(t')\, \mathrm{d}t'\right\} \\ \boldsymbol{D}'(t) = \epsilon_0 \left\{\boldsymbol{E}'(t) + \displaystyle\int_{-\infty}^{t} \chi(t,t')\boldsymbol{E}'(t')\, \mathrm{d}t'\right\} \end{cases}$$

where $\boldsymbol{E}'(t) = \boldsymbol{E}(t-\tau)$. Time invariance, Property 3, implies $\boldsymbol{D}'(t) = \boldsymbol{D}(t-\tau)$. If the first integral is evaluated at time $t-\tau$, the two integrals are

$$\begin{cases} \boldsymbol{D}(t-\tau) = \epsilon_0 \left\{\boldsymbol{E}(t-\tau) + \displaystyle\int_{-\infty}^{t-\tau} \chi(t-\tau,t')\boldsymbol{E}(t')\, \mathrm{d}t'\right\} \\ \boldsymbol{D}(t-\tau) = \epsilon_0 \left\{\boldsymbol{E}(t-\tau) + \displaystyle\int_{-\infty}^{t} \chi(t,t')\boldsymbol{E}(t'-\tau)\, \mathrm{d}t'\right\} \end{cases}$$

Equating the two expressions implies

$$\int_{-\infty}^{t-\tau} \chi(t-\tau,t')\boldsymbol{E}(t')\, \mathrm{d}t' = \int_{-\infty}^{t} \chi(t,t')\boldsymbol{E}(t'-\tau)\, \mathrm{d}t' = \int_{-\infty}^{t-\tau} \chi(t,t'+\tau)\boldsymbol{E}(t')\, \mathrm{d}t'$$

where we have made a change of variables in the last equality. We get

$$\int_{-\infty}^{t-\tau} (\chi(t-\tau,t') - \chi(t,t'+\tau))\boldsymbol{E}(t')\, \mathrm{d}t' = \boldsymbol{0}$$

The field $\boldsymbol{E}$ is here arbitrary, which implies

$$\chi(t-\tau,t') = \chi(t,t'+\tau)$$

for all t, τ, and t', or equivalently (take e.g., $t'=0$ and exchange $\tau \to t'$)

$$\chi(t,t') = \chi(t-t',0)$$

We observe that the function $\chi(t,t')$ is only a function of the difference in time, $t-t'$, and causality implies that the susceptibility function $\chi(t,0)=0$ is zero for $t<0$. Therefore, the constitutive relations of the material can be written[19]

$$\begin{cases} \boldsymbol{D}(t) = \epsilon_0 \left\{\boldsymbol{E}(t) + \displaystyle\int_{-\infty}^{t} \chi(t-t')\boldsymbol{E}(t')\, \mathrm{d}t'\right\} \\ \boldsymbol{B}(t) = \mu_0 \boldsymbol{H}(t) \end{cases} \tag{1.18}$$

[19] We prefer to keep the notation χ even if it formally is not the same function as above—the second argument is ignored.

The function $\chi(t)$ is called the susceptibility function of the material (unit frequency), and it gives the polarization of the medium excited by a delta function (for definition, see Appendix I). To see this, let $\boldsymbol{E}(t) = \boldsymbol{E}_0\delta(t)$. We then have

$$\boldsymbol{D}(t) = \underbrace{\epsilon_0\delta(t)\boldsymbol{E}_0}_{\text{Instantaneous response}} + \underbrace{\epsilon_0\chi(t)\boldsymbol{E}_0}_{\text{Transient field}}$$

The susceptibility function $\chi(t)$ is the mathematical model of the memory properties of the medium and its dispersive effects.

1.2.1.1 Optical Response

If there are several physical processes that contribute to the electrical properties of the material, there are often several different time scales involved. For example, the interaction of the electric field with the low mass electrons in an atom is a fast process compared to the slower processes that occur when heavier atoms or molecules interact with the field. Usually, the polarization of the material contains a contribution which originates from very fast processes in the material. This contribution is usually called the optical response, and can be modeled if we divide the susceptibility function $\chi(t)$ in a sum of two terms $\chi_1(t)$ and $\chi_2(t)$, i.e.,

$$\chi(t) = \chi_1(t) + \chi_2(t)$$

where $\chi_1(t)$ varies with a more rapid time scale compared to $\chi_2(t)$, which models the slower variations in the polarization effects of the material. We visualize this division in Figure 1.3. If the electric field only varies slowly compared to the variations in $\chi_1(t)$, it is convenient to include the effects of $\chi_1(t)$ as an instant contribution, similar to the first term in (1.18). The electric field $\boldsymbol{E}$ is then—compared to the variations in $\chi_1(t)$—approximately constant. Therefore, we can take the electric field $\boldsymbol{E}$ outside the integrations and we obtain:

$$\begin{aligned}
\boldsymbol{D}(t) &= \epsilon_0 \left\{\boldsymbol{E}(t) + \int_{-\infty}^{t} \left[\chi_1(t-t') + \chi_2(t-t')\right] \boldsymbol{E}(t')\, \mathrm{d}t'\right\} \\
&= \epsilon_0 \left\{\boldsymbol{E}(t) + \int_{-\infty}^{t} \chi_1(t-t')\, \mathrm{d}t' \boldsymbol{E}(t) + \int_{-\infty}^{t} \chi_2(t-t')\boldsymbol{E}(t')\, \mathrm{d}t'\right\} \\
&= \epsilon_0 \left\{\boldsymbol{E}(t) + \int_{0}^{\infty} \chi_1(t')\, \mathrm{d}t' \boldsymbol{E}(t) + \int_{-\infty}^{t} \chi_2(t-t')\boldsymbol{E}(t')\, \mathrm{d}t'\right\}
\end{aligned}$$

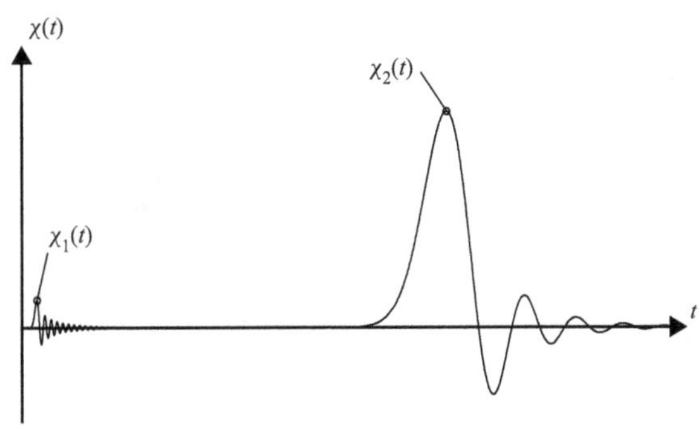

Figure 1.3 The susceptibility function $\chi(t)$ divided in two terms which exemplifies the optical response of the material. The time scale is arbitrary.

It is convenient to introduce the following notation, which defines the optical response:

$$\epsilon = 1 + \int_0^\infty \chi_1(t)\, \mathrm{d}t$$

which implies

$$\begin{cases} \boldsymbol{D}(t) = \epsilon_0 \left\{ \epsilon \boldsymbol{E}(t) + \int_{-\infty}^{t} \chi(t - t')\boldsymbol{E}(t')\, \mathrm{d}t' \right\} \\ \boldsymbol{B}(t) = \mu_0 \boldsymbol{H}(t) \end{cases} \tag{1.19}$$

where the index 2 on the susceptibility function is dropped for convenience.

Another equivalent way of introducing the optical response is to model the fast processes as a delta function contribution. To this end, we model the susceptibility function as

$$[\epsilon - 1]\, \delta(t) + \chi(t)$$

which in (1.18) also gives (1.19), but more directly. The first term in this expression is a model of the momentaneous response of the medium to an electric field excitation. In most situations, the origin of this term is due to charges with small inertia.

Comment 1.1
There is an inherent conflict with the concept of optical response in the sense that all macroscopic material models break down at the length scale associated with very fast transients. When the length scale of the fields becomes of the order of the size of the constitutive part of the material, e.g., nanoscale, the material does not behave as a bulk material any longer, but the fields interact with the constitutive parts of the material as individual scatterers and not as a homogeneous material. Nevertheless, the concept of optical response is appropriate when used and employed with care. At higher frequencies still, the entire classical modeling of materials becomes inaccurate—this is the realm of quantum phenomena, and outside the scope of this textbook.

1.2.1.2 Conductivity

The difference between bound and free charges is apparent for static (time independent) fields. Bound charges polarize the medium, and free charges contribute to the currents in the medium. This difference is wiped out for general time-dependent fields as the analysis in this section shows.

In general, there are two kinds of current densities, $\boldsymbol{J}_{\mathrm{imp}}$ and $\boldsymbol{J}_{\mathrm{ind}}$—the impressed current density and the induced current density, respectively. The impressed current density is supplied by external sources, which are controlled externally, and they are not caused by the existing electromagnetic fields. The induced currents, on the other hand, are caused by the present electromagnetic fields. It is the latter, the induced current density that we now address and model.

Several media, especially many metals, have mobile charges. The Ohm's law[20] with conductivity σ

$$\boldsymbol{J} = \sigma \boldsymbol{E} \tag{1.20}$$

is often used as a model of charge transport in these media. The conductivity σ is a material parameter, and can depend on the spatial coordinate if the medium is inhomogeneous. The limit case $\sigma \to \infty$ corresponds to a PEC.[21]

The Ohm law is an instantaneous response, and a model that is analog to the optical response discussed above. A somewhat more general set of constitutive relations than (1.19) and (1.20) is

$$\begin{cases} \boldsymbol{D}(t) = \epsilon_0 \left\{ \epsilon \boldsymbol{E}(t) + \int_{-\infty}^{t} \chi(t-t') \boldsymbol{E}(t') \, \mathrm{d}t' \right\} \\ \boldsymbol{J}(t) = \sigma \boldsymbol{E}(t) + \epsilon_0 \int_{-\infty}^{t} \Sigma(t-t') \boldsymbol{E}(t') \, \mathrm{d}t' \\ \boldsymbol{B}(t) = \mu_0 \mu \boldsymbol{H}(t) \end{cases} \tag{1.21}$$

In addition to the instantaneous term σ between the current density $\boldsymbol{J}$ and the electric field $\boldsymbol{E}$, the extended Ohm's law contains a term that models memory effects or dispersion effects. This model has, in addition to the dispersive effects described by the susceptibility function $\chi(t)$ (bound charges), also dispersive effects of the free charges in the function $\Sigma(t)$. The coupling between the $\boldsymbol{B}$- and the $\boldsymbol{H}$-fields is also more general than in (1.19). The real constant μ is a measure of the momentaneous magnetic properties of the medium.

In this section, we prove that the effect of mobile charges, which we model by the conductivity σ and the function $\Sigma(t)$, in fact can be included in the susceptibility function $\chi(t)$. Physically, this means that we classify the mobile charges as bound charges. An arrangement of this kind has no effect on the physical quantities, such as electric and magnetic fields. Conversely, we can prove that all dispersive effects, which are modeled by the susceptibility function $\chi(t)$, can be transferred to an effective conductivity and a dispersive term. In this case, we classify the bound charges modeled by the susceptibility $\chi(t)$, as mobile charges with the Ohm's law. This is also possible, provided that the physical quantities are unaffected by this rearrangement. To see this, we prove that there is non-uniqueness in the constitutive relations as they are formulated in (1.21).

The constitutive relations in (1.21) are not uniquely determined, but every choice of the constitutive relations on the form

$$\begin{cases} \boldsymbol{D}(t) = \epsilon_0 \left\{ \epsilon \boldsymbol{E}(t) + \int_{-\infty}^{t} \left(\chi(t-t') + f(t-t')\right) \boldsymbol{E}(t') \, \mathrm{d}t' \right\} \\ \boldsymbol{J}(t) = \left(\sigma - \epsilon_0 f(0^+)\right) \boldsymbol{E}(t) + \epsilon_0 \int_{-\infty}^{t} \left(\Sigma(t-t') - \frac{\partial f(t-t')}{\partial t} \right) \boldsymbol{E}(t') \, \mathrm{d}t' \\ \boldsymbol{B}(t) = \mu_0 \mu \boldsymbol{H}(t) \end{cases}$$

[20]Georg Simon Ohm (1789–1854), German physicist and mathematician.

[21]To be more precise, the PEC material in the dynamic case also has to generate zero magnetic flux density, and therefore we also have $\mu \to 0$ for a PEC material.

where $f(t)$ is an arbitrary (differentiable) real-valued function ($f(t) = 0,\ t < 0$ and $f(0^+) = \lim_{t\to 0, t>0} f(t)$), gives the same right-hand side in the Ampère law. We see this immediately by insertion in the Ampère law

$$\begin{aligned}\nabla \times \boldsymbol{H}(t) &= \boldsymbol{J} + \frac{\partial \boldsymbol{D}(t)}{\partial t}\\ &= \left(\sigma - \epsilon_0 f(0^+)\right)\boldsymbol{E}(t) + \epsilon_0 \int_{-\infty}^{t} \left(\Sigma(t-t') - \frac{\partial f(t-t')}{\partial t}\right)\boldsymbol{E}(t')\,\mathrm{d}t'\\ &\quad + \epsilon_0 \left\{\epsilon \frac{\partial \boldsymbol{E}(t)}{\partial t} + \frac{\partial}{\partial t}\int_{-\infty}^{t} \left(\chi(t-t') + f(t-t')\right)\boldsymbol{E}(t')\,\mathrm{d}t'\right\}\end{aligned}$$

Differentiation of the last integral leads to

$$\frac{\partial}{\partial t}\int_{-\infty}^{t} f(t-t')\boldsymbol{E}(t')\,\mathrm{d}t' = f(0^+)\boldsymbol{E}(t) + \int_{-\infty}^{t} \frac{\partial f(t-t')}{\partial t}\boldsymbol{E}(t')\,\mathrm{d}t'$$

and we obtain

$$\begin{aligned}\nabla \times \boldsymbol{H}(t) &= \sigma \boldsymbol{E}(t) + \epsilon_0 \int_{-\infty}^{t} \Sigma(t-t')\boldsymbol{E}(t')\,\mathrm{d}t'\\ &\quad + \epsilon_0 \left\{\epsilon \frac{\partial \boldsymbol{E}(t)}{\partial t} + \frac{\partial}{\partial t}\int_{-\infty}^{t} \chi(t-t')\boldsymbol{E}(t')\,\mathrm{d}t'\right\}\end{aligned}$$

We observe that the function $f(t)$ does not affect the Ampère law. Every choice of $f(t)$ gives the same right-hand side in the Ampère law.

Each choice of $f(t)$ is a reclassification of bound charges (modeled by the electric flux density $\boldsymbol{D}$) and the free charges (modeled by the charge density $\boldsymbol{J}$). To illustrate this reclassification and the non-uniqueness in the constitutive relations, we show two special cases in the next two examples.

Example 1.1

The dispersion model: In this example, we show that the constitutive relations given by (1.21) can be transformed such that $\boldsymbol{J} = \boldsymbol{0}$, i.e., the entire contribution from the charge density $\boldsymbol{J}$ is included in the electric flux density $\boldsymbol{D}$. To obtain this, choose the function $f(t)$ as

$$f(t) = \mathrm{H}(t)\left[\frac{\sigma}{\epsilon_0} + \int_0^t \Sigma(t')\,\mathrm{d}t'\right]$$

where $\mathrm{H}(t)$ is the Heaviside[22] step function. This implies

$$\begin{cases} f(0^+) = \dfrac{\sigma}{\epsilon_0}\\ f'(t) = \Sigma(t), \quad t > 0\end{cases}$$

[22] Oliver Heaviside (1850–1925), English electrical engineer, mathematician, and physicist.

The constitutive relations then become

$$\begin{cases} \boldsymbol{D}(t) = \epsilon_0 \left\{ \epsilon \boldsymbol{E}(t) + \int_{-\infty}^{t} \left(\chi(t-t') + f(t-t')\right)\boldsymbol{E}(t')\,\mathrm{d}t' \right\} \\ \boldsymbol{J}(t) = \boldsymbol{0} \\ \boldsymbol{B}(t) = \mu_0\mu \boldsymbol{H}(t) \end{cases}$$

This set of constitutive relations is called the dispersion model, since the contribution from the current density $\boldsymbol{J}$ is zero. These constitutive relations are equivalent to the ones in (1.21), in that they give the same right-hand side expression in the Ampère law.

Therefore, there is no loss of generality to let the contribution of the free charges, i.e., the Ohm's law, (1.20), be included in the susceptibility function $\chi_{\text{new}}(t)$

$$\chi_{\text{new}}(t) = \mathrm{H}(t)\frac{\sigma}{\epsilon_0} + \chi(t)$$

and the effects of the mobile charges have been absorbed in the susceptibility function. This new susceptibility function is a rearrangement of the free charges such that they are now included in the contribution of the bound charges. ■

Example 1.2

The conductivity model: We can also choose the function $f(t)$ such that another extreme is obtained, the conductivity model. Choose the function $f(t)$ as

$$f(t) = -\chi(t)$$

We get the following constitutive relations:

$$\begin{cases} \boldsymbol{D}(t) = \epsilon_0\epsilon \boldsymbol{E}(t) \\ \boldsymbol{J}(t) = \left(\sigma + \epsilon_0\chi(0^+)\right)\boldsymbol{E}(t) + \epsilon_0 \int_{-\infty}^{t} \left(\Sigma(t-t') + \frac{\partial\chi(t-t')}{\partial t}\right)\boldsymbol{E}(t')\,\mathrm{d}t' \\ \boldsymbol{B}(t) = \mu_0\mu \boldsymbol{H}(t) \end{cases} \tag{1.22}$$

In this particular set of constitutive relations, we have a susceptibility function $\chi(t)$ that is zero. All effects of dispersion are here collected into the current density $\boldsymbol{J}$. These constitutive relations are equivalent with the ones given in (1.21). ■

Of course, other choices of the function $f(t)$ are possible, which are a mixture of these two extreme cases. In this book, the dispersion model is usually used.

We end the analysis of isotropic materials by inverting the relation between the electric flux density $\boldsymbol{D}(\boldsymbol{r}, t)$ and the electric field $\boldsymbol{E}(\boldsymbol{r}, t)$. The relation (1.19), i.e.,

$$\boldsymbol{D}(t) = \epsilon_0 \left\{ \epsilon \boldsymbol{E}(t) + \int_{-\infty}^{t} \chi(t-t')\boldsymbol{E}(t')\,\mathrm{d}t' \right\}$$

can be inverted by applying the resolvent of the susceptibility function $\chi(t)$. The resolvent kernel $\Upsilon(t)$ satisfies the resolvent equation

$$\frac{1}{\epsilon}\chi(t)+\epsilon\Upsilon(t)+\int_0^t \chi(t-t')\Upsilon(t')\,\mathrm{d}t'=0 \tag{1.23}$$

This equation is uniquely soluble for $t>0$, and for $t<0$, $\Upsilon(t)=0$, see Appendix F.8 for more details. In terms of the resolvent kernel, the inverse is

$$\boldsymbol{E}(t)=\frac{1}{\epsilon_0}\left\{\frac{1}{\epsilon}\boldsymbol{D}(t)+\int_{-\infty}^t \Upsilon(t-t')\boldsymbol{D}(t')\,\mathrm{d}t'\right\} \tag{1.24}$$

This relation is easily verified by inserting the constitutive relation into (1.24) and using the resolvent equation (1.23).

1.2.2 Examples

We exemplify the constitutive relations in this subsection. For convenience, the dispersion model is used. In a first example, we describe the Debye model or the relaxation model, which is used to model interaction of the electromagnetic fields in polar liquids, i.e., liquids with molecules with permanent electric dipole moment. In a second example, the Lorentz model or the resonance model is described. This model is often used for electromagnetic interaction in solids.

1.2.2.1 The Debye Model

The first example is the Debye model[23] for a dispersive material. This model is adequate for polar liquids, which have molecules with permanent electric dipole moment.

We assume the molecules (or the atoms if that is appropriate) have a permanent electric dipole moment $\boldsymbol{p}$, which normally is randomly oriented due to thermal fluctuations, see Figure 1.4 (left). The polarization of the material $\boldsymbol{P}$ is defined as the total electric dipole moment per unit volume, i.e.,

$$\boldsymbol{P}=\frac{\sum_i \boldsymbol{p}_i}{\Delta V} \quad \text{for a sufficiently small volume } \Delta V$$

where the sum over the index i is over all molecules located inside the volume ΔV. In an undisturbed state without electric field we have $\boldsymbol{P}=\boldsymbol{0}$, due to random orientation of the molecules. The polarization $\boldsymbol{P}$ changes due to two competing processes. The intensity of these processes depends strongly on many exterior parameters, e.g., the temperature T. In this treatment, we

[23] Peter Joseph William Debye (1884–1966), Dutch physicist and physical chemist.

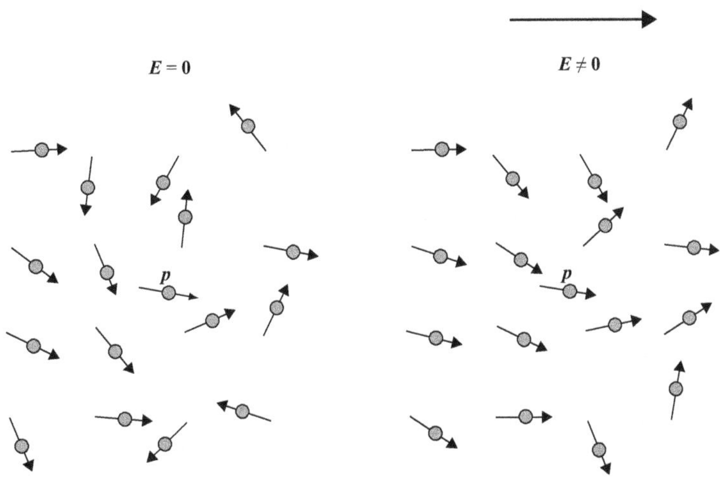

Figure 1.4 Molecules (circles) with permanent electric dipole moment (arrows) without (left) and with (right) an aligning exterior electric field $\boldsymbol{E}$.

assume that these exterior parameters are fixed. The temperature dependence is investigated in Example 1.11 on page 51. The processes are:

1. One process is striving to align the polarization $\boldsymbol{P}$ parallel to the total electric field $\boldsymbol{E}$.[24] We assume that the rate of changes in $\boldsymbol{P}$ due to this process is proportional to the electric field $\boldsymbol{E}$, more precisely to $\epsilon_0 \alpha \boldsymbol{E}$. The frequency $\alpha > 0$ is a measure of this change.
2. A second competing process is trying to disorient the polarization by collisions. If $\tau > 0$ is the relaxation time[25] for this process, the rate of changes in $\boldsymbol{P}$ is assumed to be proportional to $-\boldsymbol{P}/\tau$. The origin of this assumption is found in thermodynamics—a change in the orientation of the dipoles has a probability per unit time $1/\tau$, and the more oriented the medium is (higher polarization), the higher probability to undergo a change. More details are found in, e.g., References 5, 225.

The Debye model for molecules with permanent electric dipole moment is illustrated in Figure 1.4.

In total, the rate of changes in the polarization $\boldsymbol{P}$ is then

$$\frac{\mathrm{d}\boldsymbol{P}(t)}{\mathrm{d}t} = \epsilon_0 \alpha \boldsymbol{E}(t) - \frac{\boldsymbol{P}(t)}{\tau} \tag{1.25}$$

[24]Strictly speaking, the electric field $\boldsymbol{E}$, which affects the dipoles, is usually not identical to the exterior field, but it is modified due to the presence of the media. Suitable corrections for the difference between this local field and the exterior field can be made, i.e., the Clausius–Mossotti (Lorenz–Lorentz) law, named after Rudolf Julius Emanuel Clausius (1822–1888), German physicist and mathematician, and Ottaviano-Fabrizio Mossotti (1791–1863), Italian physicist.

[25]Typical values of the relaxation time τ lie in the interval 10^{-9}–10^{-11} s.

This equation models the interaction of the electromagnetic fields with the charges of the material—in this case the permanent electric dipoles of the material. We also call this equation the dynamics of the charges in the material.

From the previous section, we know that the dispersive effects of the material can be written as, see (1.19) and (1.8) (notice that we are using the dispersion model)

$$\boldsymbol{P}(t) = \boldsymbol{D}(t) - \epsilon_0 \boldsymbol{E}(t) = \epsilon_0 \left\{ (\epsilon - 1)\boldsymbol{E}(t) + \int_{-\infty}^{t} \chi(t - t')\boldsymbol{E}(t')\, \mathrm{d}t' \right\}$$

The time derivative of this expression is

$$\frac{1}{\epsilon_0}\frac{\mathrm{d}\boldsymbol{P}(t)}{\mathrm{d}t} = (\epsilon - 1)\frac{\mathrm{d}\boldsymbol{E}(t)}{\mathrm{d}t} + \chi(0^+)\boldsymbol{E}(t) + \int_{-\infty}^{t} \frac{\partial \chi(t - t')}{\partial t}\boldsymbol{E}(t')\, \mathrm{d}t'$$

where $\chi'(t)$ denotes the time derivative of $\chi(t)$, and $\chi(0^+) = \lim_{t\to 0, t>0} \chi(t)$, which might be non-zero, i.e., $\chi(t)$ is discontinuous at zero. Insert in (1.25) and collect terms. We get

$$\begin{aligned} &(\epsilon - 1)\frac{\mathrm{d}\boldsymbol{E}(t)}{\mathrm{d}t} + \left(\chi(0^+) - \alpha + \frac{1}{\tau}(\epsilon - 1)\right)\boldsymbol{E}(t) \\ &\quad + \int_{-\infty}^{t} \left(\frac{\partial \chi(t - t')}{\partial t} + \frac{1}{\tau}\chi(t - t')\right)\boldsymbol{E}(t')\, \mathrm{d}t' = \boldsymbol{0} \end{aligned}$$

The field $\boldsymbol{E}$ is arbitrary, which implies that the coefficient in front of each term must be zero, i.e.,

$$\begin{cases} \epsilon - 1 = 0 \\ \chi(0^+) - \alpha + \dfrac{1}{\tau}(\epsilon - 1) = 0 \\ \chi'(t) + \dfrac{1}{\tau}\chi(t) = 0 \end{cases}$$

If we simplify, we get

$$\begin{cases} \epsilon = 1 \\ \chi(0^+) = \alpha \\ \chi'(t) + \dfrac{1}{\tau}\chi(t) = 0 \end{cases}$$

The first condition, $\epsilon = 1$, implies that there is no optical response, and from the last two conditions on $\chi(t)$, we easily get

$$\chi(t) = \alpha \mathrm{e}^{-t/\tau}, \qquad t \geq 0$$

The final expressions for the Debye model are

$$\begin{cases} \boldsymbol{D}(t) = \epsilon_0 \left\{ \boldsymbol{E}(t) + \displaystyle\int_{-\infty}^{t} \chi(t - t')\boldsymbol{E}(t')\, \mathrm{d}t' \right\} \\ \chi(t) = \mathrm{H}(t)\alpha \mathrm{e}^{-t/\tau} \end{cases} \tag{1.26}$$

The susceptibility function $\chi(t)$ for this model is an exponentially decreasing function.

Example 1.3
The constitutive relations in (1.26) are given in the dispersion model, see Example 1.1, i.e.,

$$\begin{cases} \boldsymbol{D}(t) = \epsilon_0 \left\{ \boldsymbol{E}(t) + \int_{-\infty}^{t} \mathrm{e}^{-(t-t')/\tau} \boldsymbol{E}(t')\, \mathrm{d}t' \right\} \\ \boldsymbol{J}(t) = \boldsymbol{0} \end{cases}$$

In Example 1.2, we showed that with a special choice of the function $f(t)$, all effects on the charges in the material could be included in the Ohm's law. To rewrite the constitutive relations (1.26) in the conductivity model, we choose $f(t) = -\chi(t)$. In this case, we get

$$f(t) = -\mathrm{H}(t)\alpha \mathrm{e}^{-t/\tau}$$

and

$$\begin{cases} f(0^+) = -\alpha \\ \dfrac{\partial f(t)}{\partial t} = \dfrac{\alpha}{\tau} \mathrm{e}^{-t/\tau}, \qquad t > 0 \end{cases}$$

Example 1.2 now gives

$$\begin{cases} \boldsymbol{D}(t) = \epsilon_0 \boldsymbol{E}(t) \\ \boldsymbol{J}(t) = \epsilon_0 \alpha \boldsymbol{E}(t) - \epsilon_0 \dfrac{\alpha}{\tau} \int_{-\infty}^{t} \mathrm{e}^{-(t-t')/\tau} \boldsymbol{E}(t')\, \mathrm{d}t' \end{cases}$$

which is the constitutive relations for a Debye material in the conductivity model, see (1.22). ■

1.2.2.2 The Lorentz Model

We assume the material has bound charges (usually electrons), which interact with the nucleus of the atom. The atoms are usually arranged in a lattice structure, but they can also be randomly oriented and have a random location like in amorphous materials.

The charges have charge q and mass m and they are assumed to be affected by three different forces:

1. An electric force $\boldsymbol{F}_1 = q\boldsymbol{E}$ from the total electric field $\boldsymbol{E}$.[26]
2. A restoring force proportional to the displacement of the charge from its equilibrium, i.e., a harmonic force, $\boldsymbol{F}_2 = -m\omega_0^2 \boldsymbol{r}$, where $\omega_0 \geq 0$ is the harmonic frequency, and $\boldsymbol{r}$ is the displacement of the charge from its equilibrium.
3. A frictional force proportional to the velocity, $\boldsymbol{r}'(t)$, of the charge $\boldsymbol{F}_3 = -m\nu \boldsymbol{r}'(t)$, where $\nu \geq 0$ is the collision frequency.[27]

---

[26] See comment under Footnote 24 on page 21. If the medium is not dense, e.g., a gas, the difference is small.
[27] Typical values of ν lie in the interval 10^7–10^9 s^{-1}.

We assume that the dynamics of the charges is described by the laws of classic mechanics. The Newton acceleration law gives

$$m\frac{\mathrm{d}^2\boldsymbol{r}}{\mathrm{d}t^2} = \boldsymbol{F}_1 + \boldsymbol{F}_2 + \boldsymbol{F}_3 = q\boldsymbol{E} - m\omega_0^2\boldsymbol{r} - m\nu\frac{\mathrm{d}\boldsymbol{r}}{\mathrm{d}t}$$

or

$$\frac{\mathrm{d}^2\boldsymbol{r}}{\mathrm{d}t^2} + \nu\frac{\mathrm{d}\boldsymbol{r}}{\mathrm{d}t} + \omega_0^2\boldsymbol{r} = \frac{q}{m}\boldsymbol{E}$$

Introduce the polarization $\boldsymbol{P}$ of the material defined by

$$\boldsymbol{P} = Nq\boldsymbol{r}$$

where N is the number charges per unit volume.[28] Rewrite the dynamics in the polarization vector as

$$\frac{\mathrm{d}^2\boldsymbol{P}(t)}{\mathrm{d}t^2} + \nu\frac{\mathrm{d}\boldsymbol{P}(t)}{\mathrm{d}t} + \omega_0^2\boldsymbol{P}(t) = \frac{Nq^2}{m}\boldsymbol{E}(t) \tag{1.27}$$

We notice that this is differential equation of the second order as compared to the Debye model which was a first-order equation. More details on the interpretations of the various terms in both these models are found in Reference 226.

In the same way as for the Debye model, we introduce the dispersion of the material as (1.19) and (1.8)

$$\boldsymbol{P}(t) = \boldsymbol{D}(t) - \epsilon_0\boldsymbol{E}(t) = \epsilon_0\left\{(\epsilon - 1)\boldsymbol{E}(t) + \int_{-\infty}^{t}\chi(t - t')\boldsymbol{E}(t')\,\mathrm{d}t'\right\}$$

The first and second derivative of this expression is

$$\frac{1}{\epsilon_0}\frac{\mathrm{d}\boldsymbol{P}(t)}{\mathrm{d}t} = (\epsilon - 1)\frac{\mathrm{d}\boldsymbol{E}(t)}{\mathrm{d}t} + \chi(0^+)\boldsymbol{E}(t) + \int_{-\infty}^{t}\chi'(t - t')\boldsymbol{E}(t')\,\mathrm{d}t'$$

$$\frac{1}{\epsilon_0}\frac{\mathrm{d}^2\boldsymbol{P}(t)}{\mathrm{d}t^2} = (\epsilon - 1)\frac{\mathrm{d}^2\boldsymbol{E}(t)}{\mathrm{d}t^2} + \chi(0^+)\frac{\mathrm{d}\boldsymbol{E}(t)}{\mathrm{d}t} + \chi'(0^+)\boldsymbol{E}(t) + \int_{-\infty}^{t}\chi''(t - t')\boldsymbol{E}(t')\,\mathrm{d}t'$$

[28] We assume this quantity is constant in time, which is an approximation.

where $\chi'(t)$ and $\chi''(t)$ are the first and second derivative of $\chi(t)$ w.r.t. time, respectively, and $\chi(0^+) = \lim_{t\to 0, t>0} \chi(t)$ and $\chi'(0^+) = \lim_{t\to 0, t>0} \chi'(t)$, which might be non-zero. Enter these expressions in the dynamics equation, (1.27), and collect the terms. We get

$$
\begin{aligned}
&(\epsilon - 1)\frac{\mathrm{d}^2\boldsymbol{E}(t)}{\mathrm{d}t^2} + \left(\chi(0^+) + \nu(\epsilon - 1)\right)\frac{\mathrm{d}\boldsymbol{E}(t)}{\mathrm{d}t} \\
&+ \left(\chi'(0^+) + \nu\chi(0^+) + \omega_0^2(\epsilon - 1) - \frac{Nq^2}{m\epsilon_0}\right)\boldsymbol{E}(t) \\
&+ \int_{-\infty}^{t} \left(\chi''(t-t') + \nu\chi'(t-t') + \omega_0^2\chi(t-t')\right)\boldsymbol{E}(t')\,\mathrm{d}t' = \boldsymbol{0}
\end{aligned}
$$

The field $\boldsymbol{E}$ is arbitrary, and therefore the coefficients in front of every term must be identically zero, i.e.,

$$
\begin{cases}
\epsilon - 1 = 0 \\
\chi(0^+) + \nu\,(\epsilon - 1) = 0 \\
\chi'(0^+) + \nu\chi(0^+) + \omega_0^2\,(\epsilon - 1) - \omega_{\mathrm{p}}^2 = 0 \\
\chi''(t) + \nu\chi'(t) + \omega_0^2\chi(t) = 0
\end{cases}
$$

where

$$
\omega_{\mathrm{p}} = \sqrt{\frac{Nq^2}{m\epsilon_0}}
$$

is the plasma frequency of the material. We get

$$
\begin{cases}
\epsilon = 1 \\
\chi(0^+) = 0 \\
\chi'(0^+) = \omega_{\mathrm{p}}^2 \\
\chi''(t) + \nu\chi'(t) + \omega_0^2\chi(t) = 0
\end{cases}
$$

The first expression, $\epsilon = 1$, indicates that we have no optical response, which is expected since there are no unresolved processes in the material. The other conditions on $\chi(t)$ are an initial value problem for a second-order ordinary differential equation (ODE) in time with the unique solution

$$
\chi(t) = \frac{\omega_{\mathrm{p}}^2}{\nu_0}\mathrm{e}^{-\frac{\nu t}{2}}\sin\nu_0 t, \qquad t \geq 0
$$

where $\nu_0^2 = \omega_0^2 - \nu^2/4$. Typical values on the collision frequency ν are in the interval 10^7–10^9 Hz.

The final expression of the constitutive relations for Lorentz model is

$$
\begin{cases}
\boldsymbol{D}(t) = \epsilon_0\left\{\boldsymbol{E}(t) + \displaystyle\int_{-\infty}^{t} \chi(t-t')\boldsymbol{E}(t')\,\mathrm{d}t'\right\} \\
\chi(t) = \mathrm{H}(t)\dfrac{\omega_{\mathrm{p}}^2}{\nu_0}\mathrm{e}^{-\frac{\nu t}{2}}\sin\nu_0 t
\end{cases}
\tag{1.28}
$$

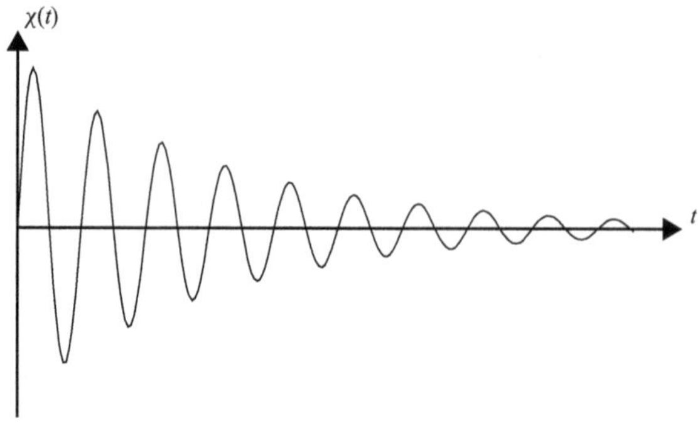

Figure 1.5 An example of a susceptibility function $\chi(t)$ for Lorentz model. The scales on the axes are in arbitrary units.

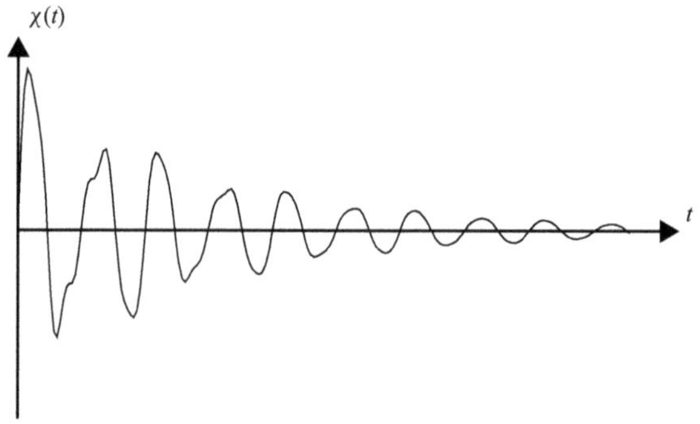

Figure 1.6 An example of a susceptibility function $\chi(t)$ for Lorentz model with three resonance frequencies. The scales on the axes are in arbitrary units.

where we explicitly have set $\chi(t) = 0$, $t < 0$ by introducing the Heaviside function $\mathrm{H}(t)$. An example of the susceptibility function for Lorentz model is given in Figure 1.5.

In a more general situation, several different processes contribute to the susceptibility function, where each contribution is a resonance model derived above. The processes have different frequencies ω_{p_i}, ω_{0i}, and ν_i. The susceptibility function for the general resonance model is a sum of all these contributions

$$\chi(t) = \mathrm{H}(t) \sum_{i=1}^{M} \frac{\omega_{\mathrm{p}_i}^2}{\sqrt{\omega_{0i}^2 - \nu_i^2/4}} \mathrm{e}^{-\frac{\nu_i t}{2}} \sin \sqrt{\omega_{0i}^2 - \nu_i^2/4}\, t$$

An example on a susceptibility function for Lorentz model with several frequencies is depicted in Figure 1.6.

If the friction on the charges can be neglected, i.e., $\nu \to 0$, then $\nu_0 = \omega_0$ and (1.28) is simplified into

$$\chi(t) = \mathrm{H}(t) \frac{\omega_{\mathrm{p}}^2}{\omega_0} \sin \omega_0 t$$

In this case, $\chi(t)$ is an undamped sinusoidal function.

On the other hand, if the restoring force can be neglected,[29] i.e., $\omega_0 \to 0$, then the frequency $\nu_0 = \mathrm{i}\nu/2$ and (1.28) becomes

$$\chi(t) = \mathrm{H}(t)\frac{\omega_\mathrm{p}^2}{\nu}\left(1 - \mathrm{e}^{-\nu t}\right)$$

Without the restoring force the electron[30] is a free conducting electron, which is not bound to any nucleus. In this case, it is possible to identify a conductivity σ of the material as in the Ohm's law, see (1.20) or more generally (1.22). We start with $\sigma = \Sigma(t) = 0$ in our conductivity model, i.e.,

$$\begin{cases} \boldsymbol{D}(t) = \epsilon_0 \epsilon \boldsymbol{E}(t) \\ \boldsymbol{J}(t) = \epsilon_0 \chi(0^+)\boldsymbol{E}(t) + \epsilon_0 \displaystyle\int_{-\infty}^{t} \frac{\partial \chi(t-t')}{\partial t} \boldsymbol{E}(t')\,\mathrm{d}t' \end{cases}$$

and we wish to identify the susceptibility function $\chi(t)$ with a conductivity in this equation. In our case, $\epsilon = 1$, $\chi(0^+) = 0$, and $\chi'(t) = \mathrm{H}(t)\omega_\mathrm{p}^2 \mathrm{e}^{-\nu t}$ and we get

$$\boldsymbol{J}(t) = \epsilon_0 \omega_\mathrm{p}^2 \int_{-\infty}^{t} \mathrm{e}^{-\nu(t-t')} \boldsymbol{E}(t')\,\mathrm{d}t'$$

If the field $\boldsymbol{E}(t')$ varies slowly on a time scale $1/\nu$, we can approximate the integral by evaluating the field $\boldsymbol{E}(t')$ at the time t and move the field outside the integral.[31] We get

$$\boldsymbol{J}(t) = \epsilon_0 \omega_\mathrm{p}^2 \boldsymbol{E}(t) \int_{-\infty}^{t} \mathrm{e}^{-\nu(t-t')}\,\mathrm{d}t' = \frac{\epsilon_0 \omega_\mathrm{p}^2}{\nu}\boldsymbol{E}(t)$$

A direct comparison in this special case gives the DC conductivity of the material

$$\sigma = \frac{\epsilon_0 \omega_\mathrm{p}^2}{\nu}$$

We immediately see that if the friction of the electrons increases (ν increases) this implies that the conductivity decreases (resistivity $= 1/\sigma$ increases), and vice versa. This observation is intuitively well motivated.

Example 1.4

Suppose a medium is described by the Lorentz model with negligible losses (collision frequency $\nu \approx 0$). The susceptibility function, which is a model of the dispersive effects in the medium, is then

$$\chi(t) = \mathrm{H}(t)\frac{\omega_p^2}{\omega_0}\sin \omega_0 t$$

---

[29]This special case is often denoted the Drude model, and it was an early model for conduction of charges in metals, see page 53.

[30]We assume that the charges in the material are electrons.

[31]This is very analogous to the argument used in Section 1.2.1.1 where we identified the optical response of the material.

In many practical situations, it is more convenient to work with constitutive relations that do not contain any convolution integrals, but only contain the field evaluated at the time t.

Integration by parts of the electric flux density $\boldsymbol{D}(t)$ can be expressed as (provided the series converges):

$$\begin{aligned} \boldsymbol{D}(t) &= \epsilon_0 \left\{ \boldsymbol{E}(t) + \int_{-\infty}^{t} \chi(t-t')\boldsymbol{E}(t')\,\mathrm{d}t' \right\} \\ &= \epsilon_0 \left\{ \boldsymbol{E}(t) + \frac{\omega_p^2}{\omega_0} \int_{-\infty}^{t} \sin\omega_0(t-t')\boldsymbol{E}(t')\,\mathrm{d}t' \right\} = \epsilon_0 \sum_{n=0}^{\infty} A_n \boldsymbol{E}^{(n)}(t) \end{aligned}$$

where

$$A_n = \begin{cases} 0, & \text{if } n \text{ is odd} \\ \delta_{n,0} + (-1)^{n/2} \dfrac{\omega_\mathrm{p}^2}{\omega_0^{n+2}}, & \text{if } n \text{ is even} \end{cases}$$

The first term $A_0 = 1 + \omega_\mathrm{p}^2/\omega_0^2 = \epsilon$ acts as an optical response. To estimate the error we have made if we replace the entire series by just the optical response, assume the electric field $\boldsymbol{E}(t)$ oscillates harmonically, say $\boldsymbol{E}(t) = \boldsymbol{E}_0 \cos(\omega t + \alpha)$, with angular frequency $\omega = 1 \times 10^{10}$ rad/s, and that the resonance frequency and the plasma frequency of the Lorentz material are $\omega_0 = 1 \times 10^{14}$ rad/s and $\omega_p = 3 \times 10^{14}$ rad/s, respectively. The relative error in the $\boldsymbol{D}(t)$ field if $\boldsymbol{D}(t) = \epsilon_0 \epsilon \boldsymbol{E}(t)$ is used instead of the complete expression is

$$\frac{\left|A_0\boldsymbol{E}(t) - \sum_{n=0}^{\infty} A_n \boldsymbol{E}^{(n)}(t)\right|}{\left|\sum_{n=0}^{\infty} A_n \boldsymbol{E}^{(n)}(t)\right|} = \frac{\frac{\omega_\mathrm{p}^2}{\omega_0^2} \sum_{n=1}^{\infty} \frac{\omega^{2n}}{\omega_0^{2n}}}{1 + \frac{\omega_\mathrm{p}^2}{\omega_0^2} + \frac{\omega_\mathrm{p}^2}{\omega_0^2} \sum_{n=1}^{\infty} \frac{\omega^{2n}}{\omega_0^{2n}}} = \frac{9\frac{10^{-8}}{1-10^{-8}}}{10 + 9\frac{10^{-8}}{1-10^{-8}}} \approx 9 \times 10^{-9}$$

The error that is made by approximating the convolution integral by an optical response $\epsilon = 10$ is negligible at this frequency ω. For a higher frequency, the error becomes much larger. Compare the result of this example with the definition and analysis of the optical response on page 15. ■

1.2.3 General Linear Media with Dispersion

So far, we have neglected magnetic effects and any coupling between electric and magnetic phenomena. If these effects are essential, we follow the same line of analysis as in Section 1.2.1. We do not give the details of this analysis here, but refer to the literature [118].

A general set of linear constitutive relations, which allows for coupling between the electric and the magnetic fields, is the following Ansatz:

$$\begin{cases} \boldsymbol{D}(t) = \epsilon_0 \left\{ \boldsymbol{\epsilon} \cdot \boldsymbol{E}(t) + \int_{-\infty}^{t} \boldsymbol{\chi}_{\mathrm{ee}}(t-t') \cdot \boldsymbol{E}(t')\, \mathrm{d}t' \right. \\ \qquad \left. + \eta_0 \int_{-\infty}^{t} \boldsymbol{\chi}_{\mathrm{em}}(t-t') \cdot \boldsymbol{H}(t')\, \mathrm{d}t' \right\} \\ \boldsymbol{B}(t) = \dfrac{1}{c_0} \left\{ \int_{-\infty}^{t} \boldsymbol{\chi}_{\mathrm{me}}(t-t') \cdot \boldsymbol{E}(t')\, \mathrm{d}t' \right. \\ \qquad \left. + \eta_0 \boldsymbol{\mu} \cdot \boldsymbol{H}(t) + \eta_0 \int_{-\infty}^{t} \boldsymbol{\chi}_{\mathrm{mm}}(t-t') \cdot \boldsymbol{H}(t')\, \mathrm{d}t' \right\} \end{cases} \tag{1.29}$$

where the optical response is modeled by the two dyadics $\boldsymbol{\epsilon}$ and $\boldsymbol{\mu}$, and the dispersive effects by the dyadic-valued susceptibility functions $\boldsymbol{\chi}_{kk'}(t)$, $k, k' = \mathrm{e}, \mathrm{m}$. In Cartesian components ($i = 1, 2, 3$), we get

$$D_i(t) = \epsilon_0 \sum_{j=1}^{3} \left\{ \epsilon_{ij} E_j(t) + \int_{-\infty}^{t} \chi_{\mathrm{ee}\,ij}(t-t') E_j(t')\, \mathrm{d}t' + \eta_0 \int_{-\infty}^{t} \chi_{\mathrm{em}\,ij}(t-t') H_j(t')\, \mathrm{d}t' \right\}$$

and

$$B_i(t) = \frac{1}{c_0} \sum_{j=1}^{3} \left\{ \int_{-\infty}^{t} \chi_{\mathrm{me}\,ij}(t-t') E_j(t')\, \mathrm{d}t' \right.$$
$$\left. + \eta_0 \mu_{ij} H_j(t) + \eta_0 \int_{-\infty}^{t} \chi_{\mathrm{mm}\,ij}(t-t') H_j(t')\, \mathrm{d}t' \right\}$$

For convenience, we have here introduced the wave impedance and the velocity of light in vacuum $\eta_0 = \sqrt{\mu_0/\epsilon_0}$ and $c_0 = \sqrt{1/\epsilon_0\mu_0}$, respectively. In this way, all susceptibility kernels $\chi_{kk'\,ij}(t)$, $k, k' = \mathrm{e}, \mathrm{m}$, $i, j = 1, 2, 3$, all have the same unit frequency, and the optical responses ϵ_{ij} and μ_{ij} are dimensionless.

The Ansatz in (1.29) is very well motivated by the analysis of isotropic media in Section 1.2.1. We are referring to this general Ansatz as the constitutive relations for dispersive bianisotropic media. The dispersion of the material is modeled by four (generalized) dyadic-valued susceptibility functions, $\boldsymbol{\chi}_{kk'}(t)$, $k, k' = \mathrm{e}, \mathrm{m}$, each, in general, having nine independent components. The susceptibility functions all have the unit frequency. The optical response of the material is modeled by $\boldsymbol{\epsilon}$ and $\boldsymbol{\mu}$, which both are dimensionless dyadics. Notice that there are no optical response between the $\boldsymbol{D}$ and $\boldsymbol{H}$ and between $\boldsymbol{B}$ and $\boldsymbol{E}$, see Reference 118.

A material is classified w.r.t. its constitutive relations as in Table 1.1, see also Figure 1.7. Media for which $\boldsymbol{\chi}_{\mathrm{em}}(t) = \boldsymbol{\chi}_{\mathrm{me}}(t) = 0$ are called dispersive anisotropic materials, and media where all quantities are proportional to the unit dyadic $\mathbf{I}_3$ are called dispersive isotropic or dispersive biisotropic materials.

In a bianisotropic material, (1.29), the electric and magnetic effects are coupled in a way that is more general than in an anisotropic material. This coupling between the electric and magnetic effects is modeled by the dyadic-valued functions $\boldsymbol{\chi}_{\mathrm{em}}(t)$ and $\boldsymbol{\chi}_{\mathrm{me}}(t)$. In this way, a magnetic field $\boldsymbol{H}$ gives rise to an electric flux density $\boldsymbol{D}$ or equivalently an electric polarization

Table 1.1 Classification of media w.r.t. their response to electromagnetic fields and their constitutive relations.

Type	$\epsilon, \chi_{ee}, \mu, \chi_{mm}$	χ_{em}, χ_{me}
Isotropic	All $\sim \mathbf{I}_3$	Both $\mathbf{0}$
Anisotropic	Some not $\sim \mathbf{I}_3$	Both $\mathbf{0}$
Biisotropic	All $\sim \mathbf{I}_3$	Both $\sim \mathbf{I}_3$
Bianisotropic	All other cases	

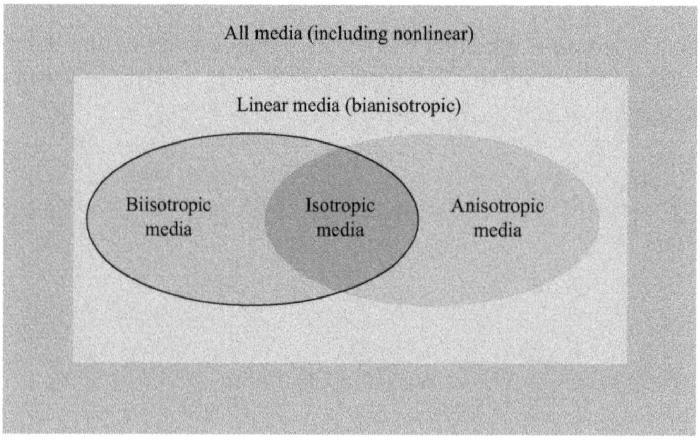

Figure 1.7 Schematic classification of media w.r.t. their constitutive relations. The area of each class is not proportional to the size of the class.

P in the material. Conversely, an electric field ***E*** gives rise to a magnetization ***M*** in the material. This is not possible in an anisotropic material.

Already during the nineteenth century, optical active media were known for optical frequencies. The properties of these media can be explained with bianisotropic or biisotropic constitutive relations.[32] There are several examples of optical active media, e.g., quartz and different sugar solutions. About 1960 Russian scientists discovered that several magnetic crystals, e.g., Cr_2O_3, have similar properties. These materials are referred to as magneto-electric material. Similar properties are desirable in the microwave regime, but then they have to be manufactured.

---

[32] Media, whose constitutive components are not identical to their mirror image, show bianisotropic effects. In nature, materials that are not invariant under mirror reflection occur both on microscopic (molecular) and macroscopic level. Interestingly, there is a dominance of right-handed-oriented natural material. On the microscopic level, we find such examples in e.g., the chromosomes, and on the macroscopic level in e.g., shells and vines.

At this stage, it is convenient to employ a six-dimensional notation, as an alternative to the more commonly used three-dimensional formulation. This notation makes the expressions very compact. To this end, define[33]

$$\mathsf{e}(\boldsymbol{r},t)=\begin{pmatrix}\boldsymbol{E}(\boldsymbol{r},t)\\ \eta_0\boldsymbol{H}(\boldsymbol{r},t)\end{pmatrix},\qquad \mathsf{d}(\boldsymbol{r},t)=c_0\begin{pmatrix}\eta_0\boldsymbol{D}(\boldsymbol{r},t)\\ \boldsymbol{B}(\boldsymbol{r},t)\end{pmatrix}\tag{1.30}$$

In this notation, the Maxwell equations without sources, (1.1) and (1.2) have the form

$$\mathsf{D}\cdot\mathsf{e}(\boldsymbol{r},t)=\frac{1}{c_0}\frac{\partial}{\partial t}\mathsf{d}(\boldsymbol{r},t)$$

where the Maxwell operator D is defined as[34]

$$\begin{aligned}\mathsf{D}\cdot\mathsf{e}(\boldsymbol{r},t)&=\begin{pmatrix}\mathbf{0} & \nabla\times\mathbf{I}_3\\ -\nabla\times\mathbf{I}_3 & \mathbf{0}\end{pmatrix}\begin{pmatrix}\boldsymbol{E}(\boldsymbol{r},t)\\ \eta_0\boldsymbol{H}(\boldsymbol{r},t)\end{pmatrix}\\ &=\begin{pmatrix}\mathbf{0} & \eta_0\nabla\times\boldsymbol{H}(\boldsymbol{r},t)\\ -\nabla\times\boldsymbol{E}(\boldsymbol{r},t) & \mathbf{0}\end{pmatrix}\end{aligned}$$

We also define two six-dimensional dyadics or matrices

$$\mathsf{A}=\begin{pmatrix}\boldsymbol{\epsilon} & \mathbf{0}\\ \mathbf{0} & \boldsymbol{\mu}\end{pmatrix},\qquad \mathsf{M}(\boldsymbol{r},t)=\begin{pmatrix}\boldsymbol{\chi}_{\mathrm{ee}}(\boldsymbol{r},t) & \boldsymbol{\chi}_{\mathrm{em}}(\boldsymbol{r},t)\\ \boldsymbol{\chi}_{\mathrm{me}}(\boldsymbol{r},t) & \boldsymbol{\chi}_{\mathrm{mm}}(\boldsymbol{r},t)\end{pmatrix}\tag{1.31}$$

where A is the six-dimensional optical response.

The constitutive relations in (1.29) now read at each fixed point $\boldsymbol{r}$ in space

$$\mathsf{d}(t)=\mathsf{A}\cdot\mathsf{e}(t)+\int_{-\infty}^{t}\mathsf{M}(t-t')\cdot\mathsf{e}(t')\,\mathrm{d}t'=\mathsf{A}\cdot\mathsf{e}(t)+\int_{0}^{\infty}\mathsf{M}(t')\cdot\mathsf{e}(t-t')\,\mathrm{d}t'\tag{1.32}$$

Example 1.5

An important, non-trivial, example of a more general constitutive relation is the cold plasma that is developed in Problem 1.5, under the influence of a static magnetic flux density $\boldsymbol{B}=B_0\hat{\boldsymbol{z}}$. The result is

$$\begin{aligned}\boldsymbol{\chi}(t)=\frac{\omega_{\mathrm{p}}^2}{\nu^2+\omega_{\mathrm{g}}^2}\mathrm{H}(t)\Big\{&\mathbf{I}_2\big(\omega_{\mathrm{g}}\mathrm{e}^{-\nu t}\sin\omega_{\mathrm{g}}t+\nu\big(1-\mathrm{e}^{-\nu t}\cos\omega_{\mathrm{g}}t\big)\big)\\ &-\mathbf{J}\left(\omega_{\mathrm{g}}\big(1-\mathrm{e}^{-\nu t}\cos\omega_{\mathrm{g}}t\big)-\nu\mathrm{e}^{-\nu t}\sin\omega_{\mathrm{g}}t\right)+\hat{\boldsymbol{z}}\hat{\boldsymbol{z}}\frac{\nu^2+\omega_{\mathrm{g}}^2}{\nu}\left(1-\mathrm{e}^{-\nu t}\right)\Big\}\end{aligned}$$

---

[33]Six-dimensional vectors and dyadics are written in sans serif boldface to distinguish them from the ordinary three-dimensional vectors, see also Appendix I for notation.

[34]The identity dyadic in three dimensions is denoted $\mathbf{I}_3$, and in two dimensions (the x–y-plane), it is denoted $\mathbf{I}_2$. The concept of a dyadic is summarized in Appendix A.

where ν is the collision frequency, $\mathbf{I}_2$ is the identity dyadic in the x–y-plane, and $\mathbf{J} = \hat{z} \times \mathbf{I}_2$, which is a rotation of $\pi/2$ along the z-axis, and the gyrotropic frequency, ω_g, and ω_p, the plasma frequency of the material, are defined by

$$\omega_g = \frac{qB_0}{m}, \qquad \omega_p = \sqrt{\frac{Nq^2}{m\epsilon_0}}$$

and N is the number of charges per unit of volume in the plasma, and m and q are the mass and the charge of the charges, respectively.

In the limit of vanishing collision frequency, $\nu \to 0$, these expressions become

$$\chi(t) = \frac{\omega_p^2}{\omega_g} \mathrm{H}(t) \left\{ \mathbf{I}_2 \sin \omega_g t - \mathbf{J} \left(1 - \cos \omega_g t\right) + \hat{z}\hat{z}\omega_g t \right\}$$

On the other hand, vanishing gyrotropic frequency, $\omega_g \to 0$, leads to a recovery of Lorentz model with vanishing restoring force, i.e.,

$$\chi(t) = \frac{\omega_p^2}{\nu} \mathrm{H}(t) \left(1 - e^{-\nu t}\right) \mathbf{I}_3$$

■

Example 1.6
In 1937, Condon[35] suggested a constitutive relation that models biisotropic (chiral or optical activity) effects [41, 42]. In this textbook, the constitutive relations for a biisotropic medium are

$$\begin{aligned} \boldsymbol{D}(t) &= \epsilon_0 \left\{ \epsilon \boldsymbol{E}(t) + \int_{-\infty}^{t} \chi_{ee}(t - t')\boldsymbol{E}(t')\,dt' + \eta_0 \int_{-\infty}^{t} \chi_{em}(t - t')\boldsymbol{H}(t')\,dt' \right\} \\ \boldsymbol{B}(t) &= \frac{1}{c_0} \left\{ \int_{-\infty}^{t} \chi_{me}(t - t')\boldsymbol{E}(t')\,dt' + \eta_0 \mu \boldsymbol{H}(t) + \eta_0 \int_{-\infty}^{t} \chi_{mm}(t - t')\boldsymbol{H}(t')\,dt' \right\} \end{aligned}$$

Condon generalized Lorentz model by adding a force proportional to the time derivative of the magnetic field $\boldsymbol{H}$. Using the notation in Section 1.2.2.2, the equation of the polarization $\boldsymbol{P}$ is

$$\frac{d^2}{dt^2}\boldsymbol{P} + \nu \frac{d}{dt}\boldsymbol{P} + \omega_0^2 \boldsymbol{P} = \epsilon_0 \left(\omega_p^2 \boldsymbol{E} + \omega_c \eta_0 \frac{\partial \boldsymbol{H}}{\partial t} \right) \tag{1.33}$$

where $\omega_p = \sqrt{\frac{Nq^2}{m\epsilon_0}}$ and ω_c are the plasma frequency and a frequency modeling the optical activity of the material, respectively.

[35] Edward Uhler Condon (1902–1974), American physicist.

Proceed as in Section 1.2.2.2 by differentiating the polarization $\boldsymbol{P}$. We get

$$\frac{1}{\epsilon_0}\frac{\mathrm{d}\boldsymbol{P}(t)}{\mathrm{d}t} = (\epsilon - 1)\frac{\mathrm{d}\boldsymbol{E}(t)}{\mathrm{d}t} + \chi_{\mathrm{ee}}(0^+)\boldsymbol{E}(t) + \int_{-\infty}^{t} \chi'_{\mathrm{ee}}(t-t')\boldsymbol{E}(t')\,\mathrm{d}t'$$
$$+ \eta_0\chi_{\mathrm{em}}(0^+)\boldsymbol{H}(t) + \eta_0\int_{-\infty}^{t} \chi'_{\mathrm{em}}(t-t')\boldsymbol{H}(t')\,\mathrm{d}t'$$

and

$$\frac{1}{\epsilon_0}\frac{\mathrm{d}^2\boldsymbol{P}(t)}{\mathrm{d}t^2} = (\epsilon - 1)\frac{\mathrm{d}^2\boldsymbol{E}(t)}{\mathrm{d}t^2} + \chi_{\mathrm{ee}}(0^+)\frac{\mathrm{d}\boldsymbol{E}(t)}{\mathrm{d}t} + \chi'_{\mathrm{ee}}(0^+)\boldsymbol{E}(t) + \int_{-\infty}^{t} \chi''_{\mathrm{ee}}(t-t')\boldsymbol{E}(t')\,\mathrm{d}t'$$
$$+ \eta_0\chi_{\mathrm{em}}(0^+)\frac{\mathrm{d}\boldsymbol{H}(t)}{\mathrm{d}t} + \eta_0\chi'_{\mathrm{em}}(0^+)\boldsymbol{H}(t) + \eta_0\int_{-\infty}^{t} \chi''_{\mathrm{em}}(t-t')\boldsymbol{H}(t')\,\mathrm{d}t'$$

Insert into the equation of dynamics, (1.33), and collect terms

$$(\epsilon - 1)\frac{\mathrm{d}^2\boldsymbol{E}(t)}{\mathrm{d}t^2} + \left(\chi_{\mathrm{ee}}(0^+) + \nu(\epsilon - 1)\right)\frac{\mathrm{d}\boldsymbol{E}(t)}{\mathrm{d}t}$$
$$+ \left(\chi'_{\mathrm{ee}}(0^+) + \nu\chi_{\mathrm{ee}}(0^+) + \omega_0^2(\epsilon - 1) - \omega_{\mathrm{p}}^2\right)\boldsymbol{E}(t)$$
$$+ \int_{-\infty}^{t} \left(\chi''_{\mathrm{ee}}(t-t') + \nu\chi'_{\mathrm{ee}}(t-t') + \omega_0^2\chi_{\mathrm{ee}}(t-t')\right)\boldsymbol{E}(t')\,\mathrm{d}t'$$
$$+ \eta_0\left(\chi_{\mathrm{em}}(0^+) - \omega_{\mathrm{c}}\right)\frac{\mathrm{d}\boldsymbol{H}(t)}{\mathrm{d}t} + \eta_0\left(\chi'_{\mathrm{em}}(0^+) + \nu\chi_{\mathrm{em}}(0^+) + \omega_0^2\right)\boldsymbol{H}(t)$$
$$+ \eta_0\int_{-\infty}^{t} \left(\chi''_{\mathrm{em}}(t-t') + \nu\chi'_{\mathrm{em}}(t-t') + \omega_0^2\chi_{\mathrm{em}}(t-t')\right)\boldsymbol{H}(t')\,\mathrm{d}t' = \boldsymbol{0}$$

Treating the fields $\boldsymbol{E}$ and $\boldsymbol{H}$ as independent fields leads to

$$\begin{cases} \epsilon - 1 = 0 \\ \chi_{\mathrm{ee}}(0^+) + \nu(\epsilon - 1) = 0 \\ \chi'_{\mathrm{ee}}(0^+) + \nu\chi_{\mathrm{ee}}(0^+) + \omega_0^2(\epsilon - 1) - \omega_{\mathrm{p}}^2 = 0 \\ \chi''_{\mathrm{ee}}(t) + \nu\chi'_{\mathrm{ee}}(t) + \omega_0^2\chi_{\mathrm{ee}}(t) = 0 \end{cases} \qquad \begin{cases} \chi_{\mathrm{em}}(0^+) - \omega_{\mathrm{c}} = 0 \\ \chi'_{\mathrm{em}}(0^+) + \nu\chi_{\mathrm{em}}(0^+) = 0 \\ \chi''_{\mathrm{em}}(t) + \nu\chi'_{\mathrm{em}}(t) + \omega_0^2\chi_{\mathrm{em}}(t) = 0 \end{cases}$$

The solutions to these problems are

$$\begin{cases} \chi_{\mathrm{ee}}(t) = \mathrm{H}(t)\dfrac{\omega_{\mathrm{p}}^2}{\nu_0}e^{-\frac{\nu t}{2}}\sin\nu_0 t \\ \chi_{\mathrm{em}}(t) = \mathrm{H}(t)\omega_{\mathrm{c}}e^{-\frac{\nu t}{2}}\left(\cos\nu_0 t - \dfrac{\nu}{2\nu_0}\sin\nu_0 t\right) \end{cases}$$

where $\nu_0^2 = \omega_0^2 - \nu^2/4$. This analysis determines the susceptibility functions $\chi_{\mathrm{ee}}(t)$ and $\chi_{\mathrm{em}}(t)$. If the material is non-magnetic $\chi_{\mathrm{mm}}(t) = 0$, and if the material satisfies additional conditions, such as the reciprocity condition developed and analyzed in Section 1.3.5 on page 70, we have also $\chi_{\mathrm{me}}(t) = -\chi_{\mathrm{em}}(t)$. ■

We end the analysis on constitutive relations by inverting the relation between the electric flux density $\boldsymbol{D}(t)$ and the electric field $\boldsymbol{E}(t)$ for an anisotropic material. The constitutive relations, (1.29), for an anisotropic material are

$$\boldsymbol{D}(t) = \epsilon_0 \left\{ \boldsymbol{\epsilon} \cdot \boldsymbol{E}(t) + \int_{-\infty}^{t} \boldsymbol{\chi}(t - t') \cdot \boldsymbol{E}(t') \, \mathrm{d}t' \right\}$$

This relation can be inverted by applying the resolvent of the susceptibility dyadic $\boldsymbol{\chi}(t)$. The resolvent kernel $\boldsymbol{\Upsilon}(t)$ satisfies the resolvent equation

$$\boldsymbol{\epsilon}^{-1} \cdot \boldsymbol{\chi}(t) + \boldsymbol{\Upsilon}(t) \cdot \boldsymbol{\epsilon} + \int_{0}^{t} \left(\boldsymbol{\Upsilon}(t - t') \cdot \boldsymbol{\epsilon} \right) \cdot \left(\boldsymbol{\epsilon}^{-1} \cdot \boldsymbol{\chi}(t') \right) \mathrm{d}t' = \boldsymbol{0} \tag{1.34}$$

This solution is uniquely soluble for $t > 0$ and for $t < 0$, $\boldsymbol{\Upsilon}(t) = \boldsymbol{0}$, see Appendix F.8 on page 679 for more details. In terms of the resolvent kernel, the inverse is

$$\boldsymbol{E}(t) = \frac{1}{\epsilon_0} \left\{ \boldsymbol{\epsilon}^{-1} \cdot \boldsymbol{D}(t) + \int_{-\infty}^{t} \boldsymbol{\Upsilon}(t - t') \cdot \boldsymbol{D}(t') \, \mathrm{d}t' \right\} \tag{1.35}$$

This relation is easily verified by inserting the constitutive relation into (1.35) and using the resolvent equation (1.34).

1.2.4 Energy and Passivity

The classification of a passive material in this section follows closely Karlsson and Kristensson [118]. A general treatment of the subject is found in Reference 305.

In Section 1.1.2, we identified the power bound in the electromagnetic fields in the volume V as, see (1.15) on page 9

$$\iiint_V \left[\boldsymbol{H} \cdot \frac{\partial \boldsymbol{B}}{\partial t} + \boldsymbol{E} \cdot \frac{\partial \boldsymbol{D}}{\partial t} \right] \mathrm{d}v$$

The integrand defines the power density of the material, and the total energy stored per volume therefore is

$$w(\boldsymbol{r}, t) = \int_{-\infty}^{t} \left[\boldsymbol{H}(t') \cdot \frac{\partial \boldsymbol{B}(t')}{\partial t'} + \boldsymbol{E}(t') \cdot \frac{\partial \boldsymbol{D}(t')}{\partial t'} \right] \mathrm{d}t'$$

To proceed, it is convenient to employ the six-dimensional notation in (1.30)–(1.32). In this notation, the Poynting theorem (1.14) (without sources, dispersive model) reads

$$\nabla \cdot \boldsymbol{S}(\boldsymbol{r}, t) + \epsilon_0 \mathrm{e}(\boldsymbol{r}, t) \cdot \frac{\partial \mathrm{d}(\boldsymbol{r}, t)}{\partial t} = 0$$

and the total energy stored per volume at a point $\boldsymbol{r}$ is

$$w(t) = \epsilon_0 \int_{-\infty}^{t} \mathrm{e}(t') \cdot \frac{\partial \mathrm{d}(t')}{\partial t'} \, \mathrm{d}t' = w_{\mathrm{em}}(t) + w_{\mathrm{d}}(t) \tag{1.36}$$

where the instantaneous part of the total energy stored is

$$
\begin{aligned}
w_{\mathrm{em}}(t) &= \epsilon_0 \int_{-\infty}^{t} \mathrm{e}(t') \cdot \left(\mathsf{A} \cdot \frac{\partial \mathrm{e}(t')}{\partial t'}\right) \mathrm{d}t' \\
&= \frac{\epsilon_0}{2} \mathrm{e}(t) \cdot (\mathsf{A} \cdot \mathrm{e}(t)) + \frac{\epsilon_0}{2} \int_{-\infty}^{t} \mathrm{e}(t') \cdot \left(\left(\mathsf{A} - \mathsf{A}^{\mathrm{t}}\right) \cdot \frac{\partial \mathrm{e}(t')}{\partial t'}\right) \mathrm{d}t'
\end{aligned}
\tag{1.37}
$$

and the dispersive part of the total energy stored is

$$
\begin{aligned}
w_{\mathrm{d}}(t) &= \epsilon_0 \int_{-\infty}^{t} \mathrm{e}(t') \cdot \int_{0}^{\infty} \mathsf{M}(t'') \cdot \frac{\partial \mathrm{e}(t' - t'')}{\partial t'} \, \mathrm{d}t'' \, \mathrm{d}t' \\
&= \epsilon_0 \int_{-\infty}^{t} \mathrm{e}(t') \cdot \frac{\partial}{\partial t'} \left(\int_{-\infty}^{t'} \mathsf{M}(t' - t'') \cdot \mathrm{e}(t'') \, \mathrm{d}t'' \right) \mathrm{d}t' \\
&= \epsilon_0 \int_{-\infty}^{t} \mathrm{e}(t') \cdot \left(\mathsf{M}(0) \cdot \mathrm{e}(t') + \int_{-\infty}^{t'} \mathsf{M}'(t' - t'') \cdot \mathrm{e}(t'') \, \mathrm{d}t'' \right) \mathrm{d}t'
\end{aligned}
\tag{1.38}
$$

where prime denotes differentiation w.r.t. time.

The medium is classified as a passive material at the fixed point $\boldsymbol{r}$, if

$$
w(t) = w_{\mathrm{em}}(t) + w_{\mathrm{d}}(t) \geq 0 \tag{1.39}
$$

for every continuously differentiable, compactly supported vector $\mathrm{e}(t)$.

The motivation behind this definition is clear from the Poynting theorem, which implies that a passive material satisfies

$$
-\int_{-\infty}^{t} \iint_{S_r} \boldsymbol{S}(\boldsymbol{r}, t') \cdot \hat{\boldsymbol{\nu}} \, \mathrm{d}S \, \mathrm{d}t' \geq 0
$$

for all electromagnetic fields $\mathrm{e}(t)$. Here, S_r is the bounding surface of an arbitrary, open neighborhood V_r of the point $\boldsymbol{r}$. This integral states that the power is always consumed inside S_r for all excitations.

We apply this definition to a continuously differentiable field e, which satisfies the condition $\mathrm{e}(t) = 0$, and moreover, each component assumes non-zero values only in a neighborhood of $\tau < t$, more precisely, $\mathrm{e}(t) = 0$, $t \notin [\tau - \epsilon, \tau + \epsilon]$, where ϵ is sufficiently small. From (1.37) and (1.38), the total energy stored per volume is

$$
\begin{aligned}
w(t) = &\frac{\epsilon_0}{2} \int_{\tau-\epsilon}^{\tau+\epsilon} \mathrm{e}(t') \cdot \left(\left(\mathsf{A} - \mathsf{A}^{\mathrm{t}}\right) \cdot \frac{\partial \mathrm{e}(t')}{\partial t'}\right) \mathrm{d}t' \\
&+ \epsilon_0 \int_{\tau-\epsilon}^{\tau+\epsilon} \mathrm{e}(t') \cdot \int_{t'-\tau-\epsilon}^{t'-\tau+\epsilon} \mathsf{M}(t'') \cdot \frac{\partial \mathrm{e}(t' - t'')}{\partial t'} \, \mathrm{d}t'' \, \mathrm{d}t' \geq 0
\end{aligned}
$$

The first integral is of the order $O(\epsilon)$, and the second integral is of order $O(\epsilon^2)$ as $\epsilon \to 0$. Therefore, the first integral dominates, and in the limit $\epsilon \to 0$, we get

$$
\frac{1}{2} \int_{\tau-\epsilon}^{\tau+\epsilon} \mathrm{e}(t') \cdot \left(\left(\mathsf{A} - \mathsf{A}^{\mathrm{t}}\right) \cdot \frac{\partial \mathrm{e}(t')}{\partial t'}\right) \mathrm{d}t' \geq 0
$$

for all fields e of this kind. Now e and its time derivative can be chosen independently, which implies that

$$\mathsf{A} = \mathsf{A}^t$$

and A has to be symmetric for a passive material, i.e.,

$$w_{em}(t) = \frac{\epsilon_0}{2}\mathsf{e}(t)\cdot(\mathsf{A}\cdot\mathsf{e}(t)) = \frac{\epsilon_0}{2}\mathsf{e}(t)\cdot\mathsf{A}\cdot\mathsf{e}(t)$$

Since this quantity is always non-negative, we conclude that the optical response A, in addition of being symmetric, also is a positive semi-definite dyadic.

The total power stored becomes

$$w(t) = \frac{\epsilon_0}{2}\mathsf{e}(t)\cdot\mathsf{A}\cdot\mathsf{e}(t) + \epsilon_0\int_{-\infty}^{t}\mathsf{e}(t')\cdot\int_0^{\infty}\mathsf{M}(t'')\cdot\frac{\partial\mathsf{e}(t'-t'')}{\partial t'}\,dt''\,dt'$$

Again, for a continuously differentiable field e, which satisfies $\mathsf{e}(t) = 0$, and which, as above, assumes non-zero values only in a neighborhood of $\tau < t$, we have

$$w(t) = \epsilon_0\int_{\tau-\epsilon}^{\tau+\epsilon}\mathsf{e}(t')\cdot\int_{t'-\tau-\epsilon}^{t'-\tau+\epsilon}\mathsf{M}(t'')\cdot\frac{\partial\mathsf{e}(t'-t'')}{\partial t'}\,dt''\,dt' \geq 0$$

which we rewrite as

$$w(t) = \epsilon_0\int_{\tau-\epsilon}^{\tau+\epsilon}\mathsf{e}(t')\cdot\left(\mathsf{M}(0)\cdot\mathsf{e}(t) + \int_{t'-\tau-\epsilon}^{t'-\tau+\epsilon}\frac{\partial\mathsf{M}(t'')}{\partial t''}\cdot\mathsf{e}(t'-t'')\,dt''\right)dt' \geq 0$$

Again, the first integral is of the order $O(\epsilon)$, and the second integral is of order $O(\epsilon^2)$ as $\epsilon \to 0$. The same argument as above implies that

$$\mathsf{a}\cdot(\mathsf{M}(0)\cdot\mathsf{a}) \geq 0$$

for all $\mathsf{a} \in \mathbb{R}^6$, i.e., M(0) is a positive semi-definite matrix.

If the early time behavior of the dispersion effects satisfies $\mathsf{M}(0) = 0$ (e.g., a Lorentz model), then a change of order of integration and a change of variables imply ($\mathsf{e}(t) = 0$)

$$\begin{aligned} w(t) &= \epsilon_0\int_{-\infty}^{t}\mathsf{e}(t')\cdot\int_{-\infty}^{t'}\frac{\partial\mathsf{M}(t'-t'')}{\partial t'}\cdot\mathsf{e}(t'')\,dt''\,dt' \\ &= \frac{\epsilon_0}{2}\int_{-\infty}^{t}\mathsf{e}(t')\cdot\int_{-\infty}^{t}\left(\frac{\partial\mathsf{M}(t'-t'')}{\partial t'} + \frac{\partial\mathsf{M}^t(t''-t')}{\partial t''}\right)\cdot\mathsf{e}(t'')\,dt''\,dt' \geq 0 \end{aligned}$$

Notice that we used the causality requirement on the dyadic-valued material function $\mathsf{M}(t)$, i.e., $\mathsf{M}(t) = 0$ for $t < 0$, and that both upper limits in the integration are the same t. We conclude that the function $K(t) = \mathsf{a}\cdot\left((\mathsf{M}'(t) + \mathsf{M}'^t(-t))\right)\cdot\mathsf{a}$ is a positive-definite function for all vectors $\mathsf{a} \in \mathbb{R}^6$, see Appendix D.4 on page 649 (prime denotes differentiation w.r.t. time).

Example 1.7

We start illustrating the theory with the Debye model, see (1.26) on page 22

$$\chi(t) = \mathrm{H}(t)\alpha \mathrm{e}^{-t/\tau}$$

We have

$$\begin{cases} \boldsymbol{\epsilon} = \mathbf{I}_3 \\ \boldsymbol{\chi}(0) = \alpha \mathbf{I}_3 \end{cases}$$

and we conclude $\boldsymbol{\chi}(0)$ is a positive-definite dyadic if $\alpha > 0$. Observe that we cannot make the conclusion that $f(t) = \boldsymbol{a} \cdot \left((\boldsymbol{\chi}'(t) + \boldsymbol{\chi}'^{\mathrm{t}}(-t))\right) \cdot \boldsymbol{a}$ is a positive-definite function, since $\boldsymbol{\chi}(0) \neq \mathbf{0}$. ■

Example 1.8
The Lorentz model, which has a susceptibility function, see (1.28) on page 25, is

$$\boldsymbol{\chi}(t) = \mathrm{H}(t)\frac{\omega_{\mathrm{p}}^2}{\nu_0}\mathrm{e}^{-\frac{\nu t}{2}} \sin \nu_0 t$$

We have

$$\begin{cases} \boldsymbol{\epsilon} = \mathbf{I}_3 \\ \boldsymbol{\chi}(0) = \mathbf{0} \end{cases} \qquad \begin{cases} \boldsymbol{\chi}'(t) = \mathrm{H}(t)\frac{\omega_{\mathrm{p}}^2}{\nu_0}\mathrm{e}^{-\frac{\nu t}{2}}\mathbf{I}_3 \left(\nu_0 \cos \nu_0 t - \frac{\nu}{2} \sin \nu_0 t\right) \\ \boldsymbol{\chi}'(0) = \omega_{\mathrm{p}}^2 \mathbf{I}_3 \end{cases}$$

Trivially, we conclude that $\boldsymbol{\chi}(0)$ is a positive-definite dyadic (more precisely positive semi-definite). Moreover, for each $\boldsymbol{a} \in \mathbb{R}^3$, the function

$$f(t) = \boldsymbol{a} \cdot \left((\boldsymbol{\chi}'(t) + \boldsymbol{\chi}'^{\mathrm{t}}(-t))\right) \cdot \boldsymbol{a} = \frac{\omega_{\mathrm{p}}^2}{\nu_0}\mathrm{e}^{-\frac{\nu|t|}{2}} |\boldsymbol{a}|^2 \left(\nu_0 \cos \nu_0 |t| - \frac{\nu}{2} \sin \nu_0 |t|\right)$$

is a positive-definite function, see Appendix D.4. To see this, engage Bochner's theorem, Theorem D.6 on page 650, and prove that $f(t)$ has a positive Fourier transform. In fact

$$\int_{-\infty}^{\infty} f(t)\mathrm{e}^{\mathrm{i}\omega t}\, \mathrm{d}t = 2|\boldsymbol{a}|^2 \frac{\omega_{\mathrm{p}}^2 \omega^2 \nu}{\left(\omega^2 - \omega_0^2\right)^2 + \omega^2 \nu^2} > 0$$

if $\nu > 0$. ■

Example 1.9
A more complex example is the model of the plasma in Example 1.5 on page 31. The constitutive relations satisfy

$$\begin{cases} \boldsymbol{\epsilon} = \mathbf{I}_3 \\ \boldsymbol{\chi}(0) = \mathbf{0} \end{cases} \qquad \begin{cases} \boldsymbol{\chi}'(t) = \omega_{\mathrm{p}}^2 \mathrm{e}^{-\nu t}\mathrm{H}(t) \left(\mathbf{I}_2 \cos \omega_{\mathrm{g}} t + \hat{z}\hat{z} - \mathbf{J} \sin \omega_{\mathrm{g}} t\right) \\ \boldsymbol{\chi}'(0) = \omega_{\mathrm{p}}^2 \mathbf{I}_3 \end{cases}$$

Note that $\boldsymbol{\chi}(0)$ is a positive-definite dyadic (more precisely positive semi-definite), and that for $\boldsymbol{a} \in \mathbb{R}^3$, the function

$$f(t) = \boldsymbol{a} \cdot \left((\boldsymbol{\chi}'(t) + \boldsymbol{\chi}^{\mathrm{t}}(-t))\right) \cdot \boldsymbol{a} = \omega_{\mathrm{p}}^2 \mathrm{e}^{-\nu|t|} \left((a_1^2 + a_2^2) \cos \omega_{\mathrm{g}} |t| + a_3^2\right)$$

is a positive-definite function, due to Bochner's theorem, since the Fourier transform of $f(t)$ is positive, i.e., In fact

$$\int_{-\infty}^{\infty} f(t)\mathrm{e}^{\mathrm{i}\omega t}\,\mathrm{d}t = 2\omega_{\mathrm{p}}^2\left(\nu\left(a_1^2+a_2^2\right)\frac{\omega^2+\omega_{\mathrm{g}}^2+\nu^2}{(\omega^2-\omega_{\mathrm{g}}^2-\nu^2)^2+4\nu^2\omega^2}+a_3^2\frac{\nu}{\omega^2+\nu^2}\right) > 0$$

if $\nu > 0$. ∎

1.3 Time-Harmonic Fields and Fourier Transform

Several important applications use time-harmonic fields. In this section, we analyze the simplification time-harmonic fields introduce.

We obtain the time-harmonic case from the general results in the previous section by either a Fourier[36] transform in the time variable of all fields (dyadic-valued, vector-valued, and scalar-valued fields) or a simple Ansatz. Both ways are looked into. The consequences of time-harmonic fields on the constitutive relations are investigated, and we introduce the concept of active, passive, and lossless media. Moreover, the concept of reciprocity is introduced, and we investigate the polarization state of a time-harmonic field, which leads to the concept of the polarization ellipse.

The Fourier transform in the time variable of a vector field, e.g., the electric field $\boldsymbol{E}(\boldsymbol{r},t)$, is defined as

$$\boldsymbol{E}(\boldsymbol{r},\omega) = \int_{-\infty}^{\infty} \boldsymbol{E}(\boldsymbol{r},t)\,\mathrm{e}^{\mathrm{i}\omega t}\,\mathrm{d}t$$

with its inverse transform

$$\boldsymbol{E}(\boldsymbol{r},t) = \frac{1}{2\pi}\int_{-\infty}^{\infty} \boldsymbol{E}(\boldsymbol{r},\omega)\,\mathrm{e}^{-\mathrm{i}\omega t}\,\mathrm{d}\omega$$

The Fourier transform for all other time-dependent fields, dyadics, and scalars is defined in a similar manner. To avoid heavy notation, we use the same symbol for the physical field $\boldsymbol{E}(\boldsymbol{r},t)$, as for the Fourier transformed field $\boldsymbol{E}(\boldsymbol{r},\omega)$—only the argument differs. This notation is conventional and well established in the literature. Moreover, note that the Fourier transformed field no longer has the same unit as the time-domain field, e.g., the physical electric field $\boldsymbol{E}(\boldsymbol{r},t)$ has the unit V/m, but the Fourier transformed field $\boldsymbol{E}(\boldsymbol{r},\omega)$ has the unit V s/m. This observation holds for all other Fourier transformed fields used in this book. In most cases, the context suggests whether it is the physical field or the Fourier transformed field that is intended. When there are doubts which field is intended, the time argument t or the angular frequency $\omega = 2\pi f$, where f is the frequency, is explicitly written out to distinguish the fields.

[36] Jean-Baptiste Joseph Fourier (1768–1830), French mathematician and physicist.

Comment 1.2
Two sign conventions for the temporal (inverse) Fourier transform exist. There is the one we use in this textbook, i.e., $e^{-i\omega t}$, which is used mostly by physicists. Electrical engineers often prefer the opposite sign in the exponential, i.e., $e^{i\omega t}$ or $e^{j\omega t}$. The choice of sign is, of course, irrelevant in the computation of all physical real quantities, but it leads to different signs in many of the complex quantities that are used in the calculations.

The choice of the electrical engineers is most appropriate when dealing with circuit applications where the dependence of the space variables is suppressed. However, using the $e^{i\omega t}$ time convention in wave propagation problems, like the scattering problems, we are dealing within this textbook, leads to an extra minus signs in front of the spatial dependence, e.g., an out-going spherical wave would be e^{-ikr}/kr with this time convention. To transfer the results in the $e^{j\omega t}$ notation to the $e^{-i\omega t}$ notation and vice versa replace all $j \leftrightarrow -i$.

All physical quantities, like the electric field $\boldsymbol{E}(\boldsymbol{r}, t)$ above, are real valued, which imply constraints on its Fourier transform. The negative values of ω are related to the positive values of ω by a complex conjugate. To see this, we write down the criterion for the field $\boldsymbol{E}(\boldsymbol{r}, t)$ to be real

$$\int_{-\infty}^{\infty} \boldsymbol{E}(\boldsymbol{r}, \omega) e^{-i\omega t} \, d\omega = \left\{ \int_{-\infty}^{\infty} \boldsymbol{E}(\boldsymbol{r}, \omega) e^{-i\omega t} \, d\omega \right\}^*$$

where the star (*) denotes the complex conjugate. For real ω, we have

$$\int_{-\infty}^{\infty} \boldsymbol{E}(\boldsymbol{r}, \omega) e^{-i\omega t} \, d\omega = \int_{-\infty}^{\infty} \boldsymbol{E}^*(\boldsymbol{r}, \omega) e^{i\omega t} \, d\omega = \int_{-\infty}^{\infty} \boldsymbol{E}^*(\boldsymbol{r}, -\omega) e^{-i\omega t} \, d\omega$$

where in the last integral, we made a change of variable $\omega \to -\omega$. Therefore, for real ω, we have

$$\boldsymbol{E}(\boldsymbol{r}, \omega) = \boldsymbol{E}^*(\boldsymbol{r}, -\omega) \tag{1.40}$$

This shows that when the physical field is constructed from its Fourier transform, it suffices to integrate over the non-negative frequencies only. By a change of variable, $\omega \to -\omega$, and the use of the condition (1.40), we have

$$\begin{aligned}
\boldsymbol{E}(\boldsymbol{r}, t) &= \frac{1}{2\pi} \int_{-\infty}^{\infty} \boldsymbol{E}(\boldsymbol{r}, \omega) \, e^{-i\omega t} \, d\omega \\
&= \frac{1}{2\pi} \int_{0}^{\infty} \left(\boldsymbol{E}(\boldsymbol{r}, \omega) e^{-i\omega t} + \boldsymbol{E}(\boldsymbol{r}, -\omega) e^{i\omega t} \right) d\omega \\
&= \frac{1}{2\pi} \int_{0}^{\infty} \left(\boldsymbol{E}(\boldsymbol{r}, \omega) e^{-i\omega t} + \boldsymbol{E}^*(\boldsymbol{r}, \omega) e^{i\omega t} \right) d\omega \\
&= \frac{1}{\pi} \operatorname{Re} \int_{0}^{\infty} \boldsymbol{E}(\boldsymbol{r}, \omega) e^{-i\omega t} \, d\omega
\end{aligned} \tag{1.41}$$

where $\operatorname{Re} z$ denotes the real part of the complex number z. A similar result holds for all other Fourier-transformed fields that we are using. From (1.40), we also conclude that the real part of $\boldsymbol{E}(\boldsymbol{r}, \omega)$ is an even function of ω and the imaginary part of $\boldsymbol{E}(\boldsymbol{r}, \omega)$ is an odd function of ω.

Table 1.2 The spectrum of the electromagnetic waves.

Band	Frequency	Wavelength	Application
ELF	<3 kHz	>100 km	
VLF	3–30 kHz	100–10 km	Navigation
LF	30–300 kHz	10–1 km	Navigation
MF	300–3 000 kHz	1 000–100 m	Radio
HF	3–30 MHz	100–10 m	Radio
VHF	30–300 MHz	10–1 m	FM, TV
UHF	300–3 000 MHz	100–30 cm	Radar, TV, mobile communication
†[a]	1–30 GHz	30–1 cm	Radar, satellite communication
†[a]	30–300 GHz	10–1 mm	Radar
	420–790 THz	0.38–0.72 μm	Visible light

[a]See also Table 1.3.

Table 1.3 Table of radar band frequencies.

Band	Frequency (GHz)
L	1–2
S	2–4
C	4–8
X	8–12
K_u	12–18
K	20–40
K_a	26.5–40
Millimeter band	30–300

Fields that are purely time harmonic are of special interest in many applications, see Table 1.2. If we concentrate on the time dependence, purely time-harmonic fields have time dependence of the form

$$\cos(\omega_0 t - \alpha)$$

Such fields are generated by the following Fourier transform:

$$\begin{aligned} \boldsymbol{E}(\boldsymbol{r}, \omega) = \pi \Big\{ &\delta(\omega - \omega_0) \left(\hat{\boldsymbol{x}} E_x(\boldsymbol{r}, \omega_0) + \hat{\boldsymbol{y}} E_y(\boldsymbol{r}, \omega_0) + \hat{\boldsymbol{z}} E_z(\boldsymbol{r}, \omega_0) \right) \\ &+ \delta(\omega + \omega_0) \left(\hat{\boldsymbol{x}} E_x^*(\boldsymbol{r}, \omega_0) + \hat{\boldsymbol{y}} E_y^*(\boldsymbol{r}, \omega_0) + \hat{\boldsymbol{z}} E_z^*(\boldsymbol{r}, \omega_0) \right) \Big\} \end{aligned}$$

The components of the field are

$$\begin{cases} E_x(\boldsymbol{r}, \omega_0) = |E_x(\boldsymbol{r}, \omega_0)| \mathrm{e}^{\mathrm{i}\alpha(\boldsymbol{r}, \omega_0)} \\ E_y(\boldsymbol{r}, \omega_0) = |E_y(\boldsymbol{r}, \omega_0)| \mathrm{e}^{\mathrm{i}\beta(\boldsymbol{r}, \omega_0)} \\ E_z(\boldsymbol{r}, \omega_0) = |E_z(\boldsymbol{r}, \omega_0)| \mathrm{e}^{\mathrm{i}\gamma(\boldsymbol{r}, \omega_0)} \end{cases}$$

where $\alpha(\boldsymbol{r},\omega_0)$, $\beta(\boldsymbol{r},\omega_0)$, and $\gamma(\boldsymbol{r},\omega_0)$ are the phase of the components, $\omega_0 \geq 0$, and where $\delta(\omega)$ denotes the delta function. Note that this Fourier transform satisfies $\boldsymbol{E}(\boldsymbol{r},\omega) = \boldsymbol{E}^*(\boldsymbol{r},-\omega)$, which is the criterion for a real-valued field, and implying that the modulus of the field is an even and the phase is an odd function in ω. The inverse Fourier transform then gives

$$\begin{aligned}\boldsymbol{E}(\boldsymbol{r},t) = \frac{1}{2\pi}\int_{-\infty}^{\infty} \boldsymbol{E}(\boldsymbol{r},\omega)\,\mathrm{e}^{-\mathrm{i}\omega t}\,\mathrm{d}\omega = \big\{\hat{\boldsymbol{x}}|E_x(\boldsymbol{r},\omega_0)|\cos(\omega_0 t - \alpha(\boldsymbol{r},\omega_0))\\ +\hat{\boldsymbol{y}}|E_y(\boldsymbol{r},\omega_0)|\cos(\omega_0 t - \beta(\boldsymbol{r},\omega_0)) + \hat{\boldsymbol{z}}|E_z(\boldsymbol{r},\omega_0)|\cos(\omega_0 t - \gamma(\boldsymbol{r},\omega_0))\big\}\end{aligned}$$

A second way of generating a purely time-harmonic field (monochromatic field) is to employ the following expression or Ansatz:

$$\boldsymbol{E}(\boldsymbol{r},t) = \mathrm{Re}\left\{\boldsymbol{E}(\boldsymbol{r},\omega)\mathrm{e}^{-\mathrm{i}\omega t}\right\} \tag{1.42}$$

where $\boldsymbol{E}(\boldsymbol{r},\omega)$ is a complex-valued vector with the same dimension as the electric field. If we write $\boldsymbol{E}(\boldsymbol{r},\omega)$ as

$$\begin{aligned}\boldsymbol{E}(\boldsymbol{r},\omega) &= \hat{\boldsymbol{x}}E_x(\boldsymbol{r},\omega) + \hat{\boldsymbol{y}}E_y(\boldsymbol{r},\omega) + \hat{\boldsymbol{z}}E_z(\boldsymbol{r},\omega)\\ &= \hat{\boldsymbol{x}}|E_x(\boldsymbol{r},\omega)|\mathrm{e}^{\mathrm{i}\alpha(\boldsymbol{r},\omega)} + \hat{\boldsymbol{y}}|E_y(\boldsymbol{r},\omega)|\mathrm{e}^{\mathrm{i}\beta(\boldsymbol{r},\omega)} + \hat{\boldsymbol{z}}|E_z(\boldsymbol{r},\omega)|\mathrm{e}^{\mathrm{i}\gamma(\boldsymbol{r},\omega)}\end{aligned}$$

we obtain the same result as in the expression above (without the subscript 0 on ω). This way of constructing purely time-harmonic waves is convenient and often used. Note that the field $\boldsymbol{E}(\boldsymbol{r},\omega)$ constructed this way has the same unit as the field $\boldsymbol{E}(\boldsymbol{r},t)$ (and also for all other purely time-harmonic fields constructed this way). This is in contrast to the Fourier transformation of the field above, but this difference seldom causes problems or misinterpretations.

1.3.1 The Maxwell Equations

As a first step in our analysis of time-harmonic fields, we Fourier transform (or more conveniently use (1.42)) the Maxwell equations (1.1) and (1.2) $\left(\frac{\partial}{\partial t} \to -\mathrm{i}\omega\right)$

$$\nabla \times \boldsymbol{E}(\boldsymbol{r},\omega) = \mathrm{i}\omega\boldsymbol{B}(\boldsymbol{r},\omega) \tag{1.43}$$

$$\nabla \times \boldsymbol{H}(\boldsymbol{r},\omega) = \boldsymbol{J}(\boldsymbol{r},\omega) - \mathrm{i}\omega\boldsymbol{D}(\boldsymbol{r},\omega) \tag{1.44}$$

The explicit harmonic time dependence $\mathrm{e}^{-\mathrm{i}\omega t}$ has been suppressed from both sides of these equations, i.e., the physical fields are

$$\boldsymbol{E}(\boldsymbol{r},t) = \mathrm{Re}\left\{\boldsymbol{E}(\boldsymbol{r},\omega)\mathrm{e}^{-\mathrm{i}\omega t}\right\}$$

This convention is applied to all purely time-harmonic fields. Note that the electromagnetic fields $\boldsymbol{E}(\boldsymbol{r},\omega)$, $\boldsymbol{B}(\boldsymbol{r},\omega)$, $\boldsymbol{D}(\boldsymbol{r},\omega)$, and $\boldsymbol{H}(\boldsymbol{r},\omega)$, and the current density $\boldsymbol{J}(\boldsymbol{r},\omega)$ in general are complex-valued vector fields.

The continuity equation (1.4) is transformed in a similar way, and we have

$$\nabla \cdot \boldsymbol{J}(\boldsymbol{r},\omega) - \mathrm{i}\omega\rho(\boldsymbol{r},\omega) = 0 \tag{1.45}$$

The two remaining equations from Section 1.1, (1.5) and (1.6), are transformed into

$$\nabla \cdot \boldsymbol{B}(\boldsymbol{r}, \omega) = 0 \tag{1.46}$$

$$\nabla \cdot \boldsymbol{D}(\boldsymbol{r}, \omega) = \rho(\boldsymbol{r}, \omega) \tag{1.47}$$

These equations are consequences of (1.43) and (1.44), and the continuity equation (1.45) (cf. Section 1.1 on page 3). In fact, take the divergence of the Maxwell equations (1.43) and (1.44) and use (1.45), which gives ($\nabla \cdot (\nabla \times \boldsymbol{A}) = 0$)

$$\begin{aligned} &\mathrm{i}\omega\nabla \cdot \boldsymbol{B}(\boldsymbol{r}, \omega) = 0 \\ &\mathrm{i}\omega\nabla \cdot \boldsymbol{D}(\boldsymbol{r}, \omega) = \nabla \cdot \boldsymbol{J}(\boldsymbol{r}, \omega) = \mathrm{i}\omega\rho(\boldsymbol{r}, \omega) \end{aligned}$$

Division by $\mathrm{i}\omega$ (provided $\omega \neq 0$) then gives (1.46) and (1.47).

To summarize, in a source-free region, the time-harmonic Maxwell equations are

$$\begin{cases} \nabla \times \boldsymbol{E}(\boldsymbol{r}, \omega) = \mathrm{i}k_0(c_0\boldsymbol{B}(\boldsymbol{r}, \omega)) \\ \nabla \times (\eta_0\boldsymbol{H}(\boldsymbol{r}, \omega)) = -\mathrm{i}k_0(c_0\eta_0\boldsymbol{D}(\boldsymbol{r}, \omega)) \end{cases} \tag{1.48}$$

where $\eta_0 = \sqrt{\mu_0/\epsilon_0}$ is the intrinsic wave impedance of vacuum, $c_0 = 1/\sqrt{\epsilon_0\mu_0}$ the speed of light in vacuum, and $k_0 = \omega/c_0$ is the wave number[37] in vacuum. In (1.48), all field quantities in parentheses have the same units, i.e., that of the electric field. This form is the standard form of the Maxwell equations that we use in this textbook.

1.3.2 Constitutive Relations

The constitutive relations introduced in Section 1.2 for general time-dependent fields contain temporal convolutions. In this section, we Fourier transform these relations and arrive at constitutive relations that hold for time-harmonic fields.

In Section 1.2.3, the general Ansatz on the constitutive relations in bianisotropic media is given in (1.29)

$$\begin{aligned} \boldsymbol{D}(t) &= \epsilon_0 \left\{ \boldsymbol{\epsilon} \cdot \boldsymbol{E}(t) + \int_{-\infty}^{t} \boldsymbol{\chi}_{\mathrm{ee}}(t-t') \cdot \boldsymbol{E}(t')\,\mathrm{d}t' + \eta_0 \int_{-\infty}^{t} \boldsymbol{\chi}_{\mathrm{em}}(t-t') \cdot \boldsymbol{H}(t')\,\mathrm{d}t' \right\} \\ \boldsymbol{B}(t) &= \frac{1}{c_0} \left\{ \int_{-\infty}^{t} \boldsymbol{\chi}_{\mathrm{me}}(t-t') \cdot \boldsymbol{E}(t')\,\mathrm{d}t' + \eta_0\boldsymbol{\mu} \cdot \boldsymbol{H}(t) + \eta_0 \int_{-\infty}^{t} \boldsymbol{\chi}_{\mathrm{mm}}(t-t') \cdot \boldsymbol{H}(t')\,\mathrm{d}t' \right\} \end{aligned}$$

or in the six-vector notation of (1.32)

$$\mathsf{d}(t) = \mathsf{A} \cdot \mathsf{e}(t) + \int_{-\infty}^{t} \mathsf{M}(t-t') \cdot \mathsf{e}(t')\,\mathrm{d}t'$$

[37] More correctly, k_0 is the angular wave number in vacuum, and f/c_0 is the wave number in vacuum.

Since the Fourier transform of a convolution between two fields is a product of their Fourier transforms, we get, see Appendix D.1 on page 639

$$\begin{aligned} \boldsymbol{D}(\omega) &= \epsilon_0 \left\{ \left(\boldsymbol{\epsilon} + \int_0^\infty \boldsymbol{\chi}_{\mathrm{ee}}(t) \mathrm{e}^{\mathrm{i}\omega t} \, \mathrm{d}t \right) \cdot \boldsymbol{E}(\omega) + \eta_0 \int_0^\infty \boldsymbol{\chi}_{\mathrm{em}}(t) \mathrm{e}^{\mathrm{i}\omega t} \, \mathrm{d}t \cdot \boldsymbol{H}(\omega) \right\} \\ \boldsymbol{B}(\omega) &= \frac{1}{c_0} \left\{ \int_0^\infty \boldsymbol{\chi}_{\mathrm{me}}(t) \mathrm{e}^{\mathrm{i}\omega t} \, \mathrm{d}t \cdot \boldsymbol{E}(\omega) + \eta_0 \left(\boldsymbol{\mu} + \int_0^\infty \boldsymbol{\chi}_{\mathrm{mm}}(t) \mathrm{e}^{\mathrm{i}\omega t} \, \mathrm{d}t \right) \cdot \boldsymbol{H}(\omega) \right\} \end{aligned}$$

It is convenient to introduce the following notation:

$$\left\{ \begin{aligned} \boldsymbol{\epsilon}(\omega) &= \boldsymbol{\epsilon} + \int_0^\infty \boldsymbol{\chi}_{\mathrm{ee}}(t) \mathrm{e}^{\mathrm{i}\omega t} \, \mathrm{d}t \\ \boldsymbol{\mu}(\omega) &= \boldsymbol{\mu} + \int_0^\infty \boldsymbol{\chi}_{\mathrm{mm}}(t) \mathrm{e}^{\mathrm{i}\omega t} \, \mathrm{d}t \end{aligned} \right. \qquad \left\{ \begin{aligned} \boldsymbol{\xi}(\omega) &= \int_0^\infty \boldsymbol{\chi}_{\mathrm{em}}(t) \mathrm{e}^{\mathrm{i}\omega t} \, \mathrm{d}t \\ \boldsymbol{\zeta}(\omega) &= \int_0^\infty \boldsymbol{\chi}_{\mathrm{me}}(t) \mathrm{e}^{\mathrm{i}\omega t} \, \mathrm{d}t \end{aligned} \right. \tag{1.49}$$

with inverses

$$\left\{ \begin{aligned} \boldsymbol{\chi}_{\mathrm{ee}}(t) &= \frac{1}{2\pi} \int_{-\infty}^\infty [\boldsymbol{\epsilon}(\omega) - \boldsymbol{\epsilon}] \, \mathrm{e}^{-\mathrm{i}\omega t} \, \mathrm{d}\omega \\ \boldsymbol{\chi}_{\mathrm{mm}}(t) &= \frac{1}{2\pi} \int_{-\infty}^\infty [\boldsymbol{\mu}(\omega) - \boldsymbol{\mu}] \, \mathrm{e}^{-\mathrm{i}\omega t} \, \mathrm{d}\omega \end{aligned} \right.$$

and

$$\left\{ \begin{aligned} \boldsymbol{\chi}_{\mathrm{em}}(t) &= \frac{1}{2\pi} \int_{-\infty}^\infty \boldsymbol{\xi}(\omega) \, \mathrm{e}^{-\mathrm{i}\omega t} \, \mathrm{d}\omega \\ \boldsymbol{\chi}_{\mathrm{me}}(t) &= \frac{1}{2\pi} \int_{-\infty}^\infty \boldsymbol{\zeta}(\omega) \, \mathrm{e}^{-\mathrm{i}\omega t} \, \mathrm{d}\omega \end{aligned} \right.$$

Note the difference between the optical responses $\boldsymbol{\epsilon}$ and $\boldsymbol{\mu}$, which are constant, real-valued dyadics, and the Fourier transform of the generalized, frequency dependent, dyadic-valued susceptibilities $\boldsymbol{\epsilon}(\omega)$ and $\boldsymbol{\mu}(\omega)$. Normally, the context reveals which of the quantities is intended. In cases where misinterpretation might appear, we explicitly write out the argument.

The generalized susceptibility functions $\boldsymbol{\chi}_{kk'}(t)$, $k, k' = \mathrm{e}, \mathrm{m}$, are real valued, which imply that the complex-valued dyadics $\boldsymbol{\epsilon}(\omega)$, $\boldsymbol{\xi}(\omega)$, $\boldsymbol{\zeta}(\omega)$, and $\boldsymbol{\mu}(\omega)$ satisfy

$$\left\{ \begin{aligned} \boldsymbol{\epsilon}(\omega) &= \boldsymbol{\epsilon}^*(-\omega) \\ \boldsymbol{\mu}(\omega) &= \boldsymbol{\mu}^*(-\omega) \end{aligned} \right. \qquad \left\{ \begin{aligned} \boldsymbol{\xi}(\omega) &= \boldsymbol{\xi}^*(-\omega) \\ \boldsymbol{\zeta}(\omega) &= \boldsymbol{\zeta}^*(-\omega) \end{aligned} \right. \tag{1.50}$$

The (angular) frequency ω in the Fourier transforms in (1.49) is assumed to be real. Nothing prevents us from extending the variable ω into the upper half-plane of the complex ω-plane. This is possible due to the fact that $\boldsymbol{\chi}_{kk'}(t)$ are causal quantities, see also Appendix D.2. We adopt $\eta = \omega + \mathrm{i}\varsigma$, where ω and ς are real, and $\varsigma \geq 0$, and extend the domain of definition in the Fourier transform to the upper half-plane in the complex variable η, i.e., the entries of the permittivity dyadic are

$$\epsilon_{ij}(\eta) = \epsilon_{ij} + \int_0^\infty \chi_{\mathrm{ee}\,ij}(t) \mathrm{e}^{\mathrm{i}\omega t - \varsigma t} \, \mathrm{d}t, \qquad \varsigma \geq 0$$

and similarly for the other dyadics in (1.49). In fact, the Fourier transform contains an extra exponentially decreasing function which ensures convergence, and this extension is an analytic extension of the function in the upper complex plane, $\varsigma \geq 0$. This is a consequence of Paley–Wiener theorem, see Theorem D.1 on page 640.

We have the identity $(-\eta^* = -\omega + \mathrm{i}\varsigma)$

$$\epsilon_{ij}(-\eta^*) = \epsilon_{ij} + \int_0^\infty \chi_{\mathrm{ee}\,ij}(t)\mathrm{e}^{-\mathrm{i}\omega t - \varsigma t}\,\mathrm{d}t = \epsilon_{ij}^*(\eta), \qquad \varsigma \geq 0$$

This relation generalizes (1.50), which holds for real frequencies ω, to complex values of the frequency in the upper half-plane. As a consequence

$$\epsilon_{ij}(\mathrm{i}\varsigma) = \epsilon_{ij} + \int_0^\infty \chi_{\mathrm{ee}\,ij}(t)\mathrm{e}^{-\varsigma t}\,\mathrm{d}t = \epsilon_{ij}^*(\mathrm{i}\varsigma)$$

is real valued.

The notation in (1.49) simplifies the constitutive relations for time-harmonic fields (or the Fourier-transformed fields). We have

$$\begin{cases} \boldsymbol{D}(\omega) = \epsilon_0\left\{\boldsymbol{\epsilon}(\omega)\cdot\boldsymbol{E}(\omega) + \eta_0\boldsymbol{\xi}(\omega)\cdot\boldsymbol{H}(\omega)\right\} \\ \boldsymbol{B}(\omega) = \dfrac{1}{c_0}\left\{\boldsymbol{\zeta}(\omega)\cdot\boldsymbol{E}(\omega) + \eta_0\boldsymbol{\mu}(\omega)\cdot\boldsymbol{H}(\omega)\right\} \end{cases} \tag{1.51}$$

or in their Cartesian components

$$\begin{cases} D_i(\omega) = \epsilon_0 \displaystyle\sum_{j=1}^{3}\left\{\epsilon_{ij}(\omega)E_j(\omega) + \eta_0\xi_{ij}(\omega)H_j(\omega)\right\} \\ B_i(\omega) = \dfrac{1}{c_0}\displaystyle\sum_{j=1}^{3}\left\{\zeta_{ij}(\omega)E_j(\omega) + \eta_0\mu_{ij}(\omega)H_j(\omega)\right\} \end{cases}$$

In the six-vector notation, we write the relations in (1.51) in a more compact form as

$$\mathsf{d}(\omega) = \left(\mathsf{A} + \int_0^\infty \mathsf{M}(t)\mathrm{e}^{\mathrm{i}\omega t}\,\mathrm{d}t\right)\cdot\mathsf{e}(\omega) = \mathsf{M}(\omega)\cdot\mathsf{e}(\omega)$$

where

$$\mathsf{e}(\omega) = \begin{pmatrix} \boldsymbol{E}(\omega) \\ \eta_0\boldsymbol{H}(\omega) \end{pmatrix}, \qquad \mathsf{d}(\omega) = c_0\begin{pmatrix} \eta_0\boldsymbol{D}(\omega) \\ \boldsymbol{B}(\omega) \end{pmatrix} \tag{1.52}$$

and

$$\mathsf{M}(\omega) = \mathsf{A} + \int_0^\infty \mathsf{M}(t)\mathrm{e}^{\mathrm{i}\omega t}\,\mathrm{d}t = \begin{pmatrix} \boldsymbol{\epsilon}(\omega) & \boldsymbol{\xi}(\omega) \\ \boldsymbol{\zeta}(\omega) & \boldsymbol{\mu}(\omega) \end{pmatrix} \tag{1.53}$$

satisfying

$$\mathsf{M}^*(\eta) = \mathsf{M}(-\eta^*), \qquad \varsigma \geq 0 \tag{1.54}$$

The dyadics (or the corresponding matrix representations) $\boldsymbol{\epsilon}$, $\boldsymbol{\xi}$, $\boldsymbol{\zeta}$, and $\boldsymbol{\mu}$ are all dimensionless dyadics as a consequence of the normalization in (1.51). The dyadics $\boldsymbol{\epsilon}$ and $\boldsymbol{\mu}$ have separate names; they are called the permittivity dyadic and the permeability dyadic, respectively.

The four dyadics $\boldsymbol{\epsilon}(\boldsymbol{r},\omega)$, $\boldsymbol{\xi}(\boldsymbol{r},\omega)$, $\boldsymbol{\zeta}(\boldsymbol{r},\omega)$, and $\boldsymbol{\mu}(\boldsymbol{r},\omega)$ depend in general of the spatial variables $\boldsymbol{r}$. For a homogeneous material, the constitutive dyadics are independent of $\boldsymbol{r}$. Notice that $\boldsymbol{\epsilon}$, $\boldsymbol{\xi}$, $\boldsymbol{\zeta}$, and $\boldsymbol{\mu}$ are generally still functions of the angular frequency ω due to (material) temporal dispersion.

1.3.2.1 Classifications

The classification of different material is made as in Table 1.4. This classification is analogous to the one we introduced for general time-dependent fields, see Table 1.1.

Anisotropic materials, are characterized by the dyadics $\boldsymbol{\epsilon}(\boldsymbol{r},\omega)$ and $\boldsymbol{\mu}(\boldsymbol{r},\omega)$, which in general contain nine independent parameters each. In an isotropic medium, the permittivity and permeability dyadics, $\boldsymbol{\epsilon}(\boldsymbol{r},\omega)$ and $\boldsymbol{\mu}(\boldsymbol{r},\omega)$, respectively, are proportional to the identity dyadic $\mathbf{I}_3$, and have only two independent parameters. In a biisotropic medium, which is the simplest complex material involving the cross-coupling terms $\boldsymbol{\xi}(\boldsymbol{r},\omega)$ and $\boldsymbol{\zeta}(\boldsymbol{r},\omega)$, all the constitutive dyadics are proportional to the identity dyadic $\mathbf{I}_3$, and have in general four independent parameters, i.e.,

$$\begin{cases} \boldsymbol{D}(\boldsymbol{r},\omega) = \epsilon_0\{\epsilon(\boldsymbol{r},\omega)\boldsymbol{E}(\boldsymbol{r},\omega) + \xi(\boldsymbol{r},\omega)\eta_0\boldsymbol{H}(\boldsymbol{r},\omega)\} \\ \boldsymbol{B}(\boldsymbol{r},\omega) = \dfrac{1}{c_0}\{\zeta(\boldsymbol{r},\omega)\boldsymbol{E}(\boldsymbol{r},\omega) + \mu(\boldsymbol{r},\omega)\eta_0\boldsymbol{H}(\boldsymbol{r},\omega)\} \end{cases} \tag{1.55}$$

The isotropic material was analyzed in Section 1.2.1, in a general time-dependent formulation. In a time-harmonic setting, these constitutive relations are

$$\begin{cases} \boldsymbol{D}(\boldsymbol{r},\omega) = \epsilon_0\epsilon(\boldsymbol{r},\omega)\boldsymbol{E}(\boldsymbol{r},\omega) \\ \boldsymbol{B}(\boldsymbol{r},\omega) = \mu_0\mu(\boldsymbol{r},\omega)\boldsymbol{H}(\boldsymbol{r},\omega) \end{cases} \tag{1.56}$$

The parameters ϵ and μ are the (relative) permittivity and permeability of the medium, respectively. The isotropic model is used frequently and is a good model for many insulation materials, e.g., glass, china, and several plastic materials.

Table 1.4 Table of classification of materials w.r.t. their constitutive relations for time-harmonic fields. The dyadic $\mathbf{I}_3$ denotes the unit dyadic in three dimensions.

Type	$\boldsymbol{\epsilon}, \boldsymbol{\mu}$	$\boldsymbol{\xi}, \boldsymbol{\zeta}$
Isotropic	Both $\sim \mathbf{I}_3$	Both $\mathbf{0}$
Anisotropic	Some not $\sim \mathbf{I}_3$	Both $\mathbf{0}$
Biisotropic	Both $\sim \mathbf{I}_3$	Both $\sim \mathbf{I}_3$
Bianisotropic	All other cases	

Table 1.5 Table of classification of anisotropic media.

Uniaxial	$\epsilon_1 = \epsilon_2 \neq \epsilon_3$
Biaxial	$\epsilon_1 \neq \epsilon_2 \neq \epsilon_3$

We also note that a material with a conductivity that satisfies the Ohm's law, (1.20) on page 17, always can be included in the constitutive relations by redefining the permittivity[38] in (1.51)

$$\boldsymbol{\epsilon}_{\mathrm{new}}(\omega) = \boldsymbol{\epsilon}_{\mathrm{old}}(\omega) + \mathrm{i}\frac{\boldsymbol{\sigma}}{\omega\epsilon_0}$$

The right-hand side in the Ampère law (1.44) is

$$\boldsymbol{J} - \mathrm{i}\omega\boldsymbol{D} = \boldsymbol{\sigma}\boldsymbol{E} - \mathrm{i}\omega\epsilon_0\left\{\boldsymbol{\epsilon}_{\mathrm{old}}\cdot\boldsymbol{E} + \eta_0\boldsymbol{\xi}\cdot\boldsymbol{H}\right\} = -\mathrm{i}\omega\epsilon_0\left\{\boldsymbol{\epsilon}_{\mathrm{new}}\cdot\boldsymbol{E} + \eta_0\boldsymbol{\xi}\cdot\boldsymbol{H}\right\}$$

where $\boldsymbol{\epsilon}_{\mathrm{new}}(\omega)$ now includes bound as well as free charges.

The anisotropic media can be classified further. The permittivity dyadic $\boldsymbol{\epsilon}$ is usually Hermitian, i.e.,

$$\boldsymbol{\epsilon} = \boldsymbol{\epsilon}^{\dagger}$$

which in the entries of the dyadic means $\epsilon_{ij}^{\dagger} = \epsilon_{ji}^{*}$. Furthermore, if the permittivity dyadic $\boldsymbol{\epsilon}$ is real, then there is a basis of real-valued orthonormal vectors such that the permittivity dyadic representation, $[\boldsymbol{\epsilon}]$, in this basis system is diagonal

$$[\boldsymbol{\epsilon}] = \begin{pmatrix} \epsilon_1 & 0 & 0 \\ 0 & \epsilon_2 & 0 \\ 0 & 0 & \epsilon_3 \end{pmatrix}$$

Anisotropic media with a diagonal representation $[\boldsymbol{\epsilon}]$ can be classified as in Table 1.5. For uniaxial media, the z-axis is a symmetry axis, which usually is called the optical axis of the material due to many applications at optical frequencies. Moreover, the uniaxial medium is positive (negative) uniaxial if $\epsilon_3 > \epsilon_1 = \epsilon_2$ ($\epsilon_3 < \epsilon_1 = \epsilon_2$).

In a medium that has a lattice structure, the symmetry of the crystal determines the values of ϵ_1, ϵ_2, and ϵ_3. The most important cases are summarized in Table 1.6.

Biisotropic media are classified w.r.t. their symmetry properties. We classify these materials with the help of Table 1.7. From this table, we see that a reciprocal,[39] biisotropic material has the following constitutive relations:

$$\begin{cases} \boldsymbol{D} = \epsilon_0\left\{\epsilon(\omega)\boldsymbol{E}(\omega) + \mathrm{i}\eta_0\chi(\omega)\boldsymbol{H}(\omega)\right\} \\ \boldsymbol{B} = \dfrac{1}{c_0}\left\{-\mathrm{i}\chi(\omega)\boldsymbol{E}(\omega) + \eta_0\mu(\omega)\boldsymbol{H}(\omega)\right\} \end{cases}$$

[38]This treatment is analogous to the one presented in Section 1.2 where the conductivity σ was included in a new susceptibility function $\chi_{\mathrm{new}}(t)$, see page 19.

[39]The concept of reciprocity is introduced in Section 1.3.5.

Table 1.6 Table of crystal symmetries and the values of ϵ_i.

Type	Diagonal elements	Crystal symmetry
Isotropic	$\epsilon_1 = \epsilon_2 = \epsilon_3$	Cubic
Uniaxial	$\epsilon_1 = \epsilon_2 \neq \epsilon_3$	Tetragonal, hexagonal trigonal
Biaxial	$\epsilon_1 \neq \epsilon_2 \neq \epsilon_3$	Orthorhombic, hexagonal monoclinic, triclinic

Table 1.7 Classification of biisotropic materials. The parameters κ and χ are: $\xi = \kappa + \mathrm{i}\chi$ and $\zeta = \kappa - \mathrm{i}\chi$, i.e., $\kappa = (\xi + \zeta)/2$ and $\chi = (\xi - \zeta)/2\mathrm{i}$.

	$\chi = 0$	$\chi = 0$
$\kappa = 0$	Isotropic	Chiral, reciprocal
$\kappa \neq 0$	Non-chiral, non-reciprocal	Chiral, non-reciprocal

These constitutive relations are often referred to as a chiral material.

Comment 1.3
In the literature [19, 66, 67], the Fedorov (Fedor Ivanovich Fedorov (1911–1994), Belorussian physicist) model

$$\begin{cases} \boldsymbol{D} = \boldsymbol{\epsilon}_{\mathrm{F}} \cdot \{\boldsymbol{E} + \boldsymbol{\alpha} \cdot (\nabla \times \boldsymbol{E})\} \\ \boldsymbol{B} = \boldsymbol{\mu}_{\mathrm{F}} \cdot \{\boldsymbol{H} + \boldsymbol{\beta} \cdot (\nabla \times \boldsymbol{H})\} \end{cases}$$

is often used, which is related to our material parameters by

$$\begin{cases} \epsilon_0 \boldsymbol{\epsilon} = \boldsymbol{\epsilon}_{\mathrm{F}} \cdot \left(\mathbf{I}_3 - \omega^2 \boldsymbol{\alpha} \cdot \boldsymbol{\mu}_{\mathrm{F}} \cdot \boldsymbol{\beta} \cdot \boldsymbol{\epsilon}_{\mathrm{F}}\right)^{-1} \\ c_0^{-1} \boldsymbol{\xi} = \mathrm{i}\omega \boldsymbol{\epsilon}_{\mathrm{F}} \cdot \boldsymbol{\alpha} \cdot \left(\mathbf{I}_3 - \omega^2 \boldsymbol{\mu}_{\mathrm{F}} \cdot \boldsymbol{\beta} \cdot \boldsymbol{\epsilon}_{\mathrm{F}} \cdot \boldsymbol{\alpha}\right)^{-1} \cdot \boldsymbol{\mu}_{\mathrm{F}} \\ c_0^{-1} \boldsymbol{\zeta} = -\mathrm{i}\omega \left(\mathbf{I}_3 - \omega^2 \boldsymbol{\mu}_{\mathrm{F}} \cdot \boldsymbol{\beta} \cdot \boldsymbol{\epsilon}_{\mathrm{F}} \cdot \boldsymbol{\alpha}\right)^{-1} \cdot \boldsymbol{\mu}_{\mathrm{F}} \cdot \boldsymbol{\beta} \cdot \boldsymbol{\epsilon}_{\mathrm{F}} \\ \mu_0 \boldsymbol{\mu} = \left(\mathbf{I}_3 - \omega^2 \boldsymbol{\mu}_{\mathrm{F}} \cdot \boldsymbol{\beta} \cdot \boldsymbol{\epsilon}_{\mathrm{F}} \cdot \boldsymbol{\alpha}\right)^{-1} \cdot \boldsymbol{\mu}_{\mathrm{F}} \end{cases}$$

1.3.2.2 Examples

Two important examples of constitutive relations were introduced in Section 1.2—the Debye and Lorentz models. We are now prepared to transform these models by (1.49).

Debye Model: The Debye model has a susceptibility function, see (1.26)

$$\chi(t) = \mathrm{H}(t)\alpha \mathrm{e}^{-t/\tau}, \qquad \epsilon = 1$$

This susceptibility function is easily transformed to the frequency domain. The result is the permittivity dyadic, $\boldsymbol{\epsilon}(\omega) = \epsilon(\omega)\mathbf{I}_3$, where

$$\epsilon(\omega) = \epsilon + \int_0^\infty \chi(t)e^{i\omega t}\,dt = 1 + \frac{\alpha\tau}{1 - i\omega\tau} \tag{1.57}$$

with real and imaginary parts

$$\begin{cases} \operatorname{Re}\epsilon(\omega) = 1 + \dfrac{\alpha\tau}{1+\omega^2\tau^2} = \dfrac{1+\omega^2\tau^2+\alpha\tau}{1+\omega^2\tau^2} \\ \operatorname{Im}\epsilon(\omega) = \dfrac{\omega\alpha\tau^2}{1+\omega^2\tau^2} \end{cases}$$

We note that $\epsilon(\omega = 0) = 1 + \alpha\tau$ and $\epsilon(\omega) = 1$ as $\omega \to \infty$. The typical behavior is depicted in Figure 1.8.

In Figure 1.9, experimental data of the permittivity for water are shown, and a fit to a Debye model is presented. The fit to the Debye model is augmented with an extra term of optical response to account for fast electronic processes in water. The Debye model is most conveniently written as

$$\epsilon(\omega) = \epsilon_\infty + \frac{\epsilon_s - \epsilon_\infty}{1 - i\omega\tau}$$

where ϵ_∞ is the permittivity for high frequencies and ϵ_s is the static value, $\omega = 0$. The real and imaginary parts of the permittivity with this notation are

$$\begin{cases} \operatorname{Re}\epsilon(\omega) = \epsilon_\infty + \dfrac{\epsilon_s - \epsilon_\infty}{1+\omega^2\tau^2} = \dfrac{\epsilon_\infty\omega^2\tau^2 + \epsilon_s}{1+\omega^2\tau^2} \\ \operatorname{Im}\epsilon(\omega) = \dfrac{\omega\tau(\epsilon_s - \epsilon_\infty)}{1+\omega^2\tau^2} \end{cases} \tag{1.58}$$

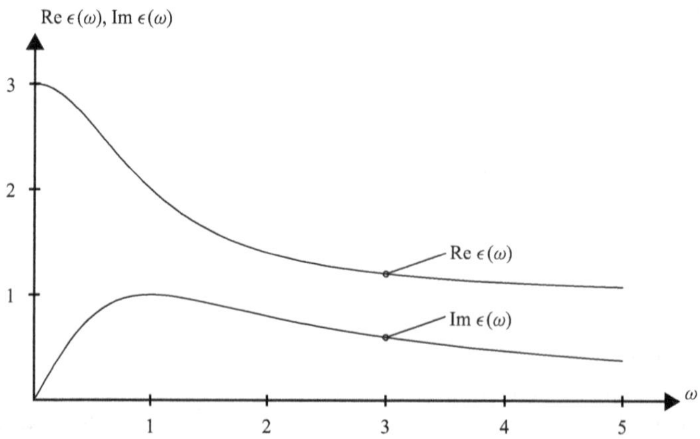

Figure 1.8 The permittivity $\epsilon(\omega)$ as a function of the angular frequency ω for the Debye model. The frequency scale is scaled by $1/\tau$ ($\alpha = 2/\tau$). The imaginary part of $\epsilon(\omega)$ assumes its maximum at $\omega = 1/\tau$ (at $\omega = 1$ with the scaling in the figure).

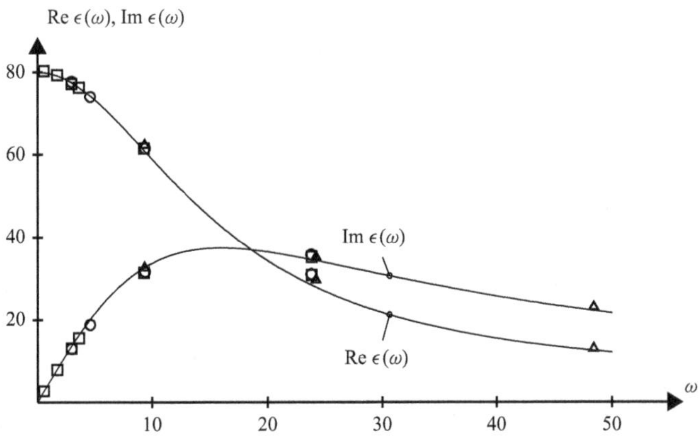

Figure 1.9 Experimental data (circles [35], squares [81], and triangles [153]) of the permittivity $\epsilon(\omega)$ as a function of the angular frequency ω for water at 20°C and frequencies up to 50 GHz. The solid lines show a fit of the experimental data to a Debye model, see text.

Table 1.8 Experimental values of the material parameters in the Debye model for water [35, 81, 153], ethanol, salt water, and fresh water in the microwave domain.

Material	ϵ_s	ϵ_∞	τ (s)	σ (S/m)	Range (GHz)
Water (20°C)	80.0	5.27	1.0×10^{-11}		0–50
Ethanol	25.1	4.4	1.2×10^{-10}		
Salt water (20°C)	80.0	5.27	1.0×10^{-11}	3–5	0–50
Fresh water (20°C)	80.0	5.27	1.0×10^{-11}	10^{-3}	0–50

Experimental values of the parameters ϵ_∞, ϵ_s, and τ for water and ethanol are listed in Table 1.8.

In water with salt ions that move freely, the permittivity is modified (modified Debye model) by an extra term of conductivity σ

$$\epsilon(\omega) = \epsilon_\infty + \frac{\epsilon_s - \epsilon_\infty}{1 - \mathrm{i}\omega\tau} + \mathrm{i}\frac{\sigma}{\omega\epsilon_0} \tag{1.59}$$

Experimental values of the parameters ϵ_∞, ϵ_s, τ, and σ for salt water and fresh water at microwave frequencies are listed in Table 1.8.

At higher frequencies, the permittivity for water has a more complicated behavior than modeled by the Debye model, due to other dominant processes. The frequency behavior of the permittivity of water at higher frequencies—up to optical frequencies—is shown in Figure 1.10.

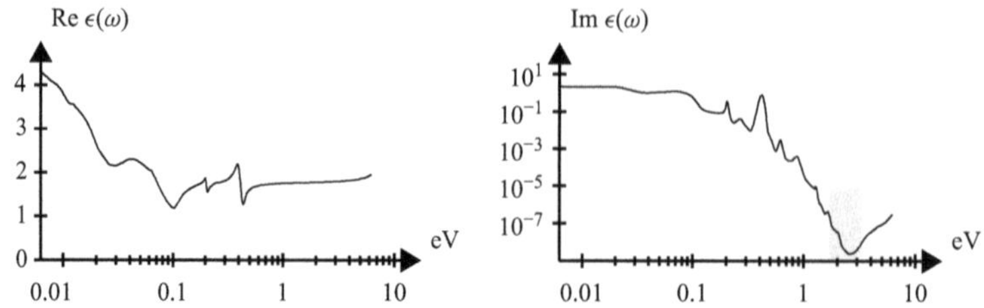

Figure 1.10 Real and imaginary parts of the permittivity for water as a function of frequency (electron volts). 1 eV corresponds to a frequency of 2.42×10^{14} Hz or a wavelength of 1.24 μm. The frequency behavior at lower frequencies is shown in Figure 1.9. Note the low imaginary part of the permittivity at the optical window, 1.7–3.3 eV, which is marked with a gray box. (Data are obtained from Hale and Querry [91] and Irvine and Pollack [98].)

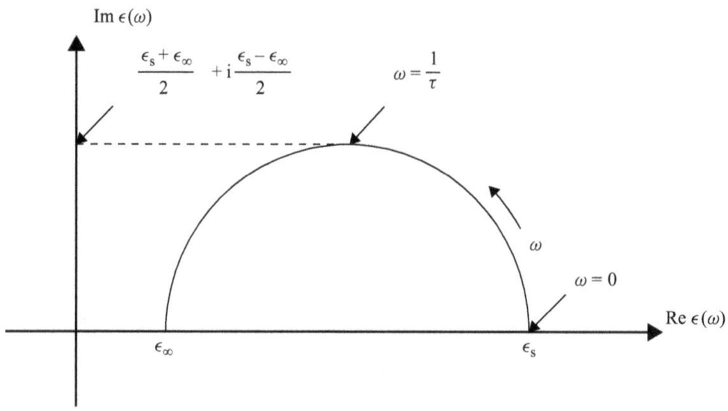

Figure 1.11 The Cole–Cole plot of the permittivity in Example 1.10 in the Debye model.

Example 1.10

From (1.58), we easily see that the real and the imaginary parts in the Debye model satisfy

$$\left(\operatorname{Re}\epsilon(\omega)-\frac{\epsilon_s+\epsilon_\infty}{2}\right)^2+(\operatorname{Im}\epsilon(\omega))^2$$
$$=(\epsilon_s-\epsilon_\infty)^2\left\{\left(\frac{1}{1+\omega^2\tau^2}-\frac{1}{2}\right)^2+\left(\frac{\omega\tau}{1+\omega^2\tau^2}\right)^2\right\}=\frac{(\epsilon_s-\epsilon_\infty)^2}{4}$$

which implies that as the angular frequency ω varies, a circle, centered at $(\epsilon_s+\epsilon_\infty)/2$ and radius[40] $(\epsilon_s-\epsilon_\infty)/2$, is traced out in the complex ϵ-plane, see Figure 1.11. The curve starts at $\epsilon=\epsilon_s$ for $\omega=0$ and ends at $\epsilon=\epsilon_\infty$ as $\omega\to\infty$. The maximum imaginary part $\epsilon=(\epsilon_s+\epsilon_\infty)/2+i(\epsilon_s-\epsilon_\infty)/2$ is obtained at $\omega=1/\tau$. This circle in the complex ϵ-plane

[40] In Section 1.3.4, we prove that the quantity $\epsilon_s-\epsilon_\infty$ is always positive for a passive material.

is the Cole–Cole plot. This feature can be used experimentally to verify that the dynamics of the charges is modeled by the Debye model, and the relevant parameters ϵ_∞, ϵ_s, and τ can be extracted from the plot. ■

Example 1.11
In this example, we investigate the temperature behavior of the permittivity[41] in a polar liquid, where we assume the dipoles are weakly interacting. The potential energy of the dipole in an electric field $\boldsymbol{E}$ depends on the orientation of the dipole. Provided the electric dipole moment of the constitutive parts of the material is $\boldsymbol{p}$, the potential energy U of each molecule of the material is

$$U = -\boldsymbol{p} \cdot \boldsymbol{E} = -pE\cos\theta$$

where the angle between the (permanent) electric dipole moment $\boldsymbol{p}$ and the electric field $\boldsymbol{E}$, directed along $\hat{z}$, is denoted θ, and $p = |\boldsymbol{p}|$ and $E = |\boldsymbol{E}|$. The polarization $\boldsymbol{P}$ then is

$$\boldsymbol{P} = Np\hat{z}\,\langle\cos\theta\rangle$$

where N is the number of dipoles per unit volume, and the ensemble average is defined as

$$\langle f(U)\rangle = \frac{\int f(U)\mathrm{e}^{-\beta U}\,\mathrm{d}U}{\int \mathrm{e}^{-\beta U}\,\mathrm{d}U}$$

where the integration is over all possible energies and $\beta = 1/(k_\mathrm{B}T)$ ($k_\mathrm{B} = 1.38062 \times 10^{-23}$ J/K is the Boltzmann constant[42] and T is the absolute temperature). In our case, we get

$$\langle\cos\theta\rangle = \frac{\int_{-1}^{1}\cos\theta\,\mathrm{e}^{\beta pE\cos\theta}\,\mathrm{d}\cos\theta}{\int_{-1}^{1}\mathrm{e}^{\beta pE\cos\theta}\,\mathrm{d}\cos\theta}$$

Introduce the dimensionless constant $x = pE\beta = pE/k_\mathrm{B}T$, and we get

$$\langle\cos\theta\rangle = \frac{\int_{-1}^{1} t\mathrm{e}^{xt}\,\mathrm{d}t}{\int_{-1}^{1}\mathrm{e}^{xt}\,\mathrm{d}t} = \frac{\mathrm{d}}{\mathrm{d}x}\ln\int_{-1}^{1}\mathrm{e}^{xt}\,\mathrm{d}t = \frac{\mathrm{d}}{\mathrm{d}x}\ln\left(\mathrm{e}^{x} - \mathrm{e}^{-x}\right) - \frac{\mathrm{d}}{\mathrm{d}x}\ln x \equiv L(x)$$

where we have introduced the Langevin[43] function $L(x) = \coth x - 1/x$, which is depicted in Figure 1.12. For small arguments, $x \ll 1$, we have $L(x) \approx x/3 + O(x^2)$, which can be used as an approximation at high temperatures T—small values of x.

Finally, we get the average polarization

$$\boldsymbol{P} = Np\hat{z}L(pE/k_\mathrm{B}T)$$

[41] More precisely, we investigate the temperature behavior of the static value of the permittivity, $\epsilon_\mathrm{s} = \epsilon(\omega = 0)$.

[42] Ludwig Eduard Boltzmann (1844–1906), Austrian physicist and philosopher.

[43] Paul Langevin (1872–1946), French physicist.

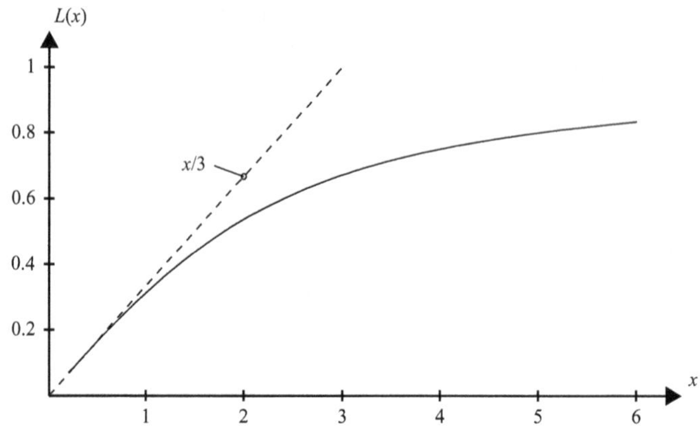

Figure 1.12 The Langevin function $L(x) = \coth x - 1/x$ (solid line) and the approximation $x/3$ (broken line).

This relation shows that the polarization of the material, in general, depends nonlinearly on the electric field. However, for large temperatures $T \gg k_B pE$, which is the most interesting case experimentally, we have a linear relation, i.e.,

$$\boldsymbol{P} = \frac{Np^2 \boldsymbol{E}}{3k_B T}$$

which implies that the static value of the permittivity ϵ_s, as a function of temperature T, is, see also Comment 1.4.

$$\epsilon_s = 1 + \frac{Np^2}{3\epsilon_0 k_B T}$$

■

Comment 1.4
There is a distinction between the local exciting field at the dipole and the macroscopic field $\boldsymbol{E}$. More precisely, the electric polarizability α is

$$\alpha = \frac{p^2}{3\epsilon_0 k_B T}$$

where the static permeability ϵ_s and the α are related by the Clausius–Mossotti (Lorenz–Lorentz) law [101]

$$\frac{\epsilon_s - 1}{\epsilon_s + 2} = \frac{N\alpha}{3}$$

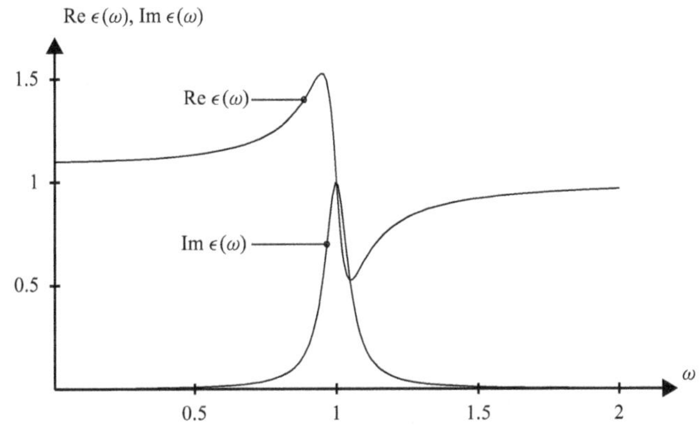

Figure 1.13 The permittivity $\epsilon(\omega)$ as a function of the angular frequency ω for the Lorentz model. The frequency scale is scaled by ω_0 ($\omega_{\rm p} = \sqrt{0.1}\omega_0$ and $\nu = 0.1\omega_0$).

Lorentz Model: Lorentz model, given by the susceptibility function, see (1.28)

$$\chi(t) = \mathrm{H}(t)\frac{\omega_{\rm p}^2}{\nu_0}\mathrm{e}^{-\frac{\nu t}{2}}\sin\nu_0 t, \qquad \epsilon = 1$$

is transformed to the frequency domain by a Fourier transform, see (1.49). The result is the permittivity dyadic, $\boldsymbol{\epsilon}(\omega) = \epsilon(\omega)\mathbf{I}_3$, where

$$\epsilon(\omega) = \epsilon + \int_0^\infty \chi(t)\mathrm{e}^{\mathrm{i}\omega t}\,\mathrm{d}t = 1 - \frac{\omega_{\rm p}^2}{\omega^2 - \omega_0^2 + \mathrm{i}\omega\nu} \tag{1.60}$$

with real and imaginary parts

$$\begin{cases} \operatorname{Re}\epsilon(\omega) = 1 - \dfrac{\omega_{\rm p}^2\left(\omega^2-\omega_0^2\right)}{\left(\omega^2-\omega_0^2\right)^2+\omega^2\nu^2} = \dfrac{\left(\omega^2-\omega_0^2\right)^2+\omega^2\nu^2-\omega_{\rm p}^2\left(\omega^2-\omega_0^2\right)}{\left(\omega^2-\omega_0^2\right)^2+\omega^2\nu^2} \\[2ex] \operatorname{Im}\epsilon(\omega) = \dfrac{\omega_{\rm p}^2\omega\nu}{\left(\omega^2-\omega_0^2\right)^2+\omega^2\nu^2} \end{cases}$$

Notice that the real part can take negative values for certain combinations of the material parameters. The scalar function $\epsilon(\omega)$ as a function of frequency for a Lorentz model is illustrated in Figure 1.13. From the expression above, we note that $\epsilon(\omega = 0) = 1 + \omega_{\rm p}^2/\omega_0^2$ and $\epsilon(\omega) = 1$ as $\omega \to \infty$.

Drude Model: The Drude model[44] corresponds to the case of vanishing restoring force, i.e., $\omega_0 \to 0$. The explicit form is

$$\epsilon(\omega) = 1 - \frac{\omega_{\rm p}^2}{\omega^2 + \mathrm{i}\omega\nu} = 1 - \frac{\omega_{\rm p}^2}{\omega(\omega + \mathrm{i}\nu)} \tag{1.61}$$

[44] Paul Karl Ludwig Drude (1863–1906), German physicist.

with real and imaginary parts

$$\begin{cases} \operatorname{Re}\epsilon(\omega) = 1 - \dfrac{\omega_{\mathrm{p}}^2}{\omega^2 + \nu^2} = -\dfrac{\omega_{\mathrm{p}}^2 - \nu^2 - \omega^2}{\omega^2 + \nu^2} \\ \operatorname{Im}\epsilon(\omega) = \dfrac{\omega_{\mathrm{p}}^2 \nu}{\omega(\omega^2 + \nu^2)} \end{cases}$$

Notice that the real part of the permittivity assumes negative values for all frequencies $\omega^2 < \omega_{\mathrm{p}}^2 - \nu^2$, and that the imaginary part has a pole at $\omega = 0$. Indeed, highly reflecting, low-loss materials, such as metals at optical frequencies show a permittivity with negative real part, see Example 10.6 on page 552. The scalar function $\epsilon(\omega)$ as a function of frequency for the Drude model is illustrated in Figure 1.14.

The Drude model is formally related to the modified Debye model in (1.59). To see this, write (1.61) as (we have added an optical response ϵ_∞ to the Drude model)

$$\epsilon(\omega) = \epsilon_\infty - \frac{\omega_{\mathrm{p}}^2/\nu^2}{1 - \mathrm{i}\omega/\nu} + \mathrm{i}\frac{\omega_{\mathrm{p}}^2/\nu}{\omega}$$

and we see that the Drude model is a special case of the modified Debye model with a conductivity term $\sigma = \epsilon_0 \omega_{\mathrm{p}}^2/\nu$ and

$$\begin{cases} \epsilon_{\mathrm{s}} = \epsilon_\infty - \omega_{\mathrm{p}}^2/\nu^2 \\ \tau = 1/\nu \\ \sigma = \epsilon_0 \omega_{\mathrm{p}}^2/\nu \end{cases}$$

Note that the modified Debye model has four free parameters (ϵ_∞, ϵ_{s}, τ, and σ), but the Drude model only has three (ϵ_∞, ω_{p}, and ν). The parameters in the modified Debye model therefore have to satisfy a constraint in order to be a Drude model. The constraint is

$$\epsilon_{\mathrm{s}} - \epsilon_\infty + \frac{\sigma\tau}{\epsilon_0} = 0$$

This constraint relates the metal data given in Table 1.9, see also Reference 75.

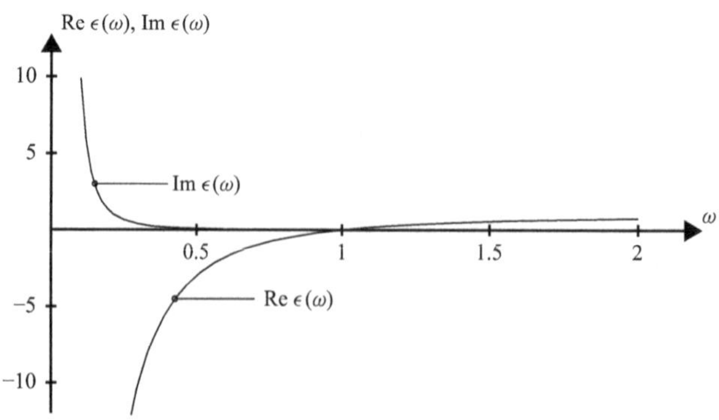

Figure 1.14 The permittivity $\epsilon(\omega)$ as a function of the angular frequency ω for the Drude model. The frequency scale is scaled by ω_{p} and $\nu = 0.01\omega_{\mathrm{p}}$.

Table 1.9 Experimental values of the material parameters in the modified Debye model for some metals [75] in the optical domain are also presented. The range of validity is given in wavelength $\lambda = c_0/f$.

Material	ϵ_s	ϵ_∞	τ (s)	σ (S/m)	Range (nm)
Gold	−15 789	11.575	8.71×10^{-15}	1.6062×10^{7}	700–1 200
Silver	−9 530.5	3.8344	7.35×10^{-15}	1.1486×10^{7}	450–1 200
Copper	−6 672.7	12.076	5.63×10^{-15}	1.0513×10^{7}	550–850
Platinum	−30.005	5.3741	3.28×10^{-16}	9.5505×10^{5}	400–1 200
Aluminum	−656.21	1.8614	1.07×10^{-15}	5.4455×10^{6}	200–700

Both the modified Debye model and the Drude model have as a special case the standard conductivity model, i.e.,

$$\epsilon(\omega) = \epsilon_{\text{reg}}(\omega) + \mathrm{i}\frac{\sigma}{\omega\epsilon_0} \tag{1.62}$$

where $\epsilon_{\text{reg}}(\omega)$ is regular at the origin. We also note that $\epsilon_{\text{reg}}(0) - \epsilon_\infty = -\omega_\text{p}^2/\nu^2 < 0$, which seems to be in conflict with passivity, see Section 1.3.4. This conclusion is, however, not correct, since the Drude model has a singularity at $\omega = 0$, which modifies the argument that makes $\epsilon(0) - \epsilon_\infty$ to be positive under the assumption of passivity, see Section 1.3.4 for more details.

Example 1.12

The plasma is an example of a gyrotropic material. The constitutive relations of a cold plasma are analyzed in Problem 1.10. The result is

$$\boldsymbol{\epsilon}(\omega) = \epsilon\mathbf{I}_2 - \mathrm{i}\mathbf{J}\epsilon_\text{g} + \hat{z}\hat{z}\epsilon_z$$

where

$$\begin{cases} \epsilon = 1 - \dfrac{\omega_\text{p}^2(\omega + \mathrm{i}\nu)}{\omega\left(\omega^2 - \omega_\text{g}^2 - \nu^2 + 2\mathrm{i}\nu\omega\right)} \\ \epsilon_\text{g} = -\dfrac{\omega_\text{p}^2\omega_\text{g}}{\omega\left(\omega^2 - \omega_\text{g}^2 - \nu^2 + 2\mathrm{i}\nu\omega\right)} \\ \epsilon_z = 1 - \dfrac{\omega_\text{p}^2}{\omega(\omega + \mathrm{i}\nu)} \end{cases}$$

and where ν is the collision frequency, $\mathbf{I}_2$ is the identity operator in the x–y-plane, and $\mathbf{J} = \hat{z} \times \mathbf{I}_2$, which is a rotation of $\pi/2$ along the z-axis. The gyrotropic frequency, ω_g, and the plasma frequency, ω_p, of the material are defined as

$$\omega_\text{g} = \frac{qB_0}{m}, \qquad \omega_\text{p} = \sqrt{\frac{Nq^2}{m\epsilon_0}}$$

where N is the number of charges per unit of volume in the plasma, and m and q are the mass and the charge of the charges, respectively.

In the limit of vanishing collision frequency, $\nu \to 0$, these expressions become

$$\begin{cases} \epsilon = 1 - \dfrac{\omega_{\mathrm{p}}^2}{\omega^2 - \omega_{\mathrm{g}}^2} \\ \epsilon_{\mathrm{g}} = -\dfrac{\omega_{\mathrm{p}}^2 \omega_{\mathrm{g}}}{\omega(\omega^2 - \omega_{\mathrm{g}}^2)} \\ \epsilon_z = 1 - \dfrac{\omega_{\mathrm{p}}^2}{\omega^2} \end{cases}$$

The other limiting case, with vanishing gyrotropic frequency, $\omega_{\mathrm{g}} \to 0$, leads to a recovery of the Drude model, (1.61), i.e.,

$$\begin{cases} \epsilon = \epsilon_z = 1 - \dfrac{\omega_{\mathrm{p}}^2}{\omega(\omega + \mathrm{i}\nu)} \\ \epsilon_{\mathrm{g}} = 0 \end{cases}$$

■

Example 1.13

Related to the Lorentz model is the Condon model for optical activity in Example 1.6 on page 32, where we derived the constitutive relations

$$\begin{cases} \chi_{\mathrm{ee}}(t) = \mathrm{H}(t)\dfrac{\omega_{\mathrm{p}}^2}{\nu_0} e^{-\frac{\nu t}{2}} \sin \nu_0 t \\ \chi_{\mathrm{em}}(t) = \mathrm{H}(t)\omega_{\mathrm{c}} e^{-\frac{\nu t}{2}} \left(\cos \nu_0 t - \dfrac{\nu}{2\nu_0} \sin \nu_0 t \right) \end{cases}$$

with Fourier transforms

$$\begin{cases} \displaystyle\int_0^\infty \chi_{\mathrm{ee}}(t) \mathrm{e}^{\mathrm{i}\omega t} \, \mathrm{d}t = -\dfrac{\omega_{\mathrm{p}}^2}{\omega^2 - \omega_0^2 + \mathrm{i}\omega\nu} \\ \displaystyle\int_0^\infty \chi_{\mathrm{em}}(t) \mathrm{e}^{\mathrm{i}\omega t} \, \mathrm{d}t = \dfrac{\mathrm{i}\omega\omega_{\mathrm{c}}}{\omega^2 - \omega_0^2 + \mathrm{i}\omega\nu} \end{cases}$$

and we can identify

$$\begin{cases} \epsilon(\omega) = 1 + \displaystyle\int_0^\infty \chi_{\mathrm{ee}}(t) \mathrm{e}^{\mathrm{i}\omega t} \, \mathrm{d}t = 1 - \dfrac{\omega_{\mathrm{p}}^2}{\omega^2 - \omega_0^2 + \mathrm{i}\omega\nu} \\ \xi(\omega) = \displaystyle\int_0^\infty \chi_{\mathrm{em}}(t) \mathrm{e}^{\mathrm{i}\omega t} \, \mathrm{d}t = \dfrac{\mathrm{i}\omega\omega_{\mathrm{c}}}{\omega^2 - \omega_0^2 + \mathrm{i}\omega\nu} \end{cases}$$

This analysis determines $\epsilon(\omega)$ and $\xi(\omega)$. If the material is non-magnetic $\mu = 1$, and if the material also is reciprocal, see Section 1.3.5 on page 70 and Table 1.7 on page 47, we have also

Table 1.10 Summary of the different models used in this book. The details of the values of the parameters are found in the book.

Model	Time domain	Time harmonic
Debye	$\chi(t) = \mathrm{H}(t)\alpha \mathrm{e}^{-t/\tau}$	$\epsilon(\omega) = \epsilon_\infty + \frac{\alpha\tau}{1-i\omega\tau}$
Modified Debye		$\epsilon(\omega) = \epsilon_\infty + \frac{\epsilon_\mathrm{s}-\epsilon_\infty}{1-\mathrm{i}\omega\tau} + \frac{\mathrm{i}\sigma}{\omega\epsilon_0}$
Lorentz	$\chi(t) = \mathrm{H}(t)\frac{\omega_\mathrm{p}^2}{\nu_0}\mathrm{e}^{-\frac{\nu t}{2}} \sin \nu_0 t$	$\epsilon(\omega) = \epsilon_\infty - \frac{\omega_\mathrm{p}^2}{\omega^2-\omega_0^2+\mathrm{i}\omega\nu}$
Drude		$\epsilon(\omega) = \epsilon_\infty - \frac{\omega_\mathrm{p}^2}{\omega(\omega+\mathrm{i}\nu)}$
Plasma	$\boldsymbol{\chi}(t) = \frac{\omega_\mathrm{p}^2}{\omega_\mathrm{g}}\mathrm{H}(t)\left\{\mathbf{I}_2 \sin \omega_\mathrm{g} t - \mathbf{J}(1-\cos \omega_\mathrm{g} t) + \hat{z}\hat{z}\omega_\mathrm{g} t\right\}$	$\boldsymbol{\epsilon}(\omega) = \epsilon\mathbf{I}_2 - \mathrm{i}\mathbf{J}\epsilon_\mathrm{g} + \hat{z}\hat{z}\epsilon_z$
Condon	$\chi_\mathrm{ee}(t) = \mathrm{H}(t)\frac{\omega_\mathrm{p}^2}{\nu_0}e^{-\frac{\nu t}{2}} \sin \nu_0 t$	$\epsilon(\omega) = \epsilon_\infty - \frac{\omega_\mathrm{p}^2}{\omega^2-\omega_0^2+\mathrm{i}\omega\nu}$
	$\chi_\mathrm{em}(t) = \mathrm{H}(t)\omega_\mathrm{c} e^{-\frac{\nu t}{2}}\left(\cos \nu_0 t - \frac{\nu}{2\nu_0} \sin \nu_0 t\right)$	$\xi(\omega) = \frac{\mathrm{i}\omega\omega_\mathrm{c}}{\omega^2-\omega_0^2+\mathrm{i}\omega\nu}$

$\zeta(\omega) = -\xi(\omega)$. Notice that $\xi(\omega) \to 0$ as $\omega \to 0$, and no biisotropic effects remain for static fields. ■

We end this long section by a summary of the different models used in the book, see Table 1.10.

1.3.3 Poynting's Theorem, Active, Passive, and Lossless Media

In Section 1.1, we derived Poynting's theorem, see (1.14) on page 9

$$\nabla \cdot \boldsymbol{S}(t) + \boldsymbol{H}(t) \cdot \frac{\partial \boldsymbol{B}(t)}{\partial t} + \boldsymbol{E}(t) \cdot \frac{\partial \boldsymbol{D}(t)}{\partial t} + \boldsymbol{E}(t) \cdot \boldsymbol{J}(t) = 0$$

The equation describes conservation of power and contains products of fields. For a product of time-harmonic fields, the most pertinent quantity is the time average over one period, see Comment 1.5. We denote the time average as $\langle\cdot\rangle$ and for Poynting's theorem, we obtain

$$\langle\nabla \cdot \boldsymbol{S}(t)\rangle + \left\langle\boldsymbol{H}(t) \cdot \frac{\partial \boldsymbol{B}(t)}{\partial t}\right\rangle + \left\langle\boldsymbol{E}(t) \cdot \frac{\partial \boldsymbol{D}(t)}{\partial t}\right\rangle + \langle\boldsymbol{E}(t) \cdot \boldsymbol{J}(t)\rangle = 0$$

The first term in this equality is the time average of the power flux density of the electromagnetic field, i.e.,

$$\langle\boldsymbol{S}(t)\rangle = \frac{1}{2}\operatorname{Re}\left\{\boldsymbol{E}(\omega) \times \boldsymbol{H}^*(\omega)\right\} \tag{1.63}$$

and similarly for the time average of the stored powers and the dissipated power (use $\frac{\partial}{\partial t} \to -\mathrm{i}\omega$)

$$\begin{cases} \left\langle \boldsymbol{H}(t) \cdot \dfrac{\partial \boldsymbol{B}(t)}{\partial t} \right\rangle = \dfrac{1}{2} \operatorname{Re} \left\{ \mathrm{i}\omega \boldsymbol{H}(\omega) \cdot \boldsymbol{B}^*(\omega) \right\} \\ \left\langle \boldsymbol{E}(t) \cdot \dfrac{\partial \boldsymbol{D}(t)}{\partial t} \right\rangle = \dfrac{1}{2} \operatorname{Re} \left\{ \mathrm{i}\omega \boldsymbol{E}(\omega) \cdot \boldsymbol{D}^*(\omega) \right\} \\ \langle \boldsymbol{E}(t) \cdot \boldsymbol{J}(t) \rangle = \dfrac{1}{2} \operatorname{Re} \left\{ \boldsymbol{E}(\omega) \cdot \boldsymbol{J}^*(\omega) \right\} \end{cases}$$

Poynting's theorem (balance of power) for time-harmonic fields, averaged over a period, becomes (notice that the time average and the differentiation w.r.t. space commute, i.e., $\langle \nabla \cdot \boldsymbol{S}(t) \rangle = \nabla \cdot \langle \boldsymbol{S}(t) \rangle$):

$$\nabla \cdot \langle \boldsymbol{S}(t) \rangle + \frac{1}{2} \operatorname{Re} \left\{ \mathrm{i}\omega \left[\boldsymbol{H}(\omega) \cdot \boldsymbol{B}^*(\omega) + \boldsymbol{E}(\omega) \cdot \boldsymbol{D}^*(\omega) \right] \right\} + \frac{1}{2} \operatorname{Re} \left\{ \boldsymbol{E}(\omega) \cdot \boldsymbol{J}^*(\omega) \right\} = 0$$

Comment 1.5

The time average of a product of two time-harmonic fields $f_1(t)$ and $f_2(t)$ is easily obtained by an average over one period $T = 2\pi/\omega$. We denote the Fourier transformed (time-harmonic) quantities by $f_1(\omega)$ and $f_2(\omega)$, respectively. We get

$$\begin{aligned} \langle f_1(t) f_2(t) \rangle &= \frac{1}{T} \int_0^T f_1(t) f_2(t) \, \mathrm{d}t = \frac{1}{T} \int_0^T \operatorname{Re} \left\{ f_1(\omega) \mathrm{e}^{-\mathrm{i}\omega t} \right\} \operatorname{Re} \left\{ f_2(\omega) \mathrm{e}^{-\mathrm{i}\omega t} \right\} \mathrm{d}t \\ &= \frac{1}{2T} \int_0^T \left\{ \operatorname{Re} \left(f_1(\omega) f_2(\omega) \mathrm{e}^{-2\mathrm{i}\omega t} \right) + \operatorname{Re} \left(f_1(\omega) f_2^*(\omega) \right) \right\} \mathrm{d}t \\ &= \frac{1}{2} \operatorname{Re} \left\{ f_1(\omega) f_2^*(\omega) \right\} \end{aligned}$$

where we used $2 \operatorname{Re}(z_1) \operatorname{Re}(z_2) = \operatorname{Re}(z_1 z_2) + \operatorname{Re}(z_1 z_2^*)$.

Of special interest is the case without currents $\boldsymbol{J} = \mathbf{0}$. Conducting currents can, as we have seen, be included in the permittivity dyadic $\boldsymbol{\epsilon}$. Poynting's theorem is then simplified to

$$\begin{aligned} \nabla \cdot \langle \boldsymbol{S}(t) \rangle &= -\frac{1}{2} \operatorname{Re} \left\{ \mathrm{i}\omega \left[\boldsymbol{H}(\omega) \cdot \boldsymbol{B}^*(\omega) + \boldsymbol{E}(\omega) \cdot \boldsymbol{D}^*(\omega) \right] \right\} \\ &= -\frac{\mathrm{i}\omega}{4} \left\{ \boldsymbol{H}(\omega) \cdot \boldsymbol{B}^*(\omega) - \boldsymbol{H}^*(\omega) \cdot \boldsymbol{B}(\omega) + \boldsymbol{E}(\omega) \cdot \boldsymbol{D}^*(\omega) - \boldsymbol{E}(\omega)^* \cdot \boldsymbol{D}(\omega) \right\} \end{aligned}$$

where we used $\operatorname{Re} z = (z + z^*)/2$.

On the right-hand side in this expression, we enter the constitutive relations from (1.51), and eliminate the fields $\boldsymbol{D}$ and $\boldsymbol{B}$. The divergence of the Poynting vector is then expressed in the fields $\boldsymbol{E}$ and $\boldsymbol{H}$

$$
\begin{aligned}
\nabla \cdot \langle \boldsymbol{S}(t) \rangle = & -\frac{\mathrm{i}\omega\epsilon_0}{4} \{\eta_0 \boldsymbol{H} \cdot (\boldsymbol{\zeta}^* \cdot \boldsymbol{E}^* + \eta_0 \boldsymbol{\mu}^* \cdot \boldsymbol{H}^*) - \eta_0 \boldsymbol{H}^* \cdot (\boldsymbol{\zeta} \cdot \boldsymbol{E} + \eta_0 \boldsymbol{\mu} \cdot \boldsymbol{H}) \\
& + \boldsymbol{E} \cdot (\boldsymbol{\epsilon}^* \cdot \boldsymbol{E}^* + \eta_0 \boldsymbol{\xi}^* \cdot \boldsymbol{H}^*) - \boldsymbol{E}^* \cdot (\boldsymbol{\epsilon} \cdot \boldsymbol{E} + \eta_0 \boldsymbol{\xi} \cdot \boldsymbol{H})\} \\
= & \frac{\mathrm{i}\omega\epsilon_0}{4} \left\{ \boldsymbol{E}^* \cdot \left(\boldsymbol{\epsilon} - \boldsymbol{\epsilon}^\dagger\right) \cdot \boldsymbol{E} + \eta_0^2 \boldsymbol{H}^* \cdot \left(\boldsymbol{\mu} - \boldsymbol{\mu}^\dagger\right) \cdot \boldsymbol{H} \right. \\
& \left. + \eta_0 \boldsymbol{E}^* \cdot \left(\boldsymbol{\xi} - \boldsymbol{\zeta}^\dagger\right) \cdot \boldsymbol{H} + \eta_0 \boldsymbol{H}^* \cdot \left(\boldsymbol{\zeta} - \boldsymbol{\xi}^\dagger\right) \cdot \boldsymbol{E} \right\}
\end{aligned}
$$

where the dagger† denotes the Hermitian adjoint of the dyadic, see Appendix I, and we have also used

$$
\boldsymbol{a} \cdot \mathbf{A}^* \cdot \boldsymbol{b}^* = \boldsymbol{b}^* \cdot \mathbf{A}^\dagger \cdot \boldsymbol{a}
$$

It is often convenient to use a combined matrix and dyadic notation. The divergence of the Poynting vector can then be written in the following compact form:

$$
\nabla \cdot \langle \boldsymbol{S}(t) \rangle = \frac{\mathrm{i}\omega\epsilon_0}{4} \begin{pmatrix} \boldsymbol{E} \\ \eta_0 \boldsymbol{H} \end{pmatrix}^\dagger \cdot \begin{pmatrix} \boldsymbol{\epsilon} - \boldsymbol{\epsilon}^\dagger & \boldsymbol{\xi} - \boldsymbol{\zeta}^\dagger \\ \boldsymbol{\zeta} - \boldsymbol{\xi}^\dagger & \boldsymbol{\mu} - \boldsymbol{\mu}^\dagger \end{pmatrix} \cdot \begin{pmatrix} \boldsymbol{E} \\ \eta_0 \boldsymbol{H} \end{pmatrix} \tag{1.64}
$$

In the six-dimensional formulation of (1.52) and (1.53), we have

$$
\nabla \cdot \langle \boldsymbol{S}(t) \rangle = \frac{\mathrm{i}\omega\epsilon_0}{4} \mathsf{e}^*(\omega) \cdot \left(\mathsf{M}(\omega) - \mathsf{M}^\dagger(\omega)\right) \cdot \mathsf{e}(\omega) = -\frac{\epsilon_0}{2} \mathsf{e}^*(\omega) \cdot \operatorname{Im} \omega \mathsf{M}(\omega) \cdot \mathsf{e}(\omega)
$$

where the imaginary part of the dyadic M is defined as

$$
\operatorname{Im} \mathsf{M} = \mathsf{M}_\mathrm{i} = \frac{1}{2\mathrm{i}} \left(\mathsf{M} - \mathsf{M}^\dagger\right)
$$

Note that the imaginary part M_i is an Hermitian dyadic, see also Appendix A.2.

The quantity $-\nabla \cdot \langle \boldsymbol{S}(t) \rangle$ gives a measure of the average power the electromagnetic field delivers to the material per unit volume. This quantity can be used to classify the material as active or passive depending on the sign of this quantity. The following definitions are introduced for a fixed (angular) frequency $\omega \neq 0$:

$$
\begin{aligned}
&\text{Passive material if} \quad \nabla \cdot \langle \boldsymbol{S}(t) \rangle \leq 0 \\
&\text{Active material if} \quad \nabla \cdot \langle \boldsymbol{S}(t) \rangle > 0
\end{aligned}
\qquad \text{for all fields } \{\boldsymbol{E}, \boldsymbol{H}\} \neq \{\mathbf{0}, \mathbf{0}\}
$$

and the notion of a lossless material (special case of passive material) is used if

$$
\text{Lossless material if} \quad \nabla \cdot \langle \boldsymbol{S}(t) \rangle = 0 \quad \text{for all fields } \{\boldsymbol{E}, \boldsymbol{H}\} \neq \{\mathbf{0}, \mathbf{0}\}
$$

These definitions have the following physical implications. Integrate $\nabla \cdot \langle \boldsymbol{S}(t) \rangle$ over a volume V bounded by S (outward directed normal $\hat{\boldsymbol{\nu}}$). The divergence theorem gives

$$\text{Passive material} \qquad \iint_S \langle \boldsymbol{S}(t) \rangle \cdot \hat{\boldsymbol{\nu}} \, \mathrm{d}S \leq 0$$

$$\text{Active material} \qquad \iint_S \langle \boldsymbol{S}(t) \rangle \cdot \hat{\boldsymbol{\nu}} \, \mathrm{d}S > 0$$

$$\text{Lossless material} \qquad \iint_S \langle \boldsymbol{S}(t) \rangle \cdot \hat{\boldsymbol{\nu}} \, \mathrm{d}S = 0$$

These definitions imply that for a passive material in V, the outward radiated power is always non-positive, $\iint_S \langle \boldsymbol{S}(t) \rangle \cdot \hat{\boldsymbol{\nu}} \, \mathrm{d}S \leq 0$. For an active material, this integral is positive, due to creation of electromagnetic energy (by non-electromagnetic sources in V). For a lossless material, the outward radiated power averaged over a period always is balanced by the power radiated inward through the surface during a period. Notice that this classification holds for time-harmonic fields. A material can be passive for one frequency and lossless for another. However, a material cannot be lossless for all frequencies.[45]

1.3.3.1 Lossless Material

In a lossless material, the dyadics $\boldsymbol{\epsilon}$, $\boldsymbol{\xi}$, $\boldsymbol{\zeta}$, and $\boldsymbol{\mu}$ in (1.51) have to satisfy a set of conditions. From (1.64), we see that

$$\begin{cases} \boldsymbol{\epsilon} = \boldsymbol{\epsilon}^\dagger \\ \boldsymbol{\mu} = \boldsymbol{\mu}^\dagger \\ \boldsymbol{\xi} = \boldsymbol{\zeta}^\dagger \end{cases} \tag{1.65}$$

since the fields $\boldsymbol{E}$ and $\boldsymbol{H}$ can be chosen arbitrarily. In an isotropic material, we immediately see that ϵ and μ in (1.56) have to be real for a lossless material. Moreover, for lossless, anisotropic material with real permittivity matrix

$$\epsilon_{ij} = \epsilon_{ji}$$

i.e., $[\epsilon]$ is a symmetric matrix, which can be diagonalized (with real coordinate axes) and a classification as in Table 1.5.

In the six-dimensional formulation of (1.52) and (1.53), the lossless materials satisfy

$$\operatorname{Im} \mathsf{M} = 0$$

[45]An isotropic material with dispersion cannot be lossless for all frequencies, since then, see (1.49)

$$0 = \operatorname{Im} \epsilon(\omega) = \int_0^\infty \chi(t) \sin \omega t \, \mathrm{d}t = \frac{1}{2\mathrm{i}} \left\{ \int_0^\infty \chi(t) e^{\mathrm{i}\omega t} \, \mathrm{d}t - \int_{-\infty}^0 \chi(-t) e^{\mathrm{i}\omega t} \, \mathrm{d}t \right\} \text{ for all } \omega$$

which implies that the function $\mathrm{H}(t)\chi(t) - \mathrm{H}(-t)\chi(-t) = 0$, and $\chi(t)$ is identically zero almost everywhere, which is a contradiction to a non-empty material.

1.3.3.2 Passive Material

We now proceed by investigating the consequences on the constitutive relations for a passive material.

In a passive material, the constitutive relations satisfy, see (1.64)

$$\begin{pmatrix} \boldsymbol{E} \\ \eta_0 \boldsymbol{H} \end{pmatrix}^{\dagger} \cdot \begin{pmatrix} \mathrm{i}\omega\left(\boldsymbol{\epsilon}-\boldsymbol{\epsilon}^{\dagger}\right) & \mathrm{i}\omega\left(\boldsymbol{\xi}-\boldsymbol{\zeta}^{\dagger}\right) \\ \mathrm{i}\omega\left(\boldsymbol{\zeta}-\boldsymbol{\xi}^{\dagger}\right) & \mathrm{i}\omega\left(\boldsymbol{\mu}-\boldsymbol{\mu}^{\dagger}\right) \end{pmatrix} \cdot \begin{pmatrix} \boldsymbol{E} \\ \eta_0 \boldsymbol{H} \end{pmatrix} \leq 0$$

for all (non-static) fields $\{\boldsymbol{E},\boldsymbol{H}\} \neq \{\boldsymbol{0},\boldsymbol{0}\}$, i.e., the Hermitian dyadic

$$2\omega \operatorname{Im} \mathsf{M} = \begin{pmatrix} \boldsymbol{\alpha} & \boldsymbol{\beta} \\ \boldsymbol{\gamma} & \boldsymbol{\delta} \end{pmatrix} = \begin{pmatrix} -\mathrm{i}\omega\left(\boldsymbol{\epsilon}-\boldsymbol{\epsilon}^{\dagger}\right) & -\mathrm{i}\omega\left(\boldsymbol{\xi}-\boldsymbol{\zeta}^{\dagger}\right) \\ -\mathrm{i}\omega\left(\boldsymbol{\zeta}-\boldsymbol{\xi}^{\dagger}\right) & -\mathrm{i}\omega\left(\boldsymbol{\mu}-\boldsymbol{\mu}^{\dagger}\right) \end{pmatrix} \geq 0$$

is a non-negative definite six-dimensional dyadic, and as a consequence, all eigenvalues of the dyadic are non-negative. Here, we have employed the six-dimensional formulation of (1.52) and (1.53), and for convenience, we have introduced the notations

$$\begin{cases} \boldsymbol{\alpha} = -\mathrm{i}\omega\left(\boldsymbol{\epsilon}-\boldsymbol{\epsilon}^{\dagger}\right) = 2\omega \operatorname{Im} \boldsymbol{\epsilon} \\ \boldsymbol{\beta} = -\mathrm{i}\omega\left(\boldsymbol{\xi}-\boldsymbol{\zeta}^{\dagger}\right) \end{cases} \qquad \begin{cases} \boldsymbol{\gamma} = -\mathrm{i}\omega\left(\boldsymbol{\zeta}-\boldsymbol{\xi}^{\dagger}\right) \\ \boldsymbol{\delta} = -\mathrm{i}\omega\left(\boldsymbol{\mu}-\boldsymbol{\mu}^{\dagger}\right) = 2\omega \operatorname{Im} \boldsymbol{\mu} \end{cases}$$

Note that $\boldsymbol{\alpha}$ and $\boldsymbol{\delta}$ are Hermitian dyadics, and that $\boldsymbol{\gamma}$ and $\boldsymbol{\delta}$ are related, i.e.,

$$\begin{cases} \boldsymbol{\alpha} = \boldsymbol{\alpha}^{\dagger} \\ \boldsymbol{\delta} = \boldsymbol{\delta}^{\dagger} \end{cases} \qquad \begin{cases} \boldsymbol{\beta} = \boldsymbol{\gamma}^{\dagger} \\ \boldsymbol{\gamma} = \boldsymbol{\beta}^{\dagger} \end{cases}$$

In a passive material, we thus have

$$\boldsymbol{E}^{*} \cdot \boldsymbol{\alpha} \cdot \boldsymbol{E} + \boldsymbol{E}^{*} \cdot \boldsymbol{\beta} \cdot \eta_0 \boldsymbol{H} + \eta_0 \boldsymbol{H}^{*} \cdot \boldsymbol{\gamma} \cdot \boldsymbol{E} + \eta_0 \boldsymbol{H}^{*} \cdot \boldsymbol{\delta} \cdot \eta_0 \boldsymbol{H} \geq 0 \tag{1.66}$$

for all (non-static) fields $\{\boldsymbol{E},\boldsymbol{H}\} \neq \{\boldsymbol{0},\boldsymbol{0}\}$. Applied to the fields $\{\boldsymbol{E},\boldsymbol{0}\}$ and $\{\boldsymbol{0},\boldsymbol{H}\}$ implies

$$\boldsymbol{\alpha} \geq 0, \qquad \boldsymbol{\delta} \geq 0$$

i.e., the three-dimensional dyadics $\boldsymbol{\alpha} = -\mathrm{i}\omega\left(\boldsymbol{\epsilon}-\boldsymbol{\epsilon}^{\dagger}\right)$ and $\boldsymbol{\delta} = -\mathrm{i}\omega\left(\boldsymbol{\mu}-\boldsymbol{\mu}^{\dagger}\right)$ are non-negative definite dyadics. Specifically, the diagonal entries and eigenvalues are non-negative.

Introduce the square root $\sqrt{\boldsymbol{\alpha}}$ of the non-negative definite dyadic $\boldsymbol{\alpha}$, i.e.,

$$\boldsymbol{\alpha} = \sqrt{\boldsymbol{\alpha}} \cdot \sqrt{\boldsymbol{\alpha}}$$

where $\sqrt{\boldsymbol{\alpha}}$ is a Hermitian dyadic. This dyadic is also non-negative definite. From the identity and $\boldsymbol{\beta}^{\dagger} = \boldsymbol{\gamma}$, we obtain

$$\begin{aligned} &\left|\sqrt{\boldsymbol{\alpha}} \cdot \boldsymbol{E} + \sqrt{\boldsymbol{\alpha}}^{-1} \cdot \boldsymbol{\beta} \cdot \eta_0 \boldsymbol{H}\right|^2 \\ &= \left(\boldsymbol{E}^{*} \cdot \sqrt{\boldsymbol{\alpha}} + \eta_0 \boldsymbol{H}^{*} \cdot \boldsymbol{\gamma} \cdot \sqrt{\boldsymbol{\alpha}}^{-1}\right) \cdot \left(\sqrt{\boldsymbol{\alpha}} \cdot \boldsymbol{E} + \sqrt{\boldsymbol{\alpha}}^{-1} \cdot \boldsymbol{\beta} \cdot \eta_0 \boldsymbol{H}\right) \\ &= \boldsymbol{E}^{*} \cdot \boldsymbol{\alpha} \cdot \boldsymbol{E} + \boldsymbol{E}^{*} \cdot \boldsymbol{\beta} \cdot \eta_0 \boldsymbol{H} + \eta_0 \boldsymbol{H}^{*} \cdot \boldsymbol{\gamma} \cdot \boldsymbol{E} + \eta_0 \boldsymbol{H}^{*} \cdot \boldsymbol{\gamma} \cdot \boldsymbol{\alpha}^{-1} \cdot \boldsymbol{\beta} \cdot \eta_0 \boldsymbol{H} \end{aligned}$$

and rewrite the identity by the use of (1.66). We get

$$\left|\sqrt{\boldsymbol{\alpha}} \cdot \boldsymbol{E} + \sqrt{\boldsymbol{\alpha}}^{-1} \cdot \boldsymbol{\beta} \cdot \eta_0 \boldsymbol{H}\right|^2 + \eta_0 \boldsymbol{H}^* \cdot \left(\boldsymbol{\delta} - \boldsymbol{\gamma} \cdot \boldsymbol{\alpha}^{-1} \cdot \boldsymbol{\beta}\right) \cdot \eta_0 \boldsymbol{H} \geq 0$$

This inequality implies that (take $\boldsymbol{E} = -\boldsymbol{\alpha}^{-1} \cdot \boldsymbol{\beta} \cdot \eta_0 \boldsymbol{H}$ and $\eta_0 \boldsymbol{H}$ arbitrary)

$$\boldsymbol{\delta} - \boldsymbol{\gamma} \cdot \boldsymbol{\alpha}^{-1} \cdot \boldsymbol{\beta} \geq 0$$

i.e., the dyadic $\boldsymbol{\delta} - \boldsymbol{\gamma} \cdot \boldsymbol{\alpha}^{-1} \cdot \boldsymbol{\beta}$ is a non-negative definite dyadic. In terms of the material dyadics $\boldsymbol{\epsilon}$, $\boldsymbol{\mu}$, $\boldsymbol{\xi}$, and $\boldsymbol{\zeta}$, this condition becomes

$$\boldsymbol{\delta} - \boldsymbol{\gamma} \cdot \boldsymbol{\alpha}^{-1} \cdot \boldsymbol{\beta} = 2\omega \operatorname{Im} \boldsymbol{\mu} + \frac{1}{2}\omega \left(\boldsymbol{\zeta} - \boldsymbol{\xi}^{\dagger}\right) \cdot (\operatorname{Im} \boldsymbol{\epsilon})^{-1} \cdot \left(\boldsymbol{\xi} - \boldsymbol{\zeta}^{\dagger}\right) \geq 0$$

In summary, the condition for a passive material for a bianisotropic material is

$$\begin{cases} \omega \operatorname{Im} \boldsymbol{\epsilon} \geq 0 \\ \omega \operatorname{Im} \boldsymbol{\mu} \geq 0 \\ \omega \left\{4 \operatorname{Im} \boldsymbol{\mu} - \left(\boldsymbol{\xi} - \boldsymbol{\zeta}^{\dagger}\right)^{\dagger} \cdot (\operatorname{Im} \boldsymbol{\epsilon})^{-1} \cdot \left(\boldsymbol{\xi} - \boldsymbol{\zeta}^{\dagger}\right)\right\} \geq 0 \end{cases} \tag{1.67}$$

Example 1.14
In an isotropic material, i.e.,

$$\begin{cases} \boldsymbol{\epsilon} = \epsilon \mathbf{I} \\ \boldsymbol{\mu} = \mu \mathbf{I} \end{cases} \qquad \begin{cases} \boldsymbol{\xi} = \mathbf{0} \\ \boldsymbol{\zeta} = \mathbf{0} \end{cases}$$

the passive condition in (1.67) implies

$$\begin{cases} \omega \operatorname{Im} \epsilon \geq 0 \\ \omega \operatorname{Im} \mu \geq 0 \end{cases}$$

This condition is true for the models by Debye and Lorentz (and Drude). ■

Example 1.15
In a biisotropic material, the constitutive relations are

$$\begin{cases} \boldsymbol{\epsilon} = \epsilon \mathbf{I} \\ \boldsymbol{\mu} = \mu \mathbf{I} \end{cases} \qquad \begin{cases} \boldsymbol{\xi} = \xi \mathbf{I} = (\kappa + \mathrm{i}\chi)\mathbf{I} \\ \boldsymbol{\zeta} = \zeta \mathbf{I} = (\kappa - \mathrm{i}\chi)\mathbf{I} \end{cases}$$

where we, for convenience, have introduced two new, complex-valued parameters, κ and χ, the reciprocity and the chirality parameters, respectively. These are expressed in the previous ones as

$$\begin{cases} \xi = \kappa + \mathrm{i}\chi \\ \zeta = \kappa - \mathrm{i}\chi \end{cases} \qquad \begin{cases} \kappa = \dfrac{\zeta + \xi}{2} \\ \chi = \mathrm{i}\dfrac{\zeta - \xi}{2} \end{cases}$$

The conditions on the material parameters in a passive biisotropic material are, see (1.67)

$$\begin{cases} \omega \operatorname{Im} \epsilon \geq 0 \\ \omega \operatorname{Im} \mu \geq 0 \\ \omega \left(4 \operatorname{Im} \epsilon \operatorname{Im} \mu - |\xi - \zeta^*|^2\right) \geq 0 \end{cases}$$

Specifically, if the material is reciprocal ($\xi = -\zeta$), we have

$$|\operatorname{Re} \xi| = |\operatorname{Re} \zeta| \leq \sqrt{\operatorname{Im} \epsilon \operatorname{Im} \mu}$$

An immediate consequence of this is that passive, reciprocal, biisotropic material cannot have vanishing imaginary parts of μ or ϵ. In particular, adopting the Condon model in Example 1.13 on page 56, and Lorentz models for ϵ and μ, i.e.,

$$\begin{cases} \epsilon(\omega) = 1 - \dfrac{\omega_{\mathrm{pe}}^2}{\omega^2 - \omega_{0\mathrm{e}}^2 + \mathrm{i}\omega\nu_{\mathrm{e}}} \\ \mu(\omega) = 1 - \dfrac{\omega_{\mathrm{pm}}^2}{\omega^2 - \omega_{0\mathrm{m}}^2 + \mathrm{i}\omega\nu_{\mathrm{m}}} \\ \xi(\omega) = -\zeta(\omega) = \dfrac{\mathrm{i}\omega\omega_{\mathrm{c}}}{\omega^2 - \omega_{0\mathrm{ce}}^2 + \mathrm{i}\omega\nu_{\mathrm{c}}} \end{cases} \qquad \begin{cases} \operatorname{Im} \epsilon(\omega) = \dfrac{\omega\nu_{\mathrm{e}}\omega_{\mathrm{pe}}^2}{\left(\omega^2 - \omega_{0\mathrm{e}}^2\right)^2 + \omega^2\nu_{\mathrm{e}}^2} \\ \operatorname{Im} \mu(\omega) = \dfrac{\omega\nu_{\mathrm{m}}\omega_{\mathrm{pm}}^2}{\left(\omega^2 - \omega_{0\mathrm{m}}^2\right)^2 + \omega^2\nu_{\mathrm{m}}^2} \\ \operatorname{Re} \xi(\omega) = \dfrac{\omega^2\nu_{\mathrm{c}}\omega_{\mathrm{c}}}{\left(\omega^2 - \omega_{0\mathrm{c}}^2\right)^2 + \omega^2\nu_{\mathrm{c}}^2} \end{cases}$$

We see that this model always becomes active at high frequencies, and therefore, the Condon model has its main potential as a model of chiral material and optic activity at low frequencies. There are many ways to remedy this deficiency. A slight modification of the model above, e.g.,

$$\begin{cases} \epsilon(\omega) = 1 - \dfrac{\omega_{\mathrm{pe}}^2}{\omega^2 - \omega_0^2 + \mathrm{i}\omega\nu_{\mathrm{e}}} + \dfrac{\alpha\tau}{1 - \mathrm{i}\omega\tau} \\ \mu(\omega) = 1 - \dfrac{\omega_{\mathrm{pm}}^2}{\omega^2 - \omega_0^2 + \mathrm{i}\omega\nu_{\mathrm{m}}} \\ \xi(\omega) = -\zeta(\omega) = \dfrac{\mathrm{i}\omega\omega_{\mathrm{c}}}{\omega^2 - \omega_0^2 + \mathrm{i}\omega\nu_{\mathrm{c}}} \end{cases}$$

makes the model passive again, since

$$\begin{cases} \operatorname{Im}\epsilon(\omega) = \dfrac{\omega\nu_{\rm e}\omega_{\rm pe}^2}{\left(\omega^2-\omega_0^2\right)^2+\omega^2\nu_{\rm e}^2} + \dfrac{\omega\alpha\tau^2}{1+\omega^2\tau^2} \\ \operatorname{Im}\mu(\omega) = \dfrac{\omega\nu_{\rm m}\omega_{\rm pm}^2}{\left(\omega^2-\omega_0^2\right)^2+\omega^2\nu_{\rm m}^2} \\ \operatorname{Re}\xi(\omega) = \dfrac{\omega^2\omega_{\rm c}\nu_{\rm c}}{\left(\omega^2-\omega_0^2\right)^2+\omega^2\nu_{\rm c}^2} \end{cases}$$

and to leading order in ω, $\operatorname{Im}\epsilon \operatorname{Im}\mu - (\operatorname{Re}\xi)^2$ behaves as $\omega \to \infty$

$$\frac{\alpha}{\omega}\frac{\nu_{\rm m}\omega_{\rm pm}^2}{\omega^3} - \frac{\omega_{\rm c}^2\nu_{\rm c}^2}{\omega^4}$$

which is non-negative provided $\alpha\nu_{\rm m}\omega_{\rm pm}^2 \geq \omega_{\rm c}^2\nu_{\rm c}^2$. ■

Example 1.16

The plasma model in Example 1.12 on page 55 shows that

$$\boldsymbol{\epsilon}(\omega) = \epsilon\mathbf{I}_2 - \mathrm{i}\mathbf{J}\epsilon_{\rm g} + \hat{z}\hat{z}\epsilon_z$$

where

$$\begin{cases} \epsilon = 1 - \dfrac{\omega_{\rm p}^2(\omega+\mathrm{i}\nu)}{\omega\left(\omega^2-\omega_{\rm g}^2-\nu^2+2\mathrm{i}\nu\omega\right)} \\ \epsilon_{\rm g} = -\dfrac{\omega_{\rm p}^2\omega_{\rm g}}{\omega\left(\omega^2-\omega_{\rm g}^2-\nu^2+2\mathrm{i}\nu\omega\right)} \\ \epsilon_z = 1 - \dfrac{\omega_{\rm p}^2}{\omega(\omega+\mathrm{i}\nu)} \end{cases}$$

To investigate whether this model is passive or not, form

$$\omega\operatorname{Im}\boldsymbol{\epsilon} = \mathbf{I}_2\omega\operatorname{Im}\epsilon - \mathrm{i}\mathbf{J}\omega\operatorname{Im}\epsilon_{\rm g} + \hat{z}\hat{z}\omega\operatorname{Im}\epsilon_z$$

The eigenvalues of $\omega\operatorname{Im}\boldsymbol{\epsilon}(\omega)$ are $\lambda = \omega\operatorname{Im}(\epsilon \pm \epsilon_{\rm g}), \omega\operatorname{Im}\epsilon_z$. Therefore, the plasma model is a passive material, since

$$\omega\operatorname{Im}\epsilon_z = \frac{\omega_{\rm p}^2\nu}{\omega^2+\nu^2} \geq 0$$

for $\nu \geq 0$, and

$$\omega\operatorname{Im}(\epsilon \pm \epsilon_{\rm g}) = -\operatorname{Im}\frac{\omega_{\rm p}^2(\omega\pm\omega_{\rm g}+\mathrm{i}\nu)}{\omega^2-\omega_{\rm g}^2-\nu^2+2\mathrm{i}\nu\omega} = \omega_{\rm p}^2\frac{(\omega\pm\omega_{\rm g})2\nu\omega - \nu(\omega^2-\omega_{\rm g}^2-\nu^2)}{(\omega^2-\omega_{\rm g}^2-\nu^2)^2+4\nu^2\omega^2}$$

or

$$\omega \operatorname{Im}(\epsilon \pm \epsilon_g) = \omega_p^2 \nu \frac{(\omega \pm \omega_g)^2 + \nu^2}{(\omega^2 - \omega_g^2 - \nu^2)^2 + 4\nu^2\omega^2} \geq 0$$

■

1.3.4 Sum Rules for the Constitutive Relations

The definition of a passive system and some of its consequences were investigated in Sections 1.2.4 and 1.3.3 in a time-domain and time-harmonic setting, respectively. In both cases, the analysis relies on the inability of the system to produce energy without external sources. It is important to start the modeling of the problem in the time-domain setting, not to miss or neglect any relevant physical properties. In this section, we derive additional consequences of the fact that the medium is passive at all frequencies. The main results of the section are the famous Kramers–Kronig[46] relations and some less famous sum rules.

We quantified the passivity of a material in the time domain in (1.36) on page 34 and (1.39) on page 35, i.e.,

$$w(t) = \epsilon_0 \int_{-\infty}^{t} \mathsf{e}(t') \cdot \frac{\partial \mathsf{d}(t')}{\partial t'} \, \mathrm{d}t' \geq 0$$

where, see (1.32) on page 31

$$\mathsf{d}(t) = \mathsf{A} \cdot \mathsf{e}(t) + \int_{-\infty}^{t} \mathsf{M}(t - t') \cdot \mathsf{e}(t') \, \mathrm{d}t'$$

In a non-magnetic, isotropic material, this relation reduces to

$$w(t) = \int_{-\infty}^{t} \boldsymbol{E}(t') \cdot \frac{\partial \boldsymbol{D}(t')}{\partial t'} \, \mathrm{d}t' + \int_{-\infty}^{t} \boldsymbol{H}(t') \cdot \frac{\partial \boldsymbol{B}(t')}{\partial t'} \, \mathrm{d}t' \geq 0$$

where

$$\boldsymbol{D}(t) = \epsilon_0 \left\{ \boldsymbol{E}(t) + \int_{-\infty}^{t} \chi(t - t') \boldsymbol{E}(t') \, \mathrm{d}t' \right\}, \quad \boldsymbol{B}(t) = \mu_0 \boldsymbol{H}(t)$$

Fascinatingly enough, the passivity criterion in the time domain implies a sequence of very strong results, some of which are presented in this section. Useful textbooks and journal articles for the details of the results below are [12, 13, 87, 123, 124, 203, 304] and references therein.

First, each component of the susceptibility kernel $\mathsf{M}(t)$, or in the isotropic case the kernel $\chi(t)$, due to causality, has a Fourier transform, $\mathsf{M}(\eta) = \mathsf{M}(\omega + i\varsigma)$ or $\epsilon(\eta) = \epsilon(\omega + i\varsigma)$, with an analytic extension in the upper complex half-plane, $\mathbb{C}_+ = \{\eta \in \mathbb{C} : \operatorname{Im} \eta = \varsigma > 0\}$. This is a consequence of Paley–Wiener theorem, see Theorem D.1 on page 640. Moreover, the extended function satisfies $\mathsf{M}^*(\eta) = \mathsf{M}(-\eta^*)$ in the closed upper half η-plane, $\varsigma \geq 0$, since $\mathsf{M}(t)$ is real valued, see also (1.54) on page 44. Consequently, the imaginary part of the entries in $\mathsf{M}(\omega)$ or $\epsilon(\omega)$ is odd for real ω. Moreover, passivity restricts the Fourier transform to $\operatorname{Im} \omega\mathsf{M}(\omega) \geq 0$ or $\operatorname{Im} \omega\epsilon(\omega) \geq 0$ for real ω, see Section 1.3.3.

[46] Hendrik Anthony Kramers (1894–1952), Dutch physicist; Ralph Kronig (1904–1995), German-American physicist.

Another very interesting result in this context is that a convolution operator for a passive system implies causality [304]. The basic assumptions in Section 1.2 included causality as one of the assumptions. For a passive system, this is not necessary. For more details, we refer to References 203, 304.

In this section, we limit ourselves to an investigation of the passivity implications in an isotropic non-magnetic material. The appropriate function to study then is

$$f(\eta) = \eta\,(\epsilon(\eta) - \epsilon_\infty)\,, \qquad \eta \in \mathbb{C}_+ \cup \mathbb{R}$$

More general results can be obtained by employing the six-dimensional matrix M, see e.g., References 62, 87 for a biisotropic application. However, the principles of the theory are more clearly illustrated with the scalar case, which we use in this section.

The passivity condition $\operatorname{Im}\omega\epsilon(\omega) \geq 0$ for real ω can be extended to the upper complex η-plane by an application of the results in Appendix D.3. Thus, we have $\operatorname{Im} f(\eta) \geq 0$ and $f^*(\eta) = -f(-\eta^*)$ in $\eta \in \mathbb{C}_+ \cup \mathbb{R}$. The imaginary part of $f(\omega)$ is thus an even function for real ω. As a consequence, $f(\eta)$ maps the upper half-plane $\mathbb{C}_+$ into the closed upper complex plane $\mathbb{C}_+ \cup \mathbb{R}$. A function with this property is called a Herglotz function, see Appendix D.5. We also assume the function $f(\eta) \to 0$ as $|\eta| \to \infty$ in $\mathbb{C}_+ \cup \mathbb{R}$.[47]

There are several ways to derive the sum rules or integral identities for the permittivity. We adopt Plemelj's formulas, see Appendix D.2, or more precisely (D.4) on page 642

$$\begin{cases} f_{\rm r}^{\rm e}(\omega) = \dfrac{2}{\pi} P \displaystyle\int_0^\infty \frac{\omega' f_{\rm i}^{\rm o}(\omega')}{\omega'^2 - \omega^2}\, d\omega' \\ f_{\rm r}^{\rm o}(\omega) = \dfrac{2\omega}{\pi} P \displaystyle\int_0^\infty \frac{f_{\rm i}^{\rm e}(\omega')}{\omega'^2 - \omega^2}\, d\omega' \end{cases} \qquad \begin{cases} f_{\rm i}^{\rm e}(\omega) = -\dfrac{2}{\pi} P \displaystyle\int_0^\infty \frac{\omega' f_{\rm r}^{\rm o}(\omega')}{\omega'^2 - \omega^2}\, d\omega' \\ f_{\rm i}^{\rm o}(\omega) = -\dfrac{2\omega}{\pi} P \displaystyle\int_0^\infty \frac{f_{\rm r}^{\rm e}(\omega')}{\omega'^2 - \omega^2}\, d\omega' \end{cases}$$

where the even and odd parts of the real and the imaginary parts of $f(\omega)$ are

$$\begin{cases} f_{\rm r,i}^{\rm e}(\omega) = \dfrac{1}{2}\left(f_{\rm r,i}(\omega) + f_{\rm r,i}(-\omega)\right) \\ f_{\rm r,i}^{\rm o}(\omega) = \dfrac{1}{2}\left(f_{\rm r,i}(\omega) - f_{\rm r,i}(-\omega)\right) \end{cases}$$

We have already concluded that the real part, $f_{\rm r}(\omega)$, of the function $f(\omega)$ is odd and the imaginary part, $f_{\rm i}(\omega)$, is even on the real axis. The Plemelj's formulas then read

$$\begin{cases} \operatorname{Re}\epsilon(\omega) = \epsilon_\infty + \dfrac{2}{\pi} P \displaystyle\int_0^\infty \frac{\omega' \operatorname{Im}\epsilon(\omega')}{\omega'^2 - \omega^2}\, d\omega' \\ \omega \operatorname{Im}\epsilon(\omega) = -\dfrac{2}{\pi} P \displaystyle\int_0^\infty \frac{\omega'^2\left(\operatorname{Re}\epsilon(\omega') - \epsilon_\infty\right)}{\omega'^2 - \omega^2}\, d\omega' \end{cases}$$

These results are known as the Kramers–Kronig relations [87, 101, 123, 124, 152, 203], where the first relation is the most used. These integrals relate the real (imaginary) part of the permittivity to the imaginary (real) part on the real axis. In principle, it suffices to know the real (imaginary) part of the function to construct the imaginary (real) part.

---

[47] Additional factors might have to be subtracted from $f(\eta)$ in order to satisfy this condition.

In general, the permittivity has a well-defined static limit and an optical response, $\epsilon_{\rm s}$ and ϵ_∞, respectively—both real valued.[48] Then in the limit $\omega \to 0$, we obtained the following sum rule:

$$\epsilon_{\rm s} - \epsilon_\infty = \frac{2}{\pi} P \int_0^\infty \frac{\operatorname{Im} \epsilon(\omega)}{\omega} \, d\omega \tag{1.68}$$

For a passive material the integrand is positive, and we conclude that $\epsilon_{\rm s} - \epsilon_\infty$ is positive for a passive material.

This result, for an isotropic material, is also given in Reference [152, page 59] and in Example 6.14, Footnote 18 on page 326.

Example 1.17
We illustrate the sum rule in (1.68) by the model of Debye, see (1.57) on page 48

$$\epsilon(\omega) = 1 + \frac{\alpha\tau}{1 - i\omega\tau}$$

This model implies

$$\eta\epsilon(\eta) = \eta + \frac{\alpha\tau\eta}{1 - i\eta\tau} = \begin{cases} \eta + \alpha\tau\eta + i\alpha\tau^2\eta^2 + \cdots, & \eta \to 0 \\ \eta + i\alpha + \cdots, & |\eta| \to \infty \end{cases}$$

The real and imaginary parts of the permittivity in this model are

$$\begin{cases} \operatorname{Re} \epsilon(\omega) - 1 = \dfrac{\alpha\tau}{1 + \omega^2\tau^2} \\ \operatorname{Im} \epsilon(\omega) = \dfrac{\omega\alpha\tau^2}{1 + \omega^2\tau^2} \end{cases}$$

with $\epsilon_\infty = 1$, $\epsilon_{\rm s} = 1 + \alpha\tau$. The sum rule in (1.68) is equivalent to the integral identity (principal value integral not necessary)[49]

$$\int_0^\infty \frac{\tau}{1 + \omega^2\tau^2} \, d\omega = \frac{\pi}{2}$$

■

[48]This is obvious from the definition of the permittivity in terms of the Fourier transform of the electric susceptibility, see (1.49), i.e.,

$$\epsilon(0) = \epsilon_\infty + \int_0^\infty \chi(t) \, dt$$

which is real.

[49]Sum rule in (1.68) is valid still, dispite the fact that $f(\eta)$ does not vanish as $|\eta| \to \infty$ in $\mathbb{C}_+ \cup \mathbb{R}$ for the Debye model.

Example 1.18
If we apply the sum rule in (1.68) to the Lorentz model in (1.60) on page 53

$$\epsilon(\omega) = 1 - \frac{\omega_{\mathrm{p}}^2}{\omega^2 - \omega_0^2 + \mathrm{i}\omega\nu}$$

we obtain a relation between the quotient $\omega_{\mathrm{p}}/\omega_0$ and the losses of the material, i.e.,

$$\frac{\omega_{\mathrm{p}}^2}{\omega_0^2} = \frac{2}{\pi} P \int_0^\infty \frac{\mathrm{Im}\,\epsilon(\omega)}{\omega}\,\mathrm{d}\omega$$

since $\epsilon_\infty = 1$, $\epsilon_{\mathrm{s}} = 1 + \omega_{\mathrm{p}}^2/\omega_0^2$. In fact, the knowledge of the losses in the material gives information of the oscillator strength $\omega_{\mathrm{p}}^2/\omega_0^2$. If the model is a sum of M different Lorentz terms, the generalization is

$$\sum_{i=1}^{M} \frac{\omega_{\mathrm{p}i}^2}{\omega_{0i}^2} = \frac{2}{\pi} P \int_0^\infty \frac{\mathrm{Im}\,\epsilon(\omega)}{\omega}\,\mathrm{d}\omega$$

■

Above, we assumed that the material did not contain a static conductivity term. Now suppose that the material has a static conductivity σ, and we have to modify the analysis above to obtain the sum rules. The permittivity now has the form

$$\epsilon(\omega) = \epsilon_{\mathrm{reg}}(\omega) + \mathrm{i}\frac{\sigma}{\omega\epsilon_0}$$

where the permittivity term $\epsilon_{\mathrm{reg}}(\omega)$ contains no singularity at $\omega = 0$. Again, use Plemelj's formulas, this time on the function $f(\eta) = \eta\,(\epsilon(\eta) - \epsilon_\infty - \mathrm{i}\sigma/\eta\epsilon_0) = \eta(\epsilon_{\mathrm{reg}}(\eta) - \epsilon_\infty)$. The result is

$$\begin{cases} \mathrm{Re}\,\epsilon_{\mathrm{reg}}(\omega) = \epsilon_\infty + \dfrac{2}{\pi} P \displaystyle\int_0^\infty \frac{\omega' \,\mathrm{Im}\,\epsilon_{\mathrm{reg}}(\omega')}{\omega'^2 - \omega^2}\,\mathrm{d}\omega' \\ \omega\,\mathrm{Im}\,\epsilon_{\mathrm{reg}}(\omega) = -\dfrac{2}{\pi} P \displaystyle\int_0^\infty \frac{\omega'^2\left(\mathrm{Re}\,\epsilon_{\mathrm{reg}}(\omega') - \epsilon_\infty\right)}{\omega'^2 - \omega^2}\,\mathrm{d}\omega' \end{cases}$$

Note that with a conductivity term, the passive material condition does *not*, in general, imply that $\omega\,\mathrm{Im}\,\epsilon_{\mathrm{reg}}(\omega) > 0$ for all ω. Passivity only implies that $\omega\,\mathrm{Im}\,\epsilon(\omega) = \omega\,\mathrm{Im}\,\epsilon_{\mathrm{reg}}(\omega) + \mathrm{Re}\,\sigma/\epsilon_0 > 0$, see the analysis in Section 1.3.3 and Example 1.19. Therefore, we cannot any longer make the conclusion that $\epsilon_{\mathrm{reg}}(0)$ is larger than the optical response.

Example 1.19
The model by Drude, see (1.61) on page 53, illustrates the theory with a static conductivity term. The model

$$\epsilon(\omega) = 1 - \frac{\omega_{\mathrm{p}}^2}{\omega^2 + \mathrm{i}\omega\nu} = 1 - \frac{\omega_{\mathrm{p}}^2}{\omega(\omega + \mathrm{i}\nu)}$$

implies

$$\epsilon_{\text{reg}}(\omega) = 1 - \frac{\omega_{\text{p}}^2}{\omega(\omega + \text{i}\nu)} - \frac{\text{i}\omega_{\text{p}}^2}{\omega\nu} = 1 - \frac{\text{i}\omega_{\text{p}}^2}{\nu(\omega + \text{i}\nu)}$$

and

$$\eta\epsilon_{\text{reg}}(\eta) = \eta - \frac{\text{i}\eta\omega_{\text{p}}^2}{\nu(\eta + \text{i}\nu)} = \begin{cases} \eta - \eta\omega_{\text{p}}^2/\nu^2 - \text{i}\eta^2\omega_{\text{p}}^2/\nu^3 + \cdots, & \eta \to 0 \\ \eta - \text{i}\omega_{\text{p}}^2/\nu + \cdots, & |\eta| \to \infty \end{cases}$$

The real and imaginary parts of the permittivity in this model are

$$\begin{cases} \operatorname{Re}\epsilon_{\text{reg}}(\omega) = 1 - \dfrac{\omega_{\text{p}}^2}{\omega^2 + \nu^2} \\ \operatorname{Im}\epsilon_{\text{reg}}(\omega) = -\dfrac{\omega_{\text{p}}^2\omega}{(\omega^2 + \nu^2)\nu} \end{cases}$$

and we see that $\omega \operatorname{Im}\epsilon_{\text{reg}}(\omega) < 0$. The sum rule for this model is (principal value integral not necessary)

$$\frac{\omega_{\text{p}}^2}{\nu^2} = \frac{2}{\pi}\int_0^\infty \frac{\omega_{\text{p}}^2}{(\omega^2 + \nu^2)\nu}\,\text{d}\omega$$

which is similar to the integral in Example 1.17. ■

Moreover, due to (1.24) on page 20, the resolvent of $\chi(t)$ is a causal quantity. In fact, the equation of the resolvent (1.23) on page 20 reads

$$\frac{\chi(t)}{\epsilon_\infty} + \epsilon_\infty\Upsilon(t) + \int_0^t \Upsilon(t - t')\chi(t')\,\text{d}t' = 0$$

The Fourier transform of the resolvent reads

$$\frac{1}{\epsilon_\infty}(\epsilon(\omega) - \epsilon_\infty) + \epsilon_\infty\Upsilon(\omega) + \Upsilon(\omega)(\epsilon(\omega) - \epsilon_\infty) = 0 \quad \Rightarrow \quad \Upsilon(\omega) = -\frac{1}{\epsilon_\infty} + \frac{1}{\epsilon(\omega)}$$

There are therefore reasons to study $f(\eta) = \eta/\epsilon(\eta) - \eta/\epsilon_\infty$. This is the Fourier transform of a causal quantity, and it has an analytic extension in $\mathbb{C}_+$ satisfying $f(-\eta^*) = -f^*(\eta)$ in $\mathbb{C}_+$. The sum rule above can be applied to this new $f(\eta)$. Following the analysis above, we get

$$\operatorname{Re} f(\omega) = \frac{2\omega}{\pi} P\int_0^\infty \frac{\operatorname{Im} f(\omega')}{\omega'^2 - \omega^2}\,\text{d}\omega'$$

or

$$\operatorname{Re}\frac{1}{\epsilon(\omega)} = \frac{1}{\epsilon_\infty} + \frac{2}{\pi} P\int_0^\infty \frac{\omega' \operatorname{Im}\{1/\epsilon(\omega')\}}{\omega'^2 - \omega^2}\,\text{d}\omega'$$

and the following sum rule holds:

$$\frac{1}{\epsilon_s} = \frac{1}{\epsilon_\infty} + \frac{2}{\pi} P \int_0^\infty \frac{\mathrm{Im}\left\{1/\epsilon(\omega')\right\}}{\omega'} \,\mathrm{d}\omega' \tag{1.69}$$

In a similar manner, an unlimited set of sum rules can be created, since the composition of Herglotz functions is Herglotz functions. We do not pursue the sum rules on the constitutive relations any further in this book, but refer to the existing literature for more details and examples [13, 123, 124]. For sum rules on the scattering properties, see Section 6.4 on page 320.

Example 1.20
Again, we illustrate this sum rule by the model of Debye (1.57) on page 48. The real and imaginary parts of the permittivity in this model are

$$\frac{1}{\epsilon(\omega)} = \frac{1 - \mathrm{i}\omega\tau}{\alpha\tau + 1 - \mathrm{i}\omega\tau} = 1 - \frac{\alpha\tau}{\alpha\tau + 1 - \mathrm{i}\omega\tau}$$

and

$$\frac{\eta}{\epsilon(\eta)} = \eta - \frac{\alpha\tau\eta}{\alpha\tau + 1 - \mathrm{i}\eta\tau} = \begin{cases} \eta - \dfrac{\alpha\tau\eta}{\alpha\tau+1} - \dfrac{\mathrm{i}\alpha\tau^2\eta^2}{(\alpha\tau+1)^2} + \cdots, & \eta \to 0 \\ \eta - \mathrm{i}\alpha + \cdots, & |\eta| \to \infty \end{cases}$$

The real and imaginary parts are

$$\begin{cases} \mathrm{Re}\, \dfrac{1}{\epsilon(\omega)} - 1 = -\dfrac{\alpha\tau(1+\alpha\tau)}{(1+\alpha\tau)^2 + \omega^2\tau^2} \\ \mathrm{Im}\, \dfrac{1}{\epsilon(\omega)} = -\dfrac{\omega\alpha\tau^2}{(1+\alpha\tau)^2 + \omega^2\tau^2} \end{cases}$$

This implies with, as in Example 1.17, $\epsilon_\infty = 1$ and $\epsilon_s = 1 + \alpha\tau$. The sum rule in (1.69) is equivalent to the integral identity (principal value integral not necessary)

$$\frac{2}{\pi} P \int_0^\infty \frac{\alpha\tau^2}{(1+\alpha\tau)^2 + \omega^2\tau^2} \,\mathrm{d}\omega' = \frac{\alpha\tau}{\alpha\tau + 1}$$

which is similar to the integral in Example 1.17. ■

1.3.5 Reciprocity

In Section 1.3.3, we studied the power dissipation in a material. Specifically, we found that the quantity $\nabla \cdot \langle \boldsymbol{S}(\boldsymbol{r}, t)\rangle$ was useful in classifying different materials. We introduced the notion of active, passive, and lossless materials, depending on whether this quantity was positive, negative, or zero, respectively. This notion was local in space, i.e., it holds at a specific point in space.

In this section, we introduce a new concept for classifying materials—the reciprocity property. The reciprocity concept compares the effects on the material from two different source

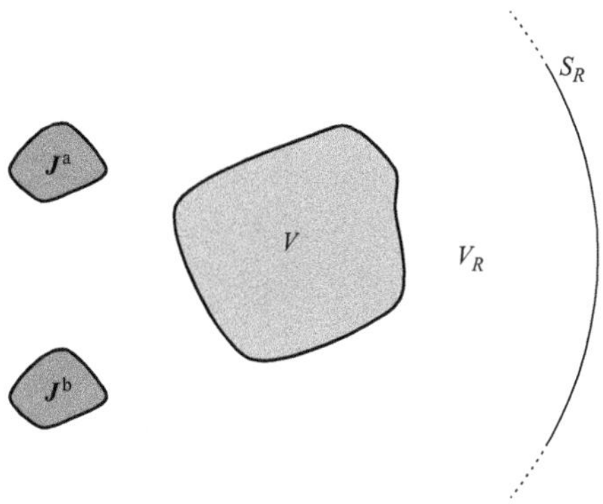

Figure 1.15 The material in the volume V is excited by two different sources $\boldsymbol{J}^{\mathrm{a}}$ and $\boldsymbol{J}^{\mathrm{b}}$, respectively.

configurations. The first set of sources, denoted $\boldsymbol{J}^{\mathrm{a}}$, gives rise to electromagnetic fields, which we denote by the superscript a, i.e., the fields are $\boldsymbol{E}^{\mathrm{a}}$, $\boldsymbol{H}^{\mathrm{a}}$, $\boldsymbol{D}^{\mathrm{a}}$, and $\boldsymbol{B}^{\mathrm{a}}$. The second set of sources, which we denote by the superscript b, has sources and fields $\boldsymbol{J}^{\mathrm{b}}$, $\boldsymbol{E}^{\mathrm{b}}$, $\boldsymbol{H}^{\mathrm{b}}$, $\boldsymbol{D}^{\mathrm{b}}$, and $\boldsymbol{B}^{\mathrm{b}}$. All sources are assumed to be located in a finite region of space outside or inside the material, which is confined to the volume V. Outside these sources and the volume V, we assume vacuum. The geometry is illustrated in Figure 1.15.

In analogy to the dissipation concept in Section 1.3.3, we make the definition of reciprocity as a local property of the material. To this end, the medium is reciprocal at $\boldsymbol{r}$ if

$$\boldsymbol{E}^{\mathrm{a}} \cdot \boldsymbol{D}^{\mathrm{b}} - \boldsymbol{E}^{\mathrm{b}} \cdot \boldsymbol{D}^{\mathrm{a}} = \boldsymbol{H}^{\mathrm{a}} \cdot \boldsymbol{B}^{\mathrm{b}} - \boldsymbol{H}^{\mathrm{b}} \cdot \boldsymbol{B}^{\mathrm{a}} \tag{1.70}$$

for all accessible fields at this point. Equivalently, in a reciprocal material the quantity $\boldsymbol{E}^{\mathrm{a}} \cdot \boldsymbol{D}^{\mathrm{b}} + \boldsymbol{B}^{\mathrm{a}} \cdot \boldsymbol{H}^{\mathrm{b}}$ is invariant under the change a $\leftrightarrow$ b.

Before investigating the consequences of this definition on the constitutive relations, we analyze the physical background to this definition. We assume the material in V is reciprocal and the surrounding region is vacuum, and integrate (1.70) over a volume V_R, a large ball of radius R, containing the material V and the sources $\boldsymbol{J}^{\mathrm{a}}$ and $\boldsymbol{J}^{\mathrm{b}}$, see Figure 1.15. We get

$$\mathrm{i}\omega \iiint_{V_R} \left\{\boldsymbol{E}^{\mathrm{a}} \cdot \boldsymbol{D}^{\mathrm{b}} - \boldsymbol{E}^{\mathrm{b}} \cdot \boldsymbol{D}^{\mathrm{a}} - \boldsymbol{H}^{\mathrm{a}} \cdot \boldsymbol{B}^{\mathrm{b}} + \boldsymbol{H}^{\mathrm{b}} \cdot \boldsymbol{B}^{\mathrm{a}}\right\} \mathrm{d}v = 0$$

Now use the Maxwell equations

$$\begin{cases} \mathrm{i}\omega \boldsymbol{D} = \boldsymbol{J} - \nabla \times \boldsymbol{H} \\ \mathrm{i}\omega \boldsymbol{B} = \nabla \times \boldsymbol{E} \end{cases}$$

and the differentiation rule $\nabla \cdot (\boldsymbol{a} \times \boldsymbol{b}) = (\nabla \times \boldsymbol{a}) \cdot \boldsymbol{b} - \boldsymbol{a} \cdot (\nabla \times \boldsymbol{b})$ to rewrite the volume integral as

$$
\begin{aligned}
0 &= \mathrm{i}\omega \iiint\limits_{V_R} \left\{ \boldsymbol{E}^{\mathrm{a}} \cdot \boldsymbol{D}^{\mathrm{b}} - \boldsymbol{E}^{\mathrm{b}} \cdot \boldsymbol{D}^{\mathrm{a}} - \boldsymbol{H}^{\mathrm{a}} \cdot \boldsymbol{B}^{\mathrm{b}} + \boldsymbol{H}^{\mathrm{b}} \cdot \boldsymbol{B}^{\mathrm{a}} \right\} \mathrm{d}v \\
&= \iiint\limits_{V_R} \left\{ \boldsymbol{E}^{\mathrm{a}} \cdot \left(\boldsymbol{J}^{\mathrm{b}} - \nabla \times \boldsymbol{H}^{\mathrm{b}} \right) - \boldsymbol{H}^{\mathrm{a}} \cdot \left(\nabla \times \boldsymbol{E}^{\mathrm{b}} \right) - \{\mathrm{a} \leftrightarrow \mathrm{b}\} \right\} \mathrm{d}v \\
&= \iiint\limits_{V_R} \{ \boldsymbol{E}^{\mathrm{a}} \cdot \boldsymbol{J}^{\mathrm{b}} - \boldsymbol{E}^{\mathrm{b}} \cdot \boldsymbol{J}^{\mathrm{a}} \} \, \mathrm{d}v + \iiint\limits_{V_R} \nabla \cdot \left(\boldsymbol{E}^{\mathrm{a}} \times \boldsymbol{H}^{\mathrm{b}} - \boldsymbol{E}^{\mathrm{b}} \times \boldsymbol{H}^{\mathrm{a}} \right) \mathrm{d}v
\end{aligned}
$$

The divergence theorem converts the last integral to a surface integral over the bounding surface S_R of the volume V_R. We get the identity

$$
\iiint\limits_{V_R} \{ \boldsymbol{E}^{\mathrm{a}} \cdot \boldsymbol{J}^{\mathrm{b}} - \boldsymbol{E}^{\mathrm{b}} \cdot \boldsymbol{J}^{\mathrm{a}} \} \, \mathrm{d}v + \iint\limits_{S_R} \left(\boldsymbol{E}^{\mathrm{a}} \times \boldsymbol{H}^{\mathrm{b}} - \boldsymbol{E}^{\mathrm{b}} \times \boldsymbol{H}^{\mathrm{a}} \right) \cdot \hat{\nu} \, \mathrm{d}S = 0
$$

This identity is usually called the Lorentz reciprocity theorem.

The domain of integration in the volume integral is only over the finite region of the sources. This integral is therefore independent of the radius R of the volume V_R. Since this integral is independent of the radius of the volume V_R, the surface integral on the right-hand side is also independent of R. In the limit as this radius goes to infinity, i.e., $V_R \to \mathbb{R}^3$, it is possible to show that the surface integral is zero.[50] For a reciprocal material, we have proved

$$
\iiint \boldsymbol{E}^{\mathrm{a}} \cdot \boldsymbol{J}^{\mathrm{b}} \, \mathrm{d}v = \iiint \boldsymbol{E}^{\mathrm{b}} \cdot \boldsymbol{J}^{\mathrm{a}} \, \mathrm{d}v
$$

If this condition is not satisfied for a set of sources or fields we have a non-reciprocal material somewhere inside V. This condition is a measure of the difference between the sources in "a" ($\boldsymbol{J}^{\mathrm{a}}$) and measuring at "b" ($\boldsymbol{E}^{\mathrm{b}}$) and vice versa. Very simplified, the reciprocity concept implies that if sources and receivers change places, the integral above remains the same.

We now continue to investigate the effects of reciprocity on the constitutive relations. Introduce the constitutive relations (1.51) in the definition of reciprocity, (1.70). We get

$$
\begin{aligned}
&\boldsymbol{E}^{\mathrm{a}} \cdot (\boldsymbol{\epsilon} \cdot \boldsymbol{E}^{\mathrm{b}} + \eta_0 \boldsymbol{\xi} \cdot \boldsymbol{H}^{\mathrm{b}}) - \boldsymbol{E}^{\mathrm{b}} \cdot (\boldsymbol{\epsilon} \cdot \boldsymbol{E}^{\mathrm{a}} + \eta_0 \boldsymbol{\xi} \cdot \boldsymbol{H}^{\mathrm{a}}) \\
&\quad - \eta_0 \boldsymbol{H}^{\mathrm{a}} \cdot (\boldsymbol{\zeta} \cdot \boldsymbol{E}^{\mathrm{b}} + \eta_0 \boldsymbol{\mu} \cdot \boldsymbol{H}^{\mathrm{b}}) + \eta_0 \boldsymbol{H}^{\mathrm{b}} \cdot (\boldsymbol{\zeta} \cdot \boldsymbol{E}^{\mathrm{a}} + \eta_0 \boldsymbol{\mu} \cdot \boldsymbol{H}^{\mathrm{a}}) = 0
\end{aligned}
$$

Using

$$
\boldsymbol{a} \cdot \mathbf{A} \cdot \boldsymbol{b} = \boldsymbol{b} \cdot \mathbf{A}^{\mathrm{t}} \cdot \boldsymbol{a}
$$

---

[50] We have assumed that the region outside the material in the volume V is vacuum. At large distances the fields satisfy the radiation condition ($\hat{\boldsymbol{r}} = \boldsymbol{r}/|\boldsymbol{r}|$, $\eta_0 = \sqrt{\mu_0/\epsilon_0}$, $k_0 = \omega/c_0$)

$$
(\hat{\boldsymbol{r}} \times \boldsymbol{E}(\boldsymbol{r})) - \eta_0 \eta \boldsymbol{H}(\boldsymbol{r}) = o((k_0 r)^{-1}) \quad \text{or} \quad \eta_0 \eta (\hat{\boldsymbol{r}} \times \boldsymbol{H}(\boldsymbol{r})) + \boldsymbol{E}(\boldsymbol{r}) = o((k_0 r)^{-1}) \quad \text{as } r \to \infty
$$

These conditions are discussed and analyzed further in Section 4.1.2, and they imply the statement.

where ^{t} denotes the transposed dyadic, we can simplify the reciprocity definition

$$\begin{aligned} & \boldsymbol{E}^{a} \cdot \left(\boldsymbol{\epsilon} - \boldsymbol{\epsilon}^{t}\right) \cdot \boldsymbol{E}^{b} - \eta_0^2 \boldsymbol{H}^{a} \cdot \left(\boldsymbol{\mu} - \boldsymbol{\mu}^{t}\right) \cdot \boldsymbol{H}^{b} \\ & \quad + \eta_0 \boldsymbol{E}^{a} \cdot \left(\boldsymbol{\xi} + \boldsymbol{\zeta}^{t}\right) \cdot \boldsymbol{H}^{b} - \eta_0 \boldsymbol{H}^{a} \cdot \left(\boldsymbol{\zeta} + \boldsymbol{\xi}^{t}\right) \cdot \boldsymbol{E}^{b} = 0 \end{aligned}$$

Since the fields are arbitrary, we obtain the following conditions on the constitutive relations for a reciprocal material:

$$\begin{cases} \boldsymbol{\epsilon} = \boldsymbol{\epsilon}^{t} \\ \boldsymbol{\mu} = \boldsymbol{\mu}^{t} \\ \boldsymbol{\xi} = -\boldsymbol{\zeta}^{t} \end{cases} \tag{1.71}$$

An immediate consequence of this result is that all isotropic materials are reciprocal. If the dielectric dyadic is real, there is a set of (real) coordinate axes that diagonalizes the dyadic and the classification in Table 1.5 on page 46 holds. This situation holds for e.g., a lossless, reciprocal material. In the six-dimensional formulation in (1.53) on page 44, the reciprocity relation reads

$$\begin{aligned} \mathsf{M}(\omega) &= \begin{pmatrix} \boldsymbol{\epsilon}(\omega) & \boldsymbol{\xi}(\omega) \\ \boldsymbol{\zeta}(\omega) & \boldsymbol{\mu}(\omega) \end{pmatrix} = \begin{pmatrix} \boldsymbol{\epsilon}^{t}(\omega) & -\boldsymbol{\zeta}^{t}(\omega) \\ -\boldsymbol{\xi}^{t}(\omega) & \boldsymbol{\mu}^{t}(\omega) \end{pmatrix} \\ &= \begin{pmatrix} \mathbf{I}_3 & \mathbf{0} \\ \mathbf{0} & -\mathbf{I}_3 \end{pmatrix} \cdot \mathsf{M}^{t}(\omega) \cdot \begin{pmatrix} \mathbf{I}_3 & \mathbf{0} \\ \mathbf{0} & -\mathbf{I}_3 \end{pmatrix} \end{aligned}$$

or that

$$\mathsf{M}(\omega) \cdot \begin{pmatrix} \mathbf{I}_3 & \mathbf{0} \\ \mathbf{0} & -\mathbf{I}_3 \end{pmatrix}$$

is a symmetric six-dimensional matrix.

1.3.6 Special Type of Solutions

There are two types of very simple solutions of the Maxwell equations in source-free region containing a homogeneous isotropic material characterized by the permittivity $\epsilon(\omega)$ and the permeability $\mu(\omega)$. These are the plane and the spherical waves, which are used frequently in the analysis in this textbook.

The plane waves are given by

$$\boldsymbol{E}(\boldsymbol{r}, \omega) = \boldsymbol{E}_0(\omega) e^{ik\hat{\boldsymbol{k}}\cdot\boldsymbol{r}}, \qquad \eta_0 \eta \boldsymbol{H}(\boldsymbol{r}, \omega) = \hat{\boldsymbol{k}} \times \boldsymbol{E}_0(\omega) e^{ik\hat{\boldsymbol{k}}\cdot\boldsymbol{r}} \tag{1.72}$$

where $\hat{\boldsymbol{k}}$ is an arbitrary real-valued unit vector, and $\boldsymbol{E}_0(\omega)$ is an arbitrary complex-valued vector, satisfying $\boldsymbol{E}_0(\omega) \cdot \hat{\boldsymbol{k}} = 0$. Here, $k = k_0(\epsilon(\omega)\mu(\omega))^{1/2}$ is the wave number of the material, $k_0 = \omega(\epsilon_0\mu_0)^{1/2}$ is the wave number of vacuum, $\eta = (\mu(\omega)/\epsilon(\omega))^{1/2}$ is the wave impedance of the material, and $\eta_0 = (\mu_0/\epsilon_0)^{1/2}$ is the wave impedance of vacuum. It is straightforward, see Problem 1.15, to show that the fields in (1.72) solve the Maxwell equations in the homogeneous isotropic material.

The spherical waves are defined in a similar manner. They are

$$\boldsymbol{E}(\boldsymbol{r},\omega) = \boldsymbol{F}(\hat{\boldsymbol{r}},\omega)\frac{\mathrm{e}^{ikr}}{r}, \qquad \eta_0\eta\boldsymbol{H}(\boldsymbol{r},\omega) = \hat{\boldsymbol{r}} \times \boldsymbol{F}(\hat{\boldsymbol{r}},\omega)\frac{\mathrm{e}^{ikr}}{r} \tag{1.73}$$

where $\boldsymbol{F}(\hat{\boldsymbol{r}},\omega)$ is an arbitrary complex-valued vector, satisfying $\boldsymbol{F}(\hat{\boldsymbol{r}},\omega) \cdot \hat{\boldsymbol{r}} = 0$. The fields in (1.73) are also easily shown to solve the Maxwell equations in the homogeneous isotropic material, see Problem 1.16.

1.3.7 Ellipse of Polarization

A time-harmonic field can be described in geometrical terms. All time-harmonic fields oscillate in a fixed plane and the field follows the trace of an ellipse in this plane. The final results in this section are coordinate free (Cartesian coordinates are used in some of the derivations, but the results are coordinate free), like most results presented in this book. This is advantageous, since the analysis can be made without referring to any specific coordinate system.

We consider the time-harmonic field $\boldsymbol{E}(t)$ (all dependence on the space coordinates $\boldsymbol{r}$ is suppressed in this section) at a fixed point in space. The time dependence of the field is

$$\boldsymbol{E}(t) = \mathrm{Re}\left\{\boldsymbol{E}_0\mathrm{e}^{-\mathrm{i}\omega t}\right\} \tag{1.74}$$

where $\boldsymbol{E}_0$ is a constant complex vector (can depend on, e.g., ω and $\boldsymbol{r}$). The Cartesian components of this field are

$$\boldsymbol{E}_0 = \hat{\boldsymbol{x}}E_{0x} + \hat{\boldsymbol{y}}E_{0y} + \hat{\boldsymbol{z}}E_{0z} = \hat{\boldsymbol{x}}|E_{0x}|\mathrm{e}^{\mathrm{i}\alpha} + \hat{\boldsymbol{y}}|E_{0y}|\mathrm{e}^{\mathrm{i}\beta} + \hat{\boldsymbol{z}}|E_{0z}|\mathrm{e}^{\mathrm{i}\gamma}$$

and α, β, and γ are the phase of the components, respectively.

First, we observe that the vector $\boldsymbol{E}(t)$ in (1.74) for all times lies in a fixed plane in space. To see this, we express the complex vector $\boldsymbol{E}_0$ in its real and imaginary parts, $\boldsymbol{E}_{0\mathrm{r}}$ and $\boldsymbol{E}_{0\mathrm{i}}$, respectively

$$\boldsymbol{E}_0 = \boldsymbol{E}_{0\mathrm{r}} + \mathrm{i}\boldsymbol{E}_{0\mathrm{i}}$$

The real vectors $\boldsymbol{E}_{0\mathrm{r}}$ and $\boldsymbol{E}_{0\mathrm{i}}$ are fixed in time, and their explicit Cartesian components are

$$\begin{aligned}\boldsymbol{E}_{0\mathrm{r}} &= \hat{\boldsymbol{x}}|E_{0x}|\cos\alpha + \hat{\boldsymbol{y}}|E_{0y}|\cos\beta + \hat{\boldsymbol{z}}|E_{0z}|\cos\gamma \\ \boldsymbol{E}_{0\mathrm{i}} &= \hat{\boldsymbol{x}}|E_{0x}|\sin\alpha + \hat{\boldsymbol{y}}|E_{0y}|\sin\beta + \hat{\boldsymbol{z}}|E_{0z}|\sin\gamma\end{aligned}$$

The vector $\boldsymbol{E}(t)$ in (1.74) is now rewritten as

$$\boldsymbol{E}(t) = \mathrm{Re}\left\{(\boldsymbol{E}_{0\mathrm{r}} + \mathrm{i}\boldsymbol{E}_{0\mathrm{i}})\,\mathrm{e}^{-\mathrm{i}\omega t}\right\} = \boldsymbol{E}_{0\mathrm{r}}\cos\omega t + \boldsymbol{E}_{0\mathrm{i}}\sin\omega t \tag{1.75}$$

from which we conclude that the vector $\boldsymbol{E}(t)$ lies in the plane spanned by the real vectors $\boldsymbol{E}_{0\mathrm{r}}$ and $\boldsymbol{E}_{0\mathrm{i}}$ for all times t. The normal to this plane is

$$\hat{\boldsymbol{\nu}} = \pm\frac{\boldsymbol{E}_{0\mathrm{r}} \times \boldsymbol{E}_{0\mathrm{i}}}{|\boldsymbol{E}_{0\mathrm{r}} \times \boldsymbol{E}_{0\mathrm{i}}|}$$

provided that $\boldsymbol{E}_{0\mathrm{r}} \times \boldsymbol{E}_{0\mathrm{i}} \neq \mathbf{0}$. In the case $\boldsymbol{E}_{0\mathrm{r}} \times \boldsymbol{E}_{0\mathrm{i}} = \mathbf{0}$, i.e., the two real vectors $\boldsymbol{E}_{0\mathrm{r}}$ and $\boldsymbol{E}_{0\mathrm{i}}$ are parallel, the field $\boldsymbol{E}$ oscillates along a fixed line in space, and no plane can be defined. Below, we classify a time-harmonic field of this type as a linearly polarized field.

In general, the real vectors $\boldsymbol{E}_{0\mathrm{r}}$ and $\boldsymbol{E}_{0\mathrm{i}}$, which span the plane in which the vector $\boldsymbol{E}(t)$ oscillates, are not orthogonal. However, it is convenient to use orthogonal vectors. To this end, we introduce two new orthogonal vectors, $\boldsymbol{a}$ and $\boldsymbol{b}$, which are linear combinations of the vectors $\boldsymbol{E}_{0\mathrm{r}}$ and $\boldsymbol{E}_{0\mathrm{i}}$. Let

$$\begin{cases} \boldsymbol{a} = \boldsymbol{E}_{0\mathrm{r}} \cos\vartheta + \boldsymbol{E}_{0\mathrm{i}} \sin\vartheta \\ \boldsymbol{b} = -\boldsymbol{E}_{0\mathrm{r}} \sin\vartheta + \boldsymbol{E}_{0\mathrm{i}} \cos\vartheta \end{cases} \tag{1.76}$$

where the angle $\vartheta \in [-\pi/4, \pi/4]$ is defined as

$$\tan 2\vartheta = \frac{2\boldsymbol{E}_{0\mathrm{r}} \cdot \boldsymbol{E}_{0\mathrm{i}}}{|\boldsymbol{E}_{0\mathrm{r}}|^2 - |\boldsymbol{E}_{0\mathrm{i}}|^2}$$

If the denominator is zero, the ϑ is $\pm\pi/4$ (the two signs correspond to a shift of the half axes, e.g., $\boldsymbol{a} \to -\boldsymbol{b}$ and $\boldsymbol{b} \to \boldsymbol{a}$). By this construction $\boldsymbol{a}$ and $\boldsymbol{b}$ are orthogonal, since

$$\begin{aligned} \boldsymbol{a} \cdot \boldsymbol{b} &= (\boldsymbol{E}_{0\mathrm{r}} \cos\vartheta + \boldsymbol{E}_{0\mathrm{i}} \sin\vartheta) \cdot (-\boldsymbol{E}_{0\mathrm{r}} \sin\vartheta + \boldsymbol{E}_{0\mathrm{i}} \cos\vartheta) \\ &= -\left(|\boldsymbol{E}_{0\mathrm{r}}|^2 - |\boldsymbol{E}_{0\mathrm{i}}|^2\right) \sin\vartheta \cos\vartheta + \boldsymbol{E}_{0\mathrm{r}} \cdot \boldsymbol{E}_{0\mathrm{i}} \left(\cos^2\vartheta - \sin^2\vartheta\right) \\ &= -\frac{1}{2}\left(|\boldsymbol{E}_{0\mathrm{r}}|^2 - |\boldsymbol{E}_{0\mathrm{i}}|^2\right) \sin 2\vartheta + \boldsymbol{E}_{0\mathrm{r}} \cdot \boldsymbol{E}_{0\mathrm{i}} \cos 2\vartheta = 0 \end{aligned}$$

by the definition of the angle ϑ.

The vectors $\boldsymbol{E}_{0\mathrm{r}}$ and $\boldsymbol{E}_{0\mathrm{i}}$ can be expressed in the vectors $\boldsymbol{a}$ and $\boldsymbol{b}$. The result is

$$\begin{cases} \boldsymbol{E}_{0\mathrm{r}} = \boldsymbol{a} \cos\vartheta - \boldsymbol{b} \sin\vartheta \\ \boldsymbol{E}_{0\mathrm{i}} = \boldsymbol{a} \sin\vartheta + \boldsymbol{b} \cos\vartheta \end{cases}$$

i.e.,

$$\boldsymbol{E}_0 = \boldsymbol{E}_{0\mathrm{r}} + \mathrm{i}\boldsymbol{E}_{0\mathrm{i}} = (\boldsymbol{a}\cos\vartheta - \boldsymbol{b}\sin\vartheta) + \mathrm{i}\,(\boldsymbol{a}\sin\vartheta + \boldsymbol{b}\cos\vartheta) = \mathrm{e}^{\mathrm{i}\vartheta}(\boldsymbol{a} + \mathrm{i}\boldsymbol{b}) \tag{1.77}$$

This representation also implies a simple form of the magnitude of the complex vector $\boldsymbol{E}_0$, i.e.,

$$|\boldsymbol{E}_0|^2 = \boldsymbol{E}_0 \cdot \boldsymbol{E}_0^* = (\boldsymbol{a} + \mathrm{i}\boldsymbol{b}) \cdot (\boldsymbol{a} - \mathrm{i}\boldsymbol{b}) = |\boldsymbol{a}|^2 + |\boldsymbol{b}|^2 = a^2 + b^2$$

We can express the physical field in terms of the vectors $\boldsymbol{a}$ and $\boldsymbol{b}$ by insertion in (1.75). We get the physical field

$$\begin{aligned} \boldsymbol{E}(t) &= \boldsymbol{E}_{0\mathrm{r}} \cos\omega t + \boldsymbol{E}_{0\mathrm{i}} \sin\omega t \\ &= (\boldsymbol{a}\cos\vartheta - \boldsymbol{b}\sin\vartheta)\cos\omega t + (\boldsymbol{a}\sin\vartheta + \boldsymbol{b}\cos\vartheta)\sin\omega t \\ &= \boldsymbol{a}\cos(\omega t - \vartheta) + \boldsymbol{b}\sin(\omega t - \vartheta) \end{aligned} \tag{1.78}$$

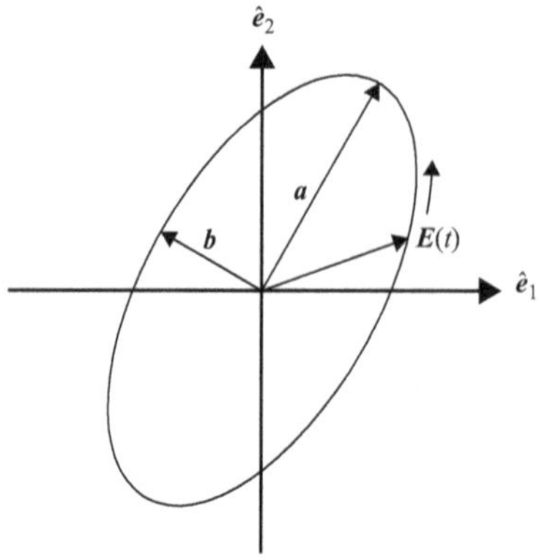

Figure 1.16 The ellipse of polarization and its half axes $\boldsymbol{a}$ and $\boldsymbol{b}$ in the plane spanned by the real vectors $\boldsymbol{E}_{0\mathrm{r}}$ and $\boldsymbol{E}_{0\mathrm{i}}$. The Cartesian coordinate axis in this plane are arbitrary.

The vectors $\boldsymbol{a}$ and $\boldsymbol{b}$ can be used as a basis in an orthogonal coordinate system in the plane where the field $\boldsymbol{E}$ oscillates. From a comparison with the equation of the ellipse in the x–y-plane (half axes a and b along the x- and the y-axes, respectively)

$$\begin{cases} x = a\cos\phi \\ y = b\sin\phi \end{cases}$$

and (1.78), we conclude that the field $\boldsymbol{E}$ traces an ellipse in the plane spanned by the vectors $\boldsymbol{a}$ and $\boldsymbol{b}$ and that these vectors are the half axes (both the direction and size) of the ellipse, see Figure 1.16. From (1.78), we also see that the field $\boldsymbol{E}$ is directed along the half axis $\boldsymbol{a}$ when $\omega t = \vartheta + 2n\pi$, and that the field $\boldsymbol{E}$ is directed along the other half axis $\boldsymbol{b}$ when $\omega t = \vartheta + \pi/2 + 2n\pi$. The angle ϑ is the parameter that marks where on the ellipse the field $\boldsymbol{E}$ is directed at $t = 0$, i.e.,

$$\boldsymbol{E}(t = 0) = \boldsymbol{a}\cos\vartheta - \boldsymbol{b}\sin\vartheta$$

and the vector $\boldsymbol{E}$ moves along the ellipse in a direction from $\boldsymbol{a}$ to $\boldsymbol{b}$ (shortest way). The vectors $\boldsymbol{a}$ and $\boldsymbol{b}$ describe the polarization state[51] of the field $\boldsymbol{E}$ completely, except for the phase angle ϑ.

We are now classifying the polarization state of the time-harmonic field $\boldsymbol{E}(t)$. This field can either be rotating along the ellipse in a clockwise or a counterclockwise direction. Without a preferred direction in space, the direction of rotation is a relative concept—depending on which side of the plane we observe the oscillations. From the direction of the power flux density of the electromagnetic fields at the point of observation, $\langle \boldsymbol{S}(t)\rangle$, we define a preferred direction in space. Let $\hat{\boldsymbol{e}}$ be the normal to the plane of polarization, such that $\langle \boldsymbol{S}(t)\rangle \cdot \hat{\boldsymbol{e}} > 0$. We use this unit vector $\hat{\boldsymbol{e}}$ as a reference direction.

The polarization of the field is now classified according to the sign of the component of

$$\mathrm{i}\boldsymbol{E}_0 \times \boldsymbol{E}_0^* = 2\boldsymbol{E}_{0\mathrm{r}} \times \boldsymbol{E}_{0\mathrm{i}} = 2\boldsymbol{a} \times \boldsymbol{b}$$

---

[51] Do not mix the concept of polarization of the material, $\boldsymbol{P}$, with the polarization of a vector field.

Table 1.11 Table of the state of polarization of a time-harmonic field, see also Comment 1.6.

$i\hat{e} \cdot (E_0 \times E_0^*)$	**Polarization**
$=0$	Linear polarization
>0	Right-handed elliptic polarization
<0	Left-handed elliptic polarization

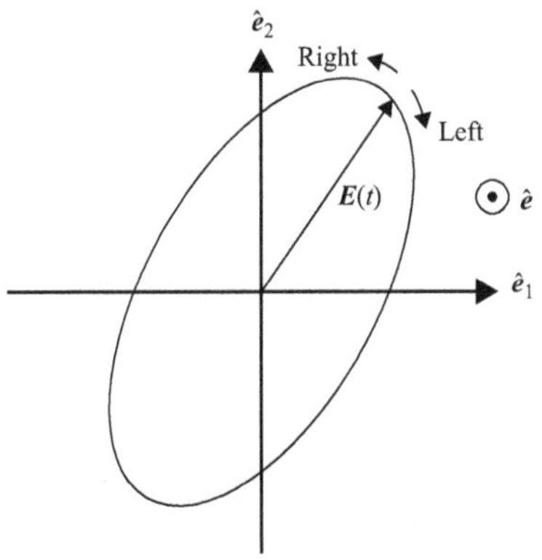

Figure 1.17 The ellipse of polarization and the definition of right- and left-handed polarizations. The unit vector $\hat{e}$ perpendicular to the plane in which the field vector $\boldsymbol{E}(t)$ oscillates and satisfies $\langle \boldsymbol{S}(t) \rangle \cdot \hat{e} > 0$. It is assumed that the vectors $\boldsymbol{a}$ and $\boldsymbol{b}$ have the position depicted in Figure 1.16.

on $\hat{e}$, see Table 1.11. The field vector either rotates counterclockwise (right-handed elliptic polarization) or clockwise (left-handed elliptic polarization) in the $\boldsymbol{a}$–$\boldsymbol{b}$-plane, see Figure 1.17, if we assume that the unit vector $\hat{e}$ is directed *toward* the observer, and that the vectors $\boldsymbol{a}$ and $\boldsymbol{b}$ have the position depicted in Figure 1.16.

The degenerated case, when the vectors $\boldsymbol{E}_{0\mathrm{r}}$ and $\boldsymbol{E}_{0\mathrm{i}}$ are parallel, implies that the field vector moves along a line through the origin—therefore, the notion of linear polarization. The linear polarization is characterized by $\boldsymbol{E}_0 \times \boldsymbol{E}_0^* = \boldsymbol{0}$. The case of a linear polarization can be viewed as a special case of an elliptic polarization, where one of the half axes is zero.

Comment 1.6
In the literature, there also occurs the opposite definition of right- and left-handed elliptic polarizations. Examples with the opposite definition are: References 101, 253, 276. In this book, we are using the same definition as, e.g., References 32, 61, 127, 130. Our definition also coincides with the IEEE standard.

One special case of elliptic polarization is particularly important. This occurs when the half axes of the ellipse, $\boldsymbol{a}$ and $\boldsymbol{b}$, have the same length, and the ellipse is a circle. We then have circular polarization. Whether the polarization is circular or not is decided by testing if

$\boldsymbol{E}_0 \cdot \boldsymbol{E}_0 = 0$. To see this, we use (1.77) and the orthogonality between the vectors $\boldsymbol{a}$ and $\boldsymbol{b}$, and we get

$$\boldsymbol{E}_0 \cdot \boldsymbol{E}_0 = \mathrm{e}^{2\mathrm{i}\vartheta} (\boldsymbol{a} + \mathrm{i}\boldsymbol{b}) \cdot (\boldsymbol{a} + \mathrm{i}\boldsymbol{b}) = \mathrm{e}^{2\mathrm{i}\vartheta} \left(|\boldsymbol{a}|^2 - |\boldsymbol{b}|^2\right) = \mathrm{e}^{2\mathrm{i}\vartheta} \left(a^2 - b^2\right)$$

The ellipse of polarization is therefore a circle, $a = |\boldsymbol{a}| = |\boldsymbol{b}| = b$, if and only if the product $\boldsymbol{E}_0 \cdot \boldsymbol{E}_0 = 0$. The direction of rotation is determined by the sign of the quantity $\mathrm{i}\hat{\boldsymbol{e}} \cdot (\boldsymbol{E}_0 \times \boldsymbol{E}_0^*)$. Right (left) circular polarization is abbreviated RCP (LCP).

Another, more convenient, way of determining whether the polarization is circular or not is motivated by the study of the quantity

$$\mathrm{i}\hat{\boldsymbol{e}} \cdot (\hat{\boldsymbol{p}}_\mathrm{e} \times \hat{\boldsymbol{p}}_\mathrm{e}^*) = \frac{2\hat{\boldsymbol{e}} \cdot (\boldsymbol{a} \times \boldsymbol{b})}{|\boldsymbol{a}|^2 + |\boldsymbol{b}|^2} = \frac{\pm 2ab}{a^2 + b^2} = \pm\left(1 - \frac{(a-b)^2}{a^2 + b^2}\right) \tag{1.79}$$

where $\hat{\boldsymbol{p}}_\mathrm{e} = \boldsymbol{E}_0 / |\boldsymbol{E}_0|$, and where we used (1.77). If this quantity is ± 1, we have RCP (upper sign) or LCP (lower sign). It is therefore convenient to define a polarization state quantity Π as

$$\Pi = \mathrm{i}\hat{\boldsymbol{e}} \cdot (\hat{\boldsymbol{p}}_\mathrm{e} \times \hat{\boldsymbol{p}}_\mathrm{e}^*) \tag{1.80}$$

This quantity is always in the interval $[-1, 1]$. $\Pi = -1$ corresponds to LCP, $\Pi = 0$ corresponds to LP, and $\Pi = 1$ corresponds to RCP. We summarize these observations in Table 1.12.

In terms of the notation above, a general polarization (unit) vector is given by

$$\hat{\boldsymbol{p}}_\mathrm{e} = \mathrm{e}^{\mathrm{i}\vartheta} \frac{(\boldsymbol{a} + \mathrm{i}\boldsymbol{b})}{\sqrt{a^2 + b^2}}$$

The ratio between the two axes $\boldsymbol{a}$ and $\boldsymbol{b}$ is also easy to compute in terms of $\hat{\boldsymbol{p}}_\mathrm{e}$. The definition of the axial ratio AR is[52]

$$\mathrm{AR} = \frac{\min\{a, b\}}{\max\{a, b\}} = \sqrt{\frac{a^2 + b^2 - |a^2 - b^2|}{a^2 + b^2 + |a^2 - b^2|}} = \sqrt{\frac{1 - |\hat{\boldsymbol{p}}_\mathrm{e} \cdot \hat{\boldsymbol{p}}_\mathrm{e}|}{1 + |\hat{\boldsymbol{p}}_\mathrm{e} \cdot \hat{\boldsymbol{p}}_\mathrm{e}|}}$$

Table 1.12 Table of the polarization state, Π, the angle χ, $\hat{\boldsymbol{e}} \cdot (\hat{\boldsymbol{a}} \times \hat{\boldsymbol{b}})$, and the axial ratio, AR, of a time-harmonic field, see also Comment 1.6.

Polarization	Π	χ	$\hat{e} \cdot (\hat{a} \times \hat{b})$	AR
Circular polarization LCP	-1	$\pi/4$	-1	1
Left-handed elliptic polarization	$(-1, 0)$	$(0, \pi/2)$	-1	$(0, 1)$
Linear polarization LP	0	$0, \pi/2$	0	0
Right-handed elliptic polarization	$(0, 1)$	$(0, \pi/2)$	1	$(0, 1)$
Circular polarization RCP	1	$\pi/4$	1	1

[52]Note that the IEEE standard of axial ratio is the reciprocal of our definition. The reason for the deviation from standard is that the linear polarization, AR $= 0$, is now numerically attainable.

since

$$\hat{\boldsymbol{p}}_e \cdot \hat{\boldsymbol{p}}_e = \frac{\boldsymbol{E}_0 \cdot \boldsymbol{E}_0}{|\boldsymbol{E}_0|^2} = \frac{e^{2i\vartheta}\left(a^2 - b^2\right)}{a^2 + b^2}$$

The axial ratio is a real number between zero and one—zero for a linearly polarized field and one for a circularly polarized field.

Define the angle χ, which also quantifies the axial ratio, by

$$\tan\chi = \frac{b}{a}, \qquad \chi \in [0, \pi/2] \quad \Rightarrow \quad \begin{cases} \cos\chi = \dfrac{a}{\sqrt{a^2+b^2}} \\ \sin\chi = \dfrac{b}{\sqrt{a^2+b^2}} \end{cases}$$

and in terms of the natural orthonormal basis $\{\hat{\boldsymbol{a}}, \hat{\boldsymbol{b}}\}$ aligned along the two half axes of the polarization ellipse, we obtain the polarization vector

$$\hat{\boldsymbol{p}}_e = e^{i\vartheta}\frac{(\boldsymbol{a} + i\boldsymbol{b})}{\sqrt{a^2+b^2}} = e^{i\vartheta}\left(\hat{\boldsymbol{a}}\cos\chi + i\hat{\boldsymbol{b}}\sin\chi\right) \tag{1.81}$$

and also the polarization state

$$\Pi = i\hat{\boldsymbol{e}} \cdot (\hat{\boldsymbol{p}}_e \times \hat{\boldsymbol{p}}_e^*) = 2\hat{\boldsymbol{e}} \cdot (\hat{\boldsymbol{a}} \times \hat{\boldsymbol{b}}) \sin\chi \cos\chi = \hat{\boldsymbol{e}} \cdot (\hat{\boldsymbol{a}} \times \hat{\boldsymbol{b}}) \sin 2\chi$$

The canonical form of an RCP field is

$$\boldsymbol{E}_0 = E_0\left(\hat{\boldsymbol{a}} + i\hat{\boldsymbol{b}}\right)$$

and the canonical form of an LCP field is

$$\boldsymbol{E}_0 = E_0\left(\hat{\boldsymbol{a}} - i\hat{\boldsymbol{b}}\right)$$

if $\{\hat{\boldsymbol{a}}, \hat{\boldsymbol{b}}, \hat{\boldsymbol{e}}\}$ forms a right-handed orthonormal basis.

To summarize, the polarization of a time-harmonic field is specified by the vector $\boldsymbol{E}_0$ which generates the two vectors $\boldsymbol{a}$ and $\boldsymbol{b}$, see (1.76). From these vectors, we can get the axial ratio

$$\text{AR} = \sqrt{\frac{1 - |\hat{\boldsymbol{p}}_e \cdot \hat{\boldsymbol{p}}_e|}{1 + |\hat{\boldsymbol{p}}_e \cdot \hat{\boldsymbol{p}}_e|}}$$

The direction of rotation is obtained by the value of $\Pi = i\hat{\boldsymbol{e}} \cdot (\hat{\boldsymbol{p}}_e \times \hat{\boldsymbol{p}}_e^*)$, and the tilt angle ψ of the vector $\boldsymbol{a}$ to the direction $\hat{\boldsymbol{e}}_1$ is, see Figure 1.21

$$\tan\psi = \frac{\hat{\boldsymbol{a}} \cdot \hat{\boldsymbol{e}}_2}{\hat{\boldsymbol{a}} \cdot \hat{\boldsymbol{e}}_1}, \qquad \psi \in [-\pi/2, \pi/2]$$

We end this section with a few examples.

Example 1.21
The most general harmonic field in the $\hat{e}_1$–$\hat{e}_2$ plane has the form (we assume $\{\hat{e}_1, \hat{e}_2, \hat{e}\}$ forms a right-handed orthonormal basis)

$$E(t) = \hat{e}_1 A \cos(\omega t - \alpha) + \hat{e}_2 B \cos(\omega t - \beta)$$

where $A \geq 0$, $B \geq 0$, and α and β are real angles. The corresponding complex vector E_0 is

$$\begin{cases} E(t) = \text{Re}\left\{E_0 e^{-i\omega t}\right\} \\ E_0 = A e^{i\alpha} \hat{e}_1 + B e^{i\beta} \hat{e}_2 \end{cases} \tag{1.82}$$

which implies

$$iE_0 \times E_0^* = iABe^{i(\alpha-\beta)}\hat{e} - iABe^{-i(\alpha-\beta)}\hat{e} = -2AB\hat{e}\sin(\alpha - \beta)$$

From this, we conclude that the field is, see Table 1.11

Left-handed polarization if $0 < \alpha - \beta < \pi$
Right-handed polarization if $\pi < \alpha - \beta < 2\pi$
Linear polarization if $\alpha = \beta$ or $\alpha = \beta + \pi$

where the inequalities are interpreted as mod 2π.
The real and imaginary parts of E_0 are

$$\begin{cases} E_{0r} = \hat{e}_1 A \cos\alpha + \hat{e}_2 B \cos\beta \\ E_{0i} = \hat{e}_1 A \sin\alpha + \hat{e}_2 B \sin\beta \end{cases}$$

From (1.76), we have

$$\begin{cases} a = \left(\hat{e}_1 A \cos\alpha + \hat{e}_2 B \cos\beta\right)\cos\vartheta + \left(\hat{e}_1 A \sin\alpha + \hat{e}_2 B \sin\beta\right)\sin\vartheta \\ b = -\left(\hat{e}_1 A \cos\alpha + \hat{e}_2 B \cos\beta\right)\sin\vartheta + \left(\hat{e}_1 A \sin\alpha + \hat{e}_2 B \sin\beta\right)\cos\vartheta \end{cases}$$

where the angle ϑ is determined by

$$\tan 2\vartheta = \frac{A^2 \sin 2\alpha + B^2 \sin 2\beta}{A^2 \cos 2\alpha + B^2 \cos 2\beta} \tag{1.83}$$

which implies that the half axes of the ellipse are

$$\begin{cases} a = A\hat{e}_1 \cos(\vartheta - \alpha) + B\hat{e}_2 \cos(\vartheta - \beta) \\ b = -A\hat{e}_1 \sin(\vartheta - \alpha) - B\hat{e}_2 \sin(\vartheta - \beta) \end{cases}$$

The length of the half axes are

$$\begin{cases} a = \sqrt{A^2 \cos^2(\vartheta - \alpha) + B^2 \cos^2(\vartheta - \beta)} \\ b = \sqrt{A^2 \sin^2(\vartheta - \alpha) + B^2 \sin^2(\vartheta - \beta)} \end{cases} \tag{1.84}$$

and the angles $\phi_a = \psi$ and ϕ_b between the $\hat{\boldsymbol{e}}_1$-axis and the half axes $\boldsymbol{a}$ and $\boldsymbol{b}$ are determined by

$$\begin{cases} \tan \phi_a = \dfrac{B\cos(\vartheta - \beta)}{A\cos(\vartheta - \alpha)} \\ \tan(\phi_b - \pi) = \dfrac{B\sin(\vartheta - \beta)}{A\sin(\vartheta - \alpha)} \end{cases} \tag{1.85}$$

respectively. ■

Example 1.22
Construct the harmonic field, oscillating in the $\hat{\boldsymbol{e}}_1$–$\hat{\boldsymbol{e}}_2$ plane, satisfying the following specifications (see also Figure 1.18):

- The field is at time $t = 0$ polarized along the $\hat{\boldsymbol{e}}_1$-axis and strength E (a given real constant), i.e., $\boldsymbol{E}(t = 0) = \hat{\boldsymbol{e}}_1 E$.
- The quotient between the axes of the ellipse is $\varepsilon = b/a$. The axis $\boldsymbol{a}$, with the length a, is located in the first quadrant, and the angle between $\boldsymbol{a}$ and the $\hat{\boldsymbol{e}}_1$-axis is ψ.
- The field has right-handed elliptic polarization ($\langle \boldsymbol{S}(t) \rangle$ is assumed to be directed along $\hat{\boldsymbol{e}} = \hat{\boldsymbol{e}}_1 \times \hat{\boldsymbol{e}}_2$).

We assume $\{\hat{\boldsymbol{e}}_1, \hat{\boldsymbol{e}}_2, \hat{\boldsymbol{e}}\}$ forms a right-handed orthonormal basis. Determine the real constants A, B, α, and β in the expression

$$\boldsymbol{E}(t) = \hat{\boldsymbol{e}}_1 A\cos(\omega t - \alpha) + \hat{\boldsymbol{e}}_2 B\cos(\omega t - \beta)$$

i.e., determine the amplitude and the phase of the $\hat{\boldsymbol{e}}_1$- and $\hat{\boldsymbol{e}}_2$-components, cf. Example 1.21.

Solution: We prefer not to follow Example 1.21, but to present an alternative solution.

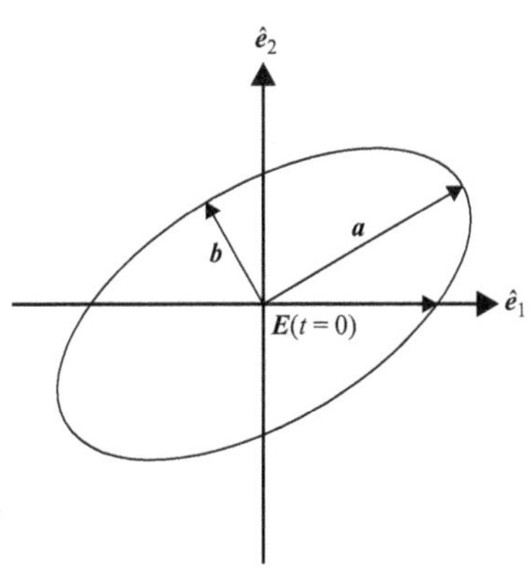

Figure 1.18 Ellipse of polarization in Example 1.22.

Construct the half axes of the ellipse, i.e., the orthogonal vectors $\boldsymbol{a}$ and $\boldsymbol{b}$ in accordance with the requirements in the example. We let

$$\begin{cases} \boldsymbol{a} = \dfrac{a}{\sqrt{1+\tan^2\psi}}\left(\hat{\boldsymbol{e}}_1 + \hat{\boldsymbol{e}}_2 \tan\psi\right) \\ \boldsymbol{b} = \dfrac{a\varepsilon}{\sqrt{1+\tan^2\psi}}\left(-\hat{\boldsymbol{e}}_1 \tan\psi + \hat{\boldsymbol{e}}_2\right) \end{cases}$$

which implies that the lengths of the vectors $\boldsymbol{a}$ and $\boldsymbol{b}$ are a and b, respectively, $\varepsilon = b/a$, the angle between $\boldsymbol{a}$ and the $\hat{\boldsymbol{e}}_1$-axis is ψ, and moreover, this choice gives a right-handed elliptic polarization of the field, since

$$(\boldsymbol{a} \times \boldsymbol{b}) \cdot \hat{\boldsymbol{e}} = a^2\varepsilon > 0$$

Now determine the angle ϑ in the expression $\boldsymbol{E}_0 = \mathrm{e}^{\mathrm{i}\vartheta}(\boldsymbol{a} + \mathrm{i}\boldsymbol{b})$

$$\boldsymbol{E}(t) = \boldsymbol{E}_{0\mathrm{r}} \cos\omega t + \boldsymbol{E}_{0\mathrm{i}} \sin\omega t = \boldsymbol{a}\cos(\omega t - \vartheta) + \boldsymbol{b}\sin(\omega t - \vartheta)$$

At time $t = 0$, we have

$$\boldsymbol{E}(0) = \boldsymbol{E}_{0\mathrm{r}} = \boldsymbol{a}\cos\vartheta - \boldsymbol{b}\sin\vartheta = E\hat{\boldsymbol{e}}_1$$

i.e., the components satisfy

$$\begin{cases} \dfrac{a}{\sqrt{1+\tan^2\psi}}(\cos\vartheta + \varepsilon\tan\psi\sin\vartheta) = E \\ \dfrac{a}{\sqrt{1+\tan^2\psi}}(\tan\psi\cos\vartheta - \varepsilon\sin\vartheta) = 0 \end{cases}$$

with solution

$$\begin{cases} \dfrac{a}{\sqrt{1+\tan^2\psi}}\cos\vartheta = \dfrac{E}{1+\tan^2\psi} \\ \dfrac{a}{\sqrt{1+\tan^2\psi}}\sin\vartheta = \dfrac{E\tan\psi}{\varepsilon\left(1+\tan^2\psi\right)} \end{cases}$$

and we have

$$\frac{a}{\sqrt{1+\tan^2\psi}}\mathrm{e}^{\mathrm{i}\vartheta} = \frac{E(\varepsilon + \mathrm{i}\tan\psi)}{\varepsilon\left(1+\tan^2\psi\right)}$$

We get

$$\begin{aligned} \boldsymbol{E}_0 = \mathrm{e}^{\mathrm{i}\vartheta}(\boldsymbol{a} + \mathrm{i}\boldsymbol{b}) &= \frac{E(\varepsilon + \mathrm{i}\tan\psi)}{\varepsilon\left(1+\tan^2\psi\right)}\left\{\left(\hat{\boldsymbol{e}}_1 + \hat{\boldsymbol{e}}_2\tan\psi\right) + \mathrm{i}\varepsilon\left(-\hat{\boldsymbol{e}}_1\tan\psi + \hat{\boldsymbol{e}}_2\right)\right\} \\ &= E\left\{\hat{\boldsymbol{e}}_1\left(1 + \mathrm{i}\frac{\tan\psi}{1+\tan^2\psi}\frac{1-\varepsilon^2}{\varepsilon}\right) + \mathrm{i}\hat{\boldsymbol{e}}_2\frac{\varepsilon^2 + \tan^2\psi}{\varepsilon\left(1+\tan^2\psi\right)}\right\} \\ &= E\left\{\hat{\boldsymbol{e}}_1\left(1 + \mathrm{i}\frac{1-\varepsilon^2}{2\varepsilon}\sin 2\psi\right) + \mathrm{i}\hat{\boldsymbol{e}}_2\left(\varepsilon\cos^2\psi + \frac{1}{\varepsilon}\sin^2\psi\right)\right\} \end{aligned}$$

From this expression of $\boldsymbol{E}_0$, we can identify the amplitudes A and B and the phases α and β in

$$\boldsymbol{E}(t) = \hat{\boldsymbol{e}}_1 A \cos(\omega t - \alpha) + \hat{\boldsymbol{e}}_2 B \cos(\omega t - \beta)$$

by rewriting the complex vector $\boldsymbol{E}_0$ in polar form

$$\boldsymbol{E}_0 = \hat{\boldsymbol{e}}_1 A e^{i\alpha} + \hat{\boldsymbol{e}}_2 B e^{i\beta}$$

i.e.,

$$\begin{cases} A = \sqrt{1 + \dfrac{\left(1-\varepsilon^2\right)^2}{4\varepsilon^2}\sin^2 2\psi} \\ B = \varepsilon \cos^2\psi + \dfrac{1}{\varepsilon}\sin^2\psi \end{cases} \qquad \begin{cases} \alpha = \arctan\left(\dfrac{1-\varepsilon^2}{2\varepsilon}\sin 2\psi\right) \\ \beta = \dfrac{\pi}{2} \end{cases}$$

■

1.4 Coherence and Degree of Polarization

In Sections 1.1 and 1.2, the focus was on general time-varying fields, and in Section 1.3, we investigated the special case of time-harmonic variations, i.e., fields oscillating with a fixed angular frequency ω. Moreover, in Section 1.3.7, we showed that all time-harmonic fields oscillate in a fixed plane and that the field traces an ellipse in this plane, i.e., the ellipse of polarization. However, all physical fields have a finite extent in time, and therefore, they are not strictly monochromatic, but a mixture of different frequencies. In addition, a mixture of polarization states often occurs in applications. This situation is particularly frequent with natural sources, e.g., electromagnetic waves that originate from the Sun or Space. In this section, we analyze fields of this kind.

A general transient field is a superposition of time-harmonic fields of different frequencies, see (1.41) on page 39

$$\boldsymbol{E}(\boldsymbol{r},t) = \frac{1}{\pi}\operatorname{Re}\int_0^\infty \boldsymbol{E}(\boldsymbol{r},\omega)e^{-i\omega t}\,d\omega$$

The fields in this section are assumed to be almost time harmonic, which implies that the temporal spectrum of the field, $\boldsymbol{E}(\boldsymbol{r},\omega)$, has a well-defined average frequency $\overline{\omega} > 0$ and half width $\Delta\omega > 0$, which is small—more precisely, $\Delta\omega/\overline{\omega} \ll 1$. As a consequence, we assume that the spectrum satisfies

$$\boldsymbol{E}(\boldsymbol{r},\omega) \approx \boldsymbol{0}, \qquad \text{for } |\omega - \overline{\omega}| \geq \frac{\Delta\omega}{2}$$

A schematic drawing of a wave and its corresponding frequency spectrum are depicted in Figure 1.19. The half width of the temporal spectrum, $\Delta\omega > 0$, introduces two new concepts, coherence time τ and coherence length l, defined by

$$\tau = \frac{2\pi}{\Delta\omega}, \qquad l = \frac{2\pi c_0}{\Delta\omega} = c_0\tau$$

The coherence length, l, is of the same order of magnitude as the length of the wave [21].

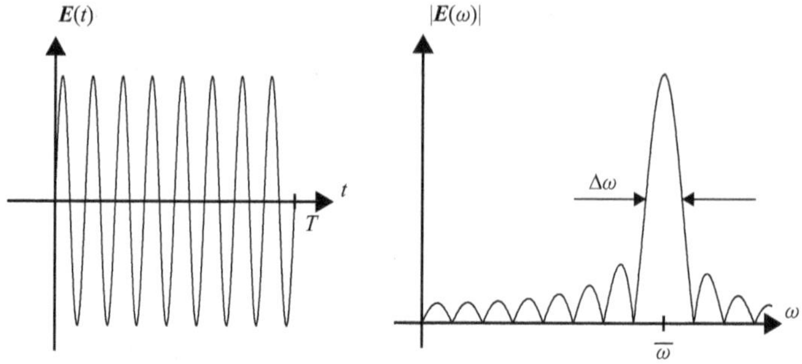

Figure 1.19 A schematic drawing of a wave and the corresponding frequency spectrum and half width $\Delta\omega$. The scales on the axes are arbitrary.

We make a model of almost monochromatic waves, and introduce a field defined by,[53] cf. (1.41) and (1.42)

$$\boldsymbol{E}(t) = \mathrm{Re}\left\{\boldsymbol{E}_0(t)\mathrm{e}^{-\mathrm{i}\overline{\omega}t}\right\}$$

where the complex-valued, time-dependent vector $\boldsymbol{E}_0(t)$ is defined by

$$\boldsymbol{E}_0(t) = \mathrm{e}^{\mathrm{i}\overline{\omega}t}\frac{1}{\pi}\int_0^\infty \boldsymbol{E}(\omega)\mathrm{e}^{-\mathrm{i}\omega t}\,\mathrm{d}\omega$$

We call this field a quasi-monochromatic field provided the assumptions for $\boldsymbol{E}(\omega)$ above hold. If the vector $\boldsymbol{E}_0(t)$ is independent of time, we have the completely time-harmonic field of Section 1.3.

The time variations of the complex-valued field $\boldsymbol{E}_0(t)$ are slow compared to a time interval of the order of $2\pi/\overline{\omega}$. To see this, we make a Fourier transform of the vector $\boldsymbol{E}_0(t)$. We have

$$\int_{-\infty}^\infty \boldsymbol{E}_0(t)\mathrm{e}^{\mathrm{i}\omega t}\,\mathrm{d}t = \frac{1}{\pi}\int_{-\infty}^\infty\int_0^\infty \boldsymbol{E}(\omega')\mathrm{e}^{\mathrm{i}(\omega+\overline{\omega}-\omega')t}\,\mathrm{d}\omega'\,\mathrm{d}t = 2\boldsymbol{E}(\omega+\overline{\omega})$$

The assumptions on the spectrum of the field imply that the Fourier transform of $\boldsymbol{E}_0(t)$ is negligible outside a frequency interval $[-\Delta\omega/2, \Delta\omega/2]$, i.e., the spectrum of $\boldsymbol{E}_0(t)$ has only low-frequency content, and since $\Delta\omega/\overline{\omega} \ll 1$, and $\boldsymbol{E}_0(t)$ is slowly varying compared to the time interval $2\pi/\overline{\omega}$. This signal is therefore a narrowband signal.

We assume that the quasi-monochromatic field oscillates in a fixed plane, $\hat{\boldsymbol{e}}_1$–$\hat{\boldsymbol{e}}_2$ plane. This is correct if the field is completely monochromatic ($\boldsymbol{E}_0$ independent of time) as in Section 1.3. For a quasi-monochromatic field this is an assumption. Under this assumption, the field $\boldsymbol{E}_0(t)$ expressed in Cartesian components $\hat{\boldsymbol{e}}_1$ and $\hat{\boldsymbol{e}}_2$ is

$$\boldsymbol{E}_0(t) = \hat{\boldsymbol{e}}_1 E_{01}(t) + \hat{\boldsymbol{e}}_2 E_{02}(t) = \hat{\boldsymbol{e}}_1|E_{01}(t)|\mathrm{e}^{\mathrm{i}\alpha(t)} + \hat{\boldsymbol{e}}_2|E_{02}(t)|\mathrm{e}^{\mathrm{i}\beta(t)}$$

and $\alpha(t)$ and $\beta(t)$ are the phases of the components, which can be functions of time.

[53]The field is analyzed at a fixed point in space, and the dependence of the spatial variables is suppressed in this section.

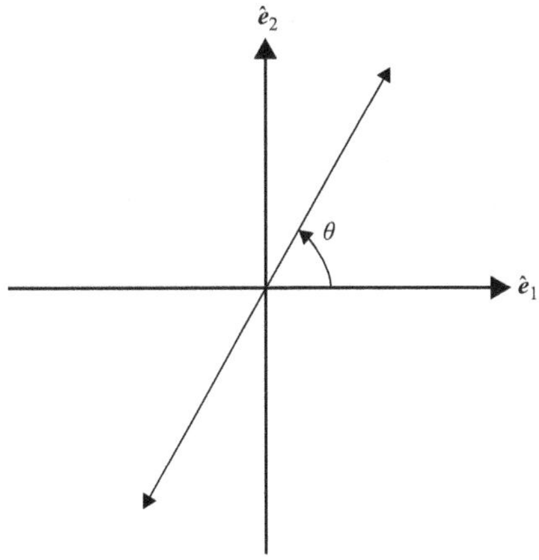

Figure 1.20 A measurement of the intensity along a fixed angle θ.

A quasi-monochromatic field is conveniently examined by investigating its properties along a fixed direction in the $\hat{e}_1$–$\hat{e}_2$ plane. The projection along a fixed direction $\hat{e}$ is

$$\hat{e} \cdot E_0(t)$$

We parameterize the direction $\hat{e}$ by the angle θ, see Figure 1.20. Explicitly, the representation is

$$\hat{e} = \hat{e}_1 \cos\theta + \hat{e}_2 e^{i\delta} \sin\theta \tag{1.86}$$

where we have introduced a phase factor δ, which gives the difference in phase between the $\hat{e}_2$-component and the $\hat{e}_1$-component of the field. Experimentally, the projection along a fixed direction is implemented by, e.g., a polarizer, and the retardation of the phase is implemented by e.g., a plate of retardation. Note that $\hat{e}$ is a complex vector, but that $\hat{e} \cdot \hat{e}^* = 1$. Only if $\delta = 0$ or π $\hat{e}$ is a real vector.

The intensity of the field, i.e., the projection of the Poynting vector (with or without retardation along the $\hat{e}_2$-component) along the direction $\hat{e}$, defined by (1.86), is proportional to the real quantity $I(\theta, \delta)$ defined by[54]

$$I(\theta, \delta) = \left\langle (\hat{e} \cdot E_0(t))(\hat{e} \cdot E_0(t))^* \right\rangle = \left\langle \left| \hat{e} \cdot E_0(t) \right|^2 \right\rangle \tag{1.87}$$

The temporal average of two complex quantities $f_1(t)$ and $f_2(t)$ over the time interval T' is given by

$$\langle f_1(t) f_2(t) \rangle = \frac{1}{2T'} \int_{-T'}^{T'} f_1(t) f_2(t) \, dt$$

The time T' is assumed to be long compared to the time over which the field varies.

[54]The intensity of the electromagnetic plane wave, see (1.72), in an isotropic material is proportional to the square of the electric field. More precisely, the intensity (the component of the Poynting vector in the propagation direction $\hat{k} = \pm\hat{e}_1 \times \hat{e}_2$) is $\left\langle |E_0(t)|^2 \right\rangle / (2\eta\eta_0)$.

The concept of intensity can be extended further by the definition of a 2×2-matrix of coherence $[\mathbf{J}]$, also called the coherency matrix. Its definition is

$$[\mathbf{J}] = \begin{pmatrix} J_{11} & J_{12} \\ J_{21} & J_{22} \end{pmatrix} = \begin{pmatrix} \langle E_{01}(t)E_{01}^*(t)\rangle & \langle E_{01}(t)E_{02}^*(t)\rangle \\ \langle E_{02}(t)E_{01}^*(t)\rangle & \langle E_{02}(t)E_{02}^*(t)\rangle \end{pmatrix} \tag{1.88}$$

By the use of row and column vectors, we rewrite the matrix of coherence as (the "dagger" (†) denotes the Hermitian adjoint)

$$[\mathbf{J}] = \left\langle \begin{pmatrix} E_{01}(t) \\ E_{02}(t) \end{pmatrix} \begin{pmatrix} E_{01}(t)^* & E_{02}(t)^* \end{pmatrix} \right\rangle = \left\langle \begin{pmatrix} E_{01}(t) \\ E_{02}(t) \end{pmatrix} \begin{pmatrix} E_{01}(t) \\ E_{02}(t) \end{pmatrix}^\dagger \right\rangle$$

where the temporal average of a matrix $[\mathbf{A}]$ has entries that are temporal averages of the entries of the matrix $[\mathbf{A}]$. The matrix of coherence quantifies the time correlation between the Cartesian components of the electric field. The diagonal elements J_{11} and J_{22} are real, positive quantities, but J_{12} and J_{21} are in general complex numbers. Note that the matrix $[\mathbf{J}]$ is Hermitian, $[\mathbf{J}] = [\mathbf{J}]^\dagger$, since

$$J_{12}^* = \langle E_{01}(t)E_{02}^*(t)\rangle^* = \langle E_{02}(t)E_{01}^*(t)\rangle = J_{21}$$

The Cauchy–Schwarz[55] inequality applied to the time integrals in the off diagonal entries of the coherency matrix implies

$$|J_{12}| = |\langle E_{01}(t)E_{02}^*(t)\rangle| \le \sqrt{\langle |E_{01}(t)|^2\rangle}\sqrt{\langle |E_{02}(t)|^2\rangle} = \sqrt{J_{11}J_{22}}$$

and consequently, the matrix $[\mathbf{J}]$ has a non-negative determinant

$$\det[\mathbf{J}] = J_{11}J_{22} - J_{12}J_{21} = J_{11}J_{22} - |J_{12}|^2 \ge 0$$

The intensity of the total field in an isotropic material is proportional to the sum of the diagonal elements of the coherency matrix, i.e., the trace of the matrix of coherence

$$\operatorname{Tr}[\mathbf{J}] = J_{11} + J_{22} = \langle |E_{01}(t)|^2\rangle + \langle |E_{02}(t)|^2\rangle = \langle |\mathbf{E}_0(t)|^2\rangle$$

The intensity along a fixed direction $\hat{\boldsymbol{e}}$, (1.86), is given by (1.87). We rewrite (1.87) in terms of the coherency matrix $[\mathbf{J}]$. The result is

$$\begin{aligned} I(\theta,\delta) &= J_{11}\cos^2\theta + J_{22}\sin^2\theta + J_{12}e^{-i\delta}\cos\theta\sin\theta + J_{21}e^{i\delta}\cos\theta\sin\theta \\ &= J_{11}\cos^2\theta + J_{22}\sin^2\theta + 2\operatorname{Re}(J_{12}e^{-i\delta})\cos\theta\sin\theta \end{aligned}$$

---

[55]Augustin-Louis Cauchy (1789–1857), French mathematician; Karl Hermann Amandus Schwarz (1843–1921), German mathematician.

since $J_{12} = J_{21}^*$. To see how this quantity varies as a function of the angle θ and the retardation δ, we rewrite the expression. We get, see Problem 1.17

$$I(\theta,\delta) = \underbrace{\frac{1}{2}(J_{11}+J_{22})}_{\langle\, |E_0(t)|^2\,\rangle} + R\cos(2\theta-\alpha) \tag{1.89}$$

where

$$\begin{cases} R = \dfrac{1}{2}\sqrt{(J_{11}-J_{22})^2 + \left(2\,\mathrm{Re}\,(J_{12}e^{-i\delta})\right)^2} \\ \tan\alpha = \dfrac{2\,\mathrm{Re}\,(J_{12}e^{-i\delta})}{J_{11}-J_{22}} \end{cases}$$

From these expressions, we see that the intensity $I(\theta,\delta)$ varies, as a function of the angles θ and δ, between

$$I_{\min} = \frac{1}{2}(J_{11}+J_{22}) - R_{\max} \le I(\theta,\delta) \le \frac{1}{2}(J_{11}+J_{22}) + R_{\max} = I_{\max}$$

where

$$R_{\max} = \frac{1}{2}\sqrt{(J_{11}-J_{22})^2 + 4|J_{12}|^2} = \frac{1}{2}(J_{11}+J_{22})\sqrt{1-\frac{4\det[\mathbf{J}]}{(J_{11}+J_{22})^2}}$$

This result can also be obtained by interpreting the intensity $I(\theta,\delta)$ as a matrix product. To see this, write the intensity $I(\theta,\delta)$ as a Hermitian form

$$I(\theta,\delta) = \begin{pmatrix}\cos\theta & \sin\theta\end{pmatrix}\begin{pmatrix} J_{11} & J_{12}e^{-i\delta} \\ J_{21}e^{i\delta} & J_{22}\end{pmatrix}\begin{pmatrix}\cos\theta \\ \sin\theta\end{pmatrix} = \boldsymbol{x}^* \cdot \tilde{\mathbf{J}} \cdot \boldsymbol{x}$$

The two eigenvalues of the matrix $[\tilde{\mathbf{J}}]$, $\lambda = \frac{1}{2}(J_{11}+J_{22}) \pm R_{\max}$, give the largest and the smallest values of the intensity.

The quantity $R_{\max}$ is a measure of the variation of the intensity as the angles θ and δ vary. This quantity is most conveniently measured by the degree of polarization P, which is a dimensionless quantity, defined by

$$\begin{aligned} P &= \frac{I_{\max}-I_{\min}}{I_{\max}+I_{\min}} = \frac{2R_{\max}}{J_{11}+J_{22}} \\ &= \sqrt{1-\frac{4\det[\mathbf{J}]}{(J_{11}+J_{22})^2}} = \sqrt{1-\frac{4\det[\mathbf{J}]}{\left(\langle |\boldsymbol{E}_0(t)|^2\rangle\right)^2}} \end{aligned} \tag{1.90}$$

This quantity varies between 0 and 1, since it assumes its smallest values when $I_{\max} = I_{\min}$ and is maximal when $I_{\min} = 0$. Therefore, the degree of polarization satisfies

$$P \in [0,1]$$

1.4.1 Unpolarized Field

The electromagnetic field from many natural sources is unpolarized or natural, which implies that the intensity $I(\theta, \delta)$ is the same in all directions $\hat{e}$, i.e., it is independent of the angle θ and the retardation δ. This implies that the quantity R_{max} is identically zero, which implies

$$\begin{cases} J_{11} = J_{22} \\ J_{12} = J_{21} = 0 \end{cases}$$

The matrix of coherence for an unpolarized field then becomes

$$[\mathbf{J}] = \begin{pmatrix} J_{11} & 0 \\ 0 & J_{22} \end{pmatrix} = \frac{\left\langle |\boldsymbol{E}_0(t)|^2 \right\rangle}{2} \begin{pmatrix} 1 & 0 \\ 0 & 1 \end{pmatrix} \tag{1.91}$$

since $\left\langle |\boldsymbol{E}_0(t)|^2 \right\rangle = J_{11} + J_{22} = 2J_{11}$.

The determinant of the matrix of coherence for an unpolarized field is

$$\det[\mathbf{J}] = J_{11}J_{22} - J_{12}J_{21} = |J_{11}|^2 = \frac{\left\langle |\boldsymbol{E}_0(t)|^2 \right\rangle^2}{4} \geq 0$$

and the degree of polarization for an unpolarized field is

$$P = \sqrt{1 - \frac{4\det[\mathbf{J}]}{(J_{11} + J_{22})^2}} = \sqrt{1 - \frac{4\det[\mathbf{J}]}{\left\langle |\boldsymbol{E}_0(t)|^2 \right\rangle^2}} = 0$$

1.4.2 Completely Polarized Field

An unpolarized field is one extreme value of the degree of polarization, $P = 0$. The other extreme value is a completely monochromatic field or a completely (fully) polarized field. The complex vector $\boldsymbol{E}_0$ is then constant in time and the matrix of coherence element is

$$[\mathbf{J}] = \begin{pmatrix} E_{01}E_{01}^* & E_{01}E_{02}^* \\ E_{02}E_{01}^* & E_{02}E_{02}^* \end{pmatrix} \tag{1.92}$$

Note that the time average now has disappeared.

The determinant of the matrix of coherence for a monochromatic field is

$$\det[\mathbf{J}] = |E_{01}|^2 |E_{02}|^2 - E_{01}E_{02}^*E_{02}E_{01}^* = 0$$

and the degree of polarization for a monochromatic field is

$$P = \sqrt{1 - \frac{4\det[\mathbf{J}]}{(J_{11} + J_{22})^2}} = 1$$

1.4.3 General Degree of Polarization

The two extreme values of the degree of polarization—unpolarized and completely polarized field—are characterized by the matrices of coherence of the following form:

$$[\mathbf{J}]_{\text{unpol}} = \begin{pmatrix} A & 0 \\ 0 & A \end{pmatrix}, \qquad [\mathbf{J}]_{\text{pol}} = \begin{pmatrix} B & D \\ D^* & C \end{pmatrix}$$

where A, B, and C are non-negative, real numbers, $A \geq 0$, $B \geq 0$, $C \geq 0$, and D is a complex number, satisfying $BC - DD^* = 0$.

We now show that every matrix of coherence, $[\mathbf{J}]$, in a unique way can be written as a sum of an unpolarized and a completely polarized field, i.e.,

$$[\mathbf{J}] = \begin{pmatrix} J_{11} & J_{12} \\ J_{21} & J_{22} \end{pmatrix} = \underbrace{\begin{pmatrix} A & 0 \\ 0 & A \end{pmatrix}}_{\text{unpol}} + \underbrace{\begin{pmatrix} B & D \\ D^* & C \end{pmatrix}}_{\text{pol}} = [\mathbf{J}]_{\text{unpol}} + [\mathbf{J}]_{\text{pol}}$$

We prove this by explicitly computing the matrix entries A, B, C, and D. The following connections hold:

$$\begin{cases} J_{11} = A + B \\ J_{12} = D \\ J_{21} = D^* \\ J_{22} = A + C \end{cases}$$

Note that the total intensity of the field is $J_{11} + J_{22} = 2A + B + C$.

Eliminate B and C in $BC - DD^* = 0$. We get

$$(J_{11} - A)(J_{22} - A) - J_{12}J_{12}^* = 0$$

The two roots, A, of this equation are

$$A = \frac{1}{2}(J_{11} + J_{22}) \pm \frac{1}{2}\sqrt{(J_{11} + J_{22})^2 - 4\det[\mathbf{J}]}$$

where $\det[\mathbf{J}] = J_{11}J_{22} - J_{12}J_{12}^*$. Both these roots are real and positive since

$$\det[\mathbf{J}] = J_{11}J_{22} - |J_{12}|^2 \leq J_{11}J_{22} \leq \frac{1}{4}(J_{11} + J_{22})^2$$

The latter inequality is easily proven by expanding $(J_{11} - J_{22})^2 \geq 0$. Only one of these roots gives positive values of B and C. The unique solution is therefore

$$\begin{cases} A = \frac{1}{2}(J_{11} + J_{22}) - \frac{1}{2}\sqrt{(J_{11} + J_{22})^2 - 4\det[\mathbf{J}]} \\ B = \frac{1}{2}(J_{11} - J_{22}) + \frac{1}{2}\sqrt{(J_{11} + J_{22})^2 - 4\det[\mathbf{J}]} \\ C = \frac{1}{2}(J_{22} - J_{11}) + \frac{1}{2}\sqrt{(J_{11} + J_{22})^2 - 4\det[\mathbf{J}]} \\ D = J_{12} \end{cases}$$

This decomposition provides us with another way of defining the degree of polarization P. The intensity of the matrix $[\mathbf{J}]_{\text{pol}}$ is given by the sum of the diagonal elements

$$\mathrm{Tr}\,[\mathbf{J}]_{\text{pol}} = B + C = \sqrt{(J_{11} + J_{22})^2 - 4\,\det\,[\mathbf{J}]}$$

The quotient $\mathrm{Tr}\,[\mathbf{J}]_{\text{pol}}\,/\mathrm{Tr}\,[\mathbf{J}]$ is

$$P = \frac{\mathrm{Tr}\,[\mathbf{J}]_{\text{pol}}}{\mathrm{Tr}\,[\mathbf{J}]} = \sqrt{1 - \frac{4\,\det\,[\mathbf{J}]}{(J_{11} + J_{22})^2}}$$

which coincides with our previous definition of the degree of polarization P. This expression shows that the degree of polarization P for a general quasi-monochromatic field is given by the quotient between the intensity of the completely polarized part, $[\mathbf{J}]_{\text{pol}}$, to the intensity of the total field.

1.4.4 The Stokes Parameters

The Stokes[56] parameters are closely related to the state of polarization of a time-harmonic or a quasi-monochromatic electromagnetic field. We denote these parameters s_i, $i = 0, 1, 2, 3$. In this section, we define these parameters for a monochromatic or a quasi-monochromatic field.

The Stokes parameters, which are real numbers, are most easily defined in terms of the entries of the matrix of coherence

$$\begin{cases} s_0 = J_{11} + J_{22} \\ s_1 = J_{11} - J_{22} \\ s_2 = J_{12} + J_{21} = 2\,\mathrm{Re}\,J_{12} \\ s_3 = \mathrm{i}\,(J_{21} - J_{12}) = 2\,\mathrm{Im}\,J_{12} \end{cases} \tag{1.93}$$

Comment 1.7
Another notation often occurring in the literature is I, Q, U, and V, defined such that $I = s_0$, $Q = s_1$, $U = \pm s_2$, and $V = s_3$. These parameters were introduced by G.G. Stokes in 1852 to describe light that was not completely polarized. Sign differences between various definitions also occur [186, 187].

Since the entries of the matrix of coherence can be interpreted as intensity quantities, $I(\theta, \delta)$, along well-defined angles θ and retardations δ, the Stokes parameters can be determined

[56] George Gabriel Stokes (1819–1903), Irish mathematician and physicist.

experimentally with a polarizer and a plate of retardation. The relationship between these quantities is, see Problem 1.18

$$\begin{cases} s_0 = I(0,0) + I(\pi/2,0) \\ s_1 = I(0,0) - I(\pi/2,0) \\ s_2 = I(\pi/4,0) - I(3\pi/4,0) \\ s_3 = I(\pi/4,\pi/2) - I(3\pi/4,\pi/2) \end{cases}$$

More details on how to perform the appropriate intensity measurements are found in the literature, see e.g., References 185, 187.

From the definitions of the parameters s_i, $i = 0, 1, 2, 3$, in (1.93), we find

$$\begin{aligned} s_1^2 + s_2^2 + s_3^2 &= (J_{11} - J_{22})^2 + (J_{12} + J_{21})^2 - (J_{12} - J_{21})^2 \\ &= (J_{11} + J_{22})^2 - 4\det[\mathbf{J}] \end{aligned}$$

and as a result of this identity, we have another way of expressing the degree of polarization P, namely, see (1.90)

$$P = \frac{\sqrt{s_1^2 + s_2^2 + s_3^2}}{s_0} \tag{1.94}$$

Since $\det[\mathbf{J}] \geq 0$ (or $P \in [0, 1]$), we also always have

$$s_1^2 + s_2^2 + s_3^2 \leq s_0^2 \tag{1.95}$$

with equality if the field is completely polarized, $P = 1$.

The unpolarized field, $P = 0$, is characterized by $s_i = 0$, $i = 1, 2, 3$, i.e.,

$$\begin{pmatrix} s_0 \\ s_1 \\ s_2 \\ s_3 \end{pmatrix} = \begin{pmatrix} \langle |\boldsymbol{E}_0(t)|^2 \rangle \\ 0 \\ 0 \\ 0 \end{pmatrix}$$

since s_0 is the total intensity of the field.

The Stokes parameters in the completely polarized field, $P = 1$, are not independent. From (1.94), we obtain

$$s_0^2 = s_1^2 + s_2^2 + s_3^2 \tag{1.96}$$

In Section 1.4.5, we interpret the completely polarized field geometrically as a point on the Poincaré[57] sphere, but before making this interpretation, we rewrite and investigate the Stokes parameters of the completely polarized field.

---

[57] Jules Henri Poincaré (1854–1912), French mathematician and theoretical physicist.

The Stokes parameters for a completely polarized field become, see (1.92)

$$\begin{cases} s_0 = |E_{01}|^2 + |E_{02}|^2 \\ s_1 = |E_{01}|^2 - |E_{02}|^2 \\ s_2 = 2\,\mathrm{Re}\,E_{01}E_{02}^* \\ s_3 = 2\,\mathrm{Im}\,E_{01}E_{02}^* \end{cases} \tag{1.97}$$

where the time averages are superfluous, since the field $\boldsymbol{E}_0$ is independent of the time.

In Section 1.3.7, we analyzed the state of polarization of an arbitrary monochromatic electromagnetic field, and we concluded that the complex vector $\boldsymbol{E}_0$ completely characterizes the state of polarization, see (1.74) on page 74. Moreover, the physical field, $\boldsymbol{E}(t)$, traces an ellipse in a fixed plane, which we take as the $\hat{\boldsymbol{e}}_1$–$\hat{\boldsymbol{e}}_2$ plane. We also introduced two real-valued, orthogonal vectors $\boldsymbol{a}$ and $\boldsymbol{b}$. These vectors are the half axes of the ellipse of the polarization. We denote the lengths of these vectors by $a = |\boldsymbol{a}|$ and $b = |\boldsymbol{b}|$, respectively, and the tilt of the ellipse is parameterized by the angle ψ (the angle between the $\hat{\boldsymbol{e}}_1$-axis and $\boldsymbol{a}$), see Figure 1.21. The relation between the complex vector $\boldsymbol{E}_0$ and the vectors $\boldsymbol{a}$ and $\boldsymbol{b}$ is given by (1.77) on page 75

$$\boldsymbol{E}_0 = e^{i\vartheta}(\boldsymbol{a} + i\boldsymbol{b})$$

and the Cartesian components, E_{01} and E_{02}, of the vector $\boldsymbol{E}_0$ are

$$\begin{cases} E_{01} = e^{i\vartheta}(a\cos\psi \mp ib\sin\psi) \\ E_{02} = e^{i\vartheta}(a\sin\psi \pm ib\cos\psi) \end{cases}$$

where the plus sign holds if the vectors $\boldsymbol{a}$ and $\boldsymbol{b}$ are positioned as in Figure 1.21, and the minus sign if the vector $\boldsymbol{b}$ has the opposite direction.

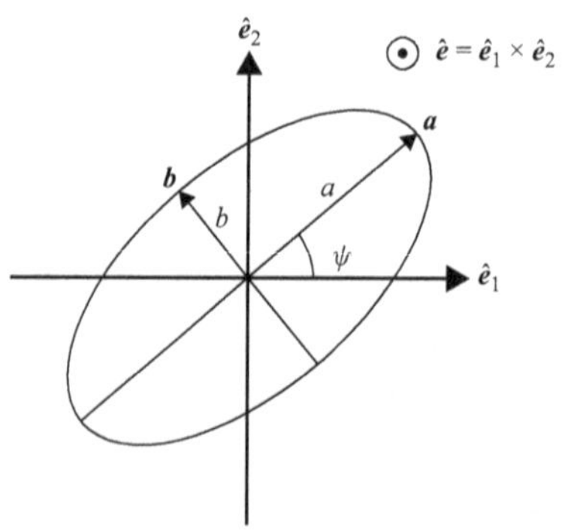

Figure 1.21 The ellipse of polarization and the definition of the tilt angle ψ.

The Stokes parameters s_i, $i = 0, 1, 2, 3$, for a completely polarized field are now written as, see (1.97)

$$\begin{cases} s_0 = a^2 + b^2 \\ s_1 = \left(a^2 - b^2\right)\cos 2\psi \\ s_2 = \left(a^2 - b^2\right)\sin 2\psi \\ s_3 = \mp 2ab \end{cases} \tag{1.98}$$

Two of the Stokes parameters, s_0 and s_3, are invariants, i.e., they do not depend on any particular coordinate system, but have the same form in every coordinate system. In Problem 1.19, it is proved

$$\begin{cases} s_0 = |\boldsymbol{a}|^2 + |\boldsymbol{b}|^2 = |\boldsymbol{E}_0|^2 = a^2 + b^2 \\ s_1^2 + s_2^2 = (|\boldsymbol{a}|^2 + |\boldsymbol{b}|^2)^2 - 4\,|\boldsymbol{a} \times \boldsymbol{b}|^2 = \left(a^2 - b^2\right)^2 \\ s_3 = -2\hat{\boldsymbol{e}} \cdot (\boldsymbol{a} \times \boldsymbol{b}) = -\mathrm{i}\hat{\boldsymbol{e}} \cdot \left(\boldsymbol{E}_0 \times \boldsymbol{E}_0^*\right) \end{cases}$$

where $\hat{\boldsymbol{e}} = \hat{\boldsymbol{e}}_1 \times \hat{\boldsymbol{e}}_2$. The sign of s_3 determines whether the field is right- or left-handed polarized, see Table 1.11 on page 77. A negative (positive) value gives a right-(left-)handed polarization.

In Section 1.4.3, we showed that a general state of polarization can be decomposed uniquely as a sum of an unpolarized field, $P = 0$, and a completely polarized field, $P = 1$. This decomposition carries over to the Stokes parameters, and we write a general set of Stokes parameters as

$$\begin{pmatrix} s_0 \\ s_1 \\ s_2 \\ s_3 \end{pmatrix} = \underbrace{\begin{pmatrix} s_0^{\text{unpol}} \\ 0 \\ 0 \\ 0 \end{pmatrix}}_{\text{unpol}} + \underbrace{\begin{pmatrix} s_0^{\text{pol}} \\ s_1 \\ s_2 \\ s_3 \end{pmatrix}}_{\text{pol}}$$

since the unpolarized field is characterized by $s_i = 0$, $i = 1, 2, 3$. The first Stokes parameters satisfy $s_0^{\text{unpol}} = s_0 - s_0^{\text{pol}}$, where the components of the completely polarized part satisfy (1.96), i.e.,

$$\left(s_0 - s_0^{\text{unpol}}\right)^2 = s_0^{\text{pol}\,2} = s_1^2 + s_2^2 + s_3^2$$

which determines the size of s_0^{unpol} and s_0^{pol}.

1.4.5 The Poincaré Sphere

The Stokes parameters for the completely polarized field, as defined in (1.98), can be represented geometrically for a completely polarized field by defining a three-dimensional vector (s_1, s_2, s_3). The relation in (1.96) shows that this vector lies on a sphere centered at the origin and with radius s_0. Two angles characterize this vector—the tilt angle 2χ and the azimuth angle 2ψ in (1.98), see Figure 1.22. The tilt angle 2χ is defined by

$$\sin 2\chi = \frac{s_3}{s_0} = -\frac{2\hat{\boldsymbol{e}} \cdot (\boldsymbol{a} \times \boldsymbol{b})}{a^2 + b^2} = \mp\frac{2ab}{a^2 + b^2}$$

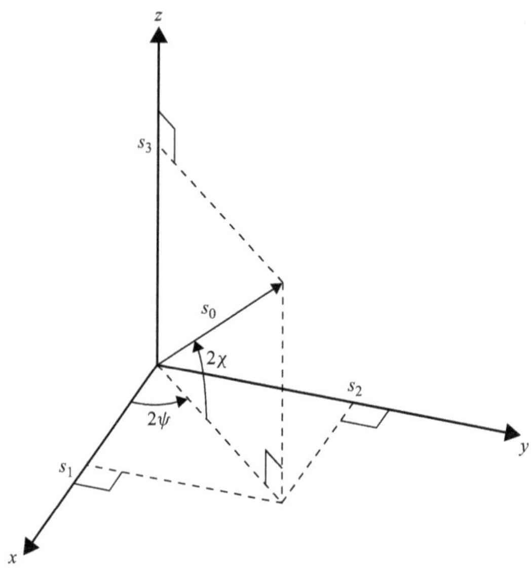

Figure 1.22 The Poincaré sphere and the angles 2ψ and 2χ.

Note that the angle $\pi/2 - 2\chi$ is the polar angle of the vector (s_1, s_2, s_3), see Figure 1.22. We also see that $\sin 2\chi$ can be written as

$$\sin 2\chi = -\mathrm{i}\hat{\boldsymbol{e}} \cdot \left(\hat{\boldsymbol{p}}_\mathrm{e} \times \hat{\boldsymbol{p}}_\mathrm{e}^*\right)$$

using the result (1.79) on page 78 in terms of the complex-valued unit vector $\hat{\boldsymbol{p}}_\mathrm{e} = \boldsymbol{E}_0 / |\boldsymbol{E}_0|$.

This geometric interpretation of the Stokes parameters shows that the upper (lower) hemisphere corresponds to a left-(right-)handed polarized field. The equator $\chi = 0$ corresponds to a linear polarized field. The poles correspond to LCP or RCP—the north pole or the south pole, respectively. This geometric interpretation is of great value in the evaluation of the state of polarization of the field.

Problems for Chapter 1

Section 1.1 The Maxwell Equations

1.1 Show the following analogous theorem:

$$\iiint_V \nabla \times \boldsymbol{A} \, \mathrm{d}v = \iint_S \hat{\boldsymbol{\nu}} \times \boldsymbol{A} \, \mathrm{d}S$$

by the use of Gauss' theorem

$$\iiint_V \nabla \cdot \boldsymbol{A} \, \mathrm{d}v = \iint_S \boldsymbol{A} \cdot \hat{\boldsymbol{\nu}} \, \mathrm{d}S$$

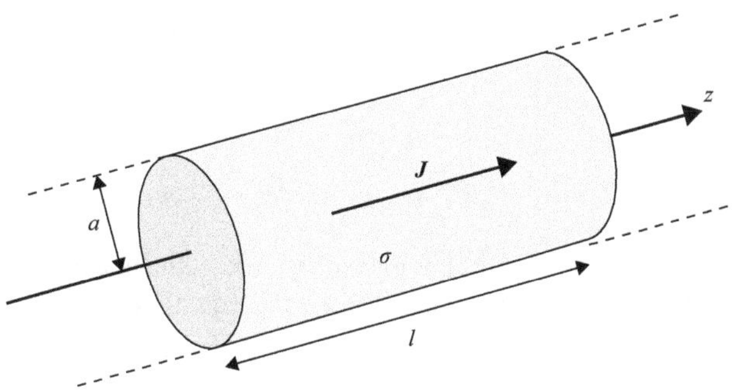

Figure 1.23 The geometry of Problem 1.2. The figure shows a finite portion (length l) of the conductor with circular cross section (radius a). The conductor consists of a material with finite conductivity σ.

1.2 An infinitely long, straight conductor of circular cross section (radius a) consists of a material with finite conductivity σ. In the conductor, a static current I is flowing. The current density $\boldsymbol{J}$ is assumed to be homogeneous over the cross section of the conductor. Compute the terms in Poynting's theorem and show that power balance holds for a volume V, which consists of a finite portion l of the conductor, see Figure 1.23.

Section 1.2 Constitutive Relations

1.3 A Debye medium with susceptibility function $\chi(t) = \alpha \exp\{-t/\tau\}$ is excited by an electric field $\boldsymbol{E} = \boldsymbol{E}_0[\mathrm{H}(t) - \mathrm{H}(t - T)]$ (abrupt on and off excitation), where $\boldsymbol{E}_0$ is a vector that is constant in time and $T > 0$. Determine the polarization $\boldsymbol{P}(t)$ in the material as a function of time t.

1.4 Denote the charge and the mass of the charge in a cold plasma by q and m, respectively. The charges move freely in a static magnetic flux density $\boldsymbol{B}$, oriented along the z-axis, i.e., $\boldsymbol{B} = B_0\hat{\boldsymbol{z}}$, under the presence of a collision frequency ν. Make the following Ansatz of the constitutive relations:

$$\boldsymbol{J}(t) = \sigma \boldsymbol{E}(t) + \epsilon_0 \int_{-\infty}^{t} \boldsymbol{\Sigma}(t - t') \cdot \boldsymbol{E}(t')\, \mathrm{d}t'$$

or in Cartesian components

$$J_i(t) = \sigma E_i(t) + \epsilon_0 \sum_{j=1}^{3} \int_{-\infty}^{t} \Sigma_{ij}(t - t') E_j(t')\, \mathrm{d}t', \qquad i = 1, 2, 3$$

Determine the constitutive relations of the plasma under the assumption that the static magnetic flux density is much stronger than the magnetic flux density generated by the charges.

Hint: Show that the current density $\boldsymbol{J}$ satisfies the following equation of motion (Lorentz force):

$$\frac{\mathrm{d}\boldsymbol{J}}{\mathrm{d}t} + \nu \boldsymbol{J} + \omega_{\mathrm{g}} \hat{\boldsymbol{z}} \times \boldsymbol{J} = \epsilon_0 \omega_{\mathrm{p}}^2 \boldsymbol{E}$$

where ω_g, the gyrotropic frequency, and ω_p, the plasma frequency of the material, are defined by

$$\omega_g = \frac{qB_0}{m}, \qquad \omega_p = \sqrt{\frac{Nq^2}{m\epsilon_0}}$$

and N is the number of charges per unit of volume in the plasma. Then show that σ and $\mathbf{\Sigma}(t)$ satisfy the following system of differential equations:

$$\begin{cases} \sigma = 0 \\ \mathbf{\Sigma}(0) = \omega_p^2 \mathbf{I}_3 \\ \dfrac{d\mathbf{\Sigma}(t)}{dt} + \nu\mathbf{\Sigma}(t) + \omega_g \hat{z} \times \mathbf{\Sigma}(t) = \mathbf{0} \end{cases}$$

1.5 Use the result of Problem 1.4 to write the constitutive relations of the plasma in the following form:

$$\boldsymbol{D}(t) = \epsilon_0 \left\{ \boldsymbol{E}(t) + \int_{-\infty}^{t} \chi(t - t') \cdot \boldsymbol{E}(t')\, dt' \right\}$$

This gives the transformation from the conductivity model, given in Problem 1.4, to the dispersion model.

Section 1.3 Time-Harmonic Fields and Fourier Transform

1.6 Find two complex vectors, $\boldsymbol{A}$ and $\boldsymbol{B}$, such that $\boldsymbol{A} \cdot \boldsymbol{B} = 0$ and

$$\boldsymbol{A}_r \cdot \boldsymbol{B}_r \neq 0$$
$$\boldsymbol{A}_i \cdot \boldsymbol{B}_i \neq 0$$

where $\boldsymbol{A}_r$ and $\boldsymbol{B}_r$ are the real parts of the vectors, respectively, and where the imaginary parts are denoted $\boldsymbol{A}_i$ and $\boldsymbol{B}_i$, respectively.

1.7 For real vectors $\boldsymbol{A}$ and $\boldsymbol{B}$, we have

$$\boldsymbol{B} \cdot (\boldsymbol{B} \times \boldsymbol{A}) = 0$$

Prove that this equality also holds for arbitrary complex vectors $\boldsymbol{A}$ and $\boldsymbol{B}$.

1.8 In some applications, a generalized Debye model is used. The susceptibility function is

$$\chi(t) = \mathrm{H}(t)(1 + \beta t)e^{-\alpha t}$$

The real constant α is assumed to be positive. What conditions must the real constant β satisfy such that $\chi(t)$ is a model of a passive material?

1.9 Consider the susceptibility function

$$\chi(t) = \mathrm{H}(t)e^{-\alpha t} \cos \beta t$$

The real constant α is assumed to be positive. For what values of the real constant β is $\chi(t)$ a model of a passive material?

1.10 Determine the constitutive relations in the frequency domain for a plasma. Use the result in Problem 1.5 to find the permittivity.

1.11 In a ferrite, the magnetization $\boldsymbol{M}$ is determined by

$$\frac{\mathrm{d}}{\mathrm{d}t}\boldsymbol{M} = g\mu_0 \boldsymbol{M} \times \boldsymbol{H}$$

where g is the gyromagnetic quotient, which for electrons is $g = -e/m \approx -1.7588 \times 10^{11}$ C/kg. Let

$$\begin{cases} \boldsymbol{H} = \hat{z}H_0 + \boldsymbol{H}_1 \\ \boldsymbol{M} = \hat{z}M_0 + \boldsymbol{M}_1 \\ \boldsymbol{B} = \hat{z}B_0 + \boldsymbol{B}_1 \end{cases}$$

where $\boldsymbol{B} = \mu_0(\boldsymbol{H} + \boldsymbol{M})$, $B_0 = \mu_0(H_0 + M_0)$, and where

$$\begin{cases} |\boldsymbol{H}_1| \ll H_0 \\ |\boldsymbol{M}_1| \ll M_0 \\ |\boldsymbol{B}_1| \ll B_0 \end{cases}$$

Determine, using the linearized equations in $\boldsymbol{H}_1$, $\boldsymbol{M}_1$, and $\boldsymbol{B}_1$, the constitutive relations in the frequency domain, i.e., find $\boldsymbol{\mu}$ in

$$B_1 = \mu_0 \boldsymbol{\mu} \cdot H_1$$

1.12 Determine the state of polarization in the following cases (a and b are real, positive constants, and α is a real constant):

(a) $\boldsymbol{E}(t) = \hat{\boldsymbol{e}}_1 a \cos(\omega t + \alpha) + \hat{\boldsymbol{e}}_2 b \cos(\omega t + \alpha)$
(b) $\boldsymbol{E}(t) = a\left(\hat{\boldsymbol{e}}_1 \cos(\omega t + \alpha) + \hat{\boldsymbol{e}}_2 \sin(\omega t + \alpha)\right)$
(c) $\boldsymbol{E}(t) = a\left(\hat{\boldsymbol{e}}_1 \cos(\omega t + \alpha) - \hat{\boldsymbol{e}}_2 \sin(\omega t + \alpha)\right)$
(d) $\boldsymbol{E}(t) = \hat{\boldsymbol{e}}_1 a \cos\omega t + \hat{\boldsymbol{e}}_2 b \sin\omega t$
(e) $\boldsymbol{E}(t) = a\left(\hat{\boldsymbol{e}}_1 \cos\omega t + \hat{\boldsymbol{e}}_2 \cos(\omega t - \pi/4)\right)$

$\langle \boldsymbol{S}(t) \rangle$ is assumed directed along $\hat{\boldsymbol{e}} = \hat{\boldsymbol{e}}_1 \times \hat{\boldsymbol{e}}_2$, and $\hat{\boldsymbol{e}}_1 \perp \hat{\boldsymbol{e}}_2$.

1.13 (a) Show that an arbitrary elliptic polarized wave can be decomposed in a superposition of an LCP and an RCP wave.

(b) Let $\boldsymbol{E}_0$ be a superposition of an LCP and an RCP wave, i.e.,

$$\boldsymbol{E}_0 = a\boldsymbol{E}_+ + b\boldsymbol{E}_-$$

where $\boldsymbol{E}_\pm = \hat{\boldsymbol{e}}_1 \pm \mathrm{i}\hat{\boldsymbol{e}}_2$. What conditions do the complex numbers a and b satisfy in order to the wave to be linearly polarized ($\langle \boldsymbol{S}(t) \rangle$ is assumed directed along $\hat{\boldsymbol{e}} = \hat{\boldsymbol{e}}_1 \times \hat{\boldsymbol{e}}_2$).

1.14 Show that a circular polarized wave satisfies

$$\boldsymbol{E}_0 = \pm \mathrm{i}\hat{\boldsymbol{e}} \times \boldsymbol{E}_0$$

where the upper (lower) sign holds for RCP (LCP).

1.15 Show that the fields in (1.72) solve the Maxwell equations in a homogeneous isotropic material.

1.16 Show that the fields in (1.73) solve the Maxwell equations in a homogeneous isotropic material at all points $r > 0$.

Section 1.4 Coherence and Degree of Polarization

1.17 Show the details in the computations of the expression in (1.89).

1.18 Show that the Stokes parameters s_i, $i = 0, 1, 2, 3$, are related to the intensity measurements $I(\theta, \delta)$ by

$$\begin{cases} s_0 = I(0,0) + I(\pi/2, 0) \\ s_1 = I(0,0) - I(\pi/2, 0) \\ s_2 = I(\pi/4, 0) - I(3\pi/4, 0) \\ s_3 = I(\pi/4, \pi/2) - I(3\pi/4, \pi/2) \end{cases}$$

1.19 Show that the Stokes parameters s_i, $i = 0, 1, 2, 3$, for a completely polarized field can be expressed in the vectors $\boldsymbol{a}$ and $\boldsymbol{b}$ or $\boldsymbol{E}_0$ in the following way:

$$\begin{cases} s_0 = |\boldsymbol{a}|^2 + |\boldsymbol{b}|^2 = |\boldsymbol{E}_0|^2 = a^2 + b^2 \\ s_1^2 + s_2^2 = (|\boldsymbol{a}|^2 + |\boldsymbol{b}|^2)^2 - 4\,|\boldsymbol{a} \times \boldsymbol{b}|^2 = \left(a^2 - b^2\right)^2 \\ s_3 = -2\hat{\boldsymbol{e}} \cdot (\boldsymbol{a} \times \boldsymbol{b}) = -\mathrm{i}\hat{\boldsymbol{e}} \cdot \left(\boldsymbol{E}_0 \times \boldsymbol{E}_0^*\right) \end{cases}$$

where $\hat{\boldsymbol{e}} = \hat{\boldsymbol{e}}_1 \times \hat{\boldsymbol{e}}_2$.

CHAPTER 2

The Green Functions and Dyadics

To analyze the electromagnetic fields, we need some results from the Green function techniques and the scalar and vector potentials. The analysis in this chapter covers both the time-domain and the frequency-domain results, but we start the investigation in the frequency domain, since this analysis is simpler. The frequency-domain results can then be used in the time domain.

The Green functions and dyadics in an isotropic media are introduced in Section 2.2. Some generalizations are found in Sections 2.3 and 2.4. As an example of how currents generate an electromagnetic field, we apply the results to the Čerenkov radiation in Section 2.5. The corresponding time-domain solution in an isotropic material is presented in Sections 2.6.

2.1 The Green Functions in Isotropic Media

In this section, we quantify the electromagnetic field in a homogeneous, isotropic material due to a given current density $\boldsymbol{J}$, which quantifies all free sources everywhere in space. The Maxwell equations are then, see (1.43) and (1.44) on page 41

$$\begin{cases} \nabla \times \boldsymbol{E}(\boldsymbol{r},\omega) = \mathrm{i}\omega \boldsymbol{B}(\boldsymbol{r},\omega) \\ \nabla \times \boldsymbol{H}(\boldsymbol{r},\omega) = \boldsymbol{J}(\boldsymbol{r},\omega) - \mathrm{i}\omega \boldsymbol{D}(\boldsymbol{r},\omega) \end{cases} \qquad \boldsymbol{r} \in \mathbb{R}^3$$

Our goal here is to solve these equations for a given current density $\boldsymbol{J}(\boldsymbol{r},\omega)$, which is assumed to vanish outside a finite volume V. We show that this is possible, provided we can determine the Green function of the problem.

The assumption of a homogeneous, isotropic material implies that the constitutive relations are

$$\begin{cases} \boldsymbol{D}(\boldsymbol{r},\omega) = \epsilon_0 \epsilon(\omega) \boldsymbol{E}(\boldsymbol{r},\omega) \\ \boldsymbol{B}(\boldsymbol{r},\omega) = \mu_0 \mu(\omega) \boldsymbol{H}(\boldsymbol{r},\omega) \end{cases} \tag{2.1}$$

where $\epsilon(\omega)$ and $\mu(\omega)$ are complex constants independent of $\boldsymbol{r}$, but they can still depend on the (angular) frequency ω. If we combine these equations, we get

$$\begin{cases} \nabla \times \boldsymbol{E}(\boldsymbol{r},\omega) = \mathrm{i}k\eta_0\eta(\omega) \boldsymbol{H}(\boldsymbol{r},\omega) \\ \eta_0\eta(\omega)\nabla \times \boldsymbol{H}(\boldsymbol{r},\omega) = \eta_0\eta(\omega)\boldsymbol{J}(\boldsymbol{r},\omega) - \mathrm{i}k\boldsymbol{E}(\boldsymbol{r},\omega) \end{cases}$$

where the wave number k is

$$k = \omega\,(\epsilon_0\mu_0\epsilon(\omega)\mu(\omega))^{1/2} = \frac{\omega}{c_0}\,(\epsilon(\omega)\mu(\omega))^{1/2}$$

and the wave impedance of vacuum, η_0, and the relative impedance, η, are defined as

$$\eta_0 = \left(\frac{\mu_0}{\epsilon_0}\right)^{1/2}, \qquad \eta(\omega) = \left(\frac{\mu(\omega)}{\epsilon(\omega)}\right)^{1/2}$$

Consequently, we get

$$\nabla\times(\nabla\times\boldsymbol{E}(\boldsymbol{r},\omega)) = \mathrm{i}k\eta_0\eta(\omega)\nabla\times\boldsymbol{H}(\boldsymbol{r},\omega) = \mathrm{i}k\eta_0\eta(\omega)\boldsymbol{J}(\boldsymbol{r},\omega) + k^2\boldsymbol{E}(\boldsymbol{r},\omega)$$

or

$$\nabla\times(\nabla\times\boldsymbol{E}(\boldsymbol{r},\omega)) - k^2\boldsymbol{E}(\boldsymbol{r},\omega) = \mathrm{i}k\eta_0\eta(\omega)\boldsymbol{J}(\boldsymbol{r},\omega) = \mathrm{i}\omega\mu_0\mu(\omega)\boldsymbol{J}(\boldsymbol{r},\omega) \tag{2.2}$$

2.1.1 Potentials and Gauge Transformations

Equation (2.2) is the partial differential equation that the electric field satisfies for a given current density distributions $\boldsymbol{J}$. The solution to this equation then determines the other fields $\boldsymbol{D}$, $\boldsymbol{B}$, and $\boldsymbol{H}$ by the Maxwell equations and the constitutive relations (2.1).

Due to the operator $\nabla\times(\nabla\times\boldsymbol{E})$, the partial differential equation (2.2) is a system of three coupled equations; one equation for each Cartesian component. The fact that it couples all components to each other makes it hard to solve the equation as it stands. We are now looking for an alternative way of writing this system of equations such that each Cartesian component decouples. One way of accomplishing this is to introduce the vector potential $\boldsymbol{A}$ defined by

$$\boldsymbol{B} = \nabla\times\boldsymbol{A} \tag{2.3}$$

The existence of a vector potential is guaranteed by the conservation of magnetic flux, $\nabla\cdot\boldsymbol{B} = 0$, see (1.46) on page 42.

The scalar potential ϕ is defined by the Faraday law

$$\nabla\times\boldsymbol{E} = \mathrm{i}\omega\boldsymbol{B} = \mathrm{i}\omega\nabla\times\boldsymbol{A}$$

We get

$$\nabla\times(\boldsymbol{E} - \mathrm{i}\omega\boldsymbol{A}) = \boldsymbol{0}$$

which implies that there is a function ϕ such that[1]

$$\boldsymbol{E} - \mathrm{i}\omega\boldsymbol{A} = -\nabla\phi$$

[1]To make the existence of ϕ clear, we assume the domain of investigation is a simply connected domain. This is the case for $\mathbb{R}^3$. The minus sign in front of the nabla operator is a convention.

or

$$E = i\omega A - \nabla\phi \tag{2.4}$$

The magnetic flux density $\boldsymbol{B}$ and the electric field $\boldsymbol{E}$ can therefore be computed from the vector potential $\boldsymbol{A}$ and the scalar potential ϕ.

There are three components of the electric field on the left-hand side of (2.4), and on the right-hand side there are four field quantities. A relation between the quantities on the right-hand side must exist. As a consequence, the vector potential $\boldsymbol{A}$ and the scalar potential ϕ are not uniquely defined. If $\boldsymbol{A}$ and ϕ are replaced by

$$\begin{cases} \boldsymbol{A}' = \boldsymbol{A} + \nabla\psi \\ \phi' = \phi + i\omega\psi \end{cases}$$

where ψ is an arbitrary (differentiable) function, the magnetic flux density and the electric field remain the same since

$$\nabla \times \boldsymbol{A}' = \nabla \times \boldsymbol{A} + \nabla \times \nabla\psi = \nabla \times \boldsymbol{A}$$

and

$$i\omega \boldsymbol{A}' - \nabla\phi' = i\omega \boldsymbol{A} + i\omega\nabla\psi - \nabla\phi - i\omega\nabla\psi = i\omega \boldsymbol{A} - \nabla\phi$$

As a consequence, we obtain the same physical fields, $\boldsymbol{E}$ and $\boldsymbol{B}$, from both the unprimed and primed potentials. This type of transformation is called a gauge transformation.[2]

The lack of uniqueness of the vector potential $\boldsymbol{A}$ and the scalar potential ϕ is now used to find a simpler differential equation for the vector potential $\boldsymbol{A}$ than (2.2) for the electric field $\boldsymbol{E}$. Therefore, we let the vector potential $\boldsymbol{A}$ and the scalar potential ϕ satisfy a constraint. Several possibilities arise, but the one that is most used is the Lorenz condition[3] or Lorenz gauge

$$\nabla \cdot \boldsymbol{A} = -\frac{k^2}{i\omega}\phi \tag{2.5}$$

It is always possible to find $\boldsymbol{A}$ and ϕ such that this condition is fulfilled. In fact, assume we have a vector potential $\boldsymbol{A}_0$ and a scalar potential ϕ_0 that satisfy

$$\begin{cases} \boldsymbol{B} = \nabla \times \boldsymbol{A}_0 \\ \boldsymbol{E} = i\omega \boldsymbol{A}_0 - \nabla\phi_0 \end{cases}$$

[2]The name gauge often raises questions. The German mathematician Hermann Weyl (1885–1955) coined the name in 1921 when he presented a theory on charged particles. In the theory, lengths could be scaled differently at different points in space. The name gauge, which has the meaning of scaling, has survived.

[3]Note that it was the Danish physicist Ludvig Valentin Lorenz (1829–1891) who first formulated this condition in a paper from 1867. The Dutch physicist Hendrik Antoon Lorentz made a similar discovery, but later. Therefore, the condition should be credited to Lorenz' name, and not Lorentz. Additional details can be found in Reference 274.

but not (2.5). Then define a new vector potential $\boldsymbol{A}$ and a new scalar potential ϕ by

$$\begin{cases} \boldsymbol{A} = \boldsymbol{A}_0 - \nabla\psi \\ \phi = \phi_0 - \mathrm{i}\omega\psi \end{cases}$$

where the function ψ is an arbitrary solution to the inhomogeneous Helmholtz equation,[4,5] i.e.,

$$\nabla^2\psi + k^2\psi = \nabla \cdot \boldsymbol{A}_0 + \frac{k^2}{\mathrm{i}\omega}\phi_0$$

Note that the right-hand side of this equation is known and non-zero, since we assumed that $\boldsymbol{A}_0$ and ϕ_0 do not satisfy the Lorenz condition. The physical fields $\boldsymbol{E}$ and $\boldsymbol{B}$ are not affected by this chance of vector potential and scalar potential since the change $\boldsymbol{A}_0 \longrightarrow \boldsymbol{A}$ and $\phi_0 \longrightarrow \phi$ is a gauge transformation.

The new potentials $\boldsymbol{A}$ and ϕ satisfy the Lorenz condition, which the following computations show

$$\begin{aligned} \nabla \cdot \boldsymbol{A} - \frac{\mathrm{i}k^2}{\omega}\phi &= \nabla \cdot (\boldsymbol{A}_0 - \nabla\psi) + \frac{k^2}{\mathrm{i}\omega}(\phi_0 - \mathrm{i}\omega\psi) \\ &= \nabla \cdot \boldsymbol{A}_0 + \frac{k^2}{\mathrm{i}\omega}\phi_0 - \nabla^2\psi - k^2\psi = 0 \end{aligned}$$

by the definition of the function ψ. It is therefore always possible to choose a vector potential $\boldsymbol{A}$ and scalar potential ϕ such that the Lorenz condition is satisfied.

Even if the vector potential $\boldsymbol{A}$ and the scalar potential ϕ satisfy the Lorenz condition, they are still not uniquely defined. This becomes clear by a similar analysis as above, but with $\boldsymbol{A}_0 = \boldsymbol{0}$ and $\phi_0 = 0$. The Lorenz condition is then satisfied from the start, but nothing prevents us from changing the potentials by choosing ψ that satisfies the homogeneous Helmholtz equation

$$\nabla^2\psi + k^2\psi = 0$$

The partial differential equation that $\boldsymbol{A}$ satisfies becomes particularly simple, if we let the Lorenz condition be satisfied. Insert (2.4) in (2.2)

$$\nabla \times (\nabla \times (\mathrm{i}\omega\boldsymbol{A} - \nabla\phi)) - k^2 (\mathrm{i}\omega\boldsymbol{A} - \nabla\phi) = \mathrm{i}\omega\mu_0\mu\boldsymbol{J}$$

Then use $\nabla \times \nabla\phi = \boldsymbol{0}$ and introduce the Lorenz condition (2.5). We get

$$\nabla \times (\nabla \times \boldsymbol{A}) - k^2\boldsymbol{A} - \nabla (\nabla \cdot \boldsymbol{A}) = \mu_0\mu\boldsymbol{J}$$

Moreover, we have

$$\nabla^2\boldsymbol{A} = \nabla (\nabla \cdot \boldsymbol{A}) - \nabla \times (\nabla \times \boldsymbol{A}) \tag{2.6}$$

---

[4]Hermann Ludwig Ferdinand von Helmholtz (1821–1894), German physician and physicist.

[5]A solution to this equation always exist (in the vector case) as shown later in this section.

which is easily seen by evaluating the identity in Cartesian components.[6] Finally, the vector potential $\boldsymbol{A}$ satisfies the following equation:

$$\nabla^2 \boldsymbol{A} + k^2 \boldsymbol{A} = -\mu_0 \mu \boldsymbol{J} \tag{2.7}$$

Provided we can solve this equation, we can determine ϕ by the use of (2.5) and the fields $\boldsymbol{E}$ and $\boldsymbol{B}$ by (2.4) and (2.3), respectively. The fields $\boldsymbol{D}$ and $\boldsymbol{H}$ are then determined by the constitutive relations. In general, (2.7) is easier to solve than (2.2), since the different Cartesian components do not couple to each other in (2.7).

2.1.2 Canonical Problem in Homogeneous Space

The solution to (2.7) is obtained by first solving a canonical problem with a simple source term. Suppose we have a solution $g(k, \boldsymbol{r}, \boldsymbol{r}')$ to the following equation:

$$\nabla^2 g(k, \boldsymbol{r}, \boldsymbol{r}') + k^2 g(k, \boldsymbol{r}, \boldsymbol{r}') = -\delta(\boldsymbol{r} - \boldsymbol{r}'), \qquad \boldsymbol{r}, \boldsymbol{r}' \in \mathbb{R}^3 \tag{2.8}$$

where the differentiations are made w.r.t. the variable $\boldsymbol{r}$ and where $\delta(\boldsymbol{r} - \boldsymbol{r}')$ denotes the three-dimensional delta distribution or delta function, i.e., its representation in the Cartesian coordinate system is

$$\delta(\boldsymbol{r} - \boldsymbol{r}') = \delta(x - x')\delta(y - y')\delta(z - z')$$

The function $g(k, \boldsymbol{r}, \boldsymbol{r}')$ is the Green function[7] of the scalar Helmholtz equation. The solution to our original problem (2.7) then is

$$\boldsymbol{A}(\boldsymbol{r}) = \mu_0 \mu \iiint_V g(k, \boldsymbol{r}, \boldsymbol{r}') \boldsymbol{J}(\boldsymbol{r}') \, \mathrm{d}v' \tag{2.9}$$

where $\mathrm{d}v'$ denotes the volume element $\mathrm{d}x' \, \mathrm{d}y' \, \mathrm{d}z'$ (integration is made over the variable $\boldsymbol{r}'$) and the domain of integration is the entire space or effectively the region in space where $\boldsymbol{J} \neq \boldsymbol{0}$.

Formally, we see that (2.9) is a solution to (2.7) by inserting the solution and interchanging the order of differentiation and integration. We have

$$\begin{aligned} \nabla^2 \boldsymbol{A}(\boldsymbol{r}) + k^2 \boldsymbol{A}(\boldsymbol{r}) &= \mu_0 \mu \iiint_V (\nabla^2 g(k, \boldsymbol{r}, \boldsymbol{r}') + k^2 g(k, \boldsymbol{r}, \boldsymbol{r}')) \boldsymbol{J}(\boldsymbol{r}') \, \mathrm{d}v' \\ &= -\mu_0 \mu \iiint_V \delta(\boldsymbol{r} - \boldsymbol{r}') \boldsymbol{J}(\boldsymbol{r}') \, \mathrm{d}v' = -\mu_0 \mu \boldsymbol{J}(\boldsymbol{r}) \end{aligned}$$

by the use of the definition of the delta distribution $\delta(\boldsymbol{r} - \boldsymbol{r}')$. Notice that these computations with the delta distribution are formal, but can be made mathematically rigorous. Here, we treat the delta distribution as an "ordinary" function.

[6] Note that this identity holds only in Cartesian coordinates. In a curvilinear coordinate system, this identity is taken as a definition of the Laplace operator on a vector field.

[7] George Green (1793–1841), British mathematician and physicist.

It remains to solve the canonical problem (2.8). This problem has spherical symmetry (the source is spherically symmetric), which implies that we are seeking a solution $g(k, \boldsymbol{r}, \boldsymbol{r}')$ that only depends on the distance $R = |\boldsymbol{r} - \boldsymbol{r}'|$. We translate the source to the origin and we get the problem

$$\nabla^2 g(k, r) + k^2 g(k, r) = -\delta(\boldsymbol{r}) \tag{2.10}$$

to solve, where $r = |\boldsymbol{r}|$. For $r \neq 0$, we have

$$\nabla^2 g(k, r) + k^2 g(k, r) = 0$$

The Laplace[8] operator ∇^2 in spherical coordinates, see Appendix H, implies

$$\frac{\mathrm{d}^2}{\mathrm{d}r^2}(r g(k, r)) + k^2 r g(k, r) = 0$$

with a general solution

$$r g(k, r) = A\mathrm{e}^{\mathrm{i}kr} + B\mathrm{e}^{-\mathrm{i}kr}$$

i.e.,

$$g(k, r) = A\frac{\mathrm{e}^{\mathrm{i}kr}}{r} + B\frac{\mathrm{e}^{-\mathrm{i}kr}}{r}$$

How to choose the constants A and B depends on the physical boundary conditions we impose at large distances r from the source at the origin.

The Green function $g(k, r)$, which represents the solution of a unit source at the origin, should, for physical reasons, represent an out-going spherical wave, see (1.73) on page 74

$$\frac{\mathrm{e}^{\mathrm{i}kr - \mathrm{i}\omega t}}{r}$$

This implies that we choose the constant $B = 0$. The remaining constant A is determined if we integrate (2.10) over a ball of radius ε centered at the origin, see Comment 2.1 for a more mathematical derivation.

$$\iiint\limits_{r \leq \varepsilon} \nabla^2 g(k, r)\, \mathrm{d}v + k^2 \iiint\limits_{r \leq \varepsilon} g(k, r)\, \mathrm{d}v = -\iiint\limits_{r \leq \varepsilon} \delta(\boldsymbol{r})\, \mathrm{d}v = -1$$

We also have the following limit:

$$\iiint\limits_{r \leq \varepsilon} g(k, r)\, \mathrm{d}v = 4\pi A \int_0^{\varepsilon} \frac{\mathrm{e}^{\mathrm{i}kr}}{r} r^2\, \mathrm{d}r = 4\pi A \int_0^{\varepsilon} \mathrm{e}^{\mathrm{i}kr} r\, \mathrm{d}r \to 0$$

[8] Pierre-Simon Laplace (1749–1827), French mathematician.

when $\varepsilon \to 0$. Moreover, the divergence theorem implies

$$\begin{aligned}\iiint\limits_{r\le\varepsilon} \nabla^2 g(k,r)\,\mathrm{d}v &= \iiint\limits_{r\le\varepsilon} \nabla\cdot\nabla g(k,r)\,\mathrm{d}v = \iint\limits_{r=\varepsilon} \hat{\boldsymbol{v}}\cdot\nabla g(k,r)\,\mathrm{d}S \\ &= \iint\limits_{r=\varepsilon} \frac{\mathrm{d}g(k,r)}{\mathrm{d}r}\,\mathrm{d}S = 4\pi\varepsilon^2 \left.\frac{\mathrm{d}g(k,r)}{\mathrm{d}r}\right|_{r=\varepsilon} \\ &= 4\pi\varepsilon^2 A\mathrm{e}^{\mathrm{i}k\varepsilon}\left\{\frac{\mathrm{i}k}{\varepsilon} - \frac{1}{\varepsilon^2}\right\} \to -4\pi A\end{aligned}$$

when $\varepsilon \to 0$. Thus, we conclude $-4\pi A = -1$, and $g(k,r) = \mathrm{e}^{\mathrm{i}kr}/4\pi r$.

Comment 2.1

A more rigorous way of determining the constant A based upon the theory of distributions is

$$\iiint g(k,r)\nabla^2\phi(\boldsymbol{r})\,\mathrm{d}v + k^2\iiint g(k,r)\phi(\boldsymbol{r})\,\mathrm{d}v = -\iiint \delta(\boldsymbol{r})\phi(\boldsymbol{r})\,\mathrm{d}v = -\phi(\boldsymbol{0})$$

where ϕ is an infinitely differentiable function with compact support, i.e., it vanishes outside a finite domain. Use the Green formula on the domain $r \ge \varepsilon$. We get

$$\begin{aligned}-\phi(\boldsymbol{0}) &= \lim_{\varepsilon\to0}\left\{\iiint\limits_{r\ge\varepsilon} g(k,r)\nabla^2\phi(\boldsymbol{r})\,\mathrm{d}v + k^2\iiint\limits_{r\ge\varepsilon} g(k,r)\phi(\boldsymbol{r})\,\mathrm{d}v\right\} \\ &= \lim_{\varepsilon\to0}\left\{\iiint\limits_{r\ge\varepsilon} \nabla^2 g(k,r)\phi(\boldsymbol{r})\,\mathrm{d}v + k^2\iiint\limits_{r\ge\varepsilon} g(k,r)\phi(\boldsymbol{r})\,\mathrm{d}v\right\} \\ &\quad + \lim_{\varepsilon\to0}\iint\limits_{r=\varepsilon}\left(\frac{\partial g(k,r)}{\partial r}\phi(\boldsymbol{r}) - \frac{\partial\phi(\boldsymbol{r})}{\partial r}g(k,r)\right)\mathrm{d}S \\ &= \lim_{\varepsilon\to0}\iint\limits_{r=\varepsilon}\frac{\partial g(k,r)}{\partial r}\phi(\boldsymbol{r})\,\mathrm{d}S = \lim_{\varepsilon\to0} A4\pi\varepsilon^2\phi(\boldsymbol{0})\frac{\mathrm{e}^{\mathrm{i}k\varepsilon}}{\varepsilon}\left(\mathrm{i}k - \frac{1}{\varepsilon}\right) = -4\pi A\phi(\boldsymbol{0})\end{aligned}$$

A translation of the source point back to $\boldsymbol{r}'$ implies

$$g(k,|\boldsymbol{r}-\boldsymbol{r}'|) = \frac{\mathrm{e}^{\mathrm{i}k|\boldsymbol{r}-\boldsymbol{r}'|}}{4\pi|\boldsymbol{r}-\boldsymbol{r}'|} \tag{2.11}$$

The solution of the vector potential $\boldsymbol{A}$ in (2.9) then is

$$\boldsymbol{A}(\boldsymbol{r}) = \mu_0\mu\iiint\limits_{V}\frac{\mathrm{e}^{\mathrm{i}k|\boldsymbol{r}-\boldsymbol{r}'|}}{4\pi|\boldsymbol{r}-\boldsymbol{r}'|}\boldsymbol{J}(\boldsymbol{r}')\,\mathrm{d}v' \tag{2.12}$$

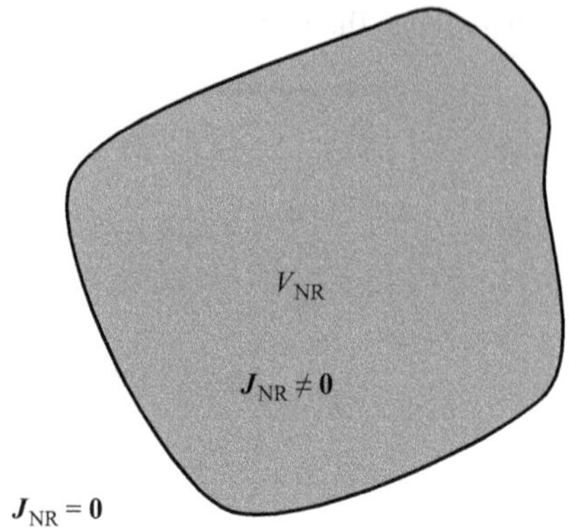

Figure 2.1 The domain V_{NR}, which contains the currents $\boldsymbol{J}_{NR}$.

The electric field $\boldsymbol{E}$ by the use of (2.4) and (2.5) is

$$\begin{aligned}\boldsymbol{E}(\boldsymbol{r}) &= \mathrm{i}\omega\left[\boldsymbol{A}(\boldsymbol{r}) + \frac{1}{k^2}\nabla(\nabla\cdot\boldsymbol{A}(\boldsymbol{r}))\right] \\ &= \mathrm{i}k\eta_0\eta\left[\mathbf{I}_3 + \frac{1}{k^2}\nabla\nabla\right]\cdot\iiint_V \frac{\mathrm{e}^{\mathrm{i}k|\boldsymbol{r}-\boldsymbol{r}'|}}{4\pi|\boldsymbol{r}-\boldsymbol{r}'|}\boldsymbol{J}(\boldsymbol{r}')\,\mathrm{d}v'\end{aligned} \tag{2.13}$$

where $\mathbf{I}_3$ is the unit dyadic in three dimensions. The magnetic field $\boldsymbol{H}$ is by the use of (2.3)

$$\boldsymbol{H}(\boldsymbol{r}) = \frac{1}{\mu_0\mu}\nabla\times\boldsymbol{A}(\boldsymbol{r}) = \nabla\times\iiint_V \frac{\mathrm{e}^{\mathrm{i}k|\boldsymbol{r}-\boldsymbol{r}'|}}{4\pi|\boldsymbol{r}-\boldsymbol{r}'|}\boldsymbol{J}(\boldsymbol{r}')\,\mathrm{d}v' \tag{2.14}$$

Notice that these expressions hold for an observation point $\boldsymbol{r}$ both inside and outside the domain V, i.e., outside the region where the currents are non-zero. The interpretation of these integrals for an observation point $\boldsymbol{r}$ inside the domain V is addressed in Section 2.2.3.

The theory presented above is illustrated in Section 2.5 with a non-trivial example of a fast-moving charged particle.

2.1.3 Non-Radiating Sources

In the previous section, we saw that the knowledge of the current density $\boldsymbol{J}$ in all space implies that the electric and magnetic fields can be determined. Equations (2.13) and (2.14) express this in a quantitative way. The question of interest in this subsection is: Are there current density distributions that give zero electric field in some part of space (not the whole space)? It soon becomes apparent, that such current distributions exist, and these current densities can be characterized in an explicit way.

If a current distribution $\boldsymbol{J}_{NR}$ in a volume V_{NR}, see Figure 2.1, gives zero electric field $\boldsymbol{E}$ outside V_{NR} (and therefore also zero magnetic field $\boldsymbol{H}$ by the Faraday law), we say that the current density $\boldsymbol{J}_{NR}$ is non-radiating, and that the field $\boldsymbol{E}$ is non-radiating in the volume outside V_{NR}. The relation between the electric field $\boldsymbol{E}$ and the current density $\boldsymbol{J}$ is generally given by (2.2)

$$\nabla \times (\nabla \times \boldsymbol{E}) - k^2 \boldsymbol{E} = \mathrm{i}\omega\mu_0\mu \boldsymbol{J}$$

The claim is now that the most general expression of a non-radiating current density $\boldsymbol{J}_{\text{NR}}$ is

$$\boldsymbol{J}_{\text{NR}}(\boldsymbol{r}) = \nabla \times (\nabla \times \boldsymbol{F}) - k^2 \boldsymbol{F} \tag{2.15}$$

where $\boldsymbol{F}$ is a differentiable vector field that is non-zero in the volume V_{NR} and it is zero in the domain outside V_{NR}. We assume the domain V_{NR} is of finite extent, i.e., the sources are located in a finite region

$$\boldsymbol{F}(\boldsymbol{r}) \begin{cases} = \boldsymbol{0}, & \boldsymbol{r} \notin V_{\text{NR}} \\ \neq \boldsymbol{0}, & \boldsymbol{r} \in V_{\text{NR}} \end{cases}$$

To prove the necessary and sufficient conditions of non-radiating sources, we first assume the current density $\boldsymbol{J}_{\text{NR}}$ is of the form (2.15) and prove that this current density generates an electric field that is zero outside V_{NR}. The electric field $\boldsymbol{E}$ that is generated by these sources satisfies

$$\nabla \times (\nabla \times \boldsymbol{E}) - k^2 \boldsymbol{E} = \mathrm{i}\omega\mu_0\mu \left(\nabla \times (\nabla \times \boldsymbol{F}) - k^2 \boldsymbol{F}\right)$$

i.e.,

$$\nabla \times (\nabla \times (\boldsymbol{E} - \mathrm{i}\omega\mu_0\mu \boldsymbol{F})) - k^2 (\boldsymbol{E} - \mathrm{i}\omega\mu_0\mu \boldsymbol{F}) = \boldsymbol{0}, \qquad \boldsymbol{r} \in \mathbb{R}^3$$

in all points in space. We assume this problem has a unique solution. The unique solvability of the equation gives that

$$\boldsymbol{E}(\boldsymbol{r}) - \mathrm{i}\omega\mu_0\mu \boldsymbol{F}(\boldsymbol{r}) = \boldsymbol{0}, \qquad \boldsymbol{r} \in \mathbb{R}^3$$

This implies that $\boldsymbol{E} = \boldsymbol{0}$ outside the volume V_{NR}, since $\boldsymbol{F}$ is zero outside V_{NR} by construction (the only non-zero values exist in the domain V_{NR}). This shows that (2.15) is a sufficient condition for a non-radiating current distribution, and that it produces a non-radiating field $\boldsymbol{E}$ outside the volume V_{NR}.

To prove the necessity, we assume the electric field $\boldsymbol{E} = \boldsymbol{0}$ outside a volume V_{NR}. Then, we have

$$\nabla \times (\nabla \times \boldsymbol{E}) - k^2 \boldsymbol{E} = \mathrm{i}\omega\mu_0\mu \boldsymbol{J}_{\text{NR}}, \qquad \boldsymbol{r} \in \mathbb{R}^3$$

This expression, however, is of the form (2.15), provided we let

$$\boldsymbol{F} = \frac{1}{\mathrm{i}\omega\mu_0\mu} \boldsymbol{E}, \qquad \boldsymbol{r} \in \mathbb{R}^3$$

Then $\boldsymbol{F} = \boldsymbol{0}$ outside V_{NR}, since $\boldsymbol{E} = \boldsymbol{0}$ there. The non-radiating current densities are therefore characterized by the condition (2.15).

Example 2.1
An example illustrates the above result. Let the current density $\boldsymbol{J}$ be a gradient of a scalar function ϕ, i.e.,

$$\boldsymbol{J}_{\mathrm{NR}} = \nabla\phi$$

where ϕ = constant outside the volume V_{NR}, but non-constant inside the volume, and let

$$\boldsymbol{F} = -\frac{1}{k^2}\nabla\phi$$

This implies that

$$\nabla \times (\nabla \times \boldsymbol{F}) - k^2\boldsymbol{F} = \boldsymbol{0} + \nabla\phi = \boldsymbol{J}_{\mathrm{NR}}$$

The vector field $\boldsymbol{F}$ is also zero outside the volume V_{NR}, since ϕ is constant in this region. Current densities of this form therefore give rise to a non-radiating field outside the volume V_{NR}. Every current density, $\boldsymbol{J}$, which is a gradient of a scalar field is a non-radiating source, and produces a non-radiating field in the region where the scalar field assumes constant values. ■

2.1.4 Generalizations

For the purpose of later use, we also analyze a slightly more general set of equations—we add a magnetic current term $\boldsymbol{J}_{\mathrm{m}}$ to the Faraday law, i.e., the set of equations then is

$$\begin{cases} \nabla \times \boldsymbol{E} = \boldsymbol{J}_{\mathrm{m}} + \mathrm{i}k\eta_0\eta\boldsymbol{H} \\ \eta_0\eta\nabla \times \boldsymbol{H} = \eta_0\eta\boldsymbol{J}_{\mathrm{e}} - \mathrm{i}k\boldsymbol{E} \end{cases} \tag{2.16}$$

where for convenience the electric current density is denoted $\boldsymbol{J}_{\mathrm{e}}$. Both $\boldsymbol{J}_{\mathrm{e}}$ and $\boldsymbol{J}_{\mathrm{m}}$ are assumed to vanish outside a finite volume V.

The continuity equations for the currents $\boldsymbol{J}_{\mathrm{e}}$ and $\boldsymbol{J}_{\mathrm{m}}$ are

$$\begin{cases} \nabla \cdot \boldsymbol{J}_{\mathrm{e}} - \mathrm{i}\omega\rho_{\mathrm{e}} = 0 \\ \nabla \cdot \boldsymbol{J}_{\mathrm{m}} - \mathrm{i}\omega\rho_{\mathrm{m}} = 0 \end{cases}$$

The equations are combined such that $\boldsymbol{H}$ is eliminated. We get

$$\begin{aligned} \nabla \times (\nabla \times \boldsymbol{E}) &= \nabla \times \boldsymbol{J}_{\mathrm{m}} + \mathrm{i}k\eta_0\eta\nabla \times \boldsymbol{H} \\ &= \nabla \times \boldsymbol{J}_{\mathrm{m}} + \mathrm{i}k(\eta_0\eta\boldsymbol{J}_{\mathrm{e}} - \mathrm{i}k\boldsymbol{E}) = \nabla \times \boldsymbol{J}_{\mathrm{m}} + \mathrm{i}k\eta_0\eta\boldsymbol{J}_{\mathrm{e}} + k^2\boldsymbol{E} \end{aligned}$$

or

$$\nabla \times (\nabla \times \boldsymbol{E}) - k^2\boldsymbol{E} = \nabla \times \boldsymbol{J}_{\mathrm{m}} + \mathrm{i}\omega\mu_0\mu\boldsymbol{J}_{\mathrm{e}} \tag{2.17}$$

This is the equation that the electric field satisfies for given current density distributions $\boldsymbol{J}_{\mathrm{e}}$ and $\boldsymbol{J}_{\mathrm{m}}$.

The analysis proceeds in an analogous way to the analysis in Section 2.1.1, with the difference that the potentials have to be defined differently. The divergence relation of interest now reads (take the divergence of the first equation in (2.16))

$$\nabla \cdot (\boldsymbol{J}_{\mathrm{m}} + \mathrm{i}\omega\mu_0\mu\boldsymbol{H}) = 0$$

Therefore, there is a vector potential $\boldsymbol{A}$ satisfying[9]

$$\frac{1}{\mathrm{i}\omega}\boldsymbol{J}_{\mathrm{m}} + \mu_0\mu\boldsymbol{H} = \nabla \times \boldsymbol{A}$$

which we reformulate as

$$\boldsymbol{H} = \frac{1}{\mu_0\mu}\nabla \times \boldsymbol{A} - \frac{1}{\mathrm{i}\omega\mu_0\mu}\boldsymbol{J}_{\mathrm{m}} \tag{2.18}$$

From this relation and (2.16), we then also conclude

$$\nabla \times (\boldsymbol{E} - \mathrm{i}\omega\boldsymbol{A}) = \boldsymbol{0}$$

which implies that there is a scalar potential ϕ such that

$$\boldsymbol{E} = \mathrm{i}\omega\boldsymbol{A} - \nabla\phi \tag{2.19}$$

Again, we can utilize the Lorenz gauge, see (2.5), and the derivation presented above. The Lorenz gauge is

$$\nabla \cdot \boldsymbol{A} = -\frac{k^2}{\mathrm{i}\omega}\phi \tag{2.20}$$

Insert the expression of the electric field into (2.17)

$$\nabla \times (\nabla \times (\mathrm{i}\omega\boldsymbol{A} - \nabla\phi)) - k^2(\mathrm{i}\omega\boldsymbol{A} - \nabla\phi) = \nabla \times \boldsymbol{J}_{\mathrm{m}} + \mathrm{i}\omega\mu_0\mu\boldsymbol{J}_{\mathrm{e}}$$

Using the Lorenz gauge, the partial differential equation that $\boldsymbol{A}$ satisfies becomes very simple, i.e.,

$$\nabla^2\boldsymbol{A} + k^2\boldsymbol{A} = -\frac{1}{\mathrm{i}\omega}\nabla \times \boldsymbol{J}_{\mathrm{m}} - \mu_0\mu\boldsymbol{J}_{\mathrm{e}}$$

where we have used $\nabla \times \nabla\phi = \boldsymbol{0}$ and $\nabla^2\boldsymbol{A} = \nabla(\nabla \cdot \boldsymbol{A}) - \nabla \times (\nabla \times \boldsymbol{A})$. The solution to this equation, using the same methods as above, is, see (2.12)

$$\boldsymbol{A}(\boldsymbol{r}) = \mu_0\mu \iiint_V \frac{\mathrm{e}^{\mathrm{i}k|\boldsymbol{r}-\boldsymbol{r}'|}}{4\pi|\boldsymbol{r}-\boldsymbol{r}'|}\left(\boldsymbol{J}_{\mathrm{e}}(\boldsymbol{r}') + \frac{1}{\mathrm{i}\omega\mu_0\mu}\nabla \times \boldsymbol{J}_{\mathrm{m}}(\boldsymbol{r}')\right)\mathrm{d}v' \tag{2.21}$$

For given current densities $\boldsymbol{J}_{\mathrm{e}}$ and $\boldsymbol{J}_{\mathrm{m}}$, we can first calculate $\boldsymbol{A}$ from (2.21), and then determine ϕ by the use of (2.20) and the fields $\boldsymbol{E}$ and $\boldsymbol{H}$ by (2.19) and (2.18), respectively.

[9] Note the difference in the definition of the vector potential compared to the one in Section 2.1.1.

The fields $\boldsymbol{D}$ and $\boldsymbol{B}$ are then determined by the constitutive relations. Explicitly, we have

$$\begin{aligned}\boldsymbol{E}(\boldsymbol{r}) &= \mathrm{i}\omega\boldsymbol{A}(\boldsymbol{r}) - \nabla\phi(\boldsymbol{r})\\ &= \mathrm{i}\omega\boldsymbol{A}(\boldsymbol{r}) + \frac{\mathrm{i}\omega}{k^2}\nabla\left(\nabla\cdot\boldsymbol{A}(\boldsymbol{r})\right) = \mathrm{i}\omega\left[\mathbf{I}_3 + \frac{1}{k^2}\nabla\nabla\right]\cdot\boldsymbol{A}(\boldsymbol{r})\end{aligned}$$

We obtain

$$\begin{aligned}\boldsymbol{E}(\boldsymbol{r}) = \mathrm{i}k\eta_0\eta\left[\mathbf{I}_3 + \frac{1}{k^2}\nabla\nabla\right]\cdot\iiint_V \frac{\mathrm{e}^{\mathrm{i}k|\boldsymbol{r}-\boldsymbol{r}'|}}{4\pi|\boldsymbol{r}-\boldsymbol{r}'|}\boldsymbol{J}_\mathrm{e}(\boldsymbol{r}')\,\mathrm{d}v'\\ + \left[\mathbf{I}_3 + \frac{1}{k^2}\nabla\nabla\right]\cdot\iiint_V \frac{\mathrm{e}^{\mathrm{i}k|\boldsymbol{r}-\boldsymbol{r}'|}}{4\pi|\boldsymbol{r}-\boldsymbol{r}'|}\nabla'\times\boldsymbol{J}_\mathrm{m}(\boldsymbol{r}')\,\mathrm{d}v'\end{aligned}$$

which we rewrite using $\nabla g(k,|\boldsymbol{r}-\boldsymbol{r}'|) = -\nabla' g(k,|\boldsymbol{r}-\boldsymbol{r}'|)$, where ∇ and ∇' denote differentiation w.r.t. the unprimed and primed variables, respectively

$$\begin{aligned}\boldsymbol{E}(\boldsymbol{r}) = \left[\mathbf{I}_3 + \frac{1}{k^2}\nabla\nabla\right]\cdot\Bigg\{\mathrm{i}k\eta_0\eta\iiint_V \frac{\mathrm{e}^{\mathrm{i}k|\boldsymbol{r}-\boldsymbol{r}'|}}{4\pi|\boldsymbol{r}-\boldsymbol{r}'|}\boldsymbol{J}_\mathrm{e}(\boldsymbol{r}')\,\mathrm{d}v'\\ + \iiint_V \nabla'\times\left(\frac{\boldsymbol{J}_\mathrm{m}(\boldsymbol{r}')\mathrm{e}^{\mathrm{i}k|\boldsymbol{r}-\boldsymbol{r}'|}}{4\pi|\boldsymbol{r}-\boldsymbol{r}'|}\right)\mathrm{d}v'\\ + \iiint_V \left(\nabla\frac{\mathrm{e}^{\mathrm{i}k|\boldsymbol{r}-\boldsymbol{r}'|}}{4\pi|\boldsymbol{r}-\boldsymbol{r}'|}\right)\times\boldsymbol{J}_\mathrm{m}(\boldsymbol{r}')\,\mathrm{d}v'\Bigg\}\end{aligned}$$

Gauss' analogous theorem and the finite support of the magnetic current density $\boldsymbol{J}_\mathrm{m}$ imply that the second volume integral vanishes, and we finally obtain

$$\begin{aligned}\boldsymbol{E}(\boldsymbol{r}) &= \left[\mathbf{I}_3 + \frac{1}{k^2}\nabla\nabla\right]\cdot\Bigg\{\mathrm{i}k\eta_0\eta\iiint_V \frac{\mathrm{e}^{\mathrm{i}k|\boldsymbol{r}-\boldsymbol{r}'|}}{4\pi|\boldsymbol{r}-\boldsymbol{r}'|}\boldsymbol{J}_\mathrm{e}(\boldsymbol{r}')\,\mathrm{d}v'\\ &\quad + \nabla\times\iiint_V \frac{\mathrm{e}^{\mathrm{i}k|\boldsymbol{r}-\boldsymbol{r}'|}}{4\pi|\boldsymbol{r}-\boldsymbol{r}'|}\boldsymbol{J}_\mathrm{m}(\boldsymbol{r}')\,\mathrm{d}v'\Bigg\}\\ &= \mathrm{i}k\eta_0\eta\left[\mathbf{I}_3 + \frac{1}{k^2}\nabla\nabla\right]\cdot\iiint_V \frac{\mathrm{e}^{\mathrm{i}k|\boldsymbol{r}-\boldsymbol{r}'|}}{4\pi|\boldsymbol{r}-\boldsymbol{r}'|}\boldsymbol{J}_\mathrm{e}(\boldsymbol{r}')\,\mathrm{d}v'\\ &\quad + \nabla\times\iiint_V \frac{\mathrm{e}^{\mathrm{i}k|\boldsymbol{r}-\boldsymbol{r}'|}}{4\pi|\boldsymbol{r}-\boldsymbol{r}'|}\boldsymbol{J}_\mathrm{m}(\boldsymbol{r}')\,\mathrm{d}v'\end{aligned}$$

Notice that one of the terms in the second integral vanishes, due to the fact that the divergence of a curl vanishes.

Similarly, for the magnetic field in (2.18), we get

$$
\begin{aligned}
H(r) &= \frac{1}{\mu_0\mu}\nabla \times A(r) - \frac{1}{\mathrm{i}\omega\mu_0\mu}J_{\mathrm{m}}(r) = \nabla \times \iiint\limits_V \frac{\mathrm{e}^{\mathrm{i}k|r-r'|}}{4\pi|r-r'|}J_{\mathrm{e}}(r')\,\mathrm{d}v' \\
&+ \frac{1}{\mathrm{i}\omega\mu_0\mu}\nabla \times \iiint\limits_V \frac{\mathrm{e}^{\mathrm{i}k|r-r'|}}{4\pi|r-r'|}\nabla' \times J_{\mathrm{m}}(r')\,\mathrm{d}v' - \frac{1}{\mathrm{i}\omega\mu_0\mu}J_{\mathrm{m}}(r) \\
&= \nabla \times \iiint\limits_V \frac{\mathrm{e}^{\mathrm{i}k|r-r'|}}{4\pi|r-r'|}J_{\mathrm{e}}(r')\,\mathrm{d}v' \\
&+ \frac{1}{\mathrm{i}k\eta_0\eta}\nabla \times \left(\nabla \times \iiint\limits_V \frac{\mathrm{e}^{\mathrm{i}k|r-r'|}}{4\pi|r-r'|}J_{\mathrm{m}}(r')\,\mathrm{d}v'\right) - \frac{1}{\mathrm{i}k\eta_0\eta}J_{\mathrm{m}}(r)
\end{aligned}
$$

which we simplify to

$$
\begin{aligned}
H(r) &= \nabla \times \iiint\limits_V \frac{\mathrm{e}^{\mathrm{i}k|r-r'|}}{4\pi|r-r'|}J_{\mathrm{e}}(r')\,\mathrm{d}v' \\
&+ \frac{k^2}{\mathrm{i}k\eta_0\eta}\left[\mathbf{I}_3 + \frac{1}{k^2}\nabla\nabla\right] \cdot \iiint\limits_V \frac{\mathrm{e}^{\mathrm{i}k|r-r'|}}{4\pi|r-r'|}J_{\mathrm{m}}(r')\,\mathrm{d}v'
\end{aligned}
$$

where we used $\nabla \times \nabla \times = \nabla\nabla \cdot -\nabla^2$ and the differential equation of the Green function (2.8). This last form of the equations can also be concluded from the symmetry between the electric and magnetic fields in the Maxwell's equations. We summarize these results

$$
\begin{aligned}
E(r) &= \mathrm{i}k\eta_0\eta\left[\mathbf{I}_3 + \frac{1}{k^2}\nabla\nabla\right] \cdot \iiint\limits_V \frac{\mathrm{e}^{\mathrm{i}k|r-r'|}}{4\pi|r-r'|}J_{\mathrm{e}}(r')\,\mathrm{d}v' \\
&+ \nabla \times \iiint\limits_V \frac{\mathrm{e}^{\mathrm{i}k|r-r'|}}{4\pi|r-r'|}J_{\mathrm{m}}(r')\,\mathrm{d}v'
\end{aligned}
\tag{2.22}
$$

and

$$
\begin{aligned}
H(r) &= \nabla \times \iiint\limits_V \frac{\mathrm{e}^{\mathrm{i}k|r-r'|}}{4\pi|r-r'|}J_{\mathrm{e}}(r')\,\mathrm{d}v' \\
&+ \frac{k^2}{\mathrm{i}k\eta_0\eta}\left[\mathbf{I}_3 + \frac{1}{k^2}\nabla\nabla\right] \cdot \iiint\limits_V \frac{\mathrm{e}^{\mathrm{i}k|r-r'|}}{4\pi|r-r'|}J_{\mathrm{m}}(r')\,\mathrm{d}v'
\end{aligned}
\tag{2.23}
$$

In summary, provided we know the currents everywhere in space, we can determine the electric and magnetic fields everywhere outside the source region. The interpretation of the integrals inside the source region is addressed in Section 2.2.3. The electric and magnetic flux densities are then determined by the constitutive relations.

2.2 The Green Dyadics in Isotropic Media

The solution of the electromagnetic problem obtained above can also be written in a compact dyadic notation employing the Green dyadics. For easy reference, we summarize our canonical problem from above, and define the Green function in free space, $g(k, |\boldsymbol{r}-\boldsymbol{r}'|)$, as the solution to, see Section 2.1.2

$$\begin{cases} \nabla^2 g(k, |\boldsymbol{r}-\boldsymbol{r}'|) + k^2 g(k, |\boldsymbol{r}-\boldsymbol{r}'|) = -\delta(\boldsymbol{r}-\boldsymbol{r}') \\ \hat{\boldsymbol{r}} \cdot \nabla g(k, |\boldsymbol{r}-\boldsymbol{r}'|) - ikg(k, |\boldsymbol{r}-\boldsymbol{r}'|) = O(r^{-2}) \end{cases} \tag{2.24}$$

The second condition is the appropriate radiation condition[10] to secure that the radiating solution is obtained. The unique solution to this problem is

$$g(k, |\boldsymbol{r}-\boldsymbol{r}'|) = \frac{e^{ik|\boldsymbol{r}-\boldsymbol{r}'|}}{4\pi |\boldsymbol{r}-\boldsymbol{r}'|}$$

2.2.1 Free-Space Green Dyadic

We start by formulating the canonical problem (2.24) in a dyadic notion. The definition of the free-space Green dyadic is

$$\mathbf{G}(k, |\boldsymbol{r}-\boldsymbol{r}'|) = \mathbf{I}_3 g(k, |\boldsymbol{r}-\boldsymbol{r}'|) = \mathbf{I}_3 \frac{e^{ik|\boldsymbol{r}-\boldsymbol{r}'|}}{4\pi |\boldsymbol{r}-\boldsymbol{r}'|} \tag{2.25}$$

and it is the unique solution of

$$\begin{cases} \nabla^2 \mathbf{G}(k, |\boldsymbol{r}-\boldsymbol{r}'|) + k^2 \mathbf{G}(k, |\boldsymbol{r}-\boldsymbol{r}'|) = -\mathbf{I}_3 \delta(\boldsymbol{r}-\boldsymbol{r}') \\ \hat{\boldsymbol{r}} \cdot \nabla \mathbf{G}(k, |\boldsymbol{r}-\boldsymbol{r}'|) - ik\mathbf{G}(k, |\boldsymbol{r}-\boldsymbol{r}'|) = O(r^{-2}) \end{cases} \tag{2.26}$$

The second condition is the appropriate radiation condition to secure that the radiating solution is obtained.[11]

The dyadic $\mathbf{G}(k, |\boldsymbol{r}-\boldsymbol{r}'|)$ is symmetric as seen from its definition. This means that

$$\hat{\boldsymbol{e}}_i \cdot \mathbf{G}(k, |\boldsymbol{r}-\boldsymbol{r}'|) \cdot \hat{\boldsymbol{e}}_j = \hat{\boldsymbol{e}}_j \cdot \mathbf{G}(k, |\boldsymbol{r}-\boldsymbol{r}'|) \cdot \hat{\boldsymbol{e}}_i$$

---

[10]The exact motivation and the definition of the radiation conditions appear in Chapter 4. Here, we merely notice that for fixed $\boldsymbol{r}'$, see also Section 4.1.1 on page 188 (more details on this limit process are found in Section 4.1.1)

$$\frac{e^{ik|\boldsymbol{r}-\boldsymbol{r}'|}}{4\pi |\boldsymbol{r}-\boldsymbol{r}'|} = \frac{e^{ikr - ik\hat{\boldsymbol{r}}\cdot\boldsymbol{r}'}}{4\pi r}\left(1 + O(r^{-1})\right), \quad \text{as } r \to \infty$$

[11]The right-hand side is a dyadic-valued quantity and the O-symbol means that each Cartesian component behaves as $O((kr)^{-2})$. For fixed $\boldsymbol{r}'$, we have

$$\mathbf{G}(k, |\boldsymbol{r}-\boldsymbol{r}'|) = \frac{e^{ikr - ik\hat{\boldsymbol{r}}\cdot\boldsymbol{r}'}}{4\pi r}\mathbf{I}_3\left(1 + O(r^{-1})\right), \quad \text{as } r \to \infty$$

for any orthonormal basis $\hat{\boldsymbol{e}}_i, i = 1, 2, 3$. However, the curl of the Green dyadic is anti-symmetric as seen from the following calculations:

$$\begin{aligned}\hat{\boldsymbol{e}}_i \cdot \left(\nabla \times \mathbf{G}(k, |\boldsymbol{r} - \boldsymbol{r}'|)\right) \cdot \hat{\boldsymbol{e}}_j &= \hat{\boldsymbol{e}}_i \cdot \left(\nabla g(k, |\boldsymbol{r} - \boldsymbol{r}'|) \times \hat{\boldsymbol{e}}_j\right) \\ &= -\hat{\boldsymbol{e}}_j \cdot \left(\nabla g(k, |\boldsymbol{r} - \boldsymbol{r}'|) \times \hat{\boldsymbol{e}}_i\right) \\ &= -\hat{\boldsymbol{e}}_j \cdot \left(\nabla \times \mathbf{G}(k, |\boldsymbol{r} - \boldsymbol{r}'|)\right) \cdot \hat{\boldsymbol{e}}_i\end{aligned}$$

This property is written as

$$\left(\nabla \times \mathbf{G}(k, |\boldsymbol{r} - \boldsymbol{r}'|)\right)^{\mathrm{t}} = -\nabla \times \mathbf{G}(k, |\boldsymbol{r} - \boldsymbol{r}'|)$$

Furthermore, due to the spatial dependence $|\boldsymbol{r} - \boldsymbol{r}'|$, we have

$$\nabla \times \mathbf{G}(k, |\boldsymbol{r} - \boldsymbol{r}'|) = -\nabla' \times \mathbf{G}(k, |\boldsymbol{r} - \boldsymbol{r}'|)$$

2.2.2 The Green Dyadic for the Electric Field in Free Space

The dyadic $\mathbf{G}(k, |\boldsymbol{r} - \boldsymbol{r}'|)$ in (2.25) is not adapted to solve electromagnetic problems. We need a Green dyadic that is associated with the equation for the electric field in (2.2). To accomplish this, we define a Green dyadic for the electric field, $\mathbf{G}_{\mathrm{e}}$, which is the unique solution of

$$\begin{cases} \nabla \times \left(\nabla \times \mathbf{G}_{\mathrm{e}}(k, \boldsymbol{r} - \boldsymbol{r}')\right) - k^2 \mathbf{G}_{\mathrm{e}}(k, \boldsymbol{r} - \boldsymbol{r}') = \mathbf{I}_3 \delta(\boldsymbol{r} - \boldsymbol{r}') \\ \nabla \times \mathbf{G}_{\mathrm{e}}(k, \boldsymbol{r} - \boldsymbol{r}') - \mathrm{i}k\hat{\boldsymbol{r}} \times \mathbf{G}_{\mathrm{e}}(k, \boldsymbol{r} - \boldsymbol{r}') = O(r^{-2}) \end{cases} \tag{2.27}$$

Notice the similarity with the equation for the electric field, see (2.2).

Comment 2.2

The radiation conditions can be justified by the use of Footnotes 10 and 11. For fixed $\boldsymbol{r}'$, we have

$$\nabla\nabla g(k, |\boldsymbol{r} - \boldsymbol{r}'|) = -k^2 \frac{\mathrm{e}^{\mathrm{i}kr - \mathrm{i}k\hat{\boldsymbol{r}}\cdot\boldsymbol{r}'}}{4\pi r} \hat{\boldsymbol{r}}\hat{\boldsymbol{r}} \left(1 + O(r^{-1})\right), \quad \text{as } r \to \infty$$

We have

$$\nabla \times \mathbf{G}_{\mathrm{e}} - \mathrm{i}k\hat{\boldsymbol{r}} \times \mathbf{G}_{\mathrm{e}} = \nabla \times \mathbf{G} - \mathrm{i}k\hat{\boldsymbol{r}} \times \mathbf{G}_{\mathrm{e}} = O\left(r^{-2}\right)$$

since

$$\nabla \times \mathbf{G} = \mathrm{i}k\hat{\boldsymbol{r}} \times \mathbf{I}_3 \frac{\mathrm{e}^{\mathrm{i}kr - \mathrm{i}k\hat{\boldsymbol{r}}\cdot\boldsymbol{r}'}}{4\pi r} \left(1 + O(r^{-1})\right)$$

and

$$\mathrm{i}k\hat{\boldsymbol{r}} \times \mathbf{G}_{\mathrm{e}} = \mathrm{i}k\hat{\boldsymbol{r}} \times \mathbf{I}_3 \frac{\mathrm{e}^{\mathrm{i}kr - \mathrm{i}k\hat{\boldsymbol{r}}\cdot\boldsymbol{r}'}}{4\pi r} \left(1 + O(r^{-1})\right)$$

The free-space Green dyadic $\mathbf{G}(k, |\boldsymbol{r} - \boldsymbol{r}'|) = \mathbf{I}_3 g(k, |\boldsymbol{r} - \boldsymbol{r}'|)$ satisfies (2.26).

$$\nabla^2 \mathbf{G}(k, |\boldsymbol{r} - \boldsymbol{r}'|) + k^2 \mathbf{G}(k, |\boldsymbol{r} - \boldsymbol{r}'|) = -\mathbf{I}_3 \delta(\boldsymbol{r} - \boldsymbol{r}')$$

Adding this equation of the free-space Green dyadic to the first equation in (2.27) and using (2.6) gives

$$\nabla \times (\nabla \times (\mathbf{G}_e - \mathbf{G})) - k^2 (\mathbf{G}_e - \mathbf{G}) = -\nabla (\nabla \cdot \mathbf{G}) = -\nabla\nabla g$$

where the dyadic $\nabla\nabla g$ is interpreted in Cartesian coordinates as

$$\nabla\nabla g|_{ij} = \frac{\partial^2 g}{\partial x_i \partial x_j}$$

We write as

$$\nabla \times \left(\nabla \times \left(\mathbf{G}_e - \mathbf{G} - \frac{1}{k^2} \nabla\nabla g \right) \right) - k^2 \left(\mathbf{G}_e - \mathbf{G} - \frac{1}{k^2} \nabla\nabla g \right) = \mathbf{0}$$

One solution of this equation is[12]

$$\mathbf{G}_e = \mathbf{G} + \frac{1}{k^2} \nabla\nabla g = \left(\mathbf{I}_3 + \frac{1}{k^2} \nabla\nabla \right) g \tag{2.28}$$

From this representation, we see that not only $\mathbf{G}$ is a symmetric dyadic but also $\mathbf{G}_e$. Furthermore, the spatial dependence of the Green dyadic for the electric field $\mathbf{G}_e$ is $\boldsymbol{r} - \boldsymbol{r}'$. Moreover,

$$\begin{aligned} \nabla \times \mathbf{G}_e(k, \boldsymbol{r} - \boldsymbol{r}') &= \nabla \times \mathbf{G}(k, |\boldsymbol{r} - \boldsymbol{r}'|) = -\nabla' \times \mathbf{G}(k, |\boldsymbol{r} - \boldsymbol{r}'|) \\ &= -\nabla' \times \mathbf{G}_e(k, \boldsymbol{r} - \boldsymbol{r}') \end{aligned}$$

Moreover, since the free-space Green dyadic is an anti-symmetric dyadic, $\nabla \times \mathbf{G}_e(k, \boldsymbol{r} - \boldsymbol{r}')$ is also an anti-symmetric dyadic. We collect the results

$$\begin{cases} \mathbf{G}_e(k, \boldsymbol{r} - \boldsymbol{r}') = \left(\mathbf{I}_3 + \dfrac{1}{k^2} \nabla\nabla \right) g(k, |\boldsymbol{r} - \boldsymbol{r}'|) \\ \qquad\qquad = \dfrac{1}{k^2} \left\{ \nabla \times (\nabla \times \mathbf{I}_3 g(k, |\boldsymbol{r} - \boldsymbol{r}'|)) - \mathbf{I}_3 \delta(\boldsymbol{r} - \boldsymbol{r}') \right\} \\ \left(\mathbf{G}_e(k, \boldsymbol{r} - \boldsymbol{r}') \right)^{\mathrm{t}} = \mathbf{G}_e(k, \boldsymbol{r} - \boldsymbol{r}') \\ \left(\nabla \times \mathbf{G}_e(k, \boldsymbol{r} - \boldsymbol{r}') \right)^{\mathrm{t}} = -\nabla \times \mathbf{G}_e(k, \boldsymbol{r} - \boldsymbol{r}') \end{cases}$$

The alternative form of $\mathbf{G}_e(k, \boldsymbol{r} - \boldsymbol{r}')$ utilizes the identity

$$\begin{aligned} \nabla \times \left(\nabla \times \mathbf{I}_3 g(k, |\boldsymbol{r} - \boldsymbol{r}'|) \right) &= \nabla \left(\nabla \cdot \mathbf{I}_3 g(k, |\boldsymbol{r} - \boldsymbol{r}'|) \right) - \mathbf{I}_3 \nabla^2 g(k, |\boldsymbol{r} - \boldsymbol{r}'|) \\ &= \nabla\nabla g(k, |\boldsymbol{r} - \boldsymbol{r}'|) - \mathbf{I}_3 \nabla^2 g(k, |\boldsymbol{r} - \boldsymbol{r}'|) \\ &= \nabla\nabla g(k, |\boldsymbol{r} - \boldsymbol{r}'|) + \mathbf{I}_3 \left(k^2 g(k, |\boldsymbol{r} - \boldsymbol{r}'|) + \delta(\boldsymbol{r} - \boldsymbol{r}') \right) \end{aligned}$$

since $\nabla \cdot \mathbf{I}_3 g = \nabla g$.

---

[12]This is the only solution provided the radiation conditions are satisfied.

2.2.3 Depolarizing Dyadic

We now investigate the distribution $\nabla\nabla g$. Of particular interest is its singular part. The effect of this distribution can be made in several different ways. Yaghjian[13] [296] starts by excluding a volume V_ϵ and proceeds in a classical manner, see also Reference 273.

In this section, we adopt a more distributional way of analysis. To attain this, multiply with a test function $\phi(\boldsymbol{r})$, and integrate over all space.[14] Furthermore, let V_ϵ be a volume centered around $\boldsymbol{r}'$ such that $V_\epsilon \to 0$ as $\epsilon \to 0$, and let S_ϵ be the bounding surface of V_ϵ with outward unit normal vector $\hat{\boldsymbol{\nu}}$. Two examples of exclusion volumes are given in Figure 2.2. The definition of distributional derivatives gives

$$\iiint_{\mathbb{R}^3} \frac{\partial^2 g}{\partial x_i \partial x_j}\phi \, \mathrm{d}v \stackrel{\text{def}}{=} \iiint_{\mathbb{R}^3} g \frac{\partial^2 \phi}{\partial x_i \partial x_j} \, \mathrm{d}v = \lim_{\epsilon\to 0} \iiint_{\mathbb{R}^3\setminus V_\epsilon} g \frac{\partial^2 \phi}{\partial x_i \partial x_j} \, \mathrm{d}v$$

where the last equality is valid, due to fact that the Green function g is a locally integrable function. Apply Gauss' theorem twice, and we get $(i,j = 1, 2, 3)$

$$\begin{aligned} \iiint_{\mathbb{R}^3} \frac{\partial^2 g}{\partial x_i \partial x_j}\phi \, \mathrm{d}v = &- \lim_{\epsilon\to 0} \iint_{S_\epsilon} \left\{ (\hat{\boldsymbol{\nu}} \cdot \hat{\boldsymbol{e}}_i) g \frac{\partial \phi}{\partial x_j} - (\hat{\boldsymbol{\nu}} \cdot \hat{\boldsymbol{e}}_j) \frac{\partial g}{\partial x_i}\phi \right\} \mathrm{d}S \\ &+ \lim_{\epsilon\to 0} \iiint_{\mathbb{R}^3\setminus V_\epsilon} \frac{\partial^2 g}{\partial x_i \partial x_j}\phi \, \mathrm{d}v \end{aligned}$$

Finally, we have to evaluate the surface integral over S_ϵ. This limit depends on the shape of the excluding volume V_ϵ. The first term in the surface integral vanishes in the limit $\epsilon \to 0$, since g is locally integrable on the surface S_ϵ, and the second can be simplified. The result is

$$\begin{aligned} &\lim_{\epsilon\to 0} \iint_{S_\epsilon} \left\{ (\hat{\boldsymbol{\nu}} \cdot \hat{\boldsymbol{e}}_i) g \frac{\partial \phi}{\partial x_j} - (\hat{\boldsymbol{\nu}} \cdot \hat{\boldsymbol{e}}_j) \frac{\partial g}{\partial x_i}\phi \right\} \mathrm{d}S \\ &\quad = \lim_{\epsilon\to 0} \iint_{S_\epsilon} \frac{(\hat{\boldsymbol{e}}_i \cdot (\boldsymbol{r} - \boldsymbol{r}'))(\hat{\boldsymbol{\nu}} \cdot \hat{\boldsymbol{e}}_j)}{4\pi |\boldsymbol{r} - \boldsymbol{r}'|^3} \phi(\boldsymbol{r}) \, \mathrm{d}S \end{aligned}$$

since

$$\nabla g(|\boldsymbol{r} - \boldsymbol{r}'|) = \frac{(\boldsymbol{r} - \boldsymbol{r}')\mathrm{e}^{\mathrm{i}k|\boldsymbol{r}-\boldsymbol{r}'|}}{4\pi |\boldsymbol{r} - \boldsymbol{r}'|^2} \left(\mathrm{i}k - \frac{1}{|\boldsymbol{r} - \boldsymbol{r}'|} \right)$$

In fact, if the surface S_ϵ is simply connected only the most singular term in ∇g survives in the limit $\epsilon \to 0$, and moreover, the exponential can be replaced by unity. Effectively, this means that all terms containing a factor k disappears (quasi-static limit). The same argument holds if

[13] Arthur D. Yaghjian (1943–), American electrical engineer.

[14] Do not confuse this function with the potential that we introduced in Section 2.1.1. The notation $\phi(\boldsymbol{r})$ is used here simply because this notation is widely used as test function.

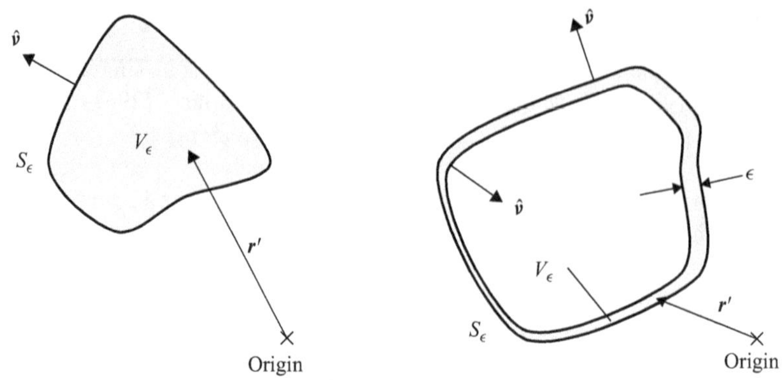

Figure 2.2 Two examples of exclusion volumes V_ϵ. In the right-hand figure, the surface S_ϵ consists of two separate parts. Notice the directions of the unit vectors.

the surface S_ϵ consists of several disconnected parts, like the right-hand drawing in Figure 2.2, since the inner and the outer surface cancel except in the vicinity of the singularity $\boldsymbol{r}'$.

The field ϕ is assumed smooth, and in the limit $\epsilon \to 0$, the surface integral defines a linear transformation (dyadic) $\mathbf{L}$, the depolarizing dyadic, defined as

$$\lim_{\epsilon\to 0} \iint_{S_\epsilon} \left\{ (\hat{\boldsymbol{\nu}} \cdot \hat{\boldsymbol{e}}_i) g \frac{\partial \phi}{\partial x_j} - (\hat{\boldsymbol{\nu}} \cdot \hat{\boldsymbol{e}}_j) \frac{\partial g}{\partial x_i} \phi \right\} \, \mathrm{d}S = \phi(\boldsymbol{r}') L_{ij}$$

where

$$L_{ij} = \lim_{\epsilon\to 0} \iint_{S_\epsilon} \frac{\left(\hat{\boldsymbol{e}}_i \cdot (\boldsymbol{r} - \boldsymbol{r}')\right) (\hat{\boldsymbol{\nu}} \cdot \hat{\boldsymbol{e}}_j)}{4\pi \, |\boldsymbol{r} - \boldsymbol{r}'|^3} \, \mathrm{d}S$$

The depolarizing dyadic $\mathbf{L}$ is a symmetric dyadic. To see this, let (i, j, k) be an even cyclic permutation of $(1, 2, 3)$. Then

$$L_{ij} - L_{ji} = \lim_{\epsilon\to 0} \iint_{S_\epsilon} \frac{(\boldsymbol{r} - \boldsymbol{r}') \times \hat{\boldsymbol{\nu}}\big|_k}{4\pi \, |\boldsymbol{r} - \boldsymbol{r}'|^3} \, \mathrm{d}S = \lim_{\epsilon\to 0} \iint_{S_\epsilon} \hat{\boldsymbol{\nu}} \times \nabla \frac{1}{4\pi \, |\boldsymbol{r} - \boldsymbol{r}'|} \bigg|_k \, \mathrm{d}S = 0$$

by an analogous theorem to Stokes' theorem, since S_ϵ is a closed surface. Moreover, the dyadic $\mathbf{L}$ has unit trace as seen by

$$\sum_{i=1}^{3} L_{ii} = \lim_{\epsilon\to 0} \iint_{S_\epsilon} \frac{(\boldsymbol{r} - \boldsymbol{r}') \cdot \hat{\boldsymbol{\nu}}}{4\pi \, |\boldsymbol{r} - \boldsymbol{r}'|^3} \, \mathrm{d}S = 1$$

since the integral is exactly the solid angle, 4π, of the surface S_ϵ as seen from the point $\boldsymbol{r}'$ inside S_ϵ, see Appendix F.5 on page 672.

The action of the electric Green dyadic therefore is

$$\iiint_{\mathbb{R}^3} \mathbf{G}_\mathrm{e} \cdot \boldsymbol{J} \,\mathrm{d}v = \lim_{\epsilon\to 0} \iiint_{\mathbb{R}^3\setminus V_\epsilon} \mathbf{G}_\mathrm{e} \cdot \boldsymbol{J} \,\mathrm{d}v - \frac{1}{k^2}\mathbf{L}\cdot \boldsymbol{J}(\boldsymbol{r}')$$

Two explicit forms of the depolarizing dyadic are used in this book. For a spherical exclusion the dyadic $\mathbf{L}$ is, see Problem 2.3

$$\mathbf{L} = \frac{1}{3}\mathbf{I}_3 \tag{2.29}$$

and therefore

$$\iiint_{\mathbb{R}^3} \phi\nabla\nabla g \,\mathrm{d}v = -\frac{1}{3}\phi(\boldsymbol{r}')\mathbf{I}_3 + \lim_{\epsilon\to 0} \iiint_{|\boldsymbol{r}-\boldsymbol{r}'|\geq\epsilon} \phi\nabla\nabla g \,\mathrm{d}v$$

Another excluding volume is used in Section 7.5, viz. the volume between two concentric spheres $r = r' - \epsilon$ and $r = r' + \epsilon$. This excluding volume is, in the limit $\epsilon \to 0$, equivalent to a disk, and the reader is encouraged to solve the details of this excluding volume in Problem 2.4. The result is

$$L_{ij} = (\hat{\boldsymbol{r}}' \cdot \hat{\boldsymbol{e}}_i)(\hat{\boldsymbol{r}}' \cdot \hat{\boldsymbol{e}}_j) \text{ or } \mathbf{L} = \hat{\boldsymbol{r}}'\hat{\boldsymbol{r}}' \tag{2.30}$$

and therefore

$$\iiint_{\mathbb{R}^3} \phi\nabla\nabla g \,\mathrm{d}v = -\phi(\boldsymbol{r}')\hat{\boldsymbol{r}}'\hat{\boldsymbol{r}}' + \lim_{\epsilon\to 0} \iiint_{V_\epsilon} \phi\nabla\nabla g \,\mathrm{d}v$$

Several other shapes of the exclusion volume are found in the literature, see e.g., References 33, 276, 296, and we refer the interested reader to these references for more details.

Expansions of the free-space Green dyadic, $\mathbf{G}(k, |\boldsymbol{r} - \boldsymbol{r}'|)$, and expansions of the electric Green dyadic, $\mathbf{G}_\mathrm{e}(k, |\boldsymbol{r} - \boldsymbol{r}'|)$, in terms of spherical and plane vector waves are presented in Chapters 7 and 10, respectively.

2.3 The Green Dyadic in Anisotropic Media

There is no known solution of the Green dyadic in a general bianisotropic media. Considerable efforts to solve the problem in complex media have been reported in the literature, see e.g., References 161–163, 166, 207–209. Our ambition in this short section is not to pursue the form of the Green dyadic in complex media in detail, but simply to formulate the appropriate equations that are used below. Thus, the purpose of this subsection is to generalize some of the equations that have been derived for an isotropic material above to a general anisotropic material.

The Maxwell equations and the appropriate constitutive relations are

$$\begin{cases} \nabla\times\boldsymbol{E}(\boldsymbol{r},\omega) = \mathrm{i}\omega\boldsymbol{B}(\boldsymbol{r},\omega) \\ \nabla\times\boldsymbol{H}(\boldsymbol{r},\omega) = \boldsymbol{J}(\boldsymbol{r},\omega) - \mathrm{i}\omega\boldsymbol{D}(\boldsymbol{r},\omega) \end{cases} \qquad \begin{cases} \boldsymbol{D}(\boldsymbol{r},\omega) = \epsilon_0\boldsymbol{\epsilon}(\boldsymbol{r},\omega)\cdot\boldsymbol{E}(\boldsymbol{r},\omega) \\ \boldsymbol{B}(\boldsymbol{r},\omega) = \dfrac{1}{c_0}\boldsymbol{\mu}(\boldsymbol{r},\omega)\cdot\eta_0\boldsymbol{H}(\boldsymbol{r},\omega) \end{cases}$$

where we have assumed inhomogeneous constitutive relations. Eliminating all fields, except the electric field $\boldsymbol{E}$, leads to

$$\nabla \times \left(\boldsymbol{\mu}^{-1} \cdot \nabla \times \boldsymbol{E}\right) = \frac{\mathrm{i}\omega\eta_0}{c_0} \nabla \times \boldsymbol{H} = \mathrm{i}k_0\eta_0 \left(\boldsymbol{J} - \mathrm{i}\omega \boldsymbol{D}\right) = \mathrm{i}k_0\eta_0 \left(\boldsymbol{J} - \mathrm{i}\omega\epsilon_0 \boldsymbol{\epsilon} \cdot \boldsymbol{E}\right)$$

or

$$\nabla \times \left(\boldsymbol{\mu}^{-1}(\boldsymbol{r},\omega) \cdot \nabla \times \boldsymbol{E}(\boldsymbol{r},\omega)\right) - k_0^2 \boldsymbol{\epsilon}(\boldsymbol{r},\omega) \cdot \boldsymbol{E}(\boldsymbol{r},\omega) = \mathrm{i}k_0\eta_0 \boldsymbol{J}(\boldsymbol{r},\omega) \tag{2.31}$$

where the wave number in vacuum is denoted $k_0 = \omega/c_0$. This equation shows that the electric field satisfies a more complex equation in an anisotropic material, and we have to abandon any attempt to find a useful scalar Green function. Instead, we develop a true dyadic approach to find the Green dyadic in the complex material.

Based on the form of the Maxwell equation, the appropriate Green dyadics satisfy

$$\begin{cases} \nabla \times \mathbf{G}_\mathrm{e}^\mathrm{an}(\boldsymbol{r},\boldsymbol{r}') = \mathrm{i}k_0 \mathbf{G}_\mathrm{b}^\mathrm{an}(\boldsymbol{r},\boldsymbol{r}') \\ \nabla \times \boldsymbol{\mu}^{-1} \cdot \mathbf{G}_\mathrm{b}^\mathrm{an}(\boldsymbol{r},\boldsymbol{r}') = \dfrac{1}{\mathrm{i}k_0} \boldsymbol{\mu}^{-1} \cdot \mathbf{I}_3 \delta(\boldsymbol{r}-\boldsymbol{r}') - \mathrm{i}k_0 \boldsymbol{\epsilon} \cdot \mathbf{G}_\mathrm{e}^\mathrm{an}(\boldsymbol{r},\boldsymbol{r}') \end{cases}$$

where the constitutive relations now have no space dependence, i.e., a homogeneous material. The dyadic field $\mathbf{G}_\mathrm{e}^\mathrm{an}$ is associated with the electric field $\boldsymbol{E}$, and $\mathbf{G}_\mathrm{b}^\mathrm{an}$ is associated with the magnetic flux density $\boldsymbol{B}$. Note that both dyadics have the same dimension m^{-1}. We can eliminate the dyadic $\mathbf{G}_\mathrm{b}^\mathrm{an}$, and we get

$$\nabla \times \boldsymbol{\mu}^{-1} \cdot \left(\nabla \times \mathbf{G}_\mathrm{e}^\mathrm{an}(\boldsymbol{r},\boldsymbol{r}')\right) - k_0^2 \boldsymbol{\epsilon} \cdot \mathbf{G}_\mathrm{e}^\mathrm{an}(\boldsymbol{r},\boldsymbol{r}') = \boldsymbol{\mu}^{-1} \cdot \mathbf{I}_3 \delta(\boldsymbol{r}-\boldsymbol{r}')$$

If the permeability of the material is homogeneous and isotropic, i.e., $\boldsymbol{\mu} = \mathbf{I}\mu$, the equation of dyadic $\mathbf{G}_\mathrm{b}^\mathrm{an}$ simplifies to

$$\nabla \times \left(\nabla \times \mathbf{G}_\mathrm{e}^\mathrm{an}(\boldsymbol{r},\boldsymbol{r}')\right) - k_0^2 \boldsymbol{\epsilon}\mu \cdot \mathbf{G}_\mathrm{e}^\mathrm{an}(\boldsymbol{r},\boldsymbol{r}') = \mathbf{I}_3 \delta(\boldsymbol{r}-\boldsymbol{r}') \tag{2.32}$$

For an isotropic material, this equation reduces to

$$\nabla \times \left(\nabla \times \mathbf{G}_\mathrm{e}^\mathrm{an}(\boldsymbol{r},\boldsymbol{r}')\right) - k^2 \mathbf{G}_\mathrm{e}^\mathrm{an}(\boldsymbol{r},\boldsymbol{r}') = \mathbf{I}_3 \delta(\boldsymbol{r}-\boldsymbol{r}')$$

where $k^2 = k_0^2 \epsilon\mu$. This is the same result for an isotropic material as derived above.

We refer to the literature for some explicit example of the Green dyadic in anisotropic materials [161–163, 166, 207–209].

2.4 The Green Dyadic in Biisotropic Media

In biisotropic media, the Green dyadic can be expressed in terms of the Green dyadic for isotropic media. How this is accomplished is analyzed in the section. Other similar treatments are found in References 151, 161.

The constitutive relations of a homogeneous biisotropic material are, see (1.55) on page 45

$$\begin{cases} \boldsymbol{D}(\boldsymbol{r},\omega) = \epsilon_0 \left\{\epsilon(\omega)\boldsymbol{E}(\boldsymbol{r},\omega) + \xi(\omega)\eta_0 \boldsymbol{H}(\boldsymbol{r},\omega)\right\} \\ \boldsymbol{B}(\boldsymbol{r},\omega) = \dfrac{1}{c_0} \left\{\zeta(\omega)\boldsymbol{E}(\boldsymbol{r},\omega) + \mu(\omega)\eta_0 \boldsymbol{H}(\boldsymbol{r},\omega)\right\} \end{cases}$$

We start by finding the appropriate equations for the fields in biisotropic media. The Maxwell equations imply

$$\begin{cases} \nabla \times \boldsymbol{E} = \mathrm{i}\omega \boldsymbol{B} = \dfrac{\mathrm{i}\omega}{c_0}\left(\zeta \boldsymbol{E} + \mu\eta_0 \boldsymbol{H}\right) \\ \nabla \times \boldsymbol{H} = \boldsymbol{J} - \mathrm{i}\omega \boldsymbol{D} = \boldsymbol{J} - \mathrm{i}\omega\epsilon_0\left(\epsilon \boldsymbol{E} + \xi\eta_0 \boldsymbol{H}\right) \end{cases}$$

which we write in a matrix form as

$$\begin{pmatrix} \nabla \times \boldsymbol{E} \\ \eta_0 \nabla \times \boldsymbol{H} \end{pmatrix} = \begin{pmatrix} \boldsymbol{0} \\ \eta_0 \boldsymbol{J} \end{pmatrix} + \mathrm{i}k_0 \begin{pmatrix} \zeta & \mu \\ -\epsilon & -\xi \end{pmatrix} \begin{pmatrix} \boldsymbol{E} \\ \eta_0 \boldsymbol{H} \end{pmatrix} \tag{2.33}$$

where we denote the wave number in free space by $k_0 = \omega/c_0$ as usual. Eliminating the magnetic field $\boldsymbol{H}$, we get

$$\begin{aligned} \nabla \times (\nabla \times \boldsymbol{E}) &= \mathrm{i}k_0 \left\{\zeta \nabla \times \boldsymbol{E} + \eta_0\mu \boldsymbol{J} - \mathrm{i}k_0\mu\left(\epsilon \boldsymbol{E} + \xi\eta_0 \boldsymbol{H}\right)\right\} \\ &= \mathrm{i}k_0 \left\{\zeta \nabla \times \boldsymbol{E} + \eta_0\mu \boldsymbol{J} - \mathrm{i}k_0\mu\epsilon \boldsymbol{E} - \xi\left(\nabla \times \boldsymbol{E} - \mathrm{i}k_0\zeta \boldsymbol{E}\right)\right\} \end{aligned}$$

This motivates the investigation of a vector-valued second-order partial differential equation of the form

$$\nabla \times (\nabla \times \boldsymbol{E}) + \alpha \nabla \times \boldsymbol{E} - \beta^2 \boldsymbol{E} = \mathrm{i}k_0\eta_0\mu \boldsymbol{J} \tag{2.34}$$

where

$$\begin{cases} \alpha = \mathrm{i}k_0(\xi - \zeta) = -2k_0\chi \\ \beta^2 = k_0^2\left(\epsilon\mu - \xi\zeta\right) = k_0^2\left(\epsilon\mu - \kappa^2 - \chi^2\right) \end{cases}$$

where we used $\xi = \kappa + \mathrm{i}\chi$ and $\zeta = \kappa - \mathrm{i}\chi$. From this equation and the Maxwell equations, (2.33), we find that the divergence satisfies

$$\begin{pmatrix} \zeta & \mu \\ -\epsilon & -\xi \end{pmatrix} \begin{pmatrix} \nabla \cdot \boldsymbol{E} \\ \eta_0 \nabla \cdot \boldsymbol{H} \end{pmatrix} = \begin{pmatrix} 0 \\ -\rho/\epsilon_0 \end{pmatrix}$$

where ρ is the charge density, and where we also have used the conservation of charge, see (1.45) on page 41. The solution then is

$$\begin{pmatrix} \nabla \cdot \boldsymbol{E} \\ \eta_0 \nabla \cdot \boldsymbol{H} \end{pmatrix} = \frac{1}{\kappa^2 + \chi^2 - \epsilon\mu} \begin{pmatrix} -\xi & -\mu \\ \epsilon & \zeta \end{pmatrix} \begin{pmatrix} 0 \\ \rho/\epsilon_0 \end{pmatrix}$$

or

$$\begin{cases} \nabla \cdot \boldsymbol{E} = \dfrac{\mu\rho}{\epsilon_0\left(\epsilon\mu - \kappa^2 - \chi^2\right)} \\ \eta_0 \nabla \cdot \boldsymbol{H} = -\dfrac{\zeta\rho}{\epsilon_0\left(\epsilon\mu - \kappa^2 - \chi^2\right)} = -\dfrac{(\kappa - \mathrm{i}\chi)\rho}{\epsilon_0\left(\epsilon\mu - \kappa^2 - \chi^2\right)} \end{cases}$$

Based on the equation of the electric field (2.34), the appropriate Green dyadic in a biisotropic material satisfies

$$\nabla \times \left(\nabla \times \mathbf{G}_\mathrm{e}^{\mathrm{bi}}(\boldsymbol{r}, \boldsymbol{r}')\right) + \alpha \nabla \times \mathbf{G}_\mathrm{e}^{\mathrm{bi}}(\boldsymbol{r}, \boldsymbol{r}') - \beta^2 \mathbf{G}_\mathrm{e}^{\mathrm{bi}}(\boldsymbol{r}, \boldsymbol{r}') = \mathbf{I}_3 \delta(\boldsymbol{r} - \boldsymbol{r}')$$

The appropriate solution is determined by requiring that the solution coincides with the Green dyadic for a non-chiral media as $\chi \to 0$.

A simple multiplication shows that the differential equation can be factorized as

$$\{\nabla \times \mathbf{I}_3 - k_+\mathbf{I}_3\} \cdot \{\nabla \times \mathbf{I}_3 + k_-\mathbf{I}_3\} \cdot \mathbf{G}_\mathrm{e}^\mathrm{bi}(\boldsymbol{r}, \boldsymbol{r}') = \mathbf{I}_3\delta(\boldsymbol{r} - \boldsymbol{r}')$$

or

$$\{\nabla \times \mathbf{I}_3 + k_-\mathbf{I}_3\} \cdot \{\nabla \times \mathbf{I}_3 - k_+\mathbf{I}_3\} \cdot \mathbf{G}_\mathrm{e}^\mathrm{bi}(\boldsymbol{r}, \boldsymbol{r}') = \mathbf{I}_3\delta(\boldsymbol{r} - \boldsymbol{r}')$$

where $k_\pm = k_0(\sqrt{\epsilon\mu - \kappa^2} \pm \chi)$, and consequently $k_- - k_+ = -2k_0\chi = \alpha$ and $k_-k_+ = \beta^2$. This decomposition allows us to write down the solution in terms of the free-space electric Green dyadics $\mathbf{G}_\mathrm{e}(k_\pm, |\boldsymbol{r} - \boldsymbol{r}'|)$, which satisfy

$$\{\nabla \times \mathbf{I}_3 - k_\pm\mathbf{I}_3\} \cdot \{\nabla \times \mathbf{I}_3 + k_\pm\mathbf{I}_3\} \cdot \mathbf{G}_\mathrm{e}(k_\pm, |\boldsymbol{r} - \boldsymbol{r}'|) = \mathbf{I}_3\delta(\boldsymbol{r} - \boldsymbol{r}')$$

or

$$\{\nabla \times \mathbf{I}_3 + k_\pm\mathbf{I}_3\} \cdot \{\nabla \times \mathbf{I}_3 - k_\pm\mathbf{I}_3\} \cdot \mathbf{G}_\mathrm{e}(k_\pm, |\boldsymbol{r} - \boldsymbol{r}'|) = \mathbf{I}_3\delta(\boldsymbol{r} - \boldsymbol{r}')$$

The solution can be formed as

$$\mathbf{G}_\mathrm{e}^\mathrm{bi}(\boldsymbol{r}, \boldsymbol{r}') = \frac{(\nabla \times \mathbf{I}_3 + k_+\mathbf{I}_3)\cdot\mathbf{G}_\mathrm{e}(k_+, |\boldsymbol{r} - \boldsymbol{r}'|) - (\nabla \times \mathbf{I}_3 - k_-\mathbf{I}_3)\cdot\mathbf{G}_\mathrm{e}(k_-, |\boldsymbol{r} - \boldsymbol{r}'|)}{k_+ + k_-} \tag{2.35}$$

since

$$\begin{aligned}
&\{\nabla \times \mathbf{I}_3 + k_-\mathbf{I}_3\} \cdot \{\nabla \times \mathbf{I}_3 - k_+\mathbf{I}_3\} \cdot \frac{(\nabla \times \mathbf{I}_3 + k_+\mathbf{I}_3) \cdot \mathbf{G}_\mathrm{e}(k_+, |\boldsymbol{r} - \boldsymbol{r}'|)}{k_+ + k_-} \\
&\quad - \{\nabla \times \mathbf{I}_3 - k_+\mathbf{I}_3\} \cdot \{\nabla \times \mathbf{I}_3 + k_-\mathbf{I}_3\} \cdot \frac{(\nabla \times \mathbf{I}_3 - k_-\mathbf{I}_3) \cdot \mathbf{G}_\mathrm{e}(k_-, |\boldsymbol{r} - \boldsymbol{r}'|)}{k_+ + k_-} \\
&\quad = \{\nabla \times \mathbf{I}_3 + k_-\mathbf{I}_3\} \cdot \frac{\mathbf{I}_3\delta(\boldsymbol{r} - \boldsymbol{r}')}{k_+ + k_-} - \{\nabla \times \mathbf{I}_3 - k_+\mathbf{I}_3\} \cdot \frac{\mathbf{I}_3\delta(\boldsymbol{r} - \boldsymbol{r}')}{k_+ + k_-} \\
&\quad = \mathbf{I}_3\delta(\boldsymbol{r} - \boldsymbol{r}')
\end{aligned}$$

This proves the explicit form of the Green dyadic in biisotropic media.

We also conclude that as the biisotropy disappears, i.e., $\xi \to 0$ and $\zeta \to 0$, the Green dyadic coincides with the Green dyadic of an isotropic material.

2.5 Čerenkov Radiation

In Section 2.1.2, we showed that knowledge of the current density everywhere in space implies that the electric and magnetic (as well as their fluxes) fields can be computed. We illustrate this result in this section by the Čerenkov radiation.

When a charged particle (e.g., an electron or ion) moves with high speed in a dielectric material, electromagnetic radiation is emitted. This was first observed experimentally in the

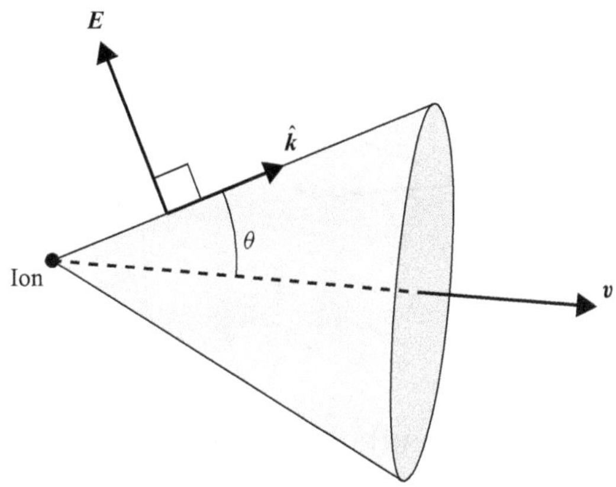

Figure 2.3 The polarization of the electric field $\boldsymbol{E}$ in Čerenkov radiation. The velocity of the charged particle is $\boldsymbol{v}$ and the direction of the radiation is $\hat{\boldsymbol{k}}$.

1930s by P.A. Čerenkov[15] and S.I. Vavilov[16] [29], and its theoretical explanation was given a few years later by I.M. Frank[17] and I.Y. Tamm[18] [73]. The radiation does not depend on any collision processes with the atoms in the material (like e.g., bremsstrahlung does), but can be explained as a purely electromagnetic phenomenon of macroscopic nature. Čerenkov radiation is utilized to detect particles in nuclear and elementary particle physics. In this section, we analyze this radiation, and, at the same time, we have an illustration of the results obtained in Section 2.1.

Experimental observations have proved that

1. The charge particle has to move very fast in the material; otherwise no radiation.
2. The direction of the radiation, $\hat{\boldsymbol{k}}$, depends on the speed of the particle.
3. The polarization of the radiation is such that the electric field $\boldsymbol{E}$ lies in the plane spanned by the velocity of the particle, $\boldsymbol{v}$, and the direction of the radiation $\hat{\boldsymbol{k}}$. Moreover, the electric field $\boldsymbol{E}$ is perpendicular to $\hat{\boldsymbol{k}}$, see Figure 2.3.

This problem is now addressed, and we investigate the electric field a fast-moving charged particle generates.

Let the particle have charge q and orient the coordinate axes so that the velocity of the particle, $\boldsymbol{v}$, is parallel to the z-axis. The surrounding material is an isotropic, homogeneous material with permittivity ϵ and permeability μ, see Figure 2.4. These material parameters may depend on ω, i.e., the material shows dispersion.

The current density $\boldsymbol{J}$, due to the movement of the particle, is

$$\boldsymbol{J}(\boldsymbol{r},t) = \hat{z}qv\delta(\boldsymbol{r} - \hat{z}vt) = \hat{z}qv\delta(x)\delta(y)\delta(z - vt)$$

---

[15] Pavel Alekseyevich Čerenkov (1904–1990), Soviet physicist.

[16] Sergey Ivanovich Vavilov (1891–1951), Soviet physicist.

[17] Ilya Mikhailovich Frank (1908–1990), Soviet physicist.

[18] Igor Yevgenyevich Tamm (1895–1971), Soviet physicist.

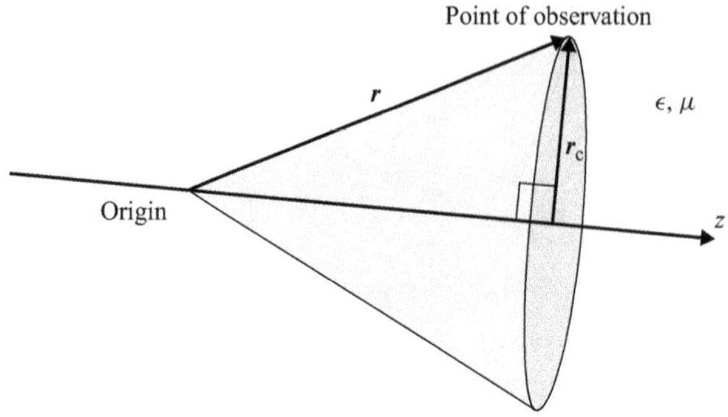

Figure 2.4 The trajectory of the particle parallel to the z-axis and the distance r_c to the trajectory.

This is the current density as a function of the spatial variables and time. A Fourier transformation w.r.t. time t gives the corresponding time-harmonic quantity[19]

$$\boldsymbol{J}(\boldsymbol{r},\omega) = \int_{-\infty}^{\infty} \boldsymbol{J}(\boldsymbol{r},t)\,\mathrm{e}^{\mathrm{i}\omega t}\,\mathrm{d}t = \hat{z}qv\delta(x)\delta(y)\int_{-\infty}^{\infty}\delta(z-vt)\,\mathrm{e}^{\mathrm{i}\omega t}\,\mathrm{d}t = \hat{z}q\delta(x)\delta(y)\mathrm{e}^{\mathrm{i}\omega z/v}$$

The vector potential is then given by (2.12) on page 105

$$\boldsymbol{A}(\boldsymbol{r},\omega) = \mu_0\mu \iiint_V \frac{\mathrm{e}^{\mathrm{i}k|\boldsymbol{r}-\boldsymbol{r}'|}}{4\pi|\boldsymbol{r}-\boldsymbol{r}'|}\boldsymbol{J}(\boldsymbol{r}',\omega)\,\mathrm{d}v'$$

where the wave number $k = k_0\,(\epsilon\mu)^{1/2}$ is assumed to be a positive number, and $k_0 = \omega/c_0$ is the wave number in vacuum. Inserting the current density, we get

$$\boldsymbol{A}(\boldsymbol{r},\omega) = \hat{z}\mu_0\mu q\int_{-\infty}^{\infty}\mathrm{d}x'\int_{-\infty}^{\infty}\mathrm{d}y'\int_{-\infty}^{\infty}\mathrm{d}z'\frac{\mathrm{e}^{\mathrm{i}kR}}{4\pi R}\delta(x')\delta(y')\mathrm{e}^{\mathrm{i}\omega z'/v}$$

where $R = |\boldsymbol{r}-\boldsymbol{r}'| = \sqrt{(x-x')^2+(y-y')^2+(z-z')^2}$. Integration over the transverse spatial variables x' and y' is easy. The result is

$$\boldsymbol{A}(\boldsymbol{r},\omega) = \hat{z}\frac{\mu_0\mu q}{4\pi}\int_{-\infty}^{\infty}\frac{\mathrm{e}^{\mathrm{i}k\left(r_c^2+(z-z')^2\right)^{1/2}+\mathrm{i}\omega z'/v}}{\left(r_c^2+(z-z')^2\right)^{1/2}}\,\mathrm{d}z'$$

[19]We assume the velocity $\boldsymbol{v}$ is independent of time t. This is an approximation and a simplification of the true physical situation.

where $r_c = \sqrt{x^2 + y^2}$ is the distance from the point of observation $\boldsymbol{r}$ to the trajectory of the particle along the z-axis, see Figure 2.4. By a change of variables, we get

$$\boldsymbol{A}(\boldsymbol{r}, \omega) = \hat{\boldsymbol{z}} \frac{\mu_0 \mu q}{4\pi} e^{i\omega z/v} \int_{-\infty}^{\infty} \frac{e^{ik\left(r_c^2 + z'^2\right)^{1/2} + i\omega z'/v}}{\left(r_c^2 + z'^2\right)^{1/2}} \, dz'$$

The remaining integral is a Hankel function of the first kind of order zero. The following integral, which is solved in Problem 2.6, is useful:

$$\int_{-\infty}^{\infty} \frac{e^{ik\left(r_c^2 + z'^2\right)^{1/2} + i\omega z'/v}}{\left(r_c^2 + z'^2\right)^{1/2}} \, dz' = i\pi H_0^{(1)}(k_{r_c} r_c)$$

where

$$\begin{cases} k_{r_c} = \left(k^2 - \dfrac{\omega^2}{v^2}\right)^{1/2} = k_0 \left(\epsilon\mu - \dfrac{c_0^2}{v^2}\right)^{1/2} \\ \operatorname{Im} k_{r_c} \geq 0 \end{cases}$$

The condition for a real k_{r_c} is

$$v \geq \frac{c_0}{\sqrt{\epsilon\mu}} \tag{2.36}$$

provided $\epsilon\mu$ is a real quantity. The square root that occurs in the right-hand side has a branch cut in the complex plane along the positive real axis. The vector potential for the charged particle then becomes

$$\boldsymbol{A}(\boldsymbol{r}, \omega) = i\hat{\boldsymbol{z}} \frac{\mu_0 \mu q}{4} e^{i\omega z/v} H_0^{(1)}(k_{r_c} r_c)$$

The electric and the magnetic fields are then calculated from (2.13) and (2.14) on page 106. We get

$$\begin{aligned} \boldsymbol{E}(\boldsymbol{r}, \omega) &= -\frac{\mu_0 \mu \omega q}{4} \left(\mathbf{I}_3 + \frac{1}{k^2} \nabla\nabla\right) \cdot \hat{\boldsymbol{z}} e^{i\omega z/v} H_0^{(1)}(k_{r_c} r_c) \\ &= -\frac{\mu_0 \mu \omega q}{4} \left(\hat{\boldsymbol{z}} e^{i\omega z/v} H_0^{(1)}(k_{r_c} r_c) + \frac{i\omega}{k^2 v} \nabla \left(e^{i\omega z/v} H_0^{(1)}(k_{r_c} r_c)\right)\right) \\ &= -\frac{q k_{r_c}}{4\epsilon_0 \epsilon \omega} e^{i\omega z/v} \left(\hat{\boldsymbol{z}} k_{r_c} H_0^{(1)}(k_{r_c} r_c) - \hat{\boldsymbol{r}}_c \frac{i\omega}{v} H_1^{(1)}(k_{r_c} r_c)\right) \end{aligned} \tag{2.37}$$

and

$$\begin{aligned} \boldsymbol{H}(\boldsymbol{r}, \omega) &= i\frac{q}{4} \nabla \times \left(\hat{\boldsymbol{z}} e^{i\omega z/v} H_0^{(1)}(k_{r_c} r_c)\right) \\ &= \frac{q}{4i} \hat{\boldsymbol{z}} \times \nabla \left(e^{i\omega z/v} H_0^{(1)}(k_{r_c} r_c)\right) = \hat{\boldsymbol{\phi}} \frac{iqk_{r_c}}{4} e^{i\omega z/v} H_1^{(1)}(k_{r_c} r_c) \end{aligned} \tag{2.38}$$

The Poynting vector, $\langle \boldsymbol{S}(t)\rangle = \mathrm{Re}\,(\boldsymbol{E}\times\boldsymbol{H}^*)\,/2$, see (1.63) on page 57, is of interest for the analysis below. We get

$$\boldsymbol{E}(\boldsymbol{r},\omega)\times\boldsymbol{H}^*(\boldsymbol{r},\omega) = \frac{|k_{r_c}|^2q^2}{16\mathrm{i}\epsilon_0\epsilon\omega}\left(\hat{\boldsymbol{r}}_c k_{r_c} H_0^{(1)}(k_{r_c}r_c)+\hat{\boldsymbol{z}}\frac{\mathrm{i}\omega}{v}H_1^{(1)}(k_{r_c}r_c)\right)\left(H_1^{(1)}(k_{r_c}r_c)\right)^*$$

The Poynting vector becomes

$$\begin{aligned}\langle \boldsymbol{S}(t)\rangle &= \mathrm{Im}\left\{\frac{|k_{r_c}|^2q^2}{32\epsilon_0\epsilon\omega}\left[\hat{\boldsymbol{r}}_c k_{r_c} H_0^{(1)}(k_{r_c}r_c)+\hat{\boldsymbol{z}}\frac{\mathrm{i}\omega}{v}H_1^{(1)}(k_{r_c}r_c)\right]\left(H_1^{(1)}(k_{r_c}r_c)\right)^*\right\}\\ &= \frac{|k_{r_c}|^2q^2}{64\mathrm{i}\epsilon_0\epsilon\omega}\hat{\boldsymbol{r}}_c\left\{k_{r_c}H_0^{(1)}(k_{r_c}r_c)\left(H_1^{(1)}(k_{r_c}r_c)\right)^*\right.\\ &\quad \left.-\left(k_{r_c}H_0^{(1)}(k_{r_c}r_c)\right)^*H_1^{(1)}(k_{r_c}r_c)\right\}+\frac{|k_{r_c}|^2q^2}{32\epsilon_0\epsilon v}\hat{\boldsymbol{z}}\left|H_1^{(1)}(k_{r_c}r_c)\right|^2\end{aligned}$$

We now focus on evaluating the expression in the parentheses in the $\hat{\boldsymbol{r}}_c$ expression, $z = k_{r_c}r_c$

$$zH_0^{(1)}(z)\left(H_1^{(1)}(z)\right)^* - \left(zH_0^{(1)}(z)\right)^*H_1^{(1)}(z)$$

Since the material is lossless, the argument z is either real $z = x$ ($v \geq c_0/\sqrt{\epsilon\mu}$) or purely imaginary $z = \mathrm{i}y$ ($v < c_0/\sqrt{\epsilon\mu}$). First, evaluate the quantity for real $z = x$, x real. We get

$$xH_0^{(1)}(x)\left(H_1^{(1)}(x)\right)^* - \left(xH_0^{(1)}(x)\right)^*H_1^{(1)}(x) = xH_0^{(1)}(x)H_1^{(2)}(x) - xH_0^{(2)}(x)H_1^{(1)}(x)$$

Evaluate this expression with the Wronskian,[20] see (B.6) on page 604

$$H_0^{(1)}(x)H_1^{(2)}(x) - H_0^{(2)}(x)H_1^{(1)}(x) = -H_0^{(1)}(x)H_0^{(2)\prime}(x) + H_0^{(2)}(x)H_0^{(1)\prime}(x) = \frac{4\mathrm{i}}{\pi x}$$

We get

$$zH_0^{(1)}(z)\left(H_1^{(1)}(z)\right)^* - \left(zH_0^{(1)}(z)\right)^*H_1^{(1)}(z) = \frac{4\mathrm{i}}{\pi},\qquad z \text{ real}$$

For a purely imaginary argument, $z = \mathrm{i}y$, y real, we have

$$\begin{aligned}&zH_0^{(1)}(z)\left(H_1^{(1)}(z)\right)^* - \left(zH_0^{(1)}(z)\right)^*H_1^{(1)}(z)\\ &= \mathrm{i}y\left\{H_0^{(1)}(\mathrm{i}y)\left(H_1^{(1)}(\mathrm{i}y)\right)^* + \left(H_0^{(1)}(\mathrm{i}y)\right)^*H_1^{(1)}(\mathrm{i}y)\right\}\\ &= \frac{4y}{\pi^2}\left(-K_0(y)K_1(y) + K_0(y)K_1(y)\right) = 0,\qquad z \text{ purely imaginary}\end{aligned}$$

where $K_0(y)$ and $K_1(y)$ are modified Bessel functions, see Appendix B.2.

---

[20] Józef Maria Hoëne-Wroński (1776–1853), Polish mathematician.

The Poynting vector simplifies to

$$\langle \boldsymbol{S}(t) \rangle = \frac{|k_{r_\mathrm{c}}|^2 q^2}{16\pi r_\mathrm{c} \epsilon_0 \epsilon \omega} \hat{\boldsymbol{r}}_\mathrm{c} + \frac{|k_{r_\mathrm{c}}|^2 q^2}{32\epsilon_0 \epsilon v} \hat{\boldsymbol{z}} \left| H_1^{(1)}(k_{r_\mathrm{c}} r_\mathrm{c}) \right|^2, \qquad v \geq c_0/\sqrt{\epsilon\mu} \tag{2.39}$$

and

$$\langle \boldsymbol{S}(t) \rangle = \frac{|k_{r_\mathrm{c}}|^2 q^2}{32\epsilon_0 \epsilon v} \hat{\boldsymbol{z}} \left| H_1^{(1)}(k_{r_\mathrm{c}} r_\mathrm{c}) \right|^2, \qquad v < c_0/\sqrt{\epsilon\mu} \tag{2.40}$$

Of special interest in this section is the field far away from the trajectory, i.e., r_c large. To this end, we let $k_{r_\mathrm{c}} r_\mathrm{c} \gg 1$ in (2.37) and (2.38), and replace the Hankel function with its asymptotic expression. For large arguments, we have, see (B.5) in Appendix B.1 on page 599 or Reference [4, page 618]

$$\begin{cases} H_0^{(1)}(z) \approx \left(\dfrac{2}{\pi z}\right)^{1/2} \mathrm{e}^{\mathrm{i}z - \mathrm{i}\pi/4} \\ H_1^{(1)}(z) \approx -\mathrm{i} \left(\dfrac{2}{\pi z}\right)^{1/2} \mathrm{e}^{\mathrm{i}z - \mathrm{i}\pi/4} \end{cases} \qquad |z| \gg 1, \qquad -\pi < \arg z < 2\pi$$

The electric and magnetic fields at a large distance from the trajectory (z-axis) are then, see (2.37) and (2.38)

$$\boldsymbol{E}(\boldsymbol{r}, \omega) = \left(\frac{2}{\pi k_{r_\mathrm{c}} r_\mathrm{c}}\right)^{1/2} \boldsymbol{E}_0 \mathrm{e}^{\mathrm{i}(k_{r_\mathrm{c}} r_\mathrm{c} + \omega z/v) - \mathrm{i}\pi/4} \tag{2.41}$$

and

$$\boldsymbol{H}(\boldsymbol{r}, \omega) = \left(\frac{2}{\pi k_{r_\mathrm{c}} r_\mathrm{c}}\right)^{1/2} \boldsymbol{H}_0 \mathrm{e}^{\mathrm{i}(k_{r_\mathrm{c}} r_\mathrm{c} + \omega z/v) - \mathrm{i}\pi/4} \tag{2.42}$$

where

$$\begin{cases} \boldsymbol{E}_0 = \dfrac{q k_{r_\mathrm{c}}}{4\omega\epsilon_0\epsilon} \left(\hat{\boldsymbol{r}}_\mathrm{c} \dfrac{\omega}{v} - \hat{\boldsymbol{z}} k_{r_\mathrm{c}}\right) \\ \boldsymbol{H}_0 = \dfrac{q k_{r_\mathrm{c}}}{4} \hat{\boldsymbol{\phi}} \end{cases}$$

and the power flux density in (2.39) and (2.40) becomes

$$\langle \boldsymbol{S}(t) \rangle = \mathrm{Re}\, \frac{|k_{r_\mathrm{c}}| q^2}{16\pi r_\mathrm{c} \epsilon_0 \epsilon \omega} \left(\hat{\boldsymbol{r}}_\mathrm{c} k_{r_\mathrm{c}} + \hat{\boldsymbol{z}} \frac{\omega}{v}\right) \mathrm{e}^{-2\,\mathrm{Im}\, k_{r_\mathrm{c}} r_\mathrm{c}} \tag{2.43}$$

Note that all neglected terms in the fields (2.41) and (2.42) vanish faster than $r_\mathrm{c}^{-1/2}$, and in (2.43) they vanish faster than r_c^{-1} as $r_\mathrm{c} \to \infty$. The expressions in (2.41) and (2.42) are cylindrical waves, i.e., the amplitude decreases with a factor $r_\mathrm{c}^{-1/2}$ as $r_\mathrm{c} \to \infty$. Notice also that the power flux density is damped exponentially in the $\hat{\boldsymbol{r}}_\mathrm{c}$-direction if k_{r_c} is not real. Therefore, assuming a lossless material, the Čerenkov radiation can only occur if the velocity of the particle is faster than $c_0/\sqrt{\epsilon\mu}$, see (2.36). Moreover, neglecting dispersion, the modulus of the Poynting vector in (2.43) is proportional to the angular frequency ω, which implies that the radiation is more intense the higher the frequency. This is also observed in water, where the Čerenkov radiation has a bluish glow.

The direction of propagation, $\hat{\boldsymbol{k}}$, of the cylindrical wave is parallel to the direction of the Poynting vector $\langle \boldsymbol{S}(t)\rangle$, i.e.,

$$\hat{\boldsymbol{k}} = \frac{\hat{\boldsymbol{r}}_c k_{r_c} + \hat{\boldsymbol{z}}\omega/v}{\left(k_{r_c}^2 + (\omega/v)^2\right)^{1/2}} = \frac{1}{k}\left(\hat{\boldsymbol{r}}_c k_{r_c} + \hat{\boldsymbol{z}}\frac{\omega}{v}\right)$$

and the polarization of the electric field, $\hat{\boldsymbol{p}}_e$, at large distances from the particle is

$$\hat{\boldsymbol{p}}_e = \frac{\hat{\boldsymbol{r}}_c \omega/v - \hat{\boldsymbol{z}} k_{r_c}}{\left(k_{r_c}^2 + (\omega/v)^2\right)^{1/2}} = \frac{1}{k}\left(\hat{\boldsymbol{r}}_c \frac{\omega}{v} - \hat{\boldsymbol{z}} k_{r_c}\right)$$

The polarization of the Čerenkov radiation, $\hat{\boldsymbol{p}}_e$, is perpendicular to the direction of propagation $\hat{\boldsymbol{k}}$, since

$$\hat{\boldsymbol{p}}_e \cdot \hat{\boldsymbol{k}} = \frac{1}{k}\left(\hat{\boldsymbol{r}}_c \frac{\omega}{v} - \hat{\boldsymbol{z}} k_{r_c}\right) \cdot \frac{1}{k}\left(\hat{\boldsymbol{r}}_c k_{r_c} + \hat{\boldsymbol{z}}\frac{\omega}{v}\right) = 0$$

Moreover, the unit vector $\hat{\boldsymbol{p}}_e$ lies in the $\hat{\boldsymbol{k}}$–$\hat{\boldsymbol{z}}$-plane, since

$$\begin{aligned}
\hat{\boldsymbol{p}}_e \cdot (\hat{\boldsymbol{k}} \times \hat{\boldsymbol{z}}) &= \frac{1}{k}\left(\hat{\boldsymbol{r}}_c \frac{\omega}{v} - \hat{\boldsymbol{z}} k_{r_c}\right) \cdot \left[\frac{1}{k}\left(\hat{\boldsymbol{r}}_c k_{r_c} + \hat{\boldsymbol{z}}\frac{\omega}{v}\right) \times \hat{\boldsymbol{z}}\right] \\
&= \frac{k_{r_c}}{k^2}\left(\hat{\boldsymbol{r}}_c \frac{\omega}{v} - \hat{\boldsymbol{z}} k_{r_c}\right) \cdot (\hat{\boldsymbol{r}}_c \times \hat{\boldsymbol{z}}) = 0
\end{aligned}$$

The angle θ between the direction of propagation, $\hat{\boldsymbol{k}}$, and the velocity of the particle, $\boldsymbol{v} = v\hat{\boldsymbol{z}}$, is

$$\cos\theta = \hat{\boldsymbol{k}} \cdot \hat{\boldsymbol{z}} = \frac{\omega}{kv} = \frac{c_0}{v\sqrt{\epsilon\mu}}$$

We notice that the larger the velocity, v, is the larger is the angle θ. Again, the condition (2.36) must be satisfied in order the angle θ to be real, see Figure 2.5.

Thus, we have proved that the experimental observations made by Čerenkov and Vavilov in the 1930s, and which are summarized in the items 1–3 on page 121 in the beginning of this section can be explained and quantified by the macroscopic theory of electromagnetics.

2.5.1 Energy Radiation

Finally, we compute the total radiated electromagnetic energy of the wave under the assumption that the material is lossless. The quantity that is of interest here is the total radiated energy per unit length of the trajectory of the particle. The energy per unit length, $\frac{\mathrm{d}E}{\mathrm{d}z}$, is by definition

$$\frac{\mathrm{d}E}{\mathrm{d}z} = \int_{-\infty}^{\infty} \iint_{S_{r_c}} \boldsymbol{S}(t) \cdot \hat{\boldsymbol{r}}_c \,\mathrm{d}S\,\mathrm{d}t = \int_{-\infty}^{\infty} \iint_{S_{r_c}} (\boldsymbol{E}(t) \times \boldsymbol{H}(t)) \cdot \hat{\boldsymbol{r}}_c \,\mathrm{d}S\,\mathrm{d}t$$

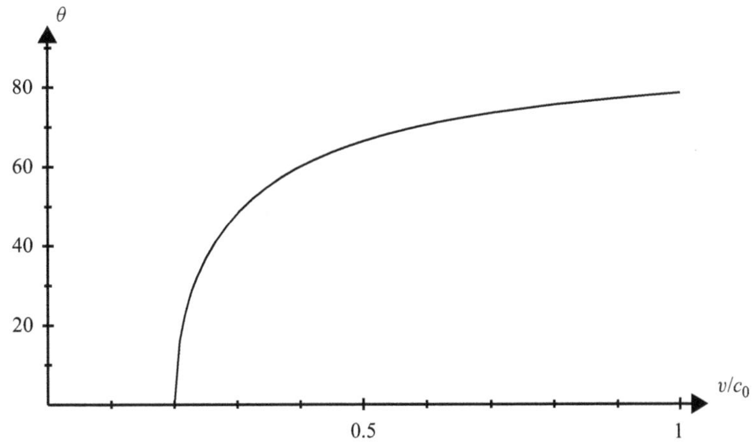

Figure 2.5 The angle θ and its variation as a function of the velocity of the particle v. The parameters of the material ϵ and μ are chosen such that $\epsilon\mu = 25$.

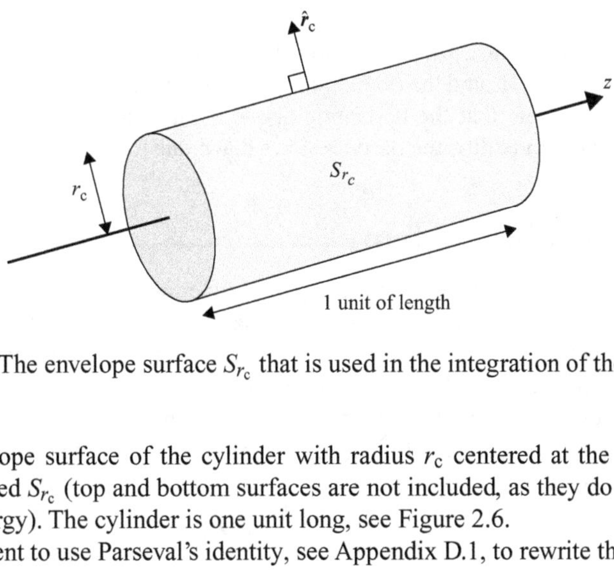

Figure 2.6 The envelope surface S_{r_c} that is used in the integration of the energy flux.

where the envelope surface of the cylinder with radius r_c centered at the trajectory of the particle is denoted S_{r_c} (top and bottom surfaces are not included, as they do not contribute to the radiated energy). The cylinder is one unit long, see Figure 2.6.

It is convenient to use Parseval's identity, see Appendix D.1, to rewrite the energy per unit length expression in terms of the Fourier transformed fields. The result is

$$\frac{\mathrm{d}E}{\mathrm{d}z} = \frac{1}{2\pi}\operatorname{Re}\iint\limits_{S_{r_c}}\int_{-\infty}^{\infty}\left(\boldsymbol{E}(\boldsymbol{r},\omega)\times\boldsymbol{H}^*(\boldsymbol{r},\omega)\right)\cdot\hat{\boldsymbol{r}}_c\,\mathrm{d}\omega\,\mathrm{d}S = \frac{1}{\pi}\iint\limits_{S_{r_c}}\int_{-\infty}^{\infty}\langle\boldsymbol{S}(t)\rangle\cdot\hat{\boldsymbol{r}}_c\,\mathrm{d}\omega\,\mathrm{d}S$$

The Poynting vector is given by (2.39) and (2.40)

$$\langle\boldsymbol{S}(t)\rangle\cdot\hat{\boldsymbol{r}}_c = \frac{|k_{r_c}|^2 q^2}{16\pi r_c\epsilon_0\epsilon\omega}, \qquad v \geq c_0/\sqrt{\epsilon\mu}$$

and zero if $v < c_0/\sqrt{\epsilon\mu}$.

The energy per unit length then is

$$\frac{\mathrm{d}E}{\mathrm{d}z} = \frac{1}{\pi} 2\pi r_{\mathrm{c}} \frac{q^2}{16\pi r_{\mathrm{c}} \epsilon_0} \int\limits_{\substack{-\infty \\ k_{r_{\mathrm{c}}} \text{ real}}}^{\infty} \frac{k_{r_{\mathrm{c}}}^2(\omega)}{\omega \epsilon(\omega)} \,\mathrm{d}\omega = \frac{q^2}{8\pi \epsilon_0} \int\limits_{\substack{-\infty \\ k_{r_{\mathrm{c}}} \text{ real}}}^{\infty} \frac{k_{r_{\mathrm{c}}}^2(\omega)}{\omega \epsilon(\omega)} \,\mathrm{d}\omega$$

The condition for $k_{r_{\mathrm{c}}}$ to be real is given by (2.36). The equation above can therefore be rewritten as

$$\frac{\mathrm{d}E}{\mathrm{d}z} = \frac{q^2 \mu_0}{8\pi} \int\limits_{\substack{-\infty \\ \epsilon(\omega)\mu(\omega) \geq (c_0/v)^2}}^{\infty} \frac{\omega}{\epsilon(\omega)} \left(\epsilon(\omega)\mu(\omega) - \left(\frac{c_0}{v}\right)^2 \right) \mathrm{d}\omega$$

or

$$\frac{\mathrm{d}E}{\mathrm{d}z} = \frac{q^2 \mu_0}{4\pi} \int\limits_{\substack{0 \\ \epsilon(\omega)\mu(\omega) \geq (c_0/v)^2}}^{\infty} \frac{\omega}{\epsilon(\omega)} \left(\epsilon(\omega)\mu(\omega) - \left(\frac{c_0}{v}\right)^2 \right) \mathrm{d}\omega$$

where $\epsilon\mu$ is real. In practice, the integration domain in ω is always finite. For very high frequencies $\epsilon(\omega)\mu(\omega) \to 1$, and the domain of integration is finite.

It is important to note that the derivation above is made under the assumption that the velocity v is constant. In reality, the particle slows down due to radiation.

2.6 Time-Domain Problem

The main objective of this section is to find the Green function, and the solution of the Maxwell equations in the time domain for a homogeneous material. We investigate the solution, given the sources in the entire space, i.e., no boundaries present, with appropriate radiation conditions at infinity. Below, the permittivity ϵ and the permeability μ are assumed real constants. These constants are the optical responses of the material, and should not be mixed with the complex permittivity $\epsilon(\omega)$ and the permeability $\mu(\omega)$ in the frequency setting. More specifically, the Maxwell equations are

$$\begin{cases} \nabla \times \boldsymbol{E}(\boldsymbol{r}, t) = -\mu_0 \mu \dfrac{\partial \boldsymbol{H}(\boldsymbol{r}, t)}{\partial t} \\ \nabla \times \boldsymbol{H}(\boldsymbol{r}, t) = \boldsymbol{J}(\boldsymbol{r}, t) + \epsilon_0 \epsilon \dfrac{\partial \boldsymbol{E}(\boldsymbol{r}, t)}{\partial t} \end{cases}$$

or combined to

$$\begin{cases} \nabla \times (\nabla \times \boldsymbol{E}(\boldsymbol{r}, t)) = -\mu_0 \mu \dfrac{\partial \boldsymbol{J}(\boldsymbol{r}, t)}{\partial t} - \dfrac{1}{c^2} \dfrac{\partial^2 \boldsymbol{E}(\boldsymbol{r}, t)}{\partial t^2} \\ \nabla \times (\nabla \times \boldsymbol{H}(\boldsymbol{r}, t)) = \nabla \times \boldsymbol{J}(\boldsymbol{r}, t) - \dfrac{1}{c^2} \dfrac{\partial^2 \boldsymbol{H}(\boldsymbol{r}, t)}{\partial t^2} \end{cases} \tag{2.44}$$

where the wave speed in the domain is $c = c_0/\sqrt{\epsilon\mu}$. The given current density $\boldsymbol{J}(\boldsymbol{r}, t)$ is assumed to vanish outside a finite volume V, and moreover, it is quiescent before some fixed time τ, i.e., $\boldsymbol{J}(\boldsymbol{r},t)=\boldsymbol{0}$ everywhere in $\mathbb{R}^3$ for $t<\tau$. Moreover, in a source-free region, both the electric and the magnetic fields satisfy the vector-valued wave equation since both fields are divergence free, $\nabla\cdot\boldsymbol{E}=0$ and $\nabla\cdot\boldsymbol{H}=0$. We have

$$\begin{cases}\nabla^2\boldsymbol{E}(\boldsymbol{r},t)-\dfrac{1}{c^2}\dfrac{\partial^2\boldsymbol{E}(\boldsymbol{r},t)}{\partial t^2}=\boldsymbol{0}\\[2mm] \nabla^2\boldsymbol{H}(\boldsymbol{r},t)-\dfrac{1}{c^2}\dfrac{\partial^2\boldsymbol{H}(\boldsymbol{r},t)}{\partial t^2}=\boldsymbol{0}\end{cases}$$

where we used $\nabla\times(\nabla\times\boldsymbol{E}(\boldsymbol{r},t))=\nabla(\nabla\cdot\boldsymbol{E}(\boldsymbol{r},t))-\nabla^2\boldsymbol{E}(\boldsymbol{r},t)$.

The time- and frequency-domain quantities are related by the Fourier transform, see Chapter 1.3. For example, the electric field $\boldsymbol{E}(\boldsymbol{r},t)$ and its Fourier transform are related as

$$\begin{cases}\boldsymbol{E}(\boldsymbol{r},\omega)=\displaystyle\int_{-\infty}^{\infty}\boldsymbol{E}(\boldsymbol{r},t)\,\mathrm{e}^{\mathrm{i}\omega t}\,\mathrm{d}t\\[2mm] \boldsymbol{E}(\boldsymbol{r},t)=\dfrac{1}{2\pi}\displaystyle\int_{-\infty}^{\infty}\boldsymbol{E}(\boldsymbol{r},\omega)\,\mathrm{e}^{-\mathrm{i}\omega t}\,\mathrm{d}\omega\end{cases}$$

and similarly for all other fields.

2.6.1 Potentials and Gauge Transformations

The solution of the time-domain problem shows large similarities to the solution in the frequency domain. We start by identifying a vector potential $\boldsymbol{A}(\boldsymbol{r},t)$, see Section 2.1.1

$$\boldsymbol{B}(\boldsymbol{r},t)=\nabla\times\boldsymbol{A}(\boldsymbol{r},t) \tag{2.45}$$

The existence of a vector potential is guaranteed by $\nabla\cdot\boldsymbol{B}(\boldsymbol{r},t)=0$. The scalar potential $\phi(\boldsymbol{r},t)$ is then defined by the use of the Faraday law

$$\nabla\times\boldsymbol{E}(\boldsymbol{r},t)=-\frac{\partial\boldsymbol{B}(\boldsymbol{r},t)}{\partial t}=-\frac{\partial}{\partial t}\nabla\times\boldsymbol{A}(\boldsymbol{r},t)$$

which we write as

$$\nabla\times\left(\boldsymbol{E}(\boldsymbol{r},t)+\frac{\partial\boldsymbol{A}(\boldsymbol{r},t)}{\partial t}\right)=\boldsymbol{0}$$

which implies that there is a function $\phi(\boldsymbol{r},t)$ such that

$$\boldsymbol{E}(\boldsymbol{r},t)=-\frac{\partial\boldsymbol{A}(\boldsymbol{r},t)}{\partial t}-\nabla\phi(\boldsymbol{r},t) \tag{2.46}$$

So far the construction of vector and scalar potentials has been identical to the frequency-domain analysis in Section 2.1.1. We proceed in a parallel manner and let the potentials satisfy the Lorenz condition

$$\nabla\cdot\boldsymbol{A}(\boldsymbol{r},t)+\frac{1}{c^2}\frac{\partial\phi(\boldsymbol{r},t)}{\partial t}=0 \tag{2.47}$$

Using the same arguments as in Section 2.1.1, we can show that it always possible to meet this condition. The partial differential equation for the vector potential is simple. Following the analysis in Section 2.1.1, we get

$$\nabla^2 \boldsymbol{A}(\boldsymbol{r},t) - \frac{1}{c^2}\frac{\partial^2 \boldsymbol{A}(\boldsymbol{r},t)}{\partial t^2} = -\mu_0\mu \boldsymbol{J}(\boldsymbol{r},t) \tag{2.48}$$

The scalar potential $\phi(\boldsymbol{r},t)$ is related to the charge density $\rho(\boldsymbol{r},t)$, which is related to the current density $\boldsymbol{J}(\boldsymbol{r},t)$ by the continuity law of charges, see (1.4) on page 3

$$\nabla \cdot \boldsymbol{J}(\boldsymbol{r},t) + \frac{\partial \rho(\boldsymbol{r},t)}{\partial t} = 0$$

which was used to derive Gauss' law (1.6), i.e.,

$$\nabla \cdot \boldsymbol{E}(\boldsymbol{r},t) = \frac{1}{\epsilon_0\epsilon}\nabla \cdot \boldsymbol{D}(\boldsymbol{r},t) = \frac{\rho(\boldsymbol{r},t)}{\epsilon_0\epsilon}$$

and using (2.46)

$$\frac{\rho(\boldsymbol{r},t)}{\epsilon_0\epsilon} = \nabla \cdot \boldsymbol{E}(\boldsymbol{r},t) = -\frac{\partial \nabla \cdot \boldsymbol{A}(\boldsymbol{r},t)}{\partial t} - \nabla^2\phi(\boldsymbol{r},t)$$

The Lorenz condition, (2.47), implies

$$\nabla^2 \phi(\boldsymbol{r},t) - \frac{1}{c^2}\frac{\partial^2 \phi(\boldsymbol{r},t)}{\partial t^2} = -\frac{\rho(\boldsymbol{r},t)}{\epsilon_0\epsilon} \tag{2.49}$$

2.6.2 Canonical Problem in Free Space

The pertinent canonical problem in the time domain is similar to the frequency-domain problem $(\boldsymbol{r},\boldsymbol{r}' \in \mathbb{R}^3, t,t' \in \mathbb{R})$

$$\nabla^2 g(|\boldsymbol{r}-\boldsymbol{r}'|,t-t') - \frac{1}{c^2}\frac{\partial^2 g(|\boldsymbol{r}-\boldsymbol{r}'|,t-t')}{\partial t^2} = -\delta(\boldsymbol{r}-\boldsymbol{r}')\delta(t-t') \tag{2.50}$$

We solve this canonical problem by the use of the corresponding result for the time-harmonic fields. A comparison with the Fourier transform of our canonical problem shows that the inverse Fourier transform is (note that this integral is not convergent, but has to be interpreted as a distribution)

$$g(|\boldsymbol{r}-\boldsymbol{r}'|,t-t') = \frac{1}{2\pi}\int_{-\infty}^{\infty} \frac{\mathrm{e}^{\mathrm{i}\omega|\boldsymbol{r}-\boldsymbol{r}'|/c}}{4\pi|\boldsymbol{r}-\boldsymbol{r}'|}\mathrm{e}^{\mathrm{i}\omega t'}\,\mathrm{e}^{-\mathrm{i}\omega t}\,\mathrm{d}\omega$$

which we easily evaluate as

$$g(|\boldsymbol{r}-\boldsymbol{r}'|,t-t') = \frac{\delta(t'-t+|\boldsymbol{r}-\boldsymbol{r}'|/c)}{4\pi|\boldsymbol{r}-\boldsymbol{r}'|} \tag{2.51}$$

This Green function is called the retarded Green function and expresses the causality principle—no effect before its cause—that is fundamental in the modeling of all physical processes, see Section 2.6.3. Notice that the Green function is only a function of the amplitude of the difference of the spatial variables $\boldsymbol{r}$ and $\boldsymbol{r}'$, and the difference between the times t and t'. The reason for this is the invariance of the wave equation, (2.50), under spatial and temporal translations.

With the solution, $g(|\boldsymbol{r}-\boldsymbol{r}'|, t-t')$, we can find the solution to (2.48). We have

$$\boldsymbol{A}(\boldsymbol{r}, t) = \mu_0\mu \int_{-\infty}^{\infty} \iiint_V g(|\boldsymbol{r}-\boldsymbol{r}'|, t-t')\boldsymbol{J}(\boldsymbol{r}', t')\,\mathrm{d}v'\,\mathrm{d}t' \tag{2.52}$$

and

$$\phi(\boldsymbol{r}, t) = \frac{1}{\epsilon_0\epsilon} \int_{-\infty}^{\infty} \iiint_V g(|\boldsymbol{r}-\boldsymbol{r}'|, t-t')\rho(\boldsymbol{r}', t')\,\mathrm{d}v'\,\mathrm{d}t' \tag{2.53}$$

which solves (2.49). Formally, we see that (2.52) is a solution to (2.48) by inserting the solution and interchanging the order of space and time differentiation and the corresponding integration in space and time. We formally have

$$\begin{aligned}
&\nabla^2 \boldsymbol{A}(\boldsymbol{r}, t) - \frac{1}{c^2}\frac{\partial^2 \boldsymbol{A}(\boldsymbol{r}, t)}{\partial t^2} \\
&= \mu_0\mu \int_{-\infty}^{\infty} \iiint_V \left(\nabla^2 g(|\boldsymbol{r}-\boldsymbol{r}'|, t-t') - \frac{1}{c^2}\frac{\partial^2 g(|\boldsymbol{r}-\boldsymbol{r}'|, t-t')}{\partial t^2}\right)\boldsymbol{J}(\boldsymbol{r}')\,\mathrm{d}v'\,\mathrm{d}t' \\
&= -\mu_0\mu \int_{-\infty}^{\infty} \iiint_V \delta(\boldsymbol{r}-\boldsymbol{r}')\delta(t-t')\boldsymbol{J}(\boldsymbol{r}', t')\,\mathrm{d}v'\,\mathrm{d}t' = -\mu_0\mu\boldsymbol{J}(\boldsymbol{r}, t)
\end{aligned}$$

by the use of the definition of the delta distributions $\delta(\boldsymbol{r}-\boldsymbol{r}')$ and $\delta(t-t')$. In an analogous way, we find that (2.53) is a solution to (2.49).

The integration in time in (2.52) and (2.53) leads to the final form of the solution to our problems

$$\boldsymbol{A}(\boldsymbol{r}, t) = \mu_0\mu \iiint_V \frac{\boldsymbol{J}(\boldsymbol{r}', t-|\boldsymbol{r}-\boldsymbol{r}'|/c)}{4\pi|\boldsymbol{r}-\boldsymbol{r}'|}\,\mathrm{d}v' = \mu_0\mu \iiint_V \frac{\boldsymbol{J}(\boldsymbol{r}', t_{\mathrm{ret}})}{4\pi|\boldsymbol{r}-\boldsymbol{r}'|}\,\mathrm{d}v'$$

and

$$\phi(\boldsymbol{r}, t) = \frac{1}{\epsilon_0\epsilon} \iiint_V \frac{\rho(\boldsymbol{r}', t-|\boldsymbol{r}-\boldsymbol{r}'|/c)}{4\pi|\boldsymbol{r}-\boldsymbol{r}'|}\,\mathrm{d}v' = \frac{1}{\epsilon_0\epsilon} \iiint_V \frac{\rho(\boldsymbol{r}', t_{\mathrm{ret}})}{4\pi|\boldsymbol{r}-\boldsymbol{r}'|}\,\mathrm{d}v'$$

The time variable in the integrand is evaluated at the retarded time

$$t_{\mathrm{ret}} = t - |\boldsymbol{r}-\boldsymbol{r}'|/c$$

This result is interpreted physically as a time delay, $|\boldsymbol{r}-\boldsymbol{r}'|/c$, due to the finite wave speed, c. Indeed, this time delay is exactly the travel time of a disturbance from the source point $\boldsymbol{r}'$ to the field point $\boldsymbol{r}$ at a speed c.

The electric and magnetic fields $\boldsymbol{E}(\boldsymbol{r}, t)$ and $\boldsymbol{H}(\boldsymbol{r}, t)$ can now be written down in terms of the sources, $\boldsymbol{J}(\boldsymbol{r}, t)$ and $\rho(\boldsymbol{r}, t)$. The electric field is, see (2.46)

$$\boldsymbol{E}(\boldsymbol{r}, t) = -\frac{\partial}{\partial t}\mu_0\mu \iiint_V \frac{\boldsymbol{J}(\boldsymbol{r}', t_{\mathrm{ret}})}{4\pi|\boldsymbol{r}-\boldsymbol{r}'|}\,\mathrm{d}v' - \nabla\frac{1}{\epsilon_0\epsilon} \iiint_V \frac{\rho(\boldsymbol{r}', t_{\mathrm{ret}})}{4\pi|\boldsymbol{r}-\boldsymbol{r}'|}\,\mathrm{d}v'$$

and the magnetic field is, see (2.45)

$$
\boldsymbol{H}(\boldsymbol{r},t) = \nabla \times \iiint_V \frac{\boldsymbol{J}(\boldsymbol{r}',t_{\text{ret}})}{4\pi|\boldsymbol{r}-\boldsymbol{r}'|}\, \mathrm{d}v'
$$

We perform the derivation under the integral sign, and we get

$$
\begin{aligned}
\boldsymbol{E}(\boldsymbol{r},t) = &-\mu_0\mu \iiint_V \frac{\partial_t \boldsymbol{J}(\boldsymbol{r}',t_{\text{ret}})}{4\pi|\boldsymbol{r}-\boldsymbol{r}'|}\, \mathrm{d}v' \\
&+ \frac{1}{\epsilon_0\epsilon} \iiint_V \left(\frac{\rho(\boldsymbol{r}',t_{\text{ret}})(\boldsymbol{r}-\boldsymbol{r}')}{4\pi|\boldsymbol{r}-\boldsymbol{r}'|^3} + \frac{\partial_t \rho(\boldsymbol{r}',t_{\text{ret}})(\boldsymbol{r}-\boldsymbol{r}')}{4\pi c|\boldsymbol{r}-\boldsymbol{r}'|^2} \right) \mathrm{d}v'
\end{aligned}
\tag{2.54}
$$

and

$$
\boldsymbol{H}(\boldsymbol{r},t) = -\iiint_V \left(\frac{(\boldsymbol{r}-\boldsymbol{r}')\times \boldsymbol{J}(\boldsymbol{r}',t_{\text{ret}})}{4\pi|\boldsymbol{r}-\boldsymbol{r}'|^3} + \frac{(\boldsymbol{r}-\boldsymbol{r}')\times \partial_t \boldsymbol{J}(\boldsymbol{r}',t_{\text{ret}})}{4\pi c|\boldsymbol{r}-\boldsymbol{r}'|^2} \right) \mathrm{d}v'
\tag{2.55}
$$

In fact

$$
\nabla \frac{1}{|\boldsymbol{r}-\boldsymbol{r}'|} = -\frac{\boldsymbol{r}-\boldsymbol{r}'}{|\boldsymbol{r}-\boldsymbol{r}'|^3}, \qquad \nabla|\boldsymbol{r}-\boldsymbol{r}'| = \frac{\boldsymbol{r}-\boldsymbol{r}'}{|\boldsymbol{r}-\boldsymbol{r}'|}
$$

and therefore

$$
\nabla \frac{\rho(\boldsymbol{r}',t_{\text{ret}})}{4\pi|\boldsymbol{r}-\boldsymbol{r}'|} = -\frac{\rho(\boldsymbol{r}',t_{\text{ret}})(\boldsymbol{r}-\boldsymbol{r}')}{4\pi|\boldsymbol{r}-\boldsymbol{r}'|^3} - \frac{\partial_t \rho(\boldsymbol{r}',t_{\text{ret}})(\boldsymbol{r}-\boldsymbol{r}')}{4\pi c|\boldsymbol{r}-\boldsymbol{r}'|^2}
$$

and

$$
\nabla \times \frac{\boldsymbol{J}(\boldsymbol{r}',t_{\text{ret}})}{4\pi|\boldsymbol{r}-\boldsymbol{r}'|} = -\frac{(\boldsymbol{r}-\boldsymbol{r}')\times \boldsymbol{J}(\boldsymbol{r}',t_{\text{ret}})}{4\pi|\boldsymbol{r}-\boldsymbol{r}'|^3} - \frac{(\boldsymbol{r}-\boldsymbol{r}')\times \partial_t \boldsymbol{J}(\boldsymbol{r}',t_{\text{ret}})}{4\pi c|\boldsymbol{r}-\boldsymbol{r}'|^2}
$$

Equations (2.54) and (2.55) are named the Jefimenko equations[21] [103, page 516].

2.6.3 Causality

Causality, which leads to propagation of disturbances and domain of dependence, is one of the fundamental features of electromagnetic wave propagation. This is a property that follows directly from the Maxwell equations, since it is a system of differential equation of hyperbolic type. Causality is essential for the understanding of propagation of waves in space, and it is also instrumental in the derivation of the sum rules in scattering below, see Section 6.4.

The first result of this section is to prove an inequality based on the energy method. This inequality not only quantifies causality, but it also proves the uniqueness of the solution of the wave equation. An alternative treatment of the problem is to formulate the problem as a symmetric hyperbolic system and to use the result for such a system. The interested reader is referred to References 65, 105, 131, 302 and to Problem 2.10.

---

[21] Oleg Dmitrovich Jefimenko (1922–2009), American physicist.

2.6.3.1 Energy Inequality

Suppose the vector field $\boldsymbol{F}(\boldsymbol{r},t)$ satisfies the vector-valued wave equation (the vector field $\boldsymbol{F}(\boldsymbol{r},t)$ can be e.g., the electric field or the magnetic field in a source-free region)

$$\nabla^2 \boldsymbol{F}(\boldsymbol{r},t) - \frac{1}{c^2}\frac{\partial^2 \boldsymbol{F}(\boldsymbol{r},t)}{\partial t^2} = \boldsymbol{0} \tag{2.56}$$

in a domain V (finite or infinite) for all time $t \geq 0$. The wave speed $c > 0$ is assumed to be constant in V.

We start by defining a few useful domains in space and space–time. For a fixed position $\boldsymbol{r}_0 \in V$, choose the time $t_0 \geq 0$ such that the ball $B(\boldsymbol{r}_0, ct_0)$ defined by

$$B(\boldsymbol{r}_0, ct_0) = \{\boldsymbol{r} : |\boldsymbol{r} - \boldsymbol{r}_0| \leq ct_0\}$$

is contained in V, i.e., $B(\boldsymbol{r}_0, ct_0) \subseteq V$. The backward light cone C_- or characteristic cone is defined as

$$C_- = \{(\boldsymbol{r},t) : 0 \leq t \leq t_0, |\boldsymbol{r} - \boldsymbol{r}_0| = c(t_0 - t)\}$$

The cone and the ball $B(\boldsymbol{r}_0, ct_0)$ are illustrated in a two-dimensional space and time setting in Figure 2.7.

We test one Cartesian component of the field $\boldsymbol{F}(\boldsymbol{r},t)$ at a time, and to simplify the notation, we denote $u(\boldsymbol{r},\tau) = F_i(\boldsymbol{r},t)$, where $i = 1,2,3$, and $\tau = ct$. The τ coordinate is a simple transform of the time variable t, and all coordinates now have the same units, i.e., length. The field $u(\boldsymbol{r},\tau)$ satisfies the scalar wave equation

$$\nabla^2 u(\boldsymbol{r},\tau) - \frac{\partial^2 u(\boldsymbol{r},\tau)}{\partial \tau^2} = 0$$

in a domain V for all time $\tau \geq 0$.

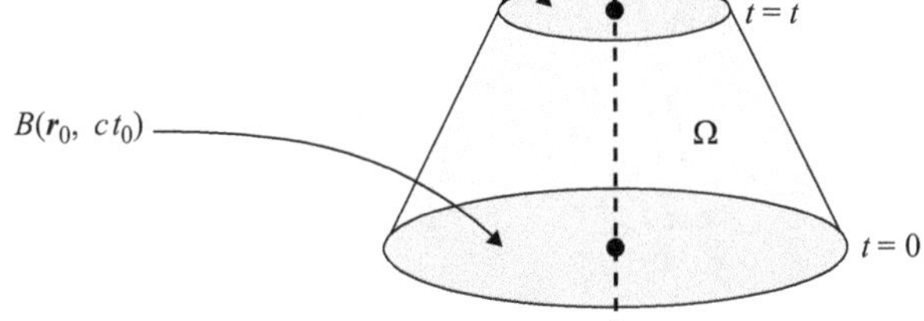

Figure 2.7 The backward light cone C_- and the ball $B(\boldsymbol{r}_0, ct_0)$ in a two-dimensional space and time setting.

Now use the identity, see Problem 2.8

$$0 = 2\frac{\partial u}{\partial \tau}\left(\nabla^2 u - \frac{\partial^2 u}{\partial \tau^2}\right) = \nabla \cdot \left(2\frac{\partial u}{\partial \tau}\nabla u\right) - \frac{\partial}{\partial \tau}\left(|\nabla u|^2 + \left(\frac{\partial u}{\partial \tau}\right)^2\right)$$

The left-hand equality is due to the fact that u satisfies the wave equation. This identity is integrated in both space and time over the part in space–time that lies inside the light cone and between the volumes $B(\boldsymbol{r}_0, \tau_0 - \tau)$ and $B(\boldsymbol{r}_0, \tau_0)$, where $\tau_0 = ct_0$. We denote this space–time domain by Ω. Integration gives

$$0 = \int_{\Omega} \left\{\nabla \cdot \left(2\frac{\partial u}{\partial \tau}\nabla u\right) - \frac{\partial}{\partial \tau}\left(|\nabla u|^2 + \left(\frac{\partial u}{\partial \tau}\right)^2\right)\right\} \, dv \, d\tau$$

Notice that the integrand in the four-dimensional space–time is the divergence of the space–time field

$$\left(2\frac{\partial u}{\partial \tau}\nabla u, -\left(|\nabla u|^2 + \left(\frac{\partial u}{\partial \tau}\right)^2\right)\right)$$

Gauss' theorem in four dimensions (space–time) implies that the four-dimensional integral can be simplified. The result is

$$0 = \int_{\partial\Omega} \left\{2\frac{\partial u}{\partial \tau}\boldsymbol{n} \cdot \nabla u - n_\tau \left(|\nabla u|^2 + \left(\frac{\partial u}{\partial \tau}\right)^2\right)\right\} \, d\sigma \tag{2.57}$$

where $\partial\Omega$ is the three-dimensional boundary of the four-dimensional domain Ω, and where $(\boldsymbol{n}, n_\tau)$ is the outward unit normal[22] vector to Ω and $d\sigma$ is the surface element on $\partial\Omega$. We evaluate this integral identity in three different parts.

First, evaluate the integral over the top and bottom domains, i.e., $B(\boldsymbol{r}_0, \tau_0 - \tau)$ and $B(\boldsymbol{r}_0, \tau_0)$, where time is $\tau = \tau$ and $\tau = 0$, respectively, and the outward unit normal vectors are $(\boldsymbol{0}, 1)$ and $(\boldsymbol{0}, -1)$, respectively. The contribution from the top domain is

$$-\iiint_{B(\boldsymbol{r}_0, \tau_0 - \tau)} \left(|\nabla u|^2 + \left(\frac{\partial u}{\partial \tau}\right)^2\right)\Bigg|_{\tau=\tau} \, dv$$

and from the bottom domain it is

$$\iiint_{B(\boldsymbol{r}_0, \tau_0)} \left(|\nabla u|^2 + \left(\frac{\partial u}{\partial \tau}\right)^2\right)\Bigg|_{\tau=0} \, dv$$

[22]This means that $|\boldsymbol{n}|^2 + n_\tau^2 = 1$.

On the third remaining part of $\partial\Omega$, the part of the light cone between $\tau = 0$ and $\tau = \tau$, which we denote by C_τ, the outward unit normal vector on C_τ is, see Problem 2.9

$$(\boldsymbol{n}, n_\tau) = \left(\frac{\boldsymbol{r}_0 - \boldsymbol{r}}{\sqrt{2}(\tau_0 - \tau)}, \frac{1}{\sqrt{2}} \right)$$

We also notice that $|\boldsymbol{n}|^2 = 1 - n_\tau^2 = 1/2 = n_\tau^2$. The integral over the part of the light cone C_τ becomes

$$\begin{aligned} &\iiint_{C_\tau} \left\{ 2\frac{\partial u}{\partial \tau}\boldsymbol{n} \cdot \nabla u - n_\tau \left(|\nabla u|^2 + \left(\frac{\partial u}{\partial \tau} \right)^2 \right) \right\} \, d\sigma \\ &= \sqrt{2} \iiint_{C_\tau} \left\{ 2n_\tau \frac{\partial u}{\partial \tau}\boldsymbol{n} \cdot \nabla u - n_\tau^2 |\nabla u|^2 - |\boldsymbol{n}|^2 \left(\frac{\partial u}{\partial \tau} \right)^2 \right\} \, d\sigma \end{aligned}$$

where we used the properties of the unit normal vector on C_τ. The integrand is identified as a square of a vector, viz.

$$\begin{aligned} \left| n_\tau \nabla u - \frac{\partial u}{\partial \tau}\boldsymbol{n} \right|^2 &= \left(n_\tau \nabla u - \frac{\partial u}{\partial \tau}\boldsymbol{n} \right) \cdot \left(n_\tau \nabla u - \frac{\partial u}{\partial \tau}\boldsymbol{n} \right) \\ &= n_\tau^2 |\nabla u|^2 + |\boldsymbol{n}|^2 \left(\frac{\partial u}{\partial \tau} \right)^2 - 2n_\tau \frac{\partial u}{\partial \tau}\boldsymbol{n} \cdot \nabla u \end{aligned}$$

and the integral on C_τ is

$$-\sqrt{2} \iiint_{C_\tau} \left| n_\tau \nabla u - \frac{\partial u}{\partial \tau}\boldsymbol{n} \right|^2 \, d\sigma$$

We now collect all three parts of the integral identity in (2.57). We get

$$\begin{aligned} 0 = &- \iiint_{B(\boldsymbol{r}_0, \tau_0 - \tau)} \left(|\nabla u|^2 + \left(\frac{\partial u}{\partial \tau} \right)^2 \right) \Bigg|_{\tau=\tau} \, dv \\ &+ \iiint_{B(\boldsymbol{r}_0, \tau_0)} \left(|\nabla u|^2 + \left(\frac{\partial u}{\partial \tau} \right)^2 \right) \Bigg|_{\tau=0} \, dv - \sqrt{2} \iiint_{C_\tau} \left| n_\tau \nabla u - \frac{\partial u}{\partial \tau}\boldsymbol{n} \right|^2 \, d\sigma \end{aligned}$$

which implies the inequality

$$\iiint_{B(\boldsymbol{r}_0, \tau_0 - \tau)} \left(|\nabla u|^2 + \left(\frac{\partial u}{\partial \tau} \right)^2 \right) \Bigg|_{\tau=\tau} \, dv \leq \iiint_{B(\boldsymbol{r}_0, \tau_0)} \left(|\nabla u|^2 + \left(\frac{\partial u}{\partial \tau} \right)^2 \right) \Bigg|_{\tau=0} \, dv$$

In terms of the components of the vector field $\boldsymbol{F}(\boldsymbol{r},t)$ in (2.56) and the original time coordinate t, this result is

$$\iiint\limits_{B(\boldsymbol{r}_0,c(t_0-t))} \left(|\nabla F_i|^2 + \frac{1}{c^2}\left(\frac{\partial F_i}{\partial t}\right)^2 \right)\Bigg|_{t=t} dv \leq \iiint\limits_{B(\boldsymbol{r}_0,ct_0)} \left(|\nabla F_i|^2 + \frac{1}{c^2}\left(\frac{\partial F_i}{\partial t}\right)^2 \right)\Bigg|_{t=0} dv \tag{2.58}$$

This is the energy integral of the wave equation, since the integrand is related to the energy density of the wave.

2.6.3.2 Uniqueness of the Initial Value Problem

The energy integral in (2.58) proves uniqueness of the initial value problem to the wave equation. Specifically, let $F_i(\boldsymbol{r},0) = 0$ and $\partial_t F_i(\boldsymbol{r},0) = 0$, for all $\boldsymbol{r} \in B(\boldsymbol{r}_0, ct_0)$, where the ball $B(\boldsymbol{r}_0, ct_0)$ is defined above. Then, due to (2.58)

$$\iiint\limits_{B(\boldsymbol{r}_0,c(t_0-t))} \left(|\nabla F_i|^2 + \frac{1}{c^2}\left(\frac{\partial F_i}{\partial t}\right)^2 \right)\Bigg|_{t=t} dv \leq 0$$

Since all terms in the integrand on the left-hand side are non-negative, we conclude that they are both zero. More precisely, we have

$$\begin{cases} \nabla F_i(\boldsymbol{r},t) = \boldsymbol{0} \\ \dfrac{\partial F_i(\boldsymbol{r},t)}{\partial t} = 0 \end{cases} \quad \text{for all } \boldsymbol{r} \in B(\boldsymbol{r}_0, c(t_0 - t)), \quad 0 \leq t \leq t_0$$

This implies that the function $F_i(\boldsymbol{r},t)$ is a constant—both in space and time—for all $(\boldsymbol{r},t)$ in the backward light cone, see Figure 2.7, and due to the initial condition this constant is zero.

Uniqueness of the initial value problem then follows in light cone, C_-, since, assume we have two different solutions, e.g., $F_i(\boldsymbol{r},t)$ and $F_i'(\boldsymbol{r},t)$, to the problem with the same initial conditions, say $F_i(\boldsymbol{r},0) = F_i'(\boldsymbol{r},0) = f(\boldsymbol{r})$ and $\partial_t F_i(\boldsymbol{r},0) = \partial_t F_i'(\boldsymbol{r},0) = g(\boldsymbol{r})$, for all $\boldsymbol{r} \in B(\boldsymbol{r}_0, ct_0)$, then the difference $F_i(\boldsymbol{r},t) - F_i'(\boldsymbol{r},t)$ satisfies homogeneous initial conditions. We have just proved that the solution to this problem is identically zero, and therefore, the solution to the initial value problem is unique, i.e., $F_i(\boldsymbol{r},t) = F_i'(\boldsymbol{r},t)$.

2.6.3.3 Domain of Dependence and Range of Influence

Another way of expressing the result above is to say that the field $F_i(\boldsymbol{r},t)$ at the point $(\boldsymbol{r}_0, t_0)$ in space–time is uniquely determined by the initial data, $f(\boldsymbol{r})$ and $g(\boldsymbol{r})$, in the ball $B(\boldsymbol{r}_0, ct_0)$ at time $t = 0$, see Figure 2.7. Initial data at any point outside $B(\boldsymbol{r}_0, ct_0)$ at time $t = 0$ has no effect on the field value $\boldsymbol{F}(\boldsymbol{r}_0, t_0)$. The domain $B(\boldsymbol{r}_0, ct_0)$ at time $t = 0$ is called the domain of dependence at the point $(\boldsymbol{r}_0, t_0)$.

We now examine the reverse situation, and ask ourselves what points in space–time, $t > 0$, that are influenced by the initial conditions at a point $\boldsymbol{r}_1$ at time $t = 0$. To study this situation, we study the forward light cone with apex at $(\boldsymbol{r}_1, 0)$, C_+, see Figure 2.8. The definition of the forward cone is

$$C_+ = \{(\boldsymbol{r},t) : t \geq 0, |\boldsymbol{r} - \boldsymbol{r}_0| = ct\}$$

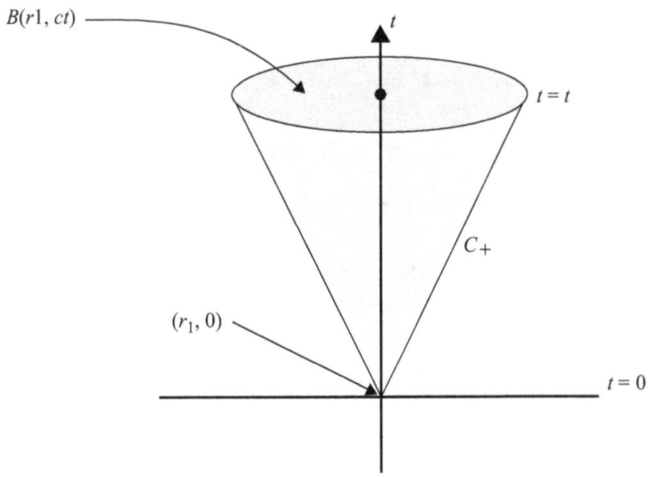

Figure 2.8 The forward light cone C_0+ in a two-dimensional space–time setting.

An immediate consequence of the domain of dependence is that any point outside the forward light cone cannot be influenced by the initial values at $(\boldsymbol{r}_1, 0)$—the backward light cone from any point outside C_+ does not include the point $(\boldsymbol{r}_1, 0)$. The forward light cone C_+ is called the range of influence to the point $(\boldsymbol{r}_1, 0)$. The forward light cone C_+ illustrates the concept of causality—the cause proceeds the effect, which is a consequence of the former. The causality property has been used already in the constitutive relations in Section 1.2.1, and it is instrumental in Section 6.4.

Problems for Chapter 2

Section 2.1 The Green Functions in Isotropic Media

2.1 The current density $\boldsymbol{J}(\boldsymbol{r}, \omega)$ is located inside a ball of radius a. The current density has only a radial component and the components depend only on the radial distance r, i.e.,

$$\boldsymbol{J}(\boldsymbol{r}, \omega) = \begin{cases} \hat{\boldsymbol{r}} f(r, \omega) & r \leq a \\ \boldsymbol{0} & r > a \end{cases}$$

Determine the electric and magnetic fields outside the source region, i.e., for $r > a$. The sources are assumed to be located in vacuum.

Note: A current density, as the one described in the problem, can be generated in a sudden separation of charges, e.g., in a nuclear explosion.

***2.2** The current density for an electric dipole with the strength p oriented along the $\hat{z}$-axis is

$$\boldsymbol{J}(\boldsymbol{r}) = -\mathrm{i}\omega p \hat{\boldsymbol{z}} \delta(\boldsymbol{r})$$

Determine the electric and the magnetic fields outside the dipole, i.e., in the region $r > 0$. Moreover, determine the power flux density (the Poynting vector) in the domain outside the dipole and the total power P that is propagating through a spherical surface centered at the dipole. The domain outside the dipole is assumed to be vacuum ($\epsilon = \mu = 1$).

Section 2.2 The Green Dyadics in Isotropic Media

2.3 Prove (2.29), i.e., show that the contribution from a spherical exclusion volume is $\phi(\boldsymbol{r}')\delta_{ij}/3$.

***2.4** Prove (2.30), i.e., show that the contribution from an excluding volume between two concentric spheres $r = r' - \epsilon$ and $r = r' + \epsilon$ is $\phi(\boldsymbol{r}')(\hat{\boldsymbol{r}}' \cdot \hat{\boldsymbol{e}}_i)(\hat{\boldsymbol{r}}' \cdot \hat{\boldsymbol{e}}_j)$.

Section 2.5 Čerenkov Radiation

2.5 How fast does a particle have to move in a homogeneous Lorentz material in order to give a Čerenkov radiation in the visible range of the spectrum $\lambda > \lambda(\text{violet}) \approx 0.4\ \mu\text{m}$ (wavelengths in vacuum)? The material is assumed to be lossless and non-magnetic $\mu = 1$, and $\omega_0 = \omega_p = 10^{16}$ rad/s.

2.6 Verify the integral

$$\int_{-\infty}^{\infty} \frac{1}{\sqrt{1+x^2}} \mathrm{e}^{\mathrm{i}a\sqrt{1+x^2}+\mathrm{i}bx}\, \mathrm{d}x = \mathrm{i}\pi H_0^{(1)}((a^2 - b^2)^{1/2})$$

where a and b are complex numbers with non-negative imaginary parts.

2.7 Show that the total radiated energy per unit length, $\frac{\mathrm{d}E}{\mathrm{d}z}$, of the Čerenkov radiation from a particle with charge q moving at a speed v in a lossless Lorentz material ($\mu = 1$) is

$$\frac{\mathrm{d}E}{\mathrm{d}z} = -\frac{q^2\omega_p^2\mu_0}{8\pi}\left\{\left(\frac{c_0}{v}\right)^2 \ln\left(1 - \left(\frac{v}{c_0}\right)^2\right) + 1\right\}$$

Show that for a slow particle, we have

$$\frac{\mathrm{d}E}{\mathrm{d}z} = \frac{q^2\omega_p^2\mu_0}{16\pi}\left(\frac{v}{c_0}\right)^2$$

Hint: Use the integral

$$\int \frac{x^2 + a}{x^2 + b} x\, \mathrm{d}x = \frac{1}{2}x^2 + \frac{1}{2}(a - b)\ln(x^2 + b)$$

Section 2.6 Time-Domain Problem

2.8 Prove the identity

$$\frac{2}{c}\frac{\partial u}{\partial t}\left(\nabla^2 u - \frac{1}{c^2}\frac{\partial^2 u}{\partial t^2}\right) = \nabla \cdot \left(\frac{2}{c}\frac{\partial u}{\partial t}\nabla u\right) - \frac{1}{c}\frac{\partial}{\partial t}\left(|\nabla u|^2 + \frac{1}{c^2}\left(\frac{\partial u}{\partial t}\right)^2\right)$$

where $u = u(\boldsymbol{r}, t)$ depends on space and time.

2.9 Prove that the outward unit normal vector on the backward light cone C_- is

$$(\boldsymbol{n}, n_\tau) = \left(\frac{-(\boldsymbol{r}-\boldsymbol{r}_0)}{\sqrt{2}(\tau_0-\tau)}, \frac{1}{\sqrt{2}}\right)$$

Hint: If $f(\boldsymbol{r},\tau) =$ constant defines the surface in space–time, the unit normal is

$$\pm\frac{\nabla_{(\boldsymbol{r},\tau)} f(\boldsymbol{r},\tau)}{\left|\nabla_{(\boldsymbol{r},\tau)} f(\boldsymbol{r},\tau)\right|}$$

where $\nabla_{(\boldsymbol{r},\tau)}$ is the nabla operator in space–time $(\boldsymbol{r},\tau)$.

2.10 Rewrite the Maxwell equations in a source-free region

$$\begin{cases} \nabla \times \boldsymbol{E}(\boldsymbol{r},t) = -\mu_0\mu \dfrac{\partial \boldsymbol{H}(\boldsymbol{r},t)}{\partial t} \\ \nabla \times \boldsymbol{H}(\boldsymbol{r},t) = \epsilon_0\epsilon \dfrac{\partial \boldsymbol{E}(\boldsymbol{r},t)}{\partial t} \end{cases}$$

as a symmetric hyperbolic system of equations, i.e.,

$$\frac{\partial \mathrm{e}(\boldsymbol{r},t)}{\partial t} = c\sum_{k=1}^{3} \mathsf{A}_k \cdot \frac{\partial \mathrm{e}(\boldsymbol{r},t)}{\partial x_k}$$

where we used the notation (notice that there is a slight difference between the notation used here and (1.30) on page 31)

$$\mathrm{e}(\boldsymbol{r},t) = \begin{pmatrix} \boldsymbol{E}(\boldsymbol{r},t) \\ \eta_0\eta\boldsymbol{H}(\boldsymbol{r},t) \end{pmatrix}$$

and where the wave speed in the medium is $c = 1/\sqrt{\epsilon_0\epsilon\mu_0\mu}$, and where the relative wave impedance $\eta = \sqrt{\mu/\epsilon}$.

Determine the three symmetric six-dimensional dyadics A_k, $k = 1, 2, 3$.

CHAPTER 3

Integral Representation of Fields

In Chapter 1, we introduced the basic equations that are used to model the electromagnetic field in vacuum, in homogeneous isotropic, and also in more complex materials. We now proceed and introduce the very useful integral representations of the electromagnetic field with the results presented in Chapters 1 and 2. These integral representations are the fundamental building blocks of the analysis of the scattering phenomena that are addressed in this textbook. We start with the integral representations for homogeneous, isotropic media, and continue with the representation in a biisotropic material. We derive the results in both time domain and frequency domain. We start, however, with the frequency-domain analysis, since this analysis is less elaborate.

The integral representations show that the electromagnetic field in a volume of an isotropic material is represented in terms of volume and surface integrals. The boundary surface is a fictitious surface, which can coincide as a limiting case with the real boundary surface between two different materials. These representations are very important and powerful tools for e.g., the development of integral equations that can be used to numerically solve electromagnetic field problems.

In this chapter, we exploit these representations in some detail. First, we start with the simpler scalar case, before we develop the vector and dyadic cases. These cases are presented in Sections 3.1 and 3.2, respectively. The surface integral representations presented in Section 3.2 are applied to the Maxwell equations in Section 3.3. The dyadic notation has the advantage of making the notation compact. This is formulated in Section 3.4, where we introduce the Green dyadic. The limit values of the integral representations as the field point approaches the surface—both at a smooth point and at a corner—are derived in Sections 3.5–3.7. We end the chapter by some generalizations to biisotropic materials given in Section 3.8, and the integral representations in the time domain in Section 3.9.

3.1 Two Scalar Fields

The integral representation of scalar fields is much easier to derive than the integral representation of vector fields, and for this reason, we begin this chapter with the scalar case. The treatment of the vector case is then easier to understand.

Let $\phi(\boldsymbol{r})$ and $\psi(\boldsymbol{r})$ be two arbitrary scalar fields defined in a domain V, which is assumed to be bounded. Moreover, assume the fields $\phi(\boldsymbol{r})$ and $\psi(\boldsymbol{r})$ are twice continuously differentiable in V. At this stage, however, they do not satisfy any specific partial differential equation.

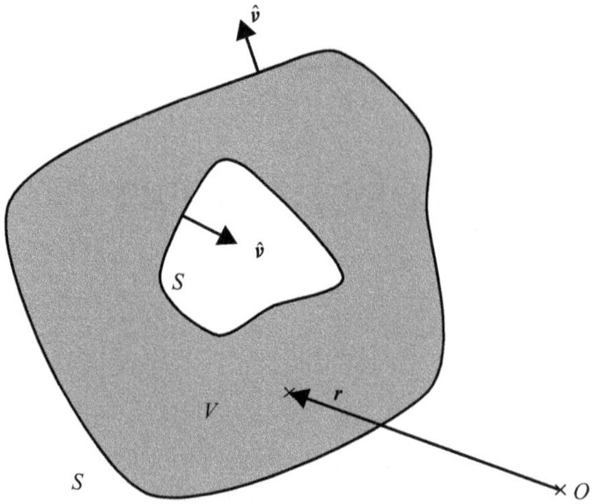

Figure 3.1 The domain of integration in Section 3.1.

The domain V is bounded by the surface S with outward unit normal vector $\hat{\boldsymbol{\nu}}$, see Figure 3.1. The surface S does not need to be simply connected, but can consist of several disjoint parts, which are piecewise smooth. Also, notice that the surface S does not have to be a surface that separates two different materials, but can be an arbitrary surface in space.

The starting point in this investigation is the following divergence, which is easily derived by the rules of nabla operator:

$$\nabla \cdot [\phi(\boldsymbol{r})\nabla\psi(\boldsymbol{r}) - \psi(\boldsymbol{r})\nabla\phi(\boldsymbol{r})] = \phi(\boldsymbol{r})\nabla^2\psi(\boldsymbol{r}) - \psi(\boldsymbol{r})\nabla^2\phi(\boldsymbol{r})$$

Integrate the identity over the volume V and use the divergence theorem. We get

$$\iint_S [\phi(\boldsymbol{r})\nabla\psi(\boldsymbol{r}) - \psi(\boldsymbol{r})\nabla\phi(\boldsymbol{r})] \cdot \hat{\boldsymbol{\nu}}(\boldsymbol{r})\, \mathrm{d}S = \iiint_V \left[\phi(\boldsymbol{r})\nabla^2\psi(\boldsymbol{r}) - \psi(\boldsymbol{r})\nabla^2\phi(\boldsymbol{r})\right] \mathrm{d}v \quad (3.1)$$

This is Green's theorem, that holds for every twice-continuously differentiable fields ϕ and ψ, and it is the foundation of the integral representation of scalar fields.

3.1.1 Integral Representation of a Scalar Field

As a first example of the use of (3.1), we derive an integral representation for a scalar field ϕ.

We let the scalar field ψ be the scalar Green function, see (2.11) on page 105

$$\psi(\boldsymbol{r}') = g(k, |\boldsymbol{r} - \boldsymbol{r}'|) = \frac{\mathrm{e}^{\mathrm{i}k|\boldsymbol{r}-\boldsymbol{r}'|}}{4\pi|\boldsymbol{r} - \boldsymbol{r}'|}$$

which satisfies, see (2.8) on page 103

$$\nabla^2 g(k, |\boldsymbol{r} - \boldsymbol{r}'|) + k^2 g(k, |\boldsymbol{r} - \boldsymbol{r}'|) = -\delta(\boldsymbol{r} - \boldsymbol{r}')$$

This function does not meet our assumption of being twice continuously differentiable at the point $r = r'$. The singularity can be treated in several ways. In this subsection, we treat the equation in a distributional sense. This is the most straightforward way of obtaining the desired result. An alternative derivation, based on classical function analysis, is presented in Section 3.1.2. The result is the same with both methods, see also Comment 2.1. on page 105.

It is worth noting that the Green function used in this chapter does not have to be the free-space Green function $g(k, |\boldsymbol{r} - \boldsymbol{r}'|)$. We only require that it is a solution to (2.8) without any specifications of its boundary values or radiation conditions.

Insert $\psi(\boldsymbol{r}') = g(k, |\boldsymbol{r} - \boldsymbol{r}'|)$ in (3.1). We get

$$\begin{aligned}
&\iint_S \left[\phi(\boldsymbol{r})\nabla g(k, |\boldsymbol{r} - \boldsymbol{r}'|) - g(k, |\boldsymbol{r} - \boldsymbol{r}'|)\nabla\phi(\boldsymbol{r})\right] \cdot \hat{\boldsymbol{\nu}}(\boldsymbol{r})\, \mathrm{d}S \\
&\quad = \iiint_V \left[\phi(\boldsymbol{r})\nabla^2 g(k, |\boldsymbol{r} - \boldsymbol{r}'|) - g(k, |\boldsymbol{r} - \boldsymbol{r}'|)\nabla^2\phi(\boldsymbol{r})\right] \mathrm{d}v \\
&\quad = -\iiint_V g(k, |\boldsymbol{r} - \boldsymbol{r}'|)\left[\nabla^2\phi(\boldsymbol{r}) + k^2\phi(\boldsymbol{r})\right] \mathrm{d}v - \iiint_V \phi(\boldsymbol{r})\delta(\boldsymbol{r} - \boldsymbol{r}')\, \mathrm{d}v \\
&\quad = -\iiint_V g(k, |\boldsymbol{r} - \boldsymbol{r}'|)\left[\nabla^2\phi(\boldsymbol{r}) + k^2\phi(\boldsymbol{r})\right] \mathrm{d}v - \begin{cases} \phi(\boldsymbol{r}'), & \boldsymbol{r}' \in V \\ 0, & \boldsymbol{r}' \in \mathbb{R}^3 \setminus \overline{V} \end{cases}
\end{aligned}$$

The bar on the volume V denotes the closed volume V, i.e., including the values on the bounding surface S, for notation see Appendix I.

We can now formulate the integral representation of a scalar field $\phi(\boldsymbol{r})$. For any twice continuously differentiable field ϕ in V, we have[1]

$$\begin{aligned}
&\iint_S \left\{g(k, |\boldsymbol{r} - \boldsymbol{r}'|)\hat{\boldsymbol{\nu}}(\boldsymbol{r}') \cdot \nabla'\phi(\boldsymbol{r}') - \phi(\boldsymbol{r}')\hat{\boldsymbol{\nu}}(\boldsymbol{r}') \cdot \nabla' g(k, |\boldsymbol{r} - \boldsymbol{r}'|)\right\} \mathrm{d}S' \\
&- \iiint_V \left\{g(k, |\boldsymbol{r} - \boldsymbol{r}'|)\left(\nabla'^2\phi(\boldsymbol{r}') + k^2\phi(\boldsymbol{r}')\right)\right\} \mathrm{d}v' = \begin{cases} \phi(\boldsymbol{r}), & \boldsymbol{r} \in V \\ 0, & \boldsymbol{r} \in \mathbb{R}^3 \setminus \overline{V} \end{cases}
\end{aligned} \tag{3.2}$$

The volume integral contains a singularity if $\boldsymbol{r} \in V$, and the integral has to be interpreted as a generalized integral.

[1] It is here more convenient to change the variable of integration to $\boldsymbol{r}'$. The integrals then depend on the variable $\boldsymbol{r}$.

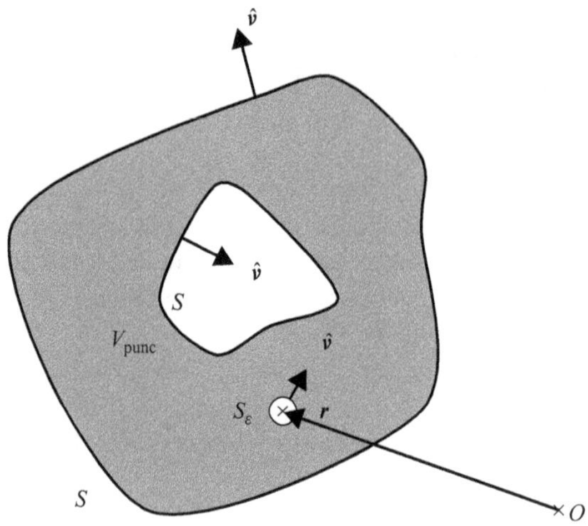

Figure 3.2 The geometry for the evaluation of the limit process of the volume integral contribution. The volume is punctured by a ball of radius ε centered at $\boldsymbol{r}$ and denoted V_{punc}. The bounding surface of this ball is S_ε.

3.1.2 Integral Representation of a Scalar Field—Alternative

It is of some pedagogical value to perform the analysis made in Section 3.1.1 without the use of the delta distribution $\delta(\boldsymbol{r}-\boldsymbol{r}')$, but instead only use classical function theory. The Green function, $\psi(\boldsymbol{r}') = g(k, |\boldsymbol{r}-\boldsymbol{r}'|)$, satisfies, see (2.8) on page 103

$$\nabla^2 g(k, |\boldsymbol{r}-\boldsymbol{r}'|) + k^2 g(k, |\boldsymbol{r}-\boldsymbol{r}'|) = 0, \qquad \boldsymbol{r} \neq \boldsymbol{r}'$$

This equation is interpreted in the classical sense. In all our computations below, we have to avoid that the point of integration coincides with the observation point. We accomplish this by excluding a small volume in the volume integral close to the observation point $\boldsymbol{r}$.

To proceed, let $\boldsymbol{r} \in V$, and let V_{punc} denote the volume V exclusive a spherical volume,[2] radius ε, with boundary S_ε, centered at the point $\boldsymbol{r}$ in V, see Figure 3.2. This is always possible for sufficiently small value of ε. Green's theorem, (3.1), applied to the volume V_{punc} then is (note that $\boldsymbol{r} \notin V_{\text{punc}}$)

$$\begin{aligned}
&\iint\limits_{S-S_\varepsilon} \left\{\phi(\boldsymbol{r}')\hat{\boldsymbol{\nu}}(\boldsymbol{r}')\cdot\nabla' g(k, |\boldsymbol{r}-\boldsymbol{r}'|) - g(k, |\boldsymbol{r}-\boldsymbol{r}'|)\hat{\boldsymbol{\nu}}(\boldsymbol{r}')\cdot\nabla'\phi(\boldsymbol{r}')\right\} \, \mathrm{d}S' \\
&\quad = \iiint\limits_{V_{\text{punc}}} \left\{\phi(\boldsymbol{r}')\nabla^2 g(k, |\boldsymbol{r}-\boldsymbol{r}'|) - g(k, |\boldsymbol{r}-\boldsymbol{r}'|)\nabla^2\phi(\boldsymbol{r}')\right\} \, \mathrm{d}v' \\
&\quad = -\iiint\limits_{V_{\text{punc}}} \left\{g(k, |\boldsymbol{r}-\boldsymbol{r}'|)\left(\nabla'^2\phi(\boldsymbol{r}') + k^2\phi(\boldsymbol{r}')\right)\right\} \, \mathrm{d}v'
\end{aligned} \tag{3.3}$$

[2]The value of the integral is independent of the shape of the surface S_ε. For a spherical exclusion, the integrals can be evaluated explicitly.

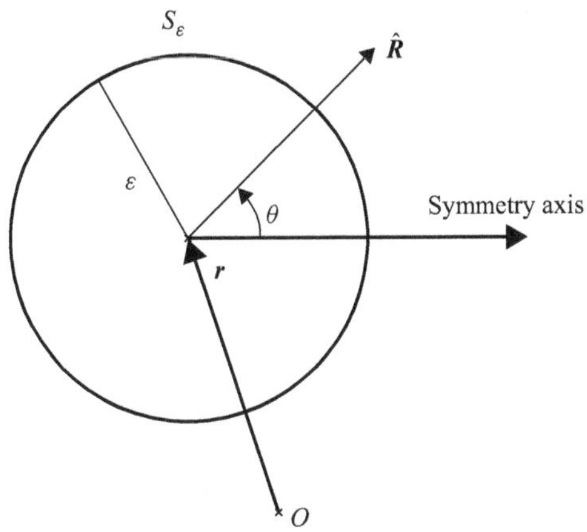

Figure 3.3 The geometry for the evaluation of integrals over the sphere S_ε.

In the integrals over the small sphere, S_ε, the unit normal vector $\hat{\nu}$ varies rapidly over the integration domain. The field ϕ and its gradient $\nabla\phi$, however, are assumed to vary more slowly, and, provided these fields are smooth (e.g., continuously differentiable), we assume that the field $\phi(\boldsymbol{r}')$ and $\nabla'\phi(\boldsymbol{r}')$ have expansions

$$\begin{cases} \phi(\boldsymbol{r}') = \phi(\boldsymbol{r}) + (\boldsymbol{r}' - \boldsymbol{r}) \cdot \nabla\phi(\boldsymbol{r}) + O(|\boldsymbol{r}' - \boldsymbol{r}|^2) \\ \nabla'\phi(\boldsymbol{r}') = \nabla\phi(\boldsymbol{r}) + O(|\boldsymbol{r}' - \boldsymbol{r}|) \end{cases} \quad \text{as } \boldsymbol{r}' \to \boldsymbol{r}$$

in a neighborhood of the point $\boldsymbol{r}$.

The following limits are useful (the surface S_ε is parameterized by the spherical angles θ and ϕ with an arbitrary fixed direction as the symmetry axis, see Figure 3.3)

$$\begin{cases} \lim_{\varepsilon \to 0} \iint_{S_\varepsilon} g(k, |\boldsymbol{r} - \boldsymbol{r}'|)\hat{\nu}(\boldsymbol{r}') \cdot \nabla'\phi(\boldsymbol{r}') \, \mathrm{d}S' = 0 \\ \lim_{\varepsilon \to 0} \iint_{S_\varepsilon} \phi(\boldsymbol{r}')\hat{\nu}(\boldsymbol{r}') \cdot \nabla' g(k, |\boldsymbol{r} - \boldsymbol{r}'|) \, \mathrm{d}S' = -\phi(\boldsymbol{r})\frac{1}{4\pi} \iint_{\Omega} \hat{\boldsymbol{R}} \cdot \hat{\boldsymbol{R}} \, \mathrm{d}\Omega = -\phi(\boldsymbol{r}) \end{cases}$$

since

$$\nabla' g(k, |\boldsymbol{r} - \boldsymbol{r}'|) = \hat{\boldsymbol{R}} \frac{\mathrm{e}^{\mathrm{i}k|\boldsymbol{r}-\boldsymbol{r}'|}}{4\pi |\boldsymbol{r} - \boldsymbol{r}'|} \left(\mathrm{i}k - \frac{1}{|\boldsymbol{r} - \boldsymbol{r}'|} \right) = \hat{\boldsymbol{R}} \frac{\mathrm{e}^{\mathrm{i}k\varepsilon}}{4\pi\varepsilon} \left(\mathrm{i}k - \frac{1}{\varepsilon} \right) \tag{3.4}$$

where $\hat{\boldsymbol{\nu}}(\boldsymbol{r}') = \hat{\boldsymbol{R}} = (\boldsymbol{r}' - \boldsymbol{r})/|\boldsymbol{r} - \boldsymbol{r}'| = (\boldsymbol{r}' - \boldsymbol{r})/\varepsilon$, Ω denotes the unit sphere, and $\mathrm{d}\Omega = \sin\theta\,\mathrm{d}\theta\,\mathrm{d}\phi$ denotes the surface measure on the unit sphere. We get as $\varepsilon \to 0$ in (3.3)

$$\begin{aligned}
&\iint\limits_S \left(g(k,|\boldsymbol{r}-\boldsymbol{r}'|)\hat{\boldsymbol{\nu}}(\boldsymbol{r}')\cdot\nabla'\phi(\boldsymbol{r}') - \phi(\boldsymbol{r}')\hat{\boldsymbol{\nu}}(\boldsymbol{r}')\cdot\nabla' g(k,|\boldsymbol{r}-\boldsymbol{r}'|)\right)\mathrm{d}S' \\
&\quad = \lim_{\varepsilon\to 0}\iint\limits_{S_\varepsilon} \left(g(k,|\boldsymbol{r}-\boldsymbol{r}'|)\hat{\boldsymbol{\nu}}(\boldsymbol{r}')\cdot\nabla'\phi(\boldsymbol{r}') - \phi(\boldsymbol{r}')\hat{\boldsymbol{\nu}}(\boldsymbol{r}')\cdot\nabla' g(k,|\boldsymbol{r}-\boldsymbol{r}'|)\right)\mathrm{d}S' \\
&\qquad + \iiint\limits_V \left\{g(k,|\boldsymbol{r}-\boldsymbol{r}'|)\left(\nabla'^2\phi(\boldsymbol{r}') + k^2\phi(\boldsymbol{r}')\right)\right\}\mathrm{d}v'
\end{aligned}$$

Only the second term in the surface integral contributes, and we obtain

$$\begin{aligned}
&\iint\limits_S \left(g(k,|\boldsymbol{r}-\boldsymbol{r}'|)\hat{\boldsymbol{\nu}}(\boldsymbol{r}')\cdot\nabla'\phi(\boldsymbol{r}') - \phi(\boldsymbol{r}')\hat{\boldsymbol{\nu}}(\boldsymbol{r}')\cdot\nabla' g(k,|\boldsymbol{r}-\boldsymbol{r}'|)\right)\mathrm{d}S' \\
&\quad = \phi(\boldsymbol{r}) + \iiint\limits_V \left\{g(k,|\boldsymbol{r}-\boldsymbol{r}'|)\left(\nabla'^2\phi(\boldsymbol{r}') + k^2\phi(\boldsymbol{r}')\right)\right\}\mathrm{d}v'
\end{aligned}$$

This result is identical to the upper part of (3.2) if $\boldsymbol{r} \in V$. For an $\boldsymbol{r} \notin V$, the singular integral is absent, and the result is identical to the lower part of (3.2).

This alternative approach is somewhat more elaborated than the δ function approach in Section 3.1.1. However, this alternative approach becomes valuable below, when dealing with the value of the integral representation, when the observation point is located on the bounding surface S, see Section 3.5.

Example 3.1

The function ϕ in (3.2) is arbitrary—the only requirement is that ϕ is twice differentiable in V. The simplest function ϕ we can choose is a constant function, i.e.,

$$\begin{cases}
\phi(\boldsymbol{r}) = 1 \\
\nabla\phi(\boldsymbol{r}) = \boldsymbol{0} \\
\nabla^2\phi(\boldsymbol{r}) = \nabla\cdot(\nabla\phi(\boldsymbol{r})) = 0
\end{cases}$$

Several terms in the integral representation vanish with this simple constant function, and (3.2) becomes

$$-\iint\limits_S \hat{\boldsymbol{\nu}}(\boldsymbol{r}')\cdot\nabla' g(k,|\boldsymbol{r}-\boldsymbol{r}'|)\,\mathrm{d}S' - k^2\iiint\limits_V g(k,|\boldsymbol{r}-\boldsymbol{r}'|)\,\mathrm{d}v' = \begin{cases} 1, & \boldsymbol{r}\in V \\ 0, & \boldsymbol{r}\in\mathbb{R}^3\setminus\overline{V}\end{cases}$$

This identity holds for all values of k, and it has a simple geometrical interpretation if we let $k = 0$. We then obtain

$$\iint\limits_S \hat{\boldsymbol{\nu}}(\boldsymbol{r}')\cdot\frac{\boldsymbol{r}'-\boldsymbol{r}}{|\boldsymbol{r}-\boldsymbol{r}'|^3}\,\mathrm{d}S' = \begin{cases} 4\pi, & \boldsymbol{r}\in V \\ 0, & \boldsymbol{r}\in\mathbb{R}^3\setminus\overline{V}\end{cases}$$

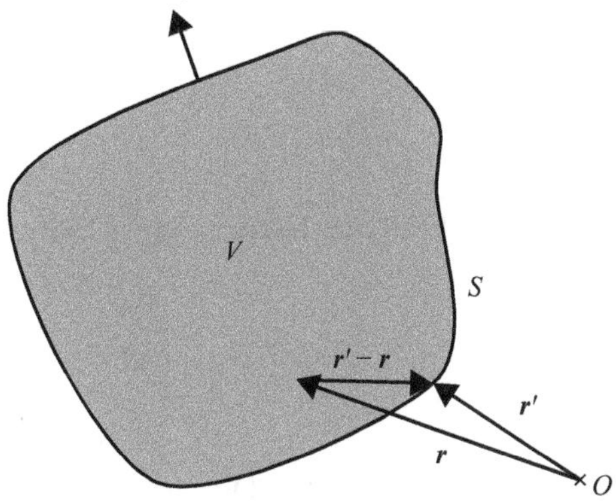

Figure 3.4 The solid angle of the surface S observed from the point $\boldsymbol{r}$.

This is the definition of solid angle of a surface S observed from a point $\boldsymbol{r}$, see also Figure 3.4 and also the definition in Appendix F.5 on page 672. ■

This example shows that many useful relations can be obtained by applying the integral representation to pertinent fields. Several other useful identities can be obtained by other choices of fields. We proceed with another, more complex, example that deals with the gradient of the scalar field ϕ.

Example 3.2
Take the gradient of (3.2). We get

$$\iint_S \left(\nabla g(k,|\boldsymbol{r}-\boldsymbol{r}'|)\left(\hat{\boldsymbol{\nu}}(\boldsymbol{r}')\cdot\nabla'\phi(\boldsymbol{r}')\right)-\phi(\boldsymbol{r}')\nabla\left(\hat{\boldsymbol{\nu}}(\boldsymbol{r}')\cdot\nabla' g(k,|\boldsymbol{r}-\boldsymbol{r}'|)\right)\right)\,\mathrm{d}S'$$
$$-\iiint_V \left\{\nabla g(k,|\boldsymbol{r}-\boldsymbol{r}'|)\left(\nabla'^2\phi(\boldsymbol{r}')+k^2\phi(\boldsymbol{r}')\right)\right\}\,\mathrm{d}v' = \begin{cases}\nabla\phi(\boldsymbol{r}), & \boldsymbol{r}\in V\\ 0, & \boldsymbol{r}\in\mathbb{R}^3\setminus\overline{V}\end{cases}$$

or, since $\nabla g(k,|\boldsymbol{r}-\boldsymbol{r}'|) = -\nabla' g(k,|\boldsymbol{r}-\boldsymbol{r}'|)$ (use the differential rule (3) on the inside of the first cover)

$$-\iint_S \left(\nabla' g(k,|\boldsymbol{r}-\boldsymbol{r}'|)\left(\hat{\boldsymbol{\nu}}(\boldsymbol{r}')\cdot\nabla'\phi(\boldsymbol{r}')\right)-\phi(\boldsymbol{r}')\left(\hat{\boldsymbol{\nu}}(\boldsymbol{r}')\cdot\nabla'\right)\nabla' g(k,|\boldsymbol{r}-\boldsymbol{r}'|)\right)\,\mathrm{d}S'$$
$$-\iiint_V \left\{\nabla g(k,|\boldsymbol{r}-\boldsymbol{r}'|)\left(\nabla'^2\phi(\boldsymbol{r}')+k^2\phi(\boldsymbol{r}')\right)\right\}\,\mathrm{d}v' = \begin{cases}\nabla\phi(\boldsymbol{r}), & \boldsymbol{r}\in V\\ 0, & \boldsymbol{r}\in\mathbb{R}^3\setminus\overline{V}\end{cases}$$

As a check, we evaluate the surface integral over an excluding sphere S_ε, see Figure 3.2, to see that this surface integral gives the correct contribution, i.e., the value $\nabla\phi(\boldsymbol{r})$ if $\boldsymbol{r}\in V$, and zero otherwise. The surface integral over the excluding surface S_ε is[3]

$$\boldsymbol{I}_\varepsilon = -\lim_{\varepsilon\to 0}\iint\limits_{S_\varepsilon}\left(\nabla' g(k,|\boldsymbol{r}-\boldsymbol{r}'|)\left(\hat{\boldsymbol{\nu}}(\boldsymbol{r}')\cdot\nabla'\phi(\boldsymbol{r}')\right) - \phi(\boldsymbol{r}')\left(\hat{\boldsymbol{\nu}}(\boldsymbol{r}')\cdot\nabla'\right)\nabla' g(k,|\boldsymbol{r}-\boldsymbol{r}'|)\right)\,\mathrm{d}S'$$

To evaluate this last integral, we make use of the following identity ($\boldsymbol{a}$ is a constant vector, and ϕ depends on the primed variables)

$$\begin{aligned}\phi\left(\hat{\boldsymbol{\nu}}\cdot\nabla'\right)\left(\nabla' g\cdot\boldsymbol{a}\right) &= \hat{\boldsymbol{\nu}}\cdot\left\{\phi\nabla'\underbrace{(\nabla' g\cdot\boldsymbol{a})}_{\nabla'\cdot(g\boldsymbol{a})}\right\} = \hat{\boldsymbol{\nu}}\cdot\left\{\phi\nabla'\times\left(\nabla' g\times\boldsymbol{a}\right)+\phi\boldsymbol{a}\nabla'^2 g\right\}\\ &= \hat{\boldsymbol{\nu}}\cdot\left\{\nabla'\times\left(\phi\left(\nabla' g\times\boldsymbol{a}\right)\right)\underbrace{-\nabla'\phi\times\left(\nabla' g\times\boldsymbol{a}\right)}_{\nabla' g(\nabla'\phi\cdot\boldsymbol{a})-\boldsymbol{a}(\nabla'\phi\cdot\nabla' g)}-k^2\boldsymbol{a}\phi g\right\}\end{aligned}$$

In this rather long calculation, we have used (6), (13), and (9) on the inside of the first cover. Moreover, we used the BAC-CAB rule (see inside of last cover) and also that $\nabla^2 g + k^2 g = 0$ if $\boldsymbol{r}\neq\boldsymbol{r}'$. For each Cartesian component of the constant vector $\boldsymbol{a}$, apply Stokes' theorem to show that the first term in the integral vanishes. The remaining terms are

$$\iint\limits_{S_\varepsilon}\phi\left(\hat{\boldsymbol{\nu}}\cdot\nabla'\right)\nabla' g\,\mathrm{d}S' = -\iint\limits_{S_\varepsilon}\left\{\left(\hat{\boldsymbol{\nu}}\cdot\nabla' g\right)\nabla'\phi - \hat{\boldsymbol{\nu}}\left(\nabla'\phi\cdot\nabla' g\right)+\hat{\boldsymbol{\nu}}k^2\phi g\right\}\,\mathrm{d}S'$$

We specialize to a spherical surface S_ε. The limit as the radius $\varepsilon\to 0$ is then easily obtained. We have ($\boldsymbol{F}(\boldsymbol{r}')$ is a smooth vector field, e.g., $\nabla'\phi(\boldsymbol{r}')$)

$$\begin{aligned}\boldsymbol{I}_\varepsilon = &-\lim_{\varepsilon\to 0}\iint\limits_{S_\varepsilon}\nabla' g(k,|\boldsymbol{r}-\boldsymbol{r}'|)\left(\hat{\boldsymbol{\nu}}(\boldsymbol{r}')\cdot\nabla'\phi(\boldsymbol{r}')\right)\,\mathrm{d}S'\\ &-\lim_{\varepsilon\to 0}\iint\limits_{S_\varepsilon}\left\{\left(\hat{\boldsymbol{\nu}}(\boldsymbol{r}')\cdot\nabla' g(k,|\boldsymbol{r}-\boldsymbol{r}'|)\right)\nabla'\phi(\boldsymbol{r}')\right.\\ &\left.-\hat{\boldsymbol{\nu}}(\boldsymbol{r}')\left(\nabla'\phi(\boldsymbol{r}')\cdot\nabla' g(k,|\boldsymbol{r}-\boldsymbol{r}'|)\right)\right\}\,\mathrm{d}S' = \nabla\phi(\boldsymbol{r})\end{aligned}$$

since, see (3.4)

$$\lim_{\varepsilon\to 0}\iint\limits_{S_\varepsilon} g(k,|\boldsymbol{r}-\boldsymbol{r}'|)\phi(\boldsymbol{r}')\hat{\boldsymbol{\nu}}(\boldsymbol{r}')\,\mathrm{d}S' = \boldsymbol{0}$$

$$\lim_{\varepsilon\to 0}\iint\limits_{S_\varepsilon}\left(\nabla' g(k,|\boldsymbol{r}-\boldsymbol{r}'|)\right)\left(\hat{\boldsymbol{\nu}}(\boldsymbol{r}')\cdot\boldsymbol{F}(\boldsymbol{r}')\right)\,\mathrm{d}S' = -\frac{1}{4\pi}\iint\limits_{\Omega}\hat{\boldsymbol{R}}(\hat{\boldsymbol{R}}\cdot\boldsymbol{F}(\boldsymbol{r}))\,\mathrm{d}\Omega = -\frac{1}{3}\boldsymbol{F}(\boldsymbol{r})$$

---

[3]In this complex example, the shape of the exclusion volume matters, cf. the analysis in Section 2.2.3 on page 115.

and

$$\lim_{\varepsilon\to 0} \iint\limits_{S_\varepsilon} \left(\hat{\boldsymbol{\nu}}(\boldsymbol{r}') \cdot \nabla' g(k, |\boldsymbol{r} - \boldsymbol{r}'|)\right) \boldsymbol{F}(\boldsymbol{r}') \, \mathrm{d}S' = -\boldsymbol{F}(\boldsymbol{r}) \frac{1}{4\pi} \iint\limits_{\Omega} \hat{\boldsymbol{R}} \cdot \hat{\boldsymbol{R}} \, \mathrm{d}\Omega = -\boldsymbol{F}(\boldsymbol{r})$$

$$\lim_{\varepsilon\to 0} \iint\limits_{S_\varepsilon} \hat{\boldsymbol{\nu}}(\boldsymbol{r}') \left(\boldsymbol{F}(\boldsymbol{r}') \cdot \nabla' g(k, |\boldsymbol{r} - \boldsymbol{r}'|)\right) \mathrm{d}S' = -\frac{1}{4\pi} \iint\limits_{\Omega} \hat{\boldsymbol{R}}(\hat{\boldsymbol{R}} \cdot \boldsymbol{F}(\boldsymbol{r})) \, \mathrm{d}\Omega = -\frac{1}{3} \boldsymbol{F}(\boldsymbol{r})$$

where we used the vector- and dyadic-valued integrals

$$\iint\limits_{\Omega} \hat{\boldsymbol{R}} \, \mathrm{d}\Omega = \mathbf{0}, \qquad \iint\limits_{\Omega} \hat{\boldsymbol{R}}\hat{\boldsymbol{R}} \, \mathrm{d}\Omega = \frac{4\pi}{3} \mathbf{I}_3 \tag{3.5}$$

The singularity can also be evaluated directly by the use of

$$\nabla'\nabla' g(k, |\boldsymbol{r} - \boldsymbol{r}'|) = \frac{\mathrm{e}^{\mathrm{i}kR}}{4\pi R} \left(\hat{\boldsymbol{R}}\hat{\boldsymbol{R}} \frac{3 - 3\mathrm{i}kR - k^2R^2}{R^2} - \mathbf{I}_3 \frac{1 - \mathrm{i}kR}{R^2} \right)$$

where

$$R = |\boldsymbol{r} - \boldsymbol{r}'|, \qquad \hat{\boldsymbol{\nu}}(\boldsymbol{r}') = \hat{\boldsymbol{R}} = \frac{\boldsymbol{r}' - \boldsymbol{r}}{|\boldsymbol{r} - \boldsymbol{r}'|}$$

This expression is obtained translating the origin in the identity

$$\partial_i \partial_j \frac{\mathrm{e}^{\mathrm{i}kr}}{4\pi r} = \frac{\mathrm{e}^{\mathrm{i}kr}}{4\pi r} \left(x_i x_j \frac{3 - 3\mathrm{i}kr - k^2 r^2}{r^4} - \delta_{ij} \frac{1 - \mathrm{i}kr}{r^2} \right)$$

The surface integral $\boldsymbol{I}_\varepsilon$ over the excluding surface S_ε then becomes

$$\begin{aligned} \boldsymbol{I}_\varepsilon &= \frac{1}{3} \nabla\phi(\boldsymbol{r}) \\ &\quad + \lim_{\varepsilon\to 0} \iint\limits_{\Omega} \left(\phi(\boldsymbol{r}) + \varepsilon\hat{\boldsymbol{R}} \cdot \nabla\phi(\boldsymbol{r}) + O(\varepsilon^2)\right) \hat{\boldsymbol{R}} \frac{\mathrm{e}^{\mathrm{i}k\varepsilon}}{4\pi\varepsilon} \left(3 - 3\mathrm{i}k\varepsilon - k^2\varepsilon^2 - 1 + \mathrm{i}k\varepsilon\right) \mathrm{d}\Omega \\ &= \frac{1}{3} \nabla\phi(\boldsymbol{r}) + \lim_{\varepsilon\to 0} \frac{1}{2\pi} \iint\limits_{\Omega} \hat{\boldsymbol{R}}(\hat{\boldsymbol{R}} \cdot \nabla\phi(\boldsymbol{r})) \, \mathrm{d}\Omega = \frac{1}{3} \nabla\phi(\boldsymbol{r}) + \frac{2}{3} \nabla\phi(\boldsymbol{r}) = \nabla\phi(\boldsymbol{r}) \end{aligned}$$

We have here used $\phi(\boldsymbol{r}') = \phi(\boldsymbol{r}) + \varepsilon\hat{\boldsymbol{R}} \cdot \nabla\phi(\boldsymbol{r}) + O(\varepsilon^2)$ in a neighborhood of the point $\boldsymbol{r}$ (Taylor's[4] theorem). This proves that the result is identical to the limit result above, and ends this rather complex example. ■

[4]Brook Taylor (1685–1731), English mathematician.

3.2 Vector and Scalar Fields

We now proceed by investigating integral representations of vector-valued fields. As we will see, several different ways of representations exist.

The result developed in Section 3.1 is exploited. Let $\phi = \boldsymbol{a} \cdot \boldsymbol{F}$, where $\boldsymbol{a}$ is an arbitrary constant vector, and $\boldsymbol{F}$ be a twice continuously differentiable vector field. The result, using (3.1), is

$$\iint_S [(\boldsymbol{a} \cdot \boldsymbol{F})\nabla\psi - \psi\nabla(\boldsymbol{a} \cdot \boldsymbol{F})] \cdot \hat{\boldsymbol{v}}\, \mathrm{d}S = \iiint_V \left[(\boldsymbol{a} \cdot \boldsymbol{F})\nabla^2\psi - \psi\nabla^2(\boldsymbol{a} \cdot \boldsymbol{F})\right] \mathrm{d}v \tag{3.6}$$

Comment 3.1

A vector-valued version of Green's theorem derived in Section 3.1, see (3.1), is easily obtained by an application of the result of Section 3.1 on each Cartesian component of the vector field. We get

$$\iint_S \left[\boldsymbol{A}(\boldsymbol{r})\frac{\partial\psi(\boldsymbol{r})}{\partial\nu} - \psi(\boldsymbol{r})\frac{\partial\boldsymbol{A}(\boldsymbol{r})}{\partial\nu}\right] \mathrm{d}S = \iiint_V \left[\boldsymbol{A}(\boldsymbol{r})\nabla^2\psi(\boldsymbol{r}) - \psi(\boldsymbol{r})\nabla^2\boldsymbol{A}(\boldsymbol{r})\right] \mathrm{d}v$$

One drawback with this expression is that the fields in the surface integral are not adapted to the boundary conditions that are natural in electromagnetic problems, i.e., a decomposition in normal and tangential components. Therefore, we develop a vector-valued integral representation, which contains these properties.

Our goal is to rewrite the equation above as an expression of the following form:

$$\boldsymbol{a} \cdot \iint_S \ldots \,\mathrm{d}S = \boldsymbol{a} \cdot \iiint_V \ldots \,\mathrm{d}v \tag{3.7}$$

This is possible, as we prove below. From this result, we are then able to extract a vector relation—Green's theorem for vector field.

Since $\nabla^2(\boldsymbol{a} \cdot \boldsymbol{F}) = \boldsymbol{a} \cdot \nabla^2\boldsymbol{F}$ and $\nabla \times (\nabla \times \boldsymbol{F}) = \nabla(\nabla \cdot \boldsymbol{F}) - \nabla^2\boldsymbol{F}$, the volume integral on the right-hand side of (3.6) is easily rewritten as

$$\begin{aligned}
&\iiint_V \left[(\boldsymbol{a} \cdot \boldsymbol{F})\nabla^2\psi + \psi\boldsymbol{a} \cdot \left(\nabla \times (\nabla \times \boldsymbol{F}) - \nabla(\nabla \cdot \boldsymbol{F})\right)\right] \mathrm{d}v \\
&\quad = \boldsymbol{a} \cdot \iiint_V \left[\boldsymbol{F}\nabla^2\psi + \psi\left(\nabla \times (\nabla \times \boldsymbol{F}) - \nabla(\nabla \cdot \boldsymbol{F})\right)\right] \mathrm{d}v
\end{aligned}$$

The surface integral in (3.6) we rewrite, with the use of the differentiation rules of the nabla operator, see (2) and (11) on the inside of the first cover, as

$$\begin{aligned}
(\boldsymbol{a} \cdot \boldsymbol{F})\nabla\psi - \psi\nabla(\boldsymbol{a} \cdot \boldsymbol{F}) &= 2(\boldsymbol{a} \cdot \boldsymbol{F})\nabla\psi - \nabla(\boldsymbol{a} \cdot (\psi\boldsymbol{F})) \\
&= 2(\boldsymbol{a} \cdot \boldsymbol{F})\nabla\psi + \nabla \times (\boldsymbol{a} \times (\psi\boldsymbol{F})) - \boldsymbol{a} \times (\nabla \times (\psi\boldsymbol{F})) - \boldsymbol{a}\,(\nabla \cdot (\psi\boldsymbol{F}))
\end{aligned}$$

Further simplifications give

$$[(\boldsymbol{a}\cdot\boldsymbol{F})\nabla\psi - \psi\nabla(\boldsymbol{a}\cdot\boldsymbol{F})]\cdot\hat{\boldsymbol{\nu}} = 2(\boldsymbol{a}\cdot\boldsymbol{F})(\hat{\boldsymbol{\nu}}\cdot\nabla\psi) + \hat{\boldsymbol{\nu}}\cdot[\nabla\times(\boldsymbol{a}\times(\psi\boldsymbol{F}))]$$
$$+\,\boldsymbol{a}\cdot\Bigg[\hat{\boldsymbol{\nu}}\times\underbrace{(\psi(\nabla\times\boldsymbol{F})+(\nabla\psi)\times\boldsymbol{F})}_{\nabla\times(\psi\boldsymbol{F})}\Bigg] - (\boldsymbol{a}\cdot\hat{\boldsymbol{\nu}})\Bigg[\underbrace{\psi(\nabla\cdot\boldsymbol{F})+(\nabla\psi)\cdot\boldsymbol{F}}_{\nabla\cdot(\psi\boldsymbol{F})}\Bigg]$$

Therefore, the surface integral on the left-hand side in (3.6) can, with the use of the BAC-CAB rule ($\boldsymbol{a}\times(\boldsymbol{b}\times\boldsymbol{c}) = \boldsymbol{b}(\boldsymbol{a}\cdot\boldsymbol{c}) - \boldsymbol{c}(\boldsymbol{a}\cdot\boldsymbol{b})$) (see inside of last cover), be written as

$$\iint_S [(\boldsymbol{a}\cdot\boldsymbol{F})\nabla\psi - \psi\nabla(\boldsymbol{a}\cdot\boldsymbol{F})]\cdot\hat{\boldsymbol{\nu}}\,\mathrm{d}S$$
$$= \boldsymbol{a}\cdot\iint_S \big[(\hat{\boldsymbol{\nu}}\cdot\nabla\psi)\boldsymbol{F} + \psi(\hat{\boldsymbol{\nu}}\times(\nabla\times\boldsymbol{F})) + (\nabla\psi)(\hat{\boldsymbol{\nu}}\cdot\boldsymbol{F})$$
$$- \hat{\boldsymbol{\nu}}\psi(\nabla\cdot\boldsymbol{F}) - \hat{\boldsymbol{\nu}}\,((\nabla\psi)\cdot\boldsymbol{F})\big]\,\mathrm{d}S + \iint_S [\nabla\times(\boldsymbol{a}\times(\psi\boldsymbol{F}))]\cdot\hat{\boldsymbol{\nu}}\,\mathrm{d}S$$

An application of the divergence theorem shows that the last surface integral on the right-hand side is zero, since $\nabla\cdot(\nabla\times\boldsymbol{A}) = \boldsymbol{0}$ for an arbitrary vector field $\boldsymbol{A}$. We have therefore rewritten both the volume and the surface integral in (3.6) on the form in (3.7). Since the constant vector $\boldsymbol{a}$ is arbitrary (let the vector $\boldsymbol{a}$ be the Cartesian unit vectors, i.e., $\boldsymbol{a} = \hat{\boldsymbol{x}}, \hat{\boldsymbol{y}}$, and $\hat{\boldsymbol{z}}$), we obtain Green's vector theorem by another use of the BAC-CAB rule

$$\iint_S \big[\psi\,(\hat{\boldsymbol{\nu}}\times(\nabla\times\boldsymbol{F})) + (\nabla\psi)(\hat{\boldsymbol{\nu}}\cdot\boldsymbol{F}) - \hat{\boldsymbol{\nu}}\psi(\nabla\cdot\boldsymbol{F}) - (\nabla\psi)\times(\hat{\boldsymbol{\nu}}\times\boldsymbol{F})\big]\,\mathrm{d}S$$
$$= \iiint_V \big[\boldsymbol{F}\nabla^2\psi + \psi\,(\nabla\times(\nabla\times\boldsymbol{F}) - \nabla(\nabla\cdot\boldsymbol{F}))\big]\,\mathrm{d}v \tag{3.8}$$

This equation is, just as Green's theorem in the scalar case, the foundation for finding integral representations of vector fields. It holds for every twice continuously differentiable vector field $\boldsymbol{F}$ and scalar ψ. The surface integral contains the vector field $\boldsymbol{F}$, decomposed in its normal and tangential components, and its derivatives—curl and divergence—evaluated at the bounding surface S. Strictly speaking, these fields are limit values of the vector field $\boldsymbol{F}$ and its derivatives as the evaluation point $\boldsymbol{r}$ approaches the surface S from the volume V (i.e., from the inside).

3.2.1 Integral Representation of a Vector Field

We now proceed with the derivation of the integral representation of the vector field $\boldsymbol{F}(\boldsymbol{r})$ by the use of the result obtained in (3.8) applied to a special scalar field $\psi(\boldsymbol{r})$. Let $\psi(\boldsymbol{r})$ be the scalar Green function, see (2.11) on page 105

$$\psi(\boldsymbol{r}) = g(k, |\boldsymbol{r}-\boldsymbol{r}'|) = \frac{\mathrm{e}^{\mathrm{i}k|\boldsymbol{r}-\boldsymbol{r}'|}}{4\pi|\boldsymbol{r}-\boldsymbol{r}'|}$$

In this expression, $\boldsymbol{r}'$ acts as a parameter. This function satisfies, see (2.8) on page 103 (differentiation w.r.t. the unprimed variables)

$$\nabla^2 g(k, |\boldsymbol{r}-\boldsymbol{r}'|) + k^2 g(k, |\boldsymbol{r}-\boldsymbol{r}'|) = -\delta(\boldsymbol{r}-\boldsymbol{r}')$$

Again, this function does not meet our assumption of being twice continuously differentiable at the point $\boldsymbol{r} = \boldsymbol{r}'$, and two ways of deriving the integral representation for vector fields are natural, just as in Sections 3.1.1 and 3.1.2. The result is the same with both methods, cf. also Comment 2.1 on page 105. We develop both methods for clarity.

Just as in scalar case above, it is here convenient to change the variable of integration in (3.8) to $\boldsymbol{r}'$. The integrals then depend on the variable $\boldsymbol{r}$, and the Green's vector theorem (3.8) becomes, provided the point $\boldsymbol{r}$ does not lie on the bounding surface S

$$\begin{aligned}
&\iint_S \big[g(k, |\boldsymbol{r}-\boldsymbol{r}'|)\left(\hat{\boldsymbol{\nu}}(\boldsymbol{r}') \times \left(\nabla' \times \boldsymbol{F}(\boldsymbol{r}')\right)\right) + \left(\nabla' g(k, |\boldsymbol{r}-\boldsymbol{r}'|)\right)\left(\hat{\boldsymbol{\nu}}(\boldsymbol{r}') \cdot \boldsymbol{F}(\boldsymbol{r}')\right) \\
&\quad - \hat{\boldsymbol{\nu}}(\boldsymbol{r}') g(k, |\boldsymbol{r}-\boldsymbol{r}'|)\left(\nabla' \cdot \boldsymbol{F}(\boldsymbol{r}')\right) - \left(\nabla' g(k, |\boldsymbol{r}-\boldsymbol{r}'|)\right) \times \left(\hat{\boldsymbol{\nu}}(\boldsymbol{r}') \times \boldsymbol{F}(\boldsymbol{r}')\right)\big] \, \mathrm{d}S' \\
&= \iiint_V \big[g(k, |\boldsymbol{r}-\boldsymbol{r}'|)\left(\nabla' \times (\nabla' \times \boldsymbol{F}(\boldsymbol{r}')) - \nabla'\left(\nabla' \cdot \boldsymbol{F}(\boldsymbol{r}')\right)\right) + \boldsymbol{F}(\boldsymbol{r}')\nabla'^2 g(k, |\boldsymbol{r}-\boldsymbol{r}'|)\big] \, \mathrm{d}v' \\
&= \iiint_V \big[g(k, |\boldsymbol{r}-\boldsymbol{r}'|)\left(\nabla' \times (\nabla' \times \boldsymbol{F}(\boldsymbol{r}')) - \nabla'(\nabla' \cdot \boldsymbol{F}(\boldsymbol{r}')) - k^2 \boldsymbol{F}(\boldsymbol{r}')\right)\big] \, \mathrm{d}v' \\
&\quad - \iiint_V \boldsymbol{F}(\boldsymbol{r}')\delta(\boldsymbol{r}-\boldsymbol{r}') \, \mathrm{d}v'
\end{aligned}$$

where we also used the differential equation that the Green function $g(k, |\boldsymbol{r}-\boldsymbol{r}'|)$ satisfy.

The value of the second volume integral on the right-hand side depends on whether the point $\boldsymbol{r}$ is contained in the volume V or not

$$\begin{aligned}
&\iiint_V \left\{g(k, |\boldsymbol{r}-\boldsymbol{r}'|)\left[\nabla' \times (\nabla' \times \boldsymbol{F}(\boldsymbol{r}')) - \nabla'(\nabla' \cdot \boldsymbol{F}(\boldsymbol{r}')) - k^2 \boldsymbol{F}(\boldsymbol{r}')\right]\right\} \mathrm{d}v' \\
&\quad - \iint_S \big[g(k, |\boldsymbol{r}-\boldsymbol{r}'|)(\hat{\boldsymbol{\nu}}(\boldsymbol{r}') \times (\nabla' \times \boldsymbol{F}(\boldsymbol{r}'))) \\
&\quad + (\nabla' g(k, |\boldsymbol{r}-\boldsymbol{r}'|))\left(\hat{\boldsymbol{\nu}}(\boldsymbol{r}') \cdot \boldsymbol{F}(\boldsymbol{r}')\right) - \hat{\boldsymbol{\nu}}(\boldsymbol{r}') g(k, |\boldsymbol{r}-\boldsymbol{r}'|)(\nabla' \cdot \boldsymbol{F}(\boldsymbol{r}')) \\
&\quad - (\nabla' g(k, |\boldsymbol{r}-\boldsymbol{r}'|)) \times (\hat{\boldsymbol{\nu}}(\boldsymbol{r}') \times \boldsymbol{F}(\boldsymbol{r}'))\big] \, \mathrm{d}S' = \begin{cases} \boldsymbol{F}(\boldsymbol{r}), & \boldsymbol{r} \in V \\ \boldsymbol{0}, & \boldsymbol{r} \in \mathbb{R}^3 \setminus \overline{V} \end{cases}
\end{aligned} \tag{3.9}$$

The field $\boldsymbol{F}(\boldsymbol{r})$ in the volume V is represented as a volume integral of the field in V, and as a surface integral of the field over the bounding surface S of V. If these integrals are evaluated at a point $\boldsymbol{r}$ that lies outside the volume V, the integrals cancel each other (extinction). It is important to point out that this does not necessarily mean that the field $\boldsymbol{F}(\boldsymbol{r})$ is identically zero outside the volume V—only the values of the integrals cancel.

3.2.2 Integral Representation of a Vector Field—Alternative

In Section 3.2.1, we derived the integral representation (3.9) using the delta distribution approach. A classical approach to derive (3.9) would be to start with the Green function $g(k, |\boldsymbol{r} - \boldsymbol{r}'|)$ satisfying (2.8) in a classical sense, i.e.,

$$\nabla^2 g(k, |\boldsymbol{r} - \boldsymbol{r}'|) + k^2 g(k, |\boldsymbol{r} - \boldsymbol{r}'|) = 0, \qquad \boldsymbol{r} \neq \boldsymbol{r}'$$

As in Section 3.1.2, let $\boldsymbol{r} \in V$, and let V_{punc} denote the volume V exclusive a spherical volume with boundary S_ε that excludes the point $\boldsymbol{r}$ in V, see Figure 3.2. The Green vector formula, (3.8), applied to the volume V_{punc} is (observe that the Green function is a regular function in the volume V_{punc})

$$\begin{aligned}
&\iiint_{V_{\text{punc}}} \left[g(k, |\boldsymbol{r} - \boldsymbol{r}'|) \left(\nabla' \times (\nabla' \times \boldsymbol{F}(\boldsymbol{r}')) - \nabla'(\nabla' \cdot \boldsymbol{F}(\boldsymbol{r}')) - k^2 \boldsymbol{F}(\boldsymbol{r}') \right) \right] \mathrm{d}v' \\
&\quad = \iint_{S - S_\varepsilon} \left[g(k, |\boldsymbol{r} - \boldsymbol{r}'|) \left(\hat{\boldsymbol{\nu}}(\boldsymbol{r}') \times (\nabla' \times \boldsymbol{F}(\boldsymbol{r}')) \right) \right. \\
&\qquad + \left(\nabla' g(k, |\boldsymbol{r} - \boldsymbol{r}'|) \right) \left(\hat{\boldsymbol{\nu}}(\boldsymbol{r}') \cdot \boldsymbol{F}(\boldsymbol{r}') \right) - \hat{\boldsymbol{\nu}}(\boldsymbol{r}') g(k, |\boldsymbol{r} - \boldsymbol{r}'|) \left(\nabla' \cdot \boldsymbol{F}(\boldsymbol{r}') \right) \\
&\qquad \left. - \left(\nabla' g(k, |\boldsymbol{r} - \boldsymbol{r}'|) \right) \times \left(\hat{\boldsymbol{\nu}}(\boldsymbol{r}') \times \boldsymbol{F}(\boldsymbol{r}') \right) \right] \mathrm{d}S'
\end{aligned} \tag{3.10}$$

In the evaluation of the contributions from the surface integrals over S_ε, several limits have to be evaluated. Useful in this respect, the following limits are computed, see (3.4) and (3.5)

$$\begin{cases}
\lim\limits_{\varepsilon \to 0} \displaystyle\iint_{S_\varepsilon} g(k, |\boldsymbol{r} - \boldsymbol{r}'|) \left(\hat{\boldsymbol{\nu}}(\boldsymbol{r}') \times (\boldsymbol{F}(\boldsymbol{r}')) \right) \mathrm{d}S' = \boldsymbol{0} \\
\lim\limits_{\varepsilon \to 0} \displaystyle\iint_{S_\varepsilon} \left(\nabla' g(k, |\boldsymbol{r} - \boldsymbol{r}'|) \right) \left(\hat{\boldsymbol{\nu}}(\boldsymbol{r}') \cdot \boldsymbol{F}(\boldsymbol{r}') \right) \mathrm{d}S' = -\frac{1}{4\pi} \iint_{\Omega} \hat{\boldsymbol{R}}(\hat{\boldsymbol{R}} \cdot \boldsymbol{F}(\boldsymbol{r})) \, \mathrm{d}\Omega = -\frac{1}{3} \boldsymbol{F}(\boldsymbol{r})
\end{cases}$$

Moreover, the following limits have also to be evaluated:

$$\begin{cases}
\lim\limits_{\varepsilon \to 0} \displaystyle\iint_{S_\varepsilon} \hat{\boldsymbol{\nu}}(\boldsymbol{r}') g(k, |\boldsymbol{r} - \boldsymbol{r}'|) \phi(\boldsymbol{r}') \, \mathrm{d}S' = \boldsymbol{0} \\
\lim\limits_{\varepsilon \to 0} \displaystyle\iint_{S_\varepsilon} \left(\nabla' g(k, |\boldsymbol{r} - \boldsymbol{r}'|) \right) \times \left(\hat{\boldsymbol{\nu}}(\boldsymbol{r}') \times \boldsymbol{F}(\boldsymbol{r}') \right) \mathrm{d}S' \\
\quad = -\dfrac{1}{4\pi} \displaystyle\iint_{\Omega} \underbrace{\hat{\boldsymbol{R}} \times (\hat{\boldsymbol{R}} \times \boldsymbol{F}(\boldsymbol{r}))}_{\hat{\boldsymbol{R}}(\hat{\boldsymbol{R}} \cdot \boldsymbol{F}(\boldsymbol{r})) - \boldsymbol{F}(\boldsymbol{r})} \mathrm{d}\Omega = \left(1 - \dfrac{1}{3} \right) \boldsymbol{F}(\boldsymbol{r}) = \dfrac{2}{3} \boldsymbol{F}(\boldsymbol{r})
\end{cases}$$

These results are now used to take the limit as $\varepsilon \to 0$ in the volume and surface integrals in (3.10). The result of this limit process becomes

$$\begin{aligned}
-\boldsymbol{F}(\boldsymbol{r}) + \iiint_V &\left[g(k,|\boldsymbol{r}-\boldsymbol{r}'|)\left(\nabla' \times \left(\nabla' \times \boldsymbol{F}(\boldsymbol{r}')\right) - \nabla'\left(\nabla' \cdot \boldsymbol{F}(\boldsymbol{r}')\right) - k^2 \boldsymbol{F}(\boldsymbol{r}')\right)\right] \mathrm{d}v' \\
= \iint_S &\left[g(k,|\boldsymbol{r}-\boldsymbol{r}'|)\left(\hat{\boldsymbol{\nu}}(\boldsymbol{r}') \times \left(\nabla' \times \boldsymbol{F}(\boldsymbol{r}')\right)\right)\right. \\
&+ \left(\nabla' g(k,|\boldsymbol{r}-\boldsymbol{r}'|)\right)\left(\hat{\boldsymbol{\nu}}(\boldsymbol{r}') \cdot \boldsymbol{F}(\boldsymbol{r}')\right) - \hat{\boldsymbol{\nu}}(\boldsymbol{r}') g(k,|\boldsymbol{r}-\boldsymbol{r}'|)\left(\nabla' \cdot \boldsymbol{F}(\boldsymbol{r}')\right) \\
&\left. - \left(\nabla' g(k,|\boldsymbol{r}-\boldsymbol{r}'|)\right) \times \left(\hat{\boldsymbol{\nu}}(\boldsymbol{r}') \times \boldsymbol{F}(\boldsymbol{r}')\right)\right] \mathrm{d}S'
\end{aligned}$$

which identical to the upper part of (3.9). For an $\boldsymbol{r} \notin V$, the singular integral does not exist, and the result identical to the lower part of (3.9).

3.3 Integral Representations of the Maxwell Equations

So far, the vector field $\boldsymbol{F}$ in this chapter has been an arbitrary vector field. We now let this field be the electric field $\boldsymbol{E}$ that satisfies the source-free Maxwell equations (1.43) and (1.44)

$$\begin{cases} \nabla \times \boldsymbol{E}(\boldsymbol{r},\omega) = \mathrm{i}\omega \boldsymbol{B}(\boldsymbol{r},\omega) \\ \nabla \times \boldsymbol{H}(\boldsymbol{r},\omega) = -\mathrm{i}\omega \boldsymbol{D}(\boldsymbol{r},\omega) \end{cases}$$

Moreover, assume the material inside V is isotropic and homogeneous, i.e.,

$$\begin{cases} \boldsymbol{D}(\boldsymbol{r},\omega) = \epsilon_0 \epsilon(\omega) \boldsymbol{E}(\boldsymbol{r},\omega) \\ \boldsymbol{B}(\boldsymbol{r},\omega) = \mu_0 \mu(\omega) \boldsymbol{H}(\boldsymbol{r},\omega) \end{cases}$$

where $\epsilon(\omega)$ and $\mu(\omega)$ may depend on the angular frequency ω, i.e., the material can be dispersive, but constants as a function of space (homogeneous material). The Maxwell equations and the constitutive relations imply

$$\begin{cases} \nabla \times \boldsymbol{E} = \mathrm{i}\omega\mu_0\mu \boldsymbol{H} = \mathrm{i}k\eta_0\eta \boldsymbol{H} \\ \nabla \times \boldsymbol{H} = -\mathrm{i}\omega\epsilon_0\epsilon \boldsymbol{E} = -\dfrac{\mathrm{i}k}{\eta_0\eta} \boldsymbol{E} \end{cases}$$

where

$$\eta = \left(\frac{\mu}{\epsilon}\right)^{1/2} \tag{3.11}$$

is the (relative) wave impedance of the material, η_0 denotes the wave impedance of vacuum, and $k = \omega\left(\epsilon_0\mu_0\epsilon\mu\right)^{1/2}$ denotes the wave number of the material.

It is now easy to eliminate the magnetic field $\boldsymbol{H}$ such that an equation in only the electric field is obtained. This has already been made in (2.2) on page 100

$$\nabla \times (\nabla \times \boldsymbol{E}) - k^2 \boldsymbol{E} = \boldsymbol{0}$$

Similarly, an equation in the magnetic field $\boldsymbol{H}$ is obtained

$$\nabla \times (\nabla \times \boldsymbol{H}) - k^2 \boldsymbol{H} = \mathbf{0}$$

Moreover, from the Maxwell equations, we get

$$\begin{cases} \nabla \cdot \boldsymbol{E} = 0 \\ \nabla \cdot \boldsymbol{H} = 0 \end{cases}$$

We apply these results to (3.9), and we obtain an integral representation of the electric field $\boldsymbol{E}$ in a surface integral over the boundary surface of the volume V. The volume integral vanishes in the integral representation and only the surface integrals remain. We have[5]

$$\begin{aligned} &-\iint_S \Big[\mathrm{i}k\eta_0\eta g(k, |\boldsymbol{r}-\boldsymbol{r}'|) \left(\hat{\boldsymbol{\nu}}(\boldsymbol{r}') \times \boldsymbol{H}(\boldsymbol{r}')\right) + \left(\nabla' g(k, |\boldsymbol{r}-\boldsymbol{r}'|)\right) \left(\hat{\boldsymbol{\nu}}(\boldsymbol{r}') \cdot \boldsymbol{E}(\boldsymbol{r}')\right) \\ &\qquad - \left(\nabla' g(k, |\boldsymbol{r}-\boldsymbol{r}'|)\right) \times \left(\hat{\boldsymbol{\nu}}(\boldsymbol{r}') \times \boldsymbol{E}(\boldsymbol{r}')\right) \Big] \, \mathrm{d}S' = \begin{cases} \boldsymbol{E}(\boldsymbol{r}), & \boldsymbol{r} \text{ inside } S \\ \mathbf{0}, & \boldsymbol{r} \text{ outside } S \end{cases} \end{aligned} \tag{3.12}$$

This result bears the name Stratton–Chu[6] formula in the literature [254]. As already pointed out, the surface S does not have to be an interface between two different materials, but it can be a general surface bounding a region in which the material parameters are homogeneous.

Similarly, we obtain an integral representation for the magnetic field $\boldsymbol{H}$ (make a dual transformation $\boldsymbol{E} \leftrightarrow \mathrm{i}\eta_0\eta\boldsymbol{H}$, since the Maxwell equations are invariant under this transformation, but the electric and the magnetic fields change places)

$$\begin{aligned} &-\iint_S \Big[-\frac{\mathrm{i}k}{\eta_0\eta} g(k, |\boldsymbol{r}-\boldsymbol{r}'|) \left(\hat{\boldsymbol{\nu}}(\boldsymbol{r}') \times \boldsymbol{E}(\boldsymbol{r}')\right) + \left(\nabla' g(k, |\boldsymbol{r}-\boldsymbol{r}'|)\right) \left(\hat{\boldsymbol{\nu}}(\boldsymbol{r}') \cdot \boldsymbol{H}(\boldsymbol{r}')\right) \\ &\qquad - \left(\nabla' g(k, |\boldsymbol{r}-\boldsymbol{r}'|)\right) \times \left(\hat{\boldsymbol{\nu}}(\boldsymbol{r}') \times \boldsymbol{H}(\boldsymbol{r}')\right) \Big] \, \mathrm{d}S' = \begin{cases} \boldsymbol{H}(\boldsymbol{r}), & \boldsymbol{r} \text{ inside } S \\ \mathbf{0}, & \boldsymbol{r} \text{ outside } S \end{cases} \end{aligned} \tag{3.13}$$

Comment 3.2

We conclude with a final comment on the relation between the normal component of the electric field and the tangential components of the magnetic field. The normal component of, e.g., the electric field, can by the Maxwell equations be written as

$$\hat{\boldsymbol{\nu}} \cdot \boldsymbol{E} = -\frac{\eta_0\eta}{\mathrm{i}k} \hat{\boldsymbol{\nu}} \cdot (\nabla \times \boldsymbol{H}) = \frac{\eta_0\eta}{\mathrm{i}k} \nabla_S \cdot \left(\hat{\boldsymbol{\nu}} \times \boldsymbol{H}\right)$$

where we used the identity [38, page 61]

$$\nabla_S \cdot \left(\hat{\boldsymbol{\nu}} \times \boldsymbol{a}\right) = -\hat{\boldsymbol{\nu}} \cdot (\nabla \times \boldsymbol{a})$$

and ∇_S denotes the surface divergence.

---

[5]The surface integral is assumed to be evaluated in a source-free region so that $\nabla \cdot \boldsymbol{E} = \nabla \cdot \boldsymbol{H} = 0$ hold on the surface S.

[6]Julius Adams Stratton (1901–1994), American electrical engineer; Lan Jen Chu (1913–1973), Chinese-American electrical engineer.

3.3.1 Elimination of Normal Component

Both the integral representations, (3.12) and (3.13), contain the normal components of the electric and the magnetic fields on S, respectively. Our goal is now to further simplify these expression such that only the tangential components of the fields occur in the surface integrals. To accomplish this, we note that

$$\nabla' g(k, |\boldsymbol{r} - \boldsymbol{r}'|) = -\nabla g(k, |\boldsymbol{r} - \boldsymbol{r}'|)$$

which leads to the following reorganization of the second term on the left-hand side in (3.13)

$$\iint_S \left(\nabla' g(k, |\boldsymbol{r} - \boldsymbol{r}'|)\right) \left(\hat{\boldsymbol{\nu}}(\boldsymbol{r}') \cdot \boldsymbol{H}(\boldsymbol{r}')\right) \mathrm{d}S' = -\nabla \iint_S g(k, |\boldsymbol{r} - \boldsymbol{r}'|) \left(\hat{\boldsymbol{\nu}}(\boldsymbol{r}') \cdot \boldsymbol{H}(\boldsymbol{r}')\right) \mathrm{d}S'$$

and that the third term on the left-hand side in (3.13) can be rewritten as

$$\begin{aligned}
&\iint_S \left(\nabla' g(k, |\boldsymbol{r} - \boldsymbol{r}'|)\right) \times \left(\hat{\boldsymbol{\nu}}(\boldsymbol{r}') \times \boldsymbol{H}(\boldsymbol{r}')\right) \mathrm{d}S' \\
&\quad = -\iint_S \left(\nabla g(k, |\boldsymbol{r} - \boldsymbol{r}'|)\right) \times \left(\hat{\boldsymbol{\nu}}(\boldsymbol{r}') \times \boldsymbol{H}(\boldsymbol{r}')\right) \mathrm{d}S' \\
&\quad = -\nabla \times \iint_S g(k, |\boldsymbol{r} - \boldsymbol{r}'|) \left(\hat{\boldsymbol{\nu}}(\boldsymbol{r}') \times \boldsymbol{H}(\boldsymbol{r}')\right) \mathrm{d}S'
\end{aligned}$$

Take the curl of (3.13), and we get, since $\nabla \times (\nabla \iint_S \ldots \mathrm{d}S') = \mathbf{0}$

$$\begin{aligned}
&-\nabla \times \left\{ \nabla \times \iint_S g(k, |\boldsymbol{r} - \boldsymbol{r}'|) \left(\hat{\boldsymbol{\nu}}(\boldsymbol{r}') \times \boldsymbol{H}(\boldsymbol{r}')\right) \mathrm{d}S' \right\} \\
&\quad + \frac{\mathrm{i}k}{\eta_0 \eta} \nabla \times \iint_S g(k, |\boldsymbol{r} - \boldsymbol{r}'|) \left(\hat{\boldsymbol{\nu}}(\boldsymbol{r}') \times \boldsymbol{E}(\boldsymbol{r}')\right) \mathrm{d}S' \\
&\quad = \begin{cases} \nabla \times \boldsymbol{H}(\boldsymbol{r}) = -\dfrac{\mathrm{i}k}{\eta_0 \eta} \boldsymbol{E}(\boldsymbol{r}), & \boldsymbol{r} \text{ inside } S \\ \mathbf{0}, & \boldsymbol{r} \text{ outside } S \end{cases}
\end{aligned}$$

where we also used $\nabla \times \boldsymbol{H}(\boldsymbol{r}) = -\mathrm{i}k\boldsymbol{E}(\boldsymbol{r})/\eta_0\eta$.

Finally, we get an integral representation of the electric field by dividing the expression by $-\mathrm{i}k/\eta_0\eta$

$$\begin{aligned}
&\frac{\eta_0 \eta}{\mathrm{i}k} \nabla \times \left\{ \nabla \times \iint_S g(k, |\boldsymbol{r} - \boldsymbol{r}'|) \left(\hat{\boldsymbol{\nu}}(\boldsymbol{r}') \times \boldsymbol{H}(\boldsymbol{r}')\right) \mathrm{d}S' \right\} \\
&\quad - \nabla \times \iint_S g(k, |\boldsymbol{r} - \boldsymbol{r}'|) \left(\hat{\boldsymbol{\nu}}(\boldsymbol{r}') \times \boldsymbol{E}(\boldsymbol{r}')\right) \mathrm{d}S' = \begin{cases} \boldsymbol{E}(\boldsymbol{r}), & \boldsymbol{r} \text{ inside } S \\ \mathbf{0}, & \boldsymbol{r} \text{ outside } S \end{cases}
\end{aligned} \tag{3.14}$$

Analogously, the magnetic field is represented as a surface integral

$$-\frac{1}{\mathrm{i}k\eta_0\eta}\nabla\times\left\{\nabla\times\iint_S g(k,|\boldsymbol{r}-\boldsymbol{r}'|)\left(\hat{\boldsymbol{\nu}}(\boldsymbol{r}')\times\boldsymbol{E}(\boldsymbol{r}')\right)\,\mathrm{d}S'\right\}$$

$$-\nabla\times\iint_S g(k,|\boldsymbol{r}-\boldsymbol{r}'|)\left(\hat{\boldsymbol{\nu}}(\boldsymbol{r}')\times\boldsymbol{H}(\boldsymbol{r}')\right)\,\mathrm{d}S'=\begin{cases}\boldsymbol{H}(\boldsymbol{r}), & \boldsymbol{r}\text{ inside } S\\ \mathbf{0}, & \boldsymbol{r}\text{ outside } S\end{cases}\tag{3.15}$$

These equations are fundamental for the formulation and the analysis of the scattering problems treated in this book.

3.4 Dyadic and Vector Fields

The integral representations in the preceding sections are based upon the scalar Green function $g(k,|\boldsymbol{r}-\boldsymbol{r}'|)$ and they have their limitations. Specifically, when we want to generalize to more complex media, it is advantageous to develop the corresponding integral representation based upon the Green dyadic $\mathbf{G}_\mathrm{e}(k,|\boldsymbol{r}-\boldsymbol{r}'|)$ for the electric field. This dyadic satisfies (2.27) on page 113

$$\nabla\times\left(\nabla\times\mathbf{G}_\mathrm{e}(k,|\boldsymbol{r}-\boldsymbol{r}'|)\right)-k^2\mathbf{G}_\mathrm{e}(k,|\boldsymbol{r}-\boldsymbol{r}'|)=\mathbf{I}_3\delta(\boldsymbol{r}-\boldsymbol{r}')$$

Before proceeding, we derive a few useful differentiation identities with dyadics. Let $\boldsymbol{F}$ and $\mathbf{D}$ be a twice differentiable vector-valued and dyadic-valued field, respectively. Furthermore, let $\boldsymbol{a}$ be an arbitrary constant vector. The differentiation rule (7) on the inside of the first cover for vector fields then gives

$$\nabla\cdot(\boldsymbol{F}\times(\nabla\times(\mathbf{D}\cdot\boldsymbol{a})))=(\nabla\times\boldsymbol{F})\cdot(\nabla\times(\mathbf{D}\cdot\boldsymbol{a}))-\boldsymbol{F}\cdot(\nabla\times(\nabla\times(\mathbf{D}\cdot\boldsymbol{a})))$$

and

$$\nabla\cdot((\nabla\times\boldsymbol{F})\times(\mathbf{D}\cdot\boldsymbol{a}))=(\nabla\times(\nabla\times\boldsymbol{F}))\cdot(\mathbf{D}\cdot\boldsymbol{a})-(\nabla\times\boldsymbol{F})\cdot(\nabla\times(\mathbf{D}\cdot\boldsymbol{a}))$$

Adding these two identities implies

$$\begin{aligned}&\nabla\cdot\{\boldsymbol{F}\times(\nabla\times(\mathbf{D}\cdot\boldsymbol{a}))+(\nabla\times\boldsymbol{F})\times(\mathbf{D}\cdot\boldsymbol{a})\}\\&=(\nabla\times(\nabla\times\boldsymbol{F}))\cdot(\mathbf{D}\cdot\boldsymbol{a})-\boldsymbol{F}\cdot(\nabla\times(\nabla\times(\mathbf{D}\cdot\boldsymbol{a})))\end{aligned}$$

The divergence theorem gives

$$\begin{aligned}&\iiint_V\{(\nabla\times(\nabla\times\boldsymbol{F}))\cdot(\mathbf{D}\cdot\boldsymbol{a})-\boldsymbol{F}\cdot(\nabla\times(\nabla\times(\mathbf{D}\cdot\boldsymbol{a})))\}\,\mathrm{d}v\\&=\iint_S\hat{\boldsymbol{\nu}}\cdot\{\boldsymbol{F}\times(\nabla\times(\mathbf{D}\cdot\boldsymbol{a}))+(\nabla\times\boldsymbol{F})\times(\mathbf{D}\cdot\boldsymbol{a})\}\,\mathrm{d}S\\&=\iint_S\left\{(\hat{\boldsymbol{\nu}}\times\boldsymbol{F})\cdot(\nabla\times(\mathbf{D}\cdot\boldsymbol{a}))+\left(\hat{\boldsymbol{\nu}}\times(\nabla\times\boldsymbol{F})\right)\cdot(\mathbf{D}\cdot\boldsymbol{a})\right\}\,\mathrm{d}S\end{aligned}$$

Since $\boldsymbol{a}$ is an arbitrary constant vector (let the vector $\boldsymbol{a}$ be the Cartesian unit vectors, i.e., $\boldsymbol{a} = \hat{\boldsymbol{x}}$, $\hat{\boldsymbol{y}}$, and $\hat{\boldsymbol{z}}$), we get

$$\iiint_V \{(\nabla \times (\nabla \times \boldsymbol{F})) \cdot \mathbf{D} - \boldsymbol{F} \cdot (\nabla \times (\nabla \times \mathbf{D}))\} \, \mathrm{d}v \cdot \boldsymbol{a}$$
$$= \iint_S \left\{(\hat{\boldsymbol{\nu}} \times \boldsymbol{F}) \cdot (\nabla \times \mathbf{D}) + \left(\hat{\boldsymbol{\nu}} \times (\nabla \times \boldsymbol{F})\right) \cdot \mathbf{D}\right\} \mathrm{d}S \cdot \boldsymbol{a}$$

and finally we have the dyadic divergence theorem

$$\iiint_V \{(\nabla \times (\nabla \times \boldsymbol{F})) \cdot \mathbf{D} - \boldsymbol{F} \cdot (\nabla \times (\nabla \times \mathbf{D}))\} \, \mathrm{d}v$$
$$= \iint_S \left\{(\hat{\boldsymbol{\nu}} \times \boldsymbol{F}) \cdot (\nabla \times \mathbf{D}) + \left(\hat{\boldsymbol{\nu}} \times (\nabla \times \boldsymbol{F})\right) \cdot \mathbf{D}\right\} \mathrm{d}S \qquad (3.16)$$

which holds for every twice continuously differentiable vector field $\boldsymbol{F}$ and dyadic field $\mathbf{D}$.

3.4.1 Integral Representation of the Electric Field—Dyadic Version

We apply the result in (3.16) to the field $\boldsymbol{F}(\boldsymbol{r}) = \boldsymbol{E}(\boldsymbol{r})$ in a source-free region V, which satisfies

$$\nabla \times (\nabla \times \boldsymbol{E}(\boldsymbol{r})) - k^2 \boldsymbol{E}(\boldsymbol{r}) = \mathbf{0}, \qquad \boldsymbol{r} \in V \qquad (3.17)$$

and $\mathbf{D}(\boldsymbol{r}) = \mathbf{G}_\mathrm{e}(k, |\boldsymbol{r} - \boldsymbol{r}'|)$, which satisfies, see (2.27) on page 113

$$\nabla \times \left(\nabla \times \mathbf{G}_\mathrm{e}(k, |\boldsymbol{r} - \boldsymbol{r}'|)\right) - k^2 \mathbf{G}_\mathrm{e}(k, |\boldsymbol{r} - \boldsymbol{r}'|) = \mathbf{I}_3 \delta(\boldsymbol{r} - \boldsymbol{r}'), \qquad \boldsymbol{r}, \boldsymbol{r}' \in \mathbb{R}^3$$

where $\mathbf{G}_\mathrm{e}$ is given by (2.28) on page 114

$$\mathbf{G}_\mathrm{e}(k, |\boldsymbol{r} - \boldsymbol{r}'|) = \left(\mathbf{I}_3 + \frac{1}{k^2}\nabla\nabla\right) g(k, |\boldsymbol{r} - \boldsymbol{r}'|) = \left(\mathbf{I}_3 + \frac{1}{k^2}\nabla'\nabla'\right) g(k, |\boldsymbol{r} - \boldsymbol{r}'|)$$

The volume integral in (3.16) simplifies to

$$\iiint_V \left\{(\nabla' \times (\nabla' \times \boldsymbol{E})) \cdot \mathbf{G}_\mathrm{e} - \boldsymbol{E} \cdot (\nabla' \times (\nabla' \times \mathbf{G}_\mathrm{e}))\right\} \mathrm{d}v'$$
$$= \iiint_V -\boldsymbol{E} \cdot \mathbf{I}_3 \delta(\boldsymbol{r} - \boldsymbol{r}') \, \mathrm{d}v' = \begin{cases} -\boldsymbol{E}(\boldsymbol{r}), & \boldsymbol{r} \in V \\ \mathbf{0}, & \boldsymbol{r} \in \mathbb{R}^3 \setminus \overline{V} \end{cases}$$

The integral representation of the electric field then becomes

$$\iint_S \left(\nabla' \times \mathbf{G}_\mathrm{e}(k, |\boldsymbol{r} - \boldsymbol{r}'|)\right) \cdot \left(\hat{\boldsymbol{\nu}}(\boldsymbol{r}') \times \boldsymbol{E}(\boldsymbol{r}')\right) \mathrm{d}S'$$
$$- \iint_S \mathbf{G}_\mathrm{e}(k, |\boldsymbol{r} - \boldsymbol{r}'|) \cdot \left(\hat{\boldsymbol{\nu}}(\boldsymbol{r}') \times \left(\nabla' \times \boldsymbol{E}(\boldsymbol{r}')\right)\right) \mathrm{d}S' = \begin{cases} \boldsymbol{E}(\boldsymbol{r}), & \boldsymbol{r} \text{ inside } S \\ \mathbf{0}, & \boldsymbol{r} \text{ outside } S \end{cases} \qquad (3.18)$$

where we also have used that $\nabla' \times \mathbf{G}_e(k, |\boldsymbol{r}-\boldsymbol{r}'|)$ is an anti-symmetric dyadic, and $\mathbf{G}_e(k, |\boldsymbol{r}-\boldsymbol{r}'|)$ is a symmetric dyadic, see Section 2.2.2 on page 113.

Comment 3.3

Notice that this result is identical to (3.14), since $\nabla \times \boldsymbol{E} = ik\eta_0\eta\boldsymbol{H}$, and

$$\begin{cases} \left(\nabla' \times \mathbf{G}_e(k, |\boldsymbol{r}-\boldsymbol{r}'|)\right) \cdot \boldsymbol{F}(\boldsymbol{r}') = -\nabla g(k, |\boldsymbol{r}-\boldsymbol{r}'|) \times \boldsymbol{F}(\boldsymbol{r}') \\ \qquad = -\nabla \times \left(g(k, |\boldsymbol{r}-\boldsymbol{r}'|)\boldsymbol{F}(\boldsymbol{r}')\right) \\ \mathbf{G}_e(k, |\boldsymbol{r}-\boldsymbol{r}'|) \cdot \boldsymbol{F}(\boldsymbol{r}') = g(k, |\boldsymbol{r}-\boldsymbol{r}'|)\boldsymbol{F}(\boldsymbol{r}') + \frac{1}{k^2}\left(\boldsymbol{F}(\boldsymbol{r}') \cdot \nabla'\right)\nabla' g(k, |\boldsymbol{r}-\boldsymbol{r}'|) \end{cases}$$

The last term is then rewritten using identities (13) and (3) on the inside of the first cover for vector fields, and the equation for the scalar Green function to get (remember $\boldsymbol{r} \neq \boldsymbol{r}'$ for the surface integrals)

$$\begin{aligned} &\nabla \times \left(\nabla \times \left(g(k, |\boldsymbol{r}-\boldsymbol{r}'|)\boldsymbol{F}(\boldsymbol{r}')\right)\right) \\ &\quad = k^2 g(k, |\boldsymbol{r}-\boldsymbol{r}'|)\boldsymbol{F}(\boldsymbol{r}') + \nabla\left(\boldsymbol{F}(\boldsymbol{r}') \cdot \nabla\left(g(k, |\boldsymbol{r}-\boldsymbol{r}'|)\right)\right) \\ &\quad = k^2 g(k, |\boldsymbol{r}-\boldsymbol{r}'|)\boldsymbol{F}(\boldsymbol{r}') + \left(\boldsymbol{F}(\boldsymbol{r}') \cdot \nabla'\right)\nabla' g(k, |\boldsymbol{r}-\boldsymbol{r}'|) \end{aligned}$$

A collection of all terms then shows that (3.18) is identical to (3.14).

The derivation of (3.18), using classical function arguments, is left to the reader, see Problem 3.4.

Example 3.3

To include a source term in the result in (3.18) is a minor generalization, and we present the result in this example. Great care of how to interpret the additional volume integral has to be made [296].

Equation (3.16) is applied to the electric field in a homogeneous isotropic material with a current density $\boldsymbol{J}$. This field satisfies, see (2.2) on page 100

$$\nabla \times (\nabla \times \boldsymbol{E}(\boldsymbol{r})) - k^2\boldsymbol{E}(\boldsymbol{r}) = i\omega\mu_0\mu\boldsymbol{J}(\boldsymbol{r})$$

Thus, let $\boldsymbol{F} = \boldsymbol{E}$ and $\mathbf{D} = \mathbf{G}_e$, which satisfies (2.27) on page 113, and integrate over the volume V. We get

$$\iiint_V \left\{i\omega\mu_0\mu\boldsymbol{J}(\boldsymbol{r}) \cdot \mathbf{G}_e(k, |\boldsymbol{r}-\boldsymbol{r}'|) - \boldsymbol{E}(\boldsymbol{r})\delta(\boldsymbol{r}-\boldsymbol{r}')\right\} \mathrm{d}v$$

$$= \iint_S \left\{\left(\hat{\boldsymbol{\nu}}(\boldsymbol{r}) \times \boldsymbol{E}(\boldsymbol{r})\right) \cdot \left(\nabla \times \mathbf{G}_e(k, |\boldsymbol{r}-\boldsymbol{r}'|)\right)\right.$$

$$\left. + \left(\hat{\boldsymbol{\nu}}(\boldsymbol{r}) \times (\nabla \times \boldsymbol{E}(\boldsymbol{r}))\right) \cdot \mathbf{G}_e(k, |\boldsymbol{r}-\boldsymbol{r}'|)\right\} \mathrm{d}S$$

As above, the integration over the delta function gives the field for a point inside V, and the extinction outside V. We get, if we exchange the primed and the unprimed variables and cyclic permutation of the last term in the surface integral

$$\iint_S \left(\hat{\boldsymbol{\nu}}(\boldsymbol{r}') \times \boldsymbol{E}(\boldsymbol{r}')\right) \cdot \left(\nabla' \times \mathbf{G}_e(k, |\boldsymbol{r} - \boldsymbol{r}'|)\right) \, dS'$$

$$- \iint_S \left(\nabla' \times \boldsymbol{E}(\boldsymbol{r}')\right) \cdot \left(\hat{\boldsymbol{\nu}}(\boldsymbol{r}') \times \mathbf{G}_e(k, |\boldsymbol{r} - \boldsymbol{r}'|)\right) \, dS'$$

$$= i\omega\mu_0\mu \iiint_V \boldsymbol{J}(\boldsymbol{r}') \cdot \mathbf{G}_e(k, |\boldsymbol{r} - \boldsymbol{r}'|) \, dv' - \begin{cases} \boldsymbol{E}(\boldsymbol{r}), & \boldsymbol{r} \text{ inside } S \\ \mathbf{0}, & \boldsymbol{r} \text{ outside } S \end{cases}$$

The volume integral has to be interpreted in the distributional sense. For more details, see Section 2.2.3 on page 115 and Reference 296. ∎

Example 3.4
A generalization to a special type of anisotropic materials can be obtained within the same theoretical framework. The electric field in a homogeneous, anisotropic material with symmetric permeability dyadic, $\boldsymbol{\epsilon}(\omega)$, and isotropic permeability, $\boldsymbol{\mu}(\omega) = \mathbf{I}\mu(\omega)$, satisfies, see (2.31) on page 118

$$\nabla \times (\nabla \times \boldsymbol{E}(\boldsymbol{r})) - k_0^2 \mu(\omega)\boldsymbol{\epsilon}(\omega) \cdot \boldsymbol{E}(\boldsymbol{r}) = i\omega\mu_0\mu(\omega)\boldsymbol{J}(\boldsymbol{r})$$

and the appropriate Green dyadic satisfies (2.32) on page 118

$$\nabla \times \left(\nabla \times \mathbf{G}_e^{an}(\boldsymbol{r}, \boldsymbol{r}')\right) - k_0^2 \mu(\omega)\boldsymbol{\epsilon}(\omega) \cdot \mathbf{G}_e^{an}(\boldsymbol{r}, \boldsymbol{r}') = \mathbf{I}_3 \delta(\boldsymbol{r} - \boldsymbol{r}')$$

In (3.16), let $\boldsymbol{F} = \boldsymbol{E}$ and $\mathbf{D} = \mathbf{G}_e^{an}$, and integrate over the volume V. Proceed as in Example 3.3, and we get for a symmetric permittivity dyadic $\boldsymbol{\epsilon}(\omega)$

$$k^2 \mu_0 \mu(\omega) \iiint_V \left(\boldsymbol{\epsilon}(\omega) \cdot \boldsymbol{E}(\boldsymbol{r})\right) \cdot \mathbf{G}_e^{an}(\boldsymbol{r}, \boldsymbol{r}') - \boldsymbol{E}(\boldsymbol{r}) \cdot \left(\boldsymbol{\epsilon}(\omega) \cdot \mathbf{G}_e^{an}(\boldsymbol{r}, \boldsymbol{r}')\right) \, dv$$

$$+ i\omega\mu_0\mu(\omega) \iiint_V \boldsymbol{J}(\boldsymbol{r}) \cdot \mathbf{G}_e^{an}(\boldsymbol{r}, \boldsymbol{r}') \, dv - \iiint_V \boldsymbol{E}(\boldsymbol{r})\delta(\boldsymbol{r} - \boldsymbol{r}') \, dv$$

$$= \iint_S \left\{\left(\hat{\boldsymbol{\nu}}(\boldsymbol{r}) \times \boldsymbol{E}(\boldsymbol{r})\right) \cdot \left(\nabla \times \mathbf{G}_e^{an}(\boldsymbol{r}, \boldsymbol{r}')\right) + \left(\hat{\boldsymbol{\nu}}(\boldsymbol{r}) \times (\nabla \times \boldsymbol{E}(\boldsymbol{r}))\right) \cdot \mathbf{G}_e^{an}(\boldsymbol{r}, \boldsymbol{r}')\right\} \, dS$$

The volume integral containing the electric field vanishes due to the symmetry of the permittivity dyadic, and we obtain

$$\iint_S \left(\hat{\nu}(\boldsymbol{r}') \times \boldsymbol{E}(\boldsymbol{r}')\right) \cdot \left(\nabla' \times \mathbf{G}_{\mathrm{e}}^{\mathrm{an}}(\boldsymbol{r}', \boldsymbol{r})\right) \mathrm{d}S' + \iint_S \left(\hat{\nu}(\boldsymbol{r}') \times (\nabla' \times \boldsymbol{E}(\boldsymbol{r}'))\right) \cdot \mathbf{G}_{\mathrm{e}}^{\mathrm{an}}(\boldsymbol{r}', \boldsymbol{r}) \, \mathrm{d}S' = \mathrm{i}\omega\mu_0\mu \iiint_V \boldsymbol{J}(\boldsymbol{r}') \cdot \mathbf{G}_{\mathrm{e}}^{\mathrm{an}}(\boldsymbol{r}', \boldsymbol{r}) \, \mathrm{d}v' - \begin{cases} \boldsymbol{E}(\boldsymbol{r}), & \boldsymbol{r} \text{ inside } S \\ \mathbf{0}, & \boldsymbol{r} \text{ outside } S \end{cases}$$

Again, the volume integral has to be interpreted in the distributional sense. ■

3.4.2 Alternative Representation of the Electric Field—Magnetic Case

An alternative representation is obtained by introducing an alternative Green dyadic $\mathbf{G}_{\mathrm{m}}(k, |\boldsymbol{r} - \boldsymbol{r}'|)$ defined as [6]

$$\mathbf{G}_{\mathrm{m}}(k, |\boldsymbol{r} - \boldsymbol{r}'|) = \mathrm{i}k\nabla \times \left\{g(k, |\boldsymbol{r} - \boldsymbol{r}'|)\mathbf{I}_3\right\} = \mathrm{i}k \sum_{j=1}^{3} \left(\nabla \times g(k, |\boldsymbol{r} - \boldsymbol{r}'|)\hat{\boldsymbol{e}}_j\right)\hat{\boldsymbol{e}}_j = \mathrm{i}k\nabla g(k, |\boldsymbol{r} - \boldsymbol{r}'|) \times \mathbf{I}_3 = -\mathrm{i}k\nabla' g(k, |\boldsymbol{r} - \boldsymbol{r}'|) \times \mathbf{I}_3$$

This dyadic is called the Green dyadic for the magnetic field in free space, and it is different, but related, to the one used above, $\mathbf{G}_{\mathrm{e}}(\boldsymbol{r}', \boldsymbol{r})$, and it satisfies the partial differential equation

$$\nabla \times \nabla \times \mathbf{G}_{\mathrm{m}}(k, |\boldsymbol{r} - \boldsymbol{r}'|) - k^2\mathbf{G}_{\mathrm{m}}(k, |\boldsymbol{r} - \boldsymbol{r}'|) = \mathrm{i}k\nabla \times \delta(\boldsymbol{r} - \boldsymbol{r}')\mathbf{I}_3$$

since $\nabla \times (\nabla \times g\hat{\boldsymbol{e}}_i) = \nabla \times (\nabla g \times \hat{\boldsymbol{e}}_i) = -\hat{\boldsymbol{e}}_i\nabla^2 g + (\hat{\boldsymbol{e}}_i \cdot \nabla)\nabla g$ where we used the vector operation (10) from the inside of the first cover of the book, $\nabla \times \left((\hat{\boldsymbol{e}}_i \cdot \nabla)\nabla g\right) = \mathbf{0}$, and $\nabla^2 g = -k^2 g - \delta(\boldsymbol{r} - \boldsymbol{r}')$.

Proceeding as above, and apply the result in (3.16) to the field $\boldsymbol{F}(\boldsymbol{r}) = \boldsymbol{E}(\boldsymbol{r})$ in a source-free region V, which satisfies

$$\nabla \times (\nabla \times \boldsymbol{E}(\boldsymbol{r})) - k^2\boldsymbol{E}(\boldsymbol{r}) = \mathbf{0}, \qquad \boldsymbol{r} \in V$$

and $\mathbf{D}(\boldsymbol{r}) = \mathbf{G}_{\mathrm{m}}(k, |\boldsymbol{r} - \boldsymbol{r}'|)$. The result is (for convenience changed the primed and unprimed coordinates)

$$\iint_S \left(\nabla' \times \mathbf{G}_{\mathrm{m}}(k, |\boldsymbol{r} - \boldsymbol{r}'|)\right) \cdot \left(\hat{\nu}(\boldsymbol{r}') \times \boldsymbol{E}(\boldsymbol{r}')\right) \mathrm{d}S' - \iint_S \mathbf{G}_{\mathrm{m}}(k, |\boldsymbol{r} - \boldsymbol{r}'|) \cdot \left(\hat{\nu}(\boldsymbol{r}') \times \left(\nabla' \times \boldsymbol{E}(\boldsymbol{r}')\right)\right) \mathrm{d}S' = \begin{cases} -\mathrm{i}k\nabla \times \boldsymbol{E}(\boldsymbol{r}), & \boldsymbol{r} \text{ inside } S \\ \mathbf{0}, & \boldsymbol{r} \text{ outside } S \end{cases}$$

since the volume integral reduces to

$$\iiint\limits_V \left\{ (\nabla' \times (\nabla' \times \boldsymbol{E}(\boldsymbol{r}'))) \cdot \mathbf{G}_{\mathrm{m}}(k, |\boldsymbol{r}-\boldsymbol{r}'|) - \boldsymbol{E}(\boldsymbol{r}') \cdot \left(\nabla' \times \left(\nabla' \times \mathbf{G}_{\mathrm{m}}(k, |\boldsymbol{r}-\boldsymbol{r}'|)\right)\right) \right\} \mathrm{d}v'$$

$$= -\iiint\limits_V \boldsymbol{E}(\boldsymbol{r}') \cdot \left(\mathrm{i}k\nabla' \times \delta(\boldsymbol{r}-\boldsymbol{r}')\mathbf{I}_3 \right) \mathrm{d}v'$$

$$= \mathrm{i}k \iiint\limits_V \nabla'\delta(\boldsymbol{r}-\boldsymbol{r}') \times \boldsymbol{E}(\boldsymbol{r}') \, \mathrm{d}v' = \begin{cases} -\mathrm{i}k\nabla \times \boldsymbol{E}(\boldsymbol{r}), & \boldsymbol{r} \in V \\ \mathbf{0}, & \boldsymbol{r} \in \mathbb{R}^3 \setminus \overline{V} \end{cases}$$

and we also have used that $\nabla' \times \mathbf{G}_{\mathrm{m}}(k, |\boldsymbol{r}-\boldsymbol{r}'|)$ is a symmetric dyadic, and that the dyadic itself, $\mathbf{G}_{\mathrm{m}}(k, |\boldsymbol{r}-\boldsymbol{r}'|)$, is an anti-symmetric dyadic.

3.5 Limit Values of the Scalar Integral Representations

In the previous sections in this chapter, we have carefully pointed out that the field point (observation point), $\boldsymbol{r}$, in the various integral representations is either inside or outside the open domain V. The bounding surface S has been excluded as a possible position of the field point $\boldsymbol{r}$. We are now ready to face the special problems that occur when we locate the field point at the surface, i.e., $\boldsymbol{r} \in S$. We first tackle the case when $\boldsymbol{r} \in S$ at a point where the tangential vectors of the surface are continuous, i.e., there is no corner or wedge at $\boldsymbol{r}$. The case with a corner or a wedge is then addressed in a separate section below, see Section 3.5.1.

The first difficulty we observe is that the surface integrals in the integral representations are not well defined—the singularity of the integrand when $\boldsymbol{r} \in S$ is proportional to $|\boldsymbol{r}-\boldsymbol{r}'|^{-n}$ where $n = 1, 2$, and even $|\boldsymbol{r}-\boldsymbol{r}'|^{-3}$ in some special cases. All surface integrals, if they exist, must then be interpreted as generalized integrals, see below for more details on this limit process. We are already familiar with this procedure in handling the volume integrals of the integral representations. In most situations, the singularity caused no problem and the volume integrals existed as generalized integrals. The introduction of the depolarizing dyadic was one exception that we have treated above, see Section 2.2.3.

To study the convergence of the generalized surface integrals, exclude a circular hole of radius ε in the surface, see Figure 3.5, and denote the punctured surface S_{punc}. We have to investigate what happens to the surface integrals as $\varepsilon \to 0$, when, in the limit, the original surface, S, is retrieved. In this section, we treat the scalar case, since the limit process in this case is easier, and come back to the vector and dyadic cases in the sections below.

The surface S_{punc} is an open surface, which we close by a spherical half cup $S_\varepsilon^{\mathrm{half}}$ of radius ε. This spherical surface can either extend into the volume, as in Figure 3.6, or extend out of the volume as in Figure 3.8. This procedure of extending the volume to a portion of space is legitimate if the surface fields ϕ and $\hat{\boldsymbol{\nu}} \cdot \nabla\phi$ are defined in a neighborhood of the surface S. We treat the case of extending into the volume first, since no extra assumption then has to be added to the field—we have already assumed the field is sufficiently regular in the volume V, which then include the spherical half cup, $S_\varepsilon^{\mathrm{half}}$, in Figure 3.6. The other case—an extension to the region outside the volume V—is treated below.

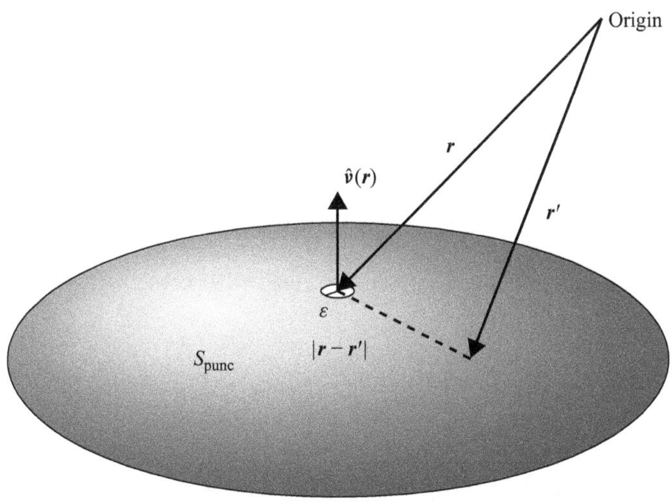

Figure 3.5 The surface S_{punc} in the limit case when $\boldsymbol{r} \in S$. The circular hole of radius ε punctures the bounding surface S.

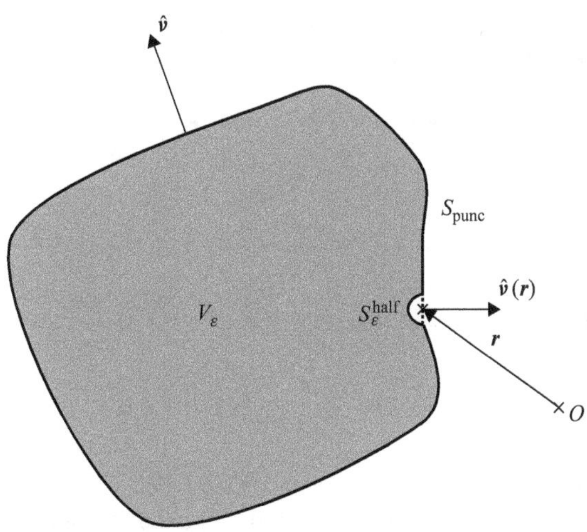

Figure 3.6 The geometry for the evaluation of the limit process of the surface integral contribution. The limit is taken from the outside of the volume V.

We employ the integral representation in (3.2) on page 143 to the volume V_ε, which has $S' = S_{\text{punc}} \cup S_\varepsilon^{\text{half}}$ as bounding surface, see Figure 3.6. For $\boldsymbol{r}$ on S, the integral representation reads

$$\iint\limits_{S_{\text{punc}} \cup S_\varepsilon^{\text{half}}} \left(g(k, |\boldsymbol{r}-\boldsymbol{r}'|)\hat{\boldsymbol{\nu}}(\boldsymbol{r}') \cdot \nabla'\phi(\boldsymbol{r}') - \phi(\boldsymbol{r}')\hat{\boldsymbol{\nu}}(\boldsymbol{r}') \cdot \nabla' g(k, |\boldsymbol{r}-\boldsymbol{r}'|) \right) \mathrm{d}S'$$
$$- \iiint\limits_{V_\varepsilon} g(k, |\boldsymbol{r}-\boldsymbol{r}'|) \left(\nabla'^2\phi(\boldsymbol{r}') + k^2\phi(\boldsymbol{r}') \right) \mathrm{d}v' = 0 \tag{3.19}$$

since the position vector $\boldsymbol{r}$ is excluded in the integration volume V_ε. In the limit $\varepsilon \to 0$, the volume integral causes no problem. This is a generalized integral with a well-defined limit, since the singularity is weak—the integrand has a singularity proportional to $\left|\boldsymbol{r}-\boldsymbol{r}'\right|^{-1}$.

We now analyze the singular surface integrals over $S_\varepsilon^{\text{half}}$ and S_{punc}. The surface integral over the punctured circular disk, S_{punc}, leads to generalized surface integrals. Provided the integrals over $S_\varepsilon^{\text{half}}$ have well-defined limits, the integral representation (3.19) proves that the generalized integrals over S_{punc} are well defined in the limit $\varepsilon \to 0$, since all other integrals are well defined.

The surface integral over hemisphere, $S_\varepsilon^{\text{half}}$, in (3.19) contains two canonical integrals. They are

$$\begin{cases} I_1(\boldsymbol{r}) = \displaystyle\iint\limits_{S_\varepsilon^{\text{half}}} g(k, |\boldsymbol{r}-\boldsymbol{r}'|)\hat{\boldsymbol{\nu}}(\boldsymbol{r}') \cdot \nabla'\phi(\boldsymbol{r}')\, \mathrm{d}S' \\ I_2(\boldsymbol{r}) = \displaystyle\iint\limits_{S_\varepsilon^{\text{half}}} \phi(\boldsymbol{r}')\hat{\boldsymbol{\nu}}(\boldsymbol{r}') \cdot \nabla' g(k, |\boldsymbol{r}-\boldsymbol{r}'|)\, \mathrm{d}S' \end{cases} \tag{3.20}$$

In these integrals, the unit normal vector $\hat{\boldsymbol{\nu}}(\boldsymbol{r}')$ varies rapidly over the integration domain. The field ϕ and its gradient $\nabla\phi$, however, are assumed to vary more slowly, and, we assume that the field $\phi(\boldsymbol{r}')$ and $\nabla'\phi(\boldsymbol{r}')$ have Taylor expansions

$$\begin{cases} \phi(\boldsymbol{r}') = \phi(\boldsymbol{r}) + (\boldsymbol{r}'-\boldsymbol{r}) \cdot \nabla\phi(\boldsymbol{r}) + O(|\boldsymbol{r}'-\boldsymbol{r}|^2) \\ \nabla'\phi(\boldsymbol{r}') = \nabla\phi(\boldsymbol{r}) + O(|\boldsymbol{r}'-\boldsymbol{r}|) \end{cases} \quad \text{as } \boldsymbol{r}' \to \boldsymbol{r} \tag{3.21}$$

These assumptions lead to

$$\begin{cases} I_1(\boldsymbol{r}) = -\displaystyle\int_0^{2\pi} \mathrm{d}\phi \int_0^{\pi/2} \varepsilon^2 \sin\theta \,\mathrm{d}\theta \frac{\mathrm{e}^{\mathrm{i}k\varepsilon}}{4\pi\varepsilon} \hat{\boldsymbol{R}} \cdot \nabla\phi(\boldsymbol{r}) \to 0, & \varepsilon \to 0 \\ I_2(\boldsymbol{r}) = -\displaystyle\int_0^{2\pi} \mathrm{d}\phi \int_0^{\pi/2} \varepsilon^2 \sin\theta \,\mathrm{d}\theta \frac{\mathrm{e}^{\mathrm{i}k\varepsilon}}{4\pi\varepsilon} \left(\mathrm{i}k - \frac{1}{\varepsilon}\right) \phi(\boldsymbol{r}) \to \frac{1}{2}\phi(\boldsymbol{r}), & \varepsilon \to 0 \end{cases}$$

where we have parameterized the integral over $S_\varepsilon^{\text{half}}$ by the spherical angles θ and ϕ with the direction $-\hat{\boldsymbol{\nu}}(\boldsymbol{r})$ (of the original surface S) as symmetry axis, see Figure 3.7, and where the unit normal on $S_\varepsilon^{\text{half}}$ is

$$\hat{\boldsymbol{\nu}}(\boldsymbol{r}') = -\hat{\boldsymbol{R}} = -\frac{\boldsymbol{r}'-\boldsymbol{r}}{|\boldsymbol{r}-\boldsymbol{r}'|}$$

Based on these limits, we see that the integrals over $S_\varepsilon^{\text{half}}$ have well-defined limits, and the limit value of the integral representation is

$$\iint\limits_S \left(g(k, |\boldsymbol{r}-\boldsymbol{r}'|)\hat{\boldsymbol{\nu}}(\boldsymbol{r}') \cdot \nabla'\phi(\boldsymbol{r}') - \phi(\boldsymbol{r}')\hat{\boldsymbol{\nu}}(\boldsymbol{r}') \cdot \nabla' g(k, |\boldsymbol{r}-\boldsymbol{r}'|)\right) \mathrm{d}S'$$
$$- \iiint\limits_V g(k, |\boldsymbol{r}-\boldsymbol{r}'|) \left(\nabla'^2\phi(\boldsymbol{r}') + k^2\phi(\boldsymbol{r}')\right) \mathrm{d}v' = \frac{1}{2}\phi(\boldsymbol{r}), \quad \boldsymbol{r} \in S \tag{3.22}$$

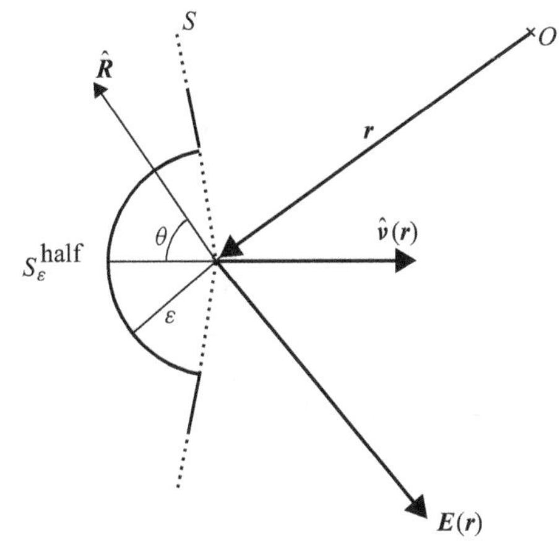

Figure 3.7 The geometry for the evaluation of integrals over the hemisphere $S_\varepsilon^{\text{half}}$.

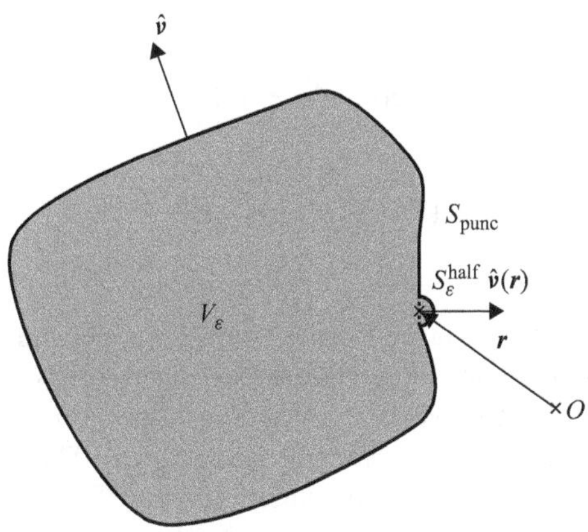

Figure 3.8 The geometry for the evaluation of the limit process of the surface integral contribution. The limit is taken from the inside of the volume V.

where the surface integral over the surface S exists as a generalized integral—the limit $\varepsilon \to 0$ over the surface S_{punc}. This limit process corresponds to the limit as the observation point approaches the surface S from outside the volume V.

We now take the limit as the observation point $\boldsymbol{r}$ approaches S from the inside by applying the integral representation to the volume in Figure 3.8. This analysis assumes that the field is defined in a neighborhood of the original volume V, such that the field is defined on the extended surface $S_\varepsilon^{\text{half}}$. The integral representation in (3.2) now gives

$$\iint_{S_{\text{punc}} \cup S_\varepsilon^{\text{half}}} \left(g(k, |\boldsymbol{r} - \boldsymbol{r}'|)\hat{\boldsymbol{\nu}}(\boldsymbol{r}') \cdot \nabla' \phi(\boldsymbol{r}') - \phi(\boldsymbol{r}')\hat{\boldsymbol{\nu}}(\boldsymbol{r}') \cdot \nabla' g(k, |\boldsymbol{r} - \boldsymbol{r}'|)\right) \, \mathrm{d}S'$$

$$- \iiint_{V_\varepsilon} g(k, |\boldsymbol{r} - \boldsymbol{r}'|) \left(\nabla'^2 \phi(\boldsymbol{r}') + k^2 \phi(\boldsymbol{r}')\right) \, \mathrm{d}v' = \phi(\boldsymbol{r})$$

Notice that the right-hand side now is $\phi(\boldsymbol{r})$, since the position vector $\boldsymbol{r}$ lies inside the domain of integration.

The evaluation of the integrals follows the previous analysis in the section. Everything is identical, except for a change in sign in the evaluation of integrals I_1 and I_2, since the unit normal vector $\hat{\boldsymbol{\nu}}$ has another direction on $S_\varepsilon^{\text{half}}$ compared to the previous situation. The final result in the limit process leads to a relation that is identical to the one in (3.22). This limit process corresponds to the limit of the observation point to the surface S taken from the inside the volume V, and we see that, in the limit, we get identical results, (3.22), independently of whether we take the limit from the inside or the outside. In this way, the results are consistent, and it does not matter if the limit is taken from the inside or the outside of S.

From the analysis above, we can also state the limit process as

$$\lim_{r\to S}\iint_S g(k,|\boldsymbol{r}-\boldsymbol{r}'|)\hat{\boldsymbol{\nu}}(\boldsymbol{r}')\cdot\nabla'\phi(\boldsymbol{r}')\,\mathrm{d}S' = \lim_{\varepsilon\to 0}\iint_{S_{\text{punc}}} g(k,|\boldsymbol{r}-\boldsymbol{r}'|)\hat{\boldsymbol{\nu}}(\boldsymbol{r}')\cdot\nabla'\phi(\boldsymbol{r}')\,\mathrm{d}S'$$

has the same limit value from both sides of the surface S (the integral is continuous across the surface), and

$$\begin{aligned}&\lim_{r\to S}\iint_S \phi(\boldsymbol{r}')\hat{\boldsymbol{\nu}}(\boldsymbol{r}')\cdot\nabla' g(k,|\boldsymbol{r}-\boldsymbol{r}'|)\,\mathrm{d}S'\\&\quad=\lim_{\varepsilon\to 0}\iint_{S_{\text{punc}}} \phi(\boldsymbol{r}')\hat{\boldsymbol{\nu}}(\boldsymbol{r}')\cdot\nabla' g(k,|\boldsymbol{r}-\boldsymbol{r}'|)\,\mathrm{d}S' \pm \frac{1}{2}\phi(\boldsymbol{r}), \qquad \boldsymbol{r}\in S\end{aligned}$$

depending on whether the limit is taken from the outside (upper sign) or the inside (lower sign) of S. The jump discontinuity of the integral on the left-hand side is exactly $\phi(\boldsymbol{r})$. Our derivation has been carried out for a bounded surface S. The jump discontinuity of the integral on the left-hand side is, however, exactly $\phi(\boldsymbol{r})$ even for an open surface S. This can be deduced by dividing the singular surface integral in two parts, where one part has no singularity, and therefore continuous in the limit process, and one singular one that contains the jump discontinuity.

3.5.1 Corners and Wedges

If the bounding surface S has a corner or a wedge at a point $\boldsymbol{r}$, with interior solid angle[7] $\Omega \neq 2\pi$ as seen from the point $\boldsymbol{r}$, the excluding surface is no longer a hemisphere, but a spherical cap or sector, S_ε^Ω, see Figure 3.9. The mathematical details of the behavior of the solutions to the Maxwell equations near corners and edges have been analyzed in the literature, see e.g., Reference 43.

The fields ϕ and $\nabla\phi$ are again assumed to be well behaved in a vicinity of the point $\boldsymbol{r}$. More explicitly, we assume the Taylor expansions in (3.21) to hold. This excludes the case of letting the bounding surface S to coincide with a material interface, since the fields in general are singular near a corner or a wedge, e.g., at a metallic corner [275]. Therefore, in this section, we assume the surface S does not coincide with an interface between two materials.

[7]Consult the definition of the solid angle in Appendix F.5.

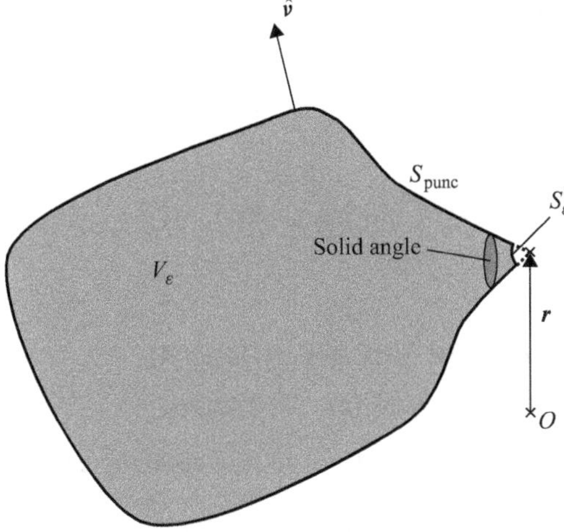

Figure 3.9 The geometry of a scatterer with a singularity on the bounding surface at $\boldsymbol{r}$. The surface S has a solid angle Ω as seen from the point $\boldsymbol{r}$.

The integrals I_1 and I_2 in (3.20) above are changed to

$$\begin{cases} I_1(\boldsymbol{r}) \to 0, & \varepsilon \to 0 \\ I_2(\boldsymbol{r}) \to \dfrac{\Omega}{4\pi}\phi(\boldsymbol{r}), & \varepsilon \to 0 \end{cases}$$

if the limit is taken from the outside, and

$$\begin{cases} I_1(\boldsymbol{r}) \to 0, & \varepsilon \to 0 \\ I_2(\boldsymbol{r}) \to -\dfrac{4\pi - \Omega}{4\pi}\phi(\boldsymbol{r}), & \varepsilon \to 0 \end{cases}$$

if the limit is taken from the inside. The integral representation in (3.2) on page 143 then implies

$$\iint_S \left(g(k,|\boldsymbol{r}-\boldsymbol{r}'|)\hat{\boldsymbol{\nu}}(\boldsymbol{r}')\cdot\nabla'\phi(\boldsymbol{r}') - \phi(\boldsymbol{r}')\hat{\boldsymbol{\nu}}(\boldsymbol{r}')\cdot\nabla' g(k,|\boldsymbol{r}-\boldsymbol{r}'|)\right)\,\mathrm{d}S'$$
$$-\iiint_V g(k,|\boldsymbol{r}-\boldsymbol{r}'|)\left(\nabla'^2\phi(\boldsymbol{r}') + k^2\phi(\boldsymbol{r}')\right)\,\mathrm{d}v' = \frac{\Omega}{4\pi}\phi(\boldsymbol{r}), \qquad \boldsymbol{r}\in S$$

both for a limit from the outside or the inside, and where all surface integrals are evaluated over the punctured surface S (generalized integral, $\varepsilon \to 0$ over the surface S_{punc}).

In particular, a wedge with interior opening angle β, the solid angle is $\Omega = 2\beta$, and we have

$$\iint_S \left(g(k,|\boldsymbol{r}-\boldsymbol{r}'|)\hat{\boldsymbol{\nu}}(\boldsymbol{r}')\cdot\nabla'\phi(\boldsymbol{r}') - \phi(\boldsymbol{r}')\hat{\boldsymbol{\nu}}(\boldsymbol{r}')\cdot\nabla' g(k,|\boldsymbol{r}-\boldsymbol{r}'|)\right)\,\mathrm{d}S'$$
$$-\iiint_V g(k,|\boldsymbol{r}-\boldsymbol{r}'|)\left(\nabla'^2\phi(\boldsymbol{r}') + k^2\phi(\boldsymbol{r}')\right)\,\mathrm{d}v' = \frac{\beta}{2\pi}\phi(\boldsymbol{r}), \qquad \boldsymbol{r}\in S$$

and for a cone with interior opening angle α (circular cross section), the solid angle is $\Omega = 2\pi(1-\cos\alpha)$, and we have

$$\iint_S \left(g(k,|\boldsymbol{r}-\boldsymbol{r}'|)\hat{\boldsymbol{\nu}}(\boldsymbol{r}')\cdot\nabla'\phi(\boldsymbol{r}') - \phi(\boldsymbol{r}')\hat{\boldsymbol{\nu}}(\boldsymbol{r}')\cdot\nabla' g(k,|\boldsymbol{r}-\boldsymbol{r}'|)\right)\,\mathrm{d}S'$$
$$-\iiint_V g(k,|\boldsymbol{r}-\boldsymbol{r}'|)\left(\nabla'^2\phi(\boldsymbol{r}') + k^2\phi(\boldsymbol{r}')\right)\,\mathrm{d}v' = \frac{1-\cos\alpha}{2}\phi(\boldsymbol{r}), \qquad \boldsymbol{r}\in S$$

3.6 Limit Values of the Vector Integral Representations—Vector Version

We now proceed with the limit of the vector-valued integral representation in (3.9) on page 152. The analysis is similar to the analysis in the scalar case in Section 3.5. We adopt the notation in Section 3.5, and apply the integral representation (3.9) at a regular point $\boldsymbol{r}\in S$, where the tangential vectors are continuous on the bounding surface to the original volume V. We first treat the case with limit from the outside, and we exclude a hemisphere from the volume V, see Figure 3.6. As above, the punctured volume is denoted V_ε. The result is

$$\iiint_{V_\varepsilon} \left[g(k,|\boldsymbol{r}-\boldsymbol{r}'|)\left(\nabla'\times\left(\nabla'\times\boldsymbol{F}(\boldsymbol{r}')\right) - \nabla'(\nabla'\cdot\boldsymbol{F}(\boldsymbol{r}')) - k^2\boldsymbol{F}(\boldsymbol{r}')\right)\right]\,\mathrm{d}v'$$
$$-\iint_{S_{\text{punc}}\cup S_\varepsilon^{\text{half}}} \left[g(k,|\boldsymbol{r}-\boldsymbol{r}'|)\left(\hat{\boldsymbol{\nu}}(\boldsymbol{r}')\times\left(\nabla'\times\boldsymbol{F}(\boldsymbol{r}')\right)\right)\right.$$
$$+\left(\nabla' g(k,|\boldsymbol{r}-\boldsymbol{r}'|)\right)\left(\hat{\boldsymbol{\nu}}(\boldsymbol{r}')\cdot\boldsymbol{F}(\boldsymbol{r}')\right) - \hat{\boldsymbol{\nu}}(\boldsymbol{r}')g(k,|\boldsymbol{r}-\boldsymbol{r}'|)\left(\nabla'\cdot\boldsymbol{F}(\boldsymbol{r}')\right)$$
$$\left.-\left(\nabla' g(k,|\boldsymbol{r}-\boldsymbol{r}'|)\right)\times\left(\hat{\boldsymbol{\nu}}(\boldsymbol{r}')\times\boldsymbol{F}(\boldsymbol{r}')\right)\right]\,\mathrm{d}S' = \boldsymbol{0}$$

since the position vector $\boldsymbol{r}$ is excluded in the integration volume V_ε. The field $\boldsymbol{F}$ is assumed sufficiently smooth in V.

To find the limit values of the surface integrals over the surface $S_\varepsilon^{\text{half}}$, we analyze the four canonical parts

$$\begin{cases} \boldsymbol{I}_1(\boldsymbol{r}) = \displaystyle\iint_{S_\varepsilon^{\text{half}}} g(k,|\boldsymbol{r}-\boldsymbol{r}'|)\left(\hat{\boldsymbol{\nu}}(\boldsymbol{r}')\times\boldsymbol{A}(\boldsymbol{r}')\right)\,\mathrm{d}S' \\ \boldsymbol{I}_2(\boldsymbol{r}) = \displaystyle\iint_{S_\varepsilon^{\text{half}}} \left(\nabla' g(k,|\boldsymbol{r}-\boldsymbol{r}'|)\right)\left(\hat{\boldsymbol{\nu}}(\boldsymbol{r}')\cdot\boldsymbol{A}(\boldsymbol{r}')\right)\,\mathrm{d}S' \end{cases}$$

and

$$\begin{cases} \boldsymbol{I}_3(\boldsymbol{r}) = \displaystyle\iint_{S_\varepsilon^{\text{half}}} g(k,|\boldsymbol{r}-\boldsymbol{r}'|)\hat{\boldsymbol{\nu}}(\boldsymbol{r}')\phi(\boldsymbol{r}')\,\mathrm{d}S' \\ \boldsymbol{I}_4(\boldsymbol{r}) = \displaystyle\iint_{S_\varepsilon^{\text{half}}} \left(\nabla' g(k,|\boldsymbol{r}-\boldsymbol{r}'|)\right)\times\left(\hat{\boldsymbol{\nu}}(\boldsymbol{r}')\times\boldsymbol{A}(\boldsymbol{r}')\right)\,\mathrm{d}S' \end{cases}$$

where $A(r)$ and $\phi(r)$ are slowly varying, smooth (continuous) canonical vector-valued and scalar fields, respectively, defined in V. Explicitly, we assume the field to be either $A = F$ or $A = \nabla \times F$, and $\phi = \nabla \cdot F$, and that the fields have the following expansions:

$$\begin{cases} A(r') = A(r) + O(|r' - r|) \\ \phi(r') = \phi(r) + O(|r' - r|) \end{cases} \quad \text{as } r' \to r$$

In these integrals, the unit normal vector $\hat{\nu}$ varies rapidly over the integration domain. Proceeding as in Section 3.5, we get

$$I_1(r) = -\int_0^{2\pi} d\phi \int_0^{\pi/2} \varepsilon^2 \sin\theta \; d\theta \frac{e^{ik\varepsilon}}{4\pi\varepsilon} \left(\hat{R} \times A(r')\right) \to 0, \qquad \varepsilon \to 0$$

$$I_2(r) = -\int_0^{2\pi} d\phi \int_0^{\pi/2} \varepsilon^2 \sin\theta \; d\theta \frac{e^{ik\varepsilon}}{4\pi\varepsilon} \left(ik - \frac{1}{\varepsilon}\right) \hat{R} \left(\hat{R} \cdot A(r')\right) \to \frac{1}{6} A(r), \qquad \varepsilon \to 0$$

$$I_3(r) = -\int_0^{2\pi} d\phi \int_0^{\pi/2} \varepsilon^2 \sin\theta \; d\theta \frac{e^{ik\varepsilon}}{4\pi\varepsilon} \hat{R}\phi(r') \to 0, \qquad \varepsilon \to 0$$

and

$$I_4(r) = -\int_0^{2\pi} d\phi \int_0^{\pi/2} \varepsilon^2 \sin\theta \; d\theta \frac{e^{ik\varepsilon}}{4\pi\varepsilon} \left(ik - \frac{1}{\varepsilon}\right) \hat{R} \times \left(\hat{R} \times A(r')\right)$$
$$\to \frac{1}{6} A(r) - \frac{1}{2} A(r) = -\frac{1}{3} A(r), \qquad \varepsilon \to 0$$

where we have used (3.4) on page 145, the BAC-CAB rule, and parameterized the integral over $S_\varepsilon^{\text{half}}$ by the spherical angles θ and ϕ with the direction $-\hat{\nu}(r)$ (of the original surface S) as the symmetry axis, see Figure 3.7, and

$$\iint_{\Omega_{\text{half}}} \hat{R}\hat{R} \; d\Omega = \frac{2\pi}{3} \mathbf{I}_3$$

where Ω_{half} is half of the unit sphere (symmetry axis arbitrary).

Based on these limits, we obtain, in analogy to the limit of the scalar fields in Section 3.5, the limit value of the integral representation of the vector fields. The result if this limit process is

$$\begin{aligned} &\iiint_V \left[g(k, |r - r'|) \left(\nabla' \times \left(\nabla' \times F(r')\right) - \nabla' \left(\nabla' \cdot F(r')\right) - k^2 F(r')\right)\right] dv' \\ &- \iint_S \left[g(k, |r - r'|) \left(\hat{\nu}(r') \times \left(\nabla' \times F(r')\right)\right) \right. \\ &+ \left(\nabla' g(k, |r - r'|)\right) \left(\hat{\nu}(r') \cdot F(r')\right) - \hat{\nu}(r') g(k, |r - r'|) \left(\nabla' \cdot F(r')\right) \\ &\left. - \left(\nabla' g(k, |r - r'|)\right) \times \left(\hat{\nu}(r') \times F(r')\right)\right] dS' = \frac{1}{2} F(r) \end{aligned} \tag{3.23}$$

where all surface integrals are evaluated over the punctured surface S (generalized integral, $\varepsilon \to 0$ over the surface S_{punc}). This limit process corresponds to taking the limit as $r \to S$ from the outside the volume V.

We now take the limit as the observation point $\boldsymbol{r}$ approaches S from the inside by applying the integral representation to the volume in Figure 3.8. This assumes that the field $\boldsymbol{F}$ can be extended outside the original volume V. We now have

$$\begin{aligned}&\iiint\limits_{V_\varepsilon} \left[g(k,|\boldsymbol{r}-\boldsymbol{r}'|)\left(\nabla'\times\left(\nabla'\times\boldsymbol{F}(\boldsymbol{r}')\right)-\nabla'\left(\nabla'\cdot\boldsymbol{F}(\boldsymbol{r}')\right)-k^2\boldsymbol{F}(\boldsymbol{r}')\right)\right] \mathrm{d}v' \\ &\quad - \iint\limits_{S_{\text{punc}}\cup S_\varepsilon^{\text{half}}} \left[g(k,|\boldsymbol{r}-\boldsymbol{r}'|)\left(\hat{\boldsymbol{\nu}}(\boldsymbol{r}')\times\left(\nabla'\times\boldsymbol{F}(\boldsymbol{r}')\right)\right)\right. \\ &\quad + \left(\nabla' g(k,|\boldsymbol{r}-\boldsymbol{r}'|)\right)\left(\hat{\boldsymbol{\nu}}(\boldsymbol{r}')\cdot\boldsymbol{F}(\boldsymbol{r}')\right) - \hat{\boldsymbol{\nu}}(\boldsymbol{r}')g(k,|\boldsymbol{r}-\boldsymbol{r}'|)\left(\nabla'\cdot\boldsymbol{F}(\boldsymbol{r}')\right) \\ &\quad \left.-\left(\nabla' g(k,|\boldsymbol{r}-\boldsymbol{r}'|)\right)\times\left(\hat{\boldsymbol{\nu}}(\boldsymbol{r}')\times\boldsymbol{F}(\boldsymbol{r}')\right)\right] \mathrm{d}S' = \boldsymbol{F}(\boldsymbol{r})\end{aligned}$$

since the position vector $\boldsymbol{r}$ is located inside the integration volume V_ε.

The evaluation of the integrals follows the previous analysis above. Everything is identical, except for a change in sign the evaluation of integrals $\boldsymbol{I}_1$–$\boldsymbol{I}_4$. The result of the limit process is identical to (3.23).

3.6.1 Maxwell Equations

Above, the field $\boldsymbol{F}(\boldsymbol{r})$ is a general differentiable field, and we do not require the field to satisfy any particular differential equation. However, if we take the field $\boldsymbol{F}(\boldsymbol{r})$ to be the electric field $\boldsymbol{E}$ that satisfies the Maxwell equations in a homogeneous, isotropic material, considerable simplifications occur in (3.23). Explicitly, the electric and the magnetic fields satisfy

$$\begin{cases}\nabla\times\boldsymbol{E} = \mathrm{i}\omega\mu_0\mu\boldsymbol{H} = \mathrm{i}k\eta_0\eta\boldsymbol{H} \\ \nabla\times\boldsymbol{H} = -\mathrm{i}\omega\epsilon_0\epsilon\boldsymbol{E} = -\dfrac{\mathrm{i}k}{\eta_0\eta}\boldsymbol{E}\end{cases}$$

which employed in (3.23) leads to

$$\begin{aligned}&-\iint\limits_{S}\left[\mathrm{i}k\eta_0\eta g(k,|\boldsymbol{r}-\boldsymbol{r}'|)\left(\hat{\boldsymbol{\nu}}(\boldsymbol{r}')\times\boldsymbol{H}(\boldsymbol{r}')\right)+\left(\nabla' g(k,|\boldsymbol{r}-\boldsymbol{r}'|)\right)\left(\hat{\boldsymbol{\nu}}(\boldsymbol{r}')\cdot\boldsymbol{E}(\boldsymbol{r}')\right)\right. \\ &\quad\left.-\left(\nabla' g(k,|\boldsymbol{r}-\boldsymbol{r}'|)\right)\times\left(\hat{\boldsymbol{\nu}}(\boldsymbol{r}')\times\boldsymbol{E}(\boldsymbol{r}')\right)\right]\mathrm{d}S' = \frac{1}{2}\boldsymbol{E}(\boldsymbol{r})\end{aligned} \tag{3.24}$$

This identity is the starting point of many numerical approaches to solve electromagnetic problems. The number of publications on this subject is extensive, and we refer the interested reader to the literature, see e.g., References 34, 38, 104.

3.6.2 Corners and Wedges

If the bounding surface S is not regular at a point $\boldsymbol{r}$, i.e., it has a corner or wedge, see Figure 3.9, we proceed as in Section 3.5.1. We define the integrals $I_i, i = 1, 2, 3, 4$ as above, and the integrals $\boldsymbol{I}_1, \boldsymbol{I}_3, \boldsymbol{I}_2 - \boldsymbol{I}_4$ have the limits

$$\boldsymbol{I}_1(\boldsymbol{r})\to\boldsymbol{0},\quad \boldsymbol{I}_3(\boldsymbol{r})\to\boldsymbol{0},\quad \boldsymbol{I}_2(\boldsymbol{r})-\boldsymbol{I}_4(\boldsymbol{r}) = -\frac{4\pi-\Omega}{4\pi}\boldsymbol{A}(\boldsymbol{r}),\qquad \varepsilon\to 0$$

from the inside, and

$$I_1(r) \to 0, \quad I_3(r) \to 0, \quad I_2(r) - I_4(r) = \frac{\Omega}{4\pi} A(r), \qquad \varepsilon \to 0$$

if the limit is taken from the outside. This leads to

$$\begin{aligned}
&\iiint_V \left[g(k, |r - r'|) \left(\nabla' \times \left(\nabla' \times F(r') \right) - \nabla' \left(\nabla' \cdot F(r') \right) - k^2 F(r') \right) \right] \mathrm{d}v' \\
&\quad - \iint_S \left[g(k, |r - r'|) \left(\hat{\nu}(r') \times \left(\nabla' \times F(r') \right) \right) \right. \\
&\quad + \left(\nabla' g(k, |r - r'|) \right) \left(\hat{\nu}(r') \cdot F(r') \right) - \hat{\nu}(r') g(k, |r - r'|) \left(\nabla' \cdot F(r') \right) \\
&\quad \left. - \left(\nabla' g(k, |r - r'|) \right) \times \left(\hat{\nu}(r') \times F(r') \right) \right] \mathrm{d}S' = \frac{\Omega}{4\pi} F(r), \qquad r \in S
\end{aligned}$$

for both a limit from the outside and the inside of V, and where all surface integrals are evaluated over the punctured surface S_{punc} (generalized integral, $\varepsilon \to 0$ over the surface S_{punc}).

In particular, a wedge with interior opening angle β, solid angle $\Omega = 2\beta$, we have

$$\begin{aligned}
&\iiint_V \left[g(k, |r - r'|) \left(\nabla' \times \left(\nabla' \times F(r') \right) - \nabla' \left(\nabla' \cdot F(r') \right) - k^2 F(r') \right) \right] \mathrm{d}v' \\
&\quad - \iint_S \left[g(k, |r - r'|) \left(\hat{\nu}(r') \times \left(\nabla' \times F(r') \right) \right) \right. \\
&\quad + \left(\nabla' g(k, |r - r'|) \right) \left(\hat{\nu}(r') \cdot F(r') \right) - \hat{\nu}(r') g(k, |r - r'|) \left(\nabla' \cdot F(r') \right) \\
&\quad \left. - \left(\nabla' g(k, |r - r'|) \right) \times \left(\hat{\nu}(r') \times F(r') \right) \right] \mathrm{d}S' = \frac{\beta}{2\pi} F(r), \qquad r \in S
\end{aligned}$$

and for a point with interior opening angle α, solid angle $\Omega = 2\pi(1 - \cos\alpha)$, we have

$$\begin{aligned}
&\iiint_V \left[g(k, |r - r'|) \left(\nabla' \times \left(\nabla' \times F(r') \right) - \nabla' \left(\nabla' \cdot F(r') \right) - k^2 F(r') \right) \right] \mathrm{d}v' \\
&\quad - \iint_S \left[g(k, |r - r'|) \left(\hat{\nu}(r') \times \left(\nabla' \times F(r') \right) \right) \right. \\
&\quad + \left(\nabla' g(k, |r - r'|) \right) \left(\hat{\nu}(r') \cdot F(r') \right) - \hat{\nu}(r') g(k, |r - r'|) \left(\nabla' \cdot F(r') \right) \\
&\quad \left. - \left(\nabla' g(k, |r - r'|) \right) \times \left(\hat{\nu}(r') \times F(r') \right) \right] \mathrm{d}S' = \frac{1 - \cos\alpha}{2} F(r), \qquad r \in S
\end{aligned}$$

3.7 Limit Values of the Vector Integral Representations—Dyadic Version

The limit value of the dyadic version in Section 3.4 has many features in common with the limit process made for the scalar and vector versions in Sections 3.5 and 3.6, respectively. In the dyadic case, we limit ourselves to fields that satisfy the Maxwell equations, see Section 3.4.1.

We adopt the notation and procedure in Sections 3.5 and 3.6, and apply the integral representation (3.18) at a regular point $\boldsymbol{r} \in S$, where the tangential vectors are continuous on the bounding surface to the original volume V.

We first treat the case with limit from the outside, and we exclude a hemisphere from the volume V, see Figure 3.6. As above, the punctured volume is denoted V_ε. Since the observation point, $\boldsymbol{r}$, lies outside the volume V_ε, we get

$$\begin{aligned}&\iint\limits_{S_{\text{punc}}\cup S_\varepsilon^{\text{half}}} \left(\nabla' \times \mathbf{G}_\text{e}(k, |\boldsymbol{r}-\boldsymbol{r}'|)\right) \cdot \left(\hat{\boldsymbol{\nu}}(\boldsymbol{r}') \times \boldsymbol{E}(\boldsymbol{r}')\right) \,\mathrm{d}S' \\ &\quad - \iint\limits_{S_{\text{punc}}\cup S_\varepsilon^{\text{half}}} \mathbf{G}_\text{e}(k, |\boldsymbol{r}-\boldsymbol{r}'|) \cdot \left(\hat{\boldsymbol{\nu}}(\boldsymbol{r}') \times \left(\nabla' \times \boldsymbol{E}(\boldsymbol{r}')\right)\right) \,\mathrm{d}S' = \mathbf{0} \end{aligned} \tag{3.25}$$

where the Green dyadic for the electric field, $\mathbf{G}_\text{e}(k, |\boldsymbol{r}-\boldsymbol{r}'|)$, is given by (2.28) on page 114

$$\mathbf{G}_\text{e}(k, |\boldsymbol{r}-\boldsymbol{r}'|) = \left(\mathbf{I}_3 + \frac{1}{k^2}\nabla\nabla\right) g(k, |\boldsymbol{r}-\boldsymbol{r}'|) = \left(\mathbf{I}_3 + \frac{1}{k^2}\nabla'\nabla'\right) g(k, |\boldsymbol{r}-\boldsymbol{r}'|)$$

and the curl of the Green dyadic for the electric field is

$$\nabla' \times \mathbf{G}_\text{e}(k, |\boldsymbol{r}-\boldsymbol{r}'|) = \nabla' g(k, |\boldsymbol{r}-\boldsymbol{r}'|) \times \mathbf{I}_3$$

The second integral in (3.25) contains a $|\boldsymbol{r}-\boldsymbol{r}'|^{-3}$ singularity, and, as the observation point approaches the boundary, this term has a singular contribution in its present form. However, this singularity is artificial, if we require that the electric field satisfies the Maxwell equations. To see this, we rewrite the singular part of the second term as (S is an arbitrary closed surface, and $\boldsymbol{r} \notin S$)

$$\begin{aligned} I_\text{sing} &= \frac{1}{k^2} \iint\limits_S \nabla'\nabla' g(k, |\boldsymbol{r}-\boldsymbol{r}'|) \cdot \left(\hat{\boldsymbol{\nu}}(\boldsymbol{r}') \times \left(\nabla' \times \boldsymbol{E}(\boldsymbol{r}')\right)\right) \,\mathrm{d}S' \\ &= \frac{1}{k^2} \nabla \iint\limits_S \left\{\nabla' g(k, |\boldsymbol{r}-\boldsymbol{r}'|) \times \left(\nabla' \times \boldsymbol{E}(\boldsymbol{r}')\right)\right\} \cdot \hat{\boldsymbol{\nu}}(\boldsymbol{r}') \,\mathrm{d}S' \end{aligned}$$

which is simplified to

$$\begin{aligned} I_\text{sing} &= \frac{1}{k^2} \nabla \iint\limits_S \nabla' \times \left\{g(k, |\boldsymbol{r}-\boldsymbol{r}'|) \left(\nabla' \times \boldsymbol{E}(\boldsymbol{r}')\right)\right\} \cdot \hat{\boldsymbol{\nu}}(\boldsymbol{r}') \,\mathrm{d}S' \\ &\quad - \frac{1}{k^2} \nabla \iint\limits_S g(k, |\boldsymbol{r}-\boldsymbol{r}'|) \left\{\nabla' \times \left(\nabla' \times \boldsymbol{E}(\boldsymbol{r}')\right)\right\} \cdot \hat{\boldsymbol{\nu}}(\boldsymbol{r}') \,\mathrm{d}S' \\ &= -\nabla \iint\limits_S g(k, |\boldsymbol{r}-\boldsymbol{r}'|) \boldsymbol{E}(\boldsymbol{r}') \cdot \hat{\boldsymbol{\nu}}(\boldsymbol{r}') \,\mathrm{d}S' \\ &= \iint\limits_S \nabla' g(k, |\boldsymbol{r}-\boldsymbol{r}'|) \left\{\boldsymbol{E}(\boldsymbol{r}') \cdot \hat{\boldsymbol{\nu}}(\boldsymbol{r}')\right\} \,\mathrm{d}S' \end{aligned}$$

due to Stokes' theorem and the equation of the electric field (3.17) on page 158. Utilizing this identity on the closed surface $S_{\text{punc}} \cup S_\varepsilon^{\text{half}}$, the identity (3.25) becomes

$$\begin{aligned}
&\iint\limits_{S_{\text{punc}} \cup S_\varepsilon^{\text{half}}} \left(\nabla' g(k, |\boldsymbol{r} - \boldsymbol{r}'|)\right) \times \left(\hat{\boldsymbol{\nu}}(\boldsymbol{r}') \times \boldsymbol{E}(\boldsymbol{r}')\right) \, \mathrm{d}S' \\
&\quad - \iint\limits_{S_{\text{punc}} \cup S_\varepsilon^{\text{half}}} g(k, |\boldsymbol{r} - \boldsymbol{r}'|)\hat{\boldsymbol{\nu}}(\boldsymbol{r}') \times \left(\nabla' \times \boldsymbol{E}(\boldsymbol{r}')\right) \, \mathrm{d}S' \\
&\quad - \iint\limits_{S_{\text{punc}} \cup S_\varepsilon^{\text{half}}} \nabla' g(k, |\boldsymbol{r} - \boldsymbol{r}'|) \left\{\boldsymbol{E}(\boldsymbol{r}') \cdot \hat{\boldsymbol{\nu}}(\boldsymbol{r}')\right\} \, \mathrm{d}S' \, \mathrm{d}S' = \boldsymbol{0}
\end{aligned}$$

In order to investigate whether the generalized surface integrals over S_{punc} exist or not in the limit $\varepsilon \to 0$, we study the limits of the surface integrals over $S_\varepsilon^{\text{half}}$ in this limit. Three integrals occur (we keep the notation of Section 3.6):

$$\begin{cases}
\boldsymbol{I}_1 = \iint\limits_{S_\varepsilon^{\text{half}}} g(k, |\boldsymbol{r} - \boldsymbol{r}'|)\hat{\boldsymbol{\nu}}(\boldsymbol{r}') \times \left(\nabla' \times \boldsymbol{E}(\boldsymbol{r}')\right) \, \mathrm{d}S' \\
\boldsymbol{I}_2 = \iint\limits_{S_\varepsilon^{\text{half}}} \nabla' g(k, |\boldsymbol{r} - \boldsymbol{r}'|) \left\{\boldsymbol{E}(\boldsymbol{r}') \cdot \hat{\boldsymbol{\nu}}(\boldsymbol{r}')\right\} \, \mathrm{d}S' \, \mathrm{d}S' \\
\boldsymbol{I}_4 = \iint\limits_{S_\varepsilon^{\text{half}}} \left(\nabla' g(k, |\boldsymbol{r} - \boldsymbol{r}'|)\right) \times \left(\hat{\boldsymbol{\nu}}(\boldsymbol{r}') \times \boldsymbol{E}(\boldsymbol{r}')\right) \, \mathrm{d}S'
\end{cases}$$

where $\boldsymbol{E}$ is a smooth field in a neighborhood of the observation point $\boldsymbol{r}$. We have already addressed the limits of these integrals in Section 3.6, and we have

$$\boldsymbol{I}_1 = \boldsymbol{0}, \quad \boldsymbol{I}_2 = \frac{1}{6}\boldsymbol{E}(\boldsymbol{r}), \quad \boldsymbol{I}_4 = -\frac{1}{3}\boldsymbol{E}(\boldsymbol{r}), \qquad \varepsilon \to 0$$

The surface integral over S_{punc} leads to a generalized surface integral. Using these limits, we finally get

$$\begin{aligned}
&\iint\limits_{S_{\text{punc}}} \nabla' g(k, |\boldsymbol{r} - \boldsymbol{r}'|) \times \left(\hat{\boldsymbol{\nu}}(\boldsymbol{r}') \times \boldsymbol{E}(\boldsymbol{r}')\right) \, \mathrm{d}S' \\
&\quad - \iint\limits_{S_{\text{punc}}} g(k, |\boldsymbol{r} - \boldsymbol{r}'|)\hat{\boldsymbol{\nu}}(\boldsymbol{r}') \times \left(\nabla' \times \boldsymbol{E}(\boldsymbol{r}')\right) \, \mathrm{d}S' \\
&\quad - \iint\limits_{S_{\text{punc}}} \nabla' g(k, |\boldsymbol{r} - \boldsymbol{r}'|)\hat{\boldsymbol{\nu}}(\boldsymbol{r}') \cdot \boldsymbol{E}(\boldsymbol{r}') \, \mathrm{d}S' = \frac{1}{2}\boldsymbol{E}(\boldsymbol{r}), \qquad \boldsymbol{r} \in S
\end{aligned}$$

where the surface integrals on the left-hand side exists as improper (generalized) integrals over S (i.e., as limits over S_{punc} as $\varepsilon \to 0$). This result is identical to the result in Section 3.7, see (3.24).

The limit from the inside is obtain in a similar way, and leads to a result which is identical to the result obtained above, i.e., the limit process from both sides gives identical result. We leave the details of this limit process to the reader to analyze.

3.8 Integral Representation for Biisotropic Materials

The field equations for a biisotropic material were developed and analyzed in Section 2.4, see (2.33) on page 119. The explicit form of these equations is

$$\begin{pmatrix} \nabla \times \boldsymbol{E} \\ \eta_0 \nabla \times \boldsymbol{H} \end{pmatrix} = \mathrm{i}k_0 \begin{pmatrix} \zeta & \mu \\ -\epsilon & -\xi \end{pmatrix} \begin{pmatrix} \boldsymbol{E} \\ \eta_0 \boldsymbol{H} \end{pmatrix} = \mathrm{i}k_0 \begin{pmatrix} \kappa - \mathrm{i}\chi & \mu \\ -\epsilon & -\kappa - \mathrm{i}\chi \end{pmatrix} \begin{pmatrix} \boldsymbol{E} \\ \eta_0 \boldsymbol{H} \end{pmatrix} \tag{3.26}$$

where we assumed the medium is source free, and we denote, as usual, the wave number in free space by $k_0 = \omega/c_0$. We have also used the complex-valued parameters, κ and χ, the reciprocity and the chirality parameters, respectively.

The integral representation of the electromagnetic field in a biisotropic material can be obtained by a modification of the results presented above. To see this, we start with the original equation, (3.26), and make a change in the dependent variables $\boldsymbol{E}$ and $\boldsymbol{H}$

$$\begin{pmatrix} \boldsymbol{E} \\ \eta_0 \boldsymbol{H} \end{pmatrix} = \begin{pmatrix} \alpha & \beta \\ \gamma & \delta \end{pmatrix} \begin{pmatrix} \boldsymbol{Q}_\mathrm{L} \\ \boldsymbol{Q}_\mathrm{R} \end{pmatrix}$$

where α, β, γ, and δ are constants, e.g., not functions of $\boldsymbol{r}$, but they can depend on ω. Now determine these constants so that (3.26) is diagonal in the new dependent variables $\boldsymbol{Q}_{\mathrm{L,R}}$. This transformation is due to Craig Bohren[8] [17]. The eigenvalues of the matrix in (3.26) are

$$\lambda_{1,2} = k_0 \left(\chi \pm \left(\epsilon\mu - \kappa^2 \right)^{1/2} \right) = \pm k_\pm$$

where

$$k_\pm = k_0 \left(\left(\epsilon\mu - \kappa^2 \right)^{1/2} \pm \chi \right)$$

denote the wave numbers of the right- and left-circularly polarized waves, respectively. The matrix that diagonalizes (3.26) is

$$\begin{pmatrix} \alpha & \beta \\ \gamma & \delta \end{pmatrix} = \begin{pmatrix} -\mathrm{i}\mu & -\mathrm{i}\mu \\ \mathrm{i}\kappa + \left(\epsilon\mu - \kappa^2 \right)^{1/2} & \mathrm{i}\kappa - \left(\epsilon\mu - \kappa^2 \right)^{1/2} \end{pmatrix}$$

with inverse

$$\begin{pmatrix} \alpha & \beta \\ \gamma & \delta \end{pmatrix}^{-1} = \frac{1}{2\mathrm{i}\mu \left(\epsilon\mu - \kappa^2 \right)^{1/2}} \begin{pmatrix} \mathrm{i}\kappa - \left(\epsilon\mu - \kappa^2 \right)^{1/2} & \mathrm{i}\mu \\ -\mathrm{i}\kappa - \left(\epsilon\mu - \kappa^2 \right)^{1/2} & -\mathrm{i}\mu \end{pmatrix}$$

---

[8] Craig F. Bohren (1940–), American atmospheric scientist and physicist.

The explicit transformations therefore are

$$\begin{pmatrix} \boldsymbol{E} \\ \eta_0 \boldsymbol{H} \end{pmatrix} = \begin{pmatrix} -\mathrm{i}\mu & -\mathrm{i}\mu \\ \mathrm{i}\kappa + \left(\epsilon\mu - \kappa^2\right)^{1/2} & \mathrm{i}\kappa - \left(\epsilon\mu - \kappa^2\right)^{1/2} \end{pmatrix} \begin{pmatrix} \boldsymbol{Q}_{\mathrm{L}} \\ \boldsymbol{Q}_{\mathrm{R}} \end{pmatrix} \tag{3.27}$$

and

$$\begin{pmatrix} \boldsymbol{Q}_{\mathrm{L}} \\ \boldsymbol{Q}_{\mathrm{R}} \end{pmatrix} = \frac{1}{2\mathrm{i}\mu\left(\epsilon\mu - \kappa^2\right)^{1/2}} \begin{pmatrix} \mathrm{i}\kappa - \left(\epsilon\mu - \kappa^2\right)^{1/2} & \mathrm{i}\mu \\ -\mathrm{i}\kappa - \left(\epsilon\mu - \kappa^2\right)^{1/2} & -\mathrm{i}\mu \end{pmatrix} \begin{pmatrix} \boldsymbol{E} \\ \eta_0 \boldsymbol{H} \end{pmatrix} \tag{3.28}$$

The new fields $\boldsymbol{Q}_{\mathrm{L,R}}$ satisfy[9]

$$\begin{aligned} \begin{pmatrix} \nabla \times \boldsymbol{Q}_{\mathrm{L}} \\ \nabla \times \boldsymbol{Q}_{\mathrm{R}} \end{pmatrix} &= \mathrm{i}k_0 \begin{pmatrix} \alpha & \beta \\ \gamma & \delta \end{pmatrix}^{-1} \begin{pmatrix} \kappa - \mathrm{i}\chi & \mu \\ -\epsilon & -\kappa - \mathrm{i}\chi \end{pmatrix} \begin{pmatrix} \alpha & \beta \\ \gamma & \delta \end{pmatrix} \begin{pmatrix} \boldsymbol{Q}_{\mathrm{L}} \\ \boldsymbol{Q}_{\mathrm{R}} \end{pmatrix} \\ &= k_0 \begin{pmatrix} -\left(\left(\epsilon\mu - \kappa^2\right)^{1/2} - \chi\right) & 0 \\ 0 & \left(\epsilon\mu - \kappa^2\right)^{1/2} + \chi \end{pmatrix} \begin{pmatrix} \boldsymbol{Q}_{\mathrm{L}} \\ \boldsymbol{Q}_{\mathrm{R}} \end{pmatrix} \\ &= \begin{pmatrix} -k_- & 0 \\ 0 & k_+ \end{pmatrix} \begin{pmatrix} \boldsymbol{Q}_{\mathrm{L}} \\ \boldsymbol{Q}_{\mathrm{R}} \end{pmatrix} \end{aligned} \tag{3.29}$$

and consequently

$$\begin{pmatrix} \nabla \times (\nabla \times \boldsymbol{Q}_{\mathrm{L}}(\boldsymbol{r})) \\ \nabla \times (\nabla \times \boldsymbol{Q}_{\mathrm{R}}(\boldsymbol{r})) \end{pmatrix} = \begin{pmatrix} k_-^2 & 0 \\ 0 & k_+^2 \end{pmatrix} \begin{pmatrix} \boldsymbol{Q}_{\mathrm{L}}(\boldsymbol{r}) \\ \boldsymbol{Q}_{\mathrm{R}}(\boldsymbol{r}) \end{pmatrix}$$

To continue, these results are now applied to the general integral representation in (3.9) on page 152 with the field $\boldsymbol{F} = \boldsymbol{Q}_{\mathrm{L}}$ and the wave number $k = k_-$. We assume that the region V is source free so that $\nabla \cdot \boldsymbol{E} = \nabla \cdot \boldsymbol{H} = 0$ hold, and consequently $\nabla \cdot \boldsymbol{Q}_{\mathrm{L}} = \nabla \cdot \boldsymbol{Q}_{\mathrm{R}} = 0$. We get

$$\begin{aligned} &-\iint_S \Big[g(k_-, |\boldsymbol{r} - \boldsymbol{r}'|)\left(\hat{\boldsymbol{\nu}}(\boldsymbol{r}') \times \left(\nabla' \times \boldsymbol{Q}_{\mathrm{L}}(\boldsymbol{r}')\right)\right) \\ &\quad + \left(\nabla' g(k_-, |\boldsymbol{r} - \boldsymbol{r}'|)\right)\left(\hat{\boldsymbol{\nu}}(\boldsymbol{r}') \cdot \boldsymbol{Q}_{\mathrm{L}}(\boldsymbol{r}')\right) \\ &\quad - \left(\nabla' g(k_-, |\boldsymbol{r} - \boldsymbol{r}'|)\right) \times \left(\hat{\boldsymbol{\nu}}(\boldsymbol{r}') \times \boldsymbol{Q}_{\mathrm{L}}(\boldsymbol{r}')\right)\Big]\, \mathrm{d}S' = \begin{cases} \boldsymbol{Q}_{\mathrm{L}}(\boldsymbol{r}), & \boldsymbol{r} \text{ inside } S \\ \boldsymbol{0}, & \boldsymbol{r} \text{ outside } S \end{cases} \end{aligned}$$

[9]Note that

$$\begin{pmatrix} \nabla \times \boldsymbol{Q}_{\mathrm{L}} \\ \nabla \times \boldsymbol{Q}_{\mathrm{R}} \end{pmatrix} = \nabla \times \left[\begin{pmatrix} \alpha & \beta \\ \gamma & \delta \end{pmatrix}^{-1} \begin{pmatrix} \boldsymbol{E} \\ \eta_0 \boldsymbol{H} \end{pmatrix}\right] = \begin{pmatrix} \alpha & \beta \\ \gamma & \delta \end{pmatrix}^{-1} \begin{pmatrix} \nabla \times \boldsymbol{E} \\ \eta_0 \nabla \times \boldsymbol{H} \end{pmatrix}$$

Similarly, we obtain an integral representation for the field $\boldsymbol{F} = \boldsymbol{Q}_\mathrm{R}$ using the wave number $k = k_+$

$$
\begin{aligned}
-\iint_S \Big[& g(k_+, |\boldsymbol{r}-\boldsymbol{r}'|)\left(\hat{\boldsymbol{\nu}}(\boldsymbol{r}') \times (\nabla' \times \boldsymbol{Q}_\mathrm{R}(\boldsymbol{r}'))\right) \\
& + (\nabla' g(k_+, |\boldsymbol{r}-\boldsymbol{r}'|))(\hat{\boldsymbol{\nu}}(\boldsymbol{r}') \cdot \boldsymbol{Q}_\mathrm{R}(\boldsymbol{r}')) \\
& - (\nabla' g(k_+, |\boldsymbol{r}-\boldsymbol{r}'|)) \times (\hat{\boldsymbol{\nu}}(\boldsymbol{r}') \times \boldsymbol{Q}_\mathrm{R}(\boldsymbol{r}'))\Big]\,\mathrm{d}S' = \begin{cases} \boldsymbol{Q}_\mathrm{R}(\boldsymbol{r}), & \boldsymbol{r} \text{ inside } S \\ \boldsymbol{0}, & \boldsymbol{r} \text{ outside } S \end{cases}
\end{aligned}
$$

Introduce the shorthand notation of the six-dimensional fields as

$$
\mathsf{e}(\boldsymbol{r}) = \begin{pmatrix} \boldsymbol{E}(\boldsymbol{r}) \\ \eta_0 \boldsymbol{H}(\boldsymbol{r}) \end{pmatrix}, \qquad \mathsf{q}(\boldsymbol{r}) = \begin{pmatrix} \boldsymbol{Q}_\mathrm{L}(\boldsymbol{r}) \\ \boldsymbol{Q}_\mathrm{R}(\boldsymbol{r}) \end{pmatrix}, \qquad \mathsf{0} = \begin{pmatrix} \boldsymbol{0} \\ \boldsymbol{0} \end{pmatrix}
$$

and

$$
\hat{\boldsymbol{\nu}} \times \mathsf{q} = \begin{pmatrix} \hat{\boldsymbol{\nu}} \times \boldsymbol{Q}_\mathrm{L} \\ \hat{\boldsymbol{\nu}} \times \boldsymbol{Q}_\mathrm{R} \end{pmatrix}, \qquad \hat{\boldsymbol{\nu}} \cdot \mathsf{q} = \begin{pmatrix} \hat{\boldsymbol{\nu}} \cdot \boldsymbol{Q}_\mathrm{L} \\ \hat{\boldsymbol{\nu}} \cdot \boldsymbol{Q}_\mathrm{R} \end{pmatrix}
$$

Using (3.29), the result is now most conveniently written as a 2×2 block system (use also $\nabla' = -\nabla$)

$$
\begin{aligned}
& -\iint_S \begin{pmatrix} -k_- g(k_-, |\boldsymbol{r}-\boldsymbol{r}'|) & 0 \\ 0 & k_+ g(k_+, |\boldsymbol{r}-\boldsymbol{r}'|) \end{pmatrix} \hat{\boldsymbol{\nu}}(\boldsymbol{r}') \times \mathsf{q}(\boldsymbol{r}')\,\mathrm{d}S' \\
& + \nabla \iint_S \begin{pmatrix} g(k_-, |\boldsymbol{r}-\boldsymbol{r}'|) & 0 \\ 0 & g(k_+, |\boldsymbol{r}-\boldsymbol{r}'|) \end{pmatrix} \hat{\boldsymbol{\nu}}(\boldsymbol{r}') \cdot \mathsf{q}(\boldsymbol{r}')\,\mathrm{d}S' \\
& - \nabla \times \iint_S \begin{pmatrix} g(k_-, |\boldsymbol{r}-\boldsymbol{r}'|) & 0 \\ 0 & g(k_+, |\boldsymbol{r}-\boldsymbol{r}'|) \end{pmatrix} \hat{\boldsymbol{\nu}}(\boldsymbol{r}') \times \mathsf{q}(\boldsymbol{r}')\,\mathrm{d}S' \\
& = \begin{cases} \mathsf{q}(\boldsymbol{r}), & \boldsymbol{r} \text{ inside } S \\ \mathsf{0}, & \boldsymbol{r} \text{ outside } S \end{cases}
\end{aligned}
$$

We proceed as in Section 3.3.1 by eliminating the normal components in these integral representations. Taking the curl of the representation to eliminate the second integral and using (3.29), we obtain

$$
\begin{aligned}
& \nabla \times \left\{ \nabla \times \iint_S \begin{pmatrix} \frac{g(k_-, |\boldsymbol{r}-\boldsymbol{r}'|)}{k_-} & 0 \\ 0 & -\frac{g(k_+, |\boldsymbol{r}-\boldsymbol{r}'|)}{k_+} \end{pmatrix} \hat{\boldsymbol{\nu}}(\boldsymbol{r}') \times \mathsf{q}(\boldsymbol{r}')\,\mathrm{d}S' \right\} \\
& - \nabla \times \iint_S \begin{pmatrix} g(k_-, |\boldsymbol{r}-\boldsymbol{r}'|) & 0 \\ 0 & g(k_+, |\boldsymbol{r}-\boldsymbol{r}'|) \end{pmatrix} \hat{\boldsymbol{\nu}}(\boldsymbol{r}') \times \mathsf{q}(\boldsymbol{r}')\,\mathrm{d}S' \\
& = \begin{cases} \mathsf{q}(\boldsymbol{r}), & \boldsymbol{r} \text{ inside } S \\ \mathsf{0}, & \boldsymbol{r} \text{ outside } S \end{cases}
\end{aligned}
$$

We are now ready to transform back to the original fields $\boldsymbol{E}$ and $\boldsymbol{H}$, by the use of (3.27) and (3.28). Since

$$\begin{pmatrix} \dfrac{-\mathrm{i}\mu}{\mathrm{i}\kappa + \left(\epsilon\mu - \kappa^2\right)^{1/2}} & \dfrac{-\mathrm{i}\mu}{\mathrm{i}\kappa - \left(\epsilon\mu - \kappa^2\right)^{1/2}} \end{pmatrix} \begin{pmatrix} a & 0 \\ 0 & b \end{pmatrix} \begin{pmatrix} \mathrm{i}\kappa - \left(\epsilon\mu - \kappa^2\right)^{1/2} & \mathrm{i}\mu \\ -\mathrm{i}\kappa - \left(\epsilon\mu - \kappa^2\right)^{1/2} & -\mathrm{i}\mu \end{pmatrix}$$
$$= \mu \begin{pmatrix} (a-b)\kappa + \mathrm{i}(a+b)\left(\epsilon\mu - \kappa^2\right)^{1/2} & \mu(a-b) \\ -\epsilon(a-b) & -(a-b)\kappa + \mathrm{i}(a+b)\left(\epsilon\mu - \kappa^2\right)^{1/2} \end{pmatrix}$$

we get

$$\nabla \times \left\{ \nabla \times \iint_S [\mathbf{A}](|\boldsymbol{r} - \boldsymbol{r}'|)\hat{\boldsymbol{v}}(\boldsymbol{r}') \times \mathrm{e}(\boldsymbol{r}')\, \mathrm{d}S' \right\}$$
$$- \nabla \times \iint_S [\mathbf{B}](|\boldsymbol{r} - \boldsymbol{r}'|)\hat{\boldsymbol{v}}(\boldsymbol{r}') \times \mathrm{e}(\boldsymbol{r}')\, \mathrm{d}S' = \begin{cases} \mathrm{e}(\boldsymbol{r}), & \boldsymbol{r} \text{ inside } S \\ 0, & \boldsymbol{r} \text{ outside } S \end{cases} \tag{3.30}$$

where the entries of the matrices $[\mathbf{A}](\cdot)$ and $[\mathbf{B}](\cdot)$ are

$$\begin{cases} A_{11}(\cdot) = \dfrac{\kappa G_+(\cdot) - \mathrm{i}G_-(\cdot)\left(\epsilon\mu - \kappa^2\right)^{1/2}}{2\mathrm{i}\left(\epsilon\mu - \kappa^2\right)^{1/2}} \\[2ex] A_{22}(\cdot) = \dfrac{-\kappa G_+(\cdot) - \mathrm{i}G_-(\cdot)\left(\epsilon\mu - \kappa^2\right)^{1/2}}{2\mathrm{i}\left(\epsilon\mu - \kappa^2\right)^{1/2}} \end{cases} \qquad \begin{cases} A_{12}(\cdot) = \dfrac{\mu G_+(\cdot)}{2\mathrm{i}\left(\epsilon\mu - \kappa^2\right)^{1/2}} \\[2ex] A_{21}(\cdot) = -\dfrac{\epsilon G_+(\cdot)}{2\mathrm{i}\left(\epsilon\mu - \kappa^2\right)^{1/2}} \end{cases}$$

where we, for convenience, introduced the collective notation

$$G_\pm(\cdot) = \frac{g(k_+, \cdot)}{k_+} \pm \frac{g(k_-, \cdot)}{k_-}$$

and where

$$\begin{cases} B_{11}(\cdot) = \dfrac{-\kappa(g(k_+,\cdot) - g(k_-,\cdot)) + \mathrm{i}\,(g(k_+,\cdot) + g(k_-,\cdot))\left(\epsilon\mu - \kappa^2\right)^{1/2}}{2\mathrm{i}(\epsilon\mu - \kappa^2)^{1/2}} \\[2ex] B_{12}(\cdot) = -\dfrac{\mu(g(k_+,\cdot) - g(k_-,\cdot))}{2\mathrm{i}\left(\epsilon\mu - \kappa^2\right)^{1/2}} \\[2ex] B_{21}(\cdot) = \dfrac{\epsilon(g(k_+,\cdot) - g(k_-,\cdot))}{2\mathrm{i}\left(\epsilon\mu - \kappa^2\right)^{1/2}} \\[2ex] B_{22}(\cdot) = \dfrac{\kappa(g(k_+,\cdot) - g(k_-,\cdot)) + \mathrm{i}(g(k_+,\cdot) + g(k_-,\cdot))\left(\epsilon\mu - \kappa^2\right)^{1/2}}{2\mathrm{i}\left(\epsilon\mu - \kappa^2\right)^{1/2}} \end{cases}$$

Equation (3.30) explicitly gives the integral representation of the coupled electric and magnetic fields in a biisotropic material. In the following examples, we give some special cases of this integral representation.

Example 3.5
The chiral material (reciprocal biisotropic material) is characterized by $\kappa = 0$, and for this special case, we obtain the following simplifications in the matrices $[\mathbf{A}](\cdot)$ and $[\mathbf{B}](\cdot)$:

$$A_{11}(\cdot) = A_{22}(\cdot) = -\frac{G_-(\cdot)}{2}, \qquad \begin{cases} A_{12}(\cdot) = \dfrac{\mu G_+(\cdot)}{2\mathrm{i}\,(\epsilon\mu)^{1/2}} \\ A_{21}(\cdot) = -\dfrac{\epsilon G_+(\cdot)}{2\mathrm{i}\,(\epsilon\mu)^{1/2}} \end{cases}$$

and

$$\begin{cases} B_{11}(\cdot) = \dfrac{(g(k_+,\cdot) + g(k_-,\cdot))}{2} \\ B_{22}(\cdot) = \dfrac{(g(k_+,\cdot) + g(k_-,\cdot))}{2} \end{cases} \qquad \begin{cases} B_{12}(\cdot) = -\dfrac{\mu(g(k_+,\cdot) - g(k_-,\cdot))}{2\mathrm{i}(\epsilon\mu)^{1/2}} \\ B_{21}(\cdot) = \dfrac{\epsilon(g(k_+,\cdot) - g(k_-,\cdot))}{2\mathrm{i}(\epsilon\mu)^{1/2}} \end{cases}$$

where $k_\pm = k_0((\epsilon\mu)^{1/2} \pm \chi)$. ■

Example 3.6
Further simplification occurs if the material is isotropic. The isotropic case is retrieved if we let $\kappa = \chi = 0$, since then $(k = k_\pm = k_0\,(\epsilon\mu)^{1/2})$

$$A_{11}(\cdot) = A_{22}(\cdot) = 0, \qquad \begin{cases} A_{12}(\cdot) = \dfrac{\mu g(k,\cdot)}{\mathrm{i}k\,(\epsilon\mu)^{1/2}} \\ A_{21}(\cdot) = -\dfrac{\epsilon g(k,\cdot)}{\mathrm{i}k\,(\epsilon\mu)^{1/2}} \end{cases}$$

and

$$\begin{cases} B_{11}(\cdot) = B_{22}(\cdot) = g(k,\cdot) \\ B_{12}(\cdot) = B_{21}(\cdot) = 0 \end{cases}$$

leading to the integral representation

$$\frac{1}{\mathrm{i}k\,(\epsilon\mu)^{1/2}} \nabla \times \left\{ \nabla \times \iint_S g(k, |\boldsymbol{r} - \boldsymbol{r}'|) \begin{pmatrix} 0 & \mu \\ -\epsilon & 0 \end{pmatrix} \hat{\boldsymbol{\nu}}(\boldsymbol{r}') \times \mathrm{e}(\boldsymbol{r}')\, \mathrm{d}S' \right\}$$

$$- \nabla \times \iint_S g(k, |\boldsymbol{r} - \boldsymbol{r}'|) \hat{\boldsymbol{\nu}}(\boldsymbol{r}') \times \mathrm{e}(\boldsymbol{r}')\, \mathrm{d}S' = \begin{cases} \mathrm{e}(\boldsymbol{r}), & \boldsymbol{r} \text{ inside } S \\ 0, & \boldsymbol{r} \text{ outside } S \end{cases}$$

which is identical to the representations in (3.14) and (3.15) in Section 3.3.1. ■

Another way to obtain an integral representation for a biisotropic material is to utilize the integral identity (3.16) on page 158, i.e.,

$$\iiint_V \left\{ (\nabla' \times (\nabla' \times \boldsymbol{F})) \cdot \mathbf{D} - \boldsymbol{F} \cdot (\nabla' \times (\nabla' \times \mathbf{D})) \right\} \mathrm{d}v'$$

$$= \iint_S \left\{ (\hat{\nu} \times \boldsymbol{F}) \cdot (\nabla' \times \mathbf{D}) + (\hat{\nu} \times (\nabla' \times \boldsymbol{F})) \cdot \mathbf{D} \right\} \mathrm{d}S'$$

We adopt, for fixed position vector $\boldsymbol{r}$, the Green dyadic $\mathbf{D}(\boldsymbol{r}') = \mathbf{G}_{\mathrm{e}}^{\mathrm{bi}}(\boldsymbol{r}', \boldsymbol{r})$ satisfying,[10] see Section 2.4 on page 118

$$\nabla \times \left(\nabla \times \mathbf{G}_{\mathrm{e}}^{\mathrm{bi}}(\boldsymbol{r}, \boldsymbol{r}') \right) + \alpha \nabla \times \mathbf{G}_{\mathrm{e}}^{\mathrm{bi}}(\boldsymbol{r}, \boldsymbol{r}') - \beta^2 \mathbf{G}_{\mathrm{e}}^{\mathrm{bi}}(\boldsymbol{r}, \boldsymbol{r}') = \mathbf{I}_3 \delta(\boldsymbol{r} - \boldsymbol{r}')$$

where

$$\begin{cases} \alpha = -2k_0\chi \\ \beta^2 = k_0^2 \left(\epsilon\mu - \kappa^2 - \chi^2 \right) \end{cases}$$

The form of $\mathbf{G}_{\mathrm{e}}^{\mathrm{bi}}$ in terms of the electric Green dyadic $\mathbf{G}_{\mathrm{e}}$ is, see (2.35) on page 120

$$\mathbf{G}_{\mathrm{e}}^{\mathrm{bi}}(\boldsymbol{r}, \boldsymbol{r}') = \frac{(\nabla \times \mathbf{I}_3 + k_+ \mathbf{I}_3) \cdot \mathbf{G}_{\mathrm{e}}(k_+, |\boldsymbol{r} - \boldsymbol{r}'|) - (\nabla \times \mathbf{I}_3 - k_- \mathbf{I}_3) \cdot \mathbf{G}_{\mathrm{e}}(k_-, |\boldsymbol{r} - \boldsymbol{r}'|)}{k_+ + k_-}$$

where, as above, the electric Green dyadic, $\mathbf{G}_{\mathrm{e}}$, in terms of the scalar Green function is, see (2.28) on page 114

$$\mathbf{G}_{\mathrm{e}}(k, |\boldsymbol{r} - \boldsymbol{r}'|) = \left(\mathbf{I}_3 + \frac{1}{k^2} \nabla\nabla \right) g(k, |\boldsymbol{r} - \boldsymbol{r}'|)$$

The electric field $\boldsymbol{F} = \boldsymbol{E}$ satisfies (2.34) on page 119

$$\nabla \times (\nabla \times \boldsymbol{E}) + \alpha \nabla \times \boldsymbol{E} - \beta^2 \boldsymbol{E} = \mathbf{0}$$

Insertion of the appropriate fields into the integral identity above gives (note that we have, for convenience, changed the role of the unprimed and the primed position vectors)

$$-\alpha \iiint_V \left\{ (\nabla' \times \boldsymbol{E}(\boldsymbol{r}')) \cdot \mathbf{G}_{\mathrm{e}}^{\mathrm{bi}}(\boldsymbol{r}', \boldsymbol{r}) - \boldsymbol{E}(\boldsymbol{r}') \cdot \left(\nabla' \times \mathbf{G}_{\mathrm{e}}^{\mathrm{bi}}(\boldsymbol{r}', \boldsymbol{r}) \right) \right\} \mathrm{d}v'$$

$$= \iint_S \left\{ (\hat{\nu} \times \boldsymbol{E}(\boldsymbol{r}')) \cdot \left(\nabla' \times \mathbf{G}_{\mathrm{e}}^{\mathrm{bi}}(\boldsymbol{r}', \boldsymbol{r}) \right) \right.$$

$$\left. + (\hat{\nu} \times (\nabla' \times \boldsymbol{E}(\boldsymbol{r}'))) \cdot \mathbf{G}_{\mathrm{e}}^{\mathrm{bi}}(\boldsymbol{r}', \boldsymbol{r}) \right\} \mathrm{d}S' + \begin{cases} \boldsymbol{E}(\boldsymbol{r}), & \boldsymbol{r} \in V \\ \mathbf{0}, & \boldsymbol{r} \in \mathbb{R}^3 \setminus \overline{V} \end{cases}$$

[10] Note the order of the arguments.

Using $(\nabla' \times \boldsymbol{E}) \cdot \mathbf{G}_\mathrm{e}^\mathrm{bi} - \boldsymbol{E} \cdot (\nabla' \times \mathbf{G}_\mathrm{e}^\mathrm{bi}) = \nabla' \cdot (\boldsymbol{E} \times \mathbf{G}_\mathrm{e}^\mathrm{bi})$, and the divergence theorem, we get

$$\iint_S \left\{ (\hat{\boldsymbol{\nu}} \times \boldsymbol{E}(\boldsymbol{r}')) \cdot \left(\nabla' \times \mathbf{G}_\mathrm{e}^\mathrm{bi}(\boldsymbol{r}', \boldsymbol{r})\right) + (\hat{\boldsymbol{\nu}} \times (\nabla' \times \boldsymbol{E}(\boldsymbol{r}'))) \cdot \mathbf{G}_\mathrm{e}^\mathrm{bi}(\boldsymbol{r}', \boldsymbol{r}) \right\} \,\mathrm{d}S'$$

$$+ \alpha \iint_S \hat{\boldsymbol{\nu}} \cdot \left(\boldsymbol{E}(\boldsymbol{r}') \times \mathbf{G}_\mathrm{e}^\mathrm{bi}(\boldsymbol{r}', \boldsymbol{r})\right) \,\mathrm{d}S' = \begin{cases} -\boldsymbol{E}(\boldsymbol{r}), & \boldsymbol{r} \text{ inside } S \\ \mathbf{0}, & \boldsymbol{r} \text{ outside } S \end{cases}$$

Finally, we eliminate the curl of the electric field by the use of (3.26), and we use the definition of α, and we rearrange terms. We get after some algebra the final expression of the integral representation of the electric field in a biisotropic media (note the cyclic permutation $\hat{\boldsymbol{\nu}} \cdot (\boldsymbol{a} \times \mathbf{A}) = (\hat{\boldsymbol{\nu}} \times \boldsymbol{a}) \cdot \mathbf{A}$ for a vector $\boldsymbol{a}$ and a dyadic $\mathbf{A}$, see Problem 3.5)

$$- \iint_S (\hat{\boldsymbol{\nu}} \times \boldsymbol{E}(\boldsymbol{r}')) \cdot \left(\mathrm{i}k_0 (\kappa + \mathrm{i}\chi) \mathbf{G}_\mathrm{e}^\mathrm{bi}(\boldsymbol{r}', \boldsymbol{r}) + \nabla' \times \mathbf{G}_\mathrm{e}^\mathrm{bi}(\boldsymbol{r}', \boldsymbol{r})\right) \,\mathrm{d}S'$$

$$- \mathrm{i}k_0\mu \iint_S \left\{ (\hat{\boldsymbol{\nu}} \times \eta_0 \boldsymbol{H}(\boldsymbol{r}')) \cdot \mathbf{G}_\mathrm{e}^\mathrm{bi}(\boldsymbol{r}', \boldsymbol{r}) \right\} \,\mathrm{d}S' = \begin{cases} \boldsymbol{E}(\boldsymbol{r}), & \boldsymbol{r} \text{ inside } S \\ \mathbf{0}, & \boldsymbol{r} \text{ outside } S \end{cases} \tag{3.31}$$

This representation is identical to the representation for the isotropic material, (3.18), if $\kappa = \chi = 0$, since $\nabla' \times \mathbf{G}_\mathrm{e}(|\boldsymbol{r} - \boldsymbol{r}'|)$ is an anti-symmetric dyadic, and $\mathbf{G}_\mathrm{e}(|\boldsymbol{r} - \boldsymbol{r}'|)$ is a symmetric dyadic. The integral representation in (3.31) and the integral representation in (3.30) are identical, see Problem 3.6.

3.9 Integral Representations in the Time Domain

The starting point of the derivation of the integral representation of a vector field in the time domain is again (3.8) on page 151, i.e.,

$$\iint_S \left[\psi(\hat{\boldsymbol{\nu}} \times (\nabla' \times \boldsymbol{F})) + (\nabla'\psi)(\hat{\boldsymbol{\nu}} \cdot \boldsymbol{F}) - \hat{\boldsymbol{\nu}}\psi(\nabla' \cdot \boldsymbol{F}) - (\nabla'\psi) \times (\hat{\boldsymbol{\nu}} \times \boldsymbol{F})\right] \,\mathrm{d}S'$$

$$= \iiint_V \left[\boldsymbol{F}\nabla'^2\psi + \psi(\nabla' \times (\nabla' \times \boldsymbol{F}) - \nabla'(\nabla' \cdot \boldsymbol{F}))\right] \,\mathrm{d}v'$$

In contrast to above, the scalar field ψ and the vector field $\boldsymbol{F}$ are now fields that depend on both the space variables $\boldsymbol{r}'$ and the time variable t'. However, the time variable t in this expression is yet only a parameter.

Now chose ψ as the Green function $g(|\boldsymbol{r} - \boldsymbol{r}'|, t - t')$ of the scalar wave equation, see (2.51) on page 130, i.e.,

$$g(|\boldsymbol{r} - \boldsymbol{r}'|, t - t') = \frac{\delta(t' - t_\mathrm{ret})}{4\pi |\boldsymbol{r} - \boldsymbol{r}'|}$$

where the retarded time $t_\mathrm{ret} = t - |\boldsymbol{r} - \boldsymbol{r}'|/c$. This Green function satisfies, see (2.50) on page 130

$$\nabla'^2 g(|\boldsymbol{r} - \boldsymbol{r}'|, t - t') - \frac{1}{c^2} \frac{\partial^2}{\partial t'^2} g(|\boldsymbol{r} - \boldsymbol{r}'|, t - t') = -\delta(\boldsymbol{r} - \boldsymbol{r}')\delta(t - t')$$

Notice that the Green function $g(r,t)$ is not a function in the classical sense, but a distribution. The computations below are made as if $g(r,t)$ is a classical function, and with this in mind, the integral representation reads

$$\begin{aligned}
&\iint_S \left[g(|\boldsymbol{r}-\boldsymbol{r}'|,t-t')\hat{\boldsymbol{\nu}}(\boldsymbol{r}')\times\left(\nabla'\times\boldsymbol{F}(\boldsymbol{r}',t')\right)+\left(\nabla' g(|\boldsymbol{r}-\boldsymbol{r}'|,t-t')\right)\left(\hat{\boldsymbol{\nu}}(\boldsymbol{r}')\cdot\boldsymbol{F}(\boldsymbol{r}',t')\right)\right.\\
&\qquad\left.-\,\hat{\boldsymbol{\nu}}(\boldsymbol{r}')g(|\boldsymbol{r}-\boldsymbol{r}'|,t-t')\left(\nabla'\cdot\boldsymbol{F}(\boldsymbol{r}',t')\right)-\left(\nabla' g(|\boldsymbol{r}-\boldsymbol{r}'|,t-t')\right)\times\left(\hat{\boldsymbol{\nu}}(\boldsymbol{r}')\times\boldsymbol{F}(\boldsymbol{r}',t')\right)\right]\mathrm{d}S'\\
&=\iiint_V \left[\frac{1}{c^2}\frac{\partial^2 g(|\boldsymbol{r}-\boldsymbol{r}'|,t-t')}{\partial t'^2}\boldsymbol{F}(\boldsymbol{r}',t')-\boldsymbol{F}(\boldsymbol{r}',t')\delta(\boldsymbol{r}-\boldsymbol{r}')\delta(t-t')\right.\\
&\qquad\left.+\,g(|\boldsymbol{r}-\boldsymbol{r}'|,t-t')\left(\nabla'\times\left(\nabla'\times\boldsymbol{F}(\boldsymbol{r}',t')\right)-\nabla'\left(\nabla'\cdot\boldsymbol{F}(\boldsymbol{r}',t')\right)\right)\right]\mathrm{d}v'
\end{aligned}$$

We now integrate this expression w.r.t. time t' over the interval $(-\infty,\infty)$, and use the explicit form of the Green function. Note that the actions of the distributions $\delta(t)$ and $\delta'(t)$, respectively, on a function $f(t)$ are

$$\int_{-\infty}^{\infty} f(t')\delta(t'-t_0)\,\mathrm{d}t' = f(t_0)$$

and

$$\int_{-\infty}^{\infty} f(t')\delta'(t'-t_0)\,\mathrm{d}t' = -f'(t_0)$$

We arrive at

$$\begin{aligned}
&\iiint_V \left[\frac{1}{4\pi|\boldsymbol{r}-\boldsymbol{r}'|}\left(\nabla'\times\left(\nabla'\times\boldsymbol{F}(\boldsymbol{r}',t_{\text{ret}})\right)-\nabla'\left(\nabla'\cdot\boldsymbol{F}(\boldsymbol{r}',t_{\text{ret}})\right)+\frac{1}{c^2}\frac{\partial^2\boldsymbol{F}(\boldsymbol{r}',t_{\text{ret}})}{\partial t'^2}\right)\right]\mathrm{d}v'\\
&\quad-\iint_S \left\{\frac{\hat{\boldsymbol{\nu}}(\boldsymbol{r}')\times\left[\nabla'\times\boldsymbol{F}(\boldsymbol{r}',t_{\text{ret}})\right]}{4\pi|\boldsymbol{r}-\boldsymbol{r}'|}+\left(\nabla'\frac{1}{4\pi|\boldsymbol{r}-\boldsymbol{r}'|}\right)\left(\hat{\boldsymbol{\nu}}(\boldsymbol{r}')\cdot\boldsymbol{F}(\boldsymbol{r}',t_{\text{ret}})\right)\right.\\
&\quad+\frac{\boldsymbol{r}-\boldsymbol{r}'}{4\pi c|\boldsymbol{r}-\boldsymbol{r}'|^2}\left(\hat{\boldsymbol{\nu}}(\boldsymbol{r}')\cdot\frac{\partial\boldsymbol{F}(\boldsymbol{r}',t_{\text{ret}})}{\partial t'}\right)-\frac{\hat{\boldsymbol{\nu}}(\boldsymbol{r}')(\nabla'\cdot\boldsymbol{F}(\boldsymbol{r}',t_{\text{ret}}))}{4\pi|\boldsymbol{r}-\boldsymbol{r}'|}\\
&\quad-\left(\nabla'\frac{1}{4\pi|\boldsymbol{r}-\boldsymbol{r}'|}\right)\times\left(\hat{\boldsymbol{\nu}}(\boldsymbol{r}')\times\boldsymbol{F}(\boldsymbol{r}',t_{\text{ret}})\right)\\
&\quad\left.-\frac{\boldsymbol{r}-\boldsymbol{r}'}{4\pi c|\boldsymbol{r}-\boldsymbol{r}'|^2}\times\left(\hat{\boldsymbol{\nu}}(\boldsymbol{r}')\times\frac{\partial\boldsymbol{F}(\boldsymbol{r}',t_{\text{ret}})}{\partial t'}\right)\right\}\mathrm{d}S' = \begin{cases}\boldsymbol{F}(\boldsymbol{r},t), & \boldsymbol{r}\in V\\ \boldsymbol{0}, & \boldsymbol{r}\in\mathbb{R}^3\setminus\overline{V}\end{cases}
\end{aligned}$$

This integral representation holds for all (sufficiently differentiable) vector fields $\boldsymbol{F}(\boldsymbol{r}',t')$. Notice that the derivation of the retarded Green function also introduces a time derivative, due to the dependence on the spatial variable in t_{ret}. In fact,

$$\nabla' g(|\boldsymbol{r}-\boldsymbol{r}'|,t-t') = \left(\nabla'\frac{1}{4\pi|\boldsymbol{r}-\boldsymbol{r}'|}\right)\delta(t'-t_{\text{ret}})-\frac{\boldsymbol{r}-\boldsymbol{r}'}{4\pi c|\boldsymbol{r}-\boldsymbol{r}'|^2}\delta'(t'-t_{\text{ret}})$$

where the last term is responsible for the time derivative that shows up. Notice also

$$\int_{-\infty}^{\infty} f(t')\delta'(t' - t_{\text{ret}})\,\mathrm{d}t' = -f'(t_{\text{ret}})$$

3.9.1 Surface Integral Representations of the Maxwell Equations

We are now ready to apply the integral representation derived above to the Maxwell equations in the time domain, see (1.1) and (1.2) on page 2, in a source-free, homogeneous, isotropic material, characterized by the permittivity ϵ and permeability μ, which both are real constants

$$\begin{cases} \nabla \times \boldsymbol{E}(\boldsymbol{r},t) = -\mu_0\mu \dfrac{\partial \boldsymbol{H}(\boldsymbol{r},t)}{\partial t} \\ \nabla \times \boldsymbol{H}(\boldsymbol{r},t) = \epsilon_0\epsilon \dfrac{\partial \boldsymbol{E}(\boldsymbol{r},t)}{\partial t} \end{cases}$$

Specifically, let $\boldsymbol{F} = \boldsymbol{E}$, and we get, see also (2.44) on page 128

$$\begin{aligned} -\iint_S \Bigg\{ & \frac{\hat{\boldsymbol{\nu}}(\boldsymbol{r}') \times \left[\nabla' \times \boldsymbol{E}(\boldsymbol{r}', t_{\text{ret}})\right]}{4\pi|\boldsymbol{r}-\boldsymbol{r}'|} + \left(\nabla' \frac{1}{4\pi|\boldsymbol{r}-\boldsymbol{r}'|}\right)\left(\hat{\boldsymbol{\nu}}(\boldsymbol{r}') \cdot \boldsymbol{E}(\boldsymbol{r}', t_{\text{ret}})\right) \\ & + \frac{\boldsymbol{r}-\boldsymbol{r}'}{4\pi c|\boldsymbol{r}-\boldsymbol{r}'|^2}\left(\hat{\boldsymbol{\nu}}(\boldsymbol{r}') \cdot \frac{\partial \boldsymbol{E}(\boldsymbol{r}', t_{\text{ret}})}{\partial t'}\right) - \left(\nabla' \frac{1}{4\pi|\boldsymbol{r}-\boldsymbol{r}'|}\right) \times \left(\hat{\boldsymbol{\nu}}(\boldsymbol{r}') \times \boldsymbol{E}(\boldsymbol{r}', t_{\text{ret}})\right) \\ & - \frac{\boldsymbol{r}-\boldsymbol{r}'}{4\pi c|\boldsymbol{r}-\boldsymbol{r}'|^2} \times \left(\hat{\boldsymbol{\nu}}(\boldsymbol{r}') \times \frac{\partial \boldsymbol{E}(\boldsymbol{r}', t_{\text{ret}})}{\partial t'}\right)\Bigg\}\,\mathrm{d}S' = \begin{cases} \boldsymbol{E}(\boldsymbol{r},t), & \boldsymbol{r} \text{ inside } S \\ \boldsymbol{0}, & \boldsymbol{r} \text{ outside } S \end{cases} \end{aligned}$$

The normal components of the field are eliminated in the same way as in the frequency domain, see Section 3.3.1. We make use of

$$\nabla' \frac{1}{4\pi|\boldsymbol{r}-\boldsymbol{r}'|} = -\nabla \frac{1}{4\pi|\boldsymbol{r}-\boldsymbol{r}'|}$$

and we get

$$\begin{aligned} & -\iint_S \frac{\hat{\boldsymbol{\nu}}(\boldsymbol{r}') \times \left[\nabla' \times \boldsymbol{E}(\boldsymbol{r}', t_{\text{ret}})\right]}{4\pi|\boldsymbol{r}-\boldsymbol{r}'|}\,\mathrm{d}S' + \nabla \iint_S \frac{\left(\hat{\boldsymbol{\nu}}(\boldsymbol{r}') \cdot \boldsymbol{E}(\boldsymbol{r}', t_{\text{ret}})\right)}{4\pi|\boldsymbol{r}-\boldsymbol{r}'|}\,\mathrm{d}S' \\ & -\nabla \times \iint_S \frac{\left(\hat{\boldsymbol{\nu}}(\boldsymbol{r}') \times \boldsymbol{E}(\boldsymbol{r}', t_{\text{ret}})\right)}{4\pi|\boldsymbol{r}-\boldsymbol{r}'|}\,\mathrm{d}S' = \begin{cases} \boldsymbol{E}(\boldsymbol{r},t), & \boldsymbol{r} \text{ inside } S \\ \boldsymbol{0}, & \boldsymbol{r} \text{ outside } S \end{cases} \end{aligned} \tag{3.32}$$

where we also used

$$\nabla \frac{f(t_{\text{ret}})}{4\pi|\boldsymbol{r}-\boldsymbol{r}'|} = -\left(\nabla' \frac{1}{4\pi|\boldsymbol{r}-\boldsymbol{r}'|}\right) f(t_{\text{ret}}) - \frac{\boldsymbol{r}-\boldsymbol{r}'}{4\pi c|\boldsymbol{r}-\boldsymbol{r}'|^2} f'(t_{\text{ret}})$$

Finally, take the curl of (3.32), use the Maxwell equations, and integrate in time[11] from $-\infty$ to t. We get

$$\frac{1}{\mu_0\mu}\nabla\times\left(\nabla\times\int_{-\infty}^{t}\iint_S \frac{\left(\hat{\boldsymbol{\nu}}(\boldsymbol{r}')\times\boldsymbol{E}(\boldsymbol{r}',t'-|\boldsymbol{r}-\boldsymbol{r}'|/c)\right)}{4\pi|\boldsymbol{r}-\boldsymbol{r}'|}\,\mathrm{d}S'\,\mathrm{d}t'\right)$$
$$-\nabla\times\iint_S \frac{\hat{\boldsymbol{\nu}}(\boldsymbol{r}')\times\boldsymbol{H}(\boldsymbol{r}',t_{\text{ret}})}{4\pi|\boldsymbol{r}-\boldsymbol{r}'|}\,\mathrm{d}S' = \begin{cases}\boldsymbol{H}(\boldsymbol{r},t), & \boldsymbol{r}\text{ inside } S\\ \boldsymbol{0}, & \boldsymbol{r}\text{ outside } S\end{cases}$$

which is put into its final form as

$$\frac{1}{\mu_0\mu}\nabla\times\left(\nabla\times\int_{-\infty}^{t_{\text{ret}}}\iint_S \frac{\left(\hat{\boldsymbol{\nu}}(\boldsymbol{r}')\times\boldsymbol{E}(\boldsymbol{r}',t')\right)}{4\pi|\boldsymbol{r}-\boldsymbol{r}'|}\,\mathrm{d}S'\,\mathrm{d}t'\right)$$
$$-\nabla\times\iint_S \frac{\hat{\boldsymbol{\nu}}(\boldsymbol{r}')\times\boldsymbol{H}(\boldsymbol{r}',t_{\text{ret}})}{4\pi|\boldsymbol{r}-\boldsymbol{r}'|}\,\mathrm{d}S' = \begin{cases}\boldsymbol{H}(\boldsymbol{r},t), & \boldsymbol{r}\text{ inside } S\\ \boldsymbol{0}, & \boldsymbol{r}\text{ outside } S\end{cases}$$

Similarly, for the electric field (replace in (3.32) $\boldsymbol{E}$ with $\boldsymbol{H}$, take the curl, use the Maxwell equations, and integrate in time from $-\infty$ to t), and we have

$$-\frac{1}{\epsilon_0\epsilon}\nabla\times\left(\nabla\times\int_{-\infty}^{t_{\text{ret}}}\iint_S \frac{\left(\hat{\boldsymbol{\nu}}(\boldsymbol{r}')\times\boldsymbol{H}(\boldsymbol{r}',t')\right)}{4\pi|\boldsymbol{r}-\boldsymbol{r}'|}\,\mathrm{d}S'\,\mathrm{d}t'\right)$$
$$-\nabla\times\iint_S \frac{\hat{\boldsymbol{\nu}}(\boldsymbol{r}')\times\boldsymbol{E}(\boldsymbol{r}',t_{\text{ret}})}{4\pi|\boldsymbol{r}-\boldsymbol{r}'|}\,\mathrm{d}S' = \begin{cases}\boldsymbol{E}(\boldsymbol{r},t), & \boldsymbol{r}\text{ inside } S\\ \boldsymbol{0}, & \boldsymbol{r}\text{ outside } S\end{cases} \tag{3.33}$$

These two integral representations are the final results of the integral representations using time-domain vector fields. Similar results can be obtained for scalar fields and dyadic fields. These results follow straightforwardly by the use of the methods in this chapter, but since we in this book only use the integral representations for vector-valued fields, we leave these details to the interested reader to complete.

Problems for Chapter 3

Section 3.1 Two Scalar Fields

3.1 Verify the dyadic-valued integrals in (3.5) in Example 3.2.

[11]The lower limit in this time integration is arbitrary as long as the magnetic field in the lower limit vanishes.

Section 3.2 Vector and Scalar Fields

3.2 Let the vector field $\boldsymbol{E}$ satisfy the electrostatic equations

$$\begin{cases} \nabla \times \boldsymbol{E} = \mathbf{0} \\ \nabla \cdot \boldsymbol{E} = 0 \end{cases}$$

in a volume V with boundary surface S and outward unit normal vector $\hat{\boldsymbol{\nu}}$. Show that this field satisfies the following integral representation:

$$\begin{aligned} &- \iint_S \nabla' g(|\boldsymbol{r} - \boldsymbol{r}'|) \left(\hat{\boldsymbol{\nu}}(\boldsymbol{r}') \cdot \boldsymbol{E}(\boldsymbol{r}')\right) \, \mathrm{d}S' \\ &+ \iint_S \nabla' g\left(|\boldsymbol{r} - \boldsymbol{r}'|\right) \times \left(\hat{\boldsymbol{\nu}}(\boldsymbol{r}') \times \boldsymbol{E}(\boldsymbol{r}')\right) \, \mathrm{d}S' = \begin{cases} \boldsymbol{E}(\boldsymbol{r}), & \boldsymbol{r} \text{ inside } S \\ \mathbf{0}, & \boldsymbol{r} \text{ outside } S \end{cases} \end{aligned} \tag{3.34}$$

where

$$g(r) = \frac{1}{4\pi r}$$

Hint: Start with the integral identity (3.8) with $\psi(\boldsymbol{r}) = g(|\boldsymbol{r} - \boldsymbol{r}'|)$ and $\boldsymbol{F} = \boldsymbol{E}$.

***3.3** Show that the result in Problem 3.2, i.e., (3.34), also follows from the integral representation results in the dynamic case, (3.14), in the limit as $k \to 0$ ($\omega \to 0$) for every $\boldsymbol{E}$- and $\boldsymbol{H}$-field that satisfy

$$\begin{cases} \nabla \times \boldsymbol{E} = \mathrm{i}k\eta_0\eta \boldsymbol{H} \\ \nabla \times \boldsymbol{H} = -\mathrm{i}\dfrac{k}{\eta_0\eta}\boldsymbol{E} \end{cases} \qquad \boldsymbol{r} \in V$$

where V is the volume bounded by the closed surface S.

Section 3.4 Dyadic and Vector Fields

***3.4** Show that the derivation of the integral representation in (3.18) can be obtained from the integral representation (3.16) by the use of the classical function arguments similar to the derivation of the vector case in Section 3.2.2.

Section 3.8 Integral Representation for Biisotropic Materials

3.5 Prove the following cyclic permutation result for two vectors $\boldsymbol{a}$ and $\boldsymbol{b}$, and a dyadic $\mathbf{A}$:

$$\boldsymbol{a} \cdot (\boldsymbol{b} \times \mathbf{A}) = (\boldsymbol{a} \times \boldsymbol{b}) \cdot \mathbf{A}$$

***3.6** Show that the dyadic representation in (3.31) is identical to (3.30).

Section 3.9 Integral Representations in the Time Domain

3.7 Derive the integral representation in the frequency domain (3.14) from the integral representation in the time domain, see (3.33).

CHAPTER 4

Introductory Scattering Theory

We have now completed the preparatory work, and we proceed with the formulation of the scattering problem. As pointed out several times above, the electromagnetic scattering problem is most naturally formulated in the time domain. However, the frequency formulation is less complicated, and for pedagogical reasons and tradition, we adopt this view also here in this textbook. Some of the results have their origins in a time-domain setting, and when this happens, we refer to the time-domain formulation that is presented in Chapter 5.

When a time-harmonic, electromagnetic wave propagates in a homogeneous, lossless, isotropic material, characterized by the real-valued parameters ϵ and μ, the wave propagates unhindered. If there is a region where the electric or magnetic material parameters have a different value, then the propagation of the wave alters—the wave is scattered. The ultimate goal of the scattering theory is to analyze this alteration of the electromagnetic fields in a quantitative way.

A typical example of a scattering geometry is depicted in Figure 4.1. The volumes $V_{\rm i}$ and $V_{\rm s}$ are assumed to be disjoint, i.e., $V_{\rm i} \cap V_{\rm s} = \emptyset$. $V_{\rm i}$ is the region or domain of the sources of the unperturbed fields, and $V_{\rm s}$ is the domain of the scatterer, which can be a single domain or consist of several disjoint regions. The bounding surface to $V_{\rm s}$ is denoted $S_{\rm s}$, and the outward unit normal vector to the surface is denoted $\hat{\boldsymbol{\nu}}$. The region exterior to $V_{\rm s}$ (and $V_{\rm i}$) is denoted $V_{\rm e}$, and its formal definition is $V_{\rm e} = \mathbb{R}^3 \setminus \left(\overline{V_{\rm s}} \cup \overline{V_{\rm i}}\right)$. Usually, the exterior region is simply connected, implying that the scattering region is "solid". A situation that occurs in many applications is that $V_{\rm i}$ contains a transmitting antenna or some other sources, and $V_{\rm s}$ contains a passive scatterer, e.g., a reflector dish or an object with different electric or magnetic material properties.

In all scattering problems, there is a controlled field denoted the incident field. The corresponding field quantities are labeled with the index i, e.g., the incident electric field is denoted $\boldsymbol{E}_{\rm i}(\boldsymbol{r}, \omega)$, and this field has its sources inside the volume $V_{\rm i}$, see Figure 4.1. Outside $V_{\rm i}$ this field is source free, i.e., the electric and magnetic fields satisfy

$$\begin{cases} \nabla \times \boldsymbol{E}_{\rm i}(\boldsymbol{r}, \omega) = {\rm i}\omega\mu_0\mu \boldsymbol{H}_{\rm i}(\boldsymbol{r}, \omega) = {\rm i}k\eta_0\eta \boldsymbol{H}_{\rm i}(\boldsymbol{r}, \omega) \\ \nabla \times \boldsymbol{H}_{\rm i}(\boldsymbol{r}, \omega) = -{\rm i}\omega\epsilon_0\epsilon \boldsymbol{E}_{\rm i}(\boldsymbol{r}, \omega) = -{\rm i}\dfrac{k}{\eta_0\eta}\boldsymbol{E}_{\rm i}(\boldsymbol{r}, \omega) \end{cases} \quad \boldsymbol{r} \notin V_{\rm i} \tag{4.1}$$

where, as above, the (relative) wave impedance of the material is $\eta = \sqrt{\mu/\epsilon}$ and the wave number of the material is $k = \omega\sqrt{\epsilon_0\mu_0\epsilon\mu}$. This is the only field present in the absence of the scatterer.

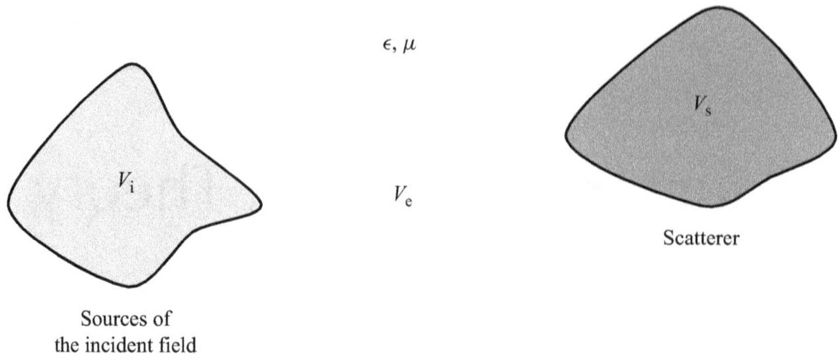

Figure 4.1 Typical example of a geometry in scattering theory. The sources of the incident field are located in V_i. The region V_s contains the passive scatterer, and V_e is the exterior region of V_s and V_i.

Scattering occurs if there is a region with different material parameters, i.e., material parameters that in some way depart from the constant values of ϵ and μ. This region is assumed to be located inside the bounded volume V_s, see Figure 4.1. These deviations in the material parameters give rise to a perturbation in the electromagnetic fields from their unperturbed valued (labeled with i), which we denote the scattered fields, and we use the index s to denote the field quantities that are associated with the scattered field, e.g., the scattered electric field is $\boldsymbol{E}_s(\boldsymbol{r},\omega)$. The scattered field has its (fictitious) sources inside the volume V_s. Outside V_s the fields are source free, i.e.,

$$\begin{cases} \nabla \times \boldsymbol{E}_s(\boldsymbol{r},\omega) = \mathrm{i}k\eta_0\eta \boldsymbol{H}_s(\boldsymbol{r},\omega) \\ \nabla \times \boldsymbol{H}_s(\boldsymbol{r},\omega) = -\mathrm{i}\dfrac{k}{\eta_0\eta}\boldsymbol{E}_s(\boldsymbol{r},\omega) \end{cases} \qquad \boldsymbol{r} \in V_e \tag{4.2}$$

The total electric field $\boldsymbol{E}(\boldsymbol{r},\omega)$, that is measured in an experiment, is the sum of these two fields, i.e.,

$$\boldsymbol{E}(\boldsymbol{r},\omega) = \boldsymbol{E}_i(\boldsymbol{r},\omega) + \boldsymbol{E}_s(\boldsymbol{r},\omega)$$

In general, the presence of the scatterer in V_s affects the (real) sources of the incident field in V_i. For example, if the sources of the incident field are generated by currents on an antenna, the scattered field modifies these currents if the scatterer is close to the antenna. This reaction on the sources complicates the solution of the scattering problem. However, in many cases this effect can be neglected. That is approximately the case, if the distance between V_i (the antenna) and V_s (scatterer) is large. Luckily, this is often the case, and, indeed, the sources of the incident field are frequently located at infinity. This effect caused by the scattered field on the sources of the incident field is always ignored in this textbook.

In this chapter, we do not specify in what way the material in V_s differs from the material outside the volume. In a general case, the material inside V_s can be a bianisotropic material. The only limitation here is that the volume V_s is bounded. As remarked above, the interaction of the scattered field on its sources is neglected, and the sources in V_i are given and not affected by the presence of the scatterer.

The integral representations of the solutions to the Maxwell equations were formulated and analyzed in Chapter 3. We make extensive use of these integral representations in this chapter. Both the surface and the volume formulations can be used to represent the electric and magnetic fields in a homogeneous material. The first section in this chapter gives a characterization of the fields at large distances from the scatterer in the far zone—utilizing both a volume and a surface formulation. We find that the surface integral formulation is more general than the volume formulation, as it contains less assumptions on the properties of the scatterer. In Sections 4.2 and 4.3, several definitions, such as the different cross sections and the scattering dyadic, are introduced, and in Section 4.4, we prove the optical theorem. Babinet's principle is treated in Section 4.5.

4.1 The Far Zone

At a large distance from the scatterer, in the far zone, the scattered electromagnetic fields have a special behavior. In this section, we investigate this special behavior in detail.

In a scattering problem, there are three different length scales—the maximum diameter of the scatterer, d, defined by

$$d = \max_{\boldsymbol{r},\boldsymbol{r}' \in V_s} |\boldsymbol{r} - \boldsymbol{r}'| \tag{4.3}$$

the wavelength, $\lambda = 2\pi/k$, and the distance to the observation or field point, r, from the origin, see Figure 4.2. The far zone is defined as the domain that satisfies the following three conditions:

$$\begin{cases} r \gg d & (a) \\ r \gg kd^2 & (b) \\ r \gg \lambda & (c) \end{cases} \tag{4.4}$$

The far zone depends on the location of the origin.

Conditions (a) and (c) imply that the far zone is characterized by an observation distance r that is much greater than the two length scales d and λ. These conditions have to be met in

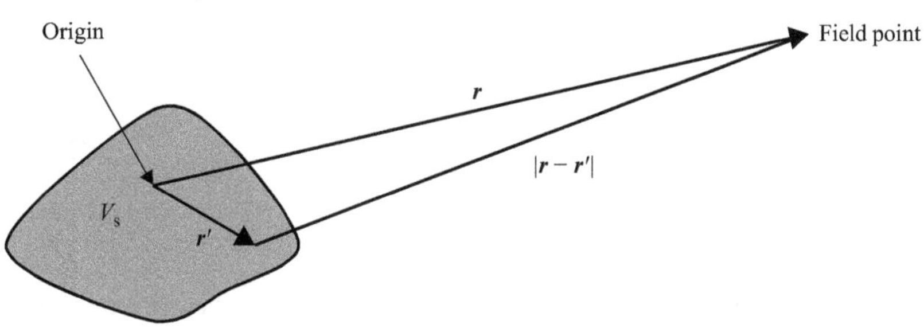

Figure 4.2 The far field from the scatterer inside the volume V_s.

order to be in the far zone, and in addition the distance has to satisfy condition (b), $r \gg kd^2$, which quantifies the Fraunhofer[1] distance.

Another way of formulating the conditions in (4.4) is to observe that the solid angle Ω of the scatterer as seen from the observation point is less than d^2/r^2, and therefore the three conditions can be reduced to

$$\begin{cases} kr \gg kd & (a) \\ kr \ll 1/\Omega & (b) \\ kr \gg 1 & (c) \end{cases}$$

There are two different ways of representing the scattered electric field of an object in the far zone. The first—less general—is formulated in terms of the induced currents set up by the incident field. This approach results in a volume formulation of the problem. The Green function results presented in Section 2.1 are appropriate here.

A more general solution of the problem is, however, to adopt the surface integral representations presented and proved in Section 3.3. This approach contains no assumptions of the material inside the scattering volume V_s, and therefore, it is very general and applies to all linear materials.

In both these cases, we find that the electric field in the far zone is

$$\boldsymbol{E}_s(\boldsymbol{r}) = \frac{e^{ikr}}{r}\boldsymbol{F}(\hat{\boldsymbol{r}})\left(1 + O(kd^2/r)\right)(1 + O(d/r))\left(1 + O((kr)^{-1})\right)$$

This result is derived in Sections 4.1.1 and 4.1.2 for the volume integral and the surface integral formulation, respectively.

4.1.1 Volume Integral Formulation

In Section 2.1, we derived the electric field, $\boldsymbol{E}$, generated by the known current density, $\boldsymbol{J}$, see (2.13) on page 106. If $\boldsymbol{J}_s$ denotes the induced current density generated by the incident field, the scattered field $\boldsymbol{E}_s$ becomes[2]

$$\boldsymbol{E}_s(\boldsymbol{r}) = i\omega\mu_0\mu\left[\mathbf{I} + \frac{1}{k^2}\nabla\nabla\right] \cdot \iiint_{V_s} \frac{e^{ik|\boldsymbol{r}-\boldsymbol{r}'|}}{4\pi|\boldsymbol{r}-\boldsymbol{r}'|}\boldsymbol{J}_s(\boldsymbol{r}')\, dv' \tag{4.5}$$

The current density $\boldsymbol{J}_s$ is generated by the incident field, but its precise form is not critical here.[3] Our goal in this section is to determine the dominant contribution of the electric field in the far zone.

---

[1] Joseph von Fraunhofer (1787–1826), German optician.

[2] Notice the distinction between the current densities $\boldsymbol{J}_s$, which is the source of the scattered field, and generated by the incident field, and $\boldsymbol{J}_S$, which is used as a notation of the surface current density at an interface, see (1.12) and (1.13) on page 8.

[3] If the scattering material in the volume V_s is an isotropic material characterized by the permittivity $\epsilon(\boldsymbol{r})$ and the permeability $\mu(\boldsymbol{r})$, then $\boldsymbol{J}_s(\boldsymbol{r}) = \frac{k_0^2}{i\omega\mu_0\mu}(\epsilon(\boldsymbol{r})\mu(\boldsymbol{r}) - \epsilon\mu)\boldsymbol{E}(\boldsymbol{r})$, where ϵ and μ are the constant permittivity and permeability, respectively, outside the scatterer, and where $\boldsymbol{E}(\boldsymbol{r})$ is the total (unknown) field inside the scatterer.

Denote the distance between the source point $\boldsymbol{r}'$ and the observation point $\boldsymbol{r}$ by $|\boldsymbol{r}-\boldsymbol{r}'|$, i.e.,

$$|\boldsymbol{r}-\boldsymbol{r}'| = \sqrt{(\boldsymbol{r}-\boldsymbol{r}')\cdot(\boldsymbol{r}-\boldsymbol{r}')} = \sqrt{r^2 + r'^2 - 2\boldsymbol{r}\cdot\boldsymbol{r}'}$$

where $r = |\boldsymbol{r}|$ and $r' = |\boldsymbol{r}'|$. Introduce the unit vector of the position vector $\hat{\boldsymbol{r}} = \boldsymbol{r}/r$, pointing away from the origin toward the observation point. With the help of the expansion $\sqrt{1+x} = 1 + x/2 + O(x^2)$, the distance $|\boldsymbol{r}-\boldsymbol{r}'|$ can be computed to leading order as

$$\begin{aligned}|\boldsymbol{r}-\boldsymbol{r}'| &= r\sqrt{1 + \left(\frac{r'}{r}\right)^2 - 2\hat{\boldsymbol{r}}\cdot\frac{\boldsymbol{r}'}{r}} = r\left\{1 + \frac{1}{2}\left[\left(\frac{r'}{r}\right)^2 - 2\hat{\boldsymbol{r}}\cdot\frac{\boldsymbol{r}'}{r}\right] + \cdots\right\} \\ &= r - \hat{\boldsymbol{r}}\cdot\boldsymbol{r}' + O(d^2/r), \qquad \text{as } r\to\infty \end{aligned} \tag{4.6}$$

where the maximal extent of the scatterer (diameter of the circumscribed sphere) d is given in (4.3).

The Green function then becomes

$$\begin{aligned}\frac{\mathrm{e}^{\mathrm{i}k|\boldsymbol{r}-\boldsymbol{r}'|}}{4\pi|\boldsymbol{r}-\boldsymbol{r}'|} &= \frac{1}{4\pi r\,(1+O(d/r))}\exp\left\{\mathrm{i}k\left(r - \hat{\boldsymbol{r}}\cdot\boldsymbol{r}' + O(d^2/r)\right)\right\} \\ &= \frac{\mathrm{e}^{\mathrm{i}kr}}{4\pi r}\mathrm{e}^{-\mathrm{i}k\hat{\boldsymbol{r}}\cdot\boldsymbol{r}'}\left(1 + O(kd^2/r)\right)(1+O(d/r))\end{aligned}$$

The dominant contribution to the scattered field in (4.5) then is

$$\boldsymbol{E}_{\mathrm{s}}(\boldsymbol{r}) = \mathrm{i}\omega\mu_0\mu\left[\mathbf{I} + \frac{1}{k^2}\nabla\nabla\right]\cdot\frac{\mathrm{e}^{\mathrm{i}kr}}{4\pi r}\iiint_{V_{\mathrm{s}}}\mathrm{e}^{-\mathrm{i}k\hat{\boldsymbol{r}}\cdot\boldsymbol{r}'}\boldsymbol{J}_{\mathrm{s}}(\boldsymbol{r}')\,\mathrm{d}v'\left(1+O(kd^2/r)\right)(1+O(d/r))$$

Note that we have used the conditions (a) and (b) in (4.4), but not condition (c) yet.

It is now convenient to introduce the vector field $\boldsymbol{K}(\hat{\boldsymbol{r}})$ defined by

$$\boldsymbol{K}(\hat{\boldsymbol{r}}) = \frac{\mathrm{i}k\eta_0\eta}{4\pi}\iiint_{V_{\mathrm{s}}}\mathrm{e}^{-\mathrm{i}k\hat{\boldsymbol{r}}\cdot\boldsymbol{r}'}\boldsymbol{J}_{\mathrm{s}}(\boldsymbol{r}')\,\mathrm{d}v'$$

Note that $\boldsymbol{K}(\hat{\boldsymbol{r}})$ is the spatial Fourier transform of the current density $\boldsymbol{J}_{\mathrm{s}}$ evaluated at the point $k\hat{\boldsymbol{r}}$. This field is a function of the direction $\hat{\boldsymbol{r}}$ only, and not on the distance r. The electric field is therefore approximated as

$$\boldsymbol{E}_{\mathrm{s}}(\boldsymbol{r}) = \left[\mathbf{I} + \frac{1}{k^2}\nabla\nabla\right]\cdot\left(\frac{\mathrm{e}^{\mathrm{i}kr}}{r}\boldsymbol{K}(\hat{\boldsymbol{r}})\right)$$

The derivation rules with the nabla operator imply

$$\nabla\cdot\left[\frac{\mathrm{e}^{\mathrm{i}kr}}{r}\boldsymbol{K}(\hat{\boldsymbol{r}})\right] = \frac{\mathrm{e}^{\mathrm{i}kr}}{r}\nabla\cdot\boldsymbol{K}(\hat{\boldsymbol{r}}) + \boldsymbol{K}(\hat{\boldsymbol{r}})\cdot\nabla\left(\frac{\mathrm{e}^{\mathrm{i}kr}}{r}\right)$$

Since the vector field $\boldsymbol{K}(\hat{\boldsymbol{r}})$ depends only on the direction $\hat{\boldsymbol{r}}$, which we parameterize with the spherical angles θ and ϕ, and not on the distance r, we get, see Appendix H

$$\nabla \cdot \boldsymbol{K}(\hat{\boldsymbol{r}}) = \frac{1}{r \sin\theta} \frac{\partial}{\partial\theta} (\sin\theta K_\theta) + \frac{1}{r \sin\theta} \frac{\partial}{\partial\phi} (K_\phi)$$

which behaves as $1/r$ as a function of r. Here, K_θ and K_ϕ are the $\hat{\boldsymbol{\theta}}$- and $\hat{\boldsymbol{\phi}}$-components of $\boldsymbol{K}$, respectively. We summarize these results

$$\nabla \cdot \left[\frac{e^{ikr}}{r} \boldsymbol{K}(\hat{\boldsymbol{r}}) \right] = ik\hat{\boldsymbol{r}} \cdot \boldsymbol{K}(\hat{\boldsymbol{r}}) \frac{e^{ikr}}{r} \left(1 + O((kr)^{-1})\right)$$

and similarly

$$\begin{aligned} \frac{1}{k^2} \nabla \left\{ \nabla \cdot \left[\frac{e^{ikr}}{r} \boldsymbol{K}(\hat{\boldsymbol{r}}) \right] \right\} &= \nabla \left\{ -\frac{1}{ik} \hat{\boldsymbol{r}} \cdot \boldsymbol{K}(\hat{\boldsymbol{r}}) \frac{e^{ikr}}{r} \left(1 + O((kr)^{-1})\right) \right\} \\ &= -\hat{\boldsymbol{r}}(\hat{\boldsymbol{r}} \cdot \boldsymbol{K}(\hat{\boldsymbol{r}})) \frac{e^{ikr}}{r} \left(1 + O((kr)^{-1})\right) \end{aligned}$$

Consequently, the leading contribution of the electric field in the far zone is

$$\boldsymbol{E}_s(\boldsymbol{r}) = \left[\boldsymbol{K}(\hat{\boldsymbol{r}}) - \hat{\boldsymbol{r}} K_r(\hat{\boldsymbol{r}}) \right] \frac{e^{ikr}}{r} = \hat{\boldsymbol{r}} \times \left(\boldsymbol{K}(\hat{\boldsymbol{r}}) \times \hat{\boldsymbol{r}} \right) \frac{e^{ikr}}{r}$$

where K_r is the $\hat{\boldsymbol{r}}$-component of $\boldsymbol{K}$, and where we also have used the BAC-CAB rule. Also note that we have used the remaining conditions for the far zone (4.4) part (*c*). The formal definition of the far field amplitude is

$$\boldsymbol{F}(\hat{\boldsymbol{r}}) = \lim_{r \to \infty} r e^{-ikr} \boldsymbol{E}_s(\boldsymbol{r})$$

Notice that the far field amplitude, $\boldsymbol{F}(\hat{\boldsymbol{r}})$, is the projection of the vector field $\boldsymbol{K}(\hat{\boldsymbol{r}})$ on the plane perpendicular to the direction $\hat{\boldsymbol{r}}$.[4] The dominant contribution to the scattered electric field in the far zone then is

$$\boldsymbol{E}_s(\boldsymbol{r}) = \frac{e^{ikr}}{r} \boldsymbol{F}(\hat{\boldsymbol{r}})$$

where

$$\boldsymbol{F}(\hat{\boldsymbol{r}}) = \hat{\boldsymbol{r}} \times \left(\boldsymbol{K}(\hat{\boldsymbol{r}}) \times \hat{\boldsymbol{r}} \right) = \frac{ik\eta_0\eta}{4\pi} \hat{\boldsymbol{r}} \times \left(\iiint_{V_s} e^{-ik\hat{\boldsymbol{r}}\cdot\boldsymbol{r}'} \boldsymbol{J}_s(\boldsymbol{r}') \, dv' \times \hat{\boldsymbol{r}} \right) \tag{4.7}$$

The scattered electric field in the far zone is a spherical wave with a polarization that is perpendicular to the direction to the observation point $\hat{\boldsymbol{r}}$, see (1.73) on page 74.

---

[4]In general, the projection of a vector $\boldsymbol{A}$ on a plane perpendicular to a unit vector $\hat{\boldsymbol{v}}$ is

$$\boldsymbol{A}_\perp = \hat{\boldsymbol{v}} \times (\boldsymbol{A} \times \hat{\boldsymbol{v}})$$

To compute the power flux density (the Poynting vector), we require knowledge of the scattered magnetic field $\boldsymbol{H}_s$ that is associated with the scattered electric field $\boldsymbol{E}_s$. The Faraday law gives the magnetic field

$$\boldsymbol{H}_s(\boldsymbol{r}) = \frac{1}{\mathrm{i}k\eta_0\eta}\nabla \times \boldsymbol{E}_s(\boldsymbol{r})$$

The leading contribution in the far zone is

$$\boldsymbol{H}_s(\boldsymbol{r}) = \frac{1}{\mathrm{i}k\eta_0\eta}\nabla \times \left\{\frac{\mathrm{e}^{\mathrm{i}kr}}{r}\boldsymbol{F}(\hat{\boldsymbol{r}})\right\} = \frac{1}{\eta_0\eta}\frac{\mathrm{e}^{\mathrm{i}kr}}{r}\hat{\boldsymbol{r}} \times \boldsymbol{F}(\hat{\boldsymbol{r}}) = \frac{1}{\eta_0\eta}\frac{\mathrm{e}^{\mathrm{i}kr}}{r}\left[\hat{\boldsymbol{\phi}}F_\theta(\hat{\boldsymbol{r}}) - \hat{\boldsymbol{\theta}}F_\phi(\hat{\boldsymbol{r}})\right]$$

The spherical components $F_\theta = \boldsymbol{F}\cdot\hat{\boldsymbol{\theta}}$ and $F_\phi = \boldsymbol{F}\cdot\hat{\boldsymbol{\phi}}$ have been introduced, and contributions that vanish faster than $1/kr$ have been neglected. Notice also that $\boldsymbol{H}_s$ is perpendicular to $\hat{\boldsymbol{r}}$ in the far zone, and that $\{\boldsymbol{E}_s, \boldsymbol{H}_s, \hat{\boldsymbol{r}}\}$ form a right-handed triple.

The power flux density of the scattered field is then computed by evaluating the Poynting vector

$$\langle \boldsymbol{S}_s(t)\rangle = \frac{1}{2}\operatorname{Re}\left\{\boldsymbol{E}_s \times \boldsymbol{H}_s^*\right\} = \frac{1}{2\eta_0\eta r^2}\boldsymbol{F}(\hat{\boldsymbol{r}}) \times \left(\hat{\boldsymbol{r}} \times \boldsymbol{F}^*(\hat{\boldsymbol{r}})\right)$$

which by the BAC-CAB rule gives

$$\langle \boldsymbol{S}_s(t)\rangle = \frac{\hat{\boldsymbol{r}}}{2\eta_0\eta r^2}|\boldsymbol{F}(\hat{\boldsymbol{r}})|^2 = \frac{\hat{\boldsymbol{r}}}{2\eta_0\eta r^2}\left[|F_\theta(\hat{\boldsymbol{r}})|^2 + |F_\phi(\hat{\boldsymbol{r}})|^2\right] \tag{4.8}$$

since $\boldsymbol{F}$ is perpendicular to $\hat{\boldsymbol{r}}$. As expected, the power flux density is radially directed along the direction of observation $\hat{\boldsymbol{r}}$.

4.1.2 Surface Integral Formulation

The formulation of the scattering problem that is presented in this section is made in terms of the tangential electric and magnetic fields evaluated exterior to the surface of the scatterer. It is in many respects more general than the previous formulation in terms of a volume integral containing information about the fields and current density inside the scatterer. The most striking difference is perhaps that in the surface integral formulation there are no explicit assumptions made on the material inside the scatterer. The material can be any linear material, i.e., a general bianisotropic material. The volume integral formulation assumes that the currents can be described inside the scatterer. In the surface integral formulation, only the tangential electric and magnetic fields from the outside are requested.

The surface integral representations of the $\boldsymbol{E}$- and the $\boldsymbol{H}$-fields in Section 3.3 were derived under the assumption that the $\boldsymbol{E}(\boldsymbol{r},\omega)$ and the $\boldsymbol{H}(\boldsymbol{r},\omega)$ fields satisfy the Maxwell equations in a source-free domain V, see Figure 3.1 on page 142. These integral representations give the $\boldsymbol{E}$- or the $\boldsymbol{H}$-field in the volume V expressed in the tangential components of the fields $\boldsymbol{E}$ and $\boldsymbol{H}$ on the bounding surface S to V (the limits on the surface S are taken from the inside of volume V), see (3.14) and (3.15), respectively. Outside the volume V, these surface integrals cancel and give the value zero. As already pointed out, the bounding surface S does not need to be an interface between two different materials, but an arbitrary surface. The only requirement is that the material inside the volume V is homogeneous and isotropic and that the field satisfies the Maxwell equations there.

We employ the integral representation on the incident field $\boldsymbol{E}_\mathrm{i}$ to the volume V_s, with bounding surface S_s, and outward unit normal vector $\hat{\boldsymbol{\nu}}$. This is possible since the incident field satisfies the source-free equations (4.1) in V_s, and since we assumed the sources of the incident field were located in V_i, which is disjunct from the volume V_s. Equation (3.14) on page 156 then gives a representation of the incident electric field in terms of the tangential electric and magnetic fields on the bounding surface S_s ($\boldsymbol{r} \notin S_\mathrm{s}$)

$$\frac{\eta_0\eta}{ik}\nabla\times\left\{\nabla\times\iint_{S_\mathrm{s}} g(k,|\boldsymbol{r}-\boldsymbol{r}'|)\left(\hat{\boldsymbol{\nu}}(\boldsymbol{r}')\times\boldsymbol{H}_\mathrm{i}(\boldsymbol{r}')\right)\,\mathrm{d}S'\right\}$$
$$-\nabla\times\iint_{S_\mathrm{s}} g(k,|\boldsymbol{r}-\boldsymbol{r}'|)\left(\hat{\boldsymbol{\nu}}(\boldsymbol{r}')\times\boldsymbol{E}_\mathrm{i}(\boldsymbol{r}')\right)\,\mathrm{d}S' = \begin{cases}\boldsymbol{E}_\mathrm{i}(\boldsymbol{r}), & \boldsymbol{r} \text{ inside } S_\mathrm{s}\\ \boldsymbol{0}, & \boldsymbol{r} \text{ outside } S_\mathrm{s}\end{cases} \tag{4.9}$$

where the unit normal vector $\hat{\boldsymbol{\nu}}$ is directed outward as depicted in Figure 4.3. Note that the fields, $\hat{\boldsymbol{\nu}}\times\boldsymbol{E}_\mathrm{i}$ and $\hat{\boldsymbol{\nu}}\times\boldsymbol{H}_\mathrm{i}$, in the integrand in (4.9) formally are the limits of the fields from the inside of S_s. However, these fields are continuous across the artificial surface S_s, and the limits can therefore be interpreted as the limits from the outside of S_s, whenever this interpretation is convenient. Moreover, we recall that the lower equation only states that the integrals on the left-hand side cancel—it does not imply that the incident field $\boldsymbol{E}_\mathrm{i}$ is zero outside S_s, see also the discussion below (3.9) on page 152.

There is also another way of utilizing the surface integral representation in our scattering problem. Apply the surface integral representation on the scattered field $\boldsymbol{E}_\mathrm{s}$ outside the volume V_s. The scattered field has no sources in this region, i.e., it satisfies the source-free Maxwell equations, (4.2). The derivation of the surface integral representation deals with a finite volume, and it is therefore convenient to start with a finite volume V exterior to V_s, but inside the spherical surface S_R, see Figure 4.3. The radius R on the spherical surface S_R is assumed large enough to enclose the scatterer V_s, but it could intersect the volume V_i as shown in Figure 4.3. It is a fictitious surface and not an interface between two different materials. On the other hand, the surface S_s is an interface to the scattering volume V_s. The values of the field at this surface are therefore limits of the fields taken from the exterior to the surface S_s (this corresponds to

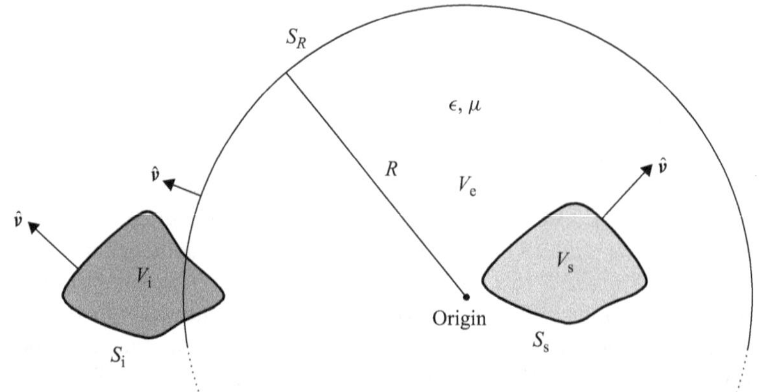

Figure 4.3 The geometry and notation used in the applications of the surface integral representation of the scattered field.

limits from the inside of the volume V). With the unit normal directions defined in Figure 4.3 (outward directed on the volume V_s and outward directed on the spherical surface S_R), we obtain the following surface integral representation of the scattered field $\boldsymbol{E}_s$, satisfying (4.2).

$$
\begin{aligned}
&\boldsymbol{I}_{S_R}(\boldsymbol{r}) - \frac{\eta_0\eta}{\mathrm{i}k}\nabla\times\left\{\nabla\times\iint_{S_s} g(k,|\boldsymbol{r}-\boldsymbol{r}'|)\left(\hat{\boldsymbol{\nu}}(\boldsymbol{r}')\times\boldsymbol{H}_s(\boldsymbol{r}')\right)\,\mathrm{d}S'\right\} \\
&+\nabla\times\iint_{S_s} g(k,|\boldsymbol{r}-\boldsymbol{r}'|)\left(\hat{\boldsymbol{\nu}}(\boldsymbol{r}')\times\boldsymbol{E}_s(\boldsymbol{r}')\right)\,\mathrm{d}S' = \begin{cases}\boldsymbol{E}_s(\boldsymbol{r}), & \boldsymbol{r} \text{ outside } S_s \\ & \text{but inside } S_R \\ \boldsymbol{0}, & \text{elsewhere}\end{cases}
\end{aligned}
\tag{4.10}
$$

where the vector-valued integral evaluated over the surface S_R is

$$
\begin{aligned}
\boldsymbol{I}_{S_R}(\boldsymbol{r}) = {}& \frac{\eta_0\eta}{\mathrm{i}k}\nabla\times\left\{\nabla\times\iint_{S_R} g(k,|\boldsymbol{r}-\boldsymbol{r}'|)\left(\hat{\boldsymbol{\nu}}(\boldsymbol{r}')\times\boldsymbol{H}_s(\boldsymbol{r}')\right)\,\mathrm{d}S'\right\} \\
&-\nabla\times\iint_{S_R} g(k,|\boldsymbol{r}-\boldsymbol{r}'|)\left(\hat{\boldsymbol{\nu}}(\boldsymbol{r}')\times\boldsymbol{E}_s(\boldsymbol{r}')\right)\,\mathrm{d}S'
\end{aligned}
$$

Note that the integral representation is applied to the region exterior to the scatterer. This implies that inside and outside S_s is reversed compared to the previous application of the integral representation on the incident field.

The integral over the surface S_R, $\boldsymbol{I}_{S_R}(\boldsymbol{r})$, is independent of the radius R, provided $\boldsymbol{r}$ is inside S_R. This is easily deduced by an application of the integral representation to a region between two concentric spheres S_{R_1} and S_{R_2} with radii $R_1 < R_2$. The spheres S_{R_1} and S_{R_2} are both assumed enclosing the volume V_s. For a position vector $\boldsymbol{r}$ inside both the surfaces S_{R_1} and S_{R_2}, i.e., at a distance $r < R_1 < R_2$ (outside the volume enclosed by the surfaces S_{R_1} and S_{R_2}), we then conclude

$$
\boldsymbol{I}_{S_{R_1}}(\boldsymbol{r}) = \boldsymbol{I}_{S_{R_2}}(\boldsymbol{r})
$$

The value of the surface integral over the surface S_R is therefore independent of the radius R for sufficiently large radius R (large enough to enclose the scatterer and the position vector $\boldsymbol{r}$). Physically, the constant value of surface integral is a contribution to the scattered field $\boldsymbol{E}_s$ from the region outside S_R, i.e., at large distances. Such a contribution violates the basic physical principles that the sources of the scattered field should be located inside the scatterer. Therefore, to obtain the correct physics, this contribution must be zero. This is equivalent to a boundary condition of the scattered field at large distances—also called the radiation condition of the scattered field. These conditions are necessary to obtain uniqueness of the scattered field and to model the correct physical behavior of the scattering problem.

We now focus on the radiation condition of the scattering problem, and we prove that a weak form of the radiation condition suffices to make the integral over the outer surface, S_R, vanish. The pertinent weak form of the radiation condition used in this book is

$$
\iint_{S_R} \left|\eta_0\eta\left(\hat{\boldsymbol{r}}'\times\boldsymbol{H}_s(\boldsymbol{r}')\right)+\boldsymbol{E}_s(\boldsymbol{r}')\right|^2\,\mathrm{d}S' \to 0, \quad \text{as } R\to\infty \tag{4.11}
$$

Sufficient conditions for this radiation condition to be fulfilled are the Silver–Müller radiation conditions.[5] We return to these often used conditions below in Comment 4.2 on page 199.

First, we show that (4.11) implies that

$$\iint_{S_R} |\boldsymbol{E}_{\rm s}(\boldsymbol{r}')|^2 \,\mathrm{d}S' \quad \text{and} \quad \iint_{S_R} |\hat{\boldsymbol{r}}' \times \boldsymbol{H}_{\rm s}(\boldsymbol{r}')|^2 \,\mathrm{d}S'$$

are both bounded in the limit $R \to \infty$. This is proved by the use of the following identity:

$$\iint_{S_R} \left|\eta_0\eta(\hat{\boldsymbol{r}}' \times \boldsymbol{H}_{\rm s}(\boldsymbol{r}')) + \boldsymbol{E}_{\rm s}(\boldsymbol{r}')\right|^2 \mathrm{d}S' = \iint_{S_R} \Big\{\eta_0^2\eta^2 \left|\hat{\boldsymbol{r}}' \times \boldsymbol{H}_{\rm s}(\boldsymbol{r}')\right|^2 + \left|\boldsymbol{E}_{\rm s}(\boldsymbol{r}')\right|^2 \\ + 2\,\mathrm{Re}\left[\eta_0\eta \boldsymbol{E}_{\rm s}^*(\boldsymbol{r}')\cdot\left(\hat{\boldsymbol{r}}' \times \boldsymbol{H}_{\rm s}(\boldsymbol{r}')\right)\right]\Big\}\,\mathrm{d}S' \tag{4.12}$$

The last integral is rewritten by an application of the divergence theorem to the volume V between S_R and $S_{\rm s}$ ($\hat{\boldsymbol{\nu}} = \hat{\boldsymbol{r}}'$ on the surface S_R)

$$\iint_{S_R} \mathrm{Re}\big[\boldsymbol{E}_{\rm s}^*(\boldsymbol{r}')\cdot(\hat{\boldsymbol{r}}' \times \boldsymbol{H}_{\rm s}(\boldsymbol{r}'))\big]\,\mathrm{d}S' = \mathrm{Re}\iint_{S_R} \big[\hat{\boldsymbol{\nu}}(\boldsymbol{r}')\cdot\big(\boldsymbol{H}_{\rm s}(\boldsymbol{r}') \times \boldsymbol{E}_{\rm s}^*(\boldsymbol{r}')\big)\big]\,\mathrm{d}S'$$

$$= \mathrm{Re}\left\{\iint_{S_{\rm s}} \big[\hat{\boldsymbol{\nu}}(\boldsymbol{r}')\cdot\big(\boldsymbol{H}_{\rm s}(\boldsymbol{r}') \times \boldsymbol{E}_{\rm s}^*(\boldsymbol{r}')\big)\big]\,\mathrm{d}S' + \iiint_V \nabla'\cdot\big(\boldsymbol{H}_{\rm s}(\boldsymbol{r}') \times \boldsymbol{E}_{\rm s}^*(\boldsymbol{r}')\big)\,\mathrm{d}v'\right\}$$

The volume integral gives zero contribution. In fact, the Maxwell equations, (4.2), in the volume integral, imply

$$\mathrm{Re}\big\{\nabla'\cdot\big(\boldsymbol{H}_{\rm s}(\boldsymbol{r}') \times \boldsymbol{E}_{\rm s}^*(\boldsymbol{r}')\big)\big\} = \mathrm{Re}\big\{\boldsymbol{E}_{\rm s}^*(\boldsymbol{r}')\cdot\big(\nabla' \times \boldsymbol{H}_{\rm s}(\boldsymbol{r}')\big) - \boldsymbol{H}_{\rm s}(\boldsymbol{r}')\cdot\big(\nabla' \times \boldsymbol{E}_{\rm s}^*(\boldsymbol{r}')\big)\big\}$$

$$= -\mathrm{Re}\left\{\frac{ik}{\eta_0\eta}|\boldsymbol{E}_{\rm s}(\boldsymbol{r}')|^2 - \mathrm{i}k\eta_0\eta|\boldsymbol{H}_{\rm s}(\boldsymbol{r}')|^2\right\} = 0$$

Comment 4.1

The identity in (4.12) also justifies the name radiation condition in (4.11). Equation (4.11) shows that the left-hand side vanishes in the limit $R \to \infty$, and since the two first integrals are non-negative, the remaining integral must be non-positive, i.e., ($\hat{\boldsymbol{\nu}} = \hat{\boldsymbol{r}}'$ on the surface S_R)

$$-\mathrm{Re}\iint_{S_R} \boldsymbol{E}_{\rm s}^*(\boldsymbol{r}')\cdot\big(\hat{\boldsymbol{r}}' \times \boldsymbol{H}_{\rm s}(\boldsymbol{r}')\big)\,\mathrm{d}S' = \mathrm{Re}\iint_{S_R} \big(\boldsymbol{E}_{\rm s}(\boldsymbol{r}') \times \boldsymbol{H}_{\rm s}^*(\boldsymbol{r}')\big)\cdot\hat{\boldsymbol{\nu}}(\boldsymbol{r}')\,\mathrm{d}S' \geq 0$$

The last integral quantifies the total radiated power out of the surface S_R, and this quantity is always non-negative, which justifies the name radiation condition.

[5] Samuel Silver (1915–1976), American electrical engineer; Claus Müller (1920–2008), German mathematician.

The radiation condition (4.11) implies

$$0 = \lim_{R\to\infty} \iint_{S_R} \left|\eta_0\eta\left(\hat{\boldsymbol{r}}' \times \boldsymbol{H}_{\rm s}(\boldsymbol{r}')\right) + \boldsymbol{E}_{\rm s}(\boldsymbol{r}')\right|^2 \,{\rm d}S'$$

$$= \lim_{R\to\infty} \iint_{S_R} \left\{\eta_0^2\eta^2 \left|\hat{\boldsymbol{r}}' \times \boldsymbol{H}_{\rm s}(\boldsymbol{r}')\right|^2 + \left|\boldsymbol{E}_{\rm s}(\boldsymbol{r}')\right|^2\right\} \,{\rm d}S'$$

$$+ 2\eta_0\eta \,{\rm Re}\left\{\iint_{S_{\rm s}} \left[\hat{\boldsymbol{\nu}}(\boldsymbol{r}')\cdot\left(\boldsymbol{H}_{\rm s}(\boldsymbol{r}') \times \boldsymbol{E}_{\rm s}^*(\boldsymbol{r}')\right)\right] \,{\rm d}S'\right\}$$

Since the integrand in the first integral over S_R on the right-hand side is non-negative and the last surface integral (over $S_{\rm s}$) is independent of R, the first integral has a finite limit as $R \to \infty$. We conclude

$$\left\{\begin{aligned} &\iint_{S_R} |\boldsymbol{E}_{\rm s}(\boldsymbol{r}')|^2 \,{\rm d}S' = O(1) \\ &\iint_{S_R} |\hat{\boldsymbol{r}}' \times \boldsymbol{H}_{\rm s}(\boldsymbol{r}')|^2 \,{\rm d}S' = O(1) \end{aligned}\right. \qquad \text{as } R \to \infty \tag{4.13}$$

i.e., the integrals are bounded in the limit $R \to \infty$. More details on these limits are provided in Section 7.8.4. In particular, it is shown that these limits are monotonic and well defined.

To proceed, we investigate the dominant contribution to the integral I_{S_R} in (4.10) as $R \to \infty$. The position vector $\boldsymbol{r}$ is assumed to be a fixed point outside or inside the volume $V_{\rm s}$ in these calculations. However, the integration variable $\boldsymbol{r}'$ (located on S_R) for large R satisfies $r' \gg r$. The dominating contributions of the gradient of the Green function and other combinations of its derivatives are essential, and we have[6] ($\boldsymbol{a}$ is a vector field independent of the position vector $\boldsymbol{r}$ and $\hat{\boldsymbol{e}} = (\boldsymbol{r}' - \boldsymbol{r})/|\boldsymbol{r} - \boldsymbol{r}'| = -\nabla|\boldsymbol{r} - \boldsymbol{r}'|$)

$$\nabla g\left(k, |\boldsymbol{r} - \boldsymbol{r}'|\right) = -\hat{\boldsymbol{e}}\frac{{\rm e}^{{\rm i}k|\boldsymbol{r}-\boldsymbol{r}'|}}{4\pi|\boldsymbol{r} - \boldsymbol{r}'|}\left({\rm i}k - \frac{1}{|\boldsymbol{r} - \boldsymbol{r}'|}\right)$$

$$\begin{aligned} \nabla \times \left(g(k, |\boldsymbol{r} - \boldsymbol{r}'|)\boldsymbol{a}\right) &= \nabla g(k, |\boldsymbol{r} - \boldsymbol{r}'|) \times \boldsymbol{a} \\ &= -{\rm i}k\left(\hat{\boldsymbol{e}} \times \boldsymbol{a}\right)\frac{{\rm e}^{{\rm i}k|\boldsymbol{r}-\boldsymbol{r}'|}}{4\pi|\boldsymbol{r} - \boldsymbol{r}'|}\left[1 + O((k|\boldsymbol{r} - \boldsymbol{r}'|)^{-1})\right] \end{aligned}$$

$$\nabla \times \left(\nabla g(k, |\boldsymbol{r} - \boldsymbol{r}'|) \times \boldsymbol{a}\right) = -k^2\frac{{\rm e}^{{\rm i}k|\boldsymbol{r}-\boldsymbol{r}'|}}{4\pi|\boldsymbol{r} - \boldsymbol{r}'|}\hat{\boldsymbol{e}} \times \left(\hat{\boldsymbol{e}} \times \boldsymbol{a}\right)\left[1 + O((k|\boldsymbol{r} - \boldsymbol{r}'|)^{-1})\right]$$

---

[6] The left-hand side is a vector and the $O(f(x))$-symbol here means that each Cartesian component of the vector vanishes as $O(f(x))$. Similar interpretation is used for the o-symbol applied to vector fields in this text.

since $k^{-1}\nabla \times (\hat{e} \times a) = O\left((k|r - r'|)^{-1}\right)$. Note that the differentiation is w.r.t. the unprimed variables. Moreover, we have, see (4.6) (change primed and unprimed variables)

$$\begin{aligned} |r - r'| &= r' - \hat{r}' \cdot r + O(r^2/r') \\ \hat{e} &= (r' - r)/|r - r'| = \hat{r}' \left(1 + O(r/r')\right) \end{aligned}$$

These estimates of the dominant contributions of the Green function at large r imply that the dominant contribution to the integral I_{S_R} in (4.10) becomes ($\hat{\nu} = \hat{r}'$ and $r' = R$ on the surface S_R)

$$\begin{aligned} I_{S_R} = \mathrm{i}k \iint\limits_{S_R} & \frac{\mathrm{e}^{\mathrm{i}k|r-r'|}}{4\pi|r - r'|} \left\{\eta_0\eta\hat{r}' \times (\hat{r}' \times (\hat{r}' \times H_s(r'))) + \hat{r}' \times (\hat{r}' \times E_s(r'))\right\} \mathrm{d}S' \\ & + \iint\limits_{S_R} \left\{O\left((kR)^{-2}\right) \hat{r}' \times (\hat{r}' \times (\hat{r}' \times H_s(r')))\right\} \mathrm{d}S' \\ & + \iint\limits_{S_R} \left\{O\left((kR)^{-2}\right) \hat{r}' \times (\hat{r}' \times E_s(r'))\right\} \mathrm{d}S' \end{aligned}$$

The Cauchy–Schwarz inequality[7] can now be applied to the last two integrals, and moreover, using (4.13), we conclude that the last two integrals vanish as $O((kR)^{-1})$. Thus, the dominant contribution to I_{S_R} is

$$\begin{aligned} I_{S_R} = \mathrm{i}k \iint\limits_{S_R} & \frac{\mathrm{e}^{\mathrm{i}k|r-r'|}}{4\pi|r - r'|} \hat{r}' \times \left\{\hat{r}' \times \left[\eta_0\eta\hat{r}' \times H_s(r') + E_s(r')\right]\right\} \mathrm{d}S' \\ & + O\left((kR)^{-1}\right) \to \mathbf{0}, \qquad \text{as } R \to \infty \end{aligned}$$

where we once again have used the Cauchy–Schwarz inequality and the radiation condition, (4.11), to prove that the entire integral vanishes in the limit $R \to \infty$.

Equation (4.10) now simplifies to

$$\begin{aligned} & -\frac{\eta_0\eta}{\mathrm{i}k}\nabla \times \left\{\nabla \times \iint\limits_{S_s} g(k, |r - r'|) \left(\hat{\nu}(r') \times H_s(r')\right) \mathrm{d}S'\right\} \\ & + \nabla \times \iint\limits_{S_s} g(k, |r - r'|) \left(\hat{\nu}(r') \times E_s(r')\right) \mathrm{d}S' = \begin{cases} E_s(r), & r \text{ outside } S_s \\ \mathbf{0}, & r \text{ inside } S_s \end{cases} \end{aligned} \tag{4.14}$$

[7] The Cauchy–Schwarz inequality for surface integrals is

$$\left|\iint\limits_S f(r)F(r)\,\mathrm{d}S\right| \leq \sqrt{\iint\limits_S |f(r)|^2\,\mathrm{d}S}\sqrt{\iint\limits_S |F(r)|^2\,\mathrm{d}S}$$

Note that integrand is expressed in the scattered tangential electric and magnetic fields. In many situations, it is advantageous to express the integrals in terms of the total tangential electric and magnetic fields, i.e., $\hat{\boldsymbol{\nu}} \times \boldsymbol{E} = \hat{\boldsymbol{\nu}} \times (\boldsymbol{E}_{\rm i} + \boldsymbol{E}_{\rm s})$. We can obtain these expressions by combining (4.9) and (4.14). This combination then involves the total tangential fields in the surface integrals over the surface of the scatterer $S_{\rm s}$

$$
\begin{aligned}
&-\frac{\eta_0\eta}{\mathrm{i}k}\nabla\times\left\{\nabla\times\iint_{S_{\rm s}} g(k,|\boldsymbol{r}-\boldsymbol{r}'|)(\hat{\boldsymbol{\nu}}(\boldsymbol{r}')\times\boldsymbol{H}(\boldsymbol{r}'))\,\mathrm{d}S'\right\}\\
&+\nabla\times\iint_{S_{\rm s}} g(k,|\boldsymbol{r}-\boldsymbol{r}'|)\left(\hat{\boldsymbol{\nu}}(\boldsymbol{r}')\times\boldsymbol{E}(\boldsymbol{r}')\right)\mathrm{d}S' = \begin{cases}\boldsymbol{E}_{\rm s}(\boldsymbol{r}), & \boldsymbol{r} \text{ outside } S_{\rm s}\\ -\boldsymbol{E}_{\rm i}(\boldsymbol{r}), & \boldsymbol{r} \text{ inside } S_{\rm s}\end{cases}
\end{aligned} \tag{4.15}
$$

Expressions (4.14) and (4.15) are the main results of this section. Both expressions give a general representation of the scattered field in terms of the boundary values of the scattered or total tangential fields on $S_{\rm s}$. No special assumptions are made on the incident field, $\boldsymbol{E}_{\rm i}$, except that its sources should be located outside the scattering volume $V_{\rm s}$, and not be affected by the scattered fields.

We conclude this section by calculating two very useful and general expressions of the far field amplitude by the use of the integral representations (4.14) and (4.15). In the far zone, see (4.4), we obtain the following dominant contribution to the terms in the integrand:

$$
g(k,|\boldsymbol{r}-\boldsymbol{r}'|) = \frac{\mathrm{e}^{\mathrm{i}kr}}{4\pi r}\mathrm{e}^{-\mathrm{i}k\hat{\boldsymbol{r}}\cdot\boldsymbol{r}'}\left(1+O(kd^2/r)\right)\left(1+O(d/r)\right)
$$

$$
\frac{1}{\mathrm{i}k}\nabla g(k,|\boldsymbol{r}-\boldsymbol{r}'|) = \hat{\boldsymbol{r}}\frac{\mathrm{e}^{\mathrm{i}kr}}{4\pi r}\mathrm{e}^{-\mathrm{i}k\hat{\boldsymbol{r}}\cdot\boldsymbol{r}'}\left(1+O(kd^2/r)\right)\left(1+O(d/r)\right)\left(1+O((kr)^{-1})\right)
$$

since $1/k\nabla f(\hat{\boldsymbol{r}}) = O((kr)^{-1})$. Moreover, we have

$$
\begin{aligned}
&\frac{1}{k^2}\nabla\times\left(\nabla g(k,|\boldsymbol{r}-\boldsymbol{r}'|)\times\boldsymbol{a}\right)\\
&\quad = \frac{\mathrm{e}^{\mathrm{i}kr}}{4\pi r}\mathrm{e}^{-\mathrm{i}k\hat{\boldsymbol{r}}\cdot\boldsymbol{r}'}\hat{\boldsymbol{r}}\times\left(\boldsymbol{a}\times\hat{\boldsymbol{r}}\right)\left(1+O(kd^2/r)\right)\left(1+O(d/r)\right)\left(1+O((kr)^{-1})\right)
\end{aligned}
$$

where $\boldsymbol{a}$ does not depend on $\boldsymbol{r}$. The far field then is, see (4.14)

$$
\boldsymbol{E}_{\rm s}(\boldsymbol{r}) = \mathrm{i}k\frac{\mathrm{e}^{\mathrm{i}kr}}{4\pi r}\hat{\boldsymbol{r}}\times\iint_{S_{\rm s}}\left[\eta_0\eta(\hat{\boldsymbol{\nu}}(\boldsymbol{r}')\times\boldsymbol{H}_{\rm s}(\boldsymbol{r}'))\times\hat{\boldsymbol{r}} + (\hat{\boldsymbol{\nu}}(\boldsymbol{r}')\times\boldsymbol{E}_{\rm s}(\boldsymbol{r}'))\right]\mathrm{e}^{-\mathrm{i}k\hat{\boldsymbol{r}}\cdot\boldsymbol{r}'}\,\mathrm{d}S'
$$

We rewrite this as, cf. the volume integral formulation in Section 4.1.1

$$
\boldsymbol{E}_{\rm s}(\boldsymbol{r}) = \frac{\mathrm{e}^{\mathrm{i}kr}}{r}\boldsymbol{F}(\hat{\boldsymbol{r}}) \tag{4.16}
$$

where the far field amplitude has the following general form:

$$\boldsymbol{F}(\hat{\boldsymbol{r}}) = \frac{\mathrm{i}k}{4\pi}\hat{\boldsymbol{r}} \times \iint_{S_s} \left[\hat{\boldsymbol{\nu}}(\boldsymbol{r}') \times \boldsymbol{E}_s(\boldsymbol{r}') - \eta_0\eta\hat{\boldsymbol{r}} \times \left(\hat{\boldsymbol{\nu}}(\boldsymbol{r}') \times \boldsymbol{H}_s(\boldsymbol{r}')\right)\right] \mathrm{e}^{-\mathrm{i}k\hat{\boldsymbol{r}}\cdot\boldsymbol{r}'}\, \mathrm{d}S' \tag{4.17}$$

This expression of the far field amplitude involves only the values of the scattered tangential electric and magnetic fields on the boundary of the scatterer S_s. No assumptions about the scatterer and the material inside S_s have been made. Only the properties outside the scatterer have been used, and no properties of the material inside S_s. In this respect, the form of this representation of the far field amplitude does not depend on the scatterer. The explicit values of the tangential fields, of course, depend on the properties inside the scatterer. This expression of the far field amplitude is therefore a much more general expression than the expression we derived by the use of the volume integral approach in Section 4.1.1.

Moreover, we directly see that the far field amplitude $\boldsymbol{F}$ in this surface integral formulation only contains $\hat{\boldsymbol{\theta}}$- and $\hat{\boldsymbol{\phi}}$-components and not $\hat{\boldsymbol{r}}$-component, i.e.,

$$\hat{\boldsymbol{r}} \cdot \boldsymbol{F}(\hat{\boldsymbol{r}}) = 0$$

If (4.15) is used as a starting point for the derivation, we obtain, in analogy to the derivation leading to (4.17), the following expression in the total fields:

$$\boldsymbol{F}(\hat{\boldsymbol{r}}) = \frac{\mathrm{i}k}{4\pi}\hat{\boldsymbol{r}} \times \iint_{S_s} \left[\hat{\boldsymbol{\nu}}(\boldsymbol{r}') \times \boldsymbol{E}(\boldsymbol{r}') - \eta_0\eta\hat{\boldsymbol{r}} \times \left(\hat{\boldsymbol{\nu}}(\boldsymbol{r}') \times \boldsymbol{H}(\boldsymbol{r}')\right)\right] \mathrm{e}^{-\mathrm{i}k\hat{\boldsymbol{r}}\cdot\boldsymbol{r}'}\, \mathrm{d}S' \tag{4.18}$$

This expression is also very useful in many situations.

The integral expression of the far field amplitude in (4.17) does not depend on which surface S it is evaluated on, as long as the surface encloses S_s. This is physically obvious—extend the scatterer to the larger domain inside S. To prove this statement mathematically is also easy. Let S be a closed surface in the exterior of the scatterer that (simply) encloses the scatterer in V_s. Then

$$\boldsymbol{F}(\hat{\boldsymbol{r}}) = \frac{\mathrm{i}k}{4\pi}\hat{\boldsymbol{r}} \times \iint_{S} \left[\hat{\boldsymbol{\nu}}(\boldsymbol{r}') \times \boldsymbol{E}_s(\boldsymbol{r}') - \eta_0\eta\hat{\boldsymbol{r}} \times \left(\hat{\boldsymbol{\nu}}(\boldsymbol{r}') \times \boldsymbol{H}_s(\boldsymbol{r}')\right)\right] \mathrm{e}^{-\mathrm{i}k\hat{\boldsymbol{r}}\cdot\boldsymbol{r}'}\, \mathrm{d}S' \tag{4.19}$$

To prove this, make an application of the analogous Gauss' theorem

$$\iiint_V \nabla \times \boldsymbol{A}\, \mathrm{d}v = \iint_S \hat{\boldsymbol{\nu}} \times \boldsymbol{A}\, \mathrm{d}S$$

to the difference between the integral in (4.19) and in (4.17). The integral becomes

$$\begin{aligned}
&\hat{\boldsymbol{r}} \times \iint_{S-S_s} \left[\hat{\boldsymbol{\nu}}(\boldsymbol{r}') \times \boldsymbol{E}_s(\boldsymbol{r}') - \eta_0\eta\hat{\boldsymbol{r}} \times \left(\hat{\boldsymbol{\nu}}(\boldsymbol{r}') \times \boldsymbol{H}_s(\boldsymbol{r}')\right)\right] \mathrm{e}^{-\mathrm{i}k\hat{\boldsymbol{r}}\cdot\boldsymbol{r}'}\, \mathrm{d}S' \\
&= \hat{\boldsymbol{r}} \times \iiint_V \left[\nabla' \times \boldsymbol{E}_s(\boldsymbol{r}')\mathrm{e}^{-\mathrm{i}k\hat{\boldsymbol{r}}\cdot\boldsymbol{r}'} - \eta_0\eta\hat{\boldsymbol{r}} \times \left(\nabla' \times \boldsymbol{H}_s(\boldsymbol{r}')\mathrm{e}^{-\mathrm{i}k\hat{\boldsymbol{r}}\cdot\boldsymbol{r}'}\right)\right] \mathrm{d}v'
\end{aligned}$$

where the volume bounded by S and S_s is denoted V. However, the integrand in the integral on the right-hand side is zero. The Maxwell equations for the scattered field and the use of the BAC-CAB rule imply

$$\begin{aligned}
&\hat{\boldsymbol{r}} \times \left\{\nabla' \times \boldsymbol{E}_s(\boldsymbol{r}')e^{-ik\hat{\boldsymbol{r}}\cdot\boldsymbol{r}'} - \eta_0\eta\hat{\boldsymbol{r}} \times \left(\nabla' \times \boldsymbol{H}_s(\boldsymbol{r}')e^{-ik\hat{\boldsymbol{r}}\cdot\boldsymbol{r}'}\right)\right\} \\
&\quad = \hat{\boldsymbol{r}} \times \left\{ik\eta\eta_0\boldsymbol{H}_s(\boldsymbol{r}') - ik\hat{\boldsymbol{r}} \times \boldsymbol{E}_s(\boldsymbol{r}') - \hat{\boldsymbol{r}} \times \left(-ik\boldsymbol{E}_s(\boldsymbol{r}') - ik\hat{\boldsymbol{r}} \times \eta_0\eta\boldsymbol{H}_s(\boldsymbol{r}')\right)\right\}e^{-ik\hat{\boldsymbol{r}}\cdot\boldsymbol{r}'} = \boldsymbol{0}
\end{aligned}$$

and (4.19) is proved. This shows that the integral representation of the far field amplitude does not depend on which surface S used for evaluation of the integral, as long as it encloses V_s. The integral, however, does depend on the fields inside V_s and the incident field. We also note that the scattered field in the expressions (4.19) can be replaced by the total fields, as long as the bounding surface S does not include any region where the sources of the incident field are located.

Comment 4.2

A sufficient condition to satisfy the radiation condition in (4.11) is one of the Silver–Müller radiation conditions, see References 197, 244. These are

$$\begin{cases} \left(\hat{\boldsymbol{r}} \times \boldsymbol{E}_s(\boldsymbol{r})\right) - \eta_0\eta\boldsymbol{H}_s(\boldsymbol{r}) = o((kr)^{-1}) \\ \eta_0\eta\left(\hat{\boldsymbol{r}} \times \boldsymbol{H}_s(\boldsymbol{r})\right) + \boldsymbol{E}_s(\boldsymbol{r}) = o((kr)^{-1}) \end{cases} \quad \text{as } r \to \infty$$

These conditions are assumed to hold uniformly in $\hat{\boldsymbol{r}}$ as $r \to \infty$. Either of the conditions is sufficient to satisfy (4.11). The name radiation condition is obvious if we calculate the power flux density, $\langle \boldsymbol{S}_s(t)\rangle$, of the scattered field

$$\begin{aligned}
\langle \boldsymbol{S}_s(t)\rangle &= \frac{1}{2}\operatorname{Re}\left\{\boldsymbol{E}_s \times \boldsymbol{H}_s^*\right\} = \frac{1}{2\eta_0\eta}\operatorname{Re}\left\{\boldsymbol{E}_s \times (\hat{\boldsymbol{r}} \times \boldsymbol{E}_s^*)\right\} + o((kr)^{-1}) \\
&= \frac{1}{2\eta_0\eta}\operatorname{Re}\left\{\hat{\boldsymbol{r}}|\boldsymbol{E}_s|^2 - \boldsymbol{E}_s^*(\hat{\boldsymbol{r}} \cdot \boldsymbol{E}_s)\right\} + o((kr)^{-1}) = \frac{\hat{\boldsymbol{r}}}{2\eta_0\eta}|\boldsymbol{E}_s|^2 + o((kr)^{-1})
\end{aligned}$$

Here, we have used the BAC-CAB rule, $\boldsymbol{a} \times (\boldsymbol{b} \times \boldsymbol{c}) = \boldsymbol{b}(\boldsymbol{a} \cdot \boldsymbol{c}) - \boldsymbol{c}(\boldsymbol{a} \cdot \boldsymbol{b})$, $\hat{\boldsymbol{r}} \cdot \boldsymbol{F}(\hat{\boldsymbol{r}}) = 0$, and the radiation condition. The dominant contribution of the power flux density at large distances from the scatterer is non-negative, i.e., $\langle \boldsymbol{S}_s(t)\rangle \cdot \hat{\boldsymbol{r}} \geq 0$, which implies that the power is radiated away from the scatterer. Notice that no plane wave satisfies the radiation conditions.

4.1.3 Translation of the Origin

We note that the far field amplitude depends on the position of the origin. Different choices of origin give different far field amplitudes. To relate the far field amplitude at two different origins, O and O', which are translated by a vector $\boldsymbol{r}_0$, see Figure 4.4, we express the position vectors, $\boldsymbol{r}$ and $\boldsymbol{r}'$, respectively, w.r.t. the two origins, i.e.,

$$\boldsymbol{r} = \boldsymbol{r}_0 + \boldsymbol{r}'$$

and the far field amplitudes w.r.t. the two origins are defined as

$$\boldsymbol{F}(\hat{\boldsymbol{r}}) = \lim_{r\to\infty} re^{-ikr}\boldsymbol{E}_s(\boldsymbol{r}), \qquad \boldsymbol{F}'(\hat{\boldsymbol{r}}) = \boldsymbol{F}'(\hat{\boldsymbol{r}}') = \lim_{r'\to\infty} r'e^{-ikr'}\boldsymbol{E}_s(\boldsymbol{r}')$$

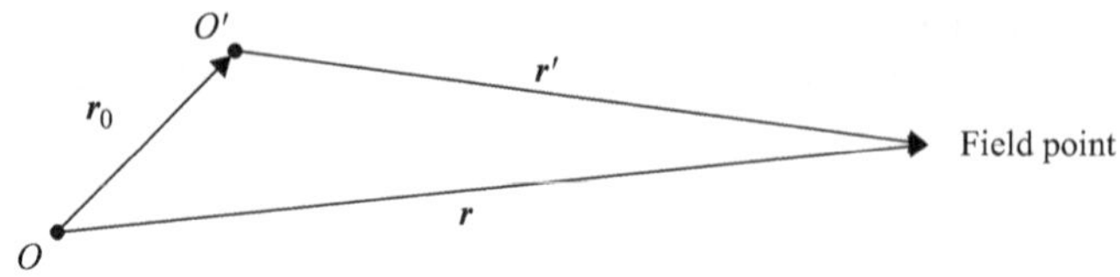

Figure 4.4 Translation of the origin and the position vectors w.r.t. the two origins O and O'.

Note that the directional vector $\hat{\boldsymbol{r}}$ is the same in the two far field amplitudes. This is a consequence of the definition of the directional vectors $\hat{\boldsymbol{r}}$ and $\hat{\boldsymbol{r}}'$, i.e.,

$$\hat{\boldsymbol{r}} = \lim_{r\to\infty} \frac{\boldsymbol{r}}{r} = \lim_{r\to\infty} \frac{\boldsymbol{r}_0 + \boldsymbol{r}'}{|\boldsymbol{r}_0 + \boldsymbol{r}'|} = \lim_{r\to\infty} \frac{\boldsymbol{r}_0/r' + \hat{\boldsymbol{r}}'}{\left|\boldsymbol{r}_0/r' + \hat{\boldsymbol{r}}'\right|} = \lim_{r'\to\infty} \frac{\boldsymbol{r}_0/r' + \hat{\boldsymbol{r}}'}{\left|\boldsymbol{r}_0/r' + \hat{\boldsymbol{r}}'\right|} = \hat{\boldsymbol{r}}'$$

For large distances we have, see (4.6)

$$r = \left|\boldsymbol{r}_0 + \boldsymbol{r}'\right| \approx r' + \hat{\boldsymbol{r}} \cdot \boldsymbol{r}_0, \qquad \text{as } r,\ r' \to \infty$$

Therefore, we get

$$\boldsymbol{F}'(\hat{\boldsymbol{r}}) = \lim_{r'\to\infty} r\mathrm{e}^{-\mathrm{i}k(r-\hat{\boldsymbol{r}}\cdot\boldsymbol{r}_0)}\boldsymbol{E}_{\mathrm{s}}(\boldsymbol{r} - \boldsymbol{r}_0) = \mathrm{e}^{\mathrm{i}k\hat{\boldsymbol{r}}\cdot\boldsymbol{r}_0} \lim_{r\to\infty} r\mathrm{e}^{-\mathrm{i}kr}\boldsymbol{E}_{\mathrm{s}}(\boldsymbol{r})$$

and consequently

$$\boldsymbol{F}'(\hat{\boldsymbol{r}}) = \mathrm{e}^{\mathrm{i}k\hat{\boldsymbol{r}}\cdot\boldsymbol{r}_0}\boldsymbol{F}(\hat{\boldsymbol{r}})$$

4.2 Cross Sections

In all scattering problems, the total field, $\boldsymbol{E}$, is divided into the undisturbed—incident field $\boldsymbol{E}_{\mathrm{i}}$—and the scattered field $\boldsymbol{E}_{\mathrm{s}}$, i.e.,

$$\boldsymbol{E} = \boldsymbol{E}_{\mathrm{i}} + \boldsymbol{E}_{\mathrm{s}}$$

The incident field, $\boldsymbol{E}_{\mathrm{i}}$, is the field over which we have control. In this section, we assume this field is a plane wave[8] with propagation direction $\hat{\boldsymbol{k}}_{\mathrm{i}}$, see (1.72) on page 73, i.e.,

$$\boldsymbol{E}_{\mathrm{i}}(\boldsymbol{r}) = \boldsymbol{E}_0\mathrm{e}^{\mathrm{i}k\hat{\boldsymbol{k}}_{\mathrm{i}}\cdot\boldsymbol{r}}$$

The strength and the polarization of the incident plane wave are specified by the complex-valued vector $\boldsymbol{E}_0$, which also is the value of the incident field at the origin. The incident

[8] A plane wave in the vicinity of V_{s} has its sources at infinity and it can be obtained by a suitable limit process of sources located at a finite distance from the scatterer.

magnetic field is, see (4.1)

$$\boldsymbol{H}_{\rm i}(\boldsymbol{r}) = \frac{1}{{\rm i}k\eta_0\eta}\nabla\times\boldsymbol{E}_{\rm i}(\boldsymbol{r}) = \frac{1}{\eta_0\eta}\left(\hat{\boldsymbol{k}}_{\rm i}\times\boldsymbol{E}_0\right){\rm e}^{{\rm i}k\hat{\boldsymbol{k}}_{\rm i}\cdot\boldsymbol{r}} = \frac{1}{\eta_0\eta}\hat{\boldsymbol{k}}_{\rm i}\times\boldsymbol{E}_{\rm i}(\boldsymbol{r})$$

The complex-valued vector $\boldsymbol{E}_0$ is orthogonal to $\hat{\boldsymbol{k}}_{\rm i}$ (the polarization is perpendicular to the direction of propagation) since, see (4.1)

$$\begin{aligned}\hat{\boldsymbol{k}}_{\rm i}\cdot\boldsymbol{E}_{\rm i} &= \frac{i\eta_0\eta}{k}\hat{\boldsymbol{k}}_{\rm i}\cdot\nabla\times\boldsymbol{H}_{\rm i} = \frac{\rm i}{k}\hat{\boldsymbol{k}}_{\rm i}\cdot\nabla\times\left[\left(\hat{\boldsymbol{k}}_{\rm i}\times\boldsymbol{E}_0\right){\rm e}^{{\rm i}k\hat{\boldsymbol{k}}_{\rm i}\cdot\boldsymbol{r}}\right]\\ &= -\hat{\boldsymbol{k}}_{\rm i}\cdot\left[\hat{\boldsymbol{k}}_{\rm i}\times\left(\hat{\boldsymbol{k}}_{\rm i}\times\boldsymbol{E}_0\right)\right]{\rm e}^{{\rm i}k\hat{\boldsymbol{k}}_{\rm i}\cdot\boldsymbol{r}} = 0\end{aligned}$$

The geometrical interpretation of orthogonality of a complex-valued vector $\boldsymbol{E}_0$ and a real-valued direction $\hat{\boldsymbol{k}}_{\rm i}$ makes only sense if the direction of the vector $\boldsymbol{E}_0$ is real valued, since only then the vector $\boldsymbol{E}_0$ can be visualized in three dimensions. If the plane wave is linearly polarized, the complex-valued vector $\boldsymbol{E}_0$ is of the form $\boldsymbol{E}_0 = \hat{\boldsymbol{p}}_0 E_0$, where E_0 is a complex number, and $\hat{\boldsymbol{p}}_0$ is real valued, and the field $\boldsymbol{E}_{\rm i}$ simplifies to

$$\boldsymbol{E}_{\rm i}(\boldsymbol{r}) = \hat{\boldsymbol{p}}_0 E_0 {\rm e}^{{\rm i}k\hat{\boldsymbol{k}}_{\rm i}\cdot\boldsymbol{r}}$$

where the real-valued unit vector $\hat{\boldsymbol{p}}_0$ satisfies $\hat{\boldsymbol{p}}_0\cdot\hat{\boldsymbol{k}}_{\rm i} = 0$.

The second field, the scattered field, $\boldsymbol{E}_{\rm s}$, is the perturbation of the electric field due to the presence of the scatterer $V_{\rm s}$. This field is zero if there is no scatterer present, i.e., if the electromagnetic properties in $V_{\rm s}$ are identical to the surrounding space. In Section 4.1, we found that the scattered field at a large distance from the scatterer $V_{\rm s}$ is a spherical wave with transverse polarization. Explicit expressions of the far field amplitude, $\boldsymbol{F}(\hat{\boldsymbol{r}})$, in the volume integral formulation is, see (4.7)

$$\boldsymbol{F}(\hat{\boldsymbol{r}}) = \frac{{\rm i}k\eta_0\eta}{4\pi}\hat{\boldsymbol{r}}\times\left(\iiint_{V_{\rm s}} {\rm e}^{-{\rm i}k\hat{\boldsymbol{r}}\cdot\boldsymbol{r}'}\boldsymbol{J}_{\rm s}(\boldsymbol{r}')\,{\rm d}v'\times\hat{\boldsymbol{r}}\right)$$

and in the surface integral formulation we have, see (4.17)

$$\boldsymbol{F}(\hat{\boldsymbol{r}}) = \frac{{\rm i}k}{4\pi}\hat{\boldsymbol{r}}\times\iint_{S_{\rm s}}\left[\hat{\boldsymbol{\nu}}(\boldsymbol{r}')\times\boldsymbol{E}_{\rm s}(\boldsymbol{r}') - \eta_0\eta\hat{\boldsymbol{r}}\times\left(\hat{\boldsymbol{\nu}}(\boldsymbol{r}')\times\boldsymbol{H}_{\rm s}(\boldsymbol{r}')\right)\right]{\rm e}^{-{\rm i}k\hat{\boldsymbol{r}}\cdot\boldsymbol{r}'}\,{\rm d}S'$$

The power flux density (power per unit area) of the scattered field radiating away from $V_{\rm s}$ was shown above, see (4.8) on page 191

$$\langle\boldsymbol{S}_{\rm s}(t)\rangle = \frac{1}{2}{\rm Re}\left\{\boldsymbol{E}_{\rm s}\times\boldsymbol{H}_{\rm s}^*\right\} = \frac{\hat{\boldsymbol{r}}}{2\eta_0\eta r^2}|\boldsymbol{F}(\hat{\boldsymbol{r}})|^2$$

where the (relative) wave impedance, η, of the exterior material is given by (3.11) on page 154.

Similarly, the power flux density of the incident field is given by

$$\langle\boldsymbol{S}_{\rm i}(t)\rangle = \frac{1}{2}{\rm Re}\left\{\boldsymbol{E}_{\rm i}\times\boldsymbol{H}_{\rm i}^*\right\} = \frac{1}{2\eta_0\eta}{\rm Re}\left\{\boldsymbol{E}_{\rm i}\times\left(\hat{\boldsymbol{k}}_{\rm i}\times\boldsymbol{E}_{\rm i}^*\right)\right\}$$

Simplifications using the BAC-CAB rule and the condition $\hat{\boldsymbol{k}}_{\mathrm{i}} \cdot \boldsymbol{E}_{\mathrm{i}} = 0$ give

$$\langle \boldsymbol{S}_{\mathrm{i}}(t) \rangle = \frac{\hat{\boldsymbol{k}}_{\mathrm{i}}}{2\eta_0 \eta} \boldsymbol{E}_0 \cdot \boldsymbol{E}_0^* = \frac{\hat{\boldsymbol{k}}_{\mathrm{i}}}{2\eta_0 \eta} |\boldsymbol{E}_0|^2 \tag{4.20}$$

A measure of the scattering capacity of the scatterer, V_{s}, is expressed by the differential scattering cross section or the bistatic radar cross section (RCS), σ_{RCS}, in technical literature. We introduce its definition[9]

$$\sigma_{\mathrm{RCS}}(\hat{\boldsymbol{r}}, \hat{\boldsymbol{k}}_{\mathrm{i}}) = \frac{\mathrm{d}\sigma}{\mathrm{d}\Omega}(\hat{\boldsymbol{r}}, \hat{\boldsymbol{k}}_{\mathrm{i}}) = 4\pi r^2 \frac{\langle \boldsymbol{S}_{\mathrm{s}}(t) \rangle \cdot \hat{\boldsymbol{r}}}{\langle \boldsymbol{S}_{\mathrm{i}}(t) \rangle \cdot \hat{\boldsymbol{k}}_{\mathrm{i}}}$$

This quantity is the quotient between the power flux density of the scattered field and the corresponding power flux density of the incident wave. The quantity is normalized with the area of the sphere of radius r, $4\pi r^2$, on which the scattered field is computed. If we introduce the expressions of the power flux density from above, we get

$$\frac{\mathrm{d}\sigma}{\mathrm{d}\Omega}(\hat{\boldsymbol{r}}, \hat{\boldsymbol{k}}_{\mathrm{i}}) = 4\pi \frac{|\boldsymbol{F}(\hat{\boldsymbol{r}})|^2}{|\boldsymbol{E}_0|^2} \tag{4.21}$$

The differential scattering cross section is a function of the observation direction, $\hat{\boldsymbol{r}}$, and the direction of the incident plane wave, $\hat{\boldsymbol{k}}_{\mathrm{i}}$. It is also a function of the frequency, but this dependence is usually suppressed. The symbol $\frac{\mathrm{d}\sigma}{\mathrm{d}\Omega}$ is not a differential in strict mathematical sense, but a quotient between two quantities. Note that the unit of the differential scattering cross section is area.

Related to the differential scattering cross section is the co-polarized bistatic cross section, $\sigma_{\mathrm{RCS,co}}$

$$\sigma_{\mathrm{RCS,co}}(\hat{\boldsymbol{r}}, \hat{\boldsymbol{k}}_{\mathrm{i}}) = 4\pi \frac{\left|\boldsymbol{E}_0^* \cdot \boldsymbol{F}(\hat{\boldsymbol{r}})\right|^2}{|\boldsymbol{E}_0|^2} \tag{4.22}$$

The total power P_{s} that the scatterer in V_{s} scatters is the integral of the radial component of the Poynting vector, $\langle \boldsymbol{S}_{\mathrm{s}}(t) \rangle \cdot \hat{\boldsymbol{r}}$, over a sphere[10] with radius r. We get

$$P_{\mathrm{s}} = \iint_{|r|=r} \langle \boldsymbol{S}_{\mathrm{s}}(t) \rangle \cdot \hat{\boldsymbol{r}} \, \mathrm{d}S = \frac{1}{2\eta_0 \eta} \iint_{\Omega} |\boldsymbol{F}(\hat{\boldsymbol{r}})|^2 \, \mathrm{d}\Omega \tag{4.23}$$

where the integration in the second integral is over the unit sphere Ω, and the surface measure of the unit sphere is $\mathrm{d}\Omega = \sin\theta \, \mathrm{d}\theta \, \mathrm{d}\phi$. The scattering cross section σ_{s} is defined as the total

---

[9]The factor 4π is sometimes excluded in the definition of the differential scattering cross section [100]. This normalization factor is a matter of convenience and related to the application. The quantity $p(\hat{\boldsymbol{r}}, \hat{\boldsymbol{k}}_{\mathrm{i}}) = \frac{\mathrm{d}\sigma}{\mathrm{d}\Omega}(\hat{\boldsymbol{r}}, \hat{\boldsymbol{k}}_{\mathrm{i}})/\sigma_{\mathrm{ext}}(\hat{\boldsymbol{k}}_{\mathrm{i}})$ is denoted the phase function [100]. Also, the definition $p(\hat{\boldsymbol{r}}, \hat{\boldsymbol{k}}_{\mathrm{i}}) = \frac{\mathrm{d}\sigma}{\mathrm{d}\Omega}(\hat{\boldsymbol{r}}, \hat{\boldsymbol{k}}_{\mathrm{i}})/\sigma_{\mathrm{s}}(\hat{\boldsymbol{k}}_{\mathrm{i}})$ occurs [187]. The cross sections $\sigma_{\mathrm{ext}}(\hat{\boldsymbol{k}}_{\mathrm{i}})$ and $\sigma_{\mathrm{s}}(\hat{\boldsymbol{k}}_{\mathrm{i}})$ are defined below.

[10]On page 228, we show that it is irrelevant which spherical surface we compute the total power P_{s} over. The only requirement is that the surface encloses V_{s}. Here, we choose the radius of the sphere large enough so that the far field amplitude of the scattered field is the dominant contribution to the scattered field.

power P_s divided by $\langle S_i(t)\rangle \cdot \hat{k}_i$, i.e.,

$$\sigma_s(\hat{k}_i) = \frac{P_s}{\langle S_i(t)\rangle \cdot \hat{k}_i} = \frac{1}{|E_0|^2} \iint_\Omega |F(\hat{r})|^2 \, d\Omega \tag{4.24}$$

or if the differential scattering cross section $\frac{d\sigma}{d\Omega}$ is introduced

$$\sigma_s(\hat{k}_i) = \frac{1}{4\pi} \iint_\Omega \frac{d\sigma}{d\Omega}(\hat{r}, \hat{k}_i) \, d\Omega$$

The scattering cross section is a function of the direction of the incident plane wave, $\hat{k}_i$, and the frequency, but this latter dependence is usually suppressed.

In most situations, the scatterer, which is confined by the surface S_s, absorbs electromagnetic energy. The total power that is absorbed by the scatterer is quantified by an integral over the Poynting vector (note the projection on the negative unit normal vector $-\hat{\nu}$)

$$P_a = -\frac{1}{2} \operatorname{Re} \iint_{S_s} \{E(r') \times H^*(r')\} \cdot \hat{\nu}(r') \, dS' \tag{4.25}$$

The integral gives the total electromagnetic power that is absorbed in the volume V_s (transformed into other energy forms). The total absorbed power defines the absorption cross section σ_a

$$\sigma_a(\hat{k}_i) = \frac{P_a}{\langle S_i(t)\rangle \cdot \hat{k}_i}$$

The absorption cross section is a function of the direction of the incident plane wave, $\hat{k}_i$, and the frequency, but this latter dependence is usually suppressed.

The scattering cross section and the absorption cross section are often combined into the extinction cross section (or total cross section), σ_{ext}, defined as

$$\sigma_{ext}(\hat{k}_i) = \sigma_a(\hat{k}_i) + \sigma_s(\hat{k}_i) = \frac{P_a + P_s}{\langle S_i(t)\rangle \cdot \hat{k}_i} \tag{4.26}$$

We return to the physical interpretation of this quantity in Section 4.4.1.

The quotient between the extinction and scattering cross sections is denoted the albedo $\alpha(\hat{k}_i) \in [0, 1]$, i.e.,

$$\alpha(\hat{k}_i) = \frac{\sigma_s(\hat{k}_i)}{\sigma_{ext}(\hat{k}_i)}$$

The efficiencies (or efficiency factors), $Q_{ext,s,a}(\hat{k}_i)$, for the three different cross sections, $\sigma_{ext,s,a}(\hat{k}_i)$, are defined as

$$Q_{ext}(\hat{k}_i) = \frac{\sigma_{ext}(\hat{k}_i)}{A(\hat{k}_i)}, \qquad Q_s(\hat{k}_i) = \frac{\sigma_s(\hat{k}_i)}{A(\hat{k}_i)}, \qquad Q_a(\hat{k}_i) = \frac{\sigma_a(\hat{k}_i)}{A(\hat{k}_i)}$$

where $A(\hat{k}_i)$ is the geometrical cross-sectional area of the scatterer, projected on the plane perpendicular to the direction of the incident plane wave, i.e., in the direction $\hat{k}_i$.

4.3 Scattering Dyadic (Matrix)

Again, we let the incident field, $\boldsymbol{E}_\mathrm{i}$, be a plane wave with propagation direction $\hat{\boldsymbol{k}}_\mathrm{i}$, i.e.,

$$\boldsymbol{E}_\mathrm{i}(\boldsymbol{r}) = \boldsymbol{E}_0 e^{ik\hat{\boldsymbol{k}}_\mathrm{i}\cdot\boldsymbol{r}}$$

In this section, the direction of incidence $\hat{\boldsymbol{k}}_\mathrm{i}$ and the direction of observation $\hat{\boldsymbol{r}}$ are assumed to be fixed and non-parallel, i.e., $\hat{\boldsymbol{k}}_\mathrm{i} \times \hat{\boldsymbol{r}} \neq \boldsymbol{0}$. The plane that these unit vectors $\hat{\boldsymbol{k}}_\mathrm{i}$ and $\hat{\boldsymbol{r}}$ span is called the plane of scattering.

Two sets of new unit vectors are defined from the unit vectors $\hat{\boldsymbol{k}}_\mathrm{i}$ and $\hat{\boldsymbol{r}}$—the triples $\{\hat{\boldsymbol{e}}_{\mathrm{i}\parallel}, \hat{\boldsymbol{e}}_{\mathrm{i}\perp}, \hat{\boldsymbol{k}}_\mathrm{i}\}$ and $\{\hat{\boldsymbol{e}}_{\mathrm{s}\parallel}, \hat{\boldsymbol{e}}_{\mathrm{s}\perp}, \hat{\boldsymbol{r}}\}$. The unit vectors, $\hat{\boldsymbol{e}}_{\mathrm{i}\perp} = \hat{\boldsymbol{e}}_{\mathrm{s}\perp}$, are both perpendicular to the plane of scattering, and the unit vectors, $\hat{\boldsymbol{e}}_{\mathrm{i}\parallel}$ and $\hat{\boldsymbol{e}}_{\mathrm{s}\parallel}$, lie in the plane of scattering, and they are perpendicular to $\hat{\boldsymbol{k}}_\mathrm{i}$ and $\hat{\boldsymbol{r}}$, respectively, see Figure 4.5. The formal definitions of these unit vectors are

$$\begin{cases} \hat{\boldsymbol{e}}_{\mathrm{i}\perp} = \dfrac{\hat{\boldsymbol{k}}_\mathrm{i} \times \hat{\boldsymbol{r}}}{|\hat{\boldsymbol{k}}_\mathrm{i} \times \hat{\boldsymbol{r}}|} \\ \hat{\boldsymbol{e}}_{\mathrm{i}\parallel} = \hat{\boldsymbol{e}}_{\mathrm{i}\perp} \times \hat{\boldsymbol{k}}_\mathrm{i} \end{cases} \qquad \begin{cases} \hat{\boldsymbol{e}}_{\mathrm{s}\perp} = \dfrac{\hat{\boldsymbol{k}}_\mathrm{i} \times \hat{\boldsymbol{r}}}{|\hat{\boldsymbol{k}}_\mathrm{i} \times \hat{\boldsymbol{r}}|} \\ \hat{\boldsymbol{e}}_{\mathrm{s}\parallel} = \hat{\boldsymbol{e}}_{\mathrm{s}\perp} \times \hat{\boldsymbol{r}} \end{cases} \tag{4.27}$$

The two sets of unit vectors, $\{\hat{\boldsymbol{e}}_{\mathrm{i}\parallel}, \hat{\boldsymbol{e}}_{\mathrm{i}\perp}, \hat{\boldsymbol{k}}_\mathrm{i}\}$ and $\{\hat{\boldsymbol{e}}_{\mathrm{s}\parallel}, \hat{\boldsymbol{e}}_{\mathrm{s}\perp}, \hat{\boldsymbol{r}}\}$, both form right-handed sets of unit vectors.

In an arbitrary polarization, $\boldsymbol{E}_0$, the incident field can be expanded in a linear combination of the two perpendicular unit vectors $\hat{\boldsymbol{e}}_{\mathrm{i}\parallel}$ and $\hat{\boldsymbol{e}}_{\mathrm{i}\perp}$

$$\boldsymbol{E}_0 = \hat{\boldsymbol{e}}_{\mathrm{i}\parallel} E_{\mathrm{i}\parallel} + \hat{\boldsymbol{e}}_{\mathrm{i}\perp} E_{\mathrm{i}\perp} \tag{4.28}$$

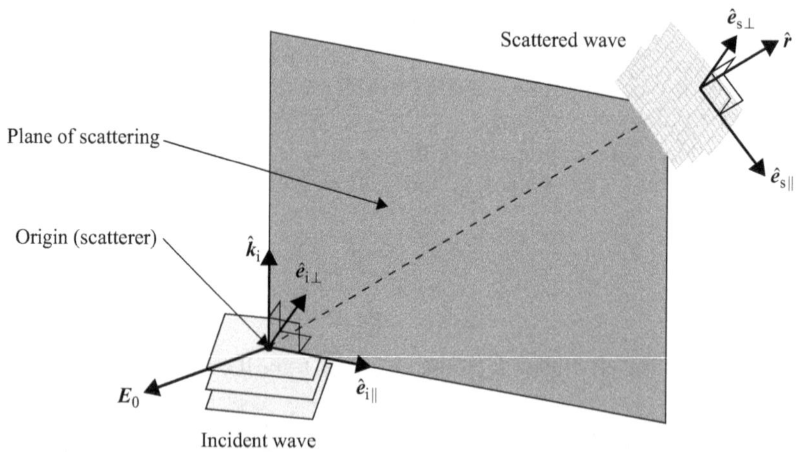

Figure 4.5 The definition of the direction of incidence, $\hat{\boldsymbol{k}}_\mathrm{i}$, the scattering direction, $\hat{\boldsymbol{r}}$, the plane of scattering, and the unit vectors $\hat{\boldsymbol{e}}_{\mathrm{i}\perp} = \hat{\boldsymbol{e}}_{\mathrm{s}\perp}$, $\hat{\boldsymbol{e}}_{\mathrm{i}\parallel}$, and $\hat{\boldsymbol{e}}_{\mathrm{s}\parallel}$. These definitions are made without any reference to a specific frame of coordinates.

since $\boldsymbol{E}_0$ is perpendicular to $\hat{\boldsymbol{k}}_{\mathrm{i}}$, $\boldsymbol{E}_0 \cdot \hat{\boldsymbol{k}}_{\mathrm{i}} = 0$. The vector $\boldsymbol{E}_0$ denotes the strength of the incident field at the origin.

Similarly, the far field amplitude, $\boldsymbol{F}(\hat{\boldsymbol{r}})$, can be expressed as a linear combination of $\hat{\boldsymbol{e}}_{\mathrm{s}\parallel}$ and $\hat{\boldsymbol{e}}_{\mathrm{s}\perp}$, i.e.,

$$\boldsymbol{F}(\hat{\boldsymbol{r}}) = \hat{\boldsymbol{e}}_{\mathrm{s}\parallel} F_{\parallel}(\hat{\boldsymbol{r}}) + \hat{\boldsymbol{e}}_{\mathrm{s}\perp} F_{\perp}(\hat{\boldsymbol{r}})$$

since $\boldsymbol{F}$ is perpendicular to $\hat{\boldsymbol{r}}$.

For a fixed pair of unit vectors, $\hat{\boldsymbol{k}}_{\mathrm{i}}$ and $\hat{\boldsymbol{r}}$, the mapping from the incident field at the origin, $\boldsymbol{E}_0$, to the far field amplitude, $\boldsymbol{F}$, of the scattered field is a linear transformation.[11] The existence of this mapping is best demonstrated by time-domain arguments, and we refer to Chapter 5 for the proper arguments. This linear transformation (or dyadic), denoted $\mathbf{S}$, acts on the complex-valued vector $\boldsymbol{E}_0$, and it is a two-dimensional mapping (maps a two-dimensional field into another two-dimensional field). We have

$$\boldsymbol{F}(\hat{\boldsymbol{r}}) = \mathbf{S}(\hat{\boldsymbol{r}}, \hat{\boldsymbol{k}}_{\mathrm{i}}) \cdot \boldsymbol{E}_0 \tag{4.29}$$

This linear transformation is called the scattering dyadic or scattering matrix or simply the S-matrix.[12] The scattering dyadic, $\mathbf{S}(\hat{\boldsymbol{r}}, \hat{\boldsymbol{k}}_{\mathrm{i}})$, depends on the unit vectors $\hat{\boldsymbol{k}}_{\mathrm{i}}$ and $\hat{\boldsymbol{r}}$, but it is independent of the particular coordinate system and the value of the incident field at the origin (only its direction of propagation). The scattering dyadic depends, of course, on the angular frequency ω—or equivalently on the wave number k—but we often suppress this dependence to avoid cumbersome notation—a more complete notation would then be $\mathbf{S}(\hat{\boldsymbol{r}}, \hat{\boldsymbol{k}}_{\mathrm{i}}, k)$.

The conditions, $\boldsymbol{F} \cdot \hat{\boldsymbol{r}} = 0$ and $\boldsymbol{E}_0 \cdot \hat{\boldsymbol{k}}_{\mathrm{i}} = 0$, imply that

$$\begin{cases} \hat{\boldsymbol{r}} \cdot \mathbf{S}(\hat{\boldsymbol{r}}, \hat{\boldsymbol{k}}_{\mathrm{i}}) = \mathbf{0} \\ \mathbf{S}(\hat{\boldsymbol{r}}, \hat{\boldsymbol{k}}_{\mathrm{i}}) \cdot \hat{\boldsymbol{k}}_{\mathrm{i}} = \mathbf{0} \end{cases} \tag{4.30}$$

so that the dyadic has only components in the plane perpendicular to $\hat{\boldsymbol{r}}$ and perpendicular to $\hat{\boldsymbol{k}}_{\mathrm{i}}$, respectively. A flag of warning is appropriate. The dyadic $\mathbf{S}(\hat{\boldsymbol{r}}, \hat{\boldsymbol{k}}_{\mathrm{i}})$ transforms a tangential vector, $\boldsymbol{E}_0$, on the unit sphere (perpendicular to the direction $\hat{\boldsymbol{k}}_{\mathrm{i}}$) into another tangential vector, $\boldsymbol{F}$ (perpendicular to the direction $\hat{\boldsymbol{r}}$) on the unit sphere. If we change the direction of observation to $\hat{\boldsymbol{r}}'$, the scattering dyadic and the plane of scattering change, so that $\hat{\boldsymbol{r}}' \cdot \mathbf{S}(\hat{\boldsymbol{r}}', \hat{\boldsymbol{k}}_{\mathrm{i}}) = \mathbf{0}$, i.e., the tangent plane changes.

The scattering dyadic $\mathbf{S}(\hat{\boldsymbol{r}}, \hat{\boldsymbol{k}}_{\mathrm{i}})$ is most conveniently represented by components in the two bases $\{\hat{\boldsymbol{e}}_{\mathrm{s}\parallel}, \hat{\boldsymbol{e}}_{\mathrm{s}\perp}\}$ and $\{\hat{\boldsymbol{e}}_{\mathrm{i}\parallel}, \hat{\boldsymbol{e}}_{\mathrm{i}\perp}\}$ defined above. In this representation, the incident field amplitude, $\boldsymbol{E}_0$, has the components $\{E_{\mathrm{i}\parallel}, E_{\mathrm{i}\perp}\}$, and the far field amplitude, $\boldsymbol{F}$, has the components $\{F_{\parallel}(\hat{\boldsymbol{r}}), F_{\perp}(\hat{\boldsymbol{r}})\}$. The transformation, written as a transformation of the components, then is

$$\begin{pmatrix} F_{\parallel}(\hat{\boldsymbol{r}}) \\ F_{\perp}(\hat{\boldsymbol{r}}) \end{pmatrix} = \begin{pmatrix} S_{\parallel\parallel}(\hat{\boldsymbol{r}}, \hat{\boldsymbol{k}}_{\mathrm{i}}) & S_{\parallel\perp}(\hat{\boldsymbol{r}}, \hat{\boldsymbol{k}}_{\mathrm{i}}) \\ S_{\perp\parallel}(\hat{\boldsymbol{r}}, \hat{\boldsymbol{k}}_{\mathrm{i}}) & S_{\perp\perp}(\hat{\boldsymbol{r}}, \hat{\boldsymbol{k}}_{\mathrm{i}}) \end{pmatrix} \begin{pmatrix} E_{\mathrm{i}\parallel} \\ E_{\mathrm{i}\perp} \end{pmatrix} = [\mathbf{S}]\,(\hat{\boldsymbol{r}}, \hat{\boldsymbol{k}}_{\mathrm{i}}) \begin{pmatrix} E_{\mathrm{i}\parallel} \\ E_{\mathrm{i}\perp} \end{pmatrix} \tag{4.31}$$

---

[11] We assume the scatterer consists of a linear, passive material.

[12] Another name for this transformation is the scattering amplitude [57, 277] or amplitude scattering matrix [18].

where

$$\begin{cases} S_{\parallel\,\parallel}(\hat{\boldsymbol{r}},\hat{\boldsymbol{k}}_{\rm i}) = \hat{\boldsymbol{e}}_{\rm s\parallel}\cdot \mathbf{S}(\hat{\boldsymbol{r}},\hat{\boldsymbol{k}}_{\rm i})\cdot \hat{\boldsymbol{e}}_{\rm i\parallel} \\ S_{\parallel\,\perp}(\hat{\boldsymbol{r}},\hat{\boldsymbol{k}}_{\rm i}) = \hat{\boldsymbol{e}}_{\rm s\parallel}\cdot \mathbf{S}(\hat{\boldsymbol{r}},\hat{\boldsymbol{k}}_{\rm i})\cdot \hat{\boldsymbol{e}}_{\rm i\perp} \\ S_{\perp\,\parallel}(\hat{\boldsymbol{r}},\hat{\boldsymbol{k}}_{\rm i}) = \hat{\boldsymbol{e}}_{\rm s\perp}\cdot \mathbf{S}(\hat{\boldsymbol{r}},\hat{\boldsymbol{k}}_{\rm i})\cdot \hat{\boldsymbol{e}}_{\rm i\parallel} \\ S_{\perp\,\perp}(\hat{\boldsymbol{r}},\hat{\boldsymbol{k}}_{\rm i}) = \hat{\boldsymbol{e}}_{\rm s\perp}\cdot \mathbf{S}(\hat{\boldsymbol{r}},\hat{\boldsymbol{k}}_{\rm i})\cdot \hat{\boldsymbol{e}}_{\rm i\perp} \end{cases} \tag{4.32}$$

The entries in the scattering matrix, $S_{\parallel\,\parallel}, S_{\parallel\,\perp}, S_{\perp\,\parallel}$, and $S_{\perp\,\perp}$, are in general all complex numbers characterizing the scattered field in the far zone, given the strength of the incident electric field at the origin, $\boldsymbol{E}_0$. The full scattering dyadic therefore has the form

$$\mathbf{S}(\hat{\boldsymbol{r}},\hat{\boldsymbol{k}}_{\rm i}) = \hat{\boldsymbol{e}}_{\rm s\parallel}S_{\parallel\,\parallel}(\hat{\boldsymbol{r}},\hat{\boldsymbol{k}}_{\rm i})\hat{\boldsymbol{e}}_{\rm i\parallel} + \hat{\boldsymbol{e}}_{\rm s\parallel}S_{\parallel\,\perp}(\hat{\boldsymbol{r}},\hat{\boldsymbol{k}}_{\rm i})\hat{\boldsymbol{e}}_{\rm i\perp} + \hat{\boldsymbol{e}}_{\rm s\perp}S_{\perp\,\parallel}(\hat{\boldsymbol{r}},\hat{\boldsymbol{k}}_{\rm i})\hat{\boldsymbol{e}}_{\rm i\parallel} + \hat{\boldsymbol{e}}_{\rm s\perp}S_{\perp\,\perp}(\hat{\boldsymbol{r}},\hat{\boldsymbol{k}}_{\rm i})\hat{\boldsymbol{e}}_{\rm i\perp}$$

The state of polarization of a wave is quantified in terms of the polarization state quantity Π defined as, see (1.80) on page 78

$$\Pi = {\rm i}\hat{\boldsymbol{e}}\cdot(\hat{\boldsymbol{p}}_{\rm e}\times\hat{\boldsymbol{p}}_{\rm e}^*)$$

The polarization state of the incident field $\boldsymbol{E}_{\rm i}(\boldsymbol{r})$ is quantified by

$$\Pi_{\rm i} = {\rm i}\hat{\boldsymbol{k}}_{\rm i}\cdot(\hat{\boldsymbol{p}}_{\rm i}\times\hat{\boldsymbol{p}}_{\rm i}^*)$$

where $\hat{\boldsymbol{p}}_{\rm i} = \boldsymbol{E}_0/|\boldsymbol{E}_0|$. In terms of the decomposition in (4.28), we get

$$\begin{aligned}\Pi_{\rm i} &= {\rm i}\frac{\hat{\boldsymbol{k}}_{\rm i}\cdot\left(\left(\hat{\boldsymbol{e}}_{\rm i\parallel}E_{\rm i\parallel}+\hat{\boldsymbol{e}}_{\rm i\perp}E_{\rm i\perp}\right)\times\left(\hat{\boldsymbol{e}}_{\rm i\parallel}E_{\rm i\parallel}^*+\hat{\boldsymbol{e}}_{\rm i\perp}E_{\rm i\perp}^*\right)\right)}{|E_{\rm i\parallel}|^2+|E_{\rm i\perp}|^2} \\ &= {\rm i}\frac{E_{\rm i\parallel}E_{\rm i\perp}^*-E_{\rm i\parallel}^*E_{\rm i\perp}}{|E_{\rm i\parallel}|^2+|E_{\rm i\perp}|^2} = 2\frac{{\rm Im}\left(E_{\rm i\parallel}^*E_{\rm i\perp}\right)}{|E_{\rm i\parallel}|^2+|E_{\rm i\perp}|^2}\end{aligned}$$

since the unit vectors $\{\hat{\boldsymbol{e}}_{\rm i\parallel},\hat{\boldsymbol{e}}_{\rm i\perp},\hat{\boldsymbol{k}}_{\rm i}\}$ form a right-handed set of unit vectors.

The polarization state of the scattered field is quantified in a similar manner by the use of the right-handed set of unit vectors $\{\hat{\boldsymbol{e}}_{\rm s\parallel},\hat{\boldsymbol{e}}_{\rm s\perp},\hat{\boldsymbol{r}}\}$ and the polarization state quantity

$$\Pi_{\rm s} = {\rm i}\hat{\boldsymbol{r}}\cdot(\hat{\boldsymbol{p}}_{\rm s}\times\hat{\boldsymbol{p}}_{\rm s}^*) = 2\frac{{\rm Im}\left(F_\parallel^*F_\perp\right)}{|F_\parallel|^2+|F_\perp|^2}$$

where $\hat{\boldsymbol{p}}_{\rm s} = \boldsymbol{F}/|\boldsymbol{F}|$. In terms of the scattering matrix, we get

$$\Pi_{\rm s} = 2\frac{{\rm Im}\left\{\left(S_{\parallel\,\parallel}^*E_{\rm i\parallel}^*+S_{\parallel\,\perp}^*E_{\rm i\perp}^*\right)\left(S_{\perp\,\parallel}E_{\rm i\parallel}+S_{\perp\,\perp}E_{\rm i\perp}\right)\right\}}{\left|S_{\parallel\,\parallel}E_{\rm i\parallel}+S_{\parallel\,\perp}E_{\rm i\perp}\right|^2+\left|S_{\perp\,\parallel}E_{\rm i\parallel}+S_{\perp\,\perp}E_{\rm i\perp}\right|^2}$$

Example 4.1

The incident field is usually linearly polarized in many applications, e.g., $\boldsymbol{E}_0 = E_0\hat{\boldsymbol{e}}$, where $\hat{\boldsymbol{e}}$ is a fixed, real-valued unit vector satisfying $\hat{\boldsymbol{e}} \cdot \hat{\boldsymbol{k}}_\mathrm{i} = 0$. If we, moreover, assume that $S_{\parallel\perp} = S_{\perp\parallel} = 0$, which e.g., is the case for spherical scatterers in Chapter 8, we get

$$\Pi_\mathrm{s} = 2\frac{\mathrm{Im}\left(S_{\parallel\parallel}^* E_{\mathrm{i}\parallel}^* S_{\perp\perp} E_{\mathrm{i}\perp}\right)}{\left|S_{\parallel\parallel} E_{\mathrm{i}\parallel}\right|^2 + |S_{\perp\perp} E_{\mathrm{i}\perp}|^2} = \sin 2\phi \frac{\mathrm{Im}\left(S_{\parallel\parallel}^* S_{\perp\perp}\right)}{\left|S_{\parallel\parallel}\cos\phi\right|^2 + |S_{\perp\perp}\sin\phi|^2}$$

since $E_{\mathrm{i}\parallel} = E_0\cos\phi$ and $E_{\mathrm{i}\perp} = E_0\sin\phi$, where ϕ is the angle between the plane spanned by $\hat{\boldsymbol{e}}$ and $\hat{\boldsymbol{k}}_\mathrm{i}$ and the scattering plane. We conclude that the scattered field is linearly polarized, $\Pi_\mathrm{s} = 0$, in the planes $\phi = 0, \pm\pi/2, \pi$.

Now let the incident field be circularly polarized, i.e., $E_{\mathrm{i}\perp} = \pm \mathrm{i} E_{\mathrm{i}\parallel}$, where RCP (LCP) is upper (lower) sign. Then, under the same assumption of symmetry as above, $S_{\parallel\perp} = S_{\perp\parallel} = 0$, the polarization state becomes

$$\Pi_\mathrm{s} = \pm 2\frac{\mathrm{Re}\left(S_{\parallel\parallel}^* S_{\perp\perp}\right)}{\left|S_{\parallel\parallel}\right|^2 + |S_{\perp\perp}|^2}$$

■

The differential scattering cross section, (4.21), is expressed in the scattering dyadic in the following way:

$$\begin{aligned}\frac{\mathrm{d}\sigma}{\mathrm{d}\Omega}(\hat{\boldsymbol{r}}, \hat{\boldsymbol{k}}_\mathrm{i}) &= 4\pi\frac{\left|\boldsymbol{F}(\hat{\boldsymbol{r}})\right|^2}{|\boldsymbol{E}_0|^2} = 4\pi\frac{\boldsymbol{E}_0^* \cdot \mathbf{S}^\dagger(\hat{\boldsymbol{r}}, \hat{\boldsymbol{k}}_\mathrm{i}) \cdot \mathbf{S}(\hat{\boldsymbol{r}}, \hat{\boldsymbol{k}}_\mathrm{i}) \cdot \boldsymbol{E}_0}{|\boldsymbol{E}_0|^2} \\ &= 4\pi\hat{\boldsymbol{p}}_\mathrm{i}^* \cdot \mathbf{S}^\dagger(\hat{\boldsymbol{r}}, \hat{\boldsymbol{k}}_\mathrm{i}) \cdot \mathbf{S}(\hat{\boldsymbol{r}}, \hat{\boldsymbol{k}}_\mathrm{i}) \cdot \hat{\boldsymbol{p}}_\mathrm{i} = 4\pi\left|\mathbf{S}(\hat{\boldsymbol{r}}, \hat{\boldsymbol{k}}_\mathrm{i}) \cdot \hat{\boldsymbol{p}}_\mathrm{i}\right|^2\end{aligned} \tag{4.33}$$

The dagger (†) denotes the adjoint of the dyadic, and the unit vector $\hat{\boldsymbol{p}}_\mathrm{i}$ is defined as

$$\hat{\boldsymbol{p}}_\mathrm{i} = \frac{\boldsymbol{E}_0}{|\boldsymbol{E}_0|} \tag{4.34}$$

Note that this unit vector in general is complex valued, e.g., for a circular polarized incident wave, and that $\hat{\boldsymbol{p}}_\mathrm{i} \cdot \hat{\boldsymbol{p}}_\mathrm{i}^* = 1$ and $\hat{\boldsymbol{p}}_\mathrm{i} \cdot \hat{\boldsymbol{p}}_\mathrm{i} = 0$.

The co-polarized bistatic cross section, $\sigma_{\mathrm{RCS,co}}$, see (4.22), can also be expressed in the scattering dyadics. The result is

$$\sigma_{\mathrm{RCS,co}}(\hat{\boldsymbol{r}}, \hat{\boldsymbol{k}}_\mathrm{i}) = 4\pi\left|\hat{\boldsymbol{p}}_\mathrm{i}^* \cdot \mathbf{S}(\hat{\boldsymbol{r}}, \hat{\boldsymbol{k}}_\mathrm{i}) \cdot \hat{\boldsymbol{p}}_\mathrm{i}\right|^2 \tag{4.35}$$

The scattering cross section, defined in (4.24), is in terms of the scattering dyadic

$$\sigma_\mathrm{s}(\hat{\boldsymbol{k}}_\mathrm{i}) = \iint_\Omega \hat{\boldsymbol{p}}_\mathrm{i}^* \cdot \mathbf{S}^\dagger(\hat{\boldsymbol{r}}, \hat{\boldsymbol{k}}_\mathrm{i}) \cdot \mathbf{S}(\hat{\boldsymbol{r}}, \hat{\boldsymbol{k}}_\mathrm{i}) \cdot \hat{\boldsymbol{p}}_\mathrm{i}\,\mathrm{d}\Omega = \iint_\Omega \left|\mathbf{S}(\hat{\boldsymbol{r}}, \hat{\boldsymbol{k}}_\mathrm{i}) \cdot \hat{\boldsymbol{p}}_\mathrm{i}\right|^2 \mathrm{d}\Omega \tag{4.36}$$

Example 4.2
In several scattering problems below, the scattering amplitude $\boldsymbol{F}(\hat{\boldsymbol{r}})$ has the canonical form

$$\boldsymbol{F}(\hat{\boldsymbol{r}}) = G\hat{\boldsymbol{r}} \times \left(\boldsymbol{E}_0 \times \hat{\boldsymbol{r}}\right) = G\left(\boldsymbol{E}_0 - \hat{\boldsymbol{r}}\left(\hat{\boldsymbol{r}} \cdot \boldsymbol{E}_0\right)\right)$$

where G is a scalar factor, which depends on $\hat{\boldsymbol{r}}$, frequency ω, and the properties of the scatterer. The scattering dyadic then is

$$\mathbf{S}(\hat{\boldsymbol{r}}, \hat{\boldsymbol{k}}_\mathrm{i}) = G\left(\mathbf{I}_3 - \hat{\boldsymbol{r}}\hat{\boldsymbol{r}}\right)$$

The entries in the scattering matrix, $S_{\parallel\parallel}$, $S_{\parallel\perp}$, $S_{\perp\parallel}$, and $S_{\perp\perp}$, are then obtained by projection, see (4.32)

$$\begin{cases} S_{\parallel\parallel} = G\hat{\boldsymbol{e}}_{\mathrm{s}\parallel} \cdot \left(\mathbf{I}_3 - \hat{\boldsymbol{r}}\hat{\boldsymbol{r}}\right) \cdot \hat{\boldsymbol{e}}_{\mathrm{i}\parallel} = G\cos\theta \\ S_{\parallel\perp} = G\hat{\boldsymbol{e}}_{\mathrm{s}\parallel} \cdot \left(\mathbf{I}_3 - \hat{\boldsymbol{r}}\hat{\boldsymbol{r}}\right) \cdot \hat{\boldsymbol{e}}_{\mathrm{i}\perp} = 0 \\ S_{\perp\parallel} = G\hat{\boldsymbol{e}}_{\mathrm{s}\perp} \cdot \left(\mathbf{I}_3 - \hat{\boldsymbol{r}}\hat{\boldsymbol{r}}\right) \cdot \hat{\boldsymbol{e}}_{\mathrm{i}\parallel} = 0 \\ S_{\perp\perp} = G\hat{\boldsymbol{e}}_{\mathrm{s}\perp} \cdot \left(\mathbf{I}_3 - \hat{\boldsymbol{r}}\hat{\boldsymbol{r}}\right) \cdot \hat{\boldsymbol{e}}_{\mathrm{i}\perp} = G \end{cases}$$

since

$$\begin{cases} \hat{\boldsymbol{e}}_{\mathrm{s}\parallel} \cdot \hat{\boldsymbol{r}} = \hat{\boldsymbol{e}}_{\mathrm{s}\perp} \cdot \hat{\boldsymbol{r}} = 0 \\ \hat{\boldsymbol{e}}_{\mathrm{s}\parallel} \cdot \hat{\boldsymbol{e}}_{\mathrm{i}\perp} = \hat{\boldsymbol{e}}_{\mathrm{s}\perp} \cdot \hat{\boldsymbol{e}}_{\mathrm{i}\parallel} = 0 \end{cases} \qquad \begin{cases} \hat{\boldsymbol{e}}_{\mathrm{s}\parallel} \cdot \hat{\boldsymbol{e}}_{\mathrm{i}\parallel} = \cos\theta \\ \hat{\boldsymbol{e}}_{\mathrm{s}\perp} \cdot \hat{\boldsymbol{e}}_{\mathrm{i}\perp} = 1 \end{cases}$$

where $\cos\theta = \hat{\boldsymbol{r}} \cdot \hat{\boldsymbol{k}}_\mathrm{i}$. In summary,

$$[\mathbf{S}]\left(\hat{\boldsymbol{r}}, \hat{\boldsymbol{k}}_\mathrm{i}\right) = \begin{pmatrix} S_{\parallel\parallel} & S_{\parallel\perp} \\ S_{\perp\parallel} & S_{\perp\perp} \end{pmatrix} = G\begin{pmatrix} \cos\theta & 0 \\ 0 & 1 \end{pmatrix}$$

■

Example 4.3
If the scatterer (both its geometry and its material parameters) is invariant under reflection in the plane of scattering, defined by the unit vectors $\hat{\boldsymbol{k}}_\mathrm{i}$ and $\hat{\boldsymbol{r}}$, then the off-diagonal elements in the scattering matrix are zero, i.e., $S_{\parallel\perp}(\hat{\boldsymbol{r}}, \hat{\boldsymbol{k}}_\mathrm{i}) = S_{\perp\parallel}(\hat{\boldsymbol{r}}, \hat{\boldsymbol{k}}_\mathrm{i}) = 0$. Another way of expressing the invariance is to say that scattering by the object and its mirror image in the plane of scattering both are possible scattering situations and that their scattered fields are identical.

In fact, invariance under reflection in the plane of scattering implies that the perpendicular components of $\boldsymbol{E}_0$ and $\boldsymbol{F}$ change sign, i.e., $E_{\mathrm{i}\perp}$ and $F_\perp$ both change sign, but the perpendicular components, $E_{\mathrm{i}\parallel}$ and $F_\parallel$, remain the same. As a consequence, the diagonal entries of the scattering matrix, $S_{\parallel\parallel}(\hat{\boldsymbol{r}}, \hat{\boldsymbol{k}}_\mathrm{i})$ and $S_{\perp\perp}(\hat{\boldsymbol{r}}, \hat{\boldsymbol{k}}_\mathrm{i})$, remain the same, and the off-diagonal elements, $S_{\parallel\perp}(\hat{\boldsymbol{r}}, \hat{\boldsymbol{k}}_\mathrm{i})$ and $S_{\perp\parallel}(\hat{\boldsymbol{r}}, \hat{\boldsymbol{k}}_\mathrm{i})$, change sign, i.e.,

$$[\mathbf{S}]_\mathrm{mirror}\left(\hat{\boldsymbol{r}}, \hat{\boldsymbol{k}}_\mathrm{i}\right) = \begin{pmatrix} S_{\parallel\parallel} & -S_{\parallel\perp} \\ -S_{\perp\parallel} & S_{\perp\perp} \end{pmatrix}$$

However, if the scatterer shows mirror symmetry in the plane of scattering, the scattering matrices entries of the scatterer and its mirror image must be identical, and, hence, the off-diagonal entries are zero.

If the scatterer is invariant under rotation about $\hat{\boldsymbol{k}}_{\mathrm{i}}$, then the off-diagonal elements in the scattering matrix are zero for all $\hat{\boldsymbol{r}}$, i.e., the off-diagonal entries of the scattering dyadic are zero for all orientations of the plane of scattering. This is, as we derive in Chapter 8, true for scattering by a dielectric sphere, but not for scattering by a biisotropic sphere, which is axially symmetric, but the material parameters lack mirror symmetry. ■

Example 4.4
If the scatterer has axial symmetry w.r.t. the direction $\hat{\boldsymbol{k}}_{\mathrm{i}}$, i.e., the scatterer is invariant under any rotation of the scatterer along the axis $\hat{\boldsymbol{k}}_{\mathrm{i}}$, the entries in the scattering matrix, $S_{\parallel\,\parallel}(\hat{\boldsymbol{r}}, \hat{\boldsymbol{k}}_{\mathrm{i}})$, $S_{\parallel\,\perp}(\hat{\boldsymbol{r}}, \hat{\boldsymbol{k}}_{\mathrm{i}})$, $S_{\perp\,\parallel}(\hat{\boldsymbol{r}}, \hat{\boldsymbol{k}}_{\mathrm{i}})$, and $S_{\perp\,\perp}(\hat{\boldsymbol{r}}, \hat{\boldsymbol{k}}_{\mathrm{i}})$, depend only on the scattering angle $\hat{\boldsymbol{r}} \cdot \hat{\boldsymbol{k}}_{\mathrm{i}} = \cos\theta$ and not on any azimuthal angle. As a consequence, $S_{\parallel\,\parallel}(\hat{\boldsymbol{r}}, \hat{\boldsymbol{k}}_{\mathrm{i}}) = S_{\parallel\,\parallel}(\hat{\boldsymbol{r}}', \hat{\boldsymbol{k}}_{\mathrm{i}})$, $S_{\parallel\,\perp}(\hat{\boldsymbol{r}}, \hat{\boldsymbol{k}}_{\mathrm{i}}) = S_{\parallel\,\perp}(\hat{\boldsymbol{r}}', \hat{\boldsymbol{k}}_{\mathrm{i}})$, $S_{\perp\,\parallel}(\hat{\boldsymbol{r}}, \hat{\boldsymbol{k}}_{\mathrm{i}}) = S_{\perp\,\parallel}(\hat{\boldsymbol{r}}', \hat{\boldsymbol{k}}_{\mathrm{i}})$, and $S_{\perp\,\perp}(\hat{\boldsymbol{r}}, \hat{\boldsymbol{k}}_{\mathrm{i}}) = S_{\perp\,\perp}(\hat{\boldsymbol{r}}', \hat{\boldsymbol{k}}_{\mathrm{i}})$, provided $\hat{\boldsymbol{r}} \cdot \hat{\boldsymbol{k}}_{\mathrm{i}} = \hat{\boldsymbol{r}}' \cdot \hat{\boldsymbol{k}}_{\mathrm{i}} = \cos\theta$. ■

Comment 4.3
There is no lack of generality to orient the coordinate system such that the incident wave impinges along the $\hat{\boldsymbol{z}}$-direction, see Figure 4.6. This makes it possible to identify the two sets of unit vectors, $\{\hat{\boldsymbol{e}}_{\mathrm{i}\parallel}, \hat{\boldsymbol{e}}_{\mathrm{i}\perp}, \hat{\boldsymbol{k}}_{\mathrm{i}}\}$ and $\{\hat{\boldsymbol{e}}_{\mathrm{s}\parallel}, \hat{\boldsymbol{e}}_{\mathrm{s}\perp}, \hat{\boldsymbol{r}}\}$, with the unit vectors of the spherical and circular cylindrical coordinate systems, i.e., $\hat{\boldsymbol{e}}_{\mathrm{i}\perp} = \hat{\boldsymbol{e}}_{\mathrm{s}\perp} = \hat{\boldsymbol{\phi}}$, $\hat{\boldsymbol{e}}_{\mathrm{s}\parallel} = \hat{\boldsymbol{\theta}}$, and $\hat{\boldsymbol{e}}_{\mathrm{i}\parallel} = \hat{\boldsymbol{r}}_{\mathrm{c}}$. In terms of the spherical angles θ and ϕ, the direction of observation $\hat{\boldsymbol{r}}$ is

$$\hat{\boldsymbol{r}} = \hat{\boldsymbol{x}} \sin\theta \cos\phi + \hat{\boldsymbol{y}} \sin\theta \sin\phi + \hat{\boldsymbol{z}} \cos\theta$$

Viewed as a linear transformation in three dimension, the scattering dyadic has a Cartesian coordinate representation

$$[\mathbf{S}] = \begin{pmatrix} S_{11} & S_{12} & S_{13} \\ S_{21} & S_{22} & S_{23} \\ S_{31} & S_{32} & S_{33} \end{pmatrix}$$

The nine elements of $[\mathbf{S}]$ are not independent. The condition, $\mathbf{S}(\hat{\boldsymbol{r}}, \hat{\boldsymbol{k}}_{\mathrm{i}}) \cdot \hat{\boldsymbol{k}}_{\mathrm{i}} = \mathbf{0}$, in (4.30) implies with $\hat{\boldsymbol{k}}_{\mathrm{i}} = \hat{\boldsymbol{z}}$ that $S_{13} = S_{23} = S_{33} = 0$, and moreover, the condition $\hat{\boldsymbol{r}} \cdot \mathbf{S}(\hat{\boldsymbol{r}}, \hat{\boldsymbol{k}}_{\mathrm{i}}) = \mathbf{0}$ implies that

$$\begin{cases} S_{11} \sin\theta \cos\phi + S_{21} \sin\theta \sin\phi + S_{31} \cos\theta = 0 \\ S_{12} \sin\theta \cos\phi + S_{22} \sin\theta \sin\phi + S_{32} \cos\theta = 0 \end{cases}$$

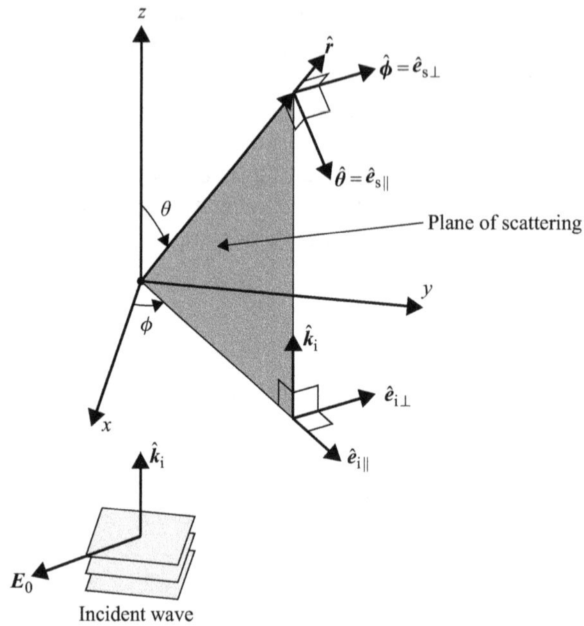

Figure 4.6 The definition of the scattering directions and unit vectors when $\hat{k}_i = \hat{z}$.

These conditions give five constraints on the entries of the scattering matrix **[S]**, and out of the nine components there are only four independent left, which agrees with the standard representation above. The appropriate projections to obtain the $S_{\parallel\parallel}$, $S_{\parallel\perp}$, $S_{\perp\parallel}$, and $S_{\perp\perp}$ components are

$$\begin{cases} S_{\parallel\parallel} = \hat{\theta} \cdot \mathbf{S}(\hat{r}, \hat{z}) \cdot \hat{r}_c \\ S_{\parallel\perp} = \hat{\theta} \cdot \mathbf{S}(\hat{r}, \hat{z}) \cdot \hat{\phi} \end{cases} \qquad \begin{cases} S_{\perp\parallel} = \hat{\phi} \cdot \mathbf{S}(\hat{r}, \hat{z}) \cdot \hat{r}_c \\ S_{\perp\perp} = \hat{\phi} \cdot \mathbf{S}(\hat{r}, \hat{z}) \cdot \hat{\phi} \end{cases}$$

4.3.1 Spherical Coordinate Representation

The directions $\{\hat{e}_{i\parallel}, \hat{e}_{i\perp}\}$ and $\{\hat{e}_{s\parallel}, \hat{e}_{s\perp}\}$ are natural, and they are defined without any reference to a coordinate system, i.e., they are defined in terms of the two vectors $\hat{k}_i$ and $\hat{r}$, see (4.27). However, numerical implementations make use of specific coordinate representations, e.g., the spherical coordinate system. In terms of the spherical angles, $\{\alpha, \beta\}$ and $\{\theta, \phi\}$, the components of the two vectors $\hat{k}_i$ and $\hat{r}$ are

$$\begin{cases} \hat{k}_i = \hat{x} \sin\alpha \cos\beta + \hat{y} \sin\alpha \sin\beta + \hat{z} \cos\alpha \\ \hat{r} = \hat{x} \sin\theta \cos\phi + \hat{y} \sin\theta \sin\phi + \hat{z} \cos\theta \end{cases}$$

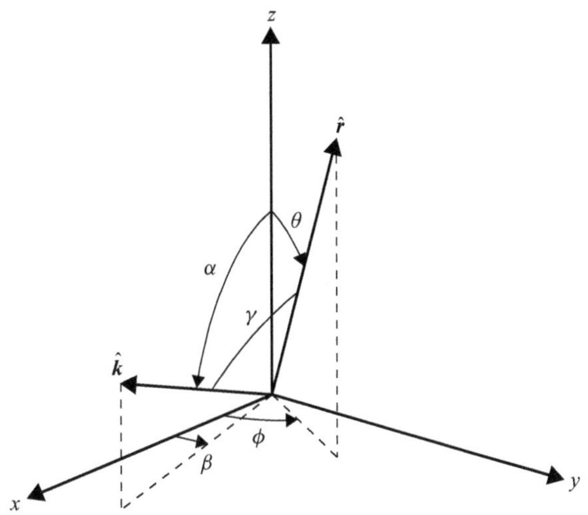

Figure 4.7 The definition of the spherical angles. The scatterer is assumed to be fixed at the origin of the Cartesian coordinate system. The unit vector $\hat{\boldsymbol{r}}$ has spherical angles (θ, ϕ), and the unit vector $\hat{\boldsymbol{k}}$ has spherical angles (α, β). The angle between the directions $\hat{\boldsymbol{r}}$ and $\hat{\boldsymbol{k}}$ is denoted γ.

The angles $\{\alpha, \beta\}$ and $\{\theta, \phi\}$ are associated with unit vectors, which explicit expressions are (cf. the unit vectors of the spherical coordinate system)

$$\begin{cases} \hat{\boldsymbol{\alpha}} = \hat{\boldsymbol{x}} \cos\alpha \cos\beta + \hat{\boldsymbol{y}} \cos\alpha \sin\beta - \hat{\boldsymbol{z}} \sin\alpha \\ \hat{\boldsymbol{\beta}} = -\hat{\boldsymbol{x}} \sin\beta + \hat{\boldsymbol{y}} \cos\beta \end{cases}$$

and

$$\begin{cases} \hat{\boldsymbol{\theta}} = \hat{\boldsymbol{x}} \cos\theta \cos\phi + \hat{\boldsymbol{y}} \cos\theta \sin\phi - \hat{\boldsymbol{z}} \sin\theta \\ \hat{\boldsymbol{\phi}} = -\hat{\boldsymbol{x}} \sin\phi + \hat{\boldsymbol{y}} \cos\phi \end{cases}$$

The two sets of unit vectors, $\{\hat{\boldsymbol{\alpha}}, \hat{\boldsymbol{\beta}}, \hat{\boldsymbol{k}}_{\mathrm{i}}\}$ and $\{\hat{\boldsymbol{\theta}}, \hat{\boldsymbol{\phi}}, \hat{\boldsymbol{r}}\}$, both form right-handed sets of unit vectors, see Figure 4.7.

The scatterer is assumed to be fixed in the Cartesian coordinate system, (x, y, z), and the incident field and the far field amplitude have components

$$\boldsymbol{E}_0(\hat{\boldsymbol{k}}_{\mathrm{i}}) = \hat{\boldsymbol{\alpha}} E_{\mathrm{i}\alpha}(\hat{\boldsymbol{k}}_{\mathrm{i}}) + \hat{\boldsymbol{\beta}} E_{\mathrm{i}\beta}(\hat{\boldsymbol{k}}_{\mathrm{i}})$$

and

$$\boldsymbol{F}(\hat{\boldsymbol{r}}) = \hat{\boldsymbol{\theta}} F_\theta(\hat{\boldsymbol{r}}) + \hat{\boldsymbol{\phi}} F_\phi(\hat{\boldsymbol{r}})$$

Our goal here in this section is to relate the components of the scattering dyadic in terms of these spherical unit vectors to the components in the system $\{\hat{\boldsymbol{e}}_{\mathrm{i}\parallel}, \hat{\boldsymbol{e}}_{\mathrm{i}\perp}\}$ and $\{\hat{\boldsymbol{e}}_{\mathrm{s}\parallel}, \hat{\boldsymbol{e}}_{\mathrm{s}\perp}\}$.

To accomplish this, we write the scattering dyadic in terms of the components of $\{\hat{\boldsymbol{\alpha}}, \hat{\boldsymbol{\beta}}\}$ and $\{\hat{\boldsymbol{\theta}}, \hat{\boldsymbol{\phi}}\}$ as

$$\boldsymbol{F}(\hat{\boldsymbol{r}}) = \left(\hat{\boldsymbol{\theta}} S_{\theta\,\alpha}\hat{\boldsymbol{\alpha}} + \hat{\boldsymbol{\theta}} S_{\theta\,\beta}\hat{\boldsymbol{\beta}} + \hat{\boldsymbol{\phi}} S_{\phi\,\alpha}\hat{\boldsymbol{\alpha}} + \hat{\boldsymbol{\phi}} S_{\phi\,\beta}\hat{\boldsymbol{\beta}}\right) \cdot \boldsymbol{E}_0(\hat{\boldsymbol{k}}_{\mathrm{i}})$$

and the representation in this basis system becomes

$$\begin{pmatrix} F_\theta(\hat{\boldsymbol{r}}) \\ F_\phi(\hat{\boldsymbol{r}}) \end{pmatrix} = \begin{pmatrix} S_{\theta\,\alpha} & S_{\theta\,\beta} \\ S_{\phi\,\alpha} & S_{\phi\,\beta} \end{pmatrix} \begin{pmatrix} E_{\mathrm{i}\alpha}(\hat{\boldsymbol{k}}_{\mathrm{i}}) \\ E_{\mathrm{i}\beta}(\hat{\boldsymbol{k}}_{\mathrm{i}}) \end{pmatrix}$$

The representations in the $\{\hat{\boldsymbol{e}}_{\mathrm{i}\,\parallel}, \hat{\boldsymbol{e}}_{\mathrm{i}\,\perp}\}$ and $\{\hat{\boldsymbol{e}}_{\mathrm{s}\,\parallel}, \hat{\boldsymbol{e}}_{\mathrm{s}\,\perp}\}$ system are then easily found. By projection, we get

$$\begin{cases} S_{\parallel\,\parallel} = \hat{\boldsymbol{e}}_{\mathrm{s}\parallel} \cdot \left(\hat{\boldsymbol{\theta}} S_{\theta\,\alpha}\hat{\boldsymbol{\alpha}} + \hat{\boldsymbol{\theta}} S_{\theta\,\beta}\hat{\boldsymbol{\beta}} + \hat{\boldsymbol{\phi}} S_{\phi\,\alpha}\hat{\boldsymbol{\alpha}} + \hat{\boldsymbol{\phi}} S_{\phi\,\beta}\hat{\boldsymbol{\beta}}\right) \cdot \hat{\boldsymbol{e}}_{\mathrm{i}\,\parallel} \\ S_{\parallel\,\perp} = \hat{\boldsymbol{e}}_{\mathrm{s}\parallel} \cdot \left(\hat{\boldsymbol{\theta}} S_{\theta\,\alpha}\hat{\boldsymbol{\alpha}} + \hat{\boldsymbol{\theta}} S_{\theta\,\beta}\hat{\boldsymbol{\beta}} + \hat{\boldsymbol{\phi}} S_{\phi\,\alpha}\hat{\boldsymbol{\alpha}} + \hat{\boldsymbol{\phi}} S_{\phi\,\beta}\hat{\boldsymbol{\beta}}\right) \cdot \hat{\boldsymbol{e}}_{\mathrm{i}\,\perp} \\ S_{\perp\,\parallel} = \hat{\boldsymbol{e}}_{\mathrm{s}\perp} \cdot \left(\hat{\boldsymbol{\theta}} S_{\theta\,\alpha}\hat{\boldsymbol{\alpha}} + \hat{\boldsymbol{\theta}} S_{\theta\,\beta}\hat{\boldsymbol{\beta}} + \hat{\boldsymbol{\phi}} S_{\phi\,\alpha}\hat{\boldsymbol{\alpha}} + \hat{\boldsymbol{\phi}} S_{\phi\,\beta}\hat{\boldsymbol{\beta}}\right) \cdot \hat{\boldsymbol{e}}_{\mathrm{i}\,\parallel} \\ S_{\perp\,\perp} = \hat{\boldsymbol{e}}_{\mathrm{s}\perp} \cdot \left(\hat{\boldsymbol{\theta}} S_{\theta\,\alpha}\hat{\boldsymbol{\alpha}} + \hat{\boldsymbol{\theta}} S_{\theta\,\beta}\hat{\boldsymbol{\beta}} + \hat{\boldsymbol{\phi}} S_{\phi\,\alpha}\hat{\boldsymbol{\alpha}} + \hat{\boldsymbol{\phi}} S_{\phi\,\beta}\hat{\boldsymbol{\beta}}\right) \cdot \hat{\boldsymbol{e}}_{\mathrm{i}\,\perp} \end{cases}$$

Simplifications imply

$$\begin{cases} S_{\parallel\,\parallel} = -S_{\theta\,\alpha}\dfrac{\hat{\boldsymbol{k}}_{\mathrm{i}} \cdot \hat{\boldsymbol{\theta}}}{\sin\gamma}\dfrac{\hat{\boldsymbol{r}} \cdot \hat{\boldsymbol{\alpha}}}{\sin\gamma} - S_{\theta\,\beta}\dfrac{\hat{\boldsymbol{k}}_{\mathrm{i}} \cdot \hat{\boldsymbol{\theta}}}{\sin\gamma}\dfrac{\hat{\boldsymbol{r}} \cdot \hat{\boldsymbol{\beta}}}{\sin\gamma} - S_{\phi\,\alpha}\dfrac{\hat{\boldsymbol{k}}_{\mathrm{i}} \cdot \hat{\boldsymbol{\phi}}}{\sin\gamma}\dfrac{\hat{\boldsymbol{r}} \cdot \hat{\boldsymbol{\alpha}}}{\sin\gamma} - S_{\phi\,\beta}\dfrac{\hat{\boldsymbol{k}}_{\mathrm{i}} \cdot \hat{\boldsymbol{\phi}}}{\sin\gamma}\dfrac{\hat{\boldsymbol{r}} \cdot \hat{\boldsymbol{\beta}}}{\sin\gamma} \\ S_{\perp\,\perp} = -S_{\theta\,\alpha}\dfrac{\hat{\boldsymbol{k}}_{\mathrm{i}} \cdot \hat{\boldsymbol{\phi}}}{\sin\gamma}\dfrac{\hat{\boldsymbol{r}} \cdot \hat{\boldsymbol{\beta}}}{\sin\gamma} + S_{\theta\,\beta}\dfrac{\hat{\boldsymbol{k}}_{\mathrm{i}} \cdot \hat{\boldsymbol{\phi}}}{\sin\gamma}\dfrac{\hat{\boldsymbol{r}} \cdot \hat{\boldsymbol{\alpha}}}{\sin\gamma} + S_{\phi\,\alpha}\dfrac{\hat{\boldsymbol{k}}_{\mathrm{i}} \cdot \hat{\boldsymbol{\theta}}}{\sin\gamma}\dfrac{\hat{\boldsymbol{r}} \cdot \hat{\boldsymbol{\beta}}}{\sin\gamma} - S_{\phi\,\beta}\dfrac{\hat{\boldsymbol{k}}_{\mathrm{i}} \cdot \hat{\boldsymbol{\theta}}}{\sin\gamma}\dfrac{\hat{\boldsymbol{r}} \cdot \hat{\boldsymbol{\alpha}}}{\sin\gamma} \end{cases}$$

and

$$\begin{cases} S_{\parallel\,\perp} = S_{\theta\,\alpha}\dfrac{\hat{\boldsymbol{k}}_{\mathrm{i}} \cdot \hat{\boldsymbol{\theta}}}{\sin\gamma}\dfrac{\hat{\boldsymbol{r}} \cdot \hat{\boldsymbol{\beta}}}{\sin\gamma} - S_{\theta\,\beta}\dfrac{\hat{\boldsymbol{k}}_{\mathrm{i}} \cdot \hat{\boldsymbol{\theta}}}{\sin\gamma}\dfrac{\hat{\boldsymbol{r}} \cdot \hat{\boldsymbol{\alpha}}}{\sin\gamma} + S_{\phi\,\alpha}\dfrac{\hat{\boldsymbol{k}}_{\mathrm{i}} \cdot \hat{\boldsymbol{\phi}}}{\sin\gamma}\dfrac{\hat{\boldsymbol{r}} \cdot \hat{\boldsymbol{\beta}}}{\sin\gamma} - S_{\phi\,\beta}\dfrac{\hat{\boldsymbol{k}}_{\mathrm{i}} \cdot \hat{\boldsymbol{\phi}}}{\sin\gamma}\dfrac{\hat{\boldsymbol{r}} \cdot \hat{\boldsymbol{\alpha}}}{\sin\gamma} \\ S_{\perp\,\parallel} = S_{\theta\,\alpha}\dfrac{\hat{\boldsymbol{k}}_{\mathrm{i}} \cdot \hat{\boldsymbol{\phi}}}{\sin\gamma}\dfrac{\hat{\boldsymbol{r}} \cdot \hat{\boldsymbol{\alpha}}}{\sin\gamma} + S_{\theta\,\beta}\dfrac{\hat{\boldsymbol{k}}_{\mathrm{i}} \cdot \hat{\boldsymbol{\phi}}}{\sin\gamma}\dfrac{\hat{\boldsymbol{r}} \cdot \hat{\boldsymbol{\beta}}}{\sin\gamma} - S_{\phi\,\alpha}\dfrac{\hat{\boldsymbol{k}}_{\mathrm{i}} \cdot \hat{\boldsymbol{\theta}}}{\sin\gamma}\dfrac{\hat{\boldsymbol{r}} \cdot \hat{\boldsymbol{\alpha}}}{\sin\gamma} - S_{\phi\,\beta}\dfrac{\hat{\boldsymbol{k}}_{\mathrm{i}} \cdot \hat{\boldsymbol{\theta}}}{\sin\gamma}\dfrac{\hat{\boldsymbol{r}} \cdot \hat{\boldsymbol{\beta}}}{\sin\gamma} \end{cases}$$

since the directions $\{\hat{\boldsymbol{e}}_{\mathrm{i}\,\parallel}, \hat{\boldsymbol{e}}_{\mathrm{i}\,\perp}\}$ and $\{\hat{\boldsymbol{e}}_{\mathrm{s}\,\parallel}, \hat{\boldsymbol{e}}_{\mathrm{s}\,\perp}\}$ by cyclic permutations become

$$\begin{cases} \hat{\boldsymbol{e}}_{\mathrm{s}\parallel} \cdot \hat{\boldsymbol{\theta}} = \hat{\boldsymbol{e}}_{\mathrm{s}\perp} \cdot \hat{\boldsymbol{\phi}} = -\dfrac{\hat{\boldsymbol{k}}_{\mathrm{i}} \cdot \hat{\boldsymbol{\theta}}}{\sin\gamma} = \dfrac{\cos\alpha\sin\theta - \sin\alpha\cos\theta\cos(\beta - \phi)}{\sin\gamma} \\ \hat{\boldsymbol{e}}_{\mathrm{s}\parallel} \cdot \hat{\boldsymbol{\phi}} = -\hat{\boldsymbol{e}}_{\mathrm{s}\perp} \cdot \hat{\boldsymbol{\theta}} = -\dfrac{\hat{\boldsymbol{k}}_{\mathrm{i}} \cdot \hat{\boldsymbol{\phi}}}{\sin\gamma} = -\dfrac{\sin\alpha\sin(\beta - \phi)}{\sin\gamma} \end{cases}$$

and

$$\begin{cases} \hat{\boldsymbol{\alpha}} \cdot \hat{\boldsymbol{e}}_{\mathrm{i}\parallel} = \hat{\boldsymbol{\beta}} \cdot \hat{\boldsymbol{e}}_{\mathrm{i}\perp} = \dfrac{\hat{\boldsymbol{r}} \cdot \hat{\boldsymbol{\alpha}}}{\sin\gamma} = \dfrac{\sin\theta\cos\alpha\cos(\beta-\phi) - \cos\theta\sin\alpha}{\sin\gamma} \\ \hat{\boldsymbol{\beta}} \cdot \hat{\boldsymbol{e}}_{\mathrm{i}\parallel} = -\hat{\boldsymbol{\alpha}} \cdot \hat{\boldsymbol{e}}_{\mathrm{i}\perp} = \dfrac{\hat{\boldsymbol{r}} \cdot \hat{\boldsymbol{\beta}}}{\sin\gamma} = -\dfrac{\sin\theta\sin(\beta-\phi)}{\sin\gamma} \end{cases}$$

where γ is the angle between the directions $\hat{\boldsymbol{k}}_{\mathrm{i}}$ and $\hat{\boldsymbol{r}}$, i.e.,

$$\cos\gamma = \sin\theta\sin\alpha\cos(\phi-\beta) + \cos\theta\cos\alpha$$

For a fixed scatterer in the Cartesian coordinate system and incidence along $\hat{\boldsymbol{k}}_{\mathrm{i}}$, the scattering matrix is conveniently obtained in the spherical system, i.e., the components

$$\begin{pmatrix} S_{\theta\,\alpha} & S_{\theta\,\beta} \\ S_{\phi\,\alpha} & S_{\phi\,\beta} \end{pmatrix}$$

are acquired. The transformation to the $\{\hat{\boldsymbol{e}}_{\mathrm{i}\parallel}, \hat{\boldsymbol{e}}_{\mathrm{i}\perp}\}$ and $\{\hat{\boldsymbol{e}}_{\mathrm{s}\parallel}, \hat{\boldsymbol{e}}_{\mathrm{s}\perp}\}$ system, we write as a matrix relation

$$\begin{pmatrix} S_{\parallel\,\parallel} \\ S_{\parallel\,\perp} \\ S_{\perp\,\parallel} \\ S_{\perp\,\perp} \end{pmatrix} = \frac{1}{\sin^2\gamma} \begin{pmatrix} -ac & -ad & -bc & -bd \\ ad & -ac & bd & -bc \\ bc & bd & -ac & -ad \\ -bd & bc & ad & -ac \end{pmatrix} \begin{pmatrix} S_{\theta\,\alpha} \\ S_{\theta\,\beta} \\ S_{\phi\,\alpha} \\ S_{\phi\,\beta} \end{pmatrix} \tag{4.37}$$

where

$$\begin{cases} a = \hat{\boldsymbol{k}}_{\mathrm{i}} \cdot \hat{\boldsymbol{\theta}} \\ b = \hat{\boldsymbol{k}}_{\mathrm{i}} \cdot \hat{\boldsymbol{\phi}} \end{cases} \qquad \begin{cases} c = \hat{\boldsymbol{r}} \cdot \hat{\boldsymbol{\alpha}} \\ d = \hat{\boldsymbol{r}} \cdot \hat{\boldsymbol{\beta}} \end{cases}$$

We can also give the inverse transformation, i.e.,

$$\begin{cases} S_{\theta\,\alpha} = \hat{\boldsymbol{\theta}} \cdot \left(\hat{\boldsymbol{e}}_{\mathrm{s}\parallel} S_{\parallel\,\parallel} \hat{\boldsymbol{e}}_{\mathrm{i}\parallel} + \hat{\boldsymbol{e}}_{\mathrm{s}\parallel} S_{\parallel\,\perp} \hat{\boldsymbol{e}}_{\mathrm{i}\perp} + \hat{\boldsymbol{e}}_{\mathrm{s}\perp} S_{\perp\,\parallel} \hat{\boldsymbol{e}}_{\mathrm{i}\parallel} + \hat{\boldsymbol{e}}_{\mathrm{s}\perp} S_{\perp\,\perp} \hat{\boldsymbol{e}}_{\mathrm{i}\perp}\right) \cdot \hat{\boldsymbol{\alpha}} \\ S_{\theta\,\beta} = \hat{\boldsymbol{\theta}} \cdot \left(\hat{\boldsymbol{e}}_{\mathrm{s}\parallel} S_{\parallel\,\parallel} \hat{\boldsymbol{e}}_{\mathrm{i}\parallel} + \hat{\boldsymbol{e}}_{\mathrm{s}\parallel} S_{\parallel\,\perp} \hat{\boldsymbol{e}}_{\mathrm{i}\perp} + \hat{\boldsymbol{e}}_{\mathrm{s}\perp} S_{\perp\,\parallel} \hat{\boldsymbol{e}}_{\mathrm{i}\parallel} + \hat{\boldsymbol{e}}_{\mathrm{s}\perp} S_{\perp\,\perp} \hat{\boldsymbol{e}}_{\mathrm{i}\perp}\right) \cdot \hat{\boldsymbol{\beta}} \\ S_{\phi\,\alpha} = \hat{\boldsymbol{\phi}} \cdot \left(\hat{\boldsymbol{e}}_{\mathrm{s}\parallel} S_{\parallel\,\parallel} \hat{\boldsymbol{e}}_{\mathrm{i}\parallel} + \hat{\boldsymbol{e}}_{\mathrm{s}\parallel} S_{\parallel\,\perp} \hat{\boldsymbol{e}}_{\mathrm{i}\perp} + \hat{\boldsymbol{e}}_{\mathrm{s}\perp} S_{\perp\,\parallel} \hat{\boldsymbol{e}}_{\mathrm{i}\parallel} + \hat{\boldsymbol{e}}_{\mathrm{s}\perp} S_{\perp\,\perp} \hat{\boldsymbol{e}}_{\mathrm{i}\perp}\right) \cdot \hat{\boldsymbol{\alpha}} \\ S_{\phi\,\beta} = \hat{\boldsymbol{\phi}} \cdot \left(\hat{\boldsymbol{e}}_{\mathrm{s}\parallel} S_{\parallel\,\parallel} \hat{\boldsymbol{e}}_{\mathrm{i}\parallel} + \hat{\boldsymbol{e}}_{\mathrm{s}\parallel} S_{\parallel\,\perp} \hat{\boldsymbol{e}}_{\mathrm{i}\perp} + \hat{\boldsymbol{e}}_{\mathrm{s}\perp} S_{\perp\,\parallel} \hat{\boldsymbol{e}}_{\mathrm{i}\parallel} + \hat{\boldsymbol{e}}_{\mathrm{s}\perp} S_{\perp\,\perp} \hat{\boldsymbol{e}}_{\mathrm{i}\perp}\right) \cdot \hat{\boldsymbol{\beta}} \end{cases}$$

or

$$\begin{pmatrix} S_{\theta\,\alpha} \\ S_{\theta\,\beta} \\ S_{\phi\,\alpha} \\ S_{\phi\,\beta} \end{pmatrix} = \frac{1}{\sin^2\gamma} \begin{pmatrix} -ac & ad & bc & -bd \\ -ad & -ac & bd & bc \\ -bc & bd & -ac & ad \\ -bd & -bc & -ad & -ac \end{pmatrix} \begin{pmatrix} S_{\parallel\,\parallel} \\ S_{\parallel\,\perp} \\ S_{\perp\,\parallel} \\ S_{\perp\,\perp} \end{pmatrix}$$

which gives the components in the $\{\hat{\boldsymbol{\theta}}, \hat{\boldsymbol{\phi}}\}$ and $\{\hat{\boldsymbol{\alpha}}, \hat{\boldsymbol{\beta}}\}$ system. Note that the matrix is the transpose of the matrix in (4.37) (orthogonal matrix).

Example 4.5
As a check, we let the incident wave impinge along the $\hat{z}$-axis, i.e., $\alpha = 0$. Then

$$\begin{cases} a = -\sin\gamma \\ b = 0 \end{cases} \qquad \begin{cases} c = \sin\gamma\cos(\phi-\beta) \\ d = \sin\gamma\sin(\phi-\beta) \end{cases}$$

since

$$\cos\gamma = \cos\theta \quad \Rightarrow \quad \gamma = \theta \in [0, \pi]$$

The result simplifies to

$$\begin{pmatrix} S_{\parallel\,\parallel} \\ S_{\parallel\,\perp} \\ S_{\perp\,\parallel} \\ S_{\perp\,\perp} \end{pmatrix} = \begin{pmatrix} \cos(\phi-\beta) & \sin(\phi-\beta) & 0 & 0 \\ -\sin(\phi-\beta) & \cos(\phi-\beta) & 0 & 0 \\ 0 & 0 & \cos(\phi-\beta) & \sin(\phi-\beta) \\ 0 & 0 & -\sin(\phi-\beta) & \cos(\phi-\beta) \end{pmatrix} \begin{pmatrix} S_{\theta\,\alpha} \\ S_{\theta\,\beta} \\ S_{\phi\,\alpha} \\ S_{\phi\,\beta} \end{pmatrix}$$

which corroborates that the difference between the two systems is a rotation of the plan of scattering an angle $\phi - \beta$ around the z-axis. If $\beta = \phi$, the two systems are identical

$$\begin{cases} S_{\parallel\,\parallel} = S_{\theta\,\alpha} \\ S_{\parallel\,\perp} = S_{\theta\,\beta} \\ S_{\perp\,\parallel} = S_{\phi\,\alpha} \\ S_{\perp\,\perp} = S_{\phi\,\beta} \end{cases}$$

■

Example 4.6
If the direction of incidence is in the x–z-plane, $\beta = 0, \pi$ then

$$\begin{cases} a = -\cos\alpha\sin\theta \pm \sin\alpha\cos\theta\cos\phi \\ b = \mp\sin\alpha\sin\phi \end{cases} \qquad \begin{cases} c = \pm\sin\theta\cos\alpha\cos\phi - \cos\theta\sin\alpha \\ d = \pm\sin\theta\sin\phi \end{cases}$$

and

$$\cos\gamma = \pm\sin\theta\sin\alpha\cos\phi + \cos\theta\cos\alpha$$

where the upper (lower) sign corresponds to $\beta = 0$ ($\beta = \pi$). ■

4.3.2 Coherency Matrix

In Section 1.4, we defined the coherency matrix for a quasi-monochromatic field. Using $\{\hat{e}_{i\parallel}, \hat{e}_{i\perp}\}$ and $\{\hat{e}_{s\parallel}, \hat{e}_{s\perp}\}$, respectively, as a basis, we define the coherency matrices for the incident and the scattered fields (far field amplitude), see (1.88) on page 86.

They are

$$[\mathbf{J}_\mathrm{i}] = \begin{pmatrix} J_{\mathrm{i}\parallel\parallel} & J_{\mathrm{i}\parallel\perp} \\ J_{\mathrm{i}\perp\parallel} & J_{\mathrm{i}\perp\perp} \end{pmatrix} = \begin{pmatrix} \left\langle E_{\mathrm{i}\parallel}E_{\mathrm{i}\parallel}^* \right\rangle & \left\langle E_{\mathrm{i}\parallel}E_{\mathrm{i}\perp}^* \right\rangle \\ \left\langle E_{\mathrm{i}\perp}E_{\mathrm{i}\parallel}^* \right\rangle & \left\langle E_{\mathrm{i}\perp}E_{\mathrm{i}\perp}^* \right\rangle \end{pmatrix} = \left\langle \begin{pmatrix} E_{\mathrm{i}\parallel}(t) \\ E_{\mathrm{i}\perp}(t) \end{pmatrix} \begin{pmatrix} E_{\mathrm{i}\parallel}(t) \\ E_{\mathrm{i}\perp}(t) \end{pmatrix}^\dagger \right\rangle$$

and

$$[\mathbf{J}_\mathrm{s}] = \begin{pmatrix} J_{\mathrm{s}\parallel\parallel} & J_{\mathrm{s}\parallel\perp} \\ J_{\mathrm{s}\perp\parallel} & J_{\mathrm{s}\perp\perp} \end{pmatrix} = \begin{pmatrix} \left\langle F_{\parallel}F_{\parallel}^* \right\rangle & \left\langle F_{\parallel}F_{\perp}^* \right\rangle \\ \left\langle F_{\perp}F_{\parallel}^* \right\rangle & \left\langle F_{\perp}F_{\perp}^* \right\rangle \end{pmatrix} = \left\langle \begin{pmatrix} F_{\parallel}(t) \\ F_{\perp}(t) \end{pmatrix} \begin{pmatrix} F_{\parallel}(t) \\ F_{\perp}(t) \end{pmatrix}^\dagger \right\rangle$$

By the use of the scattering matrix, (4.31), we get

$$\begin{aligned} [\mathbf{J}_\mathrm{s}] &= \left\langle [\mathbf{S}] \begin{pmatrix} E_{\mathrm{i}\parallel}(t) \\ E_{\mathrm{i}\perp}(t) \end{pmatrix} \left([\mathbf{S}] \begin{pmatrix} E_{\mathrm{i}\parallel}(t) \\ E_{\mathrm{i}\perp}(t) \end{pmatrix} \right)^\dagger \right\rangle = \left\langle [\mathbf{S}] \begin{pmatrix} E_{\mathrm{i}\parallel}(t) \\ E_{\mathrm{i}\perp}(t) \end{pmatrix} \begin{pmatrix} E_{\mathrm{i}\parallel}(t) \\ E_{\mathrm{i}\perp}(t) \end{pmatrix}^\dagger [\mathbf{S}]^\dagger \right\rangle \\ &= [\mathbf{S}] \left\langle \begin{pmatrix} E_{\mathrm{i}\parallel}(t) \\ E_{\mathrm{i}\perp}(t) \end{pmatrix} \begin{pmatrix} E_{\mathrm{i}\parallel}(t) \\ E_{\mathrm{i}\perp}(t) \end{pmatrix}^\dagger \right\rangle [\mathbf{S}]^\dagger = [\mathbf{S}]\,[\mathbf{J}_\mathrm{i}]\,[\mathbf{S}]^\dagger \end{aligned} \tag{4.38}$$

since the scattering matrix [S] is assumed independent of time. We can also write this relation in a matrix notation as

$$\begin{pmatrix} \left\langle F_{\parallel}F_{\parallel}^* \right\rangle \\ \left\langle F_{\parallel}F_{\perp}^* \right\rangle \\ \left\langle F_{\perp}F_{\parallel}^* \right\rangle \\ \left\langle F_{\perp}F_{\perp}^* \right\rangle \end{pmatrix} = [\mathbf{Z}_\mathrm{J}] \begin{pmatrix} \left\langle E_{\mathrm{i}\parallel}E_{\mathrm{i}\parallel}^* \right\rangle \\ \left\langle E_{\mathrm{i}\parallel}E_{\mathrm{i}\perp}^* \right\rangle \\ \left\langle E_{\mathrm{i}\perp}E_{\mathrm{i}\parallel}^* \right\rangle \\ \left\langle E_{\mathrm{i}\perp}E_{\mathrm{i}\perp}^* \right\rangle \end{pmatrix}$$

where the coherency phase matrix $[\mathbf{Z}_\mathrm{J}]$ is defined by

$$[\mathbf{Z}_\mathrm{J}] = \begin{pmatrix} S_{\parallel\parallel}S_{\parallel\parallel}^* & S_{\parallel\parallel}S_{\parallel\perp}^* & S_{\parallel\perp}S_{\parallel\parallel}^* & S_{\parallel\perp}S_{\parallel\perp}^* \\ S_{\parallel\parallel}S_{\perp\parallel}^* & S_{\parallel\parallel}S_{\perp\perp}^* & S_{\parallel\perp}S_{\perp\parallel}^* & S_{\parallel\perp}S_{\perp\perp}^* \\ S_{\perp\parallel}S_{\parallel\parallel}^* & S_{\perp\parallel}S_{\parallel\perp}^* & S_{\perp\perp}S_{\parallel\parallel}^* & S_{\perp\perp}S_{\parallel\perp}^* \\ S_{\perp\parallel}S_{\perp\parallel}^* & S_{\perp\parallel}S_{\perp\perp}^* & S_{\perp\perp}S_{\perp\parallel}^* & S_{\perp\perp}S_{\perp\perp}^* \end{pmatrix} \tag{4.39}$$

Example 4.7

If the polarization of the incident field is linearly polarized with an angle α to the plane of scattering, see Figure 4.8, and amplitude E_0, then the amplitude of the incident field is

$$\boldsymbol{E}_0 = E_0 \left(\hat{\boldsymbol{e}}_{\mathrm{i}\parallel} \cos\alpha + \hat{\boldsymbol{e}}_{\mathrm{i}\perp} \sin\alpha \right)$$

The angle α can vary and a suitable average of the range of angles can finally be made. The coherency matrix in this special case becomes

$$[\mathbf{J}_\mathrm{i}] = |E_0|^2 \begin{pmatrix} \cos^2\alpha & \cos\alpha\sin\alpha \\ \sin\alpha\cos\alpha & \sin^2\alpha \end{pmatrix}$$

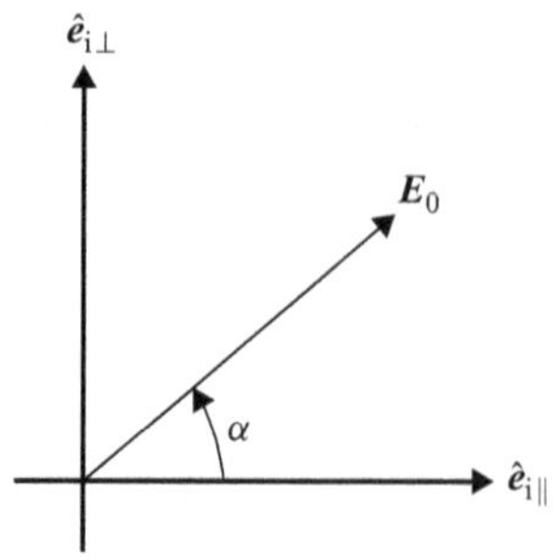

Figure 4.8 Definition of the angle α.

The corresponding matrix for the scattered field (far field amplitude) is

$$\begin{pmatrix} F_{\parallel}(\hat{r}) \\ F_{\perp}(\hat{r}) \end{pmatrix} = E_0 \begin{pmatrix} S_{\parallel\parallel} & S_{\parallel\perp} \\ S_{\perp\parallel} & S_{\perp\perp} \end{pmatrix} \begin{pmatrix} \cos\alpha \\ \sin\alpha \end{pmatrix} = E_0 \begin{pmatrix} S_{\parallel\parallel}\cos\alpha + S_{\parallel\perp}\sin\alpha \\ S_{\perp\parallel}\cos\alpha + S_{\perp\perp}\sin\alpha \end{pmatrix}$$

and the differential scattering cross section, (4.21), becomes

$$\begin{aligned} \frac{\mathrm{d}\sigma}{\mathrm{d}\Omega}(\hat{r}) &= 4\pi \frac{|\boldsymbol{F}(\hat{r})|^2}{|\boldsymbol{E}_0|^2} = 4\pi \frac{|F_{\parallel}(\hat{r})|^2 + |F_{\perp}(\hat{r})|^2}{|\boldsymbol{E}_0|^2} \\ &= 4\pi \left\{ \left[|S_{\parallel\parallel}|^2 + |S_{\perp\parallel}|^2 \right] \cos^2\alpha + \left[|S_{\parallel\perp}|^2 + |S_{\perp\perp}|^2 \right] \sin^2\alpha \right. \\ &\quad \left. + 2\,\mathrm{Re}\left[S_{\parallel\parallel} S_{\parallel\perp}^* + S_{\perp\parallel} S_{\perp\perp}^* \right] \sin\alpha\cos\alpha \right\} \end{aligned} \tag{4.40}$$

■

Example 4.8
In many applications, the incident wave has no definite polarization. The coherency matrix of the incident field for this unpolarized field is diagonal, see (1.91) on page 88. We have

$$[\mathbf{J}_\mathrm{i}] = \frac{\left\langle |\boldsymbol{E}_0(t)|^2 \right\rangle}{2} [\mathbf{I}]$$

and the corresponding coherency matrix of the scattered field (far field amplitude) then becomes, see (4.38)

$$[\mathbf{J}_\mathrm{s}] = \frac{\left\langle |\boldsymbol{E}_0(t)|^2 \right\rangle}{2} [\mathbf{S}]\,[\mathbf{S}]^\dagger = \frac{\left\langle |\boldsymbol{E}_0(t)|^2 \right\rangle}{2} \begin{pmatrix} |S_{\parallel\parallel}|^2 + |S_{\parallel\perp}|^2 & S_{\parallel\parallel} S_{\perp\parallel}^* + S_{\parallel\perp} S_{\perp\perp}^* \\ S_{\perp\parallel} S_{\parallel\parallel}^* + S_{\perp\perp} S_{\parallel\perp}^* & |S_{\perp\parallel}|^2 + |S_{\perp\perp}|^2 \end{pmatrix}$$

The differential scattering cross section of an unpolarized incident plane wave then is

$$\left.\frac{\mathrm{d}\sigma}{\mathrm{d}\Omega}\right|_{\mathrm{unpol}}(\hat{r}) = 4\pi \frac{\mathrm{Tr}\,[\mathbf{J}_\mathrm{s}]}{\left\langle |\boldsymbol{E}_0(t)|^2 \right\rangle} = 2\pi \left\{ |S_{\parallel\parallel}|^2 + |S_{\parallel\perp}|^2 + |S_{\perp\parallel}|^2 + |S_{\perp\perp}|^2 \right\}$$

Notice that this result is also possible to obtain by an average over the angle α in (4.40). The degree of polarization of the scattered field by an unpolarized incident field is, see (1.90) on page 87

$$P|_{\text{unpol}} = \sqrt{1 - \frac{4 \det [\mathbf{J}_s]}{(\operatorname{Tr}[\mathbf{J}_s])^2}} = \sqrt{1 - \frac{4\,|\det[\mathbf{S}]|^2}{\left(|S_{\parallel\parallel}|^2 + |S_{\parallel\perp}|^2 + |S_{\perp\parallel}|^2 + |S_{\perp\perp}|^2\right)^2}}$$

$$= \sqrt{1 - \frac{4\left|S_{\parallel\parallel}S_{\perp\perp} - S_{\parallel\perp}S_{\perp\parallel}\right|^2}{\left(|S_{\parallel\parallel}|^2 + |S_{\parallel\perp}|^2 + |S_{\perp\parallel}|^2 + |S_{\perp\perp}|^2\right)^2}}$$

When $S_{\parallel\perp} = S_{\perp\parallel} = 0$, as they are for a PEC or a dielectric spherical scatterer, see Chapter 8, we get

$$\left.\frac{\mathrm{d}\sigma}{\mathrm{d}\Omega}\right|_{\text{unpol}}(\hat{\boldsymbol{r}}) = 2\pi\left\{|S_{\parallel\parallel}|^2 + |S_{\perp\perp}|^2\right\} \tag{4.41}$$

and

$$P|_{\text{unpol}} = \frac{\left||S_{\parallel\parallel}|^2 - |S_{\perp\perp}|^2\right|}{|S_{\parallel\parallel}|^2 + |S_{\perp\perp}|^2} \tag{4.42}$$

■

4.3.3 Mueller Matrix or Phase Matrix

In Chapter 1, we introduced the Stokes parameters as combinations of different components of the coherency matrix [J], see (1.93) on page 90

$$\begin{cases} s_0 = J_{11} + J_{22} \\ s_1 = J_{11} - J_{22} \\ s_2 = J_{12} + J_{21} = 2\operatorname{Re} J_{12} \\ s_3 = \mathrm{i}\,(J_{21} - J_{12}) = 2\operatorname{Im} J_{12} \end{cases}$$

This set of real numbers characterizes the intensity and the polarization of a time-harmonic or a quasi-monochromatic field. It is convenient to characterize both the incident field and the scattered field in terms of their Stokes parameters, i.e.,

$$\begin{pmatrix} s_0^{\mathrm{i}} \\ s_1^{\mathrm{i}} \\ s_2^{\mathrm{i}} \\ s_3^{\mathrm{i}} \end{pmatrix} \quad \begin{pmatrix} s_0^{\mathrm{s}} \\ s_1^{\mathrm{s}} \\ s_2^{\mathrm{s}} \\ s_3^{\mathrm{s}} \end{pmatrix}$$

Above, we have seen that the scattered and the incident fields are related by the scattering dyadic, and, thus, there is a linear relation between their Stokes parameters. The relation between the incident and the scattered Stokes parameters is a 4 × 4 matrix, called the Mueller matrix[13] or the phase matrix [**M**]. The form of this matrix is derived in Problem 4.5. The result is

$$\begin{pmatrix} s_0^{\mathrm{s}} \\ s_1^{\mathrm{s}} \\ s_2^{\mathrm{s}} \\ s_3^{\mathrm{s}} \end{pmatrix} = \begin{pmatrix} M_{11} & M_{12} & M_{13} & M_{14} \\ M_{21} & M_{22} & M_{23} & M_{24} \\ M_{31} & M_{32} & M_{33} & M_{34} \\ M_{41} & M_{42} & M_{43} & M_{44} \end{pmatrix} \begin{pmatrix} s_0^{\mathrm{i}} \\ s_1^{\mathrm{i}} \\ s_2^{\mathrm{i}} \\ s_3^{\mathrm{i}} \end{pmatrix} \tag{4.43}$$

where the matrix [**M**] has four natural subparts

$$\begin{cases} M_{11} = \dfrac{1}{2}\left\{\left|S_{\parallel\parallel}\right|^2 + \left|S_{\parallel\perp}\right|^2 + \left|S_{\perp\parallel}\right|^2 + \left|S_{\perp\perp}\right|^2\right\} \\[2ex] M_{12} = \dfrac{1}{2}\left\{\left|S_{\parallel\parallel}\right|^2 - \left|S_{\parallel\perp}\right|^2 + \left|S_{\perp\parallel}\right|^2 - \left|S_{\perp\perp}\right|^2\right\} \\[2ex] M_{21} = \dfrac{1}{2}\left\{\left|S_{\parallel\parallel}\right|^2 + \left|S_{\parallel\perp}\right|^2 - \left|S_{\perp\parallel}\right|^2 - \left|S_{\perp\perp}\right|^2\right\} \\[2ex] M_{22} = \dfrac{1}{2}\left\{\left|S_{\parallel\parallel}\right|^2 - \left|S_{\parallel\perp}\right|^2 - \left|S_{\perp\parallel}\right|^2 + \left|S_{\perp\perp}\right|^2\right\} \end{cases}$$

$$\begin{cases} M_{33} = \mathrm{Re}\left\{S_{\parallel\parallel}S_{\perp\perp}^* + S_{\parallel\perp}S_{\perp\parallel}^*\right\} \\[1ex] M_{34} = -\mathrm{Im}\left\{S_{\parallel\parallel}S_{\perp\perp}^* - S_{\parallel\perp}S_{\perp\parallel}^*\right\} \\[1ex] M_{43} = \mathrm{Im}\left\{S_{\parallel\parallel}S_{\perp\perp}^* + S_{\parallel\perp}S_{\perp\parallel}^*\right\} \\[1ex] M_{44} = \mathrm{Re}\left\{S_{\parallel\parallel}S_{\perp\perp}^* - S_{\parallel\perp}S_{\perp\parallel}^*\right\} \end{cases}$$

$$\begin{cases} M_{13} = \mathrm{Re}\left\{S_{\parallel\parallel}S_{\parallel\perp}^* + S_{\perp\parallel}S_{\perp\perp}^*\right\} \\[1ex] M_{14} = -\mathrm{Im}\left\{S_{\parallel\parallel}S_{\parallel\perp}^* + S_{\perp\parallel}S_{\perp\perp}^*\right\} \\[1ex] M_{23} = \mathrm{Re}\left\{S_{\parallel\parallel}S_{\parallel\perp}^* - S_{\perp\parallel}S_{\perp\perp}^*\right\} \\[1ex] M_{24} = -\mathrm{Im}\left\{S_{\parallel\parallel}S_{\parallel\perp}^* - S_{\perp\parallel}S_{\perp\perp}^*\right\} \end{cases}$$

[13] Hans Mueller (1900–1965), American physicist.

and

$$
\begin{cases}
M_{31} = \mathrm{Re}\left\{S_{\parallel\parallel} S_{\perp\parallel}^* + S_{\parallel\perp} S_{\perp\perp}^*\right\} \\
M_{32} = \mathrm{Re}\left\{S_{\parallel\parallel} S_{\perp\parallel}^* - S_{\parallel\perp} S_{\perp\perp}^*\right\} \\
M_{41} = \mathrm{Im}\left\{S_{\parallel\parallel} S_{\perp\parallel}^* + S_{\parallel\perp} S_{\perp\perp}^*\right\} \\
M_{42} = \mathrm{Im}\left\{S_{\parallel\parallel} S_{\perp\parallel}^* - S_{\parallel\perp} S_{\perp\perp}^*\right\}
\end{cases}
$$

In Section 1.4.4, we showed that the Stokes parameters satisfied, see (1.95) on page 91

$$
s_1^2 + s_2^2 + s_3^2 \le s_0^2
$$

The Cauchy–Schwarz inequality,[14] see Appendix F.9 on page 680, implies

$$
\sum_{k=0}^{3} s_k^{\mathrm{s}} s_k^{\mathrm{i}} = s_0^{\mathrm{s}} s_0^{\mathrm{i}} + \sum_{k=1}^{3} s_k^{\mathrm{s}} s_k^{\mathrm{i}} \ge s_0^{\mathrm{s}} s_0^{\mathrm{i}} - \sqrt{\sum_{k=1}^{3} s_k^{\mathrm{s}\,2}} \sqrt{\sum_{k=1}^{3} s_k^{\mathrm{i}\,2}}
$$

and therefore

$$
\sum_{k=0}^{3} s_k^{\mathrm{s}} s_k^{\mathrm{i}} \ge s_0^{\mathrm{s}} s_0^{\mathrm{i}} - \sqrt{s_0^{\mathrm{s}\,2}} \sqrt{s_0^{\mathrm{i}\,2}} = 0
$$

A consequence of this relation is that

$$
\begin{pmatrix} s_0^{\mathrm{i}} \\ s_1^{\mathrm{i}} \\ s_2^{\mathrm{i}} \\ s_3^{\mathrm{i}} \end{pmatrix}^{\mathrm{t}}
\begin{pmatrix} M_{11} & M_{12} & M_{13} & M_{14} \\ M_{21} & M_{22} & M_{23} & M_{24} \\ M_{31} & M_{32} & M_{33} & M_{34} \\ M_{41} & M_{42} & M_{43} & M_{44} \end{pmatrix}
\begin{pmatrix} s_0^{\mathrm{i}} \\ s_1^{\mathrm{i}} \\ s_2^{\mathrm{i}} \\ s_3^{\mathrm{i}} \end{pmatrix}
= \sum_{k=0}^{3} s_k^{\mathrm{s}} s_k^{\mathrm{i}} \ge 0
$$

Unfortunately, the Stokes parameters do not span $\mathbb{R}^4$, so we cannot make the conclusion that the Mueller matrix is positive semi-definite. Important inequalities of its entries are, however, found in Reference 97.

[14] Cauchy–Schwarz inequality is used as

$$
\sum_{k=1}^{3} s_k^{\mathrm{s}} s_k^{\mathrm{i}} \ge - \sqrt{\sum_{k=1}^{3} s_k^{\mathrm{s}\,2}} \sqrt{\sum_{k=1}^{3} s_k^{\mathrm{i}\,2}}
$$

Example 4.9

In many applications, the off-diagonal entries in the scattering matrix are zero, $S_{\parallel\perp} = S_{\perp\parallel} = 0$. This is, for example, the case when the scatterer has axial symmetry, and this symmetry leads to considerable simplifications in the Mueller matrix. The relation between the scattered and the incident Stokes parameters simplifies to (the notation in the matrix on the right-hand side follows standard)

$$\begin{pmatrix} s_0^{\mathrm{s}} \\ s_1^{\mathrm{s}} \\ s_2^{\mathrm{s}} \\ s_3^{\mathrm{s}} \end{pmatrix} = \begin{pmatrix} a_1 & b_1 & 0 & 0 \\ b_1 & a_1 & 0 & 0 \\ 0 & 0 & a_3 & b_2 \\ 0 & 0 & -b_2 & a_3 \end{pmatrix} \begin{pmatrix} s_0^{\mathrm{i}} \\ s_1^{\mathrm{i}} \\ s_2^{\mathrm{i}} \\ s_3^{\mathrm{i}} \end{pmatrix}$$

where the only non-zero components of the Mueller matrix are

$$\begin{cases} a_1 = M_{11} = M_{22} = \dfrac{1}{2}\left\{|S_{\parallel\parallel}|^2 + |S_{\perp\perp}|^2\right\} \\[2ex] b_1 = M_{12} = M_{21} = \dfrac{1}{2}\left\{|S_{\parallel\parallel}|^2 - |S_{\perp\perp}|^2\right\} \end{cases}$$

and

$$\begin{cases} a_3 = M_{33} = M_{44} = \operatorname{Re}\left\{S_{\parallel\parallel}S_{\perp\perp}^*\right\} \\[1ex] b_2 = M_{34} = -M_{43} = -\operatorname{Im}\left\{S_{\parallel\parallel}S_{\perp\perp}^*\right\} \end{cases}$$

The Stokes parameter of an unpolarized incident field is, see (1.91) on page 88

$$\begin{pmatrix} s_0^{\mathrm{i}} \\ s_1^{\mathrm{i}} \\ s_2^{\mathrm{i}} \\ s_3^{\mathrm{i}} \end{pmatrix} = \begin{pmatrix} \left\langle |\boldsymbol{E}_0(t)|^2 \right\rangle \\ 0 \\ 0 \\ 0 \end{pmatrix}$$

which implies that the degree of polarization $P|_{\mathrm{unpol}}$ of the scattered field becomes, see (1.94) on page 91[15]

$$P|_{\mathrm{unpol}} = \frac{\sqrt{s_1^{\mathrm{s}\,2} + s_2^{\mathrm{s}\,2} + s_3^{\mathrm{s}\,2}}}{s_0^{\mathrm{s}}} = \frac{|s_1^{\mathrm{s}}|}{s_0^{\mathrm{s}}} = \frac{|b_1|}{a_1}$$

In Figure 4.9, we illustrate these results for a PEC and a non-magnetic dielectric sphere with an incident unpolarized plane wave. The methods presented in Chapter 8 are used to compute these results. The material parameters of the dielectric sphere are $\epsilon_1/\epsilon = 1.33^2$ and $\mu_1/\mu = 1$. Notice that the scattered field is completely polarized, $P = 1$, at low frequencies, $ka = 0$, at $\theta = 60°$ for the PEC sphere, and at $\theta = 90°$ for the dielectric sphere. ■

---

[15]The negative of the right-hand side, without absolute value in numerator, $-b_1/a_1$, is called the degree of linear depolarization. This is a factor in the interval $[-1, 1]$—positive if the vibrations of the field perpendicular to the plane of scattering dominate over the vibrations in the scattering plane; otherwise it is negative.

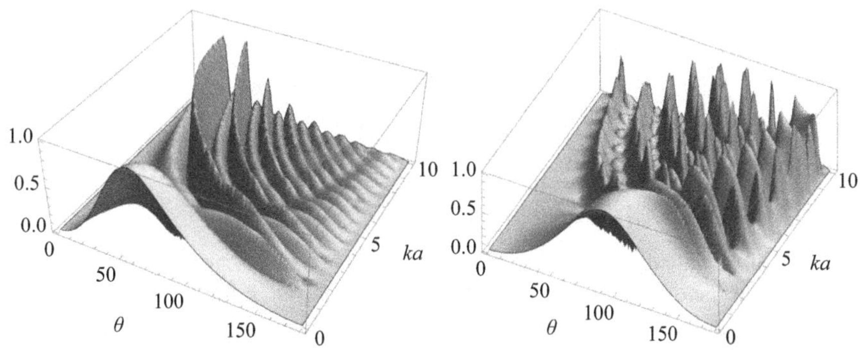

Figure 4.9 The degree of polarization $P|_{\text{unpol}}$ for a PEC sphere (left) and a non-magnetic dielectric sphere (right) as a function of the electric size ka and the scattering angle θ. The incident plane wave is unpolarized, and the material parameters of the dielectric sphere are $\epsilon_1/\epsilon = 1.33^2$ and $\mu_1/\mu = 1$.

4.3.4 Superposition

An incident field with a single incident direction $\hat{\boldsymbol{k}}_{\text{i}}$ has the form

$$\boldsymbol{E}_{\text{i}}(\boldsymbol{r}) = \boldsymbol{E}_0(\hat{\boldsymbol{k}}_{\text{i}})\text{e}^{\text{i}k\hat{\boldsymbol{k}}_{\text{i}}\cdot\boldsymbol{r}}$$

and the corresponding electric field in the far zone is

$$\boldsymbol{E}_{\text{s}}(\boldsymbol{r}) = \boldsymbol{F}(\hat{\boldsymbol{r}}, \hat{\boldsymbol{k}}_{\text{i}})\frac{\text{e}^{\text{i}kr}}{r}$$

The far field amplitude $\boldsymbol{F}(\hat{\boldsymbol{r}}, \hat{\boldsymbol{k}}_{\text{i}})$ is related to the incident field amplitude $\boldsymbol{E}_0(\hat{\boldsymbol{k}}_{\text{i}})$ and the scattering dyadic $\mathbf{S}(\hat{\boldsymbol{r}}, \hat{\boldsymbol{k}}_{\text{i}})$ as

$$\boldsymbol{F}(\hat{\boldsymbol{r}}, \hat{\boldsymbol{k}}_{\text{i}}) = \mathbf{S}(\hat{\boldsymbol{r}}, \hat{\boldsymbol{k}}_{\text{i}}) \cdot \boldsymbol{E}_0(\hat{\boldsymbol{k}}_{\text{i}})$$

We now let the incident field be a superposition of plane waves. The general form of this incident field is a weighted integral over the different incident directions

$$\boldsymbol{E}_{\text{i}}(\boldsymbol{r}) = \iint_{\Omega} \boldsymbol{E}_0(\hat{\boldsymbol{k}}_{\text{i}})\text{e}^{\text{i}k\hat{\boldsymbol{k}}_{\text{i}}\cdot\boldsymbol{r}}\, d\Omega_{\hat{\boldsymbol{k}}_{\text{i}}}$$

where the integration is the direction of the incident field direction $\hat{\boldsymbol{k}}_{\text{i}}$ over the unit sphere $\Omega_{\hat{\boldsymbol{k}}_{\text{i}}}$.[16] The weight function $\boldsymbol{E}_0(\hat{\boldsymbol{k}}_{\text{i}})$ describes and quantifies the superposition. Then, due to superposition, the far field amplitude for the superposed incident field is

$$\boldsymbol{E}_{\text{s}}(\boldsymbol{r}) = \boldsymbol{F}(\hat{\boldsymbol{r}})\frac{\text{e}^{\text{i}kr}}{r}$$

[16]This domain can be generalized to include points outside the unit sphere of directions $\hat{\boldsymbol{k}}_{\text{i}}$. In fact, the domain of integration can assume complex spherical angles of $\hat{\boldsymbol{k}}_{\text{i}}$. This is used in Chapter 10 to solve the scattering problem in planar geometries by means of the Fourier transform.

where

$$\boldsymbol{F}(\hat{\boldsymbol{r}}) = \iint_{\Omega} \mathbf{S}(\hat{\boldsymbol{r}}, \hat{\boldsymbol{k}}_{\mathrm{i}}) \cdot \boldsymbol{E}_0(\hat{\boldsymbol{k}}_{\mathrm{i}})\, d\Omega_{\hat{\boldsymbol{k}}_{\mathrm{i}}}$$

4.3.5 Translation of the Origin

The far field amplitude depends on the choice of origin, and the scattering dyadic, which maps the value of the incident plane wave at the origin to the far field amplitude, therefore also depends on the choice of the origin. This section analyzes the relation between the scattering dyadic at two different positions of the origin.

In Section 4.1.3, we found that translating the origin, the far field amplitude changed, i.e.,

$$\boldsymbol{F}'(\hat{\boldsymbol{r}}) = \mathrm{e}^{\mathrm{i}k\hat{\boldsymbol{r}}\cdot\boldsymbol{r}_0}\boldsymbol{F}(\hat{\boldsymbol{r}})$$

where the translation is expressed as, see also Figure 4.4

$$\boldsymbol{r} = \boldsymbol{r}_0 + \boldsymbol{r}'$$

The translation has also consequences on the scattering dyadic. From (4.29), the far field amplitude is

$$\boldsymbol{F}(\hat{\boldsymbol{r}}) = \mathbf{S}(\hat{\boldsymbol{r}}, \hat{\boldsymbol{k}}_{\mathrm{i}}) \cdot \boldsymbol{E}_0$$

and, correspondingly, the far field amplitude with reference to the translated origin is

$$\boldsymbol{F}'(\hat{\boldsymbol{r}}) = \mathbf{S}'(\hat{\boldsymbol{r}}, \hat{\boldsymbol{k}}_{\mathrm{i}}) \cdot \boldsymbol{E}_0\mathrm{e}^{\mathrm{i}k\hat{\boldsymbol{k}}_{\mathrm{i}}\cdot\boldsymbol{r}_0}$$

since the incident field at the origin O' is $\boldsymbol{E}_0\mathrm{e}^{\mathrm{i}k\hat{\boldsymbol{k}}_{\mathrm{i}}\cdot\boldsymbol{r}_0}$. These expressions imply

$$\mathbf{S}'(\hat{\boldsymbol{r}}, \hat{\boldsymbol{k}}_{\mathrm{i}}) \cdot \boldsymbol{E}_0\mathrm{e}^{\mathrm{i}k\hat{\boldsymbol{k}}_{\mathrm{i}}\cdot\boldsymbol{r}_0} = \mathrm{e}^{\mathrm{i}k\hat{\boldsymbol{r}}\cdot\boldsymbol{r}_0}\mathbf{S}(\hat{\boldsymbol{r}}, \hat{\boldsymbol{k}}_{\mathrm{i}}) \cdot \boldsymbol{E}_0$$

leading to

$$\mathbf{S}'(\hat{\boldsymbol{r}}, \hat{\boldsymbol{k}}_{\mathrm{i}}) = \mathrm{e}^{\mathrm{i}k(\hat{\boldsymbol{r}}-\hat{\boldsymbol{k}}_{\mathrm{i}})\cdot\boldsymbol{r}_0}\mathbf{S}(\hat{\boldsymbol{r}}, \hat{\boldsymbol{k}}_{\mathrm{i}}) \tag{4.44}$$

which relates the scattering dyadic evaluated at the two origins. We notice that the two scattering dyadics are identical in the forward direction, $\hat{\boldsymbol{r}} = \hat{\boldsymbol{k}}_{\mathrm{i}}$, independently of the translation $\boldsymbol{r}_0$.

The translation property of the scattering dyadic in (4.44) also implies that the coherency phase matrix $[\mathbf{Z}_{\mathrm{J}}]$ in (4.39) and the Mueller matrix or phase matrix $[\mathbf{M}]$ in (4.43) are independent of the location of the origin for any directions $\hat{\boldsymbol{r}}$ and $\hat{\boldsymbol{k}}_{\mathrm{i}}$, i.e., the location of the scatterer is immaterial to the entries of the coherency phase matrix and the Mueller matrix.

4.3.6 Reciprocity of the Scattering Dyadic

Scatterers consisting of reciprocal materials lead to certain symmetries in the scattering dyadic. This symmetry is discussed in this section.

Let $\{\boldsymbol{E}^{\mathrm{a}}, \boldsymbol{H}^{\mathrm{a}}\}$ and $\{\boldsymbol{E}^{\mathrm{b}}, \boldsymbol{H}^{\mathrm{b}}\}$ be two sets of solutions to the Maxwell equations in the exterior region V_{e} corresponding to two different incident directions $\hat{\boldsymbol{k}}_{\mathrm{i}}^{\mathrm{a}}$ and $\hat{\boldsymbol{k}}_{\mathrm{i}}^{\mathrm{b}}$, respectively. The incident fields and the dominant contributions of the scattered electric fields in the far zone are denoted

$$\begin{cases} \boldsymbol{E}_{\mathrm{i}}^{\mathrm{a}}(\boldsymbol{r}) = \boldsymbol{E}_0^{\mathrm{a}} \mathrm{e}^{\mathrm{i}k\hat{\boldsymbol{k}}_{\mathrm{i}}^{\mathrm{a}}\cdot\boldsymbol{r}} \\ \boldsymbol{E}_{\mathrm{i}}^{\mathrm{b}}(\boldsymbol{r}) = \boldsymbol{E}_0^{\mathrm{b}} \mathrm{e}^{\mathrm{i}k\hat{\boldsymbol{k}}_{\mathrm{i}}^{\mathrm{b}}\cdot\boldsymbol{r}} \end{cases} \qquad \begin{cases} \boldsymbol{E}_{\mathrm{s}}^{\mathrm{a}}(\boldsymbol{r}) = \boldsymbol{F}^{\mathrm{a}}(\hat{\boldsymbol{r}}, \hat{\boldsymbol{k}}_{\mathrm{i}}^{\mathrm{a}}) \dfrac{\mathrm{e}^{\mathrm{i}kr}}{r} \\ \boldsymbol{E}_{\mathrm{s}}^{\mathrm{b}}(\boldsymbol{r}) = \boldsymbol{F}^{\mathrm{b}}(\hat{\boldsymbol{r}}, \hat{\boldsymbol{k}}_{\mathrm{i}}^{\mathrm{b}}) \dfrac{\mathrm{e}^{\mathrm{i}kr}}{r} \end{cases}$$

and the magnetic counterparts are

$$\begin{cases} \eta\eta_0 \boldsymbol{H}_{\mathrm{i}}^{\mathrm{a}}(\boldsymbol{r}) = \hat{\boldsymbol{k}}_{\mathrm{i}}^{\mathrm{a}} \times \boldsymbol{E}_0^{\mathrm{a}} \mathrm{e}^{\mathrm{i}k\hat{\boldsymbol{k}}_{\mathrm{i}}^{\mathrm{a}}\cdot\boldsymbol{r}} \\ \eta\eta_0 \boldsymbol{H}_{\mathrm{i}}^{\mathrm{b}}(\boldsymbol{r}) = \hat{\boldsymbol{k}}_{\mathrm{i}}^{\mathrm{b}} \times \boldsymbol{E}_0^{\mathrm{b}} \mathrm{e}^{\mathrm{i}k\hat{\boldsymbol{k}}_{\mathrm{i}}^{\mathrm{b}}\cdot\boldsymbol{r}} \end{cases} \qquad \begin{cases} \eta\eta_0 \boldsymbol{H}_{\mathrm{s}}^{\mathrm{a}}(\boldsymbol{r}) = \hat{\boldsymbol{r}} \times \boldsymbol{F}^{\mathrm{a}}(\hat{\boldsymbol{r}}, \hat{\boldsymbol{k}}_{\mathrm{i}}^{\mathrm{a}}) \dfrac{\mathrm{e}^{\mathrm{i}kr}}{r} \\ \eta\eta_0 \boldsymbol{H}_{\mathrm{s}}^{\mathrm{b}}(\boldsymbol{r}) = \hat{\boldsymbol{r}} \times \boldsymbol{F}^{\mathrm{b}}(\hat{\boldsymbol{r}}, \hat{\boldsymbol{k}}_{\mathrm{i}}^{\mathrm{b}}) \dfrac{\mathrm{e}^{\mathrm{i}kr}}{r} \end{cases}$$

The far field amplitudes are related to the corresponding incident fields and the scattering dyadics as

$$\begin{cases} \boldsymbol{F}^{\mathrm{a}}(\hat{\boldsymbol{r}}, \hat{\boldsymbol{k}}_{\mathrm{i}}^{\mathrm{a}}) = \mathbf{S}(\hat{\boldsymbol{r}}, \hat{\boldsymbol{k}}_{\mathrm{i}}^{\mathrm{a}}) \cdot \boldsymbol{E}_0^{\mathrm{a}} \\ \boldsymbol{F}^{\mathrm{b}}(\hat{\boldsymbol{r}}, \hat{\boldsymbol{k}}_{\mathrm{i}}^{\mathrm{b}}) = \mathbf{S}(\hat{\boldsymbol{r}}, \hat{\boldsymbol{k}}_{\mathrm{i}}^{\mathrm{b}}) \cdot \boldsymbol{E}_0^{\mathrm{b}} \end{cases}$$

The following reciprocity result is proved:

$$\mathbf{S}(-\hat{\boldsymbol{k}}_{\mathrm{i}}, -\hat{\boldsymbol{r}}) = \mathbf{S}^{\mathrm{t}}(\hat{\boldsymbol{r}}, \hat{\boldsymbol{k}}_{\mathrm{i}}) \tag{4.45}$$

In particular, reciprocity implies that the scattering dyadic is symmetric in the back scattering direction, $\hat{\boldsymbol{r}} = -\hat{\boldsymbol{k}}_{\mathrm{i}}$, i.e.,

$$\mathbf{S}(-\hat{\boldsymbol{k}}_{\mathrm{i}}, \hat{\boldsymbol{k}}_{\mathrm{i}}) = \mathbf{S}^{\mathrm{t}}(-\hat{\boldsymbol{k}}_{\mathrm{i}}, \hat{\boldsymbol{k}}_{\mathrm{i}})$$

This result holds for all scatterers for which the integral

$$\iint_{S_{\mathrm{s}}} \left\{ \boldsymbol{E}^{\mathrm{a}} \times \boldsymbol{H}^{\mathrm{b}} - \boldsymbol{E}^{\mathrm{b}} \times \boldsymbol{H}^{\mathrm{a}} \right\} \cdot \hat{\boldsymbol{\nu}} \, \mathrm{d}S = 0$$

which holds for e.g., a perfectly conducting object, due to the boundary condition on the PEC scatterer. At the end of this section, we also prove that this integral is identically zero for a reciprocal medium inside V_{S}.

The result in (4.45) shows that the transpose of the scattering dyadic for one set of incident and observation directions is identical to the scattering dyadic if the incident and the observation directions are interchanged and reversed.

The starting point of the proof is the following identity:

$$\iint_{S_{\mathrm{s}}} \left\{ \boldsymbol{E}^{\mathrm{a}} \times \boldsymbol{H}^{\mathrm{b}} - \boldsymbol{E}^{\mathrm{b}} \times \boldsymbol{H}^{\mathrm{a}} \right\} \cdot \hat{\boldsymbol{\nu}} \, \mathrm{d}S = \iint_{S_R} \left\{ \boldsymbol{E}^{\mathrm{a}} \times \boldsymbol{H}^{\mathrm{b}} - \boldsymbol{E}^{\mathrm{b}} \times \boldsymbol{H}^{\mathrm{a}} \right\} \cdot \hat{\boldsymbol{\nu}} \, \mathrm{d}S \tag{4.46}$$

where $\{\boldsymbol{E}^{\mathrm{a}}, \boldsymbol{H}^{\mathrm{a}}\}$ and $\{\boldsymbol{E}^{\mathrm{b}}, \boldsymbol{H}^{\mathrm{b}}\}$ are any two sets of solutions to the Maxwell equations in the region outside S_{s}, but inside any spherical surface S_R, which encloses the scatter. The identity is a simple consequence of the Maxwell equations and an application of the divergence theorem to the identity

$$\begin{aligned}\nabla \cdot &\left\{\boldsymbol{E}^{\mathrm{a}} \times \boldsymbol{H}^{\mathrm{b}} - \boldsymbol{E}^{\mathrm{b}} \times \boldsymbol{H}^{\mathrm{a}}\right\} \\ &= \mathrm{i}k\eta_0\eta \boldsymbol{H}^{\mathrm{a}} \cdot \boldsymbol{H}^{\mathrm{b}} + \mathrm{i}\frac{k}{\eta_0\eta}\boldsymbol{E}^{\mathrm{a}} \cdot \boldsymbol{E}^{\mathrm{b}} - \mathrm{i}k\eta_0\eta \boldsymbol{H}^{\mathrm{b}} \cdot \boldsymbol{H}^{\mathrm{a}} - \mathrm{i}\frac{k}{\eta_0\eta}\boldsymbol{E}^{\mathrm{b}} \cdot \boldsymbol{E}^{\mathrm{a}} = 0\end{aligned}$$

Notice that the integrand in the integral on the left-hand side of (4.46), due to the boundary conditions on S_{s}, is continuous over the boundary surface S_{s}, provided no surface charges or surface currents are present. In fact

$$\begin{aligned}(\boldsymbol{E}_+^{\mathrm{a}} \times \boldsymbol{H}_+^{\mathrm{b}}) \cdot \hat{\boldsymbol{\nu}} &= (\hat{\boldsymbol{\nu}} \times \boldsymbol{E}_+^{\mathrm{a}}) \cdot \boldsymbol{H}_+^{\mathrm{b}} = (\hat{\boldsymbol{\nu}} \times \boldsymbol{E}_-^{\mathrm{a}}) \cdot \boldsymbol{H}_+^{\mathrm{b}} \\ &= -(\hat{\boldsymbol{\nu}} \times \boldsymbol{H}_+^{\mathrm{b}}) \cdot \boldsymbol{E}_-^{\mathrm{a}} = -(\hat{\boldsymbol{\nu}} \times \boldsymbol{H}_-^{\mathrm{b}}) \cdot \boldsymbol{E}_-^{\mathrm{a}} = (\boldsymbol{E}_-^{\mathrm{a}} \times \boldsymbol{H}_-^{\mathrm{b}}) \cdot \hat{\boldsymbol{\nu}}\end{aligned}$$

and similarly for the second term.

Using (4.46), we have to prove that (4.45) is a consequence of the assumption

$$I_1 + I_2 + I_3 = \iint\limits_{S_R} \left\{\boldsymbol{E}^{\mathrm{a}} \times \boldsymbol{H}^{\mathrm{b}} - \boldsymbol{E}^{\mathrm{b}} \times \boldsymbol{H}^{\mathrm{a}}\right\} \cdot \hat{\boldsymbol{\nu}}\, \mathrm{d}S = 0$$

where

$$\begin{cases} I_1 = \displaystyle\iint\limits_{S_R} \left\{\boldsymbol{E}_{\mathrm{i}}^{\mathrm{a}} \times \boldsymbol{H}_{\mathrm{i}}^{\mathrm{b}} - \boldsymbol{E}_{\mathrm{i}}^{\mathrm{b}} \times \boldsymbol{H}_{\mathrm{i}}^{\mathrm{a}}\right\} \cdot \hat{\boldsymbol{\nu}}\, \mathrm{d}S \\ I_2 = \displaystyle\iint\limits_{S_R} \left\{\boldsymbol{E}_{\mathrm{s}}^{\mathrm{a}} \times \boldsymbol{H}_{\mathrm{s}}^{\mathrm{b}} - \boldsymbol{E}_{\mathrm{s}}^{\mathrm{b}} \times \boldsymbol{H}_{\mathrm{s}}^{\mathrm{a}}\right\} \cdot \hat{\boldsymbol{\nu}}\, \mathrm{d}S \\ I_3 = \displaystyle\iint\limits_{S_R} \left\{\boldsymbol{E}_{\mathrm{i}}^{\mathrm{a}} \times \boldsymbol{H}_{\mathrm{s}}^{\mathrm{b}} - \boldsymbol{E}_{\mathrm{i}}^{\mathrm{b}} \times \boldsymbol{H}_{\mathrm{s}}^{\mathrm{a}} + \boldsymbol{E}_{\mathrm{s}}^{\mathrm{a}} \times \boldsymbol{H}_{\mathrm{i}}^{\mathrm{b}} - \boldsymbol{E}_{\mathrm{s}}^{\mathrm{b}} \times \boldsymbol{H}_{\mathrm{i}}^{\mathrm{a}}\right\} \cdot \hat{\boldsymbol{\nu}}\, \mathrm{d}S \end{cases}$$

and where the sum of the integrals, $I_1 + I_2 + I_3$, is independent of the radius R. The explicit value of the different parts I_i, $i = 1, 2, 3$, is evaluated as $R \to \infty$.

The first integral is easily evaluated by using the divergence theorem. We get

$$\begin{aligned}\eta\eta_0 I_1 &= \iint\limits_{S_R} \left\{\boldsymbol{E}_0^{\mathrm{a}} \times (\hat{\boldsymbol{k}}_{\mathrm{i}}^{\mathrm{b}} \times \boldsymbol{E}_0^{\mathrm{b}}) - \boldsymbol{E}_0^{\mathrm{b}} \times (\hat{\boldsymbol{k}}_{\mathrm{i}}^{\mathrm{a}} \times \boldsymbol{E}_0^{\mathrm{a}})\right\} \cdot \hat{\boldsymbol{\nu}} \mathrm{e}^{\mathrm{i}k(\hat{\boldsymbol{k}}_{\mathrm{i}}^{\mathrm{a}} + \hat{\boldsymbol{k}}_{\mathrm{i}}^{\mathrm{b}}) \cdot \boldsymbol{r}}\, \mathrm{d}S \\ &= \mathrm{i}k \iiint\limits_{V_R} (\hat{\boldsymbol{k}}_{\mathrm{i}}^{\mathrm{a}} + \hat{\boldsymbol{k}}_{\mathrm{i}}^{\mathrm{b}}) \cdot \left\{\boldsymbol{E}_0^{\mathrm{a}} \times (\hat{\boldsymbol{k}}_{\mathrm{i}}^{\mathrm{b}} \times \boldsymbol{E}_0^{\mathrm{b}}) - \boldsymbol{E}_0^{\mathrm{b}} \times (\hat{\boldsymbol{k}}_{\mathrm{i}}^{\mathrm{a}} \times \boldsymbol{E}_0^{\mathrm{a}})\right\} \mathrm{e}^{\mathrm{i}k(\hat{\boldsymbol{k}}_{\mathrm{i}}^{\mathrm{a}} + \hat{\boldsymbol{k}}_{\mathrm{i}}^{\mathrm{b}}) \cdot \boldsymbol{r}}\, \mathrm{d}v\end{aligned}$$

where V_R is the ball with center at the origin and radius R. Using the BAC-CAB rule, $\boldsymbol{a}\times(\boldsymbol{b}\times\boldsymbol{c}) = \boldsymbol{b}(\boldsymbol{a}\cdot\boldsymbol{c}) - \boldsymbol{c}(\boldsymbol{a}\cdot\boldsymbol{b})$ and the two orthogonal properties $\boldsymbol{E}_0^{\mathrm{a}}\cdot\hat{\boldsymbol{k}}_{\mathrm{i}}^{\mathrm{a}} = 0$ and $\boldsymbol{E}_0^{\mathrm{b}}\cdot\hat{\boldsymbol{k}}_{\mathrm{i}}^{\mathrm{b}} = 0$, we get for all values of R

$$\eta\eta_0 I_1 = \mathrm{i}k \iiint_{V_R} \Big\{(1+\hat{\boldsymbol{k}}_{\mathrm{i}}^{\mathrm{a}}\cdot\hat{\boldsymbol{k}}_{\mathrm{i}}^{\mathrm{b}})(\boldsymbol{E}_0^{\mathrm{a}}\cdot\boldsymbol{E}_0^{\mathrm{b}}) - (\hat{\boldsymbol{k}}_{\mathrm{i}}^{\mathrm{a}}\cdot\boldsymbol{E}_0^{\mathrm{b}})(\hat{\boldsymbol{k}}_{\mathrm{i}}^{\mathrm{b}}\cdot\boldsymbol{E}_0^{\mathrm{a}})$$
$$- (1+\hat{\boldsymbol{k}}_{\mathrm{i}}^{\mathrm{a}}\cdot\hat{\boldsymbol{k}}_{\mathrm{i}}^{\mathrm{b}})(\boldsymbol{E}_0^{\mathrm{a}}\cdot\boldsymbol{E}_0^{\mathrm{b}}) + (\hat{\boldsymbol{k}}_{\mathrm{i}}^{\mathrm{b}}\cdot\boldsymbol{E}_0^{\mathrm{a}})(\hat{\boldsymbol{k}}_{\mathrm{i}}^{\mathrm{a}}\cdot\boldsymbol{E}_0^{\mathrm{b}})\Big\}\,\mathrm{e}^{\mathrm{i}k(\hat{\boldsymbol{k}}_{\mathrm{i}}^{\mathrm{a}}+\hat{\boldsymbol{k}}_{\mathrm{i}}^{\mathrm{b}})\cdot\boldsymbol{r}}\,\mathrm{d}v = 0$$

The second integral I_2 is seen to vanish in the limit $R\to\infty$, since the integrand is proportional to

$$\frac{\mathrm{e}^{2\mathrm{i}kR}}{R^2}\left\{\boldsymbol{F}^{\mathrm{a}}(\hat{\boldsymbol{r}},\hat{\boldsymbol{k}}_{\mathrm{i}}^{\mathrm{a}})\times(\hat{\boldsymbol{r}}\times\boldsymbol{F}^{\mathrm{b}}(\hat{\boldsymbol{r}},\hat{\boldsymbol{k}}_{\mathrm{i}}^{\mathrm{b}})) - \boldsymbol{F}^{\mathrm{b}}(\hat{\boldsymbol{r}},\hat{\boldsymbol{k}}_{\mathrm{i}}^{\mathrm{b}})\times(\hat{\boldsymbol{r}}\times\boldsymbol{F}^{\mathrm{a}}(\hat{\boldsymbol{r}},\hat{\boldsymbol{k}}_{\mathrm{i}}^{\mathrm{a}}))\right\}\cdot\hat{\boldsymbol{r}}$$
$$+ o\left((kR)^{-2}\right) = o\left((kR)^{-2}\right)$$

where we have utilized the two orthogonality relations $\hat{\boldsymbol{r}}\cdot\boldsymbol{F}^{\mathrm{b}}(\hat{\boldsymbol{r}},\hat{\boldsymbol{k}}_{\mathrm{i}}^{\mathrm{b}}) = 0$ and $\hat{\boldsymbol{r}}\cdot\boldsymbol{F}^{\mathrm{a}}(\hat{\boldsymbol{r}},\hat{\boldsymbol{k}}_{\mathrm{i}}^{\mathrm{a}}) = 0$.

The remaining integral has to be evaluated more explicitly. The dominant part of the integral as $R\to\infty$ is

$$\eta\eta_0 I_3 = R^2\frac{\mathrm{e}^{\mathrm{i}kR}}{R}\iint_{\Omega}\Big\{\boldsymbol{E}_0^{\mathrm{a}}\times(\hat{\boldsymbol{r}}\times\boldsymbol{F}^{\mathrm{b}}(\hat{\boldsymbol{r}},\hat{\boldsymbol{k}}_{\mathrm{i}}^{\mathrm{b}}))\mathrm{e}^{\mathrm{i}kR\hat{\boldsymbol{k}}_{\mathrm{i}}^{\mathrm{a}}\cdot\hat{\boldsymbol{r}}} - \boldsymbol{E}_0^{\mathrm{b}}\times(\hat{\boldsymbol{r}}\times\boldsymbol{F}^{\mathrm{a}}(\hat{\boldsymbol{r}},\hat{\boldsymbol{k}}_{\mathrm{i}}^{\mathrm{a}}))\mathrm{e}^{\mathrm{i}kR\hat{\boldsymbol{k}}_{\mathrm{i}}^{\mathrm{b}}\cdot\hat{\boldsymbol{r}}}$$
$$+ \boldsymbol{F}^{\mathrm{a}}(\hat{\boldsymbol{r}},\hat{\boldsymbol{k}}_{\mathrm{i}}^{\mathrm{a}})\times(\hat{\boldsymbol{k}}_{\mathrm{i}}^{\mathrm{b}}\times\boldsymbol{E}_0^{\mathrm{b}})\mathrm{e}^{\mathrm{i}kR\hat{\boldsymbol{k}}_{\mathrm{i}}^{\mathrm{b}}\cdot\hat{\boldsymbol{r}}} - \boldsymbol{F}^{\mathrm{b}}(\hat{\boldsymbol{r}},\hat{\boldsymbol{k}}_{\mathrm{i}}^{\mathrm{b}})\times(\hat{\boldsymbol{k}}_{\mathrm{i}}^{\mathrm{a}}\times\boldsymbol{E}_0^{\mathrm{a}})\mathrm{e}^{\mathrm{i}kR\hat{\boldsymbol{k}}_{\mathrm{i}}^{\mathrm{a}}\cdot\hat{\boldsymbol{r}}}\Big\}\cdot\hat{\boldsymbol{r}}\,\mathrm{d}\Omega$$

or

$$\eta\eta_0 I_3 = R\mathrm{e}^{\mathrm{i}kR}\iint_{\Omega}\Big\{(\boldsymbol{E}_0^{\mathrm{a}}\cdot\boldsymbol{F}^{\mathrm{b}}(\hat{\boldsymbol{r}},\hat{\boldsymbol{k}}_{\mathrm{i}}^{\mathrm{b}}))(1-\hat{\boldsymbol{r}}\cdot\hat{\boldsymbol{k}}_{\mathrm{i}}^{\mathrm{a}})\mathrm{e}^{\mathrm{i}kR\hat{\boldsymbol{k}}_{\mathrm{i}}^{\mathrm{a}}\cdot\hat{\boldsymbol{r}}} - (\boldsymbol{E}_0^{\mathrm{b}}\cdot\boldsymbol{F}^{\mathrm{a}}(\hat{\boldsymbol{r}},\hat{\boldsymbol{k}}_{\mathrm{i}}^{\mathrm{a}}))(1-\hat{\boldsymbol{r}}\cdot\hat{\boldsymbol{k}}_{\mathrm{i}}^{\mathrm{b}})\mathrm{e}^{\mathrm{i}kR\hat{\boldsymbol{k}}_{\mathrm{i}}^{\mathrm{b}}\cdot\hat{\boldsymbol{r}}}$$
$$- (\boldsymbol{E}_0^{\mathrm{b}}\cdot\hat{\boldsymbol{r}})(\boldsymbol{F}^{\mathrm{a}}(\hat{\boldsymbol{r}},\hat{\boldsymbol{k}}_{\mathrm{i}}^{\mathrm{a}})\cdot\hat{\boldsymbol{k}}_{\mathrm{i}}^{\mathrm{b}})\mathrm{e}^{\mathrm{i}kR\hat{\boldsymbol{k}}_{\mathrm{i}}^{\mathrm{b}}\cdot\hat{\boldsymbol{r}}} + \left(\boldsymbol{E}_0^{\mathrm{a}}\cdot\hat{\boldsymbol{r}}\right)\left(\boldsymbol{F}^{\mathrm{b}}(\hat{\boldsymbol{r}},\hat{\boldsymbol{k}}_{\mathrm{i}}^{\mathrm{b}})\cdot\hat{\boldsymbol{k}}_{\mathrm{i}}^{\mathrm{a}}\right)\mathrm{e}^{\mathrm{i}kR\hat{\boldsymbol{k}}_{\mathrm{i}}^{\mathrm{a}}\cdot\hat{\boldsymbol{r}}}\Big\}\,\mathrm{d}\Omega$$

Here, $d\Omega$ denotes the measure on the unit sphere Ω. We have also used the results that $\hat{\boldsymbol{r}}\cdot\boldsymbol{F}^{\mathrm{b}}(\hat{\boldsymbol{r}},\hat{\boldsymbol{k}}_{\mathrm{i}}^{\mathrm{b}}) = 0$ and $\hat{\boldsymbol{r}}\cdot\boldsymbol{F}^{\mathrm{a}}(\hat{\boldsymbol{r}},\hat{\boldsymbol{k}}_{\mathrm{i}}^{\mathrm{a}}) = 0$.

The generic integral that occurs here has two stationary points, $\hat{\boldsymbol{r}} = \pm\hat{\boldsymbol{k}}_{\mathrm{i}}$, and the dominant contribution as $R\to\infty$ is, see also Comment 4.4

$$\iint_{\Omega} f(\hat{\boldsymbol{r}})\mathrm{e}^{\mathrm{i}kR\hat{\boldsymbol{k}}\cdot\hat{\boldsymbol{r}}}\,\mathrm{d}\Omega = \frac{2\pi}{\mathrm{i}kR}\mathrm{e}^{-\mathrm{i}kR}\left(f(\hat{\boldsymbol{r}}=\hat{\boldsymbol{k}})\mathrm{e}^{2\mathrm{i}kR} - f(\hat{\boldsymbol{r}}=-\hat{\boldsymbol{k}})\right) + O((kR)^{-2})$$

Comment 4.4

This result above can be obtained by either using the result in Appendix G or by integration by parts in

$$\iint_{\Omega} f(\hat{\boldsymbol{r}}) \mathrm{e}^{\mathrm{i}kR\hat{\boldsymbol{k}}\cdot\hat{\boldsymbol{r}}} \, \mathrm{d}\Omega = \int_0^{2\pi} \int_0^{\pi} f(\theta, \phi) \mathrm{e}^{\mathrm{i}kR\cos\theta} \sin\theta \, \mathrm{d}\theta \, \mathrm{d}\phi$$
$$= -\frac{1}{\mathrm{i}kR} \int_0^{2\pi} \left\{ f(\theta, \phi) \mathrm{e}^{\mathrm{i}kR\cos\theta} \Big|_{\theta=0}^{\theta=\pi} - \int_0^{\pi} \frac{\partial f(\theta, \phi)}{\partial \theta} \mathrm{e}^{\mathrm{i}kR\cos\theta} \, \mathrm{d}\theta \right\} \mathrm{d}\phi$$
$$= -\frac{2\pi}{\mathrm{i}kR} \left(f(-\hat{\boldsymbol{k}}) \mathrm{e}^{-\mathrm{i}kR} - f(\hat{\boldsymbol{k}}) \mathrm{e}^{\mathrm{i}kR} \right) + O((kR)^{-2})$$

This result is also named Jones' lemma (Douglas S. Jones (1922–2013), British mathematician) [107].

The I_3 integral then is (use $\boldsymbol{F}^{\mathrm{a}}(\pm\hat{\boldsymbol{k}}_{\mathrm{i}}^{\mathrm{b}}, \hat{\boldsymbol{k}}_{\mathrm{i}}^{\mathrm{a}}) \cdot \hat{\boldsymbol{k}}_{\mathrm{i}}^{\mathrm{b}} = 0$ and $\boldsymbol{F}^{\mathrm{a}}(\pm\hat{\boldsymbol{k}}_{\mathrm{i}}^{\mathrm{a}}, \hat{\boldsymbol{k}}_{\mathrm{i}}^{\mathrm{b}}) \cdot \hat{\boldsymbol{k}}_{\mathrm{i}}^{\mathrm{a}} = 0$)

$$\eta\eta_0 I_3 = -\frac{4\pi}{\mathrm{i}k} \left\{ \boldsymbol{E}_0^{\mathrm{a}} \cdot \boldsymbol{F}^{\mathrm{b}}(-\hat{\boldsymbol{k}}_{\mathrm{i}}^{\mathrm{a}}, \hat{\boldsymbol{k}}_{\mathrm{i}}^{\mathrm{b}}) - \boldsymbol{E}_0^{\mathrm{b}} \cdot \boldsymbol{F}^{\mathrm{a}}(-\hat{\boldsymbol{k}}_{\mathrm{i}}^{\mathrm{b}}, \hat{\boldsymbol{k}}_{\mathrm{i}}^{\mathrm{a}}) \right\} + o(1)$$

As a consequence of $I_1 + I_2 + I_3 = 0$, the reciprocity relation of the scattering matrix $\mathbf{S}$ is obtained. We get

$$\boldsymbol{E}_0^{\mathrm{a}} \cdot \mathbf{S}(-\hat{\boldsymbol{k}}_{\mathrm{i}}^{\mathrm{a}}, \hat{\boldsymbol{k}}_{\mathrm{i}}^{\mathrm{b}}) \cdot \boldsymbol{E}_0^{\mathrm{b}} = \boldsymbol{E}_0^{\mathrm{b}} \cdot \mathbf{S}(-\hat{\boldsymbol{k}}_{\mathrm{i}}^{\mathrm{b}}, \hat{\boldsymbol{k}}_{\mathrm{i}}^{\mathrm{a}}) \cdot \boldsymbol{E}_0^{\mathrm{a}}, \quad \text{for all } \boldsymbol{E}_0^{\mathrm{a,b}}$$

which is equivalent to (4.45), and the proof is completed.

It remains to prove that

$$I = \iint_{S_{\mathrm{s}}} \left\{ \boldsymbol{E}^{\mathrm{a}} \times \boldsymbol{H}^{\mathrm{b}} - \boldsymbol{E}^{\mathrm{b}} \times \boldsymbol{H}^{\mathrm{a}} \right\} \cdot \hat{\boldsymbol{\nu}} \, \mathrm{d}S = 0$$

for a reciprocal medium inside S_{s}. The field values in this integral are formally taken from the outside, i.e., in the region V_{e}. However, we have already concluded that the integrand is the same if the limit taken from the inside, i.e., in the region V_{s}. Use Gauss' theorem on the surface integral, and we get

$$I = \iiint_{V_{\mathrm{s}}} \nabla \cdot \left\{ \boldsymbol{E}^{\mathrm{a}} \times \boldsymbol{H}^{\mathrm{b}} - \boldsymbol{E}^{\mathrm{b}} \times \boldsymbol{H}^{\mathrm{a}} \right\} \mathrm{d}v$$

The integrand is

$$\nabla \cdot \left\{ \boldsymbol{E}^{\mathrm{a}}(\boldsymbol{r}) \times \boldsymbol{H}^{\mathrm{b}}(\boldsymbol{r}) - \boldsymbol{E}^{\mathrm{b}}(\boldsymbol{r}) \times \boldsymbol{H}^{\mathrm{a}}(\boldsymbol{r}) \right\} = (\nabla \times \boldsymbol{E}^{\mathrm{a}}(\boldsymbol{r})) \cdot \boldsymbol{H}^{\mathrm{b}}(\boldsymbol{r}) - \boldsymbol{E}^{\mathrm{a}}(\boldsymbol{r}) \cdot (\nabla \times \boldsymbol{H}^{\mathrm{b}}(\boldsymbol{r}))$$
$$- (\nabla \times \boldsymbol{E}^{\mathrm{b}}(\boldsymbol{r})) \cdot \boldsymbol{H}^{\mathrm{a}}(\boldsymbol{r}) + \boldsymbol{E}^{\mathrm{b}}(\boldsymbol{r}) \cdot (\nabla \times \boldsymbol{H}^{\mathrm{a}}(\boldsymbol{r}))$$

Use the Maxwell equations and the constitutive relations

$$\begin{cases} \boldsymbol{D}(\boldsymbol{r}) = \epsilon_0\left\{\boldsymbol{\epsilon}(\boldsymbol{r})\cdot\boldsymbol{E}(\boldsymbol{r}) + \eta_0\boldsymbol{\xi}(\boldsymbol{r})\cdot\boldsymbol{H}(\boldsymbol{r})\right\} \\ \boldsymbol{B}(\boldsymbol{r}) = \dfrac{1}{c_0}\left\{\boldsymbol{\zeta}(\boldsymbol{r})\cdot\boldsymbol{E}(\boldsymbol{r}) + \eta_0\boldsymbol{\mu}(\boldsymbol{r})\cdot\boldsymbol{H}(\boldsymbol{r})\right\} \end{cases}$$

to obtain

$$\begin{aligned} &\nabla\cdot\left\{\boldsymbol{E}^{\mathrm{a}}(\boldsymbol{r})\times\boldsymbol{H}^{\mathrm{b}}(\boldsymbol{r}) - \boldsymbol{E}^{\mathrm{b}}(\boldsymbol{r})\times\boldsymbol{H}^{\mathrm{a}}(\boldsymbol{r})\right\} \\ &\quad= \frac{\mathrm{i}\omega}{c_0}\boldsymbol{H}^{\mathrm{b}}(\boldsymbol{r})\cdot\left\{\boldsymbol{\zeta}(\boldsymbol{r})\cdot\boldsymbol{E}^{\mathrm{a}}(\boldsymbol{r}) + \eta_0\boldsymbol{\mu}(\boldsymbol{r})\cdot\boldsymbol{H}^{\mathrm{a}}(\boldsymbol{r})\right\} + \mathrm{i}\omega\epsilon_0\boldsymbol{E}^{\mathrm{a}}(\boldsymbol{r})\cdot\left\{\boldsymbol{\epsilon}(\boldsymbol{r})\cdot\boldsymbol{E}^{\mathrm{b}}(\boldsymbol{r}) + \eta_0\boldsymbol{\xi}(\boldsymbol{r})\cdot\boldsymbol{H}^{\mathrm{b}}(\boldsymbol{r})\right\} \\ &\quad\quad - \frac{\mathrm{i}\omega}{c_0}\boldsymbol{H}^{\mathrm{a}}(\boldsymbol{r})\cdot\left\{\boldsymbol{\zeta}(\boldsymbol{r})\cdot\boldsymbol{E}^{\mathrm{b}}(\boldsymbol{r}) + \eta_0\boldsymbol{\mu}(\boldsymbol{r})\cdot\boldsymbol{H}^{\mathrm{b}}(\boldsymbol{r})\right\} - \mathrm{i}\omega\epsilon_0\boldsymbol{E}^{\mathrm{b}}(\boldsymbol{r})\cdot\left\{\boldsymbol{\epsilon}(\boldsymbol{r})\cdot\boldsymbol{E}^{\mathrm{a}}(\boldsymbol{r}) + \eta_0\boldsymbol{\xi}(\boldsymbol{r})\cdot\boldsymbol{H}^{\mathrm{a}}(\boldsymbol{r})\right\} \\ &\quad= \mathrm{i}\omega\epsilon_0\left\{\eta_0\boldsymbol{H}^{\mathrm{b}}(\boldsymbol{r})\cdot(\boldsymbol{\zeta}(\boldsymbol{r}) + \boldsymbol{\xi}^{\mathrm{t}}(\boldsymbol{r}))\cdot\boldsymbol{E}^{\mathrm{a}}(\boldsymbol{r}) + \eta_0^2\boldsymbol{H}^{\mathrm{b}}(\boldsymbol{r})\cdot\left(\boldsymbol{\mu}(\boldsymbol{r}) - \boldsymbol{\mu}^{\mathrm{t}}(\boldsymbol{r})\right)\cdot\boldsymbol{H}^{\mathrm{a}}(\boldsymbol{r})\right. \\ &\quad\quad \left. + \boldsymbol{E}^{\mathrm{a}}(\boldsymbol{r})\cdot(\boldsymbol{\epsilon}(\boldsymbol{r}) - \boldsymbol{\epsilon}^{\mathrm{t}}(\boldsymbol{r}))\cdot\boldsymbol{E}^{\mathrm{b}}(\boldsymbol{r}) - \eta_0\boldsymbol{H}^{\mathrm{a}}(\boldsymbol{r})\cdot(\boldsymbol{\zeta}(\boldsymbol{r}) + \boldsymbol{\xi}^{\mathrm{t}}(\boldsymbol{r}))\cdot\boldsymbol{E}^{\mathrm{b}}(\boldsymbol{r})\right\} \end{aligned}$$

In Section 1.3.5, we found that the material parameters in a reciprocal material satisfy, see (1.71) on page 73

$$\begin{cases} \boldsymbol{\epsilon}(\boldsymbol{r}) = \boldsymbol{\epsilon}^{\mathrm{t}}(\boldsymbol{r}) \\ \boldsymbol{\mu}(\boldsymbol{r}) = \boldsymbol{\mu}^{\mathrm{t}}(\boldsymbol{r}) \\ \boldsymbol{\xi}(\boldsymbol{r}) = -\boldsymbol{\zeta}^{\mathrm{t}}(\boldsymbol{r}) \end{cases}$$

This symmetry implies that the integral $I = 0$, and we have proved that a reciprocal material in V_{s} implies that the scattering dyadic satisfies (4.45).

4.4 Optical Theorem

The derivation of the optical theorem for electromagnetic waves is the main goal of this section. A historical exposition is given by Newton [200]. The optical theorem is a very useful result and some of its important consequences are also exploited in this section. There is a generalized optical theorem, but we postpone the analysis of this concept to Section 7.8.3.

The incident field, $\boldsymbol{E}_{\mathrm{i}}$, is assumed, as in Section 4.2, to be a plane wave, and as usual, the scatterer is confined to the volume V_{s} with bounding surface S_{s}, see Figure 4.1. The extinction cross section σ_{ext} was defined in (4.26) on page 203 as

$$\sigma_{\mathrm{ext}} = \sigma_{\mathrm{a}} + \sigma_{\mathrm{s}} = \frac{P_{\mathrm{a}} + P_{\mathrm{s}}}{\langle \boldsymbol{S}_{\mathrm{i}}(t)\rangle\cdot\hat{\boldsymbol{k}}_{\mathrm{i}}}$$

where P_{a} and P_{s} are the total absorbed and scattered powers, respectively, see (4.25) and (4.23). The power flux density of the incident field is $\langle \boldsymbol{S}_{\mathrm{i}}(t)\rangle\cdot\hat{\boldsymbol{k}}_{\mathrm{i}}$, see (4.20) on page 202.

The scattered power P_{s} is quantified as the surface integral of the normal component of $\mathrm{Re}\left(\boldsymbol{E}_{\mathrm{s}}\times\boldsymbol{H}_{\mathrm{s}}^*\right)/2$ over any closed surface enclosing the scatterer in V_{s}. To see this, apply the

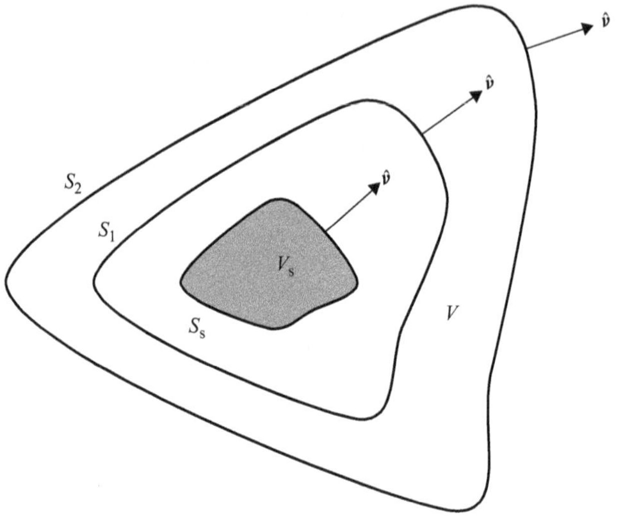

Figure 4.10 Two surfaces, S_1 and S_2, both enclosing V_s. The volume V denotes the region between two surfaces S_1 and S_2.

divergence theorem on a volume V between two surfaces S_1 and S_2 both enclosing V_s, see Figure 4.10. We get (compare the derivation on page 194)

$$\begin{aligned}\iint_{S_2} \left(\boldsymbol{E}_s \times \boldsymbol{H}_s^*\right) \cdot \hat{\boldsymbol{\nu}}\, dS - \iint_{S_1} \left(\boldsymbol{E}_s \times \boldsymbol{H}_s^*\right) \cdot \hat{\boldsymbol{\nu}}\, dS \\ = \iiint_V \nabla \cdot \left(\boldsymbol{E}_s \times \boldsymbol{H}_s^*\right) dv = \iiint_V \left\{(\nabla \times \boldsymbol{E}_s) \cdot \boldsymbol{H}_s^* - \boldsymbol{E}_s \cdot (\nabla \times \boldsymbol{H}_s^*)\right\} dv \\ = \iiint_V \left\{ik\eta_0\eta\, |\boldsymbol{H}_s|^2 - \frac{ik}{\eta_0\eta}\, |\boldsymbol{E}_s|^2\right\} dv\end{aligned}$$

Take the real part of this expression and we conclude (remember we have a lossless material outside V_s)

$$\iint_{S_2} \langle \boldsymbol{S}_s(t)\rangle \cdot \hat{\boldsymbol{\nu}}\, dS = \iint_{S_1} \langle \boldsymbol{S}_s(t)\rangle \cdot \hat{\boldsymbol{\nu}}\, dS$$

From this result, we infer that P_s can be obtained as an integral over the bounding surface S_s (limit values from the outside) instead as the limit value over a large spherical surface. With (4.25), we obtain

$$P_a + P_s = \frac{1}{2} \operatorname{Re} \iint_{S_s} \left\{\boldsymbol{E}_s \times \boldsymbol{H}_s^* - \boldsymbol{E} \times \boldsymbol{H}^*\right\} \cdot \hat{\boldsymbol{\nu}}\, dS$$

Moreover, we have

$$\operatorname{Re}\iint_{S_s} \left(\boldsymbol{E}_i \times \boldsymbol{H}_i^*\right) \cdot \hat{\boldsymbol{\nu}} \, dS = 0$$

which states that the net power flux of the incident field through the surface S_s is zero. In fact, use the divergence theorem on the surface integral, and that the incident field is source free in V_s, see (4.1), i.e.,

$$\begin{aligned}
\operatorname{Re}\iint_{S_s} \left(\boldsymbol{E}_i \times \boldsymbol{H}_i^*\right) \cdot \hat{\boldsymbol{\nu}} \, dS &= \operatorname{Re}\iiint_{V_s} \nabla \cdot \left(\boldsymbol{E}_i \times \boldsymbol{H}_i^*\right) dv \\
&= \operatorname{Re}\iiint_{V_s} \left\{\boldsymbol{H}_i^* \cdot (\nabla \times \boldsymbol{E}_i) - \boldsymbol{E}_i \cdot \left(\nabla \times \boldsymbol{H}_i^*\right)\right\} dv \\
&= \operatorname{Re}\iiint_{V_s} \left\{ik\eta_0\eta \left|\boldsymbol{H}_i\right|^2 - \frac{ik}{\eta_0\eta}\left|\boldsymbol{E}_i\right|^2\right\} dv = 0
\end{aligned}$$

The expression for $P_a + P_s$ above can then be rewritten by adding this surface integral over the incident field as

$$\begin{aligned}
P_a + P_s &= \frac{1}{2}\operatorname{Re}\iint_{S_s} \left\{\boldsymbol{E}_s \times \boldsymbol{H}_s^* - \boldsymbol{E} \times \boldsymbol{H}^*\right\} \cdot \hat{\boldsymbol{\nu}} \, dS \\
&= \frac{1}{2}\operatorname{Re}\iint_{S_s} \left\{\boldsymbol{E}_s \times \boldsymbol{H}_s^* - \boldsymbol{E} \times \boldsymbol{H}_s^* - \boldsymbol{E} \times \boldsymbol{H}_i^* + \boldsymbol{E}_i \times \boldsymbol{H}_i^*\right\} \cdot \hat{\boldsymbol{\nu}} \, dS \\
&= -\frac{1}{2}\operatorname{Re}\iint_{S_s} \left\{\boldsymbol{E}_i \times \boldsymbol{H}_s^* + \boldsymbol{E}_s \times \boldsymbol{H}_i^*\right\} \cdot \hat{\boldsymbol{\nu}} \, dS \\
&= -\frac{1}{2}\operatorname{Re}\iint_{S_s} \left\{\boldsymbol{E}_i^* \times \boldsymbol{H}_s + \boldsymbol{E}_s \times \boldsymbol{H}_i^*\right\} \cdot \hat{\boldsymbol{\nu}} \, dS
\end{aligned}$$

The explicit form of the incident field is now used

$$\begin{cases}
\boldsymbol{E}_i(\boldsymbol{r}) = \boldsymbol{E}_0 e^{ik\hat{\boldsymbol{k}}_i \cdot \boldsymbol{r}} \\
\boldsymbol{H}_i(\boldsymbol{r}) = \dfrac{1}{ik\eta_0\eta}\nabla \times \boldsymbol{E}_i = \dfrac{1}{\eta_0\eta}(\hat{\boldsymbol{k}}_i \times \boldsymbol{E}_0) e^{ik\hat{\boldsymbol{k}}_i \cdot \boldsymbol{r}}
\end{cases}$$

and we get

$$P_a + P_s = -\frac{1}{2}\operatorname{Re}\iint_{S_s} \left\{\boldsymbol{E}_0^* \times \boldsymbol{H}_s + \frac{1}{\eta_0\eta}\boldsymbol{E}_s \times (\hat{\boldsymbol{k}}_i \times \boldsymbol{E}_0^*)\right\} \cdot \hat{\boldsymbol{\nu}} \, e^{-ik\hat{\boldsymbol{k}}_i \cdot \boldsymbol{r}} \, dS$$

A cyclic permutation, the BAC-CAB rule, and $\boldsymbol{E}_0 \cdot \hat{\boldsymbol{k}}_\mathrm{i} = 0$ give

$$\left[\boldsymbol{E}_\mathrm{s} \times (\hat{\boldsymbol{k}}_\mathrm{i} \times \boldsymbol{E}_0^*)\right] \cdot \hat{\boldsymbol{\nu}} = (\hat{\boldsymbol{\nu}} \times \boldsymbol{E}_\mathrm{s}) \cdot (\hat{\boldsymbol{k}}_\mathrm{i} \times \boldsymbol{E}_0^*) = -\boldsymbol{E}_0^* \cdot \left[\hat{\boldsymbol{k}}_\mathrm{i} \times (\hat{\boldsymbol{\nu}} \times \boldsymbol{E}_\mathrm{s})\right]$$

$$\left(\boldsymbol{E}_0^* \times \boldsymbol{H}_\mathrm{s}\right) \cdot \hat{\boldsymbol{\nu}} = -\boldsymbol{E}_0^* \cdot (\hat{\boldsymbol{\nu}} \times \boldsymbol{H}_\mathrm{s}) = \boldsymbol{E}_0^* \cdot \left\{\hat{\boldsymbol{k}}_\mathrm{i} \times \left[\hat{\boldsymbol{k}}_\mathrm{i} \times (\hat{\boldsymbol{\nu}} \times \boldsymbol{H}_\mathrm{s})\right]\right\}$$

The expression for $P_\mathrm{a} + P_\mathrm{s}$ then is

$$P_\mathrm{a} + P_\mathrm{s} = \frac{1}{2\eta_0\eta} \operatorname{Re}\left\{\boldsymbol{E}_0^* \cdot \iint_{S_\mathrm{s}} \hat{\boldsymbol{k}}_\mathrm{i} \times \left\{\hat{\boldsymbol{\nu}} \times \boldsymbol{E}_\mathrm{s}(\boldsymbol{r}) - \eta_0\eta\hat{\boldsymbol{k}}_\mathrm{i} \times \left(\hat{\boldsymbol{\nu}} \times \boldsymbol{H}_\mathrm{s}(\boldsymbol{r})\right)\right\} \mathrm{e}^{-\mathrm{i}k\hat{\boldsymbol{k}}_\mathrm{i}\cdot\boldsymbol{r}} \, \mathrm{d}S\right\}$$

This expression can now be identified as the far field amplitude $\boldsymbol{F}$ in (4.17) on page 198 evaluated in the forward direction $\hat{\boldsymbol{r}} = \hat{\boldsymbol{k}}_\mathrm{i}$. The result is

$$P_\mathrm{a} + P_\mathrm{s} = \frac{1}{2\eta_0\eta} \operatorname{Re}\left\{\frac{4\pi}{\mathrm{i}k} \boldsymbol{E}_0^* \cdot \boldsymbol{F}(\hat{\boldsymbol{k}}_\mathrm{i})\right\}$$

and we finally get

$$\begin{aligned}\sigma_\mathrm{ext} &= \frac{P_\mathrm{a} + P_\mathrm{s}}{\frac{|\boldsymbol{E}_0|^2}{2\eta_0\eta}} = \frac{4\pi}{k} \operatorname{Im}\left\{\frac{\boldsymbol{E}_0^* \cdot \boldsymbol{F}(\hat{\boldsymbol{k}}_\mathrm{i})}{|\boldsymbol{E}_0|^2}\right\} = \frac{4\pi}{k} \operatorname{Im}\left\{\frac{\boldsymbol{E}_0^* \cdot \mathbf{S}(\hat{\boldsymbol{k}}_\mathrm{i}, \hat{\boldsymbol{k}}_\mathrm{i}) \cdot \boldsymbol{E}_0}{|\boldsymbol{E}_0|^2}\right\} \\ &= \frac{4\pi}{k} \operatorname{Im}\left\{\hat{\boldsymbol{p}}_\mathrm{e}^* \cdot \mathbf{S}(\hat{\boldsymbol{k}}_\mathrm{i}, \hat{\boldsymbol{k}}_\mathrm{i}) \cdot \hat{\boldsymbol{p}}_\mathrm{e}\right\}\end{aligned} \tag{4.47}$$

where $\hat{\boldsymbol{p}}_\mathrm{i} = \boldsymbol{E}_0 / |\boldsymbol{E}_0|$, and where we used (4.29) to rewrite the far field amplitude in the scattering dyadic $\mathbf{S}$. The relation (4.47) is called the optical theorem. Notice that the assumptions on the scatterer inside V_s are very weak.

As a final note, we observe that the argument of the imaginary part on the right-hand side in (4.47), $\boldsymbol{E}_0^* \cdot \mathbf{S}(\hat{\boldsymbol{k}}_\mathrm{i}, \hat{\boldsymbol{k}}_\mathrm{i}) \cdot \boldsymbol{E}_0$, defines a quadratic form generated by the scattering matrix in the forward direction, $\mathbf{S}(\hat{\boldsymbol{k}}_\mathrm{i}, \hat{\boldsymbol{k}}_\mathrm{i})$. This quadratic form has two, in general complex, eigenvalues. The imaginary part of these eigenvalues gives bounds on the extinction cross section on the left-hand side of (4.47). In Section 7.8.2, we explore these bounds in more detail for a spherical scatterer.

4.4.1 Extinction

The concept of extinction was introduced in Section 4.2. In this section, we explore this subject further [11, 18]. Consider a single scatterer (or several isolated scatterers) with bounding surface S_s, and let a plane wave impinge on the scatterer in the direction $\hat{\boldsymbol{k}}_\mathrm{i}$. The incident field is

$$\begin{cases}\boldsymbol{E}_\mathrm{i}(\boldsymbol{r}) = \boldsymbol{E}_0 \mathrm{e}^{\mathrm{i}k\hat{\boldsymbol{k}}_\mathrm{i}\cdot\boldsymbol{r}} \\ \boldsymbol{H}_\mathrm{i}(\boldsymbol{r}) = \dfrac{1}{\eta\eta_0} \hat{\boldsymbol{k}}_\mathrm{i} \times \boldsymbol{E}_\mathrm{i}(\boldsymbol{r})\end{cases}$$

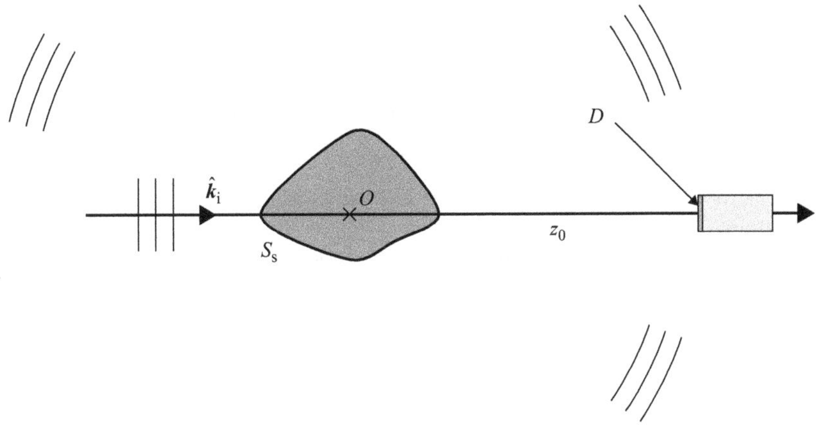

Figure 4.11 The scatterer S_s and the detector with detection area D at a distance z_0 from the scatterer.

The far field amplitude is, as above, denoted $\boldsymbol{F}(\hat{\boldsymbol{r}})$, and the scattered field in the far zone is to leading order

$$\begin{cases} \boldsymbol{E}_s(\boldsymbol{r}) = \dfrac{e^{ikr}}{r}\boldsymbol{F}(\hat{\boldsymbol{r}}) \\ \boldsymbol{H}_s(\boldsymbol{r}) = \dfrac{1}{\eta\eta_0}\hat{\boldsymbol{r}} \times \boldsymbol{E}_s(\boldsymbol{r}) \end{cases}$$

Let D denote the active area of a detector located in the far field region of scatterer facing the incident direction, i.e., its absorbing direction is $\hat{\boldsymbol{k}}_i$, see Figure 4.11. The distance from the origin to the detector is denoted z_0.

The total power absorbed by the detector is (we assume the absorption is 100% effective)

$$\begin{aligned} P &= \frac{1}{2}\operatorname{Re}\iint_D (\boldsymbol{E} \times \boldsymbol{H}^*) \cdot \hat{\boldsymbol{k}}_i \, dS \\ &= \frac{1}{2}\operatorname{Re}\iint_D (\boldsymbol{E}_i \times \boldsymbol{H}_i^* + \boldsymbol{E}_s \times \boldsymbol{H}_s^* + \boldsymbol{E}_i \times \boldsymbol{H}_s^* + \boldsymbol{E}_s \times \boldsymbol{H}_i^*) \cdot \hat{\boldsymbol{k}}_i \, dS \end{aligned}$$

We identify a sum of three different contributions, $P = P_i + P_s + P_{ext}$. The first two contributions are:

$$\begin{cases} P_i = \dfrac{1}{2}\operatorname{Re}\displaystyle\iint_D (\boldsymbol{E}_i \times \boldsymbol{H}_i^*) \cdot \hat{\boldsymbol{k}}_i \, dS = \dfrac{1}{2\eta\eta_0}|\boldsymbol{E}_0|^2 |D| \\ P_s = \dfrac{1}{2}\operatorname{Re}\displaystyle\iint_D (\boldsymbol{E}_s \times \boldsymbol{H}_s^*) \cdot \hat{\boldsymbol{k}}_i \, dS \end{cases}$$

where P_i is the power received by the detector from the incident wave, and $|D|$ is the area of the detector. This is the total contribution to the power P, if the scatterer is absent. P_s is the power received by the detector due to the scattered wave in the direction of the detector $\hat{\boldsymbol{r}} = \hat{\boldsymbol{k}}_{\mathrm{i}}$. The third contribution is

$$P_{\mathrm{ext}} = \frac{1}{2}\,\mathrm{Re} \iint\limits_{D} (\boldsymbol{E}_{\mathrm{i}} \times \boldsymbol{H}_{\mathrm{s}}^* + \boldsymbol{E}_{\mathrm{s}} \times \boldsymbol{H}_{\mathrm{i}}^*) \cdot \hat{\boldsymbol{k}}_{\mathrm{i}} \, \mathrm{d}S$$

where P_{ext} is the extinction power, which is a consequence of the interaction of the incident and the scattered fields.

If the detector is located in the far zone of the scatterer, then to leading order, we have, see (4.23) on page 202

$$\begin{aligned} P_{\mathrm{s}} &= \frac{1}{2\eta\eta_0}\,\mathrm{Re} \iint\limits_{D} |\boldsymbol{F}(\hat{\boldsymbol{r}})|^2 \frac{\hat{\boldsymbol{r}} \cdot \hat{\boldsymbol{k}}_{\mathrm{i}}}{r^2} \, \mathrm{d}S = \frac{1}{2\eta\eta_0} \left|\boldsymbol{F}(\hat{\boldsymbol{r}}_D)\right|^2 \iint\limits_{D} \frac{\hat{\boldsymbol{r}} \cdot \hat{\boldsymbol{k}}_{\mathrm{i}}}{r^2} \, \mathrm{d}S \\ &= \frac{P_{\mathrm{i}}}{|D|} \left|\mathbf{S}(\hat{\boldsymbol{r}}_D, \hat{\boldsymbol{k}}_{\mathrm{i}}) \cdot \hat{\boldsymbol{p}}_{\mathrm{e}}\right|^2 \Omega_D(\mathbf{0}) \end{aligned}$$

where $\hat{\boldsymbol{p}}_{\mathrm{i}} = \boldsymbol{E}_0 / |\boldsymbol{E}_0|$, and where $\Omega_D(\mathbf{0})$ is the solid angle of the detector D as seen from the origin, see Appendix F.5 on page 672, and where we also have used the mean value theorem for the surface integral over D.

We proceed with the extinction power and replace the scattered fields with their far field expressions

$$\begin{aligned} P_{\mathrm{ext}} &= \frac{1}{2\eta\eta_0}\,\mathrm{Re} \iint\limits_{D} \frac{1}{r} \Big(\boldsymbol{E}_0 \times (\hat{\boldsymbol{r}} \times \boldsymbol{F}^*(\hat{\boldsymbol{r}})) \mathrm{e}^{\mathrm{i}k\hat{\boldsymbol{k}}_{\mathrm{i}} \cdot \boldsymbol{r} - \mathrm{i}kr} + \boldsymbol{F}(\hat{\boldsymbol{r}}) \times (\hat{\boldsymbol{k}}_{\mathrm{i}} \times \boldsymbol{E}_0^*) \mathrm{e}^{-\mathrm{i}k\hat{\boldsymbol{k}}_{\mathrm{i}} \cdot \boldsymbol{r} + \mathrm{i}kr} \Big) \cdot \hat{\boldsymbol{k}}_{\mathrm{i}} \, \mathrm{d}S \\ &= \frac{1}{2\eta\eta_0}\,\mathrm{Re} \iint\limits_{D} \frac{1}{r} \Big(\big[(\boldsymbol{E}_0 \cdot \boldsymbol{F}^*(\hat{\boldsymbol{r}}))(\hat{\boldsymbol{r}} \cdot \hat{\boldsymbol{k}}_{\mathrm{i}}) - (\boldsymbol{F}^*(\hat{\boldsymbol{r}}) \cdot \hat{\boldsymbol{k}}_{\mathrm{i}})(\boldsymbol{E}_0 \cdot \hat{\boldsymbol{r}}) \big] \mathrm{e}^{\mathrm{i}k\hat{\boldsymbol{k}}_{\mathrm{i}} \cdot \boldsymbol{r} - \mathrm{i}kr} \\ &\qquad + (\boldsymbol{F}(\hat{\boldsymbol{r}}) \cdot \boldsymbol{E}_0^*) \mathrm{e}^{-\mathrm{i}k\hat{\boldsymbol{k}}_{\mathrm{i}} \cdot \boldsymbol{r} + \mathrm{i}kr} \Big) \, \mathrm{d}S \end{aligned}$$

since $\boldsymbol{E}_0 \cdot \hat{\boldsymbol{k}}_{\mathrm{i}} = 0$. To leading order, the extinction power for large distances z_0 becomes

$$\begin{aligned} P_{\mathrm{ext}} &= \frac{1}{2\eta\eta_0}\,\mathrm{Re} \frac{2\pi}{\mathrm{i}k} \Big((\boldsymbol{E}_0 \cdot \boldsymbol{F}(\hat{\boldsymbol{k}}_{\mathrm{i}})^*) - (\boldsymbol{F}(\hat{\boldsymbol{k}}_{\mathrm{i}}) \cdot \boldsymbol{E}_0^*) \Big) \\ &= -\frac{2\pi}{k\eta\eta_0} \mathrm{Im}(\boldsymbol{F}(\hat{\boldsymbol{k}}_{\mathrm{i}}) \cdot \boldsymbol{E}_0^*) = -P_{\mathrm{i}} \frac{\sigma_{\mathrm{ext}}}{|D|} \end{aligned}$$

where we used the optical theorem in (4.47) on page 230. To obtain this asymptotic evaluation, the following integral is useful (for detail see Example G.2 on page 686)

$$\iint\limits_{D} G(x,y) \mathrm{e}^{\mathrm{i}kz_0(1 - r/z_0)} \, \mathrm{d}x \, \mathrm{d}y = \frac{2\pi z_0}{\mathrm{i}k} G(0,0) + o\left(1/(kz_0)\right), \qquad \text{as } kz_0 \to \infty$$

The detected power, P, then becomes

$$\begin{aligned} P &= \frac{P_{\mathrm{i}}}{|D|} \left(|D| + \left|\mathbf{S}(\hat{\boldsymbol{r}}_D, \hat{\boldsymbol{k}}_{\mathrm{i}}) \cdot \hat{\boldsymbol{p}}_{\mathrm{e}}\right|^2 \Omega_D(\mathbf{0}) - \sigma_{\mathrm{ext}} \right) \\ &= \frac{P_{\mathrm{i}}}{|D|} \left(|D| + \left|\mathbf{S}(\hat{\boldsymbol{r}}_D, \hat{\boldsymbol{k}}_{\mathrm{i}}) \cdot \hat{\boldsymbol{p}}_{\mathrm{e}}\right|^2 \Omega_D(\mathbf{0}) - \sigma_{\mathrm{s}} - \sigma_{\mathrm{a}} \right) \end{aligned}$$

The first term in this expression is the contribution from the incident field, and the second term is the scattered power that is detected by the detector. The last two terms are due to the power that is scattered in all directions, σ_s, and the power that is absorbed by the scatterer, σ_a, respectively. These last two contributions, the extinction power, lower the total power received by the detector—hence the notion extinction.

4.5 Plane Interface Case and Babinet's Principle

A scatterer which is flat and part of a plane interface deserves special attention, since several additional conclusions can be made on the form of the scattered field. This is due to the symmetry properties of the scattered fields caused by a plane scatterer. This geometry enters in many electromagnetic applications, e.g., electromagnetic scattering by disks and—for the complementary geometry—penetration of electromagnetic fields through holes, cracks, and joints,

Consider the plane $z = 0$. Part of this surface consists of metal, S_M, and the rest is an aperture in the plane. The aperture part of the plane is denoted S_A. We assume for simplicity that the metal part is simply connected and finite, but modifications of this assumption are easy to make. A possible extension of the analysis presented in this section is to treat a periodic pattern of apertures or metal parts. The two parts, S_M and S_A, are complementary domains, i.e., $S_M \cap S_A = \emptyset$ and $\overline{S_M} \cup \overline{S_A} = \{\boldsymbol{r} \in \mathbb{R}^3 : z = 0\}$, see Figure 4.12.

The sources are assumed to be located in the domain to the left of the interface, i.e., in the region $z < 0$. We assume the sources are located at a finite distance from

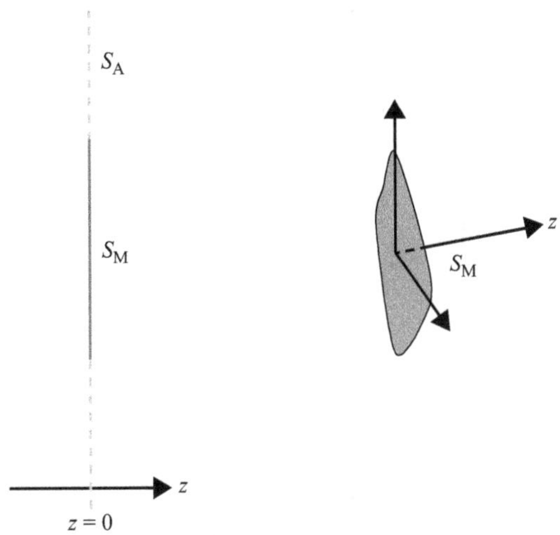

Figure 4.12 Geometry of the plane metallic scatterer, S_M, and its complementary surface, S_A, in the plane $z = 0$. The left figure shows the geometry in a side view with metal in solid line and complementary surface in dashed line. The right figure shows a three-dimensional view of the finite metal part S_M. The z-direction is normal to the plane metallic scatterer.

the origin.[17] The integral representations in (3.14) and (3.15) on pages 156 and 157, respectively, applied to the incident fields in the region $z > 0$ are

$$
\begin{aligned}
&-\frac{\eta_0\eta}{\mathrm{i}k}\nabla\times\left(\nabla\times\iint\limits_{z'=0} g(k,|\boldsymbol{r}-\boldsymbol{r}'|)\left(\hat{z}\times\boldsymbol{H}_{\mathrm{i}}(\boldsymbol{r}')\right)\,\mathrm{d}S'\right)\\
&+\nabla\times\iint\limits_{z'=0} g(k,|\boldsymbol{r}-\boldsymbol{r}'|)\left(\hat{z}\times\boldsymbol{E}_{\mathrm{i}}(\boldsymbol{r}')\right)\,\mathrm{d}S' = \begin{cases}\boldsymbol{E}_{\mathrm{i}}(\boldsymbol{r}), & z>0\\ \boldsymbol{0}, & z<0\end{cases}
\end{aligned}
\tag{4.48}
$$

Similarly, we have for the incident magnetic field:

$$
\begin{aligned}
&\frac{1}{\mathrm{i}k\eta_0\eta}\nabla\times\left(\nabla\times\iint\limits_{z'=0} g(k,|\boldsymbol{r}-\boldsymbol{r}'|)\left(\hat{z}\times\boldsymbol{E}_{\mathrm{i}}(\boldsymbol{r}')\right)\,\mathrm{d}S'\right)\\
&+\nabla\times\iint\limits_{z'=0} g(k,|\boldsymbol{r}-\boldsymbol{r}'|)\left(\hat{z}\times\boldsymbol{H}_{\mathrm{i}}(\boldsymbol{r}')\right)\,\mathrm{d}S' = \begin{cases}\boldsymbol{H}_{\mathrm{i}}(\boldsymbol{r}), & z>0\\ \boldsymbol{0}, & z<0\end{cases}
\end{aligned}
\tag{4.49}
$$

We have here also assumed that, due to radiation condition, the integrals from the large hemisphere vanish in the limit.

Notice that it is irrelevant which side of the surface $z = 0$ the field in the integrand is evaluated, since the incident field is continuous over $z = 0$.

We can rewrite the integral representation in (4.48) as

$$
\boldsymbol{I}_1(\boldsymbol{r}) + \boldsymbol{I}_2(\boldsymbol{r}) = \begin{cases}\boldsymbol{E}_{\mathrm{i}}(\boldsymbol{r}), & z>0\\ \boldsymbol{0}, & z<0\end{cases}
\tag{4.50}
$$

where the two integrals, $\boldsymbol{I}_1(\boldsymbol{r})$ and $\boldsymbol{I}_2(\boldsymbol{r})$, are

$$
\boldsymbol{I}_1(\boldsymbol{r}) = -\frac{\eta_0\eta}{\mathrm{i}k}\nabla\times\left(\nabla\times\iint\limits_{z'=0} g(k,|\boldsymbol{r}-\boldsymbol{r}'|)\left(\hat{z}\times\boldsymbol{H}_{\mathrm{i}}(\boldsymbol{r}')\right)\,\mathrm{d}S'\right), \qquad z\neq 0
$$

and

$$
\boldsymbol{I}_2(\boldsymbol{r}) = \nabla\times\iint\limits_{z'=0} g(k,|\boldsymbol{r}-\boldsymbol{r}'|)\left(\hat{z}\times\boldsymbol{E}_{\mathrm{i}}(\boldsymbol{r}')\right)\,\mathrm{d}S', \qquad z\neq 0
$$

Both integrals contain vector fields with components only in the x–y-plane, and moreover, since $g(k, |\boldsymbol{r} - \boldsymbol{r}'|)$ is even in z, the integrands are also even in z. After the application of the

[17] To handle plane wave incidence with sources at infinity, subtract the plane wave contribution (specularly reflected and transmitted plane waves) and analyze the remaining part.

curl operations, this symmetry conditions imply that, see Problem 4.6:

$$\begin{cases} \hat{z} \times \boldsymbol{I}_1(\boldsymbol{r}) \text{ even in } z \text{ (tangential components)} \\ \hat{z} \cdot \boldsymbol{I}_1(\boldsymbol{r}) \text{ odd in } z \text{ (normal component)} \\ \hat{z} \times \boldsymbol{I}_2(\boldsymbol{r}) \text{ odd in } z \text{ (tangential components)} \\ \hat{z} \cdot \boldsymbol{I}_2(\boldsymbol{r}) \text{ even in } z \text{ (normal component)} \end{cases} \tag{4.51}$$

We denote the mirror point to a point $\boldsymbol{r}$ ($z > 0$) in the right-hand side domain by $\boldsymbol{r}^* = \mathbf{R} \cdot \boldsymbol{r}$ ($z \to -z$), where the reflection dyadic $\mathbf{R} = \mathbf{I}_3 - 2\hat{z}\hat{z}$. The extinction part of the integral representation of the incident electric field, (4.50), shows that

$$\boldsymbol{I}_1(\boldsymbol{r}^*) = -\boldsymbol{I}_2(\boldsymbol{r}^*), \qquad z > 0$$

which due to the symmetry conditions of the integrals $\boldsymbol{I}_1(\boldsymbol{r})$ and $\boldsymbol{I}_2(\boldsymbol{r})$ imply that

$$\hat{z} \times \boldsymbol{I}_1(\boldsymbol{r}) = \hat{z} \times \boldsymbol{I}_1(\boldsymbol{r}^*) = -\hat{z} \times \boldsymbol{I}_2(\boldsymbol{r}^*) = \hat{z} \times \boldsymbol{I}_2(\boldsymbol{r}), \qquad z > 0$$

and

$$\hat{z} \cdot \boldsymbol{I}_1(\boldsymbol{r}) = -\hat{z} \cdot \boldsymbol{I}_1(\boldsymbol{r}^*) = \hat{z} \cdot \boldsymbol{I}_2(\boldsymbol{r}^*) = \hat{z} \cdot \boldsymbol{I}_2(\boldsymbol{r}), \qquad z > 0$$

The integral representation of the incident electric field in the region $z > 0$ then implies two alternative integral formulations of the incident electric field in the region $z > 0$, i.e.,

$$\hat{z} \times \boldsymbol{E}_{\mathrm{i}}(\boldsymbol{r}) = \hat{z} \times \boldsymbol{I}_1(\boldsymbol{r}) + \hat{z} \times \boldsymbol{I}_2(\boldsymbol{r}) = \begin{cases} 2\hat{z} \times \boldsymbol{I}_1(\boldsymbol{r}) \\ 2\hat{z} \times \boldsymbol{I}_2(\boldsymbol{r}) \end{cases} \qquad z > 0$$

and

$$\hat{z} \cdot \boldsymbol{E}_{\mathrm{i}}(\boldsymbol{r}) = \hat{z} \cdot \boldsymbol{I}_1(\boldsymbol{r}) + \hat{z} \cdot \boldsymbol{I}_2(\boldsymbol{r}) = \begin{cases} 2\hat{z} \cdot \boldsymbol{I}_1(\boldsymbol{r}) \\ 2\hat{z} \cdot \boldsymbol{I}_2(\boldsymbol{r}) \end{cases} \qquad z > 0$$

Therefore, in the domain $z > 0$, we have

$$\boldsymbol{E}_{\mathrm{i}}(\boldsymbol{r}) = \begin{cases} -\dfrac{2\eta_0\eta}{\mathrm{i}k} \nabla \times \left(\nabla \times \displaystyle\iint_{z'=0} g(k, |\boldsymbol{r} - \boldsymbol{r}'|) \left(\hat{z} \times \boldsymbol{H}_{\mathrm{i}}(\boldsymbol{r}')\right) \mathrm{d}S' \right) \\ 2\nabla \times \displaystyle\iint_{z'=0} g(k, |\boldsymbol{r} - \boldsymbol{r}'|) \left(\hat{z} \times \boldsymbol{E}_{\mathrm{i}}(\boldsymbol{r}')\right) \mathrm{d}S' \end{cases} \qquad z > 0$$

and similarly from (4.49), we obtain two alternative formulations of the incident magnetic field

$$\boldsymbol{H}_{\mathrm{i}}(\boldsymbol{r}) = \begin{cases} \dfrac{2}{\mathrm{i}k\eta_0\eta} \nabla \times \left(\nabla \times \displaystyle\iint_{z'=0} g(k, |\boldsymbol{r} - \boldsymbol{r}'|) \left(\hat{z} \times \boldsymbol{E}_{\mathrm{i}}(\boldsymbol{r}')\right) \mathrm{d}S' \right) \\ 2\nabla \times \displaystyle\iint_{z'=0} g(k, |\boldsymbol{r} - \boldsymbol{r}'|) \left(\hat{z} \times \boldsymbol{H}_{\mathrm{i}}(\boldsymbol{r}')\right) \mathrm{d}S' \end{cases} \qquad z > 0$$

Notice that these representations only hold in the region $z > 0$, and not in the region $z < 0$, where the incident field has its sources.

The scattered fields, $\boldsymbol{E}_{\mathrm{s}}$ and $\boldsymbol{H}_{\mathrm{s}}$, are the additional fields due to the presence of the metallic parts on the interface. The total fields, $\boldsymbol{E} = \boldsymbol{E}_{\mathrm{i}} + \boldsymbol{E}_{\mathrm{s}}$ and $\boldsymbol{H} = \boldsymbol{H}_{\mathrm{i}} + \boldsymbol{H}_{\mathrm{s}}$, satisfy the appropriate boundary conditions at the interface $z = 0$. Similar symmetries as for the incident field can be obtained from the integral representation of the scattered fields applied to the domain $z > 0$ (we assume that, due to radiation condition, the integrals from the large sphere vanish in the limit):

$$\boldsymbol{E}_{\mathrm{s}}(\boldsymbol{r}) = \begin{cases} -\dfrac{2\eta_0\eta}{\mathrm{i}k}\nabla \times \left(\nabla \times \displaystyle\iint_{z'=0^+} g(k, |\boldsymbol{r}-\boldsymbol{r}'|)\,(\hat{z} \times \boldsymbol{H}_{\mathrm{s}}(\boldsymbol{r}'))\,\mathrm{d}S'\right) \\ 2\nabla \times \displaystyle\iint_{z'=0^+} g(k, |\boldsymbol{r}-\boldsymbol{r}'|)\,(\hat{z} \times \boldsymbol{E}_{\mathrm{s}}(\boldsymbol{r}'))\,\mathrm{d}S' \end{cases} \qquad z > 0$$

and

$$\boldsymbol{H}_{\mathrm{s}}(\boldsymbol{r}) = \begin{cases} \dfrac{2}{\mathrm{i}k\eta_0\eta}\nabla \times \left(\nabla \times \displaystyle\iint_{z'=0^+} g(k, |\boldsymbol{r}-\boldsymbol{r}'|)\,(\hat{z} \times \boldsymbol{E}_{\mathrm{s}}(\boldsymbol{r}'))\,\mathrm{d}S'\right) \\ 2\nabla \times \displaystyle\iint_{z'=0^+} g(k, |\boldsymbol{r}-\boldsymbol{r}'|)\,(\hat{z} \times \boldsymbol{H}_{\mathrm{s}}(\boldsymbol{r}'))\,\mathrm{d}S' \end{cases} \qquad z > 0$$

Notice that the tangential surface fields have to be evaluated at $z = 0^+$, since these fields are not in general continuous over the interface $z = 0$.

However, more symmetries can be deduced for the scattered field if we apply the integral representation to the domain $z < 0$. This application was not possible for the incident field, since the incident field has its sources in this region. The result is

$$\frac{\eta_0\eta}{\mathrm{i}k}\nabla \times \left(\nabla \times \iint_{z'=0^-} g(k, |\boldsymbol{r}-\boldsymbol{r}'|)\,(\hat{z} \times \boldsymbol{H}_{\mathrm{s}}(\boldsymbol{r}'))\,\mathrm{d}S'\right) - \nabla \times \iint_{z'=0^-} g(k, |\boldsymbol{r}-\boldsymbol{r}'|)\,(\hat{z} \times \boldsymbol{E}_{\mathrm{s}}(\boldsymbol{r}'))\,\mathrm{d}S' = \begin{cases} \boldsymbol{0}, & z > 0 \\ \boldsymbol{E}_{\mathrm{s}}(\boldsymbol{r}), & z < 0 \end{cases}$$

and

$$-\frac{1}{\mathrm{i}k\eta_0\eta}\nabla \times \left(\nabla \times \iint_{z'=0^-} g(k, |\boldsymbol{r}-\boldsymbol{r}'|)\,(\hat{z} \times \boldsymbol{E}_{\mathrm{s}}(\boldsymbol{r}'))\,\mathrm{d}S'\right) - \nabla \times \iint_{z'=0^-} g(k, |\boldsymbol{r}-\boldsymbol{r}'|)\,(\hat{z} \times \boldsymbol{H}_{\mathrm{s}}(\boldsymbol{r}'))\,\mathrm{d}S' = \begin{cases} \boldsymbol{0}, & z > 0 \\ \boldsymbol{H}_{\mathrm{s}}(\boldsymbol{r}), & z < 0 \end{cases}$$

Notice that the tangential fields in the integrals are limits taken from the left-hand side of the interface, i.e., $z = 0^-$. With similar arguments as above, we get

$$E_s(r) = \begin{cases} \dfrac{2\eta_0\eta}{ik}\nabla \times \left(\nabla \times \displaystyle\iint_{z'=0^-} g(k, |r - r'|)\left(\hat{z} \times H_s(r')\right) dS'\right) & \\ & z < 0 \\ -2\nabla \times \displaystyle\iint_{z'=0^-} g(k, |r - r'|)\left(\hat{z} \times E_s(r')\right) dS' & \end{cases}$$

and

$$H_s(r) = \begin{cases} -\dfrac{2}{ik\eta_0\eta}\nabla \times \left(\nabla \times \displaystyle\iint_{z'=0^-} g(k, |r - r'|)\left(\hat{z} \times E_s(r')\right) dS'\right) & \\ & z < 0 \\ -2\nabla \times \displaystyle\iint_{z'=0^-} g(k, |r - r'|)\left(\hat{z} \times H_s(r')\right) dS' & \end{cases}$$

From the boundary conditions, we know that $\hat{z} \times E_s = -\hat{z} \times E_i$ on both sides of the metallic part, S_M, at the interface $z = 0$. Since the incident field is continuous at the interface $z = 0$, $\hat{z} \times E_s$ is also continuous at S_M. On the other hand, in the aperture, S_A, the tangential electric field, and consequently, the tangential scattered electric field has to be continuous. Thus, we conclude that the tangential scattered electric field is continuous everywhere on the interface, $z = 0$, both on the metal part and on the aperture part. Therefore, it does not matter which surface field we enter in the following integral representation of the scattered electric field:

$$E_s(r) = \pm 2\nabla \times \iint_{z=0} g(k, |r - r'|)\left(\hat{z} \times E_s(r')\right) dS', \qquad z \gtrless 0$$

The symmetry properties of the integral I_2 defined above in (4.51) implies that $\hat{z} \times E_s(r)$ is even in $z \to -z$ and $\hat{z} \cdot E_s(r)$ is odd in $z \to -z$. Further use of the representations of the scattered fields above and the properties of the tangential electric field on both sides of the interface $z = 0$ make it possible to conclude that $\hat{z} \times H_s(r)$ is odd in $z \to -z$ and $\hat{z} \cdot H_s(r)$ is even in $z \to -z$. We summarize the symmetries properties (they are also consistent with the limiting values at the interface $z = 0$):

$$\begin{cases} \hat{z} \times E_s(r) \text{ even in } z \to -z \\ \hat{z} \cdot E_s(r) \text{ odd in } z \to -z \end{cases} \qquad \begin{cases} \hat{z} \times H_s(r) \text{ odd in } z \to -z \\ \hat{z} \cdot H_s(r) \text{ even in } z \to -z \end{cases}$$

Notice that these symmetries do not imply that the normal components of the scattered electric field and the tangential components of the scattered magnetic field are zero at $z = 0$. The fields in general have jump discontinuities at the interface. A more compact way of writing these symmetries are

$$\begin{cases} E_s(r^*) = \mathbf{R} \cdot E_s(r) = E_s(r) - 2\hat{z}\left(\hat{z} \cdot E_s(r)\right) \\ H_s(r^*) = -\mathbf{R} \cdot H_s(r) = -H_s(r) + 2\hat{z}\left(\hat{z} \cdot H_s(r)\right) \end{cases} \tag{4.52}$$

where the point $r^* = \mathbf{R} \cdot r$ is the point r reflected in $z = 0$, and the reflection dyadic $\mathbf{R} = \mathbf{I}_3 - 2\hat{z}\hat{z}$.

As concluded above, on the metallic part, the tangential components of the total electric field are zero. Moreover, the tangential parts of the scattered magnetic field are, due to continuity and the fact that $\hat{z} \times \boldsymbol{H}_{\mathrm{s}}(\boldsymbol{r})$ is odd in z, zero in the aperture, S_{A}, i.e., the boundary conditions on the screen can be written as

$$\begin{cases} \hat{z} \times \boldsymbol{E}(\boldsymbol{r}) = \boldsymbol{0}, & \boldsymbol{r} \text{ on } S_{\mathrm{M}} \\ \hat{z} \times \boldsymbol{H}_{\mathrm{s}}(\boldsymbol{r}) = \boldsymbol{0}, & \boldsymbol{r} \text{ on } S_{\mathrm{A}} \end{cases} \tag{4.53}$$

Therefore, from the integral representations above, we have for all $z \neq 0$

$$\boldsymbol{E}_{\mathrm{s}}(\boldsymbol{r}) = -\frac{2\eta_0\eta}{\mathrm{i}k} \nabla \times \left(\nabla \times \iint_{S_{\mathrm{M}}} g(k, |\boldsymbol{r} - \boldsymbol{r}'|) \left(\hat{z} \times \boldsymbol{H}_{\mathrm{s}}(\boldsymbol{r}') \right)\big|_{+} \, \mathrm{d}S' \right)$$

or

$$\boldsymbol{E}_{\mathrm{s}}(\boldsymbol{r}) = 2 \operatorname{sign}(z) \nabla \times \iint_{z'=0} g(k, |\boldsymbol{r} - \boldsymbol{r}'|) \left(\hat{z} \times \boldsymbol{E}_{\mathrm{s}}(\boldsymbol{r}') \right) \mathrm{d}S'$$

and

$$\boldsymbol{H}_{\mathrm{s}}(\boldsymbol{r}) = \frac{2 \operatorname{sign}(z)}{\mathrm{i}k\eta_0\eta} \nabla \times \left(\nabla \times \iint_{z'=0} g(k, |\boldsymbol{r} - \boldsymbol{r}'|) \left(\hat{z} \times \boldsymbol{E}_{\mathrm{s}}(\boldsymbol{r}') \right) \mathrm{d}S' \right)$$

or

$$\boldsymbol{H}_{\mathrm{s}}(\boldsymbol{r}) = 2\nabla \times \iint_{S_{\mathrm{M}}} g(k, |\boldsymbol{r} - \boldsymbol{r}'|) \left(\hat{z} \times \boldsymbol{H}_{\mathrm{s}}(\boldsymbol{r}') \right)\big|_{+} \, \mathrm{d}S'$$

where the plus sign on the field in the integrand indicates that the field is limit value taken at $z = 0^{+}$.

To these integral representations of the scattered fields, we add the integral representation of the incident field from above, and use the boundary conditions. We then get our final result

$$\begin{cases} \boldsymbol{E}(\boldsymbol{r}) = 2\nabla \times \displaystyle\iint_{S_{\mathrm{A}}} g(k, |\boldsymbol{r} - \boldsymbol{r}'|) \left(\hat{z} \times \boldsymbol{E}(\boldsymbol{r}') \right) \mathrm{d}S' \\ \boldsymbol{H}_{\mathrm{s}}(\boldsymbol{r}) = 2\nabla \times \displaystyle\iint_{S_{\mathrm{M}}} g(k, |\boldsymbol{r} - \boldsymbol{r}'|) \left(\hat{z} \times \boldsymbol{H}_{\mathrm{s}}(\boldsymbol{r}') \right)\big|_{+} \, \mathrm{d}S' \end{cases} \quad z > 0 \tag{4.54}$$

and

$$\begin{cases} \boldsymbol{E}_{\mathrm{s}}(\boldsymbol{r}) = -\dfrac{2\eta_0\eta}{\mathrm{i}k} \nabla \times \left(\nabla \times \displaystyle\iint_{S_{\mathrm{M}}} g(k, |\boldsymbol{r} - \boldsymbol{r}'|) \left(\hat{z} \times \boldsymbol{H}_{\mathrm{s}}(\boldsymbol{r}') \right)\big|_{+} \, \mathrm{d}S' \right) \\ \boldsymbol{H}(\boldsymbol{r}) = \dfrac{2}{\mathrm{i}k\eta_0\eta} \nabla \times \left(\nabla \times \displaystyle\iint_{S_{\mathrm{A}}} g(k, |\boldsymbol{r} - \boldsymbol{r}'|) \left(\hat{z} \times \boldsymbol{E}(\boldsymbol{r}') \right) \mathrm{d}S' \right) \end{cases} \quad z > 0$$

These latter two integral representations add no additional information since they follow from (4.54) and the Maxwell equations.

The far zone has, in general, no meaning for the plane interface case, if the metallic part is infinite in extent. However, if the metallic part is finite, we can define a far field amplitude, and we obtain an expression directly from the equations above. The result is (formally replace $\nabla \to \mathrm{i}k\hat{\boldsymbol{r}}$)

$$\boldsymbol{F}(\hat{\boldsymbol{r}}) = -\frac{\mathrm{i}k\eta_0\eta}{2\pi}\hat{\boldsymbol{r}} \times \left(\hat{\boldsymbol{r}} \times \iint_{S_\mathrm{M}} \left(\hat{z} \times \boldsymbol{H}_\mathrm{s}(\boldsymbol{r}')\right)\big|_+ \mathrm{e}^{-\mathrm{i}k\hat{\boldsymbol{r}}\cdot\boldsymbol{r}'}\,\mathrm{d}S'\right)$$

From (4.52), we also conclude that the far field amplitude, $\boldsymbol{F}(\hat{\boldsymbol{r}})$, and the scattering dyadic, $\mathbf{S}(\hat{\boldsymbol{r}}, \hat{\boldsymbol{k}}_\mathrm{i})$, in the direction $\hat{\boldsymbol{r}}$ and its mirror direction $\hat{\boldsymbol{r}}^* = \mathbf{R}\cdot\hat{\boldsymbol{r}}$ are related as

$$\begin{cases} \boldsymbol{F}(\hat{\boldsymbol{r}}^*) = \mathbf{R}\cdot\boldsymbol{F}(\hat{\boldsymbol{r}}) \\ \mathbf{S}(\hat{\boldsymbol{r}}^*, \hat{\boldsymbol{k}}_\mathrm{i}) = \mathbf{R}\cdot\mathbf{S}(\hat{\boldsymbol{r}}, \hat{\boldsymbol{k}}_\mathrm{i}) \end{cases} \tag{4.55}$$

In particular, in the normal directions, $\hat{\boldsymbol{r}} = \pm\hat{z}$, the scattering dyadic is identical, i.e., $\mathbf{S}(-\hat{z}, \hat{\boldsymbol{k}}_\mathrm{i}) = \mathbf{S}(\hat{z}, \hat{\boldsymbol{k}}_\mathrm{i})$.

4.5.1 Babinet's Principle

Babinet's principle[18] relates the scattered field due to a plane metallic scatterer, S_M, as depicted in Figure 4.13 to the transmitted field through an aperture of the complementary geometry. In the complementary geometry, the metal part and the aperture part change places.

Babinet's principle is a consequence of the symmetry of the problem and the boundary conditions (4.53)

$$\begin{cases} \hat{z} \times \boldsymbol{E}(\boldsymbol{r}) = \boldsymbol{0}, & \boldsymbol{r} \text{ on the metal part} \\ \hat{z} \times \boldsymbol{H}_\mathrm{s}(\boldsymbol{r}) = \boldsymbol{0}, & \boldsymbol{r} \text{ on the aperture part} \end{cases}$$

Original Problem: Denote the incident and the scattered fields of the original scattering problem by

$$\begin{cases} \boldsymbol{E}_\mathrm{i}(\boldsymbol{r}) = \boldsymbol{A}(\boldsymbol{r}) \\ \eta_0\eta\boldsymbol{H}_\mathrm{i}(\boldsymbol{r}) \end{cases} \text{(incident fields)} \qquad \begin{cases} \boldsymbol{E}_\mathrm{s}(\boldsymbol{r}) \\ \eta_0\eta\boldsymbol{H}_\mathrm{s}(\boldsymbol{r}) \end{cases} \text{(scattered fields)}$$

where the field $\boldsymbol{A}(\boldsymbol{r})$ is a prescribed incident electric field with sources in the region $z < 0$. This scattering problem satisfies the following boundary conditions on the metal part, S_M, and the aperture part, S_A, see (4.53):

$$\begin{cases} \hat{z} \times \boldsymbol{E}_\mathrm{s}(\boldsymbol{r}) = -\hat{z} \times \boldsymbol{A}(\boldsymbol{r}), & \boldsymbol{r} \text{ on } S_\mathrm{M} \\ \hat{z} \times \boldsymbol{H}_\mathrm{s}(\boldsymbol{r}) = \boldsymbol{0}, & \boldsymbol{r} \text{ on } S_\mathrm{A} \end{cases} \tag{4.56}$$

We assume this problem has a unique solution in $z > 0$.

[18] Jacques Babinet (1794–1872), French physicist, mathematician, and astronomer.

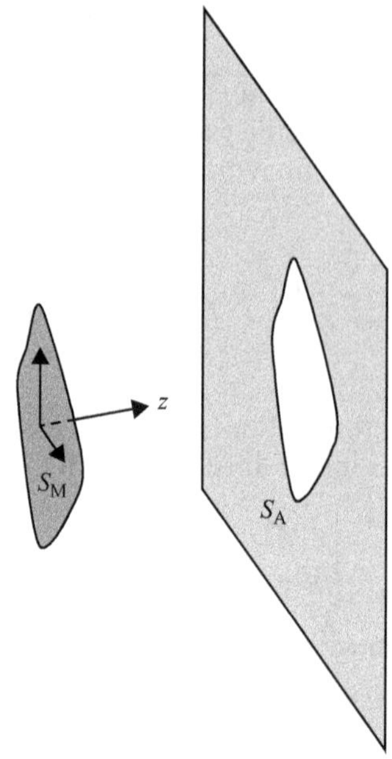

Figure 4.13 Geometry of the plane metallic scatterer, S_M, and its complementary surface, S_A, in the plane $z = 0$. The aperture part S_A is laterally displaced from the metal part S_M for clarity. In the complementary geometry, S_A is the metallic part.

Complementary Problem: The corresponding fields of the complementary problem are

$$\begin{cases} \boldsymbol{E}_i^c(\boldsymbol{r}) \\ \eta_0\eta\boldsymbol{H}_i^c(\boldsymbol{r}) = \boldsymbol{A}(\boldsymbol{r}) \end{cases} \quad \text{(incident fields)} \qquad \begin{cases} \boldsymbol{E}_s^c(\boldsymbol{r}) \\ \eta_0\eta\boldsymbol{H}_s^c(\boldsymbol{r}) \end{cases} \quad \text{(scattered fields)}$$

Notice that in the complementary problem, the metal part now corresponds to S_A of the original problem, and the aperture part to S_M of the original problem. Also notice that the field $\boldsymbol{A}(\boldsymbol{r})$ now quantifies the incident magnetic field (change in polarization of the incident field).

The total fields of the complementary problems satisfy the following boundary conditions, see (4.53):

$$\begin{cases} -\eta_0\eta\hat{z} \times \boldsymbol{H}^c(\boldsymbol{r}) = -\eta_0\eta\hat{z} \times \boldsymbol{H}_i^c(\boldsymbol{r}) = -\hat{z} \times \boldsymbol{A}(\boldsymbol{r}), & \boldsymbol{r} \text{ on } S_M \\ \hat{z} \times \boldsymbol{E}^c(\boldsymbol{r}) = \boldsymbol{0}, & \boldsymbol{r} \text{ on } S_A \end{cases} \tag{4.57}$$

We also assume this problem has a unique solution in $z > 0$.

Comparing (4.56) and (4.57), the duality principle (if $\boldsymbol{E}$ and $\eta_0\eta\boldsymbol{H}$ is a solution to the Maxwell equations, then the exchange $\boldsymbol{E} \to -\eta_0\eta\boldsymbol{H}$ and $\eta_0\eta\boldsymbol{H} \to \boldsymbol{E}$ also is a solution) and the unique solvability of the scattering problems imply

$$\boldsymbol{E}_s(\boldsymbol{r}) = -\eta_0\eta\boldsymbol{H}^c(\boldsymbol{r}), \qquad z > 0$$

This is the Babinet's principle. Notice that this conclusion can only be made in the region $z > 0$, since only there the two fields have no sources, and therefore satisfy the same source-free Maxwell equations. We can also formulate the Babinet's principle in terms of the total fields. The result is

$$\boldsymbol{E}(\boldsymbol{r}) + \eta_0\eta\boldsymbol{H}^c(\boldsymbol{r}) = \boldsymbol{A}(\boldsymbol{r}), \qquad z > 0$$

Problems for Chapter 4

Section 4.1 The Far Zone

4.1 The current density $\boldsymbol{J}(\boldsymbol{r},\omega)$ in a straight wire antenna oriented along the z-axis is assumed to be

$$\boldsymbol{J}(\boldsymbol{r},\omega) = \begin{cases} I_0\hat{\boldsymbol{z}}\sin k\left(\frac{l}{2} - |z|\right)\delta(x)\delta(y), & |z| \le \frac{l}{2} \\ \boldsymbol{0}, & \text{elsewhere} \end{cases}$$

where δ is the delta distribution and $k = \omega/c_0$. The antenna is located in vacuum and has the total length l, see Figure 4.14. Calculate the far field amplitude $\boldsymbol{F}(\hat{\boldsymbol{r}})$ produced by this current density.

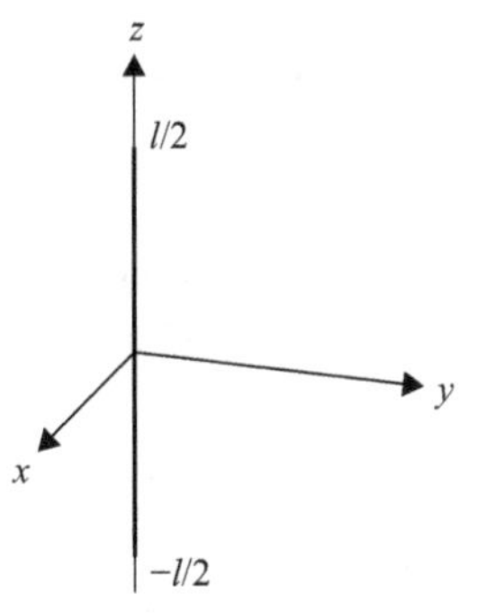

Figure 4.14 Geometry for Problem 4.1.

*__4.2__ Compute the far field amplitude $\boldsymbol{F}(\hat{\boldsymbol{r}})$ from a plane circular loop in the x–y-plane that carries the total current I. The loop is located in vacuum and has a radius a, see Figure 4.15. The current density in spherical coordinates is

$$\boldsymbol{J}_\mathrm{s}(\boldsymbol{r},\omega) = \boldsymbol{J}_\mathrm{s}(r,\theta,\phi,\omega) = \frac{I\hat{\boldsymbol{\phi}}}{a}\delta(r-a)\delta(\theta-\pi/2)$$

where δ is the delta function. Calculate also the long wavelength contribution of the far field amplitude for this configuration.

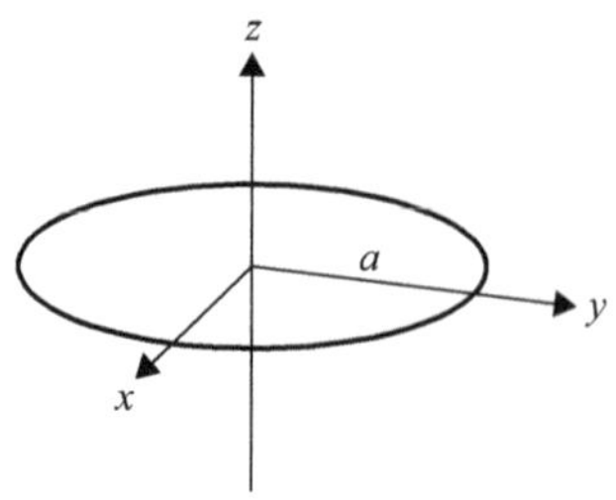

Figure 4.15 Geometry for Problem 4.2.

Relevant integrals:

$$\begin{cases} \displaystyle\int_0^{2\pi} \sin\alpha \mathrm{e}^{iz\cos(\alpha-\phi)}\,\mathrm{d}\alpha = \mathrm{i}2\pi J_1(z)\sin\phi \\ \displaystyle\int_0^{2\pi} \cos\alpha \mathrm{e}^{iz\cos(\alpha-\phi)}\,\mathrm{d}\alpha = \mathrm{i}2\pi J_1(z)\cos\phi \end{cases}$$

where $J_1(z)$ is the Bessel function of order 1.

4.3 Equations (4.17) and (4.18) give two different expressions of the far field amplitude with the integral representation. The difference between the integrals is the following integral:

$$\hat{\boldsymbol{r}} \times \iint_{S_\mathrm{s}} \left[\hat{\boldsymbol{\nu}} \times \boldsymbol{E}_\mathrm{i}(\boldsymbol{r}') - \eta_0\eta\hat{\boldsymbol{r}} \times (\hat{\boldsymbol{\nu}} \times \boldsymbol{H}_\mathrm{i}(\boldsymbol{r}'))\right]\mathrm{e}^{-\mathrm{i}k\hat{\boldsymbol{r}}\cdot\boldsymbol{r}'}\,\mathrm{d}S'$$

where the $\boldsymbol{E}_\mathrm{i}$ and the $\boldsymbol{H}_\mathrm{i}$ fields denote the incident field.

Show that this integral is zero for every $\boldsymbol{E}_\mathrm{i}$ and $\boldsymbol{H}_\mathrm{i}$ fields that satisfy

$$\begin{cases} \nabla \times \boldsymbol{E}_\mathrm{i} = \mathrm{i}k\eta_0\eta\boldsymbol{H}_\mathrm{i} \\ \nabla \times \boldsymbol{H}_\mathrm{i} = -\mathrm{i}\dfrac{k}{\eta_0\eta}\boldsymbol{E}_\mathrm{i} \end{cases} \qquad \boldsymbol{r} \in V_\mathrm{s}$$

where V_s is the volume bounded by the surface S_s.

***4.4** Show the following alternative form of the far field amplitude:

$$\boldsymbol{F}(\hat{\boldsymbol{r}}) = \frac{k^2}{4\pi}\left\{\hat{\boldsymbol{r}} \times \left(\hat{\boldsymbol{r}} \times \iint_{S_\mathrm{s}} \mathrm{e}^{-\mathrm{i}k\hat{\boldsymbol{r}}\cdot\boldsymbol{r}'}\boldsymbol{r}'\left[\eta_0\eta\hat{\boldsymbol{r}} \cdot (\hat{\boldsymbol{\nu}}(\boldsymbol{r}') \times \boldsymbol{H}(\boldsymbol{r}')) - \hat{\boldsymbol{\nu}}(\boldsymbol{r}') \cdot \boldsymbol{E}(\boldsymbol{r}')\right]\mathrm{d}S'\right)\right.$$

$$\left. -\hat{\boldsymbol{r}} \times \iint_{S_\mathrm{s}} \mathrm{e}^{-\mathrm{i}k\hat{\boldsymbol{r}}\cdot\boldsymbol{r}'}\boldsymbol{r}'\left[\hat{\boldsymbol{r}} \cdot (\hat{\boldsymbol{\nu}}(\boldsymbol{r}') \times \boldsymbol{E}(\boldsymbol{r}')) + \eta_0\eta\hat{\boldsymbol{\nu}}(\boldsymbol{r}') \cdot \boldsymbol{H}(\boldsymbol{r}')\right]\mathrm{d}S'\right\}$$

Section 4.3 Scattering Dyadic (Matrix)

4.5 Determine the transformation (Mueller matrix) [**M**] between the scattered and the incident Stokes parameters in terms of the elements of the scattering matrix [**S**], i.e., determine the entries M_{ij}, $i,j = 1, 2, 3, 4$, in

$$\begin{pmatrix} s_0^{\mathrm{s}} \\ s_1^{\mathrm{s}} \\ s_2^{\mathrm{s}} \\ s_3^{\mathrm{s}} \end{pmatrix} = \begin{pmatrix} M_{11} & M_{12} & M_{13} & M_{14} \\ M_{21} & M_{22} & M_{23} & M_{24} \\ M_{31} & M_{32} & M_{33} & M_{34} \\ M_{41} & M_{42} & M_{43} & M_{44} \end{pmatrix} \begin{pmatrix} s_0^{\mathrm{i}} \\ s_1^{\mathrm{i}} \\ s_2^{\mathrm{i}} \\ s_3^{\mathrm{i}} \end{pmatrix}$$

Section 4.5 Plane Interface Case and Babinet's Principle

4.6 Show that the vector-valued integrals

$$\begin{cases} \boldsymbol{C}_1(\boldsymbol{r}) = \nabla \times \iint\limits_{z'=0} g(k, |\boldsymbol{r} - \boldsymbol{r}'|)\boldsymbol{A}_{xy}(\boldsymbol{r}')\, \mathrm{d}S' \\ \boldsymbol{C}_2(\boldsymbol{r}) = \nabla \times \left(\nabla \times \iint\limits_{z'=0} g(k, |\boldsymbol{r} - \boldsymbol{r}'|)\boldsymbol{A}_{xy}(\boldsymbol{r}')\, \mathrm{d}S' \right) \end{cases} \qquad z \neq 0$$

where $\boldsymbol{A}_{xy}$ is a vector satisfying $\hat{z} \cdot \boldsymbol{A}_{xy} = 0$, i.e., it is tangential to the x–y-plane, have the symmetries

$$\begin{cases} \hat{z} \times \boldsymbol{C}_1(\boldsymbol{r}) \text{ odd in } z \to -z \text{ (tangential components)} \\ \hat{z} \cdot \boldsymbol{C}_1(\boldsymbol{r}) \text{ even in } z \to -z \text{ (normal component)} \\ \hat{z} \times \boldsymbol{C}_2(\boldsymbol{r}) \text{ even in } z \to -z \text{ (tangential components)} \\ \hat{z} \cdot \boldsymbol{C}_2(\boldsymbol{r}) \text{ odd in } z \to -z \text{ (normal component)} \end{cases}$$

4.7 Show that the Poynting vectors of the scattered fields at mirror points are related as

$$\boldsymbol{P}(\boldsymbol{r}^*) = \mathbf{R} \cdot \boldsymbol{P}(\boldsymbol{r})$$

where $\boldsymbol{P} = \mathrm{Re}\,(\boldsymbol{E}_{\mathrm{s}} \times \boldsymbol{H}_{\mathrm{s}}^*)/2$, and where the point $\boldsymbol{r}^* = \mathbf{R} \cdot \boldsymbol{r}$ is the point $\boldsymbol{r}$ reflected in the plane $z = 0$, and the reflection dyadic $\mathbf{R} = \mathbf{I}_3 - 2\hat{z}\hat{z}$.

CHAPTER 5

Scattering in the Time Domain

The motivation behind some of the scattering results in Chapter 4, which were derived under the assumption of fixed (angular) frequency ω, has its origin in a time-domain setting of the scattering problem. This chapter treats the scattering problem as a time-domain problem. Following the logics of the analysis of material interaction in Chapter 1, where we started with a space–time description, this chapter should have been presented first, followed by the time-harmonic analysis in Chapter 4. However, the physically motivated time-domain analysis contains more complex mathematics than the time-harmonic analysis, and we follow the tradition and keep the present order of the chapters.[1]

All fields in this chapter vary as functions of both time t and the three spatial variables $\boldsymbol{r}$, and moreover, all field quantities are real valued in contrast to the fields in Chapter 4, which usually assume complex values. Outside the scattering region, the fields propagate in a non-dispersive material with constant permittivity ϵ and permeability μ. Both these quantities are positive real numbers (optical responses). The scatterer (obstacle), however, can consist of an arbitrary, dispersive bianisotropic material, with constitutive relations described in a time-domain setting, see Chapter 1. In contrast to the frequency formulation in Chapter 4, where we only allowed linear material properties of the scatterer, the properties of the scatterer in this chapter can be more general, e.g., show nonlinear field response. The formulation presented in this chapter is therefore very general.

The basic model equations of the fields in the time domain were derived in Section 2.6, and the integral representation of time-dependent fields was treated in Section 3.9. We exploit these results in this chapter, and we start by formulating the scattering problem for time-dependent fields. This result is found in Section 5.1. We continue by deriving an appropriate expression of energy balance for time-dependent fields in Section 5.2. The relations between the fields, the far field amplitude, and the scattering dyadic in the time-domain formulation and the corresponding time-harmonic results in Chapter 4 are presented in Section 5.3. The optical theorem in the time domain is analyzed in Section 5.4, and we end this chapter with Section 5.5, where we illustrate some of the consequences of the optical theorem in the time domain.

[1] It is possible to read Chapters 4 and 5 in reverse order with some minor gaps in the logics and an initially less detailed description and presentation of the scattering problem and its setup, which is presented in the introduction of Chapter 4.

5.1 The Scattering Problem

A scattering problem is a comparison between two different, but related, wave propagation situations. The underlying wave propagation problem deals with propagation of field quantities in space and time in a volume with undisturbed electromagnetic properties—the propagation of the incident fields, $\boldsymbol{E}_i(\boldsymbol{r}, t)$ and $\boldsymbol{H}_i(\boldsymbol{r}, t)$. The sources of the incident fields are located in a region V_i, see Figure 5.1. These fields propagate in a domain of constant, real-valued permittivity ϵ and permeability μ (positive real numbers) everywhere in space outside the volume V_i, see Section 5.1.1 for more details and analysis of the propagation of these fields.

The second wave propagation problem is identical to the first, except that in the vicinity of the origin there is a domain, V_s, with altered electromagnetic properties. These fields are denoted as $\boldsymbol{E}(\boldsymbol{r}, t)$ and $\boldsymbol{H}(\boldsymbol{r}, t)$. As above in Chapter 4, the domain that contains the scattering object, V_s, has a closed bounding surface S_s, and outward unit normal vector $\hat{\boldsymbol{\nu}}$, see Figure 5.1. The differences between the fields in this problem and the incident fields are denoted the scattered fields, $\boldsymbol{E}_s(\boldsymbol{r}, t) = \boldsymbol{E}(\boldsymbol{r}, t) - \boldsymbol{E}_i(\boldsymbol{r}, t)$ and $\boldsymbol{H}_s(\boldsymbol{r}, t) = \boldsymbol{H}(\boldsymbol{r}, t) - \boldsymbol{H}_i(\boldsymbol{r}, t)$, see Section 5.1.3.

5.1.1 The Incident Field

In the analysis of the scattering problem in the time-harmonic setting in Chapter 4, we frequently used the integral representations of the incident and the scattered fields. In this time-domain setting, we make no exception, and employ the integral representations of time-domain fields. These integral representations were developed and derived in Section 3.9 on page 180.

The incident field has sources that are located in a region V_i outside the scattering domain V_s. To be explicit, we assume that the domains V_i and V_s are separated by a plane. The normal to this plane defines a fixed direction, which we call the z-direction (the z-axis points to the right), oriented so that V_i is located to the left and V_s to the right of this plane, see e.g., $z = z_0$ in Figure 5.1. The definition of this separating plane is

$$z = z_0 = \min_{\boldsymbol{r}\in S_s} \hat{\boldsymbol{z}} \cdot \boldsymbol{r}$$

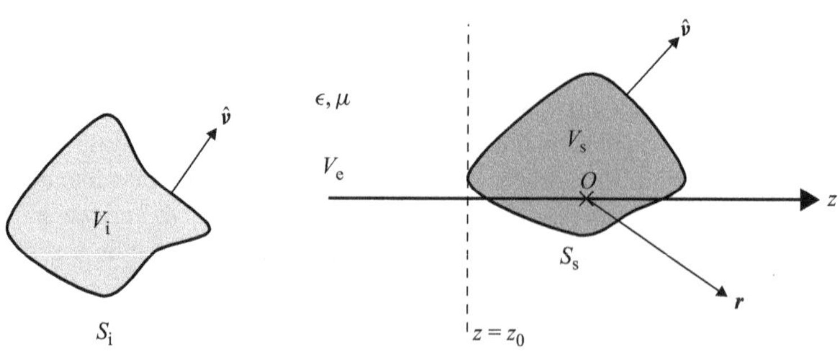

Figure 5.1 The scattering domain, V_s, with the origin O, and the domain V_i where the sources of the problem are located. The region to the left of the separating plane $z = z_0 = \min_{\boldsymbol{r}'\in S_s} \hat{\boldsymbol{z}} \cdot \boldsymbol{r}'$ contains all sources of the incident field, but no sources of the scattered field.

and in the volume V_s, the incident electric and magnetic fields satisfy

$$\begin{cases} \nabla \times \boldsymbol{E}_i(\boldsymbol{r},t) = -\mu_0\mu \dfrac{\partial \boldsymbol{H}_i(\boldsymbol{r},t)}{\partial t} \\ \nabla \times \boldsymbol{H}_i(\boldsymbol{r},t) = \epsilon_0\epsilon \dfrac{\partial \boldsymbol{E}_i(\boldsymbol{r},t)}{\partial t} \end{cases} \quad \boldsymbol{r} \in V_s$$

Equation (3.33) on page 183 gives an integral representation of the incident field in terms of its boundary values on the bounding surface S_s of V_s. We have

$$-\frac{1}{\epsilon_0\epsilon}\nabla \times \left(\nabla \times \iint_{S_s} \int_{-\infty}^{t_{ret}} \frac{(\hat{\boldsymbol{\nu}}(\boldsymbol{r}') \times \boldsymbol{H}_i(\boldsymbol{r}',t'))}{4\pi |\boldsymbol{r}-\boldsymbol{r}'|} \, dt' \, dS' \right)$$
$$- \nabla \times \iint_{S_s} \frac{\hat{\boldsymbol{\nu}}(\boldsymbol{r}') \times \boldsymbol{E}_i(\boldsymbol{r}',t_{ret})}{4\pi |\boldsymbol{r}-\boldsymbol{r}'|} \, dS' = \begin{cases} \boldsymbol{0}, & \boldsymbol{r} \text{ outside } S_s \\ \boldsymbol{E}_i(\boldsymbol{r},t), & \boldsymbol{r} \text{ inside } S_s \end{cases} \tag{5.1}$$

where the retarded time $t_{ret} = t - |\boldsymbol{r}-\boldsymbol{r}'|/c$. The wave speed in the undisturbed region is $c = c_0/\sqrt{\epsilon\mu}$.

We restrict ourselves in the analysis below to a special type of incident field—the plane wave. The analysis can be generalized to more general incident fields, but the plane wave suffices for the treatment in this chapter. The incident plane wave propagates in the positive z-direction, and we adopt the vector-valued function $\boldsymbol{E}_0(t)$ defined by

$$\boldsymbol{E}_i(\boldsymbol{r},t) = \boldsymbol{E}_0(t - \hat{\boldsymbol{z}} \cdot \boldsymbol{r}/c) = \boldsymbol{E}_0(t - z/c)$$

where $\boldsymbol{E}_0(t) \cdot \hat{\boldsymbol{z}} = 0$, since we require $\nabla \cdot \boldsymbol{E}_i(\boldsymbol{r},t) = 0$ and a sharp wave front of the wave, see below. The source region V_i is assumed to be located far enough to the left of the scatterer to make the plane wave approximation in the vicinity of the scatterer valid. Moreover, we assume the incident field has a sharp wave front, which reaches the origin at $t = t_0$. Causality implies that the incident field is quiescent to the right of the wave front, and we conclude, see Figure 5.2

$$\boldsymbol{E}_i(\boldsymbol{r},t_0) = \boldsymbol{E}_0(t_0 - z/c) = \boldsymbol{0}, \quad z > 0 \quad \Leftrightarrow \quad \boldsymbol{E}_0(t) = \boldsymbol{0}, \quad t < t_0$$

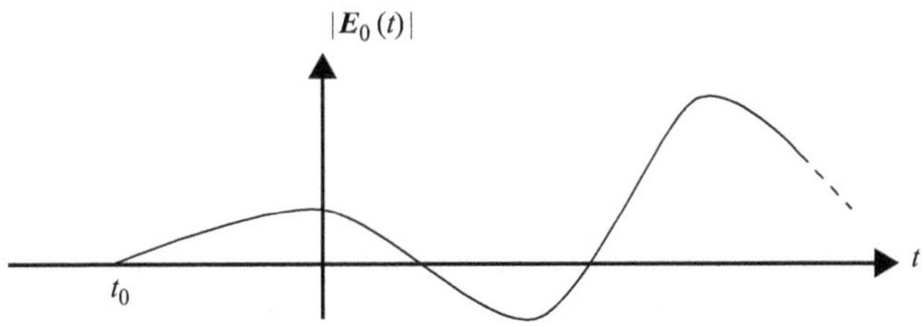

Figure 5.2 Typical amplitude of the incident field. The time t_0 denotes the time when the wave front reaches the origin. The field eventually goes to zero at large times.

Moreover, we assume the incident field vanishes in the limit $t \to \infty$ for fixed values of z. This assumption also covers the case where incident field has finite duration, i.e., $\boldsymbol{E}_0(t) = \boldsymbol{0}$ if $t \notin [t_0, t_1]$ for some constant time t_1.[2]

The plane wave relation in (1.72) on page 73 holds also in the time domain. This is seen from the Faraday law

$$\eta_0\eta \boldsymbol{H}_{\mathrm{i}}(\boldsymbol{r},t) = -c\int_{-\infty}^{t} \nabla \times \boldsymbol{E}_{\mathrm{i}}(\boldsymbol{r},t')\,\mathrm{d}t' = \hat{\boldsymbol{z}} \times \boldsymbol{E}_0(t - z/c)$$

as $\boldsymbol{E}_0(t - z/c) = \boldsymbol{0}$ as $t \to -\infty$. The Poynting vector of the incident is

$$\boldsymbol{P}_{\mathrm{i}}(z,t) = \boldsymbol{E}_{\mathrm{i}}(\boldsymbol{r},t) \times \boldsymbol{H}_{\mathrm{i}}(\boldsymbol{r},t) = \frac{1}{\eta_0\eta}\hat{\boldsymbol{z}}\,|\boldsymbol{E}_0(t - z/c)|^2$$

Moreover, if $\boldsymbol{E}_0 \in \boldsymbol{L}^2(\mathbb{R})$, see Appendix F.9.1 on page 681 for definitions on function spaces, the total incident energy per unit area is finite, i.e.,

$$P_{\mathrm{i}} = \int_{-\infty}^{\infty} \boldsymbol{P}_{\mathrm{i}}(z,t)\cdot\hat{\boldsymbol{z}}\,\mathrm{d}t = \frac{1}{\eta_0\eta}\int_{-\infty}^{\infty} |\boldsymbol{E}_0(t - z/c)|^2\,\mathrm{d}t = \frac{1}{\eta_0\eta}\int_{-\infty}^{\infty} |\boldsymbol{E}_0(t)|^2\,\mathrm{d}t < \infty$$

Note that the total incident energy per unit area is independent of z.

5.1.2 Scattering Problem—Formulation

The complement of the scattering region V_{s} (and V_{i}) is assumed to be connected[3] and denoted $V_{\mathrm{e}} = \mathbb{R}^3 \setminus \left(\overline{V_{\mathrm{s}}} \cup \overline{V_{\mathrm{i}}}\right)$. In this domain, the permittivity ϵ and the permeability μ are independent of the spatial variables, $\boldsymbol{r}$, i.e., a homogeneous, dispersion-free region. The wave speed in this region is $c = c_0/\sqrt{\epsilon\mu}$.

Let $z = z_0 = \min_{\boldsymbol{r}' \in S_{\mathrm{s}}} \hat{\boldsymbol{z}} \cdot \boldsymbol{r}'$ be the plane at which the wave front of the incident wave first hits the scatterer, see Figure 5.1. This first contact happens at $T = t_0 + z_0/c$. Due to causality, see Section 2.6.3, the total field is equal to the incident field everywhere in $\mathbb{R}^3$ for $t < T$, i.e.,

$$\boldsymbol{E}(\boldsymbol{r},t) = \boldsymbol{E}_{\mathrm{i}}(\boldsymbol{r},t), \quad \text{for all } \boldsymbol{r} \in \mathbb{R}^3, \quad t < T = t_0 + z_0/c$$

This motivates us to state the appropriate scattering problem in terms of the total electric and magnetic fields. The electric and magnetic fields satisfy the Maxwell equations in free space

$$\begin{cases} \nabla \times \boldsymbol{E}(\boldsymbol{r},t) = -\mu_0\mu\dfrac{\partial \boldsymbol{H}(\boldsymbol{r},t)}{\partial t} \\ \nabla \times \boldsymbol{H}(\boldsymbol{r},t) = \epsilon_0\epsilon\dfrac{\partial \boldsymbol{E}(\boldsymbol{r},t)}{\partial t} \end{cases} \quad \boldsymbol{r} \in V_{\mathrm{e}}, \quad t > T \tag{5.2}$$

[2]Under these assumptions, the time integral of the vector field $\boldsymbol{E}_0(t)$ must be zero, i.e., each component has zero time average. This conclusion can be justified by the observation that the plane wave has its sources far out to the left, and it is realized by a far field amplitude generated by these sources and evaluated in the vicinity of the origin. The use of the result in Problem 5.4 then completes the argument.

[3]The assumption on connectivity of the complement excludes voids in the scatterer. This assumption makes the analysis below simpler, but extensions are possible to formulate.

subject to the initial conditions

$$\begin{cases} \boldsymbol{E}(\boldsymbol{r},t)|_{t=T} = \boldsymbol{E}_0(T - z/c) \\ \eta_0\eta\, \boldsymbol{H}(\boldsymbol{r},t)|_{t=T} = \hat{\boldsymbol{z}} \times \boldsymbol{E}_0(T - z/c) \end{cases} \quad \boldsymbol{r} \in V_{\mathrm{e}} \quad \text{(initial conditions)} \tag{5.3}$$

The scattered fields are defined as the perturbations to the incident fields due to the presence of the scatterer, i.e.,

$$\boldsymbol{E}_{\mathrm{s}}(\boldsymbol{r},t) = \boldsymbol{E}(\boldsymbol{r},t) - \boldsymbol{E}_{\mathrm{i}}(\boldsymbol{r},t) \quad \text{and} \quad \boldsymbol{H}_{\mathrm{s}}(\boldsymbol{r},t) = \boldsymbol{H}(\boldsymbol{r},t) - \boldsymbol{H}_{\mathrm{i}}(\boldsymbol{r},t)$$

Moreover, due to causality and the specific form of the incident field, the total field $\boldsymbol{E}(\boldsymbol{r},t) = \boldsymbol{E}_{\mathrm{i}}(\boldsymbol{r},t)$ at all points in the set $V_{\mathrm{e}}(t)$, defined for $t > T$ as, see Figure 5.3

$$V_{\mathrm{e}}(t) = \left\{\boldsymbol{r} : \boldsymbol{r} \in V_{\mathrm{e}} \text{ and } c(t - t_0) < \hat{\boldsymbol{z}} \cdot \boldsymbol{r}' + |\boldsymbol{r} - \boldsymbol{r}'| \text{ for all } \boldsymbol{r}' \in S_{\mathrm{s}}\right\} \tag{5.4}$$

provided the optical response in the medium generates a phase (wave front) velocity that is smaller than the velocity c of the exterior region. This to ensure that the wave front of the interior field does not precede the wave front of the incident field. In most applications, this is the most common situation. One way to derive the set $V_{\mathrm{e}}(t)$ above is to rewrite the inequality as $|\boldsymbol{r} - \boldsymbol{r}'| > c(t - T) + z_0 - \hat{\boldsymbol{z}} \cdot \boldsymbol{r}'$ for all $\boldsymbol{r}' \in S_{\mathrm{s}}$. The set is characterized as the domain outside every sphere with center at the boundary S_{s}, say at $\boldsymbol{r}'$, and radius that touches the plane

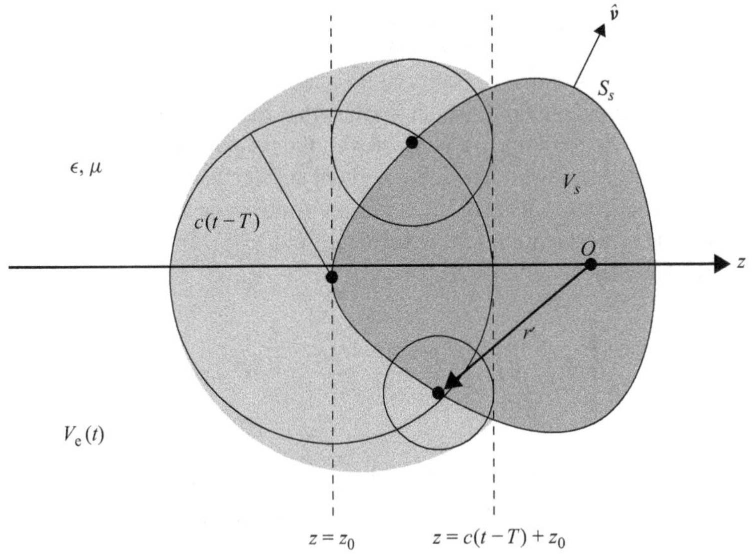

Figure 5.3 The set of points $V_{\mathrm{e}}(t)$ of undisturbed fields, $\boldsymbol{E}(\boldsymbol{r})$ and $\boldsymbol{H}(\boldsymbol{r})$, in space–time is the domain outside the light gray volume in V_{e}. The incident wave touches the scatterer at $z = z_0$ at time $t = T$, and at time $t > T$ the incident wave front has reached the plane $z = z_0 + c(t - T)$.

$c(t-T)+z_0$, i.e., the radius of the sphere is $c(t-T)+z_0-\hat{z}\cdot\boldsymbol{r}'$. The causality property can also be phrased as

$$\boldsymbol{E}_s(\boldsymbol{r},t)=\boldsymbol{0}, \quad \text{for all } \boldsymbol{r}\in V_e(t) \tag{5.5}$$

Note that this condition implies that at a fixed time t, the scattered electric and magnetic fields are zero sufficiently far away from the scatterer. This causality condition replaces the radiation condition we used for the time-harmonic results in Chapter 4.

5.1.3 Scattered Field

To analyze the problem in (5.2) and (5.3), we apply the integral representation (3.33) on page 183 to the scattered field. Since the scattered field has its (fictitious) sources inside the volume V_s, we get

$$\frac{1}{\epsilon_0\epsilon}\nabla\times\left(\nabla\times\iint_{S_s}\int_{-\infty}^{t_{\text{ret}}}\frac{\hat{\boldsymbol{\nu}}(\boldsymbol{r}')\times\boldsymbol{H}_s(\boldsymbol{r}',t')}{4\pi|\boldsymbol{r}-\boldsymbol{r}'|}\,dt'\,dS'\right)$$
$$+\nabla\times\iint_{S_s}\frac{\hat{\boldsymbol{\nu}}(\boldsymbol{r}')\times\boldsymbol{E}_s(\boldsymbol{r}',t_{\text{ret}})}{4\pi|\boldsymbol{r}-\boldsymbol{r}'|}\,dS'=\begin{cases}\boldsymbol{E}_s(\boldsymbol{r},t), & \boldsymbol{r}\text{ outside } S_s\\ \boldsymbol{0}, & \boldsymbol{r}\text{ inside } S_s\end{cases} \tag{5.6}$$

where, as above, the retarded time $t_{\text{ret}}=t-|\boldsymbol{r}-\boldsymbol{r}'|/c$. The integral representation above also contains a surface integral over a large spherical surface, S_R, that gives zero contribution, see Figure 4.3 on page 192. In fact, for time t in a fixed finite time interval, the contribution from this surface vanishes due to causality of the scattered field. This is a consequence of the specific form of the incident field, which implies that the scattered field $\boldsymbol{E}_s(\boldsymbol{r},t)=\boldsymbol{0}$ at all points in $V_e(t)$, see (5.5) and Figure 5.3. Moreover, the tangential components of the fields in the integrals are limits from the outside, but without any surface current density present at the surface, the tangential components of the fields are continuous at the surface S_s.

We combine the representation for the incident field, (5.1), and the integral representation for the scattered field above. The result is

$$\frac{1}{\epsilon_0\epsilon}\nabla\times\left(\nabla\times\iint_{S_s}\int_{-\infty}^{t_{\text{ret}}}\frac{\hat{\boldsymbol{\nu}}(\boldsymbol{r}')\times\boldsymbol{H}(\boldsymbol{r}',t')}{4\pi|\boldsymbol{r}-\boldsymbol{r}'|}\,dt'\,dS'\right)$$
$$+\nabla\times\iint_{S_s}\frac{\hat{\boldsymbol{\nu}}(\boldsymbol{r}')\times\boldsymbol{E}(\boldsymbol{r}',t_{\text{ret}})}{4\pi|\boldsymbol{r}-\boldsymbol{r}'|}\,dS'=\begin{cases}\boldsymbol{E}_s(\boldsymbol{r},t), & \boldsymbol{r}\text{ outside } S_s\\ -\boldsymbol{E}_i(\boldsymbol{r},t), & \boldsymbol{r}\text{ inside } S_s\end{cases} \tag{5.7}$$

These two expressions, (5.6) and (5.7), are the final expressions of the analysis in this chapter. Below, we explore the consequences of these integral representations in some detail. In particular, we find an expression of the form of the scattered electric and magnetic fields far away from the scatterer.

5.1.4 Far Field Amplitude

The structure of the time-harmonic far field electric and magnetic fields was characterized by the far field amplitude $\boldsymbol{F}(\hat{\boldsymbol{r}},\omega)$. Not surprisingly, this is also the case in the time-domain scattering analysis, but, whereas in the time-harmonic case, the far field amplitude characterizes the fields far away from the scatterer, the far field amplitude in the time-domain case characterizes the fields both far away from the scatterer and far away in the future. The details of this limit procedure are given in this section.

The field outside the scatterer has the form, see (5.7)

$$\begin{aligned}\boldsymbol{E}_s(\boldsymbol{r},t) &= \frac{1}{\epsilon_0\epsilon}\nabla\times\left(\nabla\times\iint_{S_s}\int_{-\infty}^{t_{\rm ret}}\frac{\hat{\boldsymbol{\nu}}(\boldsymbol{r}')\times\boldsymbol{H}(\boldsymbol{r}',t')}{4\pi\,|\boldsymbol{r}-\boldsymbol{r}'|}\,{\rm d}t'\,{\rm d}S'\right)\\ &\quad+\nabla\times\iint_{S_s}\frac{\hat{\boldsymbol{\nu}}(\boldsymbol{r}')\times\boldsymbol{E}(\boldsymbol{r}',t_{\rm ret})}{4\pi\,|\boldsymbol{r}-\boldsymbol{r}'|}\,{\rm d}S',\qquad \boldsymbol{r}\in V_e\end{aligned}$$

or in terms of the scattered fields alone, see (5.6)

$$\begin{aligned}\boldsymbol{E}_s(\boldsymbol{r},t) &= \frac{1}{\epsilon_0\epsilon}\nabla\times\left(\nabla\times\iint_{S_s}\int_{-\infty}^{t_{\rm ret}}\frac{\hat{\boldsymbol{\nu}}(\boldsymbol{r}')\times\boldsymbol{H}_s(\boldsymbol{r}',t')}{4\pi\,|\boldsymbol{r}-\boldsymbol{r}'|}\,{\rm d}t'\,{\rm d}S'\right)\\ &\quad+\nabla\times\iint_{S_s}\frac{\hat{\boldsymbol{\nu}}(\boldsymbol{r}')\times\boldsymbol{E}_s(\boldsymbol{r}',t_{\rm ret})}{4\pi\,|\boldsymbol{r}-\boldsymbol{r}'|}\,{\rm d}S',\qquad \boldsymbol{r}\in V_e\end{aligned}$$

Our next task is to prove that the dominant contribution of the electric field at large distance (and sufficiently large time t) from the scatterer decays to leading order as $1/r$. To see this, we evaluate the scattered electric field far away from the scatterer and keep only the dominant terms in an expansion in powers of $1/r$.

The scattered electric field depends on the spatial variable $\boldsymbol{r}$ in two ways—by the denominator $|\boldsymbol{r}-\boldsymbol{r}'|$ and by the retarded time variable $t_{\rm ret}$. At large distances from the scatterer, derivation w.r.t. the $\boldsymbol{r}$ variable in the denominator leads to factors that decay as

$$\nabla\frac{1}{|\boldsymbol{r}-\boldsymbol{r}'|} = -\frac{\boldsymbol{r}-\boldsymbol{r}'}{|\boldsymbol{r}-\boldsymbol{r}'|^3} = -\frac{\boldsymbol{r}}{r^3}\left(1+O(d/r)\right),\qquad \text{as } r\to\infty$$

where d is the diameter of the scatterer, i.e., $d=\max_{\boldsymbol{r},\boldsymbol{r}'\in V_s}|\boldsymbol{r}-\boldsymbol{r}'|$, see (4.3) on page 187, and since $|\boldsymbol{r}-\boldsymbol{r}'|=r-\hat{\boldsymbol{r}}\cdot\boldsymbol{r}'+O(d^2/r)=r\left(1+O(d/r)\right)$, see (4.6) on page 189. Similarly, differentiation of the retarded time implies a main contribution of the form

$$\nabla t_{\rm ret} = -\frac{\nabla|\boldsymbol{r}-\boldsymbol{r}'|}{c} = -\frac{\boldsymbol{r}-\boldsymbol{r}'}{c|\boldsymbol{r}-\boldsymbol{r}'|} = -\frac{\hat{\boldsymbol{r}}}{c}\left(1+O(d/r)\right)$$

At large distances from the scatterer, two canonical integrals appear. The first integral contains an integration w.r.t. time. This integral has the form

$$\nabla\times\iint_{S_s}\int_{-\infty}^{t_{\text{ret}}}\frac{\boldsymbol{A}(\boldsymbol{r}',t')}{4\pi|\boldsymbol{r}-\boldsymbol{r}'|}\,\mathrm{d}t'\,\mathrm{d}S' = -\frac{\hat{\boldsymbol{r}}}{4\pi rc}\times\iint_{S_s}\boldsymbol{A}\big(\boldsymbol{r}',t-(r-\hat{\boldsymbol{r}}\cdot\boldsymbol{r}')/c\big)\,\mathrm{d}S'\,(1+O(d/r))$$

The second canonical integral lacks time integration, and it has the following form:

$$\nabla\times\iint_{S_s}\frac{\boldsymbol{A}(\boldsymbol{r}',t_{\text{ret}})}{4\pi|\boldsymbol{r}-\boldsymbol{r}'|}\,\mathrm{d}S' = -\frac{\hat{\boldsymbol{r}}}{4\pi rc}\times\iint_{S_s}\frac{\partial\boldsymbol{A}(\boldsymbol{r}',t-(r-\hat{\boldsymbol{r}}\cdot\boldsymbol{r}')/c)}{\partial t}\,\mathrm{d}S'\,(1+O(d/r))$$

where $\boldsymbol{A}(\boldsymbol{r},t)$ is an arbitrary smooth vector field. To leading order, these canonical integrals lead to the following expansion of the scattered electric field at large distance from the scatterer:

$$\begin{aligned}\lim_{r\to\infty} r\boldsymbol{E}_s(\boldsymbol{r},t+r/c) &= \frac{\mu_0\mu}{4\pi}\hat{\boldsymbol{r}}\times\left(\hat{\boldsymbol{r}}\times\iint_{S_s}\hat{\boldsymbol{\nu}}(\boldsymbol{r}')\times\frac{\partial}{\partial t}\boldsymbol{H}(\boldsymbol{r}',t+\hat{\boldsymbol{r}}\cdot\boldsymbol{r}'/c)\,\mathrm{d}S'\right)\\ &\quad-\frac{1}{4\pi c}\hat{\boldsymbol{r}}\times\iint_{S_s}\hat{\boldsymbol{\nu}}(\boldsymbol{r}')\times\frac{\partial}{\partial t}\boldsymbol{E}(\boldsymbol{r}',t+\hat{\boldsymbol{r}}\cdot\boldsymbol{r}'/c)\,\mathrm{d}S'\end{aligned}$$

where we also have used, see Problem 5.1

$$\nabla\times\left(\hat{\boldsymbol{r}}\times\iint_{S_s}\frac{\boldsymbol{A}(\boldsymbol{r}',t_{\text{ret}})}{4\pi|\boldsymbol{r}-\boldsymbol{r}'|}\,\mathrm{d}S'\right) = -\frac{1}{4\pi rc}\hat{\boldsymbol{r}}\times\left(\hat{\boldsymbol{r}}\times\iint_{S_s}\frac{\partial}{\partial t}\boldsymbol{A}(\boldsymbol{r}',t_{\text{ret}})\,\mathrm{d}S'\right)(1+O(d/r)) \tag{5.8}$$

From the analysis presented above, we conclude that the scattered fields have the following limits as $r\to\infty$:

$$\begin{aligned}\lim_{\substack{r\to\infty\\ t-r/c=\tau}} r\boldsymbol{E}_s(\boldsymbol{r},t) &= \lim_{r\to\infty} r\boldsymbol{E}_s(r\hat{\boldsymbol{r}},\tau+r/c) = \boldsymbol{F}(\hat{\boldsymbol{r}},\tau)\\ \eta_0\eta\lim_{\substack{r\to\infty\\ t-r/c=\tau}} r\boldsymbol{H}_s(\boldsymbol{r},t) &= \hat{\boldsymbol{r}}\times\boldsymbol{F}(\hat{\boldsymbol{r}},\tau)\end{aligned}$$

where $\tau = t - r/c$ is constant during the limit process, and where the derivation of the expression of the magnetic field is referred to Problem 5.2. The dominant field contributions in the far zone, expressed in the far field amplitude, $\boldsymbol{F}(\hat{\boldsymbol{r}},\tau)$, are

$$\begin{cases}\boldsymbol{E}_s(\boldsymbol{r},t)\approx\dfrac{\boldsymbol{F}(\hat{\boldsymbol{r}},t-r/c)}{r}\\[2ex] \eta_0\eta\boldsymbol{H}_s(\boldsymbol{r},t)\approx\dfrac{\hat{\boldsymbol{r}}\times\boldsymbol{F}(\hat{\boldsymbol{r}},t-r/c)}{r}\end{cases} \tag{5.9}$$

The explicit expressions of the far field amplitude, $\boldsymbol{F}(\hat{\boldsymbol{r}}, \tau)$, are obtained from the surface integral representations above. In terms of the total tangential fields on the surface, we have

$$\boldsymbol{F}(\hat{\boldsymbol{r}}, \tau) = \frac{\mu_0 \mu}{4\pi} \frac{\partial}{\partial \tau} \hat{\boldsymbol{r}} \times \left(\hat{\boldsymbol{r}} \times \iint_{S_s} \hat{\boldsymbol{\nu}}(\boldsymbol{r}') \times \boldsymbol{H}(\boldsymbol{r}', \tau + \hat{\boldsymbol{r}} \cdot \boldsymbol{r}'/c) \, \mathrm{d}S' \right.$$
$$\left. - \frac{1}{\eta_0 \eta} \iint_{S_s} \hat{\boldsymbol{\nu}}(\boldsymbol{r}') \times \boldsymbol{E}(\boldsymbol{r}', \tau + \hat{\boldsymbol{r}} \cdot \boldsymbol{r}'/c) \, \mathrm{d}S' \right) \tag{5.10}$$

and in terms of the tangential scattered fields, we also have

$$\boldsymbol{F}(\hat{\boldsymbol{r}}, \tau) = \frac{\mu_0 \mu}{4\pi} \frac{\partial}{\partial \tau} \hat{\boldsymbol{r}} \times \left(\hat{\boldsymbol{r}} \times \iint_{S_s} \hat{\boldsymbol{\nu}}(\boldsymbol{r}') \times \boldsymbol{H}_s(\boldsymbol{r}', \tau + \hat{\boldsymbol{r}} \cdot \boldsymbol{r}'/c) \, \mathrm{d}S' \right.$$
$$\left. - \frac{1}{\eta_0 \eta} \iint_{S_s} \hat{\boldsymbol{\nu}}(\boldsymbol{r}') \times \boldsymbol{E}_s(\boldsymbol{r}', \tau + \hat{\boldsymbol{r}} \cdot \boldsymbol{r}'/c) \, \mathrm{d}S' \right) \tag{5.11}$$

Either representations of the far field amplitude $\boldsymbol{F}(\hat{\boldsymbol{r}}, \tau)$ satisfy

$$\hat{\boldsymbol{r}} \cdot \boldsymbol{F}(\hat{\boldsymbol{r}}, \tau) = 0, \qquad \text{for all } \tau \in \mathbb{R}, \quad \text{and } \hat{\boldsymbol{r}} \in \Omega$$

where Ω denotes the unit sphere, and $\boldsymbol{F}(\hat{\boldsymbol{r}}, \tau)$ is zero for sufficiently large negative times τ. More precisely, by causality, the far field amplitude $\boldsymbol{F}(\hat{\boldsymbol{r}}, \tau)$ is identically zero in a specific time interval for a given direction $\hat{\boldsymbol{r}}$. Define the time interval $Z(\hat{\boldsymbol{r}})$ by

$$Z(\hat{\boldsymbol{r}}) = \left\{ \tau : c\tau < ct_0 + \boldsymbol{r}' \cdot (\hat{\boldsymbol{z}} - \hat{\boldsymbol{r}}) \quad \text{for all } \boldsymbol{r}' \in S_s \right\} \tag{5.12}$$

For given $\hat{\boldsymbol{r}}$, this set determines the time interval $Z(\hat{\boldsymbol{r}})$ where $\boldsymbol{F}(\hat{\boldsymbol{r}}, \tau) = \boldsymbol{0}$. In fact, the scattered electric field is zero in $V_e(t)$ defined in (5.4). In the far zone, the dominant contribution in the criterion in $V_e(t)$ is

$$c(t - t_0) < \hat{\boldsymbol{z}} \cdot \boldsymbol{r}' + r - \hat{\boldsymbol{r}} \cdot \boldsymbol{r}'$$

which is identical to the criterion in $Z(\hat{\boldsymbol{r}})$. Specifically, in the forward direction $\hat{\boldsymbol{r}} = \hat{\boldsymbol{z}}$, $Z(\hat{\boldsymbol{z}}) = (-\infty, t_0]$, and in the backward direction, $\hat{\boldsymbol{r}} = -\hat{\boldsymbol{z}}$, the interval is $Z(-\hat{\boldsymbol{z}}) = (-\infty, t_0 + 2z_0/c]$.

We end this section by connecting the results in the time domain with the previously obtained results in the frequency domain in Chapter 4. The far field amplitudes in the time and the frequency domains are related by

$$\begin{cases} \boldsymbol{F}(\hat{\boldsymbol{r}}, \omega) = \displaystyle\int_{-\infty}^{\infty} \boldsymbol{F}(\hat{\boldsymbol{r}}, \tau) \, \mathrm{e}^{\mathrm{i}\omega\tau} \, \mathrm{d}\tau \\ \boldsymbol{F}(\hat{\boldsymbol{r}}, \tau) = \dfrac{1}{2\pi} \displaystyle\int_{-\infty}^{\infty} \boldsymbol{F}(\hat{\boldsymbol{r}}, \omega) \, \mathrm{e}^{-\mathrm{i}\omega\tau} \, \mathrm{d}\omega \end{cases}$$

More details on this connection between the time-domain and the time-harmonic results are found in Section 5.3.

5.1.5 Scattering Dyadic

The far field amplitude, $\boldsymbol{F}(\hat{\boldsymbol{r}}, \tau)$, and the incident field, $\boldsymbol{E}_0(t)$, depend linearly on each other by a convolution integral in time, i.e.,

$$\boldsymbol{F}(\hat{\boldsymbol{r}}, \tau) = \int_{-\infty}^{\infty} \mathbf{S}(\hat{\boldsymbol{r}}, \hat{\boldsymbol{k}}_\mathrm{i}, \tau - t) \cdot \boldsymbol{E}_0(t)\, \mathrm{d}t$$

This is a consequence of linearity and time invariance of the scattering problem[4]—consult the analysis of the constitutive relations in Section 1.2 for a similar treatment. The dyadic-valued kernel $\mathbf{S}(\hat{\boldsymbol{r}}, \hat{\boldsymbol{k}}_\mathrm{i}, t)$ is the scattering dyadic in the time domain.[5] Its connection to the scattering dyadic $\mathbf{S}(\hat{\boldsymbol{r}}, \hat{\boldsymbol{k}}_\mathrm{i})$ in Section 4.3 is given in Section 5.3.

We now investigate the consequences of causality on the scattering dyadic evaluated in the forward direction. Non-forward direction can also be treated, but with increasing complexity. Some details are found in Reference 201.

From the assumptions above, $\boldsymbol{E}_0(t) = \mathbf{0}$ for $t < t_0$. In the forward direction, $\hat{\boldsymbol{r}} = \hat{\boldsymbol{k}}_\mathrm{i}$, causality implies that, $\boldsymbol{F}(\hat{\boldsymbol{k}}_\mathrm{i}, \tau) = \mathbf{0}$ for $\tau \in (-\infty, t_0]$, see the definition of $Z(\hat{\boldsymbol{r}})$ in (5.12). Explicitly, the fields in the forward direction are

$$\boldsymbol{E}_\mathrm{i}(r\hat{\boldsymbol{z}}, t) = \boldsymbol{E}_0(t - r/c) \quad \text{and} \quad \boldsymbol{E}_\mathrm{s}(r\hat{\boldsymbol{z}}, t) \approx \frac{\boldsymbol{F}(\hat{\boldsymbol{z}}, t - r/c)}{r}$$

Causality implies that

$$\boldsymbol{F}(\hat{\boldsymbol{k}}_\mathrm{i}, \tau) = \int_{-\infty}^{\tau} \mathbf{S}(\hat{\boldsymbol{k}}_\mathrm{i}, \hat{\boldsymbol{k}}_\mathrm{i}, \tau - t) \cdot \boldsymbol{E}_0(t)\, \mathrm{d}t \tag{5.13}$$

since the scattered field cannot precede the incident field in the forward direction (note the same time delay in both the forward scattering amplitude $\boldsymbol{F}$ and the incident field $\boldsymbol{E}_0$). In other words, in the forward direction $\hat{\boldsymbol{r}} = \hat{\boldsymbol{k}}_\mathrm{i}$, the scattering dyadic is causal, i.e.,

$$\mathbf{S}(\hat{\boldsymbol{k}}_\mathrm{i}, \hat{\boldsymbol{k}}_\mathrm{i}, t) = \mathbf{0}, \qquad t < 0 \tag{5.14}$$

This result is concluded under the assumption that the material inside V_s does not have an optical response that produces a wave front velocity inside V_s that exceeds the speed c in V_e. This is the usual situation if the surrounding material is air or vacuum, since the wave speed in vacuum is an upper limit of the wave speeds that can exist in nature.

---

[4]This connection to the time-harmonic case is not valid if the constitutive relations of the material in the scatterer do not satisfy the requirements of the constitutive relations made in Section 1.2, i.e., nonlinear materials are now excluded.

[5]To make the notation in the time-domain case more similar to the notation in the time-harmonic case, we denote the incident direction of the plane wave by $\hat{\boldsymbol{k}}_\mathrm{i}$.

5.2 Energy Balance in the Time Domain

In a source-free domain V, where the permittivity ϵ and the permeability μ are independent of the spatial variables, $\boldsymbol{r}$, i.e., a homogeneous, non-dispersive material, the Poynting's theorem, (1.15) on page 9, reads

$$\iint_S [\boldsymbol{E}(\boldsymbol{r},t) \times \boldsymbol{H}(\boldsymbol{r},t)] \cdot \hat{\boldsymbol{\nu}}(\boldsymbol{r})\, \mathrm{d}S = -\frac{\epsilon_0\epsilon}{2}\frac{\mathrm{d}}{\mathrm{d}t} \iiint_V \left[|\eta_0\eta \boldsymbol{H}(\boldsymbol{r},t)|^2 + |\boldsymbol{E}(\boldsymbol{r},t)|^2\right] \mathrm{d}v$$

where S is the bounding surface of V, and $\hat{\boldsymbol{\nu}}$ is the outward unit normal vector. This identity is a simple consequence of the fact that in an isotropic material the fields satisfy $\boldsymbol{H} \cdot \partial_t \boldsymbol{B} = \mu_0\mu\partial_t (\boldsymbol{H} \cdot \boldsymbol{H})/2$ and $\boldsymbol{E} \cdot \partial_t \boldsymbol{D} = \epsilon_0\epsilon\partial_t (\boldsymbol{E} \cdot \boldsymbol{E})/2$.

We integrate this identity in time from $\tilde{t}$ to t. The result is

$$\int_{\tilde{t}}^{t} \iint_S \left[\boldsymbol{E}(\boldsymbol{r},t') \times \boldsymbol{H}(\boldsymbol{r},t')\right] \cdot \hat{\boldsymbol{\nu}}(\boldsymbol{r})\, \mathrm{d}S\, \mathrm{d}t' = W(\tilde{t}) - W(t) \tag{5.15}$$

where the stored electromagnetic energy in the domain V at time t is denoted

$$W(t) = \frac{\epsilon_0\epsilon}{2} \iiint_V \left[|\eta_0\eta \boldsymbol{H}(\boldsymbol{r},t)|^2 + |\boldsymbol{E}(\boldsymbol{r},t)|^2\right] \mathrm{d}v$$

If the surface integral on the left-hand side is positive for all time t, which is the case if energy is radiated out of S, i.e.,

$$\iint_S [\boldsymbol{E}(\boldsymbol{r},t) \times \boldsymbol{H}(\boldsymbol{r},t)] \cdot \hat{\boldsymbol{\nu}}(\boldsymbol{r})\, \mathrm{d}S > 0, \qquad \text{for all } t$$

then the energy balance shows that the stored energy $W(t)$ is a decreasing quantity as a function of time. This statement is nothing else but conservation of energy. If energy is radiating out of the bounding surface S, the total stored energy in V has to decrease with time.

5.3 Connection to the Time-Harmonic Results

We are now ready to relate the results obtained in this chapter to the results in the previous chapter for time-harmonic fields.

The time-domain quantities are related to the corresponding frequency-domain quantities by a temporal Fourier transform as we have already exploited in Section 1.3. Explicitly, we have for the incident electric field

$$\begin{cases} \boldsymbol{E}_{\mathrm{i}}(\boldsymbol{r},\omega) = \displaystyle\int_{-\infty}^{\infty} \boldsymbol{E}_{\mathrm{i}}(\boldsymbol{r},\tau)\, \mathrm{e}^{\mathrm{i}\omega t}\, \mathrm{d}t \\ \boldsymbol{E}_{\mathrm{i}}(\boldsymbol{r},t) = \dfrac{1}{2\pi} \displaystyle\int_{-\infty}^{\infty} \boldsymbol{E}_{\mathrm{i}}(\boldsymbol{r},\omega)\, \mathrm{e}^{-\mathrm{i}\omega t}\, \mathrm{d}\omega \end{cases}$$

and for the scattered electric field

$$\begin{cases} \boldsymbol{E}_{\mathrm{s}}(\boldsymbol{r},\omega) = \displaystyle\int_{-\infty}^{\infty} \boldsymbol{E}_{\mathrm{s}}(\boldsymbol{r},\tau)\,\mathrm{e}^{\mathrm{i}\omega t}\,\mathrm{d}t \\ \boldsymbol{E}_{\mathrm{s}}(\boldsymbol{r},t) = \displaystyle\frac{1}{2\pi}\int_{-\infty}^{\infty} \boldsymbol{E}_{\mathrm{s}}(\boldsymbol{r},\omega)\,\mathrm{e}^{-\mathrm{i}\omega t}\,\mathrm{d}\omega \end{cases}$$

Similarly, the far field amplitude and the scattering dyadic are related by a temporal Fourier transform pair

$$\begin{cases} \boldsymbol{F}(\hat{\boldsymbol{r}},\omega) = \displaystyle\int_{-\infty}^{\infty} \boldsymbol{F}(\hat{\boldsymbol{r}},\tau)\,\mathrm{e}^{\mathrm{i}\omega\tau}\,\mathrm{d}\tau \\ \boldsymbol{F}(\hat{\boldsymbol{r}},\tau) = \displaystyle\frac{1}{2\pi}\int_{-\infty}^{\infty} \boldsymbol{F}(\hat{\boldsymbol{r}},\omega)\,\mathrm{e}^{-\mathrm{i}\omega\tau}\,\mathrm{d}\omega \end{cases}$$

and

$$\begin{cases} \mathbf{S}(\hat{\boldsymbol{r}},\hat{\boldsymbol{k}}_{\mathrm{i}},\omega) = \displaystyle\int_{-\infty}^{\infty} \mathbf{S}(\hat{\boldsymbol{r}},\hat{\boldsymbol{k}}_{\mathrm{i}},\tau)\,\mathrm{e}^{\mathrm{i}\omega\tau}\,\mathrm{d}\tau \\ \mathbf{S}(\hat{\boldsymbol{r}},\hat{\boldsymbol{k}}_{\mathrm{i}},\tau) = \displaystyle\frac{1}{2\pi}\int_{-\infty}^{\infty} \mathbf{S}(\hat{\boldsymbol{r}},\hat{\boldsymbol{k}}_{\mathrm{i}},\omega)\,\mathrm{e}^{-\mathrm{i}\omega\tau}\,\mathrm{d}\omega \end{cases}$$

All time-domain quantities are real valued and therefore the Fourier coefficients for negative and positive frequencies are related. Specifically, we have for the Fourier transform of the incident and scattered electric fields, the far field amplitude, and the scattering dyadic, see (1.40) on page 39

$$\begin{cases} \boldsymbol{E}_{\mathrm{i}}(\boldsymbol{r},-\omega) = \boldsymbol{E}_{\mathrm{i}}^{*}(\boldsymbol{r},\omega) \\ \boldsymbol{E}_{\mathrm{s}}(\boldsymbol{r},-\omega) = \boldsymbol{E}_{\mathrm{s}}^{*}(\boldsymbol{r},\omega) \end{cases} \qquad \begin{cases} \boldsymbol{F}(\hat{\boldsymbol{r}},-\omega) = \boldsymbol{F}^{*}(\hat{\boldsymbol{r}},\omega) \\ \mathbf{S}(\hat{\boldsymbol{r}},\hat{\boldsymbol{k}}_{\mathrm{i}},-\omega) = \mathbf{S}^{*}(\hat{\boldsymbol{r}},\hat{\boldsymbol{k}}_{\mathrm{i}},\omega) \end{cases}$$

The time-domain formulation is a perfect tool to guarantee that the mathematical model satisfies the important causality property. In particular, the scattering dyadic is causal in the forward direction $\hat{\boldsymbol{r}} = \hat{\boldsymbol{k}}_{\mathrm{i}}$, see (5.14), and therefore it has an analytical extension into the upper complex η-plane. This is a consequence of Paley–Wiener theorem, see Theorem D.1 on page 640. Following the notation in Section 1.3.2, the components of the scattering dyadic satisfy the following relation in the upper complex frequency plane:

$$S_{ij}(-\eta^{*}) = S_{ij}^{*}(\eta), \qquad i,j = 1,2,3 \tag{5.16}$$

where we have adopted $\eta = \omega + \mathrm{i}\varsigma$, where ω and ς are real, and $\varsigma \geq 0$.

5.4 Optical Theorem

We are now prepared to exploit the energy conservation law in Section 5.2, which together with causality arguments leads to the optical theorem in the time domain. This optical theorem is a time-domain counterpart of the optical theorem for time-harmonic fields in Section 4.4.

However, notice that the energy concept is different for time-harmonic fields (time averages over one period) and for time-domain fields (total energy for all times). The presentation in this section follows closely the investigations in References 51, 52, 115, 116.

We start by defining the total absorbed energy, W_a, and the total scattered energy, W_s, by the scatterer from time T, before which all fields are quiescent, to infinity.[6] These definitions are:

$$\begin{cases} W_a = -\int_T^\infty \iint_{S_s} [\boldsymbol{E}(\boldsymbol{r},t) \times \boldsymbol{H}(\boldsymbol{r},t)] \cdot \hat{\boldsymbol{\nu}}(\boldsymbol{r})\, dS\, dt \\ W_s = \int_T^\infty \iint_{S_s} [\boldsymbol{E}_s(\boldsymbol{r},t) \times \boldsymbol{H}_s(\boldsymbol{r},t)] \cdot \hat{\boldsymbol{\nu}}(\boldsymbol{r})\, dS\, dt \end{cases} \tag{5.17}$$

The total absorbed energy contains the total fields on the surface S_s. If we rewrite the total electric and magnetic fields in terms of the incident and the scattered fields, we obtain

$$\begin{aligned} W_a &= -\int_T^\infty \iint_{S_s} [(\boldsymbol{E}_i(\boldsymbol{r},t) + \boldsymbol{E}_s(\boldsymbol{r},t)) \times (\boldsymbol{H}_i(\boldsymbol{r},t) + \boldsymbol{H}_s(\boldsymbol{r},t))] \cdot \hat{\boldsymbol{\nu}}(\boldsymbol{r})\, dS\, dt \\ &= -W_s - \int_T^\infty \iint_{S_s} [\boldsymbol{E}_i(\boldsymbol{r},t) \times \boldsymbol{H}_i(\boldsymbol{r},t)] \cdot \hat{\boldsymbol{\nu}}(\boldsymbol{r})\, dS\, dt \\ &\quad - \int_T^\infty \iint_{S_s} [\boldsymbol{E}_i(\boldsymbol{r},t) \times \boldsymbol{H}_s(\boldsymbol{r},t) + \boldsymbol{E}_s(\boldsymbol{r},t) \times \boldsymbol{H}_i(\boldsymbol{r},t)] \cdot \hat{\boldsymbol{\nu}}(\boldsymbol{r})\, dS\, dt \end{aligned}$$

The two remaining surface integrals are now analyzed. The integral over the incident field is always zero under the assumptions made on the form of the incident field. In fact, use (5.15)

$$\begin{aligned} &\int_T^\infty \iint_{S_s} [\boldsymbol{E}_i(\boldsymbol{r},t) \times \boldsymbol{H}_i(\boldsymbol{r},t)] \cdot \hat{\boldsymbol{\nu}}(\boldsymbol{r})\, dS\, dt \\ &\quad = \frac{1}{\eta_0\eta} \int_T^\infty \iint_{S_s} \left[\boldsymbol{E}_0(t' - z/c) \times \left(\hat{\boldsymbol{z}} \times \boldsymbol{E}_0(t' - z/c)\right)\right] \cdot \hat{\boldsymbol{\nu}}(\boldsymbol{r})\, dS\, dt' \\ &\quad = W_i(T) - W_i(\infty) \end{aligned}$$

where

$$W_i(t) = \epsilon_0\epsilon \iiint_{V_s} |\boldsymbol{E}_0(t - z/c)|^2\, dv$$

[6] If the incident pulse is of finite duration, the upper limit in the time integration is finite by causality. The examples in Section 5.5 focus on this case.

However, both $W_i(T)$ and $W_i(\infty)$ are zero, due to the assumptions made on the incident field. Thus, we can simplify the identity above, and write the sum of the total absorbed and scattered energy as

$$W_a + W_s = -\int_T^\infty \iint_{S_s} \Big[\boldsymbol{E}_0(t - z/c) \times \boldsymbol{H}_s(\boldsymbol{r}, t) + \frac{1}{\eta_0\eta} \boldsymbol{E}_s(\boldsymbol{r}, t) \times \big(\hat{\boldsymbol{z}} \times \boldsymbol{E}_0(t - z/c)\big) \Big] \cdot \hat{\boldsymbol{\nu}}(\boldsymbol{r})\, \mathrm{d}S\, \mathrm{d}t$$

of by changing the time variable

$$W_a + W_s = -\int_{t_0}^\infty \iint_{S_s} \Big[\boldsymbol{E}_0(t) \times \boldsymbol{H}_s(\boldsymbol{r}, t + z/c) + \frac{1}{\eta_0\eta} \boldsymbol{E}_s(\boldsymbol{r}, t + z/c) \times \big(\hat{\boldsymbol{z}} \times \boldsymbol{E}_0(t)\big) \Big] \cdot \hat{\boldsymbol{\nu}}(\boldsymbol{r})\, \mathrm{d}S\, \mathrm{d}t \tag{5.18}$$

where $t_0 = T - \min_{\boldsymbol{r}\in S_s} \hat{\boldsymbol{z}} \cdot \boldsymbol{r}/c = T - z_0/c$.

A surface integral expression of the far field amplitude of the scattered field was developed in Section 5.1.4. In particular, in the forward direction $\hat{\boldsymbol{r}} = \hat{\boldsymbol{z}}$, the far field amplitude, $\boldsymbol{F}(\hat{\boldsymbol{z}}, \tau)$, is, see (5.11)

$$\boldsymbol{F}(\hat{\boldsymbol{z}}, \tau) = \frac{\mu_0\mu}{4\pi} \frac{\partial}{\partial\tau} \hat{\boldsymbol{z}} \times \left(\hat{\boldsymbol{z}} \times \iint_{S_s} \hat{\boldsymbol{\nu}}(\boldsymbol{r}) \times \boldsymbol{H}_s(\boldsymbol{r}, \tau + z/c)\, \mathrm{d}S - \frac{1}{\eta_0\eta} \iint_{S_s} \hat{\boldsymbol{\nu}}(\boldsymbol{r}) \times \boldsymbol{E}_s(\boldsymbol{r}, \tau + z/c)\, \mathrm{d}S \right)$$

Use this expression to calculate the following integral:

$$\begin{aligned}
\frac{4\pi}{\mu_0\mu} \boldsymbol{E}_0(\tau) \cdot \int_{t_0}^{\tau} \boldsymbol{F}(\hat{\boldsymbol{z}}, \tau')\, \mathrm{d}\tau'
&= \boldsymbol{E}_0(\tau) \cdot \left\{ \hat{\boldsymbol{z}} \times \left(\hat{\boldsymbol{z}} \times \iint_{S_s} \hat{\boldsymbol{\nu}}(\boldsymbol{r}) \times \boldsymbol{H}_s(\boldsymbol{r}, \tau + z/c)\, \mathrm{d}S - \frac{1}{\eta_0\eta} \iint_{S_s} \hat{\boldsymbol{\nu}}(\boldsymbol{r}) \times \boldsymbol{E}_s(\boldsymbol{r}, \tau + z/c)\, \mathrm{d}S \right) \right\} \\
&= -\boldsymbol{E}_0(\tau) \cdot \iint_{S_s} \hat{\boldsymbol{\nu}}(\boldsymbol{r}) \times \boldsymbol{H}_s(\boldsymbol{r}, \tau + z/c)\, \mathrm{d}S \\
&\quad + \frac{1}{\eta_0\eta} \iint_{S_s} \big(\hat{\boldsymbol{\nu}}(\boldsymbol{r}) \times \boldsymbol{E}_s(\boldsymbol{r}, \tau + z/c)\big) \cdot \big(\hat{\boldsymbol{z}} \times \boldsymbol{E}_0(\tau)\big)\, \mathrm{d}S
\end{aligned}$$

since $\hat{\boldsymbol{z}} \cdot \boldsymbol{E}_0(\tau) = 0$ and $\boldsymbol{F}(\hat{\boldsymbol{z}}, t_0) = \boldsymbol{0}$ by causality (see above $Z(\hat{\boldsymbol{z}}) = (-\infty, t_0]$).

By cyclic permutations of the vectors in this equality, we identify this expression in (5.18), and finally we obtain the time-domain version of the optical theorem

$$W_{\mathrm{T}} = W_{\mathrm{a}} + W_{\mathrm{s}} = -\frac{4\pi}{\mu_0 \mu} \int_{t_0}^{\infty} \boldsymbol{E}_0(\tau) \cdot \left(\int_{t_0}^{\tau} \boldsymbol{F}(\hat{z}, \tau')\, \mathrm{d}\tau' \right) \mathrm{d}\tau \tag{5.19}$$

where the total energy W_{T} denotes the sum of the total absorbed and scattered energy by the scatterer. If the scatterer is lossless, the total energy reduces to the total scattered energy W_{s}.

We observe that the total energy is uniquely determined by the far field amplitude in the forward direction during the time interval $[t_0, \infty)$. If the incident field has a finite time duration then the total energy is uniquely determined by the far field amplitude in the forward direction during the time interval $[t_0, t_1]$, where t_1 is the time when the trailing edge of the incident field leaves the origin. In the next section, we exploit this conclusion in some detail.

5.5 Some Applications of the Optical Theorem

We conclude this chapter by deriving and illustrating some interesting results from the optical theorem in the time domain, which was obtained in Section 5.4. In particular, we are going to calculate the total energy by the scatterer in a very efficient way by the use of (5.19). We refer the reader to References 115, 116 for more details and additional interesting examples and results.

In the remainder of this chapter, we assume that the incident plane wave has a sharp wave front that hits the scatterer at time $T = t_0 + z_0/c$, and arrives at the origin at t_0. Moreover, the incident wave has finite length, i.e., $\boldsymbol{E}_0(t)$ is zero outside some time interval $t_0 \le t \le t_1$, where t_1 denotes the time when the trailing edge leaves the origin. We have

$$\boldsymbol{E}_0(t) = \boldsymbol{0}, \qquad t \notin [t_0, t_1]$$

The optical theorem in (5.19) states that only the far field amplitude during the time interval $[t_0, t_1]$ determines the total energy of the scattering problem. As a consequence, if the time duration of the incident plane wave is short enough, no multiple scattering effects can influence the total energy. This is a direct consequence of causality, and to make this clear, we illustrate this statement with a few examples in the two sections below. Several additional results of the optical theorem in the time domain are found in Reference 116.

5.5.1 Several Scatterers

Two scatterers illustrate the fact that the interaction effects have no impact on the total energy, if the time duration of the incident plane wave is short enough. The materials of the scatterers are not essential—it is the separation of the scatterers in space that matters. In fact, the scatterers can consist of arbitrary nonlinear materials.

Denote the total energy of each individual scatterer, in the absence of the other scatterer, by W_{T1} and W_{T2}, respectively. In general, the total energy of the two scatterers configuration is $W_{\mathrm{T}} \neq W_{\mathrm{T1}} + W_{\mathrm{T2}}$. However, if the incident pulse is short enough, no interaction effects influence the total energy due to the finite wave propagation between the scatterers. This is illustrated in Figure 5.4, where the path $a \to b \to c \to d$ denotes the shortest wave path between the scatterers (including also a possible path with interaction of scatterer 2 first).

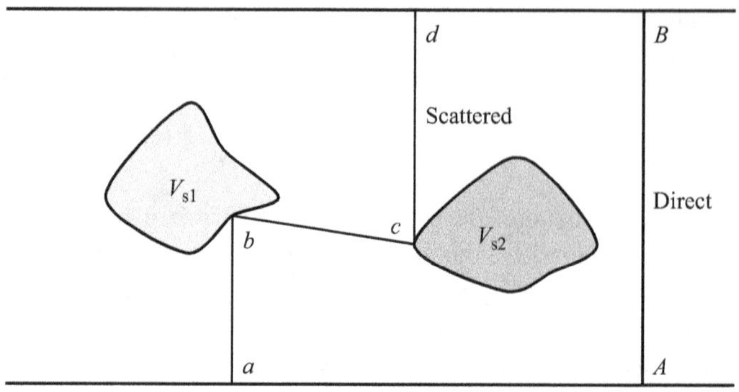

Figure 5.4 Multiple scattering by two separate scatterers. The path $a \to b \to c \to d$ denotes the shortest wave path between the scatterers (including also a possible path with interaction of scatterer 2 first). The direct path is denoted $A \to B$.

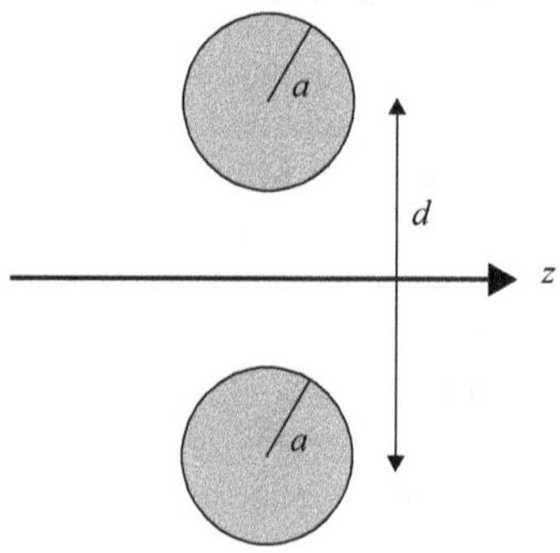

Figure 5.5 The geometry of the two spheres in Example 5.1.

If the difference between this path and the direct path $A \to B$ in Figure 5.4 is longer than the duration of the incident field, there is equality in the expression above, i.e.,

$$W_{\mathrm{T}} = W_{\mathrm{T}1} + W_{\mathrm{T}2}$$

since no interaction or multiple scattering effects can occur during the time duration of the incident field $[t_0, t_1]$. Only if duration of the incident wave is longer than the difference path $a \to b \to c \to d$ and $A \to B$, multiple scattering effects have any influence on the total energy. The next example quantifies this analysis by a simple example with two homogeneous, dielectric spheres.

Example 5.1
Two identical, homogeneous spheres of radius a are separated by a distance d. The incident wave impinges perpendicularly to the line that connects the two spheres, see Figure 5.5.

The incident plane wave in this example is assumed to have the form, see the inset in Figure 5.8:

$$\boldsymbol{E}_0(t) = \hat{\boldsymbol{x}} \frac{T - t}{t_0} \mathrm{e}^{-(T-t)^2/t_0^2}$$

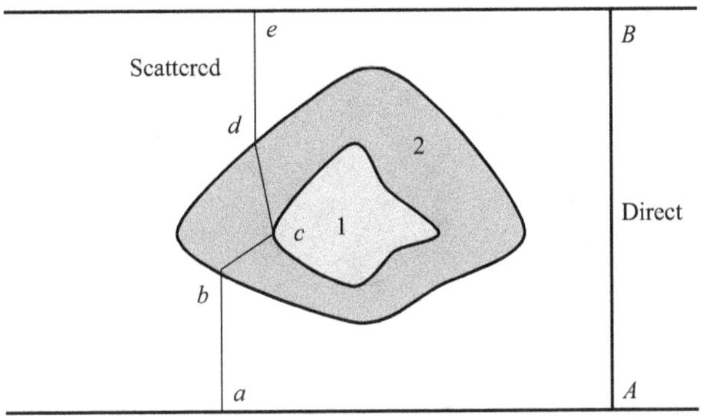

Figure 5.6 The geometry of a layered object consisting of two layers 1 and 2, and the two interaction paths, the direct path $A \to B$ and the shortest interaction path $a \to b \to c \to d \to e$.

Then, with the data, $T = 5$ ns and $t_0 = 1$ ns, the incident field is approximately zero outside the interval $[2, 8]$ ns. The time duration of the incident field (6 ns) corresponds to a distance $6 \times 10^{-9} c_0 \approx 1.8$ m in vacuum, which implies that if the spheres with radii $a = 1$ m are separated by a distance $d > 4$ m, the multiple scattering effects are delayed by time factor larger than 6 ns and no multiple effects have any impact on the total energy. With material parameters of the spheres, $\epsilon_1/\epsilon = 4$ and $\mu_1/\mu = 1$, the total energy for one sphere is $W_{T1} = W_{T2} = 7.1$ pNm, see Example 5.2, and therefore, the total energy of the two spheres is $W_T = 14.2$ pNm. ■

5.5.2 Layered Scatterers

Another geometry that also illustrates this causality phenomenon clearly is the layered object (homogeneous layers or not). The geometry of a layered scatterer is depicted in Figure 5.6, where the path $a \to b \to c \to d \to e$ denotes the shortest interaction path between the layers. Again, if the duration of the incident field is shorter than the time difference between the paths $a \to b \to c \to d \to e$ and $A \to B$, the effect of the inner layer does not show up in the total energy. We quantify this observation by a layered sphere in the example below.

Example 5.2

We illustrate the conclusions made above by three different scattering objects. The first scatterer is a homogeneous sphere of radius b. The second and third objects are layered scatterers—more precisely, homogeneous, dielectric shells with the same outer radius b, but with different inner radii a. The geometry and the material parameters are depicted in Figure 5.7.

The incident plane wave in this example is assumed to have the form, see the inset in Figure 5.8:

$$\boldsymbol{E}_0(t) = \hat{\boldsymbol{x}} \frac{T - t}{t_0} \mathrm{e}^{-(T-t)^2/t_0^2}$$

With the data in the figure ($T = 5$ ns and $t_0 = 1$ ns), the incident field is approximately zero outside the interval $[2, 8]$ ns. The time duration of the incident field (6 ns) corresponds to a

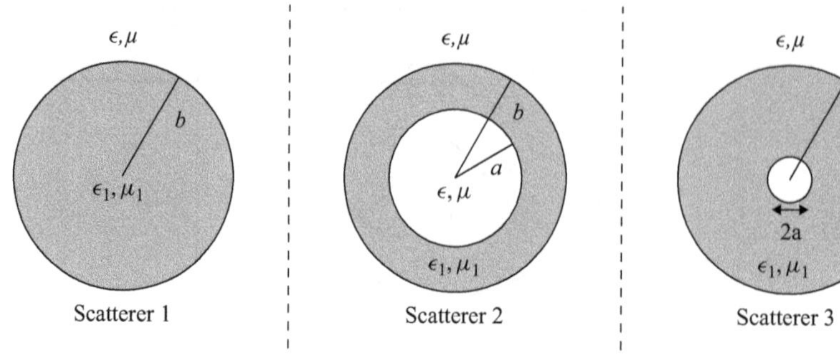

Figure 5.7 The geometry in Example 5.2 with one homogeneous sphere (left) and two different dielectric shells (middle and right). The explicit values used in this example are: $\epsilon_1/\epsilon = 4$, $\mu_1/\mu = 1$, $a = 0.6\,\mathrm{m}$ (middle), $a = 0.2\,\mathrm{m}$ (right), and $b = 1\,\mathrm{m}$.

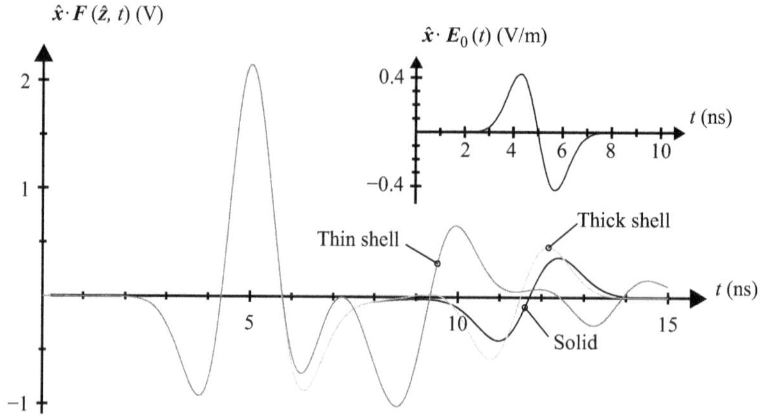

Figure 5.8 The far field amplitude as a function of time for two dielectric shells and a solid sphere in the forward direction for data in Example 5.2. Two of the curves (solid and thick shell) coincide in the interval [0, 8] ns, and the curve of the thin shell starts to deviate already at 6 ns. The inset shows the incident field with data: $T = 5$ ns and $t_0 = 1$ ns.

distance $6 \times 10^{-9}c \approx 0.9\,\mathrm{m}$ in the material of the sphere ($c = c_0/2$ in the layer). This field has the Fourier transform

$$\boldsymbol{E}_0(\omega) = -\hat{\boldsymbol{x}}\frac{\mathrm{i}\sqrt{\pi}t_0^2\omega}{2}\mathrm{e}^{\mathrm{i}T\omega - t_0^2\omega^2/4}$$

The far field amplitude in the forward direction $\boldsymbol{F}(\hat{z}, t)$ then is (use (5.13), the Parseval's formula in Appendix D.1, results in Chapter 8, and (1.41) on page 39)

$$\begin{aligned}\boldsymbol{F}(\hat{z}, t) &= \int_{-\infty}^{t} \mathbf{S}(\hat{z}, \hat{z}, t - t') \cdot \boldsymbol{E}_0(t')\,\mathrm{d}t' \\ &= \frac{1}{2\pi}\int_{-\infty}^{\infty} \mathbf{S}(\hat{z}, \hat{z}, \omega) \cdot \boldsymbol{E}_0(\omega)\mathrm{e}^{-\mathrm{i}\omega t}\,\mathrm{d}\omega = \frac{1}{\pi}\,\mathrm{Re}\int_{0}^{\infty} \mathbf{S}(\hat{z}, \hat{z}, \omega) \cdot \boldsymbol{E}_0(\omega)\mathrm{e}^{-\mathrm{i}\omega t}\,\mathrm{d}\omega\end{aligned}$$

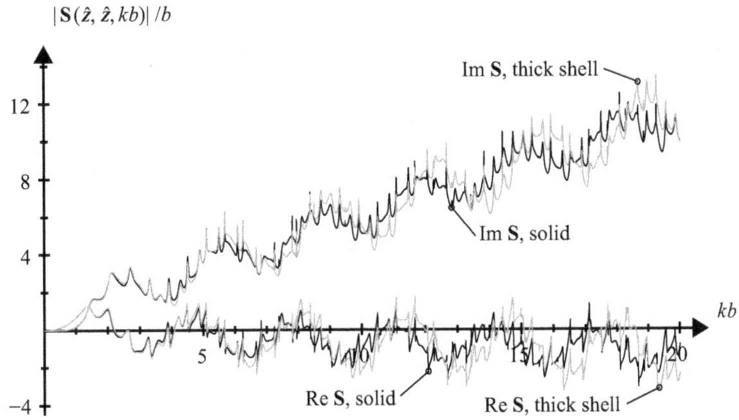

Figure 5.9 The scattering dyadic in the forward direction (solid sphere and thick shell) for data in Example 5.2. The scattering dyadic is scaled by the outer radius b.

The result is illustrated in Figure 5.8. It is hard to see any causality consequences in the frequency-domain results in Figure 5.9, except for coincident curves at low frequencies. However, in the time domain, the effect is obvious, see Figure 5.8.

Since the far field amplitude of the solid sphere and the thick shell coincide during the support of the incident field [2, 8] ns, the total energy of these two scattering examples is identical. Numerical calculation shows that both have $W_s = 7.1$ pNm. However, the far field amplitude of the thin shell starts to deviate from the other two curves at 6 ns, which is in the support of the incident field. Consequently, the total energy of the thin shell is slightly different from the other two. In fact, we have $W_s = 7.3$ pNm for the thin shell. These energy calculations are in agreement with a direct computation of the scattered energy using the result of Problem 5.5. ■

Problems for Chapter 5

Section 5.1 The Scattering Problem

5.1 Show the estimate (5.8), i.e.,

$$\nabla \times (\hat{\boldsymbol{r}} \times \boldsymbol{A}(t_{\text{ret}})) = -\frac{1}{c}\hat{\boldsymbol{r}} \times \left(\hat{\boldsymbol{r}} \times \frac{\mathrm{d}}{\mathrm{d}t}\boldsymbol{A}(t_{\text{ret}})\right)(1 + O(1/r))$$

where $\boldsymbol{A}(t)$ is a smooth vector field and $t_{\text{ret}} = t - |\boldsymbol{r} - \boldsymbol{r}'|/c$.

5.2 Verify that the leading contribution to the scattered magnetic field satisfies

$$\eta_0 \eta \lim_{\substack{r\to\infty \\ t-r/c=\tau}} r\boldsymbol{H}_s(\boldsymbol{r}, t) = \hat{\boldsymbol{r}} \times \boldsymbol{F}(\hat{\boldsymbol{r}}, \tau)$$

where the far field amplitude $\boldsymbol{F}\left(\hat{\boldsymbol{r}}, \tau\right)$ is given by (5.10) or (5.11).

5.3 The Jefimenko equation for the electric field was derived in (2.54) on page 132. Assume the fictitious current density and charge density of the scattered field are denoted $\boldsymbol{J}_{\rm s}(\boldsymbol{r}, t)$ and $\rho_{\rm s}(\boldsymbol{r}, t)$, respectively, i.e.,

$$\boldsymbol{E}_{\rm s}(\boldsymbol{r}, t) = -\mu_0\mu \iiint_{V_{\rm s}} \frac{\partial_t \boldsymbol{J}_{\rm s}\left(\boldsymbol{r}', t_{\rm ret}\right)}{4\pi \left|\boldsymbol{r} - \boldsymbol{r}'\right|} \, {\rm d}v'$$

$$+ \frac{1}{\epsilon_0\epsilon} \iiint_{V_{\rm s}} \left\{ \frac{\rho_{\rm s}\left(\boldsymbol{r}', t_{\rm ret}\right)\left(\boldsymbol{r} - \boldsymbol{r}'\right)}{4\pi \left|\boldsymbol{r} - \boldsymbol{r}'\right|^3} + \frac{\partial_t \rho_{\rm s}\left(\boldsymbol{r}', t_{\rm ret}\right)\left(\boldsymbol{r} - \boldsymbol{r}'\right)}{4\pi c \left|\boldsymbol{r} - \boldsymbol{r}'\right|^2} \right\} {\rm d}v'$$

Find an expression of the far field amplitude in the time domain $\boldsymbol{F}(\hat{\boldsymbol{r}}, \tau)$ using this expression.

5.4 Show that

$$\int_{-\infty}^{\infty} \boldsymbol{F}(\hat{\boldsymbol{r}}, \tau) \, {\rm d}\tau = \boldsymbol{0}, \qquad \text{for all } \hat{\boldsymbol{r}} \in \Omega$$

under the assumptions made on the incident pulse in this chapter. Ω denotes the unit sphere.

Section 5.4 Optical Theorem

5.5 Show that an alternative expression of the total scattered energy in (5.17) is

$$W_{\rm s} = \frac{1}{\eta_0\eta} \iint_{\Omega} \int_{\tau \notin Z(\hat{\boldsymbol{r}})} \left|\boldsymbol{F}(\hat{\boldsymbol{r}}, \tau)\right|^2 \, {\rm d}\tau \, {\rm d}\Omega$$

where Ω is the unit sphere and ${\rm d}\Omega$ is the surface element on Ω and the interval $Z(\hat{\boldsymbol{r}})$ is defined in (5.12).

CHAPTER 6

Approximations and Applications

The scattering problem that we formulated and analyzed in Chapter 4 was general and without any major approximations. To find explicit solutions to a scattering problem, a boundary value problem or an integral equation has to be solved. These solutions are often very hard to find indeed. It is therefore highly motivated to find approximate solutions to the problem. In this chapter, we introduce some of the most popular approximations used to solve scattering problems. An overview of the different approximations and their domain of usage is given in Figure 6.1, where a is the typical length scale of the scatterer.

These approximate solutions are of great importance in many applications, e.g., wave propagation in the ionosphere and in optics. We first address approximations that are pertinent if the wavelength is long compared to characteristic lengths of the scatterer. This is accomplished in Section 6.1. In a short section, Section 6.2, we treat approximations for weak scatterers. Approximations that are adequate for short wavelengths are then introduced in Section 6.3. The use of sum rules is exploited on scattering problems in Section 6.4, and some of the appropriate approximations that involve many scatterers are analyzed in Section 6.5.

6.1 Long Wavelength Approximation

The scatterer is, as usual, assumed to be a passive scatterer, and it is located in the volume V_s. The complementary volume, i.e., $V_e = \mathbb{R}^3 \setminus \overline{V_s}$, is assumed to be simply connected. Moreover, in this section, the wavelength λ is large compared to the radius $a = \max_{\boldsymbol{r},\boldsymbol{r}' \in V_s} |\boldsymbol{r} - \boldsymbol{r}'|/2$ of the scatterer. This assumption leads to useful approximations of the far field amplitude in the long wavelength limit—the Rayleigh[1] limit or Rayleigh scattering.[2]

6.1.1 Near Field Approximation

Exterior to the scatterer ($\boldsymbol{r} \in V_e$), the total electric and magnetic fields, $\boldsymbol{E}$ and $\boldsymbol{H}$, are assumed to have a series expansion in powers of the exterior wave number k (the exterior material

[1] John William Strutt, 3rd Baron Rayleigh (1842–1919), English physicist.

[2] Rayleigh scattering explains a number of scattering phenomena, e.g., the blue color of the sky and the red sun at sunset. This subject and its history are presented in Reference 300.

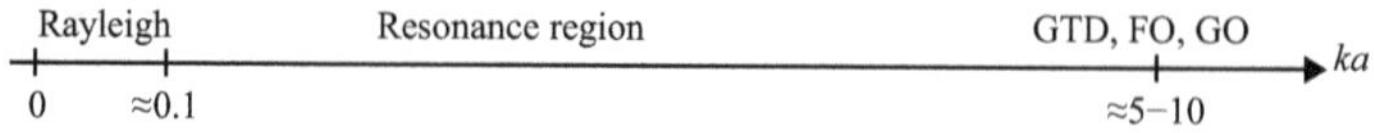

Figure 6.1 Different frequency domains and the appropriate approximations. The typical length scale of the scatterer is a and $k = 2\pi/\lambda$.

parameters, ϵ and μ, are assumed to be independent of frequency). More precisely,[3]

$$\begin{cases} \boldsymbol{E}(\boldsymbol{r}) = \sum_{n=0}^{\infty} (\mathrm{i}k)^n \boldsymbol{E}^n(\boldsymbol{r}) \\ \boldsymbol{H}(\boldsymbol{r}) = \sum_{n=0}^{\infty} (\mathrm{i}k)^n \boldsymbol{H}^n(\boldsymbol{r}) \end{cases} \quad \boldsymbol{r} \in V_{\mathrm{e}} \tag{6.1}$$

where the expansion coefficients $\boldsymbol{E}^n$ and $\boldsymbol{H}^n$, $n = 0, 1, 2, \ldots$, are independent of the wave number k. At this stage, there is no guarantee that these series converge, but the hope is that the first terms in such a series give fair representations of the fields at low frequencies.

The electric and magnetic fields satisfy the Maxwell equations in the region exterior to the scatterer, i.e.,

$$\begin{cases} \nabla \times \boldsymbol{E} = \mathrm{i}k\eta_0\eta\boldsymbol{H} \\ \eta_0\eta\nabla \times \boldsymbol{H} = -\mathrm{i}k\boldsymbol{E} \end{cases} \quad \boldsymbol{r} \in V_{\mathrm{e}}$$

If we insert the expansion (6.1) into the Maxwell equations and identify terms of equal powers in k, the lowest order terms have to satisfy

$$\begin{cases} \nabla \times \boldsymbol{E}^0 = \boldsymbol{0} \\ \nabla \times \boldsymbol{H}^0 = \boldsymbol{0} \end{cases} \quad \boldsymbol{r} \in V_{\mathrm{e}}$$

which imply that the zero-order terms can be written as potential fields (static-like fields), i.e.,

$$\begin{cases} \boldsymbol{E}^0(\boldsymbol{r}) = -\nabla\phi(\boldsymbol{r}) \\ \eta_0\eta\boldsymbol{H}^0(\boldsymbol{r}) = -\nabla\psi(\boldsymbol{r}) \end{cases} \quad \boldsymbol{r} \in V_{\mathrm{e}} \tag{6.2}$$

The higher order terms satisfy

$$\begin{cases} \nabla \times \boldsymbol{E}^{n+1} = \eta_0\eta\boldsymbol{H}^n \\ \eta_0\eta\nabla \times \boldsymbol{H}^{n+1} = -\boldsymbol{E}^n \end{cases} \quad \boldsymbol{r} \in V_{\mathrm{e}}, \quad n = 0, 1, 2, \ldots \tag{6.3}$$

and

$$\begin{cases} \nabla \cdot \boldsymbol{E}^n = 0 \\ \nabla \cdot \boldsymbol{H}^n = 0 \end{cases} \quad \boldsymbol{r} \in V_{\mathrm{e}}, \quad n = 0, 1, 2, \ldots$$

[3]The derivation presented in this section follows closely [47, 126].

The $n = 0$ case and (6.2) imply that the potentials ϕ and ψ satisfy

$$\begin{cases} \nabla^2 \phi(\boldsymbol{r}) = 0 \\ \nabla^2 \psi(\boldsymbol{r}) = 0 \end{cases} \quad \boldsymbol{r} \in V_{\mathrm{e}} \tag{6.4}$$

Our scattering problem consists of finding the scattered electric field $\boldsymbol{E}_{\mathrm{s}}(\boldsymbol{r})$ for an incident plane wave. To solve this problem, the incident plane wave is expanded in powers of k. We get[4]

$$\begin{cases} \boldsymbol{E}_{\mathrm{i}}(\boldsymbol{r}) = \boldsymbol{E}_0 \mathrm{e}^{\mathrm{i}k\hat{\boldsymbol{k}}_{\mathrm{i}}\cdot\boldsymbol{r}} = \boldsymbol{E}_0 \sum_{n=0}^{\infty} (\mathrm{i}k)^n \dfrac{(\hat{\boldsymbol{k}}_{\mathrm{i}}\cdot\boldsymbol{r})^n}{n!} \\ \eta_0\eta \boldsymbol{H}_{\mathrm{i}}(\boldsymbol{r}) = \hat{\boldsymbol{k}}_{\mathrm{i}} \times \boldsymbol{E}_0 \mathrm{e}^{\mathrm{i}k\hat{\boldsymbol{k}}_{\mathrm{i}}\cdot\boldsymbol{r}} = \hat{\boldsymbol{k}}_{\mathrm{i}} \times \boldsymbol{E}_0 \sum_{n=0}^{\infty} (\mathrm{i}k)^n \dfrac{(\hat{\boldsymbol{k}}_{\mathrm{i}}\cdot\boldsymbol{r})^n}{n!} \end{cases}$$

The first term in this expansion is

$$\begin{cases} \lim_{k\to 0} \boldsymbol{E}_{\mathrm{i}}(\boldsymbol{r}) = \boldsymbol{E}_0 = \nabla\left(\boldsymbol{E}_0\cdot\boldsymbol{r} + c_1\right) \\ \eta_0\eta \lim_{k\to 0} \boldsymbol{H}_{\mathrm{i}}(\boldsymbol{r}) = \hat{\boldsymbol{k}}_{\mathrm{i}} \times \boldsymbol{E}_0 = \nabla\left((\hat{\boldsymbol{k}}_{\mathrm{i}} \times \boldsymbol{E}_0)\cdot\boldsymbol{r} + c_2\right) \end{cases}$$

where c_1 and c_2 are arbitrary constants.

6.1.2 Far Field Amplitude

The main goal in this section is to find the dominant contribution to the far field amplitude $\boldsymbol{F}$ defined by scattered electric field $\boldsymbol{E}_{\mathrm{s}}(\boldsymbol{r})$ in the far field zone

$$\boldsymbol{E}_{\mathrm{s}}(\boldsymbol{r}) \approx \frac{\mathrm{e}^{\mathrm{i}kr}}{r}\boldsymbol{F}(\hat{\boldsymbol{r}}) \quad \text{as } r \to \infty$$

We start with the general expression of the far field amplitude in (4.18) on page 198:

$$\boldsymbol{F}(\hat{\boldsymbol{r}}) = \frac{\mathrm{i}k}{4\pi}\hat{\boldsymbol{r}} \times \iint_{S_{\mathrm{s}}} \left[\hat{\boldsymbol{\nu}}(\boldsymbol{r}') \times \boldsymbol{E}(\boldsymbol{r}') - \eta_0\eta\hat{\boldsymbol{r}} \times \left(\hat{\boldsymbol{\nu}}(\boldsymbol{r}') \times \boldsymbol{H}(\boldsymbol{r}')\right)\right]\mathrm{e}^{-\mathrm{i}k\hat{\boldsymbol{r}}\cdot\boldsymbol{r}'}\,\mathrm{d}S'$$

The exponential in the integrand is expanded in a power series in k, i.e.,

$$\mathrm{e}^{-\mathrm{i}k\hat{\boldsymbol{r}}\cdot\boldsymbol{r}'} = \sum_{l=0}^{\infty} (\mathrm{i}k)^l \frac{\left(-\hat{\boldsymbol{r}}\cdot\boldsymbol{r}'\right)^l}{l!}$$

This gives

$$\boldsymbol{F}(\hat{\boldsymbol{r}}) = \frac{\mathrm{i}k}{4\pi}\sum_{l=0}^{\infty} (\mathrm{i}k)^l\,\hat{\boldsymbol{r}} \times \iint_{S_{\mathrm{s}}} \left[\hat{\boldsymbol{\nu}} \times \boldsymbol{E}(\boldsymbol{r}') - \eta_0\eta\hat{\boldsymbol{r}} \times \left(\hat{\boldsymbol{\nu}} \times \boldsymbol{H}(\boldsymbol{r}')\right)\right]\frac{\left(-\hat{\boldsymbol{r}}\cdot\boldsymbol{r}'\right)^l}{l!}\,\mathrm{d}S'$$

[4] Notice the difference between the lowest order field, $\boldsymbol{E}^0$, and the complex polarization vector of the incident plane wave, $\boldsymbol{E}_0$.

which, using (6.1), can be written as a power series expansion as

$$\begin{aligned}
\boldsymbol{F}(\hat{\boldsymbol{r}}) &= \frac{\mathrm{i}k}{4\pi}\sum_{l=0}^{\infty}\sum_{n=0}^{\infty}(\mathrm{i}k)^{n+l}\,\hat{\boldsymbol{r}}\times\iint_{S_s}\left[\hat{\boldsymbol{\nu}}\times\boldsymbol{E}^n(\boldsymbol{r}')-\eta_0\eta\hat{\boldsymbol{r}}\times\left(\hat{\boldsymbol{\nu}}\times\boldsymbol{H}^n(\boldsymbol{r}')\right)\right]\frac{\left(-\hat{\boldsymbol{r}}\cdot\boldsymbol{r}'\right)^l}{l!}\,\mathrm{d}S'\\
&= \frac{\mathrm{i}k}{4\pi}\sum_{m=0}^{\infty}(\mathrm{i}k)^m\sum_{n=0}^{m}\hat{\boldsymbol{r}}\times\iint_{S_s}\left[\hat{\boldsymbol{\nu}}\times\boldsymbol{E}^n(\boldsymbol{r}')-\eta_0\eta\hat{\boldsymbol{r}}\times\left(\hat{\boldsymbol{\nu}}\times\boldsymbol{H}^n(\boldsymbol{r}')\right)\right]\frac{\left(-\hat{\boldsymbol{r}}\cdot\boldsymbol{r}'\right)^{m-n}}{(m-n)!}\,\mathrm{d}S'
\end{aligned}$$

or

$$\boldsymbol{F}(\hat{\boldsymbol{r}}) = \sum_{m=0}^{\infty}(\mathrm{i}k)^m\,\boldsymbol{F}^m(\hat{\boldsymbol{r}}) = \sum_{m=1}^{\infty}(\mathrm{i}k)^m\,\boldsymbol{F}^m(\hat{\boldsymbol{r}})$$

where

$$\boldsymbol{F}^m(\hat{\boldsymbol{r}}) = \frac{\mathrm{i}k}{4\pi}\sum_{n=0}^{m}\hat{\boldsymbol{r}}\times\iint_{S_s}\left[\hat{\boldsymbol{\nu}}\times\boldsymbol{E}^n(\boldsymbol{r}')-\eta_0\eta\hat{\boldsymbol{r}}\times\left(\hat{\boldsymbol{\nu}}\times\boldsymbol{H}^n(\boldsymbol{r}')\right)\right]\frac{\left(-\hat{\boldsymbol{r}}\cdot\boldsymbol{r}'\right)^{m-n}}{(m-n)!}\,\mathrm{d}S'$$

since the first terms

$$\begin{cases}
\displaystyle\iint_{S_s}\left[\hat{\boldsymbol{\nu}}(\boldsymbol{r}')\times\nabla\phi(\boldsymbol{r}')\right]\mathrm{d}S' = \int_C \phi(\boldsymbol{r}')\,\mathrm{d}\boldsymbol{r}' = \boldsymbol{0}\\
\displaystyle\iint_{S_s}\left[\hat{\boldsymbol{\nu}}(\boldsymbol{r}')\times\nabla\psi(\boldsymbol{r}')\right]\mathrm{d}S' = \int_C \psi(\boldsymbol{r}')\,\mathrm{d}\boldsymbol{r}' = \boldsymbol{0}
\end{cases}$$

due to Stokes' analogous theorem on a closed surface S_s.

In the long wavelength limit, we expect that only the first few terms in the power series expansion in k contribute. The leading contribution to the far field amplitude in the long wavelength limit therefore is

$$\begin{aligned}
\boldsymbol{F}(\hat{\boldsymbol{r}}) &\approx \mathrm{i}k\boldsymbol{F}^1(\hat{\boldsymbol{r}})\\
&= -\frac{k^2}{4\pi}\hat{\boldsymbol{r}}\times\iint_{S_s}\left[\hat{\boldsymbol{\nu}}\times\boldsymbol{E}^0-\eta_0\eta\hat{\boldsymbol{r}}\times\left(\hat{\boldsymbol{\nu}}\times\boldsymbol{H}^0\right)\right]\left(-\hat{\boldsymbol{r}}\cdot\boldsymbol{r}'\right)\mathrm{d}S'\\
&\quad -\frac{k^2}{4\pi}\hat{\boldsymbol{r}}\times\iint_{S_s}\left[\hat{\boldsymbol{\nu}}\times\boldsymbol{E}^1-\eta_0\eta\hat{\boldsymbol{r}}\times\left(\hat{\boldsymbol{\nu}}\times\boldsymbol{H}^1\right)\right]\mathrm{d}S'
\end{aligned}$$

This representation is expressed in the lowest order field amplitudes, $\boldsymbol{E}^0$ and $\boldsymbol{H}^0$, together with the next-order terms, $\boldsymbol{E}^1$ and $\boldsymbol{H}^1$.

Our ambition now is to rewrite this expression in the lowest order terms $\boldsymbol{E}^0$ and $\boldsymbol{H}^0$ only. To accomplish this, use (6.3) and the Stokes' theorem in following vector identity:

$$\begin{aligned}\iint_{S_s} x_i'\hat{\boldsymbol{\nu}} \cdot \eta_0\eta\boldsymbol{H}^0 \, \mathrm{d}S' &= \iint_{S_s} x_i'\hat{\boldsymbol{\nu}} \cdot \left(\nabla' \times \boldsymbol{E}^1\right) \mathrm{d}S' \\ &= \iint_{S_s} \hat{\boldsymbol{\nu}} \cdot \left(\nabla' \times \left(x_i'\boldsymbol{E}^1\right)\right) \mathrm{d}S' + \hat{\boldsymbol{e}}_i \cdot \iint_{S_s} \left(\hat{\boldsymbol{\nu}} \times \boldsymbol{E}^1\right) \mathrm{d}S' \\ &= \hat{\boldsymbol{e}}_i \cdot \iint_{S_s} \left(\hat{\boldsymbol{\nu}} \times \boldsymbol{E}^1\right) \mathrm{d}S'\end{aligned}$$

Similarly,

$$\begin{aligned}\iint_{S_s} x_i'\hat{\boldsymbol{\nu}} \cdot \boldsymbol{E}^0 \, \mathrm{d}S' &= -\iint_{S_s} x_i'\hat{\boldsymbol{\nu}} \cdot \left(\nabla' \times \eta_0\eta\boldsymbol{H}^1\right) \mathrm{d}S' \\ &= -\iint_{S_s} \hat{\boldsymbol{\nu}} \cdot \left(\nabla' \times \left(x_i'\eta_0\eta\boldsymbol{H}^1\right)\right) \mathrm{d}S' - \hat{\boldsymbol{e}}_i \cdot \iint_{S_s} \left(\hat{\boldsymbol{\nu}} \times \eta_0\eta\boldsymbol{H}^1\right) \mathrm{d}S' \\ &= -\hat{\boldsymbol{e}}_i \cdot \iint_{S_s} \left(\hat{\boldsymbol{\nu}} \times \eta_0\eta\boldsymbol{H}^1\right) \mathrm{d}S'\end{aligned}$$

The lowest order far field contribution can therefore be written in a form that involves the lowest order fields, $\boldsymbol{E}^0$ and $\boldsymbol{H}^0$, only

$$\begin{aligned}\boldsymbol{F}(\hat{\boldsymbol{r}}) = &\frac{k^2}{4\pi}\hat{\boldsymbol{r}} \times \iint_{S_s} \left[\hat{\boldsymbol{\nu}} \times \boldsymbol{E}^0 - \eta_0\eta\hat{\boldsymbol{r}} \times \left(\hat{\boldsymbol{\nu}} \times \boldsymbol{H}^0\right)\right] \left(\hat{\boldsymbol{r}} \cdot \boldsymbol{r}'\right) \mathrm{d}S' \\ &- \frac{k^2}{4\pi}\hat{\boldsymbol{r}} \times \iint_{S_s} \left[\boldsymbol{r}'\left(\hat{\boldsymbol{\nu}} \cdot \eta_0\eta\boldsymbol{H}^0\right) + \hat{\boldsymbol{r}} \times \boldsymbol{r}'\left(\hat{\boldsymbol{\nu}} \cdot \boldsymbol{E}^0\right)\right] \mathrm{d}S'\end{aligned}$$

We use (6.2) and the Stokes' alternative theorem, $\iint_{S_s} \hat{\boldsymbol{\nu}} \times \nabla'\phi \, \mathrm{d}S' = \boldsymbol{0}$, since the surface S_s is closed, and get

$$\begin{aligned}\boldsymbol{F}(\hat{\boldsymbol{r}}) = &-\frac{k^2}{4\pi}\hat{\boldsymbol{r}} \times \iint_{S_s} \left[\hat{\boldsymbol{\nu}} \times \nabla'\phi - \hat{\boldsymbol{r}} \times \left(\hat{\boldsymbol{\nu}} \times \nabla'\psi\right)\right] \left(\hat{\boldsymbol{r}} \cdot \boldsymbol{r}'\right) \mathrm{d}S' \\ &+ \frac{k^2}{4\pi}\hat{\boldsymbol{r}} \times \iint_{S_s} \left[\boldsymbol{r}'\frac{\partial\psi}{\partial\nu} + \hat{\boldsymbol{r}} \times \boldsymbol{r}'\frac{\partial\phi}{\partial\nu}\right] \mathrm{d}S' \\ = &\frac{k^2}{4\pi}\hat{\boldsymbol{r}} \times \iint_{S_s} \left[\phi\hat{\boldsymbol{\nu}} \times \hat{\boldsymbol{r}} - \hat{\boldsymbol{r}} \times \left(\psi\hat{\boldsymbol{\nu}} \times \hat{\boldsymbol{r}}\right) + \boldsymbol{r}'\frac{\partial\psi}{\partial\nu} + \hat{\boldsymbol{r}} \times \boldsymbol{r}'\frac{\partial\phi}{\partial\nu}\right] \mathrm{d}S'\end{aligned}$$

The dominant contribution contains one electric and one magnetic contribution.

Finally, we write the far field amplitude $\boldsymbol{F}(\hat{\boldsymbol{r}})$ to leading order as

$$\boldsymbol{F}(\hat{\boldsymbol{r}}) = -\frac{k^2}{4\pi\epsilon_0\epsilon}\left[\hat{\boldsymbol{r}} \times (\hat{\boldsymbol{r}} \times \boldsymbol{p}) + \frac{\hat{\boldsymbol{r}} \times \boldsymbol{m}}{c} + O(kd)\right] \tag{6.5}$$

where the electric and magnetic dipole moments, $\boldsymbol{p}$ and $\boldsymbol{m}$, respectively, are

$$\begin{cases} \boldsymbol{p} = \epsilon_0\epsilon \iint\limits_{S_s} \left[\hat{\boldsymbol{\nu}}\phi - \boldsymbol{r}\frac{\partial\phi}{\partial\nu}\right] \mathrm{d}S \\ \boldsymbol{m} = \frac{1}{\eta_0\eta} \iint\limits_{S_s} \left[\hat{\boldsymbol{\nu}}\psi - \boldsymbol{r}\frac{\partial\psi}{\partial\nu}\right] \mathrm{d}S \end{cases} \tag{6.6}$$

and where $c = c_0/\sqrt{\epsilon\mu}$, and $d = \max_{\boldsymbol{r},\boldsymbol{r}'\in V_s} |\boldsymbol{r} - \boldsymbol{r}'|$ is the diameter of V_s.

We note that in the long wavelength limit, the scattering properties are characterized by the electric and magnetic dipole moments of the scatterer, $\boldsymbol{p}$ and $\boldsymbol{m}$, respectively. We also notice that the far field amplitude $\boldsymbol{F}(\hat{\boldsymbol{r}})$ is proportional to k^2, i.e., the frequency to the second power.

The electric and magnetic dipole moments are linearly related to the excitation of scatterer, $\boldsymbol{E}_0$. These dyadic-valued transformations are denoted the electric and the magnetic polarizability dyadics, $\boldsymbol{\gamma}_\mathrm{e}$ and $\boldsymbol{\gamma}_\mathrm{m}$, respectively. They are defined by

$$\begin{cases} \boldsymbol{p} = \epsilon_0\epsilon\boldsymbol{\gamma}_\mathrm{e} \cdot \boldsymbol{E}_0 \\ \boldsymbol{m} = \frac{1}{\eta_0\eta}\boldsymbol{\gamma}_\mathrm{m} \cdot (\hat{\boldsymbol{k}}_\mathrm{i} \times \boldsymbol{E}_0) \end{cases} \tag{6.7}$$

The electric and the magnetic polarizability dyadics, $\boldsymbol{\gamma}_\mathrm{e}$ and $\boldsymbol{\gamma}_\mathrm{m}$, both have the unit volume (m^3). The polarizability dyadics are independent of the excitation of the scatterer, and depend only on the material of the scatterer and its geometry. For symmetric permittivity and permeability dyadics, $\boldsymbol{\gamma}_\mathrm{e}$ and $\boldsymbol{\gamma}_\mathrm{m}$, are symmetric, which are proved in Section 6.1.6.

It remains to compute the potentials ϕ and ψ to find the polarizability dyadics. For several types of scatterers, these potentials are solutions of classical boundary value problems. In the next example, one important case is investigated.

Example 6.1

For a perfectly conducting, non-magnetic scatterer, the following boundary value problem for the electric field has to be solved:

$$\begin{cases} \nabla^2\phi(\boldsymbol{r}) = 0, & \boldsymbol{r} \in V_\mathrm{e} \\ \phi(\boldsymbol{r}) = C, & \boldsymbol{r} \in S_\mathrm{s} \\ \iint\limits_{S_s} \frac{\partial\phi(\boldsymbol{r})}{\partial\nu}\,\mathrm{d}S = 0 \\ \phi(\boldsymbol{r}) = -\boldsymbol{E}_0 \cdot \boldsymbol{r} + C + O(r^{-2}), & \text{as } r \to \infty \end{cases}$$

The surface integral of the normal derivative of ϕ on the surface S_s has to be zero, since there is no net charge on the scatterer. The last condition guaranties that $-\nabla\phi(\boldsymbol{r}) = \boldsymbol{E}_0$ at large distances from the origin, which is the field without scatterer. The solution ϕ is uniquely determined up to a constant C.

In a dynamic case—low frequency, but not zero—not only the electric field vanishes inside the perfectly conducting scatterer, but also the magnetic field $\eta_0\eta\boldsymbol{H} = -\nabla\psi$ and the corresponding magnetic flux density. This leads to the following modified magnetic boundary value problem:

$$\begin{cases} \nabla^2\psi(\boldsymbol{r}) = 0, \quad \boldsymbol{r} \in V_{\rm e} \\ \dfrac{\partial\psi(\boldsymbol{r})}{\partial\nu} = 0, \quad \boldsymbol{r} \in S_{\rm s} \\ \psi(\boldsymbol{r}) = -(\hat{\boldsymbol{k}}_{\rm i} \times \boldsymbol{E}_0) \cdot \boldsymbol{r} + C + O(r^{-2}), \qquad \text{as } r \to \infty \end{cases}$$

The normal magnetic field on the surface $\eta_0\eta\hat{\boldsymbol{r}} \cdot \boldsymbol{H} = -\frac{\partial\psi}{\partial\nu} = 0$ on the surface, since the normal component of the magnetic flux density is continuous over the surface $S_{\rm s}$. The solution ψ is uniquely determined up to a constant C.

As a particular explicit example, take a perfectly conducting, non-magnetic sphere of radius a. The electric potential ϕ is

$$\phi(\boldsymbol{r}) = -\boldsymbol{E}_0 \cdot \boldsymbol{r}\left(1 - \frac{a^3}{r^3}\right) + C, \qquad \boldsymbol{r} \in V_{\rm e}$$

which gives, see (6.6)

$$\boldsymbol{p} = \epsilon_0\epsilon \iint_{r=a} \left[C\hat{\boldsymbol{r}} + 3a\hat{\boldsymbol{r}}\left(\boldsymbol{E}_0 \cdot \hat{\boldsymbol{r}}\right)\right] \,{\rm d}S = \epsilon_0\epsilon \iint_{r=a} 3a\hat{\boldsymbol{r}}\left(\boldsymbol{E}_0 \cdot \hat{\boldsymbol{r}}\right) \,{\rm d}S = 4\pi a^3 \epsilon_0\epsilon \boldsymbol{E}_0$$

The electric polarizability dyadic then becomes

$$\boldsymbol{\gamma}_{\rm e} = 4\pi a^3 \mathbf{I}_3$$

The magnetic potential ψ is

$$\psi(\boldsymbol{r}) = -(\hat{\boldsymbol{k}}_{\rm i} \times \boldsymbol{E}_0) \cdot \boldsymbol{r}\left(1 + \frac{a^3}{2r^3}\right) + C, \qquad \boldsymbol{r} \in V_{\rm e}$$

which gives, see (6.6)

$$\begin{aligned} \boldsymbol{m} &= \frac{1}{\eta_0\eta} \iint_{r=a} \left[-\hat{\boldsymbol{r}}\left((\hat{\boldsymbol{k}}_{\rm i} \times \boldsymbol{E}_0) \cdot \hat{\boldsymbol{r}}\right)\frac{3a}{2} + C\hat{\boldsymbol{r}}\right] {\rm d}S \\ &= -\frac{1}{\eta_0\eta} \iint_{r=a} \frac{3a}{2}\hat{\boldsymbol{r}}\left((\hat{\boldsymbol{k}}_{\rm i} \times \boldsymbol{E}_0) \cdot \hat{\boldsymbol{r}}\right) {\rm d}S = -2\pi a^3 \frac{1}{\eta_0\eta}(\hat{\boldsymbol{k}}_{\rm i} \times \boldsymbol{E}_0) \end{aligned}$$

The magnetic polarizability dyadic then becomes

$$\boldsymbol{\gamma}_{\rm m} = -2\pi a^3 \mathbf{I}_3$$

■

6.1.3 The Scattering Dyadic

The far field amplitude in the long wavelength limit can now be written in an explicit form as, see (6.5) and (6.7)

$$\boldsymbol{F}(\hat{\boldsymbol{r}}) = -\frac{k^2}{4\pi}\left[\hat{\boldsymbol{r}} \times \left(\hat{\boldsymbol{r}} \times \boldsymbol{\gamma}_{\mathrm{e}} \cdot \boldsymbol{E}_0\right) + \hat{\boldsymbol{r}} \times \boldsymbol{\gamma}_{\mathrm{m}} \cdot (\hat{\boldsymbol{k}}_{\mathrm{i}} \times \boldsymbol{E}_0)\right] = \mathbf{S}(\hat{\boldsymbol{r}}, \hat{\boldsymbol{k}}_{\mathrm{i}}) \cdot \boldsymbol{E}_0$$

where the scattering dyadic is

$$\mathbf{S}(\hat{\boldsymbol{r}}, \hat{\boldsymbol{k}}_{\mathrm{i}}) = -\frac{k^2}{4\pi}\left[\hat{\boldsymbol{r}} \times (\hat{\boldsymbol{r}} \times \boldsymbol{\gamma}_{\mathrm{e}}) + \hat{\boldsymbol{r}} \times \boldsymbol{\gamma}_{\mathrm{m}} \cdot \left(\hat{\boldsymbol{k}}_{\mathrm{i}} \times \mathbf{I}_3\right)\right] \tag{6.8}$$

Specifically, the scattering dyadic in the forward direction $\hat{\boldsymbol{r}} = \hat{\boldsymbol{k}}_{\mathrm{i}}$ is

$$\begin{aligned}
\boldsymbol{E}_0^* \cdot \mathbf{S}(\hat{\boldsymbol{k}}_{\mathrm{i}}, \hat{\boldsymbol{k}}_{\mathrm{i}}) \cdot \boldsymbol{E}_0 &= -\frac{k^2}{4\pi}\boldsymbol{E}_0^* \cdot \left[\hat{\boldsymbol{k}}_{\mathrm{i}} \times \left(\hat{\boldsymbol{k}}_{\mathrm{i}} \times (\boldsymbol{\gamma}_{\mathrm{e}} \cdot \boldsymbol{E}_0)\right) + \hat{\boldsymbol{k}}_{\mathrm{i}} \times \left(\boldsymbol{\gamma}_{\mathrm{m}} \cdot (\hat{\boldsymbol{k}}_{\mathrm{i}} \times \boldsymbol{E}_0)\right)\right] \\
&= \frac{k^2}{4\pi}\left[\boldsymbol{E}_0^* \cdot \boldsymbol{\gamma}_{\mathrm{e}} \cdot \boldsymbol{E}_0 + \left(\hat{\boldsymbol{k}}_{\mathrm{i}} \times \boldsymbol{E}_0^*\right) \cdot \boldsymbol{\gamma}_{\mathrm{m}} \cdot (\hat{\boldsymbol{k}}_{\mathrm{i}} \times \boldsymbol{E}_0)\right]
\end{aligned} \tag{6.9}$$

since $\hat{\boldsymbol{k}}_{\mathrm{i}} \cdot \boldsymbol{E}_0 = 0$. Similarly, in the backscattering direction $\hat{\boldsymbol{r}} = -\hat{\boldsymbol{k}}_{\mathrm{i}}$, we get

$$\boldsymbol{E}_0^* \cdot \mathbf{S}(-\hat{\boldsymbol{k}}_{\mathrm{i}}, \hat{\boldsymbol{k}}_{\mathrm{i}}) \cdot \boldsymbol{E}_0 = \frac{k^2}{4\pi}\left[\boldsymbol{E}_0^* \cdot \boldsymbol{\gamma}_{\mathrm{e}} \cdot \boldsymbol{E}_0 - \left(\hat{\boldsymbol{k}}_{\mathrm{i}} \times \boldsymbol{E}_0^*\right) \cdot \boldsymbol{\gamma}_{\mathrm{m}} \cdot (\hat{\boldsymbol{k}}_{\mathrm{i}} \times \boldsymbol{E}_0)\right]$$

6.1.4 Cross Sections

In the long wavelength limit, it is possible to find closed-form expressions of the differential scattering cross section, $\frac{\mathrm{d}\sigma}{\mathrm{d}\Omega}(\hat{\boldsymbol{r}}, \hat{\boldsymbol{k}}_{\mathrm{i}})$, and the scattering cross section, $\sigma_{\mathrm{s}}(\hat{\boldsymbol{k}}_{\mathrm{i}})$.

We employ (4.33) on page 207 and (6.8) to get

$$\frac{\mathrm{d}\sigma}{\mathrm{d}\Omega}(\hat{\boldsymbol{r}}, \hat{\boldsymbol{k}}_{\mathrm{i}}) = \frac{k^4}{4\pi}\left|\hat{\boldsymbol{r}} \times (\hat{\boldsymbol{r}} \times (\boldsymbol{\gamma}_{\mathrm{e}} \cdot \hat{\boldsymbol{p}}_{\mathrm{e}})) + \hat{\boldsymbol{r}} \times (\boldsymbol{\gamma}_{\mathrm{m}} \cdot \hat{\boldsymbol{p}}_{\mathrm{m}})\right|^2$$

where we have specified the polarization as

$$\hat{\boldsymbol{p}}_{\mathrm{e}} = \frac{\boldsymbol{E}_0}{|\boldsymbol{E}_0|}, \qquad \hat{\boldsymbol{p}}_{\mathrm{m}} = \hat{\boldsymbol{k}}_{\mathrm{i}} \times \hat{\boldsymbol{p}}_{\mathrm{e}} \tag{6.10}$$

Notice that the differential scattering cross section behaves as k^4 at low frequencies. We simplify this expression[5]

$$\frac{\mathrm{d}\sigma}{\mathrm{d}\Omega}(\hat{\boldsymbol{r}}, \hat{\boldsymbol{k}}_\mathrm{i}) = \frac{k^4}{4\pi}\{ \left|\boldsymbol{\gamma}_\mathrm{e} \cdot \hat{\boldsymbol{p}}_\mathrm{e}\right|^2 - \left|\hat{\boldsymbol{r}} \cdot (\boldsymbol{\gamma}_\mathrm{e} \cdot \hat{\boldsymbol{p}}_\mathrm{e})\right|^2 + \left|\boldsymbol{\gamma}_\mathrm{m} \cdot \hat{\boldsymbol{p}}_\mathrm{m}\right|^2 - \left|\hat{\boldsymbol{r}} \cdot (\boldsymbol{\gamma}_\mathrm{m} \cdot \hat{\boldsymbol{p}}_\mathrm{m})\right|^2 + 2\hat{\boldsymbol{r}} \cdot \mathrm{Re}\left((\boldsymbol{\gamma}_\mathrm{e} \cdot \hat{\boldsymbol{p}}_\mathrm{e}) \times (\boldsymbol{\gamma}_\mathrm{m} \cdot \hat{\boldsymbol{p}}_\mathrm{m})^*\right)\}$$

The scattering cross section is obtained by an integration of $|\mathbf{S}(\hat{\boldsymbol{r}}, \hat{\boldsymbol{k}}_\mathrm{i}) \cdot \hat{\boldsymbol{p}}_\mathrm{e}|^2$ over the unit sphere Ω, see (4.36) on page 207. The following integrals are pertinent to use:

$$\iint_\Omega \mathrm{d}\Omega = 4\pi, \qquad \iint_\Omega \hat{\boldsymbol{r}}\, \mathrm{d}\Omega = \boldsymbol{0}, \qquad \iint_\Omega \left|\hat{\boldsymbol{r}} \cdot \boldsymbol{a}\right|^2 \mathrm{d}\Omega = \frac{4\pi}{3} |\boldsymbol{a}|^2$$

and we get

$$\sigma_\mathrm{s}(\hat{\boldsymbol{k}}_\mathrm{i}) = \iint \left|\mathbf{S}(\hat{\boldsymbol{r}}, \hat{\boldsymbol{k}}_\mathrm{i}) \cdot \hat{\boldsymbol{p}}_\mathrm{e}\right|^2 \mathrm{d}\Omega = \frac{k^4}{6\pi}\left\{\left|\boldsymbol{\gamma}_\mathrm{e} \cdot \hat{\boldsymbol{p}}_\mathrm{e}\right|^2 + \left|\boldsymbol{\gamma}_\mathrm{m} \cdot \hat{\boldsymbol{p}}_\mathrm{m}\right|^2\right\} \tag{6.11}$$

Example 6.2
If the polarization dyadics are proportional to the identity dyadic, i.e., $\boldsymbol{\gamma}_\mathrm{e} = \gamma_\mathrm{e}\mathbf{I}_3$ and $\boldsymbol{\gamma}_\mathrm{m} = \gamma_\mathrm{m}\mathbf{I}_3$, then[6]

$$\begin{aligned}\frac{\mathrm{d}\sigma}{\mathrm{d}\Omega}(\hat{\boldsymbol{r}}, \hat{\boldsymbol{k}}_\mathrm{i}) &= \frac{k^4}{4\pi}\left\{|\gamma_\mathrm{e}|^2 \left|\hat{\boldsymbol{r}} \times \hat{\boldsymbol{p}}_\mathrm{e}\right|^2 + |\gamma_\mathrm{m}|^2 \left|\hat{\boldsymbol{r}} \times \hat{\boldsymbol{p}}_\mathrm{m}\right|^2 + 2\hat{\boldsymbol{r}} \cdot \mathrm{Re}\left(\gamma_\mathrm{e}\gamma_\mathrm{m}^*\hat{\boldsymbol{p}}_\mathrm{e} \times \hat{\boldsymbol{p}}_\mathrm{m}^*\right)\right\} \\ &= \frac{k^4}{4\pi}\left\{|\gamma_\mathrm{e}|^2 \left|\hat{\boldsymbol{r}} \times \hat{\boldsymbol{p}}_\mathrm{e}\right|^2 + |\gamma_\mathrm{m}|^2 \left|\hat{\boldsymbol{r}} \times \hat{\boldsymbol{p}}_\mathrm{m}\right|^2 + 2\hat{\boldsymbol{r}} \cdot \hat{\boldsymbol{k}}_\mathrm{i}\, \mathrm{Re}\left(\gamma_\mathrm{e}\gamma_\mathrm{m}^*\right)\right\}\end{aligned}$$

since $\hat{\boldsymbol{p}}_\mathrm{e} \cdot \hat{\boldsymbol{k}}_\mathrm{i} = 0$, and

$$\sigma_\mathrm{s}(\hat{\boldsymbol{k}}_\mathrm{i}) = \frac{k^4}{6\pi}\left\{|\gamma_\mathrm{e}|^2 + |\gamma_\mathrm{m}|^2\right\}$$

■

[5]Useful in this simplification are

$$|\boldsymbol{a} + \boldsymbol{b}|^2 = (\boldsymbol{a} + \boldsymbol{b}) \cdot \left(\boldsymbol{a}^* + \boldsymbol{b}^*\right) = |\boldsymbol{a}|^2 + |\boldsymbol{b}|^2 + 2\,\mathrm{Re}\left(\boldsymbol{a} \cdot \boldsymbol{b}^*\right)$$

and

$$\left|\hat{\boldsymbol{r}} \times \boldsymbol{a}\right|^2 = (\hat{\boldsymbol{r}} \times \boldsymbol{a}) \cdot (\hat{\boldsymbol{r}} \times \boldsymbol{a}^*) = |\boldsymbol{a}|^2 - |\hat{\boldsymbol{r}} \cdot \boldsymbol{a}|^2$$

[6]There are several geometries where the polarizability dyadics are proportional to the identity dyadic, e.g., the sphere, the cube, see References 243, 248.

Example 6.3
The electric and magnetic polarizability dyadics, $\boldsymbol{\gamma}_{\text{e}}$ and $\boldsymbol{\gamma}_{\text{m}}$, for a perfectly conducting, non-magnetic spherical scatterer of radius a are found in Example 6.1

$$\boldsymbol{\gamma}_{\text{e}} = 4\pi a^3 \mathbf{I}_3, \qquad \boldsymbol{\gamma}_{\text{m}} = -2\pi a^3 \mathbf{I}_3$$

The scattering cross section then becomes, see Example 6.2

$$\sigma_{\text{s}}(\hat{\boldsymbol{k}}_{\text{i}}) = \frac{10\pi k^4 a^6}{3}$$

■

If there is no power absorbed in the scatterer, the absorption cross section $\sigma_{\text{a}} = 0$, and the optical theorem (4.47) on page 230 implies

$$\frac{k^4}{6\pi}\left\{\left|\boldsymbol{\gamma}_{\text{e}}\cdot\hat{\boldsymbol{p}}_{\text{e}}\right|^2 + \left|\boldsymbol{\gamma}_{\text{m}}\cdot\hat{\boldsymbol{p}}_{\text{m}}\right|^2\right\} = \sigma_{\text{ext}} = \frac{4\pi}{k}\operatorname{Im}\left\{\hat{\boldsymbol{p}}_{\text{e}}^*\cdot\mathbf{S}(\hat{\boldsymbol{k}}_{\text{i}},\hat{\boldsymbol{k}}_{\text{i}})\cdot\hat{\boldsymbol{p}}_{\text{e}}\right\}$$

From (6.9), we have that the real part of $\hat{\boldsymbol{p}}_{\text{e}}^*\cdot\mathbf{S}(\hat{\boldsymbol{k}}_{\text{i}},\hat{\boldsymbol{k}}_{\text{i}})\cdot\hat{\boldsymbol{p}}_{\text{e}}$ behaves as k^2 as $k \to 0$. In the non-absorbing case, and in situations where the optical theorem can be applied, we see that the imaginary part of $\hat{\boldsymbol{p}}_{\text{e}}^*\cdot\mathbf{S}(\hat{\boldsymbol{k}}_{\text{i}},\hat{\boldsymbol{k}}_{\text{i}})\cdot\hat{\boldsymbol{p}}_{\text{e}}$ behaves as k^5 as $k \to 0$. Indeed, the low-frequency behavior for the PEC disk, see (9.8) on page 487, verifies this statement.

6.1.5 Internal Field

In Section 6.1.2, we analyzed the far field amplitude of the scatterer and found that, in the long wavelength regime, it was characterized by the electric and magnetic polarizability dyadics, $\boldsymbol{\gamma}_{\text{e}}$ and $\boldsymbol{\gamma}_{\text{m}}$, respectively. We now focus on the internal fields that are needed to compute these dyadics.

Assume the scatterer in V_{s} consists of a reciprocal, anisotropic material with relative permittivity, $\boldsymbol{\epsilon}_1(\boldsymbol{r},\omega)$, and relative permeability, $\boldsymbol{\mu}_1(\boldsymbol{r},\omega)$. The Maxwell equations are[7]

$$\begin{cases}\nabla\times\boldsymbol{E}(\boldsymbol{r},\omega) = \mathrm{i}k\eta_0\eta\dfrac{\boldsymbol{\mu}_1(\boldsymbol{r},\omega)}{\mu}\cdot\boldsymbol{H}(\boldsymbol{r},\omega) \\ \nabla\times\boldsymbol{H}(\boldsymbol{r},\omega) = -\dfrac{\mathrm{i}k}{\eta_0\eta}\dfrac{\boldsymbol{\epsilon}_1(\boldsymbol{r},\omega)}{\epsilon}\cdot\boldsymbol{E}(\boldsymbol{r},\omega)\end{cases} \quad \boldsymbol{r}\in V_{\text{s}} \tag{6.12}$$

where we scale the material parameters with the exterior permittivity,[8] ϵ, and permeability, μ, and where the relative wave impedance and wave number of the exterior medium are $\eta = (\mu/\epsilon)^{1/2}$ and $k = k_0(\epsilon\mu)^{1/2}$, respectively. We assume the exterior material parameters have well-defined limits as $\omega \to 0$.

We proceed the analysis as in Section 6.1.1 for the near field approximation, i.e., we assume a series expansion of the fields in powers of k, which is the wave number of the exterior domain.

---

[7]We emphasize the frequency behavior dependence on ω in all quantities explicitly in this section, since we later are going to the static limit $\omega \to 0$.

[8]In most applications, the constants are 1.

Keeping the same notation, we have

$$\boldsymbol{E}(\boldsymbol{r}) = \sum_{n=0}^{\infty} (\mathrm{i}k)^n \boldsymbol{E}^n(\boldsymbol{r}), \quad \boldsymbol{H}(\boldsymbol{r}) = \sum_{n=0}^{\infty} (\mathrm{i}k)^n \boldsymbol{H}^n(\boldsymbol{r}), \qquad \boldsymbol{r} \in V_{\mathrm{s}}$$

where the expansion coefficients $\boldsymbol{E}^n(\boldsymbol{r})$ and $\boldsymbol{H}^n(\boldsymbol{r})$, $n = 0, 1, 2, \ldots$, are independent of the wave number k, but they depend on the location $\boldsymbol{r}$. Following the same steps of analysis as in Section 6.1.1, we obtain the following results:

$$\begin{cases} \nabla \times \boldsymbol{E}^0 = \boldsymbol{0} \\ \nabla \times \boldsymbol{H}^0 = \boldsymbol{0} \end{cases} \qquad \boldsymbol{r} \in V_{\mathrm{s}}$$

which implies that the zero-order terms can be written as potential fields, i.e.,

$$\begin{cases} \boldsymbol{E}^0(\boldsymbol{r}) = -\nabla\phi(\boldsymbol{r}) \\ \eta_0\eta\boldsymbol{H}^0(\boldsymbol{r}) = -\nabla\psi(\boldsymbol{r}) \end{cases} \qquad \boldsymbol{r} \in V_{\mathrm{s}}$$

Moreover, the divergence equations

$$\begin{cases} \nabla \cdot (\boldsymbol{\epsilon}_1(\boldsymbol{r}, \omega) \cdot \boldsymbol{E}(\boldsymbol{r}, \omega)) = 0 \\ \nabla \cdot (\boldsymbol{\mu}_1(\boldsymbol{r}, \omega) \cdot \boldsymbol{H}(\boldsymbol{r}, \omega)) = 0 \end{cases} \qquad \boldsymbol{r} \in V_{\mathrm{s}}$$

have to be satisfied. To the lowest order in k, the potentials ϕ and ψ satisfy

$$\begin{cases} \nabla \cdot (\boldsymbol{\epsilon}_1(\boldsymbol{r}, 0) \cdot \nabla\phi(\boldsymbol{r})) = 0 \\ \nabla \cdot (\boldsymbol{\mu}_1(\boldsymbol{r}, 0) \cdot \nabla\psi(\boldsymbol{r})) = 0 \end{cases} \qquad \boldsymbol{r} \in V_{\mathrm{s}}$$

We connect this result to the potential problem of the exterior domain, see (6.4), and we get

$$\begin{cases} \nabla \cdot (\boldsymbol{\epsilon}_1(\boldsymbol{r}, 0) \cdot \nabla\phi(\boldsymbol{r})) = 0, & \boldsymbol{r} \in \mathbb{R}^3 \\ \phi(\boldsymbol{r})|_- = \phi(\boldsymbol{r})|_+, & \boldsymbol{r} \in S_{\mathrm{s}} \\ \hat{\boldsymbol{\nu}} \cdot \boldsymbol{\epsilon}_1(\boldsymbol{r}, 0) \cdot \nabla\phi(\boldsymbol{r})\big|_- = \epsilon\hat{\boldsymbol{\nu}} \cdot \nabla\phi(\boldsymbol{r})\big|_+, & \boldsymbol{r} \in S_{\mathrm{s}} \\ \phi(\boldsymbol{r}) = -\boldsymbol{E}_0 \cdot \boldsymbol{r} + C + O(r^{-2}), & \text{as } r \to \infty \end{cases} \tag{6.13}$$

and similarly for the magnetic potential ψ. Once the potentials ϕ and ψ are determined, the polarization and magnetization are determined by (6.6), i.e.,

$$\begin{cases} \boldsymbol{p} = \epsilon_0\epsilon \displaystyle\iint_{S_{\mathrm{s}}} \left[\hat{\boldsymbol{\nu}}\phi - \boldsymbol{r}\frac{\partial\phi}{\partial\nu}\right] \mathrm{d}S \\ \boldsymbol{m} = \dfrac{1}{\eta_0\eta} \displaystyle\iint_{S_{\mathrm{s}}} \left[\hat{\boldsymbol{\nu}}\psi - \boldsymbol{r}\frac{\partial\psi}{\partial\nu}\right] \mathrm{d}S \end{cases}$$

Notice that the values of the normal derivatives in these integrals are taken from the outside of S_{s}.

If the material parameters, $\boldsymbol{\epsilon}_1(\boldsymbol{r}, 0) = \boldsymbol{\epsilon}_1(0)$ and/or $\boldsymbol{\mu}_1(\boldsymbol{r}, 0) = \boldsymbol{\mu}_1(0)$, are constant in V_{s}, the potential satisfies $\nabla \cdot (\boldsymbol{\epsilon}_1(0) \cdot \nabla\phi(\boldsymbol{r})) = 0$ in V_{s}, and we can simplify the second part of the integrals by repeated use of Gauss' theorem. We get ($i = 1, 2, 3$)

$$\begin{aligned}\epsilon \iint\limits_{S_s} x_i \left.\frac{\partial \phi}{\partial \nu}\right|_+ \mathrm{d}S &= \iint\limits_{S_s} x_i \hat{\boldsymbol{\nu}} \cdot \boldsymbol{\epsilon}_1(0) \cdot \nabla \phi|_- \, \mathrm{d}S \\ &= \iiint\limits_{V_s} \nabla \cdot (x_i \boldsymbol{\epsilon}_1(0) \cdot \nabla \phi) \, \mathrm{d}v = \hat{\boldsymbol{e}}_i \cdot \boldsymbol{\epsilon}_1(0) \cdot \iiint\limits_{V_s} \nabla \phi \, \mathrm{d}v \\ &= \hat{\boldsymbol{e}}_i \cdot \boldsymbol{\epsilon}_1(0) \cdot \iint\limits_{S_s} \hat{\boldsymbol{\nu}} \phi \, \mathrm{d}S\end{aligned}$$

and

$$\begin{cases} \boldsymbol{p} = \epsilon_0 \epsilon \left(\mathbf{I}_3 - \dfrac{\boldsymbol{\epsilon}_1(0)}{\epsilon} \right) \cdot \displaystyle\iint\limits_{S_s} \hat{\boldsymbol{\nu}} \phi \, \mathrm{d}S \\ \boldsymbol{m} = \dfrac{1}{\eta_0 \eta} \left(\mathbf{I}_3 - \dfrac{\boldsymbol{\mu}_1(0)}{\mu} \right) \cdot \displaystyle\iint\limits_{S_s} \hat{\boldsymbol{\nu}} \psi \, \mathrm{d}S \end{cases} \tag{6.14}$$

Example 6.4

As a simple example, we choose a homogeneous, isotropic spherical scatterer of radius a. For a homogeneous spherical scatterer (radius a, real-valued permittivity $\epsilon_1(\omega)$, and real-valued permeability $\mu_1(\omega)$), the pertinent potential problem to solve is stated in (6.13). The solution to this problem is

$$\phi(\boldsymbol{r}) = \begin{cases} -\boldsymbol{E}_0 \cdot \boldsymbol{r} \dfrac{3\epsilon}{\epsilon_1(0) + 2\epsilon} + C, & 0 \le r < a \\ -\boldsymbol{E}_0 \cdot \boldsymbol{r} \left(1 - \dfrac{\epsilon_1(0) - \epsilon}{\epsilon_1(0) + 2\epsilon} \dfrac{a^3}{r^3} \right) + C, & r > a \end{cases}$$

where C is an arbitrary constant.

The electric and magnetic polarizability dyadics, $\boldsymbol{\gamma}_{\mathrm{e}}$ and $\boldsymbol{\gamma}_{\mathrm{m}}$, can be computed analytically, see (6.14)

$$\begin{aligned}\boldsymbol{p} &= \epsilon_0 \epsilon \left(1 - \frac{\epsilon_1(0)}{\epsilon} \right) \iint\limits_{S_s} \hat{\boldsymbol{r}} \phi \, \mathrm{d}S' \\ &= -\epsilon_0 \epsilon \left(1 - \frac{\epsilon_1(0)}{\epsilon} \right) \frac{3\epsilon}{\epsilon_1(0) + 2\epsilon} a^3 \iint\limits_{\Omega} \hat{\boldsymbol{r}} \hat{\boldsymbol{r}} \, \mathrm{d}\Omega \cdot \boldsymbol{E}_0 = 4\pi a^3 \epsilon_0 \epsilon \frac{\epsilon_1(0) - \epsilon}{\epsilon_1(0) + 2\epsilon} \boldsymbol{E}_0\end{aligned}$$

since

$$\iint\limits_{\Omega} \hat{\boldsymbol{r}} \, \mathrm{d}\Omega = \mathbf{0}, \qquad \iint\limits_{\Omega} \hat{\boldsymbol{r}} \hat{\boldsymbol{r}} \, \mathrm{d}\Omega = \frac{4\pi}{3} \mathbf{I}_3$$

This leads to

$$
\begin{cases}
\boldsymbol{\gamma}_{\mathrm{e}} = 4\pi a^3 \dfrac{\epsilon_1(0) - \epsilon}{\epsilon_1(0) + 2\epsilon} \mathbf{I}_3 \\
\boldsymbol{\gamma}_{\mathrm{m}} = 4\pi a^3 \dfrac{\mu_1(0) - \mu}{\mu_1(0) + 2\mu} \mathbf{I}_3
\end{cases}
$$

The scattering cross section then becomes, see Example 6.2

$$
\sigma_{\mathrm{s}}(\hat{\boldsymbol{k}}_{\mathrm{i}}) = \frac{8\pi k^4 a^6}{3} \left\{ \left| \frac{\epsilon_1(0) - \epsilon}{\epsilon_1(0) + 2\epsilon} \right|^2 + \left| \frac{\mu_1(0) - \mu}{\mu_1(0) + 2\mu} \right|^2 \right\}
$$

■

The static permittivity and permeability dyadics, $\boldsymbol{\epsilon}_1(\boldsymbol{r}, 0)$ and $\boldsymbol{\mu}_1(\boldsymbol{r}, 0)$, have been assumed real valued in our analysis above. If the material is modeled with a conductivity term, e.g.,

$$
\boldsymbol{\epsilon}_1(\boldsymbol{r}, \omega) = \boldsymbol{\epsilon}(\boldsymbol{r}, 0) + \frac{\mathrm{i}\boldsymbol{\sigma}(\boldsymbol{r}, \omega)}{\epsilon_0 \omega}
$$

the static limit shows great similarities with the perfectly conducting case, see Example 6.1.[9] The reader is encouraged to consult in Reference 126 for more details on the analysis of this conductivity model.

6.1.5.1 Integral Equation Formulation

The internal field can also be obtained in terms of the solution of an integral equation. We rewrite the Maxwell equations in (6.12) as

$$
\begin{cases}
\nabla \times \boldsymbol{E}(\boldsymbol{r}, \omega) = \mathrm{i}k\eta_0\eta \left(\dfrac{\boldsymbol{\mu}_1(\boldsymbol{r}, \omega)}{\mu} - \mathbf{I}_3 \right) \cdot \boldsymbol{H}(\boldsymbol{r}, \omega) + \mathrm{i}k\eta_0\eta \boldsymbol{H}(\boldsymbol{r}, \omega) \\
\nabla \times \boldsymbol{H}(\boldsymbol{r}, \omega) = -\dfrac{\mathrm{i}k}{\eta_0\eta} \left(\dfrac{\boldsymbol{\epsilon}_1(\boldsymbol{r}, \omega)}{\epsilon} - \mathbf{I}_3 \right) \cdot \boldsymbol{E}(\boldsymbol{r}, \omega) - \dfrac{\mathrm{i}k}{\eta_0\eta} \boldsymbol{E}(\boldsymbol{r}, \omega)
\end{cases} \quad \boldsymbol{r} \in \mathbb{R}^3
$$

and we identify an electric and a magnetic current density

$$
\begin{cases}
\boldsymbol{J}_{\mathrm{e}}(\boldsymbol{r}, \omega) = -\dfrac{\mathrm{i}k}{\eta_0\eta} \left(\dfrac{\boldsymbol{\epsilon}_1(\boldsymbol{r}, \omega)}{\epsilon} - \mathbf{I}_3 \right) \cdot \boldsymbol{E}(\boldsymbol{r}, \omega) \\
\boldsymbol{J}_{\mathrm{m}}(\boldsymbol{r}, \omega) = \mathrm{i}k\eta_0\eta \left(\dfrac{\boldsymbol{\mu}_1(\boldsymbol{r}, \omega)}{\mu} - \mathbf{I}_3 \right) \cdot \boldsymbol{H}(\boldsymbol{r}, \omega)
\end{cases} \quad \boldsymbol{r} \in \mathbb{R}^3
$$

These current densities vanish outside the scatterer V_{s}, and they are the sources of the scattered fields. The (scattered) electric and magnetic fields can now be obtained, or rather represented

[9] This might at first seem contradicting, since all real materials show conductivity, even if it is very small for many insulators. However, the typical time scale, after which we have stationary conditions, is ϵ_0/σ, which for a good insulator can be of the order 1–10^5 s.

as a volume integral, see (2.22) and (2.23) on page 111. We get for all $\boldsymbol{r} \in \mathbb{R}^3$

$$\begin{aligned} \boldsymbol{E}(\boldsymbol{r},\omega) = \boldsymbol{E}_{\mathrm{i}}(\boldsymbol{r},\omega) + k^2 \left[\mathbf{I}_3 + \frac{1}{k^2}\nabla\nabla\right] \cdot \iiint_{V_{\mathrm{s}}} \frac{\mathrm{e}^{\mathrm{i}k|\boldsymbol{r}-\boldsymbol{r}'|}}{4\pi|\boldsymbol{r}-\boldsymbol{r}'|} \left(\frac{\boldsymbol{\epsilon}_1(\boldsymbol{r}',\omega)}{\epsilon} - \mathbf{I}_3\right) \cdot \boldsymbol{E}(\boldsymbol{r}',\omega)\,\mathrm{d}v' \\ + \mathrm{i}k\eta_0\eta\nabla \times \iiint_{V_{\mathrm{s}}} \frac{\mathrm{e}^{\mathrm{i}k|\boldsymbol{r}-\boldsymbol{r}'|}}{4\pi|\boldsymbol{r}-\boldsymbol{r}'|} \left(\frac{\boldsymbol{\mu}_1(\boldsymbol{r}',\omega)}{\mu} - \mathbf{I}_3\right) \cdot \boldsymbol{H}(\boldsymbol{r}',\omega)\,\mathrm{d}v' \end{aligned} \tag{6.15}$$

and

$$\begin{aligned} \boldsymbol{H}(\boldsymbol{r},\omega) = \boldsymbol{H}_{\mathrm{i}}(\boldsymbol{r},\omega) - \frac{\mathrm{i}k}{\eta_0\eta}\nabla \times \iiint_{V_{\mathrm{s}}} \frac{\mathrm{e}^{\mathrm{i}k|\boldsymbol{r}-\boldsymbol{r}'|}}{4\pi|\boldsymbol{r}-\boldsymbol{r}'|} \left(\frac{\boldsymbol{\epsilon}_1(\boldsymbol{r}',\omega)}{\epsilon} - \mathbf{I}_3\right) \cdot \boldsymbol{E}(\boldsymbol{r}',\omega)\,\mathrm{d}v' \\ + k^2 \left[\mathbf{I}_3 + \frac{1}{k^2}\nabla\nabla\right] \cdot \iiint_{V_{\mathrm{s}}} \frac{\mathrm{e}^{\mathrm{i}k|\boldsymbol{r}-\boldsymbol{r}'|}}{4\pi|\boldsymbol{r}-\boldsymbol{r}'|} \left(\frac{\boldsymbol{\mu}_1(\boldsymbol{r}',\omega)}{\mu} - \mathbf{I}_3\right) \cdot \boldsymbol{H}(\boldsymbol{r}',\omega)\,\mathrm{d}v' \end{aligned}$$

The far field amplitude then is ($\nabla \to \mathrm{i}k\hat{\boldsymbol{r}}$)

$$\begin{aligned} \boldsymbol{F}(\hat{\boldsymbol{r}}) = k^2 \left[\mathbf{I}_3 - \hat{\boldsymbol{r}}\hat{\boldsymbol{r}}\right] \cdot \iiint_{V_{\mathrm{s}}} \frac{\mathrm{e}^{-\mathrm{i}k\hat{\boldsymbol{r}}\cdot\boldsymbol{r}'}}{4\pi} \left(\frac{\boldsymbol{\epsilon}_1(\boldsymbol{r}',\omega)}{\epsilon} - \mathbf{I}_3\right) \cdot \boldsymbol{E}(\boldsymbol{r}',\omega)\,\mathrm{d}v' \\ - k^2\eta_0\eta\hat{\boldsymbol{r}} \times \iiint_{V_{\mathrm{s}}} \frac{\mathrm{e}^{-\mathrm{i}k\hat{\boldsymbol{r}}\cdot\boldsymbol{r}'}}{4\pi} \left(\frac{\boldsymbol{\mu}_1(\boldsymbol{r}',\omega)}{\mu} - \mathbf{I}_3\right) \cdot \boldsymbol{H}(\boldsymbol{r}',\omega)\,\mathrm{d}v' \end{aligned} \tag{6.16}$$

We return to the general integral expressions of the field, and take the limit $k \to 0$. These limits are ($\boldsymbol{r} \in \mathbb{R}^3$)

$$\begin{cases} \boldsymbol{E}^0(\boldsymbol{r}) = \boldsymbol{E}_0 + \nabla\nabla \cdot \displaystyle\iiint_{V_{\mathrm{s}}} \frac{1}{4\pi|\boldsymbol{r}-\boldsymbol{r}'|} \left(\frac{\boldsymbol{\epsilon}_1(\boldsymbol{r}',0)}{\epsilon} - \mathbf{I}_3\right) \cdot \boldsymbol{E}^0(\boldsymbol{r}')\,\mathrm{d}v' \\ \boldsymbol{H}^0(\boldsymbol{r}) = \boldsymbol{H}_0 + \nabla\nabla \cdot \displaystyle\iiint_{V_{\mathrm{s}}} \frac{1}{4\pi|\boldsymbol{r}-\boldsymbol{r}'|} \left(\frac{\boldsymbol{\mu}_1(\boldsymbol{r}',0)}{\mu} - \mathbf{I}_3\right) \cdot \boldsymbol{H}^0(\boldsymbol{r}')\,\mathrm{d}v' \end{cases} \tag{6.17}$$

where we explicitly have assumed that the incident field is a plane wave with constant polarization vectors, $\boldsymbol{E}_0$ and $\boldsymbol{H}_0$, respectively. If the position vector $\boldsymbol{r}$ is restricted to the scatterer, i.e., $\boldsymbol{r} \in V_{\mathrm{s}}$, these integral relations become integral equations, which can be solved for the unknown fields $\boldsymbol{E}^0$ and $\boldsymbol{H}^0$. From these expressions, we again conclude that the electric and magnetic fields can be written as a gradient of a potential, e.g., the electric potential $\phi(\boldsymbol{r})$ and $\boldsymbol{E}^0 = -\nabla\phi$. The explicit expression of the electric potential is

$$\phi(\boldsymbol{r}) = -\boldsymbol{E}_0 \cdot \boldsymbol{r} - \nabla \cdot \iiint_{V_{\mathrm{s}}} \frac{1}{4\pi|\boldsymbol{r}-\boldsymbol{r}'|} \left(\frac{\boldsymbol{\epsilon}_1(\boldsymbol{r}',0)}{\epsilon} - \mathbf{I}_3\right) \cdot \boldsymbol{E}^0(\boldsymbol{r}')\,\mathrm{d}v' + C$$

To leading order in k, we also have the asymptotic behavior of the scattered field far away from the scatterer, see (6.16)

$$\begin{aligned} \boldsymbol{E}^0(\boldsymbol{r}) - \boldsymbol{E}_0 = \frac{k^2}{4\pi r} \Bigg\{ &[\mathbf{I}_3 - \hat{\boldsymbol{r}}\hat{\boldsymbol{r}}] \cdot \iiint_{V_s} \left(\frac{\boldsymbol{\epsilon}_1(\boldsymbol{r}',0)}{\epsilon} - \mathbf{I}_3 \right) \cdot \boldsymbol{E}^0(\boldsymbol{r}')\, \mathrm{d}v' \\ &- \eta_0\eta \hat{\boldsymbol{r}} \times \iiint_{V_s} \left(\frac{\boldsymbol{\mu}_1(\boldsymbol{r}',0)}{\mu} - \mathbf{I}_3 \right) \cdot \boldsymbol{H}^0(\boldsymbol{r}')\, \mathrm{d}v' \Bigg\} \end{aligned}$$

from which we conclude by comparison with (6.5)

$$\boldsymbol{F}(\hat{\boldsymbol{r}}) = -\frac{k^2}{4\pi\epsilon_0\epsilon} \left[\hat{\boldsymbol{r}} \times (\hat{\boldsymbol{r}} \times \boldsymbol{p}) + \frac{\hat{\boldsymbol{r}} \times \boldsymbol{m}}{c} + O(kd) \right]$$

that the electric and magnetic dipole moments are

$$\begin{cases} \boldsymbol{p} = \epsilon_0\epsilon \iiint_{V_s} \left(\dfrac{\boldsymbol{\epsilon}_1(\boldsymbol{r}',0)}{\epsilon} - \mathbf{I}_3 \right) \cdot \boldsymbol{E}^0(\boldsymbol{r}')\, \mathrm{d}v' \\ \boldsymbol{m} = \iiint_{V_s} \left(\dfrac{\boldsymbol{\mu}_1(\boldsymbol{r}',0)}{\mu} - \mathbf{I}_3 \right) \cdot \boldsymbol{H}^0(\boldsymbol{r}')\, \mathrm{d}v' \end{cases} \tag{6.18}$$

Moreover, we notice that the electric and the magnetic equations are identical in form—they differ by an exchange of $\boldsymbol{\epsilon}_1(\boldsymbol{r},0)/\epsilon$ and $\boldsymbol{\mu}_1(\boldsymbol{r},0)/\mu$.

6.1.6 Polarizability Dyadics

It is intuitively clear that a local increase of the permittivity or permeability increases the polarization. Moreover, an increase of the volume of the object also increases the polarization—a statement that is consistent with the local increase of the material parameters. We devote this section to a mathematical proof of these statements.

Jones [112] has published a similar result for isotropic materials, and we here generalize his result to anisotropic materials. We concentrate on estimates of the electric polarizability—the magnetic properties follow by similar arguments. The permittivity dyadic $\boldsymbol{\epsilon}(\boldsymbol{r})$ is assumed to be symmetric and positive definite throughout this section.

The starting point of the estimates is the integral representation of the static electric field, see (6.17)

$$\boldsymbol{E}^0(\boldsymbol{r}) = \boldsymbol{E}_0 + \nabla\nabla \cdot \iiint_{V_s} \frac{\boldsymbol{\chi}(\boldsymbol{r}') \cdot \boldsymbol{E}^0(\boldsymbol{r}')}{4\pi|\boldsymbol{r}-\boldsymbol{r}'|}\, \mathrm{d}v' = \boldsymbol{E}_0 + \nabla\nabla \cdot \boldsymbol{Q}(\boldsymbol{r}), \qquad \boldsymbol{r} \in \mathbb{R}^3 \tag{6.19}$$

where the real-valued[10] susceptibility function $\chi(\boldsymbol{r})$ is

$$\chi(\boldsymbol{r}) = \frac{\epsilon(\boldsymbol{r})}{\epsilon} - \mathbf{I}_3$$

and the vector $\boldsymbol{Q}(\boldsymbol{r})$ is

$$\boldsymbol{Q}(\boldsymbol{r}) = \iiint_{V_s} \frac{\chi(\boldsymbol{r}') \cdot \boldsymbol{E}^0(\boldsymbol{r}')}{4\pi |\boldsymbol{r} - \boldsymbol{r}'|} \, dv'$$

The susceptibility $\chi(\boldsymbol{r})$ is zero outside the scattering region V_s, and the constant ϵ is real valued. The electric dipole moment, $\boldsymbol{p}$, and the electric polarizability dyadic, $\boldsymbol{\gamma}_e$, are, see (6.18) and (6.7)

$$\begin{cases} \boldsymbol{p} = \epsilon_0 \epsilon \iiint_{V_s} \chi(\boldsymbol{r}) \cdot \boldsymbol{E}^0(\boldsymbol{r}) \, dv' \\ \boldsymbol{p} = \epsilon_0 \epsilon \boldsymbol{\gamma}_e \cdot \boldsymbol{E}_0 \end{cases}$$

Equation (6.19) is an integral equation provided the position vector is restricted to the scattering region V_s. The equivalent partial differential equation to (6.19) is found by rewriting the equation as

$$\boldsymbol{E}^0(\boldsymbol{r}) = \boldsymbol{E}_0 + \nabla^2 \boldsymbol{Q}(\boldsymbol{r}) + \nabla \times (\nabla \times \boldsymbol{Q}(\boldsymbol{r}))$$

which implies

$$\boldsymbol{E}^0(\boldsymbol{r}) = \boldsymbol{E}_0 - \chi(\boldsymbol{r}) \cdot \boldsymbol{E}^0(\boldsymbol{r}) + \nabla \times (\nabla \times \boldsymbol{Q}(\boldsymbol{r}))$$

since (let $k \to 0$ in (2.8) on page 103)

$$\nabla^2 \frac{1}{4\pi |\boldsymbol{r} - \boldsymbol{r}'|} = -\delta(\boldsymbol{r} - \boldsymbol{r}')$$

or

$$\frac{\epsilon(\boldsymbol{r})}{\epsilon} \cdot \boldsymbol{E}^0(\boldsymbol{r}) = \boldsymbol{E}_0 + \nabla \times (\nabla \times \boldsymbol{Q}(\boldsymbol{r}))$$

and consequently

$$\nabla \cdot \left(\frac{\epsilon(\boldsymbol{r})}{\epsilon} \cdot \boldsymbol{E}^0(\boldsymbol{r}) - \boldsymbol{E}_0 \right) = \nabla \cdot \left(\frac{\epsilon(\boldsymbol{r})}{\epsilon} \cdot \boldsymbol{E}^0(\boldsymbol{r}) \right) = 0$$

The aim is now to find an appropriate expression for the electric polarizability dyadic $\boldsymbol{\gamma}_e$. To accomplish this, we introduce a set of orthonormal basis vectors, $\hat{\boldsymbol{e}}_i$, $i = 1, 2, 3$. Denote the solution to the integral equation[11] in (6.19) for an exciting field $\boldsymbol{E}_0 = \hat{\boldsymbol{e}}_j$, $j = 1, 2, 3$,

---

[10]The effect of a conductivity term is discussed in Reference 126, see also Footnote 9 on page 277.

[11]To obtain an integral equation, the position vector has to be restricted to the scattering domain, i.e., $\boldsymbol{r} \in V_s$.

by $\boldsymbol{E}_j^0$, i.e.,

$$\boldsymbol{E}_j^0(\boldsymbol{r}) = \hat{\boldsymbol{e}}_j + \nabla\nabla \cdot \iiint_{V_s} \frac{\boldsymbol{\chi}(\boldsymbol{r}') \cdot \boldsymbol{E}_j^0(\boldsymbol{r}')}{4\pi|\boldsymbol{r}-\boldsymbol{r}'|} \,\mathrm{d}v', \qquad \boldsymbol{r} \in \mathbb{R}^3, \quad j = 1,2,3 \tag{6.20}$$

Then

$$\boldsymbol{\gamma}_\mathrm{e} \cdot \hat{\boldsymbol{e}}_j = \iiint_{V_s} \boldsymbol{\chi}(\boldsymbol{r}') \cdot \boldsymbol{E}_j^0(\boldsymbol{r}') \,\mathrm{d}v', \qquad j = 1,2,3$$

with Cartesian components

$$\hat{\boldsymbol{e}}_i \cdot \boldsymbol{\gamma}_\mathrm{e} \cdot \hat{\boldsymbol{e}}_j = \iiint_{V_s} \hat{\boldsymbol{e}}_i \cdot \boldsymbol{\chi}(\boldsymbol{r}) \cdot \boldsymbol{E}_j^0(\boldsymbol{r}) \,\mathrm{d}v, \qquad i,j = 1,2,3$$

From (6.20), we conclude that

$$\boldsymbol{E}_j^0 = -\nabla V_j(\boldsymbol{r}), \qquad j = 1,2,3$$

where V_j satisfies[12]

$$V_j(\boldsymbol{r}) = -x_j + \nabla \cdot \iiint_{V_s} \frac{\boldsymbol{\chi}(\boldsymbol{r}') \cdot \nabla' V_j(\boldsymbol{r}')}{4\pi|\boldsymbol{r}-\boldsymbol{r}'|} \,\mathrm{d}v', \qquad j = 1,2,3$$

or in differential form

$$\nabla \cdot \left(\boldsymbol{\epsilon}(\boldsymbol{r}) \cdot \nabla V_j(\boldsymbol{r})\right) = 0, \qquad j = 1,2,3$$

We introduce the notation $W_j = x_j + V_j, j = 1,2,3$, where W_j satisfies

$$W_j(\boldsymbol{r}) = -\nabla \cdot \iiint_{V_s} \frac{\boldsymbol{\chi}(\boldsymbol{r}') \cdot \hat{\boldsymbol{e}}_j}{4\pi|\boldsymbol{r}-\boldsymbol{r}'|} \,\mathrm{d}v' + \nabla \cdot \iiint_{V_s} \frac{\boldsymbol{\chi}(\boldsymbol{r}') \cdot \nabla' W_j(\boldsymbol{r}')}{4\pi|\boldsymbol{r}-\boldsymbol{r}'|} \,\mathrm{d}v'$$

This expression shows that $W_j = O(r^{-2})$ as $r \to \infty$. The analogous differential form is

$$\nabla \cdot \left(\boldsymbol{\epsilon}(\boldsymbol{r}) \cdot \nabla W_i(\boldsymbol{r}) - \boldsymbol{\epsilon}(\boldsymbol{r}) \cdot \hat{\boldsymbol{e}}_i\right) = 0, \qquad i = 1,2,3 \tag{6.21}$$

A typical element of the electric polarization dyadic is then

$$\begin{aligned}\hat{\boldsymbol{e}}_i \cdot \boldsymbol{\gamma}_\mathrm{e} \cdot \hat{\boldsymbol{e}}_j &= -\iiint_{V_s} \hat{\boldsymbol{e}}_i \cdot \boldsymbol{\chi}(\boldsymbol{r}) \cdot \nabla V_j(\boldsymbol{r}) \,\mathrm{d}v \\ &= \iiint_{V_s} \hat{\boldsymbol{e}}_i \cdot \boldsymbol{\chi}(\boldsymbol{r}) \cdot \hat{\boldsymbol{e}}_j \,\mathrm{d}v - \iiint_{V_s} \hat{\boldsymbol{e}}_i \cdot \boldsymbol{\chi}(\boldsymbol{r}) \cdot \nabla W_j(\boldsymbol{r}) \,\mathrm{d}v, \qquad i,j = 1,2,3\end{aligned}$$

[12] An irrelevant constant is ignored, i.e., the constant is set to zero.

The electric polarization dyadic is symmetric, provided the electric susceptibility dyadic χ (permittivity dyadic ϵ) is symmetric. In fact, using the results above, we obtain by an application of Gauss' theorem (assuming the permittivity dyadic $\epsilon(\boldsymbol{r})$ smooth[13])

$$\iiint_{V_s} \hat{e}_i \cdot (\epsilon(\boldsymbol{r}) - \epsilon \mathbf{I}_3) \cdot \nabla W_j(\boldsymbol{r})\, dv = \iiint_{\mathbb{R}^3} \hat{e}_i \cdot (\epsilon(\boldsymbol{r}) - \epsilon \mathbf{I}_3) \cdot \nabla W_j(\boldsymbol{r})\, dv$$
$$= -\iiint_{\mathbb{R}^3} \nabla \cdot \left\{\hat{e}_i \cdot (\epsilon(\boldsymbol{r}) - \epsilon \mathbf{I}_3)\right\} W_j(\boldsymbol{r})\, dv$$

Further simplifications employing the symmetry of the permittivity dyadic give

$$\iiint_{V_s} \hat{e}_i \cdot (\epsilon(\boldsymbol{r}) - \epsilon \mathbf{I}_3) \cdot \nabla W_j(\boldsymbol{r})\, dv = -\iiint_{\mathbb{R}^3} \nabla \cdot \left\{\epsilon(\boldsymbol{r}) \cdot \hat{e}_i\right\} W_j(\boldsymbol{r})\, dv$$
$$= -\iiint_{\mathbb{R}^3} \nabla \cdot \{\epsilon(\boldsymbol{r}) \cdot \nabla W_i(\boldsymbol{r})\} W_j(\boldsymbol{r})\, dv$$
$$= \iiint_{\mathbb{R}^3} \nabla W_i(\boldsymbol{r}) \cdot \epsilon(\boldsymbol{r}) \cdot \nabla W_j(\boldsymbol{r})\, dv$$

where the differential equation (6.21) has been used.

The polarizability dyadic can therefore be written as

$$\gamma_e = \iiint_{V_s} \chi(\boldsymbol{r})\, dv + \mathbf{D}$$

where the symmetric dyadic $\mathbf{D}$ is

$$\hat{e}_i \cdot \mathbf{D} \cdot \hat{e}_j = -\iiint_{\mathbb{R}^3} \hat{e}_i \cdot \chi(\boldsymbol{r}) \cdot \nabla W_j(\boldsymbol{r})\, dv$$
$$= -\iiint_{\mathbb{R}^3} \nabla W_i(\boldsymbol{r}) \cdot \frac{\epsilon(\boldsymbol{r})}{\epsilon} \cdot \nabla W_j(\boldsymbol{r})\, dv \qquad i,j = 1, 2, 3 \tag{6.22}$$

From (6.22), we conclude that the diagonal elements of $\mathbf{D}$ are all non-positive, since the permittivity dyadic ϵ is positive definite. Moreover, by the Cauchy–Schwarz inequality, we get, since the integrand is positive definite

$$\left|\hat{e}_i \cdot \mathbf{D} \cdot \hat{e}_j\right|^2 \le \left(\hat{e}_i \cdot \mathbf{D} \cdot \hat{e}_i\right)\left(\hat{e}_j \cdot \mathbf{D} \cdot \hat{e}_j\right), \qquad i,j = 1, 2, 3$$

[13] If the permittivity shows a finite jump discontinuity at a regular surface, this case can be treated by a limit procedure of this smooth case.

and upper bounds on the diagonal elements

$$\hat{\boldsymbol{e}}_i \cdot \boldsymbol{\gamma}_{\mathrm{e}} \cdot \hat{\boldsymbol{e}}_i \leq \iiint\limits_{V_{\mathrm{s}}} \hat{\boldsymbol{e}}_i \cdot \boldsymbol{\chi}(\boldsymbol{r}) \cdot \hat{\boldsymbol{e}}_i \, \mathrm{d}v, \qquad i = 1, 2, 3$$

Notice that the right-hand side of this inequality is the arithmetic mean of $\hat{\boldsymbol{e}}_i \cdot \boldsymbol{\chi}(\boldsymbol{r}) \cdot \hat{\boldsymbol{e}}_i$ over the scatterer in V_{S}.

It is also possible to derive a lower bound of the diagonal elements by the use of (6.22). We have

$$\begin{aligned}
\hat{\boldsymbol{e}}_i \cdot \boldsymbol{\gamma}_{\mathrm{e}} \cdot \hat{\boldsymbol{e}}_i &= \iiint\limits_{\mathbb{R}^3} \hat{\boldsymbol{e}}_i \cdot \boldsymbol{\chi}(\boldsymbol{r}) \cdot \hat{\boldsymbol{e}}_i \, \mathrm{d}v \\
&\quad - 2 \iiint\limits_{\mathbb{R}^3} \hat{\boldsymbol{e}}_i \cdot \boldsymbol{\chi}(\boldsymbol{r}) \cdot \nabla W_i(\boldsymbol{r}) \, \mathrm{d}v + \iiint\limits_{\mathbb{R}^3} \nabla W_i(\boldsymbol{r}) \cdot \frac{\boldsymbol{\epsilon}(\boldsymbol{r})}{\epsilon} \cdot \nabla W_i(\boldsymbol{r}) \, \mathrm{d}v \\
&= \iiint\limits_{\mathbb{R}^3} \hat{\boldsymbol{e}}_i \cdot \left(\mathbf{I}_3 - \epsilon \boldsymbol{\epsilon}^{-1}(\boldsymbol{r})\right) \cdot \hat{\boldsymbol{e}}_i \, \mathrm{d}v + \iiint\limits_{\mathbb{R}^3} \boldsymbol{f}(\boldsymbol{r}) \cdot \frac{\boldsymbol{\epsilon}(\boldsymbol{r})}{\epsilon} \cdot \boldsymbol{f}(\boldsymbol{r}) \, \mathrm{d}v \\
&\geq \iiint\limits_{\mathbb{R}^3} \hat{\boldsymbol{e}}_i \cdot \left(\mathbf{I}_3 - \epsilon \boldsymbol{\epsilon}^{-1}(\boldsymbol{r})\right) \cdot \hat{\boldsymbol{e}}_i \, \mathrm{d}v = \iiint\limits_{V_{\mathrm{s}}} \hat{\boldsymbol{e}}_i \cdot \left(\mathbf{I}_3 - \epsilon \boldsymbol{\epsilon}^{-1}(\boldsymbol{r})\right) \cdot \hat{\boldsymbol{e}}_i \, \mathrm{d}v
\end{aligned}$$

where $\boldsymbol{f}(\boldsymbol{r}) = \nabla W_i(\boldsymbol{r}) - \left(\mathbf{I}_3 - \epsilon \boldsymbol{\epsilon}^{-1}(\boldsymbol{r})\right) \cdot \hat{\boldsymbol{e}}_i$. In combination, we have

$$\iiint\limits_{V_{\mathrm{s}}} \hat{\boldsymbol{e}}_i \cdot \left(\mathbf{I}_3 - \epsilon \boldsymbol{\epsilon}^{-1}(\boldsymbol{r})\right) \cdot \hat{\boldsymbol{e}}_i \, \mathrm{d}v \leq \hat{\boldsymbol{e}}_i \cdot \boldsymbol{\gamma}_{\mathrm{e}} \cdot \hat{\boldsymbol{e}}_i \leq \iiint\limits_{V_{\mathrm{s}}} \hat{\boldsymbol{e}}_i \cdot \boldsymbol{\chi}(\boldsymbol{r}) \cdot \hat{\boldsymbol{e}}_i \, \mathrm{d}v, \qquad i = 1, 2, 3$$

The entries of the polarizability dyadic, $\boldsymbol{\gamma}_{\mathrm{e}}$, depend on the geometry and the material parameter, $\boldsymbol{\epsilon}(\boldsymbol{r})$, of the obstacle. Intuitively, we expect that the eigenvalues of the polarizability dyadic are increasing functions of the material parameter. This is now given a more precise meaning, and to this end, we denote the solutions to the problems with material parameters $\boldsymbol{\epsilon}(\boldsymbol{r})$ and $\boldsymbol{\epsilon}'(\boldsymbol{r})$ by $W_i(\boldsymbol{r})$ and $W_i'(\boldsymbol{r})$, respectively, i.e., they satisfy

$$\nabla \cdot \left(\boldsymbol{\epsilon}'(\boldsymbol{r}) \cdot \nabla W_i'(\boldsymbol{r}) - \boldsymbol{\epsilon}'(\boldsymbol{r}) \cdot \hat{\boldsymbol{e}}_i\right) = 0, \qquad i = 1, 2, 3$$

and

$$\nabla \cdot \left(\boldsymbol{\epsilon}(\boldsymbol{r}) \cdot \nabla W_i(\boldsymbol{r}) - \boldsymbol{\epsilon}(\boldsymbol{r}) \cdot \hat{\boldsymbol{e}}_i\right) = 0, \qquad i = 1, 2, 3$$

By the use of a similar technique as above and (6.22), we derive the following identity:

$$\begin{aligned}
&\iiint\limits_{\mathbb{R}^3} \hat{e}_i \cdot \left(\epsilon'(r) - \epsilon(r)\right) \cdot \nabla W_j'(r)\, \mathrm{d}v \\
&= -\iiint\limits_{\mathbb{R}^3} \nabla \cdot \left\{\left(\epsilon'(r) - \epsilon(r)\right) \cdot \hat{e}_i\right\} W_j'(r)\, \mathrm{d}v \\
&= -\iiint\limits_{\mathbb{R}^3} \left\{\nabla \cdot \left\{\epsilon'(r) \cdot \nabla W_i'(r)\right\} W_j'(r) - \nabla \cdot \left\{\epsilon(r) \cdot \nabla W_i(r)\right\} W_j'(r)\right\} \mathrm{d}v \\
&= \iiint\limits_{\mathbb{R}^3} \nabla W_i'(r) \cdot \epsilon'(r) \cdot \nabla W_j'(r)\, \mathrm{d}v - \iiint\limits_{\mathbb{R}^3} \nabla W_j'(r) \cdot \epsilon(r) \cdot \nabla W_i(r)\, \mathrm{d}v \\
&= -\epsilon \hat{e}_i \cdot \mathbf{D}' \cdot \hat{e}_j - \iiint\limits_{\mathbb{R}^3} \nabla W_j'(r) \cdot \epsilon(r) \cdot \nabla W_i(r)\, \mathrm{d}v
\end{aligned}$$

where

$$\hat{e}_i \cdot \mathbf{D}' \cdot \hat{e}_j = -\iiint\limits_{\mathbb{R}^3} \nabla W_i'(r) \cdot \frac{\epsilon'(r)}{\epsilon} \cdot \nabla W_j'(r)\, \mathrm{d}v \quad i,j = 1,2,3$$

Similar analysis gives

$$\iiint\limits_{\mathbb{R}^3} \hat{e}_i \cdot \left(\epsilon'(r) - \epsilon(r)\right) \cdot \nabla W_j(r)\, \mathrm{d}v = \epsilon \hat{e}_i \cdot \mathbf{D} \cdot \hat{e}_j + \iiint\limits_{\mathbb{R}^3} \nabla W_j(r) \cdot \epsilon'(r) \cdot \nabla W_i'(r)\, \mathrm{d}v$$

The difference in the components of the dyadic $\mathbf{D}$ then is

$$\begin{aligned}
\epsilon \hat{e}_i \cdot \left(\mathbf{D}' - \mathbf{D}\right) \cdot \hat{e}_j = &\iiint\limits_{\mathbb{R}^3} \nabla W_j(r) \cdot \epsilon'(r) \cdot \nabla W_i'(r)\, \mathrm{d}v - \iiint\limits_{\mathbb{R}^3} \nabla W_j'(r) \cdot \epsilon(r) \cdot \nabla W_i(r)\, \mathrm{d}v \\
&- \iiint\limits_{\mathbb{R}^3} \hat{e}_i \cdot \left(\epsilon'(r) - \epsilon(r)\right) \cdot \left(\nabla W_j'(r) + \nabla W_j(r)\right) \mathrm{d}v
\end{aligned}$$

Denote $\epsilon'(r) = \epsilon(r) + \varepsilon\epsilon\eta(r)$, where $\eta(r)$ is a symmetric, positive-definite dyadic, and let eventually $\varepsilon \to 0$. Then

$$\begin{aligned}
\hat{e}_i \cdot \left(\gamma_e' - \gamma_e\right) \cdot \hat{e}_j = \varepsilon &\iiint\limits_{\mathbb{R}^3} \nabla W_j(r) \cdot \eta(r) \cdot \nabla W_i'(r)\, \mathrm{d}v \\
&+ \iiint\limits_{\mathbb{R}^3} \nabla W_j(r) \cdot \frac{\epsilon(r)}{\epsilon} \cdot \nabla W_i'(r)\, \mathrm{d}v \\
&- \iiint\limits_{\mathbb{R}^3} \nabla W_j'(r) \cdot \frac{\epsilon(r)}{\epsilon} \cdot \nabla W_i(r)\, \mathrm{d}v \\
&- \varepsilon \iiint\limits_{\mathbb{R}^3} \hat{e}_i \cdot \eta(r) \cdot \left(\nabla W_j'(r) + \nabla W_j(r) - \hat{e}_j\right) \mathrm{d}v
\end{aligned}$$

To prove that the dyadic $\boldsymbol{\gamma}'_{\rm e}-\boldsymbol{\gamma}_{\rm e}$ is positive semi-definite, multiply with arbitrary real constants a_i and a_j, and sum over i and j. If we denote

$$\begin{cases} \boldsymbol{a}(\boldsymbol{r}) = \sum_{i=1}^{3} a_i \nabla W_i(\boldsymbol{r}) \\ \boldsymbol{a}'(\boldsymbol{r}) = \sum_{i=1}^{3} a_i \nabla W'_i(\boldsymbol{r}) \end{cases} \qquad \boldsymbol{c} = \sum_{i=1}^{3} a_i \hat{\boldsymbol{e}}_i$$

we get (use the symmetry of the permittivity dyadic $\boldsymbol{\epsilon}$)

$$\begin{aligned} \boldsymbol{c}\cdot\left(\boldsymbol{\gamma}'_{\rm e}-\boldsymbol{\gamma}_{\rm e}\right)\cdot\boldsymbol{c} &= \varepsilon \iiint_{\mathbb{R}^3} \boldsymbol{a}(\boldsymbol{r})\cdot\boldsymbol{\eta}(\boldsymbol{r})\cdot\boldsymbol{a}'(\boldsymbol{r})\,{\rm d}v + \iiint_{\mathbb{R}^3} \boldsymbol{a}(\boldsymbol{r})\cdot\frac{\boldsymbol{\epsilon}(\boldsymbol{r})}{\epsilon}\cdot\boldsymbol{a}'(\boldsymbol{r})\,{\rm d}v \\ &- \iiint_{\mathbb{R}^3} \boldsymbol{a}'(\boldsymbol{r})\cdot\frac{\boldsymbol{\epsilon}(\boldsymbol{r})}{\epsilon}\cdot\boldsymbol{a}(\boldsymbol{r})\,{\rm d}v - \varepsilon \iiint_{\mathbb{R}^3} \boldsymbol{c}\cdot\boldsymbol{\eta}(\boldsymbol{r})\cdot\left(\boldsymbol{a}'(\boldsymbol{r})+\boldsymbol{a}(\boldsymbol{r})-\boldsymbol{c}\right){\rm d}v \\ &= \varepsilon \iiint_{\mathbb{R}^3} \boldsymbol{a}(\boldsymbol{r})\cdot\boldsymbol{\eta}(\boldsymbol{r})\cdot\boldsymbol{a}'(\boldsymbol{r})\,{\rm d}v - \varepsilon \iiint_{\mathbb{R}^3} \boldsymbol{c}\cdot\boldsymbol{\eta}(\boldsymbol{r})\cdot\left(\boldsymbol{a}'(\boldsymbol{r})+\boldsymbol{a}(\boldsymbol{r})-\boldsymbol{c}\right){\rm d}v \end{aligned}$$

The permittivities $\boldsymbol{\epsilon}'$ and $\boldsymbol{\epsilon}$ differ by the order ε. To first order, the solutions also differ by ε, i.e., $\boldsymbol{a}'(\boldsymbol{r}) = \boldsymbol{a}(\boldsymbol{r}) + O(\varepsilon)$, and

$$\boldsymbol{c}\cdot\left(\boldsymbol{\gamma}'_{\rm e}-\boldsymbol{\gamma}_{\rm e}\right)\cdot\boldsymbol{c} = \underbrace{\varepsilon \iiint_{\mathbb{R}^3} \left(\boldsymbol{a}(\boldsymbol{r})-\boldsymbol{c}\right)\cdot\boldsymbol{\eta}(\boldsymbol{r})\cdot\left(\boldsymbol{a}(\boldsymbol{r})-\boldsymbol{c}\right){\rm d}v}_{\geq 0} + O(\varepsilon^2)$$

Provided $\boldsymbol{\eta}(\boldsymbol{r}) \geq 0$, the dyadic $\boldsymbol{\gamma}'_{\rm e}-\boldsymbol{\gamma}_{\rm e}$ is positive semi-definite and it has positive eigenvalues, i.e., the eigenvalues of polarizability dyadic increase. Thus, any local increase of the permittivity in the obstacle leads to an increase in the eigenvalues. We summarize this result in

$$\boldsymbol{\epsilon}'(\boldsymbol{r}) \geq \boldsymbol{\epsilon}(\boldsymbol{r}) \quad \Longrightarrow \quad \boldsymbol{\gamma}'_{\rm e} \geq \boldsymbol{\gamma}_{\rm e}$$

6.1.6.1 Variational Result

We conclude this section by proving a variational result that is related to the result above. The result is general, and we choose to apply the result to the permittivity dyadic $\boldsymbol{\epsilon}$, but it can be applied to any dyadic that fulfill the assumptions made.

Let $\boldsymbol{\epsilon}$ be a symmetric, positive semi-definite dyadic, that assumes a constant value $\epsilon\mathbf{I}_3$ outside a finite domain. All continuous functions u with piecewise continuous first partial derivatives and whose behavior at infinity is $u = O(1/r)$ and $\nabla u = O(1/r^2)$ then satisfies, $i = 1, 2, 3$

$$\hat{\boldsymbol{e}}_i\cdot\mathbf{D}\cdot\hat{\boldsymbol{e}}_i \leq \iiint_{\mathbb{R}^3} \left(-2\hat{\boldsymbol{e}}_i\cdot\boldsymbol{\chi}(\boldsymbol{r})\cdot\nabla u(\boldsymbol{r}) + \nabla u(\boldsymbol{r})\cdot\frac{\boldsymbol{\epsilon}(\boldsymbol{r})}{\epsilon}\cdot\nabla u(\boldsymbol{r})\right){\rm d}v$$

where $\mathbf{D}$ is defined in (6.22). Equality holds, if and only if $u = W_i$, where W_i is the solution to (6.21).

The proof starts with the observation that equality holds if $u = W_i$. This is seen from (6.22). To proceed with the proof, let $W_i = u + h$. The function h satisfies the condition at infinity, since W_i does. Then, since $\nabla \cdot (\boldsymbol{\epsilon} \cdot \nabla W_i) = \nabla \cdot (\boldsymbol{\epsilon} \cdot \hat{e}_i) = \epsilon \nabla \cdot (\boldsymbol{\chi} \cdot \hat{e}_i)$, we obtain by Gauss' theorem

$$\begin{aligned}
\iiint\limits_{\mathbb{R}^3} \hat{e}_i \cdot \boldsymbol{\chi}(\boldsymbol{r}) \cdot \nabla u(\boldsymbol{r})\, \mathrm{d}v &= -\iiint\limits_{\mathbb{R}^3} \nabla \cdot \left(\boldsymbol{\chi}(\boldsymbol{r}) \cdot \hat{e}_i\right) u(\boldsymbol{r})\, \mathrm{d}v \\
&= -\iiint\limits_{\mathbb{R}^3} \nabla \cdot \left(\frac{\boldsymbol{\epsilon}(\boldsymbol{r})}{\epsilon} \cdot \nabla W_i(\boldsymbol{r})\right) u(\boldsymbol{r})\, \mathrm{d}v \\
&= \iiint\limits_{\mathbb{R}^3} \nabla u(\boldsymbol{r}) \cdot \frac{\boldsymbol{\epsilon}(\boldsymbol{r})}{\epsilon} \cdot \nabla W_i(\boldsymbol{r})\, \mathrm{d}v \\
&= \iiint\limits_{\mathbb{R}^3} \nabla u(\boldsymbol{r}) \cdot \frac{\boldsymbol{\epsilon}(\boldsymbol{r})}{\epsilon} \cdot \nabla u(\boldsymbol{r})\, \mathrm{d}v + \iiint\limits_{\mathbb{R}^3} \nabla u(\boldsymbol{r}) \cdot \frac{\boldsymbol{\epsilon}(\boldsymbol{r})}{\epsilon} \cdot \nabla h(\boldsymbol{r})\, \mathrm{d}v
\end{aligned}$$

and similarly

$$\begin{aligned}
\iiint\limits_{\mathbb{R}^3} \hat{e}_i \cdot \boldsymbol{\chi}(\boldsymbol{r}) \cdot \nabla h(\boldsymbol{r})\, \mathrm{d}v &= \iiint\limits_{\mathbb{R}^3} \nabla h(\boldsymbol{r}) \cdot \frac{\boldsymbol{\epsilon}(\boldsymbol{r})}{\epsilon} \cdot \nabla W_i(\boldsymbol{r})\, \mathrm{d}v \\
&= \iiint\limits_{\mathbb{R}^3} \nabla h(\boldsymbol{r}) \cdot \frac{\boldsymbol{\epsilon}(\boldsymbol{r})}{\epsilon} \cdot \nabla h(\boldsymbol{r})\, \mathrm{d}v + \iiint\limits_{\mathbb{R}^3} \nabla h(\boldsymbol{r}) \cdot \frac{\boldsymbol{\epsilon}(\boldsymbol{r})}{\epsilon} \cdot \nabla u(\boldsymbol{r})\, \mathrm{d}v
\end{aligned}$$

This leads to

$$\begin{aligned}
\hat{e}_i \cdot \mathbf{D} \cdot \hat{e}_i = -\iiint\limits_{\mathbb{R}^3} \hat{e}_i \cdot \boldsymbol{\chi}(\boldsymbol{r}) \cdot \nabla W_i(\boldsymbol{r})\, \mathrm{d}v &= -2\iiint\limits_{\mathbb{R}^3} \hat{e}_i \cdot \boldsymbol{\chi}(\boldsymbol{r}) \cdot \nabla u(\boldsymbol{r})\, \mathrm{d}v \\
&+ \iiint\limits_{\mathbb{R}^3} \hat{e}_i \cdot \boldsymbol{\chi}(\boldsymbol{r}) \cdot \nabla u(\boldsymbol{r})\, \mathrm{d}v - \iiint\limits_{\mathbb{R}^3} \hat{e}_i \cdot \boldsymbol{\chi}(\boldsymbol{r}) \cdot \nabla h(\boldsymbol{r})\, \mathrm{d}v
\end{aligned}$$

and

$$\begin{aligned}
\hat{e}_i \cdot \mathbf{D} \cdot \hat{e}_i &= -2\iiint\limits_{\mathbb{R}^3} \hat{e}_i \cdot \boldsymbol{\chi}(\boldsymbol{r}) \cdot \nabla u(\boldsymbol{r})\, \mathrm{d}v + \iiint\limits_{\mathbb{R}^3} \nabla u(\boldsymbol{r}) \cdot \frac{\boldsymbol{\epsilon}(\boldsymbol{r})}{\epsilon} \cdot \nabla u(\boldsymbol{r})\, \mathrm{d}v \\
&- \iiint\limits_{\mathbb{R}^3} \nabla h(\boldsymbol{r}) \cdot \frac{\boldsymbol{\epsilon}(\boldsymbol{r})}{\epsilon} \cdot \nabla h(\boldsymbol{r})\, \mathrm{d}v \\
&\leq -2\iiint\limits_{\mathbb{R}^3} \hat{e}_i \cdot \boldsymbol{\chi}(\boldsymbol{r}) \cdot \nabla u(\boldsymbol{r})\, \mathrm{d}v + \iiint\limits_{\mathbb{R}^3} \nabla u(\boldsymbol{r}) \cdot \frac{\boldsymbol{\epsilon}(\boldsymbol{r})}{\epsilon} \cdot \nabla u(\boldsymbol{r})\, \mathrm{d}v
\end{aligned}$$

with equality if and only if $h =$ constant. This constant has to be zero due to the conditions at infinity, which proves the variational result.

6.2 Weak-Scatterer Approximation

In this section, we investigate some of the simplifications that can be made if the material parameters of the scatterer deviate very little from the parameters of the surrounding material.

During the investigation of the internal fields in Section 6.1.5, we encountered an integral relation of the electric field in the scatterer. We have for all $\boldsymbol{r} \in \mathbb{R}^3$, see (6.15)

$$\boldsymbol{E}(\boldsymbol{r}) = \boldsymbol{E}_{\mathrm{i}}(\boldsymbol{r}) + k^2 \left[\mathbf{I}_3 + \frac{1}{k^2}\nabla\nabla \right] \cdot \iiint\limits_{V_s} \frac{\mathrm{e}^{\mathrm{i}k|\boldsymbol{r}-\boldsymbol{r}'|}}{4\pi |\boldsymbol{r}-\boldsymbol{r}'|} \left(\frac{\epsilon_1(\boldsymbol{r}')}{\epsilon} - \mathbf{I}_3 \right) \cdot \boldsymbol{E}(\boldsymbol{r}') \, \mathrm{d}v'$$

where we have assumed that the scatterer has the same magnetic properties as the exterior region, i.e., $\mu_1(\boldsymbol{r}) = \mu$ everywhere. If we restrict the position vector to the scatterer, V_s, we obtain an integral equation in the unknown electric field $\boldsymbol{E}(\boldsymbol{r})$ in V_s. With the solution of this integral equation, the far field amplitude is obtained by integration, see (6.16)

$$\boldsymbol{F}(\hat{\boldsymbol{r}}) = k^2 \left[\mathbf{I}_3 - \hat{\boldsymbol{r}}\hat{\boldsymbol{r}} \right] \cdot \iiint\limits_{V_s} \frac{\mathrm{e}^{-\mathrm{i}k\hat{\boldsymbol{r}}\cdot\boldsymbol{r}'}}{4\pi} \left(\frac{\epsilon_1(\boldsymbol{r}')}{\epsilon} - \mathbf{I}_3 \right) \cdot \boldsymbol{E}(\boldsymbol{r}') \, \mathrm{d}v'$$

or

$$\boldsymbol{F}(\hat{\boldsymbol{r}}) = \hat{\boldsymbol{r}} \times \left(\boldsymbol{K}(\hat{\boldsymbol{r}}) \times \hat{\boldsymbol{r}} \right)$$

where

$$\boldsymbol{K}(\hat{\boldsymbol{r}}) = \frac{k^2}{4\pi} \iiint\limits_{V_s} \mathrm{e}^{-\mathrm{i}k\hat{\boldsymbol{r}}\cdot\boldsymbol{r}'} \left(\frac{\epsilon_1(\boldsymbol{r}')}{\epsilon} - \mathbf{I}_3 \right) \cdot \boldsymbol{E}(\boldsymbol{r}') \, \mathrm{d}v'$$

So far, no approximations are made. However, if the scatterer is weak, considerable simplifications can be made. This is done in the section below.

6.2.1 Born Approximation

We now investigate what simplifications that can be made if the material parameters differ just little from the surrounding material parameters.

By a weak scatterer, we mean

$$\left\| \frac{\epsilon_1(\boldsymbol{r}')}{\epsilon} - \mathbf{I}_3 \right\| \ll 1$$

where $\| \cdot \|$ denotes the norm of the dyadic or matrix. This suggest that we can approximate the unknown field inside the integral of the far field amplitude above with $\boldsymbol{E}(\boldsymbol{r}) \approx \boldsymbol{E}_{\mathrm{i}}(\boldsymbol{r})$. This approximation is called the Born[14] approximation or the weak-scatterer approximation, and it very useful in many applications.

[14]Max Born (1882–1970), German physicist and mathematician.

The far field amplitude is now explicit. In fact, we have

$$\boldsymbol{K}(\hat{\boldsymbol{r}}) = \frac{k^2}{4\pi} \iiint\limits_{V_s} \mathrm{e}^{-\mathrm{i}k\hat{\boldsymbol{r}}\cdot\boldsymbol{r}'} \left(\frac{\epsilon_1(\boldsymbol{r}')}{\epsilon} - \mathbf{I}_3 \right) \cdot \boldsymbol{E}_\mathrm{i}(\boldsymbol{r}')\, \mathrm{d}v'$$

where $\boldsymbol{E}_\mathrm{i}(\boldsymbol{r})$ denotes the known incident field.

Example 6.5

A simple example that illustrates the Born approximation is scattering of a plane wave by a non-magnetic, homogeneous dielectric sphere (permittivity ϵ_1) of radius a.

The far field amplitude in Born approximation is given by

$$\boldsymbol{K}(\hat{\boldsymbol{r}}) = \frac{k^2 \boldsymbol{E}_0}{4\pi} \left(\frac{\epsilon_1}{\epsilon} - 1 \right) \iiint\limits_{V_s} \mathrm{e}^{\mathrm{i}k(\hat{\boldsymbol{k}}_\mathrm{i} - \hat{\boldsymbol{r}})\cdot\boldsymbol{r}'}\, \mathrm{d}v'$$

where the plane wave excitation is

$$\boldsymbol{E}_\mathrm{i}(\boldsymbol{r}) = \boldsymbol{E}_0 \mathrm{e}^{\mathrm{i}k\hat{\boldsymbol{k}}_\mathrm{i}\cdot\boldsymbol{r}}$$

The integration is easily done in spherical coordinates with the z-axis oriented along the direction $\hat{\boldsymbol{k}}_\mathrm{i} - \hat{\boldsymbol{r}}$. We get

$$\begin{aligned}
\boldsymbol{K}(\hat{\boldsymbol{r}}) &= \frac{k^2 \boldsymbol{E}_0}{2} \left(\frac{\epsilon_1}{\epsilon} - 1 \right) \int_0^\pi \sin\theta'\, \mathrm{d}\theta' \int_0^a r'^2\, \mathrm{d}r' \mathrm{e}^{\mathrm{i}k_b r' \cos\theta'} \\
&= \frac{k^2 \boldsymbol{E}_0}{k_b} \left(\frac{\epsilon_1}{\epsilon} - 1 \right) \int_0^a r' \sin k_b r'\, \mathrm{d}r'
\end{aligned}$$

where $\boldsymbol{k}_b = k(\hat{\boldsymbol{k}}_\mathrm{i} - \hat{\boldsymbol{r}})$, $k_b = |\boldsymbol{k}_b|$, and θ' is the angle between $\boldsymbol{k}_b$ and $\boldsymbol{r}'$. Evaluation of the r' integral gives

$$\boldsymbol{K}(\hat{\boldsymbol{r}}) = \frac{k^2 \boldsymbol{E}_0}{k_b^3} \left(\frac{\epsilon_1}{\epsilon} - 1 \right) (\sin k_b a - k_b a \cos k_b a)$$

The entries of the scattering matrix then are, see Example 4.2 on page 208

$$[\mathbf{S}]\,(\hat{\boldsymbol{r}}, \hat{\boldsymbol{k}}_\mathrm{i}) = \begin{pmatrix} S_{\parallel\parallel} & S_{\parallel\perp} \\ S_{\perp\parallel} & S_{\perp\perp} \end{pmatrix} = G \begin{pmatrix} \cos\theta & 0 \\ 0 & 1 \end{pmatrix}$$

where θ is the angle between $\hat{\boldsymbol{k}}_\mathrm{i}$ and $\hat{\boldsymbol{r}}$, and

$$G = \frac{k^2}{k_b^3} \left(\frac{\epsilon_1}{\epsilon} - 1 \right) (\sin k_b a - k_b a \cos k_b a)$$

■

6.3 High-Frequency Approximation

In Section 6.1, we introduced the long wavelength approximation of the scattering problem. We now focus on approximations in another part of the spectrum, and assume that the length scale of the scatterer is large compared with the wavelength—high-frequency approximations. This high-frequency limit is a difficult problem indeed, and with an extensive literature. Examples of early and later attempts to solve this difficult problem are [26, 31, 44, 108–110, 121, 122, 128, 168, 181].

We start by finding methods that are useful in solving scattering problems at high frequencies. Physical arguments are often used to motivate some of the methods presented in this section. These arguments make by no means the methods less accurate or applicable. In fact, some of the methods are widely used in many high-frequency methods. As in the previous sections, the analysis in not restricted to an incident plane wave, unless this is pointed out in the text. We prefer to introduce these high-frequency methods by the analysis of an explicit antenna example—the reflector antenna.

The starting point for representation of the scattered field is the exact surface integral representation of the total field, see (4.15) on page 197

$$
\begin{aligned}
&-\frac{\eta_0\eta}{\mathrm{i}k}\nabla\times\left\{\nabla\times\iint_{S_\mathrm{s}} g(k,|\boldsymbol{r}-\boldsymbol{r}'|)\left(\hat{\boldsymbol{\nu}}(\boldsymbol{r}')\times\boldsymbol{H}(\boldsymbol{r}')\right)\,\mathrm{d}S'\right\}\\
&+\nabla\times\iint_{S_\mathrm{s}} g(k,|\boldsymbol{r}-\boldsymbol{r}'|)\left(\hat{\boldsymbol{\nu}}(\boldsymbol{r}')\times\boldsymbol{E}(\boldsymbol{r}')\right)\,\mathrm{d}S'=\begin{cases}\boldsymbol{E}_\mathrm{s}(\boldsymbol{r}),\ \boldsymbol{r}\text{ outside }S_\mathrm{s}\\-\boldsymbol{E}_\mathrm{i}(\boldsymbol{r}),\ \boldsymbol{r}\text{ inside }S_\mathrm{s}\end{cases}
\end{aligned}
$$

High-frequency approximation is often used in antenna applications, and a typical geometry (passive reflector antenna) is depicted in Figure 6.2. The incident field is here generated by a feeding horn. The scatterer consists of a metallic reflector. The unit normal $\hat{\boldsymbol{\nu}}$ of the scatterer is defined in Figure 6.2. The boundary condition on the metallic reflector is, see (1.13) on page 8

$$
\hat{\boldsymbol{\nu}}\times\boldsymbol{E}=\mathbf{0}
$$

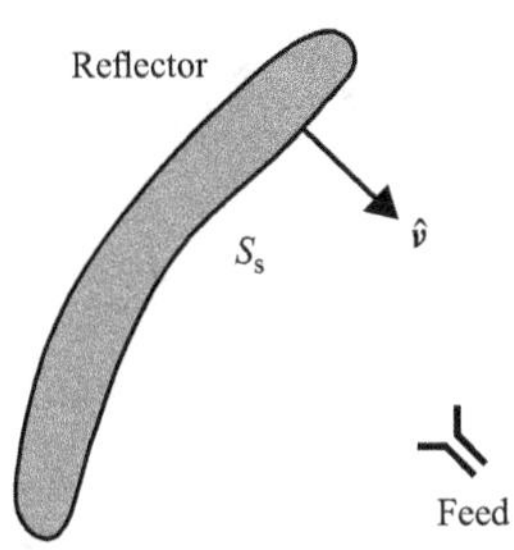

Figure 6.2 Example of a geometry in an antenna application.

This boundary condition implies that the scattered field (outside the scatterer) is

$$\boldsymbol{E}_{\mathrm{s}}(\boldsymbol{r}) = -\frac{\eta_0\eta}{\mathrm{i}k}\nabla \times \left\{ \nabla \times \iint_{S_{\mathrm{s}}} g(k, |\boldsymbol{r}-\boldsymbol{r}'|)\left(\hat{\boldsymbol{\nu}}(\boldsymbol{r}') \times \boldsymbol{H}(\boldsymbol{r}')\right) \mathrm{d}S' \right\}$$

We identify the unknown field $\hat{\boldsymbol{\nu}} \times \boldsymbol{H}$ on the scatterer as the surface current density $\boldsymbol{J}_S$, see (1.13) on page 8. Another way of writing the scattered field then is

$$\boldsymbol{E}_{\mathrm{s}}(\boldsymbol{r}) = -\frac{\eta_0\eta}{\mathrm{i}k}\nabla \times \left\{ \nabla \times \iint_{S_{\mathrm{s}}} g(k, |\boldsymbol{r}-\boldsymbol{r}'|)\boldsymbol{J}_S(\boldsymbol{r}') \,\mathrm{d}S' \right\}$$

The feed horn, which is the source, generates an incident field, $\boldsymbol{E}_{\mathrm{i}}$, that induces a surface current density, $\boldsymbol{J}_S$, on the reflector so that the boundary conditions are fulfilled. At short wavelengths, the scatterer can be divided into two zones or regions—an illuminated (lit) zone and a shadow zone. By experience, the high-frequency approximation methods give the best results if the geometry of the scatterer has a clear division between the lit and the shadow zones. Geometries with deep concavities give less accurate results.

The border between the lit and shadow zones is constructed by geometrical optics, i.e., straight rays as in optics. In Figure 6.3, the shadow zone is denoted S_{s}^-, and the lit zone S_{s}^+. These surfaces are in general open surfaces. The unit normal vectors of the two surfaces are defined in Figure 6.3.

In the shadow zone, the electromagnetic fields are weak, and we approximate them by zero fields. This approximation holds better and better as frequency gets higher and higher. Therefore, the scattered field can be approximated as a surface integral over the lit surface S_{s}^+

$$\boldsymbol{E}_{\mathrm{s}}(\boldsymbol{r}) \approx -\frac{\eta_0\eta}{\mathrm{i}k}\nabla \times \left\{ \nabla \times \iint_{S_{\mathrm{s}}^+} g(k, |\boldsymbol{r}-\boldsymbol{r}'|)\boldsymbol{J}_S(\boldsymbol{r}') \,\mathrm{d}S' \right\} \tag{6.23}$$

The scattered field $\boldsymbol{E}_{\mathrm{s}}$ can be determined by integration as soon as we know the surface current density, $\boldsymbol{J}_S$, on the illuminated zone of the scatterer. The surface current density, $\boldsymbol{J}_S$, is in most situations very complicated to find and the solution typically involves a boundary value problem or integral equation to solve. As a consequence, one prefers to use accurate

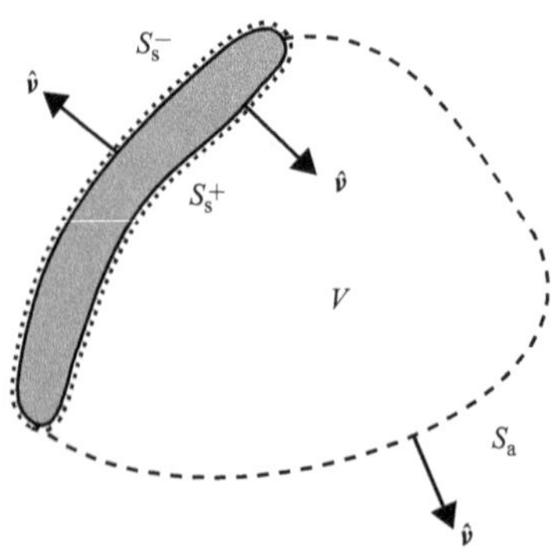

Figure 6.3 The illuminated (lit) zone S_{s}^+, the shadow zone S_{s}^-, and the aperture surface S_{a}.

approximation of the surface current. If these approximations reflect the underlying physics well, it is likely that also the scattered field is accurately modeled.

6.3.1 Aperture Formulation

Some useful and often used approximations at high frequencies are presented below, but first, we present an alternative surface integral representation of the scattered field over a surface partly different from S_s. We obtain this alternative representation by an application of the integral representation (3.14) on page 156 on the volume V, which is bounded by the surfaces S_a and S_s^+, see Figure 6.3. Notice that the sources of the incident field are assumed to be located outside the volume V so that the Maxwell equations are source free in V. With the unit normal vectors defined in Figure 6.3 (note that the unit normal $\hat{\boldsymbol{\nu}}$ of S_s^+ points into the volume V), we get[15]

$$
\begin{aligned}
&\frac{\eta_0\eta}{\mathrm{i}k}\nabla\times\left\{\nabla\times\iint_{S_a} g(k,|\boldsymbol{r}-\boldsymbol{r}'|)\left(\hat{\boldsymbol{\nu}}(\boldsymbol{r}')\times\boldsymbol{H}(\boldsymbol{r}')\right)\,\mathrm{d}S'\right\}\\
&-\nabla\times\iint_{S_a} g(k,|\boldsymbol{r}-\boldsymbol{r}'|)\left(\hat{\boldsymbol{\nu}}(\boldsymbol{r}')\times\boldsymbol{E}(\boldsymbol{r}')\right)\,\mathrm{d}S'\\
&-\frac{\eta_0\eta}{\mathrm{i}k}\nabla\times\left\{\nabla\times\iint_{S_s^+} g(k,|\boldsymbol{r}-\boldsymbol{r}'|)\left(\hat{\boldsymbol{\nu}}(\boldsymbol{r}')\times\boldsymbol{H}(\boldsymbol{r}')\right)\,\mathrm{d}S'\right\}\\
&+\nabla\times\iint_{S_s^+} g(k,|\boldsymbol{r}-\boldsymbol{r}'|)\left(\hat{\boldsymbol{\nu}}(\boldsymbol{r}')\times\boldsymbol{E}(\boldsymbol{r}')\right)\,\mathrm{d}S'=\begin{cases}\boldsymbol{E}(\boldsymbol{r}), & \boldsymbol{r}\in V\\ \boldsymbol{0}, & \boldsymbol{r}\notin V\end{cases}
\end{aligned}
$$

If the position vector $\boldsymbol{r}$ is outside V, we get

$$
\begin{aligned}
&-\frac{\eta_0\eta}{\mathrm{i}k}\nabla\times\left\{\nabla\times\iint_{S_s^+} g(k,|\boldsymbol{r}-\boldsymbol{r}'|)\boldsymbol{J}_S(\boldsymbol{r}')\,\mathrm{d}S'\right\}\\
&=-\frac{\eta_0\eta}{\mathrm{i}k}\nabla\times\left\{\nabla\times\iint_{S_a} g(k,|\boldsymbol{r}-\boldsymbol{r}'|)\left(\hat{\boldsymbol{\nu}}(\boldsymbol{r}')\times\boldsymbol{H}(\boldsymbol{r}')\right)\,\mathrm{d}S'\right\}\\
&\quad+\nabla\times\iint_{S_a} g(k,|\boldsymbol{r}-\boldsymbol{r}'|)\left(\hat{\boldsymbol{\nu}}(\boldsymbol{r}')\times\boldsymbol{E}(\boldsymbol{r}')\right)\,\mathrm{d}S'
\end{aligned}
$$

where the boundaries $\hat{\boldsymbol{\nu}}\times\boldsymbol{E}=\boldsymbol{0}$ and $\hat{\boldsymbol{\nu}}\times\boldsymbol{H}=\boldsymbol{J}_S$ have been employed on the surface S_s^+. This implies that the scattered field in (6.23) can be represented in an alternative way ($\boldsymbol{r}$ outside V)

$$
\begin{aligned}
\boldsymbol{E}_s(\boldsymbol{r})\approx&-\frac{\eta_0\eta}{\mathrm{i}k}\nabla\times\left\{\nabla\times\iint_{S_a} g(k,|\boldsymbol{r}-\boldsymbol{r}'|)\left(\hat{\boldsymbol{\nu}}(\boldsymbol{r}')\times\boldsymbol{H}(\boldsymbol{r}')\right)\,\mathrm{d}S'\right\}\\
&+\nabla\times\iint_{S_a} g(k,|\boldsymbol{r}-\boldsymbol{r}'|)\left(\hat{\boldsymbol{\nu}}(\boldsymbol{r}')\times\boldsymbol{E}(\boldsymbol{r}')\right)\,\mathrm{d}S'
\end{aligned}
$$

[15]We follow closely the analysis presented by Yaghjian [297].

An obvious advantage with this alternative formula is that the aperture surface S_a is arbitrary. The potential choice of the aperture surface might be utilized in such a way that the fields $\boldsymbol{E}$ and $\boldsymbol{H}$ on the aperture surface S_a are easy to approximate.

Here, we are particularly interested in the far field amplitude of the scattering by the reflector, and we let $\boldsymbol{r} \to \infty$. Similar to the derivation of (4.18) on page 198, we get

$$\boldsymbol{F}(\hat{\boldsymbol{r}}) = -\frac{\mathrm{i}k\eta_0\eta}{4\pi}\hat{\boldsymbol{r}} \times \left[\hat{\boldsymbol{r}} \times \iint_{S_s^+} \boldsymbol{J}_S(\boldsymbol{r}')\mathrm{e}^{-\mathrm{i}k\hat{\boldsymbol{r}}\cdot\boldsymbol{r}'}\,\mathrm{d}S'\right] \tag{6.24}$$

and

$$\boldsymbol{F}(\hat{\boldsymbol{r}}) = \frac{\mathrm{i}k}{4\pi}\hat{\boldsymbol{r}} \times \iint_{S_a} \left[\hat{\boldsymbol{\nu}}(\boldsymbol{r}') \times \boldsymbol{E}(\boldsymbol{r}') - \eta_0\eta\hat{\boldsymbol{r}} \times \left(\hat{\boldsymbol{\nu}}(\boldsymbol{r}') \times \boldsymbol{H}(\boldsymbol{r}')\right)\right]\mathrm{e}^{-\mathrm{i}k\hat{\boldsymbol{r}}\cdot\boldsymbol{r}'}\,\mathrm{d}S' \tag{6.25}$$

for the two alternative surface integral representations above. Which representation is preferred depends on which approximation of the fields that is used.

6.3.2 Reflection at a Metallic Surface

In this section, we derive a coordinate-free expression of reflection by a plane perfectly conducting surface. The canonical field for the incident field is a plane wave. This reflection problem constitutes the basis of several high-frequency approximations that are treated in this book, and it motivates us to study this scattering process in some detail.

Let $\hat{\boldsymbol{k}}_i$ and $\hat{\boldsymbol{k}}_s$ be propagation directions of the incident and the scattered (reflected), respectively, and let $\hat{\boldsymbol{\nu}}$ denote the unit normal of the surface, see Figure 6.4. We assume that $\hat{\boldsymbol{\nu}} \cdot \hat{\boldsymbol{k}}_i < 0$, so that the incident plane wave actually is incident on the surface with the direction of the surface normal in Figure 6.4.

The incident fields $\boldsymbol{E}_i$ and $\boldsymbol{H}_i$, and the scattered fields $\boldsymbol{E}_s$ and $\boldsymbol{H}_s$ are related to each other by the plane wave relations, cf. (1.72) on page 73

$$\begin{cases} \boldsymbol{E}_i = -\eta_0\eta\hat{\boldsymbol{k}}_i \times \boldsymbol{H}_i \\ \eta_0\eta\boldsymbol{H}_i = \hat{\boldsymbol{k}}_i \times \boldsymbol{E}_i \end{cases} \qquad \begin{cases} \boldsymbol{E}_s = -\eta_0\eta\hat{\boldsymbol{k}}_s \times \boldsymbol{H}_s \\ \eta_0\eta\boldsymbol{H}_s = \hat{\boldsymbol{k}}_s \times \boldsymbol{E}_s \end{cases} \tag{6.26}$$

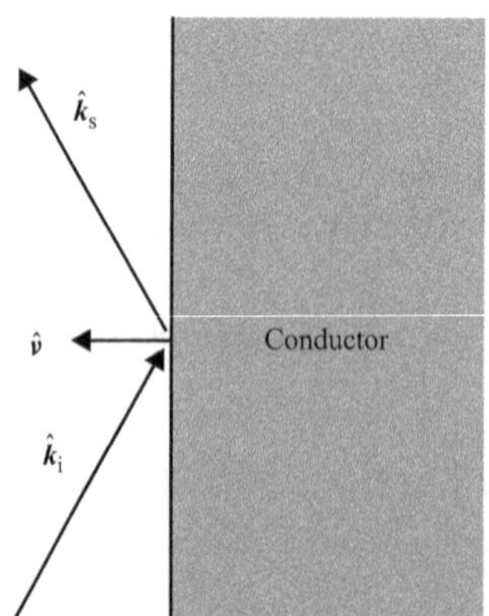

Figure 6.4 Plane wave reflection by a plane perfectly conductor.

The direction of the reflected (scattered) field in terms of the incident direction is given by

$$\hat{\boldsymbol{k}}_{\mathrm{s}} = \hat{\boldsymbol{k}}_{\mathrm{i}} - 2\hat{\boldsymbol{\nu}}\left(\hat{\boldsymbol{\nu}} \cdot \hat{\boldsymbol{k}}_{\mathrm{i}}\right) \tag{6.27}$$

which means a change in sign of the normal component compared with the direction of the incident field $\hat{\boldsymbol{k}}_{\mathrm{i}}$. Note that $\hat{\boldsymbol{k}}_{\mathrm{s}}$ is of unit length by construction. In fact, coordinate independent calculations give

$$\begin{aligned}
\hat{\boldsymbol{k}}_{\mathrm{s}} \cdot \hat{\boldsymbol{k}}_{\mathrm{s}} &= \left(\hat{\boldsymbol{k}}_{\mathrm{i}} - 2\hat{\boldsymbol{\nu}}\left(\hat{\boldsymbol{\nu}} \cdot \hat{\boldsymbol{k}}_{\mathrm{i}}\right)\right) \cdot \left(\hat{\boldsymbol{k}}_{\mathrm{i}} - 2\hat{\boldsymbol{\nu}}\left(\hat{\boldsymbol{\nu}} \cdot \hat{\boldsymbol{k}}_{\mathrm{i}}\right)\right) \\
&= \hat{\boldsymbol{k}}_{\mathrm{i}} \cdot \hat{\boldsymbol{k}}_{\mathrm{i}} - 2\left(\hat{\boldsymbol{\nu}} \cdot \hat{\boldsymbol{k}}_{\mathrm{i}}\right)^2 - 2\left(\hat{\boldsymbol{\nu}} \cdot \hat{\boldsymbol{k}}_{\mathrm{i}}\right)^2 + 4\left(\hat{\boldsymbol{\nu}} \cdot \hat{\boldsymbol{k}}_{\mathrm{i}}\right)^2 = \hat{\boldsymbol{k}}_{\mathrm{i}} \cdot \hat{\boldsymbol{k}}_{\mathrm{i}} = 1
\end{aligned}$$

The representation of the direction of the reflected field in (6.27) is a coordinate-free representation, which is advantageous is many calculations. We can also express the direction of the reflected field by the use of the reflection dyadic $\mathbf{R}$, defined by

$$\mathbf{R} = \mathbf{I}_3 - 2\hat{\boldsymbol{\nu}}\hat{\boldsymbol{\nu}}$$

The direction of the reflected field then is

$$\hat{\boldsymbol{k}}_{\mathrm{s}} = \mathbf{R} \cdot \hat{\boldsymbol{k}}_{\mathrm{i}} = \left(\mathbf{I}_3 - 2\hat{\boldsymbol{\nu}}\hat{\boldsymbol{\nu}}\right) \cdot \hat{\boldsymbol{k}}_{\mathrm{i}} = \hat{\boldsymbol{k}}_{\mathrm{i}} - 2\hat{\boldsymbol{\nu}}\left(\hat{\boldsymbol{\nu}} \cdot \hat{\boldsymbol{k}}_{\mathrm{i}}\right)$$

We proceed by finding the polarization of the reflected field. On the conducting surface, the total field, $\boldsymbol{E}_{\mathrm{i}} + \boldsymbol{E}_{\mathrm{s}}$, satisfies

$$\hat{\boldsymbol{\nu}} \times \left(\boldsymbol{E}_{\mathrm{i}} + \boldsymbol{E}_{\mathrm{s}}\right) = \mathbf{0}$$

This condition implies that the total field $\boldsymbol{E}_{\mathrm{i}} + \boldsymbol{E}_{\mathrm{s}}$ is directed along the surface normal $\hat{\boldsymbol{\nu}}$, i.e.,

$$\boldsymbol{E}_{\mathrm{i}} + \boldsymbol{E}_{\mathrm{s}} = A\hat{\boldsymbol{\nu}}$$

The unknown scalar A is easily determined by the use of (6.26) and (6.27) and the condition that $\hat{\boldsymbol{k}}_{\mathrm{i}} \cdot \boldsymbol{E}_{\mathrm{i}} = \hat{\boldsymbol{k}}_{\mathrm{s}} \cdot \boldsymbol{E}_{\mathrm{s}} = 0$. We get

$$\begin{aligned}
0 = \hat{\boldsymbol{k}}_{\mathrm{s}} \cdot \boldsymbol{E}_{\mathrm{s}} &= \left(\hat{\boldsymbol{k}}_{\mathrm{i}} - 2\hat{\boldsymbol{\nu}}\left(\hat{\boldsymbol{\nu}} \cdot \hat{\boldsymbol{k}}_{\mathrm{i}}\right)\right) \cdot \left(-\boldsymbol{E}_{\mathrm{i}} + A\hat{\boldsymbol{\nu}}\right) \\
&= 2\left(\hat{\boldsymbol{\nu}} \cdot \hat{\boldsymbol{k}}_{\mathrm{i}}\right)\left(\hat{\boldsymbol{\nu}} \cdot \boldsymbol{E}_{\mathrm{i}}\right) + A\left(\hat{\boldsymbol{\nu}} \cdot \hat{\boldsymbol{k}}_{\mathrm{i}}\right) - 2A\left(\hat{\boldsymbol{\nu}} \cdot \hat{\boldsymbol{k}}_{\mathrm{i}}\right)
\end{aligned}$$

from which, we infer (note that we have assumed that $\hat{\boldsymbol{\nu}} \cdot \hat{\boldsymbol{k}}_{\mathrm{i}} \neq 0$)

$$A = 2\hat{\boldsymbol{\nu}} \cdot \boldsymbol{E}_{\mathrm{i}}$$

and the coordinate-free representation of the reflected field $\boldsymbol{E}_{\mathrm{s}}$ is

$$\boldsymbol{E}_{\mathrm{s}} = -\boldsymbol{E}_{\mathrm{i}} + 2\hat{\boldsymbol{\nu}}\left(\hat{\boldsymbol{\nu}} \cdot \boldsymbol{E}_{\mathrm{i}}\right) \tag{6.28}$$

Similar coordinate-free representation of the reflected magnetic field is obtained from (6.26)–(6.28)

$$\begin{aligned}\eta_0\eta \boldsymbol{H}_\mathrm{s} &= \hat{\boldsymbol{k}}_\mathrm{s} \times \boldsymbol{E}_\mathrm{s} = \left(\hat{\boldsymbol{k}}_\mathrm{i} - 2\hat{\boldsymbol{\nu}}\left(\hat{\boldsymbol{\nu}}\cdot\hat{\boldsymbol{k}}_\mathrm{i}\right)\right) \times \left(-\boldsymbol{E}_\mathrm{i} + 2\hat{\boldsymbol{\nu}}\left(\hat{\boldsymbol{\nu}}\cdot\boldsymbol{E}_\mathrm{i}\right)\right)\\ &= -\eta_0\eta\boldsymbol{H}_\mathrm{i} + 2\left(\hat{\boldsymbol{k}}_\mathrm{i}\times\hat{\boldsymbol{\nu}}\right)\left(\hat{\boldsymbol{\nu}}\cdot\boldsymbol{E}_\mathrm{i}\right) + 2\left(\hat{\boldsymbol{\nu}}\times\boldsymbol{E}_\mathrm{i}\right)\left(\hat{\boldsymbol{\nu}}\cdot\hat{\boldsymbol{k}}_\mathrm{i}\right)\\ &= -\eta_0\eta\boldsymbol{H}_\mathrm{i} + 2\hat{\boldsymbol{\nu}}\times\left(\boldsymbol{E}_\mathrm{i}\left(\hat{\boldsymbol{\nu}}\cdot\hat{\boldsymbol{k}}_\mathrm{i}\right) - \hat{\boldsymbol{k}}_\mathrm{i}\left(\hat{\boldsymbol{\nu}}\cdot\boldsymbol{E}_\mathrm{i}\right)\right)\\ &= -\eta_0\eta\boldsymbol{H}_\mathrm{i} + 2\hat{\boldsymbol{\nu}}\times\left(\hat{\boldsymbol{\nu}}\times\left(\boldsymbol{E}_\mathrm{i}\times\hat{\boldsymbol{k}}_\mathrm{i}\right)\right) = -\eta_0\eta\boldsymbol{H}_\mathrm{i} - 2\eta_0\eta\hat{\boldsymbol{\nu}}\times\left(\hat{\boldsymbol{\nu}}\times\boldsymbol{H}_\mathrm{i}\right)\end{aligned}$$

The reflected magnetic field $\boldsymbol{H}_\mathrm{s}$ therefore is

$$\boldsymbol{H}_\mathrm{s} = -\boldsymbol{H}_\mathrm{i} - 2\hat{\boldsymbol{\nu}}\times\left(\hat{\boldsymbol{\nu}}\times\boldsymbol{H}_\mathrm{i}\right) = \boldsymbol{H}_\mathrm{i} - 2\hat{\boldsymbol{\nu}}\left(\hat{\boldsymbol{\nu}}\cdot\boldsymbol{H}_\mathrm{i}\right)$$

In summary, the coordinate-free representations are

$$\begin{cases}\hat{\boldsymbol{k}}_\mathrm{s} = \hat{\boldsymbol{k}}_\mathrm{i} - 2\hat{\boldsymbol{\nu}}\left(\hat{\boldsymbol{\nu}}\cdot\hat{\boldsymbol{k}}_\mathrm{i}\right)\\ \boldsymbol{E}_\mathrm{s} = -\boldsymbol{E}_\mathrm{i} + 2\hat{\boldsymbol{\nu}}\left(\hat{\boldsymbol{\nu}}\cdot\boldsymbol{E}_\mathrm{i}\right)\\ \boldsymbol{H}_\mathrm{s} = \boldsymbol{H}_\mathrm{i} - 2\hat{\boldsymbol{\nu}}\left(\hat{\boldsymbol{\nu}}\cdot\boldsymbol{H}_\mathrm{i}\right)\end{cases}$$

or with the reflection dyadic **R**

$$\begin{cases}\hat{\boldsymbol{k}}_\mathrm{s} = \mathbf{R}\cdot\hat{\boldsymbol{k}}_\mathrm{i}\\ \boldsymbol{E}_\mathrm{s} = -\mathbf{R}\cdot\boldsymbol{E}_\mathrm{i}\\ \boldsymbol{H}_\mathrm{s} = \mathbf{R}\cdot\boldsymbol{H}_\mathrm{i}\end{cases} \tag{6.29}$$

These expressions immediately give the surface current density on the conductor

$$\boldsymbol{J}_S = \hat{\boldsymbol{\nu}}\times\boldsymbol{H} = \hat{\boldsymbol{\nu}}\times\left(\boldsymbol{H}_\mathrm{i} + \boldsymbol{H}_\mathrm{s}\right) = 2\hat{\boldsymbol{\nu}}\times\boldsymbol{H}_\mathrm{i} \tag{6.30}$$

This result is used in the physical optics approximation in Sections 6.3.3 and 6.3.4.

6.3.3 Physical Optics Approximation

The physical optics (PO) approximation can be used to calculate the scattered field by a perfectly conducting scatterer, if the radius of curvature of the scatterer is large compared with the wavelength. Under these circumstances, we assume scattering occurs locally and in the same way as when a plane wave reflects from an infinite plane perfectly conductor. This scattering case was under investigation in Section 6.3.2. We found that the surface current density on the perfectly conducting surface S_s was, see (6.30), $\boldsymbol{J}_S = 2\hat{\boldsymbol{\nu}}\times\boldsymbol{H}_\mathrm{i}$. The far field amplitude, (6.24), in the physical optics approximation then becomes

$$\boldsymbol{F}_\mathrm{PO}(\hat{\boldsymbol{r}}) = -\frac{\mathrm{i}k\eta_0\eta}{2\pi}\hat{\boldsymbol{r}}\times\left[\hat{\boldsymbol{r}}\times\iint_{S_\mathrm{s}^+}\hat{\boldsymbol{\nu}}(\boldsymbol{r}')\times\boldsymbol{H}_\mathrm{i}(\boldsymbol{r}')\mathrm{e}^{-\mathrm{i}k\hat{\boldsymbol{r}}\cdot\boldsymbol{r}'}\,\mathrm{d}S'\right] \tag{6.31}$$

This expression is an explicit integral as soon as the incident (feeding) field is known on the surface S_s^+.

The expression of the far field amplitude in the physical optics approximation, see (6.31), is defined even if the incident field is not a plane wave. If it is used for more general incident fields, the incident field is approximated locally as a plane wave close to the surface. Only then the surface current density is related to the incident field as in (6.30). This approximation holds best for short wavelengths, and it assumes that the radius of curvature of the scatterer is large compared to the wavelengths. We expect less accurate results if the scatterer has wedges or corners. Near a wedge or a corner, the electromagnetic fields show a singular behavior. One way to account for these effects is to include wedge and corner effects obtained by the geometrical theory of diffraction (GTD). This is out of the scope of the treatment in this book. Details on this topic can be found in, e.g., References 102, 129, 180.

We conclude the section by explicitly assuming the incident field is a plane wave with amplitude $\boldsymbol{E}_0$ and incident direction $\hat{\boldsymbol{k}}_{\mathrm{i}}$. In such a case, we can define a backscattering direction of the far field amplitude, i.e., $\hat{\boldsymbol{r}} = -\hat{\boldsymbol{k}}_{\mathrm{i}}$. In many technical applications, e.g., radar, it is only possible to observe in the backscattered direction. In the backscattered direction, the far field amplitude simplifies. Use the BAC-CAB rule on the integrand above when $\hat{\boldsymbol{r}} = -\hat{\boldsymbol{k}}_{\mathrm{i}}$. We get

$$\begin{aligned}\hat{\boldsymbol{r}} \times \left[\hat{\boldsymbol{r}} \times \left(\hat{\boldsymbol{\nu}}(\boldsymbol{r}') \times \boldsymbol{H}_{\mathrm{i}}(\boldsymbol{r}')\right)\right] &= \hat{\boldsymbol{k}}_{\mathrm{i}} \times \left[\hat{\boldsymbol{k}}_{\mathrm{i}} \times \left(\hat{\boldsymbol{\nu}}(\boldsymbol{r}') \times \boldsymbol{H}_{\mathrm{i}}(\boldsymbol{r}')\right)\right] \\ &= -\left(\hat{\boldsymbol{k}}_{\mathrm{i}} \times \boldsymbol{H}_{\mathrm{i}}(\boldsymbol{r}')\right)\left(\hat{\boldsymbol{k}}_{\mathrm{i}} \cdot \hat{\boldsymbol{\nu}}(\boldsymbol{r}')\right) = \frac{1}{\eta_0\eta}\boldsymbol{E}_{\mathrm{i}}(\boldsymbol{r}')\left(\hat{\boldsymbol{k}}_{\mathrm{i}} \cdot \hat{\boldsymbol{\nu}}(\boldsymbol{r}')\right)\end{aligned}$$

since $\hat{\boldsymbol{k}}_{\mathrm{i}} \cdot \boldsymbol{H}_{\mathrm{i}}(\boldsymbol{r}') = 0$ and $\eta_0\eta\hat{\boldsymbol{k}}_{\mathrm{i}} \times \boldsymbol{H}_{\mathrm{i}}(\boldsymbol{r}') = -\boldsymbol{E}_{\mathrm{i}}(\boldsymbol{r}')$ for a plane wave in an isotropic material. With the explicit form of the incident wave, viz.

$$\boldsymbol{E}_{\mathrm{i}}(\boldsymbol{r}) = \boldsymbol{E}_0 \mathrm{e}^{\mathrm{i}k\hat{\boldsymbol{k}}_{\mathrm{i}}\cdot\boldsymbol{r}}$$

we obtain the following expression of the far field amplitude in the backscattered direction:

$$\boldsymbol{F}_{\mathrm{PO}}(-\hat{\boldsymbol{k}}_{\mathrm{i}}) = -\frac{\mathrm{i}k}{2\pi}\boldsymbol{E}_0 \iint_{S_{\mathrm{s}}^+} \left(\hat{\boldsymbol{k}}_{\mathrm{i}} \cdot \hat{\boldsymbol{\nu}}(\boldsymbol{r}')\right) \mathrm{e}^{2\mathrm{i}k\hat{\boldsymbol{k}}_{\mathrm{i}}\cdot\boldsymbol{r}'} \, \mathrm{d}S'$$

We immediately note that in this approximation, the far field amplitude is parallel to the polarization of the incident field, $\boldsymbol{E}_0$, i.e., no cross polarization effects occur in the backscattered direction in the physical optics approximation. The differential scattering cross section, (4.21) on page 202, evaluated in the backscattering direction $\hat{\boldsymbol{r}} = -\hat{\boldsymbol{k}}_{\mathrm{i}}$, becomes

$$\frac{\mathrm{d}\sigma}{\mathrm{d}\Omega}(-\hat{\boldsymbol{k}}_{\mathrm{i}}, \hat{\boldsymbol{k}}_{\mathrm{i}}) = \frac{k^2}{\pi}\left|\iint_{S_{\mathrm{s}}^+} \left(\hat{\boldsymbol{k}}_{\mathrm{i}} \cdot \hat{\boldsymbol{\nu}}(\boldsymbol{r}')\right) \mathrm{e}^{2\mathrm{i}k\hat{\boldsymbol{k}}_{\mathrm{i}}\cdot\boldsymbol{r}'} \, \mathrm{d}S'\right|^2$$

We end this section with a series of rather elaborate examples that illustrate the usefulness of the physical optics approximation.

Example 6.6
In this example, we compute the extinction cross section σ_{ext} in the physical optics approximation by the use of the optical theorem. As the starting point, we take (6.31) and an incident plane wave

$$\boldsymbol{E}_{\mathrm{i}}(\boldsymbol{r}) = \boldsymbol{E}_0 \mathrm{e}^{\mathrm{i}k\hat{\boldsymbol{k}}_{\mathrm{i}}\cdot\boldsymbol{r}}$$

We evaluate this expression in the forward direction in order to use later it as an input in the optical theorem. Since

$$\eta_0\eta \boldsymbol{H}_{\rm i}(\boldsymbol{r}) = \hat{\boldsymbol{k}}_{\rm i} \times \boldsymbol{E}_{\rm i}(\boldsymbol{r}) = \hat{\boldsymbol{k}}_{\rm i} \times \boldsymbol{E}_0 {\rm e}^{{\rm i}k\hat{\boldsymbol{k}}_{\rm i}\cdot\boldsymbol{r}}$$

holds for plane waves, we get

$$\boldsymbol{F}_{\rm PO}(\hat{\boldsymbol{k}}_{\rm i}) = -\frac{{\rm i}k}{2\pi}\hat{\boldsymbol{k}}_{\rm i} \times \left[\hat{\boldsymbol{k}}_{\rm i} \times \iint\limits_{S_{\rm s}^+} \hat{\boldsymbol{\nu}}(\boldsymbol{r}') \times (\hat{\boldsymbol{k}}_{\rm i} \times \boldsymbol{E}_0)\, {\rm d}S'\right]$$

We simplify the integrand by the use of the BAC-CAB rule

$$\hat{\boldsymbol{k}}_{\rm i} \times \Big\{\hat{\boldsymbol{k}}_{\rm i} \times \underbrace{\Big[\hat{\boldsymbol{\nu}} \times \Big(\hat{\boldsymbol{k}}_{\rm i} \times \boldsymbol{E}_0\Big)\Big]}_{\hat{\boldsymbol{k}}_{\rm i}(\hat{\boldsymbol{\nu}}\cdot\boldsymbol{E}_0)-\boldsymbol{E}_0\left(\hat{\boldsymbol{\nu}}\cdot\hat{\boldsymbol{k}}_{\rm i}\right)}\Big\} = -\underbrace{\hat{\boldsymbol{k}}_{\rm i} \times \Big(\hat{\boldsymbol{k}}_{\rm i} \times \boldsymbol{E}_0\Big)}_{\hat{\boldsymbol{k}}_{\rm i}(\hat{\boldsymbol{k}}_{\rm i}\cdot\boldsymbol{E}_0)-\boldsymbol{E}_0}\Big(\hat{\boldsymbol{\nu}} \cdot \hat{\boldsymbol{k}}_{\rm i}\Big) = \boldsymbol{E}_0\Big(\hat{\boldsymbol{\nu}} \cdot \hat{\boldsymbol{k}}_{\rm i}\Big)$$

since $\hat{\boldsymbol{k}}_{\rm i} \cdot \boldsymbol{E}_0 = 0$. We get

$$\boldsymbol{F}_{\rm PO}(\hat{\boldsymbol{k}}_{\rm i}) = -\frac{{\rm i}k}{2\pi}\boldsymbol{E}_0 \iint\limits_{S_{\rm s}^+} \hat{\boldsymbol{\nu}}(\boldsymbol{r}') \cdot \hat{\boldsymbol{k}}_{\rm i}\, {\rm d}S'$$

The integral expresses the projected cross-sectional area, $A(\hat{\boldsymbol{k}}_{\rm i})$, of the illuminated part of the scatterer, see Figure 6.5. By definition, we have

$$A(\hat{\boldsymbol{k}}_{\rm i}) = -\iint\limits_{S_{\rm s}^+} \hat{\boldsymbol{\nu}}(\boldsymbol{r}') \cdot \hat{\boldsymbol{k}}_{\rm i}\, {\rm d}S' \tag{6.32}$$

and

$$\boldsymbol{F}_{\rm PO}(\hat{\boldsymbol{k}}_{\rm i}) = \frac{{\rm i}k}{2\pi}\boldsymbol{E}_0 A(\hat{\boldsymbol{k}}_{\rm i})$$

or

$$\mathbf{S}_{\rm PO}(\hat{\boldsymbol{k}}_{\rm i}, \hat{\boldsymbol{k}}_{\rm i}) = \frac{{\rm i}k}{2\pi}A(\hat{\boldsymbol{k}}_{\rm i})\mathbf{I}_3 \tag{6.33}$$

The optical theorem, (4.47) on page 230, now gives

$$\sigma_{\rm ext} = \frac{4\pi}{k}\,{\rm Im}\left\{\frac{\boldsymbol{E}_0^* \cdot \boldsymbol{F}_{\rm PO}(\hat{\boldsymbol{k}}_{\rm i})}{|\boldsymbol{E}_0|^2}\right\} = 2A(\hat{\boldsymbol{k}}_{\rm i})$$

This result holds for a large class of scatterers, even though it has been derived under the assumption that the physical optics approximation is valid. The result is often called the extinction paradox, since the extinction cross section $\sigma_{\rm ext}$ is twice the total geometrical cross-sectional area $A(\hat{\boldsymbol{k}}_{\rm i})$. The intuitive result would be just the total geometrical cross-sectional area $A(\hat{\boldsymbol{k}}_{\rm i})$. Geometrical optics (rays) and reflection of these rays, account for a factor $A(\hat{\boldsymbol{k}}_{\rm i})$—corresponding to the total geometrical cross section of the scatterer—to the extinction cross section $\sigma_{\rm ext}$. It is tempting to believe that this result should hold better and better for higher frequencies. Instead,

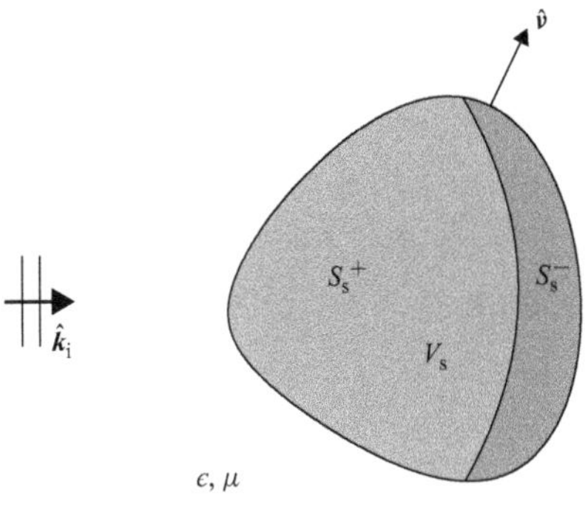

Figure 6.5 The scatterer and the illuminated and shadow regions.

as the physical optics approximation result shows, the extinction cross section σ_{ext} approaches twice this value. The explanation to this phenomena is still under debate, see References 10, 24, 89, 150, 201, 277. ■

Example 6.7
A perfectly conducting reflector has the shape of a parabolic cylinder. The orientation of the cylindrical reflector is illustrated in Figure 6.6. The reflector is supposed to generate a main lobe with a lobe width (defined as the angle between the directions in the main lobe where the power flux density has decreased to half of its maximum value) less than 4° both in the x–z-plane (H-plane) and in y–z-plane (E-plane) at the frequency 4 GHz.

We are going to use the physical optics approximation to solve this problem. The analogous computation with the geometrical optic approximation is presented in Example 6.11 on page 314.

The parabolic reflector is oriented so that its surface is described by $x^2 = 4F\,(z + F)$, where the focal length of the parabolic cylinder is $F > 0$. The projected surface of the parabolic cylinder in the x–y-plane is a rectangle, which can be used as a parametrization of the reflector, i.e., $-b/2 \le x \le b/2$, $-h/2 \le y \le h/2$. We define the vector-valued function $\boldsymbol{S}(x, y) = \boldsymbol{r}'$ that describes the lit surface S_s^+ by

$$\boldsymbol{S}(x,y) = \hat{\boldsymbol{x}}x + \hat{\boldsymbol{y}}y + \hat{\boldsymbol{z}}\left(\frac{x^2}{4F} - F\right), \qquad \begin{cases} -b/2 \le x \le b/2 \\ -h/2 \le y \le h/2 \end{cases}$$

For every (x, y)-value, the vector $\boldsymbol{S}(x, y)$ describes a unique point on the reflector surface.

The reflector is fed by a line source located at the focal line of the parabolic reflector. This source can be realized as a large number of short, straight antennas aligned in the focal line of the reflector, see Figure 6.6. The antennas are fed with the same phase and a cylindrical

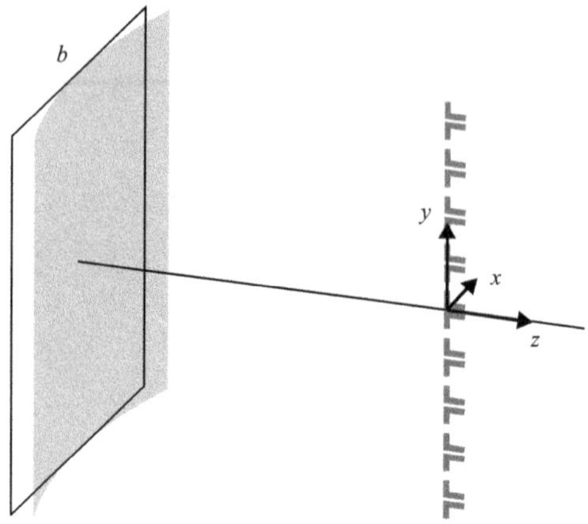

Figure 6.6 The geometry of the parabolic cylinder in Example 6.7.

wave is generated. The electric field of the feed is directed in the y-direction. The electric and magnetic fields from the feed are

$$\begin{cases} \boldsymbol{E}_{\mathrm{i}}(\boldsymbol{r}) = E_0 H_0^{(1)}\left(k\sqrt{x^2+z^2}\right)\hat{\boldsymbol{y}} \\ \boldsymbol{H}_{\mathrm{i}}(\boldsymbol{r}) = \dfrac{E_0}{\mathrm{i}k\eta_0\eta}\nabla\times\left(H_0^{(1)}\left(k\sqrt{x^2+z^2}\right)\hat{\boldsymbol{y}}\right) = \mathrm{i}\dfrac{E_0}{\eta_0\eta}H_0^{(1)\prime}\left(k\sqrt{x^2+z^2}\right)\dfrac{\hat{\boldsymbol{x}}z-\hat{\boldsymbol{z}}x}{\sqrt{x^2+z^2}} \end{cases}$$

where the function $H_0^{(1)}$ is the Hankel function of the first kind, see Appendix B.1. The far field of the dipoles ($kr \gg 1$) is given by (see Appendix B.1 and the asymptotic expansions of the Hankel functions of the first kind)

$$\begin{cases} \boldsymbol{E}_{\mathrm{i}}(\boldsymbol{r}) = E_0\sqrt{\dfrac{2}{\mathrm{i}k\pi\sqrt{x^2+z^2}}}\mathrm{e}^{\mathrm{i}k\sqrt{x^2+z^2}}\hat{\boldsymbol{y}} \\ \boldsymbol{H}_{\mathrm{i}}(\boldsymbol{r}) = -\dfrac{E_0}{\eta_0\eta}\sqrt{\dfrac{2}{\mathrm{i}k\pi\sqrt{x^2+z^2}}}\mathrm{e}^{\mathrm{i}k\sqrt{x^2+z^2}}\dfrac{\hat{\boldsymbol{x}}z-\hat{\boldsymbol{z}}x}{\sqrt{x^2+z^2}} \end{cases} \tag{6.34}$$

We let $F = 1$ m. Then at the frequency $f = 4$ GHz, $kF = 2\pi fF/c_0 \approx 84 \gg 1$, i.e., the approximation is accurate.

Differentiating $\boldsymbol{S}(x,y)$ w.r.t. the parameters x and y, we obtain the tangent vectors on the reflector surface, i.e.,

$$\begin{cases} \tau_1(x,y) = \dfrac{\partial\boldsymbol{S}}{\partial x}(x,y) = \hat{\boldsymbol{x}} + \hat{\boldsymbol{z}}\dfrac{x}{2F} \\ \tau_2(x,y) = \dfrac{\partial\boldsymbol{S}}{\partial y}(x,y) = \hat{\boldsymbol{y}} \end{cases}$$

Normalizing the vector product, we obtain the unit normal vector to the reflector surface. Specifically, we have

$$\hat{\boldsymbol{\nu}}(x,y) = \frac{\boldsymbol{\tau}_1(x,y) \times \boldsymbol{\tau}_2(x,y)}{|\boldsymbol{\tau}_1(x,y) \times \boldsymbol{\tau}_2(x,y)|} = \frac{-\hat{\boldsymbol{x}}x + \hat{\boldsymbol{z}}2F}{\sqrt{x^2+4F^2}} \tag{6.35}$$

The distance from the feed to a point on the reflector surface is given by

$$r_{\rm c}(x) = |\boldsymbol{S}(x,0)| = \sqrt{x^2 + \left(\frac{x^2}{4F} - F\right)^2} = \frac{x^2+4F^2}{4F} \tag{6.36}$$

Moreover, the surface element of the reflector $\mathrm{d}S$ is

$$\mathrm{d}S = \left|\frac{\partial \boldsymbol{S}}{\partial x}(x,y) \times \frac{\partial \boldsymbol{S}}{\partial y}(x,y)\right| \mathrm{d}x\,\mathrm{d}y = \frac{\sqrt{x^2+4F^2}}{2F}\,\mathrm{d}x\,\mathrm{d}y \tag{6.37}$$

The incident direction, $\hat{\boldsymbol{k}}_{\rm i}(x)$, that hits a point on the surface, parameterized by (x,y), is

$$\hat{\boldsymbol{k}}_{\rm i}(x) = \frac{\boldsymbol{S}(x,0)}{|\boldsymbol{S}(x,0)|} = \frac{\hat{\boldsymbol{x}}x + \hat{\boldsymbol{z}}\left(\frac{x^2}{4F} - F\right)}{\sqrt{\left(\frac{x^2}{4F} - F\right)^2 + x^2}} = \frac{\hat{\boldsymbol{x}}4Fx + \hat{\boldsymbol{z}}\left(x^2 - 4F^2\right)}{x^2+4F^2} \tag{6.38}$$

which is independent of y.

We investigate if the reflector antenna can be designed, i.e., choose b and h, so that the beam width of the main lobe is less than 4° in both the x–z and the y–z-planes. We use the physical optics approximation given by (6.31) ($\eta = 1$)

$$\boldsymbol{F}_{\rm PO}(\hat{\boldsymbol{r}}) = -\frac{\mathrm{i}k\eta_0}{2\pi}\hat{\boldsymbol{r}} \times \left[\hat{\boldsymbol{r}} \times \iint_{S_{\rm s}^+} \hat{\boldsymbol{\nu}}(\boldsymbol{r}') \times \boldsymbol{H}_{\rm i}(\boldsymbol{r}')\mathrm{e}^{-\mathrm{i}k\hat{\boldsymbol{r}}\cdot\boldsymbol{r}'}\,\mathrm{d}S'\right] \tag{6.39}$$

where the integration is over the illuminated part of the reflector surface. We denote the surface integral in (6.39) by $\boldsymbol{I}$. From (6.37), we get

$$\begin{aligned}\boldsymbol{I} &= \iint_{S_{\rm s}^+} \hat{\boldsymbol{\nu}}(\boldsymbol{r}') \times \boldsymbol{H}_{\rm i}(\boldsymbol{r}')\mathrm{e}^{-\mathrm{i}k\hat{\boldsymbol{r}}\cdot\boldsymbol{r}'}\,\mathrm{d}S' \\ &= \int_{-b/2}^{b/2} \mathrm{d}x \int_{-h/2}^{h/2} \mathrm{d}y \frac{\sqrt{x^2+4F^2}}{2F}\left(\hat{\boldsymbol{\nu}}(x,y) \times \boldsymbol{H}_{\rm i}\left(\boldsymbol{S}(x,y)\right)\right)\mathrm{e}^{-\mathrm{i}k\hat{\boldsymbol{r}}\cdot\boldsymbol{S}(x,y)}\end{aligned}$$

From (6.34) and (6.35), we get

$$\begin{aligned}\hat{\boldsymbol{\nu}}(x,y) \times \boldsymbol{H}_{\rm i}\left(\boldsymbol{S}(x,y)\right) &= -\frac{E_0}{\eta_0\eta}\sqrt{\frac{2}{\mathrm{i}k\pi r_{\rm c}(x)}}\mathrm{e}^{\mathrm{i}kr_{\rm c}(x)}\frac{-\hat{\boldsymbol{x}}x + \hat{\boldsymbol{z}}2F}{\sqrt{x^2+4F^2}} \times \frac{\hat{\boldsymbol{x}}\left(x^2/4F - F\right) - \hat{\boldsymbol{z}}x}{x^2/4F + F} \\ &= \frac{E_0}{\eta_0\eta}\sqrt{\frac{2}{\mathrm{i}k\pi r_{\rm c}(x)}}\mathrm{e}^{\mathrm{i}kr_{\rm c}(x)}\frac{2F}{\sqrt{x^2+4F^2}}\hat{\boldsymbol{y}}\end{aligned}$$

The argument in the exponential function remains to be analyzed. It is

$$\hat{\boldsymbol{r}} \cdot \boldsymbol{S}(x, y) = x \sin\theta \cos\phi + y \sin\theta \sin\phi + \left(\frac{x^2}{4F} - F\right)\cos\theta$$

where the direction of observation is parameterized by the spherical angles, i.e., the unit vector $\hat{\boldsymbol{r}} = \hat{\boldsymbol{x}} \sin\theta \cos\phi + \hat{\boldsymbol{y}} \sin\theta \sin\phi + \hat{\boldsymbol{z}} \cos\theta$.

The surface integral $\boldsymbol{I}$ is computed with the help of (6.36)

$$\boldsymbol{I} = \hat{\boldsymbol{y}}\frac{E_0}{\eta_0\eta}\int_{-b/2}^{b/2} \mathrm{d}x \int_{-h/2}^{h/2} \mathrm{d}y \sqrt{\frac{8F}{ik\pi(x^2 + 4F^2)}} \cdot \mathrm{e}^{ik\left(\left(\frac{x^2}{4F}+F\right) - x\sin\theta\cos\phi - y\sin\theta\sin\phi - \left(\frac{x^2}{4F}-F\right)\cos\theta\right)}$$

and the far field amplitude in the physical optics approximation, (6.39), becomes

$$\begin{aligned}\boldsymbol{F}_{\mathrm{PO}}(\hat{\boldsymbol{r}}) = {} & -\frac{ikE_0}{2\pi}\mathrm{e}^{ikF(1+\cos\theta)}\hat{\boldsymbol{r}} \times (\hat{\boldsymbol{r}} \times \hat{\boldsymbol{y}}) \\ & \cdot \int_{-b/2}^{b/2} \mathrm{d}x \int_{-h/2}^{h/2} \mathrm{d}y \sqrt{\frac{8F}{ik\pi(x^2 + 4F^2)}}\,\mathrm{e}^{ik\left(\frac{x^2}{4F}(1-\cos\theta) - x\sin\theta\cos\phi - y\sin\theta\sin\phi\right)}\end{aligned}$$

The integration in the y-variable is easily done analytically with

$$\int_{-h/2}^{h/2} \mathrm{e}^{-i\alpha y}\,\mathrm{d}y = h\frac{\sin\frac{h\alpha}{2}}{\frac{h\alpha}{2}} \tag{6.40}$$

Then, with the simplification $\hat{\boldsymbol{r}} \times (\hat{\boldsymbol{r}} \times \hat{\boldsymbol{y}}) = -\hat{\boldsymbol{\theta}} \cos\theta \sin\phi - \hat{\boldsymbol{\phi}} \cos\phi$, we get

$$\begin{aligned}\boldsymbol{F}_{\mathrm{PO}}(\hat{\boldsymbol{r}}) = {} & \frac{ikE_0h}{2\pi}\mathrm{e}^{ikF(1+\cos\theta)}\left(\hat{\boldsymbol{\theta}}\cos\theta\sin\phi + \hat{\boldsymbol{\phi}}\cos\phi\right)\frac{\sin\left(\frac{kh}{2}\sin\theta\sin\phi\right)}{\frac{kh}{2}\sin\theta\sin\phi} \\ & \cdot \int_{-b/2}^{b/2} \sqrt{\frac{8F}{ik\pi(x^2 + 4F^2)}}\,\mathrm{e}^{ik\left(\frac{x^2}{4F}(1-\cos\theta) - x\sin\theta\cos\phi\right)}\,\mathrm{d}x\end{aligned}$$

In the forward direction of the reflector, $\theta = 0$, the far field amplitude can be computed analytically. The result is

$$\begin{aligned}\boldsymbol{F}_{\mathrm{PO}}(\hat{\boldsymbol{z}}) &= \hat{\boldsymbol{y}}\frac{ikE_0h}{2\pi}\mathrm{e}^{2ikF}\sqrt{\frac{8F}{ik\pi}}\int_{-kb/2}^{kb/2}\frac{\mathrm{d}t}{\sqrt{t^2 + 4k^2F^2}} \\ &= \hat{\boldsymbol{y}}\frac{ikE_0h}{2\pi}\mathrm{e}^{2ikF}\sqrt{\frac{8F}{ik\pi}}\ln\frac{\sqrt{k^2b^2 + 16k^2F^2} + kb}{\sqrt{k^2b^2 + 16k^2F^2} - kb} \\ &= \hat{\boldsymbol{y}}\frac{ikE_0h}{\pi}\mathrm{e}^{2ikF}\sqrt{\frac{8F}{ik\pi}}\ln\left(\frac{b}{4F} + \sqrt{\frac{b^2}{16F^2} + 1}\right)\end{aligned}$$

For an arbitrary direction, the integral in the x-variable has to be computed numerically. The square of the amplitude of the far field amplitude is

$$|\boldsymbol{F}_{\mathrm{PO}}(\hat{\boldsymbol{r}})|^2 = \frac{2k|E_0|^2h^2F}{\pi^3}\left(\cos^2\theta\sin^2\phi+\cos^2\phi\right)\left|\frac{\sin\left(\frac{kh}{2}\sin\theta\sin\phi\right)}{\frac{kh}{2}\sin\theta\sin\phi}\right|^2 \cdot\left|\int_{-kb/2}^{kb/2}\frac{\mathrm{d}t}{\sqrt{t^2+4k^2F^2}}\mathrm{e}^{\mathrm{i}\left(\frac{t^2}{4kF}(1-\cos\theta)-t\sin\theta\cos\phi\right)}\right|^2$$

In the forward direction, $\theta = 0$, we get

$$|\boldsymbol{F}_{\mathrm{PO}}(\hat{\boldsymbol{z}})|^2 = \frac{8k|E_0|^2h^2F}{\pi^3}\left|\ln\left(\frac{b}{4F}+\sqrt{\frac{b^2}{16F^2}+1}\right)\right|^2$$

The width of the main lobe is investigated numerically. In the H-plane (x–z-plane), the half width, θ_{hv}, is determined by solving the equation

$$|\boldsymbol{F}_{\mathrm{PO}}(\theta_{\mathrm{hv}},\phi=0)|^2 = 0.5\,|\boldsymbol{F}_{\mathrm{PO}}(\hat{\boldsymbol{z}})|^2$$

and analogously, in the E-plane (y–z-plane), the half width is determined by

$$|\boldsymbol{F}_{\mathrm{PO}}(\theta_{\mathrm{hv}},\phi=\pi/2)|^2 = 0.5\,|\boldsymbol{F}_{\mathrm{PO}}(\hat{\boldsymbol{z}})|^2$$

The results are illustrated in Figure 6.7, and the figure shows that $b = h = 1$ m give a lobe width of approximately 4° at the frequency $f = 4$ GHz. Note that the difference between the lobe widths in the E- and H-planes is very small. In Figure 6.8, the square of the amplitude of the far field amplitude (normalized with the amplitude in the forward direction) is depicted as a function of θ in both the E- and H-planes. The level of the first side lobe is approximately -13 dB in both the E- and the H-planes. ■

Example 6.8

A circular, perfectly conducting reflector has the shape of an axially symmetric parabola, see Figure 6.9. The projected circular surface in the x–y-plane has the radius a. The parabola is illuminated by a short electric dipole antenna located at the focal point of the parabola (origin of the coordinate system). The electric dipole is oriented along $\hat{\boldsymbol{y}}$. The media outside the reflector is air, which we approximate by vacuum, $\epsilon = \mu = 1$. We determine the far field amplitude of the reflector with the physical optics approximation. The corresponding computations with the geometrical optics approximation is presented in Example 6.12 on page 318.

The coordinate system is oriented so that the equation of the surface is $x^2+y^2 = 4F(z+F)$, where $F > 0$ is the focal length of the parabola. We define the vector-valued function $\boldsymbol{S}(r_{\mathrm{c}},\alpha) = \boldsymbol{r}'$ as

$$\boldsymbol{S}(r_{\mathrm{c}},\alpha) = \hat{\boldsymbol{r}}_{\mathrm{c}}r_{\mathrm{c}} + \hat{\boldsymbol{z}}\left(r_{\mathrm{c}}^2/4F - F\right)$$

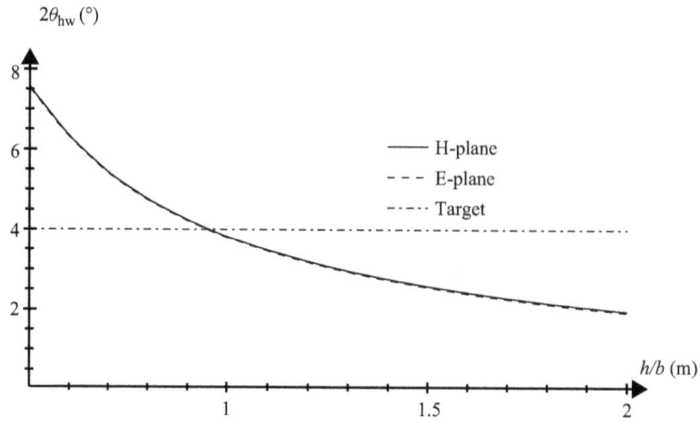

Figure 6.7 The width of the main lobe in degrees as a function of b and h for the cylindrical reflector in Example 6.7. The computations are made with the physical optics approximation. In the H-plane (x–z-plane), b is determined with $h = F = 1$ m, and in the E-plane (y–z-plane), h is determined with $b = F = 1$ m. The frequency is $f = 4$ GHz. Compare with the computations in Figure 6.15, where the computations are made with the geometrical optics approximation.

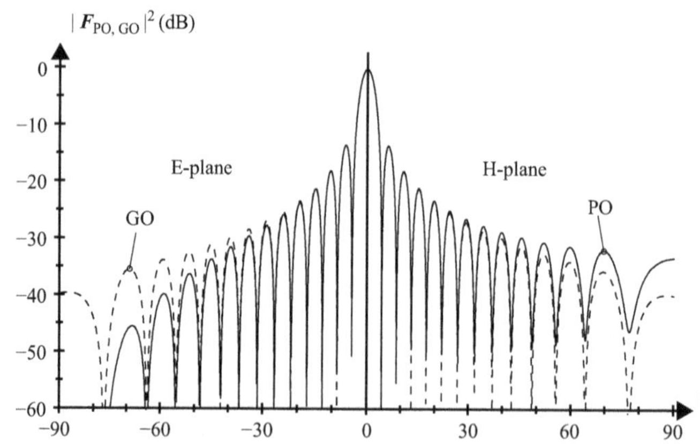

Figure 6.8 Comparison between the far field amplitude of the cylindrical reflector computed with the physical optics approximation and the geometrical optics approximation in the E-plane (y–z-plane, left side of the figure) and the H-plane (x–z-plane, right side of the figure), respectively. Solid line represents the physical optics approximation, see Example 6.7, and the broken line represents the geometrical optics approximation, see Example 6.11. $h = b = F = 1$ m and the frequency $f = 4$ GHz.

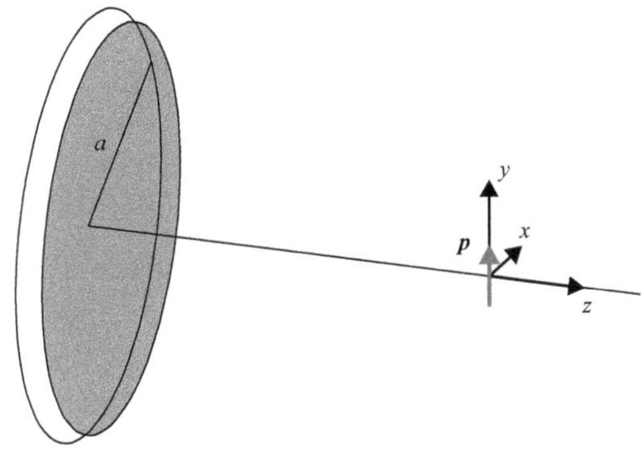

Figure 6.9 The geometry of the parabolic reflector in Example 6.8. The vertical electric dipole $\boldsymbol{p}$ is located at the origin.

where we have introduced the cylindrical coordinates r_c and α in the x–y-plane where the angle α is measured positive from the x-axis toward the y-axis, i.e.,

$$\begin{cases} \hat{\boldsymbol{r}}_c r_c = \hat{\boldsymbol{x}} x + \hat{\boldsymbol{y}} y \\ \hat{\boldsymbol{r}}_c = \hat{\boldsymbol{x}} \cos\alpha + \hat{\boldsymbol{y}} \sin\alpha \\ \hat{\boldsymbol{\alpha}} = -\hat{\boldsymbol{x}} \sin\alpha + \hat{\boldsymbol{y}} \cos\alpha \end{cases}$$

This vector-valued function $\boldsymbol{S}(r_c, \alpha)$ describes the reflector surface. Every value of (r_c, α) defines a unique point on the reflector surface.

Differentiate $\boldsymbol{S}(r_c, \alpha)$ w.r.t. r_c and α to obtain the tangential vectors of the surface.

$$\begin{cases} \tau_1(r_c, \alpha) = \dfrac{\partial \boldsymbol{S}(r_c, \alpha)}{\partial r_c} = \hat{\boldsymbol{r}}_c + \hat{\boldsymbol{z}} r_c / 2F \\ \tau_2(r_c, \alpha) = \dfrac{\partial \boldsymbol{S}(r_c, \alpha)}{\partial \alpha} = \hat{\boldsymbol{\alpha}} r_c \end{cases}$$

The normalized vector product of the tangent vectors defines the unit normal vector of the surface, i.e.,

$$\hat{\boldsymbol{\nu}}(r_c, \alpha) = \frac{\frac{\partial \boldsymbol{S}}{\partial r_c} \times \frac{\partial \boldsymbol{S}}{\partial \alpha}}{\left| \frac{\partial \boldsymbol{S}}{\partial r_c} \times \frac{\partial \boldsymbol{S}}{\partial \alpha} \right|} = \frac{\hat{\boldsymbol{z}} r_c - \hat{\boldsymbol{r}}_c r_c^2 / 2F}{\sqrt{r_c^2 + r_c^4 / 4F^2}} = \frac{-\hat{\boldsymbol{r}}_c r_c + \hat{\boldsymbol{z}} 2F}{\sqrt{r_c^2 + 4F^2}} \tag{6.41}$$

The distance from the dipole to the reflector surface is

$$r'(r_c) = |\boldsymbol{S}(r_c, \alpha)| = \sqrt{r_c^2 + \left(r_c^2 / 4F - F\right)^2} = r_c^2 / 4F + F \tag{6.42}$$

which is independent of α.

The far field from the exciting dipole is, see Problem 2.2 on page 137

$$\boldsymbol{E}_i = \frac{pk^2}{\epsilon_0} \frac{e^{ikr}}{4\pi r} \left(\hat{\boldsymbol{r}}_0 \times \left(\hat{\boldsymbol{y}} \times \hat{\boldsymbol{r}}_0\right)\right) \tag{6.43}$$

where p denotes the dipole moment of the dipole, $k = \omega/c_0$, and $\hat{\boldsymbol{r}}_0$ is the direction of observation measured from the location of the antenna (the origin). The magnetic field in the far zone is then

$$\boldsymbol{H}_{\mathrm{i}} = k\omega p \frac{\mathrm{e}^{ikr}}{4\pi r} \left(\hat{\boldsymbol{r}}_0 \times \hat{\boldsymbol{y}}\right) \tag{6.44}$$

The physical optics approximation of the far field amplitude is given by (6.31) ($\eta = 1$)

$$\boldsymbol{F}_{\mathrm{PO}}(\hat{\boldsymbol{r}}) = -\frac{ik\eta_0}{2\pi} \hat{\boldsymbol{r}} \times \left[\hat{\boldsymbol{r}} \times \iint_{S_{\mathrm{s}}^+} \hat{\boldsymbol{\nu}}(\boldsymbol{r}') \times \boldsymbol{H}_{\mathrm{i}}(\boldsymbol{r}')\mathrm{e}^{-ik\hat{\boldsymbol{r}}\cdot\boldsymbol{r}'}\, \mathrm{d}S'\right] \tag{6.45}$$

where the integration is over the illuminated part of the reflector surface. We denote the surface integral in (6.45) by $\boldsymbol{I}(\hat{\boldsymbol{r}})$, i.e.,

$$\begin{aligned}\boldsymbol{I}(\hat{\boldsymbol{r}}) &= \iint_{S_{\mathrm{s}}^+} \hat{\boldsymbol{\nu}}(\boldsymbol{r}') \times \boldsymbol{H}_{\mathrm{i}}(\boldsymbol{r}')\mathrm{e}^{-ik\hat{\boldsymbol{r}}\cdot\boldsymbol{r}'}\, \mathrm{d}S' \\ &= \int_0^a \mathrm{d}r_{\mathrm{c}} \int_0^{2\pi} \mathrm{d}\alpha\, \hat{\boldsymbol{\nu}}(r_{\mathrm{c}}, \alpha) \times \boldsymbol{H}_{\mathrm{i}}\left(\boldsymbol{S}\left(r_{\mathrm{c}}, \alpha\right)\right) \mathrm{e}^{-ik\hat{\boldsymbol{r}}\cdot\boldsymbol{S}(r_{\mathrm{c}},\alpha)} \left|\frac{\partial \boldsymbol{S}}{\partial r_{\mathrm{c}}} \times \frac{\partial \boldsymbol{S}}{\partial \alpha}\right|\end{aligned}$$

The surface element of the parabola is

$$\left|\frac{\partial \boldsymbol{S}}{\partial r_{\mathrm{c}}} \times \frac{\partial \boldsymbol{S}}{\partial \alpha}\right| = \sqrt{r_{\mathrm{c}}^2 + r_{\mathrm{c}}^4/4F^2} = r_{\mathrm{c}}\sqrt{r'(r_{\mathrm{c}})/F}$$

The magnetic field is, see (6.44)

$$\boldsymbol{H}_{\mathrm{i}}\left(\boldsymbol{S}\left(r_{\mathrm{c}}, \alpha\right)\right) = k\omega p \frac{\mathrm{e}^{ikr'(r_{\mathrm{c}})}}{4\pi r'(r_{\mathrm{c}})} \left(\hat{\boldsymbol{k}}_{\mathrm{i}} \times \hat{\boldsymbol{y}}\right) = k\omega p \frac{\mathrm{e}^{ikr'(r_{\mathrm{c}})}}{4\pi r'(r_{\mathrm{c}})} \left(\hat{\boldsymbol{k}}_{\mathrm{i}} \times \left(\hat{\boldsymbol{r}}_{\mathrm{c}} \sin\alpha + \hat{\boldsymbol{\alpha}} \cos\alpha\right)\right)$$

where the direction of the incident field is

$$\hat{\boldsymbol{k}}_{\mathrm{i}}\left(r_{\mathrm{c}}, \alpha\right) = \frac{\boldsymbol{S}\left(r_{\mathrm{c}}, \alpha\right)}{\left|\boldsymbol{S}\left(r_{\mathrm{c}}, \alpha\right)\right|} = \frac{4\hat{\boldsymbol{r}}_{\mathrm{c}} F r_{\mathrm{c}} + \hat{\boldsymbol{z}}\left(r_{\mathrm{c}}^2 - 4F^2\right)}{r_{\mathrm{c}}^2 + 4F^2} \tag{6.46}$$

It is important to distinguish between the two unit vectors $\hat{\boldsymbol{r}}$ and $\hat{\boldsymbol{k}}_{\mathrm{i}}$. The unit vector $\hat{\boldsymbol{r}}$ denotes the direction from the origin (antenna feed) to the observation point, but $\hat{\boldsymbol{k}}_{\mathrm{i}}$ is the direction from the origin (antenna feed) to a point on the reflector (reflection point). The magnetic field is written as

$$\boldsymbol{H}_{\mathrm{i}}\left(\boldsymbol{S}\left(r_{\mathrm{c}}, \alpha\right)\right) = k\omega p \frac{\mathrm{e}^{ikr'(r_{\mathrm{c}})}}{4\pi r'(r_{\mathrm{c}})} \frac{4\hat{\boldsymbol{z}} F r_{\mathrm{c}} \cos\alpha - \hat{\boldsymbol{x}}\left(r_{\mathrm{c}}^2 - 4F^2\right)}{r_{\mathrm{c}}^2 + 4F^2} \tag{6.47}$$

An expansion of the vector product $\hat{\boldsymbol{\nu}} \times \boldsymbol{H}_{\mathrm{i}}$ by using (6.41) implies

$$\begin{aligned}\hat{\boldsymbol{\nu}}(r_{\mathrm{c}}, \alpha) \times \boldsymbol{H}_{\mathrm{i}}\left(\boldsymbol{S}\left(r_{\mathrm{c}}, \alpha\right)\right) = \frac{k\omega p}{(4F)^{3/2}} \frac{\mathrm{e}^{ikr'(r_{\mathrm{c}})}}{4\pi \left(r'(r_{\mathrm{c}})\right)^{5/2}} &\left[\hat{\boldsymbol{y}}\left(4Fr_{\mathrm{c}}^2 \cos^2\alpha - 2F\left(r_{\mathrm{c}}^2 - 4F^2\right)\right)\right. \\ &\left. -\hat{\boldsymbol{x}} 4Fr_{\mathrm{c}}^2 \sin\alpha \cos\alpha - \hat{\boldsymbol{z}} r_{\mathrm{c}} \sin\alpha \left(r_{\mathrm{c}}^2 - 4F^2\right)\right]\end{aligned}$$

The surface integral $\boldsymbol{I}(\hat{\boldsymbol{r}})$ can now be written as

$$\boldsymbol{I}(\hat{\boldsymbol{r}}) = \frac{k\omega p}{8F^2}\int_0^a r_c\, dr_c \int_0^{2\pi} d\alpha \frac{e^{ik\left(r'(r_c)-\hat{\boldsymbol{r}}\cdot\boldsymbol{S}(r_c,\alpha)\right)}}{4\pi\,(r'(r_c))^2}\left[\hat{\boldsymbol{y}}\left(2Fr_c^2\cos 2\alpha + 8F^3\right)\right.$$
$$\left. -\hat{\boldsymbol{x}}2Fr_c^2\sin 2\alpha - \hat{\boldsymbol{z}}r_c\sin\alpha\left(r_c^2 - 4F^2\right)\right]$$

To simplify the complex-valued integral $\boldsymbol{I}(\hat{\boldsymbol{r}})$, we evaluate the α-integrals analytically. Useful integrals are

$$\begin{cases}\displaystyle\int_0^{2\pi} e^{iz\cos(\alpha-\phi)}\, d\alpha = \int_0^{2\pi} e^{iz\cos\alpha}\, d\alpha = 2\pi J_0(z)\\ \displaystyle\int_0^{2\pi} e^{iz\cos(\alpha-\phi)}\sin\alpha\, d\alpha = \int_0^{2\pi} e^{iz\cos\alpha}\sin(\alpha+\phi)\, d\alpha = 2\pi i J_1(z)\sin\phi\end{cases}$$

and

$$\begin{cases}\displaystyle\int_0^{2\pi} e^{iz\cos(\alpha-\phi)}\cos 2\alpha\, d\alpha = \int_0^{2\pi} e^{iz\cos\alpha}\cos 2(\alpha+\phi)\, d\alpha = -2\pi J_2(z)\cos 2\phi\\ \displaystyle\int_0^{2\pi} e^{iz\cos(\alpha-\phi)}\sin 2\alpha\, d\alpha = \int_0^{2\pi} e^{iz\cos\alpha}\sin 2(\alpha+\phi)\, d\alpha = -2\pi J_2(z)\sin 2\phi\end{cases}$$

which are obtained by using different combinations of (B.3) on page 599. For a given direction of observation $\hat{\boldsymbol{r}} = \hat{\boldsymbol{x}}\sin\theta\cos\phi + \hat{\boldsymbol{y}}\sin\theta\sin\phi + \hat{\boldsymbol{z}}\cos\theta$, we obtain

$$\int_0^{2\pi} e^{-ikr_c\sin\theta\cos(\alpha-\phi)}\left[\hat{\boldsymbol{y}}\left(2Fr_c^2\cos 2\alpha + 8F^3\right) - \hat{\boldsymbol{x}}2Fr_c^2\sin 2\alpha - \hat{\boldsymbol{z}}r_c\sin\alpha\left(r_c^2-4F^2\right)\right] d\alpha$$
$$= 2\pi\Big[\hat{\boldsymbol{y}}\left(8F^3J_0(kr_c\sin\theta) - 2Fr_c^2J_2(kr_c\sin\theta)\cos 2\phi\right)$$
$$+\hat{\boldsymbol{x}}2Fr_c^2J_2(kr_c\sin\theta)\sin 2\phi + i\hat{\boldsymbol{z}}r_c\left(r_c^2-4F^2\right)J_1(kr_c\sin\theta)\sin\phi\Big]$$

The remaining integrals in the r_c-variable are then determined by numerical integration. We get

$$\boldsymbol{I}(\hat{\boldsymbol{r}}) = e^{ikF(1+\cos\theta)}k\omega p\int_0^a r_c\, dr_c \frac{e^{ikr_c^2(1-\cos\theta)/4F}}{\left(r_c^2+4F^2\right)^2}$$
$$\cdot\Big[8F^3\hat{\boldsymbol{y}}J_0(kr_c\sin\theta) + 2Fr_c^2J_2(kr_c\sin\theta)\left(\hat{\boldsymbol{x}}\sin 2\phi - \hat{\boldsymbol{y}}\cos 2\phi\right)$$
$$+ i\hat{\boldsymbol{z}}r_c\left(r_c^2-4F^2\right)J_1(kr_c\sin\theta)\sin\phi\Big]$$

The surface integral $\boldsymbol{I}(\hat{\boldsymbol{r}})$ then determines the far field amplitude by (6.45). In particular, the integral simplifies in the H-plane (x–z-plane or $\phi = 0$-plane) to

$$\boldsymbol{I}(\hat{\boldsymbol{r}}) = 2kF\omega p\hat{\boldsymbol{y}}e^{ikF(1+\cos\theta)}$$
$$\cdot\int_0^a r_c\, dr_c\frac{e^{ikr_c^2(1-\cos\theta)/4F}}{\left(r_c^2+4F^2\right)^2}\left[4F^2J_0(kr_c\sin\theta) - r_c^2J_2(kr_c\sin\theta)\right]$$

The final expression of the far field amplitude is, see (6.45)

$$\boldsymbol{F}_{\text{PO}}(\hat{\boldsymbol{r}}) = \frac{\mathrm{i}k^2 F\omega p\eta_0\hat{\boldsymbol{y}}}{\pi}\mathrm{e}^{\mathrm{i}kF(1+\cos\theta)} \cdot \int_0^a r_\mathrm{c}\,\mathrm{d}r_\mathrm{c}\frac{\mathrm{e}^{\mathrm{i}kr_\mathrm{c}^2(1-\cos\theta)/4F}}{\left(r_\mathrm{c}^2+4F^2\right)^2}\left[4F^2 J_0(kr_\mathrm{c}\sin\theta) - r_\mathrm{c}^2 J_2(kr_\mathrm{c}\sin\theta)\right]$$

since $\hat{\boldsymbol{r}}\times(\hat{\boldsymbol{r}}\times\hat{\boldsymbol{y}}) = -\hat{\boldsymbol{y}}$ if $\hat{\boldsymbol{r}} = \hat{\boldsymbol{x}}\sin\theta + \hat{\boldsymbol{z}}\cos\theta$.

In the E-plane (y–z-plane or $\phi = \pi/2$-plane) the integral $\boldsymbol{I}(\hat{\boldsymbol{r}})$ simplifies to

$$\begin{aligned}\boldsymbol{I}(\hat{\boldsymbol{r}}) &= \mathrm{e}^{\mathrm{i}kF(1+\cos\theta)}k\omega p\int_0^a r_\mathrm{c}\,\mathrm{d}r_\mathrm{c}\frac{\mathrm{e}^{\mathrm{i}kr_\mathrm{c}^2(1-\cos\theta)/4F}}{\left(r_\mathrm{c}^2+4F^2\right)^2} \\ &\quad\cdot\left[2F\hat{\boldsymbol{y}}\left(4F^2J_0(kr_\mathrm{c}\sin\theta)+r_\mathrm{c}^2J_2(kr_\mathrm{c}\sin\theta)\right)+\mathrm{i}\hat{\boldsymbol{z}}r_\mathrm{c}\left(r_\mathrm{c}^2-4F^2\right)J_1(kr_\mathrm{c}\sin\theta)\right] \\ &= I_y(\hat{\boldsymbol{r}})\hat{\boldsymbol{y}} + I_z(\hat{\boldsymbol{r}})\hat{\boldsymbol{z}}\end{aligned}$$

The final expression of the far field amplitude is, see (6.45)

$$\boldsymbol{F}_{\text{PO}}(\hat{\boldsymbol{r}}) = -\frac{\mathrm{i}k\eta_0}{2\pi}\left(-\hat{\boldsymbol{y}}\cos\theta+\hat{\boldsymbol{z}}\sin\theta\right)\left(I_y(\hat{\boldsymbol{r}})\cos\theta - I_z(\hat{\boldsymbol{r}})\sin\theta\right)$$

since if $\hat{\boldsymbol{r}} = \hat{\boldsymbol{y}}\sin\theta + \hat{\boldsymbol{z}}\cos\theta$

$$\hat{\boldsymbol{r}}\times(\hat{\boldsymbol{r}}\times(I_y\hat{\boldsymbol{y}}+I_z\hat{\boldsymbol{z}})) = \left(-\hat{\boldsymbol{y}}\cos\theta+\hat{\boldsymbol{z}}\sin\theta\right)\left(I_y\cos\theta - I_z\sin\theta\right)$$

and therefore $\left|\hat{\boldsymbol{r}}\times(\hat{\boldsymbol{r}}\times(I_y\hat{\boldsymbol{y}}+I_z\hat{\boldsymbol{z}}))\right|^2 = \left|I_y\cos\theta - I_z\sin\theta\right|^2$.

A numerical example of the squared absolute value of the far field amplitude, $|\boldsymbol{F}_{\text{PO}}(\hat{\boldsymbol{r}})|^2$, (suitably normalized) with $a = 0.4$ m and $F = 0.5$ m at the frequency 11 GHz as a function of the angle θ, is depicted in Figure 6.10. The amplitude in the forward direction can be calculated analytically, see Problem 6.11. ■

6.3.4 Geometrical Optics Approximation

We are now going to utilize the alternative representation of the far field, see (6.25) on page 292, over the aperture surface S_a. The construction of the field on the aperture surface is done with ray optics—hence the name geometrical optics (GO) approximation. As in the physical optics approximation, the reflection on the surface of the scatterer is assumed to obey the plane wave relations (ray optics). In the physical optics approximation, the surface integral representation is used to compute the propagation of the fields from the surface to the measurement point. In the geometrical optics approximation, the fields are assumed to propagate along straight lines from the scatterer to the aperture surface S_a. From the aperture surface S_a, the fields at the measurement point are then computed by the surface integral representation. The construction of the ray optics approximation can be found in the literature, e.g., References 102, 180.

We illustrate the method by an incident plane wave or an antenna horn that feeds a reflector antenna. The field from the antenna horn feed at a distance R is approximated by

$$\boldsymbol{E}_\mathrm{i} \approx \boldsymbol{E}_0(\hat{\boldsymbol{R}})\frac{\mathrm{e}^{\mathrm{i}kR}}{R}$$

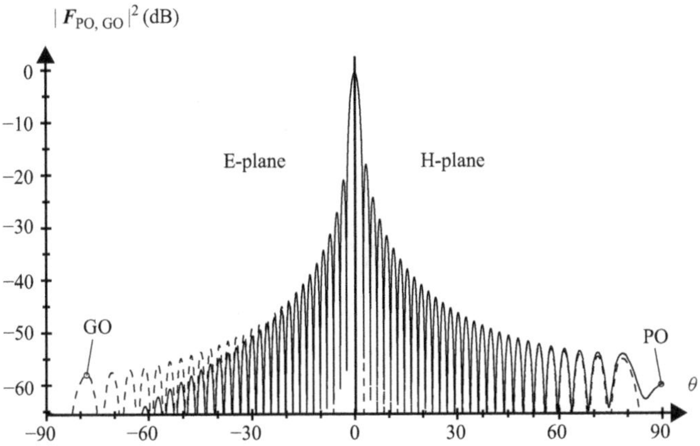

Figure 6.10 Comparison between the physical optics approximation and the geometrical optics approximation of $|\boldsymbol{F}_{PO}(\hat{\boldsymbol{r}})|^2$ in the H-plane (x–z-plane or $\phi = 0$-plane) and the E-plane (y–z-plane or $\phi = \pi/2$-plane) of the parabolic reflector. Solid (broken) curve represents the physical (geometrical) optics approximation, see Example 6.8 (Example 6.12). $a = 0.4$ m, $F = 0.5$ m, and the frequency $f = 11$ GHz.

where $\boldsymbol{E}_0(\hat{\boldsymbol{R}})$ is the angular distribution of the radiation of the feed. The incident field at the reflector propagates as a spherical wave.[16] The scattered field is assumed to satisfy the plane wave relations

$$\boldsymbol{E}_{sGO} = -\eta_0\eta\hat{\boldsymbol{k}}_{GO} \times \boldsymbol{H}_{sGO}$$
$$\boldsymbol{H}_{sGO} = \frac{1}{\eta_0\eta}\hat{\boldsymbol{k}}_{GO} \times \boldsymbol{E}_{sGO}$$

and they are obtained by geometrical construction from the laws of ray optics (the angle of incidence = angle of reflection, and $\hat{\boldsymbol{\nu}} \times (\boldsymbol{E}_i + \boldsymbol{E}_{sGO}) = \boldsymbol{0}$ on the surface S_s^+), see Figure 6.11. Notice that no integration has to be performed to construct the scattered field.

Moreover, we assume that the contribution of the incident field in the surface integral in (6.25) can be neglected in comparison to the contribution of the scattered field. The incident field interacts in a destructive way in the forward direction of the reflector, $\hat{\boldsymbol{k}}_{GO}$, if $\hat{\boldsymbol{k}}_{GO}$ is in the opposite direction to the incident direction $\hat{\boldsymbol{k}}_i$, see Example 6.9 for further details in the case of an incident plane wave. The scattered field, however, interacts constructively in the forward direction of the reflector. The far field in (6.25) then only contains an integral over the scattered field

$$\boldsymbol{F}_{GO}(\hat{\boldsymbol{r}}) = \frac{ik}{4\pi}\hat{\boldsymbol{r}} \times \iint_{S_a} \left[\hat{\boldsymbol{\nu}}(\boldsymbol{r}') \times \boldsymbol{E}_{sGO}(\boldsymbol{r}') - \eta_0\eta\hat{\boldsymbol{r}} \times \left(\hat{\boldsymbol{\nu}}(\boldsymbol{r}') \times \boldsymbol{H}_{sGO}(\boldsymbol{r}')\right)\right] e^{-ik\hat{\boldsymbol{r}}\cdot\boldsymbol{r}'} \, dS' \tag{6.48}$$

[16]This is a simplification. Geometrical optics approximation allows for a wave front with two different radii of curvatures.

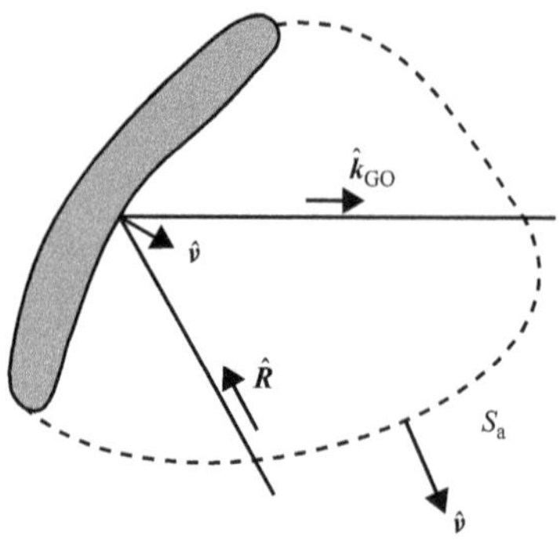

Figure 6.11 Construction of the scattered field in the geometrical optics approximation.

This expression is simplified by introducing additional approximations. The reflector is designed in such a way that the propagation direction of the scattered field, $\hat{\boldsymbol{k}}_{GO}$, which is constructed by the ray optics laws, varies very little. We can, therefore, let $\hat{\boldsymbol{k}}_{GO}$ be a constant vector in the integrand. Moreover, the field strength is most intensive in this direction, i.e., in the direction $\hat{\boldsymbol{r}} = \hat{\boldsymbol{k}}_{GO}$. Then, we have

$$\boldsymbol{H}_{sGO} \approx \frac{1}{\eta_0 \eta} \hat{\boldsymbol{r}} \times \boldsymbol{E}_{sGO}$$

Repeated use of the BAC-CAB rule implies ($\hat{\boldsymbol{r}} \cdot \boldsymbol{E}_{sGO} \approx 0$)

$$\begin{aligned}
\hat{\boldsymbol{r}} \times \left\{\hat{\boldsymbol{r}} \times \left[\hat{\boldsymbol{\nu}} \times (\hat{\boldsymbol{r}} \times \boldsymbol{E}_{sGO})\right]\right\} &= \hat{\boldsymbol{r}} \times \left\{\hat{\boldsymbol{r}} \times \left[\hat{\boldsymbol{r}}\left(\hat{\boldsymbol{\nu}} \cdot \boldsymbol{E}_{sGO}\right) - \boldsymbol{E}_{sGO}\left(\hat{\boldsymbol{\nu}} \cdot \hat{\boldsymbol{r}}\right)\right]\right\} \\
&= -\hat{\boldsymbol{r}} \times \left\{\hat{\boldsymbol{r}} \times \boldsymbol{E}_{sGO}\left(\hat{\boldsymbol{\nu}} \cdot \hat{\boldsymbol{r}}\right)\right\} = -\left\{\hat{\boldsymbol{r}}\left(\hat{\boldsymbol{r}} \cdot \boldsymbol{E}_{sGO}\right) - \boldsymbol{E}_{sGO}\right\}\left(\hat{\boldsymbol{\nu}} \cdot \hat{\boldsymbol{r}}\right) \\
&= \boldsymbol{E}_{sGO}\left(\hat{\boldsymbol{\nu}} \cdot \hat{\boldsymbol{r}}\right) = -\hat{\boldsymbol{r}} \times \left(\hat{\boldsymbol{\nu}} \times \boldsymbol{E}_{sGO}\right)
\end{aligned}$$

The far field amplitude can then be approximated as

$$\boldsymbol{F}_{GO}(\hat{\boldsymbol{r}}) = \frac{ik}{2\pi} \hat{\boldsymbol{r}} \times \iint_{S_a} \hat{\boldsymbol{\nu}}(\boldsymbol{r}') \times \boldsymbol{E}_{sGO}(\boldsymbol{r}') e^{-ik\hat{\boldsymbol{r}}\cdot\boldsymbol{r}'} \, dS', \qquad \hat{\boldsymbol{r}} \approx \hat{\boldsymbol{k}}_{GO} \tag{6.49}$$

The advantage with the expressions (6.48) and (6.49) is that the aperture surface can be chosen with some freedom, such that the integral becomes simple to compute. In general, the geometrical optic approximation requires higher frequencies than the physical optics approximation for the same accuracy, since the approximations are coarser. Note also that these expressions of the far field amplitude cannot be used in the optical theorem, since these expressions on $\boldsymbol{F}_{GO}(\hat{\boldsymbol{r}})$ are not valid in the direction of the incident direction, but only for $\hat{\boldsymbol{r}} = \hat{\boldsymbol{k}}_{GO}$, which often coincides with the direction opposite to the direction of the incident field.

We illustrate the use of the geometrical optics approximation with a series of examples.

Example 6.9
In this example, we show that the contribution of an incident plane wave in the integral (6.25) over the aperture surface S_a is negligible, except in the forward direction $\hat{r} = \hat{k}_i$. More precisely, we show that

$$I(\hat{r}) = \frac{ik}{4\pi}\hat{r} \times \iint_{S_a} \left[\hat{\nu}(r') \times E_i(r') - \eta_0\eta\hat{r} \times \left(\hat{\nu}(r') \times H_i(r')\right)\right] e^{-ik\hat{r}\cdot r'}\, dS'$$

where

$$\begin{cases} E_i(r) = E_0 e^{ik\hat{k}_i\cdot r} \\ \eta_0\eta H_i(r) = \hat{k}_i \times E_i(r) \end{cases}$$

is negligible except when $\hat{r} \approx \hat{k}_i$. Introduction of the fields E_i and H_i gives

$$I(\hat{r}) = \frac{ik}{4\pi}\hat{r} \times \iint_{S_a} \left[\hat{\nu}(r') \times E_0 - \hat{r} \times \left(\hat{\nu}(r') \times \left[\hat{k}_i \times E_0\right]\right)\right] e^{ik(\hat{k}_i - \hat{r})\cdot r'}\, dS'$$

At high frequencies, $kd \gg 1$ (d is the diameter of the scatterer), the integrand oscillates rapidly, due to the exponential function, except when $\hat{r} \approx \hat{k}_i$, which is the forward direction. The oscillations imply that the contribution of the integrand cancel, except when $\hat{r} \approx \hat{k}_i$. The higher the frequency, the better the cancellations. Thus, the integral $I(\hat{r})$ is negligible except in the forward direction $\hat{r} = \hat{k}_i$. More details on the asymptotic evaluation of integrals with oscillating terms are found in Appendix G, and the reader is referred to this appendix for an explanation of the mathematical details.

The value of the integral $I(\hat{r})$ in the forward direction is of interest, and we rewrite the integrand by the use of the BAC-CAB rule. Moreover, use $\hat{k}_i \cdot E_0 = 0$, and simplify the integrand in the following way:

$$\begin{aligned} &\underbrace{\hat{k}_i \times (\hat{\nu} \times E_0)}_{-E_0(\hat{k}_i\cdot\hat{\nu})} - \hat{k}_i \times \left(\hat{k}_i \times \underbrace{\left(\hat{\nu} \times \left[\hat{k}_i \times E_0\right]\right)}_{\hat{k}_i(\hat{\nu}\cdot E_0) - E_0(\hat{k}_i\cdot\hat{\nu})}\right) \\ &= -E_0(\hat{k}_i \cdot \hat{\nu}) + \hat{k}_i \times (\hat{k}_i \times E_0)(\hat{k}_i \cdot \hat{\nu}) = -2E_0(\hat{k}_i \cdot \hat{\nu}) \end{aligned}$$

The projected area of the cross section, see (6.32), can, by the use of the divergence theorem, be written as (see Figure 6.3 for the definition of the unit normals of the surfaces S_s^+ and S_a)

$$A(\hat{k}_i) = -\iint_{S_s^+} \hat{\nu}(r') \cdot \hat{k}_i\, dS' = -\iint_{S_a} \hat{\nu}(r') \cdot \hat{k}_i\, dS'$$

since $\nabla' \cdot \hat{k}_i = 0$. We get

$$I(\hat{k}_i) = -\frac{ik}{2\pi} E_0 \iint_{S_a} \hat{k}_i \cdot \hat{\nu}(r')\, dS' = \frac{ik}{2\pi} E_0 A(\hat{k}_i)$$

In a neighborhood of the forward direction, $\hat{\boldsymbol{r}} = \hat{\boldsymbol{k}}_{\mathrm{i}}$, we approximate the integrand by

$$\hat{\boldsymbol{r}} \times \left[\hat{\boldsymbol{\nu}} \times \boldsymbol{E}_0 - \hat{\boldsymbol{r}} \times \left(\hat{\boldsymbol{\nu}} \times \left[\hat{\boldsymbol{k}}_{\mathrm{i}} \times \boldsymbol{E}_0\right]\right)\right] \approx -2\boldsymbol{E}_0(\hat{\boldsymbol{k}}_{\mathrm{i}} \cdot \hat{\boldsymbol{\nu}})$$

according to the calculations above. The argument of the exponential function is rewritten by introducing a transverse component to the direction $\hat{\boldsymbol{k}}_{\mathrm{i}}$, i.e.,

$$\boldsymbol{r}' = \boldsymbol{r}'_{\perp} + z'\hat{\boldsymbol{k}}_{\mathrm{i}}, \qquad \hat{\boldsymbol{k}}_{\mathrm{i}} \cdot \boldsymbol{r}'_{\perp} = 0$$

The following approximation is pertinent:

$$(\hat{\boldsymbol{k}}_{\mathrm{i}} - \hat{\boldsymbol{r}}) \cdot \boldsymbol{r}' = z' - \hat{\boldsymbol{r}} \cdot \boldsymbol{r}'_{\perp} - z' \underbrace{\hat{\boldsymbol{r}} \cdot \hat{\boldsymbol{k}}_{\mathrm{i}}}_{\approx 1} \approx -\hat{\boldsymbol{r}} \cdot \boldsymbol{r}'_{\perp}$$

The following approximation, close to the forward direction, $\hat{\boldsymbol{r}} \approx \hat{\boldsymbol{k}}_{\mathrm{i}}$, is obtained:

$$\boldsymbol{I}(\hat{\boldsymbol{r}}) = -\frac{\mathrm{i}k}{2\pi}\boldsymbol{E}_0 \iint_{S_{\mathrm{a}}} (\hat{\boldsymbol{k}}_{\mathrm{i}} \cdot \hat{\boldsymbol{\nu}}(\boldsymbol{r}'))\mathrm{e}^{-\mathrm{i}k\hat{\boldsymbol{r}}\cdot\boldsymbol{r}'_{\perp}}\, \mathrm{d}S'$$

Note that the integral on the right-hand side gives the same value for all scatterers with the same illuminated surface S_{s}^{+}, since $\nabla' \cdot (\hat{\boldsymbol{k}}_{\mathrm{i}}\mathrm{e}^{-\mathrm{i}k\hat{\boldsymbol{r}}\cdot\boldsymbol{r}'_{\perp}}) = 0$. ■

Example 6.10

In this example, the geometrical optics approximation is applied to a dihedral corner reflector (right angle between the two planes), which is illuminated by a plane wave along its symmetry axis (along the negative z-axis). The geometry and the coordinate axes are given in Figure 6.12. The height of the dihedral corner reflector is denoted h and its projected width is b, see Figure 6.12.

In this example, the geometrical optics field is easily constructed by simple reflection in the plane surfaces of the corner reflector, but to obtain some systematic treatment of the reflection, it is suitable to introduce some general procedure that also can be applied to more complex geometries.

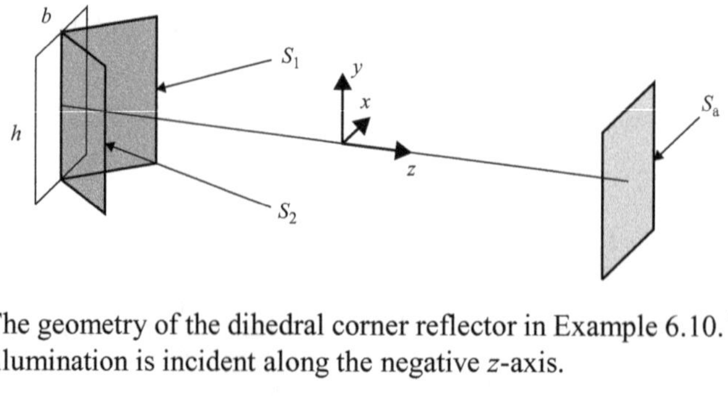

Figure 6.12 The geometry of the dihedral corner reflector in Example 6.10. The plane wave illumination is incident along the negative z-axis.

We start by noting that all waves remain plane waves after a reflection in the plane surfaces of the corner reflector. Moreover, introduce the two dihedral surfaces S_1 and S_2 with coordinate representations

$$\begin{cases} \boldsymbol{S}_1(x,y) = \hat{\boldsymbol{x}}x + \hat{\boldsymbol{y}}y + \hat{\boldsymbol{z}}(l+x), & 0 \le x \le b/2, \quad -h/2 \le y \le h/2 \\ \boldsymbol{S}_2(x,y) = \hat{\boldsymbol{x}}x + \hat{\boldsymbol{y}}y + \hat{\boldsymbol{z}}(l-x), & -b/2 \le x \le 0, \quad -h/2 \le y \le h/2 \end{cases}$$

where l is a parameter that gives the position of the corner reflector relative to the origin (in the figure $l < 0$). The normal vectors of the surfaces are

$$\begin{cases} \hat{\boldsymbol{\nu}}_1 = \dfrac{-\hat{\boldsymbol{x}} + \hat{\boldsymbol{z}}}{\sqrt{2}} \\ \hat{\boldsymbol{\nu}}_2 = \dfrac{\hat{\boldsymbol{x}} + \hat{\boldsymbol{z}}}{\sqrt{2}} \end{cases}$$

To use the geometrical optics approximation, we must first determine the scattered field on the aperture surface S_a by ray (geometrical) optics. We chose the plane $z = \xi$ as the aperture surface (all other parts of the aperture surface give zero contribution).

In our case, the incident field is ($\hat{\boldsymbol{k}}_\text{i} = -\hat{\boldsymbol{z}}$, $\boldsymbol{E}_0 \cdot \hat{\boldsymbol{k}}_\text{i} = 0$)

$$\boldsymbol{E}_\text{i}(\boldsymbol{r}) = \boldsymbol{E}_0 \mathrm{e}^{\mathrm{i}k\hat{\boldsymbol{k}}_\text{i}\cdot\boldsymbol{r}} = \boldsymbol{E}_0 \mathrm{e}^{-\mathrm{i}kz}$$

The reflected field is constructed by the reflection relations in Section 6.3.2, see (6.29)

$$\begin{cases} \hat{\boldsymbol{k}}_\text{s} = \mathbf{R} \cdot \hat{\boldsymbol{k}}_\text{i} \\ \boldsymbol{E}_\text{s} = -\mathbf{R} \cdot \boldsymbol{E}_\text{i} \\ \boldsymbol{H}_\text{s} = \mathbf{R} \cdot \boldsymbol{H}_\text{i} \end{cases}$$

The reflected field splits into two parts, depending on whether the reflection takes place in surface S_1 or in surface S_2 first. The reflection dyadic differs for the two surfaces. We have

$$\mathbf{R}_i = \mathbf{I}_3 - 2\hat{\boldsymbol{\nu}}_i\hat{\boldsymbol{\nu}}_i, \qquad i = 1, 2$$

The reflected (scattered) field on the aperture surface, S_a, is now constructed. In the first case, we let the incident field impinge on surface S_1. The following sequence of reflections leads to the construction of fields on the aperture surface:

$$\text{Incident field} \longrightarrow \boldsymbol{S}_1 \longrightarrow \boldsymbol{S}_2 \longrightarrow \text{Aperture field at } S_\text{a}$$

For each reflection holds:

1. The incident field propagates to the surface S_1 where the incident field on the surface S_1 is

$$\boldsymbol{E}_0 \mathrm{e}^{\mathrm{i}k\hat{\boldsymbol{k}}_\text{i}\cdot\boldsymbol{S}_1(x,y)}, \qquad 0 \le x \le b/2, \quad -h/2 \le y \le h/2$$

2. This field is reflected at the surface S_1 by the reflection dyadic $\mathbf{R}_1$. The result is

$$(-\mathbf{R}_1) \cdot \boldsymbol{E}_0 \mathrm{e}^{\mathrm{i}k\hat{\boldsymbol{k}}_\text{i}\cdot\boldsymbol{S}_1(x,y)}$$

3. This field propagates in the direction $\hat{\boldsymbol{k}}_1 = \mathbf{R}_1 \cdot \hat{\boldsymbol{k}}_{\rm i} = -\hat{\boldsymbol{x}}$ to the surface S_2 (reflection point $S_2(-x, y)$) where its value is

$$\mathrm{e}^{ik\hat{\boldsymbol{k}}_1\cdot(\boldsymbol{S}_2(-x,y)-\boldsymbol{S}_1(x,y))}\cdot(-\mathbf{R}_1)\cdot \boldsymbol{E}_0\mathrm{e}^{ik\hat{\boldsymbol{k}}_{\rm i}\cdot\boldsymbol{S}_1(x,y)}$$

where $\hat{\boldsymbol{k}}_1\cdot(\boldsymbol{S}_2(-x,y)-\boldsymbol{S}_1(x,y))$ is the phase difference between the rays evaluated on each surface, respectively.

4. The reflected field at the surface S_2 is obtained by an application of the reflection dyadic $\mathbf{R}_2$

$$(-\mathbf{R}_2)\cdot\mathrm{e}^{ik\hat{\boldsymbol{k}}_1\cdot(\boldsymbol{S}_2(-x,y)-\boldsymbol{S}_1(x,y))}\cdot(-\mathbf{R}_1)\cdot \boldsymbol{E}_0\mathrm{e}^{ik\hat{\boldsymbol{k}}_{\rm i}\cdot\boldsymbol{S}_1(x,y)}$$

5. The final propagation to the point $\boldsymbol{r}$ on the aperture surface $S_{\rm a}$ is performed in the direction $\hat{\boldsymbol{k}}_{\rm GO} = \mathbf{R}_2\cdot\hat{\boldsymbol{k}}_1 = \mathbf{R}_2\cdot\mathbf{R}_1\cdot\hat{\boldsymbol{k}}_{\rm i} = \hat{\boldsymbol{z}}$

$$\mathrm{e}^{ik\hat{\boldsymbol{k}}_{\rm GO}\cdot(\boldsymbol{r}-\boldsymbol{S}_2(-x,y))}\cdot(-\mathbf{R}_2)\cdot\mathrm{e}^{ik\hat{\boldsymbol{k}}_1\cdot(\boldsymbol{S}_2(-x,y)-\boldsymbol{S}_1(x,y))}\cdot(-\mathbf{R}_1)\cdot \boldsymbol{E}_0\mathrm{e}^{ik\hat{\boldsymbol{k}}_{\rm i}\cdot\boldsymbol{S}_1(x,y)}$$

The field at the aperture surface $\boldsymbol{E}_{\rm a}(\boldsymbol{r}) = \boldsymbol{E}_{\rm sGO}(\boldsymbol{r})$, $\boldsymbol{r} = x\hat{\boldsymbol{x}} + y\hat{\boldsymbol{y}} + \xi\hat{\boldsymbol{z}}$, is

$$\boldsymbol{E}_{\rm a}(\boldsymbol{r}) = \mathrm{e}^{ik\hat{\boldsymbol{k}}_{\rm GO}\cdot(\boldsymbol{r}-\boldsymbol{S}_2(-x,y))}\cdot\mathbf{R}_2\cdot\mathrm{e}^{ik\hat{\boldsymbol{k}}_1\cdot(\boldsymbol{S}_2(-x,y)-\boldsymbol{S}_1(x,y))}\cdot\mathbf{R}_1\cdot \boldsymbol{E}_0\mathrm{e}^{ik\hat{\boldsymbol{k}}_{\rm i}\cdot\boldsymbol{S}_1(x,y)}$$

The vectors simplify to

$$\mathbf{R}_2\cdot\mathbf{R}_1\cdot\boldsymbol{E}_0 = \boldsymbol{E}_0 - 2\hat{\boldsymbol{x}}(\boldsymbol{E}_0\cdot\hat{\boldsymbol{x}}) = \boldsymbol{E}_0 - 2\hat{\boldsymbol{x}}E_{0x}$$

since $\mathbf{R}_2\cdot\mathbf{R}_1 = \mathbf{I}_3 - 2\hat{\boldsymbol{x}}\hat{\boldsymbol{x}} - 2\hat{\boldsymbol{z}}\hat{\boldsymbol{z}}$ and $\boldsymbol{E}_0\cdot\hat{\boldsymbol{z}} = 0$.

The phase difference in the exponent is

$$\begin{aligned}
&\hat{\boldsymbol{k}}_{\rm GO}\cdot(\boldsymbol{r}-\boldsymbol{S}_2(-x,y)) + \hat{\boldsymbol{k}}_1\cdot(\boldsymbol{S}_2(-x,y)-\boldsymbol{S}_1(x,y)) + \hat{\boldsymbol{k}}_{\rm i}\cdot\boldsymbol{S}_1(x,y)\\
&\quad = \xi - (l+x) - (-x-x) - (l+x) = \xi - 2l
\end{aligned}$$

The entire reflection contribution from the reflection in S_1 then becomes

$$\boldsymbol{E}_{\rm a}(x,y) = \left(\boldsymbol{E}_0 - 2\hat{\boldsymbol{x}}E_{0x}\right)\mathrm{e}^{ik(\xi-2l)}, \qquad 0\le x\le b/2, \quad -h/2\le y\le h/2$$

The construction of the second part of the aperture field, where the reflection first takes place at the surface S_2, is a consequence of the following the chain of reflections (interchange indices 1 and 2):

$$\text{Incident field} \longrightarrow \boldsymbol{S}_2 \longrightarrow \boldsymbol{S}_1 \longrightarrow \text{Aperture field at } S_{\rm a}$$

which is identical to the first case since the reflections dyadics $\mathbf{R}_i$, $i = 1, 2$, commute (holds since $\hat{\boldsymbol{\nu}}_1\cdot\hat{\boldsymbol{\nu}}_2 = 0$). The entire reflection contribution from the reflection in S_2 is

$$\boldsymbol{E}_{\rm a}(x,y) = \left(\boldsymbol{E}_0 - 2\hat{\boldsymbol{x}}E_{0x}\right)\mathrm{e}^{ik(\xi-2l)}, \qquad -b/2\le x\le 0, \quad -h/2\le y\le h/2$$

The expression of the far field amplitude for geometrical optics approximation, $\boldsymbol{F}_{\mathrm{GO}}(\hat{\boldsymbol{r}})$, in (6.48) ($\eta = 1$)

$$\boldsymbol{F}_{\mathrm{GO}}(\hat{\boldsymbol{r}}) = \frac{\mathrm{i}k}{4\pi}\hat{\boldsymbol{r}} \times \iint_{S_\mathrm{a}} \left[\hat{\boldsymbol{\nu}}(\boldsymbol{r}') \times \boldsymbol{E}_{\mathrm{sGO}}(\boldsymbol{r}') - \eta_0 \hat{\boldsymbol{r}} \times \left(\hat{\boldsymbol{\nu}}(\boldsymbol{r}') \times \boldsymbol{H}_{\mathrm{sGO}}(\boldsymbol{r}')\right)\right] \mathrm{e}^{-\mathrm{i}k\hat{\boldsymbol{r}}\cdot\boldsymbol{r}'}\, \mathrm{d}S'$$

simplifies for the dihedral corner reflector. The integrand is rewritten as (note that on S_a holds $z = \xi$ and $\hat{\boldsymbol{\nu}} = \hat{\boldsymbol{z}}$)

$$\hat{\boldsymbol{\nu}}(\boldsymbol{r}') \times \boldsymbol{E}_{\mathrm{sGO}}(\boldsymbol{r}') - \eta_0 \hat{\boldsymbol{r}} \times \left(\hat{\boldsymbol{\nu}}(\boldsymbol{r}') \times \boldsymbol{H}_{\mathrm{sGO}}(\boldsymbol{r}')\right) = (\hat{\boldsymbol{z}} + \hat{\boldsymbol{r}}) \times \boldsymbol{E}_\mathrm{a}(x, y)$$

since $\eta_0 \hat{\boldsymbol{z}} \times \boldsymbol{H}_{\mathrm{sGO}} = -\boldsymbol{E}_{\mathrm{sGO}} = -\boldsymbol{E}_\mathrm{a}$. We get

$$\begin{aligned}\boldsymbol{F}_{\mathrm{GO}}(\hat{\boldsymbol{r}}) = {} & \frac{ik}{4\pi}\mathrm{e}^{ik(\xi(1-\cos\theta)-2l)}\hat{\boldsymbol{r}} \times \left[(\hat{\boldsymbol{z}} + \hat{\boldsymbol{r}}) \times \left(\boldsymbol{E}_0 - 2\hat{\boldsymbol{x}}E_{0x}\right)\right] \\ & \cdot \int_{-b/2}^{b/2} \mathrm{d}x \int_{-h/2}^{h/2} \mathrm{d}y\, \mathrm{e}^{-ik(x\sin\theta\cos\phi + y\sin\theta\sin\phi)}\end{aligned}$$

where the spherical angles θ and ϕ are defined in the usual way by the direction of the position vector $\hat{\boldsymbol{r}}$

$$\hat{\boldsymbol{r}} = \hat{\boldsymbol{x}}\sin\theta\cos\phi + \hat{\boldsymbol{y}}\sin\theta\sin\phi + \hat{\boldsymbol{z}}\cos\theta$$

By the use of (6.40), we get

$$\begin{aligned}\boldsymbol{F}_{\mathrm{GO}}(\hat{\boldsymbol{r}}) = {} & \frac{ikbh}{4\pi}\mathrm{e}^{ik(\xi(1-\cos\theta)-2l)}\hat{\boldsymbol{r}} \times \left[(\hat{\boldsymbol{z}} + \hat{\boldsymbol{r}}) \times \left(\boldsymbol{E}_0 - 2\hat{\boldsymbol{x}}E_{0x}\right)\right] \\ & \cdot \frac{\sin\left(\frac{kb}{2}\sin\theta\cos\phi\right)}{\frac{kb}{2}\sin\theta\cos\phi} \frac{\sin\left(\frac{kh}{2}\sin\theta\sin\phi\right)}{\frac{kh}{2}\sin\theta\sin\phi}\end{aligned} \tag{6.50}$$

Compare this expression with the simplified expression, we obtained close to the forward direction, see (6.49)

$$\begin{aligned}\boldsymbol{F}_{\mathrm{GO}}(\hat{\boldsymbol{r}}) = {} & \frac{ikbh}{2\pi}\mathrm{e}^{ik(\xi(1-\cos\theta)-2l)}\hat{\boldsymbol{r}} \times \left[\hat{\boldsymbol{z}} \times \left(\boldsymbol{E}_0 - 2\hat{\boldsymbol{x}}E_{0x}\right)\right] \\ & \cdot \frac{\sin\left(\frac{kb}{2}\sin\theta\cos\phi\right)}{\frac{kb}{2}\sin\theta\cos\phi} \frac{\sin\left(\frac{kh}{2}\sin\theta\sin\phi\right)}{\frac{kh}{2}\sin\theta\sin\phi}, \qquad \hat{\boldsymbol{r}} \approx \hat{\boldsymbol{z}}\end{aligned}$$

By the use of (6.50) and the result in Problem 6.10, we obtain

$$\left|\boldsymbol{F}_{\mathrm{GO}}(\hat{\boldsymbol{r}})\right|^2 = \frac{k^2b^2h^2}{16\pi^2}(1+\cos\theta)^2\left|\boldsymbol{E}_0\right|^2 \left|\frac{\sin\left(\frac{kb}{2}\sin\theta\cos\phi\right)}{\frac{kb}{2}\sin\theta\cos\phi}\right|^2 \left|\frac{\sin\left(\frac{kh}{2}\sin\theta\sin\phi\right)}{\frac{kh}{2}\sin\theta\sin\phi}\right|^2$$

which is independent of the polarization of the incident field.

The half width of the main lobe as a function of frequency is shown in Figure 6.13. The differential scattering cross section (antenna diagram) is shown in Figure 6.14. ■

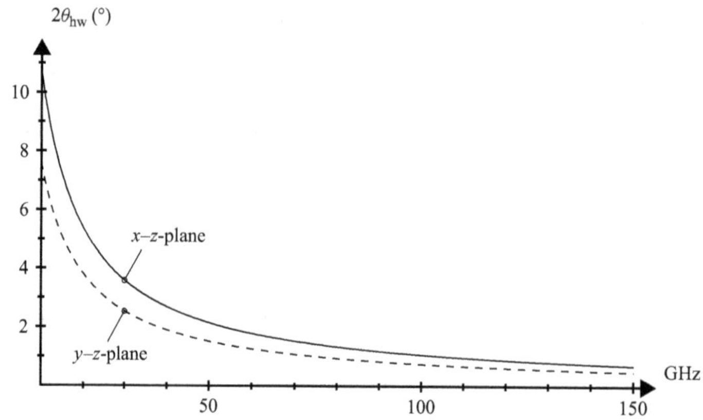

Figure 6.13 Half width of the main lobe for the dihedral corner reflector in Example 6.10. The computations are made with the geometrical optics approximation in the x–z-plane ($\phi = 0$-plane) and in the y–z-plane ($\phi = \pi/2$-plane). Numerical data are $b = 20/\sqrt{2}$ cm, and $h = 20$ cm.

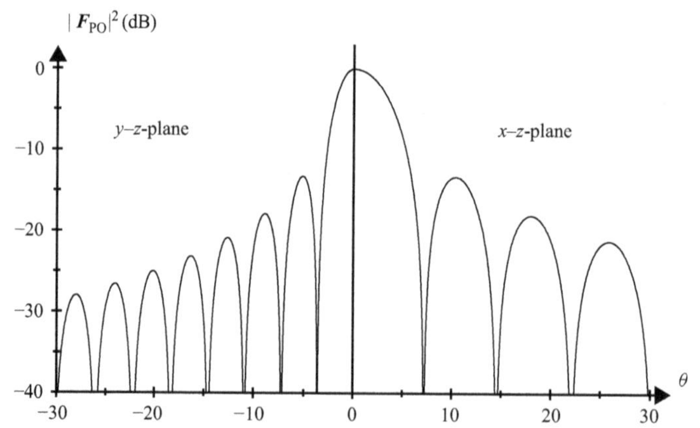

Figure 6.14 The far field amplitude $\left|\boldsymbol{F}_{GO}(\hat{\boldsymbol{r}})\right|^2$ of the dihedral corner reflector in Example 6.10 normalized by $(kbh/2\pi)^2 \, |\boldsymbol{E}_0|^2$, $kb = 50$, $kh = 100$.

Example 6.11

In this example, we carry out the calculations for the cylindrical parabolic reflector in Example 6.7 on page 297 with the geometrical optical approximation. The definitions of the geometry, etc. as well as several vector calculations, we collect from Example 6.7.

The general expression of the far field amplitude in the geometrical optics approximation is, see (6.48) ($\eta = 1$)

$$\boldsymbol{F}_{\mathrm{GO}}(\hat{\boldsymbol{r}}) = \frac{\mathrm{i}k}{4\pi}\hat{\boldsymbol{r}} \times \iint_{S_\mathrm{a}} \left[\hat{\boldsymbol{\nu}}(\boldsymbol{r}') \times \boldsymbol{E}_{\mathrm{sGO}}(\boldsymbol{r}') - \eta_0 \hat{\boldsymbol{r}} \times (\hat{\boldsymbol{\nu}}(\boldsymbol{r}') \times \boldsymbol{H}_{\mathrm{sGO}}(\boldsymbol{r}'))\right] \mathrm{e}^{-\mathrm{i}k\hat{\boldsymbol{r}}\cdot\boldsymbol{r}'} \, \mathrm{d}S' \quad (6.51)$$

where the integration is carried out over the aperture surface S_a.

To determine the far field from the reflector, we first compute the electric field in the aperture plane, which is $z = \xi > b^2/16F - F$ (more precisely the rectangular surface $z = \xi$, $-b/2 \leq x \leq b/2$, $-h/2 \leq y \leq h/2$) in front of the reflector (all other parts of the aperture surface give zero contribution). At each point in the aperture plane, there is a unique ray to the origin, where the source (antenna) is located.

The reflected (scattered) direction is given by

$$\hat{\boldsymbol{k}}_\mathrm{s}(x,y) = \hat{\boldsymbol{k}}_\mathrm{i}(x) - 2\hat{\boldsymbol{\nu}}(x,y)\left(\hat{\boldsymbol{\nu}}(x,y) \cdot \hat{\boldsymbol{k}}_\mathrm{i}(x)\right)$$

where $\hat{\boldsymbol{k}}_\mathrm{i}(x)$ is given by (6.38) and $\hat{\boldsymbol{\nu}}(x,y)$ by (6.35). Simple calculations give

$$\hat{\boldsymbol{k}}_\mathrm{s}(x,y) = \hat{\boldsymbol{k}}_\mathrm{s} = \frac{\hat{\boldsymbol{x}}4Fx + \hat{\boldsymbol{z}}\left(x^2 - 4F^2\right)}{x^2 + 4F^2} + 2\frac{-\hat{\boldsymbol{x}}x + \hat{\boldsymbol{z}}2F}{\sqrt{x^2+4F^2}}\frac{2F}{\sqrt{x^2+4F^2}} = \hat{\boldsymbol{z}}$$

i.e., the field is a plane wave after the reflection in the parabolic reflector, just as prescribed by the construction.

The reflected (scattered) fields $\boldsymbol{E}_\mathrm{s}(\boldsymbol{S}(x,y))$ and $\boldsymbol{H}_\mathrm{s}(\boldsymbol{S}(x,y))$ on the reflector surface are easily determined by the plane wave conditions in Section 6.3.2, see (6.29). We obtain

$$\begin{cases} \boldsymbol{E}_\mathrm{s}(\boldsymbol{S}(x,y)) = -\boldsymbol{E}_\mathrm{i}(\boldsymbol{S}(x,y)) + 2\hat{\boldsymbol{\nu}}(x,y)\left(\hat{\boldsymbol{\nu}}(x,y) \cdot \boldsymbol{E}_\mathrm{i}(\boldsymbol{S}(x,y))\right) \\ \boldsymbol{H}_\mathrm{s}(\boldsymbol{S}(x,y)) = \boldsymbol{H}_\mathrm{i}(\boldsymbol{S}(x,y)) - 2\hat{\boldsymbol{\nu}}(x,y)\left(\hat{\boldsymbol{\nu}}(x,y) \cdot \boldsymbol{H}_\mathrm{i}(\boldsymbol{S}(x,y))\right) \end{cases}$$

where $\boldsymbol{E}_\mathrm{i}(\boldsymbol{S}(x,y))$ and $\boldsymbol{H}_\mathrm{i}(\boldsymbol{S}(x,y))$ are given by (6.34). The distance from the feed to the reflector, $r_\mathrm{c}(x)$, is given by (6.36). The result is

$$\begin{cases} \boldsymbol{E}_\mathrm{s}(\boldsymbol{S}(x,y)) = -E_0\sqrt{\dfrac{8F}{\mathrm{i}k\pi\left(x^2+4F^2\right)}}\mathrm{e}^{\mathrm{i}k\left(x^2+4F^2\right)/4F}\hat{\boldsymbol{y}} \\ \boldsymbol{H}_\mathrm{s}(\boldsymbol{S}(x,y)) = \dfrac{E_0}{\eta_0}\sqrt{\dfrac{8F}{\mathrm{i}k\pi\left(x^2+4F^2\right)}}\mathrm{e}^{\mathrm{i}k\left(x^2+4F^2\right)/4F}\hat{\boldsymbol{x}} \end{cases} \quad (6.52)$$

since $\hat{\boldsymbol{\nu}}(x,y) \cdot \hat{\boldsymbol{y}} = 0$, and

$$\frac{\hat{\boldsymbol{x}}\left(x^2-4F^2\right) - \hat{\boldsymbol{z}}4Fx}{x^2+4F^2} - 2\frac{-\hat{\boldsymbol{x}}x + \hat{\boldsymbol{z}}2F}{\sqrt{x^2+4F^2}}\left(\frac{-\hat{\boldsymbol{x}}x + \hat{\boldsymbol{z}}2F}{\sqrt{x^2+4F^2}} \cdot \frac{\hat{\boldsymbol{x}}\left(x^2-4F^2\right) - \hat{\boldsymbol{z}}4Fx}{x^2+4F^2}\right) = -\hat{\boldsymbol{x}}$$

It remains to determine the distance from the point of reflection to the corresponding point at the aperture plane. This distance is $\xi - \left(x^2/4F - F\right)$ and the fields at the aperture plane,

$\boldsymbol{E}_{\mathrm{a}} = \boldsymbol{E}_{\mathrm{sGO}}$ and $\boldsymbol{H}_{\mathrm{a}} = \boldsymbol{H}_{\mathrm{sGO}}$, are

$$\begin{cases} \boldsymbol{E}_{\mathrm{a}}(x,y) = \boldsymbol{E}_{\mathrm{s}}(\boldsymbol{S}(x,y))\mathrm{e}^{\mathrm{i}k(\xi-(x^2/4F-F))} \\ \boldsymbol{H}_{\mathrm{a}}(x,y) = \boldsymbol{H}_{\mathrm{s}}(\boldsymbol{S}(x,y))\mathrm{e}^{\mathrm{i}k(\xi-(x^2/4F-F))} \end{cases}$$

where $\boldsymbol{E}_{\mathrm{s}}(\boldsymbol{S}(x,y))$ and $\boldsymbol{H}_{\mathrm{s}}(\boldsymbol{S}(x,y))$ are given by (6.52).

At the aperture surface S_{a} in (6.51), the unit normal vector is $\hat{\boldsymbol{\nu}} = \hat{\boldsymbol{z}}$. Moreover, we have

$$\eta_0 \hat{\boldsymbol{z}} \times \boldsymbol{H}_{\mathrm{a}}(x,y) = -\boldsymbol{E}_{\mathrm{a}}(x,y)$$

which gives simplifications in (6.51).

$$\boldsymbol{F}_{\mathrm{GO}}(\hat{\boldsymbol{r}}) = \frac{\mathrm{i}k}{4\pi}\hat{\boldsymbol{r}} \times \int_{-b/2}^{b/2} \mathrm{d}x \int_{-h/2}^{h/2} \mathrm{d}y \left[(\hat{\boldsymbol{z}} + \hat{\boldsymbol{r}}) \times \boldsymbol{E}_{\mathrm{a}}(x,y)\right] \mathrm{e}^{-\mathrm{i}k(x\sin\theta\cos\phi + y\sin\theta\sin\phi + \xi\cos\theta)}$$

where we also introduced the spherical angles θ and ϕ for the direction of observation, i.e., $\hat{\boldsymbol{r}} = \hat{\boldsymbol{x}}\sin\theta\cos\phi + \hat{\boldsymbol{y}}\sin\theta\sin\phi + \hat{\boldsymbol{z}}\cos\theta$. Insertion of all these expressions gives

$$\begin{aligned} \boldsymbol{F}_{\mathrm{GO}}(\hat{\boldsymbol{r}}) = {} & -E_0\frac{\mathrm{i}k}{4\pi}\mathrm{e}^{\mathrm{i}k\xi(1-\cos\theta)+2\mathrm{i}kF}\hat{\boldsymbol{r}} \times \left[(\hat{\boldsymbol{z}} + \hat{\boldsymbol{r}}) \times \hat{\boldsymbol{y}}\right] \\ & \cdot \int_{-b/2}^{b/2} \mathrm{d}x \int_{-h/2}^{h/2} \mathrm{d}y \sqrt{\frac{8F}{\mathrm{i}k\pi\left(x^2+4F^2\right)}}\mathrm{e}^{-\mathrm{i}k(x\sin\theta\cos\phi + y\sin\theta\sin\phi)} \end{aligned}$$

The integral in the y variable can be made analytically, see (6.40). The result is

$$\begin{aligned} \boldsymbol{F}_{\mathrm{GO}}(\hat{\boldsymbol{r}}) = {} & -E_0\frac{\mathrm{i}kh}{4\pi}\mathrm{e}^{\mathrm{i}k\xi(1-\cos\theta)+2\mathrm{i}kF}\frac{\sin\left(\frac{kh}{2}\sin\theta\sin\phi\right)}{\frac{kh}{2}\sin\theta\sin\phi}\hat{\boldsymbol{r}} \times \left[(\hat{\boldsymbol{z}} + \hat{\boldsymbol{r}}) \times \hat{\boldsymbol{y}}\right] \\ & \cdot \int_{-b/2}^{b/2} \sqrt{\frac{8F}{\mathrm{i}k\pi\left(x^2+4F^2\right)}}\mathrm{e}^{-\mathrm{i}kx\sin\theta\cos\phi}\,\mathrm{d}x \end{aligned} \tag{6.53}$$

The integration w.r.t. the x variable has to be made numerically. Just as in the physical optics approximation case, the value in the forward direction can be solved analytically. The result is identical to the result in the physical optics approximation

$$\begin{aligned} \boldsymbol{F}_{\mathrm{GO}}(\hat{\boldsymbol{z}}) &= \hat{\boldsymbol{y}}E_0\frac{\mathrm{i}kh}{2\pi}\sqrt{\frac{8F}{\mathrm{i}k\pi}}\mathrm{e}^{2\mathrm{i}kF}\int_{-kb/2}^{kb/2}\frac{\mathrm{d}t}{\sqrt{t^2+4k^2F^2}} \\ &= \hat{\boldsymbol{y}}\frac{\mathrm{i}kE_0h}{\pi}\sqrt{\frac{8F}{\mathrm{i}k\pi}}\mathrm{e}^{2\mathrm{i}kF}\ln\left(\frac{b}{4F} + \sqrt{\frac{b^2}{16F^2}+1}\right) \end{aligned}$$

The square of the magnitude of the far field amplitude (6.53) is

$$\left|\boldsymbol{F}_{\mathrm{GO}}(\hat{\boldsymbol{r}})\right|^2 = \frac{k|E_0|^2h^2F}{2\pi^3}\left|\frac{\sin\left(\frac{kh}{2}\sin\theta\sin\phi\right)}{\frac{kh}{2}\sin\theta\sin\phi}\right|^2(1+\cos\theta)^2$$

$$\cdot\left|\int_{-kb/2}^{kb/2}\frac{\mathrm{d}t}{\sqrt{t^2+4k^2F^2}}\mathrm{e}^{-it\sin\theta\cos\phi}\right|^2$$

where we used the simplification, $\left|\hat{\boldsymbol{r}}\times\left[(\hat{\boldsymbol{z}}+\hat{\boldsymbol{r}})\times\hat{\boldsymbol{y}}\right]\right|^2 = (1+\cos\theta)^2$, see Problem 6.10. In the forward direction of the reflector, $\theta = 0$, we get

$$\left|\boldsymbol{F}_{\mathrm{GO}}(\hat{\boldsymbol{z}})\right|^2 = \frac{8k|E_0|^2h^2F}{\pi^3}\left|\ln\left(\frac{b}{4F}+\sqrt{\frac{b^2}{16F^2}+1}\right)\right|^2$$

Computations of the width of the main lobe with the geometrical optics approximation are presented in Figure 6.15, which can be compared with the corresponding computations made with the physical optics approximation in Figure 6.7. The absolute square of the far field amplitude is illustrated in Figure 6.8. The results show that both methods are comparable. The results differ appreciably only when the angle between the forward direction and the direction of observation is greater than 45°. ∎

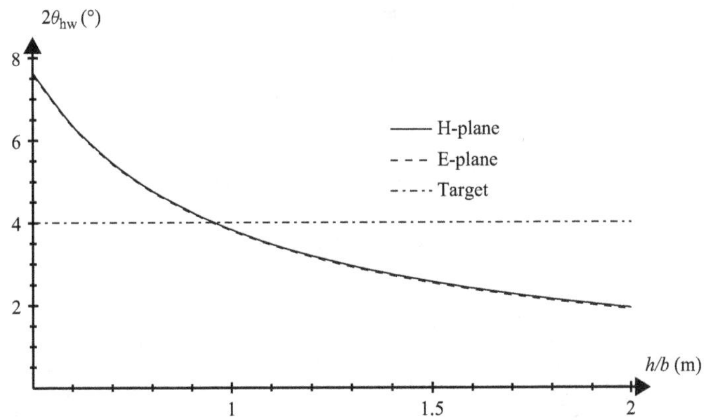

Figure 6.15 The width of the main lobe in degrees as a function of b and h for the cylindrical parabolic reflector in Example 6.11. The computations are made with the geometrical optics approximation. In the H-plane (x–z-plane), b is determined with $h = F = 1$ m, and in the E-plane (y–z-plane), h is determined with $b = F = 1$ m. The frequency is $f = 4$ GHz. Compare with Figure 6.7 where the computations are made with the physical optics approximation.

Example 6.12
In this example, we make the computations in Example 6.8 on page 301 with the geometrical optics approximation. Notation and computations are found in Example 6.8.

The far field from the parabolic reflector is determined in the geometrical optics approximation by first making an approximation of the electric field on the aperture surface, which is taken as a flat, circular surface of radius a in front of the reflector (all other parts of the aperture surface give zero contribution). Every point on the aperture surface is connected by a unique ray to the origin, where the antenna is located. The aperture surface is parameterized by $z = \xi > a^2/4F - F$.

We know both the incident electric and magnetic fields and the unit normal vector of the reflector surface, see (6.43), (6.47), and (6.41)

$$\begin{cases} \boldsymbol{E}_\mathrm{i}\left(\boldsymbol{S}\left(r_\mathrm{c},\alpha\right)\right) = -\dfrac{k}{\epsilon_0\omega}\hat{\boldsymbol{k}}_\mathrm{i} \times \boldsymbol{H}_\mathrm{i}\left(\boldsymbol{S}\left(r_\mathrm{c},\alpha\right)\right) \\ \boldsymbol{H}_\mathrm{i}\left(\boldsymbol{S}\left(r_\mathrm{c},\alpha\right)\right) = k\omega p\dfrac{\mathrm{e}^{\mathrm{i}kr'(r_\mathrm{c})}}{4\pi r'(r_\mathrm{c})}\dfrac{\hat{\boldsymbol{z}}4Fr_\mathrm{c}\cos\alpha - \hat{\boldsymbol{x}}\left(r_\mathrm{c}^2 - 4F^2\right)}{r_\mathrm{c}^2 + 4F^2} \\ \hat{\boldsymbol{\nu}}(r_\mathrm{c},\alpha) = \dfrac{-\hat{\boldsymbol{r}}_\mathrm{c}r_\mathrm{c} + \hat{\boldsymbol{z}}2F}{\sqrt{r_\mathrm{c}^2 + 4F^2}} \end{cases}$$

where $\hat{\boldsymbol{k}}_\mathrm{i}$ and $r'(r_\mathrm{c})$ are, see (6.46) and (6.42)

$$\begin{cases} \hat{\boldsymbol{k}}_\mathrm{i}\left(r_\mathrm{c},\alpha\right) = \dfrac{\hat{\boldsymbol{r}}_\mathrm{c}4Fr_\mathrm{c} + \hat{\boldsymbol{z}}\left(r_\mathrm{c}^2 - 4F^2\right)}{r_\mathrm{c}^2 + 4F^2} \\ r'(r_\mathrm{c}) = r_\mathrm{c}^2/4F + F \end{cases}$$

The direction of the scattered field is, see (6.29)

$$\begin{aligned} \hat{\boldsymbol{k}}_\mathrm{s} &= \hat{\boldsymbol{k}}_\mathrm{i}\left(r_\mathrm{c},\alpha\right) - 2\hat{\boldsymbol{\nu}}(r_\mathrm{c},\alpha)\left(\hat{\boldsymbol{\nu}}(r_\mathrm{c},\alpha)\cdot\hat{\boldsymbol{k}}_\mathrm{i}\left(r_\mathrm{c},\alpha\right)\right) \\ &= \frac{\hat{\boldsymbol{r}}_\mathrm{c}4Fr_\mathrm{c} + \hat{\boldsymbol{z}}\left(r_\mathrm{c}^2 - 4F^2\right)}{r_\mathrm{c}^2 + 4F^2} + 4F\frac{-\hat{\boldsymbol{r}}_\mathrm{c}r_\mathrm{c} + \hat{\boldsymbol{z}}2F}{r_\mathrm{c}^2 + 4F^2} = \hat{\boldsymbol{z}} \end{aligned}$$

The explicit value of the incident electric field is

$$\begin{aligned} \boldsymbol{E}_\mathrm{i}\left(\boldsymbol{S}\left(r_\mathrm{c},\alpha\right)\right) &= -k^2p\frac{\mathrm{e}^{\mathrm{i}kr'(r_\mathrm{c})}}{4\pi\epsilon_0 r'(r_\mathrm{c})}\frac{\hat{\boldsymbol{r}}_\mathrm{c}4Fr_\mathrm{c} + \hat{\boldsymbol{z}}\left(r_\mathrm{c}^2 - 4F^2\right)}{r_\mathrm{c}^2 + 4F^2} \times \frac{\hat{\boldsymbol{z}}4Fr_\mathrm{c}\cos\alpha - \hat{\boldsymbol{x}}\left(r_\mathrm{c}^2 - 4F^2\right)}{r_\mathrm{c}^2 + 4F^2} \\ &= k^2p\frac{\mathrm{e}^{\mathrm{i}kr'(r_\mathrm{c})}}{4\pi\epsilon_0 r'(r_\mathrm{c})}\frac{\hat{\boldsymbol{\alpha}}(4Fr_\mathrm{c})^2\cos\alpha - \hat{\boldsymbol{z}}4Fr_\mathrm{c}\sin\alpha\left(r_\mathrm{c}^2 - 4F^2\right) + \hat{\boldsymbol{y}}\left(r_\mathrm{c}^2 - 4F^2\right)^2}{\left(r_\mathrm{c}^2 + 4F^2\right)^2} \end{aligned}$$

The reflected (scattered) field, $\boldsymbol{E}_\mathrm{s}$, on the reflector surface is easily determined by the relation, see (6.29)

$$\boldsymbol{E}_\mathrm{s}(\boldsymbol{S}(r_\mathrm{c},\alpha)) = -\boldsymbol{E}_\mathrm{i}(\boldsymbol{S}(r_\mathrm{c},\alpha)) + 2\hat{\boldsymbol{\nu}}(r_\mathrm{c},\alpha)\left(\hat{\boldsymbol{\nu}}(r_\mathrm{c},\alpha)\cdot\boldsymbol{E}_\mathrm{i}(\boldsymbol{S}(r_\mathrm{c},\alpha))\right)$$

Straightforward algebra gives

$$E_s(S(r_c,\alpha)) = k^2 p \frac{e^{ikr'(r_c)}}{4\pi\epsilon_0 r'(r_c)} \frac{1}{r_c^2+4F^2}\left\{\hat{x} r_c^2 \sin 2\alpha - \hat{y}\left(4F^2 + r_c^2\cos 2\alpha\right)\right\}$$

It remains to find the distance between the point of reflection to the aperture plane. This distance is $\xi - \left(r_c^2/4F - F\right)$, and the electric field at the aperture plane, E_a, is

$$E_a(r_c,\alpha) = E_s(S(r_c,\alpha))e^{ik\left(\xi - (r_c^2/4F - F)\right)}$$

The geometrical optics approximation of the far field amplitude is given by, see (6.48)

$$F_{GO}(\hat{r}) = \frac{ik}{4\pi}\hat{r}\times\left[(\hat{z}+\hat{r})\times\iint_{S_a} E_a(r')e^{-ik\hat{r}\cdot r'}\,dS'\right] \tag{6.54}$$

since $\eta_0\hat{z}\times H_a(r_c,\alpha) = -E_a(r_c,\alpha)$, and where the integration is over the aperture surface S_a. We introduce the spherical angles θ and ϕ for the direction of observation. The integration in (6.54) has to be made numerically. We observe that E_a does not have any z-component in the aperture plane, and define $I(\hat{r})$ as

$$I(\hat{r}) = \iint_{S_a} E_a(r')e^{-ik\hat{r}\cdot r'}\,dS'$$

where the complex-valued vector $I(\hat{r})$ is determined by numerical integration for every value of θ and ϕ. We have, see Problem 6.10 and (6.54)

$$\left|F_{GO}(\hat{r})\right|^2 = \left(\frac{k}{4\pi}\right)^2 \left|I(\hat{r})\right|^2 (1+\cos\theta)^2$$

From Example 6.8 on page 301, we collect the following integrals:

$$\begin{cases} \displaystyle\int_0^{2\pi} e^{iz\cos(\alpha-\phi)}\,d\alpha = 2\pi J_0(z) \\ \displaystyle\int_0^{2\pi} e^{iz\cos(\alpha-\phi)}\cos 2\alpha\,d\alpha = -2\pi J_2(z)\cos 2\phi \\ \displaystyle\int_0^{2\pi} e^{iz\cos(\alpha-\phi)}\sin 2\alpha\,d\alpha = -2\pi J_2(z)\sin 2\phi \end{cases}$$

Integration in the α variable gives

$$\hat{x}\cdot I(\hat{r}) = -\frac{2Fk^2 p}{\epsilon_0}e^{ik(\xi(1-\cos\theta)+2F)}\sin 2\phi \int_0^a \frac{J_2(kr_c\sin\theta)\,r_c^3}{\left(r_c^2+4F^2\right)^2}\,dr_c$$

and

$$\hat{y}\cdot I(\hat{r}) = \frac{2Fk^2 p}{\epsilon_0}e^{ik(\xi(1-\cos\theta)+2F)}\int_0^a \frac{r_c\left(J_2(kr_c\sin\theta)\,r_c^2\cos 2\phi - 4F^2 J_0(kr_c\sin\theta)\right)}{\left(r_c^2+4F^2\right)^2}\,dr_c$$

In the H-plane (x–z-plane or $\phi = 0$-plane), the integrals simplify to

$$I(\hat{r}) = \frac{2Fk^2 p\hat{y}}{\epsilon_0} e^{ik(\xi(1-\cos\theta)+2F)} \int_0^a \frac{r_c \left(J_2(kr_c \sin\theta) r_c^2 - 4F^2 J_0(kr_c \sin\theta)\right)}{\left(r_c^2 + 4F^2\right)^2} \, dr_c$$

In the E-plane (y–z-plane or $\phi = \pi/2$-plane), the integrals simplify to

$$I(\hat{r}) = -\frac{2Fk^2 p\hat{y}}{\epsilon_0} e^{ik(\xi(1-\cos\theta)+2F)} \int_0^a \frac{r_c \left(J_2(kr_c \sin\theta) r_c^2 + 4F^2 J_0(kr_c \sin\theta)\right)}{\left(r_c^2 + 4F^2\right)^2} \, dr_c$$

Numerical calculations of the far field amplitude are depicted in Figure 6.10 on page 307. The results show that the two methods are equivalent for small angles of observations, i.e., in the forward direction. The amplitude in the forward direction can be computed analytically, see Problem 6.12. ■

6.4 Sum Rule for the Extinction Cross Section

This section exploits the analytic properties of the scattering quantities. In particular, we work out the details on how the overall frequency behavior of the scattering cross section is related to the low-frequency properties of the scatterer. These are called sum rules. They are very useful in many applications, and also as tests and benchmarks to numerical code. Some of the results presented in this section are found in References 88, 90, 143, 154, 246–249, 271. An analysis of the sum rules using time-domain arguments is found in Reference 86, and an investigation on the sum rules for transmission through an aperture (Babinet's principle) is reported in Reference 85.

We take Plemelj's formulas as the starting point of this analysis. A short review of these formulas is given in Appendix D.2. To satisfy Plemelj's formulas, see (D.2) on page 642, the function $f(k)$ has to be analytic in the upper complex plane, $\mathbb{C}_+ = \{k \in \mathbb{C} : \operatorname{Im} k > 0\}$, regular on the real axis, and to vanish at infinity in the closed upper complex plane, see Appendix D.2

$$\begin{cases} f_r(k) = \dfrac{1}{\pi} P \displaystyle\int_{-\infty}^{\infty} \frac{f_i(k')}{k'-k} \, dk' \\ f_i(k) = -\dfrac{1}{\pi} P \displaystyle\int_{-\infty}^{\infty} \frac{f_r(k')}{k'-k} \, dk' \end{cases} \qquad k \in \mathbb{R}$$

where the real and imaginary parts of the complex-valued function $f(k)$ are denoted $f_r(k)$ and $f_i(k)$, respectively, i.e., $f(k) = f_r(k) + i f_i(k)$. If we decompose the real-valued functions $f_r(k)$ and $f_i(k)$ in their even and odd parts as

$$f_{r,i}(k) = f_{r,i}^{e}(k) + f_{r,i}^{o}(k), \qquad k \in \mathbb{R}$$

where

$$\begin{cases} f_{r,i}^{e}(k) = \dfrac{1}{2} \left(f_{r,i}(k) + f_{r,i}(-k)\right) \\ f_{r,i}^{o}(k) = \dfrac{1}{2} \left(f_{r,i}(k) - f_{r,i}(-k)\right) \end{cases}$$

Plemelj's formulas become, see (D.4) on page 642

$$\begin{cases} f_{\mathrm{r}}^{\mathrm{e}}(k) = \dfrac{2}{\pi} P\displaystyle\int_0^\infty \dfrac{k' f_{\mathrm{i}}^{\mathrm{o}}(k')}{k'^2 - k^2}\, \mathrm{d}k' \\ f_{\mathrm{r}}^{\mathrm{o}}(k) = \dfrac{2k}{\pi} P\displaystyle\int_0^\infty \dfrac{f_{\mathrm{i}}^{\mathrm{e}}(k')}{k'^2 - k^2}\, \mathrm{d}k' \end{cases} \qquad \begin{cases} f_{\mathrm{i}}^{\mathrm{e}}(k) = -\dfrac{2}{\pi} P\displaystyle\int_0^\infty \dfrac{k' f_{\mathrm{r}}^{\mathrm{o}}(k')}{k'^2 - k^2}\, \mathrm{d}k' \\ f_{\mathrm{i}}^{\mathrm{o}}(k) = -\dfrac{2k}{\pi} P\displaystyle\int_0^\infty \dfrac{f_{\mathrm{r}}^{\mathrm{e}}(k')}{k'^2 - k^2}\, \mathrm{d}k' \end{cases} \tag{6.55}$$

where $k \in \mathbb{R}$.

We apply Plemelj's formulas to a function $f(k)$ related to the scattering dyadic in the forward direction, $\mathbf{S}(\hat{\boldsymbol{k}}_{\mathrm{i}}, \hat{\boldsymbol{k}}_{\mathrm{i}}, k)$. This dyadic-valued field is, due to causality, analytic in the upper half complex κ-plane, see (5.16) on page 256, i.e.,

$$S_{ij}(\hat{\boldsymbol{k}}_{\mathrm{i}}, \hat{\boldsymbol{k}}_{\mathrm{i}}, -\kappa^*) = S_{ij}^*(\hat{\boldsymbol{k}}_{\mathrm{i}}, \hat{\boldsymbol{k}}_{\mathrm{i}}, \kappa), \qquad i,j = 1,2,3, \quad \operatorname{Im}\kappa \geq 0$$

This is a consequence of Paley–Wiener theorem, see Theorem D.1 on page 640. Note that we make a distinction between the real argument k, used in Plemelj's formulas, and complex argument κ in the upper complex plane.

A suitable choice of the function $f(\kappa)$ is ($\kappa \in \mathbb{R} \cup \mathbb{C}_+$)

$$f(\kappa) = \frac{\hat{\boldsymbol{p}}_{\mathrm{e}}^* \cdot \mathbf{S}(\hat{\boldsymbol{k}}_{\mathrm{i}}, \hat{\boldsymbol{k}}_{\mathrm{i}}, \kappa) \cdot \hat{\boldsymbol{p}}_{\mathrm{e}}}{\kappa^2} \tag{6.56}$$

where $\hat{\boldsymbol{p}}_{\mathrm{e}}$ is an arbitrary complex-valued unit vector (independent of κ), which is associated with the polarization of the incident electric field, see (6.10) on page 272

$$\hat{\boldsymbol{p}}_{\mathrm{e}} = \frac{\boldsymbol{E}_0}{|\boldsymbol{E}_0|}$$

We have seen, using the physical optics approximation, the asymptotic behavior for large (real) values of k is, see (6.33) on page 296

$$\mathbf{S}(\hat{\boldsymbol{k}}_{\mathrm{i}}, \hat{\boldsymbol{k}}_{\mathrm{i}}, k) = \frac{\mathrm{i}k}{2\pi} A(\hat{\boldsymbol{k}}_{\mathrm{i}})\mathbf{I}_3 + O(1), \qquad \text{as } k \to \infty$$

where the projected area, $A(\hat{\boldsymbol{k}}_{\mathrm{i}})$, is real valued and independent of k. We assume this relation holds in general for all complex values of κ in the upper complex plane, i.e.,

$$f(\kappa) = -\frac{A(\hat{\boldsymbol{k}}_{\mathrm{i}})}{2\pi \mathrm{i}\kappa} + O(\kappa^{-2}), \qquad \text{as } |\kappa| \to \infty \tag{6.57}$$

In summary, the function $f(\kappa)$ is analytic in the upper complex plane, due to causality, and its assumed behavior for large arguments implies that it vanishes as $|\kappa| \to \infty$ in the upper complex plane, i.e., it satisfies the requirements of Plemelj's formulas.

We now investigate the properties of the function, $f(\kappa)$, as $\kappa \to 0$, a little closer. The behavior of the function $f(k)$ at low frequencies is found by using the results of Section 6.1, where the low-frequency limit of the scattering dyadic is investigated, which is most conveniently expressed as, see (6.9) on page 272

$$f(k) = a_0 + O(k), \qquad \text{as } \mathbb{R} \ni k \to 0$$

where

$$a_0 = \frac{1}{4\pi}\left(\hat{\boldsymbol{p}}_e^* \cdot \boldsymbol{\gamma}_e \cdot \hat{\boldsymbol{p}}_e + (\hat{\boldsymbol{k}}_i \times \hat{\boldsymbol{p}}_e^*) \cdot \boldsymbol{\gamma}_m \cdot (\hat{\boldsymbol{k}}_i \times \hat{\boldsymbol{p}}_e)\right)$$

The electric and magnetic polarizability dyadics are denoted $\boldsymbol{\gamma}_e$ and $\boldsymbol{\gamma}_m$, respectively. They are independent of the excitation, and depend only on the geometry and material parameters of the scatterer. Note that a_0 is a real number if the polarizability dyadics are symmetric and real valued.

We start with a special incident polarization. For a linearly polarized incident field, $\hat{\boldsymbol{p}}_e$ has the form $\hat{\boldsymbol{p}}_e = e^{i\vartheta}\hat{\boldsymbol{e}}$, where $\hat{\boldsymbol{e}}$ is a real-valued unit vector. Under this assumption, $f(\kappa)$ satisfies $f(\kappa) = f^*(-\kappa^*)$, implying that the real (imaginary) part of $f(k)$ is an even (odd) function for real k, i.e., $f_r^o(k) = f_i^e(k) = 0$. Plemelj's formulas read (linear polarization, $\hat{\boldsymbol{p}}_e = e^{i\vartheta}\hat{\boldsymbol{e}}$)

$$\begin{cases} \operatorname{Re}\dfrac{\hat{\boldsymbol{e}} \cdot \mathbf{S}(\hat{\boldsymbol{k}}_i, \hat{\boldsymbol{k}}_i, k) \cdot \hat{\boldsymbol{e}}}{k^2} = \dfrac{2}{\pi} P\displaystyle\int_0^\infty \dfrac{\operatorname{Im}\hat{\boldsymbol{e}} \cdot \mathbf{S}(\hat{\boldsymbol{k}}_i, \hat{\boldsymbol{k}}_i, k') \cdot \hat{\boldsymbol{e}}}{k'\left(k'^2 - k^2\right)}\, dk' \\[2ex] \operatorname{Im}\dfrac{\hat{\boldsymbol{e}} \cdot \mathbf{S}(\hat{\boldsymbol{k}}_i, \hat{\boldsymbol{k}}_i, k) \cdot \hat{\boldsymbol{e}}}{k^2} = -\dfrac{2k}{\pi} P\displaystyle\int_0^\infty \dfrac{\operatorname{Re}\hat{\boldsymbol{e}} \cdot \mathbf{S}(\hat{\boldsymbol{k}}_i, \hat{\boldsymbol{k}}_i, k') \cdot \hat{\boldsymbol{e}}}{k'^2\left(k'^2 - k^2\right)}\, dk' \end{cases}$$

In particular, as $k \to 0$, we get

$$\int_0^\infty \frac{\operatorname{Im}\hat{\boldsymbol{e}} \cdot \mathbf{S}(\hat{\boldsymbol{k}}_i, \hat{\boldsymbol{k}}_i, k) \cdot \hat{\boldsymbol{e}}}{k^3}\, dk = \frac{\pi a_0}{2}, \qquad \text{LP} \tag{6.58}$$

We leave the special case of linearly polarized incident fields for a moment, and analyze the general polarization in more detail. The unit vector $\hat{\boldsymbol{p}}_e$ is now generally complex valued, as it is for an incident field that is not linearly polarized. Now, in general, both the real and the imaginary parts of $f(k)$ contain even and odd parts. An explicit example of this property can be viewed in Comment 8.4 on page 473.

The optical theorem, see (4.47) on page 230, states that, for an incident plane wave impinging along $\hat{\boldsymbol{k}}_i$ and electrically polarized along $\hat{\boldsymbol{p}}_e$, the extinction cross section is related to the far field amplitude in the forward direction, i.e.,

$$0 \le \sigma_{\text{ext}}(\hat{\boldsymbol{k}}_i, \hat{\boldsymbol{p}}_e, k) = \frac{4\pi}{k} \operatorname{Im}\left\{\hat{\boldsymbol{p}}_e^* \cdot \mathbf{S}(\hat{\boldsymbol{k}}_i, \hat{\boldsymbol{k}}_i, k) \cdot \hat{\boldsymbol{p}}_e\right\} = 4\pi k \operatorname{Im} f(k) \tag{6.59}$$

This relation shows that $k \operatorname{Im} f(k) \ge 0$ for all real k (positive or negative). Since $kf_i^o(k)$ is an even function, and $kf_i^e(k)$ is an odd function for real k, we get for real k (positive or negative)

$$0 \le \operatorname{Im} kf(k) = kf_i^e(k) + kf_i^o(k) \quad \Rightarrow \quad \left|kf_i^e(k)\right| \le kf_i^o(k) \tag{6.60}$$

Below, we use this estimate on $kf_i^o(k)$ on the real k-axis.

From (1.81) on page 79, we see that the general polarization vector, $\hat{\boldsymbol{p}}_e$, of the incident field can be decomposed in two, real-valued, orthogonal unit vectors $\hat{\boldsymbol{a}}$ and $\hat{\boldsymbol{b}} = \pm\hat{\boldsymbol{k}}_i \times \hat{\boldsymbol{a}}$—plus sign for right-handed elliptic polarization, and minus sign for left-handed elliptic polarization—and where $\hat{\boldsymbol{k}}_i \cdot \hat{\boldsymbol{a}} = 0$. Both unit vectors $\boldsymbol{a}$ and $\boldsymbol{b}$ are independent of κ. We have

$$\hat{\boldsymbol{p}}_e = e^{i\vartheta}\left(\hat{\boldsymbol{a}}\cos\chi + i\hat{\boldsymbol{b}}\sin\chi\right)$$

and

$$kf(k) = \frac{\hat{\boldsymbol{p}}_e^* \cdot \mathbf{S}(\hat{\boldsymbol{k}}_i, \hat{\boldsymbol{k}}_i, k) \cdot \hat{\boldsymbol{p}}_e}{k} = \frac{1}{k}\left(S_{aa}\cos^2\chi + S_{bb}\sin^2\chi\right) \pm \frac{i\Pi}{2k}\left(S_{ab} - S_{ba}\right)$$

where the polarization state $\Pi = \hat{\boldsymbol{k}}_i \cdot (\hat{\boldsymbol{a}} \times \hat{\boldsymbol{b}}) \sin 2\chi$, and

$$\begin{cases} S_{aa} = \hat{\boldsymbol{a}} \cdot \mathbf{S}(\hat{\boldsymbol{k}}_i, \hat{\boldsymbol{k}}_i, k) \cdot \hat{\boldsymbol{a}} \\ S_{bb} = \hat{\boldsymbol{b}} \cdot \mathbf{S}(\hat{\boldsymbol{k}}_i, \hat{\boldsymbol{k}}_i, k) \cdot \hat{\boldsymbol{b}} \\ S_{ab} = \hat{\boldsymbol{a}} \cdot \mathbf{S}(\hat{\boldsymbol{k}}_i, \hat{\boldsymbol{k}}_i, k) \cdot \hat{\boldsymbol{b}} \\ S_{ba} = \hat{\boldsymbol{b}} \cdot \mathbf{S}(\hat{\boldsymbol{k}}_i, \hat{\boldsymbol{k}}_i, k) \cdot \hat{\boldsymbol{a}} \end{cases}$$

Note that the elements S_{aa}, S_{bb}, S_{ab}, and S_{ba} are independent of the angle χ, but depend on k and the incident direction $\hat{\boldsymbol{k}}_i$. Due to the optical theorem, $\operatorname{Im} S_{aa}/k$ and $\operatorname{Im} S_{bb}/k$ are both positive (corresponding to real polarization vectors $\hat{\boldsymbol{p}}_e = \hat{\boldsymbol{a}}$ and $\hat{\boldsymbol{p}}_e = \hat{\boldsymbol{b}}$, respectively). The imaginary part of $kf(k)$ satisfies

$$0 \le kf_i^e(k) + kf_i^o(k) = \frac{1}{k}\left(\operatorname{Im}\left(S_{aa}\cos^2\chi + S_{bb}\sin^2\chi\right)\right) \pm \frac{\Pi}{2k}\operatorname{Re}\left(S_{ab} - S_{ba}\right)$$

Consequently, since $\operatorname{Re} S_{\alpha\alpha'}$ is an even function in k on the real axis, and $\operatorname{Im} S_{\alpha\alpha'}$ is odd for $\alpha, \alpha' = a, b$, we get

$$\begin{cases} kf_i^o(k) = \dfrac{1}{k}\operatorname{Im}\left(S_{aa}\cos^2\chi + S_{bb}\sin^2\chi\right) \\ kf_i^e(k) = \pm\dfrac{\Pi}{2k}\operatorname{Re}\left(S_{ab} - S_{ba}\right) \end{cases}$$

Since the inequality in (6.60) holds for all polarizations states χ, we conclude (take e.g., $\chi = \pi/4$)

$$\left|\frac{1}{k}\operatorname{Re}\left(S_{ab} - S_{ba}\right)\right| \le \frac{1}{k}\operatorname{Im}\left(S_{aa} + S_{bb}\right)$$

For a general polarization, we have derived the estimate

$$\left|kf_i^e(k)\right| = \frac{|\Pi|}{2}\left|\frac{1}{k}\operatorname{Re}\left(S_{ab} - S_{ba}\right)\right| \le \frac{|\Pi|}{2k}\operatorname{Im}\left(S_{aa} + S_{bb}\right) \tag{6.61}$$

We now have all properties and estimates that we need to apply Plemelj's formulas, (6.55), to a general polarization $\hat{\boldsymbol{p}}_e$. An immediate consequence of the two formulas on the right-hand side of (6.55) and the optical theorem (6.59) is

$$\begin{aligned} \sigma_{\text{ext}}(k) &= 4\pi kf_i^e(k) + 4\pi kf_i^o(k) \\ &= -8kP\int_0^\infty \frac{k' f_r^o(k')}{k'^2 - k^2}\,dk' - 8k^2 P\int_0^\infty \frac{f_r^e(k')}{k'^2 - k^2}\,dk' \end{aligned}$$

From the high-frequency response of the extinction cross section, (6.57), we get as $k \to \infty$

$$\int_0^\infty f_r^e(k)\,dk = \frac{A(\hat{\boldsymbol{k}}_i)}{4}$$

Moreover, Plemelj's formulas, (6.55), can be evaluated in the limit $k \to 0$. The only non-trivial identities are

$$\begin{cases} a_0 = \dfrac{2}{\pi}\displaystyle\int_0^\infty \dfrac{f_{\mathrm{i}}^{\mathrm{o}}(k)}{k}\,\mathrm{d}k \\ 0 = \dfrac{2}{\pi}\displaystyle\int_0^\infty \dfrac{f_{\mathrm{r}}^{\mathrm{o}}(k)}{k}\,\mathrm{d}k \end{cases}$$

A consequence of the second identity is that $f_{\mathrm{r}}^{\mathrm{o}}(k)$ must assume both positive and negative values in the interval $k \in [0, \infty)$. The first relation we write as an inequality by adding an integral term with integrand $f_{\mathrm{i}}^{\mathrm{e}}(k)/k$

$$-\int_0^\infty \frac{\left|f_{\mathrm{i}}^{\mathrm{e}}(k)\right|}{k}\,\mathrm{d}k \le \int_0^\infty \frac{f_{\mathrm{i}}(k)}{k}\,\mathrm{d}k - \frac{\pi a_0}{2} \le \int_0^\infty \frac{\left|f_{\mathrm{i}}^{\mathrm{e}}(k)\right|}{k}\,\mathrm{d}k$$

or using the optical theorem, see (6.59)

$$-4\pi\int_0^\infty \frac{\left|f_{\mathrm{i}}^{\mathrm{e}}(k)\right|}{k}\,\mathrm{d}k \le \int_0^\infty \frac{\sigma_{\mathrm{ext}}(k)}{k^2}\,\mathrm{d}k - 2\pi^2 a_0 \le 4\pi\int_0^\infty \frac{\left|f_{\mathrm{i}}^{\mathrm{e}}(k)\right|}{k}\,\mathrm{d}k$$

The integral on the left- and the right-hand sides is related to the polarization contribution for a circular polarization. To see this, define the circular polarized field, $\chi = \pi/4$, and $\hat{\boldsymbol{b}} = \pm\hat{\boldsymbol{k}}_{\mathrm{i}} \times \hat{\boldsymbol{a}}$

$$\begin{cases} \hat{\boldsymbol{p}}_{\mathrm{e}}^{\mathrm{c}} = \mathrm{e}^{\mathrm{i}\vartheta}\left(\hat{\boldsymbol{a}} + \mathrm{i}\hat{\boldsymbol{b}}\right)/\sqrt{2} = \mathrm{e}^{\mathrm{i}\vartheta}\left(\hat{\boldsymbol{a}} \pm \mathrm{i}\hat{\boldsymbol{k}}_{\mathrm{i}} \times \hat{\boldsymbol{a}}\right)/\sqrt{2} \\ \hat{\boldsymbol{p}}_{\mathrm{m}}^{\mathrm{c}} = \hat{\boldsymbol{k}}_{\mathrm{i}} \times \hat{\boldsymbol{p}}_{\mathrm{e}}^{\mathrm{c}} = \mathrm{e}^{\mathrm{i}\vartheta}\left(\hat{\boldsymbol{k}}_{\mathrm{i}} \times \hat{\boldsymbol{a}} \mp \mathrm{i}\hat{\boldsymbol{a}}\right)/\sqrt{2} = \mp\mathrm{i}\mathrm{e}^{\mathrm{i}\vartheta}\left(\hat{\boldsymbol{a}} + \mathrm{i}\hat{\boldsymbol{b}}\right)/\sqrt{2} \end{cases}$$

and

$$\begin{aligned} a_0^{\mathrm{c}} &= \frac{1}{4\pi}\left(\hat{\boldsymbol{p}}_{\mathrm{e}}^{\mathrm{c}*} \cdot \boldsymbol{\gamma}_{\mathrm{e}} \cdot \hat{\boldsymbol{p}}_{\mathrm{e}}^{\mathrm{c}} + \hat{\boldsymbol{p}}_{\mathrm{m}}^{\mathrm{c}*} \cdot \boldsymbol{\gamma}_{\mathrm{m}} \cdot \hat{\boldsymbol{p}}_{\mathrm{m}}^{\mathrm{c}}\right) \\ &= \frac{1}{8\pi}\left(\hat{\boldsymbol{a}} \cdot \boldsymbol{\gamma}_{\mathrm{e}} \cdot \hat{\boldsymbol{a}} + (\hat{\boldsymbol{k}}_{\mathrm{i}} \times \hat{\boldsymbol{a}}) \cdot \boldsymbol{\gamma}_{\mathrm{m}} \cdot (\hat{\boldsymbol{k}}_{\mathrm{i}} \times \hat{\boldsymbol{a}})\right) + \frac{1}{8\pi}\left(\hat{\boldsymbol{b}} \cdot \boldsymbol{\gamma}_{\mathrm{e}} \cdot \hat{\boldsymbol{b}} + (\hat{\boldsymbol{k}}_{\mathrm{i}} \times \hat{\boldsymbol{b}}) \cdot \boldsymbol{\gamma}_{\mathrm{m}} \cdot (\hat{\boldsymbol{k}}_{\mathrm{i}} \times \hat{\boldsymbol{b}})\right) \end{aligned}$$

where $\hat{\boldsymbol{b}} = \pm\hat{\boldsymbol{k}}_{\mathrm{i}} \times \hat{\boldsymbol{a}}$, and the last equality is due to the symmetry of the polarization dyadics. Then, due to (6.61) and (6.58), we get

$$\begin{aligned} \int_0^\infty \frac{\left|f_{\mathrm{i}}^{\mathrm{e}}(k)\right|}{k}\,\mathrm{d}k &\le \frac{|\Pi|}{2}\int_0^\infty \frac{\operatorname{Im}(S_{aa} + S_{bb})}{k^3}\,\mathrm{d}k \\ &= \frac{|\Pi|}{2}\frac{\pi}{2}\left\{\frac{1}{4\pi}\left(\hat{\boldsymbol{a}} \cdot \boldsymbol{\gamma}_{\mathrm{e}} \cdot \hat{\boldsymbol{a}} + (\hat{\boldsymbol{k}}_{\mathrm{i}} \times \hat{\boldsymbol{a}}) \cdot \boldsymbol{\gamma}_{\mathrm{m}} \cdot (\hat{\boldsymbol{k}}_{\mathrm{i}} \times \hat{\boldsymbol{a}})\right)\right. \\ &\quad \left. + \frac{1}{4\pi}\left(\hat{\boldsymbol{b}} \cdot \boldsymbol{\gamma}_{\mathrm{e}} \cdot \hat{\boldsymbol{b}} + (\hat{\boldsymbol{k}}_{\mathrm{i}} \times \hat{\boldsymbol{b}}) \cdot \boldsymbol{\gamma}_{\mathrm{m}} \cdot (\hat{\boldsymbol{k}}_{\mathrm{i}} \times \hat{\boldsymbol{b}})\right)\right\} = \frac{|\Pi|}{2}\pi a_0^{\mathrm{c}} \end{aligned}$$

and the sum rule becomes

$$-2\pi^2 a_0^{\mathrm{c}}\,|\Pi| \le \int_0^\infty \frac{\sigma_{\mathrm{ext}}(k)}{k^2}\,\mathrm{d}k - 2\pi^2 a_0 \le 2\pi^2 a_0^{\mathrm{c}}\,|\Pi|$$

For a linear polarization, $\Pi = 0$, we get an equality, i.e.,

$$\int_0^\infty \frac{\sigma_{\text{ext}}(k)}{k^2}\,\mathrm{d}k = \frac{\pi}{2}\left(\hat{\boldsymbol{p}}_\text{e}^* \cdot \boldsymbol{\gamma}_\text{e} \cdot \hat{\boldsymbol{p}}_\text{e} + \hat{\boldsymbol{p}}_\text{m}^* \cdot \boldsymbol{\gamma}_\text{m} \cdot \hat{\boldsymbol{p}}_\text{m}\right), \qquad \text{LP} \tag{6.62}$$

where $\hat{\boldsymbol{p}}_\text{m} = \hat{\boldsymbol{k}}_\text{i} \times \hat{\boldsymbol{p}}_\text{e}$, or

$$\int_0^\infty \sigma_{\text{ext}}(\lambda)\,\mathrm{d}\lambda = \pi^2\left(\hat{\boldsymbol{p}}_\text{e}^* \cdot \boldsymbol{\gamma}_\text{e} \cdot \hat{\boldsymbol{p}}_\text{e} + \hat{\boldsymbol{p}}_\text{m}^* \cdot \boldsymbol{\gamma}_\text{m} \cdot \hat{\boldsymbol{p}}_\text{m}\right), \qquad \text{LP} \tag{6.63}$$

where we have made a change in variable $k \to \lambda = 2\pi/k$ in the integral on the left-hand side. The identities also hold for a general polarization, provided the scattering dyadic is symmetric in the forward direction, $\mathbf{S}^\text{t}(\hat{\boldsymbol{k}}_\text{i}, \hat{\boldsymbol{k}}_\text{i}, k) = \mathbf{S}(\hat{\boldsymbol{k}}_\text{i}, \hat{\boldsymbol{k}}_\text{i}, k)$, since then $S_{ab} = S_{ba}$. These equalities are identical to the one obtained in (6.58).

The sum rule in (6.62) limits the amount of power—in a broadband sense—an obstacle can scatter and absorb. The left-hand side is an integral over a non-negative quantity over all positive frequencies, and the right-hand side quantifies the low-frequency behavior of the scatterer. In Section 6.1.6, we found that the polarizability dyadics grow if the volume or the material parameters, permittivity or permeability, grow. Note that this observation does not mean that the extinction cross section grows as the volume or the material parameters grow—only that the integral over all frequencies grows. This type of identity is an excellent benchmark of evaluating numerical code of scattering problems.

It remains to estimate the limit on the right-hand side of (6.62) in terms of the eigenvalues of the polarizability dyadics. We immediately see that the largest eigenvalues of the polarizability dyadics bound the right-hand side of (6.62). If the two real eigenvalues of $\boldsymbol{\gamma}_\text{e}$ are denoted $\lambda^\text{e}_{1,2}$, and the two real eigenvalues of $\boldsymbol{\gamma}_\text{m}$ are denoted $\lambda^\text{m}_{1,2}$, respectively, we get

$$\frac{\pi}{2}\left(\lambda_1^\text{e} + \lambda_1^\text{m}\right) \leq \int_0^\infty \frac{\sigma_{\text{ext}}(k)}{k^2}\,\mathrm{d}k \leq \frac{\pi}{2}\left(\lambda_2^\text{e} + \lambda_2^\text{m}\right)$$

provided the eigenvalues are ordered properly. This expression provides an upper and a lower estimate of the extinction (scattering and absorption) capacity of the obstacle. We illustrate the results in this section by a few useful examples.

Example 6.13

The simplest example concerns a perfectly conducting sphere of radius a. The electric and magnetic polarizability dyadics for this geometry are $\boldsymbol{\gamma}_\text{e} = 4\pi a^3\mathbf{I}_3$ and $\boldsymbol{\gamma}_\text{m} = -2\pi a^3\mathbf{I}_3$, respectively, see Example 6.1 on page 270. The explicit expression of the sum rule, (6.62) and (6.63), for a generally[17] polarized incident wave then is

$$\int_0^\infty \frac{\sigma_{\text{ext}}(k)}{k^2}\,\mathrm{d}k = \pi^2 a^3 \tag{6.64}$$

or

$$\int_0^\infty \sigma_{\text{ext}}(\lambda)\,\mathrm{d}\lambda = 2\pi^3 a^3$$

[17] The polarizability dyadics are proportional to the identity dyadic, so that $\mathbf{S}(\hat{\boldsymbol{k}}_\text{i}, \hat{\boldsymbol{k}}_\text{i}, k)$ is symmetric.

This shows that the integrated extinction cross section cannot grow faster than a^3, i.e., the volume. ■

Example 6.14
Another simple example is a homogeneous, dielectric spherical scatterer of radius a. For a homogeneous spherical scatterer (radius a, real-valued permittivity $\epsilon(\omega = 0)$, and real-valued permeability $\mu(\omega = 0)$) surrounded by vacuum, the electric and magnetic polarizability dyadics, $\boldsymbol{\gamma}_{\mathrm{e}}$ and $\boldsymbol{\gamma}_{\mathrm{m}}$, can be computed analytically, see Example 6.4 on page 276

$$\boldsymbol{\gamma}_{\mathrm{e}} = 4\pi a^3 \frac{\epsilon(0) - 1}{\epsilon(0) + 2} \mathbf{I}_3, \qquad \boldsymbol{\gamma}_{\mathrm{m}} = 4\pi a^3 \frac{\mu(0) - 1}{\mu(0) + 2} \mathbf{I}_3$$

The explicit expression of the sum rule, (6.62) and (6.63), for a generally polarized incident wave then is,[18] see also (6.9) on page 272

$$\int_0^\infty \frac{\sigma_{\mathrm{ext}}(k)}{k^2}\, \mathrm{d}k = 2\pi^2 a^3 \left(\frac{\epsilon(0) - 1}{\epsilon(0) + 2} + \frac{\mu(0) - 1}{\mu(0) + 2} \right)$$

or

$$\int_0^\infty \sigma_{\mathrm{ext}}(\lambda)\, \mathrm{d}\lambda = 4\pi^3 a^3 \left(\frac{\epsilon(0) - 1}{\epsilon(0) + 2} + \frac{\mu(0) - 1}{\mu(0) + 2} \right)$$

This shows that the integrated extinction cross section cannot grow faster than a^3, i.e., the volume. ■

Example 6.15
As an extension of Example 6.14, we treat the case of non-zero conductivity. This is the situation if we e.g., have a Drude–Lorentz model, see the Drude model on page 53

$$\epsilon(\omega) = 1 - \frac{\omega_{\mathrm{pL}}^2}{\omega^2 - \omega_0^2 + \mathrm{i}\omega\nu_{\mathrm{L}}} - \frac{\omega_{\mathrm{pD}}^2}{\omega^2 + \mathrm{i}\omega\nu_{\mathrm{D}}}$$

which at low frequencies behaves as

$$\epsilon(\omega) \approx \mathrm{i}\frac{\omega_{\mathrm{pD}}^2}{\omega\nu_{\mathrm{D}}}$$

[18]This result also provides an indirect proof that $\epsilon(\omega)$ has to be greater or equal to 1 in the limit $\omega \to 0$ for a non-magnetic material, since the extinction cross section on the left-hand side always is a positive quantity (there is also another possibility $\epsilon(0) < -2$), see also Reference [152, page 59].

and a DC conductivity of $\sigma = \epsilon_0\omega_{\mathrm{pD}}^2/\nu_{\mathrm{D}}$. This model leads to the polarizabilities (non-magnetic materials)

$$\begin{cases} \boldsymbol{\gamma}_{\mathrm{e}} = 4\pi a^3 \mathbf{I}_3 \\ \boldsymbol{\gamma}_{\mathrm{m}} = \mathbf{0} \end{cases}$$

irrespectively of the value of the conductivity. The sum rule for a homogeneous sphere with DC conductivity, therefore is

$$\int_0^\infty \frac{\sigma_{\mathrm{ext}}(k)}{k^2}\,\mathrm{d}k = 2\pi^2 a^3$$

and the result is illustrated in Figure 6.16. The method used for computing these results is presented in Section 8.2. All curves in the figure have the same weighted area as integral of the variable ka (with the normalization in the figure, the value is π for all curves). ■

Example 6.16
The sum rule presented above quantifies how the overall scattering strength and its bandwidth are related. In fact, a simple example illustrates this relation and the practical use of the sum rules.

A relevant question to ask would be: How large is the maximum scattering/absorption strength (absorption and total scattering in all directions) of an obstacle at a certain bandwidth, if the obstacle is confined to a volume $|V|$?

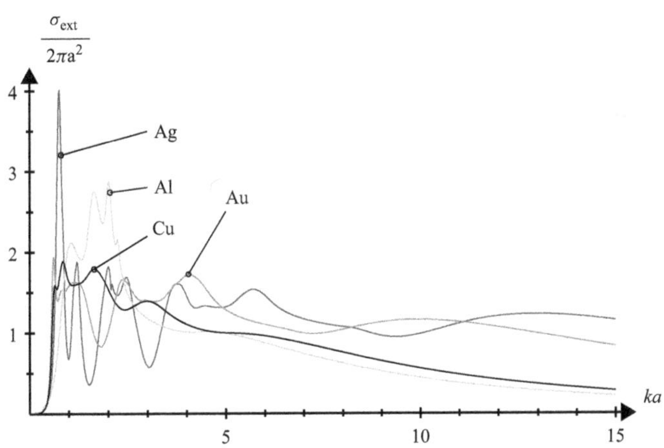

Figure 6.16 The extinction cross section for four different metallic spheres. All curves in the figure have the same weighted area (with the normalization in the figure, the value is π for all curves). The data of the metallic spheres are given in Reference 221.

In Figure 6.17, we show the normalized extinction cross section σ_{ext} for a sphere of radius a filled with a homogeneous, non-magnetic, Lorentz material. The normalized material parameters of the Lorentz material used in this example are

$$\begin{cases} \epsilon_\infty = 1 \\ \omega_{\text{p}} = c/a \\ \omega_0 = c/a \end{cases} \qquad \begin{cases} \nu = 0.05c/a \\ \mu = 1 \end{cases}$$

where c is the speed of light in the lossless material outside the scatterer. In this example, the electric polarizability $\gamma_{\text{e}} = 0.75|V|$, where $|V| = 4\pi a^3/3$ is the volume of the sphere.

The extinction integral in this example is, see Example 6.14

$$\frac{1}{2\pi a^3}\int_0^\infty \frac{\sigma_{\text{ext}}(k)}{k^2}\,\mathrm{d}k = \int_0^\infty \frac{\sigma_{\text{ext}}(ka)}{2\pi (ka)^2}\,\mathrm{d}ka = \frac{\pi}{4}$$

The two dotted boxes in Figure 6.17 both have the same value of the extinction integral as the solid line. The boxes have different scattering strength, which affects the bandwidth. These boxes quantify the bandwidth versus the maximum scattering strength of the scatterer. We also conclude that a curve that is everywhere outside a box of this kind cannot be realized with a passive material—given the polarizability of the object. ■

Example 6.17

A more advanced example illustrating the sum rule is the perfectly conducting disk of radius a. The polarizability dyadics for the PEC disk oriented in the x–y-plane are, see Example 7.5 on page 400

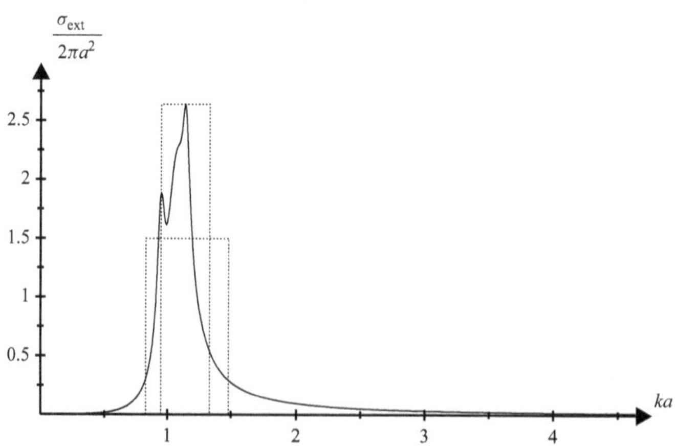

Figure 6.17 The normalized extinction cross section as a function of frequency ka for a Lorentz sphere of radius (solid line). The explicit data are given in the text. The two dotted boxes have the same extinction integral as the solid line.

$$\begin{cases} \gamma_e = \dfrac{16a^3}{3}(\hat{x}\hat{x} + \hat{y}\hat{y}) \\ \gamma_m = -\dfrac{8a^3}{3}\hat{z}\hat{z} \end{cases}$$

The sum rule for the disk becomes, see Problem 6.14 (the incidence angle is α)

$$\int_0^\infty \frac{\sigma_{ext}(k)}{k^2}\, dk = \frac{4\pi a^3}{3} \begin{cases} 1 + \cos^2\alpha & \text{(TE)} \\ 2\cos^2\alpha & \text{(TM)} \end{cases}$$

The extinction cross section, $\sigma_{ext}(\hat{\boldsymbol{k}}_i)$, for scattering by a perfectly conducting disk at different directions of incidence as a function of frequency is depicted in Figure 9.4 on page 490. ■

Example 6.18
In Section 4.5 on page 233, we examined thin metallic (PEC) scatterers. Let S_M be a flat, thin scatterer located in the plane $z = 0$. We then know that the scattering dyadic satisfies $\mathbf{S}(-\hat{z}, \hat{\boldsymbol{k}}_i) = \mathbf{S}(\hat{z}, \hat{\boldsymbol{k}}_i)$, see (4.55) on page 239, and in particular, we get for $\hat{\boldsymbol{k}}_i = \hat{z}$

$$\mathbf{S}(-\hat{z}, \hat{z}) = \mathbf{S}(\hat{z}, \hat{z})$$

Moreover, reciprocity is satisfied so (4.45) on page 223 implies

$$\mathbf{S}(-\hat{z}, \hat{z}) = \mathbf{S}^t(-\hat{z}, \hat{z})$$

From these results, we can derive that the scattering dyadic in the forward direction is symmetric for normal incidence. In fact

$$\mathbf{S}^t(\hat{z}, \hat{z}) = \mathbf{S}^t(-\hat{z}, \hat{z}) = \mathbf{S}(-\hat{z}, \hat{z}) = \mathbf{S}(\hat{z}, \hat{z})$$

and by (6.62)

$$\int_0^\infty \frac{\sigma_{ext}(k)}{k^2}\, dk = \frac{\pi}{2}\left(\hat{\boldsymbol{p}}_e^* \cdot \gamma_e \cdot \hat{\boldsymbol{p}}_e + \hat{\boldsymbol{p}}_m^* \cdot \gamma_m \cdot \hat{\boldsymbol{p}}_m\right)$$

holds for any polarization, since $S_{ab} = S_{ba}$, and consequently, $f_i^e(k) = 0$. ■

6.4.1 Additional Sum Rules

The sum rule presented above was obtained by an application of Plemelj's formulas. The usefulness of this sum rule depends on the fact that the imaginary part of $f(k)$ was non-negative on the positive real axis. This allows us to estimate the scattering ability of the object. This subject is developed in Reference 248.

We start by extending the sum rule in (6.62). For simplicity and in order to get explicit expressions, we assume the incident plane wave is linearly polarized in this section. The general low-frequency expansion behavior of the function $f(k)$, defined in (6.56) above, is

$$f(k) = \frac{\hat{\boldsymbol{p}}_{\mathrm{e}}^* \cdot \mathbf{S}(\hat{\boldsymbol{k}}_{\mathrm{i}}, \hat{\boldsymbol{k}}_{\mathrm{i}}, k) \cdot \hat{\boldsymbol{p}}_{\mathrm{e}}}{k^2} = a_0 + O(k), \qquad \text{as } \mathbb{R} \ni k \to 0$$

where the real-valued constant a_0 is

$$a_0 = \frac{1}{4\pi} \left(\hat{\boldsymbol{p}}_{\mathrm{e}}^* \cdot \boldsymbol{\gamma}_{\mathrm{e}} \cdot \hat{\boldsymbol{p}}_{\mathrm{e}} + \hat{\boldsymbol{p}}_{\mathrm{m}}^* \cdot \boldsymbol{\gamma}_{\mathrm{m}} \cdot \hat{\boldsymbol{p}}_{\mathrm{m}} \right)$$

This holds both for lossy and lossless obstacles ($\sigma_{\mathrm{a}} \geq 0$). As a consequence of this low-frequency behavior, we derived the sum rule in (6.62).

However, if the obstacle is lossless, we can do better. In that case, $\sigma_{\mathrm{a}} = 0$, and the extinction cross section and the scattering cross section coincide, i.e., $\sigma_{\mathrm{ext}} = \sigma_{\mathrm{s}}$. Since, by the optical theorem, $\sigma_{\mathrm{ext}}(k) = 4\pi \,\mathrm{Im}\, (\hat{\boldsymbol{p}}_{\mathrm{e}}^* \cdot \mathbf{S}(\hat{\boldsymbol{k}}_{\mathrm{i}}, \hat{\boldsymbol{k}}_{\mathrm{i}}, k) \cdot \hat{\boldsymbol{p}}_{\mathrm{e}})/k$, and, see (4.36) on page 207

$$\sigma_{\mathrm{s}}(k, \hat{\boldsymbol{k}}_{\mathrm{i}}) = \iint \left| \mathbf{S}(\hat{\boldsymbol{r}}, \hat{\boldsymbol{k}}_{\mathrm{i}}, k) \cdot \hat{\boldsymbol{p}}_{\mathrm{e}} \right|^2 \mathrm{d}\Omega$$

we conclude that $|\mathbf{S}(\hat{\boldsymbol{r}}, \hat{\boldsymbol{k}}_{\mathrm{i}}) \cdot \hat{\boldsymbol{p}}_{\mathrm{e}}|^2$ and $\mathrm{Im}\, (\hat{\boldsymbol{p}}_{\mathrm{e}}^* \cdot \mathbf{S}(\hat{\boldsymbol{k}}_{\mathrm{i}}, \hat{\boldsymbol{k}}_{\mathrm{i}}) \cdot \hat{\boldsymbol{p}}_{\mathrm{e}})/k$ have similar leading power expansion in k at the origin. Therefore, for a lossless scatterer the quantity $\mathrm{Im}\, (\hat{\boldsymbol{p}}_{\mathrm{e}}^* \cdot \mathbf{S}(\hat{\boldsymbol{k}}_{\mathrm{i}}, \hat{\boldsymbol{k}}_{\mathrm{i}}) \cdot \hat{\boldsymbol{p}}_{\mathrm{e}})/k$ is expected to behave as k^4 at low frequencies.[19] The following low-frequency expansion of the scattering dyadic is assumed to hold

$$f(\kappa) = \frac{\hat{\boldsymbol{p}}_{\mathrm{e}}^* \cdot \mathbf{S}(\hat{\boldsymbol{k}}_{\mathrm{i}}, \hat{\boldsymbol{k}}_{\mathrm{i}}, \kappa) \cdot \hat{\boldsymbol{p}}_{\mathrm{e}}}{\kappa^2} = a_0 + a_2\kappa^2 + O(\kappa^3), \qquad \text{as } \kappa \to 0$$

where a_0 and a_2 are real valued. Moreover, for large arguments in the upper complex plane it behaves as, see (6.57)

$$f(\kappa) = O(\kappa^{-1}), \qquad \text{as } |\kappa| \to \infty$$

We apply Theorem D.3 on page 643 to the function $f(\kappa)$. In this expansion, $N = 3, M = 1$, and $b_0 = b_1 = 0$. We get

$$P\int_{-\infty}^{\infty} \frac{\mathrm{Im} f(k)}{k^p} \,\mathrm{d}k = \pi \left(a_{p-1} - b_{p-1} \right), \qquad p = 1, 2, 3$$

or with the optical theorem (6.59) the only non-vanishing integrals are[20]

$$\int_0^{\infty} \frac{\sigma_{\mathrm{ext}}(k)}{k^{2+p}} \,\mathrm{d}k = 2\pi^2 a_p, \qquad p = 0, 2$$

---

[19] In general, by the use of Cauchy–Schwarz inequality, we have

$$\left|\mathrm{Im}\left(\hat{\boldsymbol{p}}_{\mathrm{e}}^* \cdot \mathbf{S}(\hat{\boldsymbol{k}}_{\mathrm{i}}, \hat{\boldsymbol{k}}_{\mathrm{i}}) \cdot \hat{\boldsymbol{p}}_{\mathrm{e}}\right)\right|^2 \leq \left|\hat{\boldsymbol{p}}_{\mathrm{e}}^* \cdot \mathbf{S}(\hat{\boldsymbol{k}}_{\mathrm{i}}, \hat{\boldsymbol{k}}_{\mathrm{i}}) \cdot \hat{\boldsymbol{p}}_{\mathrm{e}}\right|^2 \leq \left|\hat{\boldsymbol{p}}_{\mathrm{e}}\right|^2 \left|\mathbf{S}(\hat{\boldsymbol{k}}_{\mathrm{i}}, \hat{\boldsymbol{k}}_{\mathrm{i}}) \cdot \hat{\boldsymbol{p}}_{\mathrm{e}}\right|^2 = \left|\mathbf{S}(\hat{\boldsymbol{k}}_{\mathrm{i}}, \hat{\boldsymbol{k}}_{\mathrm{i}}) \cdot \hat{\boldsymbol{p}}_{\mathrm{e}}\right|^2$$

[20] With an even integrand, the Cauchy principal value integral becomes a generalized integral at the origin.

provided the integrals exist as generalized integrals. The index $p = 0$ duplicates the result in (6.58), and $p = 2$ reads

$$\int_0^\infty \frac{\sigma_{\text{ext}}(k)}{k^4}\,\mathrm{d}k = 2\pi^2 a_2 \tag{6.65}$$

Before we proceed with the construction of additional sum rules, we make a useful observation of the function $f(\kappa)$. The function $f(\kappa)$ is analytic in the upper complex plane with appropriate behavior at infinity, satisfies the cross symmetry $f(\kappa) = f^*(-\kappa^*)$, and moreover, satisfies $\operatorname{Im} f(k) > 0$ for real $k > 0$. Meĭman's theorem, see Appendix D.3 on page 645, then shows that $f(\kappa)$ has no roots in the upper complex plane.[21] As a consequence, $1/f(\kappa)$ is an analytic function in the upper complex plane.

As soon becomes apparent, there is an infinite amount of sum rules we can construct, but the usefulness of most of these is limited. When we now develop additional sum rules, our aim is to preserve the property of a non-negative integrand.

New sum rules can be obtained by a composition of two analytic functions, such that the composition is analytic in the upper complex plane and that a non-negative imaginary part of the function is preserved. This is exactly the inherent properties of the class of functions called Herglotz functions, see Appendix D.5. A composition of Herglotz functions preserves the property that the upper complex plane is mapped into itself.

As a first example, let $f(\kappa)$ be defined as in (6.56), and compose this function with $-1/\kappa$. From above, the function $F(\kappa) = -1/f(\kappa)$ is an analytic function in the upper complex plane, and moreover, $\operatorname{Im} F(k) > 0$, for real $k > 0$. If, for a lossless obstacle

$$f(\kappa) = a_0 + a_2\kappa^2 + O(\kappa^3) = a_0\left(1 + \frac{a_2}{a_0}\kappa^2 + O(\kappa^3)\right), \qquad \text{as } \kappa \to 0$$

then

$$\begin{aligned} F(\kappa) &= -\frac{1}{f(\kappa)} = -\frac{1}{a_0}\left(1 - \frac{a_2}{a_0}\kappa^2 + O(\kappa^3)\right) \\ &= -\frac{1}{a_0} + \frac{a_2}{a_0^2}\kappa^2 + O(\kappa^3), \qquad \text{as } \kappa \to 0 \end{aligned}$$

At large arguments, we have

$$F(\kappa) = -\frac{1}{f(\kappa)} = O(\kappa), \qquad \text{as } |\kappa| \to \infty$$

Furthermore, for real k, we have

$$\operatorname{Im} F(k) = \frac{\operatorname{Im} f(k)}{|f(k)|^2} = \frac{4\pi \operatorname{Im} kf(k)}{4\pi k|f(k)|^2} = \frac{k^3\sigma_{\text{ext}}(k)}{\sigma_{\text{RCS,co}}(k)}$$

where the co-polarized bistatic cross section, see (4.35) on page 207, in the forward direction of the radar cross section is $\sigma_{\text{RCS,co}}(k) = 4\pi k^4 |f(k)|^2$ (linear polarization is assumed).

[21] In fact, $\kappa f(\kappa)$ is a Herglotz function, see Appendix D.5.

Theorem D.3 on page 643 applied to $F(k)$ then gives ($N = 3$, $M = -1$, and formally all $b_m = 0$, notice that $N + M = 2$ so the theorem still applies)

$$P\int_0^\infty \frac{\operatorname{Im} F(k)}{k^p}\,\mathrm{d}k = \frac{\pi}{2}a_{p-1}, \qquad p = 3$$

or

$$\int_0^\infty \frac{\sigma_{\mathrm{ext}}(k)}{\sigma_{\mathrm{RCS,co}}(k)}\,\mathrm{d}k = \frac{a_2\pi}{2a_0^2} \tag{6.66}$$

A more advanced example of a sum rule would be to form the function

$$F(\kappa) = \frac{\mathrm{i}}{\kappa}\ln\left(1 - \mathrm{i}\alpha\kappa f(\kappa)\right), \qquad \kappa \in \mathbb{C}_+$$

Given the expansion

$$f(\kappa) = a_0 + O(\kappa)$$

the function $F(\kappa)$ has the following low-frequency expansion:

$$F(\kappa) = \frac{\mathrm{i}}{\kappa}\ln\left(1 - \mathrm{i}\alpha\kappa f(\kappa)\right) = \alpha f(\kappa) + O(\kappa) = \alpha a_0 + O(\kappa)$$

and for large arguments, we have

$$F(\kappa) = O(\kappa^{-1})$$

since $\kappa f(\kappa) = O(1)$ for large arguments.

The function $F(\kappa)$ is an analytic function in the upper complex plane for all $\alpha > 0$. In fact, $\operatorname{Re}(1 - \mathrm{i}\alpha\kappa f(\kappa)) = 1 + \alpha \operatorname{Im}(\kappa f(\kappa)) > 0$ for all κ in the upper complex plane, and the argument of the logarithm lies in the right-half complex plane. Moreover, the function satisfies $\operatorname{Im} F(k) > 0$, for $k > 0$, which can be concluded by the use of the result from above. We get

$$\begin{aligned}
\operatorname{Im} F(k) &= \operatorname{Re}\frac{1}{k}\ln\left(1 - \mathrm{i}\alpha k f(k)\right) \\
&= \frac{1}{2k}\ln\left((1 + \alpha k \operatorname{Im} f(k))^2 + \alpha^2 (k \operatorname{Re} f(k))^2\right) \\
&= \frac{1}{2k}\ln\left(1 + 2\alpha k \operatorname{Im} f(k) + \alpha^2 |k f(k)|^2\right) \\
&= \frac{1}{2k}\ln\left(1 + \alpha\sigma_{\mathrm{ext}}(k)/2\pi + \alpha^2\sigma_{\mathrm{RCS,co}}(k)/4\pi k^2\right)
\end{aligned}$$

Theorem D.3 on page 643 applied to $F(k)$, then gives (use $N = 1$, $M = 1$, and $b_0 = b_1 = 0$)

$$\int_0^\infty \ln\left(1 + \alpha\sigma_{\mathrm{ext}}(k)/2\pi + \alpha^2\sigma_{\mathrm{RCS,co}}(k)/4\pi k^2\right)\frac{\mathrm{d}k}{k^2} = \pi\alpha a_0 \tag{6.67}$$

We can also differentiate this result w.r.t. α. We get

$$\int_0^\infty \frac{\sigma_{\mathrm{ext}}(k) + \alpha\sigma_{\mathrm{RCS,co}}(k)/k^2}{2\pi + \alpha\sigma_{\mathrm{ext}}(k) + \alpha^2\sigma_{\mathrm{RCS,co}}(k)/2k^2}\frac{\mathrm{d}k}{k^2} = \pi a_0 \tag{6.68}$$

Note that the right-hand side is independent of α, and that the right-hand side limit $\alpha \to 0$ of the identity gives an identity identical to (6.58).

We verify these new sum rules in a series of examples.

Example 6.19
We illustrate the result in (6.65) with the perfectly conducting sphere. From Problem 8.3, we get

$$f(k) = \frac{a^3}{2} + \frac{113a^5}{180}k^2 + O(k^3)$$

The sum rule in (6.65) reads

$$\int_0^\infty \frac{\sigma_{\text{ext}}(k)}{k^4}\,\mathrm{d}k = \frac{113\pi^2 a^5}{90} \tag{6.69}$$

Similarly, (6.66)–(6.68) give

$$\int_0^\infty \frac{\sigma_{\text{ext}}(k)}{\sigma_{\text{RCS,co}}(k)}\,\mathrm{d}k = \frac{113\pi}{90a} \tag{6.70}$$

$$\begin{aligned}\int_0^\infty \ln\left(1 + \alpha\sigma_{\text{ext}}(k)/2\pi + \alpha^2\sigma_{\text{RCS,co}}(k)/4\pi k^2\right)\frac{\mathrm{d}k}{k^2} &= \frac{\pi\alpha a^3}{2}\\ \int_0^\infty \frac{\sigma_{\text{ext}}(k) + \alpha\sigma_{\text{RCS,co}}(k)/k^2}{2\pi + \alpha\sigma_{\text{ext}}(k) + \alpha^2\sigma_{\text{RCS,co}}(k)/2k^2}\frac{\mathrm{d}k}{k^2} &= \frac{\pi a^3}{2}\end{aligned} \tag{6.71}$$

The integrands of some of these sum rules in this example are depicted in Figure 6.18. ■

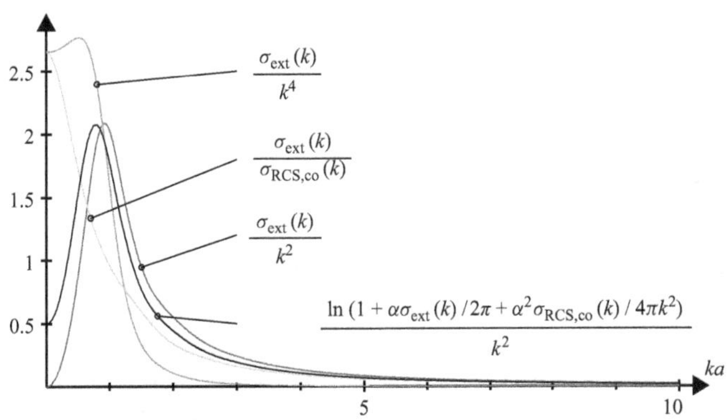

Figure 6.18 The integrands in the sum rules of (6.64), (6.69), (6.70), and (6.71). The integrands are scaled appropriately, so the areas under all curves are π when integrated w.r.t. ka.

Example 6.20
Several interesting sum rules can be obtained for the perfectly conducting disk of radius a. Introduce $f_{\perp,\parallel}(k) = \hat{\boldsymbol{p}}^*_{e\perp,\parallel} \cdot \mathbf{S}(\hat{\boldsymbol{k}}_\mathrm{i}, \hat{\boldsymbol{k}}_\mathrm{i}) \cdot \hat{\boldsymbol{p}}_{e\perp,\parallel}/k^2$, which depend on the polarization—transverse electric (magnetic) field TE (TM). The low-frequency behavior of these functions are, see (9.8) on page 487 (θ is angle of incidence)

$$f_\perp(k) = \frac{2a^3}{3\pi}\left(1+\cos^2\theta\right) + \frac{2k^2a^5}{45\pi}\left(16 - 5\sin^2\theta + \sin^4\theta\right) + O(k^3)$$

and

$$f_\parallel(k) = \frac{4a^3}{3\pi}\cos^2\theta + \frac{2k^2a^5}{45\pi}\cos^2\theta\left(16 - 3\sin^2\theta\right) + O(k^3)$$

and (6.65) becomes

$$\int_0^\infty \frac{\sigma_\mathrm{ext}(k)}{k^4}\,\mathrm{d}k = \frac{4\pi a^5}{45}\begin{cases}16 - 5\sin^2\theta + \sin^4\theta & \text{(TE)}\\ \cos^2\theta\left(16 - 3\sin^2\theta\right) & \text{(TM)}\end{cases}$$

The extinction cross section, $\sigma_\mathrm{ext}(\hat{\boldsymbol{k}}_\mathrm{i})$, for scattering by a perfectly conducting disk at different directions of incidence as a function of frequency is depicted in Figure 9.4 on page 490.

Similarly, we get using (6.66)

$$\int_0^\infty \frac{\sigma_\mathrm{ext}(k)}{\sigma_\mathrm{RCS,co}(k)}\,\mathrm{d}k = \frac{\pi^2}{20a}\begin{cases}\dfrac{16 - 5\sin^2\theta + \sin^4\theta}{\left(1+\cos^2\theta\right)^2} & \text{(TE)}\\[2ex] \dfrac{16 - 3\sin^2\theta}{4\cos^2\theta} & \text{(TM)}\end{cases}$$

Using (6.67) and (6.68), we get ($\alpha > 0$)

$$\int_0^\infty \ln\left(1 + \alpha\sigma_\mathrm{ext}(k)/2\pi + \alpha^2\sigma_\mathrm{RCS,co}(k)/4\pi k^2\right)\frac{\mathrm{d}k}{k^2} = \frac{2a^3\alpha}{3}\begin{cases}1+\cos^2\theta & \text{(TE)}\\ 2\cos^2\theta & \text{(TM)}\end{cases}$$

and

$$\int_0^\infty \frac{\sigma_\mathrm{ext}(k) + \alpha\sigma_\mathrm{RCS,co}(k)/k^2}{2\pi + \alpha\sigma_\mathrm{ext}(k) + \alpha^2\sigma_\mathrm{RCS,co}(k)/2k^2}\frac{\mathrm{d}k}{k^2} = \frac{2a^3}{3}\begin{cases}1+\cos^2\theta & \text{(TE)}\\ 2\cos^2\theta & \text{(TM)}\end{cases}$$

We see that for $\alpha \to 0$, we recover the sum rule above. ■

Additional sum rules can be obtained by multiplying $f(k)$ with an exponential function $\mathrm{e}^{\mathrm{i}ka}$ where $a > 0$. Note that the multiplication by a factor $\mathrm{e}^{\mathrm{i}ka}$ implies a time shift $t \to t - a$ in the physical quantities, i.e., we are not in conflict with causality. Since the function $F(\kappa) = f(\kappa)\mathrm{e}^{\mathrm{i}\kappa a}$ is analytic in the upper complex plane, if $f(\kappa)$ is analytic in the upper complex plane, we get in analogy with (6.58)

$$\int_0^\infty \frac{\operatorname{Im} F(k)}{k}\,\mathrm{d}k = \frac{\pi a_0}{2}$$

or

$$\int_0^\infty \frac{\operatorname{Im} f(k)\cos ka + \operatorname{Re} f(k)\sin ka}{k}\, dk = \frac{\pi a_0}{2}$$

Notice that the right-hand side is independent of $a > 0$. In the limit $a \to 0$, we retrieve the sum rule in (6.58). This sum rule has limited use, since the integrand is not strictly positive.

6.5 Scattering by Many Scatterers—Multiple Scattering

This section addresses the special problem of scattering by an ensemble of separate objects—each described by their individual scattering characteristics. This problem is exceptionally difficult to solve, since the field that impinges on one particular scatterer in the ensemble consists not only of the globally incident field—usually a plane wave—but also the scattered field from all other scatterers in the ensemble, which are unknown. This is called multiple scattering. If the number of individual scatterers becomes large this problem becomes exceedingly difficult to solve. The literature on the subject is vast—for some of the early works see e.g., References 70, 71, 155, 156, 267, 287. Several of the results in this section are verified in a more quantifiable treatment in Section 9.2, and several general textbooks about the subject have been published, see e.g., References 174, 187.

In this section, we make a general formulation of the multi-particle scattering problem, and exploit the special formulation that occurs if the scatterers are located far from each other. The far field approximation of each scattering process can then be employed. We then conclude the session by the limiting case where multiple scattering effects can be neglected, which is a good approximation if the scatterers are sparsely distributed and/or weak. The field that impinges on a scatterer in the ensemble is then approximated by the global incident field, $\boldsymbol{E}_{\mathrm{i}}$, generated by the external sources of the problem. This approximation is referred to as single scattering process.

Assume there are N different scatterers in the ensemble, each occupying a volume $V_{\mathrm{s}p}$, bounded by a surface $S_{\mathrm{s}p}$, $p = 1, 2, \ldots, N$, and that the ensemble is excited by an incident field $\boldsymbol{E}_{\mathrm{i}}(\boldsymbol{r})$, whose sources are located outside the ensemble. All scatterers are confined to the volume V_{s}, and we assume no individual scatterers are overlapping, i.e., $V_{\mathrm{s}p} \cap V_{\mathrm{s}q} = \emptyset, p \neq q$. The geometry of the ensemble is shown in Figure 6.19. As above, the material parameters of the medium surrounding (outside) the scatterers $V_{\mathrm{s}p}$ are ϵ and μ, both real valued.

Comment 6.1

Formally, this decomposition of the scattering volume V_{s} in several constitutive parts also can be applied to a single scatterer. In such a decomposition there is no surrounding material between the "individual scatterers." This decomposition is the starting point of several numerical algorithms to solve the scattering problems.

We focus on an arbitrary scatterer, which is labeled p, with a local origin at $\boldsymbol{r}_p$, $p = 1, 2, \ldots, N$, relative to the common origin, O, and we essentially follow the formulation of the scattering problem as it is presented in Reference 287. Each scatterer has its own material properties, which do not have to be the same for all scatterers. Three different fields in the vicinity of a scatterer are relevant—the incident field, $\boldsymbol{E}_{\mathrm{i}}(\boldsymbol{r})$, the field that excites the scatterer, $\boldsymbol{E}_{\mathrm{exc}p}(\boldsymbol{r})$, and the scattered field by the scatterer, $\boldsymbol{E}_{\mathrm{s}p}(\boldsymbol{r})$. The total electric field outside

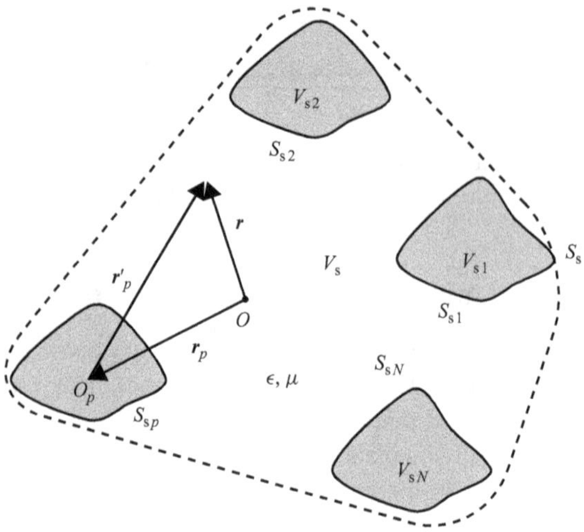

Figure 6.19 Geometry for multiple scatterers, V_{sp}, with common origin O and local origin O_p. The dashed contour confines the total volume V_s with bounding surface S_s.

the scatterers consists of the incident field $\boldsymbol{E}_i(\boldsymbol{r})$ and the sum of all the scattered fields from the individual scatterers, i.e.,

$$\boldsymbol{E}(\boldsymbol{r}) = \boldsymbol{E}_i(\boldsymbol{r}) + \sum_{p=1}^{N} \boldsymbol{E}_{sp}(\boldsymbol{r}), \qquad \boldsymbol{r} \notin \bigcup_{p=1}^{N} V_{sp}$$

This statement is proved by the integral representation in Chapter 9. We simply accept this physically well-motivated statement here, and return to the problem in Chapter 9.

The scattered and the exciting fields are not independent of each other. The scattered field depends *linearly* on the excited field and we formally write as

$$\boldsymbol{E}_{sp}(\boldsymbol{r}) = (T_p \boldsymbol{E}_{excp})(\boldsymbol{r}), \qquad \boldsymbol{r} \notin V_{sp}, \quad p = 1, 2, \ldots, N$$

where the transition operator of the pth scatterer is denoted T_p, which acts on the excited field in the vicinity of the pth scatterer. More precisely, the position vector is assumed to be located outside the scatterer, i.e., outside S_p ($\boldsymbol{r} \notin V_{sp}$). In Chapter 9, an explicit algorithm to compute the transition operator is developed. The formal definition of the exciting field $\boldsymbol{E}_{excp}(\boldsymbol{r})$ is done below, see (6.74).

The total electric field at a field point $\boldsymbol{r}$ then becomes

$$\boldsymbol{E}(\boldsymbol{r}) = \boldsymbol{E}_i(\boldsymbol{r}) + \sum_{p=1}^{N} \boldsymbol{E}_{sp}(\boldsymbol{r}) = \boldsymbol{E}_i(\boldsymbol{r}) + \sum_{p=1}^{N} (T_p \boldsymbol{E}_{excp})(\boldsymbol{r}), \qquad \boldsymbol{r} \notin \bigcup_{p=1}^{N} V_{sp} \tag{6.72}$$

Similarly, the field inside each of the scatterers is related to the exciting field by a linear operator T_p^I. We define

$$E(r) = (T_p^I E_{\mathrm{exc}p})(r), \qquad r \in V_{\mathrm{s}p}, \quad p = 1, 2, \ldots, N$$

We collect these expressions into one single equation for the total field regardless of whether the observation point, r, lies inside or outside a scatterer by the use of the function (characteristic function of the scatterer p) [287][22]

$$\Theta_p(r) = \begin{cases} 1, & r \in V_{\mathrm{s}p} \\ 0, & r \notin V_{\mathrm{s}p} \end{cases} \qquad p = 1, 2, \ldots, N$$

The result is

$$\begin{aligned} E(r) = \prod_{p=1}^{N} \left(1 - \Theta_p(r)\right) \left\{ E_{\mathrm{i}}(r) + \sum_{q=1}^{N} (T_q E_{\mathrm{exc}q})(r) \right\} \\ + \sum_{p=1}^{N} \Theta_p(r) (T_p^I E_{\mathrm{exc}p})(r), \qquad r \in \mathbb{R}^3 \end{aligned} \tag{6.73}$$

which also is called Foldy–Lax equations.[23] Note that the factor $\prod_{p=1}^{N} \left(1 - \Theta_p(r)\right)$ is 0 for r inside one of the scatterers, and 1 elsewhere.

The exciting field, $E_{\mathrm{exc}p}(r)$, which depends on the position of all scatterers, is assumed to satisfy the equations of the unperturbed media (no scatterers present), and has to satisfy the appropriate consistency relation at the field point r. This set of relations can be considered as the definition of the exciting field, and we have [176, 287]

$$E_{\mathrm{exc}p}(r) = E_{\mathrm{i}}(r) + \sum_{\substack{q=1 \\ q \neq p}}^{N} (T_q E_{\mathrm{exc}q})(r), \qquad r \in V_{\mathrm{s}p}, \quad p = 1, 2, \ldots, N \tag{6.74}$$

or

$$\sum_{p=1}^{N} \Theta_p(r) \left\{ E_{\mathrm{exc}p}(r) - E_{\mathrm{i}}(r) - \sum_{\substack{q=1 \\ q \neq p}}^{N} (T_q E_{\mathrm{exc}q})(r) \right\} = \mathbf{0}, \qquad r \in \mathbb{R}^3 \tag{6.75}$$

[22] More precisely, the function $\Theta_p(r)$ in Reference 287 is $1 - \Theta_p(r)$.

[23] Leslie Lawrance Foldy (1919–2001), American theoretical physicist; Melvin Lax (1922–2002), American physicist.

2

In Chapter 9, this consistency relation is proved by the use of the integral representation of the field. Combine (6.73) and (6.75) to obtain

$$\begin{aligned}\boldsymbol{E}(\boldsymbol{r}) &= \boldsymbol{E}_{\rm i}(\boldsymbol{r}) + \prod_{p=1}^{N}\left(1-\Theta_p(\boldsymbol{r})\right)\sum_{q=1}^{N}(T_q\boldsymbol{E}_{{\rm exc}q})(\boldsymbol{r}) + \sum_{p=1}^{N}\Theta_p(\boldsymbol{r})\sum_{\substack{q=1\\ q\neq p}}^{N}(T_q\boldsymbol{E}_{{\rm exc}q})(\boldsymbol{r})\\ &+\sum_{p=1}^{N}\Theta_p(\boldsymbol{r})\left\{(T_p^I\boldsymbol{E}_{{\rm exc}p})(\boldsymbol{r}) - \boldsymbol{E}_{{\rm exc}p}(\boldsymbol{r})\right\}, \qquad \boldsymbol{r}\in\mathbb{R}^3\end{aligned}$$

where we used the identity, see Problem 6.15

$$\prod_{p=1}^{N}\left(1-\Theta_p(\boldsymbol{r})\right) + \sum_{p=1}^{N}\Theta_p(\boldsymbol{r}) = 1, \qquad \boldsymbol{r}\in\mathbb{R}^3 \tag{6.76}$$

Rewrite the expression above as

$$\begin{aligned}\boldsymbol{E}(\boldsymbol{r}) &= \boldsymbol{E}_{\rm i}(\boldsymbol{r}) + \sum_{p=1}^{N}\left(1-\Theta_p(\boldsymbol{r})\right)(T_p\boldsymbol{E}_{{\rm exc}p})(\boldsymbol{r})\\ &+\sum_{p=1}^{N}\Theta_p(\boldsymbol{r})\left\{(T_p^I\boldsymbol{E}_{{\rm exc}p})(\boldsymbol{r}) - \boldsymbol{E}_{{\rm exc}p}(\boldsymbol{r})\right\}, \qquad \boldsymbol{r}\in\mathbb{R}^3\end{aligned}$$

since, by the use of (6.76), we have

$$\begin{aligned}&\prod_{p=1}^{N}\left(1-\Theta_p(\boldsymbol{r})\right)\sum_{q=1}^{N}(T_q\boldsymbol{E}_{{\rm exc}q})(\boldsymbol{r}) + \sum_{p=1}^{N}\Theta_p(\boldsymbol{r})\sum_{\substack{q=1\\ q\neq p}}^{N}(T_q\boldsymbol{E}_{{\rm exc}q})(\boldsymbol{r})\\ &= \prod_{p=1}^{N}\left(1-\Theta_p(\boldsymbol{r})\right)\sum_{q=1}^{N}(T_q\boldsymbol{E}_{{\rm exc}q})(\boldsymbol{r}) + \sum_{p=1}^{N}\Theta_p(\boldsymbol{r})\sum_{q=1}^{N}(T_q\boldsymbol{E}_{{\rm exc}q})(\boldsymbol{r})\\ &-\sum_{p=1}^{N}\Theta_p(\boldsymbol{r})(T_p\boldsymbol{E}_{{\rm exc}p})(\boldsymbol{r}) = \sum_{p=1}^{N}\left(1-\Theta_p(\boldsymbol{r})\right)(T_p\boldsymbol{E}_{{\rm exc}p})(\boldsymbol{r}), \qquad \boldsymbol{r}\in\mathbb{R}^3\end{aligned}$$

6.5.1 Far Field Approximation

Provided all distances between the scatterers are large, such that all scatterers are located in the far zone of all other objects, the scattered field can be approximated by the far field contributions by the scatterers. This approximation simplifies the form of the transition operator T_p above. We explore this approximation in this section.

We assume the exterior sources generate a plane wave impinging along the direction $\hat{\boldsymbol{k}}_{\rm i}$, i.e.,

$$\boldsymbol{E}_{\rm i}(\boldsymbol{r}) = \boldsymbol{E}_0 e^{{\rm i}k\hat{\boldsymbol{k}}_{\rm i}\cdot\boldsymbol{r}}$$

where $\boldsymbol{E}_0$ is the amplitude of the plane wave at the common origin. The incident field at the local origin of the pth scatterer has the amplitude ($\boldsymbol{r} = \boldsymbol{r}_p$)

$$\boldsymbol{E}_{\mathrm{i}}(\boldsymbol{r}_p) = \boldsymbol{E}_0 \mathrm{e}^{\mathrm{i}k\hat{\boldsymbol{k}}_{\mathrm{i}}\cdot\boldsymbol{r}_p}$$

The functional behavior of the scattered field on the exciting field is in general a complex mapping (operator equation), but in the far field this operation is a simple multiplication. We have, see (4.16) on page 197

$$(T_p\boldsymbol{E}_{\mathrm{exc}p})(\boldsymbol{r}) = (T_p\boldsymbol{E}_{\mathrm{exc}p})(\boldsymbol{r}_p + \boldsymbol{r}'_p) = \frac{\mathrm{e}^{\mathrm{i}kr'_p}}{r'_p}\boldsymbol{F}_p(\hat{\boldsymbol{r}}'_p) \tag{6.77}$$

where $\boldsymbol{r}'_p = \boldsymbol{r} - \boldsymbol{r}_p$ is the location vector from the local origin of scatterer p to the point of observation $\boldsymbol{r}$, see Figure 6.19, and $r'_p = |\boldsymbol{r}'_p|$, $\hat{\boldsymbol{r}}'_p = \boldsymbol{r}'_p/r'_p$.

Outside all scatterers (in the far field zone of all scatterers), we then obtain from (6.72)

$$\boldsymbol{E}(\boldsymbol{r}) = \boldsymbol{E}_0 \mathrm{e}^{\mathrm{i}k\hat{\boldsymbol{k}}_{\mathrm{i}}\cdot\boldsymbol{r}} + \sum_{p=1}^{N} \frac{\mathrm{e}^{\mathrm{i}kr'_p}}{r'_p}\boldsymbol{F}_p(\hat{\boldsymbol{r}}'_p), \qquad \boldsymbol{r} \notin \bigcup_{p=1}^{N} V_{\mathrm{s}p} \tag{6.78}$$

The far field amplitude $\boldsymbol{F}_p(\hat{\boldsymbol{r}}'_p)$ is still unknown, but once these are known, we can calculate the field $\boldsymbol{E}(\boldsymbol{r})$. If the exciting field at the pth scatterer is a plane wave with incident direction $\hat{\boldsymbol{e}}_p$ and polarization $\boldsymbol{E}_p$, the far field amplitude can be expressed in terms of the scattering dyadic $\mathbf{S}_p(\hat{\boldsymbol{r}}'_p, \hat{\boldsymbol{e}}_p)$, see (4.29) on page 205

$$\boldsymbol{F}_p(\hat{\boldsymbol{r}}'_p) = \mathbf{S}_p(\hat{\boldsymbol{r}}'_p, \hat{\boldsymbol{e}}_p) \cdot \boldsymbol{E}_p$$

Under these assumptions the scattering properties of the pth scatterer are characterized by its scattering dyadic $\mathbf{S}_p(\hat{\boldsymbol{r}}'_p, \hat{\boldsymbol{e}}_p)$, but the excitation of the scatterer is unknown. This excitation we now determine.

The far field amplitude $\boldsymbol{F}_p(\hat{\boldsymbol{r}}'_p)$ is linearly related to the excitation of the pth scatterer evaluated at $\boldsymbol{r}_p$, which is the $\boldsymbol{E}_{\mathrm{exc}p}(\boldsymbol{r}_p)$. This field is determined by evaluating (6.74) at $\boldsymbol{r}_q$, which by the approximation in (6.77) reads

$$\boldsymbol{E}_{\mathrm{exc}p}(\boldsymbol{r}_p) = \boldsymbol{E}_0 \mathrm{e}^{\mathrm{i}k\hat{\boldsymbol{k}}_{\mathrm{i}}\cdot\boldsymbol{r}_p} + \sum_{\substack{r=1\\ r\neq p}}^{N} \frac{\mathrm{e}^{\mathrm{i}kR_{pr}}}{R_{pr}}\boldsymbol{F}_r(\hat{\boldsymbol{R}}_{pr}), \qquad p = 1, 2, \ldots, N \tag{6.79}$$

where $\boldsymbol{R}_{pr} = \boldsymbol{r}_p - \boldsymbol{r}_r$, $R_{pr} = \left|\boldsymbol{R}_{pr}\right|$, and $\hat{\boldsymbol{R}}_{pr} = \boldsymbol{R}_{pr}/R_{pr}$, the unit vector between the local origins O_r and O_p—pointing toward O_p. The far field amplitude $\boldsymbol{F}_r(\hat{\boldsymbol{R}}_{pr})$ does not depend on any distances, only directions, and it satisfies $\boldsymbol{F}_r(\hat{\boldsymbol{R}}_{pr})\cdot\hat{\boldsymbol{R}}_{pr} = 0$. The right-hand side of (6.79) is a superposition of plane waves with incident directions $\hat{\boldsymbol{k}}_{\mathrm{i}}$ and $\hat{\boldsymbol{R}}_{pr}$ and polarizations $\boldsymbol{E}_{\mathrm{i}}(\boldsymbol{r}_p)$ and $\boldsymbol{F}_r(\hat{\boldsymbol{R}}_{pr})$, respectively.[24] This proves that the exciting field at the pth scatterer is a superposition of plane waves, and it is possible to use the scattering dyadic to determine its scattered field.

[24] In the far zone, the spherical waves in the sum are approximated by plane waves with incident directions $\hat{\boldsymbol{R}}_{pr}$ and polarization $\boldsymbol{F}_r(\hat{\boldsymbol{R}}_{pr})$.

To proceed, operate with T_p on (6.79) to transform it to the far field contribution, and evaluate the final expression at $\boldsymbol{r}_q$, see (6.77). We get

$$\frac{\mathrm{e}^{\mathrm{i}kR_{qp}}}{R_{qp}}\boldsymbol{F}_p(\hat{\boldsymbol{R}}_{qp}) = \frac{\mathrm{e}^{\mathrm{i}kR_{qp}}}{R_{qp}}\mathbf{S}_p(\hat{\boldsymbol{R}}_{qp}, \hat{\boldsymbol{k}}_\mathrm{i}) \cdot \boldsymbol{E}_0\mathrm{e}^{\mathrm{i}k\hat{\boldsymbol{k}}_\mathrm{i}\cdot\boldsymbol{r}_p}$$
$$+ \frac{\mathrm{e}^{\mathrm{i}kR_{qp}}}{R_{qp}}\sum_{\substack{r=1\\ r\neq p}}^{N}\frac{\mathrm{e}^{\mathrm{i}kR_{pr}}}{R_{pr}}\mathbf{S}_p(\hat{\boldsymbol{R}}_{qp}, \hat{\boldsymbol{R}}_{pr}) \cdot \boldsymbol{F}_r(\hat{\boldsymbol{R}}_{pr}), \qquad p \neq q = 1, 2, \ldots, N$$

With known scattering properties, i.e., $\mathbf{S}_p$, this equation is now a system of $N(N-1)$ linear algebraic equations in the unknown $\boldsymbol{E}_{qp}$ defined by[25]

$$\boldsymbol{E}_{qp} = \frac{\mathrm{e}^{\mathrm{i}kR_{qp}}}{R_{qp}}\boldsymbol{F}_p(\hat{\boldsymbol{R}}_{qp}), \qquad p \neq q = 1, 2, \ldots, N$$

The system of equation then reads ($p \neq q = 1, 2, \ldots, N$)

$$\boldsymbol{E}_{qp} = \frac{\mathrm{e}^{\mathrm{i}kR_{qp}}}{R_{qp}}\mathbf{S}_p(\hat{\boldsymbol{R}}_{qp}, \hat{\boldsymbol{k}}_\mathrm{i}) \cdot \boldsymbol{E}_0\mathrm{e}^{\mathrm{i}k\hat{\boldsymbol{k}}_\mathrm{i}\cdot\boldsymbol{r}_p} + \frac{\mathrm{e}^{\mathrm{i}kR_{qp}}}{R_{qp}}\sum_{\substack{r=1\\ r\neq p}}^{N}\mathbf{S}_p(\hat{\boldsymbol{R}}_{qp}, \hat{\boldsymbol{R}}_{pr}) \cdot \boldsymbol{E}_{pr} \tag{6.80}$$

For a given configuration of scatterers, this set of equations can, in principle, be used to solve for the unknown excitation $\boldsymbol{E}_{qp}$. For consistency, define $\boldsymbol{E}_{pp} = \mathbf{0}$.

Outside all scatterer (in the far field zone of all scatterers), we obtain from (6.78) and the far field approximation in (6.79)

$$\boldsymbol{E}(\boldsymbol{r}) = \boldsymbol{E}_0\mathrm{e}^{\mathrm{i}k\hat{\boldsymbol{k}}_\mathrm{i}\cdot\boldsymbol{r}} + \sum_{p=1}^{N}\frac{\mathrm{e}^{\mathrm{i}kr'_p}}{r'_p}\left\{\mathbf{S}_p(\hat{\boldsymbol{r}}'_p, \hat{\boldsymbol{k}}_\mathrm{i}) \cdot \boldsymbol{E}_0\mathrm{e}^{\mathrm{i}k\hat{\boldsymbol{k}}_\mathrm{i}\cdot\boldsymbol{r}_p} + \sum_{\substack{q=1\\ q\neq p}}^{N}\mathbf{S}_p(\hat{\boldsymbol{r}}'_p, \hat{\boldsymbol{R}}_{pq}) \cdot \boldsymbol{E}_{pq}\right\}$$

and the scattered field becomes

$$\boldsymbol{E}_\mathrm{s}(\boldsymbol{r}) = \sum_{p=1}^{N}\frac{\mathrm{e}^{\mathrm{i}kr'_p}}{r'_p}\left\{\mathbf{S}_p(\hat{\boldsymbol{r}}'_p, \hat{\boldsymbol{k}}_\mathrm{i}) \cdot \boldsymbol{E}_0\mathrm{e}^{\mathrm{i}k\hat{\boldsymbol{k}}_\mathrm{i}\cdot\boldsymbol{r}_p} + \sum_{\substack{q=1\\ q\neq p}}^{N}\mathbf{S}_p(\hat{\boldsymbol{r}}'_p, \hat{\boldsymbol{R}}_{pq}) \cdot \boldsymbol{E}_{pq}\right\} \tag{6.81}$$

We conclude that, at least formally, we have solved the multiple scattering problem provided all scatterers lie in the far zone to each other. First, solve the linear system of equations in (6.80) for $\boldsymbol{E}_{pq}$, and then calculate the scattered field by (6.81).

---

[25]An alternative set of unknowns is $\boldsymbol{F}_p(\hat{\boldsymbol{R}}_{qp})$. It is a matter of choice which set of unknowns to use.

6.5.2 Single Scattering

The consistency equations can formally be inserted in the expression of the total electric field above. Repeated insertion of (6.74) in (6.72) gives

$$\begin{aligned}E(r) &= E_{\rm i}(r) + \sum_{p=1}^{N}(T_pE_{\rm i})(r) + \sum_{p=1}^{N}\sum_{\substack{q=1\\q\neq p}}^{N}(T_p(T_qE_{{\rm exc}q}))(r)\\ &= E_{\rm i}(r) + \sum_{p=1}^{N}(T_pE_{\rm i})(r) + \sum_{p=1}^{N}\sum_{\substack{q=1\\q\neq p}}^{N}(T_p(T_qE_{\rm i}))(r) + \cdots\end{aligned}$$

where we have used the linear dependence of the transition operator T_p. The first term is, of course, the incident field (no scattering at all), and the second is referred to as the single scattering term. The excited field in each term is the incident field at the position r_p, $p = 1, 2, \ldots, N$. The third term is the first contribution to multiple scattering (double scattering). There are several procedures to terminate this infinite sequence of operations. We point out that this iteration procedure is a mathematical construction, and that the real scattering effects cannot be decomposed in this way. We now proceed making several restrictions on the fields. A more complete treatment of this problem is found in Section 9.2.1.

Neglecting the interaction (multiple scattering effects) between all the scatterers, the single scattering approximation is obtained, and the total scattered field is the sum of all the scattered fields from the individual scatterers (far zone contributions)

$$E_{\rm s}(r) \approx \sum_{p=1}^{N}\frac{e^{ikr'_p}}{r'_p}F_p(\hat{r}'_p) = \sum_{p=1}^{N}\frac{e^{ikr'_p}}{r'_p}e^{ik\hat{k}_{\rm i}\cdot r_p}\mathbf{S}_p(\hat{r}'_p, \hat{k}_{\rm i})\cdot E_0$$

Now assume the field point is far away from all scatterers, so that we are in the far zone of all scatterers, i.e., $r \gg r_p$, $p = 1, 2, \ldots, N$. We approximate as usual ($d = \max_p \max_{\{r_p, r'_p\}\in V_{{\rm s}p}} |r_p - r'_p|$)

$$r'_p = |r - r_p| = r\left\{1 - \hat{r}\cdot\frac{r_p}{r} + O(d^2/r^2)\right\} = r - \hat{r}\cdot r_p + O(d^2/r), \text{ as } r \to \infty$$

and we get ($\hat{r}'_p \approx \hat{r}$)

$$E_{\rm s}(r) = \frac{e^{ikr}}{r}F(\hat{r})$$

where the total far field amplitude $F(\hat{r})$ is given by

$$F(\hat{r}) = \sum_{p=1}^{N}e^{-ik\hat{r}\cdot r_p}F_p(\hat{r}) = \sum_{p=1}^{N}e^{ikq\cdot r_p}\mathbf{S}_p(\hat{r}, \hat{k}_{\rm i})\cdot E_0 \tag{6.82}$$

Here, the vector q is defined as

$$q = \hat{k}_{\rm i} - \hat{r}$$

In general, the vector $\boldsymbol{q}$ is not of unit length. This vector can be interpreted as the change in the direction of the wave due to scattering. The scattering dyadic of the ensemble of scatterers $\mathbf{S}(\hat{\boldsymbol{r}}, \hat{\boldsymbol{k}}_{\rm i})$ can be identified from these expressions

$$\boldsymbol{F}(\hat{\boldsymbol{r}}) = \mathbf{S}(\hat{\boldsymbol{r}}, \hat{\boldsymbol{k}}_{\rm i}) \cdot \boldsymbol{E}_0$$

where

$$\mathbf{S}(\hat{\boldsymbol{r}}, \hat{\boldsymbol{k}}_{\rm i}) = \sum_{p=1}^{N} e^{ik\boldsymbol{q}\cdot\boldsymbol{r}_p} \mathbf{S}_p(\hat{\boldsymbol{r}}, \hat{\boldsymbol{k}}_{\rm i})$$

Specifically, in the forward scattering direction, $\hat{\boldsymbol{r}} = \hat{\boldsymbol{k}}_{\rm i}$, we have

$$\mathbf{S}(\hat{\boldsymbol{k}}_{\rm i}, \hat{\boldsymbol{k}}_{\rm i}) = \sum_{p=1}^{N} \mathbf{S}_p(\hat{\boldsymbol{k}}_{\rm i}, \hat{\boldsymbol{k}}_{\rm i})$$

In the forward direction, all phase factors in the scattering dyadic are identical, and the scattering contributions add coherently.

The optical theorem, (4.47) on page 230, can be used to evaluate the extinction cross section $\sigma_{\rm ext}$ for the entire ensemble of scatterers expressed in the extinction cross section of each scatterer $\sigma_{{\rm ext}p}$. We get

$$\sigma_{\rm ext} = \frac{4\pi}{k} \operatorname{Im}\left\{\hat{\boldsymbol{p}}_{\rm e}^* \cdot \mathbf{S}(\hat{\boldsymbol{k}}_{\rm i}, \hat{\boldsymbol{k}}_{\rm i}) \cdot \hat{\boldsymbol{p}}_{\rm e}\right\} = \sum_{p=1}^{N} \frac{4\pi}{k} \operatorname{Im}\left\{\hat{\boldsymbol{p}}_{\rm e}^* \cdot \mathbf{S}_p(\hat{\boldsymbol{k}}_{\rm i}, \hat{\boldsymbol{k}}_{\rm i}) \cdot \hat{\boldsymbol{p}}_{\rm e}\right\} = \sum_{p=1}^{N} \sigma_{{\rm ext}p}$$

where the individual extinction cross sections of the scatterers $\sigma_{{\rm ext}p}$ are

$$\sigma_{{\rm ext}p} = \frac{4\pi}{k} \operatorname{Im}\left\{\hat{\boldsymbol{p}}_{\rm e}^* \cdot \mathbf{S}_p(\hat{\boldsymbol{k}}_{\rm i}, \hat{\boldsymbol{k}}_{\rm i}) \cdot \hat{\boldsymbol{p}}_{\rm e}\right\}$$

Example 6.21

If the scatterer is small compared to the wavelength (long wavelength approximation), we have a closed-form expression of the scattering dyadic. From (6.8) on page 272

$$\mathbf{S}_p(\hat{\boldsymbol{r}}, \hat{\boldsymbol{k}}_{\rm i}) = -\frac{k^2}{4\pi}\left[\hat{\boldsymbol{r}} \times \left(\hat{\boldsymbol{r}} \times \boldsymbol{\gamma}_{{\rm e}p}\right) + \hat{\boldsymbol{r}} \times \boldsymbol{\gamma}_{{\rm m}p} \cdot \left(\hat{\boldsymbol{k}}_{\rm i} \times \mathbf{I}_3\right)\right]$$

we obtain

$$\mathbf{S}(\hat{\boldsymbol{r}}, \hat{\boldsymbol{k}}_{\rm i}) = \sum_{p=1}^{N} e^{ik\boldsymbol{q}\cdot\boldsymbol{r}_p} \mathbf{S}_p(\hat{\boldsymbol{r}}, \hat{\boldsymbol{k}}_{\rm i})$$

■

If all scatterers are equal, i.e., $\mathbf{S}_p(\hat{\boldsymbol{r}}, \hat{\boldsymbol{k}}_{\mathrm{i}}) = \mathbf{s}(\hat{\boldsymbol{r}}, \hat{\boldsymbol{k}}_{\mathrm{i}})$, $n = 1, 2, \ldots, N$, the far field amplitude $\boldsymbol{F}(\hat{\boldsymbol{r}})$ in (6.82) is simplified.

$$\boldsymbol{F}(\hat{\boldsymbol{r}}) = \mathbf{s}(\hat{\boldsymbol{r}}, \hat{\boldsymbol{k}}_{\mathrm{i}}) \cdot \boldsymbol{E}_0 \sum_{p=1}^{N} \mathrm{e}^{\mathrm{i}k\boldsymbol{q}\cdot\boldsymbol{r}_p}$$

In this special case, we observe that the far field amplitude consists of the far field amplitude of an individual scatterer multiplied by a complex number in the form of a sum. In particular, for the differential scattering cross section, see (4.21) on page 202, we get

$$\frac{\mathrm{d}\sigma}{\mathrm{d}\Omega}(\hat{\boldsymbol{r}}, \hat{\boldsymbol{k}}_{\mathrm{i}}) = 4\pi \frac{|\boldsymbol{F}(\hat{\boldsymbol{r}})|^2}{|\boldsymbol{E}_0|^2} = 4\pi \frac{\left|\mathbf{s}(\hat{\boldsymbol{r}}, \hat{\boldsymbol{k}}_{\mathrm{i}}) \cdot \boldsymbol{E}_0\right|^2}{|\boldsymbol{E}_0|^2} \mathcal{F}(\hat{\boldsymbol{r}})$$

where the phase factor $\mathcal{F}(\hat{\boldsymbol{r}})$ is

$$\mathcal{F}(\hat{\boldsymbol{r}}) = \left|\sum_{p=1}^{N} \mathrm{e}^{\mathrm{i}k\boldsymbol{q}\cdot\boldsymbol{r}_p}\right|^2$$

This quantity is also called the structure factor and it can be expressed as

$$\mathcal{F}(\hat{\boldsymbol{r}}) = \sum_{p,p'=1}^{N} \mathrm{e}^{\mathrm{i}k\boldsymbol{q}\cdot(\boldsymbol{r}_p - \boldsymbol{r}_{p'})}$$

In particular, in the forward direction, when $\hat{\boldsymbol{r}} = \hat{\boldsymbol{k}}_{\mathrm{i}}$ or $\boldsymbol{q} = \mathbf{0}$, the structure factor is

$$\mathcal{F}(\hat{\boldsymbol{k}}_{\mathrm{i}}) = N^2$$

and the differential scattering cross section becomes

$$\frac{\mathrm{d}\sigma}{\mathrm{d}\Omega}(\hat{\boldsymbol{k}}_{\mathrm{i}}, \hat{\boldsymbol{k}}_{\mathrm{i}}) = 4\pi N^2 \frac{|\mathbf{s}(\hat{\boldsymbol{k}}_{\mathrm{i}}, \hat{\boldsymbol{k}}_{\mathrm{i}}) \cdot \boldsymbol{E}_0|^2}{|\boldsymbol{E}_0|^2}$$

Problems for Chapter 6

Section 6.1 Long Wavelength Approximation

6.1 Calculate the scattering dyadic $\mathbf{S}$ in the long wavelength limit for scattering of an incident plane wave by a dielectric sphere of radius a, permittivity ϵ, and permeability $\mu = 1$ ($\mu = \epsilon = 1$ outside the sphere). The incident plane wave is

$$\boldsymbol{E}_{\mathrm{i}}(\boldsymbol{r}) = \boldsymbol{E}_0 \mathrm{e}^{\mathrm{i}k\hat{\boldsymbol{k}}_{\mathrm{i}}\cdot\boldsymbol{r}}$$

Calculate also the scattering cross section $\sigma_{\mathrm{s}}|_{\mathrm{unpol}}$ and the degree of polarization $P|_{\mathrm{unpol}}$ for an unpolarized incident plane wave.

Hint: First, calculate the equivalent electric dipole moment of the sphere, $\boldsymbol{p}$, for the incident excitation $\boldsymbol{E}_0$

$$\boldsymbol{p} = 4\pi\epsilon_0 a^3 \frac{\epsilon - 1}{\epsilon + 2}\boldsymbol{E}_0$$

6.2 Calculate the scattering dyadic $\mathbf{S}$ in the long wavelength limit for scattering of an incident plane wave by a perfectly conducting sphere of radius a ($\mu = \epsilon = 1$ outside the sphere). The incident plane wave is

$$\boldsymbol{E}_{\mathrm{i}}(\boldsymbol{r}) = \boldsymbol{E}_0 \mathrm{e}^{\mathrm{i}k\hat{\boldsymbol{k}}_{\mathrm{i}}\cdot\boldsymbol{r}}$$

Calculate also the total scattering cross section $\sigma_{\mathrm{s}}|_{\mathrm{unpol}}$ and the degree of polarization $P|_{\mathrm{unpol}}$ for an unpolarized incident plane wave.
Hint: First, calculate the equivalent electric and magnetic dipole moments of the sphere, $\boldsymbol{p}$ and $\boldsymbol{m}$, respectively, for the incident excitation $\boldsymbol{E}_0$

$$\boldsymbol{p} = 4\pi\epsilon_0 a^3 \boldsymbol{E}_0$$

$$\boldsymbol{m} = -\frac{2\pi}{\eta_0} a^3 \hat{\boldsymbol{k}}_0 \times \boldsymbol{E}_0$$

Section 6.3 High-Frequency Approximation

6.3 A metallic reflector has the shape of a thin, rectangular screen (sides a and b), see Figure 6.20. The reflector is fed a plane wave at normal incidence and the electric field polarized parallel to the side a, i.e.,

$$\boldsymbol{E}_{\mathrm{i}}(\boldsymbol{r}) = \hat{\boldsymbol{x}} E_0 \mathrm{e}^{-\mathrm{i}kz}$$

Compute the far field amplitude $\boldsymbol{F}(\hat{\boldsymbol{r}})$ with the physical optics approximation, the extinction cross section σ_{ext} by the use of the optical theorem, and the total scattering cross section σ_{s}.

***6.4** A metallic reflector has the shape of a plane, thin, circular disk (radius a), see Figure 6.21. A plane wave impinges normally on the reflector, i.e.,

$$\boldsymbol{E}_{\mathrm{i}}(\boldsymbol{r}) = \boldsymbol{E}_0 \mathrm{e}^{-\mathrm{i}kz}$$

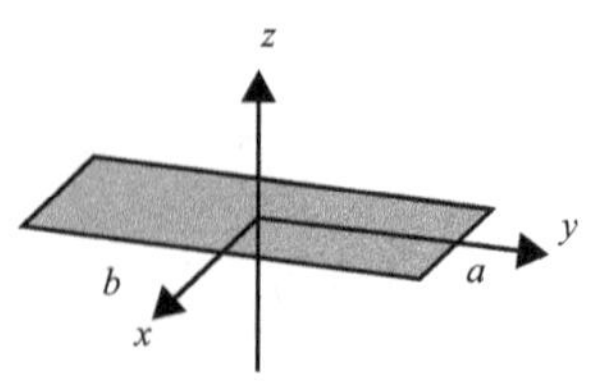

Figure 6.20 Geometry for Problem 6.3.

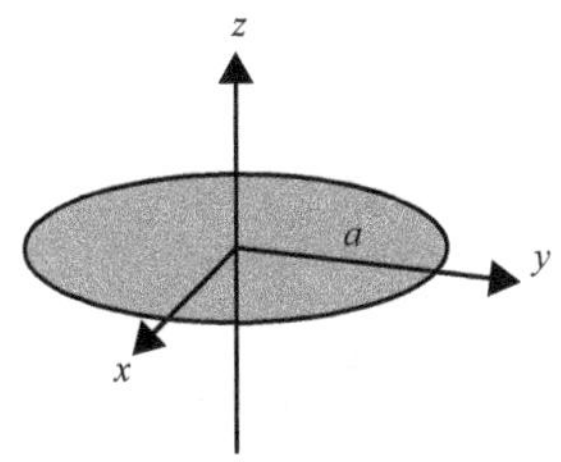

Figure 6.21 Geometry in Problem 6.4.

Compute the far field amplitude $\boldsymbol{F}(\hat{\boldsymbol{r}})$ by the physical optics approximation, the extinction cross section σ_{ext} by the use of the optical theorem, and the total scattering cross section σ_{s}.

6.5 Compute the far field amplitude $\boldsymbol{F}(\hat{\boldsymbol{r}})$ in the forward direction of the reflector $\hat{\boldsymbol{r}} = \hat{\boldsymbol{z}}$ for the same geometry and incident field as in Problem 6.4, but with the geometrical optics approximation.

*__6.6__ The far field amplitude in Example 6.7 (PO approximation) is

$$\boldsymbol{F}_{\text{PO}}\left(\hat{\boldsymbol{r}}\right) = \mathrm{i}\frac{kE_0 h}{2\pi}\mathrm{e}^{\mathrm{i}kF(1+\cos\theta)}\left(\hat{\boldsymbol{\theta}}\cos\theta\sin\phi + \hat{\boldsymbol{\phi}}\cos\phi\right)\frac{\sin\left(\frac{kh}{2}\sin\theta\sin\phi\right)}{\frac{kh}{2}\sin\theta\sin\phi} \cdot \int_{-b/2}^{b/2}\sqrt{\frac{8F}{\mathrm{i}k\pi\left(x^2+4F^2\right)}}\mathrm{e}^{\mathrm{i}k\left(\frac{x^2}{4F}(1-\cos\theta)-x\sin\theta\cos\phi\right)}\,\mathrm{d}x$$

Evaluate the remaining integral asymptotically under the assumption that the width b is large compared to the wavelength and that $\theta \neq 0, \pi$ or $\phi = \pi/2$, i.e., under the assumption that $kb\sin\theta\cos\phi \gg 1$.

*__6.7__ The far field amplitude in Example 6.11 (GO approximation) is

$$\boldsymbol{F}_{\text{GO}}\left(\hat{\boldsymbol{r}}\right) = -\mathrm{i}E_0\frac{kh}{4\pi}\mathrm{e}^{\mathrm{i}k\xi(1-\cos\theta)+2\mathrm{i}kF}\frac{\sin\left(\frac{kh}{2}\sin\theta\sin\phi\right)}{\frac{kh}{2}\sin\theta\sin\phi}\hat{\boldsymbol{r}}\times\left[\left(\hat{\boldsymbol{z}}+\hat{\boldsymbol{r}}\right)\times\hat{\boldsymbol{y}}\right] \cdot \int_{-b/2}^{b/2}\sqrt{\frac{8F}{\mathrm{i}k\pi\left(x^2+4F^2\right)}}\mathrm{e}^{-\mathrm{i}kx\sin\theta\cos\phi}\,\mathrm{d}x$$

Evaluate the remaining integral asymptotically under the assumption that the width b is large compared to the wavelength and that $\theta \neq 0, \pi$ or $\phi = \pi/2$, i.e., under the assumption that $kb\sin\theta\cos\phi \gg 1$.

*__6.8__ Evaluate the far field amplitude for the corner reflector in Example 6.10 with the physical optics approximation.

6.9 Compute the far field amplitude in the backscattering direction using the physical optics approximation for an incident plane wave impinging on a perfectly conducting sphere

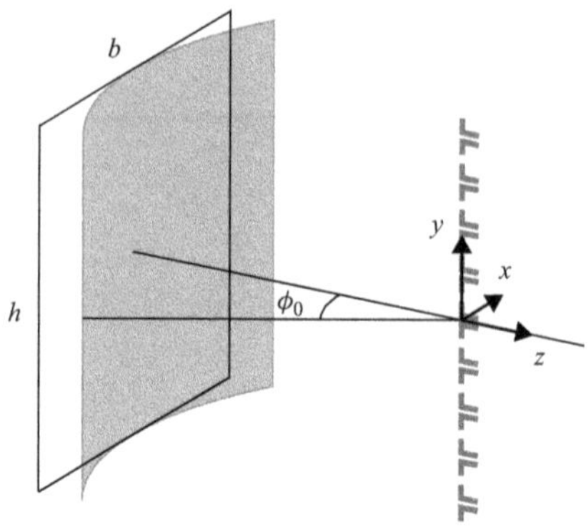

Figure 6.22 The geometry of the circular reflector (a sector of a circular cylinder) in Problem 6.13. The cylinder has the opening angle ϕ_0.

of radius a ($\mu = \epsilon = 1$ in the surrounding material). The incident plane wave is

$$\boldsymbol{E}_\mathrm{i}(\boldsymbol{r}) = \boldsymbol{E}_0 \mathrm{e}^{\mathrm{i}k\hat{\boldsymbol{k}}_\mathrm{i}\cdot\boldsymbol{r}}$$

6.10 Show that

$$\left|\hat{\boldsymbol{r}} \times \left[(\hat{\boldsymbol{z}} \pm \hat{\boldsymbol{r}}) \times \boldsymbol{A}\right]\right|^2 = (1 \pm \cos\theta)^2 |\boldsymbol{A}|^2$$

where $\hat{\boldsymbol{r}} \cdot \hat{\boldsymbol{z}} = \cos\theta$ and $\hat{\boldsymbol{z}} \cdot \boldsymbol{A} = 0$.

6.11 Compute the far field amplitude $\boldsymbol{F}_\mathrm{PO}(\hat{\boldsymbol{z}})$ in the forward scattering direction, $\theta = 0$, using the physical optics approximation for the parabolic reflector described in Example 6.8.

6.12 Compute the far field amplitude $\boldsymbol{F}_\mathrm{GO}(\hat{\boldsymbol{z}})$ in the forward scattering direction, $\theta = 0$, using the geometrical optics approximation for the parabolic reflector described in Example 6.12.

6.13 Use the physical optics approximation to analyze the far field amplitude of a perfectly conducting circular cylindrical reflector with radius a, height h, and opening angle ϕ_0. The reflector is fed by a line source located at the focal line of the circular reflector, see Figure 6.22. Make a comparison with the results obtained for the perfectly conducting cylindrical parabolic reflector in Example 6.7.

Section 6.4 Sum Rule for the Extinction Cross Section

6.14 Show that for TE and TM polarizations

$$\hat{\boldsymbol{p}}_\mathrm{e}^* \cdot \boldsymbol{\gamma}_\mathrm{e} \cdot \hat{\boldsymbol{p}}_\mathrm{e} + \hat{\boldsymbol{p}}_\mathrm{m}^* \cdot \boldsymbol{\gamma}_\mathrm{m} \cdot \hat{\boldsymbol{p}}_\mathrm{m} = \frac{8a^3}{3} \begin{cases} 1 + \cos^2\alpha & \text{(TE)} \\ 2\cos^2\alpha & \text{(TM)} \end{cases}$$

for a perfectly conducting disk of radius a located in the x–y-plane. The angle of incidence is denoted α.

Section 6.5 Scattering by Many Scatterers

6.15 Prove the identity (6.76)

$$\prod_{p=1}^{N}\left(1-\Theta_p(\boldsymbol{r})\right)+\sum_{p=1}^{N}\Theta_p(\boldsymbol{r})=1,\qquad \boldsymbol{r}\in\mathbb{R}^3$$

where

$$\Theta_p(\boldsymbol{r})=\begin{cases}1, & \boldsymbol{r}\in V_{\text{s}p}\\ 0, & \boldsymbol{r}\notin V_{\text{s}p}\end{cases}\qquad p=1,2,\ldots,N$$

CHAPTER 7

Spherical Vector Waves

There is a special interest in finding solutions of the Maxwell equations in source-free, homogeneous, isotropic materials in the spherical coordinate system (r, θ, ϕ). The reason for this is at least twofold: (1) we aim at developing efficient tools to solve scattering problems with spherical symmetries, and (2) outside the circumscribed sphere of the scatterer, the fields are naturally expanded in vector waves with spherical symmetries—we have already seen in Chapter 4 that the scattered field in the far zone is a spherical wave. For these reasons, we study solutions to the source-free Maxwell equations

$$\begin{cases} \nabla \times \boldsymbol{E}(\boldsymbol{r}, \omega) = \mathrm{i}k\eta_0\eta \boldsymbol{H}(\boldsymbol{r}, \omega) \\ \eta_0\eta\nabla \times \boldsymbol{H}(\boldsymbol{r}, \omega) = -\mathrm{i}k\boldsymbol{E}(\boldsymbol{r}, \omega) \end{cases}$$

in the spherical coordinate system (r, θ, ϕ). The material parameters, explicitly given by the wave number k and the wave impedance η, are assumed to be constants in terms of the spatial variables (although they may depend on the angular frequency ω). In the general case, these constants can be complex numbers, but in many applications, they are real valued. This latter situation occurs when finding solutions outside the circumscribed sphere of the scatterer in a lossless surrounding material.

We eliminate the field $\boldsymbol{H}$, and we get

$$\nabla \times (\nabla \times \boldsymbol{E}(\boldsymbol{r})) - k^2\boldsymbol{E}(\boldsymbol{r}) = \boldsymbol{0} \tag{7.1}$$

where, as above, $k^2 = k_0^2\epsilon\mu$. This equation is the starting point for the development of the special solutions that we are aiming at in this chapter.

In Section 7.1, we motivate the particular form of the vector-valued solutions to the vector Helmholtz equation used in this textbook. These solutions—spherical vector waves—are formally defined in Section 7.2. Some of the properties of the solutions are analyzed in Sections 7.3 and 7.4, and the expansion of the electric Green dyadic in spherical vector waves is derived in Section 7.5. The Null-field equations are presented in Section 7.6. The remaining sections in this chapter contain some results and consequences that are used in the subsequent chapters. Specifically, Sections 7.7 and 7.8 contain the expansion of the sources and the far field amplitude in the spherical vector waves, respectively. We conclude with an investigation of the relation between transition matrix entries and the dipole moments of a single scatterer in Section 7.9.

7.1 Preparatory Discussions

Our goal in this section is to construct a set of vector-valued solutions to the Maxwell equations that can serve as a basis for the construction of more general solutions to electromagnetic scattering problems. We start from scratch in our quest for suitable solutions, which gives a somewhat detailed analysis. The more traditional route is found in Comment 7.1.

To find a solution in a spherical region, we utilize the vector spherical harmonics. They are, see Appendix C.4

$$
\boldsymbol{A}_{\tau\sigma ml}(\hat{\boldsymbol{r}}) = \boldsymbol{A}_{\tau n}(\hat{\boldsymbol{r}}) = \boldsymbol{A}_n(\hat{\boldsymbol{r}}) \qquad \begin{cases} \tau = 1, 2, 3 \\ \sigma = \mathrm{e}, \mathrm{o} \\ m = 0, 1, \dots, l-1, l \\ l = 0, 1, 2, 3, \dots \end{cases}
$$

where the unit vector $\hat{\boldsymbol{r}}$ in spherical angles θ and ϕ is

$$
\hat{\boldsymbol{r}} = \hat{\boldsymbol{x}} \sin\theta \cos\phi + \hat{\boldsymbol{y}} \sin\theta \sin\phi + \hat{\boldsymbol{z}} \cos\theta
$$

The index n is a multi-index that consists of three or four different indices, i.e., $n = \sigma ml$ or $n = \tau\sigma ml$, depending on the context, where $\tau = 1, 2, 3$, $\sigma = \mathrm{e}, \mathrm{o}$, $m = 0, 1, 2, \dots, l$, and $l = 0, 1, 2, \dots$. The functions are defined and described in Appendix C.4, and we assume the reader is familiar with these functions—if not, please consult the literature [4, 195]. Their definitions in terms of the spherical harmonics, $Y_n(\hat{\boldsymbol{r}})$, are, see (C.6) on page 630

$$
\begin{cases}
\boldsymbol{A}_{1n}(\hat{\boldsymbol{r}}) = \dfrac{1}{\sqrt{l(l+1)}} \nabla \times \left(\boldsymbol{r} Y_n(\hat{\boldsymbol{r}})\right) = \dfrac{1}{\sqrt{l(l+1)}} \nabla Y_n(\hat{\boldsymbol{r}}) \times \boldsymbol{r} \\
\boldsymbol{A}_{2n}(\hat{\boldsymbol{r}}) = \dfrac{1}{\sqrt{l(l+1)}} r \nabla Y_n(\hat{\boldsymbol{r}}) \\
\boldsymbol{A}_{3n}(\hat{\boldsymbol{r}}) = \hat{\boldsymbol{r}} Y_n(\hat{\boldsymbol{r}})
\end{cases}
$$

For $l = 0$ the vector spherical harmonics, $\boldsymbol{A}_{1\sigma 00}(\hat{\boldsymbol{r}}) = \boldsymbol{A}_{2\sigma 00}(\hat{\boldsymbol{r}}) = \boldsymbol{0}$, by definition, and only $\boldsymbol{A}_{3\mathrm{e}00}(\hat{\boldsymbol{r}}) \neq \boldsymbol{0}$. The vector spherical harmonics are orthonormal on the unit sphere, i.e.,

$$
\iint_{\Omega} \boldsymbol{A}_n(\hat{\boldsymbol{r}}) \cdot \boldsymbol{A}_{n'}(\hat{\boldsymbol{r}})\, \mathrm{d}\Omega = \iint_{\Omega} \boldsymbol{A}_{\tau\sigma ml}(\hat{\boldsymbol{r}}) \cdot \boldsymbol{A}_{\tau'\sigma'm'l'}(\hat{\boldsymbol{r}})\, \mathrm{d}\Omega = \delta_{\tau\tau'}\delta_{\sigma\sigma'}\delta_{mm'}\delta_{ll'} = \delta_{nn'}
$$

where the surface measure on the unit sphere Ω is $\mathrm{d}\Omega = \sin\theta\, \mathrm{d}\theta\, \mathrm{d}\phi$. Other important properties of the vector spherical harmonics are:

$$
\begin{cases}
\hat{\boldsymbol{r}} \cdot \boldsymbol{A}_{\tau n}(\hat{\boldsymbol{r}}) = 0, \quad \tau = 1, 2 \\
\hat{\boldsymbol{r}} \times \boldsymbol{A}_{3n}(\hat{\boldsymbol{r}}) = \boldsymbol{0}
\end{cases}
\qquad
\begin{cases}
\boldsymbol{A}_{1n}(\hat{\boldsymbol{r}}) = \boldsymbol{A}_{2n}(\hat{\boldsymbol{r}}) \times \hat{\boldsymbol{r}} \\
\boldsymbol{A}_{2n}(\hat{\boldsymbol{r}}) = \hat{\boldsymbol{r}} \times \boldsymbol{A}_{1n}(\hat{\boldsymbol{r}})
\end{cases}
\tag{7.2}
$$

More details on the vector spherical harmonics, $\boldsymbol{A}_{\tau\sigma ml}(\hat{\boldsymbol{r}})$, are collected in Appendix C.4. Moreover, every square integrable vector-valued function, $\boldsymbol{F}(\hat{\boldsymbol{r}})$, defined on the unit sphere has a convergent generalized Fourier expansion in the vector spherical harmonics

$$\boldsymbol{F}(\hat{\boldsymbol{r}}) = \sum_{\tau=1}^{3} \sum_{\sigma=\mathrm{e,o}} \sum_{l=0}^{\infty} \sum_{m=0}^{l} a_{\tau\sigma ml} \boldsymbol{A}_{\tau\sigma ml}(\hat{\boldsymbol{r}}) = \sum_{n} a_n \boldsymbol{A}_n(\hat{\boldsymbol{r}}), \qquad \theta \in [0, \pi], \ \phi \in [0, 2\pi)$$

where the (Fourier) coefficients a_n are determined by the integral

$$a_n = \iint_{\Omega} \boldsymbol{F}(\hat{\boldsymbol{r}}) \cdot \boldsymbol{A}_n(\hat{\boldsymbol{r}}) \, \mathrm{d}\Omega$$

Comment 7.1

The construction of the spherical vector waves in this section follows a somewhat unorthodox route. The reason for this is to give a more comprehensive analysis compared to the traditional way of analysis presentation in e.g., Reference 196, where the construction of the spherical vector waves starts by the construction of the first spherical vector wave, which is

$$\boldsymbol{F}(\boldsymbol{r}) = \nabla \times (\boldsymbol{r}\psi(\boldsymbol{r})) = \nabla\psi(\boldsymbol{r}) \times \boldsymbol{r}$$

where $\psi(\boldsymbol{r})$ satisfies the scalar Helmholtz equation

$$\nabla^2\psi(\boldsymbol{r}) + k^2\psi(\boldsymbol{r}) = 0$$

In fact, use e.g., rule (11) for the nabla operator on the inside of the first cover, to obtain

$$\begin{aligned} \nabla \times (\nabla \times \boldsymbol{F}) &= -\nabla \times \{\nabla \times (\boldsymbol{r} \times \nabla\psi)\} \\ &= -\nabla \times \Big\{ -\nabla(\boldsymbol{r} \cdot \nabla\psi) + 2(\nabla\psi \cdot \nabla)\boldsymbol{r} + \boldsymbol{r} \times \underbrace{(\nabla \times \nabla\psi)}_{=0} \\ &\qquad + \psi \times \underbrace{(\nabla \times \boldsymbol{r})}_{=0} + \boldsymbol{r}\nabla^2\psi - \nabla\psi\underbrace{(\nabla \cdot \boldsymbol{r})}_{=3} \Big\} \end{aligned}$$

where we also used the rules (14), (18), and (20) for the nabla operator. The first term disappears due to the rule (14), and we obtain

$$\nabla \times (\nabla \times \boldsymbol{F}) = -\nabla \times \left(2(\nabla\psi \cdot \nabla)\boldsymbol{r} + \boldsymbol{r}\nabla^2\psi\right)$$

Finally, use (23) and the differential equation for ψ above to get

$$\nabla \times (\nabla \times \boldsymbol{F}) = -\nabla \times \left(2\nabla\psi - k^2\boldsymbol{r}\psi\right) = k^2\nabla\psi \times \boldsymbol{r} = k^2\boldsymbol{F}$$

A second spherical vector wave, $\boldsymbol{G}(\boldsymbol{r})$, is then constructed by a repeated use of the curl, i.e.,

$$\boldsymbol{G}(\boldsymbol{r}) = \frac{1}{k}\nabla \times \boldsymbol{F}(\boldsymbol{r}) \quad \Rightarrow \quad \nabla \times (\nabla \times \boldsymbol{G}(\boldsymbol{r})) = k^2\boldsymbol{G}(\boldsymbol{r})$$

We return to the original problem of this section, and suppose the Maxwell equations are satisfied in a region in space bounded by two concentric spheres of radii $r_1 < r_2$, i.e., in the region $r_1 \leq r \leq r_2$, where the radii $0 \leq r_1 < r_2 \leq \infty$. If $r_1 = 0$, the region is a sphere of radius r_2, and if $r_2 = \infty$, then the region is the volume outside the sphere of radius r_1. To find a solution in this region, we utilize the completeness of the vector spherical harmonics $\boldsymbol{A}_{\tau n}(\hat{\boldsymbol{r}})$, and expand the electric field for a fixed radius $r_1 \leq r \leq r_2$ as

$$\boldsymbol{E}(\boldsymbol{r}) = \sum_n \left\{ f_{1n}(r)\boldsymbol{A}_{1n}(\hat{\boldsymbol{r}}) + f_{2n}(r)\boldsymbol{A}_{2n}(\hat{\boldsymbol{r}}) + f_{3n}(r)\boldsymbol{A}_{3n}(\hat{\boldsymbol{r}}) \right\}$$

The expansion coefficients $f_{\tau n}$ depend in general on the radial coordinate r, since we expect different expansion coefficients for different r, but there is no dependence on the direction $\hat{\boldsymbol{r}}$.

Assuming the series converges uniformly in the angular variables, we can differentiate inside the summation sign. We are helped by the curls of the vector spherical harmonics, see (C.15) on page 633

$$\begin{cases} \nabla \times \boldsymbol{A}_{1n}(\hat{\boldsymbol{r}}) = \dfrac{\boldsymbol{A}_{2n}(\hat{\boldsymbol{r}}) + \sqrt{l(l+1)}\boldsymbol{A}_{3n}(\hat{\boldsymbol{r}})}{r} \\ \nabla \times \boldsymbol{A}_{2n}(\hat{\boldsymbol{r}}) = -\dfrac{\boldsymbol{A}_{1n}(\hat{\boldsymbol{r}})}{r} \\ \nabla \times \boldsymbol{A}_{3n}(\hat{\boldsymbol{r}}) = \dfrac{\sqrt{l(l+1)}\boldsymbol{A}_{1n}(\hat{\boldsymbol{r}})}{r} \end{cases}$$

The use of (7.2), then implies that the curl of the expansion of the electric field has the following form:

$$\begin{aligned} \nabla \times \boldsymbol{E}(\boldsymbol{r}) &= \sum_n \left(\nabla \times \left(f_{1n}(r)\boldsymbol{A}_{1n}(\hat{\boldsymbol{r}}) + f_{2n}(r)\boldsymbol{A}_{2n}(\hat{\boldsymbol{r}}) + f_{3n}(r)\boldsymbol{A}_{3n}(\hat{\boldsymbol{r}}) \right) \right) \\ &= \sum_n \left(f'_{1n}(r)\boldsymbol{A}_{2n}(\hat{\boldsymbol{r}}) - f'_{2n}(r)\boldsymbol{A}_{1n}(\hat{\boldsymbol{r}}) + f_{1n}(r)\frac{\boldsymbol{A}_{2n}(\hat{\boldsymbol{r}}) + \sqrt{l(l+1)}\boldsymbol{A}_{3n}(\hat{\boldsymbol{r}})}{r} \right. \\ &\quad \left. - f_{2n}(r)\frac{\boldsymbol{A}_{1n}(\hat{\boldsymbol{r}})}{r} + f_{3n}(r)\frac{\sqrt{l(l+1)}\boldsymbol{A}_{1n}(\hat{\boldsymbol{r}})}{r} \right) \end{aligned}$$

We collect terms

$$\begin{aligned} \nabla \times \boldsymbol{E}(\boldsymbol{r}) = \sum_n \Bigg(&\frac{(rf_{1n}(r))'}{r}\boldsymbol{A}_{2n}(\hat{\boldsymbol{r}}) - \frac{(rf_{2n}(r))' - f_{3n}(r)\sqrt{l(l+1)}}{r}\boldsymbol{A}_{1n}(\hat{\boldsymbol{r}}) \\ &+ \frac{f_{1n}(r)\sqrt{l(l+1)}}{r}\boldsymbol{A}_{3n}(\hat{\boldsymbol{r}}) \Bigg) \end{aligned}$$

Another application of the curl operator implies

$$
\begin{aligned}
\nabla\times(\nabla\times\boldsymbol{E}(\boldsymbol{r})) = \sum_n \Bigg\{ &-\frac{\mathrm{d}}{\mathrm{d}r}\left(\frac{(rf_{1n}(r))'}{r}\right)\boldsymbol{A}_{1n}(\hat{\boldsymbol{r}}) - \frac{(rf_{1n}(r))'}{r^2}\boldsymbol{A}_{1n}(\hat{\boldsymbol{r}}) \\
&- \frac{\mathrm{d}}{\mathrm{d}r}\left(\frac{(rf_{2n}(r))' - f_{3n}(r)\sqrt{l(l+1)}}{r}\right)\boldsymbol{A}_{2n}(\hat{\boldsymbol{r}}) + l(l+1)\frac{f_{1n}(r)}{r^2}\boldsymbol{A}_{1n}(\hat{\boldsymbol{r}}) \\
&- \frac{(rf_{2n}(r))' - f_{3n}(r)\sqrt{l(l+1)}}{r^2}\left(\boldsymbol{A}_{2n}(\hat{\boldsymbol{r}}) + \sqrt{l(l+1)}\boldsymbol{A}_{3n}(\hat{\boldsymbol{r}})\right)\Bigg\}
\end{aligned}
$$

Collecting terms, simplifying, and insertion in (7.1) give

$$
\begin{aligned}
&\nabla\times(\nabla\times\boldsymbol{E}(\boldsymbol{r})) - k^2\boldsymbol{E}(\boldsymbol{r}) \\
&= \sum_n \Bigg\{ -\frac{1}{r}\frac{\mathrm{d}}{\mathrm{d}r}((rf_{1n}(r))')\boldsymbol{A}_{1n}(\hat{\boldsymbol{r}}) + l(l+1)\frac{f_{1n}(r)}{r^2}\boldsymbol{A}_{1n}(\hat{\boldsymbol{r}}) \\
&\quad - \frac{1}{r}\frac{\mathrm{d}}{\mathrm{d}r}\left((rf_{2n}(r))' - f_{3n}(r)\sqrt{l(l+1)}\right)\boldsymbol{A}_{2n}(\hat{\boldsymbol{r}}) \\
&\quad - \frac{(rf_{2n}(r))' - f_{3n}(r)\sqrt{l(l+1)}}{r^2}\sqrt{l(l+1)}\boldsymbol{A}_{3n}(\hat{\boldsymbol{r}})\Bigg\} \\
&\quad - k^2\sum_n \left\{f_{1n}(r)\boldsymbol{A}_{1n}(\hat{\boldsymbol{r}}) + f_{2n}(r)\boldsymbol{A}_{2n}(\hat{\boldsymbol{r}}) + f_{3n}(r)\boldsymbol{A}_{3n}(\hat{\boldsymbol{r}})\right\} = \boldsymbol{0}
\end{aligned}
$$

Since the vector spherical harmonics are orthogonal, the radial functions have to satisfy (suppress the index n)

$$
\begin{cases}
r(rf_1(r))'' - l(l+1)f_1(r) + k^2r^2f_1(r) = 0 \\
r(rf_2(r))'' - \sqrt{l(l+1)}\,rf_3'(r) + k^2r^2f_2(r) = 0 \\
\sqrt{l(l+1)}(rf_2(r))' - l(l+1)f_3(r) + k^2r^2f_3(r) = 0
\end{cases}
\tag{7.3}
$$

We observe that the first function f_1 does not couple to the other two functions, f_2 and f_3. However, f_2 and f_3 couple to each other, and they depend on each other. In fact, the following relation between $f_2(r)$ and $f_3(r)$ in (7.3) eliminates $f_2(r)$ in the second and third equations

$$
f_2(r) = \frac{\left(r^2f_3(r)\right)'}{r\sqrt{l(l+1)}}
\tag{7.4}
$$

since then the second and third equations become

$$
\begin{cases}
r\left(r^2f_3(r)\right)''' - l(l+1)\,rf_3'(r) + k^2r^2\dfrac{\left(r^2f_3(r)\right)'}{r} = 0 \\
\left(r^2f_3(r)\right)'' - l(l+1)f_3(r) + k^2r^2f_3(r) = 0
\end{cases}
$$

or

$$\begin{cases} r\dfrac{\mathrm{d}}{\mathrm{d}r}\left(\left(r^2 f_3(r)\right)'' - l(l+1)f_3(r) + k^2r^2 f_3(r)\right) = 0 \\ \left(r^2 f_3(r)\right)'' - l(l+1)f_3(r) + k^2r^2 f_3(r) = 0 \end{cases}$$

and the last equation implies the first.

The first equation in (7.3) is the differential equation for spherical Bessel functions, i.e.,

$$\frac{\mathrm{d}}{\mathrm{d}z}\left(z^2\frac{\mathrm{d}Z_l(z)}{\mathrm{d}z}\right) + (z^2 - l(l+1))Z_l(z) = 0$$

Details about these functions are listed in Appendix B.3. It is also convenient to identify the differential equation for a spherical Bessel function times a power, i.e., $F_{nl}(z) = z^n Z_l(z)$, where $n = 0, \pm 1, \pm 2, \ldots$ is an integer. The function $F_{nl}(z)$ satisfies, see (B.24) on page 617

$$\frac{\mathrm{d}}{\mathrm{d}z}\left(z^2\frac{\mathrm{d}F_{nl}(z)}{\mathrm{d}z}\right) - 2nz\frac{\mathrm{d}F_{nl}(z)}{\mathrm{d}z} + (z^2 - l(l+1) + n(n-1))F_{nl}(z) = 0$$

From this equation, we see that $f_3(r) = F_{-1l}(kr)$.

We summarize, the radial functions have to be a combination of[1]

$$\begin{cases} f_{1n}(r) = j_l(kr) \text{ and/or } h_l^{(1)}(kr) \\ f_{2n}(r) = \dfrac{(krj_l(kr))'}{kr} \text{ and/or } \dfrac{(krh_l^{(1)}(kr))'}{kr} \\ f_{3n}(r) = \sqrt{l(l+1)}\dfrac{j_l(kr)}{kr} \text{ and/or } \sqrt{l(l+1)}\dfrac{h_l^{(1)}(kr)}{kr} \end{cases}$$

where the solutions f_{2n} and f_{3n} are tied together by (7.4). Notice that the prime here denotes differentiation w.r.t. the whole argument of the functions, i.e., kr.

Comment 7.2

Under certain circumstances, it is convenient to group the first two functions into a common radial function, $V_{\tau l}(kr)$ defined as, see Reference 92:

$$V_{\tau l}(kr) = \begin{cases} j_l(kr), & \tau = 1 \\ \dfrac{(krj_l(kr))'}{kr}, & \tau = 2 \end{cases}$$

In some circumstances, this notation leads to more compact formulas.

An alternative set of notation for the radial function, that is commonly used in the literature, and also used frequently in this textbook, is the Riccati–Bessel functions, $\psi_l(kr)$ and $\xi_l(kr)$,

[1]Another alternative is to use $h_l^{(1)}(kr)$ and $h_l^{(2)}(kr)$ as the two independent solutions.

see Appendix B.3.3 on page 617 and below. The radial functions $V_{\tau l}(kr)$ expressed in these functions are

$$V_{\tau l}(kr) = \begin{cases} \dfrac{\psi_l(kr)}{kr}, & \tau = 1 \\ \dfrac{\psi_l'(kr)}{kr}, & \tau = 2 \end{cases} = \frac{1}{kr}\frac{\mathrm{d}^{\tau-1}\psi_l(kr)}{\mathrm{d}(kr)^{\tau-1}}$$

In this textbook, depending on the case, we use different notations, with the ambition to obtain compact expressions. To make this set of radial function complete, also define the out-going or radiation radial functions.

$$U_{\tau l}(kr) = \begin{cases} \dfrac{\xi_l(kr)}{kr}, & \tau = 1 \\ \dfrac{\xi_l'(kr)}{kr}, & \tau = 2 \end{cases} = \frac{1}{kr}\frac{\mathrm{d}^{\tau-1}\xi_l(kr)}{\mathrm{d}(kr)^{\tau-1}}$$

We are now ready to define two sets of spherical vector solutions to the Maxwell equations in spherical regions. This is done in the next section.

7.2 Definition of Spherical Vector Waves

Following the results in Section 7.1, we define the out-going, or radiating, spherical vector waves, $\boldsymbol{u}_{\tau\sigma ml}(k\boldsymbol{r})$ as

$$\begin{cases} \boldsymbol{u}_{1n}(k\boldsymbol{r}) = h_l^{(1)}(kr)\boldsymbol{A}_{1n}(\hat{\boldsymbol{r}}) \\ \boldsymbol{u}_{2n}(k\boldsymbol{r}) = \dfrac{(krh_l^{(1)}(kr))'}{kr}\boldsymbol{A}_{2n}(\hat{\boldsymbol{r}}) + \sqrt{l(l+1)}\dfrac{h_l^{(1)}(kr)}{kr}\boldsymbol{A}_{3n}(\hat{\boldsymbol{r}}) \end{cases} \tag{7.5}$$

Below, in Section 7.8, the rationale for the name out-going or radiating spherical vector waves is provided.

The regular spherical vector waves $\boldsymbol{v}_{\tau\sigma ml}(k\boldsymbol{r})$ are defined as

$$\begin{cases} \boldsymbol{v}_{1n}(k\boldsymbol{r}) = j_l(kr)\boldsymbol{A}_{1n}(\hat{\boldsymbol{r}}) \\ \boldsymbol{v}_{2n}(k\boldsymbol{r}) = \dfrac{(krj_l(kr))'}{kr}\boldsymbol{A}_{2n}(\hat{\boldsymbol{r}}) + \sqrt{l(l+1)}\dfrac{j_l(kr)}{kr}\boldsymbol{A}_{3n}(\hat{\boldsymbol{r}}) \end{cases} \tag{7.6}$$

For the convenience in many scattering applications, these definitions are often expressed in terms of the Riccati–Bessel functions, i.e.,

$$\begin{cases} \boldsymbol{u}_{1n}(k\boldsymbol{r}) = \dfrac{\xi_l(kr)}{kr}\boldsymbol{A}_{1n}(\hat{\boldsymbol{r}}) \\ \boldsymbol{u}_{2n}(k\boldsymbol{r}) = \dfrac{\xi_l'(kr)}{kr}\boldsymbol{A}_{2n}(\hat{\boldsymbol{r}}) + \sqrt{l(l+1)}\dfrac{\xi_l(kr)}{(kr)^2}\boldsymbol{A}_{3n}(\hat{\boldsymbol{r}}) \end{cases}$$

and

$$\begin{cases} \boldsymbol{v}_{1n}(k\boldsymbol{r}) = \dfrac{\psi_l(kr)}{kr}\boldsymbol{A}_{1n}(\hat{\boldsymbol{r}}) \\ \boldsymbol{v}_{2n}(k\boldsymbol{r}) = \dfrac{\psi_l'(kr)}{kr}\boldsymbol{A}_{2n}(\hat{\boldsymbol{r}}) + \sqrt{l(l+1)}\dfrac{\psi_l(kr)}{(kr)^2}\boldsymbol{A}_{3n}(\hat{\boldsymbol{r}}) \end{cases}$$

where the Riccati–Bessel functions, $\psi_l(z)$ and $\xi_l(z)$, are defined in Appendix B.3.3—the explicit definitions are given in (B.26) on page 618.

The regular and out-going spherical vector waves are very similar—they only differ by the spherical Bessel or Hankel functions. Note that for $l = 0$, $\boldsymbol{u}_{1\sigma 00}(k\boldsymbol{r}) = \boldsymbol{u}_{2\sigma 00}(k\boldsymbol{r}) = \boldsymbol{v}_{1\sigma 00}(k\boldsymbol{r}) = \boldsymbol{v}_{2\sigma 00}(k\boldsymbol{r}) = \boldsymbol{0}$, so in both cases the indices l, m, and σ assume the following values:

$$l = 1, 2, 3, \ldots, \qquad m = 0, 1, \ldots, l-1, l, \qquad \sigma = \mathrm{e}, \mathrm{o}$$

Both these sets satisfy

$$\begin{cases} \nabla \times (\nabla \times \boldsymbol{u}_{\tau n}(k\boldsymbol{r})) = k^2 \boldsymbol{u}_{\tau n}(k\boldsymbol{r}) \\ \nabla \times (\nabla \times \boldsymbol{v}_{\tau n}(k\boldsymbol{r})) = k^2 \boldsymbol{v}_{\tau n}(k\boldsymbol{r}) \end{cases}$$

and, as a consequence of these equations

$$\begin{cases} \nabla \cdot \boldsymbol{u}_{\tau n}(k\boldsymbol{r}) = 0 \\ \nabla \cdot \boldsymbol{v}_{\tau n}(k\boldsymbol{r}) = 0 \end{cases}$$

Moreover, we find by direct calculations

$$\nabla \times \boldsymbol{v}_{1n}(k\boldsymbol{r}) = kj_l'(kr)\boldsymbol{A}_{2n}(\hat{\boldsymbol{r}}) + j_l(kr)\frac{\boldsymbol{A}_{2n}(\hat{\boldsymbol{r}}) + \sqrt{l(l+1)}\boldsymbol{A}_{3n}(\hat{\boldsymbol{r}})}{r} = k\boldsymbol{v}_{2n}(k\boldsymbol{r})$$

and

$$\nabla \times \boldsymbol{v}_{2n}(k\boldsymbol{r}) = \frac{1}{k}\nabla \times (\nabla \times \boldsymbol{v}_{1n}(k\boldsymbol{r})) = k\boldsymbol{v}_{1n}(k\boldsymbol{r})$$

This way we have shown that

$$\nabla \times \boldsymbol{v}_{\tau n}(k\boldsymbol{r}) = k\boldsymbol{v}_{\overline{\tau} n}(k\boldsymbol{r}) \tag{7.7}$$

where we have introduced the dual index to τ, defined by $\overline{1} = 2$ and $\overline{2} = 1$.

These results lead to an alternative definition of the regular spherical vector waves, $\boldsymbol{v}_{\tau\sigma ml}(k\boldsymbol{r})$, in terms of the Riccati–Bessel functions, $\psi_l(z)$, i.e.,

$$\begin{cases} \boldsymbol{v}_{1n}(k\boldsymbol{r}) = \dfrac{\psi_l(kr)}{kr}\boldsymbol{A}_{1n}(\hat{\boldsymbol{r}}) \\ \boldsymbol{v}_{2n}(k\boldsymbol{r}) = \dfrac{1}{k}\nabla \times \left(\dfrac{\psi_l(kr)}{kr}\boldsymbol{A}_{1n}(\hat{\boldsymbol{r}})\right) \end{cases} \tag{7.8}$$

A similar relation holds for the out-going or radiating spherical vector waves in terms of the Riccati–Bessel functions, $\xi_l(z)$. We get

$$\left\{\begin{aligned} \boldsymbol{u}_{1n}(k\boldsymbol{r}) &= \frac{\xi_l(kr)}{kr}\boldsymbol{A}_{1n}(\hat{\boldsymbol{r}}) \\ \boldsymbol{u}_{2n}(k\boldsymbol{r}) &= \frac{1}{k}\nabla\times\left(\frac{\xi_l(kr)}{kr}\boldsymbol{A}_{1n}(\hat{\boldsymbol{r}})\right) \end{aligned}\right. \tag{7.9}$$

The functions in (7.5) and (7.6) have zero divergence as stated above. These functions are often augmented by a third spherical vector wave, $\tau = 3$, that has non-zero divergence, and therefore do not enter in the expansions of the electric field in an isotropic, homogeneous, source-free region:

$$\left\{\begin{aligned} \boldsymbol{u}_{3n}(k\boldsymbol{r}) &= \frac{1}{k}\nabla\left(h_l^{(1)}(kr)Y_n(\hat{\boldsymbol{r}})\right) = \frac{1}{k}\nabla u_n(\boldsymbol{r}) \\ \boldsymbol{v}_{3n}(k\boldsymbol{r}) &= \frac{1}{k}\nabla\left(j_l(kr)Y_n(\hat{\boldsymbol{r}})\right) = \frac{1}{k}\nabla v_n(\boldsymbol{r}) \end{aligned}\right. \tag{7.10}$$

where the spherical scalar waves are defined as

$$\left\{\begin{aligned} u_n(k\boldsymbol{r}) &= h_l^{(1)}(kr)Y_n(\hat{\boldsymbol{r}}) \\ v_n(k\boldsymbol{r}) &= j_l(kr)Y_n(\hat{\boldsymbol{r}}) \end{aligned}\right. \tag{7.11}$$

This type of wave is called the longitudinal (irrotational) spherical vector waves, as opposed to the ones defined above, $\tau = 1, 2$, in (7.5) and (7.6), which are called the transverse spherical vector waves. These alternative definitions also imply a compact way of expressing the spherical vector waves. The result for the radiating spherical vector waves is

$$\left\{\begin{aligned} \boldsymbol{u}_{1n}(k\boldsymbol{r}) &= \frac{1}{\sqrt{l(l+1)}}\nabla\times(\boldsymbol{r}u_n(k\boldsymbol{r})) \\ \boldsymbol{u}_{2n}(k\boldsymbol{r}) &= \frac{1}{k\sqrt{l(l+1)}}\nabla\times(\nabla\times(\boldsymbol{r}u_n(k\boldsymbol{r}))) \\ \boldsymbol{u}_{3n}(k\boldsymbol{r}) &= \frac{1}{k}\nabla u_n(k\boldsymbol{r}) \end{aligned}\right.$$

and regular spherical vector waves $\boldsymbol{v}_{\tau n}(k\boldsymbol{r})$ is

$$\left\{\begin{aligned} \boldsymbol{v}_{1n}(k\boldsymbol{r}) &= \frac{1}{\sqrt{l(l+1)}}\nabla\times(\boldsymbol{r}v_n(k\boldsymbol{r})) \\ \boldsymbol{v}_{2n}(k\boldsymbol{r}) &= \frac{1}{k\sqrt{l(l+1)}}\nabla\times(\nabla\times(\boldsymbol{r}v_n(k\boldsymbol{r}))) \\ \boldsymbol{v}_{3n}(k\boldsymbol{r}) &= \frac{1}{k}\nabla v_n(k\boldsymbol{r}) \end{aligned}\right.$$

These relations are consequences of the differentiation rules of the nabla operator and the identity $\nabla \times \boldsymbol{r} = \boldsymbol{0}$

$$\nabla \times (\boldsymbol{r}u_n(kr)) = \nabla u_n(kr) \times \boldsymbol{r} = h_l^{(1)}(kr)\nabla Y_n(\hat{\boldsymbol{r}}) \times \boldsymbol{r} = \sqrt{l(l+1)}h_l^{(1)}(kr)\boldsymbol{A}_{1n}(\hat{\boldsymbol{r}})$$

To make the list complete, we repeat the definitions above, and get

$$\begin{cases} \boldsymbol{u}_{1n}(k\boldsymbol{r}) = h_l^{(1)}(kr)\boldsymbol{A}_{1n}(\hat{\boldsymbol{r}}) \\ \boldsymbol{u}_{2n}(k\boldsymbol{r}) = \dfrac{(krh_l^{(1)}(kr))'}{kr}\boldsymbol{A}_{2n}(\hat{\boldsymbol{r}}) + \sqrt{l(l+1)}\dfrac{h_l^{(1)}(kr)}{kr}\boldsymbol{A}_{3n}(\hat{\boldsymbol{r}}) \\ \boldsymbol{u}_{3n}(k\boldsymbol{r}) = \dfrac{1}{k}\nabla\left(h_l^{(1)}(kr)Y_n(\hat{\boldsymbol{r}})\right) = h_l^{(1)\prime}(kr)\boldsymbol{A}_{3n}(\hat{\boldsymbol{r}}) + \sqrt{l(l+1)}\dfrac{h_l^{(1)}(kr)}{kr}\boldsymbol{A}_{2n}(\hat{\boldsymbol{r}}) \end{cases}$$

and regular spherical vector waves $\boldsymbol{v}_{\tau n}(k\boldsymbol{r})$ as

$$\begin{cases} \boldsymbol{v}_{1n}(k\boldsymbol{r}) = j_l(kr)\boldsymbol{A}_{1n}(\hat{\boldsymbol{r}}) \\ \boldsymbol{v}_{2n}(k\boldsymbol{r}) = \dfrac{(krj_l(kr))'}{kr}\boldsymbol{A}_{2n}(\hat{\boldsymbol{r}}) + \sqrt{l(l+1)}\dfrac{j_l(kr)}{kr}\boldsymbol{A}_{3n}(\hat{\boldsymbol{r}}) \\ \boldsymbol{v}_{3n}(k\boldsymbol{r}) = \dfrac{1}{k}\nabla\left(j_l(kr)Y_n(\hat{\boldsymbol{r}})\right) = j_l'(kr)\boldsymbol{A}_{3n}(\hat{\boldsymbol{r}}) + \sqrt{l(l+1)}\dfrac{j_l(kr)}{kr}\boldsymbol{A}_{2n}(\hat{\boldsymbol{r}}) \end{cases}$$

These two sets of regular and out-going spherical vector waves are each complete on the sphere, $r = a$, due to the completeness of the vector spherical harmonics, provided the spherical vector waves are linearly independent—we need three linearly independent spherical vector waves to represent three different vector components.

To prove linear independence, we form a linear combination of the three spherical vector waves equal to the zero vector, and prove that the coefficients must be zero, i.e.,

$$\sum_{\tau=1}^{3}\sum_{\sigma=\mathrm{e,o}}\sum_{l=0}^{L}\sum_{m=0}^{l} c_{\tau\sigma ml}\boldsymbol{u}_{\tau\sigma ml}(k\boldsymbol{r}) = \boldsymbol{0}$$

for all truncations $L \geq 0$, see Appendix F.9.1.3. This implies, due to the orthogonality of the vector spherical harmonics, $\boldsymbol{A}_n(\hat{\boldsymbol{r}})$, on the unit sphere, that

$$\begin{pmatrix} h_l^{(1)}(kr) & 0 & 0 \\ 0 & \dfrac{(krh_l^{(1)}(kr))'}{kr} & \sqrt{l(l+1)}\dfrac{h_l^{(1)}(kr)}{kr} \\ 0 & \sqrt{l(l+1)}\dfrac{h_l^{(1)}(kr)}{kr} & h_l^{(1)\prime}(kr) \end{pmatrix}\begin{pmatrix} c_{1n} \\ c_{2n} \\ c_{3n} \end{pmatrix} = \begin{pmatrix} 0 \\ 0 \\ 0 \end{pmatrix}$$

Non-trivial solutions to this system of equations exist only if

$$
\begin{aligned}
0 &= \frac{h_l^{(1)}(kr)}{(kr)^2}\left(kr(krh_l^{(1)}(kr))'h_l^{(1)\prime}(kr) - l(l+1)\left(h_l^{(1)}(kr)\right)^2\right) \\
&= \frac{h_l^{(1)}(kr)}{(kr)^2}\left[krh_l^{(1)\prime}(kr) + (l+1)h_l^{(1)}(kr)\right]\left[krh_l^{(1)\prime}(kr) - lh_l^{(1)}(kr)\right] \\
&= -h_l^{(1)}(kr)h_{l-1}^{(1)}(kr)h_{l+1}^{(1)}(kr)
\end{aligned}
$$

where we used the recursion relations of the spherical Bessel and the Hankel functions (B.18) on page 611 in Appendix B.3. For a fixed value of $z = kr$, the condition above can never be fulfilled, since there are no zeros of $h_l^{(1)}(z)$ (l integer) in the upper half of the complex z-plane including the real axis, i.e., $\operatorname{Im} z \geq 0$, see Reference [1, Section 10.1.66].[2] Notice that the statement $\operatorname{Im} kr \geq 0$ is identical to the requirement that the medium is passive, since $r > 0$. Therefore, all coefficients $c_{\tau\sigma ml} = 0$, and the out-going spherical vector waves are independent. Similarly, the completeness of the regular spherical vector waves, $\boldsymbol{v}_n(k\boldsymbol{r})$, on a spherical surface is proved.[3] The analysis above then proves that the spherical vector waves on a spherical surface are linearly independent.

Comment 7.3
There are many different definitions of the spherical vector waves presented in this section. The connection to the ones in Morse and Feshbach is [196]

$$
\begin{cases}
\boldsymbol{v}_{1n}(k\boldsymbol{r}) = \boldsymbol{M}_n^1(\boldsymbol{r}) \\
\boldsymbol{v}_{2n}(k\boldsymbol{r}) = \boldsymbol{N}_n^1(\boldsymbol{r}) \\
\boldsymbol{v}_{3n}(k\boldsymbol{r}) = \boldsymbol{L}_n^1(\boldsymbol{r})
\end{cases}
\qquad
\begin{cases}
\boldsymbol{u}_{1n}(k\boldsymbol{r}) = \boldsymbol{M}_n^3(\boldsymbol{r}) \\
\boldsymbol{u}_{2n}(k\boldsymbol{r}) = \boldsymbol{N}_n^3(\boldsymbol{r}) \\
\boldsymbol{u}_{3n}(k\boldsymbol{r}) = \boldsymbol{L}_n^3(\boldsymbol{r})
\end{cases}
$$

Waterman uses the notation [285]

$$
\begin{cases}
\boldsymbol{v}_n(k\boldsymbol{r}) = \operatorname{Re} \boldsymbol{\psi}_n(\boldsymbol{r}) \\
\boldsymbol{u}_n(k\boldsymbol{r}) = \boldsymbol{\psi}_n(\boldsymbol{r})
\end{cases}
$$

7.2.1 Expansions of the Fields

The electric field in a source-free region satisfies

$$
\nabla \times (\nabla \times \boldsymbol{E}(\boldsymbol{r}, \omega)) = k^2 \boldsymbol{E}(\boldsymbol{r}, \omega)
$$

[2] The absence of real roots is proved in Lemma B.1 on page 611.

[3] In the case of regular spherical vector waves, there are, for each fixed value of kr, a finite number of l-values, corresponding to the real zeros of $j_l(kr)$.

The electric field outside a sphere that circumscribes the scatterer, therefore, due to completeness of the spherical vector waves, has a general expansion

$$\boldsymbol{E}(\boldsymbol{r},\omega)=\sum_{\substack{n\\ \tau=1,2}}\left(a_n\boldsymbol{v}_n(k\boldsymbol{r})+f_n\boldsymbol{u}_n(k\boldsymbol{r})\right) \tag{7.12}$$

Notice that the summation only includes $\tau = 1, 2$, since the electric field is divergence free in free space—only if we deal with a region containing sources of the electromagnetic fields, the functions $\boldsymbol{v}_{3\sigma ml}$ and $\boldsymbol{u}_{3\sigma ml}$ are needed. The formal proof of this statement is postponed until Section 7.6. Note that we, in general, need an expansion in both regular and out-going spherical vector waves. Note that the notation $\sum_n$ for this multiple sum extends over $\tau = 1, 2$ and it starts at $l = 1$. The expansion of the corresponding magnetic field then is by the use of the Faraday law

$$\boldsymbol{H}(\boldsymbol{r},\omega)=\frac{1}{\mathrm{i}\eta_0\eta}\sum_{\substack{n\\ \tau=1,2}}\left(a_n\boldsymbol{v}_{\overline{n}}(k\boldsymbol{r})+f_n\boldsymbol{u}_{\overline{n}}(k\boldsymbol{r})\right) \tag{7.13}$$

The collective dual index $\overline{n} = \{\overline{\tau}\sigma ml\}$ is also used. In Sections 7.7 and 7.8, we prove that in the solution of a scattering problem, the coefficients $a_{\tau\sigma ml}$ are expansion coefficients of the incident field, and that $f_{\tau\sigma ml}$ are expansion coefficients of the scattered field.

The expansions of the electric and the magnetic fields in (7.12) and (7.13) are the partial wave or multipole expansions of the fields. Every l, m, σ-term in the sum defines a partial wave or multipole (l, m, σ) of degree l. From the alternative representations, (7.6) and (7.5), we see that the multipole with $\tau = 1$ has no radial component ($\hat{\boldsymbol{r}}$-component), since $\hat{\boldsymbol{r}} \cdot \boldsymbol{A}_{1n}(\hat{\boldsymbol{r}}) = 0$. This τ-index generates a transverse electric (TE) multipole of degree (l, m, σ). Similarly, $\tau = 2$ is a transverse magnetic (TM) multipole of degree (l, m, σ).

7.3 Orthogonality and Reciprocity Relations

The transverse, divergence-free (solenoidal) spherical vector waves were defined in (7.8) and (7.9), or (7.5) and (7.6), and the longitudinal (irrotational) spherical vector wave is defined in (7.10). In this chapter, we use, as above, a multi-index n for the indices $\tau\sigma ml$. For $\tau = 1, 2$, the dual index $\overline{n}$ is defined as $\overline{n} = \{\overline{\tau}\sigma ml\}$, where, as before, $\overline{1} = 2$ and $\overline{2} = 1$.

In this section, we derive the orthogonality (or reciprocity) relations[4] for these waves. For the regular spherical vector waves, the relation is

$$\iint_S \left\{\boldsymbol{v}_n \times (\nabla \times \boldsymbol{v}_{n'}) - \boldsymbol{v}_{n'} \times (\nabla \times \boldsymbol{v}_n)\right\} \cdot \hat{\boldsymbol{\nu}}\, \mathrm{d}S = 0, \qquad \tau, \tau' = 1, 2 \tag{7.14}$$

[4]Whether the relations should be called orthogonality or reciprocity relations are the matter of taste. The relations show resemblance with the reciprocity relation developed in Section 1.3.5, but they rely on orthogonality of the vector spherical harmonics.

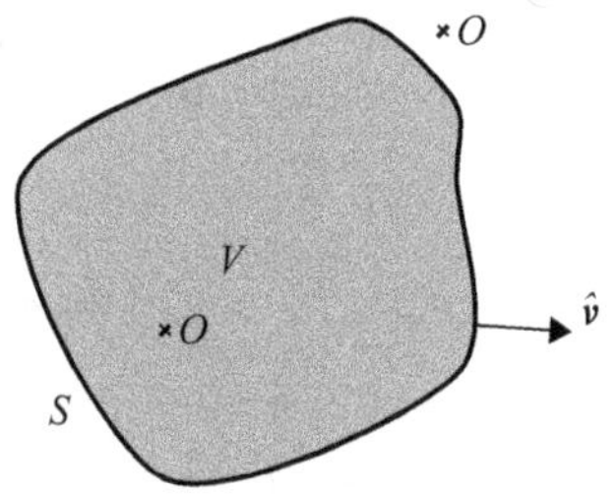

Figure 7.1 The geometry used in the proof of the orthogonality relations in Section 7.3. Two different positions of the origin O—one inside the volume V, and one outside the volume—are displayed.

for all closed, regular surfaces S, see Figure 7.1. To prove this orthogonality relation, we make use of the divergence theorem, or Gauss' theorem, and then expand the integrand using the differentiation rule $\nabla \cdot (\boldsymbol{a} \times \boldsymbol{b}) = \boldsymbol{b} \cdot (\nabla \times \boldsymbol{a}) - \boldsymbol{a} \cdot (\nabla \times \boldsymbol{b})$. We get

$$\begin{aligned}
&\iint_S \{\boldsymbol{v}_n \times (\nabla \times \boldsymbol{v}_{n'}) - \boldsymbol{v}_{n'} \times (\nabla \times \boldsymbol{v}_n)\} \cdot \hat{\boldsymbol{\nu}} \, \mathrm{d}S \\
&\quad = \iiint_V \nabla \cdot \{\boldsymbol{v}_n \times (\nabla \times \boldsymbol{v}_{n'}) - \boldsymbol{v}_{n'} \times (\nabla \times \boldsymbol{v}_n)\} \, \mathrm{d}v \\
&\quad = \iiint_V \Big\{(\nabla \times \boldsymbol{v}_n) \cdot (\nabla \times \boldsymbol{v}_{n'}) - \boldsymbol{v}_n \cdot (\nabla \times (\nabla \times \boldsymbol{v}_{n'})) \\
&\qquad - (\nabla \times \boldsymbol{v}_{n'}) \cdot (\nabla \times \boldsymbol{v}_n) + \boldsymbol{v}_{n'} \cdot (\nabla \times (\nabla \times \boldsymbol{v}_n))\Big\} \, \mathrm{d}v = 0
\end{aligned}$$

since $\nabla \times (\nabla \times \boldsymbol{v}_n) = k^2 \boldsymbol{v}_n$, and where V is the volume enclosed by the surface S, see Figure 7.1.

The identity for a combination of regular and out-going waves is derived in a similar way. The analysis depends on whether the origin of the coordinate system is enclosed by the surface S or not.

If the surface S does not include the origin $\boldsymbol{r} = \boldsymbol{0}$ then, as in the regular wave case

$$\iint_S \{\boldsymbol{v}_n \times (\nabla \times \boldsymbol{u}_{n'}) - \boldsymbol{u}_{n'} \times (\nabla \times \boldsymbol{v}_n)\} \cdot \hat{\boldsymbol{\nu}} \, \mathrm{d}S = 0, \qquad \tau, \tau' = 1, 2, \ \boldsymbol{0} \notin V$$

for all closed, regular surfaces S not including the origin. If, however, the surface S includes the origin, the divergence theorem gives (we assume ϵ is small enough to ensure that the sphere, $r = \epsilon$, centered at the origin, lies inside S)

$$\begin{aligned}
&\iint_S \{\boldsymbol{v}_n \times (\nabla \times \boldsymbol{u}_{n'}) - \boldsymbol{u}_{n'} \times (\nabla \times \boldsymbol{v}_n)\} \cdot \hat{\boldsymbol{\nu}} \, \mathrm{d}S \\
&\quad = \iint_{r=\epsilon} \{\boldsymbol{v}_n \times (\nabla \times \boldsymbol{u}_{n'}) - \boldsymbol{u}_{n'} \times (\nabla \times \boldsymbol{v}_n)\} \cdot \hat{\boldsymbol{r}} \, \mathrm{d}S \\
&\quad = k \iint_{r=\epsilon} \Big\{\boldsymbol{v}_n \cdot (\boldsymbol{u}_{\overline{n}'} \times \hat{\boldsymbol{r}}) - \boldsymbol{u}_{n'} \cdot (\boldsymbol{v}_{\overline{n}} \times \hat{\boldsymbol{r}})\Big\} \, \mathrm{d}S
\end{aligned}$$

The integral over the sphere, $r = \epsilon$, is zero if $\tau \neq \tau'$, since the integrand is proportional to $\boldsymbol{A}_{\tau\sigma ml} \cdot \boldsymbol{A}_{\tau'\sigma' m' l'}$. Orthogonality then gives the result. For $\tau = \tau' = 1$, we explicitly compute the surface integral

$$\iint_S \{\boldsymbol{v}_n \times (\nabla \times \boldsymbol{u}_{n'}) - \boldsymbol{u}_{n'} \times (\nabla \times \boldsymbol{v}_n)\} \cdot \hat{\boldsymbol{\nu}}\, \mathrm{d}S$$
$$= k\epsilon^2 \delta_{\sigma\sigma'} \delta_{mm'} \delta_{ll'} \left\{ j_l(k\epsilon) \frac{(k\epsilon h_l^{(1)}(k\epsilon))'}{k\epsilon} - h_l^{(1)}(k\epsilon) \frac{(k\epsilon j_l(k\epsilon))'}{k\epsilon} \right\}$$

or

$$\iint_S \{\boldsymbol{v}_n \times (\nabla \times \boldsymbol{u}_{n'}) - \boldsymbol{u}_{n'} \times (\nabla \times \boldsymbol{v}_n)\} \cdot \hat{\boldsymbol{\nu}}\, \mathrm{d}S$$
$$= k\epsilon^2 \delta_{\sigma\sigma'} \delta_{mm'} \delta_{ll'} \left\{ j_l(k\epsilon) h_l^{(1)\prime}(k\epsilon) - h_l^{(1)}(k\epsilon) j_l'(k\epsilon) \right\} = -\delta_{\sigma\sigma'} \delta_{mm'} \delta_{ll'} \frac{1}{\mathrm{i}k} = -\delta_{nn'} \frac{1}{\mathrm{i}k}$$

where we used the Wronskian for the spherical Bessel and Hankel functions, see Appendix B.3

$$j_l(z) h_l^{(1)\prime}(z) - h_l^{(1)}(z) j_l'(z) = \frac{\mathrm{i}}{z^2}$$

Notice that the orthogonality relation is independent of the value of ϵ as expected. Similarly, for $\tau = \tau' = 2$, we have

$$\iint_S \{\boldsymbol{v}_n \times (\nabla \times \boldsymbol{u}_{n'}) - \boldsymbol{u}_{n'} \times (\nabla \times \boldsymbol{v}_n)\} \cdot \hat{\boldsymbol{\nu}}\, \mathrm{d}S$$
$$= k\epsilon^2 \delta_{\sigma\sigma'} \delta_{mm'} \delta_{ll'} \left\{ -\frac{(k\epsilon j_l(k\epsilon))'}{k\epsilon} h_l^{(1)}(k\epsilon) + j_l(k\epsilon) \frac{(k\epsilon h_l^{(1)}(k\epsilon))'}{k\epsilon} \right\} = -\delta_{nn'} \frac{1}{\mathrm{i}k}$$

In summary $(\tau, \tau' = 1, 2)$

$$\iint_S \{\boldsymbol{v}_n \times (\nabla \times \boldsymbol{u}_{n'}) - \boldsymbol{u}_{n'} \times (\nabla \times \boldsymbol{v}_n)\} \cdot \hat{\boldsymbol{\nu}}\, \mathrm{d}S = -\delta_{nn'} \frac{1}{\mathrm{i}k}, \qquad \mathbf{0} \in V \tag{7.15}$$

if the surface S encloses the origin.

The out-going waves have an orthogonality relation similar to the regular one in (7.14)

$$\iint_S \{\boldsymbol{u}_n \times (\nabla \times \boldsymbol{u}_{n'}) - \boldsymbol{u}_{n'} \times (\nabla \times \boldsymbol{u}_n)\} \cdot \hat{\boldsymbol{\nu}}\, \mathrm{d}S$$
$$= k\epsilon^2 \delta_{nn'} \left\{ \frac{(k\epsilon h_l^{(1)}(k\epsilon))'}{k\epsilon} h_l^{(1)}(k\epsilon) - h_l^{(1)}(k\epsilon) \frac{(k\epsilon h_l^{(1)}(k\epsilon))'}{k\epsilon} \right\} = 0 \tag{7.16}$$

for all regular surfaces S, regardless of whether S includes the origin or not

$$\iint_S \{\boldsymbol{u}_n \times (\nabla \times \boldsymbol{u}_{n'}) - \boldsymbol{u}_{n'} \times (\nabla \times \boldsymbol{u}_n)\} \cdot \hat{\boldsymbol{\nu}} \, dS = 0$$

7.3.1 Spherical Scalar Waves

To make the analysis complete, we also give the analogous results for the spherical scalar waves defined in (7.11). The divergence theorem implies

$$\iint_S \{v_n \nabla v_{n'} - v_{n'} \nabla v_n\} \cdot \hat{\boldsymbol{\nu}} \, dS = \iiint_V \nabla \cdot \{v_n \nabla v_{n'} - v_{n'} \nabla v_n\} \, dv$$
$$= \iiint_V \left\{v_n \nabla^2 v_{n'} - v_{n'} \nabla^2 v_n\right\} dv = -\iiint_V \left\{v_n k^2 v_{n'} - v_{n'} k^2 v_n\right\} dv = 0$$

where V is the volume enclosed by the surface S. To summarize,

$$\iint_S \{v_n \nabla v_{n'} - v_{n'} \nabla v_n\} \cdot \hat{\boldsymbol{\nu}} \, dS = 0$$

Similarly, for the radiating spherical wave, if the origin is enclosed by the surface S, $\mathbf{0} \in V$, we get

$$\iint_S \{v_n \nabla u_{n'} - u_{n'} \nabla v_n\} \cdot \hat{\boldsymbol{\nu}} \, dS = \iint_{r=\epsilon} \{v_n \nabla u_{n'} - u_{n'} \nabla v_n\} \cdot \hat{\boldsymbol{r}} \, dS$$

Orthogonality of the spherical harmonics, $Y_n(\hat{\boldsymbol{r}})$, then implies

$$\iint_S \{v_n \nabla u_{n'} - u_{n'} \nabla v_n\} \cdot \hat{\boldsymbol{\nu}} \, dS = k\epsilon^2 \delta_{nn'} \left\{j_l(k\epsilon) h_l^{(1)\prime}(k\epsilon) - h_l^{(1)}(k\epsilon) j_l'(k\epsilon)\right\} = -\delta_{nn'} \frac{1}{ik}$$

where we again used the Wronskian for the spherical Bessel and Hankel functions. Notice again that the orthogonality relation is independent of the value of ϵ as expected. We have

$$\iint_S \{v_n \nabla u_{n'} - u_{n'} \nabla v_n\} \cdot \hat{\boldsymbol{\nu}} \, dS = -\delta_{nn'} \frac{1}{ik}, \qquad \mathbf{0} \in V \tag{7.17}$$

We also easily prove

$$\iint_S \{v_n \nabla u_{n'} - u_{n'} \nabla v_n\} \cdot \hat{\boldsymbol{\nu}} \, dS = 0, \qquad \mathbf{0} \notin V$$

and

$$\iint_S \{u_n \nabla u_{n'} - u_{n'} \nabla u_n\} \cdot \hat{\boldsymbol{\nu}} \, dS = 0$$

7.3.2 Power Transport

The orthogonality relations derived above are useful in the evaluation of the power radiating out of an arbitrary surface S containing the origin and the circumscribed sphere of the scatterer S_s. The electric and the magnetic fields outside the circumscribed sphere of the scatterer S_s have the following expansions, see Section 7.2.1

$$\begin{cases} \boldsymbol{E}(\boldsymbol{r},\omega) = \sum_n (a_n \boldsymbol{v}_n(k\boldsymbol{r}) + f_n \boldsymbol{u}_n(k\boldsymbol{r})) \\ \boldsymbol{H}(\boldsymbol{r},\omega) = \dfrac{1}{\mathrm{i}k\eta_0\eta} \sum_n (a_n \nabla \times \boldsymbol{v}_n(k\boldsymbol{r}) + f_n \nabla \times \boldsymbol{u}_n(k\boldsymbol{r})) \end{cases}$$

and the radiated power radiating through S is

$$\begin{aligned} P &= \frac{1}{2} \operatorname{Re} \iint_S \{\boldsymbol{E} \times \boldsymbol{H}^*\} \cdot \hat{\boldsymbol{\nu}} \, \mathrm{d}S \\ &= -\frac{1}{2k\eta_0\eta} \sum_{n,n'} \operatorname{Im} \iint_S \{(a_n \boldsymbol{v}_n + f_n \boldsymbol{u}_n) \times (a_{n'}^* \nabla \times \boldsymbol{v}_{n'} + f_{n'}^* \nabla \times \boldsymbol{u}_{n'}^*)\} \cdot \hat{\boldsymbol{\nu}} \, \mathrm{d}S \end{aligned}$$

There are three types of terms in this expression, and we examine each one separately.

The first type of integral is

$$\begin{aligned} 2 \operatorname{Im} \sum_{n,n'} a_n a_{n'}^* \iint_S \boldsymbol{v}_n \times (\nabla \times \boldsymbol{v}_{n'}) \cdot \hat{\boldsymbol{\nu}} \, \mathrm{d}S \\ = -\mathrm{i} \sum_{n,n'} a_n a_{n'}^* \iint_S \{\boldsymbol{v}_n \times (\nabla \times \boldsymbol{v}_{n'}) - \boldsymbol{v}_{n'} \times (\nabla \times \boldsymbol{v}_n)\} \cdot \hat{\boldsymbol{\nu}} \, \mathrm{d}S = 0 \end{aligned}$$

due to the orthogonality relation (7.14).

The second type is analyzed in an analogous manner. We get

$$\begin{aligned} &2 \operatorname{Im} \sum_{n,n'} f_n f_{n'}^* \iint_S \boldsymbol{u}_n \times \left(\nabla \times \boldsymbol{u}_n^*\right) \cdot \hat{\boldsymbol{\nu}} \, \mathrm{d}S \\ &= -\mathrm{i} \sum_{n,n'} f_n f_{n'}^* \iint_S \left(\boldsymbol{u}_n \times \left(\nabla \times \boldsymbol{u}_{n'}^*\right) - \boldsymbol{u}_{n'}^* \times (\nabla \times \boldsymbol{u}_n)\right) \cdot \hat{\boldsymbol{\nu}} \, \mathrm{d}S = -\frac{2}{k} \sum_n |f_n|^2 \end{aligned}$$

We have here used the identity $\boldsymbol{u}_n^* = 2\boldsymbol{v}_n - \boldsymbol{u}_n$ (the argument kr is a real number), (7.15) and (7.16), since the origin is located inside the surface S.

The final term is, again using (7.15)

$$2\,\mathrm{Im}\sum_{n,n'} f_n a_{n'}^* \iint_S \boldsymbol{u}_n \times (\nabla \times \boldsymbol{v}_{n'}) \cdot \hat{\boldsymbol{\nu}}\, \mathrm{d}S + 2\,\mathrm{Im}\sum_{n,n'} a_n f_{n'}^* \iint_S \boldsymbol{v}_n \times \left(\nabla \times \boldsymbol{u}_{n'}^*\right) \cdot \hat{\boldsymbol{\nu}}\, \mathrm{d}S$$

$$= -\mathrm{i}\sum_{n,n'} f_n a_{n'}^* \iint_S \left(\boldsymbol{u}_n \times (\nabla \times \boldsymbol{v}_{n'}) - \boldsymbol{v}_{n'} \times (\nabla \times \boldsymbol{u}_n)\right) \cdot \hat{\boldsymbol{\nu}}\, \mathrm{d}S$$

$$+\mathrm{i}\sum_{n,n'} f_n^* a_{n'} \iint_S \left(\boldsymbol{u}_n^* \times (\nabla \times \boldsymbol{v}_{n'}) - \boldsymbol{v}_{n'} \times \left(\nabla \times \boldsymbol{u}_n^*\right)\right) \cdot \hat{\boldsymbol{\nu}}\, \mathrm{d}S$$

$$= -\frac{1}{k}\sum_n \left(f_n a_n^* + f_n^* a_n\right) = -\frac{2}{k}\sum_n \mathrm{Re}\left(f_n a_n^*\right)$$

Finally, we get

$$P = \frac{1}{2k^2\eta_0\eta}\sum_n \left(|f_n|^2 + \mathrm{Re}\left(f_n a_n^*\right)\right) \tag{7.18}$$

This statement is equivalent to the optical theorem, see Problem 7.2.

7.4 Some Properties of the Spherical Vector Waves

In this section, we collect a few useful properties of the spherical vector waves that are used in the chapters below. More explicitly, we investigate the linear independence of the tangential components of the radiating spherical vector waves, and the general translation properties of the spherical vector waves.

7.4.1 Linear Independence

The concept of linear independence is defined in Appendix F.9.1.3 on page 682. We have proved, see page 358, that the out-going spherical vector waves, $\tau = 1, 2, 3$, are linearly independent on a spherical surface. The same result was obtained for the regular spherical vector waves with the exception of a discrete set of frequencies. In this section, we show a similar result for the tangential components of the out-going spherical vector waves for $\tau = 1, 2$.

The following analysis shows that the tangential components of the radiating spherical vector waves on a simply connected surface S—not necessarily a spherical surface—are linearly independent. More explicitly, the family

$$\left\{\hat{\boldsymbol{\nu}}(\boldsymbol{r}) \times \boldsymbol{u}_n(k\boldsymbol{r})\right\}_n, \qquad \tau = 1, 2$$

is linearly independent on $\boldsymbol{L}_{\mathrm{t}}^2(S)$. The function space $\boldsymbol{L}_{\mathrm{t}}^2(S)$ consists of all tangential complex vector-valued function defined on the surface S with finite $\boldsymbol{L}^2$ norm, see Appendix F.9.1 on page 681, i.e.,

$$\boldsymbol{L}_{\mathrm{t}}^2(S) = \left\{\boldsymbol{u} : \|\boldsymbol{u}\|_{\boldsymbol{L}_{\mathrm{t}}^2(S)} < \infty; \boldsymbol{u} \cdot \hat{\boldsymbol{\nu}} = 0\right\}$$

To prove this statement, consider

$$\sum_{\tau=1}^{2}\sum_{\sigma=\mathrm{e,o}}\sum_{l=0}^{L}\sum_{m=0}^{l} c_{\tau\sigma ml}\hat{\boldsymbol{\nu}}(\boldsymbol{r})\times\boldsymbol{u}_{\tau\sigma ml}(k\boldsymbol{r})=\boldsymbol{0},\qquad \boldsymbol{r}\in S$$

To prove linear independence of the family $\{\hat{\boldsymbol{\nu}}(\boldsymbol{r})\times\boldsymbol{u}_n(k\boldsymbol{r})\}_{n(\tau=1,2)}$, we have to prove that $c_{\tau\sigma ml}=0$ for all indices and all $L\geq 0$.

Outside the surface S, we define the function

$$\boldsymbol{u}(\boldsymbol{r})=\sum_{\tau=1}^{2}\sum_{\sigma=\mathrm{e,o}}\sum_{l=0}^{L}\sum_{m=0}^{l} c_{\tau\sigma ml}\boldsymbol{u}_{\tau\sigma ml}(k\boldsymbol{r}),\qquad \boldsymbol{r}\text{ outside } S$$

This function is well defined and it satisfies the electric field equation in region outside S, and moreover, it satisfies the Silver–Müller radiations conditions (no convergence problems since the sum is finite). By construction its tangential components, $\hat{\boldsymbol{\nu}}\times\boldsymbol{u}$, vanish on S. Due to uniqueness of the exterior problem, $\boldsymbol{u}=\boldsymbol{0}$ everywhere outside S [40]. In particular, it vanishes on a spherical surface of radius R, S_R, that encloses the surface S. Orthogonality of the vector spherical harmonics on the sphere with radius R implies

$$\begin{aligned}0=\iint_{S_R}|\boldsymbol{u}(\boldsymbol{r})|^2\,\mathrm{d}S&=\sum_{\sigma=\mathrm{e,o}}\sum_{l=0}^{L}\sum_{m=0}^{l}|c_{1\sigma ml}|^2\left|h_l^{(1)}(kR)\right|^2\\&+\sum_{\sigma=\mathrm{e,o}}\sum_{l=0}^{L}\sum_{m=0}^{l}|c_{2\sigma ml}|^2\left|\frac{(kRh_l^{(1)}(kR))'}{kR}\right|^2\\&+\sum_{\sigma=\mathrm{e,o}}\sum_{l=0}^{L}\sum_{m=0}^{l}|c_{2\sigma ml}|^2 l(l+1)\left|\frac{h_l^{(1)}(kR)}{kR}\right|^2\end{aligned}$$

Each term in the sum is positive and must therefore be identically zero, and since the spherical Hankel functions, $h_l^{(1)}(z)$ and $(zh_l^{(1)}(z))'$, have no real zeros, see Lemma B.1 on page 611, we conclude that $c_{\tau\sigma ml}=0$ for all indices. The proof of the statement is completed.

A similar result can be obtained for the regular spherical vector waves with the exception of a discrete set of frequencies.

7.4.2 The Translation Matrices

The translation properties of the spherical vector waves are instrumental for the formulation and the solution of the scattering problem of many individual scatterers, see Section 9.2. These translation properties are well known, see Reference 174, and we refer to, e.g., Reference 23 for details.

The spherical vector waves $\boldsymbol{u}_n(k\boldsymbol{r})$ and $\boldsymbol{v}_n(k\boldsymbol{r})$ solve (7.1) everywhere in space with the exception of the origin for the radiating vector waves $\boldsymbol{u}_n(k\boldsymbol{r})$, where they are undefined. However, (7.1) has identical form in a translated, primed, Cartesian coordinate system, see Figure 7.2. The connection between the primed and the unprimed systems, origin O' and O, respectively, is $\boldsymbol{r}'=\boldsymbol{r}+\boldsymbol{d}$. The completeness properties of the spherical vector waves imply

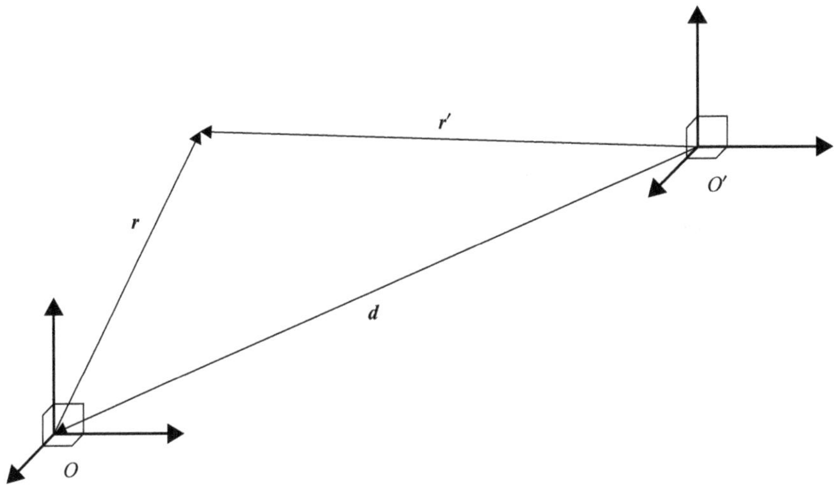

Figure 7.2 The relation between the translated origins O and O' and their position vectors $\boldsymbol{r}$ and $\boldsymbol{r}'$.

that the solutions expressed in the primed coordinate system must have an expansion in the solutions in the unprimed system. This section summarizes these translation properties.

Let $\boldsymbol{r}' = \boldsymbol{r} + \boldsymbol{d}$, see Figure 7.2. Then

$$\begin{cases} \boldsymbol{v}_n(k\boldsymbol{r}') = \sum_{n'} \mathcal{R}_{nn'}(k\boldsymbol{d})\boldsymbol{v}_{n'}(k\boldsymbol{r}), & \text{for all } \boldsymbol{d} \\ \boldsymbol{u}_n(k\boldsymbol{r}') = \sum_{n'} \mathcal{R}_{nn'}(k\boldsymbol{d})\boldsymbol{u}_{n'}(k\boldsymbol{r}), & r > d \\ \boldsymbol{u}_n(k\boldsymbol{r}') = \sum_{n'} \mathcal{P}_{nn'}(k\boldsymbol{d})\boldsymbol{v}_{n'}(k\boldsymbol{r}), & r < d \end{cases} \tag{7.19}$$

The explicit expressions of the matrices $\mathcal{R}_{nn'}(k\boldsymbol{d})$ and $\mathcal{P}_{nn'}(k\boldsymbol{d})$ are given in Appendix F.7 on page 675.

We refer the reader to Appendix F.7 for useful identities of these translation matrices. One of these identities is translation in the opposite direction, which is identical to the transpose of the translation matrices, i.e.,

$$\mathcal{R}^{\mathrm{t}}(k\boldsymbol{d}) = \mathcal{R}(-k\boldsymbol{d}), \qquad \mathcal{P}^{\mathrm{t}}(k\boldsymbol{d}) = \mathcal{P}(-k\boldsymbol{d})$$

7.5 Expansion of the Green Dyadic

The completeness of the spherical vector waves can be used to find an expansion of the Green dyadic for the electric field in free space in terms of spherical vector waves. This, very useful expansion, is accomplished in this section.

7.5.1 The Green Function in Free Space

Before we approach the main problem of finding the expansion of the Green dyadic for the electric field in free space in spherical vector waves, we undertake the task of finding the expansion of the Green function $g(k, |\boldsymbol{r} - \boldsymbol{r}'|)$, see (2.11) on page 105, in terms of spherical scalar waves $v_n(k\boldsymbol{r})$ and $u_n(k\boldsymbol{r})$ defined in (7.11) on page 357.

To prove the expansion of the Green function, we employ the integral representation, see (3.2) on page 143

$$\phi(\boldsymbol{r}) = \iint_S \left\{ g(k, |\boldsymbol{r} - \boldsymbol{r}'|)\nabla'\phi(\boldsymbol{r}') - \phi(\boldsymbol{r}')\nabla' g(k, |\boldsymbol{r} - \boldsymbol{r}'|) \right\} \cdot \hat{\boldsymbol{\nu}}(\boldsymbol{r}')\, \mathrm{d}S' \tag{7.20}$$

where we explicitly assumed that $\boldsymbol{r}$ is located inside S, and that the function $\phi(\boldsymbol{r})$ satisfies the Helmholtz equation

$$\nabla^2\phi(\boldsymbol{r}) + k^2\phi(\boldsymbol{r}) = 0$$

Examples of such functions are the regular spherical scalar waves $v_n(k\boldsymbol{r})$.

For each fixed value of $\boldsymbol{r}$, $g(k, |\boldsymbol{r} - \boldsymbol{r}'|)$ is a function of $\boldsymbol{r}'$ in the domain $r < r'$ that satisfies the radiation conditions as $r' \to \infty$, cf. (2.24) on page 112, and the Helmholtz equation, see (2.8) on page 103

$$\nabla'^2 g(k, |\boldsymbol{r} - \boldsymbol{r}'|) + k^2 g(k, |\boldsymbol{r} - \boldsymbol{r}'|) = 0, \qquad r < r'$$

The completeness of the spherical harmonics on the sphere $r' =$ constant and the assumed radiation condition for $r < r'$ imply that the Green function has an expansion in terms of $u_n(k\boldsymbol{r}')$, e.g.,

$$g(k, |\boldsymbol{r} - \boldsymbol{r}'|) = \sum_n a_n(\boldsymbol{r}) u_n(k\boldsymbol{r}'), \qquad r < r'$$

To determine the unknown coefficients $a_n(\boldsymbol{r})$, we use (7.20) for a surface S that has a inscribed sphere of radius R that satisfies $r < R$. Then for $\phi(\boldsymbol{r}) = v_n(k\boldsymbol{r})$, we get (assume summation and integration can be interchanged)

$$v_n(k\boldsymbol{r}) = \sum_{n'} a_{n'}(\boldsymbol{r}) \iint_S \left\{ u_{n'}(k\boldsymbol{r}')\nabla' v_n(k\boldsymbol{r}') - v_n(k\boldsymbol{r}')\nabla' u_{n'}(k\boldsymbol{r}') \right\} \cdot \hat{\boldsymbol{\nu}}(\boldsymbol{r}')\, \mathrm{d}S'$$

The orthogonality results in (7.17) imply

$$v_n(k\boldsymbol{r}) = \frac{a_n(\boldsymbol{r})}{\mathrm{i}k}$$

The coefficients $a_n(\boldsymbol{r})$ are now determined and we conclude that the scalar Green function has the expansion

$$g(k, |\boldsymbol{r} - \boldsymbol{r}'|) = \mathrm{i}k \sum_n v_n(k\boldsymbol{r}) u_n(k\boldsymbol{r}'), \qquad r < r'$$

and similarly for a change of arguments. To summarize, we have

$$g(k, |\boldsymbol{r} - \boldsymbol{r}'|) = \mathrm{i}k \sum_n v_n(k\boldsymbol{r}_<) u_n(k\boldsymbol{r}_>), \qquad r \neq r' \tag{7.21}$$

where, if $r' > r$, $\boldsymbol{r}_> = \boldsymbol{r}'$ and $\boldsymbol{r}_< = \boldsymbol{r}$, and if $r' < r$, $\boldsymbol{r}_> = \boldsymbol{r}$ and $\boldsymbol{r}_< = \boldsymbol{r}'$. This expansion is uniformly convergent in compact (finite and closed) domains, provided $r \neq r'$ in the domain [40].

7.5.2 The Green Dyadic for the Electric Field in Free Space

Our aim is to find an expansion in spherical vector waves of the electric Green dyadic $\mathbf{G}_\mathrm{e}(k, |\boldsymbol{r} - \boldsymbol{r}'|)$ in free space. This dyadic satisfies (2.27) on page 113, i.e.,

$$\nabla \times \left(\nabla \times \mathbf{G}_\mathrm{e}(k, |\boldsymbol{r} - \boldsymbol{r}'|)\right) - k^2 \mathbf{G}_\mathrm{e}(k, |\boldsymbol{r} - \boldsymbol{r}'|) = \mathbf{I}\delta(\boldsymbol{r} - \boldsymbol{r}')$$

Following (3.18) on page 158, we easily get for an $\boldsymbol{r}$ inside S ($\nabla \times \mathbf{G}_\mathrm{e}$ is an anti-symmetric dyadic and $\mathbf{G}_\mathrm{e}$ is a symmetric dyadic)

$$\boldsymbol{A}(\boldsymbol{r}) = -\iint_S \left\{(\hat{\boldsymbol{\nu}} \times \boldsymbol{A}) \cdot \left(\nabla' \times \mathbf{G}_\mathrm{e}\right) + (\hat{\boldsymbol{\nu}} \times (\nabla' \times \boldsymbol{A})) \cdot \mathbf{G}_\mathrm{e}\right\} \mathrm{d}S'$$

for each vector field $\boldsymbol{A}$ that satisfies

$$\nabla \times (\nabla \times \boldsymbol{A}(\boldsymbol{r})) - k^2 \boldsymbol{A}(\boldsymbol{r}) = \mathbf{0}, \quad \boldsymbol{r} \text{ inside } S$$

Make a scalar multiplication with a constant vector $\boldsymbol{a}$

$$\boldsymbol{A}(\boldsymbol{r}) \cdot \boldsymbol{a} = -\iint_S \left\{(\hat{\boldsymbol{\nu}} \times \boldsymbol{A}) \cdot \left(\nabla' \times (\mathbf{G}_\mathrm{e} \cdot \boldsymbol{a})\right) + (\hat{\boldsymbol{\nu}} \times (\nabla' \times \boldsymbol{A})) \cdot (\mathbf{G}_\mathrm{e} \cdot \boldsymbol{a})\right\} \mathrm{d}S' \tag{7.22}$$

For a fixed value of $\boldsymbol{r}$, and each fixed, constant vector $\boldsymbol{a}$, the vector-valued function $\boldsymbol{g}(\boldsymbol{r}') = \mathbf{G}_\mathrm{e}\left(k, |\boldsymbol{r} - \boldsymbol{r}'|\right) \cdot \boldsymbol{a}$ is a function of $\boldsymbol{r}'$ that satisfies

$$\nabla' \times (\nabla' \times \boldsymbol{g}(\boldsymbol{r}')) - k^2 \boldsymbol{g}(\boldsymbol{r}') = \mathbf{0}, \qquad r < r'$$

and moreover, as a function of the primed variables, it satisfies the radiation conditions in the domain $r < r'$. Thus, in the region $r < r'$, due to the results above, $\boldsymbol{g}(\boldsymbol{r}') = \mathbf{G}_\mathrm{e}(k, |\boldsymbol{r} - \boldsymbol{r}'|) \cdot \boldsymbol{a}$ has an expansion

$$\boldsymbol{g}(\boldsymbol{r}') = \mathbf{G}_\mathrm{e}(k, |\boldsymbol{r} - \boldsymbol{r}'|) \cdot \boldsymbol{a} = \sum_{\substack{n \\ \tau=1,2}} a_{\tau n}(\boldsymbol{r}) \boldsymbol{u}_{\tau n}(k\boldsymbol{r}'), \qquad r < r' \tag{7.23}$$

Notice that only $\tau = 1, 2$ enter into the sum, since the function $\boldsymbol{g}(\boldsymbol{r}')$ is solenoidal in the region $r < r'$. The expansion coefficients $a_{\tau n}(\boldsymbol{r})$ are unknown and have to be determined. This is done by the orthogonality relations derived in Section 7.3.

Let S be a surface with an inscribed sphere that encloses $\boldsymbol{r}$ and, furthermore, let $\boldsymbol{A}(\boldsymbol{r}) = \boldsymbol{v}_{\tau'n'}(k\boldsymbol{r})$, $\tau' = 1, 2$. For all points $\boldsymbol{r}'$ on S, we then have $r < r'$. Then (7.22) implies with the use of the expansion (7.23)

$$\boldsymbol{v}_{\tau'n'}(k\boldsymbol{r}) \cdot \boldsymbol{a} = -\sum_{\substack{n \\ \tau=1,2}} a_{\tau n}(\boldsymbol{r}) \iint_S \left\{(\hat{\boldsymbol{\nu}} \times \boldsymbol{v}_{\tau'n'}) \cdot \left(\nabla' \times \boldsymbol{u}_{\tau n}\right) + (\hat{\boldsymbol{\nu}} \times (\nabla' \times \boldsymbol{v}_{\tau'n'})) \cdot \boldsymbol{u}_{\tau n}\right\} \mathrm{d}S'$$

Use the orthogonality relation (7.15), and we get

$$\mathrm{i}k\boldsymbol{v}_{\tau n}(k\boldsymbol{r}) \cdot \boldsymbol{a} = a_{\tau n}(\boldsymbol{r})$$

From (7.23), we have

$$\mathbf{G}_{\mathrm{e}}(k, |\boldsymbol{r} - \boldsymbol{r}'|) \cdot \boldsymbol{a} = \mathrm{i}k \sum_{\substack{n \\ \tau=1,2}} \boldsymbol{u}_{\tau n}(k\boldsymbol{r}')\boldsymbol{v}_{\tau n}(k\boldsymbol{r}) \cdot \boldsymbol{a}, \qquad r < r'$$

for all constant vectors $\boldsymbol{a}$. This implies that

$$\mathbf{G}_{\mathrm{e}}(k, |\boldsymbol{r} - \boldsymbol{r}'|) = \mathrm{i}k \sum_{\substack{n \\ \tau=1,2}} \boldsymbol{u}_{\tau n}(k\boldsymbol{r}')\boldsymbol{v}_{\tau n}(k\boldsymbol{r}), \qquad r < r'$$

and similarly for a change of arguments. Since $\mathbf{G}_{\mathrm{e}}$ is a symmetric dyadic, we have

$$\mathbf{G}_{\mathrm{e}}(k, |\boldsymbol{r} - \boldsymbol{r}'|) = \mathrm{i}k \sum_{\substack{n \\ \tau=1,2}} \boldsymbol{v}_{\tau n}(k\boldsymbol{r}_<)\boldsymbol{u}_{\tau n}(k\boldsymbol{r}_>) = \mathrm{i}k \sum_{\substack{n \\ \tau=1,2}} \boldsymbol{u}_{\tau n}(k\boldsymbol{r}_>)\boldsymbol{v}_{\tau n}(k\boldsymbol{r}_<), \qquad r \neq r' \tag{7.24}$$

where $\boldsymbol{r}_<$ ($\boldsymbol{r}_>$) is the position vector with the smallest (largest) distance to the origin, i.e., if $r < r'$ then $\boldsymbol{r}_< = \boldsymbol{r}$ and $\boldsymbol{r}_> = \boldsymbol{r}'$. This expansion is uniformly convergent in compact (finite and closed) domains, provided $r \neq r'$ in the domain [133, 195].

Notice that the singularity of the Green dyadic is not included in this expansion. This singularity is located on the surface $r = r'$, which is excluded in the domain of validity. For further details, see Reference 37.

7.5.3 Free-Space Green Dyadic

It is also of interest for us to expand the free-space Green dyadic $\mathbf{G}$ in spherical vector waves. The definition of the Green dyadic is

$$\mathbf{G}(k, |\boldsymbol{r} - \boldsymbol{r}'|) = \mathbf{I}g(k, |\boldsymbol{r} - \boldsymbol{r}'|) = \mathbf{I}\frac{\mathrm{e}^{\mathrm{i}k|\boldsymbol{r}-\boldsymbol{r}'|}}{4\pi|\boldsymbol{r} - \boldsymbol{r}'|}$$

and it is related to the Green dyadic for the electric field $\mathbf{G}_{\mathrm{e}}$ by (2.28) on page 114

$$\mathbf{G} = \mathbf{G}_{\mathrm{e}} - \frac{1}{k^2}\nabla\nabla g$$

We easily rewrite $\nabla\nabla g$ as, use (7.21)

$$\nabla\nabla g(k,|\boldsymbol{r}-\boldsymbol{r}'|)=\mathrm{i}k\nabla\nabla\sum_n v_n(k\boldsymbol{r}_<)u_n(k\boldsymbol{r}_>)=-\mathrm{i}k\nabla\nabla'\sum_n v_n(k\boldsymbol{r}_<)u_n(k\boldsymbol{r}_>)$$

$$=-\mathrm{i}k^3\sum_{\substack{n\\ \tau=3}}\boldsymbol{v}_n(k\boldsymbol{r}')\boldsymbol{u}_n(k\boldsymbol{r}),\qquad r\neq r'$$

Thus, we have

$$\mathbf{G}(k,|\boldsymbol{r}-\boldsymbol{r}'|)=\mathrm{i}k\sum_{\substack{n\\ \tau=1,2,3}}\boldsymbol{v}_{\tau n}(k\boldsymbol{r}_<)\boldsymbol{u}_{\tau n}(k\boldsymbol{r}_>)=\mathrm{i}k\sum_{\substack{n\\ \tau=1,2,3}}\boldsymbol{u}_{\tau n}(k\boldsymbol{r}_>)\boldsymbol{v}_{\tau n}(k\boldsymbol{r}_<),\qquad r\neq r'$$

This completes the expansions of the Green dyadics in terms of spherical vector waves. These expansions are instrumental in the development of the Null-field equations in Section 7.6 and the Null-field approach in Chapter 9.

7.6 Null-Field Equations

We are now in a position to utilize the result in Section 7.5.2 and the integral representation of the scattered and the incident fields in (4.15) on page 197, i.e.,

$$-\frac{\eta_0\eta}{\mathrm{i}k}\nabla\times\left\{\nabla\times\iint_{S_\mathrm{s}}\mathbf{G}_\mathrm{e}(k,|\boldsymbol{r}-\boldsymbol{r}'|)\cdot(\hat{\boldsymbol{\nu}}(\boldsymbol{r}')\times\boldsymbol{H}(\boldsymbol{r}'))\,\mathrm{d}S'\right\}$$

$$+\nabla\times\iint_{S_\mathrm{s}}\mathbf{G}_\mathrm{e}(k,|\boldsymbol{r}-\boldsymbol{r}'|)\cdot(\hat{\boldsymbol{\nu}}(\boldsymbol{r}')\times\boldsymbol{E}(\boldsymbol{r}'))\,\mathrm{d}S'=\begin{cases}\boldsymbol{E}_\mathrm{s}(\boldsymbol{r}),\ \boldsymbol{r}\text{ outside }S_\mathrm{s}\\ -\boldsymbol{E}_\mathrm{i}(\boldsymbol{r}),\ \boldsymbol{r}\text{ inside }S_\mathrm{s}\end{cases}$$

where we have used the result in Section 7.5.2 to rewrite the integral representation in terms of the Green dyadic for the electric field in free space, $\mathbf{G}_\mathrm{e}(k,|\boldsymbol{r}-\boldsymbol{r}'|)$. In fact,

$$\nabla\times\iint_{S_\mathrm{s}}g(k,|\boldsymbol{r}-\boldsymbol{r}'|)\boldsymbol{F}(\boldsymbol{r}')\,\mathrm{d}S'=\nabla\times\iint_{S_\mathrm{s}}\mathbf{I}g(k,|\boldsymbol{r}-\boldsymbol{r}'|)\cdot\boldsymbol{F}(\boldsymbol{r}')\,\mathrm{d}S'$$

$$=\nabla\times\iint_{S_\mathrm{s}}\mathbf{G}_\mathrm{e}(k,|\boldsymbol{r}-\boldsymbol{r}'|)\cdot\boldsymbol{F}(\boldsymbol{r}')\,\mathrm{d}S'$$

since $\nabla'=-\nabla$ and $\nabla\times\nabla\big(\nabla'g(k,|\boldsymbol{r}-\boldsymbol{r}'|)\cdot\boldsymbol{F}(\boldsymbol{r}')\big)=\mathbf{0}$.

For an argument outside the smallest circumscribed sphere, $|\boldsymbol{r}|>\max_{\boldsymbol{r}'\in S_\mathrm{s}}r'=R$, we obtain

$$\boldsymbol{E}_\mathrm{s}(\boldsymbol{r},\omega)=-\frac{\eta_0\eta}{\mathrm{i}k}\nabla\times\left\{\nabla\times\iint_{S_\mathrm{s}}\mathbf{G}_\mathrm{e}(k,|\boldsymbol{r}-\boldsymbol{r}'|)\cdot(\hat{\boldsymbol{\nu}}(\boldsymbol{r}')\times\boldsymbol{H}(\boldsymbol{r}'))\,\mathrm{d}S'\right\}$$

$$+\nabla\times\iint_{S_\mathrm{s}}\mathbf{G}_\mathrm{e}(k,|\boldsymbol{r}-\boldsymbol{r}'|)\cdot(\hat{\boldsymbol{\nu}}(\boldsymbol{r}')\times\boldsymbol{E}(\boldsymbol{r}'))\,\mathrm{d}S',\qquad r>R$$

The expansion of the Green dyadic, $\mathbf{G}_e(k,|\boldsymbol{r}-\boldsymbol{r}'|)$, in terms of spherical vector waves from Section 7.5.2, see (7.24), reads

$$\mathbf{G}_e(k,|\boldsymbol{r}-\boldsymbol{r}'|) = \mathrm{i}k \sum_{\substack{n\\ \tau=1,2}} \boldsymbol{u}_{\tau n}(k\boldsymbol{r})\boldsymbol{v}_{\tau n}(k\boldsymbol{r}'), \qquad \boldsymbol{r}' \in S_s$$

since $\boldsymbol{r}_< = \boldsymbol{r}'$ and $\boldsymbol{r}_> = \boldsymbol{r}$ on all points $\boldsymbol{r}' \in S_s$.

Insert the expansion of the Green dyadics in the integral above and interchange the integration and summation, which is legitimate, since the sum is uniformly convergent on compact domains. We get

$$\boldsymbol{E}_s(\boldsymbol{r},\omega) = \sum_{\substack{n\\ \tau=1,2}} f_{\tau n}\boldsymbol{u}_{\tau n}(k\boldsymbol{r}) = \sum_n f_n\boldsymbol{u}_n(k\boldsymbol{r}), \qquad |\boldsymbol{r}| > \max_{\boldsymbol{r}'\in S_s} r' = R$$

where

$$f_{\tau n} = -\eta_0\eta k^2 \iint_{S_s} \boldsymbol{v}_{\tau n}(k\boldsymbol{r}') \cdot (\hat{\boldsymbol{\nu}}(\boldsymbol{r}') \times \boldsymbol{H}(\boldsymbol{r}'))\, \mathrm{d}S' + \mathrm{i}k^2 \iint_{S_s} \boldsymbol{v}_{\bar{\tau} n}(k\boldsymbol{r}') \cdot (\hat{\boldsymbol{\nu}}(\boldsymbol{r}') \times \boldsymbol{E}(\boldsymbol{r}'))\, \mathrm{d}S' \tag{7.25}$$

since $\nabla \times \boldsymbol{v}_{\tau n}(k\boldsymbol{r}') = k\boldsymbol{v}_{\bar{\tau} n}(k\boldsymbol{r}')$. This proves the existence of an expansion of the electric field in terms of radiating spherical vector waves outside the smallest circumscribed sphere of S_s.

The coefficients $f_{\tau n}$ can also be written in terms of the tangential scattered field on S_s as

$$f_{\tau n} = -\eta_0\eta k^2 \iint_{S_s} \boldsymbol{v}_{\tau n}(k\boldsymbol{r}') \cdot \left(\hat{\boldsymbol{\nu}}(\boldsymbol{r}') \times \boldsymbol{H}_s(\boldsymbol{r}')\right) \mathrm{d}S' + \mathrm{i}k^2 \iint_{S_s} \boldsymbol{v}_{\bar{\tau} n}(k\boldsymbol{r}') \cdot \left(\hat{\boldsymbol{\nu}}(\boldsymbol{r}') \times \boldsymbol{E}_s(\boldsymbol{r}')\right) \mathrm{d}S'$$

since, by the divergence theorem (Gauss' theorem)

$$\begin{aligned}
&\iint_{S_s} \left\{\mathrm{i}\eta_0\eta k\boldsymbol{v}_{\tau n} \cdot (\hat{\boldsymbol{\nu}} \times \boldsymbol{H}_i) + k\boldsymbol{v}_{\bar{\tau} n} \cdot \left(\hat{\boldsymbol{\nu}} \times \boldsymbol{E}_i\right)\right\} \mathrm{d}S \\
&\quad = -\iint_{S_s} \hat{\boldsymbol{\nu}} \cdot \{(\boldsymbol{v}_{\tau n} \times (\nabla \times \boldsymbol{E}_i)) + ((\nabla \times \boldsymbol{v}_{\tau n}) \times \boldsymbol{E}_i)\}\, \mathrm{d}S \\
&\quad = -\iiint_{V_s} \nabla \cdot \{(\boldsymbol{v}_{\tau n} \times (\nabla \times \boldsymbol{E}_i)) + ((\nabla \times \boldsymbol{v}_{\tau n}) \times \boldsymbol{E}_i)\}\, \mathrm{d}v = 0
\end{aligned}$$

where we used

$$\begin{cases} \mathrm{i}k\eta\eta_0\boldsymbol{H}_i(\boldsymbol{r}) = \nabla \times \boldsymbol{E}_i(\boldsymbol{r}) \\ k\boldsymbol{v}_{\bar{\tau} n}(k\boldsymbol{r}) = \nabla \times \boldsymbol{v}_{\tau n}(k\boldsymbol{r}) \end{cases} \quad \begin{cases} \nabla \times (\nabla \times \boldsymbol{E}_i(\boldsymbol{r})) = k^2\boldsymbol{E}_i(\boldsymbol{r}) \\ \nabla \times (\nabla \times \boldsymbol{v}_{\tau n}(k\boldsymbol{r})) = k^2\boldsymbol{v}_{\tau n}(k\boldsymbol{r}) \end{cases} \quad \boldsymbol{r} \in V_s$$

and for all $\boldsymbol{r} \in V_s$

$$\begin{aligned}
\nabla \cdot \{(\boldsymbol{v}_{\tau n} \times (\nabla \times \boldsymbol{E}_i)) + ((\nabla \times \boldsymbol{v}_{\tau n}) \times \boldsymbol{E}_i)\} &= (\nabla \times \boldsymbol{v}_{\tau n}) \times (\nabla \times \boldsymbol{E}_i) - k^2\boldsymbol{v}_{\tau n} \cdot \boldsymbol{E}_i \\
&\quad + k^2\boldsymbol{v}_{\tau n} \cdot \boldsymbol{E}_i - (\nabla \times \boldsymbol{v}_{\tau n}) \times (\nabla \times \boldsymbol{E}_i) = 0
\end{aligned}$$

Similarly, for an argument inside the largest inscribed sphere, $|\boldsymbol{r}| < \min_{r'\in S_s} r'$, we obtain

$$\boldsymbol{E}_i(\boldsymbol{r}) = \frac{\eta_0\eta}{ik}\nabla\times\left\{\nabla\times\iint_{S_s}\mathbf{G}_e(k,|\boldsymbol{r}-\boldsymbol{r}'|)\cdot(\hat{\boldsymbol{\nu}}(\boldsymbol{r}')\times\boldsymbol{H}(\boldsymbol{r}'))\,dS'\right\}$$
$$-\nabla\times\iint_{S_s}\mathbf{G}_e(k,|\boldsymbol{r}-\boldsymbol{r}'|)\cdot\left(\hat{\boldsymbol{\nu}}(\boldsymbol{r}')\times\boldsymbol{E}(\boldsymbol{r}')\right)\,dS',\qquad |\boldsymbol{r}| < \min_{\boldsymbol{r}'\in S_s} r'$$

and the expansion of $\mathbf{G}_e(k,|\boldsymbol{r}-\boldsymbol{r}'|)$ now is

$$\mathbf{G}_e(k,|\boldsymbol{r}-\boldsymbol{r}'|) = ik\sum_{\substack{n\\ \tau=1,2}}\boldsymbol{v}_{\tau n}(k\boldsymbol{r})\boldsymbol{u}_{\tau n}(k\boldsymbol{r}'),\qquad \boldsymbol{r}'\in S_s$$

since $\boldsymbol{r}_< = \boldsymbol{r}$ and $\boldsymbol{r}_> = \boldsymbol{r}'$ on all points $\boldsymbol{r}'\in S_s$. Insertion and interchanging the summation and integration give

$$\boldsymbol{E}_i(\boldsymbol{r},\omega) = \sum_{\substack{n\\ \tau=1,2}} a_{\tau n}\boldsymbol{v}_{\tau n}(k\boldsymbol{r}) = \sum_n a_n\boldsymbol{v}_n(k\boldsymbol{r}),\qquad |\boldsymbol{r}| < \min_{\boldsymbol{r}'\in S_s} r'$$

where

$$a_{\tau n} = \eta_0\eta k^2\iint_{S_s}\boldsymbol{u}_{\tau n}(k\boldsymbol{r}')\cdot(\hat{\boldsymbol{\nu}}(\boldsymbol{r}')\times\boldsymbol{H}(\boldsymbol{r}'))\,dS' - ik^2\iint_{S_s}\boldsymbol{u}_{\bar{\tau}n}(k\boldsymbol{r}')\cdot\left(\hat{\boldsymbol{\nu}}(\boldsymbol{r}')\times\boldsymbol{E}(\boldsymbol{r}')\right)\,dS'$$

This set of equations is called the Null-field equations. These equations are uniquely solvable, and we refer the reader to the literature regarding details (mostly on the corresponding acoustic problem) on this topic [39, 172, 173, 255].

The coefficients $f_{\tau n}$ in (7.25) do not depend on which surface S_s they are evaluated on in the following respect.

Let S_1 and S_2 be two closed surfaces in the exterior of the scatterer that both enclose the scatterer in V_s. The volume bounded by S_1 and S_2 is denoted V, see Figure 4.10 on page 228. Then

$$ik\iint_{S_1}\left\{i\eta_0\eta k\boldsymbol{v}_{\tau n}(k\boldsymbol{r})\cdot(\hat{\boldsymbol{\nu}}(\boldsymbol{r})\times\boldsymbol{H}_s(\boldsymbol{r})) + k\boldsymbol{v}_{\bar{\tau}n}(k\boldsymbol{r})\cdot\left(\hat{\boldsymbol{\nu}}(\boldsymbol{r})\times\boldsymbol{E}_s(\boldsymbol{r})\right)\right\}\,dS$$
$$= ik\iint_{S_2}\left\{i\eta_0\eta k\boldsymbol{v}_{\tau n}(k\boldsymbol{r})\cdot(\hat{\boldsymbol{\nu}}(\boldsymbol{r})\times\boldsymbol{H}_s(\boldsymbol{r})) + k\boldsymbol{v}_{\bar{\tau}n}(k\boldsymbol{r})\cdot\left(\hat{\boldsymbol{\nu}}(\boldsymbol{r})\times\boldsymbol{E}_s(\boldsymbol{r})\right)\right\}\,dS \quad (7.26)$$

To prove this, we start by noting that ($\boldsymbol{r}\in V_e$)

$$\begin{cases} ik\eta\eta_0\boldsymbol{H}_s(\boldsymbol{r}) = \nabla\times\boldsymbol{E}_s(\boldsymbol{r})\\ k\boldsymbol{v}_{\bar{\tau}n}(k\boldsymbol{r}) = \nabla\times\boldsymbol{v}_{\tau n}(k\boldsymbol{r})\end{cases}\qquad \begin{cases}\nabla\times(\nabla\times\boldsymbol{E}_s(\boldsymbol{r})) = k^2\boldsymbol{E}_s(\boldsymbol{r})\\ \nabla\times(\nabla\times\boldsymbol{v}_{\tau n}(k\boldsymbol{r})) = k^2\boldsymbol{v}_{\tau n}(k\boldsymbol{r})\end{cases}$$

By the divergence theorem (Gauss' theorem), the difference between the integrals in (7.26) is

$$
\begin{aligned}
\mathrm{i}k & \iint\limits_{S_2-S_1} \left\{\mathrm{i}\eta_0\eta k \boldsymbol{v}_{\tau n}(k\boldsymbol{r}) \cdot \left(\hat{\boldsymbol{\nu}}(\boldsymbol{r}) \times \boldsymbol{H}_\mathrm{s}(\boldsymbol{r})\right) + k \boldsymbol{v}_{\bar{\tau} n}(k\boldsymbol{r}) \cdot \left(\hat{\boldsymbol{\nu}}(\boldsymbol{r}) \times \boldsymbol{E}_\mathrm{s}(\boldsymbol{r})\right)\right\} \mathrm{d}S \\
&= -\mathrm{i}k \iint\limits_{S_2-S_1} \hat{\boldsymbol{\nu}}(\boldsymbol{r}) \cdot \{\boldsymbol{v}_{\tau n}(k\boldsymbol{r}) \times (\nabla \times \boldsymbol{E}_\mathrm{s}(\boldsymbol{r})) + (\nabla \times \boldsymbol{v}_{\tau n}(k\boldsymbol{r})) \times \boldsymbol{E}_\mathrm{s}(\boldsymbol{r})\} \, \mathrm{d}S \\
&= -\mathrm{i}k \iiint_V \nabla \cdot (\boldsymbol{v}_{\tau n}(k\boldsymbol{r}) \times (\nabla \times \boldsymbol{E}_\mathrm{s}(\boldsymbol{r})) + (\nabla \times \boldsymbol{v}_{\tau n}(k\boldsymbol{r})) \times \boldsymbol{E}_\mathrm{s}(\boldsymbol{r})) \, \mathrm{d}v
\end{aligned}
$$

However, the integrand in the last integral is zero, since

$$
\begin{aligned}
\nabla \cdot (\boldsymbol{v}_{\tau n} \times (\nabla \times \boldsymbol{E}_\mathrm{s}) + (\nabla \times \boldsymbol{v}_{\tau n}) \times \boldsymbol{E}_\mathrm{s}) &= (\nabla \times \boldsymbol{v}_{\tau n}) \cdot (\nabla \times \boldsymbol{E}_\mathrm{s}) - k^2 \boldsymbol{v}_{\tau n} \cdot \boldsymbol{E}_\mathrm{s} \\
&+ k^2 \boldsymbol{v}_{\tau n} \cdot \boldsymbol{E}_\mathrm{s} - (\nabla \times \boldsymbol{v}_{\tau n}) \cdot (\nabla \times \boldsymbol{E}_\mathrm{s}) = 0
\end{aligned}
$$

and (7.26) is proved. This shows that the coefficients $f_{\tau n}$ do not depend on which surface S is used for evaluation of the coefficients, as long as it encloses V_s. The coefficients, however, do depend on the material inside V_s and the incident field. We also note that the scattered field in the expressions (7.26) can be replaced by the total fields, as long as the bounding surface S does not include any region where the sources of the incident field are located.

The results in this section show that there exists a series expansion in terms of the spherical vector waves of the electric (and magnetic) field inside the inscribed and outside the circumscribed spheres of the scatterer, respectively. The explicit form of the expansion coefficients is also given in terms of the tangential electric and magnetic fields on the scatterer. In Chapter 9, these equations are exploited in more detail.

7.7 Expansion of Sources

The expansions of two different kinds of sources—the plane wave and the vertical electric dipole—are presented in this section.

7.7.1 Expansion of a Plane Wave

The incident field is commonly taken as a plane wave, which is a well-defined vector field everywhere in space. Therefore, the plane wave has an expansion in terms of the regular spherical vector waves $\boldsymbol{v}_n(k\boldsymbol{r})$, i.e.,

$$
\boldsymbol{E}_\mathrm{i}(\boldsymbol{r}, \omega) = \boldsymbol{E}_0 \mathrm{e}^{\mathrm{i}k\hat{\boldsymbol{k}}_\mathrm{i} \cdot \boldsymbol{r}} = \sum_{\tau=1}^{3} \sum_{\sigma=\mathrm{e,o}} \sum_{l=0}^{\infty} \sum_{m=0}^{l} a_{\tau\sigma ml} \boldsymbol{v}_{\tau\sigma ml}(k\boldsymbol{r}) = \sum_n a_n \boldsymbol{v}_n(k\boldsymbol{r})
$$

Notice that at this stage, we have made no assumption on the complex vector $\boldsymbol{E}_0$. For the moment it is arbitrary. The expansion coefficients $a_{\tau n}$ can be computed, see Problem 7.5. The result is

$$
\begin{cases}
a_{1n} = 4\pi \mathrm{i}^l \boldsymbol{A}_{1n}(\hat{\boldsymbol{k}}_\mathrm{i}) \cdot \boldsymbol{E}_0 = 4\pi \mathrm{i}^l \boldsymbol{A}_{2n}(\hat{\boldsymbol{k}}_\mathrm{i}) \cdot \left(\hat{\boldsymbol{k}}_\mathrm{i} \times \boldsymbol{E}_0\right) \\
a_{2n} = -4\pi \mathrm{i}^{l+1} \boldsymbol{A}_{2n}(\hat{\boldsymbol{k}}_\mathrm{i}) \cdot \boldsymbol{E}_0 = 4\pi \mathrm{i}^{l+1} \boldsymbol{A}_{1n}(\hat{\boldsymbol{k}}_\mathrm{i}) \cdot \left(\hat{\boldsymbol{k}}_\mathrm{i} \times \boldsymbol{E}_0\right) \\
a_{3n} = -4\pi \mathrm{i}^{l+1} \boldsymbol{E}_0 \cdot \boldsymbol{A}_{3n}(\hat{\boldsymbol{k}}_\mathrm{i})
\end{cases}
\tag{7.27}
$$

The first two are conveniently collected into one expression

$$a_{\tau n} = 4\pi \mathrm{i}^{l+1-\tau} \boldsymbol{A}_{\tau n}(\hat{\boldsymbol{k}}_\mathrm{i}) \cdot \boldsymbol{E}_0 = 4\pi \mathrm{i}^{l-1+\tau} \boldsymbol{A}_{\bar{\tau} n}(\hat{\boldsymbol{k}}_\mathrm{i}) \cdot \left(\hat{\boldsymbol{k}}_\mathrm{i} \times \boldsymbol{E}_0\right), \qquad \tau = 1,2 \tag{7.28}$$

or with the use of the complex-valued vector spherical harmonics defined below in (7.32)

$$a_{\tau n} = 4\pi \boldsymbol{a}_{\tau n}(\hat{\boldsymbol{k}}_\mathrm{i}) \cdot \boldsymbol{E}_0, \qquad \tau = 1,2$$

The plane wave has to satisfy the Maxwell equations, which implies that the complex vector $\boldsymbol{E}_0$ and the direction of the plane wave $\hat{\boldsymbol{k}}_\mathrm{i}$ have to be orthogonal, i.e.,

$$\boldsymbol{E}_0 \cdot \hat{\boldsymbol{k}}_\mathrm{i} = 0$$

Directly, we notice $a_{3n} = 0$ in the expansion, since $\boldsymbol{A}_{3n}(\hat{\boldsymbol{k}}_\mathrm{i})$ is proportional to $\hat{\boldsymbol{k}}_\mathrm{i}$, and consequently, the summation in the τ-index is only over 1 and 2. An alternative way of expressing this condition is to require that the plane wave has zero divergence. Since only $\tau = 1,2$ contribute to the sum, the l-summation starts on $l = 1$ ($l = 0$ gives no contribution), i.e.,

$$\boldsymbol{E}_\mathrm{i}(\boldsymbol{r}, \omega) = \boldsymbol{E}_0 \mathrm{e}^{\mathrm{i}k\hat{\boldsymbol{k}}_\mathrm{i}\cdot\boldsymbol{r}} = \sum_{\substack{n \\ \tau=1,2}} a_{\tau n} \boldsymbol{v}_{\tau n}(k\boldsymbol{r})$$

In particular, if the incident direction is along the positive z-direction, i.e., $\hat{\boldsymbol{k}}_\mathrm{i} = \hat{\boldsymbol{z}}$, we get, see (C.9) on page 631

$$\begin{cases} a_{1n} = \mathrm{i}^l \delta_{m1} \sqrt{2\pi(2l+1)} \boldsymbol{E}_0 \cdot \left(\delta_{\sigma\mathrm{o}} \hat{\boldsymbol{x}} - \delta_{\sigma\mathrm{e}} \hat{\boldsymbol{y}}\right) \\ a_{2n} = -\mathrm{i}^{l+1} \delta_{m1} \sqrt{2\pi(2l+1)} \boldsymbol{E}_0 \cdot \left(\delta_{\sigma\mathrm{e}} \hat{\boldsymbol{x}} + \delta_{\sigma\mathrm{o}} \hat{\boldsymbol{y}}\right) \\ a_{3n} = 0 \end{cases} \qquad \hat{\boldsymbol{k}}_\mathrm{i} = \hat{\boldsymbol{z}} \tag{7.29}$$

7.7.2 Expansion of a Vertical Electric Dipole

The electric field generated by a vertical electric dipole located at $\boldsymbol{r}_\mathrm{s}$ is proportional to $\boldsymbol{u}_{2\mathrm{e}01}(k(\boldsymbol{r} - \boldsymbol{r}_\mathrm{s}))$. This field can easily be expanded in regular spherical vector waves, $\boldsymbol{v}_n(k\boldsymbol{r})$, close to the origin. We obtain this expansion by the use of the translation matrices presented in Section 7.4.2. The explicit expression is, see (7.19)

$$\begin{aligned} \boldsymbol{u}_{2\mathrm{e}01}(k(\boldsymbol{r} - \boldsymbol{r}_\mathrm{s})) &= \sum_n \mathcal{P}_{2\mathrm{e}01,n}(-k\boldsymbol{r}_\mathrm{s}) \boldsymbol{v}_n(k\boldsymbol{r}) \\ &= \sum_n \mathcal{P}_{n,2\mathrm{e}01}(k\boldsymbol{r}_\mathrm{s}) \boldsymbol{v}_n(k\boldsymbol{r}) = \sum_n a_n \boldsymbol{v}_n(k\boldsymbol{r}), \qquad r < r_\mathrm{s} \end{aligned}$$

where the expansion coefficients a_n are

$$a_n = \mathcal{P}_{n,2\mathrm{e}01}(k\boldsymbol{r}_\mathrm{s})$$

Other orientations of the electric dipole (involve other m values and σ values) and the magnetic dipole of arbitrary orientation (involve $\tau = 1$ and $l = 1$) are found by a similar application of the translation matrices, see Problem 7.6.

7.8 Far Field Amplitude and the Transition Matrix

The scattered field has its (fictitious) sources inside the scatterer. Outside the smallest circumscribed sphere of the scatterer, radius R, the scattered field has an expansion in out-going spherical vector waves, i.e.,

$$\boldsymbol{E}_{\rm s}(\boldsymbol{r},\omega)=\sum_{\tau=1}^{2}\sum_{\sigma=\rm e,o}\sum_{l=1}^{\infty}\sum_{m=0}^{l}f_{\tau\sigma ml}\boldsymbol{u}_{\tau\sigma ml}(k\boldsymbol{r})=\sum_{\substack{n\\ \tau=1,2}}f_{\tau n}\boldsymbol{u}_{\tau n}(k\boldsymbol{r}),\qquad r>R \tag{7.30}$$

where $f_{\tau n}$ is given by (7.25). Note that the sum is over $\tau = 1,2$ and that the l-sum starts at $l = 1$.

The corresponding magnetic field has the expansion

$$\boldsymbol{H}_{\rm s}(\boldsymbol{r},\omega)=\frac{1}{\mathrm{i}\eta_0\eta}\sum_{\substack{n\\ \tau=1,2}}f_{\tau n}\boldsymbol{u}_{\bar{\tau}n}(k\boldsymbol{r}),\qquad r>R$$

The spherical Hankel functions, $h_l^{(1)}(z)$, have the asymptotic expansion valid for large arguments ($|z|\gg 1$)

$$\begin{cases} h_l^{(1)}(z)=\dfrac{\mathrm{i}^{-l-1}}{z}\mathrm{e}^{\mathrm{i}z}+O(z^{-2})\\[2ex] h_l^{(1)\prime}(z)=\dfrac{\mathrm{i}^{-l}}{z}\mathrm{e}^{\mathrm{i}z}+O(z^{-2})\end{cases}$$

and at a large distance from the origin, the out-going spherical vector waves have the following asymptotic expansions

$$\boldsymbol{u}_{\tau n}(k\boldsymbol{r})=\mathrm{i}^{-l-2+\tau}\frac{\mathrm{e}^{\mathrm{i}kr}}{kr}\boldsymbol{A}_{\tau n}(\hat{\boldsymbol{r}})+o((kr)^{-1}),\qquad \tau=1,2$$

Note that this approximation only holds for $\tau = 1,2$, which are the appropriate τ-index of the electric and the magnetic fields.

From these asymptotic expansions, we conjecture that the far field amplitude $\boldsymbol{F}(\hat{\boldsymbol{r}})$, in a series expansion of vector spherical harmonics, is

$$\boldsymbol{F}(\hat{\boldsymbol{r}})=\frac{1}{\mathrm{i}k}\sum_{\substack{n\\ \tau=1,2}}\mathrm{i}^{-l+\tau-1}f_{\tau n}\boldsymbol{A}_{\tau n}(\hat{\boldsymbol{r}})=\frac{1}{\mathrm{i}k}\sum_{\substack{n\\ \tau=1,2}}f_{\tau n}\boldsymbol{a}_{\tau n}^{*}(\hat{\boldsymbol{r}}) \tag{7.31}$$

where the complex vector spherical harmonics, $\boldsymbol{a}_n(\hat{\boldsymbol{r}})$, are

$$\boldsymbol{a}_n(\hat{\boldsymbol{r}})=\mathrm{i}^{l-\tau+1}\boldsymbol{A}_n(\hat{\boldsymbol{r}}) \tag{7.32}$$

With the parity relation (C.8) on page 631, $\boldsymbol{A}_n(-\hat{\boldsymbol{r}})=(-1)^{l+\tau-1}\boldsymbol{A}_n(\hat{\boldsymbol{r}})$, we get

$$\boldsymbol{a}_n(-\hat{\boldsymbol{r}})=(-1)^{l+\tau-1}\boldsymbol{a}_n(\hat{\boldsymbol{r}})=\boldsymbol{a}_n^{*}(\hat{\boldsymbol{r}}) \tag{7.33}$$

One way to verify this conjecture in (7.31) is to prove that this series is identical to (4.18) on page 198

$$F(\hat{r}) = \frac{\mathrm{i}k}{4\pi}\hat{r} \times \iint_{S_s} \left[\hat{\nu}(r') \times E_s(r') - \eta_0\eta\hat{r} \times \left(\hat{\nu}(r') \times H_s(r')\right)\right]\mathrm{e}^{-\mathrm{i}k\hat{r}\cdot r'}\,\mathrm{d}S'$$

It is not essential to evaluate this integral on the surface S_s. In fact, any surface that circumscribes S_s gives the same answer, see (4.19) on page 198. We choose a spherical surface of radius $R_1 > R$ that circumscribes the surface S_s. On this spherical surface, we know that the expansions (7.30)

$$E_s(r,\omega) = \sum_{\substack{n\\ \tau=1,2}} f_{\tau n} u_{\tau n}(kr), \qquad r \geq R_1$$

and

$$H_s(r,\omega) = \frac{1}{\mathrm{i}\eta_0\eta} \sum_{\substack{n\\ \tau=1,2}} f_{\tau n} u_{\overline{\tau} n}(kr), \qquad r \geq R_1$$

hold. The integral representation of the far field is then evaluated as (interchange of summation and integration is valid since the series are uniformly convergent on compact domains)

$$\begin{aligned} F(\hat{r}) &= \frac{\mathrm{i}kR_1^2}{4\pi}\hat{r} \times \iint_{\Omega} \left[\hat{r}' \times E_s(R_1\hat{r}') - \eta_0\eta\hat{r} \times \left(\hat{r}' \times H_s(R_1\hat{r}')\right)\right]\mathrm{e}^{-\mathrm{i}kR_1\hat{r}\cdot\hat{r}'}\,\mathrm{d}\Omega' \\ &= \frac{\mathrm{i}kR_1^2}{4\pi}\hat{r} \times \sum_{\substack{n\\ \tau=1,2}} f_{\tau n} \left\{\iint_{\Omega} \left[\hat{r}' \times u_{\tau n}(kR_1\hat{r}') + \mathrm{i}\hat{r} \times \left(\hat{r}' \times u_{\overline{\tau} n}(kR_1\hat{r}')\right)\right]\mathrm{e}^{-\mathrm{i}kR_1\hat{r}\cdot\hat{r}'}\,\mathrm{d}\Omega'\right\} \end{aligned}$$

This expression suggests that we need to evaluate integrals of the vector spherical harmonics over the unit sphere. From Problem 7.4, we collect

$$\iint_{\Omega} A_{\tau n}(\hat{r}')\mathrm{e}^{\mathrm{i}kR_1\hat{r}\cdot\hat{r}'}\,\mathrm{d}\Omega' = 4\pi\mathrm{i}^{l-\tau+1} v_{\tau n}(kR_1\hat{r}), \qquad \tau = 1,2$$

and due to (C.7) on page 631

$$\hat{r} \times A_{\tau n}(\hat{r}) = (-1)^{\tau-1} A_{\overline{\tau} n}(\hat{r}), \qquad \tau = 1,2$$

and (C.8) on page 631

$$A_{\tau n}(-\hat{r}) = (-1)^{l+\tau-1} A_{\tau n}(\hat{r}), \qquad \tau = 1,2$$

we obtain

$$\hat{r} \times \iint_{\Omega} \hat{r}' \times A_{\tau n}(\hat{r}')\mathrm{e}^{-\mathrm{i}k\hat{k}\cdot r'}\,\mathrm{d}\Omega' = 4\pi\mathrm{i}^{-l-\tau} V_{\overline{\tau} l}(r) A_{\tau n}(\hat{r}), \qquad \tau = 1,2$$

where we temporarily have used the radial functions $V_{\tau l}(kr)$ and $U_{\tau l}(kr)$, see Comment 7.2

$$\begin{cases} V_{1l}(kr) = j_l(kr) \\ V_{2l}(kr) = \dfrac{(krj_l(kr))'}{kr} \end{cases} \qquad \begin{cases} U_{1l}(kr) = h_l^{(1)}(kr) \\ U_{2l}(kr) = \dfrac{(krh_l^{(1)}(kr))'}{kr} \end{cases}$$

Collecting these results, we get

$$\boldsymbol{F}(\hat{\boldsymbol{r}}) = \mathrm{i}kR_1^2 \sum_{\substack{n \\ \tau=1,2}} f_{\tau n} \mathrm{i}^{-l-\tau} \left\{ U_{\tau l}(kR_1)V_{\overline{\tau} l}(kR_1) - U_{\overline{\tau} l}(kR_1)V_{\tau l}(kR_1) \right\} \boldsymbol{A}_{\tau n}(\hat{\boldsymbol{r}})$$

The Wronskian of the radial functions implies

$$\begin{aligned} U_{1l}(kR_1)V_{2l}(kR_1) - U_{2l}(kR_1)V_{1l}(kR_1) &= h_l^{(1)}(kR_1)\frac{(kR_1j_l(kR_1))'}{kR_1} - \frac{(kR_1h_l^{(1)}(kR_1))'}{kR_1}j_l(kR_1) \\ &= h_l^{(1)}(kR_1)j_l'(kR_1) - h_l^{(1)'}(kR_1)j_l(kR_1) = -\frac{\mathrm{i}}{(kR_1)^2} \end{aligned}$$

The final expression of the far field amplitude is

$$\boldsymbol{F}(\hat{\boldsymbol{r}}) = \frac{1}{\mathrm{i}k} \sum_{\substack{n \\ \tau=1,2}} f_{\tau n} \mathrm{i}^{-l+\tau-1} \boldsymbol{A}_{\tau n}(\hat{\boldsymbol{r}})$$

which proves the conjecture in (7.31).

For a given far field amplitude $\boldsymbol{F}(\hat{\boldsymbol{r}})$, the coefficients f_{1n} and f_{2n} are determined by the orthogonality of the vector spherical harmonics. We get

$$f_{\tau n} = \mathrm{i}^{l+2-\tau} k \iint_{\Omega} \boldsymbol{A}_{\tau n}(\hat{\boldsymbol{r}}) \cdot \boldsymbol{F}(\hat{\boldsymbol{r}}) \, \mathrm{d}\Omega$$

Traditionally, two types of scattering problems are identified. The determination of the scattered field, with the knowledge of the incident field and the geometry and material properties of the scatterer, is called a direct scattering problem. The other type of scattering problem is the inverse scattering problem—from the knowledge of the incident and the scattered fields, determine the scatterer.

In a direct scattering problem, the incident field, which is determined by the expansion coefficients a_n, is known, and we seek the expansion coefficients of the scattered field, f_n, which determine the scattered field. Provided the scatterer consists of a linear material, there is a linear relation between the two sets of expansion coefficients a_n and f_n. We write this relation formally as

$$f_n = \sum_{n'} T_{nn'} a_{n'} \tag{7.34}$$

The infinite dimensional matrix $T_{nn'}$ is called the transition matrix or the T-matrix for the scatterer. The T-matrix is independent of the incident field and it is determined only by the geometry and the material properties of the scatterer. One way of solving the direct scattering problem is to find the transition matrix of the scatterer. In Chapter 8, the transition matrix for same simple geometries is calculated (spherical cases when the transition matrix is a diagonal matrix). A more general method to calculate the transition matrix for bounded scatterers is presented in Chapter 9. We have investigated the sum rules for the extinction cross section in Section 6.4. Sum rules for the transition matrix are investigated in Reference 12.

7.8.1 Scattering Dyadic

The multipole expansion of a plane wave was developed in Section 7.7.1, i.e.,

$$\boldsymbol{E}_{\mathrm{i}}(\boldsymbol{r},\omega)=\boldsymbol{E}_0\mathrm{e}^{\mathrm{i}k\hat{\boldsymbol{k}}_{\mathrm{i}}\cdot\boldsymbol{r}}=\sum_n a_n\boldsymbol{v}_n(k\boldsymbol{r})$$

where

$$a_n=4\pi\mathrm{i}^{l-\tau+1}\boldsymbol{E}_0\cdot\boldsymbol{A}_n(\hat{\boldsymbol{k}}_{\mathrm{i}})=4\pi\boldsymbol{E}_0\cdot\boldsymbol{a}_n(\hat{\boldsymbol{k}}_{\mathrm{i}}),\qquad\tau=1,2$$

and where the vector $\boldsymbol{a}_n(\hat{\boldsymbol{r}})$ is defined in (7.32). Note that the complex vector spherical harmonics $\boldsymbol{a}_n(\hat{\boldsymbol{r}})$ do not depend on the frequency k, and they form a complete set of vectors on the unit sphere.

From the expansion of the far field amplitude, (7.31), we get in terms of the transition matrix of the scatterer

$$\boldsymbol{F}(\hat{\boldsymbol{r}})=\frac{1}{\mathrm{i}k}\sum_{n,n'}\boldsymbol{a}_n^*(\hat{\boldsymbol{r}})T_{nn'}a_{n'}=\frac{4\pi}{\mathrm{i}k}\sum_{n,n'}\boldsymbol{a}_n^*(\hat{\boldsymbol{r}})T_{nn'}\boldsymbol{a}_{n'}(\hat{\boldsymbol{k}}_{\mathrm{i}})\cdot\boldsymbol{E}_0$$

The general expression of the scattering dyadic $\mathbf{S}(\hat{\boldsymbol{r}},\hat{\boldsymbol{k}}_{\mathrm{i}})$ then becomes [285]

$$\mathbf{S}(\hat{\boldsymbol{r}},\hat{\boldsymbol{k}}_{\mathrm{i}})=\frac{4\pi}{\mathrm{i}k}\sum_{n,n'}\boldsymbol{a}_n^*(\hat{\boldsymbol{r}})T_{nn'}\boldsymbol{a}_{n'}(\hat{\boldsymbol{k}}_{\mathrm{i}})\tag{7.35}$$

and

$$\mathbf{S}^{\dagger}(\hat{\boldsymbol{r}},\hat{\boldsymbol{k}}_{\mathrm{i}})=-\frac{4\pi}{\mathrm{i}k}\sum_{n,n'}\boldsymbol{a}_n^*(\hat{\boldsymbol{k}}_{\mathrm{i}})T_{nn'}^{\dagger}\boldsymbol{a}_{n'}(\hat{\boldsymbol{r}})$$

In particular, in the forward direction, $\hat{\boldsymbol{r}}=\hat{\boldsymbol{k}}_{\mathrm{i}}$, we obtain

$$\mathbf{S}(\hat{\boldsymbol{k}}_{\mathrm{i}},\hat{\boldsymbol{k}}_{\mathrm{i}})=\frac{4\pi}{\mathrm{i}k}\sum_{n,n'}\boldsymbol{a}_n^*(\hat{\boldsymbol{k}}_{\mathrm{i}})T_{nn'}\boldsymbol{a}_{n'}(\hat{\boldsymbol{k}}_{\mathrm{i}})\tag{7.36}$$

and the backscattering direction, $\hat{\boldsymbol{r}}=-\hat{\boldsymbol{k}}_{\mathrm{i}}$, we have, see (7.33)

$$\mathbf{S}(-\hat{\boldsymbol{k}}_{\mathrm{i}},\hat{\boldsymbol{k}}_{\mathrm{i}})=\frac{4\pi}{\mathrm{i}k}\sum_{n,n'}\boldsymbol{a}_n(\hat{\boldsymbol{k}}_{\mathrm{i}})T_{nn'}\boldsymbol{a}_{n'}(\hat{\boldsymbol{k}}_{\mathrm{i}})$$

The relation in (7.35) can also be used in reverse. If we know the scattering dyadic, $\mathbf{S}(\hat{\boldsymbol{r}}, \hat{\boldsymbol{k}}_\text{i})$, for all directions $\hat{\boldsymbol{r}}$ and incident directions $\hat{\boldsymbol{k}}_\text{i}$, then by orthogonality of the complex vector spherical harmonics, we obtain the transition matrix as

$$T_{nn'} = \text{i}^{l-l'-\tau+\tau'} \frac{\text{i}k}{4\pi} \iint\limits_{\Omega} \left(\iint\limits_{\Omega} \boldsymbol{A}_n(\hat{\boldsymbol{r}}) \cdot \mathbf{S}(\hat{\boldsymbol{r}}, \hat{\boldsymbol{r}}') \cdot \boldsymbol{A}_{n'}(\hat{\boldsymbol{r}}')\, \text{d}\Omega' \right) \text{d}\Omega$$

Alternative approaches to find the transition matrix using far field expressions are presented in References 76–78.

If the scattering dyadic satisfies the reciprocity condition (4.45) on page 223, i.e.,

$$\mathbf{S}(-\hat{\boldsymbol{k}}_\text{i}^\text{a}, \hat{\boldsymbol{k}}_\text{i}^\text{b}) = \mathbf{S}^\text{t}(-\hat{\boldsymbol{k}}_\text{i}^\text{b}, \hat{\boldsymbol{k}}_\text{i}^\text{a})$$

we obtain from (7.35)

$$\sum_{n,n'} \boldsymbol{a}_n^*(-\hat{\boldsymbol{k}}_\text{i}^\text{a}) T_{nn'} \boldsymbol{a}_{n'}(\hat{\boldsymbol{k}}_\text{i}^\text{b}) = \sum_{n,n'} \boldsymbol{a}_n(\hat{\boldsymbol{k}}_\text{i}^\text{a}) T_{nn'}^\text{t} \boldsymbol{a}_{n'}^*(-\hat{\boldsymbol{k}}_\text{i}^\text{b})$$

Use (7.33), and insert in above. We obtain

$$\sum_{n,n'} \boldsymbol{a}_n^*(-\hat{\boldsymbol{k}}_\text{i}^\text{a}) T_{nn'} \boldsymbol{a}_{n'}(\hat{\boldsymbol{k}}_\text{i}^\text{b}) = \sum_{n,n'} \boldsymbol{a}_n^*(-\hat{\boldsymbol{k}}_\text{i}^\text{a}) T_{nn'}^\text{t} \boldsymbol{a}_{n'}(\hat{\boldsymbol{k}}_\text{i}^\text{b})$$

For this relation to be true for all unit vectors $\hat{\boldsymbol{k}}_\text{i}^\text{a}$ and $\hat{\boldsymbol{k}}_\text{i}^\text{b}$, the transition matrix must be symmetric, i.e.,

$$T = T^\text{t}$$

Example 7.1

Let the direction of incidence be $\hat{\boldsymbol{k}}_\text{i} = \hat{\boldsymbol{z}}$. The scattering dyadic in the forward scattering direction then becomes

$$\mathbf{S}(\hat{\boldsymbol{z}}, \hat{\boldsymbol{z}}) = \frac{4\pi}{\text{i}k} \sum_{n,n'} \text{i}^{(l'-l)-(\tau'-\tau)} \boldsymbol{A}_n(\hat{\boldsymbol{z}}) T_{nn'} \boldsymbol{A}_{n'}(\hat{\boldsymbol{z}})$$

We use (C.9) on page 631 and get

$$\mathbf{S}(\hat{\boldsymbol{z}}, \hat{\boldsymbol{z}}) = \frac{1}{2\text{i}k} \sum_{l,l'=1}^{\infty} \sum_{\sigma,\sigma'} \sum_{\tau,\tau'} \text{i}^{(l'-l)-(\tau'-\tau)} \sqrt{(2l+1)(2l'+1)}\, \boldsymbol{b}_{\tau\sigma} T_{\tau\sigma 1l,\tau'\sigma' 1l'} \boldsymbol{b}_{\tau'\sigma'}$$

where

$$\boldsymbol{b}_{\tau\sigma} = \left(-\hat{\boldsymbol{y}}\delta_{\sigma\text{e}} + \hat{\boldsymbol{x}}\delta_{\sigma\text{o}}\right)\delta_{\tau 1} + \left(\hat{\boldsymbol{x}}\delta_{\sigma\text{e}} + \hat{\boldsymbol{y}}\delta_{\sigma\text{o}}\right)\delta_{\tau 2}$$

The components of the scattering dyadic are

$$\mathbf{S}(\hat{\boldsymbol{z}}, \hat{\boldsymbol{z}}) = \hat{\boldsymbol{x}}\hat{\boldsymbol{x}} S_{xx} + \hat{\boldsymbol{x}}\hat{\boldsymbol{y}} S_{xy} + \hat{\boldsymbol{y}}\hat{\boldsymbol{x}} S_{yx} + \hat{\boldsymbol{y}}\hat{\boldsymbol{y}} S_{yy}$$

where

$$S_{xx} = \frac{1}{2ik} \sum_{l,l'=1}^{\infty} \mathrm{i}^{l'-l} \sqrt{(2l+1)(2l'+1)} \big(T_{1\mathrm{o}1l,1\mathrm{o}1l'} - \mathrm{i} T_{1\mathrm{o}1l,2\mathrm{e}1l'} + \mathrm{i} T_{2\mathrm{e}1l,1\mathrm{o}1l'} + T_{2\mathrm{e}1l,2\mathrm{e}1l'} \big)$$

$$S_{xy} = \frac{1}{2ik} \sum_{l,l'=1}^{\infty} \mathrm{i}^{l'-l} \sqrt{(2l+1)(2l'+1)} \big(-T_{1\mathrm{o}1l,1\mathrm{e}1l'} - \mathrm{i} T_{1\mathrm{o}1l,2\mathrm{o}1l'} - \mathrm{i} T_{2\mathrm{e}1l,1\mathrm{e}1l'} + T_{2\mathrm{e}1l,2\mathrm{o}1l'} \big)$$

$$S_{yx} = \frac{1}{2ik} \sum_{l,l'=1}^{\infty} \mathrm{i}^{l'-l} \sqrt{(2l+1)(2l'+1)} \big(-T_{1\mathrm{e}1l,1\mathrm{o}1l'} + \mathrm{i} T_{1\mathrm{e}1l,2\mathrm{e}1l'} + \mathrm{i} T_{2\mathrm{o}1l,1\mathrm{o}1l'} + T_{2\mathrm{o}1l,2\mathrm{e}1l'} \big)$$

$$S_{yy} = \frac{1}{2ik} \sum_{l,l'=1}^{\infty} \mathrm{i}^{l'-l} \sqrt{(2l+1)(2l'+1)} \big(T_{1\mathrm{e}1l,1\mathrm{e}1l'} + \mathrm{i} T_{1\mathrm{e}1l,2\mathrm{o}1l'} - \mathrm{i} T_{2\mathrm{o}1l,1\mathrm{e}1l'} + T_{2\mathrm{o}1l,2\mathrm{o}1l'} \big)$$

■

Example 7.2
If we let the direction of incidence be $\hat{\boldsymbol{k}}_{\mathrm{i}} = \hat{z}$, then the scattering dyadic in the backscattering direction simplifies to

$$\mathbf{S}(-\hat{z}, \hat{z}) = -\frac{4\pi}{\mathrm{i}k} \sum_{n,n'} \mathrm{i}^{(l+l')-(\tau+\tau')} \boldsymbol{A}_n(\hat{z}) T_{nn'} \boldsymbol{A}_{n'}(\hat{z})$$

Using (C.9) on page 631 implies

$$\mathbf{S}(-\hat{z}, \hat{z}) = -\frac{1}{2ik} \sum_{l,l'=1}^{\infty} \sum_{\sigma,\sigma'} \sum_{\tau,\tau'} \mathrm{i}^{(l+l')-(\tau+\tau')} \sqrt{(2l+1)(2l'+1)} \boldsymbol{b}_{\tau\sigma} T_{\tau\sigma 1l,\tau'\sigma' 1l'} \boldsymbol{b}_{\tau'\sigma'}$$

where

$$\boldsymbol{b}_{\tau\sigma} = \big(-\hat{\boldsymbol{y}} \delta_{\sigma\mathrm{e}} + \hat{\boldsymbol{x}} \delta_{\sigma\mathrm{o}} \big) \delta_{\tau 1} + \big(\hat{\boldsymbol{x}} \delta_{\sigma\mathrm{e}} + \hat{\boldsymbol{y}} \delta_{\sigma\mathrm{o}} \big) \delta_{\tau 2}$$

The components of the scattering dyadic are

$$\mathbf{S}(-\hat{z}, \hat{z}) = \hat{\boldsymbol{x}}\hat{\boldsymbol{x}} S_{xx} + \hat{\boldsymbol{x}}\hat{\boldsymbol{y}} S_{xy} + \hat{\boldsymbol{y}}\hat{\boldsymbol{x}} S_{yx} + \hat{\boldsymbol{y}}\hat{\boldsymbol{y}} S_{yy}$$

where

$$S_{xx} = -\frac{1}{2ik}\sum_{l,l'=1}^{\infty} \mathrm{i}^{l+l'}\sqrt{(2l+1)(2l'+1)}\left(-T_{1\mathrm{o}1l,1\mathrm{o}1l'} + \mathrm{i}T_{1\mathrm{o}1l,2\mathrm{e}1l'} + \mathrm{i}T_{2\mathrm{e}1l,1\mathrm{o}1l'} + T_{2\mathrm{e}1l,2\mathrm{e}1l'}\right)$$

$$S_{xy} = -\frac{1}{2ik}\sum_{l,l'=1}^{\infty} \mathrm{i}^{l+l'}\sqrt{(2l+1)(2l'+1)}\left(T_{1\mathrm{o}1l,1\mathrm{e}1l'} + \mathrm{i}T_{1\mathrm{o}1l,2\mathrm{o}1l'} - \mathrm{i}T_{2\mathrm{e}1l,1\mathrm{e}1l'} + T_{2\mathrm{e}1l,2\mathrm{o}1l'}\right)$$

$$S_{yx} = -\frac{1}{2ik}\sum_{l,l'=1}^{\infty} \mathrm{i}^{l+l'}\sqrt{(2l+1)(2l'+1)}\left(T_{1\mathrm{e}1l,1\mathrm{o}1l'} - \mathrm{i}T_{1\mathrm{e}1l,2\mathrm{e}1l'} + \mathrm{i}T_{2\mathrm{o}1l,1\mathrm{o}1l'} + T_{2\mathrm{o}1l,2\mathrm{e}1l'}\right)$$

$$S_{yy} = -\frac{1}{2ik}\sum_{l,l'=1}^{\infty} \mathrm{i}^{l+l'}\sqrt{(2l+1)(2l'+1)}\left(-T_{1\mathrm{e}1l,1\mathrm{e}1l'} - \mathrm{i}T_{1\mathrm{e}1l,2\mathrm{o}1l'} - \mathrm{i}T_{2\mathrm{o}1l,1\mathrm{e}1l'} + T_{2\mathrm{o}1l,2\mathrm{o}1l'}\right)$$

■

To find the components of the scattering dyadic in the basis $\{\hat{e}_{s\parallel}, \hat{e}_{s\perp}\}$ and $\{\hat{e}_{i\parallel}, \hat{e}_{i\perp}\}$, see (4.27) on page 204, we need to evaluate the projections of $\boldsymbol{a}_n(\hat{\boldsymbol{r}})$ on $\{\hat{e}_{s\parallel}, \hat{e}_{s\perp}\}$ and, similarly, the projections of $\boldsymbol{a}_n(\hat{\boldsymbol{k}}_i)$ on $\{\hat{e}_{i\parallel}, \hat{e}_{i\perp}\}$. The degenerate case is treated in Examples 7.1 and 7.2. We therefore assume that the vectors $\hat{\boldsymbol{k}}_i$ and $\hat{\boldsymbol{r}}$ are not co-linear. Denote the angle between the directions $\hat{\boldsymbol{k}}_i$ and $\hat{\boldsymbol{r}}$ by γ. We then have

$$\boldsymbol{a}_n(\hat{\boldsymbol{r}}) \cdot \hat{\boldsymbol{e}}_{s\perp} = \mathrm{i}^{l-\tau+1}\boldsymbol{A}_n(\hat{\boldsymbol{r}}) \cdot \frac{\hat{\boldsymbol{k}}_i \times \hat{\boldsymbol{r}}}{\sin\gamma} = \mathrm{i}^{l+\tau-1}\frac{\boldsymbol{A}_{\overline{n}}(\hat{\boldsymbol{r}}) \cdot \hat{\boldsymbol{k}}_i}{\sin\gamma} = \mathrm{i}\frac{\boldsymbol{a}_{\overline{n}}(\hat{\boldsymbol{r}}) \cdot \hat{\boldsymbol{k}}_i}{\sin\gamma}$$

where we have used (C.7) on page 631, and $\hat{\boldsymbol{r}} \times \boldsymbol{A}_n(\hat{\boldsymbol{r}}) = (-1)^{\tau+1}\boldsymbol{A}_{\overline{n}}(\hat{\boldsymbol{r}})$ and $\boldsymbol{a}_{\overline{n}}(\hat{\boldsymbol{r}}) = \mathrm{i}^{l+\tau-2}\boldsymbol{A}_{\overline{n}}(\hat{\boldsymbol{r}})$. Moreover, we have

$$\boldsymbol{a}_n(\hat{\boldsymbol{r}}) \cdot \hat{\boldsymbol{e}}_{s\parallel} = \boldsymbol{a}_n(\hat{\boldsymbol{r}}) \cdot \frac{\hat{\boldsymbol{r}}(\hat{\boldsymbol{k}}_i \cdot \hat{\boldsymbol{r}}) - \hat{\boldsymbol{k}}_i}{\sin\gamma} = -\frac{\boldsymbol{a}_n(\hat{\boldsymbol{r}}) \cdot \hat{\boldsymbol{k}}_i}{\sin\gamma}$$

Similar projections give

$$\begin{cases} \boldsymbol{a}_n(\hat{\boldsymbol{k}}_i) \cdot \hat{\boldsymbol{e}}_{i\perp} = -\mathrm{i}\dfrac{\boldsymbol{a}_{\overline{n}}(\hat{\boldsymbol{k}}_i) \cdot \hat{\boldsymbol{r}}}{\sin\gamma} \\[2ex] \boldsymbol{a}_n(\hat{\boldsymbol{k}}_i) \cdot \hat{\boldsymbol{e}}_{i\parallel} = \dfrac{\boldsymbol{a}_n(\hat{\boldsymbol{k}}_i) \cdot \hat{\boldsymbol{r}}}{\sin\gamma} \end{cases}$$

and we obtain

$$\begin{cases} S_{\parallel\,\parallel}(\hat{\boldsymbol{r}},\hat{\boldsymbol{k}}_{\rm i}) = \hat{\boldsymbol{e}}_{\rm s\parallel}\cdot \mathbf{S}(\hat{\boldsymbol{r}},\hat{\boldsymbol{k}}_{\rm i})\cdot \hat{\boldsymbol{e}}_{\rm i\parallel} = \dfrac{4\pi {\rm i}}{k\sin^2\gamma}\displaystyle\sum_{n,n'} \hat{\boldsymbol{k}}_{\rm i}\cdot \boldsymbol{a}_n^*(\hat{\boldsymbol{r}})T_{nn'}\boldsymbol{a}_{n'}(\hat{\boldsymbol{k}}_{\rm i})\cdot \hat{\boldsymbol{r}} \\[2ex] S_{\parallel\,\perp}(\hat{\boldsymbol{r}},\hat{\boldsymbol{k}}_{\rm i}) = \hat{\boldsymbol{e}}_{\rm s\parallel}\cdot \mathbf{S}(\hat{\boldsymbol{r}},\hat{\boldsymbol{k}}_{\rm i})\cdot \hat{\boldsymbol{e}}_{\rm i\perp} = \dfrac{4\pi}{k\sin^2\gamma}\displaystyle\sum_{n,n'} \hat{\boldsymbol{k}}_{\rm i}\cdot \boldsymbol{a}_n^*(\hat{\boldsymbol{r}})T_{n\overline{n}'}\boldsymbol{a}_{n'}(\hat{\boldsymbol{k}}_{\rm i})\cdot \hat{\boldsymbol{r}} \\[2ex] S_{\perp\,\parallel}(\hat{\boldsymbol{r}},\hat{\boldsymbol{k}}_{\rm i}) = \hat{\boldsymbol{e}}_{\rm s\perp}\cdot \mathbf{S}(\hat{\boldsymbol{r}},\hat{\boldsymbol{k}}_{\rm i})\cdot \hat{\boldsymbol{e}}_{\rm i\parallel} = -\dfrac{4\pi}{k\sin^2\gamma}\displaystyle\sum_{n,n'} \hat{\boldsymbol{k}}_{\rm i}\cdot \boldsymbol{a}_n^*(\hat{\boldsymbol{r}})T_{\overline{n}n'}\boldsymbol{a}_{n'}(\hat{\boldsymbol{k}}_{\rm i})\cdot \hat{\boldsymbol{r}} \\[2ex] S_{\perp\,\perp}(\hat{\boldsymbol{r}},\hat{\boldsymbol{k}}_{\rm i}) = \hat{\boldsymbol{e}}_{\rm s\perp}\cdot \mathbf{S}(\hat{\boldsymbol{r}},\hat{\boldsymbol{k}}_{\rm i})\cdot \hat{\boldsymbol{e}}_{\rm i\perp} = \dfrac{4\pi {\rm i}}{k\sin^2\gamma}\displaystyle\sum_{n,n'} \hat{\boldsymbol{k}}_{\rm i}\cdot \boldsymbol{a}_n^*(\hat{\boldsymbol{r}})T_{\overline{n}\overline{n}'}\boldsymbol{a}_{n'}(\hat{\boldsymbol{k}}_{\rm i})\cdot \hat{\boldsymbol{r}} \end{cases} \tag{7.37}$$

7.8.2 Cross Sections

The general expression of the differential scattering cross section is, see the expression in (4.33) on page 207

$$\frac{{\rm d}\sigma}{{\rm d}\Omega}(\hat{\boldsymbol{r}},\hat{\boldsymbol{k}}_{\rm i}) = 4\pi \left|\mathbf{S}(\hat{\boldsymbol{r}},\hat{\boldsymbol{k}}_{\rm i})\cdot \hat{\boldsymbol{p}}_{\rm e}\right|^2 \tag{7.38}$$

where the complex-valued unit vector $\hat{\boldsymbol{p}}_{\rm e} = \boldsymbol{E}_0/|\boldsymbol{E}_0|$. We get, in terms of the transition matrix

$$\frac{{\rm d}\sigma}{{\rm d}\Omega}(\hat{\boldsymbol{r}},\hat{\boldsymbol{k}}_{\rm i}) = \frac{64\pi^3}{k^2}\left|\sum_{n,n'} \boldsymbol{a}_n^*(\hat{\boldsymbol{r}})T_{nn'}\boldsymbol{a}_{n'}(\hat{\boldsymbol{k}}_{\rm i})\cdot \hat{\boldsymbol{p}}_{\rm e}\right|^2$$

In particular, in the backward direction, $\hat{\boldsymbol{r}} = -\hat{\boldsymbol{k}}_{\rm i}$, we obtain using the parity properties of the vector spherical harmonics, see (7.33)

$$\frac{{\rm d}\sigma}{{\rm d}\Omega}(-\hat{\boldsymbol{k}}_{\rm i},\hat{\boldsymbol{k}}_{\rm i}) = \frac{64\pi^3}{k^2}\left|\sum_{n,n'} \boldsymbol{a}_n(\hat{\boldsymbol{k}}_{\rm i})T_{nn'}\boldsymbol{a}_{n'}(\hat{\boldsymbol{k}}_{\rm i})\cdot \hat{\boldsymbol{p}}_{\rm e}\right|^2$$

The scattering cross section, $\sigma_{\rm s}(\hat{\boldsymbol{k}}_{\rm i})$, is, due to the orthonormal properties of the vector spherical harmonics

$$\begin{aligned} \sigma_{\rm s}(\hat{\boldsymbol{k}}_{\rm i}) &= \frac{1}{4\pi}\iint_{\Omega} \frac{{\rm d}\sigma}{{\rm d}\Omega}(\hat{\boldsymbol{r}},\hat{\boldsymbol{k}}_{\rm i})\,{\rm d}\Omega = \frac{16\pi^2}{k^2}\sum_n \left|\sum_{n'} T_{nn'}\boldsymbol{a}_{n'}(\hat{\boldsymbol{k}}_{\rm i})\cdot \hat{\boldsymbol{p}}_{\rm e}\right|^2 \\ &= \frac{1}{k^2|\boldsymbol{E}_0|^2}\sum_{n,n',n''} a_{n''}^* T_{n''n}^\dagger T_{nn'}a_{n'} = \frac{1}{k^2|\boldsymbol{E}_0|^2}\sum_n |f_n|^2 \end{aligned} \tag{7.39}$$

since, see (7.28)

$$a_n = 4\pi \boldsymbol{a}_n(\hat{\boldsymbol{k}}_{\rm i})\cdot \hat{\boldsymbol{p}}_{\rm e}|\boldsymbol{E}_0|$$

and where we also have used (7.34). For convenience, we write this relation in a shorthand notation, using obvious matrix multiplication, as

$$\sigma_{\rm s}(\hat{\boldsymbol{k}}_{\rm i}) = \frac{1}{k^2|\boldsymbol{E}_0|^2} a^* T^\dagger T a$$

We now use the expression of the scattering dyadic in the forward direction $\hat{\boldsymbol{r}} = \hat{\boldsymbol{k}}_{\rm i}$, see (7.36). We get by the use of the optical theorem, see (4.47) on page 230, the extinction cross section $\sigma_{\rm ext}(\hat{\boldsymbol{k}}_{\rm i})$, i.e.,

$$\begin{aligned}
\sigma_{\rm ext}(\hat{\boldsymbol{k}}_{\rm i}) &= \frac{4\pi}{k} \operatorname{Im} \left\{ \hat{\boldsymbol{p}}_{\rm e}^* \cdot \mathbf{S}(\hat{\boldsymbol{k}}_{\rm i}, \hat{\boldsymbol{k}}_{\rm i}) \cdot \hat{\boldsymbol{p}}_{\rm e} \right\} \\
&= -\frac{16\pi^2}{k^2} \operatorname{Re} \left\{ \sum_{n,n'} \hat{\boldsymbol{p}}_{\rm e}^* \cdot \boldsymbol{a}_n^*(\hat{\boldsymbol{k}}_{\rm i}) T_{nn'} \boldsymbol{a}_{n'}(\hat{\boldsymbol{k}}_{\rm i}) \cdot \hat{\boldsymbol{p}}_{\rm e} \right\} \\
&= -\frac{8\pi^2}{k^2} \sum_{n,n'} \hat{\boldsymbol{p}}_{\rm e}^* \cdot \boldsymbol{a}_n^*(\hat{\boldsymbol{k}}_{\rm i}) \left(T_{nn'} + T_{nn'}^\dagger \right) \boldsymbol{a}_{n'}(\hat{\boldsymbol{k}}_{\rm i}) \cdot \hat{\boldsymbol{p}}_{\rm e} \\
&= -\frac{1}{2k^2|\boldsymbol{E}_0|^2} \sum_{n,n'} a_n^* \left(T_{nn'} + T_{nn'}^\dagger \right) a_{n'} \\
&= -\frac{1}{k^2|\boldsymbol{E}_0|^2} \operatorname{Re} \sum_{n,n'} a_n^* T_{nn'} a_{n'} = -\frac{1}{k^2|\boldsymbol{E}_0|^2} \operatorname{Re} \sum_{n} a_n^* f_n
\end{aligned} \tag{7.40}$$

where we have used (7.34). We write this relation in a shorthand notation as

$$\sigma_{\rm ext}(\hat{\boldsymbol{k}}_{\rm i}) = -\frac{1}{2k^2|\boldsymbol{E}_0|^2} a^* \left(T + T^\dagger \right) a = -\frac{1}{k^2|\boldsymbol{E}_0|^2} a^* \operatorname{Re} T a \tag{7.41}$$

The absorption cross section then is given by $\sigma_{\rm a}(\hat{\boldsymbol{k}}_{\rm i})$

$$\sigma_{\rm a}(\hat{\boldsymbol{k}}_{\rm i}) = \sigma_{\rm ext}(\hat{\boldsymbol{k}}_{\rm i}) - \sigma_{\rm s}(\hat{\boldsymbol{k}}_{\rm i})$$

Since the absorption cross section, $\sigma_{\rm a}(\hat{\boldsymbol{k}}_{\rm i})$, is a positive quantity for a passive scatterer, we get

$$-\frac{1}{k^2|\boldsymbol{E}_0|^2} a^* \left(\operatorname{Re} T + T^\dagger T \right) a = \sigma_{\rm ext}(\hat{\boldsymbol{k}}_{\rm i}) - \sigma_{\rm s}(\hat{\boldsymbol{k}}_{\rm i}) = \sigma_{\rm a}(\hat{\boldsymbol{k}}_{\rm i}) \geq 0 \tag{7.42}$$

Moreover, since $a^* T^\dagger T a \geq 0$

$$-a^* \left(T + T^\dagger \right) a \geq 0$$

The T-matrix characterizes the scattering properties of the obstacle completely, and it maps the expansion coefficients of the incident field, a_n, to the expansion coefficients of the scattered field, f_n. It is sometimes convenient to characterize the scattering in terms of the S-matrix defined as

$$S = 2T + I \quad \text{or more explicitly} \quad S_{nn'} = 2T_{nn'} + \delta_{nn'}$$

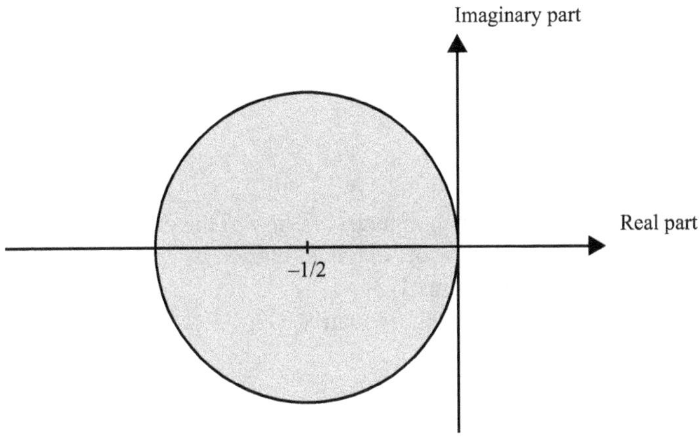

Figure 7.3 The location of the eigenvalues and diagonal entries of the transition matrix T.

This S-matrix maps the expansion coefficients of the in-going waves[5] to the expansion coefficients of the scattered field, f_n. From (7.42), we get

$$a^*\left(I - S^\dagger S\right) a \geq 0$$

and the eigenvalues of $I - S^\dagger S$ are non-negative, which implies that the eigenvalues of S, λ_n, lie inside the unit disk in the complex plane centered at the origin, i.e., $|\lambda_n| \leq 1$, and the eigenvalues of T are $(\lambda_n - 1)/2$, and they lie inside the disk in the complex plane centered at $-1/2$ with radius $1/2$, see Figure 7.3.

This relation also tells us that the matrices $I - S^\dagger S$ and $-\operatorname{Re} T - T^\dagger T$ are a non-negative matrices, see Appendix D.4 on page 649. In particular, the entries of the S matrix satisfy

$$\begin{cases} \sum_{n'} |S_{nn'}|^2 \leq 1, \quad \text{for all } n \\ \left| \delta_{nn'} - \sum_{n''} S_{nn''} S^*_{n'n''} \right|^2 \\ \quad \leq \left(1 - \sum_{n''} |S_{nn''}|^2\right)\left(1 - \sum_{n''} |S_{n'n''}|^2\right), \quad \text{for all } n, n' \end{cases}$$

The first inequality implies that

$$|S_{nn'}| \leq 1 \quad \Rightarrow \quad \left| T_{nn'} + \frac{1}{2}\delta_{nn'} \right| \leq \frac{1}{2}, \quad \text{for all } n, n'$$

[5]We have refrained from defining these waves here, but they are identical to the definition of $\boldsymbol{u}_n$ except that the spherical Hankel function of the first kind, $h_l^{(1)}(kr)$, is replaced by a spherical Hankel function of the second kind, $h_l^{(2)}(kr)$.

Therefore, all matrix entries of the S-matrix lie inside the unit disk, and we can identify complex numbers, $\theta_{nn'}$, called the phase shifts,[6] in the upper half complex plane, i.e., $\operatorname{Im}\theta_{nn'} \geq 0$, that satisfy

$$S_{nn'} = \mathrm{e}^{2\mathrm{i}\theta_{nn'}} \quad \Rightarrow \quad T_{nn'} = -\frac{1}{2}\delta_{nn'} + \frac{1}{2}\mathrm{e}^{2\mathrm{i}\theta_{nn'}} \tag{7.43}$$

and the diagonal entries of the transition matrix lie inside the complex disk centered at $-1/2$ with radius $1/2$, see Figure 7.3. The off-diagonal entries lie inside the disk in the complex plane centered at the origin with radius $1/2$.

If the scatterer is lossless, S is a unitary matrix, i.e.,

$$S^{\dagger}S = I \quad \text{or} \quad \operatorname{Re} T + T^{\dagger}T = 0 \tag{7.44}$$

and the eigenvalues of S lie on the unit disk, $\lambda_n = \exp 2\mathrm{i}\delta_n$, where δ_n is real.

The following example is indispensable in the analysis presented for spherical scatterers in Chapter 8.

Example 7.3

For a transition matrix that is diagonal in l, m, and σ, and moreover, independent of the indices m and σ, i.e.,

$$T_{nn'} = t_{\tau\tau' l}\delta_{ll'}\delta_{mm'}\delta_{\sigma\sigma'}$$

considerable simplifications in the results above can be obtained. This special case is relevant in scattering by obstacles with spherical symmetry, see Chapter 8.

The expression of the scattering dyadic in (7.35) simplifies when we have a transition matrix with these symmetries. Assuming incidence along the $\hat{\boldsymbol{k}}_{\mathrm{i}} = \hat{\boldsymbol{z}}$, we get using (7.37) and $\gamma = \theta$

$$\begin{cases} S_{\parallel\,\parallel}(\hat{\boldsymbol{r}},\hat{\boldsymbol{z}}) = \dfrac{4\pi\mathrm{i}}{k\sin^2\theta}\displaystyle\sum_{n,n'}\hat{\boldsymbol{z}}\cdot\boldsymbol{a}_n^*(\hat{\boldsymbol{r}})T_{nn'}\boldsymbol{a}_{n'}(\hat{\boldsymbol{z}})\cdot\hat{\boldsymbol{r}} \\[2ex] S_{\parallel\,\perp}(\hat{\boldsymbol{r}},\hat{\boldsymbol{z}}) = \dfrac{4\pi}{k\sin^2\theta}\displaystyle\sum_{n,n'}\hat{\boldsymbol{z}}\cdot\boldsymbol{a}_n^*(\hat{\boldsymbol{r}})T_{n\overline{n'}}\boldsymbol{a}_{n'}(\hat{\boldsymbol{z}})\cdot\hat{\boldsymbol{r}} \\[2ex] S_{\perp\,\parallel}(\hat{\boldsymbol{r}},\hat{\boldsymbol{z}}) = -\dfrac{4\pi}{k\sin^2\theta}\displaystyle\sum_{n,n'}\hat{\boldsymbol{z}}\cdot\boldsymbol{a}_n^*(\hat{\boldsymbol{r}})T_{\overline{n}n'}\boldsymbol{a}_{n'}(\hat{\boldsymbol{z}})\cdot\hat{\boldsymbol{r}} \\[2ex] S_{\perp\,\perp}(\hat{\boldsymbol{r}},\hat{\boldsymbol{z}}) = \dfrac{4\pi\mathrm{i}}{k\sin^2\theta}\displaystyle\sum_{n,n'}\hat{\boldsymbol{z}}\cdot\boldsymbol{a}_n^*(\hat{\boldsymbol{r}})T_{\overline{n}\overline{n'}}\boldsymbol{a}_{n'}(\hat{\boldsymbol{z}})\cdot\hat{\boldsymbol{r}} \end{cases}$$

[6]The more commonly used notation $\delta_{nn'}$ is avoided since it can be misinterpreted as the Kronecker delta (function) symbol. Zero matrix entries correspond to an infinite imaginary part in $\theta_{nn'}$.

For an incidence along the z-axis the basis vectors $\{\hat{\boldsymbol{e}}_{\mathrm{s}\parallel}, \hat{\boldsymbol{e}}_{\mathrm{s}\perp}\} = \{\hat{\boldsymbol{\theta}}, \hat{\boldsymbol{\phi}}\}$ and $\{\hat{\boldsymbol{e}}_{\mathrm{i}\parallel}, \hat{\boldsymbol{e}}_{\mathrm{i}\perp}\} = \{\hat{\boldsymbol{r}}_{\mathrm{c}}, \hat{\boldsymbol{\phi}}\}$, in terms of the spherical and cylindrical unit vectors, respectively, see Figure 4.5 on page 204.

From (C.9) on page 631, we see that only $m = 1$ contributes with projections

$$\hat{\boldsymbol{r}} \cdot \boldsymbol{A}_{\tau\sigma ml}(\hat{z}) = \delta_{m1} \sin\theta \sqrt{\frac{2l+1}{8\pi}} \begin{cases} \begin{Bmatrix} -\sin\phi \\ \cos\phi \end{Bmatrix} & \tau = 1, \begin{Bmatrix} \sigma = \mathrm{e} \\ \sigma = \mathrm{o} \end{Bmatrix} \\ \begin{Bmatrix} \cos\phi \\ \sin\phi \end{Bmatrix} & \tau = 2, \begin{Bmatrix} \sigma = \mathrm{e} \\ \sigma = \mathrm{o} \end{Bmatrix} \end{cases}$$

Similarly,

$$\hat{z} \cdot \boldsymbol{A}_{\tau\sigma 1l}(\hat{\boldsymbol{r}}) = -\frac{\sin\theta}{l(l+1)} \sqrt{\frac{2l+1}{2\pi}} \begin{cases} \pi_l(\cos\theta) \begin{Bmatrix} -\sin\phi \\ \cos\phi \end{Bmatrix} & \tau = 1, \begin{Bmatrix} \sigma = \mathrm{e} \\ \sigma = \mathrm{o} \end{Bmatrix} \\ \tau_l(\cos\theta) \begin{Bmatrix} \cos\phi \\ \sin\phi \end{Bmatrix} & \tau = 2, \begin{Bmatrix} \sigma = \mathrm{e} \\ \sigma = \mathrm{o} \end{Bmatrix} \end{cases}$$

where we have used the abbreviations, see Appendix C.1.1

$$\begin{cases} \pi_l(x) = P_l'(x) \\ \tau_l(x) = l(l+1)P_l(x) - xP_l'(x) \end{cases} \tag{7.45}$$

and (C.13) on page 632. An alternative representation of the functions $\pi_l(x)$ and $\tau_l(x)$ is presented in Problem 7.7.

The entries in the scattering dyadic then become

$$\begin{pmatrix} S_{\parallel\parallel}(\hat{\boldsymbol{r}}, \hat{z}) & S_{\parallel\perp}(\hat{\boldsymbol{r}}, \hat{z}) \\ S_{\perp\parallel}(\hat{\boldsymbol{r}}, \hat{z}) & S_{\perp\perp}(\hat{\boldsymbol{r}}, \hat{z}) \end{pmatrix} = \frac{1}{k} \begin{pmatrix} -\mathrm{i}S_1 & -S_2 \\ S_3 & -\mathrm{i}S_4 \end{pmatrix} \tag{7.46}$$

where the four sums, S_i, $i = 1, 2, 3, 4$, are defined as

$$\begin{cases} S_1 = \sum_{l=1}^{\infty} \frac{2l+1}{l(l+1)} \left(\pi_l(\cos\theta) t_{11l} + \tau_l(\cos\theta) t_{22l}\right) \\ S_2 = \sum_{l=1}^{\infty} \frac{2l+1}{l(l+1)} \left(\pi_l(\cos\theta) t_{12l} + \tau_l(\cos\theta) t_{21l}\right) \end{cases}$$

and

$$\begin{cases} S_3 = \sum_{l=1}^{\infty} \frac{2l+1}{l(l+1)} \left(\pi_l(\cos\theta) t_{21l} + \tau_l(\cos\theta) t_{12l}\right) \\ S_4 = \sum_{l=1}^{\infty} \frac{2l+1}{l(l+1)} \left(\pi_l(\cos\theta) t_{22l} + \tau_l(\cos\theta) t_{11l}\right) \end{cases}$$

The scattering dyadic becomes

$$\mathbf{S}(\hat{\boldsymbol{r}},\hat{\boldsymbol{z}}) = \hat{\boldsymbol{\theta}}\left(S_{\parallel\parallel}(\hat{\boldsymbol{r}},\hat{\boldsymbol{z}})\hat{\boldsymbol{r}}_{\mathrm{c}} + S_{\parallel\perp}(\hat{\boldsymbol{r}},\hat{\boldsymbol{z}})\hat{\boldsymbol{\phi}}\right) + \hat{\boldsymbol{\phi}}\left(S_{\perp\parallel}(\hat{\boldsymbol{r}},\hat{\boldsymbol{z}})\hat{\boldsymbol{r}}_{\mathrm{c}} + S_{\perp\perp}(\hat{\boldsymbol{r}},\hat{\boldsymbol{z}})\hat{\boldsymbol{\phi}}\right)$$

In the forward direction, $\hat{\boldsymbol{r}} = \hat{\boldsymbol{z}}$, $(\theta = 0)$, the entries become

$$\begin{cases} S_{\parallel\parallel}(\hat{\boldsymbol{z}},\hat{\boldsymbol{z}}) = S_{\perp\perp}(\hat{\boldsymbol{z}},\hat{\boldsymbol{z}}) = \dfrac{1}{2\mathrm{i}k}\displaystyle\sum_{l=1}^{\infty}(2l+1)(t_{11l}+t_{22l}) = \dfrac{s_1}{2\mathrm{i}k} \\ S_{\parallel\perp}(\hat{\boldsymbol{z}},\hat{\boldsymbol{z}}) = -S_{\perp\parallel}(\hat{\boldsymbol{z}},\hat{\boldsymbol{z}}) = -\dfrac{1}{2k}\displaystyle\sum_{l=1}^{\infty}(2l+1)(t_{12l}+t_{21l}) = -\dfrac{s_2}{2k} \end{cases}$$

since

$$\begin{cases} \pi_l(1) = P_l'(1) = \dfrac{l(l+1)}{2} \\ \tau_l(1) = l(l+1)P_l(1) - P_l'(1) = \dfrac{l(l+1)}{2} \end{cases}$$

and where the sums s_1 and s_2 are

$$\begin{cases} s_1 = \displaystyle\sum_{l=1}^{\infty}(2l+1)\left(t_{11l}+t_{22l}\right) \\ s_2 = \displaystyle\sum_{l=1}^{\infty}(2l+1)\left(t_{12l}+t_{21l}\right) \end{cases}$$

In particular, in the forward direction, the scattering matrix in (7.46)

$$\begin{pmatrix} S_{\parallel\parallel}(\hat{\boldsymbol{z}},\hat{\boldsymbol{z}}) & S_{\parallel\perp}(\hat{\boldsymbol{z}},\hat{\boldsymbol{z}}) \\ S_{\perp\parallel}(\hat{\boldsymbol{z}},\hat{\boldsymbol{z}}) & S_{\perp\perp}(\hat{\boldsymbol{z}},\hat{\boldsymbol{z}}) \end{pmatrix} = \frac{1}{2\mathrm{i}k}\begin{pmatrix} s_1 & -\mathrm{i}s_2 \\ \mathrm{i}s_2 & s_1 \end{pmatrix}$$

has the eigenvalues $\lambda_{1,2} = (s_1 \pm s_2)/(2\mathrm{i}k)$.

In the backward direction, $\hat{\boldsymbol{r}} = -\hat{\boldsymbol{z}}$, $(\theta = \pi)$, we have similar expressions:

$$\begin{cases} S_{\parallel\parallel}(-\hat{\boldsymbol{z}},\hat{\boldsymbol{z}}) = -S_{\perp\perp}(-\hat{\boldsymbol{z}},\hat{\boldsymbol{z}}) = \dfrac{\mathrm{i}}{2k}\displaystyle\sum_{l=1}^{\infty}(2l+1)(-1)^l\left(t_{11l}-t_{22l}\right) = \dfrac{\mathrm{i}}{2k}s_1' \\ S_{\parallel\perp}(-\hat{\boldsymbol{z}},\hat{\boldsymbol{z}}) = S_{\perp\parallel}(-\hat{\boldsymbol{z}},\hat{\boldsymbol{z}}) = \dfrac{1}{2k}\displaystyle\sum_{l=1}^{\infty}(2l+1)(-1)^l\left(t_{12l}-t_{21l}\right) = \dfrac{1}{2k}s_2' \end{cases}$$

since

$$\begin{cases} \pi_l(-1) = P_l'(-1) = -(-1)^l\dfrac{l(l+1)}{2} \\ \tau_l(-1) = l(l+1)P_l(-1) + P_l'(-1) = (-1)^l\dfrac{l(l+1)}{2} \end{cases}$$

and where the sums s_1' and s_2' are

$$\begin{cases} s_1' = \sum_{l=1}^{\infty} (2l+1)(-1)^l \left(t_{11l} - t_{22l}\right) \\ s_2' = \sum_{l=1}^{\infty} (2l+1)(-1)^l \left(t_{12l} - t_{21l}\right) \end{cases}$$

This leads to the dyadic expressions[7]

$$\begin{cases} \mathbf{S}(\hat{z}, \hat{z}) = \frac{1}{2ik} \left\{ \mathbf{I}_2 s_1 + i\mathbf{J} s_2 \right\} \\ \mathbf{S}(-\hat{z}, \hat{z}) = \frac{1}{2ik} \left\{ \mathbf{I}_2 s_1' + i\mathbf{J} s_2' \right\} \end{cases} \tag{7.47}$$

where $\mathbf{I}_2 = \hat{e}_{i\parallel}\hat{e}_{i\parallel} + \hat{e}_{i\perp}\hat{e}_{i\perp} = \hat{x}\hat{x} + \hat{y}\hat{y}$ is the unit dyadic in the x–y-plane, and $\mathbf{J} = \hat{z} \times \mathbf{I}_2 = \hat{e}_{i\perp}\hat{e}_{i\parallel} - \hat{e}_{i\parallel}\hat{e}_{i\perp} = \hat{y}\hat{x} - \hat{x}\hat{y}$ denotes the dyadic that rotates the vector $\pi/2$ around the positive z-axis.

The differential scattering cross section becomes

$$\begin{aligned} \frac{d\sigma}{d\Omega}(\hat{r}, \hat{z}) &= 4\pi \left| \mathbf{S}(\hat{r}, \hat{z}) \cdot \hat{p}_e \right|^2 \\ &= \frac{4\pi}{k^2} \left\{ \left| -iS_1 \hat{r}_c \cdot \hat{p}_e - S_2 \hat{\boldsymbol{\phi}} \cdot \hat{p}_e \right|^2 + \left| S_3 \hat{r}_c \cdot \hat{p}_e - iS_4 \hat{\boldsymbol{\phi}} \cdot \hat{p}_e \right|^2 \right\} \end{aligned} \tag{7.48}$$

If the transition matrix is diagonal also in the τ index, i.e., $t_{12l} = t_{21l} = 0$, then the differential scattering cross section has an alternative simple structure. The result of Problem 7.8 shows that the differential scattering cross section can be expressed as a sum of Legendre polynomials $P_l(\cos\theta)$

$$\frac{d\sigma}{d\Omega}(\hat{r}, \hat{z}) = \sum_{l=0}^{\infty} \omega_l P_l(\cos\theta)$$

where the coefficients are given in the answers to the problems.

The differential scattering cross section in the forward direction can also be written down explicitly in terms of the transition matrix entries. The result is

$$\begin{aligned} \frac{d\sigma}{d\Omega}(\hat{z}, \hat{z}) &= 4\pi \left| \mathbf{S}(\hat{z}, \hat{z}) \cdot \hat{p}_e \right|^2 = \frac{\pi}{k^2} \left| s_1 \hat{p}_e + i s_2 \hat{z} \times \hat{p}_e \right|^2 \\ &= \frac{\pi}{k^2} \left\{ |s_1|^2 + |s_2|^2 - 2\,\mathrm{Re}\left(i s_1 s_2^* \hat{p}_e \cdot \left(\hat{z} \times \hat{p}_e^* \right) \right) \right\} \\ &= \frac{\pi}{k^2} \left\{ |s_1|^2 + |s_2|^2 + 2\Pi\,\mathrm{Re}\left(s_1 s_2^* \right) \right\} \end{aligned}$$

[7] Note that $\hat{e}_{s\parallel} = -\hat{e}_{i\parallel}$ and $\hat{e}_{s\perp} = \hat{e}_{i\perp}$ when $\hat{r} = -\hat{z}$.

since $\Pi = \mathrm{i}\hat{z} \cdot (\hat{\boldsymbol{p}}_\mathrm{e} \times \hat{\boldsymbol{p}}_\mathrm{e}^*) \in [-1, 1]$ is a real number, see Section 1.3.7.[8] The last term vanishes if either the sum $\mathrm{Re}\,(s_1 s_2^*)$ is zero, or if the incident wave is linearly polarized, i.e., $\Pi = \mathrm{i}\hat{z} \cdot (\hat{\boldsymbol{p}}_\mathrm{e} \times \hat{\boldsymbol{p}}_\mathrm{e}^*) = 0$, see Table 1.12 on page 78. Similarly, in the backward direction, we have

$$\frac{\mathrm{d}\sigma}{\mathrm{d}\Omega}(-\hat{z}, \hat{z}) = 4\pi \left|\mathbf{S}(-\hat{z}, \hat{z}) \cdot \hat{\boldsymbol{p}}_\mathrm{e}\right|^2 = \frac{\pi}{k^2}\left\{\left|s_1'\right|^2 + \left|s_2'\right|^2 + 2\Pi\,\mathrm{Re}\left(s_1' s_2'^*\right)\right\}$$

The scattering cross section is obtain from (7.39)

$$\sigma_\mathrm{s}(\hat{z}) = \frac{16\pi^2}{k^2} \sum_{\tau\sigma m l} \left|t_{\tau 1 l}\boldsymbol{a}_{1\sigma m l}(\hat{z}) \cdot \hat{\boldsymbol{p}}_\mathrm{e} + t_{\tau 2 l}\boldsymbol{a}_{2\sigma m l}(\hat{z}) \cdot \hat{\boldsymbol{p}}_\mathrm{e}\right|^2$$

and using (C.9) on page 631

$$\hat{\boldsymbol{p}}_\mathrm{e} \cdot \boldsymbol{A}_{\tau\sigma m l}(\hat{z}) = \delta_{m1}\sqrt{\frac{2l+1}{8\pi}} \begin{cases} \begin{Bmatrix} -\hat{y} \cdot \hat{\boldsymbol{p}}_\mathrm{e} \\ \hat{x} \cdot \hat{\boldsymbol{p}}_\mathrm{e} \end{Bmatrix} & \tau = 1, \begin{Bmatrix} \sigma = \mathrm{e} \\ \sigma = \mathrm{o} \end{Bmatrix} \\ \begin{Bmatrix} \hat{x} \cdot \hat{\boldsymbol{p}}_\mathrm{e} \\ \hat{y} \cdot \hat{\boldsymbol{p}}_\mathrm{e} \end{Bmatrix} & \tau = 2, \begin{Bmatrix} \sigma = \mathrm{e} \\ \sigma = \mathrm{o} \end{Bmatrix} \end{cases}$$

the result is

$$\begin{aligned} \sigma_\mathrm{s}(\hat{z}) &= \frac{2\pi}{k^2} \sum_{\tau l} (2l+1) \left\{\left|-t_{\tau 1 l}\hat{y} \cdot \hat{\boldsymbol{p}}_\mathrm{e} - \mathrm{i}t_{\tau 2 l}\hat{x} \cdot \hat{\boldsymbol{p}}_\mathrm{e}\right|^2 + \left|t_{\tau 1 l}\hat{x} \cdot \hat{\boldsymbol{p}}_\mathrm{e} - \mathrm{i}t_{\tau 2 l}\hat{y} \cdot \hat{\boldsymbol{p}}_\mathrm{e}\right|^2\right\} \\ &= \frac{2\pi}{k^2} \sum_{\tau l} (2l+1) \Big\{ |t_{\tau 1 l}|^2 + |t_{\tau 2 l}|^2 \\ &\quad - 2\,\mathrm{Re}\left[\mathrm{i}t_{\tau 1 l}t_{\tau 2 l}^* \left(\left(\hat{y} \cdot \hat{\boldsymbol{p}}_\mathrm{e}\right)\left(\hat{x} \cdot \hat{\boldsymbol{p}}_\mathrm{e}^*\right) - \left(\hat{x} \cdot \hat{\boldsymbol{p}}_\mathrm{e}\right)\left(\hat{y} \cdot \hat{\boldsymbol{p}}_\mathrm{e}^*\right)\right)\right] \Big\} \end{aligned}$$

since $\left|\hat{x} \cdot \hat{\boldsymbol{p}}_\mathrm{e}\right|^2 + \left|\hat{y} \cdot \hat{\boldsymbol{p}}_\mathrm{e}\right|^2 = \hat{\boldsymbol{p}}_\mathrm{e}^* \cdot \mathbf{I}_2 \cdot \hat{\boldsymbol{p}}_\mathrm{e} = 1$. Further simplifications give

$$\sigma_\mathrm{s}(\hat{z}) = \frac{2\pi}{k^2} \sum_{\tau l} (2l+1) \left\{|t_{\tau 1 l}|^2 + |t_{\tau 2 l}|^2 + 2\Pi\,\mathrm{Re}\left(t_{\tau 1 l}t_{\tau 2 l}^*\right)\right\} \tag{7.49}$$

since $\left(\hat{x} \cdot \hat{\boldsymbol{p}}_\mathrm{e}\right)\left(\hat{y} \cdot \hat{\boldsymbol{p}}_\mathrm{e}^*\right) - \left(\hat{y} \cdot \hat{\boldsymbol{p}}_\mathrm{e}\right)\left(\hat{x} \cdot \hat{\boldsymbol{p}}_\mathrm{e}^*\right) = \hat{\boldsymbol{p}}_\mathrm{e}^* \cdot \left(\hat{z} \times \mathbf{I}_2\right) \cdot \hat{\boldsymbol{p}}_\mathrm{e} = \hat{z} \cdot \left(\hat{\boldsymbol{p}}_\mathrm{e} \times \hat{\boldsymbol{p}}_\mathrm{e}^*\right)$. Notice that the last term vanishes if either the transition matrix is diagonal in the τ-index, or if the incident wave is linearly polarized, i.e., $\Pi = \mathrm{i}\hat{z} \cdot (\hat{\boldsymbol{p}}_\mathrm{e} \times \hat{\boldsymbol{p}}_\mathrm{e}^*) = 0$, see Table 1.12 on page 78.

By the optical theorem, the extinction cross section of the scatterer is now determined. Of interest in this derivation is the evaluation of the following quantity:

$$\hat{\boldsymbol{p}}_\mathrm{e}^* \cdot \mathbf{S}(\hat{z}, \hat{z}) \cdot \hat{\boldsymbol{p}}_\mathrm{e} = \frac{1}{2\mathrm{i}k}\left\{s_1 + \mathrm{i}\hat{\boldsymbol{p}}_\mathrm{e}^* \cdot (\hat{z} \times \hat{\boldsymbol{p}}_\mathrm{e}) s_2\right\}$$

---

[8] $\Pi = -1$ indicates LCP, $\Pi = 0$ LP, and $\Pi = 1$ corresponds to RCP.

where we used (7.47). The optical theorem then implies

$$\sigma_{\text{ext}}(\hat{z}) = \frac{4\pi}{k}\,\text{Im}\left\{\hat{\boldsymbol{p}}_{\text{e}}^{*} \cdot \mathbf{S}(\hat{z},\hat{z}) \cdot \hat{\boldsymbol{p}}_{\text{e}}\right\} = -\frac{2\pi}{k^2}\,\text{Re}(s_1 + \Pi s_2) \tag{7.50}$$

Again, we observe that the last term vanishes if either the sum $\text{Re}\, s_2$ is zero, or if the incident wave is linearly polarized, i.e., $\Pi = \mathrm{i}\hat{z} \cdot \left(\hat{\boldsymbol{p}}_{\text{e}} \times \hat{\boldsymbol{p}}_{\text{e}}^{*}\right) = 0$, see Table 1.12 on page 78. We have already concluded that the eigenvalues of the scattering matrix in the forward direction are $\lambda_{1,2} = (s_1 \pm s_2)/(2\mathrm{i}k)$. In a final note in Section 4.4 on page 227, we concluded that these eigenvalues bound the extinction cross section. This in complete agreement with result obtained in this example. An illustration of this result is found in Figure 8.30 on page 471.

Finally, the absorption cross section becomes $\sigma_{\text{a}}(\hat{z})$

$$\sigma_{\text{a}}(\hat{z}) = \sigma_{\text{ext}}(\hat{z}) - \sigma_{\text{s}}(\hat{z})$$

For a lossless scatterer, the expressions in (7.49) and (7.50) are identical. The reader is encouraged to prove this in Problem 7.9. ∎

Example 7.4

If the transition matrix is diagonal, like in Example 7.3, but also in the τ index, i.e., $t_{12l} = t_{21l} = 0$, there is an interesting interpretation of the phase shifts $\theta_{nn'} = \delta_{nn'}\theta_n$ defined above.[9] Outside the circumscribed sphere of the scatterer, we have a representation of the total electric and magnetic fields, $\boldsymbol{E}$ and $\boldsymbol{H}$, expanded in terms of the spherical vector waves, see (7.12) and (7.13), i.e.,

$$\begin{cases} \boldsymbol{E}(\boldsymbol{r},\omega) = \displaystyle\sum_n \left(a_n \boldsymbol{v}_n(k\boldsymbol{r}) + f_n \boldsymbol{u}_n(k\boldsymbol{r})\right) \\ \boldsymbol{H}(\boldsymbol{r},\omega) = \dfrac{1}{\mathrm{i}\eta_0\eta} \displaystyle\sum_n \left(a_n \boldsymbol{v}_{\overline{n}}(k\boldsymbol{r}) + f_n \boldsymbol{u}_{\overline{n}}(k\boldsymbol{r})\right) \end{cases} \qquad r > \max_{\boldsymbol{r}' \in S_{\text{s}}} |\boldsymbol{r}'| = R$$

Introduce the asymptotic expansion for the spherical Bessel and Hankel functions for real arguments x, see Appendix B.3 on page 607

$$\begin{cases} j_l(x) = \dfrac{1}{x}\sin\left(x - \dfrac{l\pi}{2}\right) + O(x^{-2}) \\ \dfrac{(xj_l(x))'}{x} = \dfrac{1}{x}\cos\left(x - \dfrac{l\pi}{2}\right) + O(x^{-2}) \end{cases} \qquad \begin{cases} h_l^{(1)}(x) = \dfrac{\mathrm{i}^{-l-1}}{x}\mathrm{e}^{\mathrm{i}x} + O(x^{-2}) \\ \dfrac{\left(xh_l^{(1)}(x)\right)'}{x} = \dfrac{\mathrm{i}^{-l}}{x}\mathrm{e}^{\mathrm{i}x} + O(x^{-2}) \end{cases}$$

[9] It suffices to characterize the phase shifts with only one index in this situation.

As $r \to \infty$, each term in the sum (each partial wave) has the following dominant contribution:

$$\frac{1}{kr}\Bigg(\delta_{\tau 1}\left(a_{1\sigma ml}\sin\left(kr-\frac{l\pi}{2}\right)+f_{1\sigma ml}\mathrm{i}^{-l-1}\mathrm{e}^{\mathrm{i}kr}\right)\boldsymbol{A}_{1\sigma ml}(\hat{\boldsymbol{r}})$$
$$+\delta_{\tau 2}\left(a_{2\sigma ml}\cos\left(kr-\frac{l\pi}{2}\right)+f_{2\sigma ml}\mathrm{i}^{-l}\mathrm{e}^{\mathrm{i}kr}\right)\boldsymbol{A}_{2\sigma ml}(\hat{\boldsymbol{r}})\Bigg)$$
$$=-\frac{1}{kr}\left(a_n\cos\left(kr-\frac{(l-\tau)\pi}{2}\right)+f_n\mathrm{e}^{\mathrm{i}\left(kr-\frac{(l-\tau)\pi}{2}\right)}\right)\boldsymbol{A}_n(\hat{\boldsymbol{r}})$$

If there is no scatterer present, each term in the sum of the incident field has the asymptotic behavior

$$-\frac{1}{kr}a_n\cos\left(kr-\frac{(l-\tau)\pi}{2}\right)\boldsymbol{A}_n(\hat{\boldsymbol{r}})$$

Introduce the transition matrix, with diagonal elements $t_n = (\exp\{2\mathrm{i}\theta_n\}-1)/2$, see the phase shifts definition in (7.43). Each partial wave then becomes ($\Theta_n = kr-(l-\tau)\pi/2$, $f_n = t_n a_n$)

$$-\frac{1}{kr}a_n\left(\cos\Theta_n+t_n\mathrm{e}^{\mathrm{i}\Theta_n}\right)\boldsymbol{A}_n(\hat{\boldsymbol{r}}) = -\frac{1}{2kr}a_n\left(\mathrm{e}^{\mathrm{i}\Theta_n}+\mathrm{e}^{-\mathrm{i}\Theta_n}+\left(\mathrm{e}^{2\mathrm{i}\theta_n}-1\right)\mathrm{e}^{\mathrm{i}\Theta_n}\right)\boldsymbol{A}_n(\hat{\boldsymbol{r}})$$
$$=-\frac{1}{2kr}a_n\mathrm{e}^{\mathrm{i}\theta_n}\left(\mathrm{e}^{\mathrm{i}(\Theta_n+\theta_n)}+\mathrm{e}^{-\mathrm{i}(\Theta_n+\theta_n)}\right)\boldsymbol{A}_n(\hat{\boldsymbol{r}})$$

or

$$-\frac{1}{kr}a_n\mathrm{e}^{\mathrm{i}\theta_n}\cos\left(\Theta_n+\theta_n\right)\boldsymbol{A}_n(\hat{\boldsymbol{r}})$$

We see that the effect of scattering is a radial shift toward the center by a distance θ_n, compared to the incident wave, in each partial wave. Moreover, each partial wave is multiplied with a phase factor $\exp\{\mathrm{i}\theta_n\}$. ■

7.8.3 Generalized Optical Theorem

The optical theorem was derived under very general assumptions of the scatterer in Section 4.4. In this section, we derive a generalization of this theorem, see also Reference 239.

Using the shorthand notation from above, we derive the identity

$$2\,\mathrm{Re}\left\{a^*(\hat{\boldsymbol{p}}_{\mathrm{e}},\hat{\boldsymbol{k}}_{\mathrm{i}})T^\dagger Ta(\hat{\boldsymbol{p}}'_{\mathrm{e}},\hat{\boldsymbol{k}}'_{\mathrm{i}})\right\} = -\left(a^*(\hat{\boldsymbol{p}}_{\mathrm{e}},\hat{\boldsymbol{k}}_{\mathrm{i}})-a^*(\hat{\boldsymbol{p}}'_{\mathrm{e}},\hat{\boldsymbol{k}}'_{\mathrm{i}})\right)T^\dagger T\left(a(\hat{\boldsymbol{p}}_{\mathrm{e}},\hat{\boldsymbol{k}}_{\mathrm{i}})-a(\hat{\boldsymbol{p}}'_{\mathrm{e}},\hat{\boldsymbol{k}}'_{\mathrm{i}})\right)$$
$$+a^*(\hat{\boldsymbol{p}}'_{\mathrm{e}},\hat{\boldsymbol{k}}'_{\mathrm{i}})T^\dagger Ta(\hat{\boldsymbol{p}}'_{\mathrm{e}},\hat{\boldsymbol{k}}'_{\mathrm{i}})+a^*(\hat{\boldsymbol{p}}_{\mathrm{e}},\hat{\boldsymbol{k}}_{\mathrm{i}})T^\dagger Ta(\hat{\boldsymbol{p}}_{\mathrm{e}},\hat{\boldsymbol{k}}_{\mathrm{i}})$$

where $a(\hat{\boldsymbol{p}}_{\mathrm{e}},\hat{\boldsymbol{k}}_{\mathrm{i}})$ is defined in (7.28), i.e.,

$$a_n(\hat{\boldsymbol{p}}_{\mathrm{e}},\hat{\boldsymbol{k}}_{\mathrm{i}}) = 4\pi\left|\boldsymbol{E}_0\right|\hat{\boldsymbol{p}}_{\mathrm{e}}\cdot\boldsymbol{a}_n(\hat{\boldsymbol{k}}_{\mathrm{i}}),\qquad \tau=1,2$$

The complex unit vectors $\hat{\boldsymbol{p}}_e$ and $\hat{\boldsymbol{p}}'_e$ are the polarization of the electric field associated with the incident directions $\hat{\boldsymbol{k}}_i$ and $\hat{\boldsymbol{k}}'_i$, respectively. From (7.35), we get by orthogonality of the vector spherical harmonics utilizing the same shorthand notation

$$2\,\mathrm{Re}\iint\limits_{\Omega} \hat{\boldsymbol{p}}_e^* \cdot \mathbf{S}^\dagger(\hat{\boldsymbol{r}},\hat{\boldsymbol{k}}_i)\cdot \mathbf{S}(\hat{\boldsymbol{r}},\hat{\boldsymbol{k}}'_i)\cdot \hat{\boldsymbol{p}}'_e \,\mathrm{d}\Omega = \frac{2}{k^2|\boldsymbol{E}_0|^2}\,\mathrm{Re}\Big\{a^*(\hat{\boldsymbol{p}}_e,\hat{\boldsymbol{k}}_i)T^\dagger Ta(\hat{\boldsymbol{p}}'_e,\hat{\boldsymbol{k}}'_i)\Big\}$$

or with the identity above

$$\begin{aligned}
&2\,\mathrm{Re}\iint\limits_{\Omega} \hat{\boldsymbol{p}}_e^* \cdot \mathbf{S}^\dagger(\hat{\boldsymbol{r}},\hat{\boldsymbol{k}}_i)\cdot \mathbf{S}(\hat{\boldsymbol{r}},\hat{\boldsymbol{k}}'_i)\cdot \hat{\boldsymbol{p}}'_e \,\mathrm{d}\Omega \\
&\quad= \frac{1}{k^2|\boldsymbol{E}_0|^2}\Big(a^*(\hat{\boldsymbol{p}}'_e,\hat{\boldsymbol{k}}'_i)T^\dagger Ta(\hat{\boldsymbol{p}}'_e,\hat{\boldsymbol{k}}'_i) + a^*(\hat{\boldsymbol{p}}_e,\hat{\boldsymbol{k}}_i)T^\dagger Ta(\hat{\boldsymbol{p}}_e,\hat{\boldsymbol{k}}_i) \\
&\qquad - \Big(a^*(\hat{\boldsymbol{p}}_e,\hat{\boldsymbol{k}}_i) - a^*(\hat{\boldsymbol{p}}'_e,\hat{\boldsymbol{k}}'_i)\Big)T^\dagger T\Big(a(\hat{\boldsymbol{p}}_e,\hat{\boldsymbol{k}}_i) - a(\hat{\boldsymbol{p}}'_e,\hat{\boldsymbol{k}}'_i)\Big)\Big)
\end{aligned}$$

We identify the two first terms on the right-hand side as scattering cross sections with different directions of incidence and polarization, see (7.40)

$$\begin{aligned}
2\,\mathrm{Re}\iint\limits_{\Omega} \hat{\boldsymbol{p}}_e^* \cdot \mathbf{S}^\dagger(\hat{\boldsymbol{r}},\hat{\boldsymbol{k}}_i)\cdot \mathbf{S}(\hat{\boldsymbol{r}}',\hat{\boldsymbol{k}}'_i)\cdot \hat{\boldsymbol{p}}'_e \,\mathrm{d}\Omega &= \sigma_s(\hat{\boldsymbol{p}}'_e,\hat{\boldsymbol{k}}'_i) + \sigma_s(\hat{\boldsymbol{p}}_e,\hat{\boldsymbol{k}}_i) \\
&\quad - \frac{1}{k^2|\boldsymbol{E}_0|^2}\Big(a^*(\hat{\boldsymbol{p}}_e,\hat{\boldsymbol{k}}_i) - a^*(\hat{\boldsymbol{p}}'_e,\hat{\boldsymbol{k}}'_i)\Big)T^\dagger T\Big(a(\hat{\boldsymbol{p}}_e,\hat{\boldsymbol{k}}_i) - a(\hat{\boldsymbol{p}}'_e,\hat{\boldsymbol{k}}'_i)\Big) \\
&= \sigma_{\mathrm{ext}}(\hat{\boldsymbol{p}}'_e,\hat{\boldsymbol{k}}'_i) + \sigma_{\mathrm{ext}}(\hat{\boldsymbol{p}}_e,\hat{\boldsymbol{k}}_i) - \sigma_a(\hat{\boldsymbol{p}}'_e,\hat{\boldsymbol{k}}'_i) - \sigma_a(\hat{\boldsymbol{p}}_e,\hat{\boldsymbol{k}}_i) \\
&\quad - \frac{1}{k^2|\boldsymbol{E}_0|^2}\Big(a^*(\hat{\boldsymbol{p}}_e,\hat{\boldsymbol{k}}_i) - a^*(\hat{\boldsymbol{p}}'_e,\hat{\boldsymbol{k}}'_i)\Big)T^\dagger T\Big(a(\hat{\boldsymbol{p}}_e,\hat{\boldsymbol{k}}_i) - a(\hat{\boldsymbol{p}}'_e,\hat{\boldsymbol{k}}'_i)\Big)
\end{aligned}$$

The last term on the right-hand side is always negative for all $a(\hat{\boldsymbol{p}}_e,\hat{\boldsymbol{k}}_i)$ and $a(\hat{\boldsymbol{p}}'_e,\hat{\boldsymbol{k}}'_i)$, which implies

$$2\,\mathrm{Re}\iint\limits_{\Omega} \hat{\boldsymbol{p}}_e^* \cdot \mathbf{S}^\dagger(\hat{\boldsymbol{r}},\hat{\boldsymbol{k}}_i)\cdot \mathbf{S}(\hat{\boldsymbol{r}},\hat{\boldsymbol{k}}'_i)\cdot \hat{\boldsymbol{p}}'_e \,\mathrm{d}\Omega \le \sigma_s(\hat{\boldsymbol{p}}'_e,\hat{\boldsymbol{k}}'_i) + \sigma_s(\hat{\boldsymbol{p}}_e,\hat{\boldsymbol{k}}_i)$$

We rewrite using (7.41), i.e.,

$$\sigma_{\mathrm{ext}}(\hat{\boldsymbol{p}}_e,\hat{\boldsymbol{k}}_i) = -\frac{1}{2k^2|\boldsymbol{E}_0|^2}a^*(\hat{\boldsymbol{p}}_e,\hat{\boldsymbol{k}}_i)\Big(T + T^\dagger\Big)a(\hat{\boldsymbol{p}}_e,\hat{\boldsymbol{k}}_i)$$

This expression leads to

$$2\,\mathrm{Re}\iint_{\Omega} \hat{\boldsymbol{p}}_{\mathrm{e}}^{*}\cdot\mathbf{S}^{\dagger}(\hat{\boldsymbol{r}},\hat{\boldsymbol{k}}_{\mathrm{i}})\cdot\mathbf{S}(\hat{\boldsymbol{r}},\hat{\boldsymbol{k}}_{\mathrm{i}}')\cdot\hat{\boldsymbol{p}}_{\mathrm{e}}'\,\mathrm{d}\Omega+\sigma_{\mathrm{a}}(\hat{\boldsymbol{p}}_{\mathrm{e}}',\hat{\boldsymbol{k}}_{\mathrm{i}}')+\sigma_{\mathrm{a}}(\hat{\boldsymbol{p}}_{\mathrm{e}},\hat{\boldsymbol{k}}_{\mathrm{i}})$$

$$=-\frac{1}{2k^2|\boldsymbol{E}_0|^2}a^*(\hat{\boldsymbol{p}}_{\mathrm{e}}',\hat{\boldsymbol{k}}_{\mathrm{i}}')\left(T+T^{\dagger}\right)a(\hat{\boldsymbol{p}}_{\mathrm{e}}',\hat{\boldsymbol{k}}_{\mathrm{i}}')-\frac{1}{2k^2|\boldsymbol{E}_0|^2}a^*(\hat{\boldsymbol{p}}_{\mathrm{e}},\hat{\boldsymbol{k}}_{\mathrm{i}})\left(T+T^{\dagger}\right)a(\hat{\boldsymbol{p}}_{\mathrm{e}},\hat{\boldsymbol{k}}_{\mathrm{i}})$$
$$-\frac{1}{k^2|\boldsymbol{E}_0|^2}\left(a^*(\hat{\boldsymbol{p}}_{\mathrm{e}},\hat{\boldsymbol{k}}_{\mathrm{i}})-a^*(\hat{\boldsymbol{p}}_{\mathrm{e}}',\hat{\boldsymbol{k}}_{\mathrm{i}}')\right)T^{\dagger}T\left(a(\hat{\boldsymbol{p}}_{\mathrm{e}},\hat{\boldsymbol{k}}_{\mathrm{i}})-a(\hat{\boldsymbol{p}}_{\mathrm{e}}',\hat{\boldsymbol{k}}_{\mathrm{i}}')\right)$$

or with aid of the scattering dyadic (7.35)

$$\hat{\boldsymbol{p}}_{\mathrm{e}}^{*}\cdot\mathbf{S}(\hat{\boldsymbol{k}}_{\mathrm{i}},\hat{\boldsymbol{k}}_{\mathrm{i}})\cdot\hat{\boldsymbol{p}}_{\mathrm{e}}=\frac{1}{4\pi\mathrm{i}k|\boldsymbol{E}_0|^2}a^*(\hat{\boldsymbol{p}}_{\mathrm{e}},\hat{\boldsymbol{k}}_{\mathrm{i}})Ta(\hat{\boldsymbol{p}}_{\mathrm{e}},\hat{\boldsymbol{k}}_{\mathrm{i}})$$

we get

$$2\,\mathrm{Re}\iint_{\Omega} \hat{\boldsymbol{p}}_{\mathrm{e}}^{*}\cdot\mathbf{S}^{\dagger}(\hat{\boldsymbol{r}},\hat{\boldsymbol{k}}_{\mathrm{i}})\cdot\mathbf{S}(\hat{\boldsymbol{r}},\hat{\boldsymbol{k}}_{\mathrm{i}}')\cdot\hat{\boldsymbol{p}}_{\mathrm{e}}'\,\mathrm{d}\Omega+\sigma_{\mathrm{a}}(\hat{\boldsymbol{p}}_{\mathrm{e}}',\hat{\boldsymbol{k}}_{\mathrm{i}}')+\sigma_{\mathrm{a}}(\hat{\boldsymbol{p}}_{\mathrm{e}},\hat{\boldsymbol{k}}_{\mathrm{i}})$$
$$=\mathrm{Im}\,\frac{4\pi}{k}\hat{\boldsymbol{p}}_{\mathrm{e}}'^{*}\cdot\mathbf{S}(\hat{\boldsymbol{k}}_{\mathrm{i}}',\hat{\boldsymbol{k}}_{\mathrm{i}}')\cdot\hat{\boldsymbol{p}}_{\mathrm{e}}'+\mathrm{Im}\,\frac{4\pi}{k}\hat{\boldsymbol{p}}_{\mathrm{e}}^{*}\cdot\mathbf{S}(\hat{\boldsymbol{k}}_{\mathrm{i}},\hat{\boldsymbol{k}}_{\mathrm{i}})\cdot\hat{\boldsymbol{p}}_{\mathrm{e}}$$
$$-\frac{1}{k^2|\boldsymbol{E}_0|^2}\left(a^*(\hat{\boldsymbol{p}}_{\mathrm{e}},\hat{\boldsymbol{k}}_{\mathrm{i}})-a^*(\hat{\boldsymbol{p}}_{\mathrm{e}}',\hat{\boldsymbol{k}}_{\mathrm{i}}')\right)T^{\dagger}T\left(a(\hat{\boldsymbol{p}}_{\mathrm{e}},\hat{\boldsymbol{k}}_{\mathrm{i}})-a(\hat{\boldsymbol{p}}_{\mathrm{e}}',\hat{\boldsymbol{k}}_{\mathrm{i}}')\right)$$

This relation is often named the generalized optical theorem. The optical theorem from above, see (4.47) on page 230, is retrieved if $\hat{\boldsymbol{k}}_{\mathrm{i}}=\hat{\boldsymbol{k}}_{\mathrm{i}}'$ and $\hat{\boldsymbol{p}}_{\mathrm{e}}=\hat{\boldsymbol{p}}_{\mathrm{e}}'$, since the scattering cross section is

$$\sigma_{\mathrm{s}}(\hat{\boldsymbol{p}}_{\mathrm{e}},\hat{\boldsymbol{k}}_{\mathrm{i}})=\iint_{\Omega}\left|\mathbf{S}(\hat{\boldsymbol{r}},\hat{\boldsymbol{k}}_{\mathrm{i}})\cdot\hat{\boldsymbol{p}}_{\mathrm{e}}\right|^2\mathrm{d}\Omega$$

If we specialize and let the scatterer be lossless, $\sigma_{\mathrm{a}}(\hat{\boldsymbol{p}}_{\mathrm{e}},\hat{\boldsymbol{k}}_{\mathrm{i}})=0$, we can derive an alternative relation, which shows similarities with the generalized optical theorem. Once again use (7.35) and (7.44)

$$\iint_{\Omega}\mathbf{S}^{\dagger}(\hat{\boldsymbol{r}},\hat{\boldsymbol{k}}_{\mathrm{i}})\cdot\mathbf{S}(\hat{\boldsymbol{r}},\hat{\boldsymbol{k}}_{\mathrm{i}}')\,\mathrm{d}\Omega=\frac{16\pi^2}{k^2}\sum_{n,n',n''}\boldsymbol{a}_n^*(\hat{\boldsymbol{p}}_{\mathrm{e}},\hat{\boldsymbol{k}}_{\mathrm{i}})T_{nn'}^{\dagger}T_{n'n''}\boldsymbol{a}_{n''}(\hat{\boldsymbol{p}}_{\mathrm{e}}',\hat{\boldsymbol{k}}_{\mathrm{i}}')$$
$$=-\frac{8\pi^2}{k^2}\sum_{n,n'}\boldsymbol{a}_n^*(\hat{\boldsymbol{p}}_{\mathrm{e}},\hat{\boldsymbol{k}}_{\mathrm{i}})\left(T_{nn'}+T_{nn'}^{\dagger}\right)\boldsymbol{a}_{n'}(\hat{\boldsymbol{p}}_{\mathrm{e}}',\hat{\boldsymbol{k}}_{\mathrm{i}}')$$
$$=\frac{2\pi}{\mathrm{i}k}\left(\mathbf{S}(\hat{\boldsymbol{k}}_{\mathrm{i}},\hat{\boldsymbol{k}}_{\mathrm{i}}')-\mathbf{S}^{\dagger}(\hat{\boldsymbol{k}}_{\mathrm{i}}',\hat{\boldsymbol{k}}_{\mathrm{i}})\right)$$

This relation expresses energy conservation for a lossless scatterer. Additional details are found in Reference 239.

7.8.4 The Decrease of the Scattered Field

In this section, we present some general results for the far field amplitude and the scattered field. In particular, we demonstrate how the average field on a spherical surface decreases as the radius of the sphere increases.

We start with a collection of the general expressions of the relevant integral quantities. Outside any circumscribed sphere of the scatterer, radius $r > R$, the scattered field can be represented in an infinite series of out-going spherical vector waves $\boldsymbol{u}_{\tau n}(k\boldsymbol{r})$. More explicitly, we have, see Sections 7.2.1 and 7.6

$$\begin{cases} \boldsymbol{E}_{\mathrm{s}}(\boldsymbol{r},\omega) = \sum\limits_{\substack{n \\ \tau=1,2}} f_{\tau n}\boldsymbol{u}_{\tau n}(k\boldsymbol{r}) \\ \boldsymbol{H}_{\mathrm{s}}(\boldsymbol{r},\omega) = \dfrac{1}{\mathrm{i}\eta\eta_0} \sum\limits_{\substack{n \\ \tau=1,2}} f_{\tau n}\boldsymbol{u}_{\bar{\tau} n}(k\boldsymbol{r}) \end{cases} \qquad |\boldsymbol{r}| > \max_{\boldsymbol{r}'\in S_{\mathrm{s}}} r' = R$$

where, as above, the index $\bar{\tau}$ is the dual index of τ, defined by $\bar{1} = 2$ and $\bar{2} = 1$. The coefficients $f_{\tau n}$ are determined by, see (7.25) on page 372

$$f_{\tau n} = \mathrm{i}k \iint\limits_{S_{\mathrm{s}}} \left\{\mathrm{i}\eta_0\eta k\boldsymbol{v}_{\tau n}(k\boldsymbol{r}) \cdot \left(\hat{\boldsymbol{\nu}}(\boldsymbol{r}) \times \boldsymbol{H}(\boldsymbol{r})\right) + k\boldsymbol{v}_{\bar{\tau} n}(k\boldsymbol{r}) \cdot \left(\hat{\boldsymbol{\nu}}(\boldsymbol{r}) \times \boldsymbol{E}(\boldsymbol{r})\right)\right\} \mathrm{d}S$$

From the expansion of the far field amplitude (7.31), we immediately get by orthogonality of the vector spherical harmonics

$$\iint\limits_{\Omega} \left|\boldsymbol{F}(\hat{\boldsymbol{r}})\right|^2 \mathrm{d}\Omega = \frac{1}{k^2} \sum_{\substack{n \\ \tau=1,2}} |f_{\tau n}|^2 = \frac{1}{k^2} \sum_{n} \left\{|f_{1n}|^2 + |f_{2n}|^2\right\}$$

Outside any circumscribed sphere, $r \geq R$, the norms of the tangential fields are (S_r is a sphere of radius r)

$$\iint\limits_{S_r} \left|\hat{\boldsymbol{r}} \times \boldsymbol{E}_{\mathrm{s}}(\boldsymbol{r})\right|^2 \mathrm{d}S = \frac{1}{k^2} \sum_{n} \left\{\left|krh_l^{(1)}(kr)\right|^2 |f_{1n}|^2 + \left|(krh_l^{(1)}(kr))'\right|^2 |f_{2n}|^2\right\}$$

and

$$\iint\limits_{S_r} \left|\eta\eta_0\hat{\boldsymbol{r}} \times \boldsymbol{H}_{\mathrm{s}}(\boldsymbol{r})\right|^2 \mathrm{d}S = \frac{1}{k^2} \sum_{n} \left\{\left|(krh_l^{(1)}(kr))'\right|^2 |f_{1n}|^2 + \left|krh_l^{(1)}(kr)\right|^2 |f_{2n}|^2\right\}$$

Similarly, we have the norms of all components of the scattered field

$$\iint\limits_{S_r} |\boldsymbol{E}_{\mathrm{s}}(\boldsymbol{r})|^2 \,\mathrm{d}S = \frac{1}{k^2} \sum_n \left\{ \left| krh_l^{(1)}(kr) \right|^2 |f_{1n}|^2 + \left(\left| (krh_l^{(1)}(kr))' \right|^2 + l(l+1) \left| h_l^{(1)}(kr) \right|^2 \right) |f_{2n}|^2 \right\}$$

and

$$\iint\limits_{S_r} |\eta\eta_0 \boldsymbol{H}_{\mathrm{s}}(\boldsymbol{r})|^2 \,\mathrm{d}S = \frac{1}{k^2} \sum_n \left\{ \left(\left| (krh_l^{(1)}(kr))' \right|^2 + l(l+1) \left| h_l^{(1)}(kr) \right|^2 \right) |f_{1n}|^2 + \left| krh_l^{(1)}(kr) \right|^2 |f_{2n}|^2 \right\}$$

The next results show that the integral quantities above are non-increasing quantities as functions of the radius r. Outside any circumscribed sphere, $r > R$, we have for any spherical surface $R < r_1 \le r_2$

$$\begin{aligned} 2 \iint\limits_{\Omega} \left| \boldsymbol{F}(\hat{\boldsymbol{r}}) \right|^2 \,\mathrm{d}\Omega &\le \iint\limits_{S_{r_2}} \left\{ \left| \hat{\boldsymbol{r}} \times \boldsymbol{E}_{\mathrm{s}}(\boldsymbol{r}) \right|^2 + \left| \eta\eta_0 \hat{\boldsymbol{r}} \times \boldsymbol{H}_{\mathrm{s}}(\boldsymbol{r}) \right|^2 \right\} \,\mathrm{d}S \\ &\le \iint\limits_{S_{r_1}} \left\{ \left| \hat{\boldsymbol{r}} \times \boldsymbol{E}_{\mathrm{s}}(\boldsymbol{r}) \right|^2 + \left| \eta\eta_0 \hat{\boldsymbol{r}} \times \boldsymbol{H}_{\mathrm{s}}(\boldsymbol{r}) \right|^2 \right\} \,\mathrm{d}S \end{aligned}$$

and

$$\iint\limits_{S_{r_2}} |\boldsymbol{E}_{\mathrm{s}}(\boldsymbol{r})|^2 \,\mathrm{d}S \le \iint\limits_{S_{r_1}} |\boldsymbol{E}_{\mathrm{s}}(\boldsymbol{r})|^2 \,\mathrm{d}S$$

$$\iint\limits_{S_{r_2}} |\eta\eta_0 \boldsymbol{H}_{\mathrm{s}}(\boldsymbol{r})|^2 \,\mathrm{d}S \le \iint\limits_{S_{r_1}} |\eta\eta_0 \boldsymbol{H}_{\mathrm{s}}(\boldsymbol{r})|^2 \,\mathrm{d}S$$

The limits

$$\begin{aligned} \lim_{r\to\infty} \iint\limits_{S_r} \left| \hat{\boldsymbol{r}} \times \boldsymbol{E}_{\mathrm{s}}(\boldsymbol{r}) \right|^2 \,\mathrm{d}S &= \lim_{r\to\infty} \iint\limits_{S_r} \left| \eta\eta_0 \hat{\boldsymbol{r}} \times \boldsymbol{H}_{\mathrm{s}}(\boldsymbol{r}) \right|^2 \,\mathrm{d}S \\ = \lim_{r\to\infty} \iint\limits_{S_r} |\boldsymbol{E}_{\mathrm{s}}(\boldsymbol{r})|^2 \,\mathrm{d}S &= \lim_{r\to\infty} \iint\limits_{S_r} |\eta\eta_0 \boldsymbol{H}_{\mathrm{s}}(\boldsymbol{r})|^2 \,\mathrm{d}S = \iint\limits_{\Omega} \left| \boldsymbol{F}(\hat{\boldsymbol{r}}) \right|^2 \,\mathrm{d}\Omega \end{aligned}$$

hold. These results follow immediately from the norms above and Lemma B.2 on page 615 in Appendix B.3.2. Additional results and estimates in other norms are presented in References 148 and 290.

7.9 Dipole Moments of a Scatterer

The aim of this section is to identify the electric and magnetic dipole moments, or, more generally, the polarizability dyadics, $\boldsymbol{\gamma}_{\rm e}$ and $\boldsymbol{\gamma}_{\rm m}$, of the scatterer in terms of the transition matrix entries $T_{nn'}$.

The far field amplitude in (7.31)

$$\boldsymbol{F}(\hat{\boldsymbol{r}}) = \frac{1}{\mathrm{i}k}\sum_n \mathrm{i}^{-l+\tau-1} f_n \boldsymbol{A}_n(\hat{\boldsymbol{r}}) \tag{7.51}$$

is the general expansion in terms of the vector spherical harmonics $\boldsymbol{A}_n(\hat{\boldsymbol{r}})$. In the long wavelength limit, the far field amplitude is expressed in the electric and magnetic dipole moments, $\boldsymbol{p}$ and $\boldsymbol{m}$, respectively, see (6.5) on page 270

$$\boldsymbol{F}(\hat{\boldsymbol{r}}) = -\frac{k^2}{4\pi\epsilon_0\epsilon}\left[\hat{\boldsymbol{r}}\times(\hat{\boldsymbol{r}}\times\boldsymbol{p}) + \frac{\hat{\boldsymbol{r}}\times\boldsymbol{m}}{c} + O(kd)\right]$$

The terms on the right-hand side are functions of the direction $\hat{\boldsymbol{r}}$ only, and can be expressed in a series of vector spherical harmonics $\boldsymbol{A}_n(\hat{\boldsymbol{r}})$. We get

$$\begin{cases} \hat{\boldsymbol{r}}\times(\hat{\boldsymbol{r}}\times\boldsymbol{p}) = \sum_n \alpha_n \boldsymbol{A}_n(\hat{\boldsymbol{r}}) \\ \hat{\boldsymbol{r}}\times\boldsymbol{m} = \sum_n \beta_n \boldsymbol{A}_n(\hat{\boldsymbol{r}}) \end{cases}$$

where due to orthonormality of the vector spherical harmonics

$$\begin{cases} \alpha_n = \iint_\Omega (\hat{\boldsymbol{r}}\times(\hat{\boldsymbol{r}}\times\boldsymbol{p}))\cdot\boldsymbol{A}_n(\hat{\boldsymbol{r}})\,\mathrm{d}\Omega = \boldsymbol{p}\cdot\iint_\Omega (\hat{\boldsymbol{r}}\times(\hat{\boldsymbol{r}}\times\boldsymbol{A}_n(\hat{\boldsymbol{r}})))\,\mathrm{d}\Omega \\ \beta_n = \iint_\Omega (\hat{\boldsymbol{r}}\times\boldsymbol{m})\cdot\boldsymbol{A}_n(\hat{\boldsymbol{r}})\,\mathrm{d}\Omega = -\boldsymbol{m}\cdot\iint_\Omega (\hat{\boldsymbol{r}}\times\boldsymbol{A}_n(\hat{\boldsymbol{r}}))\,\mathrm{d}\Omega \end{cases}$$

With the use of the result in Problem 7.4

$$\begin{cases} \iint_\Omega \boldsymbol{A}_{1n}(\hat{\boldsymbol{r}})\,\mathrm{d}\Omega = 4\pi\mathrm{i}^l\boldsymbol{v}_{1n}(\boldsymbol{0}) = \boldsymbol{0} \\ \iint_\Omega \boldsymbol{A}_{2n}(\hat{\boldsymbol{r}})\,\mathrm{d}\Omega = -4\pi\mathrm{i}^{l+1}\boldsymbol{v}_{2n}(\boldsymbol{0}) \\ \iint_\Omega \boldsymbol{A}_{3n}(\hat{\boldsymbol{r}})\,\mathrm{d}\Omega = -4\pi\mathrm{i}^{l+1}\boldsymbol{v}_{3n}(\boldsymbol{0}) \end{cases}$$

or, with the explicit form of the spherical vector waves, we get

$$\iint_\Omega \boldsymbol{A}_n(\hat{\boldsymbol{r}})\,\mathrm{d}\Omega = \delta_{l1}\left(\hat{\boldsymbol{x}}\delta_{\sigma\mathrm{e}}\delta_{m1} + \hat{\boldsymbol{y}}\delta_{\sigma\mathrm{o}}\delta_{m1} + \hat{\boldsymbol{z}}\delta_{\sigma\mathrm{e}}\delta_{m0}\right)\left(\delta_{\tau2}\sqrt{\frac{8\pi}{3}} + \delta_{\tau3}\sqrt{\frac{4\pi}{3}}\right)$$

Moreover, we use, see (C.7) on page 631

$$\begin{cases} \boldsymbol{A}_{1n}(\hat{\boldsymbol{r}}) = \boldsymbol{A}_{2n}(\hat{\boldsymbol{r}}) \times \hat{\boldsymbol{r}} \\ \boldsymbol{A}_{2n}(\hat{\boldsymbol{r}}) = \hat{\boldsymbol{r}} \times \boldsymbol{A}_{1n}(\hat{\boldsymbol{r}}) \end{cases}$$

which imply

$$\begin{cases} \alpha_n = -\delta_{\tau 2}\delta_{l1}\left(\boldsymbol{p}\cdot\hat{\boldsymbol{x}}\delta_{\sigma\mathrm{e}}\delta_{m1} + \boldsymbol{p}\cdot\hat{\boldsymbol{y}}\delta_{\sigma\mathrm{o}}\delta_{m1} + \boldsymbol{p}\cdot\hat{\boldsymbol{z}}\delta_{\sigma\mathrm{e}}\delta_{m0}\right)\sqrt{\dfrac{8\pi}{3}} \\ \beta_n = -\delta_{\tau 1}\delta_{l1}\left(\boldsymbol{m}\cdot\hat{\boldsymbol{x}}\delta_{\sigma\mathrm{e}}\delta_{m1} + \boldsymbol{m}\cdot\hat{\boldsymbol{y}}\delta_{\sigma\mathrm{o}}\delta_{m1} + \boldsymbol{m}\cdot\hat{\boldsymbol{z}}\delta_{\sigma\mathrm{e}}\delta_{m0}\right)\sqrt{\dfrac{8\pi}{3}} \end{cases}$$

Therefore

$$\begin{cases} \hat{\boldsymbol{r}}\times(\hat{\boldsymbol{r}}\times\boldsymbol{p}) = -\sqrt{\dfrac{8\pi}{3}}\left\{(\boldsymbol{p}\cdot\hat{\boldsymbol{x}})\,\boldsymbol{A}_{2\mathrm{e}11}(\hat{\boldsymbol{r}}) + (\boldsymbol{p}\cdot\hat{\boldsymbol{y}})\,\boldsymbol{A}_{2\mathrm{o}11}(\hat{\boldsymbol{r}}) + (\boldsymbol{p}\cdot\hat{\boldsymbol{z}})\,\boldsymbol{A}_{2\mathrm{e}01}(\hat{\boldsymbol{r}})\right\} \\ \hat{\boldsymbol{r}}\times\boldsymbol{m} = -\sqrt{\dfrac{8\pi}{3}}\left\{(\boldsymbol{m}\cdot\hat{\boldsymbol{x}})\,\boldsymbol{A}_{1\mathrm{e}11}(\hat{\boldsymbol{r}}) + (\boldsymbol{m}\cdot\hat{\boldsymbol{y}})\,\boldsymbol{A}_{1\mathrm{o}11}(\hat{\boldsymbol{r}}) + (\boldsymbol{m}\cdot\hat{\boldsymbol{z}})\,\boldsymbol{A}_{1\mathrm{e}01}(\hat{\boldsymbol{r}})\right\} \end{cases}$$

and the far field amplitude in the far zone becomes

$$\begin{aligned} \boldsymbol{F}(\hat{\boldsymbol{r}}) = \frac{k^2}{\epsilon_0\epsilon\sqrt{6\pi}}\Big[&(\boldsymbol{p}\cdot\hat{\boldsymbol{x}})\,\boldsymbol{A}_{2\mathrm{e}11}(\hat{\boldsymbol{r}}) + (\boldsymbol{p}\cdot\hat{\boldsymbol{y}})\,\boldsymbol{A}_{2\mathrm{o}11}(\hat{\boldsymbol{r}}) + (\boldsymbol{p}\cdot\hat{\boldsymbol{z}})\,\boldsymbol{A}_{2\mathrm{e}01}(\hat{\boldsymbol{r}}) \\ &+ \frac{1}{c}\left[(\boldsymbol{m}\cdot\hat{\boldsymbol{x}})\,\boldsymbol{A}_{1\mathrm{e}11}(\hat{\boldsymbol{r}}) + (\boldsymbol{m}\cdot\hat{\boldsymbol{y}})\,\boldsymbol{A}_{1\mathrm{o}11}(\hat{\boldsymbol{r}}) + (\boldsymbol{m}\cdot\hat{\boldsymbol{z}})\,\boldsymbol{A}_{1\mathrm{e}01}(\hat{\boldsymbol{r}})\right] + O(kd)\Big] \end{aligned}$$

From the expansion of the far field amplitude in vector spherical harmonics in (7.51), we can identify the electric and the magnetic dipole moments of the scatterer in terms of the expansion coefficients f_n of the scattered field. The result is

$$\begin{cases} \boldsymbol{p}\cdot\hat{\boldsymbol{x}} = \lim\limits_{k\to 0}\dfrac{-\mathrm{i}\epsilon_0\epsilon\sqrt{6\pi}}{k^3}f_{2\mathrm{e}11} \\ \boldsymbol{p}\cdot\hat{\boldsymbol{y}} = \lim\limits_{k\to 0}\dfrac{-\mathrm{i}\epsilon_0\epsilon\sqrt{6\pi}}{k^3}f_{2\mathrm{o}11} \\ \boldsymbol{p}\cdot\hat{\boldsymbol{z}} = \lim\limits_{k\to 0}\dfrac{-\mathrm{i}\epsilon_0\epsilon\sqrt{6\pi}}{k^3}f_{2\mathrm{e}01} \end{cases} \qquad \begin{cases} \boldsymbol{m}\cdot\hat{\boldsymbol{x}} = \lim\limits_{k\to 0}\dfrac{-\sqrt{6\pi}}{k^3\eta_0\eta}f_{1\mathrm{e}11} \\ \boldsymbol{m}\cdot\hat{\boldsymbol{y}} = \lim\limits_{k\to 0}\dfrac{-\sqrt{6\pi}}{k^3\eta_0\eta}f_{1\mathrm{o}11} \\ \boldsymbol{m}\cdot\hat{\boldsymbol{z}} = \lim\limits_{k\to 0}\dfrac{-\sqrt{6\pi}}{k^3\eta_0\eta}f_{1\mathrm{e}01} \end{cases}$$

The expansion coefficients of the far field amplitude, f_n, can also be expressed in terms of the transition matrix $T_{nn'}$

$$\begin{cases} \boldsymbol{p}\cdot\hat{\boldsymbol{x}} = \lim_{k\to 0} \dfrac{-\mathrm{i}\epsilon_0\epsilon\sqrt{6\pi}}{k^3}\sum_n T_{2\mathrm{e}11n}a_n \\ \boldsymbol{p}\cdot\hat{\boldsymbol{y}} = \lim_{k\to 0} \dfrac{-\mathrm{i}\epsilon_0\epsilon\sqrt{6\pi}}{k^3}\sum_n T_{2\mathrm{o}11n}a_n \\ \boldsymbol{p}\cdot\hat{\boldsymbol{z}} = \lim_{k\to 0} \dfrac{-\mathrm{i}\epsilon_0\epsilon\sqrt{6\pi}}{k^3}\sum_n T_{2\mathrm{e}01n}a_n \end{cases} \qquad \begin{cases} \boldsymbol{m}\cdot\hat{\boldsymbol{x}} = \lim_{k\to 0} \dfrac{-\sqrt{6\pi}}{k^3\eta_0\eta}\sum_n T_{1\mathrm{e}11n}a_n \\ \boldsymbol{m}\cdot\hat{\boldsymbol{y}} = \lim_{k\to 0} \dfrac{-\sqrt{6\pi}}{k^3\eta_0\eta}\sum_n T_{1\mathrm{o}11n}a_n \\ \boldsymbol{m}\cdot\hat{\boldsymbol{z}} = \lim_{k\to 0} \dfrac{-\sqrt{6\pi}}{k^3\eta_0\eta}\sum_n T_{1\mathrm{e}01n}a_n \end{cases}$$

For an incident plane wave, the expansion coefficients are given by (7.28)

$$a_n = 4\pi\mathrm{i}^{l+1-\tau}\boldsymbol{A}_n(\hat{\boldsymbol{k}}_\mathrm{i})\cdot\boldsymbol{E}_0 = 4\pi\mathrm{i}^{l-1+\tau}\boldsymbol{A}_{\overline{n}}(\hat{\boldsymbol{k}}_\mathrm{i})\cdot\left(\hat{\boldsymbol{k}}_\mathrm{i}\times\boldsymbol{E}_0\right), \qquad \tau = 1, 2$$

and we obtain expressions of the electric and magnetic polarizability dyadics, $\boldsymbol{\gamma}_\mathrm{e}$ and $\boldsymbol{\gamma}_\mathrm{m}$, respectively, defined in (6.7) on page 270

$$\begin{cases} \boldsymbol{p} = \epsilon_0\epsilon\boldsymbol{\gamma}_\mathrm{e}\cdot\boldsymbol{E}_0 \\ \boldsymbol{m} = \boldsymbol{\gamma}_\mathrm{m}\cdot\boldsymbol{H}_0 = \dfrac{1}{\eta_0\eta}\boldsymbol{\gamma}_\mathrm{m}\cdot\left(\hat{\boldsymbol{k}}_\mathrm{i}\times\boldsymbol{E}_0\right) \end{cases}$$

The result is

$$\begin{cases} \hat{\boldsymbol{x}}\cdot\boldsymbol{\gamma}_\mathrm{e}\cdot\boldsymbol{E}_0 = \lim_{k\to 0}\dfrac{4\pi\sqrt{6\pi}}{k^3}\sum_n \mathrm{i}^{l-\tau}T_{2\mathrm{e}11n}\boldsymbol{A}_n(\hat{\boldsymbol{k}}_\mathrm{i})\cdot\boldsymbol{E}_0 \\ \hat{\boldsymbol{y}}\cdot\boldsymbol{\gamma}_\mathrm{e}\cdot\boldsymbol{E}_0 = \lim_{k\to 0}\dfrac{4\pi\sqrt{6\pi}}{k^3}\sum_n \mathrm{i}^{l-\tau}T_{2\mathrm{o}11n}\boldsymbol{A}_n(\hat{\boldsymbol{k}}_\mathrm{i})\cdot\boldsymbol{E}_0 \\ \hat{\boldsymbol{z}}\cdot\boldsymbol{\gamma}_\mathrm{e}\cdot\boldsymbol{E}_0 = \lim_{k\to 0}\dfrac{4\pi\sqrt{6\pi}}{k^3}\sum_n \mathrm{i}^{l-\tau}T_{2\mathrm{e}01n}\boldsymbol{A}_n(\hat{\boldsymbol{k}}_\mathrm{i})\cdot\boldsymbol{E}_0 \end{cases} \tag{7.52}$$

and

$$\begin{cases} \hat{\boldsymbol{x}}\cdot\boldsymbol{\gamma}_\mathrm{m}\cdot\left(\hat{\boldsymbol{k}}_\mathrm{i}\times\boldsymbol{E}_0\right) = \lim_{k\to 0}\dfrac{4\pi\sqrt{6\pi}}{k^3}\sum_n \mathrm{i}^{l+\tau+1}T_{1\mathrm{e}11n}\boldsymbol{A}_{\overline{n}}(\hat{\boldsymbol{k}}_\mathrm{i})\cdot\left(\hat{\boldsymbol{k}}_\mathrm{i}\times\boldsymbol{E}_0\right) \\ \hat{\boldsymbol{y}}\cdot\boldsymbol{\gamma}_\mathrm{m}\cdot\left(\hat{\boldsymbol{k}}_\mathrm{i}\times\boldsymbol{E}_0\right) = \lim_{k\to 0}\dfrac{4\pi\sqrt{6\pi}}{k^3}\sum_n \mathrm{i}^{l+\tau+1}T_{1\mathrm{o}11n}\boldsymbol{A}_{\overline{n}}(\hat{\boldsymbol{k}}_\mathrm{i})\cdot\left(\hat{\boldsymbol{k}}_\mathrm{i}\times\boldsymbol{E}_0\right) \\ \hat{\boldsymbol{z}}\cdot\boldsymbol{\gamma}_\mathrm{m}\cdot\left(\hat{\boldsymbol{k}}_\mathrm{i}\times\boldsymbol{E}_0\right) = \lim_{k\to 0}\dfrac{4\pi\sqrt{6\pi}}{k^3}\sum_n \mathrm{i}^{l+\tau+1}T_{1\mathrm{e}01n}\boldsymbol{A}_{\overline{n}}(\hat{\boldsymbol{k}}_\mathrm{i})\cdot\left(\hat{\boldsymbol{k}}_\mathrm{i}\times\boldsymbol{E}_0\right) \end{cases} \tag{7.53}$$

Example 7.5

A non-trivial example of the result above is the dipole moment of a perfectly conducting disk of radius a. If the disk is located in the x–y-plane, the lowest order T-matrix elements, $T_{\tau\sigma ml,\tau'\sigma'm'l'}$, are [147], see also Example 9.2 on page 486

$$\begin{cases} T_{1\mathrm{e}01,1\mathrm{e}01} = -\dfrac{4\mathrm{i}}{9\pi}k^3a^3 \\ T_{2\mathrm{e}11,2\mathrm{e}11} = \dfrac{8\mathrm{i}}{9\pi}k^3a^3 \\ T_{2\mathrm{o}11,2\mathrm{o}11} = \dfrac{8\mathrm{i}}{9\pi}k^3a^3 \end{cases}$$

Evaluating (7.52) and (7.53), only the following elements survive:

$$\begin{cases} \hat{\boldsymbol{x}} \cdot \boldsymbol{\gamma}_{\mathrm{e}} \cdot \boldsymbol{E}_0 = \dfrac{16a^3}{3}\hat{\boldsymbol{x}} \cdot \boldsymbol{E}_0 \\ \hat{\boldsymbol{y}} \cdot \boldsymbol{\gamma}_{\mathrm{e}} \cdot \boldsymbol{E}_0 = \dfrac{16a^3}{3}\hat{\boldsymbol{y}} \cdot \boldsymbol{E}_0 \\ \hat{\boldsymbol{z}} \cdot \boldsymbol{\gamma}_{\mathrm{e}} \cdot \boldsymbol{E}_0 = 0 \end{cases}$$

and

$$\begin{cases} \hat{\boldsymbol{x}} \cdot \boldsymbol{\gamma}_{\mathrm{m}} \cdot \left(\hat{\boldsymbol{k}}_{\mathrm{i}} \times \boldsymbol{E}_0\right) = 0 \\ \hat{\boldsymbol{y}} \cdot \boldsymbol{\gamma}_{\mathrm{m}} \cdot \left(\hat{\boldsymbol{k}}_{\mathrm{i}} \times \boldsymbol{E}_0\right) = 0 \\ \hat{\boldsymbol{z}} \cdot \boldsymbol{\gamma}_{\mathrm{m}} \cdot \left(\hat{\boldsymbol{k}}_{\mathrm{i}} \times \boldsymbol{E}_0\right) = -\dfrac{8a^3}{3}\hat{\boldsymbol{z}} \cdot \left(\hat{\boldsymbol{k}}_{\mathrm{i}} \times \boldsymbol{E}_0\right) \end{cases}$$

These expressions imply that

$$\begin{cases} \boldsymbol{\gamma}_{\mathrm{e}} = \dfrac{16a^3}{3}(\hat{\boldsymbol{x}}\hat{\boldsymbol{x}} + \hat{\boldsymbol{y}}\hat{\boldsymbol{y}}) \\ \boldsymbol{\gamma}_{\mathrm{m}} = -\dfrac{8a^3}{3}\hat{\boldsymbol{z}}\hat{\boldsymbol{z}} \end{cases}$$

The scattering cross section, correct to lowest order in ka, for an incident plane wave that impinges along $\hat{\boldsymbol{k}}_{\mathrm{i}} = \hat{\boldsymbol{x}}\sin\alpha + \hat{\boldsymbol{z}}\cos\alpha$ and with polarization

$$\hat{\boldsymbol{p}}_{\mathrm{e}} = \hat{\boldsymbol{x}}\cos\alpha\cos\gamma + \hat{\boldsymbol{y}}\sin\gamma - \hat{\boldsymbol{z}}\sin\alpha\cos\gamma$$

where $\gamma = \pi/2$ corresponds to a TE-wave, and $\gamma = 0$ to a TM-wave, then is, see (6.11) on page 273

$$\frac{\sigma_{\mathrm{s}}(\hat{\boldsymbol{k}}_{\mathrm{i}})}{2\pi a^2} = \frac{k^4}{12\pi^2a^2}\left\{\left|\boldsymbol{\gamma}_{\mathrm{e}} \cdot \hat{\boldsymbol{p}}_{\mathrm{e}}\right|^2 + \left|\boldsymbol{\gamma}_{\mathrm{m}} \cdot \hat{\boldsymbol{p}}_{\mathrm{m}}\right|^2\right\} = \frac{64k^4a^4}{27\pi^2}\left(1 + \sin^2\alpha\left(\frac{5}{4}\sin^2\gamma - 1\right)\right)$$

More generally, the scattering dyadic is, see (6.8) on page 272

$$\begin{aligned}\mathbf{S}(\hat{\boldsymbol{r}},\hat{\boldsymbol{k}}_\mathrm{i})\cdot\boldsymbol{E}_0&=-\frac{k^2}{4\pi}\left[\hat{\boldsymbol{r}}\times\left(\hat{\boldsymbol{r}}\times\boldsymbol{\gamma}_\mathrm{e}\right)+\hat{\boldsymbol{r}}\times\boldsymbol{\gamma}_\mathrm{m}\cdot\left(\hat{\boldsymbol{k}}_\mathrm{i}\times\mathbf{I}_3\right)\right]\cdot\boldsymbol{E}_0\\&=-\frac{2k^2a^3}{3\pi}\left[2\hat{\boldsymbol{r}}\times\left(\hat{\boldsymbol{r}}\times\left(\boldsymbol{E}_0-\hat{\boldsymbol{z}}\hat{\boldsymbol{z}}\cdot\boldsymbol{E}_0\right)\right)-\left(\hat{\boldsymbol{r}}\times\hat{\boldsymbol{z}}\right)\hat{\boldsymbol{z}}\cdot\left(\hat{\boldsymbol{k}}_\mathrm{i}\times\boldsymbol{E}_0\right)\right]\end{aligned}$$

■

Example 7.6
For a transition matrix that is diagonal in l, m, and σ, cf. Example 7.3, and moreover, independent of the indices m and σ, i.e.,

$$T_{nn'}=t_{\tau\tau' l}\delta_{ll'}\delta_{mm'}\delta_{\sigma\sigma'}$$

we get simplifications, see (C.10) and (C.11) in Appendix C.4

$$\sum_n \mathrm{i}^{l-\tau}T_{2\mathrm{e}11n}\boldsymbol{A}_n(\hat{\boldsymbol{k}}_\mathrm{i})\cdot\boldsymbol{E}_0=\sqrt{\frac{3}{8\pi}}\left(-t_{211}\left(\hat{\boldsymbol{k}}_\mathrm{i}\times\hat{\boldsymbol{x}}\right)-\mathrm{i}t_{221}\hat{\boldsymbol{x}}\right)\cdot\boldsymbol{E}_0$$

$$\sum_n \mathrm{i}^{l-\tau}T_{2\mathrm{o}11n}\boldsymbol{A}_n(\hat{\boldsymbol{k}}_\mathrm{i})\cdot\boldsymbol{E}_0=\sqrt{\frac{3}{8\pi}}\left(-t_{211}\left(\hat{\boldsymbol{k}}_\mathrm{i}\times\hat{\boldsymbol{y}}\right)-\mathrm{i}t_{221}\hat{\boldsymbol{y}}\right)\cdot\boldsymbol{E}_0$$

$$\sum_n \mathrm{i}^{l-\tau}T_{2\mathrm{e}01n}\boldsymbol{A}_n(\hat{\boldsymbol{k}}_\mathrm{i})\cdot\boldsymbol{E}_0=\sqrt{\frac{3}{8\pi}}\left(-t_{211}\left(\hat{\boldsymbol{k}}_\mathrm{i}\times\hat{\boldsymbol{z}}\right)-\mathrm{i}t_{221}\hat{\boldsymbol{z}}\right)\cdot\boldsymbol{E}_0$$

and similarly for the magnetic case, i.e.,

$$\sum_n \mathrm{i}^{l+\tau+1}T_{1\mathrm{e}11n}\boldsymbol{A}_{\overline{n}}(\hat{\boldsymbol{k}}_\mathrm{i})\cdot\left(\hat{\boldsymbol{k}}_\mathrm{i}\times\boldsymbol{E}_0\right)=\sqrt{\frac{3}{8\pi}}\left(-\mathrm{i}t_{111}\hat{\boldsymbol{x}}-t_{121}\left(\hat{\boldsymbol{k}}_\mathrm{i}\times\hat{\boldsymbol{x}}\right)\right)\cdot\left(\hat{\boldsymbol{k}}_\mathrm{i}\times\boldsymbol{E}_0\right)$$

$$\sum_n \mathrm{i}^{l+\tau+1}T_{1\mathrm{o}11n}\boldsymbol{A}_{\overline{n}}(\hat{\boldsymbol{k}}_\mathrm{i})\cdot\left(\hat{\boldsymbol{k}}_\mathrm{i}\times\boldsymbol{E}_0\right)=\sqrt{\frac{3}{8\pi}}\left(-\mathrm{i}t_{111}\hat{\boldsymbol{y}}-t_{121}\left(\hat{\boldsymbol{k}}_\mathrm{i}\times\hat{\boldsymbol{y}}\right)\right)\cdot\left(\hat{\boldsymbol{k}}_\mathrm{i}\times\boldsymbol{E}_0\right)$$

$$\sum_n \mathrm{i}^{l+\tau+1}T_{1\mathrm{e}01n}\boldsymbol{A}_{\overline{n}}(\hat{\boldsymbol{k}}_\mathrm{i})\cdot\left(\hat{\boldsymbol{k}}_\mathrm{i}\times\boldsymbol{E}_0\right)=\sqrt{\frac{3}{8\pi}}\left(-\mathrm{i}t_{111}\hat{\boldsymbol{z}}-t_{121}\left(\hat{\boldsymbol{k}}_\mathrm{i}\times\hat{\boldsymbol{z}}\right)\right)\cdot\left(\hat{\boldsymbol{k}}_\mathrm{i}\times\boldsymbol{E}_0\right)$$

The components of the polarizability dyadics then are

$$\begin{cases} \gamma_{\mathrm{e}} = \lim_{k\to 0} \dfrac{6\pi t_{221}\mathbf{I}_3}{\mathrm{i}k^3} + \lim_{k\to 0} \dfrac{6\pi t_{211}\hat{\boldsymbol{k}}_{\mathrm{i}} \times \mathbf{I}_3}{k^3} \\ \gamma_{\mathrm{m}} = \lim_{k\to 0} \dfrac{6\pi t_{111}\mathbf{I}_3}{\mathrm{i}k^3} + \lim_{k\to 0} \dfrac{6\pi t_{121}\hat{\boldsymbol{k}}_{\mathrm{i}} \times \mathbf{I}_3}{k^3} \end{cases} \tag{7.54}$$

These are the general expressions of the polarization dyadics for a diagonal transition matrix. ■

Example 7.7
We exemplify the result in Example 7.6 by the perfectly conducting sphere. In Section 8.1.1, the lowest order entries of the T-matrix are derived, i.e.,

$$\begin{cases} t_{11} = -\mathrm{i}\dfrac{k^3a^3}{3} \\ t_{21} = \mathrm{i}\dfrac{2k^3a^3}{3} \end{cases}$$

where $t_{\tau l} = \delta_{\tau\tau'} t_{\tau\tau' l}$. Equation (7.54) then implies

$$\begin{cases} \gamma_{\mathrm{e}} = 4\pi a^3 \mathbf{I}_3 \\ \gamma_{\mathrm{m}} = -2\pi a^3 \mathbf{I}_3 \end{cases}$$

■

Example 7.8
A second example of the use of the result in Example 7.6 is the dielectric sphere in Section 8.2.2. From the transition matrix elements in (8.13) on page 427, we get

$$\begin{cases} t_{11} = \dfrac{2\mathrm{i}k^3a^3}{3} \dfrac{\mu_1 - \mu}{\mu_1 + 2\mu} \\ t_{21} = \dfrac{2\mathrm{i}k^3a^3}{3} \dfrac{\epsilon_1 - \epsilon}{\epsilon_1 + 2\epsilon} \end{cases}$$

Equation (7.54) then implies

$$\begin{cases} \gamma_{\mathrm{e}} = 4\pi a^3 \dfrac{\epsilon_1 - \epsilon}{\epsilon_1 + 2\epsilon} \mathbf{I}_3 \\ \gamma_{\mathrm{m}} = 4\pi a^3 \dfrac{\mu_1 - \mu}{\mu_1 + 2\mu} \mathbf{I}_3 \end{cases}$$

provided the values of ϵ_1 and μ_1 are real numbers at zero frequency. The perfectly conducting case is obtained by $\epsilon_1 \to \infty$ and $\mu_1 \to 0$, and the polarizability dyadics coincide with the result of Example 7.7. ■

Problems for Chapter 7

Section 7.2 Definition of Spherical Vector Waves

7.1 Show that the longitudinal wave, $\tau = 3$, in (7.10)

$$\begin{cases} \boldsymbol{u}_{3n}(k\boldsymbol{r}) = \frac{1}{k}\nabla\left(h_l^{(1)}(kr)Y_n(\hat{\boldsymbol{r}})\right) \\ \boldsymbol{v}_{3n}(k\boldsymbol{r}) = \frac{1}{k}\nabla\left(j_l(kr)Y_n(\hat{\boldsymbol{r}})\right) \end{cases}$$

also can be written as

$$\begin{cases} \boldsymbol{u}_{3n}(k\boldsymbol{r}) = h_l^{(1)\prime}(kr)\boldsymbol{A}_{3n}(\hat{\boldsymbol{r}}) + \sqrt{l(l+1)}\frac{h_l^{(1)}(kr)}{kr}\boldsymbol{A}_{2n}(\hat{\boldsymbol{r}}) \\ \boldsymbol{v}_{3n}(k\boldsymbol{r}) = j_l'(kr)\boldsymbol{A}_{3n}(\hat{\boldsymbol{r}}) + \sqrt{l(l+1)}\frac{j_l(kr)}{kr}\boldsymbol{A}_{2n}(\hat{\boldsymbol{r}}) \end{cases}$$

Section 7.3 Orthogonality and Reciprocity Relations

7.2 Show that (7.18) is equivalent to the optical theorem.

Section 7.6 Null-Field Equations

7.3 The sources of the incident field are located in the volume $V_{\rm i}$ (bounding surface $S_{\rm i}$), see Figure 7.4. The origin O is chosen such that the region inside the sphere, centered at the origin with radius R, is free space with material parameters ϵ and μ. Use the integral representation to find an expression of the incident field inside this sphere, $r < R$. Show that the incident field has an expansion in the vicinity of the origin O in terms of the regular spherical vector waves, i.e.,

$$\boldsymbol{E}_{\rm i}(\boldsymbol{r}) = \sum_n a_n \boldsymbol{v}_n(\boldsymbol{r})$$

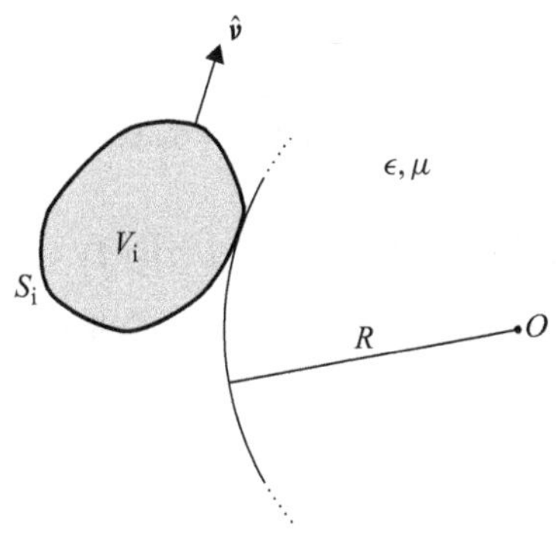

Figure 7.4 The geometry of the sources of the incident field in Problem 7.3.

Find an expression of the expansion coefficients a_n in terms of the tangential electric and magnetic fields on the bounding surface $S_{\rm i}$.

Section 7.7 Expansion of Sources

***7.4** Show the following vector identities for the vector spherical harmonics $\boldsymbol{A}_{\tau\sigma ml}$:

$$\begin{cases} \iint\limits_{\Omega} \boldsymbol{A}_{1n}(\hat{\boldsymbol{r}})\mathrm{e}^{\mathrm{i}k\hat{\boldsymbol{k}}\cdot\boldsymbol{r}}\, \mathrm{d}\Omega = 4\pi\mathrm{i}^l j_l(kr)\boldsymbol{A}_{1n}(\hat{\boldsymbol{k}}) = 4\pi\mathrm{i}^l \boldsymbol{v}_{1n}(\boldsymbol{k}r) \\ \iint\limits_{\Omega} \boldsymbol{A}_{2n}(\hat{\boldsymbol{r}})\mathrm{e}^{\mathrm{i}k\hat{\boldsymbol{k}}\cdot\boldsymbol{r}}\, \mathrm{d}\Omega = -\dfrac{4\pi\mathrm{i}^{l+1}}{r}\nabla_k \times \left(j_l(kr)\boldsymbol{A}_{1n}(\hat{\boldsymbol{k}})\right) = -4\pi\mathrm{i}^{l+1}\boldsymbol{v}_{2n}(\boldsymbol{k}r) \\ \iint\limits_{\Omega} \boldsymbol{A}_{3n}(\hat{\boldsymbol{r}})\mathrm{e}^{\mathrm{i}k\hat{\boldsymbol{k}}\cdot\boldsymbol{r}}\, \mathrm{d}\Omega = -\dfrac{4\pi\mathrm{i}^{l+1}}{r}\nabla_k \left(j_l(kr)Y_n(\hat{\boldsymbol{k}})\right) = -4\pi\mathrm{i}^{l+1}\boldsymbol{v}_{3n}(\boldsymbol{k}r) \end{cases}$$

where $\boldsymbol{k} = k\hat{\boldsymbol{k}}$ and

$$\nabla_k = \hat{\boldsymbol{x}}\frac{\partial}{\partial k_x} + \hat{\boldsymbol{y}}\frac{\partial}{\partial k_y} + \hat{\boldsymbol{z}}\frac{\partial}{\partial k_z}$$

7.5 Show that a plane wave $\boldsymbol{E}_0\mathrm{e}^{\mathrm{i}k\hat{\boldsymbol{k}}_{\rm i}\cdot\boldsymbol{r}}$ has the following expansion in regular spherical vector waves $\boldsymbol{v}_{\tau\sigma ml}(k\boldsymbol{r})$:

$$\boldsymbol{E}_0\mathrm{e}^{\mathrm{i}k\hat{\boldsymbol{k}}_{\rm i}\cdot\boldsymbol{r}} = \sum_{l=0}^{\infty}\sum_{m=0}^{l}\sum_{\sigma=\mathrm{e,o}}\sum_{\tau=1}^{3} a_{\tau\sigma ml}\boldsymbol{v}_{\tau\sigma ml}(k\boldsymbol{r})$$

where

$$\begin{cases} a_{1n} = 4\pi\mathrm{i}^l\boldsymbol{E}_0\cdot\boldsymbol{A}_{1n}(\hat{\boldsymbol{k}}_{\rm i}) \\ a_{2n} = -4\pi\mathrm{i}^{l+1}\boldsymbol{E}_0\cdot\boldsymbol{A}_{2n}(\hat{\boldsymbol{k}}_{\rm i}) \\ a_{3n} = -4\pi\mathrm{i}^{l+1}\boldsymbol{E}_0\cdot\boldsymbol{A}_{3n}(\hat{\boldsymbol{k}}_{\rm i}) \end{cases}$$

Hint: Use the result of Problem 7.4.

7.6 Determine the expansion coefficients a_n of the electric field generated by a horizontal electric dipole located at $\boldsymbol{r}_{\rm s}$. Determine also the expansion coefficients a_n of the electric field for a horizontal magnetic dipole located at $\boldsymbol{r}_{\rm s}$.

Section 7.8 Far Field Amplitude and the Transition Matrix

7.7 Show that the functions $\pi_l(x)$ and $\tau_l(x)$ in (7.45) have the following alternative representation

$$\begin{cases} \pi_l(\cos\theta) = \dfrac{1}{\sin\theta}P_l^1(\cos\theta) \\ \tau_l(x) = \dfrac{\mathrm{d}}{\mathrm{d}\theta}P_l^1(\cos\theta) \end{cases}$$

*7.8 For a dielectric spherical scatterer, the transition matrix has the form

$$T_{nn'} = t_{\tau l}\delta_{nn'} = t_{\tau l}\delta_{\tau\tau'}\delta_{ll'}\delta_{mm'}\delta_{\sigma\sigma'}$$

and the differential scattering cross section is, see (7.48) on page 389

$$\frac{d\sigma}{d\Omega}(\hat{\boldsymbol{r}},\hat{\boldsymbol{z}}) = 4\pi\left\{\left|S_{\parallel\parallel}(\hat{\boldsymbol{r}},\hat{\boldsymbol{z}})\right|^2\left|\hat{\boldsymbol{r}}_{\mathrm{c}}\cdot\hat{\boldsymbol{p}}_{\mathrm{e}}\right|^2 + \left|S_{\perp\perp}(\hat{\boldsymbol{r}},\hat{\boldsymbol{z}})\right|^2\left|\hat{\boldsymbol{\phi}}\cdot\hat{\boldsymbol{p}}_{\mathrm{e}}\right|^2\right\}$$

where

$$S_{\parallel\parallel}(\hat{\boldsymbol{r}},\hat{\boldsymbol{z}}) = \frac{1}{ik}\sum_{l=1}^{\infty}\frac{2l+1}{l(l+1)}(\pi_l(\cos\theta)t_{1l} + \tau_l(\cos\theta)t_{2l})$$

and

$$S_{\perp\perp}(\hat{\boldsymbol{r}},\hat{\boldsymbol{z}}) = \frac{1}{ik}\sum_{l=1}^{\infty}\frac{2l+1}{l(l+1)}(\pi_l(\cos\theta)t_{2l} + \tau_l(\cos\theta)t_{1l})$$

Show that the differential scattering cross section can be written as

$$\frac{d\sigma}{d\Omega}(\hat{\boldsymbol{r}},\hat{\boldsymbol{z}}) = \sum_{l=0}^{\infty}\omega_l P_l(\cos\theta)$$

and determine the expansion coefficients ω_l.

7.9 Show that the expressions in (7.49) and (7.50) are identical for a lossless scatterer.

CHAPTER 8

Scattering by Spherical Objects

The usefulness of spherical vector waves becomes obvious when we solve scattering problems by obstacles that have spherical symmetries. In Sections 8.1–8.5, we solve the scattering by perfectly conducting, dielectric, layered, anisotropic, and biisotropic sphere, respectively. These solutions are called the Mie solution[1] or partial wave solution, see e.g., References 54, 58, 94. A method to compute scattering by non-spherical shapes is presented in Chapter 9. Scattering by more complex materials has also been investigated in the literature, e.g., scattering by a gyrotropic sphere [72, 160].

8.1 Scattering by a Perfectly Conducting Sphere

One of the simplest cases of electromagnetic wave scattering is scattering by a perfectly conducting sphere. The sphere is assumed to have radius a and its center is located at the origin of the coordinate system. Let a plane wave impinge on the sphere along the positive z-axis. The geometry is depicted in Figure 8.1. The polarization of the plane wave is characterized by the complex vector $\boldsymbol{E}_0$, which lies in the x–y-plane. The expansion of the plane wave in spherical vector waves is, see (7.29) on page 375

$$\boldsymbol{E}_{\rm i}(\boldsymbol{r},\omega) = \boldsymbol{E}_0 e^{ikz} = \sum_n a_n \boldsymbol{v}_n(k\boldsymbol{r})$$

$$\begin{cases} a_{1n} = i^l \delta_{m1}\sqrt{2\pi(2l+1)}\boldsymbol{E}_0 \cdot \left(\delta_{\sigma o}\hat{\boldsymbol{x}} - \delta_{\sigma e}\hat{\boldsymbol{y}}\right) \\ a_{2n} = -i^{l+1}\delta_{m1}\sqrt{2\pi(2l+1)}\boldsymbol{E}_0 \cdot \left(\delta_{\sigma e}\hat{\boldsymbol{x}} + \delta_{\sigma o}\hat{\boldsymbol{y}}\right) \end{cases} \qquad \hat{\boldsymbol{k}}_{\rm i} = \hat{\boldsymbol{z}}$$

The expansion of the scattered field outside the sphere in out-going spherical vector waves is

$$\boldsymbol{E}_{\rm s}(\boldsymbol{r},\omega) = \sum_n f_n \boldsymbol{u}_n(k\boldsymbol{r}), \qquad r > a$$

The goal now is to determine the unknown expansion coefficients f_n in terms of the known expansion coefficients a_n. This is accomplished by forcing the total electric field

[1]Gustav Adolf Feodor Wilhelm Ludwig Mie (1869–1957), German physicist.

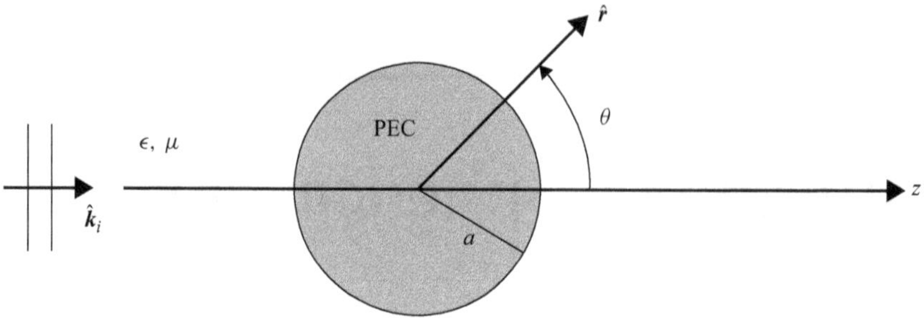

Figure 8.1 The geometry of scattering by a perfectly conducting sphere.

to satisfy the boundary condition on the surface of the sphere. The total field outside the sphere is

$$E(r,\omega) = \sum_n (a_n v_n(kr) + f_n u_n(kr)), \qquad r > a$$

The boundary condition on the surface of the sphere $r = a$ is, see (1.13) on page 8

$$\hat{r} \times E(r,\omega)|_{r=a} = 0$$

This condition implies, due to the orthogonality conditions of the vector spherical harmonics, $A_{\tau n}(\hat{r}) = A_{\tau\sigma ml}(\hat{r})$, that the expansion coefficients must satisfy

$$\begin{cases} a_{1n}\psi_l(ka) + f_{1n}\xi_l(ka) = 0 \\ a_{2n}\psi_l'(ka) + f_{2n}\xi_l'(ka) = 0 \end{cases}$$

where we have used the Riccati–Bessel functions $\psi_l(z)$ and $\xi_l(z)$ defined in Appendix B.3.3, see (B.26) on page 618. The coefficient in front of $A_{3n}(\hat{r})$ gives no condition, since this vector spherical harmonic is proportional to $\hat{r}$. We rewrite the coefficients as

$$f_{\tau\sigma ml} = t_{\tau l} a_{\tau\sigma ml}$$

The matrix entries $t_{\tau l}$ are independent of the indices m and σ, and they are explicitly given by

$$\begin{cases} t_{1l} = -\dfrac{\psi_l(ka)}{\xi_l(ka)} = -\dfrac{V_{1l}(ka)}{U_{1l}(ka)} \\[2ex] t_{2l} = -\dfrac{\psi_l'(ka)}{\xi_l'(ka)} = -\dfrac{V_{2l}(ka)}{U_{2l}(ka)} \end{cases} \tag{8.1}$$

where we also have used the radial functions $V_{\tau l}(kr)$ and $U_{\tau l}(kr)$ defined in Comment 7.2 on page 354. The direct scattering problem is now solved.

Once the transition matrix entries, $t_{\tau l}$, are found, all relevant scattering quantities—like the different cross sections—can be expressed using the results in Section 7.8.2 on page 383.

Specifically, these matrix elements and the results in Example 7.3 on page 386 now give the scattering dyadic, see (7.46) on page 387. We have

$$\begin{cases} S_{\parallel\parallel} = \dfrac{1}{ik}\displaystyle\sum_{l=1}^{\infty}\dfrac{2l+1}{l(l+1)}\left\{t_{1l}\pi_l(\cos\theta) + t_{2l}\tau_l(\cos\theta)\right\} \\ S_{\parallel\perp} = S_{\perp\parallel} = 0 \\ S_{\perp\perp} = \dfrac{1}{ik}\displaystyle\sum_{l=1}^{\infty}\dfrac{2l+1}{l(l+1)}\left\{t_{1l}\tau_l(\cos\theta) + t_{2l}\pi_l(\cos\theta)\right\} \end{cases} \tag{8.2}$$

where, see Appendix C.1.1

$$\begin{cases} \pi_l(x) = P_l'(x) \\ \tau_l(x) = l(l+1)P_l(x) - xP_l'(x) \end{cases}$$

and the scattering dyadic in the forward direction, see (7.47) on page 389

$$\mathbf{S}(\hat{z},\hat{z}) = \frac{1}{2ik}\mathbf{I}_2\sum_{l=1}^{\infty}(2l+1)\left(t_{1l}+t_{2l}\right)$$

where $\mathbf{I}_2$ is the unit dyadic in the x–y-plane, implies

$$\boldsymbol{F}(\hat{z}) = \frac{\boldsymbol{E}_0}{2ik}\sum_{l=1}^{\infty}(2l+1)\left(t_{1l}+t_{2l}\right)$$

The differential scattering cross section for the perfectly conducting sphere is obtained from (7.48) on page 389. The result is

$$\frac{\mathrm{d}\sigma}{\mathrm{d}\Omega}(\hat{\boldsymbol{r}},\hat{z}) = 4\pi\left|\mathbf{S}(\hat{\boldsymbol{r}},\hat{z})\cdot\hat{\boldsymbol{p}}_{\mathrm{e}}\right|^2 = 4\pi\left\{\left|S_{\parallel\parallel}\right|^2\left|\hat{\boldsymbol{r}}_{\mathrm{c}}\cdot\hat{\boldsymbol{p}}_{\mathrm{e}}\right|^2 + \left|S_{\perp\perp}\right|^2\left|\hat{\boldsymbol{\phi}}\cdot\hat{\boldsymbol{p}}_{\mathrm{e}}\right|^2\right\}$$

and the scattering cross section becomes, see (7.49) on page 390

$$\sigma_{\mathrm{s}} = \frac{2\pi}{k^2}\sum_{l=1}^{\infty}(2l+1)\left(|t_{1l}|^2 + |t_{2l}|^2\right) \tag{8.3}$$

independently of the polarization of the incident wave, i.e., it holds for all elliptic polarizations as well as for all linearly polarization states. Several expressions above contain an infinite summation over the l-index. An efficient way to truncate the series is determined in Reference 294.[2]

---

[2]These criteria are explicitly given by [294]

$$l_{\max} = \begin{cases} ka + 4(ka)^{1/3} + 1, & 0.02 \le ka \le 8 \\ ka + 4.05(ka)^{1/3} + 2, & 8 \le ka \le 4\,200 \\ ka + 4(ka)^{1/3} + 2, & 4\,200 \le ka \le 20\,000 \end{cases}$$

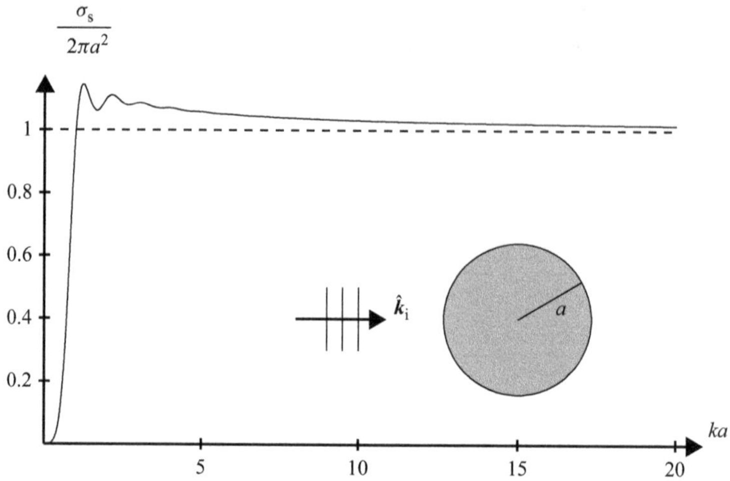

Figure 8.2 The scattering cross section σ_s for a perfectly conducting sphere as a function of the electrical size ka. Note that the scattering cross section is normalized by $2\pi a^2$.

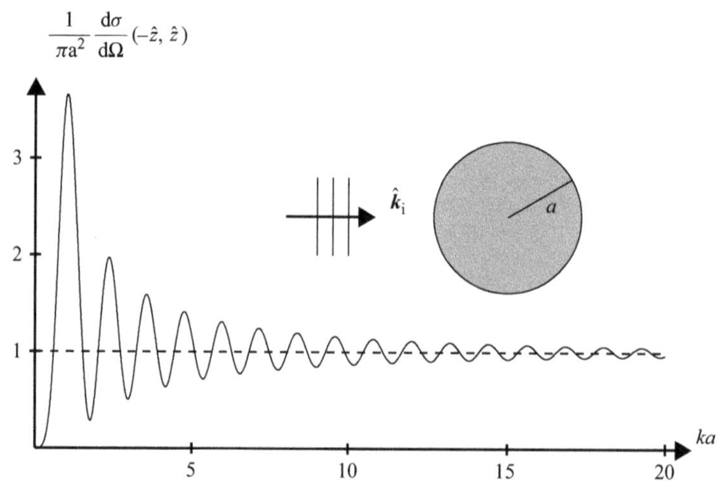

Figure 8.3 The differential scattering cross section $\frac{d\sigma}{d\Omega}(\hat{r}, \hat{z})$ in the backward direction, $\hat{r} = -\hat{z}$, for a perfectly conducting sphere as a function of the electrical size ka. Note the normalization πa^2.

In Figure 8.2, the scattering cross section is depicted as a function of the electrical size, ka, of the sphere. The scattering cross section is normalized with twice the geometrical cross section of the sphere, $2\pi a^2$, cf. the extinction paradox in Example 6.6 on page 295. The differential scattering cross section in the backward direction, $\hat{r} = -\hat{z}$, is depicted in Figure 8.3 as a function of the electrical size ka (normalized with πa^2), see also Problem 8.6. The components scattering dyadic, $S_{\parallel}(ka)$ or $S_{\perp}(ka)$, in the forward direction in the complex plane is illustrated as a function of the electrical size ka in Figure 8.4.

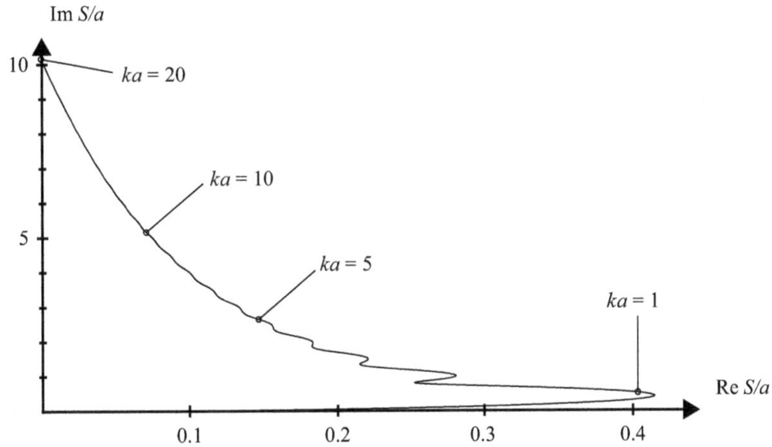

Figure 8.4 The components scattering dyadic, $S_{\parallel}(ka)$ or $S_{\perp}(ka)$, in the forward direction for a perfectly conducting sphere in the complex plane as a function of the electrical size ka. The components of the scattering dyadic are scaled by the radius a. Note the two different scales on the horizontal and vertical axes.

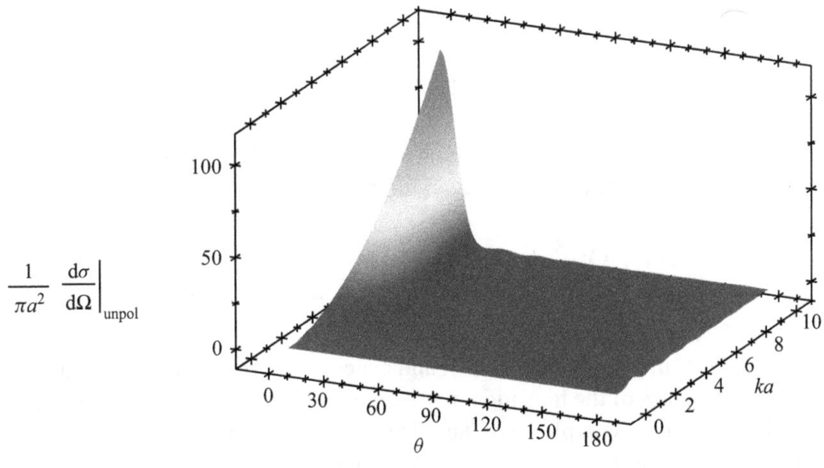

Figure 8.5 The differential scattering cross section for a PEC sphere as a function of the electric size ka and the scattering angle θ. The incident plane wave is unpolarized, and the amplitude is scaled with πa^2.

In Figure 8.5, the differential scattering cross section is depicted for an unpolarized incident plane wave as a function of both electric size, ka, and scattering angle, θ, see (4.41) on page 217. The right-hand side of Figure 8.5 (backward direction, $\theta = 180°$) coincides with Figure 8.3, and the large amplitudes on the left-hand side of the drawing illustrate the enhancement of the differential scattering cross section in the forward direction ($\theta = 0°$)—more pronounced, the larger the electric size of the scatterer.

The denominator in (8.1) has no zeros on the real ka-axis, which is concluded by Lemma B.3 on page 618, but there are zeros in the lower complex ka-plane, i.e., $\operatorname{Im} ka < 0$. The lack of poles in the transition matrix elements in the upper complex half-plane is due to causality requirements of the scattered field. The poles in the lower complex half-plane are symmetrically located around the imaginary axis, since the following relations of the Riccati–Bessel functions hold, see Problem 8.2:

$$\begin{cases} \psi_l(-z^*) = (-1)^{l+1}\,(\psi_l(z))^* \\ \xi_l(-z^*) = (-1)^{l+1}\,(\xi_l(z))^* \end{cases}$$

which show that if ka is a root to the denominator in (8.1), then $-(ka)^*$ is also a root to the same equation.

From Appendix B.3.3, we then find that the zeros, $z = ka$, are determined by the complex zeros of the polynomials, see (B.28) and (B.29)

$$\begin{cases} p_l(ka) + \mathrm{i}q_l(ka) = 0 \\ P_l(ka) + \mathrm{i}Q_l(ka) = 0 \end{cases}$$

for $\tau = 1, 2$, respectively. More explicitly

$$\sum_{k=0}^{l} \frac{(l+k)!}{k!(l-k)!}\frac{1}{(-2\mathrm{i}z)^k} = 0, \qquad \tau = 1$$

This is a polynomial in $1/z$ of degree l, and hence, there are l zeros. Similarly, for $\tau = 2$ (use the recursion relations for the Riccati-Bessel functions)

$$\frac{1}{z}\sum_{k=0}^{l} \frac{(l+1)(l+k)!}{k!(l-k)!}\frac{1}{(-2\mathrm{i}z)^k} - \sum_{k=0}^{l+1} \frac{(l+1+k)!}{k!(l+1-k)!}\frac{1}{(-2\mathrm{i}z)^k} = 0, \qquad \tau = 2$$

This is a polynomial in $1/z$ of degree $l+1$, and hence, there are $l+1$ zeros. Figure 8.6 shows the location of the poles of the transition matrix, or equivalently, the zeros of $\xi_l(ka)$ and $\xi_l'(ka)$ in a limited region of the complex ka-plane. These poles are conveniently found numerically by an implementation of the root-finding algorithm in Appendix D.7.

The scattering cross section in Figure 8.2 has no clear resonance structure of the kind a dielectric body in Section 8.2 shows. The reason for this is that the poles of the transition matrix, given by the zeros of $\xi_l(ka)$ and $\xi_l'(ka)$, respectively, see (8.1), all lie relatively far away from the real axis, see Figure 8.6. A more detailed treatment of the resonance phenomena for a dielectric scatterer is presented in Section 8.2.3.

The extinction cross section σ_{ext} is identical to σ_{s}, since the absorption cross section $\sigma_{\text{a}} = 0$ for perfectly conducting obstacles. This is an immediate consequence of the boundary conditions on the surface of the sphere, since the total power absorbed by the sphere is

$$P_{\text{a}} = -\frac{1}{2}\operatorname{Re}\iint_{S_{\text{s}}} \left(\boldsymbol{E}\times\boldsymbol{H}^*\right)\cdot\hat{\boldsymbol{\nu}}\,\mathrm{d}S' = -\frac{1}{2}\operatorname{Re}\iint_{S_{\text{s}}} \left(\hat{\boldsymbol{\nu}}\times\boldsymbol{E}\right)\cdot\boldsymbol{H}^*\,\mathrm{d}S' = 0$$

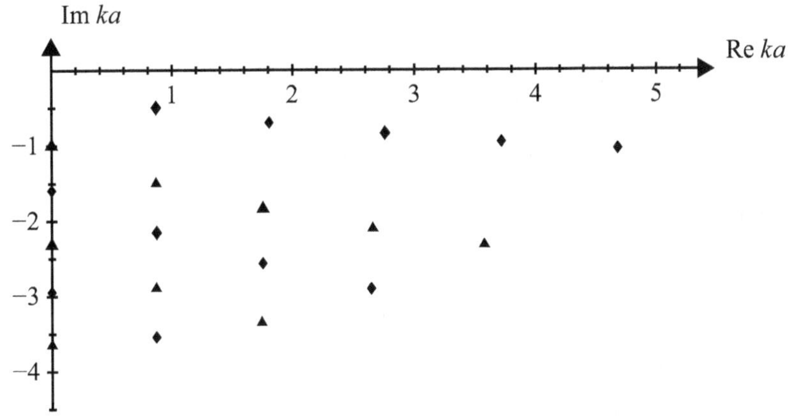

Figure 8.6 The location of the poles of the transition matrix for the perfectly conducing sphere in the complex ka-plane. The triangles indicate the location of the poles corresponding to $\tau = 1$ and the diamonds the poles of $\tau = 2$. The l values, $l = 1, 2, 3, 4, 5$, are shown.

The optical theorem then implies an alternative expression of the scattering cross section, see (7.50) on page 391:

$$\sigma_s = \sigma_{ext} = -\frac{2\pi}{k^2} \operatorname{Re} \sum_{l=1}^{\infty} (2l+1)(t_{1l} + t_{2l})$$

Note the similarities and the differences in this expression and the one in (8.3).

The T-matrix elements are defined in (8.1). These elements can be quantified in terms of a phase shift $\theta_{\tau l}(ka)$, see Example 7.4 on page 391. To see this, we rewrite the T-matrix elements as

$$\begin{cases} t_{1l} = -\dfrac{j_l(ka)}{h_l^{(1)}(ka)} = -\dfrac{h_l^{(2)}(ka) + h_l^{(1)}(ka)}{2h_l^{(1)}(ka)} = \dfrac{1}{2}\left(e^{2i\theta_{1l}(ka)} - 1\right) \\ t_{2l} = -\dfrac{(kaj_l(ka))'}{(kah_l^{(1)}(ka))'} = -\dfrac{(kah_l^{(2)}(ka))' + (kah_l^{(1)}(ka))'}{2(kah_l^{(1)}(ka))'} \\ \quad = \dfrac{1}{2}\left(e^{2i\theta_{2l}(ka)} - 1\right) \end{cases}$$

Using (B.13) and (B.15) in Appendix B.3, we obtain the phase shifts, $\theta_{\tau l}(ka)$, as

$$\begin{cases} e^{2i\theta_{1l}(ka)} = -\dfrac{h_l^{(2)}(ka)}{h_l^{(1)}(ka)} = e^{-2ika+il\pi}\dfrac{p_l(ka) - iq_l(ka)}{p_l(ka) + iq_l(ka)} = e^{-2ika+il\pi+2i\vartheta_{1l}(ka)} \\ e^{2i\theta_{2l}(ka)} = -\dfrac{(kah_l^{(2)}(ka))'}{(kah_l^{(1)}(ka))'} = e^{-2ika+il\pi}\dfrac{P_l(ka) - iQ_l(ka)}{P_l(ka) + iQ_l(ka)} \\ \quad = e^{-2ika+il\pi+2i\vartheta_{2l}(ka)} \end{cases} \tag{8.4}$$

where the quotients

$$\frac{p_l(ka) - \mathrm{i}q_l(ka)}{p_l(ka) + \mathrm{i}q_l(ka)} \quad \text{and} \quad \frac{P_l(ka) - \mathrm{i}Q_l(ka)}{P_l(ka) + \mathrm{i}Q_l(ka)}$$

both have modulus one, and we denote their phases by $2\vartheta_{\tau l}(ka)$, $\tau = 1, 2$, respectively. Therefore, the transition matrix elements, $t_{\tau l}$, for each value of the index τl lie on a circle in the complex plane centered at $-1/2$ and with radius 1/2, see Figure 7.3 on page 385. This is a form that is suited for the high-frequency asymptotic analysis in Section 8.1.2. We pursue this analysis below.

8.1.1 Long Wavelength Approximation

If the radius a is small compared to the wavelength λ, or equivalently $ka \ll 1$, we can obtain useful approximate expressions of the quantities above. In Section 6.1, we investigated the long wavelength limit in general, and in this section, we address the perfectly conducting sphere in some detail.

The Rayleigh approximation means that $ka \ll 1$. Under this assumption, we can find approximative expressions for the entries of the transition matrix, $t_{\tau l}$, in (8.1). In Appendix B.3.3, the dominant contributions for small arguments of the Riccati–Bessel functions are listed, see (B.27) on page 618

$$\begin{cases} \psi_l(z) = \dfrac{2^l l! z^{l+1}}{(2l+1)!} + O(z^{l+3}) \\ \xi_l(z) = -\mathrm{i}\dfrac{(2l)!}{2^l l! z^l} + O(z^{-l+2}) \end{cases} \quad \text{as } z \to 0$$

These dominant contributions imply that the dominant contributions to the transition matrix in the Rayleigh limit are

$$\begin{cases} t_{1l}(ka) = -\dfrac{\psi_l(ka)}{\xi_l(ka)} = -\mathrm{i}\dfrac{2^{2l}(l!)^2(ka)^{2l+1}}{(2l)!(2l+1)!} + O((ka)^{2l+3}) \\ t_{2l}(ka) = -\dfrac{\psi_l'(ka)}{\xi_l'(ka)} = \mathrm{i}\dfrac{2^{2l}(l+1)!(l-1)!(ka)^{2l+1}}{(2l)!(2l+1)!} + O((ka)^{2l+3}) \end{cases}$$

The dominant contributions come from $l = 1$, which are

$$\begin{cases} t_{11} = -\mathrm{i}\dfrac{k^3a^3}{3} + O((ka)^5) \\ t_{21} = \mathrm{i}\dfrac{2k^3a^3}{3} + O((ka)^5) \end{cases} \quad l = 1$$

with corresponding phase shifts $\theta_{\tau 1}(ka)$

$$e^{2\mathrm{i}\theta_{\tau l}} = 2t_{\tau l} + 1 \quad \Longrightarrow \quad \begin{cases} \theta_{11}(ka) = -\dfrac{k^3a^3}{3} + O((ka)^5) \\ \theta_{21}(ka) = \dfrac{2k^3a^3}{3} + O((ka)^5) \end{cases} \quad \text{as } ka \to 0$$

The dominating contributions to the entries of the scattering dyadic are, see (8.2)

$$\begin{cases} S_{\parallel\parallel} = k^2a^3\left(\cos\theta - \dfrac{1}{2}\right) \\ S_{\parallel\perp} = S_{\perp\parallel} = 0 \\ S_{\perp\perp} = k^2a^3\left(1 - \dfrac{1}{2}\cos\theta\right) \end{cases}$$

and the far field amplitude becomes

$$\boldsymbol{F}(\hat{\boldsymbol{r}}) = k^2a^3\left\{\hat{\boldsymbol{e}}_{\mathrm{s}\parallel}\left(\cos\theta - \frac{1}{2}\right)E_{\mathrm{i}\parallel} + \hat{\boldsymbol{e}}_{\mathrm{s}\perp}\left(1 - \frac{1}{2}\cos\theta\right)E_{\mathrm{i}\perp}\right\}$$

where $\hat{\boldsymbol{e}}_{\mathrm{s}\parallel}$ and $\hat{\boldsymbol{e}}_{\mathrm{s}\perp}$ are defined in Section 4.3.

We are now able to compare this result obtained by the use of the spherical vector wave, and the corresponding general expression obtained in Section 6.1.2. Specifically, from (6.8) on page 272

$$\boldsymbol{F}(\hat{\boldsymbol{r}}) = \mathbf{S}(\hat{\boldsymbol{r}}, \hat{z}) \cdot \boldsymbol{E}_0 = -\frac{k^2}{4\pi}\left[\hat{\boldsymbol{r}} \times (\hat{\boldsymbol{r}} \times \boldsymbol{\gamma}_{\mathrm{e}}) + \hat{\boldsymbol{r}} \times \boldsymbol{\gamma}_{\mathrm{m}} \cdot (\hat{z} \times \mathbf{I})\right] \cdot \boldsymbol{E}_0$$

Use the results in Example 7.7 on page 402, $\boldsymbol{\gamma}_{\mathrm{e}} = 4\pi a^3\mathbf{I}$ and $\boldsymbol{\gamma}_{\mathrm{m}} = -2\pi a^3\mathbf{I}$, and we obtain

$$\boldsymbol{F}(\hat{\boldsymbol{r}}) = -\frac{k^2a^3}{2}\left[2\hat{\boldsymbol{r}} \times (\hat{\boldsymbol{r}} \times \boldsymbol{E}_0) - \hat{\boldsymbol{r}} \times (\hat{z} \times \boldsymbol{E}_0)\right]$$

which is identical to the expression found above, viz.

$$\begin{cases} F_{\perp}(\hat{\boldsymbol{r}}) = \hat{\boldsymbol{e}}_{\mathrm{s}\perp} \cdot \boldsymbol{F}(\hat{\boldsymbol{r}}) = \dfrac{k^2a^3}{2}\left(\hat{\boldsymbol{e}}_{\mathrm{i}\perp} \cdot \boldsymbol{E}_0\right)(2 - \cos\theta) \\ F_{\parallel}(\hat{\boldsymbol{r}}) = \hat{\boldsymbol{e}}_{\mathrm{s}\parallel} \cdot \boldsymbol{F}(\hat{\boldsymbol{r}}) = \dfrac{k^2a^3}{2}\left(\hat{\boldsymbol{e}}_{\mathrm{i}\parallel} \cdot \boldsymbol{E}_0\right)(2\cos\theta - 1) \end{cases}$$

where the decomposition $\hat{\boldsymbol{r}} = \hat{\boldsymbol{k}}_{\mathrm{i}}(\hat{\boldsymbol{k}}_{\mathrm{i}} \cdot \hat{\boldsymbol{r}}) + \hat{\boldsymbol{e}}_{\mathrm{i}\parallel}(\hat{\boldsymbol{e}}_{\mathrm{i}\parallel} \cdot \hat{\boldsymbol{r}})$ is useful in the derivation of this expression.

The differential scattering cross section in the long wavelength approximation is then easily found

$$\begin{aligned} \frac{\mathrm{d}\sigma}{\mathrm{d}\Omega}(\hat{\boldsymbol{r}}) &= 4\pi\frac{|\boldsymbol{F}(\hat{\boldsymbol{r}})|^2}{|\boldsymbol{E}_0|^2} \\ &= 4\pi\frac{k^4a^6}{|\boldsymbol{E}_0|^2}\left(\left(\cos\theta - \frac{1}{2}\right)^2|E_{\mathrm{i}\parallel}|^2 + \left(1 - \frac{1}{2}\cos\theta\right)^2|E_{\mathrm{i}\perp}|^2\right) \end{aligned}$$

The scattering cross section for the perfectly conducting sphere, see (8.3), in the Rayleigh limit becomes (cf. also the result in Problem 4.2)

$$\sigma_{\mathrm{s}} = \frac{6\pi}{k^2}\left(|t_{11}|^2 + |t_{21}|^2\right) = \frac{6\pi}{k^2}\left(\frac{k^6a^6}{9} + \frac{4k^6a^6}{9}\right) = \frac{10\pi k^4a^6}{3}$$

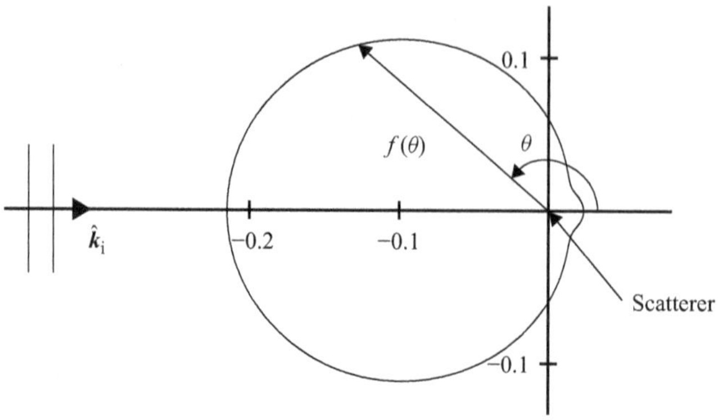

Figure 8.7 The normalized radiation diagram $f(\theta) = \frac{3}{10\pi}\left(\frac{5}{8}(1+\cos^2\theta) - \cos\theta\right)$ for scattering of an unpolarized incident plane wave by a perfectly conducting sphere in the Rayleigh limit. The normalization is chosen so that $\iint_\Omega f(\theta)\, d\Omega = 1$.

With unpolarized incident field, the differential scattering cross section simplifies, since $|E_{i\perp}|^2 = |E_{i\parallel}|^2 = |\boldsymbol{E}_0|^2/2$, and we get, see (4.41) on page 217

$$\left.\frac{d\sigma}{d\Omega}\right|_{\text{unpol}}(\hat{\boldsymbol{r}}) = 4\pi k^4 a^6 \left(\frac{5}{8}(1+\cos^2\theta) - \cos\theta\right)$$

The radiation diagram is depicted in Figure 8.7.

Finally, the degree of polarization of the scattered field for an unpolarized incident plane wave is calculated, see (4.42) on page 217:

$$P|_{\text{unpol}} = \frac{\left||S_{\parallel\parallel}|^2 - |S_{\perp\perp}|^2\right|}{|S_{\parallel\parallel}|^2 + |S_{\perp\perp}|^2} = \frac{3\sin^2\theta}{5\cos^2\theta + 5 - 8\cos\theta}$$

This function is depicted in Figure 8.8. Note that the scattered field is totally polarized, $P|_{\text{unpol}} = 1$, at $\theta = 60°$, even if the incident field is unpolarized. The degree of polarization $P|_{\text{unpol}}$ in the resonance region, $ka = 10$, as a function of the scattering angle θ is also illustrated in Figure 8.8.

8.1.2 High-Frequency Asymptotics

The traditional way of analyzing the high-frequency behavior of scattering by a PEC sphere is by employing the Watson transformation, see Appendix D.6, and the use of residue calculus and saddle-point methods. We pursue a different approach here with the phase shifts, and refer the interested reader to References 202, 204, 241, 242, 272 for a suitable overview of the approach by the Watson transformation, see also Appendix D.6 on page 655.

We limit ourselves in this section to find high-frequency approximations to the transition matrix entries for fixed l.

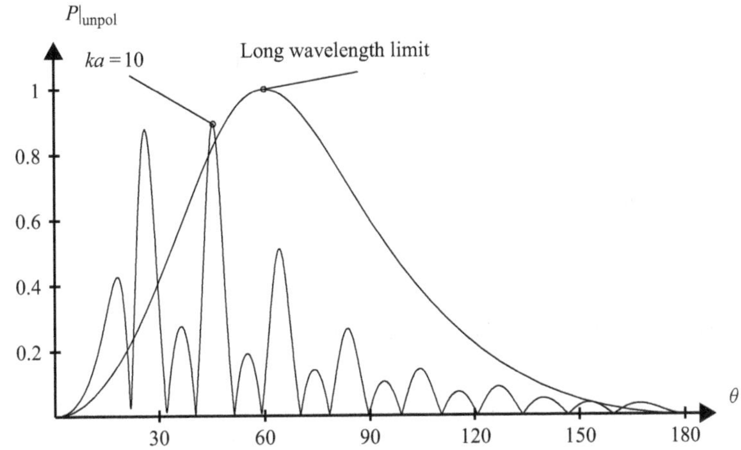

Figure 8.8 The degree of polarization $P|_{\mathrm{unpol}}$ for a perfectly conducting sphere as a function of the scattering angle θ. Confer also Figure 4.9 on page 221.

The phase shifts, $\theta_{\tau l}$, for a perfectly conducting sphere have the general form, see (8.4)

$$\begin{cases} e^{2i\theta_{1l}} = e^{-2ika+il\pi} \dfrac{p_l(ka) - iq_l(ka)}{p_l(ka) + iq_l(ka)} \\ e^{2i\theta_{2l}} = e^{-2ika+il\pi} \dfrac{P_l(ka) - iQ_l(ka)}{P_l(ka) + iQ_l(ka)} \end{cases}$$

To determinate the dominant contribution to the angles $\theta_{\tau l}$, we need to study the quotients for large ka. For fixed l value, we get, see (B.14) and (B.17) in Appendix B.3

$$\begin{cases} \dfrac{p_l(ka) - iq_l(ka)}{p_l(ka) + iq_l(ka)} = 1 - i\dfrac{l(l+1)}{ka} + O\big((ka)^{-2}\big) \\ \dfrac{P_l(ka) - iQ_l(ka)}{P_l(ka) + iQ_l(ka)} = -\left(1 - i\dfrac{l(l+1)}{ka} + O\big((ka)^{-2}\big)\right) \end{cases}$$

which implies that the phase shifts $\theta_{\tau l}$ have a high-frequency asymptotic

$$e^{2i\theta_{\tau l}} = e^{-2ika+i(l+\tau-1)\pi} \left(1 - i\frac{l(l+1)}{ka} + O\big((ka)^{-2}\big)\right), \qquad \tau = 1, 2$$

Then, the transition matrix becomes to leading order in ka, see also Example 7.4 on page 391 ($\gamma = ka - (l + \tau - 1)\pi/2$)

$$\begin{aligned} t_{\tau l} &= \frac{1}{2}\left(e^{2i\theta_{\tau l}} - 1\right) = \frac{1}{2} e^{-i\gamma} \left(e^{-i\gamma}\left(1 - i\frac{l(l+1)}{ka} + O\big((ka)^{-2}\big)\right) - e^{i\gamma}\right) \\ &= -ie^{-i\gamma} \sin\gamma - ie^{-2i\gamma} \frac{l(l+1)}{2ka} + O\big((ka)^{-2}\big), \qquad \tau = 1, 2 \end{aligned}$$

or

$$
\begin{aligned}
t_{\tau l} = &-\mathrm{i}^{l+\tau}\mathrm{e}^{-\mathrm{i}ka}\sin\left(ka - \frac{(l+\tau-1)\pi}{2}\right) \\
&+\mathrm{i}(-1)^{l+\tau}\mathrm{e}^{-2\mathrm{i}ka}\frac{l(l+1)}{2ka} + O\big((ka)^{-2}\big), \qquad \tau = 1,2
\end{aligned}
$$

These expressions give the dominant contribution of the transition matrix entries as the frequency goes to infinity, i.e., $ka \to \infty$.

8.2 Scattering by a Dielectric Sphere

A scattering problem with many interesting and important applications is scattering by a homogeneous, dielectric sphere. The sphere is modeled by a (complex) permittivity $\epsilon_1(\omega)$ and permeability $\mu_1(\omega)$, and the radius of the sphere is a. We adopt a coordinate system in which the center of the sphere coincides with the origin of the coordinate system. The surrounding (background) material is, as above, modeled by the real permittivity ϵ and the real permeability μ. The incident field is, as above, a plane wave impinging along the positive z-axis. Geometry and material parameters are shown in Figure 8.9. The complex vector $\boldsymbol{E}_0$ lies in the x–y-plane. Its expansion in regular spherical vector waves is

$$
\boldsymbol{E}_{\mathrm{i}}(\boldsymbol{r},\omega) = \boldsymbol{E}_0\mathrm{e}^{\mathrm{i}kz} = \sum_n a_n \boldsymbol{v}_n(k\boldsymbol{r})
$$

where the expansion coefficients a_n are given in (7.29) on page 375.

The multipole expansion of the scattered electric field is as above

$$
\boldsymbol{E}_{\mathrm{s}}(\boldsymbol{r},\omega) = \sum_n f_n \boldsymbol{u}_n(k\boldsymbol{r}), \qquad r > a
$$

The total electric and magnetic fields outside the sphere then become

$$
\left\{
\begin{aligned}
\boldsymbol{E}(\boldsymbol{r},\omega) &= \sum_n \left(a_n \boldsymbol{v}_n(k\boldsymbol{r}) + f_n \boldsymbol{u}_n(k\boldsymbol{r})\right) \\
\boldsymbol{H}(\boldsymbol{r},\omega) &= \frac{1}{\mathrm{i}\eta_0\eta}\sum_n \left(a_n \boldsymbol{v}_{\overline{n}}(k\boldsymbol{r}) + f_n \boldsymbol{u}_{\overline{n}}(k\boldsymbol{r})\right)
\end{aligned}
\right. \qquad r > a
$$

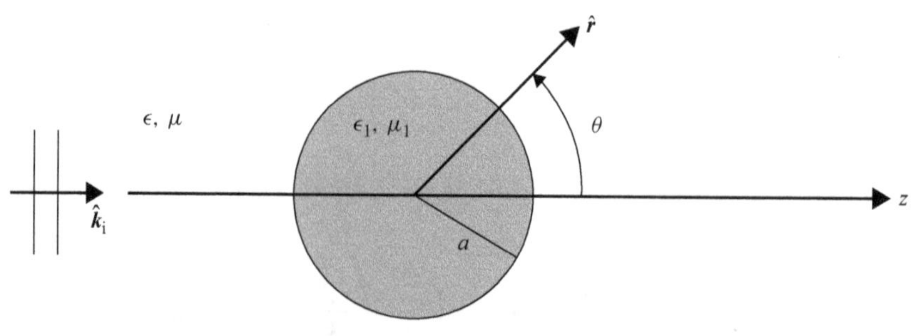

Figure 8.9 The geometry and the material parameters of the homogeneous, dielectric sphere in Section 8.2.

where $\overline{n}$ is the dual index of n, and where the relative wave impedance of the surrounding material is $\eta = \sqrt{\mu/\epsilon}$.

To match the boundary conditions on the surface of the sphere, we expand the total electric and magnetic fields inside the sphere. Since these fields are well behaved everywhere inside the sphere, we expand in regular spherical vector waves, i.e.,

$$\begin{cases} \boldsymbol{E}_1(\boldsymbol{r},\omega) = \sum_n \alpha_n \boldsymbol{v}_n(k_1\boldsymbol{r}) \\ \boldsymbol{H}_1(\boldsymbol{r},\omega) = \dfrac{1}{\mathrm{i}\eta_0\eta_1} \sum_n \alpha_n \boldsymbol{v}_{\overline{n}}(k_1\boldsymbol{r}) \end{cases} \qquad r < a \tag{8.5}$$

where we have introduced the relative wave impedance of the material inside the sphere $\eta_1 = (\mu_1/\epsilon_1)^{1/2}$. Note that we are using the wave number of the material inside the sphere, $k_1 = \omega(\epsilon_1\mu_1)^{1/2}/c_0$, in the regular spherical vector waves.

The boundary conditions on the spherical surface $r = a$ are, see (1.12) on page 8 (no surface currents $\boldsymbol{J}_S$ are assumed present)

$$\begin{cases} \hat{\boldsymbol{r}} \times \boldsymbol{E}_1(\boldsymbol{r},\omega)|_{r=a} = \hat{\boldsymbol{r}} \times \boldsymbol{E}(\boldsymbol{r},\omega)|_{r=a} \\ \hat{\boldsymbol{r}} \times \boldsymbol{H}_1(\boldsymbol{r},\omega)|_{r=a} = \hat{\boldsymbol{r}} \times \boldsymbol{H}(\boldsymbol{r},\omega)|_{r=a} \end{cases}$$

These conditions imply, due to the orthogonality conditions of the vector spherical harmonics, $\boldsymbol{A}_n(\hat{\boldsymbol{r}})$, that

$$\begin{cases} \alpha_{1n} j_l(k_1a) = a_{1n} j_l(ka) + f_{1n} h_l^{(1)}(ka) \\ \alpha_{2n} \dfrac{(k_1a j_l(k_1a))'}{k_1a} = a_{2n} \dfrac{(ka j_l(ka))'}{ka} + f_{2n} \dfrac{(ka h_l^{(1)}(ka))'}{ka} \\ \dfrac{1}{\eta_1}\alpha_{1n} \dfrac{(k_1a j_l(k_1a))'}{k_1a} = \dfrac{1}{\eta} a_{1n} \dfrac{(ka j_l(ka))'}{ka} + \dfrac{1}{\eta} f_{1n} \dfrac{(ka h_l^{(1)}(ka))'}{ka} \\ \dfrac{1}{\eta_1}\alpha_{2n} j_l(k_1a) = \dfrac{1}{\eta}\left(a_{2n} j_l(ka) + f_{2n} h_l^{(1)}(ka)\right) \end{cases}$$

The coefficient in front of $\boldsymbol{A}_{3n}(\hat{\boldsymbol{r}})$ gives no condition, since this vector spherical harmonics is proportional to $\hat{\boldsymbol{r}}$. Rewrite these equations as

$$\begin{cases} \alpha_{1n} j_l(k_1a) = a_{1n}\left(j_l(ka) + t_{1l} h_l^{(1)}(ka)\right) \\ \dfrac{1}{\mu_1}\alpha_{1n}(k_1a j_l(k_1a))' = \dfrac{1}{\mu} a_{1n}\left((ka j_l(ka))' + t_{1l}(ka h_l^{(1)}(ka))'\right) \\ \alpha_{2n} \dfrac{(k_1a j_l(k_1a))'}{k_1a} = a_{2n}\left(\dfrac{(ka j_l(ka))'}{ka} + t_{2l} \dfrac{(ka h_l^{(1)}(ka))'}{ka}\right) \\ \dfrac{1}{\eta_1}\alpha_{2n} j_l(k_1a) = \dfrac{1}{\eta} a_{2n}\left(j_l(ka) + t_{2l} h_l^{(1)}(ka)\right) \end{cases}$$

where we have written the solution as a diagonal transition matrix, i.e.,

$$f_{\tau l} = t_{\tau l} a_{\tau l}$$

The entries $t_{\tau l}$ are independent of the index m, and they can be obtained from the system of equations above. The result is

$$t_{\tau l} = -\frac{j_l(ka)(k_1 a j_l(k_1 a))' - \gamma_\tau (ka j_l(ka))' j_l(k_1 a)}{h_l^{(1)}(ka)(k_1 a j_l(k_1 a))' - \gamma_\tau (ka h_l^{(1)}(ka))' j_l(k_1 a)} \tag{8.6}$$

where

$$\gamma_\tau = \delta_{\tau 1}\frac{\mu_1}{\mu} + \delta_{\tau 2}\frac{\epsilon_1}{\epsilon}$$

Notice the complete symmetry between the electric multipoles, $\tau = 2$, and the magnetic ones, $\tau = 1$.

An alternative form of the transition matrix entries, in terms of the Riccati–Bessel functions, $\psi_n(z) = z j_n(z)$ and $\xi_n(z) = z h_n^{(1)}(z)$, see Appendix B.3.3, is

$$t_{\tau l} = -\frac{\psi_l(ka)\psi_l'(k_1 a) - m_\tau \psi_l'(ka)\psi_l(k_1 a)}{\xi_l(ka)\psi_l'(k_1 a) - m_\tau \xi_l'(ka)\psi_l(k_1 a)} \tag{8.7}$$

where

$$m_\tau = \delta_{\tau 1}\frac{\eta_1}{\eta} + \delta_{\tau 2}\frac{\eta}{\eta_1}$$

The numerical implementation is more efficient if we rewrite the entries in terms of the logarithmic derivatives $D_l(x) = \psi_l'(x)/\psi_l(x)$ and $G_l(x) = \xi_l'(x)/\xi_l(x)$ as [119]

$$t_{\tau l} = -\frac{\psi_l(ka)D_l(k_1 a) - m_\tau \psi_l'(ka)}{\xi_l(ka)D_l(k_1 a) - m_\tau \xi_l'(ka)} = -\frac{\psi_l(ka)}{\xi_l(ka)}\frac{D_l(k_1 a) - m_\tau D_l(ka)}{D_l(k_1 a) - m_\tau G_l(ka)} \tag{8.8}$$

This formulation of the transition matrix elements allows for calculations of scattering by a dielectric sphere at very high frequencies, i.e., $ka \gg 1$, see References 58, 281 and more generally for the rainbow problem [2].

Example 7.3 on page 386 now gives the entries of the scattering dyadic, see (7.46) on page 387

$$\begin{cases} S_{\|\|} = \dfrac{1}{ik}\displaystyle\sum_{l=1}^{\infty}\frac{2l+1}{l(l+1)}\{t_{1l}\pi_l(\cos\theta) + t_{2l}\tau_l(\cos\theta)\} \\ S_{\|\perp} = S_{\perp\|} = 0 \\ S_{\perp\perp} = \dfrac{1}{ik}\displaystyle\sum_{l=1}^{\infty}\frac{2l+1}{l(l+1)}\{t_{1l}\tau_l(\cos\theta) + t_{2l}\pi_l(\cos\theta)\} \end{cases} \tag{8.9}$$

where

$$\begin{cases} \pi_l(x) = P_l'(x) \\ \tau_l(x) = l(l+1)P_l(x) - xP_l'(x) \end{cases}$$

The scattering dyadic in the forward direction, see (7.47)

$$\mathbf{S}\left(\hat{z},\hat{z}\right)=\frac{1}{2ik}\mathbf{I}_2\sum_{l=1}^{\infty}(2l+1)\left(t_{1l}+t_{2l}\right)$$

and consequently

$$\boldsymbol{F}(\hat{z})=\frac{\boldsymbol{E}_0}{2ik}\sum_{l=1}^{\infty}(2l+1)\left(t_{1l}+t_{2l}\right)$$

where $\mathbf{I}_2$ is the unit dyadic in the x–y-plane.

The differential scattering cross section for the dielectric sphere is obtained from the general results, see (4.33) on page 207 or (7.48) on page 389. The result is

$$\frac{\mathrm{d}\sigma}{\mathrm{d}\Omega}(\hat{\boldsymbol{r}},\hat{z})=4\pi\left|\mathbf{S}(\hat{\boldsymbol{r}},\hat{z})\cdot\hat{\boldsymbol{p}}_{\mathrm{e}}\right|^2=4\pi\left\{\left|S_{\parallel\parallel}\right|^2\left|\hat{\boldsymbol{r}}_{\mathrm{c}}\cdot\hat{\boldsymbol{p}}_{\mathrm{e}}\right|^2+\left|S_{\perp\perp}\right|^2\left|\hat{\boldsymbol{\phi}}\cdot\hat{\boldsymbol{p}}_{\mathrm{e}}\right|^2\right\}$$

and the scattering cross section becomes, see (7.49) on page 390

$$\sigma_{\mathrm{s}}=\frac{2\pi}{k^2}\sum_{l=1}^{\infty}(2l+1)\left(|t_{1l}|^2+|t_{2l}|^2\right)$$

As for the perfectly conducting sphere, the scattering cross section is independent of the polarization of the incident field.

The optical theorem implies that the extinction cross section, σ_{ext}, is, see (7.50) on page 391

$$\sigma_{\mathrm{ext}}=\frac{4\pi}{k}\operatorname{Im}\left\{\hat{\boldsymbol{p}}_{\mathrm{e}}^{*}\cdot\mathbf{S}(\hat{z},\hat{z})\cdot\hat{\boldsymbol{p}}_{\mathrm{e}}\right\}=-\frac{2\pi}{k^2}\operatorname{Re}\sum_{l=1}^{\infty}(2l+1)\left(t_{1l}+t_{2l}\right)$$

and the absorption cross section is $\sigma_{\mathrm{a}}=\sigma_{\mathrm{ext}}-\sigma_{\mathrm{s}}$. The infinite sums above are conveniently truncated by the criterion given in Reference 294, see Footnote 2 on page 409.

Example 8.1

In Figure 8.10, the theory is illustrated by a computation of three different cross sections, the extinction cross section σ_{ext}, absorption cross section σ_{a}, and the scattering cross section σ_{s} for a dielectric sphere with material parameters $\epsilon_1/\epsilon=(1.5+\mathrm{i}0.01)^2$ and $\mu_1/\mu=1$. All scattering cross sections are normalized with twice the geometrical cross section of the sphere, $2\pi a^2$. The components of the scattering dyadic, $S_{\perp}(ka)$ or $S_{\perp}(ka)$, in the forward direction are depicted in Figure 8.11, and in Figure 8.12, the differential scattering cross section is depicted for an unpolarized incident plane wave as a function of both electric size and scattering angle. ■

Example 8.2

The material parameters, ϵ_1 and μ_1, are in general functions of the frequency. We illustrate this frequency behavior by computing the three cross sections for a drop of water of radius

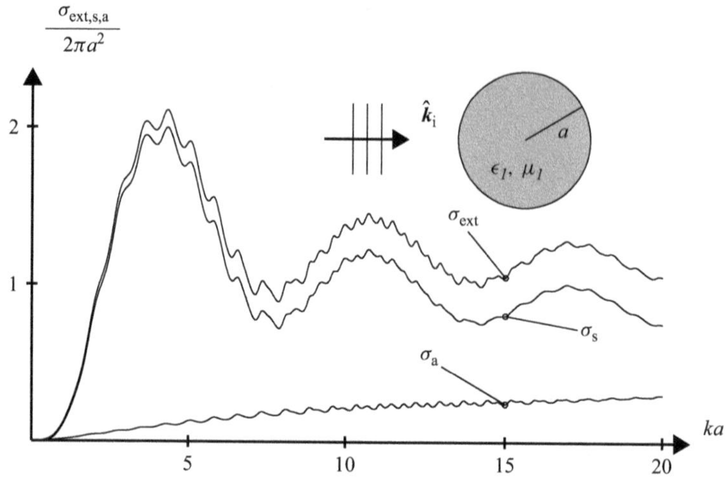

Figure 8.10 The extinction cross section, the scattering cross section, and the absorption cross section for a dielectric sphere as a function of the electrical size ka. The material parameters are $\epsilon_1/\epsilon = (1.5 + \text{i}0.01)^2$ and $\mu_1/\mu = 1$. All scattering cross sections are normalized with $2\pi a^2$.

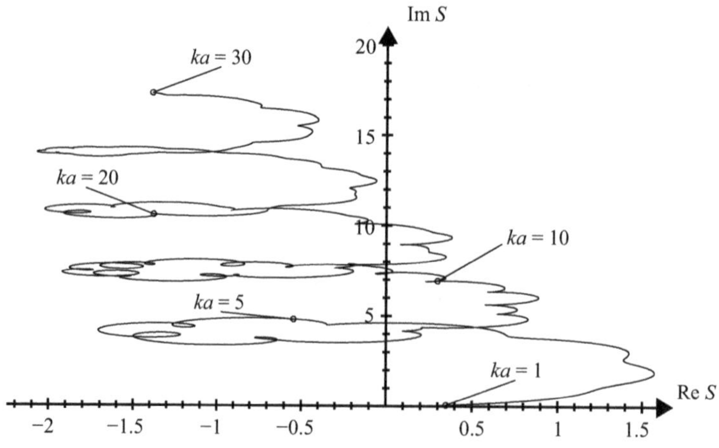

Figure 8.11 The components of the scattering dyadic, $S_\perp(ka)$ or $S_\perp(ka)$, for a dielectric sphere in the complex plane as a function of the electrical size ka. The material parameters are $\epsilon_1/\epsilon = (1.5 + \text{i}0.01)^2$ and $\mu_1/\mu = 1$. The components of the scattering dyadic are scaled by the radius a. Note the two different scales on the horizontal and vertical axes.

$a = 1$ mm in the frequency interval 0–100 GHz. The permittivity is assumed to depend on the frequency as (the Debye model)

$$\epsilon(\omega) = \epsilon_\infty + \frac{\epsilon_s - \epsilon_\infty}{1 - \text{i}\omega\tau}$$

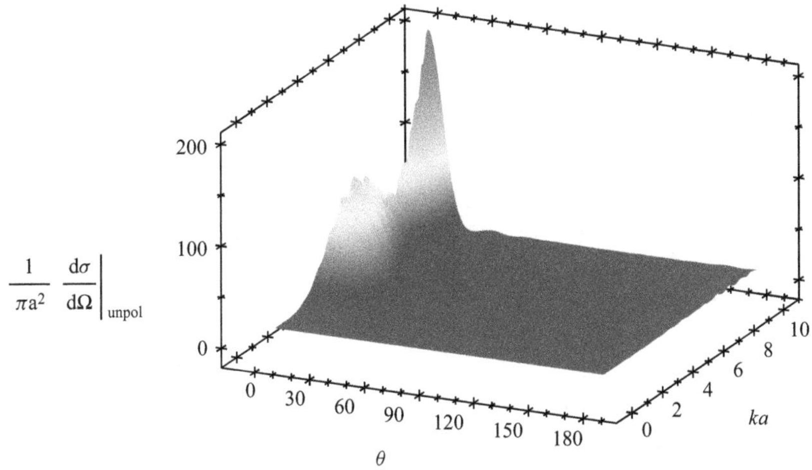

Figure 8.12 The differential scattering cross section for a dielectric sphere as a function of the electric size ka and the scattering angle θ. The material parameters are $\epsilon_1/\epsilon = (1.5 + i0.01)^2$ and $\mu_1/\mu = 1$. The incident plane wave is unpolarized, and the amplitude is scaled with πa^2.

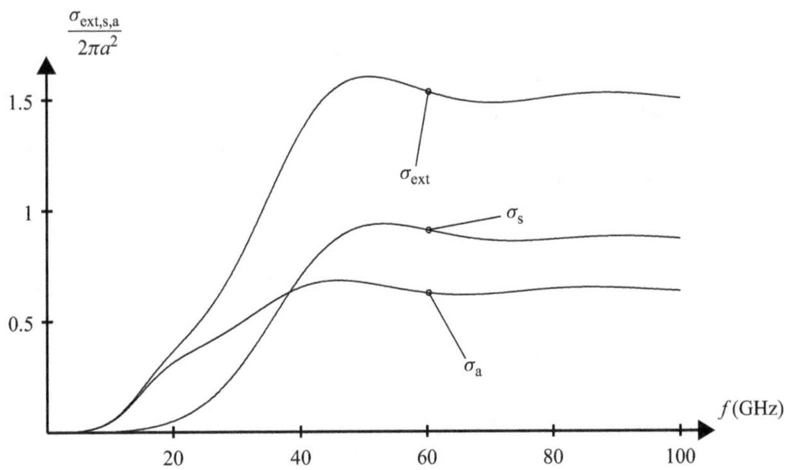

Figure 8.13 The extinction cross section, the scattering cross section, and the absorption cross section for a spherical drop of water ($a = 1$ mm) as a function of frequency f (GHz). All cross sections are normalized with $2\pi a^2$.

where ϵ_∞ is the permittivity for high frequencies, and ϵ_s is the value for $\omega = 0$ (static value). Explicit values are: $\tau = 1.0 \times 10^{-11}$ s, $\epsilon_\infty = 5.27$, and $\epsilon_s = 80.0$, cf. Table 1.8 on page 49. The results are depicted in Figure 8.13.

Notice that at low frequencies absorption dominates, while at higher frequencies scattering effects are larger than the absorption—electrically small particles mostly absorb power and electrically large particles mostly scatter power. ■

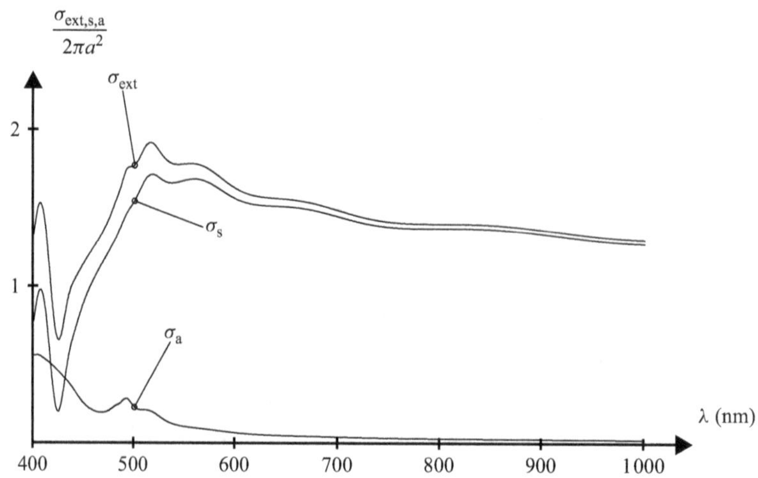

Figure 8.14 The extinction cross section, the scattering cross section, and the absorption cross section, normalized with $2\pi a^2$ as a function of the wavelength $\lambda = c/f$ in nm, for a gold sphere of radius $a = 400$ nm with data given in Table 1.9 on page 55.

Example 8.3

At optical frequencies, metals are modeled by the Drude model (or modified Debye model), (1.61) on page 53. In Figure 8.14, we illustrate the extinction, scattering, and absorption cross sections as a function of the wavelength $\lambda = c/f$ ($c = c_0/\sqrt{\epsilon\mu}$ is the speed of light in the surrounding material, and f is the frequency) for a homogeneous gold sphere with material data given in Table 1.9 on page 55. ■

Example 8.4

The perfectly conducting transition matrix entries, given by (8.1), are retrieved by letting the permittivity $\epsilon_1/\epsilon \to \infty$ and letting permeability μ_1/μ be bounded. This is obtained by the use of the conductivity model, see (1.62) on page 55, viz.

$$\epsilon_1(\omega) = \widetilde{\epsilon}(\omega) + \mathrm{i}\frac{\sigma}{\omega\epsilon_0}$$

where $\widetilde{\epsilon}(\omega)$ is regular function at the origin. In this model, we eventually let $\sigma \to \infty$.

To see the result of this limit process, we adopt the asymptotic behavior of the Riccati–Bessel functions for large arguments and fixed l value, see (B.11) and (B.15) in Appendix B.3, i.e.,

$$\psi_l(z) = zj_l(z) = \sin\left(z - \frac{l\pi}{2}\right) + O\left(\mathrm{e}^{\mathrm{Im}\, z}/z\right), \qquad z \in \mathbb{C}, \quad \text{as } z \to \infty$$

and

$$\psi_l'(z) = (zj_l(z))' = \cos\left(z - \frac{l\pi}{2}\right) + O\left(\mathrm{e}^{\mathrm{Im}\, z}/z\right), \qquad z \in \mathbb{C}, \quad \text{as } z \to \infty$$

Notice that the sine and the cosine functions grow equally fast in the complex plane.

These asymptotic relations inserted in (8.7) lead to (the l value and ka are fixed, and we adopt the notation $z = k_1 a$)

$$t_{\tau l} = -\frac{\psi_l(ka)\cos\left(z - \frac{l\pi}{2}\right) - m_\tau \psi_l'(ka)\sin\left(z - \frac{l\pi}{2}\right) + (1 + m_\tau)\, O\left(e^{\operatorname{Im} z}/z\right)}{\xi_l(ka)\cos\left(z - \frac{l\pi}{2}\right) - m_\tau \xi_l'(ka)\sin\left(z - \frac{l\pi}{2}\right) + (1 + m_\tau)\, O\left(e^{\operatorname{Im} z}/z\right)}$$

as $z \to \infty e^{i\pi/4}$, where

$$m_\tau = \delta_{\tau 1}\frac{\eta_1}{\eta} + \delta_{\tau 2}\frac{\eta}{\eta_1}$$

The limits are

$$t_{1l} = -\frac{\psi_l(ka)}{\xi(ka)}, \qquad t_{2l} = -\frac{\psi_l'(ka)}{\xi'(ka)}, \qquad \text{as } z \to \infty e^{i\pi/4}$$

since

$$\begin{cases} \dfrac{\eta_1}{\eta} = \left(\dfrac{\epsilon\mu_1}{\epsilon_1\mu}\right)^{1/2} \to 0 \\ \dfrac{\eta}{\eta_1} = \left(\dfrac{\epsilon_1\mu}{\epsilon\mu_1}\right)^{1/2} \to \infty e^{i\pi/4} \\ \dfrac{k_1 a}{ka} = \left(\dfrac{\epsilon_1\mu_1}{\epsilon\mu}\right)^{1/2} \to \infty e^{i\pi/4} \end{cases} \qquad \text{as } \sigma \to \infty$$

and we see that in the limit $\sigma \to \infty$, the transition matrix elements are identical to (8.1). Notice that the frequency is fixed in this limit process, i.e. ka fixed. It is therefore not a high frequency result. ■

8.2.1 Internal Field

The internal field is determined by the expansion in (8.5). To find an explicit solution of this field, we need to find the coefficients $\alpha_{\tau n}$. They are explicitly given in terms of the Riccati–Bessel functions, see Appendix B.3.3, by

$$\begin{cases} \alpha_{1n} = \left(\psi_l(ka) + t_{1n}\xi_l(ka)\right)\dfrac{k_1 a_{1n}}{k\psi_l(k_1 a)} \\ \alpha_{2n} = \left(\psi_l'(ka) + t_{2n}\xi_l'(ka)\right)\dfrac{k_1 a_{2n}}{k\psi_l'(k_1 a)} \end{cases}$$

which, with the use of the entries of the transition matrix in (8.7), are simplified to

$$\begin{cases} \alpha_{1n} = \dfrac{\eta_1}{\eta}\dfrac{k_1}{k}\dfrac{\psi_l'(ka)\xi_l(ka) - \psi_l(ka)\xi_l'(ka)}{\xi_l(ka)\psi_l'(k_1 a) - \frac{\eta_1}{\eta}\xi_l'(ka)\psi_l(k_1 a)} a_{1n} \\ \alpha_{2n} = \dfrac{k_1}{k}\dfrac{\psi_l'(ka)\xi_l(ka) - \psi_l(ka)\xi_l'(ka)}{\xi_l(ka)\psi_l'(k_1 a) - \frac{\eta}{\eta_1}\xi_l'(ka)\psi_l(k_1 a)} a_{2n} \end{cases}$$

or using the Wronskian for the Riccati–Bessel functions, see (B.30) on page 618

$$\psi_l(z)\xi_l'(z) - \psi_l'(z)\xi_l(z) = \mathrm{i}$$

we get

$$\alpha_{\tau n} = -\mathrm{i}\frac{\delta_{\tau 1}\frac{\mu_1}{\mu} + \delta_{\tau 2}\frac{k_1}{k}}{\xi_l(ka)\psi_l'(k_1 a) - m_\tau \xi_l'(ka)\psi_l(k_1 a)} a_{\tau n} \tag{8.10}$$

where

$$m_\tau = \delta_{\tau 1}\frac{\eta_1}{\eta} + \delta_{\tau 2}\frac{\eta}{\eta_1}$$

8.2.2 Long Wavelength Approximation

If the radius of the sphere is small compared to the wavelength λ, i.e., $ka \ll 1$, we approximate the spherical Bessel functions by keeping only the leading contribution in ka, see (B.10) on page 609. We have

$$\begin{cases} \psi_l(z) = \dfrac{2^l l! z^{l+1}}{(2l+1)!} + O(z^{l+2}) \\ \psi_l'(z) = \dfrac{2^l (l+1)! z^l}{(2l+1)!} + O(z^{l+2}) \end{cases} \qquad \begin{cases} \xi_l^{(1)}(z) = -\mathrm{i}\dfrac{(2l)!}{2^l l! z^l} + O(z^{-l+1}) \\ \xi_l'(z) = \mathrm{i}\dfrac{(2l)!}{2^l (l-1)! z^{l+1}} + O(z^{-l+1}) \end{cases}$$

as $z \to 0$. The transition matrix entries $t_{\tau l}$ in (8.7) then have the leading contribution

$$t_{\tau l} = -\mathrm{i}\frac{(ka)^{2l+1} 2^{2l} (l!)^2}{(2l+1)!(2l)!} \frac{ka\psi_l'(k_1 a) - m_\tau (l+1)\psi_l(k_1 a)}{ka\psi_l'(k_1 a) + m_\tau l \psi_l(k_1 a)} + O\left((ka)^{2l+3}\right)$$

The leading contribution in powers of ka is for $l = 1$. Higher values of l contribute with $O\left((ka)^5\right)$. We have

$$t_{\tau 1} = -\mathrm{i}\frac{(ka)^3}{3} \frac{ka\psi_1'(k_1 a) - 2m_\tau \psi_1(k_1 a)}{ka\psi_1'(k_1 a) + m_\tau \psi_l(k_1 a)} + O((ka)^5) \tag{8.11}$$

The field inside the scatterer has expansion coefficients $\alpha_{\tau n}$, which to lowest order in ka become, see (8.10)

$$\alpha_{\tau n} = \frac{2^l l! (ka)^{l+1}}{(2l)!} \frac{\delta_{\tau 1}\frac{\mu_1}{\mu} + \delta_{\tau 2}\frac{k_1}{k}}{ka\psi_l'(k_1 a) + m_\tau l \psi_l(k_1 a)} a_{\tau n} + O\left((ka)^{l+2}\right) \tag{8.12}$$

Depending on the material parameters in the sphere, we obtain different approximations. We identify the following two important approximations:

1. Small values of ϵ_1 and μ_1, or more precisely, $k_1 a \ll 1$ (Rayleigh scattering)
2. Weak scatterer, i.e., $\epsilon_1 - \epsilon$ and $\mu_1 - \mu$ small, or $|k_1 a - ka| \ll 1$ (Rayleigh–Gans scattering)

8.2.2.1 Rayleigh Scattering—Far Field

If the radius of the sphere a is small compared to the wavelength λ and simultaneously ϵ_1 and μ_1 are not too large, more precisely, $ka \ll 1$ and $k_1 a \ll 1$, we can then find useful approximations of the entries of the transition matrix, $t_{\tau l}$. We get the dominant contribution in the Rayleigh limit from $l = 1$, see (8.11)

$$\begin{cases} t_{11} = \dfrac{2ik^3a^3}{3}\dfrac{\mu_1-\mu}{\mu_1+2\mu} + O\big((ka)^5\big) \\ t_{21} = \dfrac{2ik^3a^3}{3}\dfrac{\epsilon_1-\epsilon}{\epsilon_1+2\epsilon} + O\big((ka)^5\big) \end{cases} \tag{8.13}$$

and the dominating contributions of the entries of the scattering dyadic $\mathbf{S}$ become, see (8.9)

$$\begin{cases} S_{\parallel\parallel} = k^2a^3\left\{\dfrac{\mu_1-\mu}{\mu_1+2\mu} + \dfrac{\epsilon_1-\epsilon}{\epsilon_1+2\epsilon}\cos\theta\right\} \\ S_{\parallel\perp} = S_{\perp\parallel} = 0 \\ S_{\perp\perp} = k^2a^3\left\{\dfrac{\mu_1-\mu}{\mu_1+2\mu}\cos\theta + \dfrac{\epsilon_1-\epsilon}{\epsilon_1+2\epsilon}\right\} \end{cases}$$

and the far field amplitude

$$\begin{aligned} \boldsymbol{F}(\hat{\boldsymbol{r}}) = k^2a^3 \Bigg\{ & \hat{\boldsymbol{e}}_{\mathrm{s}\parallel}\left(\frac{\mu_1-\mu}{\mu_1+2\mu} + \frac{\epsilon_1-\epsilon}{\epsilon_1+2\epsilon}\cos\theta\right)E_{\mathrm{i}\parallel} \\ & + \hat{\boldsymbol{e}}_{\mathrm{s}\perp}\left(\frac{\mu_1-\mu}{\mu_1+2\mu}\cos\theta + \frac{\epsilon_1-\epsilon}{\epsilon_1+2\epsilon}\right)E_{\mathrm{i}\perp}\Bigg\} \end{aligned}$$

We observe that the corresponding result of the perfectly conducting sphere can be obtained as a limit case of the dielectric sphere by letting $\epsilon_1 \to \infty$ and simultaneously $\mu_1 \to 0$.

The differential scattering cross section becomes

$$\begin{aligned} \frac{\mathrm{d}\sigma}{\mathrm{d}\Omega}(\hat{\boldsymbol{r}}) = 4\pi\frac{|\boldsymbol{F}(\hat{\boldsymbol{r}})|^2}{|\boldsymbol{E}_0|^2} &= 4\pi\frac{k^4a^6}{|\boldsymbol{E}_0|^2}\left|\frac{\mu_1-\mu}{\mu_1+2\mu} + \frac{\epsilon_1-\epsilon}{\epsilon_1+2\epsilon}\cos\theta\right|^2 |E_{\mathrm{i}\parallel}|^2 \\ &+ 4\pi\frac{k^4a^6}{|\boldsymbol{E}_0|^2}\left|\frac{\mu_1-\mu}{\mu_1+2\mu}\cos\theta + \frac{\epsilon_1-\epsilon}{\epsilon_1+2\epsilon}\right|^2 |E_{\mathrm{i}\perp}|^2 \end{aligned}$$

and the scattering cross section for the dielectric sphere is

$$\sigma_{\mathrm{s}} = \frac{6\pi}{k^2}\left(|t_{11}|^2 + |t_{21}|^2\right) = \frac{8\pi k^4a^6}{3}\left(\left|\frac{\mu_1-\mu}{\mu_1+2\mu}\right|^2 + \left|\frac{\epsilon_1-\epsilon}{\epsilon_1+2\epsilon}\right|^2\right)$$

The extinction cross section, σ_{ext}, is computed by the use of the optical theorem. We get

$$\sigma_{\mathrm{ext}} = 4\pi ka^3\,\mathrm{Im}\left\{\frac{\mu_1-\mu}{\mu_1+2\mu} + \frac{\epsilon_1-\epsilon}{\epsilon_1+2\epsilon}\right\}$$

and the absorption cross section $\sigma_{\mathrm{a}} = \sigma_{\mathrm{ext}} - \sigma_{\mathrm{s}}$.

If the incident plane wave is unpolarized, i.e., $|E_{i\perp}|^2 = |E_{i\parallel}|^2 = |\boldsymbol{E}_0|^2/2$, the differential scattering cross section simplifies

$$\left.\frac{d\sigma}{d\Omega}\right|_{\text{unpol}}(\hat{\boldsymbol{r}}) = 2\pi k^4 a^6 \left\{ \left| \frac{\mu_1 - \mu}{\mu_1 + 2\mu} + \frac{\epsilon_1 - \epsilon}{\epsilon_1 + 2\epsilon} \cos\theta \right|^2 + \left| \frac{\mu_1 - \mu}{\mu_1 + 2\mu} \cos\theta + \frac{\epsilon_1 - \epsilon}{\epsilon_1 + 2\epsilon} \right|^2 \right\}$$

A common situation in many applications occurs when the sphere has magnetic properties that are identical to the surrounding material, i.e., $\mu_1 = \mu$. The differential scattering cross section, the scattering cross section, and the extinction cross section then are

$$\frac{d\sigma}{d\Omega}(\hat{\boldsymbol{r}}) = 4\pi \frac{k^4 a^6}{|\boldsymbol{E}_0|^2} \left| \frac{\epsilon_1 - \epsilon}{\epsilon_1 + 2\epsilon} \right|^2 \left(\cos^2\theta \, |E_{i\parallel}|^2 + |E_{i\perp}|^2 \right)$$

and (cf. also the result in Problem 6.1)

$$\sigma_s = \frac{8\pi k^4 a^6}{3} \left| \frac{\epsilon_1 - \epsilon}{\epsilon_1 + 2\epsilon} \right|^2$$

and

$$\sigma_{\text{ext}} = 4\pi k a^3 \, \text{Im} \, \frac{\epsilon_1 - \epsilon}{\epsilon_1 + 2\epsilon}$$

respectively.

The definition $\sigma_{\text{ext}} = \sigma_a + \sigma_s$ is used to compute the total absorption cross section to lowest order in powers of ka. The result is

$$\sigma_a = \frac{12\pi k a^3 \epsilon \, \text{Im} \, \epsilon_1}{(\text{Re} \, \epsilon_1 + 2\epsilon)^2 + (\text{Im} \, \epsilon_1)^2}$$

Notice that the absorption cross section contributes with a lower power in frequency, ka, than the scattering cross section in the low-frequency limit. Therefore, at low frequencies, absorption in general dominates over scattering, cf. Figure 8.13.

The differential scattering cross section and the degree of polarization of the scattered field for an unpolarized incident plane wave when $\mu_1 = \mu$ are of some interest. These results are

$$\left.\frac{d\sigma}{d\Omega}\right|_{\text{unpol}}(\hat{\boldsymbol{r}}) = 2\pi k^4 a^6 \left| \frac{\epsilon_1 - \epsilon}{\epsilon_1 + 2\epsilon} \right|^2 \left(1 + \cos^2\theta\right)$$

and, see (4.42) on page 217

$$P|_{\text{unpol}} = \frac{\sin^2\theta}{1 + \cos^2\theta}$$

The radiation diagram for this case is depicted in Figure 8.15, and the degree of polarization for a dielectric sphere, with material parameters $\epsilon_1/\epsilon = 1.33^2$ and $\mu_1/\mu = 1$, is illustrated in Figure 8.16 for the electric size $ka = 10$ and in the Rayleigh limit. The degree of polarization shows a complex pattern with rapid variation for the electric size $ka = 10$. Note that

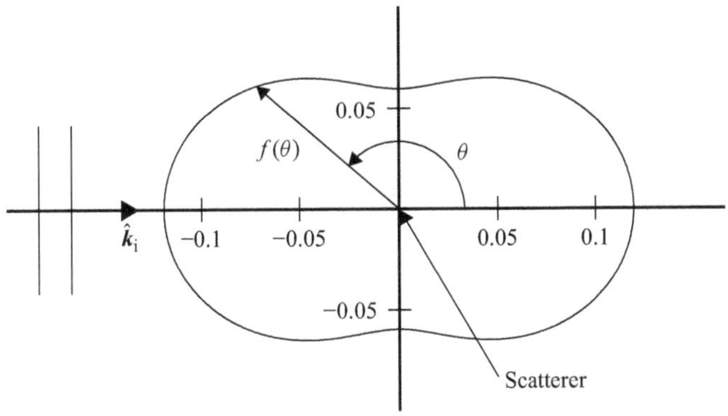

Figure 8.15 Radiation diagram for a plane wave incident on a dielectric sphere in the Rayleigh limit, $f(\theta) = \frac{3}{16\pi}\left(1 + \cos^2\theta\right)$. The normalization is chosen so that $\iint_\Omega f(\theta)\, \mathrm{d}\Omega = 1$.

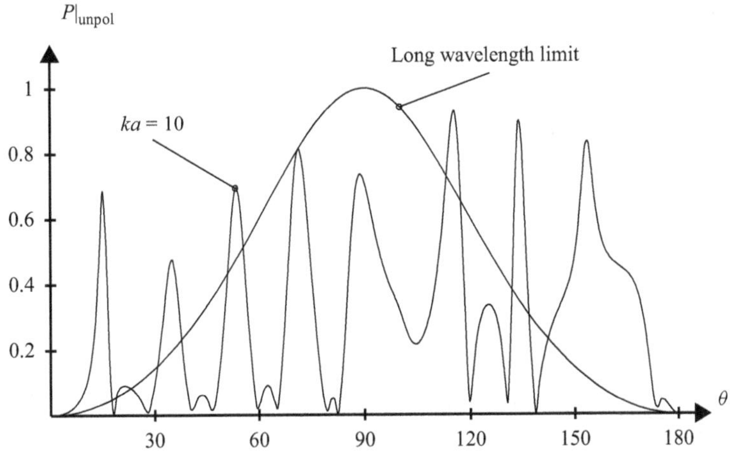

Figure 8.16 The degree of polarization $P|_{\text{unpol}}$ for a dielectric sphere as a function of the scattering angle θ. Material parameters are $\epsilon_1/\epsilon = 1.33^2$ and $\mu_1/\mu = 1$. Confer also Figure 4.9 on page 221.

the scattered field is completely polarized, $P|_{\text{unpol}} = 1$, in the Rayleigh limit in directions orthogonal to the incident direction, $\theta = \pi/2$, even if the incident field is unpolarized.

8.2.2.2 Rayleigh Scattering—Internal Field

Not only the far field properties of the scattered field by a dielectric scatterer are of interest, but also the total internal fields of the scatterer are relevant e.g., for understanding resonance phenomena.

As above, we let the incident plane wave impinge along the positive z-direction. Then only $m = 1$ contributes to the internal field, and, furthermore, keeping only terms up to order kr, only $l = 1$ and 2 contribute. The field inside the scatterer has expansion coefficients $\alpha_{\tau n}$, which to lowest order in ka become, see (8.12) and (B.10) on page 609

$$\alpha_{\tau n} = (2l+1)\left(\frac{k}{k_1}\right)^l \frac{\delta_{\tau 1}\frac{\mu_1}{\mu} + \delta_{\tau 2}\frac{k_1}{k}}{(l+1) + l\gamma_\tau} a_{\tau n} + O\left(\left(\frac{k}{k_1}\right)^{l+1}\right)$$

where $\gamma_\tau = \delta_{\tau 1}\mu_1/\mu + \delta_{\tau 2}\epsilon_1/\epsilon$. The explicit form of the coefficients is

$$\alpha_{\tau\sigma 11} = 3\frac{k}{k_1}\frac{\delta_{\tau 1}\frac{\mu_1}{\mu} + \delta_{\tau 2}\frac{k_1}{k}}{2+\gamma_\tau} a_{\tau\sigma 11}, \qquad \alpha_{\tau\sigma 12} = 5\left(\frac{k}{k_1}\right)^2 \frac{\delta_{\tau 1}\frac{\mu_1}{\mu} + \delta_{\tau 2}\frac{k_1}{k}}{3+2\gamma_\tau} a_{\tau\sigma 12}$$

The internal field, correct up to order kr, then becomes

$$\begin{aligned} \boldsymbol{E}(\boldsymbol{r}) &= \sum_{\substack{n \\ \tau=1,2}} \alpha_{\tau n}\boldsymbol{v}_{\tau n}(k_1\boldsymbol{r}) = \frac{3\epsilon}{\epsilon_1+2\epsilon}\boldsymbol{E}_0 - \frac{ik}{2}\frac{3\mu_1}{\mu_1+2\mu}\boldsymbol{r}\times(\hat{z}\times\boldsymbol{E}_0) \\ &+ \frac{ik}{2}\frac{5\epsilon}{2\epsilon_1+3\epsilon}\left\{\boldsymbol{E}_0\left(\boldsymbol{r}\cdot\hat{z}\right) + \hat{z}\left(\boldsymbol{r}\cdot\boldsymbol{E}_0\right)\right\} + O((kr)^2) \end{aligned} \tag{8.14}$$

We have here used (7.29) on page 375 and (C.11) on page 632 to obtain

$$\begin{cases} \sum\limits_{\sigma=\mathrm{e,o}} a_{1\sigma 11}\boldsymbol{A}_{1\sigma 11}(\hat{\boldsymbol{r}}) = \dfrac{3\mathrm{i}}{2}\hat{\boldsymbol{r}}\times\left(\boldsymbol{E}_0\times\hat{z}\right) \\ \sum\limits_{\sigma=\mathrm{e,o}} a_{2\sigma 11}\boldsymbol{A}_{2\sigma 11}(\hat{\boldsymbol{r}}) = \dfrac{3}{2}\hat{\boldsymbol{r}}\times\left(\boldsymbol{E}_0\times\hat{\boldsymbol{r}}\right) \\ \sum\limits_{\sigma=\mathrm{e,o}} a_{2\sigma 11}\boldsymbol{A}_{3\sigma 11}(\hat{\boldsymbol{r}}) = \dfrac{3}{\sqrt{2}}\hat{\boldsymbol{r}}\left(\boldsymbol{E}_0\cdot\hat{\boldsymbol{r}}\right) \end{cases}$$

and (C.12) on page 632 to obtain

$$\begin{cases} \sum\limits_{\sigma=\mathrm{e,o}} a_{2\sigma 12}\boldsymbol{A}_{2\sigma 12}(\hat{\boldsymbol{r}}) = \mathrm{i}\dfrac{5}{2}\left\{\cos\theta\boldsymbol{E}_0 - \left(\hat{\boldsymbol{r}}\cos\theta + \hat{\boldsymbol{\theta}}\sin\theta\right)\left(\boldsymbol{E}_0\cdot\hat{\boldsymbol{r}}\right)\right\} \\ \sum\limits_{\sigma=\mathrm{e,o}} a_{2\sigma 12}\boldsymbol{A}_{3\sigma 12}(\hat{\boldsymbol{r}}) = 5\mathrm{i}\sqrt{\dfrac{3}{2}}\hat{\boldsymbol{r}}\cos\theta\left(\boldsymbol{E}_0\cdot\hat{\boldsymbol{r}}\right) \end{cases}$$

The first term in (8.14) is the homogeneous internal field of a dipole [101, 276]. We also identify the second term as the magnetic dipole contribution since

$$\eta_0\eta_1\boldsymbol{H}(\boldsymbol{r}) = \frac{1}{ik_1}\nabla\times\boldsymbol{E}(\boldsymbol{r}) = \frac{k}{k_1}\frac{3\mu_1}{\mu_1+2\mu}\left(\hat{z}\times\boldsymbol{E}_0\right)$$

where we used $\nabla\times(\boldsymbol{a}\times\boldsymbol{r}) = 2\boldsymbol{a}$ and $\nabla\times(\boldsymbol{a}(\boldsymbol{r}\cdot\boldsymbol{b})) = \boldsymbol{b}\times\boldsymbol{a}$ for constant vectors $\boldsymbol{a}$ and $\boldsymbol{b}$. Another way of writing this result is

$$\boldsymbol{H}(\boldsymbol{r}) = \frac{3\mu}{\mu_1+2\mu}\boldsymbol{H}_0$$

where $\eta_0\eta\boldsymbol{H}_0 = \hat{z}\times\boldsymbol{E}_0$, which shows relevant symmetry between the electric and magnetic cases.

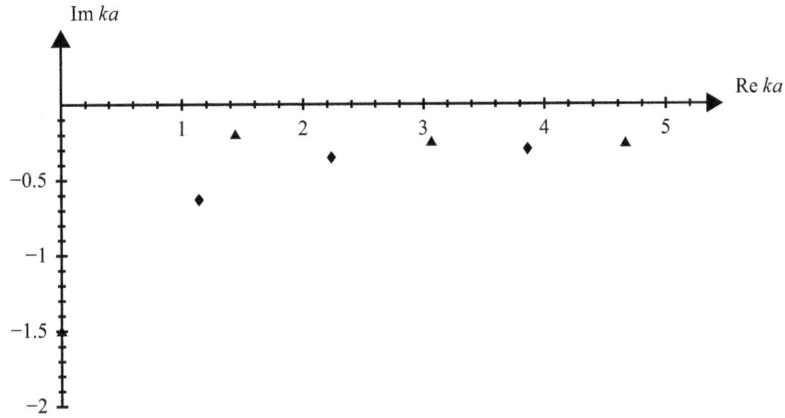

Figure 8.17 The poles in the complex ka-plane for the transition matrix $t_{\tau l}$ for $l = 1$. In this example $\epsilon_1/\epsilon = 4$ and $\mu_1/\mu = 1$. The triangles mark the poles for $\tau = 1$ and the diamonds correspond to $\tau = 2$.

8.2.3 Resonances

Scattering quantities for dielectric scatterers show sharp peaks, which correspond to resonances in the structure [63, 237]. The connection between these peaks in, e.g., the scattering cross section, and the singularities of the transition matrix entries is explained in the excellent book by Taylor [258, Chapter 13], see also Comment 8.1. Another often used name for these regular peaks is ripple structure. The subject is also closely related to the concept of plasmons [28, 212, 234–236]. The resonances are particularly obvious at high frequencies. We take a closer look at these resonance effects in this section.

For a homogeneous, dielectric sphere of radius a and material parameters ϵ_1 and μ_1, we denote, as above, the transition matrix by $t_{\tau l}$, see (8.6) on page 420. Surrounding material is assumed to be homogeneous with material parameters ϵ and μ. The matrix elements $t_{l\tau}$ are independent of the indices m and σ are

$$t_{\tau l} = -\frac{j_l(ka)(k_1 a j_l(k_1 a))' - \gamma_\tau (ka j_l(ka))' j_l(k_1 a)}{h_l^{(1)}(ka)(k_1 a j_l(k_1 a))' - \gamma_\tau (ka h_l^{(1)}(ka))' j_l(k_1 a)}$$

where $\gamma_\tau = \delta_{\tau 1}\mu_1/\mu + \delta_{\tau 2}\epsilon_1/\epsilon$. This expression has poles at the zeros of the denominator, i.e., at points ka in the complex ka-plane for which

$$h_l^{(1)}(ka)(k_1 a j_l(k_1 a))' - \gamma_\tau (ka h_l^{(1)}(ka))' j_l(k_1 a) = 0 \tag{8.15}$$

In Figure 8.17, these poles are shown in the complex ka-plane for $l = 1$ in a region close to the origin. The poles are symmetrically located around the imaginary axis, and there are poles only in the lower half-plane, i.e., $\text{Im}\, ka < 0$. The lack of poles in the upper half-plane is due to causality requirements of the scattered field. The symmetry properties are found from the following relations of the spherical Bessel and Hankel functions:

$$\begin{cases} j_l(-z) = (-1)^l j_l(z) \\ n_l(-z) = (-1)^{l+1} n_l(z) \end{cases} \qquad \begin{cases} h_l^{(1)}(-z) = (-1)^l h_l^{(2)}(z) \\ h_l^{(2)}(-z) = (-1)^l h_l^{(1)}(z) \end{cases} \qquad l = 0, 1, 2, \ldots$$

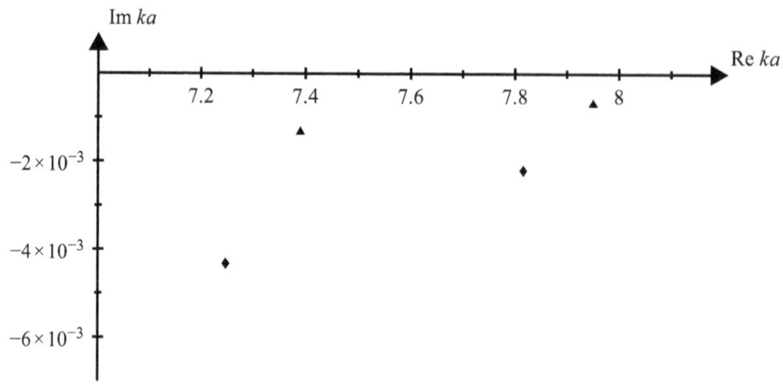

Figure 8.18 The poles in the complex ka-plane of the transition matrix $t_{\tau l}$. In this example $\epsilon_1/\epsilon = 4$ and $\mu_1/\mu = 1$. The triangles mark the poles for $\tau = 1$ and the diamonds correspond to $\tau = 2$. Note the scale on the imaginary axis.

and

$$\begin{cases} j_l(z^*) = (j_l(z))^* \\ h_l^{(1)}(z^*) = (h_l^{(2)}(z))^* \\ h_l^{(2)}(z^*) = (h_l^{(1)}(z))^* \end{cases} \qquad l = 0, 1, 2, \ldots$$

These relations show that

$$\begin{cases} j_l(-z^*) = (-1)^l j_l(z^*) = (-1)^l \, (j_l(z))^* \\ h_l^{(1)}(-z^*) = (-1)^l h_l^{(2)}(z^*) = (-1)^l \left(h_l^{(1)}(z)\right)^* \end{cases}$$

which show that if ka is a root to (8.15), then $-(ka)^*$ is also a root to the same equation provided the exterior material parameters are real valued.

The pole structure gives rise to the resonances in the scattering data if the poles are located close enough to the real axis—the closer the more sharp the resonance is. With the data of Figure 8.17, there are four poles very close to the real axis in the interval Re $ka \in [7, 8]$, see Figure 8.18. These poles give sharp resonances in the scattering cross section σ_s, as seen in Figure 8.19. In this particular example, there is a 30% increase of the scattering cross section at these resonance frequencies. Note that the peaks close to $ka = 7.4$ and $ka = 8.0$ are sharper, since the corresponding poles lie closer to the real axis than the other two poles. These poles correspond to $\tau = 1$.

Comment 8.1

The connection between the singularities of the transition matrix $t_{\tau l}$ and the peaks in the scattering cross section is described in the literature, see e.g., Reference 258. In this comment, we review some of the details.

Denote the location of a pole in a transition matrix entry by $z_0 = x_0 + \mathrm{i}y_0$, where $y_0 < 0$, and denote for simplicity $ka = x \in \mathbb{R}$. In a neighborhood of the pole on the real axis, the

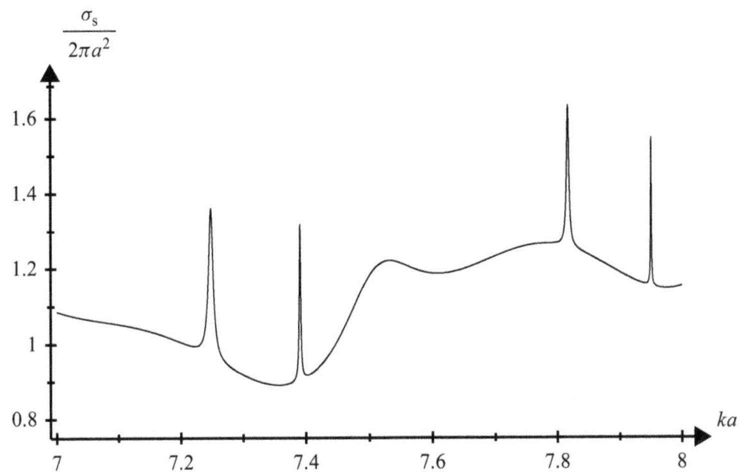

Figure 8.19 The scattering cross section $\sigma_s/2\pi a^2$ for a dielectric sphere. The material parameters of the sphere are $\epsilon_1/\epsilon = 4$ and $\mu_1/\mu = 1$. Note the normalization.

transition matrix entry has the form

$$t_{\tau l}(x) = \frac{t_{bg}(x)}{x - z_0}$$

where the background transition matrix $t_{bg}(x)$ is a slowly varying function, see below. Close to the pole z_0, the dominant contribution in the scattering cross section $\sigma_s(x)$ is the term

$$\frac{2\pi}{k^2}(2l+1)\left|t_{\tau l}(x)\right|^2 = \frac{2\pi}{k^2}(2l+1)\left|\frac{t_{bg}(x)}{x - z_0}\right|^2$$

The maximum occurs at

$$|x - z_0|^{-2}\left(\frac{d\left|t_{bg}(x)\right|^2}{dx} - 2\left|\frac{t_{bg}(x)}{x - z_0}\right|^2 (x - x_0)\right) = 0$$

or (assume $\left|t_{bg}(x)\right| > 0$)

$$\left((x - x_0)^2 + y_0^2\right)\left|t_{bg}(x)\right|' - \left|t_{bg}(x)\right|(x - x_0) = 0$$

where prime denotes differentiation w.r.t. x. If $\left|t_{bg}(x)\right|' = 0$ (constant background), then the maximum occurs at $x = x_0$. If $\left|t_{bg}(x)\right|' \neq 0$, then the maximum is slightly shifted away from x_0. This root x satisfies the approximate equation (use a linear approximation $\left|t_{bg}(x)\right| \approx \left|t_{bg}(x_0)\right| + \left|t_{bg}(x_0)\right|'(x - x_0)$)

$$\left((x - x_0)^2 + y_0^2\right)\left|t_{bg}(x_0)\right|' - \left(\left|t_{bg}(x_0)\right| + \left|t_{bg}(x_0)\right|'(x - x_0)\right)(x - x_0) = 0$$

with solution

$$x \approx x_0 + y_0^2 \frac{|t_{bg}(x_0)|'}{|t_{bg}(x_0)|}$$

which quantifies the small shift of the resonance due to non-constant background, i.e., $|t_{bg}(x_0)|' \neq 0$. Note that a decrease in the value of y_0^2 decreases the shift.

8.2.4 Interference Structure

The resonances in the scattering quantities are determined by the locations of the zeros of the denominator, and these effects were analyzed in Section 8.2.3. In this section, we concentrate on another phenomenon that has its origin in the numerator of the transition matrix entries [18]. The transition matrix elements in (8.6) have the generic form

$$t_{\tau l} = -\frac{j_l(ka)(k_1 a j_l(k_1 a))' - \gamma_\tau (ka j_l(ka))' j_l(k_1 a)}{h_l^{(1)}(ka)(k_1 a j_l(k_1 a))' - \gamma_\tau (ka h_l^{(1)}(ka))' j_l(k_1 a)}$$

where γ_τ is either μ_1/μ or ϵ_1/ϵ. We concentrate on the *numerator* and utilize the expansions of the spherical Bessel and Hankel functions, see (B.11) and (B.13) in Appendix B.3. The numerator then contains products of the trigonometric functions, i.e.,

$$\begin{cases} 2\sin\left(ka - \frac{l\pi}{2}\right)\sin\left(k_1 a - \frac{l\pi}{2}\right) = \cos(ka - k_1 a) - \cos(ka + k_1 a - l\pi) \\ 2\sin\left(ka - \frac{l\pi}{2}\right)\cos\left(k_1 a - \frac{l\pi}{2}\right) = \sin(ka + k_1 a - l\pi) + \sin(ka - k_1 a) \\ 2\cos\left(ka - \frac{l\pi}{2}\right)\sin\left(k_1 a - \frac{l\pi}{2}\right) = \sin(ka + k_1 a - l\pi) - \sin(ka - k_1 a) \\ 2\cos\left(ka - \frac{l\pi}{2}\right)\cos\left(k_1 a - \frac{l\pi}{2}\right) = \cos(ka - k_1 a) + \cos(ka + k_1 a - l\pi) \end{cases}$$

and we observe that there are parts of the numerator of t_l that contain a common factor $\sin(ka - k_1 a)$ or $\cos(ka - k_1 a)$. These factors are independent of the summation index l. Each element of the transition matrix $t_{\tau l}$, therefore, has the form

$$t_{\tau l} = A_{\tau l}(ka) + B_{\tau l}(ka)\sin(ka - k_1 a) + C_{\tau l}(ka)\cos(ka - k_1 a)$$

Thus, the transition matrix elements exhibit a periodicity[3] of $ka - k_1 a = 2\pi$, and since the scattering cross section contains the sum of $|t_{\tau l}|^2$, the scattering cross section shows extremal points at frequencies separated by an interval determined by $2(ka - k_1 a) = 2\pi$, i.e.,

$$\Delta(ka) = \frac{\pi}{\left|1 - \frac{k_1}{k}\right|} = \frac{\pi}{\left|\sqrt{\frac{\epsilon_1 \mu_1}{\epsilon \mu}} - 1\right|}$$

[3]This is not a proper periodicity, since the coefficients $A_{\tau l}(ka)$, $B_{\tau l}(ka)$, and $C_{\tau l}(ka)$ also change.

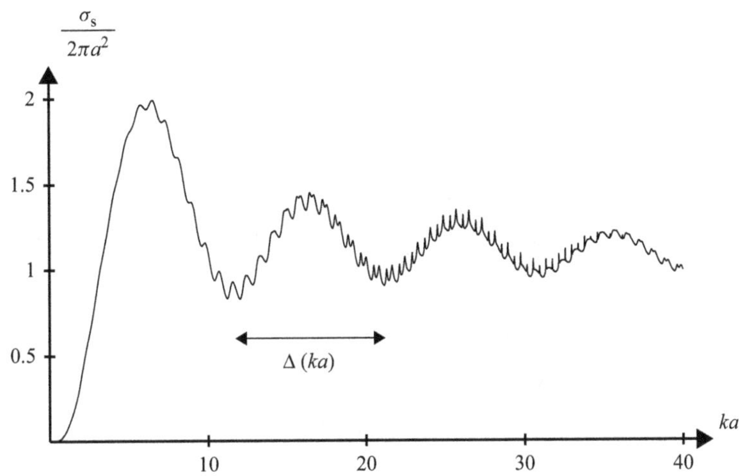

Figure 8.20 The scattering cross section $\sigma_s/2\pi a^2$ as a function of ka for a lossless dielectric sphere with permittivity $\epsilon_1/\epsilon = 1.33^2$ and permeability $\mu_1/\mu = 1$.

Example 8.5

The interference phenomenon is illustrated in Figure 8.10, where, if we ignore losses, the explicit numbers are

$$\Delta(ka) = \frac{\pi}{|1.5 - 1|} = 2\pi$$

The agreement with this simple estimate is quite good. ■

Example 8.6

Another illustration of the interference phenomenon is presented in Figure 8.20. The dielectric sphere is lossless with permittivity $\epsilon_1/\epsilon = 1.33^2$ and permeability $\mu_1/\mu = 1$. In this particular example, the interference period is

$$\Delta(ka) = \frac{\pi}{(1.33 - 1)} \approx 9.5$$

which agrees with the result in the figure. ■

8.3 Scattering by Layered Spherical Objects

Scattering by a layered spherical object is a straightforward generalization of the results presented in Section 8.2. A layered spherical sphere also serves as a staircase approximation of a continuous radial variation of the material parameters. We restrict ourselves to the case of isotropic layers, and leave the analysis of even more general scattering situations to Section 8.4.

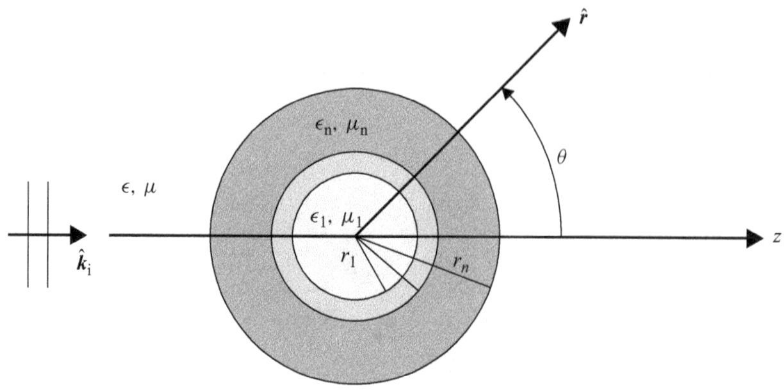

Figure 8.21 The geometry of the layered sphere and its material constants.

The subject has been treated extensively in the literature, see e.g., References 3, 64, 69, 74, 106, 170, 211, 227, 262, 299.

Let $r_1 \leq r_2 \leq \cdots \leq r_n$ be the radii of the n layers of the sphere, and moreover, let ϵ_i and μ_i, $i = 1, 2, \ldots, n$, be the (relative) permittivity and permeability, respectively, of the layers, see Figure 8.21. The outmost radius is r_n, and outside this sphere we have a homogeneous, isotropic material with lossless (real-valued) material parameters ϵ and μ. In the iteration scheme below, it is convenient to extend the set of radii of the layers defined above by denoting the origin by $r_0 = 0$, and also the exterior region by defining $r_{n+1} = \infty$.

The material parameters in each layer are constant in space, but can be dispersive, i.e., functions of the (angular) frequency ω. The corresponding wave numbers are denoted $k_i = k_0 \left(\epsilon_i \mu_i\right)^{1/2}$, $i = 1, 2, \ldots, n$, and we adopt the notation $k_{n+1} = k_0 \sqrt{\epsilon \mu}$ for the medium outside the obstacle. Notice that the electrical radii[4] $\operatorname{Re}(k_i r_i)$, $i = 1, 2, \ldots, n$, need not an ordered set. It may very well be that e.g., $\operatorname{Re}(k_2 r_2) > \operatorname{Re}(k_3 r_3)$ even if $r_2 \leq r_3$.

In each region, $r_{i-1} < r < r_i$, $i = 1, 2, \ldots, n+1$, the electric field has an expansion in the regular and the out-going spherical vector waves $\boldsymbol{v}_n$ and $\boldsymbol{u}_n$ with appropriate wave number k_i, i.e., ($i = 1, 2, \ldots, n+1$)

$$\boldsymbol{E}^{(i)}(\boldsymbol{r}, \omega) = \sum_n A_n^{(i)} \left(\boldsymbol{v}_n(k_i \boldsymbol{r}) + t_{\tau l}^{(i-1)} \boldsymbol{u}_n(k_i \boldsymbol{r}) \right), \qquad r_{i-1} < r < r_i \tag{8.16}$$

for some constants $A_n^{(i)}$, $i = 1, 2, \ldots, n+1$, and $t_{\tau l}^{(i)}$, $i = 0, 1, \ldots, n$, $\tau = 1, 2$. We include the innermost layer, $i = 1$, by defining $t_{\tau l}^{(0)} = 0$—a consequence of the fact that the electric field is non-singular at the origin $r_0 = 0$. The corresponding magnetic fields are ($i = 1, 2, \ldots, n+1$)

$$\boldsymbol{H}^{(i)}(\boldsymbol{r}, \omega) = \frac{1}{\mathrm{i}\eta_0 \eta_i} \sum_n A_n^{(i)} \left(\boldsymbol{v}_{\overline{n}}(k_i \boldsymbol{r}) + t_{\tau l}^{(i-1)} \boldsymbol{u}_{\overline{n}}(k_i \boldsymbol{r}) \right), \qquad r_{i-1} < r < r_i$$

[4]The electrical radius of an object is the wave number times the geometrical radius.

where $\overline{n}$ is the dual index of n, and where the relative wave impedance of the material is $\eta_i = (\mu_i/\epsilon_i)^{1/2}$, $i = 1, 2, \ldots, n$, and the surrounding material $\eta_{n+1} = \sqrt{\mu/\epsilon}$. The final goal is to identify the transition matrix entries of the entire layered scatterer $t_{\tau l} = t_{\tau l}^{(n)}$.

The boundary conditions at each interface—continuous tangential electric and magnetic fields—imply, due to the orthogonality conditions of the vector spherical harmonics, $\boldsymbol{A}_n(\hat{\boldsymbol{r}})$, that

$$\begin{cases} A_{1\sigma ml}^{(i)}\left(j_l(k_ir_i) + t_{1l}^{(i-1)}h_l^{(1)}(k_ir_i)\right) = A_{1\sigma ml}^{(i+1)}\left(j_l(k_{i+1}r_i) + t_{1l}^{(i)}h_l^{(1)}(k_{i+1}r_i)\right) \\ A_{2\sigma ml}^{(i)}\left(\dfrac{(k_ir_ij_l(k_ir_i))'}{k_ir_i} + t_{2l}^{(i-1)}\dfrac{(k_ir_ih_l^{(1)}(k_ir_i))'}{k_ir_i}\right) \\ \quad = A_{2\sigma ml}^{(i+1)}\left(\dfrac{(k_{i+1}r_ij_l(k_{i+1}r_i))'}{k_{i+1}r_i} + t_{2l}^{(i)}\dfrac{(k_{i+1}r_ih_l^{(1)}(k_{i+1}r_i))'}{k_{i+1}r_i}\right) \end{cases}$$

due to continuity of the tangential electric field, and

$$\begin{cases} A_{1\sigma ml}^{(i)}\dfrac{1}{\eta_i}\left(\dfrac{(k_ir_ij_l(k_ir_i))'}{k_ir_i} + t_{1l}^{(i-1)}\dfrac{(k_ir_ih_l^{(1)}(k_ir_i))'}{k_ir_i}\right) \\ \quad = A_{1\sigma ml}^{(i+1)}\dfrac{1}{\eta_{i+1}}\left(\dfrac{(k_{i+1}r_ij_l(k_{i+1}r_i))'}{k_{i+1}r_i} + t_{1l}^{(i)}\dfrac{(k_{i+1}r_ih_l^{(1)}(k_{i+1}r_i))'}{k_{i+1}r_i}\right) \\ A_{2\sigma ml}^{(i)}\dfrac{1}{\eta_i}\left(j_l(k_ir_i) + t_{2l}^{(i-1)}h_l^{(1)}(k_ir_i)\right) \\ \quad = A_{2\sigma ml}^{(i+1)}\dfrac{1}{\eta_{i+1}}\left(j_l(k_{i+1}r_i) + t_{2l}^{(i)}h_l^{(1)}(k_{i+1}r_i)\right) \end{cases}$$

due to the continuity of the tangential magnetic field. We proceed by eliminating the coefficients $A_n^{(i)}$ and $A_n^{(i+1)}$ by dividing the relations with each other, and we also introduce the Riccati–Bessel functions and their derivatives, $\psi_l(z) = zj_l(z)$ and $\xi_l(z) = zh_l^{(1)}(z)$. The result is

$$m_\tau^{(i)}\frac{\psi_l(k_ir_i) + t_{\tau l}^{(i-1)}\xi_l(k_ir_i)}{\psi_l'(k_ir_i) + t_{\tau l}^{(i-1)}\xi_l'(k_ir_i)} = \frac{\psi_l(k_{i+1}r_i) + t_{\tau l}^{(i)}\xi_l^{(1)}(k_{i+1}r_i)}{\psi_l'(k_{i+1}r_i) + t_{\tau l}^{(i)}\xi_l'(k_{i+1}r_i)}$$

where

$$m_\tau^{(i)} = \delta_{\tau 1}\frac{\eta_i}{\eta_{i+1}} + \delta_{\tau 2}\frac{\eta_{i+1}}{\eta_i}$$

We now solve for $t_{\tau l}^{(i)}$. This leads to a recursion formula—a Möbius[5] transformation—for the constants $t_{\tau l}^{(i)}$, $i = 1, \ldots, n$. The result is ($i = 1, 2, \ldots, n$)

$$t_{\tau l}^{(i)} = -\frac{m_\tau^{(i)}\left(\psi_l(x_i) + t_{\tau l}^{(i-1)}\xi_l(x_i)\right)\psi_l'(y_i) - \left(\psi_l'(x_i) + t_{\tau l}^{(i-1)}\xi_l'(x_i)\right)\psi_l(y_i)}{m_\tau^{(i)}\left(\psi_l(x_i) + t_{\tau l}^{(i-1)}\xi_l(x_i)\right)\xi_l'(y_i) - \left(\psi_l'(x_i) + t_{\tau l}^{(i-1)}\xi_l'(x_i)\right)\xi_l(y_i)}$$

[5] August Ferdinand Möbius (1790–1868), German mathematician and theoretical astronomer.

where $x_i = k_i r_i$ and $y_i = k_{i+1} r_i$—the electrical radius in terms of the material parameters on the inside and the outside of the interface r_i, respectively. We rewrite the recursion formula above as

$$t_{\tau l}^{(i)} = -\frac{a_{\tau l}^{(i)} t_{\tau l}^{(i-1)} + b_{\tau l}^{(i)}}{c_{\tau l}^{(i)} t_{\tau l}^{(i-1)} + d_{\tau l}^{(i)}}, \qquad i = 1, 2, \ldots, n$$

where

$$\begin{cases} a_{\tau l}^{(i)} = m_\tau^{(i)} \xi_l(x_i) \psi_l'(y_i) - \xi_l'(x_i) \psi_l(y_i) \\ b_{\tau l}^{(i)} = m_\tau^{(i)} \psi_l(x_i) \psi_l'(y_i) - \psi_l'(x_i) \psi_l(y_i) \\ c_{\tau l}^{(i)} = m_\tau^{(i)} \xi_l(x_i) \xi_l'(y_i) - \xi_l'(x_i) \xi_l(y_i) \\ d_{\tau l}^{(i)} = m_\tau^{(i)} \psi_l(x_i) \xi_l'(y_i) - \psi_l'(x_i) \xi_l(y_i) \end{cases}$$

The iteration starts either at $i = 0$ with $t_{\tau l}^{(0)} = 0$, or equivalently, at $i = 1$ and $t_{\tau l}^{(1)}$ as

$$t_{\tau l}^{(1)} = -\frac{m_\tau^{(1)} \psi_l(x_1) \psi_l'(y_1) - \psi_l'(x_1) \psi_l(y_1)}{m_\tau^{(1)} \psi_l(x_1) \xi_l'(y_1) - \psi_l'(x_1) \xi_l(y_1)}$$

If the innermost layer is a perfectly conducting material, the iteration starts at $i = 2$, and it is initialized by

$$t_{1l}^{(1)} = -\frac{\psi_l(k_2 r_1)}{\xi_l(k_2 r_1)} = -\frac{\psi_l(y_1)}{\xi_l(y_1)}, \qquad t_{2l}^{(1)} = -\frac{\psi_l'(k_2 r_1)}{\xi_l'(k_2 r_1)} = -\frac{\psi_l'(y_1)}{\xi_l'(y_1)}$$

This initialization corresponds to $m_1^{(1)} = 0$ and $m_2^{(1)} = \infty$, respectively. Equivalently, we start at $i = 1$ and $t_{\tau l}^{(0)} = 0$, and[6]

$$\begin{cases} a_{1l}^{(1)} = b_{1l}^{(1)} = \psi_l(y_i) \\ c_{1l}^{(1)} = d_{1l}^{(1)} = \xi_l(y_i) \end{cases} \qquad \begin{cases} a_{2l}^{(1)} = b_{2l}^{(1)} = \psi_l'(y_i) \\ c_{2l}^{(1)} = d_{2l}^{(1)} = \xi_l'(y_i) \end{cases}$$

Each transformation is uniquely defined by the constants $a_{\tau l}^{(i)}$, $b_{\tau l}^{(i)}$, $c_{\tau l}^{(i)}$, and $d_{\tau l}^{(i)}$. We make the identification, see Appendix F.4 on page 670, with the coefficients in a Möbius transformation, i.e.,

$$t_{\tau l}^{(i)} \leftrightarrow \begin{pmatrix} a_{\tau l}^{(i)} & b_{\tau l}^{(i)} \\ c_{\tau l}^{(i)} & d_{\tau l}^{(i)} \end{pmatrix}, \qquad i = 1, 2, \ldots, n$$

---

[6]In fact, the values of $a_{\tau l}^{(1)}$ and $c_{\tau l}^{(1)}$ are irrelevant in this case.

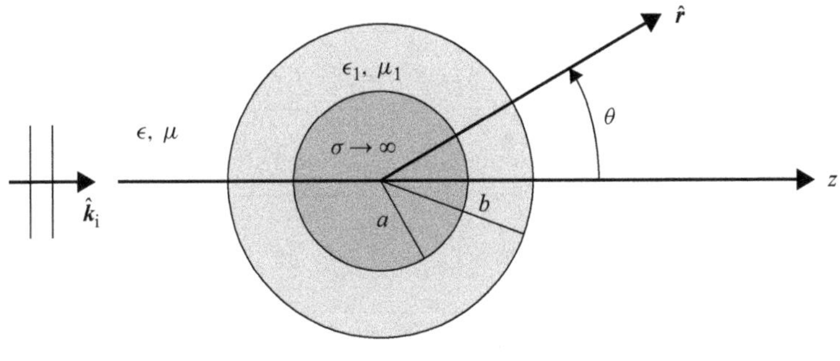

Figure 8.22 Geometry of the dielectric-coated, perfectly conducting sphere in Example 8.7.

From Appendix F.4, the coefficients in the final transition matrix $t_{\tau l} = t_{\tau l}^{(n)}$ are obtained by matrix multiplication, i.e.,

$$t_{\tau l} \leftrightarrow \begin{pmatrix} a_{\tau l}^{(n)} & b_{\tau l}^{(n)} \\ c_{\tau l}^{(n)} & d_{\tau l}^{(n)} \end{pmatrix} = \begin{pmatrix} a_{\tau l}^{(n-1)} & b_{\tau l}^{(n-1)} \\ c_{\tau l}^{(n-1)} & d_{\tau l}^{(n-1)} \end{pmatrix} \cdots \begin{pmatrix} a_{\tau l}^{(1)} & b_{\tau l}^{(1)} \\ c_{\tau l}^{(1)} & d_{\tau l}^{(1)} \end{pmatrix}$$

acting on the complex number 0. This identification with a Möbius transformation makes the numerical implementation of the transition matrix easy and systematic.

Example 8.7

The dielectric-coated, perfectly conducting sphere of radius a, shown in Figure 8.22, is an interesting example. From results above (the coating has material parameters ϵ_1 and μ_1, and outer radius b), the transition matrix entries are

$$t_{\tau l} = -\frac{m_\tau \left(\psi_l(k_1 b) + t_{\tau l}^{\mathrm{PEC}} \xi_l(k_1 b)\right) \psi_l'(kb) - \left(\psi_l'(k_1 b) + t_{\tau l}^{\mathrm{PEC}} \xi_l'(k_1 b)\right) \psi_l(kb)}{m_\tau \left(\psi_l(k_1 b) + t_{\tau l}^{\mathrm{PEC}} \xi_l(k_1 b)\right) \xi_l'(kb) - \left(\psi_l'(k_1 b) + t_{\tau l}^{\mathrm{PEC}} \xi_l'(k_1 b)\right) \xi_l(kb)}$$

where

$$t_{1l}^{\mathrm{PEC}} = -\frac{\psi_l(k_1 a)}{\xi_l(k_1 a)}, \qquad t_{2l}^{\mathrm{PEC}} = -\frac{\psi_l'(k_1 a)}{\xi_l'(k_1 a)}$$

and

$$m_\tau = \delta_{\tau 1} \frac{\eta_1}{\eta} + \delta_{\tau 2} \frac{\eta}{\eta_1}$$

and the wave number and wave impedance are $k_1 = k_0 \left(\epsilon_1 \mu_1\right)^{1/2}$ and $\eta_1 = \left(\mu_1/\epsilon_1\right)^{1/2}$, respectively.

To compute the polarizability dyadic for this scatterer, we specialize to $l = 1$ and keep only the lowest order terms in a and b. We obtain from the small argument expansions of $\psi_1(z)$ and

$\xi_1(z)$ in Appendix B.3 and the low-frequency results in Section 8.1.1

$$\begin{cases} t_{11} = -\dfrac{\mathrm{i}(kb)^3}{3}\dfrac{\mu A_{11} - 2\mu_1 B_{11}}{\mu A_{11} + \mu_1 B_{11}} \\ t_{21} = -\dfrac{\mathrm{i}(kb)^3}{3}\dfrac{\epsilon A_{21} - 2\epsilon_1 B_{21}}{\epsilon A_{21} + \epsilon_1 B_{21}} \end{cases}$$

where ($\xi = a/b \in [0, 1]$)

$$\begin{cases} A_{11} = 2 + \xi^3 \\ B_{11} = 1 - \xi^3 \end{cases} \qquad \begin{cases} A_{21} = 2 - 2\xi^3 \\ B_{21} = 1 + 2\xi^3 \end{cases}$$

The polarizability dyadics, see (7.54) on page 402, are

$$\begin{cases} \gamma_{\mathrm{e}} = \lim\limits_{k\to 0} \dfrac{6\pi t_{21}\mathbf{I}_3}{\mathrm{i}k^3} = 4\pi b^3 \dfrac{\epsilon_1\left(1 + 2\xi^3\right) - \epsilon\left(1 - \xi^3\right)}{\epsilon_1\left(1 + 2\xi^3\right) + 2\epsilon\left(1 - \xi^3\right)}\mathbf{I}_3 \\ \gamma_{\mathrm{m}} = \lim\limits_{k\to 0} \dfrac{6\pi t_{11}\mathbf{I}_3}{\mathrm{i}k^3} = -2\pi b^3 \dfrac{\mu\left(2 + \xi^3\right) - 2\mu_1\left(1 - \xi^3\right)}{\mu\left(2 + \xi^3\right) + \mu_1\left(1 - \xi^3\right)}\mathbf{I}_3 \end{cases} \tag{8.17}$$

We observe that both polarizabilities are identically zero if

$$\epsilon_1 = \frac{1 - \xi^3}{1 + 2\xi^3}\epsilon, \qquad \mu_1 = \frac{2 + \xi^3}{2\left(1 - \xi^3\right)}\mu$$

For such a material, the far field amplitude vanishes faster than k^2 at low frequencies, cf. (6.8) on page 272. Notice also that this choice of material parameters satisfies

$$\frac{\epsilon\mu}{\epsilon_1\mu_1} = 2\frac{1 + 2\xi^3}{2 + \xi^3} \in (1, 4)$$

for all $\xi > 0$, corresponding to a phase velocity in the coating that is larger than the phase velocity of the surrounding material, i.e., $c_1/c \in (1, 2)$ in the coating. By an application of the sum rule in Section 6.4, it is tempting to draw the conclusion that this scatterer neither scatter nor absorb anything at any frequency, since zero polarizability coefficients imply that $\sigma_{\mathrm{ext}} = 0$ for all frequencies. This conclusion, however, is erroneous, since the causality condition, that is crucial in the derivation of the sum rule, does not hold here. In fact, the speed of light in the material is larger than in the surrounding material. This makes the use of the sum rules faulty.

If the coating is thin, ξ is a number close to one, and we introduce $\xi^3 = 1 - \eta$, where $\eta > 0$ is a small number. Keeping only leading-order contribution in η in (8.17), we get an approximative expression of the electric and the magnetic polarizability dyadics. The result is

$$\begin{aligned} \gamma_{\mathrm{e}} &= 4\pi b^3 \frac{\epsilon_1(3 - 2\eta) - \epsilon\eta}{\epsilon_1(3 - 2\eta) + 2\epsilon\eta} = 4\pi b^3\left(3\epsilon_1 - (2\epsilon_1 + \epsilon)\eta\right)\left(\frac{1}{3\epsilon_1} - 2\frac{\epsilon - \epsilon_1}{9\epsilon_1^2}\eta + O(\eta^2)\right) \\ &= 4\pi b^3\left(1 - \frac{\epsilon}{\epsilon_1}\eta\right) + O(\eta^2) \end{aligned}$$

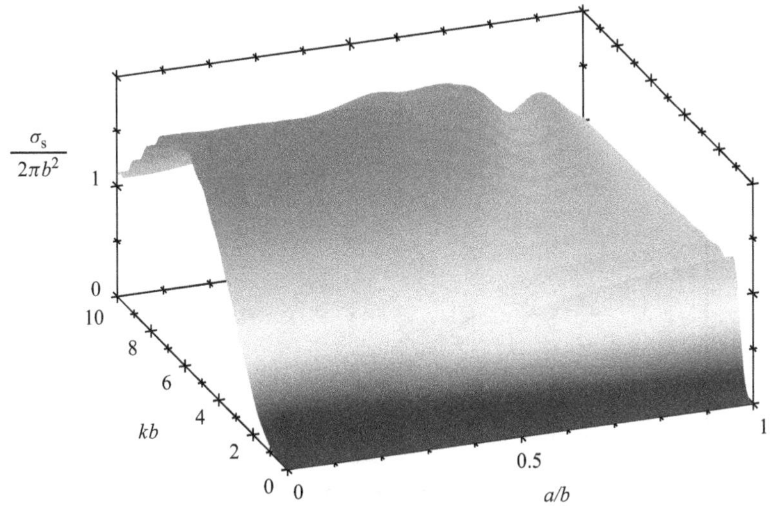

Figure 8.23 The scattering cross section σ_s for a dielectric-coated PEC sphere as a function of the electric size kb and the quotient of the radii a/b, where a is the radius of the inner PEC sphere, and b is the outer radius of the dielectric coating, see Figure 8.22. The material parameters are $\epsilon_1/\epsilon = 1.33^2$ and $\mu_1/\mu = 1$, and the cross section is scaled with $2\pi b^2$.

and

$$\begin{aligned}\gamma_m &= -2\pi b^3 \frac{\mu(3-\eta) - 2\mu_1\eta}{\mu(3-\eta) + \mu_1\eta} \\ &= -2\pi b^3 (3\mu - (2\mu_1 + \mu)\eta) \left(\frac{1}{3\mu} + \frac{\mu - \mu_1}{9\mu^2}\eta + O(\eta^2) \right) \\ &= -2\pi b^3 \left(1 - \frac{\mu_1}{\mu}\eta \right) + O(\eta^2)\end{aligned}$$

■

Example 8.8
It is interesting to see how the scattering properties vary as the material parameters continuously change from a PEC sphere to a homogeneous dielectric sphere.

In Figure 8.23, we depict the scattering cross section, σ_s, for a non-magnetic dielectric-coated PEC sphere as a function of the electric size kb and the quotient of the radii a/b, where a is the radius of the inner PEC sphere and b is the outer radius of the dielectric coating, see Figure 8.22. The material parameters are $\epsilon_1/\epsilon = 1.33^2$ and $\mu_1/\mu = 1$. The amplitude is scaled with $2\pi b^2$. The values of the left-hand side, $a/b = 0$, a homogeneous dielectric sphere, are identical to the values in Figure 8.20, and the values of the right-hand side, $a/b = 1$, a PEC sphere, in the figure are identical to the result in Figure 8.2. ■

Example 8.9
To illustrate the use of a staircase approximation of a radially, continuously varying permittivity profile, we assume the material parameters of the sphere of outer radius a are

$$\epsilon_1(r)/\epsilon = 2 - \left(\frac{r}{a}\right)^2, \quad \mu_1/\mu = 1, \qquad r \in [0, a]$$

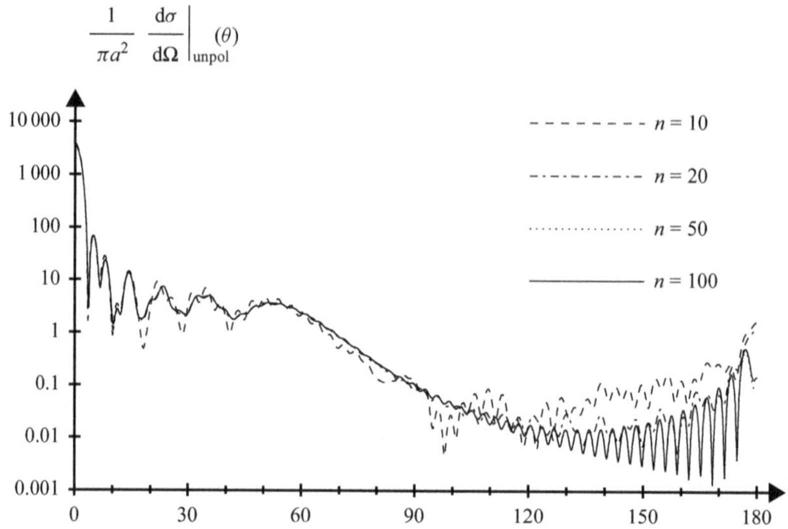

Figure 8.24 The differential scattering cross section for the Lüneburg lens in Example 8.9. The incident wave is an unpolarized plane wave and $ka = 60$. The differential scattering cross section is normalized with πa^2.

This canonical example is the permittivity of a lens antenna (Lüneburg[7] lens) [106, 211, 299].

The continuously varying permittivity profile is approximated by

$$\epsilon_i/\epsilon = 2 - \left(\frac{r_{i-1} + r_i}{2a}\right)^2, \quad \mu_1/\mu = 1, \qquad i = 1, 2, \ldots, n$$

where $r_i = ia/n, i = 1, 2, \ldots, n$. In Figure 8.24, the differential scattering cross section is shown as a function of the scattering angle θ for a series of staircase approximations $n = 10, 20, 50, 100$ and $ka = 60$. We see that the curves corresponding to $n = 50$ and $n = 100$ hardly differ on the resolution in the figure. ■

Example 8.10

The sum rules of Section 6.4 can be illustrated in a very striking way by a layered sphere, where the inner core is a lossless dielectric, and the outer layer has losses. For a sphere of constant outer radius a, the sum rule is independent of the thickness d of the outer layer. This is due to the fact that all spheres have the same electric and magnetic polarizability dyadics, see below.

This phenomenon is illustrated in Figure 8.25 for series of nanoshells, where the core is made of silicon dioxide (SiO_2), and the outer layers are made of gold (Au). Material data are obtained from the literature [125, 270], see also Reference 12. The four curves show the extinction cross section of the layered sphere as a function of the wavelength λ. In all four cases, the following sum rule holds, see (6.63) on page 325:

$$\int_0^\infty \sigma_{\text{ext}}(\lambda)\, d\lambda = \pi^2 \left(\hat{\boldsymbol{p}}_{\text{e}}^* \cdot \boldsymbol{\gamma}_{\text{e}} \cdot \hat{\boldsymbol{p}}_{\text{e}} + \hat{\boldsymbol{p}}_{\text{m}}^* \cdot \boldsymbol{\gamma}_{\text{m}} \cdot \hat{\boldsymbol{p}}_{\text{m}}\right) = 4\pi^3 a^3$$

[7] Rudolf Karl Lüneburg (1903–1949), German-American mathematician and optician.

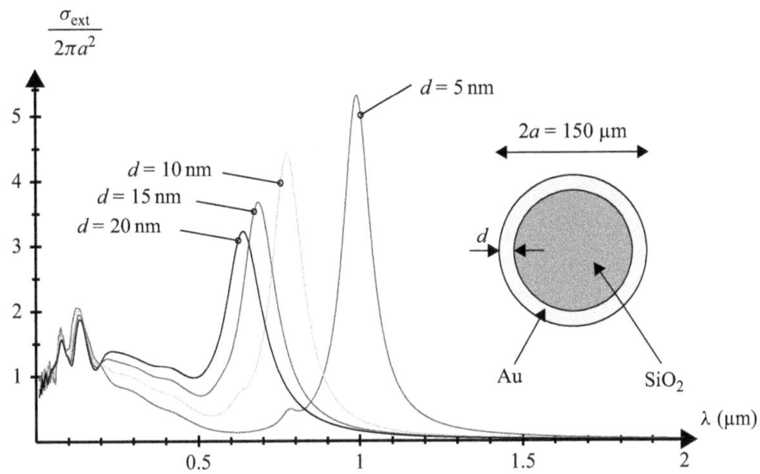

Figure 8.25 The normalized extinction cross section $\sigma_{ext}/2\pi a^2$ for four nanoshells as a function of the wavelength λ. The core is made of SiO_2, and the outer layer is made of gold. The outer radius $a = 75$ nm in all four cases, and the thickness of the outer layer is $d = 5, 10, 15, 20$ nm, respectively. Data of the materials are obtained from References 125, 270. The area under all four curves is, with the normalization in the illustration, $2\pi^2 a$.

since the polarizability dyadics of the sphere are $\boldsymbol{\gamma}_e = 4\pi a^3 \mathbf{I}_3$ and $\boldsymbol{\gamma}_m = \mathbf{0}$ in all four cases, see also Example 6.15 on page 326. All four curves have, with the normalization in the illustration, an area of $2\pi^2 a$, which is verified by numerical integration. Due to the symmetric scattering dyadic in the forward direction, the result is not restricted to a linearly polarized incident plane wave. ■

8.3.1 Resonance Frequencies in a Spherical Cavity

The determination of the resonance frequencies of a spherical cavity is a problem closely related to the scattering problem by a layered spherical object, and therefore included in this section as a special case.

Let $a < b$ be the inner and outer radii of the spherical cavity, see Figure 8.26. The cavity contains a lossless material with relative permittivity ϵ and permeability μ, and we denote the wave number $k = k_0\sqrt{\epsilon\mu}$. Inside the cavity, the electric field has an expansion similar to (8.16)

$$\boldsymbol{E}(\boldsymbol{r},\omega) = \sum_n A_n \left(\boldsymbol{v}_n(k\boldsymbol{r}) + t_{\tau l}^{\rm PEC} \boldsymbol{u}_n(k\boldsymbol{r})\right), \qquad a < r < b$$

where

$$t_{1l}^{\rm PEC} = -\frac{\psi_l(ka)}{\xi_l(ka)}, \qquad t_{2l}^{\rm PEC} = -\frac{\psi_l'(ka)}{\xi_l'(ka)} \tag{8.18}$$

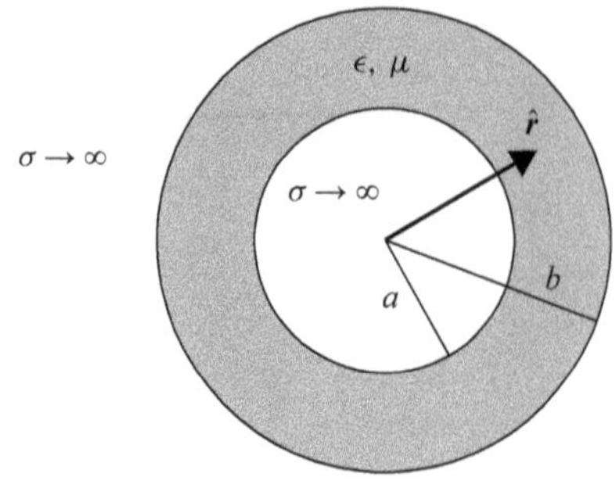

Figure 8.26 Geometry of the spherical cavity in Section 8.3.1.

This expansion of the electric field satisfies the appropriate boundary condition—vanishing tangential electric field—at $r = a$. To satisfy the boundary condition on the outer surface, $r = b$, only certain frequencies are allowed. The boundary conditions—vanishing tangential electric field—at $r = b$, lead to

$$\begin{cases} \psi_l(kb)\xi_l(ka) - \psi_l(ka)\xi_l(kb) = 0, & \tau = 1 \\ \psi_l'(kb)\xi_l'(ka) - \psi_l'(ka)\xi_l'(kb) = 0, & \tau = 2 \end{cases}$$

Denote the roots of these transcendental equations by $k_{\tau ln}$, $n = 1, 2, 3, \ldots$, i.e.,

$$\begin{cases} \psi_l(k_{1ln}b)\xi_l(k_{1ln}a) - \psi_l(k_{1ln}a)\xi_l(k_{1ln}b) = 0 \\ \psi_l'(k_{2ln}b)\xi_l'(k_{2ln}a) - \psi_l'(k_{2ln}a)\xi_l'(k_{2ln}b) = 0 \end{cases} \tag{8.19}$$

The roots to these equations determine the resonance frequencies of the spherical cavity.

The complete system of vector waves (eigenfunctions) for the spherical cavity is

$$\begin{cases} \boldsymbol{\Psi}_{1ln}(k_{1ln}\boldsymbol{r}) = \left(\dfrac{\psi_l(k_{1ln}r)}{k_{1ln}r} + t_{1l}^{\text{PEC}} \dfrac{\xi_l(k_{1ln}r)}{k_{1ln}r} \right) \boldsymbol{A}_{1n}(\hat{\boldsymbol{r}}) \\ \boldsymbol{\Psi}_{2ln}(k_{2ln}\boldsymbol{r}) = \left(\dfrac{\psi_l'(k_{2ln}r)}{k_{2ln}r} + t_{2l}^{\text{PEC}} \dfrac{\xi_l'(k_{2ln}r)}{k_{2ln}r} \right) \boldsymbol{A}_{2n}(\hat{\boldsymbol{r}}) \\ \qquad + \sqrt{l(l+1)} \left(\dfrac{\psi_l(k_{2ln}r)}{k_{2ln}^2 r^2} + t_{2l}^{\text{PEC}} \dfrac{\xi_l(k_{2ln}r)}{k_{2ln}^2 r^2} \right) \boldsymbol{A}_{3n}(\hat{\boldsymbol{r}}) \end{cases}$$

where the coefficients $t_{\tau l}^{\text{PEC}}$ are given by (8.18) (with $k = k_{\tau ln}$) and the wave numbers $k_{\tau ln}$ are determined by (8.19). These vector waves satisfy the boundary condition in the cavity, and they are natural expansion functions in the cavity. The completeness of the functions $\boldsymbol{\Psi}_{\tau ln}(k_{\tau ln}\boldsymbol{r})$ is proved in the literature [113].

Example 8.11

Very low frequency resonance phenomena can be observed with the Earth and the ionosphere viewed as a resonance cavity. Experimental verifications of these resonances have been

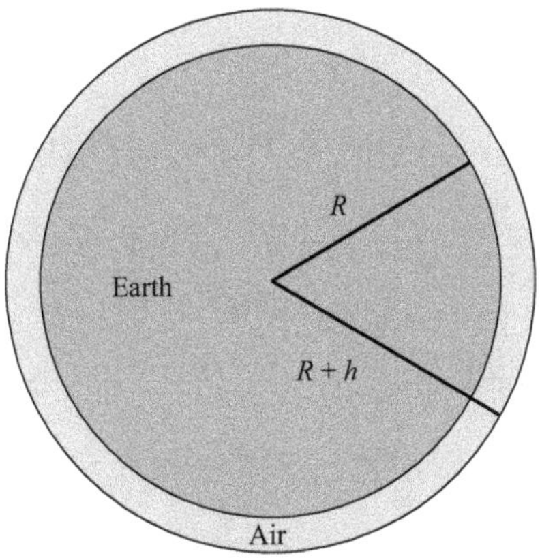

Figure 8.27 Geometry of the spherical resonance cavity for the Schumann resonances. The height of the air layer, h is exaggerated compared to R.

made [60, 205, 218, 219]. These resonances—the Schumann[8] resonances—are generated by global lightning activity. Since the Earth and the ionosphere are relatively good conductors ($\sigma \approx 10^{-1}$ mhos/m and $\sigma \approx 10^{-7}$–10^{-4} mhos/m, respectively), a resonance cavity is formed by the Earth and the ionosphere, see Figure 8.27. A crude model of this cavity is obtained if we approximate the boundary conditions on the Earth $r = R \approx 6\,378$ km and the lower surface of the ionosphere $r = R+h$ by perfectly conducting surfaces. A more advanced model is analyzed in Problem 8.9. A useful average value of the height of the lower surface of the ionosphere is $h \approx 100$ km. The wave numbers $k_{\tau ln}$, and therefore the resonance frequencies, are determined by the equations in (8.19) with $a = R$ and $b = R + h$.

Since $h/R \ll 1$, we expand the Riccati–Bessel functions in a Taylor series around $r = R$

$$\begin{cases} \psi_l(x+\varepsilon) = \psi_l(x) + \varepsilon\psi_l'(x) + O(\varepsilon^2) \\ \xi_l(x+\varepsilon) = \xi_l(x) + \varepsilon\xi_l'(x) + O(\varepsilon^2) \end{cases}$$

and

$$\begin{cases} \psi_l'(x+\varepsilon) = \psi_l'(x) + \varepsilon\psi_l''(x) + O(\varepsilon^2) = \varepsilon\left(\dfrac{l(l+1)}{x^2} - 1\right)\psi_l(x) + \psi_l'(x) + O(\varepsilon^2) \\ \xi_l'(x+\varepsilon) = \xi_l'(x) + \varepsilon\xi_l''(x) + O(\varepsilon^2) = \varepsilon\left(\dfrac{l(l+1)}{x^2} - 1\right)\xi_l(x) + \xi_l'(x) + O(\varepsilon^2) \end{cases}$$

where we have used the differential equation of the Riccati–Bessel functions, viz.

$$Z_l''(x) = \left(\frac{l(l+1)}{x^2} - 1\right)Z_l(x)$$

[8]Winfried Otto Schumann (1888–1974), German physicist.

Saving terms up to order $\varepsilon = k_{\tau ln}h$ and neglecting higher terms, from (8.19), we get for $\tau = 1$ ($x = k_{1ln}R$ and $\varepsilon = k_{1ln}h$)

$$\left(\psi_l(x) + \varepsilon\psi_l'(x)\right)\xi_l(x) - \psi_l(x)\left(\xi_l(x) + \varepsilon\xi_l'(x)\right) + O(\varepsilon^2) = 0$$

which can be simplified to

$$-\mathrm{i}\varepsilon + O(\varepsilon^2) = 0$$

where we have used the Wronskian

$$\psi_l(x)\xi_l'(x) - \psi_l'(x)\xi_l(x) = \mathrm{i}$$

The corresponding calculations for $\tau = 2$ give ($x = k_{2ln}R$ and $\varepsilon = k_{2ln}h$)

$$\begin{aligned}&\left(\varepsilon\left(\frac{l(l+1)}{x^2} - 1\right)\psi(x) + \psi_l'(x)\right)\xi_l'(x)\\&\quad - \psi_l'(x)\left(\varepsilon\left(\frac{l(l+1)}{x^2} - 1\right)\xi(x) + \xi_l'(x)\right) + O(\varepsilon^2) = 0\end{aligned}$$

which we simplify with the Wronskian to

$$\mathrm{i}\varepsilon\left(\frac{l(l+1)}{x^2} - 1\right) + O(\varepsilon^2) = 0$$

The equation for the eigenfrequencies, correct up to first order, for $\tau = 2$ is ($\tau = 1$ gives no equation with terms up to this order)

$$k_{2ln}^2 R^2 = l(l+1)$$

or (here we drop the index n)

$$f_{2l} = \frac{\sqrt{l(l+1)}c_0}{2\pi R}, \qquad l = 1, 2, 3, \ldots$$

Insert the constants $R = 6\,378$ km and $c_0 = 2.99792 \times 10^8$ m/s, and we have

$$f_{2l} = 7.5\sqrt{l(l+1)} = 10.6, 18.3, 25.9, 33.5, 41.0, 48.5, 56.0, \ldots \text{ Hz}$$

The first roots to (8.19) can be found numerically. For $\tau = 1$, the lowest frequency is 1 499 Hz, while for $\tau = 2$, they are 10.5 and 18.2 Hz. This shows that the approximations we made above are quite accurate.

The average values obtained experimentally are $f = 8, 14, 20, 26, 32$ Hz [218]. The discrepancies between the experimental values and the calculated ones are not surprising considering the crude approximations that we have made. In reality, the Earth and the ionosphere are not perfectly conducting surfaces, see also Problem 8.9, and the conductivity varies between land and sea. Furthermore, the height and the density of the ionosphere vary, and the ionosphere is more developed on the lit side of the Earth. These effects are not taken into account in our simple model. The index of refraction varies in the atmosphere (decreases approximately linearly with height). We can compensate for these imperfections by replacing the radius of the Earth R by an effective radius R_{eff}. A commonly used value is $R_{\text{eff}} = 4R/3$, which gives a very good agreement to the experimental data.

Finally, the Schumann resonances have another very interesting implication in elementary particle physics. The lowest Schumann resonance, $f = 8$ Hz, is the lowest resonance frequency observed in nature. This value gives an upper limit on the rest mass of the photon, m_γ, through Einstein's relation

$$\frac{h\omega}{2\pi} \geq m_\gamma c_0^2$$

This relation gives

$$m_\gamma \leq 5.9 \times 10^{-50} \text{ kg}$$

This very small mass supports the existing theory of a massless photon. ∎

8.4 Scattering by an Anisotropic Sphere

The successful analysis in the previous sections of this chapter encourages us to proceed with a more advanced scattering problem. This section treats scattering by an anisotropic sphere where the material parameters vary with the radial coordinate r. Both the perfectly conducting sphere and the homogeneous dielectric sphere in Sections 8.1 and 8.2 are special cases of the analysis in this section. In fact, the result in this section covers scattering by any radially layered spherical object, both isotropic, as in Section 8.3, and the special kind of anisotropy that is modeled in (8.21).

The form of the spherical vector waves that we developed in Chapter 7 is not useful for the more general scattering problem we now address. We start our analysis by finding the relevant differential equation and the appropriate expansion functions in an anisotropic material.

An anisotropic material is characterized by the constitutive relations, i.e.,

$$\begin{cases} \boldsymbol{D}(\boldsymbol{r},\omega) = \epsilon_0 \boldsymbol{\epsilon}(\boldsymbol{r},\omega) \cdot \boldsymbol{E}(\boldsymbol{r},\omega) \\ \boldsymbol{B}(\boldsymbol{r},\omega) = \dfrac{1}{c_0} \boldsymbol{\mu}(\boldsymbol{r},\omega) \cdot \eta_0 \boldsymbol{H}(\boldsymbol{r},\omega) \end{cases}$$

where the dyadics $\boldsymbol{\epsilon}(\boldsymbol{r},\omega)$ and $\boldsymbol{\mu}(\boldsymbol{r},\omega)$, in addition of being dependent on the frequency (dispersion), now also can depend on the position vector (inhomogeneous). The fields in a source-free domain have to satisfy

$$\begin{cases} \nabla \times \boldsymbol{E}(\boldsymbol{r},\omega) = \mathrm{i}k_0 \boldsymbol{\mu}(\boldsymbol{r},\omega) \cdot \eta_0 \boldsymbol{H}(\boldsymbol{r},\omega) \\ \nabla \times \eta_0 \boldsymbol{H}(\boldsymbol{r},\omega) = -\mathrm{i}k_0 \boldsymbol{\epsilon}(\boldsymbol{r},\omega) \cdot \boldsymbol{E}(\boldsymbol{r},\omega) \end{cases}$$

where, as usual, $k_0 = \omega/c_0$. We eliminate the field $\boldsymbol{H}(\boldsymbol{r},\omega)$, and we get, cf. (2.31) on page 118

$$\nabla \times \left(\boldsymbol{\mu}^{-1}(\boldsymbol{r},\omega) \cdot (\nabla \times \boldsymbol{E}(\boldsymbol{r},\omega))\right) - k_0^2 \boldsymbol{\epsilon}(\boldsymbol{r},\omega) \cdot \boldsymbol{E}(\boldsymbol{r},\omega) = \boldsymbol{0} \tag{8.20}$$

As a consequence of this equation, the electric field satisfies ($\omega \neq 0$)

$$\nabla \cdot (\boldsymbol{\epsilon}(\boldsymbol{r},\omega) \cdot \boldsymbol{E}(\boldsymbol{r},\omega)) = 0$$

Equation (8.20) is the basic equation for the electric field in a general anisotropic material. The solution depends on the permittivity and permeability dyadics $\boldsymbol{\epsilon}(\boldsymbol{r},\omega)$ and $\boldsymbol{\mu}(\boldsymbol{r},\omega)$. The specific form of these dyadics determines the features of the solution. In the next section, we specialize the anisotropy, such that it becomes relevant to a spherical scatterer with spherical symmetry.

8.4.1 Radial Expansion Functions

In this section, we investigate suitable expansion functions of the electromagnetic field in a special type of radial anisotropy. We assume that the permittivity and the permeability dyadics, $\boldsymbol{\epsilon}$ and $\boldsymbol{\mu}$, respectively, have the form (radially inhomogeneous and uniaxial w.r.t. the radial direction)[9]

$$\begin{cases} \boldsymbol{\epsilon}(\boldsymbol{r}) = \epsilon_r(r)\hat{\boldsymbol{r}}\hat{\boldsymbol{r}} + \epsilon_\perp(r)\left(\mathbf{I}_3 - \hat{\boldsymbol{r}}\hat{\boldsymbol{r}}\right) \\ \boldsymbol{\mu}(\boldsymbol{r}) = \mu_r(r)\hat{\boldsymbol{r}}\hat{\boldsymbol{r}} + \mu_\perp(r)\left(\mathbf{I}_3 - \hat{\boldsymbol{r}}\hat{\boldsymbol{r}}\right) \end{cases} \tag{8.21}$$

In the radial direction, the permittivity and the permeability are denoted $\epsilon_r(r)$ and $\mu_r(r)$, respectively, and orthogonal to the radial direction of the sphere, material parameters are denoted $\epsilon_\perp(r)$ and $\mu_\perp(r)$, respectively. The material parameters can vary w.r.t. the radial coordinate r. These constitutive relations cover a large class of scattering problems by spherical objects, e.g., inhomogeneous isotropic objects and multi-layered isotropic spherical objects. Below, we consider a spherical scatterer, centered at the origin, filled with this material, i.e., $0 \leq r \leq a$, where a is the radius of the spherical scatterer.

Encouraged by the success in Chapter 7, we proceed by the following Ansatz of the electric field

$$\boldsymbol{E}(\boldsymbol{r}) = \sum_n \left\{ g_{1n}(r)\boldsymbol{A}_{1n}(\hat{\boldsymbol{r}}) + g_{2n}(r)\boldsymbol{A}_{2n}(\hat{\boldsymbol{r}}) + g_{3n}(r)\boldsymbol{A}_{3n}(\hat{\boldsymbol{r}}) \right\}$$

This field is assumed to satisfy (8.20), and our aim is to find the ODE that the radial functions $g_{\tau n}(r)$ in the anisotropic material have to satisfy. This approach gives[10] (use (C.7) and (C.15) in Appendix C.4, or alternatively, consult Chapter 7 with $f_{\tau n} \leftrightarrow g_{\tau n}$)

$$\nabla \times \boldsymbol{E}(\boldsymbol{r}) = \sum_n \left\{ \frac{(rg_1(r))'}{r}\boldsymbol{A}_{2n}(\hat{\boldsymbol{r}}) - \frac{(rg_2(r))' - g_3(r)\sqrt{l(l+1)}}{r}\boldsymbol{A}_{1n}(\hat{\boldsymbol{r}}) + \frac{g_1(r)\sqrt{l(l+1)}}{r}\boldsymbol{A}_{3n}(\hat{\boldsymbol{r}}) \right\}$$

where the prime denotes differentiation w.r.t. the radial coordinate r. Another application of the curl operator implies

$$\begin{aligned} \nabla\times\left(\boldsymbol{\mu}^{-1}(\boldsymbol{r})\cdot(\nabla\times\boldsymbol{E}(\boldsymbol{r}))\right) = \sum_n \Bigg\{ &-\left(\frac{(rg_1(r))'}{\mu_\perp(r)r}\right)'\boldsymbol{A}_{1n}(\hat{\boldsymbol{r}}) \\ &- \left(\frac{(rg_2(r))' - g_3(r)\sqrt{l(l+1)}}{\mu_\perp(r)r}\right)'\boldsymbol{A}_{2n}(\hat{\boldsymbol{r}}) - \frac{(rg_1(r))'}{\mu_\perp(r)r^2}\boldsymbol{A}_{1n}(\hat{\boldsymbol{r}}) \\ &- \frac{(rg_2(r))' - g_3(r)\sqrt{l(l+1)}}{\mu_\perp(r)r^2}\left(\boldsymbol{A}_{2n}(\hat{\boldsymbol{r}}) + \sqrt{l(l+1)}\boldsymbol{A}_{3n}(\hat{\boldsymbol{r}})\right) \\ &+ \frac{l(l+1)}{\mu_r(r)}\frac{g_1(r)}{r^2}\boldsymbol{A}_{1n}(\hat{\boldsymbol{r}}) \Bigg\} \end{aligned}$$

---

[9]The dependence of the (angular) frequency ω is suppressed by convenience.

[10]For convenience, we suppress the index n, i.e., l, on the radial functions.

We treat each term in the sum separately, since $A_{\tau n}$ are orthogonal (linearly independent) at each direction in space. Identifying each factor containing the vector spherical harmonics in (8.20), we get

$$\begin{cases} \dfrac{\left(\frac{1}{\mu_\perp(r)}(rg_1(r))'\right)'}{r} - \dfrac{l(l+1)}{\mu_r(r)}\dfrac{g_1(r)}{r^2} + k_0^2\epsilon_\perp(r)g_1(r) = 0 \\ \dfrac{\left\{\frac{1}{\mu_\perp(r)}\left((rg_2(r))' - g_3(r)\sqrt{l(l+1)}\right)\right\}'}{r} + k_0^2\epsilon_\perp(r)g_2(r) = 0 \\ \dfrac{(rg_2(r))' - g_3(r)\sqrt{l(l+1)}}{\mu_\perp(r)r^2}\sqrt{l(l+1)} + k_0^2\epsilon_r(r)g_3(r) = 0 \end{cases}$$

The first equation involves only the function $g_1(r)$, while the other two equations couple the functions $g_2(r)$ and $g_3(r)$. We observe that a certain combination of $g_2(r)$ and $g_3(r)$ reduces the problem to one equation. To see this, temporarily denote (remember we have suppressed the index n, i.e., l, on the radial functions)

$$\begin{cases} g(r) = rg_1(r) \\ h(r) = -\dfrac{(rg_2(r))' - g_3(r)\sqrt{l(l+1)}}{\mu_\perp(r)} \end{cases}$$

and we obtain by rewriting the first equation, and by combining the last two equations

$$\begin{cases} \mu_\perp(r)\left(\dfrac{g'(r)}{\mu_\perp(r)}\right)' + \left(k_0^2\epsilon_\perp(r)\mu_\perp(r) - \dfrac{l(l+1)\mu_\perp(r)}{r^2\mu_r(r)}\right)g(r) = 0 \\ \epsilon_\perp(r)\left(\dfrac{h'(r)}{\epsilon_\perp(r)}\right)' + \left(k_0^2\epsilon_\perp(r)\mu_\perp(r) - \dfrac{l(l+1)\epsilon_\perp(r)}{r^2\epsilon_r(r)}\right)h(r) = 0 \end{cases} \tag{8.22}$$

and the three radial functions are given in terms of the two functions $g(r)$ and $h(r)$, i.e.,

$$\begin{cases} g_1(r) = \dfrac{g(r)}{r} \\ g_2(r) = \dfrac{h'(r)}{k_0^2 r\epsilon_\perp(r)} \\ g_3(r) = \dfrac{h(r)\sqrt{l(l+1)}}{k_0^2 r^2\epsilon_r(r)} \end{cases}$$

The appropriate expansion functions of the electric field for a radially anisotropic and inhomogeneous material then become (the first is scaled with k_0 in order to have all vector waves dimensionless, and we reintroduce the l index on the radial functions)

$$\begin{cases} \boldsymbol{w}_{1n}(\boldsymbol{r}) = \dfrac{g_l(r)}{k_0 r}\boldsymbol{A}_{1n}(\hat{\boldsymbol{r}}) \\ \boldsymbol{w}_{2n}(\boldsymbol{r}) = \dfrac{h_l'(r)}{k_0^2 r\epsilon_\perp(r)}\boldsymbol{A}_{2n}(\hat{\boldsymbol{r}}) + \dfrac{h_l(r)\sqrt{l(l+1)}}{k_0^2 r^2\epsilon_r(r)}\boldsymbol{A}_{3n}(\hat{\boldsymbol{r}}) \end{cases} \tag{8.23}$$

where $g_l(r)$ and $h_l(r)$ solve (8.22), and where the prime denotes differentiation w.r.t. the radial coordinate r. Useful are also the curls of these two expansion functions. These are

$$\begin{cases} \nabla \times \boldsymbol{w}_{1n}(\boldsymbol{r}) = \dfrac{g_l'(r)}{k_0 r}\boldsymbol{A}_{2n}(\hat{\boldsymbol{r}}) + \dfrac{g_l(r)}{k_0 r}\dfrac{\sqrt{l(l+1)}}{r}\boldsymbol{A}_{3n}(\hat{\boldsymbol{r}}) \\ \nabla \times \boldsymbol{w}_{2n}(\boldsymbol{r}) = \dfrac{h_l(r)\mu_\perp(r)}{r}\boldsymbol{A}_{1n}(\hat{\boldsymbol{r}}) \end{cases} \tag{8.24}$$

and the divergence is by construction zero, i.e. (see also (C.14) on page 633)

$$\nabla \cdot (\boldsymbol{\epsilon} \cdot \boldsymbol{w}_{\tau n}(\boldsymbol{r})) = 0, \qquad \tau = 1, 2$$

To proceed, we need to determine the radial functions $g_l(r)$ and $h_l(r)$, and especially the behavior of these functions at the origin.

8.4.1.1 Behavior at the Origin

For convenience, we rewrite the equations in (8.22) in the standard form as [138]

$$\begin{cases} \dfrac{\mathrm{d}^2 g(r)}{\mathrm{d}r^2} + p_1(r)\dfrac{\mathrm{d}g(r)}{\mathrm{d}r} + q_1(r)g(r) = 0 \\ \dfrac{\mathrm{d}^2 h(r)}{\mathrm{d}r^2} + p_2(r)\dfrac{\mathrm{d}h(r)}{\mathrm{d}r} + q_2(r)h(r) = 0 \end{cases}$$

where

$$\begin{cases} p_1(r) = -\dfrac{\mu_\perp'(r)}{\mu_\perp(r)} \\ q_1(r) = k_0^2\epsilon_\perp(r)\mu_\perp(r) - \dfrac{l(l+1)\mu_\perp(r)}{r^2\mu_r(r)} \end{cases}$$

and

$$\begin{cases} p_2(r) = -\dfrac{\epsilon_\perp'(r)}{\epsilon_\perp(r)} \\ q_2(r) = k_0^2\epsilon_\perp(r)\mu_\perp(r) - \dfrac{l(l+1)\epsilon_\perp(r)}{r^2\epsilon_r(r)} \end{cases}$$

Notice the symmetry between the two differential equations ($\epsilon \leftrightarrow \mu$). The solutions divide into two classes or modes, a magnetic, $\tau = 1$, and an electric, $\tau = 2$, mode.

If the permittivity and the permeability are well behaved at the origin, e.g., analytic and non-zero, the point $r = 0$ is a regular singular point of the ODE [138]. We can then use the method of Frobenius[11] to find the behavior of the solution near the origin, see e.g., Reference 138. Assuming

$$\begin{cases} rp(r) = \alpha_0 + \alpha_1 r + O(r^2) \\ r^2 q(r) = \beta_0 + \beta_1 r + O(r^2) \end{cases}$$

[11] Ferdinand Georg Frobenius (1849–1917), German mathematician.

the least singular solution to the differential equations

$$\frac{d^2f(r)}{dr^2} + p(r)\frac{df(r)}{dr} + q(r)f(r) = 0$$

is [138]

$$f(r) = r^{\nu}\left(1 + a_1 r + O(r^2)\right)$$

where ν is the root with the largest real part of (the indicial equation)

$$\nu^2 + (\alpha_0 - 1)\nu + \beta_0 = 0$$

and

$$a_1 = -\frac{\alpha_1\nu + \beta_1}{2\nu + \alpha_0}$$

The case $l = 0$ is special, but it does not enter into the analysis of electromagnetism, and we can safely assume $l = 1, 2, \ldots$.

To proceed, we have to find the first expansion coefficients α_0, α_1, β_0, and β_1. Relevant expansions in our application are (we assume the permittivity and the permeability are well-behaved functions at the origin, i.e., they are smooth functions, such that a Taylor expansion exists)

$$\begin{cases} \epsilon_r(r) = \epsilon_r(0) + \epsilon_r'(0)r + O(r^2) \\ \epsilon_\perp(r) = \epsilon_\perp(0) + \epsilon_\perp'(0)r + O(r^2) \end{cases} \qquad \begin{cases} \mu_r(r) = \mu_r(0) + \mu_r'(0)r + O(r^2) \\ \mu_\perp(r) = \mu_\perp(0) + \mu_\perp'(0)r + O(r^2) \end{cases}$$

where we assume $\epsilon_r(0) \neq 0$, $\epsilon_\perp(0) \neq 0$, $\mu_r(0) \neq 0$, and $\mu_\perp(0) \neq 0$. Due to the assumed behavior of the material parameters at the origin, the differential equation of $g(r)$ has expansion coefficients

$$\begin{cases} \alpha_0 = \lim_{r\to 0} r p_1(r) = -\lim_{r\to 0} r\frac{\mu_\perp'(r)}{\mu_\perp(r)} = 0 \\ \alpha_1 = \lim_{r\to 0}\left(p_1(r) - \frac{\alpha_0}{r}\right) = \lim_{r\to 0} p_1(r) = -\frac{\mu_\perp'(0)}{\mu_\perp(0)} \end{cases}$$

and by series expansion of the factor in $q_1(r)$ at the origin, we get

$$\begin{cases} \beta_0 = \lim_{r\to 0} r^2 q_1(r) = -l(l+1)\frac{\mu_\perp(0)}{\mu_r(0)} \\ \beta_1 = \lim_{r\to 0} r\left(q_1(r) - \frac{\beta_0}{r^2}\right) = l(l+1)\frac{\mu_r'(0)\mu_\perp(0) - \mu_\perp'(0)\mu_r(0)}{(\mu_r(0))^2} \end{cases}$$

which lead to a singular behavior of $g(r)$ at the origin as

$$g(r) = r^{\nu_1}\left(1 + \frac{r\mu_\perp'(0)}{2\mu_\perp(0)} - l(l+1)r\frac{\mu_r'(0)\mu_\perp(0) - \mu_\perp'(0)\mu_r(0)}{2\nu_1(\mu_r(0))^2} + O(r^2)\right)$$

where (assume $\mu_\perp(0)/\mu_r(0) > 0$)

$$\nu_1 = \frac{1}{2} + \sqrt{\frac{1}{4} + \frac{l(l+1)\mu_\perp(0)}{\mu_r(0)}}$$

Similarly, the result for $h(r)$ is

$$h(r) = r^{\nu_2}\left(1 + \frac{r\epsilon'_\perp(0)}{2\epsilon_\perp(0)} - l(l+1)r\frac{\epsilon'_r(0)\epsilon_\perp(0) - \epsilon'_\perp(0)\epsilon_r(0)}{2\nu_2(\epsilon_r(0))^2} + O(r^2)\right)$$

where (assume $\epsilon_\perp(0)/\epsilon_r(0) > 0$)

$$\nu_2 = \frac{1}{2} + \sqrt{\frac{1}{4} + \frac{l(l+1)\epsilon_\perp(0)}{\epsilon_r(0)}}$$

We illustrate the potential of the results in a series of examples.

Example 8.12

The simplest example is when the material is homogeneous and isotropic, i.e., $\epsilon_\perp(r) = \epsilon_r(r) = \epsilon_1$, $\mu_\perp(r) = \mu_r(r) = \mu_1$. Then, letting $k_1^2 = k_0^2\epsilon_1\mu_1$, we obtain from (8.22)

$$\begin{cases} g''(r) + \left(k_1^2 - \dfrac{l(l+1)}{r^2}\right) g(r) = 0 \\ h''(r) + \left(k_1^2 - \dfrac{l(l+1)}{r^2}\right) h(r) = 0 \end{cases}$$

and

$$\begin{cases} g_1(r) = \dfrac{g(r)}{r} \\ g_2(r) = \dfrac{h'(r)}{k_0^2 r\epsilon_1} = \dfrac{\mu_1}{k_1 r}\dfrac{\mathrm{d}h(r)}{\mathrm{d}(k_1 r)} \\ g_3(r) = \dfrac{h(r)\sqrt{l(l+1)}}{k_0^2 r^2\epsilon_1} = \mu_1\dfrac{h(r)\sqrt{l(l+1)}}{k_1^2 r^2} \end{cases}$$

First, we identify the pertinent solution to the differential equations as spherical Bessel functions times the argument, i.e., $g(r) = h(r) = k_1 r j_l(k_1 r) = \psi_l(k_1 r)$, see (B.25) on page 617. We thus have (a constant factor k_1 has no importance in $g_1(r)$, and a constant factor of μ has no importance in $g_2(r)$ and $g_3(r)$)

$$\begin{cases} g_1(r) = \dfrac{\psi_l(k_1 r)}{k_1 r} = j_l(k_1 r) \\ g_2(r) = \dfrac{\psi'_l(k_1 r)}{k_1 r} = \dfrac{(k_1 r j_l(k_1 r))'}{k_1 r} \\ g_3(r) = \sqrt{l(l+1)}\dfrac{\psi_l(k_1 r)}{(k_1 r)^2} = \sqrt{l(l+1)}\dfrac{j_l(k_1 r)}{k_1 r} \end{cases}$$

where the prime denotes differentiation w.r.t. the argument $k_1 r$. This is in agreement with the result obtained in Section 7.1. We also conclude that the indices $\nu_1 = \nu_2 = l + 1$, leading to the correct small argument behavior of the radial functions $\psi_l(k_1 r)/(k_1 r)$ and $\psi_l'(k_1 r)/(k_1 r)$, i.e., $(k_1 r)^l$. ■

Example 8.13
A slightly more complex example appears if the material is isotropic but inhomogeneous, i.e., $\epsilon_\perp(r) = \epsilon_r(r) = \epsilon_1(r)$, $\mu_\perp(r) = \mu_r(r) = \mu_1(r)$. Adopt $k_1^2(r) = k_0^2 \epsilon_1(r)\mu_1(r)$, and we get from (8.22)

$$\begin{cases} g''(r) - \dfrac{\mu_1'(r)}{\mu_1(r)} g'(r) + \left(k_1^2(r) - \dfrac{l(l+1)}{r^2} \right) g(r) = 0 \\ h''(r) - \dfrac{\epsilon_1'(r)}{\epsilon_1(r)} h'(r) + \left(k_1^2(r) - \dfrac{l(l+1)}{r^2} \right) h(r) = 0 \end{cases}$$

and

$$\begin{cases} g_1(r) = \dfrac{g(r)}{r} \\ g_2(r) = \dfrac{h'(r)}{k_0^2 r \epsilon_1(r)} = \mu_1(r) \dfrac{h'(r)}{k_1^2(r) r} \\ g_3(r) = \dfrac{h(r)\sqrt{l(l+1)}}{k_0^2 r^2 \epsilon_1(r)} = \mu_1(r) \dfrac{h(r)\sqrt{l(l+1)}}{k_1^2(r) r^2} \end{cases}$$

The behavior at the origin is determined by

$$\nu_1 = \nu_2 = \frac{1}{2} + \sqrt{\frac{1}{4} + l(l+1)} = l + 1$$

and

$$\begin{cases} g(r) = r^{l+1} \left(1 + \dfrac{r\mu_1'(0)}{2\mu_1(0)} + O(r^2) \right) \\ h(r) = r^{l+1} \left(1 + \dfrac{r\epsilon_1'(0)}{2\epsilon_1(0)} + O(r^2) \right) \end{cases}$$

■

Example 8.14
As an example with piecewise constant material parameters, let the material be isotropic and homogeneous in two intervals, i.e.,

$$\epsilon(r) = \begin{cases} \epsilon_1, & r \in [0, r_0] \\ \epsilon_2, & r \in [r_0, a] \end{cases} \qquad \mu(r) = \begin{cases} \mu_1, & r \in [0, r_0] \\ \mu_2, & r \in [r_0, a] \end{cases}$$

Adopt $k_n^2 = k_0^2 \epsilon_n \mu_n$, $n = 1, 2$, and from Example 8.12, we get the possible radial solutions. We have

$$\begin{cases} g(r) = h(r) = \psi_l(k_1 r), & r \in [0, r_0] \\ g(r) = h(r) = \psi_l(k_2 r) \text{ or } \xi_l(k_2 r), & r \in [r_0, a] \end{cases}$$

In the core, $[0, r_0]$, the solution must be regular, i.e., $\psi_l(k_1 r)$ is appropriate. However, in the shell, $[r_0, a]$, the solution can be a combination of the Riccati–Bessel functions $\psi_l(k_2 r)$ and $\xi_l(k_2 r)$. Therefore, the general solution is

$$\begin{cases} g(r) = A_{1l} \psi_l(k_1 r), & r \in [0, r_0] \\ g(r) = B_{1l} \left(\psi_l(k_2 r) + t_{1l} \xi_l^{(1)}(k_2 r) \right), & r \in [r_0, a] \end{cases}$$

and

$$\begin{cases} h(r) = A_{2l} \psi_l(k_1 r), & r \in [0, r_0] \\ h(r) = B_{2l} \left(\psi_l(k_2 r) + t_{2l} \xi_l^{(1)}(k_2 r) \right), & r \in [r_0, a] \end{cases}$$

for some constants $A_{\tau l}$, $B_{\tau l}$, and $t_{\tau l}$.

From the differential equations in (8.22), we conclude that at the point $r = r_0$, where the material parameters are discontinuous, we require that $g(r)$ and $g'(r)/\mu(r)$, and $h(r)$ and $h'(r)/\epsilon(r)$ are all continuous at $r = r_0$. In this example, we get

$$\begin{cases} A_{1l} \psi_l(k_1 r_0) = B_{1l} \left(\psi_l(k_2 r_0) + t_{1l} \xi_l(k_2 r_0) \right) \\ A_{1l} \dfrac{k_1 \psi_l'(k_1 r_0)}{\mu_1} = B_{1l} \left(\dfrac{k_2 \psi_l'(k_2 r_0)}{\mu_2} + t_{1l} \dfrac{k_2 \xi_l'(k_2 r_0)}{\mu_2} \right) \end{cases}$$

and

$$\begin{cases} A_{2l} \psi_l(k_1 r_0) = B_{2l} \left(\psi_l(k_2 r_0) + t_{2l} \xi_l(k_2 r_0) \right) \\ A_{2l} \dfrac{k_1 \psi_l'(k_1 r_0)}{\epsilon_1} = B_{2l} \left(\dfrac{k_2 \psi_l'(k_2 r_0)}{\epsilon_2} + t_{2l} \dfrac{k_2 \xi_l'(k_2 r_0)}{\epsilon_2} \right) \end{cases}$$

Division of the equations gives the $t_{\tau l}$ constants, cf. (8.7)

$$t_{\tau l} = -\frac{\psi_l(k_2 r_0) \psi_l'(k_1 r_0) - m_\tau \psi_l'(k_2 r_0) \psi_l(k_1 r_0)}{\xi_l(k_2 r_0) \psi_l'(k_1 r_0) - m_\tau \xi_l'(k_2 r_0) \psi_l(k_1 r_0)}$$

where

$$m_\tau = \delta_{\tau 1} \frac{\eta_1}{\eta_2} + \delta_{\tau 2} \frac{\eta_2}{\eta_1}$$

where we have adopted the wave impedance of the layers $\eta_i = (\mu_i/\epsilon_i)^{1/2}$, $i = 1, 2$.

Of interest below, let $\epsilon_2 = \epsilon$ and $\mu_2 = \mu$, i.e., same material in the interval $[r_0, a]$ as the surrounding material. Then

$$t_{\tau l} = -\frac{\psi_l(k r_0) \psi_l'(k_1 r_0) - m_\tau \psi_l'(k r_0) \psi_l(k_1 r_0)}{\xi_l(k r_0) \psi_l'(k_1 r_0) - m_\tau \xi_l'(k r_0) \psi_l(k_1 r_0)}$$

where

$$m_\tau = \delta_{\tau 1}\frac{\eta_1}{\eta} + \delta_{\tau 2}\frac{\eta}{\eta_1}$$

with solutions

$$\begin{cases} g(r) = A_{1l}\psi_l(k_1 r), & r \in [0, r_0] \\ g(r) = B_{1l}\left(\psi_l(kr) + t_{1l}\xi_l^{(1)}(kr)\right), & r \in [r_0, a] \end{cases}$$

and

$$\begin{cases} h(r) = A_{2l}\psi_l(k_1 r), & r \in [0, r_0] \\ h(r) = B_{2l}\left(\psi_l(kr) + t_{2l}\xi_l^{(1)}(kr)\right), & r \in [r_0, a] \end{cases}$$

This is the solution to a scattering problem of a dielectric sphere of radius $r = r_0$, since both the exterior solution and the entries of the transition matrix are those of a dielectric sphere of radius $r = r_0$. ■

The radial functions $g(r)$ and $h(r)$ satisfying (8.22) constitute the building blocks in the calculation of the scattering properties of the anisotropic sphere. In particular, the transition matrix elements of the anisotropic sphere are needed. These matrix elements are the topic of the next section.

8.4.2 Transition Matrix

The previous section investigated the form of the appropriate spherical vector waves inside the anisotropic material. We now address the solution of the scattering problem of the anisotropic sphere.

The scattering problem of an anisotropic sphere of radius a (homogeneous or inhomogeneous) is solved using the same technique as employed in Section 8.2. The sphere is modeled by the constitutive relations in (8.21). We adopt a coordinate system in which the center of the sphere coincides with the origin of the coordinate system. The surrounding (background) material is, as above, modeled by the real permittivity ϵ and permeability μ. The incident field is, as above, a plane wave impinging along the positive z-axis

$$\boldsymbol{E}_{\mathrm{i}}(\boldsymbol{r}, \omega) = \boldsymbol{E}_0 \mathrm{e}^{\mathrm{i}kz} = \sum_n a_n \boldsymbol{v}_n(k\boldsymbol{r})$$

with coefficient a_n given in (7.29) on page 375.

As in Section 8.2, the multipole expansion of the scattered field outside the sphere is

$$\boldsymbol{E}_{\mathrm{s}}(\boldsymbol{r}, \omega) = \sum_n f_n \boldsymbol{u}_n(k\boldsymbol{r}), \qquad r > a$$

and the corresponding total magnetic field outside the sphere is

$$\eta_0 \boldsymbol{H}(\boldsymbol{r}, \omega) = \frac{1}{\mathrm{i}\eta}\sum_n \left(a_n \boldsymbol{v}_{\overline{n}}(k\boldsymbol{r}) + f_n \boldsymbol{u}_{\overline{n}}(k\boldsymbol{r})\right), \qquad r > a$$

where $\overline{n}$ is the dual index of n.

We expand the electric field inside the anisotropic sphere of radius $r = a$ in (as usual, n is a multi-index $n = \tau\sigma ml$)

$$\boldsymbol{E}_1(\boldsymbol{r}) = \sum_n \alpha_n \boldsymbol{w}_n(\boldsymbol{r}), \qquad r < a$$

where the expansion functions are defined in (8.23) (we use the convention with a single multi-index n instead of the index τn). Then, the magnetic field is

$$\eta_0 \boldsymbol{H}_1(\boldsymbol{r}) = \frac{1}{\mathrm{i}k_0} \boldsymbol{\mu}^{-1}(\boldsymbol{r}) \cdot \nabla \times \boldsymbol{E}_1(\boldsymbol{r}) = \frac{1}{\mathrm{i}k_0} \sum_n \alpha_n \boldsymbol{\mu}^{-1}(\boldsymbol{r}) \cdot (\nabla \times \boldsymbol{w}_n(\boldsymbol{r})), \qquad r < a$$

Matching of the tangential electric and magnetic fields at the interface $r = a$ implies, see (8.24)

$$\begin{cases} \alpha_{1n} \dfrac{g_l(a)}{k_0 a} = \dfrac{a_{1n}\psi_l(ka) + f_{1n}\xi_l(ka)}{ka} \\ \alpha_{2n} \dfrac{h_l'(a)}{k_0^2 a \epsilon_\perp(a)} = \dfrac{a_{2n}\psi_l'(ka) + f_{2n}\xi_l'(ka)}{ka} \end{cases} \tag{8.25}$$

$$\begin{cases} \alpha_{1n} \dfrac{g_l'(a)}{k_0^2 a \mu_\perp(a)} = \dfrac{1}{\eta} \dfrac{a_{1n}\psi_l'(ka) + f_{1n}\xi_l'(ka)}{ka} \\ \alpha_{2n} \dfrac{h_l(a)}{k_0 a} = \dfrac{1}{\eta} \dfrac{a_{2n}\psi_l(ka) + f_{2n}\xi_l(ka)}{ka} \end{cases} \tag{8.26}$$

where $g_l(r)$ and $h_l(r)$ are the regular solutions[12] at the origin to the differential equations in (8.22). We have also expressed the spherical vector waves outside the scatterer in terms of the Riccati–Bessel functions, see (B.26) on page 618.

The diagonal transition matrix elements are, as usual, denoted as $t_{\tau l}$, i.e.,

$$f_{\tau\sigma ml} = t_{\tau l} a_{\tau\sigma ml}$$

The transition matrix elements are easily computed. They are

$$\begin{cases} t_{1l} = -\dfrac{\psi_l(ka)g_l'(a)/k - \gamma_1\psi_l'(ka)g_l(a)}{\xi_l(ka)g_l'(a)/k - \gamma_1\xi_l'(ka)g_l(a)} \\ t_{2l} = -\dfrac{\psi_l(ka)h_l'(a)/k - \gamma_2\psi_l'(ka)h_l(a)}{\xi_l(ka)h_l'(a)/k - \gamma_2\xi_l'(ka)h_l(a)} \end{cases} \tag{8.27}$$

where

$$\gamma_\tau = \delta_{\tau 1}\frac{\mu_\perp(a)}{\mu} + \delta_{\tau 2}\frac{\epsilon_\perp(a)}{\epsilon}$$

---

[12]The normalization of the functions $g(r)$ and $h(r)$ is irrelevant, since only the quotient of the function and its derivative enters.

Notice that the radial derivatives on $g_l(r)$ and $h_l(r)$ are scaled with the wave number of the exterior domain, k, in order to get a dimensionless quantity.

The knowledge of the transition matrix solves the scattering problem, and all relevant scattering quantities can be computed, see Example 7.3 on page 386.

Example 8.15

As a simple example of the results above, let the material in the sphere be perfectly conducting. This corresponds to $g(a) = h'(a) = 0$ and therefore from (8.27)

$$\begin{cases} t_{1l} = -\dfrac{\psi_l(ka)}{\xi_l(ka)} \\ t_{2l} = -\dfrac{\psi_l'(ka)}{\xi_l'(ka)} \end{cases}$$

These results agree with the result in Section 8.1. ■

Example 8.16

We can check the result in (8.27) with the result in Section 8.2. If the material in the sphere is homogeneous and isotropic, $\epsilon_\perp(r) = \epsilon_r(r) = \epsilon_1$, $\mu_\perp(r) = \mu_r(r) = \mu_1$. Then, letting $k_1^2 = k_0^2\epsilon_1\mu_1$ and using

$$g_l(r) = h_l(r) = \psi_l(k_1 r) = k_1 r j_l(k_1 r)$$

the transition matrix entries in (8.27) simplify to

$$t_{\tau l} = -\frac{\psi_l(ka)\psi_l'(k_1 a) - m_1\psi_l'(ka)\psi_l(k_1 a)}{\xi_l(ka)\psi_l'(k_1 a) - m_1\xi_l'(ka)\psi_l(k_1 a)}$$

where

$$m_\tau = \left(\frac{\epsilon\mu}{\epsilon_1\mu_1}\right)^{1/2} \gamma_\tau = \delta_{\tau 1}\frac{\eta_1}{\eta} + \delta_{\tau 2}\frac{\eta}{\eta_1}$$

This is in agreement with the result obtained in Section 8.2. ■

Example 8.17

If the material in the sphere is isotropic and homogeneous in two intervals, see Example 8.14, i.e.,

$$\epsilon(r) = \begin{cases} \epsilon_1, & r \in [0, r_0] \\ \epsilon_2, & r \in [r_0, a] \end{cases} \qquad \mu(r) = \begin{cases} \mu_1, & r \in [0, r_0] \\ \mu_2, & r \in [r_0, a] \end{cases}$$

Then with the notation $k_n^2 = k_0^2\epsilon_n\mu_n$, $n = 1, 2$, the relevant solutions are

$$\begin{cases} g(r) = A_{1l}\psi_l(k_1 r), & r \in [0, r_0] \\ g(r) = B_{1l}\left(\psi_l(k_2 r) + t_{1l}(r_0)\xi_l(k_2 r)\right), & r \in [r_0, a] \end{cases}$$

and

$$\begin{cases} h(r) = A_{2l}\psi_l(k_1 r), & r \in [0, r_0] \\ h(r) = B_{2l}\left(\psi_l(k_2 r) + t_{2l}(r_0)\xi_l(k_2 r)\right), & r \in [r_0, a] \end{cases}$$

where $t_{\tau l}(r_0)$ constants are, see Example 8.14

$$t_{\tau l}(r_0) = -\frac{\psi_l(k_2 r_0)\psi_l'(k_1 r_0) - m_\tau^{(1)}\psi_l'(k_2 r_0)\psi_l(k_1 r_0)}{\xi_l(k_2 r_0)\psi_l'(k_1 r_0) - m_\tau^{(1)}\xi_l'(k_2 r_0)\psi_l(k_1 r_0)}$$

where

$$m_\tau^{(1)} = \delta_{\tau 1}\frac{\eta_1}{\eta_2} + \delta_{\tau 2}\frac{\eta_2}{\eta_1}$$

and where wave impedance of the layers is $\eta_i = (\mu_i/\epsilon_i)^{1/2}$, $i = 1, 2$.

The total transition matrix is given by (8.27). Insert the values of the functions $g(a)$ and $h(a)$ and their derivatives, which implies

$$t_{\tau l} = -\frac{\psi_l(ka)\left(\psi_l'(k_2 a) + t_{\tau l}(r_0)\xi_l'(k_2 a)\right) - m_\tau^{(2)}\psi_l'(ka)\left(\psi_l(k_2 a) + t_{\tau l}(r_0)\xi_l(k_2 a)\right)}{\xi_l(ka)\left(\psi_l'(k_2 a) + t_{\tau l}(r_0)\xi_l'(k_2 a)\right) - m_\tau^{(2)}\xi_l'(ka)\left(\psi_l(k_2 a) + t_{\tau l}(r_0)\xi_l(k_2 a)\right)}$$

where

$$m_\tau^{(2)} = \delta_{\tau 1}\frac{\eta_2}{\eta} + \delta_{\tau 2}\frac{\eta}{\eta_2}$$

which is in agreement with the results obtained for a layered structure in Section 8.3.

As a special case, let $\epsilon_2 = \epsilon$ and $\mu_2 = \mu$, i.e., same as the surrounding material in the interval $[r_0, a]$. The transition matrix entries then simplify to

$$t_{\tau l} = t_{\tau l}(r_0), \qquad \tau = 1, 2$$

We see that the transition matrix is independent of the outer radius, $r = a$, provided the material outside the radius, $r = r_0$, is identical to the surrounding material. This conclusion is, of course, expected, but our analysis also shows that this conclusion is correct. ■

8.4.3 Non-Uniqueness of the Scattering Problem

The transition matrix contains all the information about the scattering properties of the object. Consequently, all objects that have identical transition matrix entries scatter the same, and hence, they cannot be distinguished by any scattering experiment. The existence of two different scatterers with the same transition matrix is closely related to the uniqueness of the inverse scattering problem, see e.g., Reference 83 and references therein. For an isotropic material, the inverse scattering problem is uniquely solvable, and therefore, two different isotropic scatterers with identical scattering properties cannot exist. For anisotropic materials, the situation is different, and in this section, we analyze this problem in some detail.

The transition matrix entries in (8.27) or the matching relations (8.25) and (8.26) show that two anisotropic spheres, both of radius a, have the same transition matrix if the quotients of the functions (remember the functions $g(r)$ and $h(r)$ depend on the index l, which we suppress in this section)

$$\frac{g'(a)}{\mu_\perp(a)g(a)} \quad \text{and} \quad \frac{h'(a)}{\epsilon_\perp(a)h(a)} \tag{8.28}$$

are the same. Under these circumstances, the two spheres are indistinguishable in a scattering experiment. In particular, a transformation of the radial coordinate, r, inside the sphere, that keeps the quotients above the same, produces such a non-uniqueness, and we investigate this case in some detail below. The goal is to create an anisotropic sphere of radius a that has the same transition matrix as an isotropic, homogeneous sphere with the same radius.

To this end, let $r = f(s)$ be a piecewise, continuously differentiable function in the interval $s \in [0, a)$, satisfying $f(0) = 0$ and $f(a) = a$. Outside the sphere, $s \geq a, f(s) = r$. These requirements give a continuous function with a possible jump discontinuity of $f'(s)$ at $r = s = a$. Examples of such functions are given below.

If the material of the sphere is homogeneous and isotropic (relative permittivity ϵ_1 and permeability μ_1), the differential equations (8.22) simplify to

$$\begin{cases} g''(r) + \left(k_1^2 - \dfrac{l(l+1)}{r^2}\right) g(r) = 0 \\ h''(r) + \left(k_1^2 - \dfrac{l(l+1)}{r^2}\right) h(r) = 0 \end{cases} \qquad r \in [0, a]$$

where $k_1 = k_0 \left(\epsilon_1 \mu_1\right)^{1/2}$. The regular solutions to these equations are[13]

$$g(r) = h(r) = \psi_l(k_1 r) = k_1 r j_l(k_1 r), \qquad r \in [0, a]$$

The change of the radial variables, $r = f(s)$ and $v(s) = g(f(s))$, produces a new set of differential equations, see Comment 8.2

$$\frac{\mathrm{d}^2 v(s)}{\mathrm{d}s^2} - \frac{f''(s)}{f'(s)} \frac{\mathrm{d}v(s)}{\mathrm{d}s} + \left(k_1^2 - \frac{l(l+1)}{(f(s))^2}\right) \left(f'(s)\right)^2 v(s) = 0, \qquad s \in [0, a] \tag{8.29}$$

We refer to this equation as a magnetic-type equation. Similarly, the other, electric-type equation, is obtained for $h(r)$ by a change of μ and ϵ below.

Comment 8.2

The change of independent variable in an ODE is a standard procedure and found in many textbooks on the subject, see e.g., Reference 138.

Let $u(z)$ be a solution of

$$\frac{\mathrm{d}^2 u(r)}{\mathrm{d}r^2} + p(r)\frac{\mathrm{d}u(r)}{\mathrm{d}r} + q(r)u(r) = 0$$

If $r = f(s)$, then $v(s) = u(f(s))$ satisfies [138, Theorem 2.1]

$$\frac{\mathrm{d}^2 v(s)}{\mathrm{d}s^2} + \underbrace{\left(p(f(s))f'(s) - \frac{f''(s)}{f'(s)}\right)}_{\overline{p}(s)} \frac{\mathrm{d}v(s)}{\mathrm{d}s} + \underbrace{q(f(s)) \left(f'(s)\right)^2}_{\overline{q}(s)} v(s) = 0$$

where $\overline{p}(s)$ and $\overline{q}(s)$ are the new functions in the differential equation.

---

[13]Again, normalization is irrelevant, see Footnote 12.

This transformation is a passive change of the radial coordinate, but it can also be viewed as an active one and a transformation of the material parameters in the original radial coordinates. This new material produces the same transition matrix, provided the side conditions in (8.28) are satisfied. At the boundary, $r = a$, we therefore impose the side conditions

$$\frac{\mathrm{d}g(a)}{\mathrm{d}r}\frac{1}{\mu_1 g(a)} = \frac{\mathrm{d}v(a)}{\mathrm{d}s}\frac{1}{\mu_\perp(a)v(a)} = \frac{\mathrm{d}g(a)}{\mathrm{d}r}\frac{\mathrm{d}f(a)}{\mathrm{d}s}\frac{1}{\mu_\perp(a)g(a)}$$

and similarly for the radial function $h(r)$. Notice the different differentiations w.r.t. the radial variables r and s, and that $g(a) = v(a)$ by construction. These relations determine the value of the derivative of $f(s)$ at $s = r = a$, which is necessary for the construction of two objects with identical scattering properties. As a consequence, it relates the values of the new material parameters $\epsilon_\perp(a)$ and $\mu_\perp(a)$ at the boundary to the original ones, i.e.,

$$f'(a) = \frac{\epsilon_\perp(a)}{\epsilon_1} = \frac{\mu_\perp(a)}{\mu_1} \tag{8.30}$$

From the differential equation in $v(s)$, (8.29), we identify the new material parameters by comparing with the coefficients in (8.22). The new material parameters $\overline{p}(s)$ and $\overline{q}(s)$ in the magnetic-type equation have to satisfy the following conditions:

$$\begin{cases} \overline{p}(s) = -\dfrac{\mu'_\perp(s)}{\mu_\perp(s)} = -\dfrac{f''(s)}{f'(s)} \\ \overline{q}(s) = k_0^2\epsilon_\perp(s)\mu_\perp(s) - \dfrac{l(l+1)\mu_\perp(s)}{s^2\mu_r(s)} = \left(k_1^2 - \dfrac{l(l+1)}{(f(s))^2}\right)(f'(s))^2 \end{cases}$$

and the conditions for the electric-type equation are

$$\begin{cases} \overline{p}(s) = -\dfrac{\epsilon'_\perp(s)}{\epsilon_\perp(s)} = -\dfrac{f''(s)}{f'(s)} \\ \overline{q}(s) = k_0^2\epsilon_\perp(s)\mu_\perp(s) - \dfrac{l(l+1)\epsilon_\perp(s)}{s^2\epsilon_r(s)} = \left(k_1^2 - \dfrac{l(l+1)}{(f(s))^2}\right)(f'(s))^2 \end{cases}$$

Integration and the employment of (8.30) give the solutions in the interior of the sphere ($s \in [0, a]$)

$$\begin{cases} \epsilon_\perp(s) = \epsilon_1 f'(s) \\ \epsilon_r(s) = \dfrac{\epsilon_1 (f(s))^2}{f'(s)s^2} \end{cases} \qquad \begin{cases} \mu_\perp(s) = \mu_1 f'(s) \\ \mu_r(s) = \dfrac{\mu_1 (f(s))^2}{f'(s)s^2} \end{cases} \tag{8.31}$$

The solutions outside the spherical region, where $f(s) = r$, are, as before

$$\begin{cases} \epsilon_\perp(s) = \epsilon_r(s) = \epsilon \\ \mu_\perp(s) = \mu_r(s) = \mu \end{cases}$$

These conditions determine the new material parameters inside the sphere. Notice that the permittivity and the permeability show anisotropy and they are in general inhomogeneous, i.e., they depend on the radial distance s.

We make one final remark before we illustrate the results above in a series of examples. If the homogeneous sphere has the same permittivity and permeability as the exterior medium, i.e., $\epsilon_1 = \epsilon$ and $\mu_1 = \mu$, the sphere does not scatter at all. The corresponding anisotropic sphere with material parameters in (8.31) also gives no scattering at any incident field and at any frequency—it is invisible. We have obtained a material that is related to the topic of a perfect cloak. We refer the reader to the vast literature on this subject [84].

Example 8.18

Let $r = f(s) = s^2/a$, $0 \le s \le a$ and $f(s) = r$ for $s \ge a$. The constraint, see (8.30), implies

$$f'(a) = 2 = \frac{\epsilon_\perp(a)}{\epsilon_1} = \frac{\mu_\perp(a)}{\mu_1}$$

A homogeneous, isotropic sphere with material parameters ϵ_1 and μ_1 has all transition matrix entries identical to a radially inhomogeneous, radially uniaxial sphere with material parameters, see (8.31), given by

$$\begin{cases} \epsilon_\perp(s) = \dfrac{2\epsilon_1 s}{a} \\ \epsilon_r(s) = \dfrac{\epsilon_1 s}{2a} \end{cases} \qquad \begin{cases} \mu_\perp(s) = \dfrac{2\mu_1 s}{a} \\ \mu_r(s) = \dfrac{\mu_1 s}{2a} \end{cases}$$

The permittivity parameters $\epsilon_\perp(s)$ and $\epsilon_r(s)$ are depicted in Figure 8.28. Notice that if $\epsilon_1 = \epsilon$ and $\mu_1 = \mu$, the anisotropic sphere is invisible.

The typical behavior of the material parameters is that at least one of the parameters is less than 1 somewhere in the interval $[0, a]$, i.e., its value is less than the surrounding media. If the exterior material is vacuum, this means that the phase velocity of the material is larger than the speed of light in vacuum. Such a (passive) material is hard to construct. For additional details on these materials, see Reference 87. ■

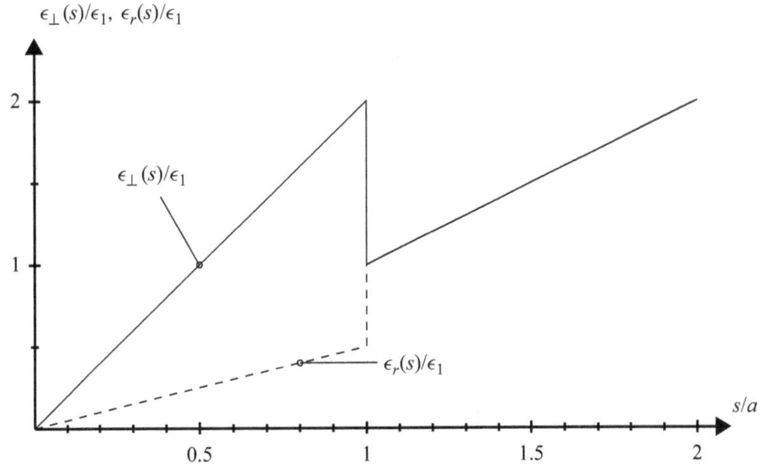

Figure 8.28 The anisotropic permittivity parameters $\epsilon_\perp(s)/\epsilon_1$ and $\epsilon_r(s)/\epsilon_1$ in Example 8.18.

Comment 8.3
As an independent check of the solution in Example 8.18, it is easy to verify that a regular solution to (this differential equation is (8.22) with the relevant permeability profile inserted)

$$g''(r) - \frac{1}{r}g'(r) + \left(\frac{4k_1^2 r^2}{a^2} - \frac{4l(l+1)}{r^2}\right) g(r) = 0$$

is

$$g(r) = \psi_l\left(k_1 r^2/a\right)$$

where $\psi_l(x)$ is the Riccati–Bessel function and $k_1 = k_0\left(\epsilon_1\mu_1\right)^{1/2}$. The value of the solution and its derivative at $r = a$ are

$$g(a) = \psi_l(k_1 a), \qquad g'(a) = \frac{2k_1 a}{a}\psi_l'(k_1 a)$$

respectively. The non-vanishing elements of the T-matrix for $\tau = 1$ therefore are, see (8.27)

$$\begin{aligned} t_{1l} &= -\frac{\mu\psi_l(ka)g_l'(a) - k\mu_\perp(a)\psi_l'(ka)g_l(a)}{\mu\xi_l(ka)g_l'(a) - k\mu_\perp(a)\xi_l'(ka)g_l(a)} \\ &= -\frac{2k_1 a\mu\psi_l(ka)\psi'(k_1 a) - 2ka\mu_1\psi_l'(ka)\psi_l(k_1 a)}{2k_1 a\mu\xi_l(ka)\psi_l'(k_1 a) - 2ka\mu_1\xi_l'(ka)\psi_l(k_1 a)} \end{aligned}$$

which is identical to the T-matrix elements, $\tau = 1$, of a homogeneous sphere. Similar results hold for the function $h(r)$ and the transition matrix entries for $\tau = 2$. The two spheres then have the same transition matrices, and therefore, they have identical scattering properties.

Example 8.19
In this example, we let $f(s)$ be a piecewise linear map in $[0, a]$, i.e., $(0 \le r_0 \le a, \alpha > 0)$

$$f(s) = \begin{cases} \alpha s/r_0 & s \in [0, r_0] \\ \{a(\alpha - r_0) + s(a - \alpha)\}/(a - r_0) & s \in [r_0, a] \\ s & s \in [a, \infty) \end{cases}$$

This map is continuous, and we augment it with a constraint in its derivative at $r = s = a$, see (8.30)

$$f'(a) = \frac{a - \alpha}{a - r_0} = \frac{\epsilon_\perp(a)}{\epsilon_1} = \frac{\mu_\perp(a)}{\mu_1}$$

This choice of function $f(s)$ implies that

$$\begin{cases} \epsilon_\perp(a) = \epsilon_1 \dfrac{a - \alpha}{a - r_0} \\ \mu_\perp(a) = \mu_1 \dfrac{a - \alpha}{a - r_0} \end{cases}$$

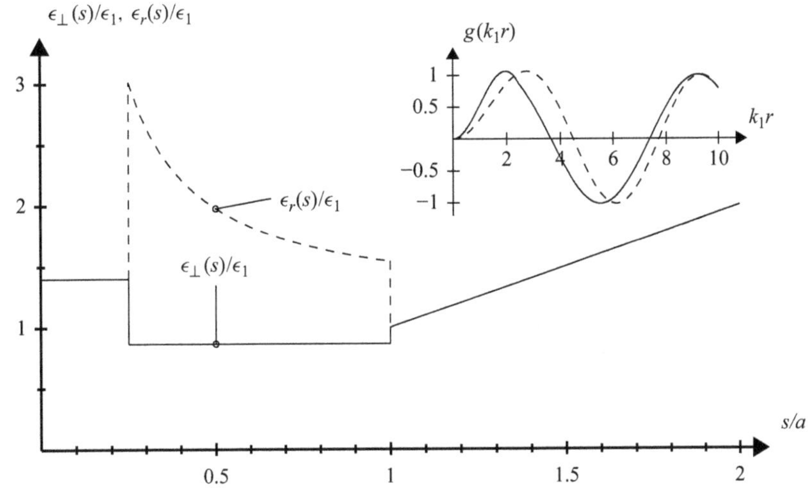

Figure 8.29 The anisotropic permittivity parameters $\epsilon_\perp(s)/\epsilon_1$ and $\epsilon_r(s)/\epsilon_1$ in Example 8.19. In this particular example, $r_0/a = 0.25$ and $\alpha/a = 0.35$. The inset shows the radial function $g(k_1 r)$ (solid curve) as function of $k_1 r$ for $l = 1$ and $k_1 a = 10$. The dashed curve shows the radial function for the corresponding homogeneous sphere.

A homogeneous, isotropic sphere with material parameters ϵ_1 and μ_1 has all transition matrix entries identical to a radially inhomogeneous, anisotropic sphere with material parameters, see (8.31), given by

$$\begin{cases} \epsilon_\perp(s) = \epsilon_1 \dfrac{\alpha}{r_0} \\ \epsilon_r(s) = \epsilon_1 \dfrac{\alpha}{r_0} \end{cases} \qquad s \in [0, r_0]$$

and

$$\begin{cases} \epsilon_\perp(s) = \epsilon_1 \dfrac{a - \alpha}{a - r_0} \\ \epsilon_r(s) = \epsilon_1 \dfrac{(a(\alpha - r_0) + s(a - \alpha))^2}{(a - \alpha)(a - r_0)s^2} \end{cases} \qquad s \in [r_0, a]$$

and similarly for the permeability parameters (replace ϵ with μ). The permittivity parameters $\epsilon_\perp(s)$ and $\epsilon_r(s)$ for a specific choice of α and r_0 are depicted in Figure 8.29. Again, if $\epsilon_1 = \epsilon$ and $\mu_1 = \mu$, the anisotropic sphere is invisible. Notice also that at least one of quotients of the material parameters is less than 1 somewhere in the interval $[0, a]$, i.e., its value is less than the surrounding media. ■

These examples can be viewed as special cases of a much more general non-uniqueness class. To investigate this, denote $\boldsymbol{r}' = s\hat{\boldsymbol{r}} = \boldsymbol{F}(\boldsymbol{r}) = g(r)\hat{\boldsymbol{r}}$, where the function $s = g(r) = f^{-1}(r)$—the inverse of the function $f(s)$ that was specified in the beginning of the section.

The Jacobian[14] of $\boldsymbol{F}(\boldsymbol{r})$ then is[15]

$$\mathbf{J}_F(\boldsymbol{r}) = \frac{g(r)}{r}\left(\mathbf{I}_3 - \hat{\boldsymbol{r}}\hat{\boldsymbol{r}}\right) + g'(r)\hat{\boldsymbol{r}}\hat{\boldsymbol{r}}$$

and the Jacobian determinant of the transformation is

$$DF(\boldsymbol{r}) = g'(r)\left(\frac{g(r)}{r}\right)^2$$

Notice that tangential part of $\mathbf{J}_F(\boldsymbol{r})$ acts as the identity on a tangential field at $r = a$.

The transformed permittivity then is [84]

$$\tilde{\boldsymbol{\epsilon}}(\boldsymbol{r}') = \left.\frac{1}{DF(\boldsymbol{r})}\mathbf{J}_F(\boldsymbol{r}) \cdot \boldsymbol{\epsilon}(\boldsymbol{r}) \cdot \mathbf{J}_F^{\mathrm{t}}(\boldsymbol{r})\right|_{\boldsymbol{r}=\boldsymbol{F}^{-1}(\boldsymbol{r}')}$$

and similarly for the transformed permeability dyadic $\tilde{\boldsymbol{\mu}}(\boldsymbol{r}')$. The two scatterers consisting of materials $\{\boldsymbol{\epsilon}(\boldsymbol{r}), \boldsymbol{\mu}(\boldsymbol{r})\}$ and $\{\tilde{\boldsymbol{\epsilon}}(\boldsymbol{r}'), \tilde{\boldsymbol{\mu}}(\boldsymbol{r}')\}$ produce the same boundary measurements, provided $\boldsymbol{F}(\boldsymbol{r}) = \boldsymbol{r}$ and the tangential part of $\mathbf{J}_F(\boldsymbol{r})$ acts as the identity at the boundary. In our case, for $\boldsymbol{\epsilon}(\boldsymbol{r}) = \epsilon_1 \mathbf{I}_3$, we have

$$\tilde{\boldsymbol{\epsilon}}(s) = \left.\frac{\epsilon_1}{g'(r)\left(\frac{g(r)}{r}\right)^2}\left(\left(\frac{g(r)}{r}\right)^2\left(\mathbf{I}_3 - \hat{\boldsymbol{r}}\hat{\boldsymbol{r}}\right) + \left(g'(r)\right)^2 \hat{\boldsymbol{r}}\hat{\boldsymbol{r}}\right)\right|_{r=f(s)}$$

which implies

$$\epsilon_r(s) = \left.\frac{\epsilon_1 g'(r)}{\left(\frac{g(r)}{r}\right)^2}\right|_{r=f(s)} = \frac{\epsilon_1 \left(f(s)\right)^2}{f'(s)s^2} \qquad \epsilon_\perp(s) = \left.\frac{\epsilon_1}{g'(r)}\right|_{r=f(s)} = \epsilon_1 f'(s)$$

and similarly for the permeability. Notice that this, more general transformation, is identical to the results above, see (8.31).

[14]Carl Gustav Jacob Jacobi (1804–1851), German mathematician.

[15]The Jacobian of the function $\boldsymbol{F}(\boldsymbol{r})$, denoted $\mathbf{J}_F(\boldsymbol{r})$, and the Jacobian determinant, denoted $DF(\boldsymbol{r})$, in three dimensions are

$$\mathbf{J}_F(\boldsymbol{r}) = \begin{pmatrix} \frac{\partial F_1(\boldsymbol{r})}{\partial x} & \frac{\partial F_1(\boldsymbol{r})}{\partial y} & \frac{\partial F_1(\boldsymbol{r})}{\partial z} \\ \frac{\partial F_2(\boldsymbol{r})}{\partial x} & \frac{\partial F_2(\boldsymbol{r})}{\partial y} & \frac{\partial F_2(\boldsymbol{r})}{\partial z} \\ \frac{\partial F_3(\boldsymbol{r})}{\partial x} & \frac{\partial F_3(\boldsymbol{r})}{\partial y} & \frac{\partial F_3(\boldsymbol{r})}{\partial z} \end{pmatrix}, \qquad DF(\boldsymbol{r}) = \begin{vmatrix} \frac{\partial F_1(\boldsymbol{r})}{\partial x} & \frac{\partial F_1(\boldsymbol{r})}{\partial y} & \frac{\partial F_1(\boldsymbol{r})}{\partial z} \\ \frac{\partial F_2(\boldsymbol{r})}{\partial x} & \frac{\partial F_2(\boldsymbol{r})}{\partial y} & \frac{\partial F_2(\boldsymbol{r})}{\partial z} \\ \frac{\partial F_3(\boldsymbol{r})}{\partial x} & \frac{\partial F_3(\boldsymbol{r})}{\partial y} & \frac{\partial F_3(\boldsymbol{r})}{\partial z} \end{vmatrix}$$

8.5 Scattering by a Biisotropic Sphere

We are now well prepared to address more complex scattering problems such as scattering by biisotropic materials. These scattering problems show many new and interesting features, e.g., polarization effects. In this section, we analyze this problem in some detail. We follow the ideas presented by Bohren [17]. The scattering problem has also been addressed in another setting using the Fedorov relations in Comment 1.3 on page 47, see References 7, 14, 151, 301 and references therein.

The field equations for a homogeneous biisotropic material were developed and analyzed in Section 2.4, see (2.33) on page 119

$$\begin{pmatrix} \nabla \times \boldsymbol{E} \\ \eta_0 \nabla \times \boldsymbol{H} \end{pmatrix} = \mathrm{i}k_0 \begin{pmatrix} \kappa - \mathrm{i}\chi & \mu_1 \\ -\epsilon_1 & -\kappa - \mathrm{i}\chi \end{pmatrix} \begin{pmatrix} \boldsymbol{E} \\ \eta_0 \boldsymbol{H} \end{pmatrix} \tag{8.32}$$

where we assume the medium is source free, and as usual, we denote the wave number in free space by $k_0 = \omega/c_0$. For convenience, we express the equations in the reciprocity and the chirality parameters, κ and χ, respectively.

8.5.1 Spherical Vector Waves in a Biisotropic Material

The spherical vector waves that were investigated in Sections 7.1 and 7.2 are actually useful for an expansion of the electromagnetic field in a biisotropic material. We have already encountered a similar problem in Section 3.8 on page 174, where the integral representation of the fields in a biisotropic material was developed. To accomplish the integral representation, we employed the transformation, due to Craig Bohren [17], see (3.27) on page 175

$$\begin{pmatrix} \boldsymbol{E} \\ \eta_0 \boldsymbol{H} \end{pmatrix} = \begin{pmatrix} -\mathrm{i}\mu_1 & -\mathrm{i}\mu_1 \\ \mathrm{i}\kappa + \left(\epsilon_1\mu_1 - \kappa^2\right)^{1/2} & \mathrm{i}\kappa - \left(\epsilon_1\mu_1 - \kappa^2\right)^{1/2} \end{pmatrix} \begin{pmatrix} \boldsymbol{Q}_{\mathrm{L}} \\ \boldsymbol{Q}_{\mathrm{R}} \end{pmatrix}$$

The new fields $\boldsymbol{Q}_{\mathrm{L,R}}$ satisfy (3.29) on page 175

$$\begin{pmatrix} \nabla \times \boldsymbol{Q}_{\mathrm{L}} \\ \nabla \times \boldsymbol{Q}_{\mathrm{R}} \end{pmatrix} = \begin{pmatrix} -k_- & 0 \\ 0 & k_+ \end{pmatrix} \begin{pmatrix} \boldsymbol{Q}_{\mathrm{L}} \\ \boldsymbol{Q}_{\mathrm{R}} \end{pmatrix}$$

where the wave numbers $k_\pm$ are

$$k_\pm = k_0 \left(\left(\epsilon_1\mu_1 - \kappa^2\right)^{1/2} \pm \chi \right)$$

This shows that we have constructed two new fields, $\boldsymbol{Q}_{\mathrm{L}}$ and $\boldsymbol{Q}_{\mathrm{R}}$, such that the differential equations that these fields satisfy do not couple. This result inspires us to proceed in a similar manner in the construction of spherical vector waves adopted to biisotropic media.

Construct a set of two new spherical vector waves from the definition of the regular spherical vector waves, $\boldsymbol{v}_{\tau n}(k\boldsymbol{r})$, in Section 7.2, viz.

$$\begin{cases} \boldsymbol{v}_{\mathrm{L}n}(\boldsymbol{r}) = \boldsymbol{v}_{1n}(k_-\boldsymbol{r}) - \boldsymbol{v}_{2n}(k_-\boldsymbol{r}) \\ \boldsymbol{v}_{\mathrm{R}n}(\boldsymbol{r}) = \boldsymbol{v}_{1n}(k_+\boldsymbol{r}) + \boldsymbol{v}_{2n}(k_+\boldsymbol{r}) \end{cases} \tag{8.33}$$

Note that the wave numbers in the spherical vector waves on the right-hand side are different. This new set of regular spherical vector waves inherits the properties of the original vector waves, e.g. the completeness.

The regular spherical vector waves $\boldsymbol{v}_{\tau n}$ in Section 7.2 for $\tau = 1, 2$ satisfy, see (7.7) on page 356

$$\begin{cases} \nabla \times \boldsymbol{v}_{1n}(k\boldsymbol{r}) = k\boldsymbol{v}_{2n}(k\boldsymbol{r}) \\ \nabla \times \boldsymbol{v}_{2n}(k\boldsymbol{r}) = k\boldsymbol{v}_{1n}(k\boldsymbol{r}) \end{cases}$$

which implies

$$\begin{pmatrix} \nabla \times \boldsymbol{v}_{\mathrm{L}n}(\boldsymbol{r}) \\ \nabla \times \boldsymbol{v}_{\mathrm{R}n}(\boldsymbol{r}) \end{pmatrix} = \begin{pmatrix} -k_- \left(\boldsymbol{v}_{1n}(k_-\boldsymbol{r}) - \boldsymbol{v}_{2n}(k_-\boldsymbol{r})\right) \\ k_+ \left(\boldsymbol{v}_{1n}(k_+\boldsymbol{r}) + \boldsymbol{v}_{2n}(k_+\boldsymbol{r})\right) \end{pmatrix} = \begin{pmatrix} -k_- & 0 \\ 0 & k_+ \end{pmatrix} \begin{pmatrix} \boldsymbol{v}_{\mathrm{L}n}(\boldsymbol{r}) \\ \boldsymbol{v}_{\mathrm{R}n}(\boldsymbol{r}) \end{pmatrix}$$

If the electric field is expanded in this set, i.e.,

$$\boldsymbol{E}(\boldsymbol{r}, \omega) = \sum_n \alpha_n \boldsymbol{v}_{\mathrm{L}n}(\boldsymbol{r}) + \beta_n \boldsymbol{v}_{\mathrm{R}n}(\boldsymbol{r}) \tag{8.34}$$

then, by the Faraday law, the corresponding magnetic field is (first row in (8.32))

$$\begin{aligned} \mathrm{i}k_0\mu_1\eta_0\boldsymbol{H}(\boldsymbol{r}, \omega) &= \nabla \times \boldsymbol{E}(\boldsymbol{r}, \omega) - \mathrm{i}k_0(\kappa - \mathrm{i}\chi)\boldsymbol{E}(\boldsymbol{r}, \omega) \\ &= k_0 \sum_n \Big\{\alpha_n \left(-\mathrm{i}\kappa - \left(\epsilon_1\mu_1 - \kappa^2\right)^{1/2}\right) \boldsymbol{v}_{\mathrm{L}n}(\boldsymbol{r}) \\ &\quad + \beta_n \left(-\mathrm{i}\kappa + \left(\epsilon_1\mu_1 - \kappa^2\right)^{1/2}\right) \boldsymbol{v}_{\mathrm{R}n}(\boldsymbol{r})\Big\} \end{aligned}$$

or

$$\mathrm{i}k_0\eta_0\boldsymbol{H}(\boldsymbol{r}, \omega) = k_0 \sum_n \left\{\alpha_n Y_- \boldsymbol{v}_{\mathrm{L}n}(\boldsymbol{r}) + \beta_n Y_+ \boldsymbol{v}_{\mathrm{R}n}(\boldsymbol{r})\right\} \tag{8.35}$$

where the material admittances $Y_\pm$ of the biisotropic material are denoted

$$Y_\pm = \frac{\pm\left(\epsilon_1\mu_1 - \kappa^2\right)^{1/2} - \mathrm{i}\kappa}{\mu_1} \tag{8.36}$$

8.5.2 The Transition Matrix for a Biisotropic Sphere

Our aim in this section is to compute the transition matrix of a homogeneous biisotropic sphere. This matrix then determines all scattering properties of the biisotropic sphere.

The sphere is modeled by a (complex) permittivity $\epsilon_1(\omega)$, permeability $\mu_1(\omega)$, chirality $\chi(\omega)$, reciprocity parameter $\kappa(\omega)$, and the radius of the sphere is a. We adopt a coordinate system where the center of the sphere coincides with the origin of the coordinate system. The surrounding (background) material is, as above, modeled by the real-valued permittivity ϵ and permeability μ. The incident field is, as above, a plane wave impinging along the positive z-axis. The complex vector $\boldsymbol{E}_0$ lies in the x–y-plane. Its expansion in spherical vector waves is

$$\boldsymbol{E}_{\mathrm{i}}(\boldsymbol{r}, \omega) = \boldsymbol{E}_0 \mathrm{e}^{\mathrm{i}kz} = \sum_n a_n \boldsymbol{v}_n(k\boldsymbol{r})$$

where the expansion coefficients a_n are given in (7.29) on page 375.

The multipole expansion of the scattered field outside the scatterer is as above

$$E_s(r,\omega) = \sum_n f_n u_n(kr)$$

and the total electric and magnetic fields outside the sphere are

$$\begin{cases} E(r,\omega) = \sum_n (a_n v_n(kr) + f_n u_n(kr)) \\ H(r,\omega) = \dfrac{1}{i\eta_0\eta} \sum_n (a_n v_{\overline{n}}(kr) + f_n u_{\overline{n}}(kr)) \end{cases} \quad r > a$$

where $\overline{n}$ is the dual index of n.

We also need, to match the boundary conditions on the surface of the sphere, the total electric and magnetic fields inside the biisotropic sphere. We expand these fields in regular spherical vector waves, see (8.33), since these fields are well behaved everywhere inside the sphere, see (8.34) and (8.35). We adopt

$$\begin{cases} E_1(r,\omega) = \sum_n \alpha_n v_{Ln}(r) + \beta_n v_{Rn}(r) \\ H_1(r,\omega) = \dfrac{1}{i\eta_0} \sum_n \{\alpha_n Y_- v_{Ln}(r) + \beta_n Y_+ v_{Rn}(r)\} \end{cases} \quad r < a$$

where the admittances $Y_\pm$ are given in (8.36). Note that we are using the wave number of the material inside the sphere, $k_\pm$, which now are two in contrast to the dielectric sphere, where we only have one wave number. Note also that the sum n only runs over the indices l, m, and σ (no τ summation, since the two modes are explicitly indexed by left and right symbols).

The boundary conditions on the spherical surface, $r = a$ are, see (1.12) on page 8 (no surface currents J_S are assumed present)

$$\begin{cases} \hat{r} \times E_1(r,\omega)|_{r=a} = \hat{r} \times E(r,\omega)|_{r=a} \\ \hat{r} \times H_1(r,\omega)|_{r=a} = \hat{r} \times H(r,\omega)|_{r=a} \end{cases}$$

These conditions imply by orthogonality of the vector spherical harmonics

$$\begin{cases} \alpha_n V_{1l}(k_- a) + \beta_n V_{1l}(k_+ a) = a_{1n} V_{1l}(ka) + f_{1n} U_{1l}(ka) \\ -\alpha_n V_{2l}(k_- a) + \beta_n V_{2l}(k_+ a) = a_{2n} V_{2l}(ka) + f_{2n} U_{2l}(ka) \end{cases}$$

and

$$\begin{cases} \alpha_n Y_- V_{1l}(k_- a) + \beta_n Y_+ V_{1l}(k_+ a) = \dfrac{1}{\eta} (a_{2n} V_{1l}(ka) + f_{2n} U_{1l}(ka)) \\ -\alpha_n Y_- V_{2l}(k_- a) + \beta_n Y_+ V_{2l}(k_+ a) = \dfrac{1}{\eta} (a_{1n} V_{2l}(ka) + f_{1n} U_{2l}(ka)) \end{cases}$$

where we for convenience have used the radial functions $V_{\tau l}(kr)$ and $U_{\tau l}(kr)$ defined in Comment 7.2 on page 354.

We write these four conditions as a system of equations, i.e.,

$$\begin{pmatrix} -U_{1l}(ka) & 0 & V_{1l}(k_-a) & V_{1l}(k_+a) \\ 0 & -U_{2l}(ka) & -V_{2l}(k_-a) & V_{2l}(k_+a) \\ 0 & -U_{1l}(ka) & \eta Y_- V_{1l}(k_-a) & \eta Y_+ V_{1l}(k_+a) \\ -U_{2l}(ka) & 0 & -\eta Y_- V_{2l}(k_-a) & \eta Y_+ V_{2l}(k_+a) \end{pmatrix} \begin{pmatrix} f_{1n} \\ f_{2n} \\ \alpha_n \\ \beta_n \end{pmatrix}$$

$$= \begin{pmatrix} a_{1n} V_{1l}(ka) \\ a_{2n} V_{2l}(ka) \\ a_{2n} V_{1l}(ka) \\ a_{1n} V_{2l}(ka) \end{pmatrix}$$

We solve for the unknown f_{1n} and f_{2n} in terms of the known a_{1n} and a_{2n}. To simplify the analysis, write the system of equations as

$$A \begin{pmatrix} f_{1n} \\ f_{2n} \end{pmatrix} + B \begin{pmatrix} \alpha_n \\ \beta_n \end{pmatrix} = X \begin{pmatrix} a_{1n} \\ a_{2n} \end{pmatrix}$$

$$C \begin{pmatrix} f_{1n} \\ f_{2n} \end{pmatrix} + D \begin{pmatrix} \alpha_n \\ \beta_n \end{pmatrix} = Y \begin{pmatrix} a_{1n} \\ a_{2n} \end{pmatrix}$$

where

$$A = \begin{pmatrix} -U_{1l}(ka) & 0 \\ 0 & -U_{2l}(ka) \end{pmatrix}, \qquad B = \begin{pmatrix} V_{1l}(k_-a) & V_{1l}(k_+a) \\ -V_{2l}(k_-a) & V_{2l}(k_+a) \end{pmatrix}$$

$$C = \begin{pmatrix} 0 & -U_{1l}(ka) \\ -U_{2l}(ka) & 0 \end{pmatrix}, \qquad D = \begin{pmatrix} \eta Y_- V_{1l}(k_-a) & \eta Y_+ V_{1l}(k_+a) \\ -\eta Y_- V_{2l}(k_-a) & \eta Y_+ V_{2l}(k_+a) \end{pmatrix}$$

and

$$X = \begin{pmatrix} V_{1l}(ka) & 0 \\ 0 & V_{2l}(ka) \end{pmatrix}, \qquad Y = \begin{pmatrix} 0 & V_{1l}(ka) \\ V_{2l}(ka) & 0 \end{pmatrix}$$

The solution then is

$$\begin{pmatrix} f_{1n} \\ f_{2n} \end{pmatrix} = \left(A - BD^{-1}C\right)^{-1} \left(X - BD^{-1}Y\right) \begin{pmatrix} a_{1n} \\ a_{2n} \end{pmatrix}$$

and

$$\begin{pmatrix} \alpha_n \\ \beta_n \end{pmatrix} = \left(D - CA^{-1}B\right)^{-1} \left(Y - CA^{-1}X\right) \begin{pmatrix} a_{1n} \\ a_{2n} \end{pmatrix}$$

which implies that the transition matrix for each l and m value has the form (the transition matrix is diagonal in the indices l, m, and σ, but not in the τ index)

$$f_{\tau\sigma ml} = \sum_{\tau'=1}^{2} t_{\tau\tau' l} a_{\tau'\sigma ml}, \qquad \tau = 1, 2$$

where

$$\begin{pmatrix} t_{11l} & t_{12l} \\ t_{21l} & t_{22l} \end{pmatrix} = \left(A - BD^{-1}C\right)^{-1} \left(X - BD^{-1}Y\right)$$

This is an explicit expression of the transition matrix entries in a 2×2 matrix form.

These T-matrix elements and the results in Example 7.3 on page 386 now give the entries of the scattering dyadic, see (7.46) on page 387. We have

$$\begin{cases} S_{\parallel\parallel} = \dfrac{1}{ik}\displaystyle\sum_l \frac{2l+1}{l(l+1)}\left(\pi_l(\cos\theta)t_{11l} + \tau_l(\cos\theta)t_{22l}\right) \\ S_{\parallel\perp} = -\dfrac{1}{k}\displaystyle\sum_l \frac{2l+1}{l(l+1)}\left(\pi_l(\cos\theta)t_{12l} + \tau_l(\cos\theta)t_{21l}\right) \\ S_{\perp\parallel} = \dfrac{1}{k}\displaystyle\sum_l \frac{2l+1}{l(l+1)}\left(\pi_l(\cos\theta)t_{21l} + \tau_l(\cos\theta)t_{12l}\right) \\ S_{\perp\perp} = \dfrac{1}{ik}\displaystyle\sum_l \frac{2l+1}{l(l+1)}\left(\pi_l(\cos\theta)t_{22l} + \tau_l(\cos\theta)t_{11l}\right) \end{cases}$$

In the forward direction, $\hat{r} = \hat{z}$ ($\theta = 0°$), the entries become, see (7.47) on page 389

$$\mathbf{S}(\hat{z}, \hat{z}) = \frac{1}{2ik}\sum_{l=1}^{\infty}(2l+1)\left\{\mathbf{I}_2\left(t_{11l} + t_{22l}\right) + i\mathbf{J}\left(t_{21l} + t_{12l}\right)\right\} \tag{8.37}$$

where $\mathbf{I}_2$ is the unit dyadic in the plane perpendicular to $\hat{z}$, and $\mathbf{J} = \hat{z} \times \mathbf{I}_2$ denotes the dyadic that rotates the vector $\pi/2$ in that plane.

The scattering cross section is, see (7.49) on page 390

$$\sigma_s = \frac{2\pi}{k^2}\sum_{\tau l}(2l+1)\left\{|t_{\tau 1l}|^2 + |t_{\tau 2l}|^2 + 2\Pi\,\mathrm{Re}\left(t_{\tau 1l}t_{\tau 2l}^*\right)\right\}$$

where $\Pi = i\hat{z} \cdot \left(\hat{p}_e \times \hat{p}_e^*\right) \in [-1, 1]$ is a real number, see Section 1.3.7.[16]

The optical theorem, see (7.50) on page 391, gives an expression of the extinction cross section, σ_{ext}

$$\sigma_{ext} = -\frac{2\pi}{k^2}\,\mathrm{Re}\sum_l(2l+1)\left\{\left(t_{11l} + t_{22l}\right) + \Pi\left(t_{12l} + t_{21l}\right)\right\}$$

If the material parameters $\chi = \kappa = 0$, we retrieve the dielectric sphere transition matrix, which we analyzed in Section 8.2. To see this, we note that, for $\chi = \kappa = 0$, $\eta Y_\pm = \pm\eta/\eta_1$ and $k_\pm a = k_1 a$, with the notation used in Section 8.2. We get

$$B = \begin{pmatrix} V_{1l}(k_1a) & V_{1l}(k_1a) \\ -V_{2l}(k_1a) & V_{2l}(k_1a) \end{pmatrix}, \qquad D = \frac{\eta}{\eta_1}\begin{pmatrix} -V_{1l}(k_1a) & V_{1l}(k_1a) \\ V_{2l}(k_1a) & V_{2l}(k_1a) \end{pmatrix}$$

[16] $\Pi = -1$ indicates LCP, $\Pi = 0$ LP, and $\Pi = 1$ corresponds to RCP.

and

$$BD^{-1} = \frac{\eta_1}{\eta} \begin{pmatrix} 0 & \dfrac{V_{1l}(k_1a)}{V_{2l}(k_1a)} \\ \dfrac{V_{2l}(k_1a)}{V_{1l}(k_1a)} & 0 \end{pmatrix}$$

Furthermore, we get

$$A - BD^{-1}C = \begin{pmatrix} -U_{1l}(ka) & 0 \\ 0 & -U_{2l}(ka) \end{pmatrix} - \frac{\eta_1}{\eta} \begin{pmatrix} -U_{2l}(ka)\dfrac{V_{1l}(k_1a)}{V_{2l}(k_1a)} & 0 \\ 0 & -U_{1l}(ka)\dfrac{V_{2l}(k_1a)}{V_{1l}(k_1a)} \end{pmatrix}$$

and

$$X - BD^{-1}Y = \begin{pmatrix} V_{1l}(ka) & 0 \\ 0 & V_{2l}(ka) \end{pmatrix} - \frac{\eta_1}{\eta} \begin{pmatrix} V_{2l}(ka)\dfrac{V_{1l}(k_1a)}{V_{2l}(k_1a)} & 0 \\ 0 & V_{1l}(ka)\dfrac{V_{2l}(k_1a)}{V_{1l}(k_1a)} \end{pmatrix}$$

The expression of the transition matrix simplifies to

$$\begin{pmatrix} t_{1l1l} & t_{1l2l} \\ t_{2l1l} & t_{2l2l} \end{pmatrix} = \left(A - BD^{-1}C\right)^{-1} \left(X - BD^{-1}Y\right)$$

$$= \begin{pmatrix} -\dfrac{\eta V_{1l}(ka)V_{2l}(k_1a) - \eta_1 V_{2l}(ka)V_{1l}(k_1a)}{\eta U_{1l}(ka)V_{2l}(k_1a) - \eta_1 U_{2l}(ka)V_{1l}(k_1a)} & 0 \\ 0 & -\dfrac{\eta V_{2l}(ka)V_{1l}(k_1a) - \eta_1 V_{1l}(ka)V_{2l}(k_1a)}{\eta U_{2l}(ka)V_{1l}(k_1a) - \eta_1 U_{1l}(ka)V_{2l}(k_1a)} \end{pmatrix}$$

$$= \begin{pmatrix} -\dfrac{\eta\psi_l(ka)\psi_l'(k_1a) - \eta_1\psi_l'(ka)\psi_1(k_1a)}{\eta\xi_l(ka)\psi_l'(k_1a) - \eta_1\xi_l'(ka)\psi_1(k_1a)} & 0 \\ 0 & -\dfrac{\eta\psi_l'(ka)\psi_l(k_1a) - \eta_1\psi_l(ka)\psi_1'(k_1a)}{\eta\xi_l'(ka)\psi_l(k_1a) - \eta_1\xi_l(ka)\psi_1'(k_1a)} \end{pmatrix}$$

which is identical to the result found in Section 8.2, see (8.7)

$$t_{\tau l} = -\frac{\psi_l(ka)\psi_l'(k_1a) - m_\tau\psi_l'(ka)\psi_l(k_1a)}{\xi_l(ka)\psi_l'(k_1a) - m_\tau\xi_l'(ka)\psi_l(k_1a)}$$

where

$$m_\tau = \delta_{\tau 1}\frac{\eta_1}{\eta} + \delta_{\tau 2}\frac{\eta}{\eta_1}$$

We illustrate the results above in two examples—one less physically relevant (with non-zero chiral properties in the static limit) and one with more realistic material parameters.

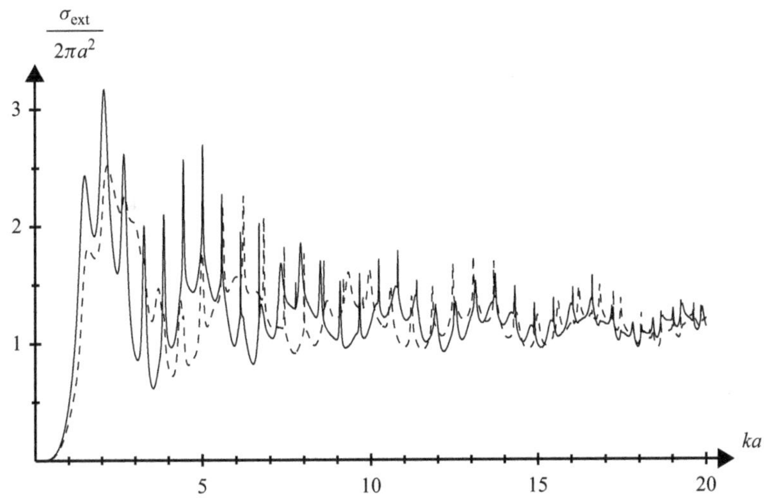

Figure 8.30 The extinction cross section, $\sigma_{\text{ext}}/2\pi a^2$, for a lossless, biisotropic sphere with $\epsilon_1/\epsilon = 4$, $\mu_1/\mu = 1$, and $\chi = \kappa = 0.1$. Two different states of polarization of the incident field are depicted, LCP (dashed curve) and RCP (solid curve). The average of these two curves gives the extinction cross section for a linearly polarized incident plane wave.

Example 8.20
In Figure 8.30, the extinction cross section of a biisotropic sphere is illustrated for two different polarization states, Π, of the incident wave. The material parameters of the lossless, biisotropic sphere are $\epsilon_1/\epsilon = 4$, $\mu_1/\mu = 1$, and $\chi = \kappa = 0.1$. Even if these material parameters are not very realistic for the frequency interval shown in Figure 8.30—a non-zero value of the chirality parameter at zero frequency is not physically correct—the computations clearly illustrate the difference in the extinction cross section for different polarization states of a biisotropic sphere. Figure 8.30 shows the LCP (dashed curve) and RCP (solid curve) incident plane waves, and the average of these two curves gives extinction cross section of a linearly polarized incident plane wave. ■

Example 8.21
The non-dispersive material model in Example 8.20 is non-physical, since we know that a biisotropic material show dispersion effects. Moreover, the parameter χ cannot assume a non-zero value at zero frequency. A more realistic model, that is passive, is given in Example 1.15 on page 62, i.e.,

$$\begin{cases} \epsilon(\omega) = 1 - \dfrac{\omega_{\text{pe}}^2}{\omega^2 - \omega_0^2 + \mathrm{i}\omega\nu_{\text{e}}} + \dfrac{\alpha\tau}{1 - \mathrm{i}\omega\tau} \\ \mu(\omega) = 1 - \dfrac{\omega_{\text{pm}}^2}{\omega^2 - \omega_0^2 + \mathrm{i}\omega\nu_{\text{m}}} \\ \xi(\omega) = -\zeta(\omega) = \dfrac{\mathrm{i}\omega\omega_{\text{c}}}{\omega^2 - \omega_0^2 + \mathrm{i}\omega\nu_{\text{c}}} \end{cases}$$

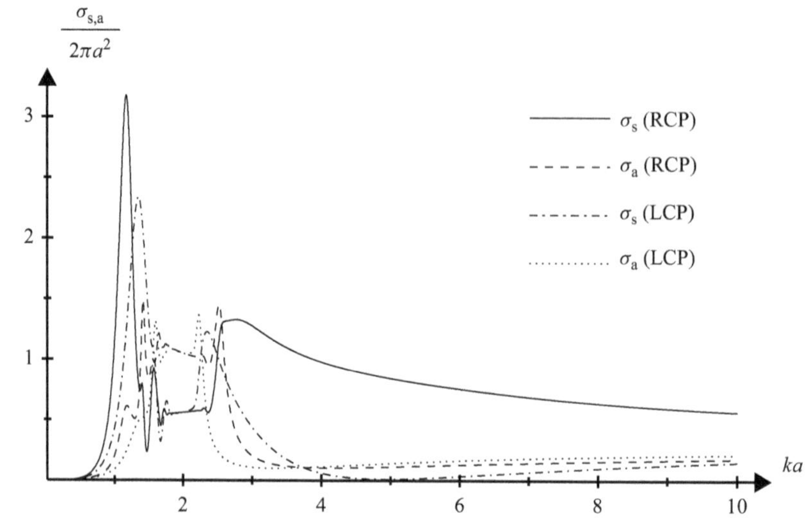

Figure 8.31 The scattering and absorption cross sections, $\sigma_{s,a}/2\pi a^2$, for a biisotropic sphere with data given in Table 8.1. Two different states of polarization of the incident field are depicted, LCP and RCP.

Table 8.1 The data of the dispersive biisotropic material in Figure 8.31. Notice that $\alpha\nu_m\omega_{pm}^2 \geq \omega_c^2\nu_c^2$, which is a requirement for a passive material.

$\omega_{pe}a/c_0 = \omega_{pm}a/c_0 = 2$
$\omega_0 a/c_0 = 2$
$\nu_e a/c_0 = \nu_m a/c_0 = \nu_c a/c_0 = 0.1$
$\omega_c a/c_0 = 0.7$
$\alpha\tau = 0.1$
$\tau c_0/a = 0.2$

In Figure 8.31, we illustrate the scattering cross section σ_s and the absorption cross section σ_a for data given in Table 8.1. From Figure 8.31, we see that in some frequency interval the scattering effects dominate, and at other frequencies the absorption is dominant. The effect also depends on the polarization of the incident wave. ■

Finally, in the next section, we analyze the low-frequency behavior of the transition matrix for a homogeneous biisotropic sphere, and we also find its electric and magnetic polarizability dyadics, $\boldsymbol{\gamma}_e$ and $\boldsymbol{\gamma}_m$, respectively.

8.5.3 Long Wavelength Approximation

We close this section with an analysis of the scattering properties of a biisotropic sphere in the low-frequency limit. Similar results have been reported in the literature, see e.g., References 164, 165. However, keep in mind that the value of the results is limited, due to the fact that most biisotropic material show no chiral effects in the limit of zero frequency.

The dominant contributions of the transition matrix in the low-frequency limit are straightforward to calculate. In Problem 8.10, the reader is encouraged to do the necessary analysis. The diagonal elements are

$$\begin{cases} t_{111} = \dfrac{2ik^3a^3\left((\epsilon_1 + 2\epsilon)(\mu_1 - \mu) - \chi^2 - \kappa^2\right)}{3\left((\epsilon_1 + 2\epsilon)(\mu_1 + 2\mu) - \chi^2 - \kappa^2\right)} \\ t_{221} = \dfrac{2ik^3a^3\left((\epsilon_1 - \epsilon)(\mu_1 + 2\mu) - \chi^2 - \kappa^2\right)}{3\left((\epsilon_1 + 2\epsilon)(\mu_1 + 2\mu) - \chi^2 - \kappa^2\right)} \end{cases}$$

and off-diagonal elements

$$\begin{cases} t_{121} = \dfrac{2k^3a^3\,(i\chi - \kappa)}{(\epsilon_1 + 2\epsilon)(\mu_1 + 2\mu) - \chi^2 - \kappa^2} \\ t_{211} = \dfrac{2k^3a^3\,(i\chi + \kappa)}{(\epsilon_1 + 2\epsilon)(\mu_1 + 2\mu) - \chi^2 - \kappa^2} \end{cases}$$

which give the low-frequency limit of the scattering dyadic in the forward direction, see (8.37)

$$\mathbf{S}(\hat{z}, \hat{z}) = \frac{k^2a^3\left\{\mathbf{I}_2\left((\epsilon_1 + 2\epsilon)(\mu_1 - \mu) + (\epsilon_1 - \epsilon)(\mu_1 + 2\mu) - 2\chi^2 - 2\kappa^2\right) + 6i\mathbf{J}\chi\right\}}{(\epsilon_1 + 2\epsilon)(\mu_1 + 2\mu) - \chi^2 - \kappa^2}$$

Comment 8.4

The sum rule of Section 6.4 can be used directly on a biisotropic sphere with result developed in this section. Use the definition of the function $f(k)$ in (6.56) on page 321, i.e.,

$$f(k) = \frac{\hat{\boldsymbol{p}}_e^* \cdot \mathbf{S}(\hat{z}, \hat{z}, k) \cdot \hat{\boldsymbol{p}}_e}{k^2}$$

Specifically, with (7.47) on page 389 and (8.37)

$$\mathbf{S}(\hat{z}, \hat{z}, k) = \frac{1}{2ik}\left\{\mathbf{I}_2 s_1(k) + i\mathbf{J} s_2(k)\right\}$$

where

$$s_1(k) = \sum_{l=1}^{\infty}(2l+1)\left(t_{11l}(k) + t_{22l}(k)\right), \qquad s_2(k) = \sum_{l=1}^{\infty}(2l+1)\left(t_{21l}(k) + t_{12l}(k)\right)$$

and we obtain with the notation from Section 6.4

$$f(k) = \frac{1}{2ik^3}\left\{s_1(k) + \Pi s_2(k)\right\}$$

since $\hat{\boldsymbol{p}}_e^* \cdot \mathbf{I} \cdot \hat{\boldsymbol{p}}_e = 1$ and $i\hat{\boldsymbol{p}}_e^* \cdot \mathbf{J} \cdot \hat{\boldsymbol{p}}_e = i\hat{\boldsymbol{p}}_e^* \cdot \left(\hat{z} \times \hat{\boldsymbol{p}}_e\right) = \Pi$, where we used, see (1.80) on page 78

$$\Pi = i\hat{z} \cdot \left(\hat{\boldsymbol{p}}_e \times \hat{\boldsymbol{p}}_e^*\right)$$

Since $\mathbf{S}(\hat{z}, \hat{z}, -k) = \mathbf{S}^*(\hat{z}, \hat{z}, k)$ for real-valued k, $\operatorname{Re} s_1(k)$ and $\operatorname{Im} s_2(k)$ are even, and $\operatorname{Im} s_1(k)$ and $\operatorname{Re} s_2(k)$ are odd functions.

The optical theorem gives, see (6.59) on page 322

$$\sigma_{\text{ext}}(k) = 4\pi k \,\text{Im} f(k) = -\frac{2\pi}{k^2} \text{Re}\{s_1(k) + \Pi s_2(k)\}$$

and Plemelj's formulas, see (6.55) on page 321

$$f_{\text{r}}^{\text{e}}(0) = \frac{2}{\pi}\int_0^\infty \frac{f_{\text{i}}^{\text{o}}(k)}{k}\,\text{d}k = -\frac{1}{\pi}\int_0^\infty \frac{\text{Re}\, s_1(k)}{k^4}\,\text{d}k$$

Moreover, the low-frequency expansion of $f(k)$ is

$$f(k) = a^3 \frac{(\epsilon_1 + 2\epsilon)(\mu_1 - \mu) + (\epsilon_1 - \epsilon)(\mu_1 + 2\mu) - 2\chi^2 - 2\kappa^2 + 6\Pi\chi}{(\epsilon_1 + 2\epsilon)(\mu_1 + 2\mu) - \chi^2 - \kappa^2} + O(k)$$

where ϵ_1, μ_1, κ, and χ are the static real values of the material parameters. The static values of the parameters κ and χ are in general zero.

The low-frequency limit of the transition matrix also gives the electric and magnetic polarizability dyadics, see (7.54) on page 402

$$\begin{aligned} \boldsymbol{\gamma}_{\text{e}} &= \lim_{k\to 0} \frac{6\pi t_{221}\mathbf{I}_2}{\text{i}k^3} + \lim_{k\to 0} \frac{6\pi t_{211}\hat{\boldsymbol{z}} \times \mathbf{I}_2}{k^3} \\ &= \frac{4\pi a^3\left((\epsilon_1 - \epsilon)(\mu_1 + 2\mu) - \chi^2 - \kappa^2\right)}{(\epsilon_1 + 2\epsilon)(\mu_1 + 2\mu) - \chi^2 - \kappa^2}\mathbf{I}_2 + \frac{12\pi a^3\,(\text{i}\chi + \kappa)}{(\epsilon_1 + 2\epsilon)(\mu_1 + 2\mu) - \chi^2 - \kappa^2}\mathbf{J} \end{aligned}$$

and

$$\begin{aligned} \boldsymbol{\gamma}_{\text{m}} &= \lim_{k\to 0} \frac{6\pi t_{111}\mathbf{I}_2}{\text{i}k^3} + \lim_{k\to 0} \frac{6\pi t_{121}\hat{\boldsymbol{z}} \times \mathbf{I}_2}{k^3} = \frac{4\pi a^3\left((\epsilon_1 + 2\epsilon)(\mu_1 - \mu) - \chi^2 - \kappa^2\right)}{(\epsilon_1 + 2\epsilon)(\mu_1 + 2\mu) - \chi^2 - \kappa^2}\mathbf{I}_2 \\ &\quad + \frac{12\pi a^3\,(\text{i}\chi - \kappa)}{(\epsilon_1 + 2\epsilon)(\mu_1 + 2\mu) - \chi^2 - \kappa^2}\mathbf{J} \end{aligned}$$

In the limit of vanishing chirality, $\chi = 0$, and reciprocity, $\kappa = 0$, we recover the electric and magnetic polarizability dyadics of a dielectric sphere, see Example 6.4 on page 276.

Introduce the polarization state $\Pi = \text{i}\hat{\boldsymbol{z}} \cdot (\hat{\boldsymbol{p}}_{\text{e}} \times \hat{\boldsymbol{p}}_{\text{e}}^*) \in [-1, 1]$, which is a real number, see Section 1.3.7. For any complex-valued polarization vector $\hat{\boldsymbol{p}}_{\text{e}}$, we get

$$\begin{aligned} \hat{\boldsymbol{p}}_{\text{e}}^* \cdot \boldsymbol{\gamma}_{\text{e}} \cdot \hat{\boldsymbol{p}}_{\text{e}} &= \frac{4\pi a^3\left((\epsilon_1 - \epsilon)(\mu_1 + 2\mu) - \chi^2 - \kappa^2\right)}{(\epsilon_1 + 2\epsilon)(\mu_1 + 2\mu) - \chi^2 - \kappa^2} \\ &\quad - \frac{12\text{i}\pi a^3\,(\text{i}\chi + \kappa)}{(\epsilon_1 + 2\epsilon)(\mu_1 + 2\mu) - \chi^2 - \kappa^2}\Pi \end{aligned}$$

and

$$(\hat{z} \times \hat{p}_e^*) \cdot \gamma_m \cdot (\hat{z} \times \hat{p}_e) = \frac{4\pi a^3 \left((\epsilon_1 + 2\epsilon)(\mu_1 - \mu) - \chi^2 - \kappa^2\right)}{(\epsilon_1 + 2\epsilon)(\mu_1 + 2\mu) - \chi^2 - \kappa^2} - \frac{12\mathrm{i}\pi a^3 (\mathrm{i}\chi - \kappa)}{(\epsilon_1 + 2\epsilon)(\mu_1 + 2\mu) - \chi^2 - \kappa^2}\Pi$$

The sum is

$$\hat{p}_e^* \cdot \gamma_e \cdot \hat{p}_e + (\hat{z} \times \hat{p}_e^*) \cdot \gamma_m \cdot (\hat{z} \times \hat{p}_e) = \frac{4\pi a^3 \left((\epsilon_1 - \epsilon)(\mu_1 + 2\mu) + (\epsilon_1 + 2)(\mu_1 - \mu) - 2\chi^2 - 2\kappa^2\right)}{(\epsilon_1 + 2)(\mu_1 + 2) - \chi^2 - \kappa^2} + \frac{24\pi a^3 \chi}{(\epsilon_1 + 2\epsilon)(\mu_1 + 2\mu) - \chi^2 - \kappa^2}\Pi$$

This quantity is relevant for the evaluation of the sum rules for the extinction cross section that were developed in Section 6.4.

Problems for Chapter 8

Section 8.1 Scattering by a Perfectly Conducting Sphere

8.1 Show that $t_{\tau l}$ in (8.1) satisfies (conservation of energy)

$$t_{\tau l} t_{\tau l}^* = -\operatorname{Re} t_{\tau l}$$

8.2 Show that

$$\begin{cases} \psi_l(-z^*) = (-1)^{l+1} \left(\psi_l(z)\right)^* \\ \xi_l(-z^*) = (-1)^{l+1} \left(\xi_l(z)\right)^* \end{cases}$$

hold for the Riccati–Bessel functions.

8.3 Show that the low-frequency expansions of the transition matrix elements for a PEC sphere of radius a, correct to the order $(ka)^5$, are ($x = ka$)

$$t_{11} = -\mathrm{i}\frac{x^3}{3} + \mathrm{i}\frac{x^5}{5} + O(x^6), \qquad t_{12} = -\mathrm{i}\frac{x^5}{45} + O(x^7)$$

and

$$t_{21} = \mathrm{i}\frac{2x^3}{3} + \mathrm{i}\frac{x^5}{5} + O(x^6), \qquad t_{22} = \mathrm{i}\frac{x^5}{30} + O(x^7)$$

Then calculate the scattering dyadic in the forward direction, $\mathbf{S}(\hat{z}, \hat{z})$, in the low-frequency limit correct to order x^5.

8.4 Determine the transition matrix elements $t_{\tau l}$ for a sphere of radius a with impedance condition on the surface $r = a$, i.e.,

$$\boldsymbol{E}_{\tan}(r = a) = Z\left(\hat{\boldsymbol{r}} \times \boldsymbol{H}(r = a)\right)$$

where Z is the surface impedance of the sphere.

Section 8.2 Scattering by a Dielectric Sphere

8.5 Show that $t_{\tau l}$ in (8.6) for real-valued ϵ_1 and μ_1 satisfies (conservation of energy)

$$t_{\tau l} t_{\tau l}^* = -\operatorname{Re} t_{\tau l}$$

8.6 Calculate the differential scattering cross section in the backward scattering direction

$$\frac{\mathrm{d}\sigma}{\mathrm{d}\Omega}(\hat{\boldsymbol{r}} = -\hat{\boldsymbol{k}}_{\mathrm{i}}) = 4\pi \frac{\left|\boldsymbol{F}(\hat{\boldsymbol{r}} = -\hat{\boldsymbol{k}}_{\mathrm{i}})\right|^2}{|\boldsymbol{E}_0|^2}$$

for a perfectly conducting sphere and a dielectric sphere expressed as a series over $t_{\tau l}$.

8.7 Calculate the electric field, $\boldsymbol{E}_1(\boldsymbol{r}, \omega)$, inside a dielectric sphere (radius a, and material parameters ϵ_1 and $\mu_1 = \mu$) in the long wavelength limit. The surrounding material has material parameters ϵ and μ.

Section 8.3 Scattering by Layered Spherical Objects

***8.8** Compute the transition matrix entries for a spherical, dielectric shell with permittivity ϵ_1 and permeability μ_1. The material parameters outside and inside the shell are ϵ and μ, and the outer and inner radii of the shell are b and a, respectively, see left drawing in Figure 8.32. If the shell is thin, $b - a \ll a$, compute the entries of the transition matrix to leading order in the thickness of the shell. Determine also the electric and magnetic polarizabilities to leading order in $b - a$.

***8.9** Derive the equation that determines the resonance frequencies of the geometry in the right drawing of Figure 8.32. The volume inside the radius a is not perfectly conducting, but has a finite conductivity σ_1, and the volume outside the radius b has conductivity σ_2. Find the lowest resonance frequencies of this resonance cavity.

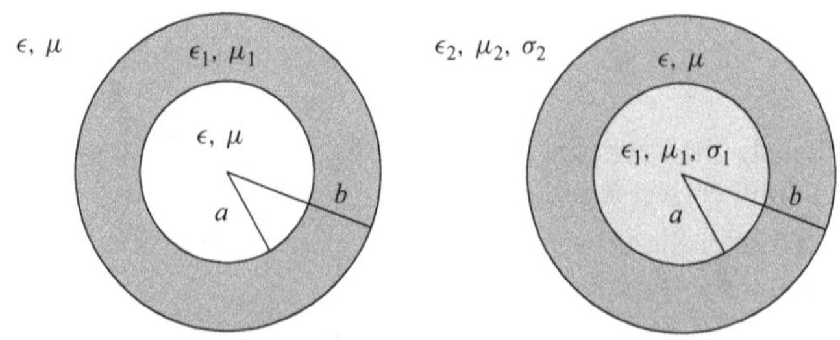

Figure 8.32 Left: Geometry of the spherical, dielectric shell in Problem 8.8. Right: Geometry of the spherical cavity in Problem 8.9.

Section 8.5 Scattering by a Biisotropic Sphere

8.10 Show that the dominant contributions to the transition matrix in long wavelength limit for a biisotropic sphere are

$$\begin{cases} t_{111} = \dfrac{2ik^3a^3\left((\epsilon_1+2\epsilon)(\mu_1-\mu)-\chi^2-\kappa^2\right)}{3\left((\epsilon_1+2\epsilon)(\mu_1+2\mu)-\chi^2-\kappa^2\right)} \\ t_{121} = \dfrac{2k^3a^3\,(i\chi-\kappa)}{(\epsilon_1+2\epsilon)(\mu_1+2\mu)-\chi^2-\kappa^2} \\ t_{211} = \dfrac{2k^3a^3\,(i\chi+\kappa)}{(\epsilon_1+2\epsilon)(\mu_1+2\mu)-\chi^2-\kappa^2} \\ t_{221} = \dfrac{2ik^3a^3\left((\epsilon_1-\epsilon)(\mu_1+2\mu)-\chi^2-\kappa^2\right)}{3\left((\epsilon_1+2\epsilon)(\mu_1+2\mu)-\chi^2-\kappa^2\right)} \end{cases}$$

8.11 Determine the scattering and the extinction cross sections, σ_s and σ_{ext}, respectively, for a biisotropic sphere of radius a in the long wavelength limit to leading order in the wave number k.

CHAPTER 9

The Null-Field Approach

The transition matrix, $T_{nn'}$, contains all information that is necessary to solve the scattering problem, such as various cross sections etc., see Chapter 7. To find the transition matrix, in the general case, is a very difficult problem. In this section, we derive a general expression of the transition matrix in terms of surface integrals over the spherical vector waves. The derivation was originally presented by Peter C. Waterman.[1] Some of his excellent scientific production is found in References 282–285, and a comprehensive database of Null-field approach is given in References 189–194, 303. Alternative approaches to find the transition matrix based on the far field are presented in References 76–78.

The derivation proceeds in a formal way, and does not address the delicate mathematical question concerning convergence, etc. These questions are addressed in e.g., References 45, 46, 56, 145, 222, 232. Several recent review articles on the method are published in the literature [188, 194, 231, 233]. We start by analyzing the single, homogeneous scatterer case in Section 9.1, and then proceed, in Section 9.2, to find the T-matrix for a collection of scatterers. The chapter ends with Section 9.3 that treats scattering by an obstacle above a perfectly conducting ground. This combination of finite structures with planar ones has many applications in the geophysical sciences.

9.1 The *T*-Matrix for a Single Homogeneous Scatterer

The single homogeneous scatterer is the simplest example, but illustrates the essence of the method very well. Therefore, it is pedagogical to start with this case. To avoid some mathematical difficulties, we assume both volumes V_s and V_e are simply connected, and that the origin of the coordinate system, O, lies in V_s.

The Null-field method or approach[2] was originally formulated by P. Waterman, and it is an efficient way of finding the T-matrix of a scatterer in terms of surface integrals over spherical vector waves. The usage of the extinction part of the integral representation is instrumental in this method, hence the Null-field approach.

---

[1] Peter C. Waterman (1928–2012), American physicist.

[2] The less suitable name Extended Boundary Condition method also occurs.

9.1.1 Perfectly Conducting Scatterer

The first, and simplest, class of scatterer is the perfectly conducting obstacle. This scatterer is characterized by the boundary condition—the total electric field $\boldsymbol{E}$ satisfies $\hat{\boldsymbol{\nu}} \times \boldsymbol{E} = \mathbf{0}$ on S_s.

Our starting point is the integral representation for the electric field (4.15) on page 197

$$-\frac{\eta_0\eta}{ik}\nabla\times\left\{\nabla\times\iint_{S_s} g(k,|\boldsymbol{r}-\boldsymbol{r}'|)\boldsymbol{J}_S(\boldsymbol{r}')\,\mathrm{d}S'\right\} = \begin{cases}\boldsymbol{E}_s(\boldsymbol{r}), & \boldsymbol{r} \text{ outside } S_s\\ -\boldsymbol{E}_i(\boldsymbol{r}), & \boldsymbol{r} \text{ inside } S_s\end{cases}$$

where the boundary condition $\hat{\boldsymbol{\nu}} \times \boldsymbol{E} = \mathbf{0}$ on S_s has been used, and we have, for convenience, denoted the surface current density $\boldsymbol{J}_S = \hat{\boldsymbol{\nu}} \times \boldsymbol{H}$.

Use the Green dyadic $\mathbf{G}_e(k,|\boldsymbol{r}-\boldsymbol{r}'|)$ of Section 2.2.2 to rewrite the integral as

$$-\frac{\eta_0\eta}{ik}\nabla\times\left\{\nabla\times\iint_{S_s} \mathbf{G}_e(k,|\boldsymbol{r}-\boldsymbol{r}'|)\cdot\boldsymbol{J}_S(\boldsymbol{r}')\,\mathrm{d}S'\right\} = \begin{cases}\boldsymbol{E}_s(\boldsymbol{r}), & \boldsymbol{r} \text{ outside } S_s\\ -\boldsymbol{E}_i(\boldsymbol{r}), & \boldsymbol{r} \text{ inside } S_s\end{cases}$$

We have here used

$$\begin{aligned}\nabla\times\iint_{S_s} g(k,|\boldsymbol{r}-\boldsymbol{r}'|)\boldsymbol{J}_S(\boldsymbol{r}')\,\mathrm{d}S' &= \nabla\times\iint_{S_s} \mathbf{I}_3 g(k,|\boldsymbol{r}-\boldsymbol{r}'|)\cdot\boldsymbol{J}_S(\boldsymbol{r}')\,\mathrm{d}S'\\ &= \nabla\times\iint_{S_s} \mathbf{G}_e(k,|\boldsymbol{r}-\boldsymbol{r}'|)\cdot\boldsymbol{J}_S(\boldsymbol{r}')\,\mathrm{d}S'\end{aligned}$$

and $\mathbf{G}_e(k,|\boldsymbol{r}-\boldsymbol{r}'|) = \mathbf{I}_3 g(k,|\boldsymbol{r}-\boldsymbol{r}'|) + k^{-2}\nabla\nabla g(k,|\boldsymbol{r}-\boldsymbol{r}'|)$.

Next step in the derivation is to decompose the Green dyadic in a set of appropriate vector waves. In this case, we employ the set of spherical vector waves developed in Chapter 7. The expansion of the Green dyadic in spherical vector waves is, see (7.24) on page 370

$$\mathbf{G}_e(k,|\boldsymbol{r}-\boldsymbol{r}'|) = \mathrm{i}k\sum_n \boldsymbol{v}_n(k\boldsymbol{r}_<)\boldsymbol{u}_n(k\boldsymbol{r}_>) \tag{9.1}$$

where $\boldsymbol{r}_<$ ($\boldsymbol{r}_>$) is the position vector with the smallest (largest) distance to the origin, i.e., if $r < r'$ then $\boldsymbol{r}_< = \boldsymbol{r}$ and $\boldsymbol{r}_> = \boldsymbol{r}'$. As everywhere in this chapter, the summation in the index τ is only over $\tau = 1, 2$. This expansion is uniformly convergent in finite domains, provided $r \neq r'$ in the domain [40].

Let $\boldsymbol{r}$ be a position vector outside the smallest circumscribed sphere of S_s, see Figure 9.1. Then $\boldsymbol{r}_> = \boldsymbol{r}$ and $\boldsymbol{r}_< = \boldsymbol{r}'$ for all points on the surface $\boldsymbol{r}' \in S_s$. A change of the order of integration and summation in the integral representation, which is legitimate due to uniform convergence of the expansion in (9.1) when $r' \neq r$, gives

$$\begin{aligned}\boldsymbol{E}_s(\boldsymbol{r}) &= -\eta_0\eta\sum_n \nabla\times\left\{\nabla\times\boldsymbol{u}_n(k\boldsymbol{r})\iint_{S_s} \boldsymbol{v}_n(k\boldsymbol{r}')\cdot\boldsymbol{J}_S(\boldsymbol{r}')\,\mathrm{d}S'\right\}\\ &= -\eta_0\eta k^2\sum_n \boldsymbol{u}_n(k\boldsymbol{r})\iint_{S_s} \boldsymbol{v}_n(k\boldsymbol{r}')\cdot\boldsymbol{J}_S(\boldsymbol{r}')\,\mathrm{d}S'\end{aligned}$$

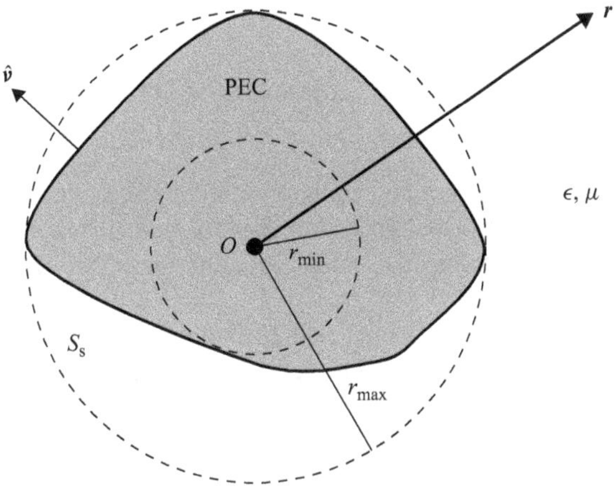

Figure 9.1 The perfectly conducting scatterer and its largest inscribed and smallest circumscribed spheres of radius r_{min} and r_{max}, respectively.

This expression verifies the expansion of the scattered field in out-going spherical vector waves outside the circumscribed sphere of the scatterer. Explicitly, we have

$$\boldsymbol{E}_{\text{s}}(\boldsymbol{r}) = \sum_n f_n \boldsymbol{u}_n(k\boldsymbol{r}), \qquad r > r_{\text{max}} \tag{9.2}$$

where the expansion coefficients f_n are

$$f_n = -\eta_0 \eta k^2 \iint_{S_{\text{s}}} \boldsymbol{v}_n(k\boldsymbol{r}) \cdot \boldsymbol{J}_S(\boldsymbol{r}) \,\mathrm{d}S \tag{9.3}$$

We now proceed in a similar way as above, but we let $\boldsymbol{r}$ be a position vector inside the largest inscribed sphere of S_{s}. Then $\boldsymbol{r}_< = \boldsymbol{r}$ and $\boldsymbol{r}_> = \boldsymbol{r}'$ for all points on the surface $\boldsymbol{r}' \in S_{\text{s}}$ and we get (again, a change of the order of integration and summation is legitimate due to uniform convergence)

$$\begin{aligned}
\boldsymbol{E}_{\text{i}}(\boldsymbol{r}) &= \eta_0 \eta \sum_n \nabla \times \left\{ \nabla \times \boldsymbol{v}_n(k\boldsymbol{r}) \iint_{S_{\text{s}}} \boldsymbol{u}_n(k\boldsymbol{r}') \cdot \boldsymbol{J}_S(\boldsymbol{r}') \,\mathrm{d}S' \right\} \\
&= \eta_0 \eta k^2 \sum_n \boldsymbol{v}_n(k\boldsymbol{r}) \iint_{S_{\text{s}}} \boldsymbol{u}_n(k\boldsymbol{r}') \cdot \boldsymbol{J}_S(\boldsymbol{r}') \,\mathrm{d}S'
\end{aligned}$$

This expression gives the expansion of the incident field in regular spherical vector waves inside the inscribed sphere of the scatterer. Explicitly, we have

$$\boldsymbol{E}_{\text{i}}(\boldsymbol{r}) = \sum_n a_n \boldsymbol{v}_n(k\boldsymbol{r}), \qquad r < r_{\text{min}} \tag{9.4}$$

where the expansion coefficients a_n are

$$a_n = \eta_0\eta k^2 \iint_{S_s} \boldsymbol{u}_n(k\boldsymbol{r}) \cdot \boldsymbol{J}_S(\boldsymbol{r})\, \mathrm{d}S \tag{9.5}$$

This expression contains infinity many equations as the index set $n = \{\tau, \sigma, m, l\}$ contains infinitely (countable) many elements. This set of equations, which has its origin in the extinction part of the integral representation, is usually called the set of Null-field equations, see also Section 7.6.

In a scattering problem, the incident field $\boldsymbol{E}_{\mathrm{i}}$ is known. This means that the expansion coefficients a_n in (9.4) are prescribed. The left-hand side of (9.5) is therefore known, and provided we, from this infinite set of equations, can extract the unknown surface current density $\boldsymbol{J}_S$, the scattering problem is solved. In fact, the expansion coefficients f_n in (9.3) are then known. The two sets of equations

$$\begin{cases} a_n = \eta_0\eta k^2 \iint_{S_s} \boldsymbol{u}_n(k\boldsymbol{r}) \cdot \boldsymbol{J}_S(\boldsymbol{r})\, \mathrm{d}S \\ f_n = -\eta_0\eta k^2 \iint_{S_s} \boldsymbol{v}_n(k\boldsymbol{r}) \cdot \boldsymbol{J}_S(\boldsymbol{r})\, \mathrm{d}S \end{cases}$$

are often viewed as pair of equations, where the first is used to eliminate the surface field, which then is inserted into the second to find the unknown coefficients f_n.

The elimination of the surface field can be made by an expansion of the surface field $\boldsymbol{J}_S(\boldsymbol{r})$ in a global, complete set of vector-valued expansion functions $\boldsymbol{\psi}_n(\boldsymbol{r})$ on the surface S_{s}. Formally, we expand the surface field as, see also Comment 9.1

$$\eta_0\eta\hat{\boldsymbol{\nu}}(\boldsymbol{r}) \times \boldsymbol{H}(\boldsymbol{r}) = \eta_0\eta\boldsymbol{J}_S(\boldsymbol{r}) = \sum_n \alpha_n \boldsymbol{\psi}_n(\boldsymbol{r}), \quad \boldsymbol{r} \in S_{\mathrm{s}} \tag{9.6}$$

where the vector-valued expansion functions $\boldsymbol{\psi}_n(\boldsymbol{r})$ are tangent to the surface S_{s}, i.e., $\hat{\boldsymbol{\nu}}(\boldsymbol{r}) \cdot \boldsymbol{\psi}_n(\boldsymbol{r}) = 0$. Several examples of complete sets of vector-valued expansion functions are available. Some examples are, see also Comment 9.2

$$\boldsymbol{\psi}_n(\boldsymbol{r}) = \begin{cases} \hat{\boldsymbol{\nu}}(\boldsymbol{r}) \times \boldsymbol{A}_n(\hat{\boldsymbol{r}}) \\ \hat{\boldsymbol{\nu}}(\boldsymbol{r}) \times (\nabla \times \boldsymbol{v}_n(k\boldsymbol{r})) = k\hat{\boldsymbol{\nu}}(\boldsymbol{r}) \times \boldsymbol{v}_{\overline{n}}(k\boldsymbol{r}) \\ \hat{\boldsymbol{\nu}}(\boldsymbol{r}) \times (\nabla \times \boldsymbol{u}_n(k\boldsymbol{r})) = k\hat{\boldsymbol{\nu}}(\boldsymbol{r}) \times \boldsymbol{u}_{\overline{n}}(k\boldsymbol{r}) \end{cases}$$

where, in the second and third sets of expansion functions, we have expressed the expansion functions $\hat{\boldsymbol{\nu}}(\boldsymbol{r}) \times \boldsymbol{v}_n(k\boldsymbol{r})$ and $\hat{\boldsymbol{\nu}}(\boldsymbol{r}) \times \boldsymbol{u}_n(k\boldsymbol{r})$ in the dual n-index. A proof of the completeness[3] of the sets $\hat{\boldsymbol{\nu}}(\boldsymbol{r}) \times \boldsymbol{v}_n(k\boldsymbol{r})$ and $\hat{\boldsymbol{\nu}}(\boldsymbol{r}) \times \boldsymbol{u}_n(k\boldsymbol{r})$ and other sets are presented in References 8, 56, 136, 198, 255. Other valuable contributions to the mathematical aspects of this elimination procedure are found in References 45, 46, 145, 222.

---

[3]A countable set of frequencies has to be excluded for the set $\hat{\boldsymbol{\nu}}(\boldsymbol{r}) \times \boldsymbol{v}_n(k\boldsymbol{r})$.

Comment 9.1
The form of the expansion of the surface field in (9.6) is somewhat misleading. The summation is always finite in practice, and the coefficients not only depend on the surface field, but also on the truncation size. In Appendix F.9.1 on page 681, the concepts of completeness and closedness are reviewed. Moreover, the equal sign has to be interpreted in the $L^2(\partial\Omega)$ sense. A more appropriate notation would be: for every $\varepsilon > 0$, there are coefficients $\alpha_n(\varepsilon)$, a truncation size $N(\varepsilon)$, and a finite sum

$$\eta_0\eta \boldsymbol{J}_S^{\varepsilon}(\boldsymbol{r}) = \sum_n^{N(\varepsilon)} \alpha_n(\varepsilon)\boldsymbol{\psi}_n(\boldsymbol{r})$$

such that

$$\left\| \boldsymbol{J}_S(\boldsymbol{r}) - \boldsymbol{J}_S^{\varepsilon}(\boldsymbol{r}) \right\| < \varepsilon$$

Completeness guarantees that such a sequence $\alpha_n(\varepsilon)$ exists. There is, however, no guarantee that the algorithm presented in this section generates a stable numerical determination scheme of this sequence.

Comment 9.2
Of the three sets of complete vector-valued expansion functions, there are indications that the first set, $\hat{\boldsymbol{\nu}}(\boldsymbol{r}) \times \boldsymbol{A}_n(\hat{\boldsymbol{r}})$, is a Riesz basis [45, 46, 145], see Appendix F.9.1. The other sets seem not to have this property. A Riesz basis shows many properties of an orthonormal basis, and such a set is suitable for expanding functions.

To facilitate the notation, introduce the two matrices R and Q defined by

$$\begin{cases} R_{nn'} = k^2 \iint\limits_{S_s} \boldsymbol{v}_n(k\boldsymbol{r}) \cdot \boldsymbol{\psi}_{n'}(\boldsymbol{r}) \, \mathrm{d}S \\ Q_{nn'} = k^2 \iint\limits_{S_s} \boldsymbol{u}_n(k\boldsymbol{r}) \cdot \boldsymbol{\psi}_{n'}(\boldsymbol{r}) \, \mathrm{d}S \end{cases} \tag{9.7}$$

Insertion of the global expansion of the surface field in (9.6) into (9.3) and (9.5) results in

$$\begin{cases} a_n = \sum_{n'} Q_{nn'}\alpha_{n'} \\ f_n = -\sum_{n'} R_{nn'}\alpha_{n'} \end{cases}$$

Notice that the entries of the matrix Q grow as the row index n increases. This is due to the properties of the out-going vector waves, $\boldsymbol{u}_n(k\boldsymbol{r})$, as the index l grows. In general, this property generates an ill-conditioned matrix.

We can formally eliminate the unknown expansion coefficients α_n in the system of equations above. We then obtain

$$f_n = -\sum_{n'} T_{nn'}a_{n'}$$

where we have identified the transition matrix $T_{nn'}$

$$T_{nn'} = -\sum_{n''} R_{nn''} Q^{-1}_{n''n'}$$

The scattering problem is formally solved, since the coefficients f_n give the scattered fields, see (9.2).

Axially symmetric obstacles (bodies of revolution) are often of special interest. In the next example, the Null-field approach is developed for this category of geometry.

Example 9.1
If the body is axially symmetric, the surface integrals in (9.7) reduce to a line integral. We choose $\boldsymbol{\psi}_n(\boldsymbol{r}) = -\mathrm{i}\hat{\boldsymbol{\nu}}(\boldsymbol{r}) \times \boldsymbol{v}_{\overline{n}}(k\boldsymbol{r})$, and get from (9.7)

$$R_{nn'} = -\mathrm{i}k^2 \iint_{S_\mathrm{s}} \boldsymbol{v}_n \cdot \left(\hat{\boldsymbol{\nu}} \times \boldsymbol{v}_{\overline{n}'}\right) \,\mathrm{d}S = \mathrm{i}k^2 \iint_{S_\mathrm{s}} (\boldsymbol{v}_n \times \boldsymbol{v}_{\overline{n}'}) \cdot \hat{\boldsymbol{\nu}} \,\mathrm{d}S$$

and

$$Q_{nn'} = -\mathrm{i}k^2 \iint_{S_\mathrm{s}} \boldsymbol{u}_n \cdot \left(\hat{\boldsymbol{\nu}} \times \boldsymbol{v}_{\overline{n}'}\right) \,\mathrm{d}S = \mathrm{i}k^2 \iint_{S_\mathrm{s}} (\boldsymbol{u}_n \times \boldsymbol{v}_{\overline{n}'}) \cdot \hat{\boldsymbol{\nu}} \,\mathrm{d}S$$

We let the symmetry axis define the z-axis, and parametrize the surface vector with the spherical angles θ and ϕ, i.e., the position vector of the surface S_s is $\boldsymbol{r} = r(\theta)\hat{\boldsymbol{r}}$. This parameterization assumes the scatterer is star shaped, i.e., the entire surface of the scatterer can be seen from the origin. The normal vector of the surface is

$$\hat{\boldsymbol{\nu}} = \frac{\frac{\partial \boldsymbol{r}}{\partial \theta} \times \frac{\partial \boldsymbol{r}}{\partial \phi}}{\left|\frac{\partial \boldsymbol{r}}{\partial \theta} \times \frac{\partial \boldsymbol{r}}{\partial \phi}\right|} = \frac{r(\theta)\hat{\boldsymbol{r}} - r'(\theta)\hat{\boldsymbol{\theta}}}{\sqrt{(r'(\theta))^2 + (r(\theta))^2}}$$

and the surface element $\mathrm{d}S$ is

$$\mathrm{d}S = r(\theta)\sin\theta \sqrt{(r'(\theta))^2 + (r(\theta))^2} \,\mathrm{d}\theta \,\mathrm{d}\phi$$

The orthogonality in the azimuthal variable ϕ shows that for $m \neq m'$ all entries of the R- and Q-matrices are zero, i.e., $R_{nn'} = Q_{nn'} = 0$ when $m \neq m'$. As a consequence, no m indices couple.

Straightforward calculations show that the structure of the Q-matrix in the τ and the σ indices is

$$Q_{nn'} = \delta_{mm'} \begin{array}{c} \\ 1\mathrm{e} \\ 2\mathrm{o} \\ 1\mathrm{o} \\ 2\mathrm{e} \end{array} \begin{array}{c} \begin{array}{cccc} 1\mathrm{e} & 2\mathrm{o} & 1\mathrm{o} & 2\mathrm{e} \end{array} \\ \begin{pmatrix} A_{ll'} & B_{ll'} & 0 & 0 \\ C_{ll'} & D_{ll'} & 0 & 0 \\ 0 & 0 & A_{ll'} & -B_{ll'} \\ 0 & 0 & -C_{ll'} & D_{ll'} \end{pmatrix} \end{array}$$

where the four functions $A_{ll'}$, $B_{ll'}$, $C_{ll'}$, and $D_{ll'}$ are (the Riccati–Bessel functions ψ_l and ξ_l defined in Appendix B.3.3 are used)

$$A_{ll'} = \mathrm{i}N_{lm}N_{l'm}\int_0^\pi \left\{\Theta_{ll'}^m(\theta)\xi_l(kr(\theta))\psi_{l'}'(kr(\theta)) + l'(l'+1)\frac{\mathrm{d}P_l^m(\cos\theta)}{\mathrm{d}\theta}P_{l'}^m(\cos\theta)kr'(\theta)h_l^{(1)}(kr(\theta))j_{l'}(kr(\theta))\right\}\sin\theta\,\mathrm{d}\theta$$

$$B_{ll'} = \mathrm{i}N_{lm}N_{l'm}m\int_0^\pi \frac{\mathrm{d}(P_l^m(\cos\theta)P_{l'}^m(\cos\theta))}{\mathrm{d}\theta}\xi_l(kr(\theta))\psi_{l'}(kr(\theta))\,\mathrm{d}\theta$$

$$C_{ll'} = -\mathrm{i}N_{lm}N_{l'm}m\int_0^\pi \left\{\frac{\mathrm{d}(P_l^m(\cos\theta)P_{l'}^m(\cos\theta))}{\mathrm{d}\theta}\xi_l'(kr(\theta))\psi_{l'}'(kr(\theta)) + l(l+1)P_l^m(\cos\theta)P_{l'}^m(\cos\theta)\frac{r'(\theta)}{r(\theta)}h_l^{(1)}(kr(\theta))\psi_{l'}'(kr(\theta)) + l'(l'+1)P_l^m(\cos\theta)P_{l'}^m(\cos\theta)\frac{r'(\theta)}{r(\theta)}\xi_l'(kr(\theta))j_{l'}(kr(\theta))\right\}\mathrm{d}\theta$$

and

$$D_{ll'} = -\mathrm{i}N_{lm}N_{l'm}\int_0^\pi \left\{\Theta_{ll'}^m(\theta)\xi_l'(kr(\theta))\psi_{l'}(kr(\theta)) + l(l+1)P_l^m(\cos\theta)\frac{\mathrm{d}P_{l'}^m(\cos\theta)}{\mathrm{d}\theta}kr'(\theta)h_l^{(1)}(kr(\theta))j_{l'}(kr(\theta))\right\}\sin\theta\,\mathrm{d}\theta$$

The normalization factor N_{lm} is defined as

$$N_{lm} = \sqrt{\frac{(2l+1)(l-m)!}{2l(l+1)(l+m)!}}$$

and the combination of associated Legendre functions

$$\Theta_{ll'}^m(\theta) = \frac{\mathrm{d}P_l^m(\cos\theta)}{\mathrm{d}\theta}\frac{\mathrm{d}P_{l'}^m(\cos\theta)}{\mathrm{d}\theta} + \frac{m^2}{\sin^2\theta}P_l^m(\cos\theta)P_{l'}^m(\cos\theta)$$

is used.

The result of Problem 9.1 shows that the functions $A_{ll'}$, $B_{ll'}$, $C_{ll'}$, and $D_{ll'}$ are related, i.e., $C_{ll'} = B_{ll'}$ and $D_{ll'} = \delta_{ll'} - A_{ll'}$, which simplifies the structure of $Q_{nn'}$

$$Q_{nn'} = \delta_{mm'}\begin{array}{c} \\ 1\mathrm{e} \\ 2\mathrm{o} \\ 1\mathrm{o} \\ 2\mathrm{e}\end{array}\begin{array}{c}\begin{array}{cccc} 1\mathrm{e} & 2\mathrm{o} & 1\mathrm{o} & 2\mathrm{e}\end{array} \\ \begin{pmatrix} A_{ll'} & B_{ll'} & 0 & 0 \\ B_{ll'} & \delta_{ll'} - A_{ll'} & 0 & 0 \\ 0 & 0 & A_{ll'} & -B_{ll'} \\ 0 & 0 & -B_{ll'} & \delta_{ll'} - A_{ll'}\end{pmatrix}\end{array}$$

The $R_{nn'}$ is identical in structure to $Q_{nn'}$, but with $h_l^{(1)}(kr(\theta))$ replaced by $j_l(kr(\theta))$ (and $\xi_l(kr(\theta))$ is replaced by $\psi_l(kr(\theta))$). For the $R_{nn'}$ matrix, we have $C_{ll'} = B_{ll'}$ and $D_{ll'} = -A_{ll'}$ (with $h_l^{(1)}(kr(\theta))$ and $\xi_l(kr(\theta))$ replaced by $j_l(kr(\theta))$ and $\psi_l(kr(\theta))$, respectively).

In general, the matrices R and Q are not symmetric in the l index, and the matrix Q has the ill-conditioned properties discussed above. However, there is a special class of geometries for which both R and Q are symmetric in the l index. These geometries are characterized by [15, 147, 285][4]

$$\frac{1}{r^2(\theta)} = r_0 + r_1 P_1(\cos\theta) + r_2 P_2(\cos\theta)$$

for some real-valued constants r_i, $i = 0, 1, 2$, and $P_l(x)$ is the Legendre polynomial of order l. This class of geometries includes the spheroid as a special case as well as some additional interesting geometries.

If the radius is constant, i.e., a sphere of radius $r = a$, then, due to orthogonality of the associate Legendre functions, the matrices in the τ index are

$$R_{nn'} = \delta_{ll'}\delta_{mm'}\delta_{\sigma\sigma'} \begin{matrix} \\ 1 \\ 2 \end{matrix} \begin{matrix} \quad 1 \quad\quad\quad\quad\quad\quad 2 \\ \begin{pmatrix} \mathrm{i}\psi_l'(ka)\psi_l(ka) & 0 \\ 0 & -\mathrm{i}\psi_l'(ka)\psi_l(ka) \end{pmatrix} \end{matrix}$$

and

$$Q_{nn'} = \delta_{ll'}\delta_{mm'}\delta_{\sigma\sigma'} \begin{matrix} \\ 1 \\ 2 \end{matrix} \begin{matrix} \quad 1 \quad\quad\quad\quad\quad\quad 2 \\ \begin{pmatrix} \mathrm{i}\psi_l'(ka)\xi_l(ka) & 0 \\ 0 & -\mathrm{i}\xi_l'(ka)\psi_l(ka) \end{pmatrix} \end{matrix}$$

This result leads to the same transition matrix entries as in Section 8.1. ■

Low-frequency analysis, such as the electric and magnetic polarizability dyadics, can efficiently be made in the setting we have presented in this section. We refer the reader to the existing literature, especially the work by Waterman [286], and to the two examples on a perfectly conducting disk of radius a below.

Example 9.2

The results in Example 9.1 are illustrated in this example. Let the surface $\boldsymbol{r} = r(\theta)\hat{\boldsymbol{r}}$ trace an oblate spheroid with half axes a and b (b along the symmetry axis), i.e.,

$$r(\theta) = \left(\frac{\cos^2\theta}{b^2} + \frac{\sin^2\theta}{a^2} \right)^{-1/2}$$

In the limit $b \to 0$, we obtain the disk limit. This limit is investigated in detail, see Reference 147, and below we give the results and some numerical examples. The extension to the perfectly conducting elliptic disk is treated in Reference 15.

---

[4]In fact, the result holds for the non-rotationally geometries if

$$\frac{1}{r^2(\hat{\boldsymbol{r}})} = \sum_{l=0}^{2}\sum_{m=0}^{l}\sum_{\sigma=\mathrm{e,o}} r_{\sigma ml} Y_{\sigma ml}(\hat{\boldsymbol{r}})$$

The transition matrix is most conveniently organized as (the indices are $\tau, \sigma, l+m$)

$$
T_{nn'} = \delta_{mm'}
\begin{array}{c|cccccccc}
 & 1ee & 2oo & 1eo & 2oe & 1oe & 2eo & 1oo & 2ee \\
\hline
1ee & 0 & 0 & 0 & 0 & 0 & 0 & 0 & 0 \\
2oo & 0 & 0 & 0 & 0 & 0 & 0 & 0 & 0 \\
1eo & 0 & 0 & T_1 & T_2 & 0 & 0 & 0 & 0 \\
2oe & 0 & 0 & T_3 & T_4 & 0 & 0 & 0 & 0 \\
1oe & 0 & 0 & 0 & 0 & 0 & 0 & 0 & 0 \\
2eo & 0 & 0 & 0 & 0 & 0 & 0 & 0 & 0 \\
1oo & 0 & 0 & 0 & 0 & 0 & 0 & T_1 & -T_2 \\
2ee & 0 & 0 & 0 & 0 & 0 & 0 & -T_3 & T_4
\end{array}
$$

The first term of transition matrix in the low-frequency expansion in power of ka is [147]

$$
\begin{aligned}
T_1 = \delta_{m0} &\left\{ \delta_{l1}\delta_{l'1}\left(-\frac{4\mathrm{i}}{9\pi}(ka)^3 + \frac{4\mathrm{i}}{25\pi}(ka)^5 - \frac{16}{81\pi^2}(ka)^6\right)\right. \\
&\left. + \left(\delta_{l,3}\delta_{l'1} + \delta_{l1}\delta_{l'3}\right)\sqrt{\frac{2}{7}}\frac{4\mathrm{i}}{225\pi}(ka)^5\right\} - \delta_{m1}\delta_{l2}\delta_{l'2}\frac{2\mathrm{i}}{225\pi}(ka)^5 + O\big((ka)^7\big) \\
T_2 = \delta_{m1}&\delta_{l2}\delta_{l'1}\frac{2\mathrm{i}}{9\sqrt{15}\pi}(ka)^5 + O\big((ka)^7\big) \\
T_3 = \delta_{m1}&\delta_{l1}\delta_{l'2}\frac{2\mathrm{i}}{9\sqrt{15}\pi}(ka)^5 + O\big((ka)^7\big) \\
T_4 = \delta_{m,0}&\delta_{l2}\delta_{l'2}\frac{4\mathrm{i}}{675\pi}(ka)^5 + \delta_{m1}\left\{\delta_{l1}\delta_{l'1}\left(\frac{8\mathrm{i}}{9\pi}(ka)^3 + \frac{22\mathrm{i}}{75\pi}(ka)^5 - \frac{64}{81\pi^2}(ka)^6\right)\right. \\
&\left. - \left(\delta_{l3}\delta_{l'1} + \delta_{l1}\delta_{l'3}\right)\frac{16\mathrm{i}}{225\sqrt{21}\pi}(ka)^5\right\} + \delta_{m2}\delta_{l2}\delta_{l'2}\frac{8\mathrm{i}}{225\pi}(ka)^5 + O\big((ka)^7\big)
\end{aligned}
$$

For a linearly polarized plane wave, the scattering dyadic in the forward direction, $\hat{\boldsymbol{r}} = \hat{\boldsymbol{k}}_{\mathrm{i}}$, is also calculated in References 147, 245. The result is (we denote the angle of incidence by α)

$$
\begin{aligned}
\mathbf{S}(\hat{\boldsymbol{k}}_{\mathrm{i}}, \hat{\boldsymbol{k}}_{\mathrm{i}}) = \hat{\boldsymbol{p}}_{\mathrm{e}\|}^{*}\hat{\boldsymbol{p}}_{\mathrm{e}\|}\frac{1}{k}&\left\{\frac{4\cos^2\alpha}{3\pi}(ka)^3 + \frac{2\cos^2\alpha}{45\pi}\left\{16 - 3\sin^2\alpha\right\}(ka)^5\right. \\
&\left. + \frac{32\mathrm{i}\cos^2\alpha}{27\pi^2}(ka)^6 + O\big((ka)^7\big)\right\} \\
+\hat{\boldsymbol{p}}_{\mathrm{e}\perp}^{*}\hat{\boldsymbol{p}}_{\mathrm{e}\perp}\frac{1}{k}&\left\{\frac{2}{3\pi}\left(1 + \cos^2\alpha\right)(ka)^3 + \frac{2}{45\pi}\left\{16 - 5\sin^2\alpha + \sin^4\alpha\right\}(ka)^5\right. \\
&\left. + \frac{8\mathrm{i}}{27\pi^2}\left(\sin^2\alpha + 4\right)(ka)^6 + O\big((ka)^7\big)\right\}
\end{aligned}
\tag{9.8}
$$

where $\hat{\boldsymbol{p}}_{\mathrm{e}\|} = \hat{\boldsymbol{e}}_{\mathrm{i}\|} \cdot \boldsymbol{E}_0 / |\boldsymbol{E}_0|$ and $\hat{\boldsymbol{p}}_{\mathrm{e}\perp} = \hat{\boldsymbol{e}}_{\mathrm{i}\perp} \cdot \boldsymbol{E}_0 / |\boldsymbol{E}_0|$. ■

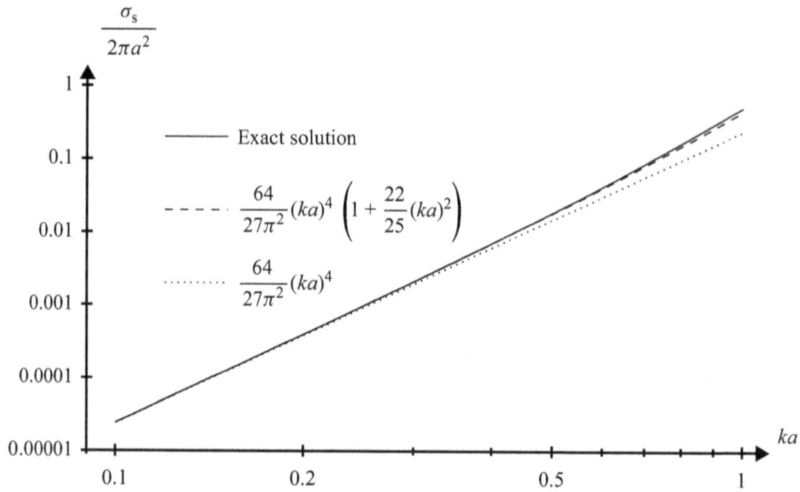

Figure 9.2 The scattering cross section $\sigma_s/2\pi a^2$ for a circular disk with radius a in the long wavelength limit. The incident plane wave impinges perpendicularly to the plane of the disk. Note the log–log scale and the normalization.

Example 9.3

The scattering cross section σ_s for a disk in the long wavelength limit at normal incidence of the plane wave is depicted in Figure 9.2. Since the disk is perfectly conducting, the scattering cross section σ_s is identical to the extinction cross section σ_{ext}. For a linearly polarized incident wave, the Rayleigh approximation is given by [147, 245]

$$\frac{\sigma_s}{2\pi a^2} = \frac{64}{27\pi^2}(ka)^4\left\{1+\frac{22}{25}(ka)^2 + O((ka)^4)\right\}$$

We see that the long wavelength approximation agrees very well for low frequencies just as the analysis suggests.

A comparison of the scattering cross section σ_s for a perfectly conducting disk and a high-frequency approximation of the scattering cross section as a function of the electric radius ka is depicted in Figure 9.3. The frequencies are now much higher than in Figure 9.2. Again, the incident plane wave impinges perpendicularly to the plane of the disk. The short wave approximation (high frequency) is given by [111, 245]

$$\begin{aligned}\frac{\sigma_s}{2\pi a^2} &= 1 - \frac{1}{ka\sqrt{\pi ka}}\sin\left(2ka-\frac{\pi}{4}\right) + \frac{1}{(ka)^2}\left(\frac{3}{4}-\frac{1}{2\pi}\cos 4ka\right)\\ &\quad - \frac{1}{4(ka)^2\sqrt{\pi ka}}\left[\frac{1}{\pi}\sin\left(6ka-\frac{3\pi}{4}\right)+\frac{27}{4}\cos\left(2ka-\frac{\pi}{4}\right)\right] + O((ka)^{-3})\end{aligned}$$

It is remarkable that the high-frequency approximation is appropriate for relatively low frequencies in this case. The high-frequency approximation contains several terms of powers

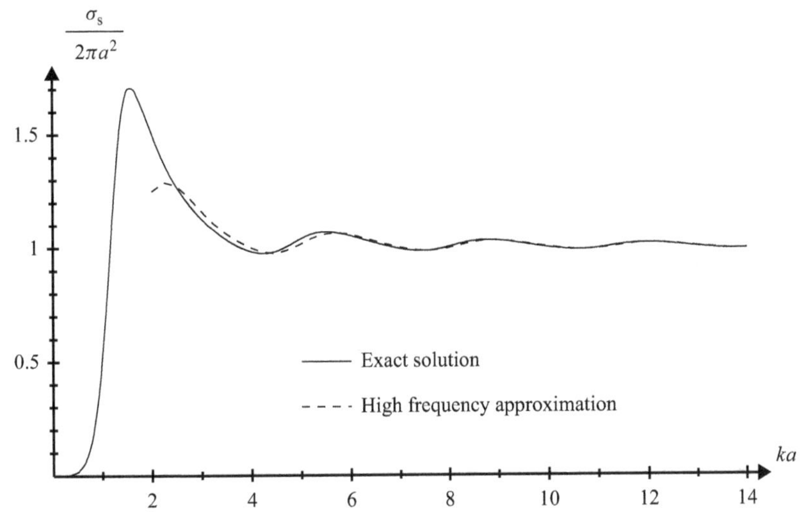

Figure 9.3 The scattering cross section $\sigma_s/2\pi a^2$ for a circular disk with radius a compared with the high-frequency approximation. The incident plane wave impinges perpendicularly to the plane of the disk. Note the normalization.

of $1/\sqrt{ka}$, computed by the geometrical theory of diffraction, and this is the cause of the good agreement down to low frequencies. In Problem 6.4, the reader is encouraged to compute the contribution obtained by the physical optics approximation, which predicts a constant scattering cross section $\sigma_s = 2\pi a^2$.

Moreover, the scattering cross section approaches the limit $2\pi a^2$ (twice the geometrical cross section of the disk) at high frequencies, which the physical optics approximation predicts, cf. Example 6.6 on page 295. We also note that the scatterer shows the largest scattering cross section in the resonance region at $ka \approx 2$. The scattering cross sections for the disk depicted at different incident angles are shown in Figure 9.4. ■

9.1.2 Dielectric Scatterer

The procedure to find the transition matrix for a perfectly conducting object can be generalized to a homogeneous dielectric scatterer. The material parameters of the scatterer are ϵ_1 and μ_1. To this end, start with the integral representation for the electric field, see (4.15) on page 197.

$$
\begin{aligned}
&-\frac{\eta_0\eta}{ik}\nabla\times\left\{\nabla\times\iint_{S_s}\mathbf{G}_e(k,|\boldsymbol{r}-\boldsymbol{r}'|)\cdot(\hat{\boldsymbol{\nu}}(\boldsymbol{r}')\times\boldsymbol{H}(\boldsymbol{r}'))\,\mathrm{d}S'\right\}\\
&+\nabla\times\iint_{S_s}\mathbf{G}_e(k,|\boldsymbol{r}-\boldsymbol{r}'|)\cdot(\hat{\boldsymbol{\nu}}(\boldsymbol{r}')\times\boldsymbol{E}(\boldsymbol{r}'))\,\mathrm{d}S'=\begin{cases}\boldsymbol{E}_s(\boldsymbol{r}), & \boldsymbol{r}\text{ outside }S_s\\-\boldsymbol{E}_i(\boldsymbol{r}), & \boldsymbol{r}\text{ inside }S_s\end{cases}
\end{aligned}
$$

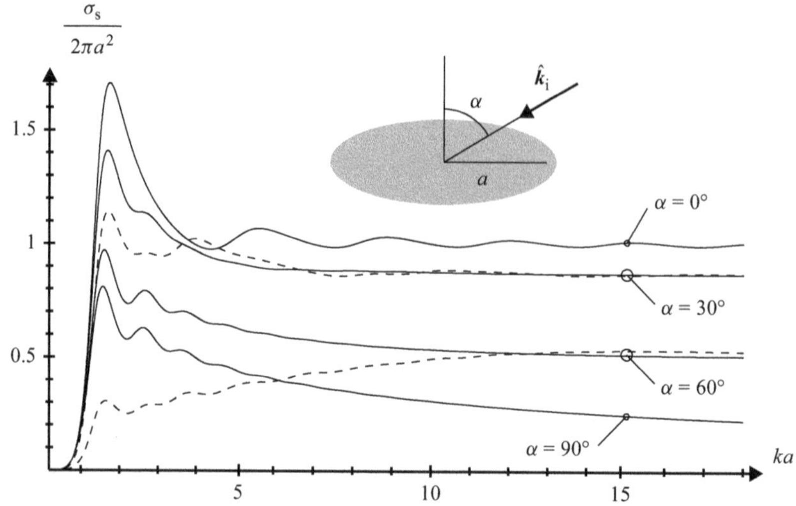

Figure 9.4 The scattering cross section of a perfectly conducting disk of radius a. The angle of incidence of the plane wave is α, and the solid lines show the TE polarization and the broken lines the TM polarization.

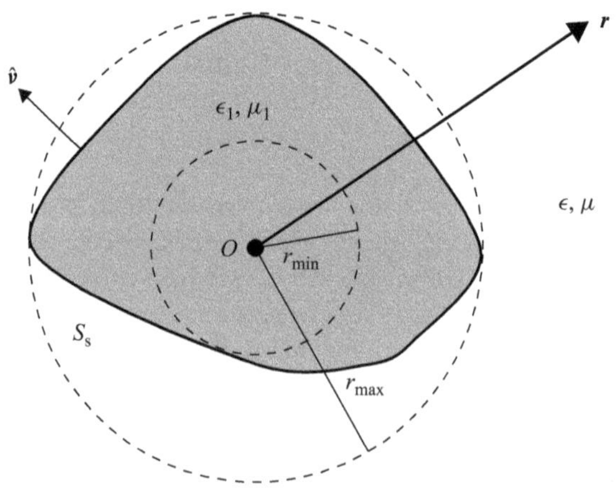

Figure 9.5 The dielectric scatterer and its inscribed and circumscribed spheres.

Let $\boldsymbol{r}$ be a position vector outside the smallest circumscribed sphere of S_s, see Figure 9.5. Then, exactly as in Section 9.1.1, $\boldsymbol{r}_> = \boldsymbol{r}$ and $\boldsymbol{r}_< = \boldsymbol{r}'$ for all points on the surface $\boldsymbol{r}' \in S_s$ in the expansion of the Green dyadic $\mathbf{G}_e(k, |\boldsymbol{r} - \boldsymbol{r}'|)$ in spherical vector waves (9.1). Insert the expansion of the Green dyadic into the integral representation above and change the order of summation and integration. The result of this operation is a general representation of the scattered field outside the smallest circumscribed sphere

$$\boldsymbol{E}_s(\boldsymbol{r}) = \sum_n f_n \boldsymbol{u}_n(k\boldsymbol{r}), \qquad r > r_{\max}$$

where

$$f_n = -k^2\eta_0\eta \iint_{S_s} \boldsymbol{v}_n(k\boldsymbol{r}) \cdot \left(\hat{\boldsymbol{v}}(\boldsymbol{r}) \times \boldsymbol{H}(\boldsymbol{r})\right) \, \mathrm{d}S + \mathrm{i}k^2 \iint_{S_s} \boldsymbol{v}_{\overline{n}}(k\boldsymbol{r}) \cdot \left(\hat{\boldsymbol{v}}(\boldsymbol{r}) \times \boldsymbol{E}(\boldsymbol{r})\right) \, \mathrm{d}S \quad (9.9)$$

where we also used (7.7) on page 356. The surface fields in this expression are the limits of the fields taken from the outside of the interface S_s. Due to the boundary conditions, these surface fields are identical to their limit values from the inside.

To find a representation of the incident field, let the position vector $\boldsymbol{r}$ be located inside the largest inscribed sphere of S_s, see Figure 9.5, so that $\boldsymbol{r}_> = \boldsymbol{r}'$ and $\boldsymbol{r}_< = \boldsymbol{r}$ for all points on the surface $\boldsymbol{r}' \in S_s$. The integral representation then implies

$$\boldsymbol{E}_i(\boldsymbol{r}) = \sum_n a_n \boldsymbol{v}_n(k\boldsymbol{r}), \qquad r < r_{\min}$$

where

$$a_n = k^2\eta_0\eta \iint_{S_s} \boldsymbol{u}_n(k\boldsymbol{r}) \cdot \left(\hat{\boldsymbol{v}}(\boldsymbol{r}) \times \boldsymbol{H}(\boldsymbol{r})\right) \, \mathrm{d}S - \mathrm{i}k^2 \iint_{S_s} \boldsymbol{u}_{\overline{n}}(k\boldsymbol{r}) \cdot \left(\hat{\boldsymbol{v}}(\boldsymbol{r}) \times \boldsymbol{E}(\boldsymbol{r})\right) \, \mathrm{d}S \quad (9.10)$$

Again, the boundary conditions show, that the surface fields are identical to their limit values from the inside.

The incident wave is known, i.e., the expansion coefficients a_n are known, and we seek the scattered field, i.e., the expansion coefficients f_n. Formally, this can be accomplished by eliminating the surface fields $\hat{\boldsymbol{v}}(\boldsymbol{r}) \times \boldsymbol{E}(\boldsymbol{r})$ and $\hat{\boldsymbol{v}}(\boldsymbol{r}) \times \boldsymbol{H}(\boldsymbol{r})$ in (9.9) and (9.10).

The elimination of the surface fields can be made by an expansion of the surface fields in a complete set of vector-valued expansion functions on the surface S_s, such as $\hat{\boldsymbol{v}}(\boldsymbol{r}) \times \boldsymbol{v}(k_1\boldsymbol{r})$, where the wave number of the material in the obstacle is $k_1 = (\epsilon_1\mu_1)^{1/2}\omega/c_0$. This is a complete system, and the expansion of the electric and the magnetic fields have the same expansion coefficients, see Comment 9.3. Formally, we have, see also Comment 9.1

$$\begin{cases} \hat{\boldsymbol{v}}(\boldsymbol{r}) \times \boldsymbol{E}(\boldsymbol{r}) = \sum_n \alpha_n \hat{\boldsymbol{v}}(\boldsymbol{r}) \times \boldsymbol{v}_n(k_1\boldsymbol{r}) \\ \eta_0 \hat{\boldsymbol{v}}(\boldsymbol{r}) \times \boldsymbol{H}(\boldsymbol{r}) = \dfrac{1}{\mathrm{i}\eta_1} \sum_n \alpha_n \hat{\boldsymbol{v}}(\boldsymbol{r}) \times \boldsymbol{v}_{\overline{n}}(k_1\boldsymbol{r}) \end{cases} \quad \boldsymbol{r} \in S_s \qquad (9.11)$$

where η_1 is the relative wave impedance of the obstacle. Note that these expansions hold on the surface S_s, and that there is no guarantee that the expansions converge or represent the field inside the scatterer. In fact, there is a convergent expansion of the electric and magnetic fields inside the largest inscribed sphere of the scatterer. This expansion is obtained by employing the integral representation to the domain inside the scatterer.

Comment 9.3

There is, in general, no guarantee that the electric and the magnetic fields in (9.11) have the same expansion coefficients. A more general approach would be two separate expansions of the following form:

$$\begin{cases} \hat{\boldsymbol{v}}(\boldsymbol{r}) \times \boldsymbol{E}(\boldsymbol{r}) = \sum_n \alpha_n \hat{\boldsymbol{v}}(\boldsymbol{r}) \times \boldsymbol{v}_n(k_1\boldsymbol{r}) \\ \mathrm{i}\eta_0\eta_1 \hat{\boldsymbol{v}}(\boldsymbol{r}) \times \boldsymbol{H}(\boldsymbol{r}) = \sum_n \beta_n \hat{\boldsymbol{v}}(\boldsymbol{r}) \times \boldsymbol{v}_{\overline{n}}(k_1\boldsymbol{r}) \end{cases}$$

However, for this particular choice of expansion functions, $\hat{\boldsymbol{\nu}}(\boldsymbol{r}) \times \boldsymbol{v}_n(k_1\boldsymbol{r})$, the expansion coefficients are identical.

Below, we give a sketch of a proof that the coefficients $\alpha_n = \beta_n$ for the expansion functions $\hat{\boldsymbol{\nu}}(\boldsymbol{r}) \times \boldsymbol{v}_n(k_1\boldsymbol{r})$, see also a similar derivation and comments in References 255, 285.

With a homogenous dielectric, we can apply the integral representation to the interior of the scatterer with the appropriate wave number. From (3.14) on page 156, we obtain

$$\begin{aligned}
&\frac{\eta_0\eta_1}{\mathrm{i}k_1}\nabla \times \left\{\nabla \times \iint_{S_s} \mathbf{G}_e(k_1, |\boldsymbol{r}-\boldsymbol{r}'|) \cdot (\hat{\boldsymbol{\nu}}(\boldsymbol{r}') \times \boldsymbol{H}(\boldsymbol{r}'))\, \mathrm{d}S'\right\} \\
&\quad -\nabla \times \iint_{S_s} \mathbf{G}_e(k_1, |\boldsymbol{r}-\boldsymbol{r}'|) \cdot \left(\hat{\boldsymbol{\nu}}(\boldsymbol{r}') \times \boldsymbol{E}(\boldsymbol{r}')\right) \mathrm{d}S' = \begin{cases} \boldsymbol{E}(\boldsymbol{r}), & \boldsymbol{r} \text{ inside } S_s \\ \boldsymbol{0}, & \boldsymbol{r} \text{ outside } S_s \end{cases}
\end{aligned}$$

Now, let $\boldsymbol{r}$ be a position vector outside the smallest circumscribed sphere of S_s. The proper expansion of the Green dyadic is now used. We obtain

$$\begin{aligned}
&\sum_n \left\{\eta_0\eta_1 k_1^2 \iint_{S_s} \boldsymbol{v}_n(k_1\boldsymbol{r}') \cdot (\hat{\boldsymbol{\nu}}(\boldsymbol{r}') \times \boldsymbol{H}(\boldsymbol{r}'))\, \mathrm{d}S' \right. \\
&\quad \left. - \mathrm{i}k_1^2 \iint_{S_s} \boldsymbol{v}_{\overline{n}}(k_1\boldsymbol{r}') \cdot \left(\hat{\boldsymbol{\nu}}(\boldsymbol{r}') \times \boldsymbol{E}(\boldsymbol{r}')\right) \mathrm{d}S' \right\} \boldsymbol{u}_n(k_1\boldsymbol{r}) = \boldsymbol{0}, \qquad r > r_{\max}
\end{aligned}$$

Orthogonality of the radiation spherical vector waves over a spherical surface $r > r_{\max}$ implies that the coefficients must be identically zero for all values of n. From the expression above, we conclude

$$\begin{aligned}
&\mathrm{i}\eta_0\eta_1 \iint_{S_s} \boldsymbol{v}_n(k_1\boldsymbol{r}') \cdot (\hat{\boldsymbol{\nu}}(\boldsymbol{r}') \times \boldsymbol{H}(\boldsymbol{r}'))\, \mathrm{d}S' \\
&\quad + \iint_{S_s} \boldsymbol{v}_{\overline{n}}(k_1\boldsymbol{r}') \cdot \left(\hat{\boldsymbol{\nu}}(\boldsymbol{r}') \times \boldsymbol{E}(\boldsymbol{r}')\right) \mathrm{d}S' = 0, \qquad \text{for all } n
\end{aligned}$$

Insert the expansions of the electric and magnetic fields. We obtain the following relations (for each level of approximation accuracy, the sum in n' is finite):

$$\begin{aligned}
&\sum_{n'} \left\{\beta_{n'} \iint_{S_s} \hat{\boldsymbol{\nu}}(\boldsymbol{r}') \cdot (\boldsymbol{v}_n(k_1\boldsymbol{r}') \times \boldsymbol{v}_{\overline{n}'}(k_1\boldsymbol{r}'))\, \mathrm{d}S' \right. \\
&\quad \left. + \alpha_{n'} \iint_{S_s} \hat{\boldsymbol{\nu}}(\boldsymbol{r}') \cdot \left(\boldsymbol{v}_{\overline{n}}(k_1\boldsymbol{r}') \times \boldsymbol{v}_{n'}(k_1\boldsymbol{r}')\right) \mathrm{d}S' \right\} = 0, \qquad \text{for all } n
\end{aligned}$$

The orthogonality relation in (7.14) on page 360 entails that this expression can be written as

$$\sum_{n'} \iint_{S_s} \left(\hat{\boldsymbol{\nu}}(\boldsymbol{r}') \times \boldsymbol{v}_{\overline{n}}(k_1\boldsymbol{r}')\right) \cdot \sum_{n'} (\alpha_{n'} - \beta_{n'})\, \boldsymbol{v}_{n'}(k_1\boldsymbol{r}')\, \mathrm{d}S' = 0, \quad \text{for all } n$$

Completeness of the system $\{\hat{\boldsymbol{\nu}}(\boldsymbol{r}) \times \boldsymbol{v}_{\overline{n}}(k_1\boldsymbol{r}')\}$, see Section F.9.1, gives

$$\sum_{n} (\alpha_n - \beta_n)\, \boldsymbol{v}_n(k_1\boldsymbol{r}) = 0, \quad \boldsymbol{r} \in S_s$$

The final step in the proof in this comment is to employ the linear independence (except at a countable set of frequencies) of the regular spherical vector functions $\boldsymbol{v}_n(k_1\boldsymbol{r})$. This property implies that $\alpha_n = \beta_n$.

Similar to the PEC obstacle in Section 9.1.1, introduce the R and Q matrices for the dielectric obstacle. These matrices are defined as

$$\begin{cases} R_{nn'} = \mathrm{i}k^2 \displaystyle\iint_{S_s} \left(\frac{\eta}{\eta_1}\boldsymbol{v}_n(k\boldsymbol{r}) \times \boldsymbol{v}_{\overline{n'}}(k_1\boldsymbol{r}) + \boldsymbol{v}_{\overline{n}}(k\boldsymbol{r}) \times \boldsymbol{v}_{n'}(k_1\boldsymbol{r})\right) \cdot \hat{\boldsymbol{\nu}}(\boldsymbol{r})\, \mathrm{d}S \\ Q_{nn'} = \mathrm{i}k^2 \displaystyle\iint_{S_s} \left(\frac{\eta}{\eta_1}\boldsymbol{u}_n(k\boldsymbol{r}) \times \boldsymbol{v}_{\overline{n'}}(k_1\boldsymbol{r}) + \boldsymbol{u}_{\overline{n}}(k\boldsymbol{r}) \times \boldsymbol{v}_{n'}(k_1\boldsymbol{r})\right) \cdot \hat{\boldsymbol{\nu}}(\boldsymbol{r})\, \mathrm{d}S \end{cases} \tag{9.12}$$

Insert the surface field expansions above, and we readily obtain the following relations:

$$\begin{cases} a_n = \displaystyle\sum_{n'} Q_{nn'}\alpha_{n'} \\ f_n = -\displaystyle\sum_{n'} R_{nn'}\alpha_{n'} \end{cases}$$

Formal elimination of the expansion coefficients $\alpha_{n'}$ gives the transition matrix $T_{nn'}$

$$T_{nn'} = -\sum_{n''} R_{nn''} Q^{-1}_{n''n'}$$

and the scattering problem is formally solved.

Example 9.4

In Example 9.1, we treated an axially symmetric body with perfectly conducting boundary conditions in some detail. In this example, we do the similar analysis for an axially symmetric, homogeneous dielectric scatterer, and just as in Section 9.1.1, the surface integrals in (9.12) reduce to line integrals for this type of geometry.

Let the z-axis define the symmetry axis and parameterize the position vector with the spherical angles θ and ϕ, i.e., $\boldsymbol{r} = r(\theta)\hat{\boldsymbol{r}}$ (star shaped scatterer). The normal vector of the surface and the surface element are identical to the ones in Example 9.1, and, again, orthogonality in the azimuthal variable shows that for $m \neq m'$ all entries of the R- and Q-matrices are zero, i.e., $R_{nn'} = Q_{nn'} = 0$ when $m \neq m'$. As a consequence, no m indices couple.

The structure of the Q matrix in the τ and the σ indices is

$$
Q_{nn'} = \delta_{mm'}
\begin{array}{c}
\\ 1\mathrm{e} \\ 2\mathrm{o} \\ 1\mathrm{o} \\ 2\mathrm{e}
\end{array}
\begin{array}{cccc}
1\mathrm{e} & 2\mathrm{o} & 1\mathrm{o} & 2\mathrm{e} \\
\end{array}
\begin{pmatrix}
\frac{\mu}{\mu_1} A_{ll'} + D_{ll'} & \frac{\mu}{\mu_1} B_{ll'} + \frac{k}{k_1} C_{ll'} & 0 & 0 \\
\frac{k}{k_1} B_{ll'} + \frac{\mu}{\mu_1} C_{ll'} & \frac{k}{k_1} A_{ll'} + \frac{\eta}{\eta_1} D_{ll'} & 0 & 0 \\
0 & 0 & \frac{\mu}{\mu_1} A_{ll'} + D_{ll'} & -\frac{\mu}{\mu_1} B_{ll'} - \frac{k}{k_1} C_{ll'} \\
0 & 0 & -\frac{k}{k_1} B_{ll'} - \frac{\mu}{\mu_1} C_{ll'} & \frac{k}{k_1} A_{ll'} + \frac{\eta}{\eta_1} D_{ll'}
\end{pmatrix}
$$

where the matrices are (the Riccati–Bessel functions ψ_l and ξ_l defined in Appendix B.3.3 are used)

$$
\begin{aligned}
A_{ll'} = \mathrm{i} N_{lm} N_{l'm} \int_0^\pi & \left\{ \Theta_{ll'}^m(\theta) \xi_l(kr(\theta)) \psi'_{l'}(k_1 r(\theta)) \right. \\
& \left. + l'(l'+1) \frac{\mathrm{d}P_l^m(\cos\theta)}{\mathrm{d}\theta} P_{l'}^m(\cos\theta) kr'(\theta) h_l^{(1)}(kr(\theta)) j_{l'}(k_1 r(\theta)) \right\} \sin\theta \, \mathrm{d}\theta
\end{aligned}
$$

$$
B_{ll'} = \mathrm{i} N_{lm} N_{l'm} m \int_0^\pi \frac{\mathrm{d}(P_l^m(\cos\theta) P_{l'}^m(\cos\theta))}{\mathrm{d}\theta} \xi_l(kr(\theta)) \psi_{l'}(k_1 r(\theta)) \, \mathrm{d}\theta
$$

$$
\begin{aligned}
C_{ll'} = -\mathrm{i} N_{lm} N_{l'm} m \int_0^\pi & \left\{ \frac{\mathrm{d}(P_l^m(\cos\theta) P_{l'}^m(\cos\theta))}{\mathrm{d}\theta} \xi'_l(kr(\theta)) \psi'_{l'}(k_1 r(\theta)) \right. \\
& + l(l+1) P_l^m(\cos\theta) P_{l'}^m(\cos\theta) \frac{r'(\theta)}{r(\theta)} h_l^{(1)}(kr(\theta)) \psi'_{l'}(k_1 r(\theta)) \\
& \left. + l'(l'+1) P_l^m(\cos\theta) P_{l'}^m(\cos\theta) \frac{r'(\theta)}{r(\theta)} \xi'_l(kr(\theta)) j_{l'}(k_1 r(\theta)) \right\} \mathrm{d}\theta
\end{aligned}
$$

and

$$
\begin{aligned}
D_{ll'} = -\mathrm{i} N_{lm} N_{l'm} \int_0^\pi & \left\{ \Theta_{ll'}^m(\theta) \xi'_l(kr(\theta)) j_{l'}(k_1 r(\theta)) kr(\theta) \right. \\
& \left. + l(l+1) P_l^m(\cos\theta) \frac{\mathrm{d}P_{l'}^m(\cos\theta)}{\mathrm{d}\theta} kr'(\theta) h_l^{(1)}(kr(\theta)) j_{l'}(k_1 r(\theta)) \right\} \sin\theta \, \mathrm{d}\theta
\end{aligned}
$$

where the normalization factor N_{lm} and the combination of associated Legendre functions $\Theta_{ll'}^m(\theta)$ are defined in Example 9.1. The $R_{nn'}$ is identical in structure to $Q'_{nn'}$, but with $h_l^{(1)}(kr(\theta))$ replaced by $j_l(kr(\theta))$ (and $\xi_l(kr(\theta))$ is replaced by $\psi_l(kr(\theta))$).

If the radius is constant, i.e., a sphere of radius $r = a$, then, as matrix in the τ index, the Q matrices, due to orthogonality, are (diagonal in the l, m, and σ indices)

$$R_{nn'} = \frac{\mathrm{i}k}{k_1} \begin{pmatrix} \frac{\eta}{\eta_1}\psi_l(ka)\psi_l'(k_1a) - \psi_l'(ka)\psi_l(k_1a) & 0 \\ 0 & \psi_l(ka)\psi_l'(k_1a) - \frac{\eta}{\eta_1}\psi_l'(ka)\psi_l(k_1a) \end{pmatrix}$$

and

$$Q_{nn'} = \frac{\mathrm{i}k}{k_1} \begin{pmatrix} \frac{\eta}{\eta_1}\xi_l(ka)\psi_l'(k_1a) - \xi_l'(ka)\psi_l(k_1a) & 0 \\ 0 & \xi_l(ka)\psi_l'(k_1a) - \frac{\eta}{\eta_1}\xi_l'(ka)\psi_l(k_1a) \end{pmatrix}$$

This result leads to the same transition matrix entries as in Section 8.2. ■

In the literature, there are numerous computer codes available, e.g., References 9, 57, 182, and the comprehensive database of Null-field approach [189–194, 303]. Some of the examples in this sections are computed using these codes.

Example 9.5

In Figure 9.6, we illustrate scattering by a spheroidal dielectric obstacle. The entries of the scattering matrix $|S_{\|\perp}|^2$ and $|S_{\perp\perp}|^2$ for a dielectric prolate spheroid are depicted. The prolate spheroid has semi-axes a and b, where the half axis a is parallel to the symmetry axis of the spheroid. In this particular example, the ratio between the semi-axes is $a/b = 2$, the material parameters are $\epsilon_1/\epsilon = 2.25$ and $\mu_1/\mu = 1$, and the electric size is $ka = 10$.

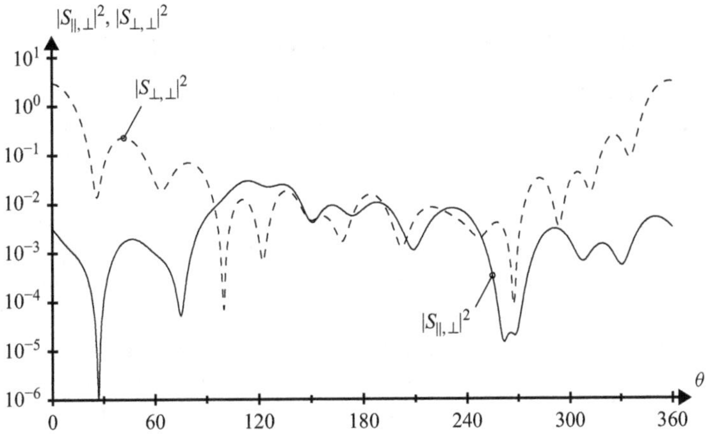

Figure 9.6 The entries of the scattering matrix $|S_{\|\perp}|^2$ and $|S_{\perp\perp}|^2$ for a dielectric prolate spheroid with data in the text.

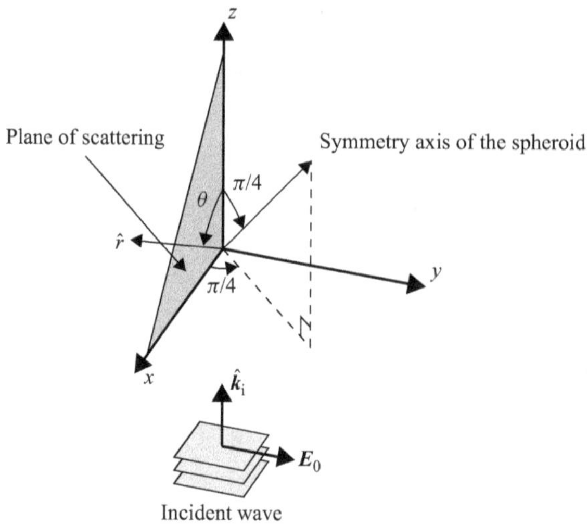

Figure 9.7 Definition of the geometry for the scatterer and the incident plane wave in Figure 9.6.

The symmetry axis of the spheroid is inclined 45° to the z-axis and 45° to the scattering plane, see Figure 9.7. The incident plane wave impinges along the positive z-direction, and the electric field of the plane wave is linearly polarized perpendicular to the scattering plane, i.e., in the y-direction. The scattering angle θ is the angle between the positive z-direction and the direction of observation $\hat{\boldsymbol{r}}$.

Note that the curves are not symmetric around $\theta = 180°$ and that the intensities, parallel and perpendicular, to the scattering plane are not equal. The scattering and the extinction cross sections are identical (lossless material parameters), $\sigma_s/2\pi a^2 = \sigma_{ext}/2\pi a^2 = 0.570$. Each maximum gives a side lobe. Note also that the number of side lobes is different for the two scattering polarizations. ■

9.2 The *T*-Matrix for a Collection of Scatterers

After the examination of scattering by a single object, we are now ready for the analysis of scattering by a collection of scatterer, which requires a much heavier machinery. Other complex geometries—multilayered and composites objects—are treated in References 214, 255–257, 306, 307.

The starting point of scattering by a collection of scatterers is the pioneer work by Foldy [71]. Several important works followed, see e.g., References 20, 25, 70, 155, 156, 174, 176, 183, 213, 259, 260, 265, 267–269, 279, 280, 287, 292, 293, and further references to the subject are found therein. The topic is also treated in several textbooks, see e.g., Reference [264, Chapter 6]. The results developed in this section are also the starting point for a treatment of randomly located scatterers, see e.g., Reference [141] and references given theirin.

We study a collection of N different scatterers, V_{sp}, with bounding surface S_{sp}, $p = 1, 2, \ldots, N$, where each scatterer is centered at the location $\boldsymbol{r}_p$, defining the position of the local origin O_p, $p = 1, 2, \ldots, N$, relative the global origin O, see Figure 9.8. The radii of the largest inscribed and smallest circumscribed spheres, both centered at the local origin, of each scatterer are denoted a_p and A_p, $p = 1, 2, \ldots, N$, respectively. The scatterers are located in

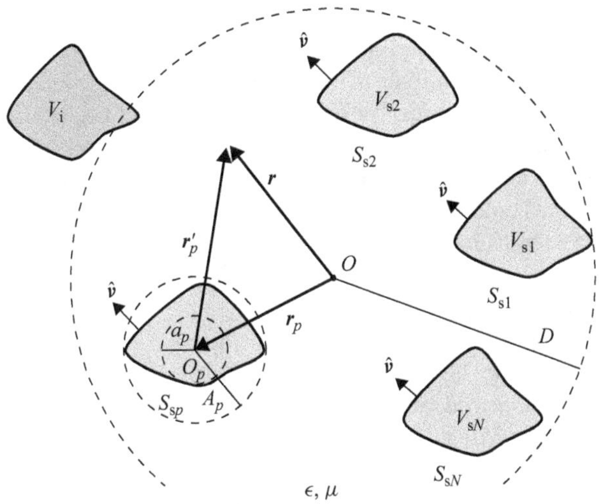

Figure 9.8 The geometry of a collection of the N scatterers and the region of prescribed sources V_i. The positions of the local origins are $\boldsymbol{r}_p, p = 1, \ldots, N$, and the radii of the largest inscribed and the smallest circumscribed spheres of each local scatterer are a_p and A_p, respectively. The entire collection of scatterers is enclosed by a sphere with radius D.

a lossless, homogeneous, isotropic media with permittivity ϵ and permeability μ. The wave number and relative wave impedance are k and η, respectively—both real numbers. We assume that no circumscribing spheres intersect. Each scatterer has its own material properties, which do not have to be the same for all scatterers. The entire collection of scatterers is enclosed by a sphere (circumscribed sphere of the entire collection) with radius D, see Figure 9.8. The prescribed sources are located in the region V_i, which is a region disjoint to all scatterers,[5] and these sources generate the field $\boldsymbol{E}_i(\boldsymbol{r})$ everywhere outside V_i.

We start with a collection of perfectly conducting obstacles, and then generalize to more complex scatterers below. As we see below, this assumption is no loss of generality. It is only to make the derivation of the results simpler. The final result holds for a collection of scatterers with arbitrary materials.

Each scatterer is characterized by the boundary condition—the total electric field, $\boldsymbol{E} = \boldsymbol{E}_i(\boldsymbol{r}) + \boldsymbol{E}_s(\boldsymbol{r})$, satisfies $\hat{\boldsymbol{\nu}} \times \boldsymbol{E} = \boldsymbol{0}$ on all $S_{sp}, p = 1, 2, \ldots, N$. Our starting point is the integral representation for the electric field, see (4.15) on page 197

$$-\frac{\eta_0\eta}{\mathrm{i}k}\nabla \times \left\{\nabla \times \sum_{p=1}^{N} \iint_{S_{sp}} \mathbf{G}_e(k, |\boldsymbol{r} - \boldsymbol{r}'|) \cdot \boldsymbol{J}_S(\boldsymbol{r}') \, \mathrm{d}S'\right\}$$

$$= \begin{cases} \boldsymbol{E}_s(\boldsymbol{r}), & \boldsymbol{r} \text{ outside all } S_{sp} \\ -\boldsymbol{E}_i(\boldsymbol{r}), & \boldsymbol{r} \text{ inside any } S_{sp} \end{cases} \quad p = 1, 2, \ldots, N \qquad (9.13)$$

[5]More precisely, the circumscribed sphere of the source region must not include any local origin $\boldsymbol{r}_p, p = 1, 2, \ldots, N$. An incident plane wave fulfill these restrictions.

where the boundary condition $\hat{\nu} \times \boldsymbol{E} = \mathbf{0}$ on each S_{sp} has been used, and we have, for convenience, denoted the surface current density $\boldsymbol{J}_S = \hat{\nu} \times \boldsymbol{H}$. The wave impedance of vacuum is denoted η_0. Note that right-hand side of the lower line contains the incident field $\boldsymbol{E}_{\rm i}(\boldsymbol{r})$, which is generated by the external sources. As seen below, this is not the total field that excites the individual scatterer. There are contributions from all other scatterers also, but in the integral representation it is the field generated by the external sources that enters.

The top row in (9.13) holds for all points outside all scatterers. This field, the scattered field $\boldsymbol{E}_{\rm s}(\boldsymbol{r})$, is due to the sources inside the scatterers, and it is source-free outside the scatterers, and it satisfies appropriate radiation conditions at infinity. This is the formal definition of the scattered field $\boldsymbol{E}_{\rm s}(\boldsymbol{r})$.

The lower row in (9.13) holds for all points inside a particular scatterer, say the pth, and represents the prescribed field (more precisely, $-\boldsymbol{E}_{\rm i}(\boldsymbol{r})$) at the location of the pth scatterer.

We decompose the Green dyadic for the electric field in free space in spherical vector waves, see (7.24) on page 370

$$\mathbf{G}_{\rm e}(k, |\boldsymbol{r} - \boldsymbol{r}'|) = {\rm i}k \sum_n \boldsymbol{v}_n(k\boldsymbol{r}_<)\boldsymbol{u}_n(k\boldsymbol{r}_>)$$

where the summation in the index τ is only over $\tau = 1, 2$, and where $\boldsymbol{r}_<$ ($\boldsymbol{r}_>$) is the position vector with the smallest (largest) distance to the origin, i.e., if $r < r'$ then $\boldsymbol{r}_< = \boldsymbol{r}$ and $\boldsymbol{r}_> = \boldsymbol{r}'$. Uniform convergence holds in finite domains, provided $r \neq r'$ in the domain [40].

The incident field is, as above, a plane wave[6] impinging along the direction $\hat{\boldsymbol{k}}_{\rm i}$ with an expansion in regular spherical vector waves, i.e.,

$$\boldsymbol{E}_{\rm i}(\boldsymbol{r}) = \boldsymbol{E}_0 {\rm e}^{{\rm i}k\hat{\boldsymbol{k}}_{\rm i}\cdot\boldsymbol{r}} = \sum_n a_n \boldsymbol{v}_n(k\boldsymbol{r})$$

where $\boldsymbol{E}_0$ is the amplitude of the plane wave at the common origin O, and the expansion coefficients, a_n, are given in (7.27) on page 374.

For an observation point $\boldsymbol{r}$ outside the circumscribed sphere (centered at the global origin O) of the entire collection of scatterers, we obtain a representation of the scattered field from the integral representation (9.13)

$$\boldsymbol{E}_{\rm s}(\boldsymbol{r}) = \sum_n f_n \boldsymbol{u}_n(k\boldsymbol{r}), \qquad r > D \tag{9.14}$$

where the expansion coefficients f_n are

$$f_n = -k^2 \eta_0 \eta \sum_{p=1}^{N} \iint_{S_{sp}} \boldsymbol{v}_n(k\boldsymbol{r}) \cdot \boldsymbol{J}_S(\boldsymbol{r})\, {\rm d}S$$

As above, the linear relation between the expansion coefficients of the scattered field, f_n, and the expansion coefficients, a_n, of the incident field is given by the transition matrix, $T_{nn'}$, of

[6] More general incident fields can be handled with the present approach, see Problem 9.3.

the entire collection of scatterers

$$f_n = \sum_{n'} T_{nn'} a_{n'}$$

The expression in (9.14) gives the scattered field in terms of the transition matrix of the entire collection of scatterers. There is, however, another procedure which emphasizes the effect of each scatterer more clearly. This analysis below resembles the analysis by Peterson and Ström [213], but emphasizes expansions w.r.t. the local origins, O_p, $p = 1, 2, \dots, N$, instead of the global origin, O.

Assume that the observation point lies outside all circumscribed spheres of the scatterers, i.e., $\boldsymbol{r}$ satisfies $|\boldsymbol{r} - \boldsymbol{r}_p| > A_p$ for all $p = 1, 2, \dots, N$. The Green's dyadic has the expansion, see (7.24) on page 370

$$\mathbf{G}_\mathrm{e}(k, |\boldsymbol{r} - \boldsymbol{r}'|) = \mathbf{G}_\mathrm{e}(k, |\boldsymbol{r} - \boldsymbol{r}_p - (\boldsymbol{r}' - \boldsymbol{r}_p)|) = \mathrm{i}k \sum_n \boldsymbol{v}_n(k(\boldsymbol{r}' - \boldsymbol{r}_p))\boldsymbol{u}_n(k(\boldsymbol{r} - \boldsymbol{r}_p)) \quad (9.15)$$

where $\boldsymbol{r}' \in S_{\mathrm{s}p}$, $p = 1, 2, \dots, N$. Notice that $|\boldsymbol{r}' - \boldsymbol{r}_p| < |\boldsymbol{r} - \boldsymbol{r}_p|$ for all $p = 1, 2, \dots, N$. The integral representation in (9.13) implies

$$\boldsymbol{E}_\mathrm{s}(\boldsymbol{r}) = \sum_{p=1}^{N} \sum_n f_n^p \boldsymbol{u}_n(k(\boldsymbol{r} - \boldsymbol{r}_p)), \qquad |\boldsymbol{r} - \boldsymbol{r}_p| > A_p, \ p = 1, 2, \dots, N \quad (9.16)$$

where the expansion coefficients, f_n^p, are

$$f_n^p = -k^2 \eta_0 \eta \iint_{S_{\mathrm{s}p}} \boldsymbol{v}_n(k(\boldsymbol{r}' - \boldsymbol{r}_p)) \cdot \boldsymbol{J}_S(\boldsymbol{r}') \, \mathrm{d}S', \qquad p = 1, 2, \dots, N$$

The expansion coefficients f_n^p depend not only on the location of the pth scatterer, $\boldsymbol{r}_p$, and its properties, but also on the locations and properties of all other scatters. To avoid cumbersome notation, however, we do not explicitly write out these variables. Thus, the scattered field, $\boldsymbol{E}_\mathrm{s}$, has a representation in terms of a contribution from each scatterer

$$\boldsymbol{E}_\mathrm{s}(\boldsymbol{r}) = \sum_{p=1}^{N} \boldsymbol{E}_{\mathrm{s}p}(\boldsymbol{r})$$

where

$$\boldsymbol{E}_{\mathrm{s}p}(\boldsymbol{r}) = \sum_n f_n^p \boldsymbol{u}_n(k(\boldsymbol{r} - \boldsymbol{r}_p)), \qquad |\boldsymbol{r} - \boldsymbol{r}_p| > A_p, \ p = 1, 2, \dots, N$$

Similarly, for an observation point inside the inscribed sphere of a particular scatterer, say the pth scatterer, we have $|\boldsymbol{r} - \boldsymbol{r}_p| < a_p$, and

$$\mathbf{G}_\mathrm{e}(k, |\boldsymbol{r} - \boldsymbol{r}'|) = \mathbf{G}_\mathrm{e}(k, |\boldsymbol{r} - \boldsymbol{r}_p - (\boldsymbol{r}' - \boldsymbol{r}_p)|) = \mathrm{i}k \sum_n \boldsymbol{u}_n(k(\boldsymbol{r}' - \boldsymbol{r}_p))\boldsymbol{v}_n(k(\boldsymbol{r} - \boldsymbol{r}_p))$$

for this particular scatterer. For all other scatterers, the expansion in (9.15) holds. The integral representation in (9.13) implies ($p = 1, 2, \ldots, N$)

$$\boldsymbol{E}_{\mathrm{i}}(\boldsymbol{r}) = \sum_n \alpha_n^p \boldsymbol{v}_n(k(\boldsymbol{r} - \boldsymbol{r}_p)) - \sum_{\substack{q=1 \\ q\neq p}}^{N} \sum_n f_n^q \boldsymbol{u}_n(k(\boldsymbol{r} - \boldsymbol{r}_q)), \qquad |\boldsymbol{r} - \boldsymbol{r}_p| < a_p \tag{9.17}$$

where the expansion coefficients, α_n^p, are defined as

$$\alpha_n^p = k^2 \eta_0 \eta \iint_{S_{\mathrm{s}p}} \boldsymbol{u}_n(k(\boldsymbol{r}' - \boldsymbol{r}_p)) \cdot \boldsymbol{J}_S(\boldsymbol{r}') \, \mathrm{d}S', \qquad p = 1, 2, \ldots, N$$

Equation (9.17) can be written as

$$\boldsymbol{E}_{\mathrm{exc}p}(\boldsymbol{r}) = \sum_n \alpha_n^p \boldsymbol{v}_n(k(\boldsymbol{r} - \boldsymbol{r}_p)) = \boldsymbol{E}_{\mathrm{i}}(\boldsymbol{r}) + \sum_{\substack{q=1 \\ q\neq p}}^{N} \boldsymbol{E}_{\mathrm{s}q}(\boldsymbol{r}), \qquad |\boldsymbol{r} - \boldsymbol{r}_p| < a_p \tag{9.18}$$

where $\boldsymbol{E}_{\mathrm{exc}p}(\boldsymbol{r})$ is the exciting field at the position of the scatterer located at $\boldsymbol{r}_p$, which consists of the incident field $\boldsymbol{E}_{\mathrm{i}}(\boldsymbol{r})$ plus all scattered fields from the scatterers except the contribution from the pth scatterer.[7]

To obtain a relation in terms of the unknown coefficients, f_n^p, we need to compare this relation using coordinates referring to the same origin, in this case O_p. The translation properties of the spherical vector waves are now employed [23, 174]. In particular, we have with the translation properties of Section 7.4.2 ($\boldsymbol{d} = \boldsymbol{r}_p - \boldsymbol{r}_q$)

$$\begin{aligned}\boldsymbol{u}_n(k(\boldsymbol{r} - \boldsymbol{r}_q)) &= \boldsymbol{u}_n(k(\boldsymbol{r} - \boldsymbol{r}_p) + k(\boldsymbol{r}_p - \boldsymbol{r}_q)) \\ &= \sum_{n'} \mathcal{P}_{nn'}(k(\boldsymbol{r}_p - \boldsymbol{r}_q)) \boldsymbol{v}_{n'}(k(\boldsymbol{r} - \boldsymbol{r}_p)), \qquad |\boldsymbol{r} - \boldsymbol{r}_p| < |\boldsymbol{r}_p - \boldsymbol{r}_q|\end{aligned}$$

The incident field has also an expansion in terms of the local origin, O_p. From above, we get by a simple translation of the origin of the incident plane wave

$$\boldsymbol{E}_{\mathrm{i}}(\boldsymbol{r}) = \mathrm{e}^{\mathrm{i}k\hat{\boldsymbol{k}}_{\mathrm{i}} \cdot \boldsymbol{r}_p} \boldsymbol{E}_0 \mathrm{e}^{\mathrm{i}k\hat{\boldsymbol{k}}_{\mathrm{i}} \cdot (\boldsymbol{r} - \boldsymbol{r}_p)} = \mathrm{e}^{\mathrm{i}k\hat{\boldsymbol{k}}_{\mathrm{i}} \cdot \boldsymbol{r}_p} \sum_n a_n \boldsymbol{v}_n(k(\boldsymbol{r} - \boldsymbol{r}_p))$$

Orthogonality of the vector spherical harmonics inside the inscribed sphere of the pth scatterer applied to (9.18) implies

$$\alpha_n^p = \mathrm{e}^{\mathrm{i}k\hat{\boldsymbol{k}}_{\mathrm{i}} \cdot \boldsymbol{r}_p} a_n + \sum_{\substack{q=1 \\ q\neq p}}^{N} \sum_{n'} f_{n'}^q \mathcal{P}_{n'n}(k(\boldsymbol{r}_p - \boldsymbol{r}_q)), \qquad p = 1, 2, \ldots, N \tag{9.19}$$

This set of equations is equivalent to the result (4.1) in Peterson and Ström [213].

---

[7]This field is not the total field at the pth scatterer with the pth scatterer removed, since the presence of the pth scatterer is needed to generate the correct sources at the locations of all other scatterers.

The transition matrix of the pth scatterer, $T^p_{nn'}$, connects the expansion coefficients α^p_n and f^p_n to each other, viz.

$$f^p_n = \sum_{n'} T^p_{nn'} \alpha^p_{n'}, \qquad p = 1, 2, \ldots, N$$

This relation is analogous to the result for a single scatterer case presented in Section 9.1. Notice that the field that excites the pth scatterer is $\boldsymbol{E}_{\text{exc}p}(\boldsymbol{r})$—not only the incident field $\boldsymbol{E}_{\text{i}}(\boldsymbol{r})$. Equation (9.19) then becomes (we also have used the relation $\mathcal{P}_{n'n}(-k\boldsymbol{r}) = \mathcal{P}_{nn'}(k\boldsymbol{r})$)

$$\alpha^p_n = \mathrm{e}^{\mathrm{i}k\hat{\boldsymbol{k}}_{\mathrm{i}}\cdot\boldsymbol{r}_p} a_n + \sum_{\substack{q=1 \\ q\neq p}}^{N} \sum_{n',n''} \mathcal{P}_{nn'}(k(\boldsymbol{r}_q - \boldsymbol{r}_p)) T^q_{n'n''} \alpha^q_{n''}, \qquad p = 1, 2, \ldots, N \tag{9.20}$$

For a given set of locations of the local origins, $\boldsymbol{r}_p$, and the individual transition matrices $T^p_{nn'}$, $p = 1, \ldots, N$, this set of linear equations determines the unknown expansion coefficients α^p_n in terms of the excitation characterized by the expansion coefficients, a_n. This relation is identical to the relation found in the literature, see e.g., References 169, 259, 278.

At this stage, we notice that if the scatterer is not a perfectly conducting scatterer as assumed above, the result can be generalized. The main reason for the assumption of perfectly conducting scatterers was to simplify the theoretical work. If a more general scatterer is present, the only change that has to be made is to replace the transition matrix of the scatterer with the appropriate one. Therefore, the results above hold for any set of scatterers—single or multiple, transparent or not, homogeneous or not—only the individual transition matrices of the scatterers (non-intersecting circumscribing spheres) are known.

An alternative form of the relation (9.20) is obtained by multiplying the relation with $T^p_{nn'}$ followed by a summation over the free index. The result of this summation is ($p = 1, 2, \ldots, N$)

$$f^p_n = \mathrm{e}^{\mathrm{i}k\hat{\boldsymbol{k}}_{\mathrm{i}}\cdot\boldsymbol{r}_p} \sum_{n'} T^p_{nn'} a_{n'} + \sum_{\substack{q=1 \\ q\neq p}}^{N} \sum_{n',n''} T^p_{nn'} \mathcal{P}_{n'n''}(k(\boldsymbol{r}_q - \boldsymbol{r}_p)) f^q_{n''} \tag{9.21}$$

Again, for a given set of locations of the local origins, $\boldsymbol{r}_p$, and the individual transition matrices $T^p_{nn'}$, $p = 1, \ldots, N$, these equations can be solved to determine the unknown expansion coefficients f^p_n in terms of the excitation characterized by the expansion coefficients, a_n.

The scattered field is therefore possible to compute by (9.16), once f^p_n are known, and these coefficients are obtained from the solution of (9.21). In principle, the field everywhere outside the scatterers (more precisely, outside the circumscribing spheres of the scatterers) can be obtained.

The coefficients of the scattered field, f_n, in terms of the expansion coefficients, f^p_n, of the local scatterers are obtained by the use of the translation matrices, see Section 7.4.2. We get, see (9.16) (we also have use $\mathcal{R}_{nn'}(-k\boldsymbol{r}_p) = \mathcal{R}_{n'n}(k\boldsymbol{r}_p)$)

$$\begin{aligned}
\boldsymbol{E}_{\text{s}}(\boldsymbol{r}) &= \sum_{p=1}^{N} \boldsymbol{E}_{\text{s}p}(\boldsymbol{r}) = \sum_{p=1}^{N} \sum_{n} f^p_n \boldsymbol{u}_n(k(\boldsymbol{r} - \boldsymbol{r}_p)) \\
&= \sum_{p=1}^{N} \sum_{n,n'} \boldsymbol{u}_{n'}(k\boldsymbol{r}) \mathcal{R}_{n'n}(k\boldsymbol{r}_p) f^p_n, \qquad |\boldsymbol{r} - \boldsymbol{r}_p| > A_p, \; p = 1, 2, \ldots, N
\end{aligned} \tag{9.22}$$

which implies, see (9.14)

$$f_n = \sum_{p=1}^{N} \sum_{n'} \mathcal{R}_{nn'}(k\boldsymbol{r}_p) f_{n'}^{p} \tag{9.23}$$

Example 9.6

The simplest system of several scatterers is a system consisting of two objects, $N = 2$. The two local origins are separated by $\boldsymbol{d} = \boldsymbol{r}_2 - \boldsymbol{r}_1$. The basic system to solve, (9.21), then reads[8]

$$\begin{cases} f_n^1 = \mathrm{e}^{\mathrm{i}k\hat{\boldsymbol{k}}_\mathrm{i}\cdot\boldsymbol{r}_1} \sum_{n'} T_{nn'}^1 a_{n'} + \sum_{n',n''} T_{nn'}^1 \mathcal{P}_{n'n''}(k\boldsymbol{d}) f_{n''}^2 \\ f_n^2 = \mathrm{e}^{\mathrm{i}k\hat{\boldsymbol{k}}_\mathrm{i}\cdot\boldsymbol{r}_2} \sum_{n'} T_{nn'}^2 a_{n'} + \sum_{n',n''} T_{nn'}^2 \mathcal{P}_{n'n''}(-k\boldsymbol{d}) f_{n''}^1 \end{cases}$$

or in a matrix notation

$$\begin{cases} f^1 = \mathrm{e}^{\mathrm{i}k\hat{\boldsymbol{k}}_\mathrm{i}\cdot\boldsymbol{r}_1} T^1 a + T^1 \mathcal{P}(k\boldsymbol{d}) f^2 \\ f^2 = \mathrm{e}^{\mathrm{i}k\hat{\boldsymbol{k}}_\mathrm{i}\cdot\boldsymbol{r}_2} T^2 a + T^2 \mathcal{P}^\mathrm{t}(k\boldsymbol{d}) f^1 \end{cases}$$

with solution, see Problem 9.4

$$\begin{cases} f^1 = \left(I - T^1\mathcal{P}(k\boldsymbol{d})T^2\mathcal{P}^\mathrm{t}(k\boldsymbol{d})\right)^{-1} \left(\mathrm{e}^{\mathrm{i}k\hat{\boldsymbol{k}}_\mathrm{i}\cdot\boldsymbol{r}_1} T^1 + \mathrm{e}^{\mathrm{i}k\hat{\boldsymbol{k}}_\mathrm{i}\cdot\boldsymbol{r}_2} T^1\mathcal{P}(k\boldsymbol{d})T^2\right) a \\ f^2 = \left(I - T^2\mathcal{P}^\mathrm{t}(k\boldsymbol{d})T^1\mathcal{P}(k\boldsymbol{d})\right)^{-1} \left(\mathrm{e}^{\mathrm{i}k\hat{\boldsymbol{k}}_\mathrm{i}\cdot\boldsymbol{r}_2} T^2 + \mathrm{e}^{\mathrm{i}k\hat{\boldsymbol{k}}_\mathrm{i}\cdot\boldsymbol{r}_1} T^2\mathcal{P}^\mathrm{t}(k\boldsymbol{d})T^1\right) a \end{cases}$$

and the total transition matrix is obtained by (9.23)

$$\begin{aligned} T = {} & \mathcal{R}(k\boldsymbol{r}_1)\left(I - T^1\mathcal{P}(k\boldsymbol{d})T^2\mathcal{P}^\mathrm{t}(k\boldsymbol{d})\right)^{-1} \left(\mathrm{e}^{\mathrm{i}k\hat{\boldsymbol{k}}_\mathrm{i}\cdot\boldsymbol{r}_1} T^1 + \mathrm{e}^{\mathrm{i}k\hat{\boldsymbol{k}}_\mathrm{i}\cdot\boldsymbol{r}_2} T^1\mathcal{P}(k\boldsymbol{d})T^2\right) \\ & + \mathcal{R}(k\boldsymbol{r}_2)\left(I - T^2\mathcal{P}^\mathrm{t}(k\boldsymbol{d})T^1\mathcal{P}(k\boldsymbol{d})\right)^{-1} \left(\mathrm{e}^{\mathrm{i}k\hat{\boldsymbol{k}}_\mathrm{i}\cdot\boldsymbol{r}_2} T^2 + \mathrm{e}^{\mathrm{i}k\hat{\boldsymbol{k}}_\mathrm{i}\cdot\boldsymbol{r}_1} T^2\mathcal{P}^\mathrm{t}(k\boldsymbol{d})T^1\right) \end{aligned}$$

This is one of many different forms of the total transition matrix of the two scatterers case. However, if the goal is to compute the scattered field numerically, the expression of the scattered field in (9.16) is more advantageous to use. An alternative derivation of the T-matrix is given in Reference [174, Section 7.10.2]. ■

Example 9.7

The two-object case in Example 9.6 is straightforwardly implemented numerically using the scattered field in (9.16) [175].

---

[8]A word of caution: The superscript 2 on T and f denotes the appropriate scatterer and not the square of the element.

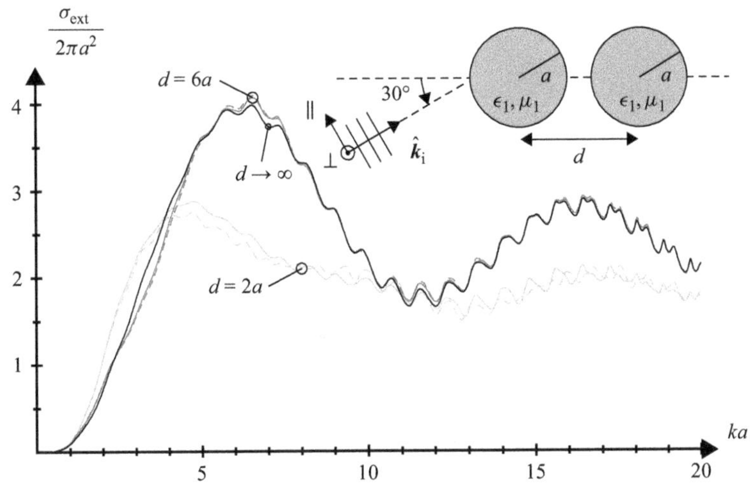

Figure 9.9 The normalized extinction cross section for two identical, dielectric spheres as a function of ka separated by a distance d. The solid curves display the parallel polarization and the dashed curves display the perpendicular polarization of the incident wave. The data of the spheres are $\epsilon_1/\epsilon = 1.33^2$ and $\mu_1/\mu = 1$.

In Figure 9.9, we show the normalized extinction cross section as a function of ka for two, identical, non-magnetic, dielectric spheres separated by a distance d. The plane wave impinges on the spheres at an angle of 30° from the symmetry axis. The data of the spheres are $\epsilon_1/\epsilon = 1.33^2$ and $\mu_1/\mu = 1$. The infinite separation of the two spheres corresponds to the sum of the extinction cross sections of the two isolated spheres, cf. Figure 8.20 (no absorption in this example, i.e., $\sigma_{\text{ext}} = \sigma_{\text{s}}$).

In Figure 9.10, the extinction cross section for a non-magnetic, dielectric sphere of radius a and a metallic (PEC) disk of radius a is depicted. The sphere has the same data as in Figure 9.9, and the symmetry axis of the disk coincides with the axis connecting the two local origins of the objects. The plane wave impinges on the sphere at an angle of 30° from the symmetry axis, see Figure 9.10. The result for infinite separation between the sphere and disk is represented by the sum of the extinction cross sections of the isolated sphere and the isolated disk (with the appropriate polarization of the incident plane wave), cf. Figures 8.20 and 9.4.

In both figures in this example, we notice that the multiple scattering effects are significant when the two objects are close to each other. On the other hand, the convergence to the sum of the extinction cross section of the isolated individual obstacles in Figure 9.9 goes quite rapidly as a function of the separation distance d. ■

9.2.1 Iterative Solution

The infinite set of equations in the unknown f_n^p, (9.21), can be solved by iteration (or repeated insertion), where each terms can be interpreted as multiple scattering contributions. We write the expansion coefficient f_n^p as a sum of different contributions ($p = 1, 2, \ldots, N$)

$$f_n^p = \sum_{l=1}^{\infty} f_n^p(l)$$

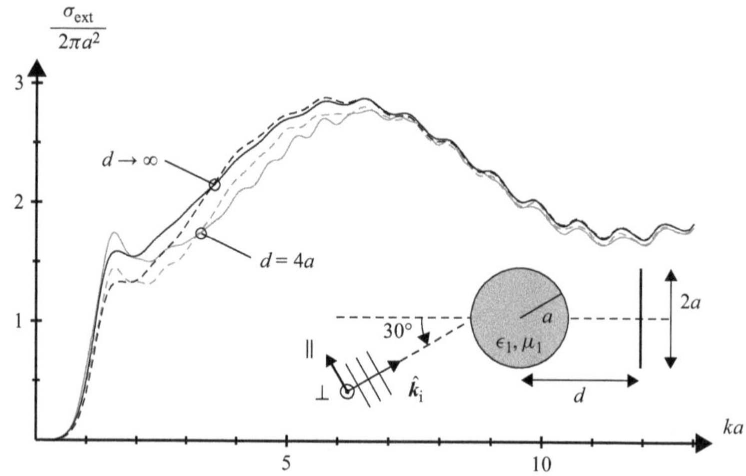

Figure 9.10 The normalized extinction cross section for a dielectric sphere and a PEC disk as a function of ka separated by a distance d. The solid curves display the parallel polarization and the dashed curves display the perpendicular polarization of the incident wave. The data of the sphere are $\epsilon_1/\epsilon = 1.33^2$ and $\mu_1/\mu = 1$.

where

$$\begin{cases} f_n^p(1) = \mathrm{e}^{\mathrm{i}k\hat{\boldsymbol{k}}_\mathrm{i}\cdot\boldsymbol{r}_p} \sum\limits_{n'} T_{nn'}^p a_{n'} \\ f_n^p(l+1) = \sum\limits_{\substack{q=1 \\ q\neq p}}^{N} \sum\limits_{n',n''} T_{nn'}^p \mathcal{P}_{n'n''}(k(\boldsymbol{r}_q - \boldsymbol{r}_p)) f_{n''}^q(l), \qquad l = 1,2,\ldots \end{cases}$$

The first term, $f_n^p(1) = \mathrm{e}^{\mathrm{i}k\hat{\boldsymbol{k}}_\mathrm{i}\cdot\boldsymbol{r}_p} \sum_{n'} T_{nn'}^p a_{n'}$, is the single scattering term, since this term produces a scattered field of the following form:

$$\boldsymbol{E}_\mathrm{s}^1(\boldsymbol{r}) = \sum_{p=1}^{N} \sum_{n} f_n^p(1) \boldsymbol{u}_n(k(\boldsymbol{r} - \boldsymbol{r}_p)) = \sum_{p=1}^{N} \mathrm{e}^{\mathrm{i}k\hat{\boldsymbol{k}}_\mathrm{i}\cdot\boldsymbol{r}_p} \sum_{n,n'} \boldsymbol{u}_n(k(\boldsymbol{r} - \boldsymbol{r}_p)) T_{nn'}^p a_{n'}$$

which is the sum of scattering contributions from each single scatterer with the incident plane wave, evaluated at the local origin, acting as the exciting field.

The next term gives a scattered field of the following form:

$$\begin{aligned}E_s^2(r) &= \sum_{p=1}^{N}\sum_{n} f_n^p(2)u_n(k(r-r_p)) \\ &= \sum_{p=1}^{N}\sum_{\substack{q=1\\ q\neq p}}^{N} e^{ik\hat{k}_i\cdot r_q}\sum_{n,n',n'',n'''} u_n(k(r-r_p))T^p_{nn''}\mathcal{P}_{n''n'}(k(r_q-r_p))T^q_{n'n'''}a_{n'''}\end{aligned}$$

This term contains a product of two transition matrices, and represents the double scattering contributions. Some aspects of the convergence of this series are presented in Reference 187.

9.2.2 Cross Sections

The extinction cross section $\sigma_{ext}(\hat{k}_i)$ for a given incident plane wave with direction $\hat{k}_i$ can be expressed in the transition matrix of the total ensemble of the scatterers. The general expression of the extinction cross section is, see the expression in (7.40) on page 384

$$\sigma_{ext}(\hat{k}_i) = -\frac{1}{k^2|E_0|^2}\,\mathrm{Re}\sum_n a_n^* T_{nn'}a_{n'} = -\frac{1}{k^2|E_0|^2}\,\mathrm{Re}\sum_n a_n^* f_n$$

Insert (9.23), and we obtain

$$\sigma_{ext}(\hat{k}_i) = -\frac{1}{k^2|E_0|^2}\sum_{p=1}^{N}\mathrm{Re}\sum_{n,n'} a_n^*\mathcal{R}_{nn'}(kr_p)f^p_{n'} = -\frac{1}{k^2|E_0|^2}\sum_{p=1}^{N}\mathrm{Re}\sum_{n} a_n^{p*}f_n^p$$

where the coefficient $a_n^p = \sum_{n'} a_{n'}\mathcal{R}_{n'n}(kr_p)$ (not that $\mathcal{R}_{n'n}(kr_p)$ is real valued) is the expansion coefficient of the incident field w.r.t. the local origin at O_p, i.e.,

$$\begin{aligned}E_i(r) &= \sum_n a_n v_n(kr) = \sum_n a_n v_n(k(r-r_p)+kr_p) \\ &= \sum_{n,n'} a_n\mathcal{R}_{nn'}(kr_p)v_{n'}(k(r-r_p)) = \sum_n a_n^p v_n(k(r-r_p))\end{aligned}$$

If we define the extinction cross section of each individual scatterer as

$$\tilde{\sigma}^p_{ext}(\hat{k}_i) = -\frac{1}{k^2|E_0|^2}\,\mathrm{Re}\sum_n a_n^{p*}f_n^p$$

we get

$$\sigma_{ext}(\hat{k}_i) = \sum_{p=1}^{N}\tilde{\sigma}^p_{ext}$$

At first glance, this result is somewhat deceptive. It does not state that the total extinction cross section is the sum of the extinction cross sections of the individual scatterer, e.g., the total extinction cross section of several identical spherical objects is not just the extinction cross sections of a single sphere times the number of spheres. The difference is that the scattered field in each term, $\widetilde{\sigma}^{p}_{\mathrm{ext}}$, contains the contribution from all other scatterers. Notice that the sum in $\widetilde{\sigma}^{p}_{\mathrm{ext}}$ contains the coefficients f^{p}_{n}, which contain all multiple scattering contributions. Another difference is that the scattering cross section for the entire ensemble of scatterers refers to a single incident plane wave, and the scattering cross sections for the individual are due to a much more complex incident field. This is also the reason for a separate notation with a tilde on the extinction cross section of the individual scatterers. For more details, we refer to References 57, 182, 187.

9.3 Obstacle above a Ground Plane

In this section, we solve a rather complex scattering problem, and we also combine the results obtained in Chapters 8 and 10 with the results obtained in this chapter. In a way, this section binds the theory in the entire book together. The solution is quite technical, but one of the simplest problem that combines planar structures and bounded scatterers. The interested reader can find more advanced problems in References 117, 136, 139, 140.

9.3.1 Formulation of the Problem

An infinite, perfectly conducting plane, S_0, separates the region V_{e} and V_{s} from the region V_- below S_0, see Figure 9.11. The material in V_{e} is assumed to be homogeneous, lossless, isotropic, characterized by the real-valued parameters ϵ and μ. The plane is parameterized by $z = z_0$, and without loss of generality, it is assumed that $z_0 < 0$, and the location of the origin O is somewhere in V_{s}. A perfectly conducting scatterer with bounding surface S_{s} defines the region V_{s}. The sources of the problem are assumed to be located in V_{i} above the surface S_0 (not shown in Figure 9.11). Moreover, the regions V_-, V_{e}, V_{i}, and V_{s} are assumed to be mutually disjoint, and $\overline{V_-} \cup \overline{V_{\mathrm{s}}} \cup V_{\mathrm{e}} \cup \overline{V_{\mathrm{i}}} = \mathbb{R}^3$.

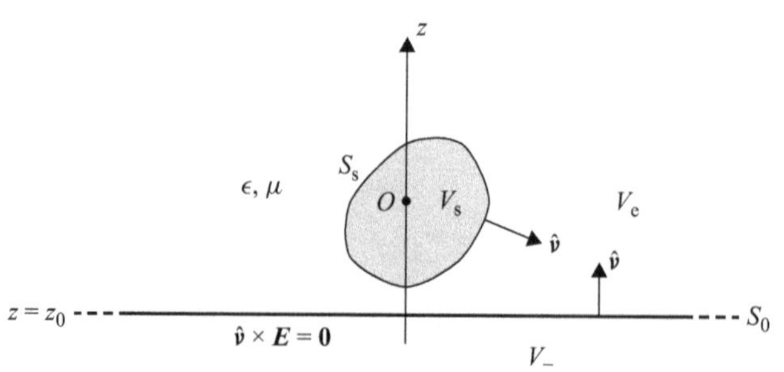

Figure 9.11 The geometry of the direct scattering problem with a perfectly conducting plane S_0 and a scatterer with bounding surface S_{s}.

The time-harmonic electric and magnetic fields in V_e satisfy the free-space Maxwell equations

$$\begin{cases} \nabla \times \boldsymbol{E}(\boldsymbol{r}) = \mathrm{i}k\eta_0\eta \boldsymbol{H}(\boldsymbol{r}) \\ \nabla \times \boldsymbol{H}(\boldsymbol{r}) = -\dfrac{\mathrm{i}k}{\eta_0\eta}\boldsymbol{E}(\boldsymbol{r}) \end{cases} \qquad \boldsymbol{r} \in V_e \tag{9.24}$$

where the (relative) wave impedance of the material is $\eta = \sqrt{\mu/\epsilon}$ and the wave number of the material is $k = \omega\sqrt{\epsilon_0\mu_0\epsilon\mu}$.

The boundary conditions on the bounding surfaces are

$$\begin{cases} \hat{z} \times \boldsymbol{E}(\boldsymbol{r}) = \boldsymbol{0}, & \boldsymbol{r} \in S_0 \\ \hat{\nu} \times \boldsymbol{E}(\boldsymbol{r}) = \boldsymbol{0}, & \boldsymbol{r} \in S_s \end{cases} \tag{9.25}$$

With an appropriate radiation condition in V_e at large distances, (9.24) has a unique solution. To proceed, we adopt a surface integral representation of the solution to this problem.

9.3.2 Integral Representation of the Solution

Let $\boldsymbol{E}_i(\boldsymbol{r})$ denote the incident electric field with sources located in V_i, and define, as above, the scattered electric field $\boldsymbol{E}_s(\boldsymbol{r}) = \boldsymbol{E}(\boldsymbol{r}) - \boldsymbol{E}_i(\boldsymbol{r})$. More details about the incident field are presented in Section 9.3.4. With the unit normals oriented as in Figure 9.11, the solution of (9.24) satisfies the surface integral representation, see (4.15) on page 197

$$-\frac{\eta_0\eta}{\mathrm{i}k}\nabla \times \left\{\nabla \times \iint_{S_0\cup S_s} g(k,|\boldsymbol{r}-\boldsymbol{r}'|)\left(\hat{\nu}(\boldsymbol{r}') \times \boldsymbol{H}(\boldsymbol{r}')\right)\,\mathrm{d}S'\right\}$$
$$+\nabla \times \iint_{S_0\cup S_s} g(k,|\boldsymbol{r}-\boldsymbol{r}'|)\left(\hat{\nu}(\boldsymbol{r}') \times \boldsymbol{E}(\boldsymbol{r}')\right)\,\mathrm{d}S' = \begin{cases} \boldsymbol{E}_s(\boldsymbol{r}), & \boldsymbol{r} \in V_e \\ -\boldsymbol{E}_i(\boldsymbol{r}), & \boldsymbol{r} \in V_- \cup V_s \end{cases}$$

where the free-space Green function is

$$g(k,|\boldsymbol{r}-\boldsymbol{r}'|) = \frac{e^{\mathrm{i}k|\boldsymbol{r}-\boldsymbol{r}'|}}{4\pi|\boldsymbol{r}-\boldsymbol{r}'|}$$

Note that the bounding surface in the surface integrals is not only the scatterer S_s but also the plane S_0. The integral representation also contains a hemispherical surface integral evaluated at large distances, but proper radiation conditions at large distances make this integral vanish [134]. Due to (9.25), the surface integral representation simplifies to

$$-\frac{\eta_0\eta}{\mathrm{i}k}\nabla \times \left(\nabla \times \iint_{S_0\cup S_s} \mathbf{G}_e(k,|\boldsymbol{r}-\boldsymbol{r}'|)\cdot \boldsymbol{J}_S(\boldsymbol{r}')\,\mathrm{d}S'\right) = \begin{cases} \boldsymbol{E}_s(\boldsymbol{r}), & \boldsymbol{r} \in V_e \\ -\boldsymbol{E}_i(\boldsymbol{r}), & \boldsymbol{r} \in V_- \cup V_s \end{cases} \tag{9.26}$$

where the surface current density is denoted $\boldsymbol{J}_S = \hat{\nu} \times \boldsymbol{H}$, and the Green dyadic for the electric field in free space is

$$\mathbf{G}_e(k,|\boldsymbol{r}-\boldsymbol{r}'|) = \left(\mathbf{I}_3 + \frac{1}{k^2}\nabla\nabla\right) g(k,|\boldsymbol{r}-\boldsymbol{r}'|)$$

The Green dyadic is now decomposed in spherical vector waves, see (7.24) on page 370

$$\mathbf{G}_e(k, |\boldsymbol{r} - \boldsymbol{r}'|) = ik \sum_n \boldsymbol{v}_n(k\boldsymbol{r}_<)\boldsymbol{u}_n(k\boldsymbol{r}_>) = ik \sum_n \boldsymbol{u}_n(k\boldsymbol{r}_>)\boldsymbol{v}_n(k\boldsymbol{r}_<) \tag{9.27}$$

where $\boldsymbol{r}_<$ ($\boldsymbol{r}_>$) is the position vector with the smallest (largest) distance to the origin, i.e., if $r < r'$ then $\boldsymbol{r}_< = \boldsymbol{r}$ and $\boldsymbol{r}_> = \boldsymbol{r}'$. The summation is, as above, over the divergence-free vector spherical vector waves, $\tau = 1, 2$. Moreover, we need the decomposition of the Green dyadic in plane planar vector waves, which is derived in Chapter 10, see (10.54) on page 574. The result is

$$\mathbf{G}_e(k, |\boldsymbol{r} - \boldsymbol{r}'|) = 2ik \sum_{j=1,2} \iint_{\mathbb{R}^2} \boldsymbol{\varphi}_j^{\pm}(\boldsymbol{k}_t; \boldsymbol{r})\boldsymbol{\varphi}_j^{\mp\dagger}(\boldsymbol{k}_t; \boldsymbol{r}')\frac{k}{k_z}\frac{dk_x\, dk_y}{k^2} \tag{9.28}$$

where the upper (lower) sign is used if $z > z'$ ($z < z'$). The definition of planar vector waves is given in (10.50) on page 573, i.e.,

$$\begin{cases} \boldsymbol{\varphi}_1^{\pm}(\boldsymbol{k}_t; \boldsymbol{r}) = \dfrac{\hat{\boldsymbol{z}} \times \boldsymbol{k}_t}{4\pi i k_t} e^{i\boldsymbol{k}_t\cdot\boldsymbol{r}_c \pm ik_z z} \\ \boldsymbol{\varphi}_2^{\pm}(\boldsymbol{k}_t; \boldsymbol{r}) = \dfrac{\mp \boldsymbol{k}_t k_z + k_t^2\hat{\boldsymbol{z}}}{4\pi k k_t} e^{i\boldsymbol{k}_t\cdot\boldsymbol{r}_c \pm ik_z z} \end{cases}$$

$$\begin{cases} \boldsymbol{\varphi}_1^{\pm\dagger}(\boldsymbol{k}_t; \boldsymbol{r}) = -\dfrac{\hat{\boldsymbol{z}} \times \boldsymbol{k}_t}{4\pi i k_t} e^{-i\boldsymbol{k}_t\cdot\boldsymbol{r}_c \pm ik_z z} \\ \boldsymbol{\varphi}_2^{\pm\dagger}(\boldsymbol{k}_t; \boldsymbol{r}) = \dfrac{\pm \boldsymbol{k}_t k_z + k_t^2\hat{\boldsymbol{z}}}{4\pi k k_t} e^{-i\boldsymbol{k}_t\cdot\boldsymbol{r}_c \pm ik_z z} \end{cases}$$

with exception that the wave number now, due to the lossless, isotropic background, is k instead of k_0, and that k_z is

$$k_z = \left(k^2 - k_t^2\right)^{1/2} = \begin{cases} \sqrt{k^2 - k_t^2} \text{ for } k_t < k \\ i\sqrt{k_t^2 - k^2} \text{ for } k_t > k \end{cases}$$

and where position vector in the x–y-plane is denoted $\boldsymbol{r}_c = \hat{\boldsymbol{x}}x + \hat{\boldsymbol{y}}y$, the transverse wave number $\boldsymbol{k}_t = \hat{\boldsymbol{x}}k_x + \hat{\boldsymbol{y}}k_y$, and $k_t = |\boldsymbol{k}_t|$. For further details on the planar vector waves, we refer to Chapter 10.

9.3.3 Transformation between Solutions

To connect the spherical vector waves and the planar vector waves, we need a transformation between the two sets of solutions. The result that is relevant in the analysis below is [23, page 183]

$$\boldsymbol{u}_n(k\boldsymbol{r}) = 2 \sum_{j=1,2} \iint_{\mathbb{R}^2} B_{nj}^{\pm}(\boldsymbol{k}_t)\boldsymbol{\varphi}_j^{\pm}(\boldsymbol{k}_t; \boldsymbol{r})\frac{k}{k_z}\frac{dk_x\, dk_y}{k^2}, \qquad z \gtrless 0 \tag{9.29}$$

where

$$B^{+}_{nj}(\boldsymbol{k}_{\mathrm{t}}) = \mathrm{i}^{-l} C_{lm} \left\{ -\mathrm{i}\delta_{\tau j} \Delta^{m}_{l}(k_z/k) \begin{Bmatrix} \cos m\beta \\ \sin m\beta \end{Bmatrix} - \delta_{\tau\bar{j}} \Pi^{m}_{l}(k_z/k) \begin{Bmatrix} -\sin m\beta \\ \cos m\beta \end{Bmatrix} \right\}$$

and

$$B^{-}_{nj}(\boldsymbol{k}_{\mathrm{t}}) = (-1)^{l+m+\tau+j+1} B^{+}_{nj}(\boldsymbol{k}_{\mathrm{t}})$$

where the dual index $\bar{j}$ is defined $\bar{1} = 2$ and $\bar{2} = 1$, and where $\boldsymbol{k}_{\mathrm{t}} = k_{\mathrm{t}}(\hat{\boldsymbol{x}} \cos\beta + \hat{\boldsymbol{y}} \sin\beta)$, and the normalization constants

$$C_{lm} = \sqrt{\frac{\varepsilon_m}{2\pi}} \sqrt{\frac{2l+1}{2} \frac{(l-m)!}{(l+m)!}}$$

and

$$\Delta^{m}_{l}(t) = -\frac{\left(1-t^2\right)^{1/2}}{\sqrt{l(l+1)}} P^{m\prime}_{l}(t), \quad \pi^{m}_{l}(t) = \frac{m}{\sqrt{l(l+1)}\left(1-t^2\right)^{1/2}} P^{m}_{l}(t)$$

There is also a useful transformation between planar vector waves and regular spherical vector waves. This transformation is [23, page 183]

$$\boldsymbol{\varphi}^{\pm}_{j}(\boldsymbol{k}_{\mathrm{t}}; \boldsymbol{r}) = \sum_{n} B^{\pm\dagger}_{nj}(\boldsymbol{k}_{\mathrm{t}}) \boldsymbol{v}_n(k\boldsymbol{r}) \tag{9.30}$$

where

$$B^{\pm\dagger}_{nj}(\boldsymbol{k}_{\mathrm{t}}) = B^{\mp}_{nj}(-\boldsymbol{k}_{\mathrm{t}})$$

This expansion is uniformly convergent in compact domains of $\mathbb{R}^3$.

9.3.4 Incident Electric Field

The sources of the incident field are assumed to be located above the plane S_0 and outside the scatterer V_{s}. More precisely, we assume they are confined to the region $V_{\mathrm{i}} \subset V_{\mathrm{e}}$, see Figure 9.12. Denote by V_{R} the largest spherical region centered at the origin not including any point in V_{i}. To find the expansion of the incident electric field in planar vector waves and the spherical vector waves, we use the integral representation of the incident electric field, see (3.14) on page 156, i.e.,

$$-\frac{\eta_0\eta}{\mathrm{i}k} \nabla \times \left\{ \nabla \times \iint_{S_{\mathrm{i}}} \mathbf{G}_{\mathrm{e}}(k, |\boldsymbol{r} - \boldsymbol{r}'|) \cdot \left(\hat{\boldsymbol{\nu}}(\boldsymbol{r}') \times \boldsymbol{H}_{\mathrm{i}}(\boldsymbol{r}')\right) \mathrm{d}S' \right\}$$

$$+ \nabla \times \iint_{S_{\mathrm{i}}} \mathbf{G}_{\mathrm{e}}(k, |\boldsymbol{r} - \boldsymbol{r}'|) \cdot \left(\hat{\boldsymbol{\nu}}(\boldsymbol{r}') \times \boldsymbol{E}_{\mathrm{i}}(\boldsymbol{r}')\right) \mathrm{d}S' = \begin{cases} \boldsymbol{E}_{\mathrm{i}}(\boldsymbol{r}), & \boldsymbol{r} \text{ outside } S_{\mathrm{i}} \\ \boldsymbol{0}, & \boldsymbol{r} \text{ inside } S_{\mathrm{i}} \end{cases}$$

where the unit normal vector, $\hat{\boldsymbol{\nu}}$, is directed outward to the bounding surface S_{i} of the volume V_{i}. Note that the surface integral is only over the surface S_{i}, since this region contains the sources

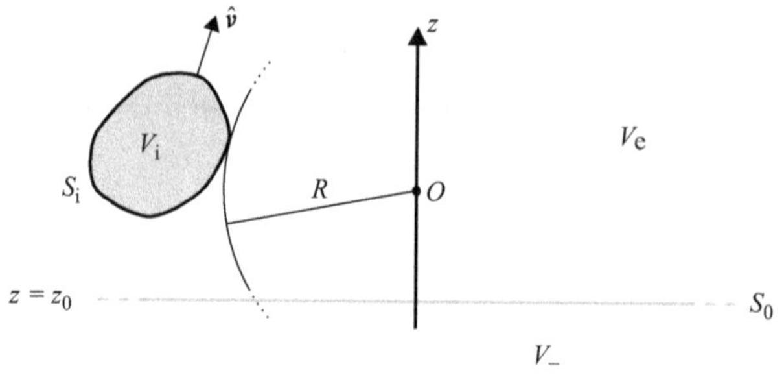

Figure 9.12 The geometry of the sources of the incident field above the perfectly conducting plane S_0. Notice that no scatterer S_s or plane surface S_0 is physically present in this geometry.

of the incident field in this problem. Any sources below the surface S_0 or inside S_s are the sources of the scattered field.

Specifically, for a position vector $\boldsymbol{r}$ in the region below S_0 ($\boldsymbol{r} \in V_-$) we get, using the Green dyadic decomposed in planar vector waves, see (9.28), and a change of order of integration

$$\boldsymbol{E}_\mathrm{i}(\boldsymbol{r}) = \sum_{j=1,2} \iint_{\mathbb{R}^2} a_j(\boldsymbol{k}_\mathrm{t}) \boldsymbol{\varphi}_j^-(\boldsymbol{k}_\mathrm{t}; \boldsymbol{r}) \frac{\mathrm{d}k_x\, \mathrm{d}k_y}{k^2}, \qquad \boldsymbol{r} \in V_- \tag{9.31}$$

where

$$a_j(\boldsymbol{k}_\mathrm{t}) = \frac{2k^3}{k_z} \iint_{S_\mathrm{i}} \left\{ \mathrm{i}\boldsymbol{\varphi}_{\bar{j}}^{+\dagger}(\boldsymbol{k}_\mathrm{t}; \boldsymbol{r}) \cdot \left(\hat{\boldsymbol{\nu}}(\boldsymbol{r}) \times \boldsymbol{E}_\mathrm{i}(\boldsymbol{r})\right) - \boldsymbol{\varphi}_j^{+\dagger}(\boldsymbol{k}_\mathrm{t}; \boldsymbol{r}) \cdot \left(\hat{\boldsymbol{\nu}}(\boldsymbol{r}) \times \eta_0 \eta \boldsymbol{H}_\mathrm{i}(\boldsymbol{r})\right) \right\} \mathrm{d}S$$

and similarly, inside V_R, in spherical vector waves, see (9.27), we get

$$\boldsymbol{E}_\mathrm{i}(\boldsymbol{r}) = \sum_n a_n \boldsymbol{v}_n(k\boldsymbol{r}), \qquad \boldsymbol{r} \in V_\mathrm{R} \tag{9.32}$$

where

$$a_n = k^2 \iint_{S_\mathrm{i}} \left\{ \mathrm{i}\boldsymbol{u}_{\bar{n}}(k\boldsymbol{r}) \cdot \left(\hat{\boldsymbol{\nu}}(\boldsymbol{r}) \times \boldsymbol{E}_\mathrm{i}(\boldsymbol{r})\right) - \boldsymbol{u}_n(k\boldsymbol{r}) \cdot \left(\hat{\boldsymbol{\nu}}(\boldsymbol{r}) \times \eta_0 \eta \boldsymbol{H}_\mathrm{i}(\boldsymbol{r})\right) \right\} \mathrm{d}S$$

where the dual index $\bar{n} = \bar{\tau}\sigma ml$.

The expansions in (9.31) and (9.32) represent the same incident field with different domains of convergence. Notice that these expansions refer to a situation where both the scatterer S_s and the plane S_0 are absent, i.e., the sources of the incident field in free space. In order to be consistent, the expansion coefficients are related to each other. Insertion of the planar vector wave expansion in terms of regular spherical vector waves, see (9.30), in (9.31) gives

$$\boldsymbol{E}_\mathrm{i}(\boldsymbol{r}) = \sum_{j=1,2} \iint_{\mathbb{R}^2} a_j(\boldsymbol{k}_\mathrm{t}) \sum_n B_{nj}^{-\dagger}(\boldsymbol{k}_\mathrm{t}) \boldsymbol{v}_n(k\boldsymbol{r}) \frac{\mathrm{d}k_x\, \mathrm{d}k_y}{k^2}$$

which is valid in the common domain of convergence of the planar vector waves and the spherical vector waves expansions. Linear independence of the regular spherical vector waves in the common domain of convergence implies

$$a_n = \sum_{j=1,2} \iint_{\mathbb{R}^2} a_j(\boldsymbol{k}_t) B_{nj}^{-\dagger}(\boldsymbol{k}_t) \frac{dk_x\, dk_y}{k^2}$$

which shows the relation between the different expansion coefficients a_n and the expansion functions $a_j(\boldsymbol{k}_t)$ of the incident field.

9.3.5 Utilizing the Surface Integral Representation

The position vector, $\boldsymbol{r}$, can take three principal positions, $\boldsymbol{r} \in V_-$, $\boldsymbol{r} \in V_s$, and $\boldsymbol{r} \in V_e$ in (9.26). We now explore these three possibilities one by one.

9.3.5.1 Below the Surfaces S_0

When the position vector $\boldsymbol{r}$ is in V_- (and outside the circumscribed sphere of S_s), the lower line of (9.26) together with (9.27)–(9.29) and (9.31) yield

$$\sum_{j=1,2} \iint_{\mathbb{R}^2} a_j(\boldsymbol{k}_t) \boldsymbol{\varphi}_j^-(\boldsymbol{k}_t;\boldsymbol{r}) \frac{dk_x\, dk_y}{k^2} = 2k^2\eta_0\eta \sum_{j=1,2} \iint_{\mathbb{R}^2} \boldsymbol{\varphi}_j^-(\boldsymbol{k}_t;\boldsymbol{r}) \left\{ \iint_{S_0} \boldsymbol{\varphi}_j^{+\dagger}(\boldsymbol{k}_t;\boldsymbol{r}') \cdot \boldsymbol{J}_S(\boldsymbol{r}')\, dS' \right.$$

$$\left. + \sum_n B_{nj}^-(\boldsymbol{k}_t) \iint_{S_s} \boldsymbol{v}_n(k\boldsymbol{r}') \cdot \boldsymbol{J}_S(\boldsymbol{r}')\, dS' \right\} \frac{k}{k_z} \frac{dk_x\, dk_y}{k^2}$$

To obtain this expression, we have used that $z < 0$, and we have also changed the order of the surface integral and the sum over n. Since the planar vector waves, $\boldsymbol{\varphi}_j^-(\boldsymbol{k}_t;\boldsymbol{r})$, are linearly independent, we have

$$a_j(\boldsymbol{k}_t) = \frac{2k^3\eta_0\eta}{k_z} \iint_{S_0} \boldsymbol{\varphi}_j^{+\dagger}(\boldsymbol{k}_t;\boldsymbol{r}') \cdot \boldsymbol{J}_S(\boldsymbol{r}')\, dS' + \frac{2k^3\eta_0\eta}{k_z} \sum_n B_{nj}^-(\boldsymbol{k}_t) \iint_{S_s} \boldsymbol{v}_n(k\boldsymbol{r}') \cdot \boldsymbol{J}_S(\boldsymbol{r}')\, dS' \tag{9.33}$$

This is an expression that links the expansion functions $a_j(\boldsymbol{k}_t)$ to the surface field $\boldsymbol{J}_S(\boldsymbol{r})$.

9.3.5.2 Inside the Largest Inscribed Sphere of V_s

When the position vector $\boldsymbol{r}$ is located inside the largest inscribed sphere in V_s, the lower line of (9.26) using (9.29) and (9.32) yields

$$\sum_n a_n \boldsymbol{v}_n(k\boldsymbol{r}) = k^2\eta_0\eta \sum_n \boldsymbol{v}_n(k\boldsymbol{r}) \left\{ \iint_{S_s} \boldsymbol{u}_n(k\boldsymbol{r}') \cdot \boldsymbol{J}_S(\boldsymbol{r}')\, dS' \right.$$

$$\left. + 2 \sum_{j=1,2} \iint_{\mathbb{R}^2} B_{nj}^-(\boldsymbol{k}_t) \iint_{S_0} \boldsymbol{\varphi}_j^-(\boldsymbol{k}_t;\boldsymbol{r}') \cdot \boldsymbol{J}_S(\boldsymbol{r}')\, dS' \frac{k}{k_z} \frac{dk_x\, dk_y}{k^2} \right\}$$

Again, we have changed the order of the surface integral and the sum over n. Since the regular spherical vector waves, $\boldsymbol{v}_n(k\boldsymbol{r})$, are linearly independent, we have

$$a_n = k^2\eta_0\eta \iint_{S_s} \boldsymbol{u}_n(k\boldsymbol{r}') \cdot \boldsymbol{J}_S(\boldsymbol{r}')\, \mathrm{d}S' + 2k^2\eta_0\eta \sum_{j=1,2} \iint_{\mathbb{R}^2} B^-_{nj}(\boldsymbol{k}_t) \iint_{S_0} \boldsymbol{\varphi}^-_j(\boldsymbol{k}_t;\boldsymbol{r}') \cdot \boldsymbol{J}_S(\boldsymbol{r}')\, \mathrm{d}S' \,\frac{k}{k_z}\frac{\mathrm{d}k_x\,\mathrm{d}k_y}{k^2} \tag{9.34}$$

This is a second expression that links the expansion coefficients a_n to the surface field $\boldsymbol{J}_S(\boldsymbol{r})$.

9.3.5.3 In V_e Outside the Smallest Sphere that Circumscribes V_s

Analogously, when the position vector $\boldsymbol{r}$ is outside the smallest circumscribed sphere of S_s in V_e, the upper line of (9.26) yields

$$\boldsymbol{E}_s(\boldsymbol{r}) = \sum_n f_n \boldsymbol{u}_n(k\boldsymbol{r}) + \sum_{j=1,2} \iint_{\mathbb{R}^2} f_j(\boldsymbol{k}_t)\boldsymbol{\varphi}^+_j(\boldsymbol{k}_t;\boldsymbol{r})\,\frac{\mathrm{d}k_x\,\mathrm{d}k_y}{k^2} \tag{9.35}$$

where

$$\begin{cases} f_n = -k^2\eta_0\eta \displaystyle\iint_{S_s} \boldsymbol{v}_n(k\boldsymbol{r}') \cdot \boldsymbol{J}_S(\boldsymbol{r}')\, \mathrm{d}S' \\ f_j(\boldsymbol{k}_t) = -\dfrac{2k^3\eta_0\eta}{k_z} \displaystyle\iint_{S_0} \boldsymbol{\varphi}^{-\dagger}_j(\boldsymbol{k}_t;\boldsymbol{r}') \cdot \boldsymbol{J}_S(\boldsymbol{r}')\, \mathrm{d}S' \end{cases} \tag{9.36}$$

This expression gives the scattered field once the expansion coefficients f_n and the expansion functions $f_j(\boldsymbol{k}_t)$ are known.

9.3.6 Expansion and Elimination of the Surface Fields

The next step is to eliminate the unknown surface current density $\boldsymbol{J}_S$ and to express this field in terms of the incident field expansion coefficients and functions.

To this end, expand the surface current density on the planar surface S_0 in planar vector waves and on the bounded scatterer S_s in a complete set of tangential vector functions, $\boldsymbol{\psi}_n$

$$\eta_0\eta \boldsymbol{J}_S(\boldsymbol{r}) = \begin{cases} \displaystyle\sum_{j=1,2} \iint_{\mathbb{R}^2} \alpha_j(\boldsymbol{k}_t)\hat{\boldsymbol{z}} \times \boldsymbol{\varphi}^+_{\bar{j}}(\boldsymbol{k}_t;\boldsymbol{r})\frac{\mathrm{d}k_x\,\mathrm{d}k_y}{k^2}, & \boldsymbol{r} \in S_0 \\ \displaystyle\sum_n \alpha_n \boldsymbol{\psi}_n(\boldsymbol{r}), & \boldsymbol{r} \in S_s \end{cases}$$

where the dual index $\bar{j}$ is defined $\bar{1} = 2$ and $\bar{2} = 1$, and introduce the two dimensionless matrices R and Q defined in (9.7), i.e.,

$$\begin{cases} R_{nn'} = k^2 \iint_{S_s} \boldsymbol{v}_n(k\boldsymbol{r}) \cdot \boldsymbol{\psi}_{n'}(\boldsymbol{r}) \, \mathrm{d}S \\ Q_{nn'} = k^2 \iint_{S_s} \boldsymbol{u}_n(k\boldsymbol{r}) \cdot \boldsymbol{\psi}_{n'}(\boldsymbol{r}) \, \mathrm{d}S \end{cases}$$

The orthogonality relations, see (10.51) on page 573 (S_0 is defined as $z = z_0 < 0$), are now used (note we use $j \to \bar{j}$)

$$\iint_{S_0} \boldsymbol{\varphi}_{j'}^{+\dagger}(\boldsymbol{k}_t; \boldsymbol{r}) \cdot \left(\hat{z} \times \boldsymbol{\varphi}_{\bar{j}}^{+}(\boldsymbol{k}_t'; \boldsymbol{r})\right) \, \mathrm{d}x \, \mathrm{d}y = (-1)^{j+1} \frac{k_z}{4\mathrm{i}k} \mathrm{e}^{2\mathrm{i}k_z z_0} \delta_{jj'} \delta(\boldsymbol{k}_t - \boldsymbol{k}_t')$$

We employ this orthogonality relation on (9.33), and we obtain

$$a_j(\boldsymbol{k}_t) = \frac{1}{2\mathrm{i}} (-1)^{j+1} \alpha_j(\boldsymbol{k}_t) \mathrm{e}^{2\mathrm{i}k_z z_0} + \frac{2k}{k_z} \sum_{n,n'} B_{nj}^{-}(\boldsymbol{k}_t) R_{nn'} \alpha_{n'} \tag{9.37}$$

Furthermore, we also use the orthogonality relation, see (10.51) on page 573 (note an extra minus sign due to $\boldsymbol{k}_t \to -\boldsymbol{k}_t$)

$$\iint_{S_0} \boldsymbol{\varphi}_{\bar{j}'}^{-}(\boldsymbol{k}_t; \boldsymbol{r}) \cdot \left(\hat{z} \times \boldsymbol{\varphi}_{\bar{j}}^{+}(\boldsymbol{k}_t'; \boldsymbol{r})\right) \, \mathrm{d}x \, \mathrm{d}y = -\frac{k_z}{4\mathrm{i}k} \delta_{jj'} \delta\left(\boldsymbol{k}_t + \boldsymbol{k}_t'\right)$$

on the second relation that connects the expansion coefficients a_n to the surface field, see (9.34). We get

$$a_n = \sum_{n'} Q_{nn'} \alpha_{n'} + c_n$$

where we define the following quantity c_n that is proved to be useful below:

$$c_n = -\frac{1}{2\mathrm{i}} \sum_{j=1,2} \iint_{\mathbb{R}^2} \alpha_j(\boldsymbol{k}_t) B_{n\bar{j}}^{-}(-\boldsymbol{k}_t) \frac{\mathrm{d}k_x \, \mathrm{d}k_y}{k^2}$$

This relation formally expresses the coefficients α_n, a_n, and c_n as

$$\alpha_n = \sum_{n'} Q_{nn'}^{-1} \left(a_{n'} - c_{n'}\right)$$

Rewrite (9.37) as

$$\alpha_j(\boldsymbol{k}_t) = 2\mathrm{i}(-1)^{j+1} a_j(\boldsymbol{k}_t) \mathrm{e}^{-2\mathrm{i}k_z z_0} + (-1)^j \frac{4\mathrm{i}k}{k_z} \mathrm{e}^{-2\mathrm{i}k_z z_0} \sum_{n,n'} B_{nj}^{-}(\boldsymbol{k}_t) R_{nn'} \alpha_{n'}$$

and insert this expression in the definition of c_n above. We obtain an infinite set of equations of the following form

$$c_n = d_n + \sum_{n', n''} A_{nn'} T_{n'n''} \left(c_{n''} - a_{n''}\right)$$

where the T-matrix of the scatterer is

$$T_{nn'} = -\sum_{n''} R_{nn''} Q^{-1}_{n''n'}$$

and where

$$d_n = \sum_{j=1,2} (-1)^j \iint_{\mathbb{R}^2} a_j(\boldsymbol{k}_\mathrm{t}) \mathrm{e}^{-2\mathrm{i}k_z z_0} B^-_{nj}(-\boldsymbol{k}_\mathrm{t}) \frac{\mathrm{d}k_x\, \mathrm{d}k_y}{k^2}$$

and

$$A_{nn'} = 2 \sum_{j=1,2} (-1)^j \iint_{\mathbb{R}^2} B^-_{nj}(-\boldsymbol{k}_\mathrm{t}) B^-_{n'j}(\boldsymbol{k}_\mathrm{t}) \mathrm{e}^{-2\mathrm{i}k_z z_0} \frac{k}{k_z} \frac{\mathrm{d}k_x\, \mathrm{d}k_y}{k^2}$$

This system of equations can be solved for c_n for every specified incident field, since the quantity d_n is known for each incident field, and the matrix entries of $A_{nn'}$ contain only geometrical data of the planar surface S_0.

Finally, using the orthogonality relation (10.51) on page 573

$$\iint_{S_0} \boldsymbol{\varphi}^{-\dagger}_{j'}(\boldsymbol{k}_\mathrm{t}; \boldsymbol{r}) \cdot \left(\hat{z} \times \boldsymbol{\varphi}^+_j(\boldsymbol{k}'_\mathrm{t}; \boldsymbol{r})\right) \mathrm{d}x\, \mathrm{d}y = \frac{k_z}{4\mathrm{i}k} \delta_{jj'} \delta(\boldsymbol{k}_\mathrm{t} - \boldsymbol{k}'_\mathrm{t})$$

we obtain the expansion coefficients f_n and the expansion functions $f_j(\boldsymbol{k}_\mathrm{t})$ of the scattered field as, see (9.36)

$$\begin{cases} f_n = -\sum_{n'} R_{nn''} \alpha_{n'} = \sum_{n'} T_{nn'} \left(a_{n'} - c_{n'}\right) \\ f_j(\boldsymbol{k}_\mathrm{t}) = -\dfrac{1}{2\mathrm{i}} \alpha_j(\boldsymbol{k}_\mathrm{t}) \\ \quad = (-1)^j a_j(\boldsymbol{k}_\mathrm{t}) \mathrm{e}^{-2\mathrm{i}k_z z_0} + \dfrac{1}{2\mathrm{i}} (-1)^{j+1} \dfrac{4\mathrm{i}k}{k_z} \mathrm{e}^{-2\mathrm{i}k_z z_0} \sum_{n,n'} B^-_{nj}(\boldsymbol{k}_\mathrm{t}) R_{nn'} \alpha_{n'} \end{cases}$$

9.3.7 Decomposition of the Scattered Field

Equation (9.35) provides an expression of the scattered electric field $\boldsymbol{E}_\mathrm{s}(\boldsymbol{r})$. It is convenient to decompose the scattered field due its origin—the surface S_0 or the bounded scatterer S_s. The relevant decomposition of the scattered field is

$$\boldsymbol{E}_\mathrm{s}(\boldsymbol{r}) = \boldsymbol{E}^\mathrm{dir}_\mathrm{s}(\boldsymbol{r}) + \boldsymbol{E}^\mathrm{anom}_\mathrm{s}(\boldsymbol{r})$$

where

$$E_s^{dir}(r) = \sum_{j=1,2} (-1)^j \iint_{\mathbb{R}^2} a_j(k_t) e^{-2ik_z z_0} \varphi_j^+(k_t; r) \frac{dk_x\, dk_y}{k^2} \tag{9.38}$$

and

$$E_s^{anom}(r) = \sum_{n,n'} T_{nn'} (a_{n'} - c_{n'}) \Bigg\{ u_n(kr) + 2 \sum_{j=1,2} (-1)^j \iint_{\mathbb{R}^2} B_{\overline{n}j}^-(k_t) e^{-2ik_z z_0} \varphi_j^+(k_t; r) \frac{k}{k_z} \frac{dk_x\, dk_y}{k^2} \Bigg\} \tag{9.39}$$

The total electric field $E(r)$ above S_0 can be decomposed in several ways. We are already familiar with the decomposition of the field in an incident field and a scattered field. In this section, it is pertinent to introduce another decomposition, i.e.,

$$E(r) = E_i(r) + E_s(r) = E^{prim}(r) + E^{sec}(r)$$

where

$$\begin{cases} E^{prim}(r) = E_i(r) + E_s^{dir}(r) \\ E^{sec}(r) = E_s^{anom}(r) \end{cases}$$

As above, the field $E_i(r)$ is the total electric field in the absence of both the surface S_0 and the scatterer S_s, and the scattered electric field $E_s(r)$ is the additional field due to the presence of the surface S_0 and the scatterer S_s. In the second decomposition, $E^{prim}(r)$ is the total electric field in the absence of the scatterer (surface S_0 present) and $E^{sec}(r)$ is the correction due to the presence of the scatterer. The explicit expressions of these fields are, see (9.38) and (9.39)

$$E^{prim}(r) = E_i(r) + \sum_{j=1,2} (-1)^j \iint_{\mathbb{R}^2} a_j(k_t) e^{-2ik_z z_0} \varphi_j^+(k_t; r) \frac{dk_x\, dk_y}{k^2}$$

and

$$E^{sec}(r) = \sum_{n,n'} T_{nn'} (a_{n'} - c_{n'}) \Bigg\{ u_n(kr) + 2 \sum_{j=1,2} (-1)^j \iint_{\mathbb{R}^2} B_{\overline{n}j}^-(k_t) e^{-2ik_z z_0} \varphi_j^+(k_t; r) \frac{k}{k_z} \frac{dk_x\, dk_y}{k^2} \Bigg\}$$

With these expressions, the entire scattering problem is solved. For a given incident field $E_i(r)$, find the expansion coefficients a_n and expansion functions $a_j(k_t)$. Then compute the quantity d_n and find c_n. The transition matrix has also to be calculated, e.g., with the method outlined in Sections 9.1.1 and 9.2. All parts to form the complete solution are then collected.

As a check, we verify that the boundary conditions on $z = z_0$ are satisfied. We have ($\boldsymbol{r}_0 = \boldsymbol{r}_\mathrm{c} + z_0\hat{\boldsymbol{z}}$)

$$\boldsymbol{E}^{\mathrm{prim}}(\boldsymbol{r}_0)\Big|_{\tan} = \boldsymbol{E}_\mathrm{i}(\boldsymbol{r}_0)|_{\tan} + \sum_{j=1,2}(-1)^j \iint_{\mathbb{R}^2} a_j(\boldsymbol{k}_\mathrm{t})\mathrm{e}^{-2\mathrm{i}k_z z_0}\, \boldsymbol{\varphi}_j^+(\boldsymbol{k}_\mathrm{t};\boldsymbol{r}_0)\Big|_{\tan} \frac{\mathrm{d}k_x\,\mathrm{d}k_y}{k^2} = \boldsymbol{0}$$

since

$$(-1)^j\mathrm{e}^{-2\mathrm{i}k_z z_0}\, \boldsymbol{\varphi}_j^+(\boldsymbol{k}_\mathrm{t};\boldsymbol{r}_0)\Big|_{\tan} = -\boldsymbol{\varphi}_j^-(\boldsymbol{k}_\mathrm{t};\boldsymbol{r}_0)\Big|_{\tan}$$

In a similar fashion, for the anomalous field, we get, see (9.29)

$$\begin{aligned}\boldsymbol{E}^{\mathrm{sec}}(\boldsymbol{r}_0)\big|_{\tan} = \sum_{n,n'} T_{nn'}(a_{n'} - c_{n'}) \Bigg\{ &\boldsymbol{u}_n(k\boldsymbol{r}_0)|_{\tan} \\ &+ 2\sum_{j=1,2}(-1)^j \iint_{\mathbb{R}^2} B_{nj}^-(\boldsymbol{k}_\mathrm{t})\mathrm{e}^{-2\mathrm{i}k_z z_0}\, \boldsymbol{\varphi}_j^+(\boldsymbol{k}_\mathrm{t};\boldsymbol{r}_0)\Big|_{\tan} \frac{k}{k_z}\frac{\mathrm{d}k_x\,\mathrm{d}k_y}{k^2}\Bigg\} = \boldsymbol{0}\end{aligned}$$

Example 9.8
An explicit incident field is a vertical electric dipole, which we locate at $\boldsymbol{r}_\mathrm{s} = \boldsymbol{r}_\mathrm{cs} + z_\mathrm{s}\hat{\boldsymbol{z}}$, where $z_\mathrm{s} > z_0$. The mirror point of the source point $\boldsymbol{r}_\mathrm{s}$ in the surface $z = z_0$ is denoted $\boldsymbol{r}_\mathrm{s}^* = \boldsymbol{r}_\mathrm{s} - 2(z_\mathrm{s} - z_0)\hat{\boldsymbol{z}} = \boldsymbol{r}_\mathrm{cs} + (2z_0 - z_\mathrm{s})\hat{\boldsymbol{z}}$. The expansion functions of the incident field $a_j(\boldsymbol{k}_\mathrm{t})$ can be found from the transformation expression in (9.29). We get for $\tau = 2$, $l = 1$, $m = 0$, and $\sigma = \mathrm{e}$

$$\boldsymbol{E}_\mathrm{i}(\boldsymbol{r}) = \boldsymbol{u}_{2\mathrm{e}01}(k(\boldsymbol{r} - \boldsymbol{r}_\mathrm{s})) = \sum_{j=1,2}\iint_{\mathbb{R}^2} a_j(\boldsymbol{k}_\mathrm{t})\boldsymbol{\varphi}_j^-(\boldsymbol{k}_\mathrm{t};\boldsymbol{r})\frac{\mathrm{d}k_x\,\mathrm{d}k_y}{k^2}, \qquad z < z_0$$

where

$$a_j(\boldsymbol{k}_\mathrm{t}) = 2\frac{k}{k_z}B_{2\mathrm{e}01j}^-(\boldsymbol{k}_\mathrm{t})\mathrm{e}^{-\mathrm{i}\boldsymbol{k}_\mathrm{t}\cdot\boldsymbol{r}_\mathrm{cs} + \mathrm{i}k_z z_\mathrm{s}} = 2\delta_{j2}\sqrt{\frac{3}{8\pi}}\frac{k_\mathrm{t}}{k_z}\mathrm{e}^{-\mathrm{i}\boldsymbol{k}_\mathrm{t}\cdot\boldsymbol{r}_\mathrm{cs} + \mathrm{i}k_z z_\mathrm{s}}$$

The direct scattered field is, see (9.38) (use $B_{2\mathrm{e}01j}^+(\boldsymbol{k}_\mathrm{t}) = B_{2\mathrm{e}01j}^-(\boldsymbol{k}_\mathrm{t})$)

$$\begin{aligned}\boldsymbol{E}_\mathrm{s}^{\mathrm{dir}}(\boldsymbol{r}) &= 2\sqrt{\frac{3}{8\pi}}\iint_{\mathbb{R}^2} \mathrm{e}^{-\mathrm{i}\boldsymbol{k}_\mathrm{t}\cdot\boldsymbol{r}_\mathrm{cs} - \mathrm{i}k_z(2z_0 - z_\mathrm{s})}\boldsymbol{\varphi}_2^+(\boldsymbol{k}_\mathrm{t};\boldsymbol{r})\frac{k_\mathrm{t}}{k_z}\frac{\mathrm{d}k_x\,\mathrm{d}k_y}{k^2} \\ &= 2\sum_{j=1,2}\iint_{\mathbb{R}^2} B_{2\mathrm{e}01j}^+(\boldsymbol{k}_\mathrm{t})\boldsymbol{\varphi}_2^+(\boldsymbol{k}_\mathrm{t};\boldsymbol{r} - \boldsymbol{r}_\mathrm{s}^*)\frac{k}{k_z}\frac{\mathrm{d}k_x\,\mathrm{d}k_y}{k^2} = \boldsymbol{u}_{2\mathrm{e}01}(k(\boldsymbol{r} - \boldsymbol{r}_\mathrm{s}^*))\end{aligned}$$

This field represents the electric field from a vertical electric dipole located at $\boldsymbol{r}_\mathrm{s}^*$. ■

Problems for Chapter 9

Section 9.1 The *T*-Matrix for a Single Homogeneous Scatterer

9.1 Show that $C_{ll'} = B_{ll'}$ and $D_{ll'} = \delta_{ll'} - A_{ll'}$ in Example 9.1.

*__9.2__ Show that the matrices $A_{ll'}$ and $B_{ll'}$ in Example 9.1 are symmetric if the parametrization of the axially symmetric scatterer is given by

$$\frac{1}{r^2(\theta)} = r_0 + r_1 P_1(\cos\theta) + r_2 P_2(\cos\theta)$$

Section 9.2 The *T*-Matrix for a Collection of Scatterers

*__9.3__ Generalize the analysis in Section 9.2 to a general incident field $\boldsymbol{E}_{\mathrm{i}}(\boldsymbol{r})$.

9.4 Verify the identity

$$\begin{aligned} &\mathrm{e}^{\mathrm{i}k\hat{\boldsymbol{k}}_{\mathrm{i}}\cdot\boldsymbol{r}_2} T^2 a + T^2 \mathcal{P}^{\mathrm{t}}(k\boldsymbol{d}) f^1 \\ &\quad = \left(I - T^2 \mathcal{P}^{\mathrm{t}}(k\boldsymbol{d}) T^1 \mathcal{P}(k\boldsymbol{d})\right)^{-1} \left(\mathrm{e}^{\mathrm{i}k\hat{\boldsymbol{k}}_{\mathrm{i}}\cdot\boldsymbol{r}_2} T^2 + \mathrm{e}^{\mathrm{i}k\hat{\boldsymbol{k}}_{\mathrm{i}}\cdot\boldsymbol{r}_1} T^2 \mathcal{P}^{\mathrm{t}}(k\boldsymbol{d}) T^1\right) a \end{aligned}$$

*__9.5__ Prove the identity

$$P_{nn'}(\hat{\boldsymbol{r}}) = -4\pi\,\mathrm{i}\boldsymbol{a}_n(-\hat{\boldsymbol{r}}) \cdot \boldsymbol{a}_{n'}(\hat{\boldsymbol{r}}) = -4\pi\,\mathrm{i}\boldsymbol{a}_n^*(\hat{\boldsymbol{r}}) \cdot \boldsymbol{a}_{n'}(\hat{\boldsymbol{r}})$$

where

$$P_{nn'}(\hat{\boldsymbol{r}}) = \lim_{kr\to\infty} kr\mathrm{e}^{-\mathrm{i}kr} \mathcal{P}_{nn'}(k\boldsymbol{r})$$

Hint: Show the identity by solving the scattering problem in Figure 9.13 in two different ways:

1. Compute the far field amplitude $\boldsymbol{F}(\hat{\boldsymbol{r}})$ of the scatterer by the use of the scattering dyadic $\mathbf{S}(\hat{\boldsymbol{r}}, \hat{\boldsymbol{k}}_{\mathrm{i}})$ of the scatterer with the appropriate incident field expressed in the far field amplitude of the source field.
2. Recompute the far field amplitude $\boldsymbol{F}(\hat{\boldsymbol{r}})$, but this time by translating the field generated by the sources to an incident field of the scatterer, and then use of the transition matrix to get the far field amplitude.

Compare the two expressions in the limit $kr_0 \to \infty$.

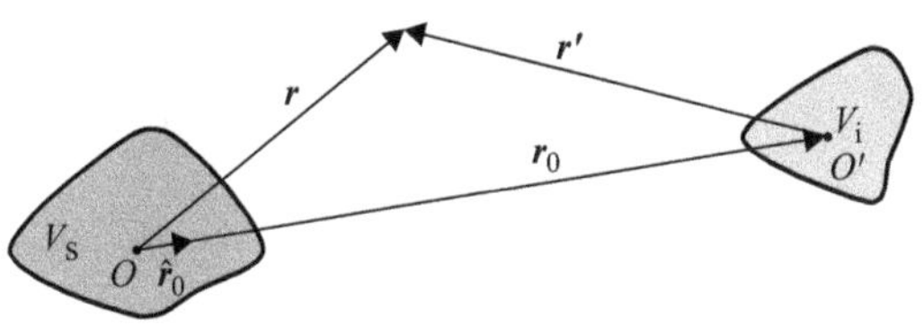

Figure 9.13 The geometry of the scatterer V_{s} and its local origin O and the sources V_{i} and their local origin O' in Problem 9.5. The distance between the origins is r_0.

CHAPTER 10

Propagation in Stratified Media

The geometry treated in this chapter differs from the geometry of the scattering problem in the chapters above. Everything up to this point has been dealing with scattering by finite obstacles. In this chapter, we treat scattering by planar structures, i.e., the geometry is large in two of its dimensions. Not surprisingly, the solution of these scattering problems employs a somewhat different approach compared to the technique used in the previous chapters. Obviously, we must master to solve these scattering problems if we aim at solving problems of the kind we addressed in e.g., Section 9.3. Scattering by planar structures is often referred to as reflection and transmission problems.

Propagation of electromagnetic waves in plane stratified media, i.e., materials that show only variation w.r.t. depth and no variation in the lateral variables, can be solved by Fourier transform technique [21, 33, 171, 210]. If we take the z-variable as the depth variable, see e.g., Figure 10.1, the material varies only w.r.t. this variable. In this technique, which is also called plane wave expansion method or spectral domain technique, all field quantities are Fourier transformed in the lateral variables x and y. The treatment in this chapter follows closely [229]. The analysis can be extended to accommodate thin frequency selective surfaces [142, 144, 215–217] and also to include the onset of surface waves [137] and patch antennas [146].

In this chapter, we develop the main tools for the analysis of wave propagation in stratified media, and we have the ambition to make a unified approach to scattering by a general bianisotropic media. The appealing part of this analysis is that all materials, from simple isotropic materials to general bianisotropic ones, are treated with the same uniform theory. This theory is built on a rather simple 2×2 matrix algebra, which is easily implemented on a computer. Of course, considerable simplifications occur if the medium is less complex, such as isotropic materials. These special cases are analyzed in detail.

The underlying equations and the decompositions used in this chapter are presented in Section 10.1, which lead to the fundamental equation introduced in Section 10.2. The concept of wave splitting is discussed in Section 10.3, and the propagator dyadic is employed in Sections 10.4 and 10.5. Several examples and numerical illustrations are given in Sections 10.6 and 10.7. The asymptotic behavior of the transmitted and reflected fields is analyzed in Section 10.8. The Green dyadic in stratified structures is analyzed in Section 10.9.

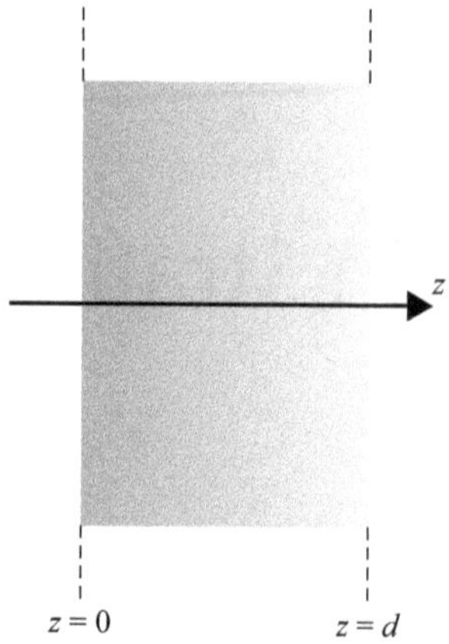

Figure 10.1 The geometry of a stratified structure. The regions outside the slab are, in general, empty half-spaces ($\epsilon = \mu = 1$).

10.1 Basic Equations

The Maxwell equations, (1.48) on page 42, model the dynamics of the fields in macroscopic media:

$$\begin{cases} \nabla \times \boldsymbol{E}(\boldsymbol{r}) = \mathrm{i}k_0 c_0 \boldsymbol{B}(\boldsymbol{r}) \\ \nabla \times (\eta_0 \boldsymbol{H}(\boldsymbol{r})) = -\mathrm{i}k_0 c_0 \eta_0 \boldsymbol{D}(\boldsymbol{r}) \end{cases} \tag{10.1}$$

where, as above, c_0, k_0, and η_0 are the speed of light, the wave number, and the wave impedance in vacuum, respectively.

The bianisotropic medium is the most general, linear, complex medium, and it is modeled with constitutive relations with at most 36 different scalar constitutive parameters—or functions of the frequency, if temporal dispersion is present, and the depth z. The time-harmonic constitutive relations of a general bianisotropic medium are, see (1.51) on page 44

$$\begin{cases} \boldsymbol{D}(\boldsymbol{r}) = \epsilon_0 \left\{ \boldsymbol{\epsilon}(z) \cdot \boldsymbol{E}(\boldsymbol{r}) + \eta_0 \boldsymbol{\xi}(z) \cdot \boldsymbol{H}(\boldsymbol{r}) \right\} \\ \boldsymbol{B}(\boldsymbol{r}) = \dfrac{1}{c_0} \left\{ \boldsymbol{\zeta}(z) \cdot \boldsymbol{E}(\boldsymbol{r}) + \eta_0 \boldsymbol{\mu}(z) \cdot \boldsymbol{H}(\boldsymbol{r}) \right\} \end{cases} \tag{10.2}$$

The four dyadics $\boldsymbol{\epsilon}(z)$, $\boldsymbol{\xi}(z)$, $\boldsymbol{\zeta}(z)$, and $\boldsymbol{\mu}(z)$ depend on the spatial variable z only, since the medium is assumed to be plane stratified (or simply stratified) in the z-direction. We suppress the dependence of the (angular) frequency to avoid cumbersome notation. For a homogeneous material, the constitutive dyadics are independent of z. Notice that the fields, e.g., $\boldsymbol{E}$, in general depend on all variables in $\boldsymbol{r}$, e.g., on x, y, and z.

10.1.1 Decomposition of Dyadics

The geometry treated in this chapter has one preferred direction. We take this direction as the z-direction in our coordinate system. The lateral directions constitute the x–y-plane.

For the purpose of studying wave propagation in layered bianisotropic structures, it is appropriate to decompose each constitutive dyadic, i.e., a general three-dimensional dyadic **A** is decomposed as

$$\mathbf{A} = \mathbf{A}_{\perp\perp} + \hat{z}\boldsymbol{A}_z + \boldsymbol{A}_{\perp}\hat{z} + \hat{z}A_{zz}\hat{z}$$

where

$$\begin{cases} \mathbf{A}_{\perp\perp} = \mathbf{I}_2 \cdot \mathbf{A} \cdot \mathbf{I}_2 \\ A_{zz} = \hat{z} \cdot \mathbf{A} \cdot \hat{z} \end{cases} \qquad \begin{cases} \boldsymbol{A}_z = \hat{z} \cdot \mathbf{A} \cdot \mathbf{I}_2 \\ \boldsymbol{A}_{\perp} = \mathbf{I}_2 \cdot \mathbf{A} \cdot \hat{z} \end{cases}$$

The dyadic $\mathbf{A}_{\perp\perp}$ is a two-dimensional dyadic in the x–y-plane, and the vectors $\boldsymbol{A}_z$ and $\boldsymbol{A}_{\perp}$ are two two-dimensional vectors in this plane. A_{zz} is a scalar. This decomposition is a simple fact of the identity

$$\mathbf{A} = \mathbf{I}_3 \cdot \mathbf{A} \cdot \mathbf{I}_3 = \left(\mathbf{I}_2 + \hat{z}\hat{z}\right) \cdot \mathbf{A} \cdot \left(\mathbf{I}_2 + \hat{z}\hat{z}\right)$$

More details on projections and the decomposition presented in this section are found in Appendix A.2.1.

10.2 The Fundamental Equation

As a preparation for the fundamental equation, we decompose all fields in a Fourier transform w.r.t. the lateral variables x and y.

10.2.1 The Fourier Transform of the Fields

In a geometry where the medium is laterally homogeneous in the variables x and y, it is natural to decompose the electromagnetic field outside the source region in a spectrum of plane waves. The plane wave decomposition amounts to a Fourier transformation of the electric and magnetic fields and flux densities w.r.t. the lateral variables x and y. The Fourier transform of a time-harmonic field, e.g., the electric field $\boldsymbol{E}(\boldsymbol{r}, \omega)$, is denoted by

$$\boldsymbol{E}(\boldsymbol{k}_t, z, \omega) = \iint_{\mathbb{R}^2} \boldsymbol{E}(\boldsymbol{r}, \omega) \mathrm{e}^{-\mathrm{i}\boldsymbol{k}_t \cdot \boldsymbol{r}_c} \, \mathrm{d}x \, \mathrm{d}y$$

where

$$\boldsymbol{k}_t = \hat{\boldsymbol{x}}k_x + \hat{\boldsymbol{y}}k_y = k_t\hat{\boldsymbol{e}}_{\parallel}, \quad \text{and} \quad \boldsymbol{r}_c = \hat{\boldsymbol{x}}x + \hat{\boldsymbol{y}}y$$

are the transverse (tangential) wave vector and the lateral position vector, respectively. The transverse (tangential) wave number is

$$k_t = \sqrt{k_x^2 + k_y^2}$$

The unit vector of the transverse wave vector in the x–y plane is $\hat{\boldsymbol{e}}_{\parallel} = \boldsymbol{k}_t / k_t$.

The inverse Fourier transform is defined by

$$\boldsymbol{E}(\boldsymbol{r}, \omega) = \frac{1}{4\pi^2} \iint_{\mathbb{R}^2} \boldsymbol{E}(\boldsymbol{k}_t, z, \omega) \mathrm{e}^{\mathrm{i}\boldsymbol{k}_t \cdot \boldsymbol{r}_c} \, \mathrm{d}k_x \, \mathrm{d}k_y \tag{10.3}$$

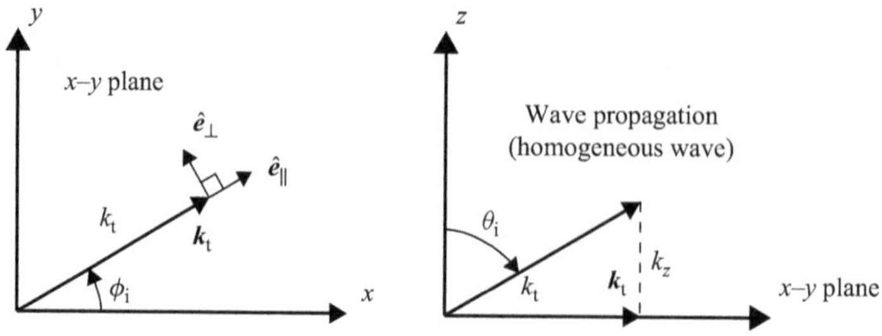

Figure 10.2 Left: The Fourier variable (transverse wave vector) $\boldsymbol{k}_t$, the azimuth angle of incidence, ϕ_i, and the two unit vectors $\hat{\boldsymbol{e}}_\parallel$ and $\hat{\boldsymbol{e}}_\perp$ in the x–y-plane. Right: The angle of incidence, θ_i, for a homogeneous plane wave.

Notice that the same notation is used to denote the Fourier transform of the field and the field itself. It is only the arguments that differ. Generally, there is no risk of confusion of what fields are intended. At rare occasion when this could happen, we explicitly write out the argument to indicate whether the field or its (spatially) Fourier transformed field is intended. There is also a difference in the dimensions between the two fields.

In vacuum,[1] the solutions can be classified as either homogeneous, obliquely propagating plane waves or inhomogeneous (evanescent) plane waves depending on whether the tangential wave number, k_t, is less or greater than the wave number in vacuum, $k_0 = \omega/c_0$. In the sections below, we show that the z dependence of $\boldsymbol{E}(\boldsymbol{k}_t, z, \omega)$ in a lossless medium has the form $\exp\{\pm \mathrm{i}k_z z\}$, where the normal (longitudinal) wave number, k_z, is defined by

$$k_z = k_0 \cos\theta_i = \left(k_0^2 - k_t^2\right)^{1/2} = \begin{cases} \sqrt{k_0^2 - k_t^2} & \text{for} \quad k_t < k_0 \\ \mathrm{i}\sqrt{k_t^2 - k_0^2} & \text{for} \quad k_t > k_0 \end{cases} \tag{10.4}$$

where we also have defined the angle of incidence, θ_i, which is real for homogeneous plane waves, see Figure 10.2. The wavelength λ_0 in vacuum is $\lambda_0 = 2\pi/k_0$.

By this definition, k_z applies to right-going waves and $-k_z$ to left-going waves, see also wave splitting in Section 10.3. Furthermore, it is convenient to introduce a set of coordinate independent orthonormal basis vectors in the x–y-plane by

$$\begin{cases} \hat{\boldsymbol{e}}_\parallel = \boldsymbol{k}_t/k_t = \hat{\boldsymbol{x}}\cos\phi_i + \hat{\boldsymbol{y}}\sin\phi_i \\ \hat{\boldsymbol{e}}_\perp = \hat{\boldsymbol{z}} \times \hat{\boldsymbol{e}}_\parallel = -\hat{\boldsymbol{x}}\sin\phi_i + \hat{\boldsymbol{y}}\cos\phi_i \end{cases}$$

where the azimuth angle of incidence, ϕ_i, is defined in Figure 10.2. The basis vectors $\{\hat{\boldsymbol{e}}_\parallel, \hat{\boldsymbol{e}}_\perp, \hat{\boldsymbol{z}}\}$ form a positively oriented orthonormal system.

All plane waves of normal and oblique incidence in vacuum are included in this description. Explicitly, oblique incidence corresponds to

$$\boldsymbol{E}(\boldsymbol{k}_t, z, \omega) = 4\pi^2 \boldsymbol{E}_0 \delta(\boldsymbol{k}_t - \boldsymbol{k}_{t0}) \exp\{\pm \mathrm{i}k_z z\}$$

[1]The analysis presented here in vacuum, $\epsilon = \mu = 1$, can easily be generalized to a homogeneous, lossless isotropic material.

where the vector $\boldsymbol{k}_{t0}$ is the projection of the direction of propagation in the x–y plane. For a normally incident wave, we have $\boldsymbol{k}_{t0} = \mathbf{0}$. Explicitly, from (10.3), we obtain the electric field

$$\boldsymbol{E}(\boldsymbol{r}, \omega) = \iint_{\mathbb{R}^2} \boldsymbol{E}_0 \delta(\boldsymbol{k}_t - \boldsymbol{k}_{t0}) e^{i\boldsymbol{k}_t \cdot \boldsymbol{r}_c} \exp\{\pm i k_z z\} \, dk_x \, dk_y = \boldsymbol{E}_0 e^{i\boldsymbol{k}_0^{\pm} \cdot \boldsymbol{r}}$$

where

$$\boldsymbol{k}_0^{\pm} = \boldsymbol{k}_{t0} \pm \hat{z}\left(k_0^2 - k_{t0}^2\right)^{1/2}$$

As a consequence, the transverse wave vector, $\boldsymbol{k}_t$, completely characterizes the plane wave, homogeneous or inhomogeneous. More details on this type of fields are found in Section 10.3.

10.2.2 Decomposition of the Maxwell Equations

Since the medium is laterally homogeneous, a Fourier transform of the Maxwell equations in a source-free region do not affect the material parameters. The Fourier transform amounts to substituting the operator identity $\nabla \longrightarrow i\boldsymbol{k}_t + \hat{z}\partial_z$ into the Maxwell field equations (10.1). This gives a system of linear, coupled ODEs in the variable z. We write this as a system of equations

$$\frac{d}{dz}\begin{pmatrix} \mathbf{J} \cdot \boldsymbol{E}(z) \\ \eta_0 \mathbf{J} \cdot \boldsymbol{H}(z) \end{pmatrix} = -i \begin{pmatrix} \boldsymbol{k}_t \times \mathbf{I}_3 & \mathbf{0} \\ \mathbf{0} & \boldsymbol{k}_t \times \mathbf{I}_3 \end{pmatrix} \cdot \begin{pmatrix} \boldsymbol{E}(z) \\ \eta_0 \boldsymbol{H}(z) \end{pmatrix} - ik_0 \begin{pmatrix} -c_0 \boldsymbol{B}(z) \\ c_0 \eta_0 \boldsymbol{D}(z) \end{pmatrix}$$

where $\mathbf{J} = \hat{z} \times \mathbf{I}_3 = \hat{z} \times \mathbf{I}_2$ is a two-dimensional rotation dyadic (rotation of $\pi/2$ in the x–y-plane). Note that $\mathbf{J} \cdot \mathbf{J} = -\mathbf{I}_2$. This reduction to a set of ODEs is, of course, not possible for a medium with variations in the variables x or y.

The ODE is now decomposed in its transverse and its longitudinal (normal) parts. We get

$$\frac{d}{dz}\begin{pmatrix} \boldsymbol{E}_{xy}(z) \\ \eta_0 \mathbf{J} \cdot \boldsymbol{H}_{xy}(z) \end{pmatrix} = i \begin{pmatrix} \boldsymbol{k}_t & \mathbf{0} \\ \mathbf{0} & \mathbf{J} \cdot \boldsymbol{k}_t \end{pmatrix} \begin{pmatrix} E_z(z) \\ \eta_0 H_z(z) \end{pmatrix} - ik_0 \begin{pmatrix} c_0 \mathbf{J} \cdot \boldsymbol{B}_{xy}(z) \\ c_0 \eta_0 \boldsymbol{D}_{xy}(z) \end{pmatrix} \tag{10.5}$$

and

$$\begin{pmatrix} c_0 \eta_0 D_z(z) \\ c_0 B_z(z) \end{pmatrix} = \frac{1}{k_0} \begin{pmatrix} \mathbf{0} & \boldsymbol{k}_t \\ \mathbf{J} \cdot \boldsymbol{k}_t & \mathbf{0} \end{pmatrix} \cdot \begin{pmatrix} \boldsymbol{E}_{xy}(z) \\ \eta_0 \mathbf{J} \cdot \boldsymbol{H}_{xy}(z) \end{pmatrix} \tag{10.6}$$

where we used the fact that $\mathbf{J} \cdot (\boldsymbol{k}_t \times \mathbf{I}_3) = \hat{z} \times (\boldsymbol{k}_t \times \mathbf{I}_3) = \boldsymbol{k}_t \hat{z}$ and $(\boldsymbol{k}_t \times \boldsymbol{E}_{xy}) \cdot \hat{z} = (\mathbf{J} \cdot \boldsymbol{k}_t) \cdot \boldsymbol{E}_{xy}$. The Fourier components of the electric and magnetic fields have also decomposed in their tangential and normal components as

$$\begin{cases} \boldsymbol{E}(\boldsymbol{k}_t, z, \omega) = \boldsymbol{E}_{xy}(\boldsymbol{k}_t, z, \omega) + \hat{z} E_z(\boldsymbol{k}_t, z, \omega) \\ \boldsymbol{H}(\boldsymbol{k}_t, z, \omega) = \boldsymbol{H}_{xy}(\boldsymbol{k}_t, z, \omega) + \hat{z} H_z(\boldsymbol{k}_t, z, \omega) \end{cases}$$

and similarly for the flux densities. Here, we have, for convenience, adopted a matrix notation for the multiplication of the different components of the fields.

Similarly, the constitutive relations, (10.2), can be decomposed in tangential and normal parts as

$$\begin{pmatrix} c_0\mathbf{J}\cdot\boldsymbol{B}_{xy}(z) \\ c_0\eta_0\boldsymbol{D}_{xy}(z) \end{pmatrix} = \begin{pmatrix} \mathbf{J}\cdot\boldsymbol{\zeta}_{\perp\perp} & -\mathbf{J}\cdot\boldsymbol{\mu}_{\perp\perp}\cdot\mathbf{J} \\ \boldsymbol{\epsilon}_{\perp\perp} & -\boldsymbol{\xi}_{\perp\perp}\cdot\mathbf{J} \end{pmatrix}\cdot\begin{pmatrix} \boldsymbol{E}_{xy}(z) \\ \eta_0\mathbf{J}\cdot\boldsymbol{H}_{xy}(z) \end{pmatrix} + \begin{pmatrix} \mathbf{J}\cdot\boldsymbol{\zeta}_{\perp} & \mathbf{J}\cdot\boldsymbol{\mu}_{\perp} \\ \boldsymbol{\epsilon}_{\perp} & \boldsymbol{\xi}_{\perp} \end{pmatrix}\begin{pmatrix} E_z(z) \\ \eta_0 H_z(z) \end{pmatrix} \tag{10.7}$$

and (remember $\mathbf{J}$ antisymmetric, i.e., $\mathbf{J}\cdot\boldsymbol{a} = -\boldsymbol{a}\cdot\mathbf{J}$)

$$\begin{pmatrix} \epsilon_{zz} & \xi_{zz} \\ \zeta_{zz} & \mu_{zz} \end{pmatrix}\begin{pmatrix} E_z(z) \\ \eta_0 H_z(z) \end{pmatrix} = \begin{pmatrix} c_0\eta_0 D_z(z) \\ c_0 B_z(z) \end{pmatrix} - \begin{pmatrix} \boldsymbol{\epsilon}_z & \mathbf{J}\cdot\boldsymbol{\xi}_z \\ \boldsymbol{\zeta}_z & \mathbf{J}\cdot\boldsymbol{\mu}_z \end{pmatrix}\cdot\begin{pmatrix} \boldsymbol{E}_{xy}(z) \\ \eta_0\mathbf{J}\cdot\boldsymbol{H}_{xy}(z) \end{pmatrix} \tag{10.8}$$

where the different projections of the constitutive dyadics are treated in detail in Section 10.1.1, i.e.,

$$\begin{cases} \boldsymbol{\epsilon} = \boldsymbol{\epsilon}_{\perp\perp} + \hat{z}\boldsymbol{\epsilon}_z + \boldsymbol{\epsilon}_{\perp}\hat{z} + \hat{z}\epsilon_{zz}\hat{z} \\ \boldsymbol{\xi} = \boldsymbol{\xi}_{\perp\perp} + \hat{z}\boldsymbol{\xi}_z + \boldsymbol{\xi}_{\perp}\hat{z} + \hat{z}\xi_{zz}\hat{z} \\ \boldsymbol{\zeta} = \boldsymbol{\zeta}_{\perp\perp} + \hat{z}\boldsymbol{\zeta}_z + \boldsymbol{\zeta}_{\perp}\hat{z} + \hat{z}\zeta_{zz}\hat{z} \\ \boldsymbol{\mu} = \boldsymbol{\mu}_{\perp\perp} + \hat{z}\boldsymbol{\mu}_z + \boldsymbol{\mu}_{\perp}\hat{z} + \hat{z}\mu_{zz}\hat{z} \end{cases}$$

Combining (10.5) and (10.7) gives

$$\begin{aligned} \frac{\mathrm{d}}{\mathrm{d}z}\begin{pmatrix} \boldsymbol{E}_{xy}(z) \\ \eta_0\mathbf{J}\cdot\boldsymbol{H}_{xy}(z) \end{pmatrix} &= -\mathrm{i}k_0\begin{pmatrix} \mathbf{J}\cdot\boldsymbol{\zeta}_{\perp\perp} & -\mathbf{J}\cdot\boldsymbol{\mu}_{\perp\perp}\cdot\mathbf{J} \\ \boldsymbol{\epsilon}_{\perp\perp} & -\boldsymbol{\xi}_{\perp\perp}\cdot\mathbf{J} \end{pmatrix}\cdot\begin{pmatrix} \boldsymbol{E}_{xy}(z) \\ \eta_0\mathbf{J}\cdot\boldsymbol{H}_{xy}(z) \end{pmatrix} \\ &\quad - \mathrm{i}k_0\begin{pmatrix} -\boldsymbol{k}_\mathrm{t}/k_0 + \mathbf{J}\cdot\boldsymbol{\zeta}_{\perp} & \mathbf{J}\cdot\boldsymbol{\mu}_{\perp} \\ \boldsymbol{\epsilon}_{\perp} & -\mathbf{J}\cdot\boldsymbol{k}_\mathrm{t}/k_0 + \boldsymbol{\xi}_{\perp} \end{pmatrix}\begin{pmatrix} E_z(z) \\ \eta_0 H_z(z) \end{pmatrix} \end{aligned} \tag{10.9}$$

Similarly, combining (10.6) and (10.8) yields

$$\begin{pmatrix} \epsilon_{zz} & \xi_{zz} \\ \zeta_{zz} & \mu_{zz} \end{pmatrix}\begin{pmatrix} E_z(z) \\ \eta_0 H_z(z) \end{pmatrix} = \begin{pmatrix} -\boldsymbol{\epsilon}_z & \boldsymbol{k}_\mathrm{t}/k_0 - \mathbf{J}\cdot\boldsymbol{\xi}_z \\ \mathbf{J}\cdot\boldsymbol{k}_\mathrm{t}/k_0 - \boldsymbol{\zeta}_z & -\mathbf{J}\cdot\boldsymbol{\mu}_z \end{pmatrix}\cdot\begin{pmatrix} \boldsymbol{E}_{xy}(z) \\ \eta_0\mathbf{J}\cdot\boldsymbol{H}_{xy}(z) \end{pmatrix} \tag{10.10}$$

which is an expression of the normal components of the electric and magnetic fields in terms of the tangential components of the fields. The last equation shows that the knowledge of the tangential components of the electric and magnetic fields imply that the longitudinal components of the two fields can be reconstructed. As a consequence, the two fields $\boldsymbol{E}_{xy}(z)$ and $\eta_0\mathbf{J}\cdot\boldsymbol{H}_{xy}(z)$ generate complete knowledge of all fields—also the flux densities $\boldsymbol{D}$ and $\boldsymbol{B}$ through the constitutive relations. Therefore, it suffices to study these two tangential fields to solve the general scattering problem in a stratified geometry.

Combining (10.9) and (10.10) gives a system of ODEs in the tangential components of the electric and magnetic fields only—the fundamental equation. The fundamental equation for one-dimensional wave propagation becomes

$$\frac{\mathrm{d}}{\mathrm{d}z}\begin{pmatrix} \boldsymbol{E}_{xy}(\boldsymbol{k}_\mathrm{t}, z, \omega) \\ \boldsymbol{L}_{xy}(\boldsymbol{k}_\mathrm{t}, z, \omega) \end{pmatrix} = \mathrm{i}k_0\mathbf{M}(\boldsymbol{k}_\mathrm{t}, z, \omega)\cdot\begin{pmatrix} \boldsymbol{E}_{xy}(\boldsymbol{k}_\mathrm{t}, z, \omega) \\ \boldsymbol{L}_{xy}(\boldsymbol{k}_\mathrm{t}, z, \omega) \end{pmatrix} \tag{10.11}$$

where for convenience a compact notation for the rotated magnetic field have been introduced $\boldsymbol{L}_{xy}(\boldsymbol{k}_t, z, \omega) = \eta_0 \mathbf{J} \cdot \boldsymbol{H}_{xy}(\boldsymbol{k}_t, z, \omega)$. The fundamental dyadic is conveniently decomposed into 2×2 blocks:

$$\mathbf{M} = \begin{pmatrix} \mathbf{M}_{11} & \mathbf{M}_{12} \\ \mathbf{M}_{21} & \mathbf{M}_{22} \end{pmatrix}$$

The fundamental dyadic is found to be

$$\mathbf{M} = \begin{pmatrix} -\mathbf{J} \cdot \boldsymbol{\zeta}_{\perp\perp} & \mathbf{J} \cdot \boldsymbol{\mu}_{\perp\perp} \cdot \mathbf{J} \\ -\boldsymbol{\epsilon}_{\perp\perp} & \boldsymbol{\xi}_{\perp\perp} \cdot \mathbf{J} \end{pmatrix} + \frac{1}{\epsilon_{zz}\mu_{zz} - \xi_{zz}\zeta_{zz}} \begin{pmatrix} \boldsymbol{k}_t/k_0 - \mathbf{J} \cdot \boldsymbol{\zeta}_{\perp} & -\mathbf{J} \cdot \boldsymbol{\mu}_{\perp} \\ -\boldsymbol{\epsilon}_{\perp} & \mathbf{J} \cdot \boldsymbol{k}_t/k_0 - \boldsymbol{\xi}_{\perp} \end{pmatrix}$$
$$\times \begin{pmatrix} -\mu_{zz}\boldsymbol{\epsilon}_z - \xi_{zz}\mathbf{J} \cdot \boldsymbol{k}_t/k_0 + \xi_{zz}\boldsymbol{\zeta}_z & \mu_{zz}\boldsymbol{k}_t/k_0 - \mu_{zz}\mathbf{J} \cdot \boldsymbol{\xi}_z + \xi_{zz}\mathbf{J} \cdot \boldsymbol{\mu}_z \\ \zeta_{zz}\boldsymbol{\epsilon}_z + \epsilon_{zz}\mathbf{J} \cdot \boldsymbol{k}_t/k_0 - \epsilon_{zz}\boldsymbol{\zeta}_z & -\zeta_{zz}\boldsymbol{k}_t/k_0 + \zeta_{zz}\mathbf{J} \cdot \boldsymbol{\xi}_z - \epsilon_{zz}\mathbf{J} \cdot \boldsymbol{\mu}_z \end{pmatrix}$$

Specifically, the block dyadics, $\mathbf{M}_{lm}$, $l, m = 1, 2$, are

$$\begin{cases} \mathbf{M}_{11} = -\mathbf{J} \cdot \boldsymbol{\zeta}_{\perp\perp} + a(\boldsymbol{k}_t/k_0 - \mathbf{J} \cdot \boldsymbol{\zeta}_{\perp}) (-\mu_{zz}\boldsymbol{\epsilon}_z - \xi_{zz}\mathbf{J} \cdot \boldsymbol{k}_t/k_0 + \xi_{zz}\boldsymbol{\zeta}_z) \\ \qquad - a(\mathbf{J} \cdot \boldsymbol{\mu}_{\perp}) (\zeta_{zz}\boldsymbol{\epsilon}_z + \epsilon_{zz}\mathbf{J} \cdot \boldsymbol{k}_t/k_0 - \epsilon_{zz}\boldsymbol{\zeta}_z) \\ \mathbf{M}_{12} = \mathbf{J} \cdot \boldsymbol{\mu}_{\perp\perp} \cdot \mathbf{J} + a(\boldsymbol{k}_t/k_0 - \mathbf{J} \cdot \boldsymbol{\zeta}_{\perp}) (\mu_{zz}\boldsymbol{k}_t/k_0 - \mu_{zz}\mathbf{J} \cdot \boldsymbol{\xi}_z + \xi_{zz}\mathbf{J} \cdot \boldsymbol{\mu}_z) \\ \qquad - a(\mathbf{J} \cdot \boldsymbol{\mu}_{\perp}) (-\zeta_{zz}\boldsymbol{k}_t/k_0 + \zeta_{zz}\mathbf{J} \cdot \boldsymbol{\xi}_z - \epsilon_{zz}\mathbf{J} \cdot \boldsymbol{\mu}_z) \\ \mathbf{M}_{21} = -\boldsymbol{\epsilon}_{\perp\perp} - a\boldsymbol{\epsilon}_{\perp}(-\mu_{zz}\boldsymbol{\epsilon}_z - \xi_{zz}\mathbf{J} \cdot \boldsymbol{k}_t/k_0 + \xi_{zz}\boldsymbol{\zeta}_z) \\ \qquad + a(\mathbf{J} \cdot \boldsymbol{k}_t/k_0 - \boldsymbol{\xi}_{\perp}) (\zeta_{zz}\boldsymbol{\epsilon}_z + \epsilon_{zz}\mathbf{J} \cdot \boldsymbol{k}_t/k_0 - \epsilon_{zz}\boldsymbol{\zeta}_z) \\ \mathbf{M}_{22} = \boldsymbol{\xi}_{\perp\perp} \cdot \mathbf{J} - a\boldsymbol{\epsilon}_{\perp}(\mu_{zz}\boldsymbol{k}_t/k_0 - \mu_{zz}\mathbf{J} \cdot \boldsymbol{\xi}_z + \xi_{zz}\mathbf{J} \cdot \boldsymbol{\mu}_z) \\ \qquad + a(\mathbf{J} \cdot \boldsymbol{k}_t/k_0 - \boldsymbol{\xi}_{\perp}) (-\zeta_{zz}\boldsymbol{k}_t/k_0 + \zeta_{zz}\mathbf{J} \cdot \boldsymbol{\xi}_z - \epsilon_{zz}\mathbf{J} \cdot \boldsymbol{\mu}_z) \end{cases} \tag{10.12}$$

where $a^{-1} = \epsilon_{zz}\mu_{zz} - \xi_{zz}\zeta_{zz}$. Notice that the result holds for all materials that are stratified, i.e., homogeneous layers (piecewise constant material parameters) or layers with material parameters that vary with depth z.

Equation (10.11) is the general equation for wave propagation in linear, laterally homogeneous media. From the solution of this equation, all pertinent electromagnetic properties can be computed, i.e., the longitudinal parts of the electric and the magnetic fields are found by (10.10), and the flux densities from the constitutive relations of the material. The fundamental dyadic depends on the tangential wave vector, $\boldsymbol{k}_t$, and the constitutive dyadics, which may or may not depend on the depth z. For homogeneous materials, $\mathbf{M}$ is independent of z. Several examples of the fundamental dyadic in different materials are given in Section 10.6.

The tangential electric and magnetic fields $\boldsymbol{E}_{xy}$ and $\boldsymbol{L}_{xy}$ are continuous everywhere in the slab, and especially over any interface of the slab due to the boundary conditions.

10.2.2.1 Lossless Material

A lossless bianisotropic material is characterized by, see (1.65) on page 60

$$\boldsymbol{\epsilon} = \boldsymbol{\epsilon}^{\dagger}, \qquad \boldsymbol{\mu} = \boldsymbol{\mu}^{\dagger}, \qquad \boldsymbol{\xi} = \boldsymbol{\zeta}^{\dagger}$$

We now investigate what implications these conditions have on the fundamental matrix $\mathbf{M}$. The Hermitian adjoint of $\mathbf{M}$ is straightforwardly found to be

$$\mathbf{M}^{\dagger} = \begin{pmatrix} \mathbf{M}_{11}^{\dagger} & \mathbf{M}_{21}^{\dagger} \\ \mathbf{M}_{12}^{\dagger} & \mathbf{M}_{22}^{\dagger} \end{pmatrix} = \begin{pmatrix} \mathbf{M}_{22} & \mathbf{M}_{21} \\ \mathbf{M}_{12} & \mathbf{M}_{11} \end{pmatrix} \tag{10.13}$$

We conclude that

$$\mathbf{M}^{\dagger} = \begin{pmatrix} \mathbf{0} & \mathbf{I}_2 \\ \mathbf{I}_2 & \mathbf{0} \end{pmatrix} \cdot \mathbf{M} \cdot \begin{pmatrix} \mathbf{0} & \mathbf{I}_2 \\ \mathbf{I}_2 & \mathbf{0} \end{pmatrix} = \begin{pmatrix} \mathbf{M}_{22} & \mathbf{M}_{21} \\ \mathbf{M}_{12} & \mathbf{M}_{11} \end{pmatrix}$$

In Section 10.5, the important use of the eigenvalues of the material dyadic $\mathbf{M}$ is demonstrated. The eigenvalues determine e.g., the speed of the wave in the z-direction. The analysis in this section proves that both $\mathbf{M}$ and $\mathbf{M}^{\dagger}$ have the same eigenvalues in the lossless case.

10.2.2.2 Fundamental Matrix in Vacuum

The fundamental dyadic in vacuum is

$$\mathbf{M}_0 = \begin{pmatrix} \mathbf{0} & -\mathbf{I}_2 + k_0^{-2}\boldsymbol{k}_{\mathrm{t}}\boldsymbol{k}_{\mathrm{t}} \\ -\mathbf{I}_2 - k_0^{-2}\mathbf{J} \cdot \boldsymbol{k}_{\mathrm{t}}\boldsymbol{k}_{\mathrm{t}} \cdot \mathbf{J} & \mathbf{0} \end{pmatrix} \tag{10.14}$$

10.3 Wave Splitting

Before facing the general propagation problem in a bianisotropic structure, we start by analyzing wave propagation in vacuum. This is the typical situation outside the stratified media, where it is assumed that the material is vacuum, except for a source region $z < z_0 < 0$, see Figure 10.3. If we have sources in several parts of space, e.g., on both sides of the slab, this assumption can easily be solved by the use of superposition.

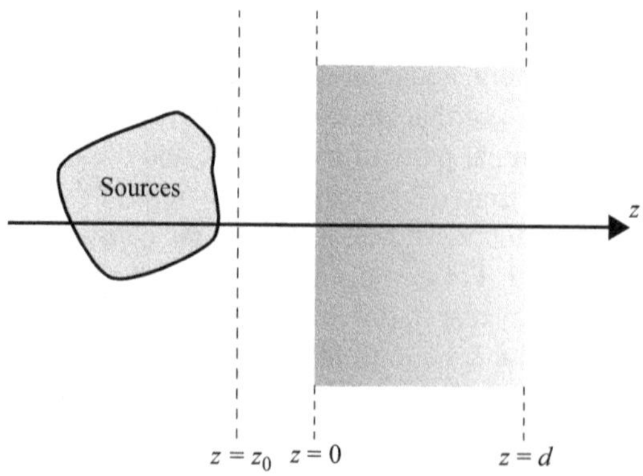

Figure 10.3 The source region and the plane, $z = z_0 < 0$, that limits its extent. Fourier transformation of the fields on any plane to the right of the source region, $z > z_0$, is well defined.

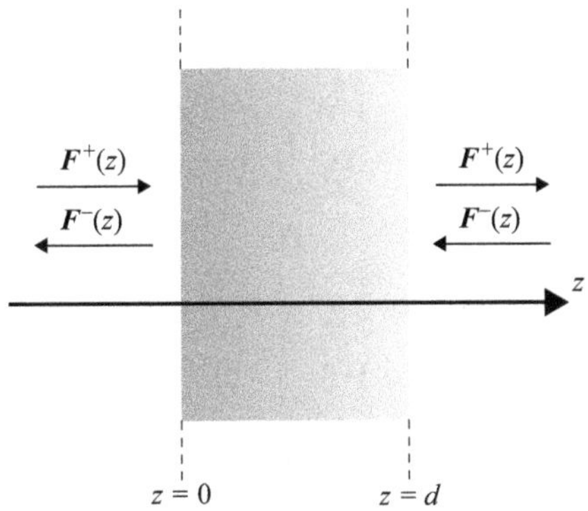

Figure 10.4 A symbolic representation of the wave splitting. The arrows represent the sign of the z-component of the power flux of the electromagnetic field.

One way to efficiently organize the solution in vacuum is to introduce a wave splitting. A wave splitting is a one-to-one transformation between the dependent vector fields, i.e., the tangential electric field and the tangential magnetic field, and two new fields—so-called split vector fields and denoted $\boldsymbol{F}^+$ and $\boldsymbol{F}^-$. These new fields, $\boldsymbol{F}^\pm$, represent the right- and left-going waves, respectively, and refer to the power flux in the preferred direction z. A symbolic representation of the wave splitting is given in Figure 10.4.

The split field vectors, $\boldsymbol{F}^\pm$, in vacuum are defined by

$$\boldsymbol{F}^\pm(\boldsymbol{k}_t, z, \omega) = \frac{1}{2}\left\{\boldsymbol{E}_{xy}(\boldsymbol{k}_t, z, \omega) \mp \mathbf{W}(\boldsymbol{k}_t)\cdot \boldsymbol{L}_{xy}(\boldsymbol{k}_t, z, \omega)\right\} \tag{10.15}$$

or equivalently

$$\begin{cases} \boldsymbol{E}_{xy}(\boldsymbol{k}_t, z, \omega) = \boldsymbol{F}^+(\boldsymbol{k}_t, z, \omega) + \boldsymbol{F}^-(\boldsymbol{k}_t, z, \omega) \\ \boldsymbol{L}_{xy}(\boldsymbol{k}_t, z, \omega) = -\mathbf{W}^{-1}(\boldsymbol{k}_t)\cdot \boldsymbol{F}^+(\boldsymbol{k}_t, z, \omega) + \mathbf{W}^{-1}(\boldsymbol{k}_t)\cdot \boldsymbol{F}^-(\boldsymbol{k}_t, z, \omega) \end{cases}$$

where the operator $\mathbf{W}(\boldsymbol{k}_t)$ and its inverse $\mathbf{W}^{-1}(\boldsymbol{k}_t)$ are defined as

$$\begin{aligned} \mathbf{W}(\boldsymbol{k}_t) &= \frac{k_z}{k_0}\left(\mathbf{I}_2 - \frac{1}{k_z^2}\boldsymbol{k}_t \times (\boldsymbol{k}_t \times \mathbf{I}_2)\right) = \hat{\boldsymbol{e}}_\parallel\hat{\boldsymbol{e}}_\parallel \frac{k_z}{k_0} + \hat{\boldsymbol{e}}_\perp\hat{\boldsymbol{e}}_\perp \frac{k_0}{k_z} \\ &= \frac{k_0}{k_z}\left(\mathbf{I}_2 - \frac{1}{k_0^2}\boldsymbol{k}_t\boldsymbol{k}_t\right) = \frac{k_0}{k_z}\mathbf{I}_2 - \frac{k_t^2}{k_0 k_z}\hat{\boldsymbol{e}}_\parallel\hat{\boldsymbol{e}}_\parallel \end{aligned} \tag{10.16}$$

and

$$\mathbf{W}^{-1}(\boldsymbol{k}_t) = \frac{k_0}{k_z}\left(\mathbf{I}_2 + \frac{1}{k_0^2}\boldsymbol{k}_t \times (\boldsymbol{k}_t \times \mathbf{I}_2)\right) = \hat{\boldsymbol{e}}_\parallel\hat{\boldsymbol{e}}_\parallel \frac{k_0}{k_z} + \hat{\boldsymbol{e}}_\perp\hat{\boldsymbol{e}}_\perp \frac{k_z}{k_0} \tag{10.17}$$

and where we used $\boldsymbol{k}_t \times (\boldsymbol{k}_t \times \mathbf{I}_2) = \mathbf{J}\cdot \boldsymbol{k}_t\boldsymbol{k}_t \cdot \mathbf{J} = -k_t^2\hat{\boldsymbol{e}}_\perp\hat{\boldsymbol{e}}_\perp$.

In a matrix notation, the wave splitting becomes

$$\begin{pmatrix} \boldsymbol{F}^+(z) \\ \boldsymbol{F}^-(z) \end{pmatrix} = \frac{1}{2}\begin{pmatrix} \mathbf{I}_2 & -\mathbf{W} \\ \mathbf{I}_2 & \mathbf{W} \end{pmatrix} \cdot \begin{pmatrix} \boldsymbol{E}_{xy}(z) \\ \boldsymbol{L}_{xy}(z) \end{pmatrix} = \mathbf{S} \cdot \begin{pmatrix} \boldsymbol{E}_{xy}(z) \\ \boldsymbol{L}_{xy}(z) \end{pmatrix} \tag{10.18}$$

with inverse

$$\begin{pmatrix} \boldsymbol{E}_{xy}(z) \\ \boldsymbol{L}_{xy}(z) \end{pmatrix} = \begin{pmatrix} \mathbf{I}_2 & \mathbf{I}_2 \\ -\mathbf{W}^{-1} & \mathbf{W}^{-1} \end{pmatrix} \cdot \begin{pmatrix} \boldsymbol{F}^+(z) \\ \boldsymbol{F}^-(z) \end{pmatrix} = \mathbf{S}^{-1} \cdot \begin{pmatrix} \boldsymbol{F}^+(z) \\ \boldsymbol{F}^-(z) \end{pmatrix} \tag{10.19}$$

where we have introduced the two four-dimensional dyadics $\mathbf{S}$ and $\mathbf{S}^{-1}$ defined by

$$\mathbf{S}(\boldsymbol{k}_t) = \frac{1}{2}\begin{pmatrix} \mathbf{I}_2 & -\mathbf{W}(\boldsymbol{k}_t) \\ \mathbf{I}_2 & \mathbf{W}(\boldsymbol{k}_t) \end{pmatrix}, \qquad \mathbf{S}^{-1}(\boldsymbol{k}_t) = \begin{pmatrix} \mathbf{I}_2 & \mathbf{I}_2 \\ -\mathbf{W}^{-1}(\boldsymbol{k}_t) & \mathbf{W}^{-1}(\boldsymbol{k}_t) \end{pmatrix}$$

At normal incidence, $\mathbf{W} = \mathbf{W}^{-1} = \mathbf{I}_2$.

Since the normal components (longitudinal parts) of the electric and magnetic fields can be expressed in terms of the tangential parts as, see (10.10)

$$\begin{pmatrix} E_z(z) \\ \eta_0 H_z(z) \end{pmatrix} = \frac{1}{k_0}\begin{pmatrix} \mathbf{0} & \boldsymbol{k}_t \\ \mathbf{J}\cdot\boldsymbol{k}_t & \mathbf{0} \end{pmatrix} \cdot \begin{pmatrix} \boldsymbol{E}_{xy}(z) \\ \boldsymbol{L}_{xy}(z) \end{pmatrix}$$

the total electric and magnetic fields are, see Problem 10.2

$$\begin{cases} \boldsymbol{E}(z) = \boldsymbol{\gamma}^+(\boldsymbol{k}_t)\cdot\boldsymbol{F}^+(z) + \boldsymbol{\gamma}^-(\boldsymbol{k}_t)\cdot\boldsymbol{F}^-(z) \\ \eta_0\boldsymbol{H}(z) = \dfrac{1}{k_0}\boldsymbol{k}^+ \times \left(\boldsymbol{\gamma}^+(\boldsymbol{k}_t)\cdot\boldsymbol{F}^+(z)\right) + \dfrac{1}{k_0}\boldsymbol{k}^- \times \left(\boldsymbol{\gamma}^-(\boldsymbol{k}_t)\cdot\boldsymbol{F}^-(z)\right) \end{cases} \tag{10.20}$$

where

$$\boldsymbol{k}^{\pm} = \boldsymbol{k}_t \pm \hat{\boldsymbol{z}}k_z \tag{10.21}$$

and where the projection dyadic $\gamma^{\pm}(\boldsymbol{k}_t)$ is defined as

$$\boldsymbol{\gamma}^{\pm}(\boldsymbol{k}_t) = \mathbf{I}_2 \mp \frac{\hat{\boldsymbol{z}}\boldsymbol{k}_t}{k_z} = \mathbf{I}_2 \mp \frac{k_t}{k_z}\hat{\boldsymbol{z}}\hat{\boldsymbol{e}}_{\parallel} \tag{10.22}$$

The relation

$$\boldsymbol{k}^{\pm}\cdot\boldsymbol{\gamma}^{\pm}(\boldsymbol{k}_t) = \mathbf{0}$$

also hold.

To see the physical implications of the wave-splitting transformation, we proceed by finding the differential equation for the split fields $\boldsymbol{F}^{\pm}$ in vacuum. The transverse fields $\boldsymbol{E}_{xy}(\boldsymbol{k}_t, z)$ and $\boldsymbol{L}_{xy}(\boldsymbol{k}_t, z)$ satisfy, see (10.11) and (10.14)

$$\frac{\mathrm{d}}{\mathrm{d}z}\begin{pmatrix} \boldsymbol{E}_{xy}(\boldsymbol{k}_t, z) \\ \boldsymbol{L}_{xy}(\boldsymbol{k}_t, z) \end{pmatrix} = \mathrm{i}k_z\begin{pmatrix} \mathbf{0} & -\mathbf{W}(\boldsymbol{k}_t) \\ -\mathbf{W}^{-1}(\boldsymbol{k}_t) & \mathbf{0} \end{pmatrix} \cdot \begin{pmatrix} \boldsymbol{E}_{xy}(\boldsymbol{k}_t, z) \\ \boldsymbol{L}_{xy}(\boldsymbol{k}_t, z) \end{pmatrix}$$

since

$$\begin{cases} \mathbf{I}_2 - \dfrac{1}{k_0^2}\boldsymbol{k}_t\boldsymbol{k}_t = \dfrac{k_z}{k_0}\mathbf{W}(\boldsymbol{k}_t) \\ \mathbf{I}_2 + \dfrac{1}{k_0^2}\boldsymbol{k}_t \times (\boldsymbol{k}_t \times \mathbf{I}_2) = \dfrac{k_z}{k_0}\mathbf{W}^{-1}(\boldsymbol{k}_t) \end{cases}$$

An application of (10.18) and (10.19) gives

$$\frac{\mathrm{d}}{\mathrm{d}z}\begin{pmatrix} \boldsymbol{F}^+(\boldsymbol{k}_t, z) \\ \boldsymbol{F}^-(\boldsymbol{k}_t, z) \end{pmatrix} = \mathrm{i}k_z \begin{pmatrix} \mathbf{I}_2 & \mathbf{0} \\ \mathbf{0} & -\mathbf{I}_2 \end{pmatrix} \cdot \begin{pmatrix} \boldsymbol{F}^+(\boldsymbol{k}_t, z) \\ \boldsymbol{F}^-(\boldsymbol{k}_t, z) \end{pmatrix}$$

From this equation, we see that the split fields, $\boldsymbol{F}^\pm$, decouple in vacuum and the solution is

$$\boldsymbol{F}^\pm(\boldsymbol{k}_t, z) = \boldsymbol{F}^\pm(\boldsymbol{k}_t, z_0)e^{\pm ik_z(z-z_0)}$$

which give

$$\begin{cases} \boldsymbol{F}^\pm(\boldsymbol{k}_t, z, \omega) = \boldsymbol{F}^\pm(\boldsymbol{k}_t, 0, \omega)\mathrm{e}^{\pm \mathrm{i}k_z z}, & z < 0 \\ \boldsymbol{F}^\pm(\boldsymbol{k}_t, z, \omega) = \boldsymbol{F}^\pm(\boldsymbol{k}_t, d, \omega)\mathrm{e}^{\pm \mathrm{i}k_z (z-d)}, & z > d \end{cases} \tag{10.23}$$

Notice that $\boldsymbol{F}^\pm(z)$ are damped in the $\pm z$-directions, respectively, if the plane waves are inhomogeneous (evanescent).

The tangential electric and magnetic fields $\boldsymbol{E}_{xy}$ and $\boldsymbol{H}_{xy}$ are continuous everywhere in the slab, and especially over any interface of the slab due to the boundary conditions. This continuity property is inherited by the split fields, i.e., $\boldsymbol{F}^\pm$ are continuous at the interfaces.

10.3.1 Power Flux Density

The average power in the z-direction is determined by the integral of Poynting vector $\boldsymbol{S}$ over a plane $z =$ constant. Since the material parameters are constant on the plane $z =$ constant, we have a Fourier representation of the fields. Using Parseval's identity, see Appendix D.1, we get

$$\begin{aligned} \iint_{\mathbb{R}^2} \hat{z} \cdot \boldsymbol{S}(\boldsymbol{r})\, \mathrm{d}x\, \mathrm{d}y &= \frac{1}{2} \operatorname{Re} \iint_{\mathbb{R}^2} \hat{z} \cdot \left(\boldsymbol{E}(\boldsymbol{r}) \times \boldsymbol{H}^*(\boldsymbol{r})\right) \mathrm{d}x\, \mathrm{d}y \\ &= -\frac{1}{8\pi^2} \operatorname{Re} \iint_{\mathbb{R}^2} \boldsymbol{E}_{xy}(\boldsymbol{k}_t, z) \cdot \left(\hat{z} \times \boldsymbol{H}_{xy}^*(\boldsymbol{k}_t, z)\right) \mathrm{d}k_x\, \mathrm{d}k_y \end{aligned}$$

We rewrite the integrand with the wave splitting, see (10.19)

$$-\operatorname{Re}\left\{\boldsymbol{E}_{xy}(z) \cdot \left(\hat{z} \times \boldsymbol{H}_{xy}^*(z)\right)\right\} = \operatorname{Re}\left\{\frac{1}{\eta_0}\left(\boldsymbol{F}^{+*}(z) + \boldsymbol{F}^{-*}(z)\right) \cdot \mathbf{W}^{-1} \cdot \left(\boldsymbol{F}^+(z) - \boldsymbol{F}^-(z)\right)\right\}$$

Two types of components occur

$$\begin{cases} \operatorname{Re}\left(\boldsymbol{F}^{+*} \cdot \mathbf{W}^{-1} \cdot \boldsymbol{F}^+ - \boldsymbol{F}^{-*} \cdot \mathbf{W}^{-1} \cdot \boldsymbol{F}^-\right) \\ \operatorname{Re}\left(\boldsymbol{F}^{-*} \cdot \mathbf{W}^{-1} \cdot \boldsymbol{F}^+ - \boldsymbol{F}^{+*} \cdot \mathbf{W}^{-1} \cdot \boldsymbol{F}^-\right) \end{cases} \tag{10.24}$$

The first term in (10.24) has the form

$$\boldsymbol{F}^{\pm}\cdot\mathbf{W}^{-1}\cdot\boldsymbol{F}^{\pm *}=\boldsymbol{F}^{\pm}\cdot\left(\hat{\boldsymbol{e}}_{\|}\hat{\boldsymbol{e}}_{\|}\frac{k_0}{k_z}+\hat{\boldsymbol{e}}_{\perp}\hat{\boldsymbol{e}}_{\perp}\frac{k_z}{k_0}\right)\cdot\boldsymbol{F}^{\pm *}=\frac{k_z}{k_0}\left(|F_{\|}^{\pm}|^2\frac{k_0^2}{k_z^2}+|F_{\perp}^{\pm}|^2\right)$$

where the projections $F_{\perp}^{\pm}$ and $F_{\|}^{\pm}$ are defined by $F_{\perp}^{\pm}=\hat{\boldsymbol{e}}_{\perp}\cdot\boldsymbol{F}^{\pm}$ and $F_{\|}^{\pm}=\hat{\boldsymbol{e}}_{\|}\cdot\boldsymbol{F}^{\pm}$.

We defined the projection dyadic $\boldsymbol{\gamma}^{\pm}(\boldsymbol{k}_t)$ in (10.22). These dyadics imply for $k_t\le k_0$

$$|\boldsymbol{\gamma}^{\pm}\cdot\boldsymbol{F}^{\pm}|^2=\left|\hat{\boldsymbol{e}}_{\|}F_{\|}^{\pm}+\hat{\boldsymbol{e}}_{\perp}F_{\perp}^{\pm}\mp\hat{\boldsymbol{z}}F_{\|}^{\pm}\frac{k_t}{k_z}\right|^2=\left(|F_{\|}^{\pm}|^2\frac{k_0^2}{k_z^2}+|F_{\perp}^{\pm}|^2\right)$$

and we have

$$\operatorname{Re}\boldsymbol{F}^{\pm}\cdot\mathbf{W}^{-1}\cdot\boldsymbol{F}^{\pm *}=\begin{cases}|\boldsymbol{\gamma}^{\pm}\cdot\boldsymbol{F}^{\pm}|^2\frac{k_z}{k_0}, & k_t\le k_0\\ 0, & k_t\ge k_0\end{cases}$$

Similarly for the second term in (10.24), we get ($\mathbf{W}^{-1}$ is a symmetric dyadic)

$$\operatorname{Re}\left(\boldsymbol{F}^{-*}\cdot\mathbf{W}^{-1}\cdot\boldsymbol{F}^{+}-\boldsymbol{F}^{+*}\cdot\mathbf{W}^{-1}\cdot\boldsymbol{F}^{-}\right)=2\operatorname{Re}\left(\mathrm{i}\boldsymbol{F}^{-*}\cdot\operatorname{Im}\left(\mathbf{W}^{-1}\right)\cdot\boldsymbol{F}^{+}\right)$$

$$=-2\left(\operatorname{Im}\frac{k_0}{k_z}\operatorname{Im}\left(F_{\|}^{+}F_{\|}^{-*}\right)+\operatorname{Im}\frac{k_z}{k_0}\operatorname{Im}\left(F_{\perp}^{+}F_{\perp}^{-*}\right)\right)$$

Finally, we get

$$\iint_{\mathbb{R}^2}\hat{\boldsymbol{z}}\cdot\boldsymbol{S}(\boldsymbol{r})\,\mathrm{d}x\,\mathrm{d}y=\frac{1}{8\pi^2\eta_0}\iint_{k_t\le k_0}\frac{k_z}{k_0}\left(|\boldsymbol{\gamma}^{+}\cdot\boldsymbol{F}^{+}|^2-|\boldsymbol{\gamma}^{-}\cdot\boldsymbol{F}^{-}|^2\right)\mathrm{d}k_x\,\mathrm{d}k_y$$

$$-\frac{1}{4\pi^2\eta_0}\iint_{k_t\ge k_0}\left(\operatorname{Im}\frac{k_0}{k_z}\operatorname{Im}\left(F_{\|}^{+}F_{\|}^{-*}\right)+\operatorname{Im}\frac{k_z}{k_0}\operatorname{Im}\left(F_{\perp}^{+}F_{\perp}^{-*}\right)\right)\mathrm{d}k_x\,\mathrm{d}k_y$$

We see that a homogeneous right-going (left-going) wave has a positive (negative) z-component of the Poynting vector, which justifies the notion right- and left-going waves in vacuum. The inhomogeneous waves give no contributions, unless the wave is a mixture of right- and left-going inhomogeneous waves.

10.3.1.1 Pure Homogeneous Waves

The analysis above holds for a general electromagnetic field. If the electromagnetic wave is a pure plane wave and homogeneous (propagating), we adopt another, more standard, route to quantify the power flux density.

The starting point for this analysis is the Fourier transformed fields in (10.20):

$$\boldsymbol{E}(\boldsymbol{k}_t,z)=\boldsymbol{\gamma}^{+}(\boldsymbol{k}_t)\cdot\boldsymbol{F}^{+}(\boldsymbol{k}_t,z)+\boldsymbol{\gamma}^{-}(\boldsymbol{k}_t)\cdot\boldsymbol{F}^{-}(\boldsymbol{k}_t,z)$$

and

$$k_0\eta_0\boldsymbol{H}(\boldsymbol{k}_t,z)=\boldsymbol{k}^{+}\times\left(\boldsymbol{\gamma}^{+}(\boldsymbol{k}_t)\cdot\boldsymbol{F}^{+}(\boldsymbol{k}_t,z)\right)+\boldsymbol{k}^{-}\times\left(\boldsymbol{\gamma}^{-}(\boldsymbol{k}_t)\cdot\boldsymbol{F}^{-}(\boldsymbol{k}_t,z)\right)$$

where $\boldsymbol{k}^{\pm}$ and $\boldsymbol{\gamma}^{\pm}(\boldsymbol{k}_t)$ are given by (10.21) and (10.22), respectively.

The relevant quantity for the power flux density is

$$\langle \boldsymbol{S}(t)\rangle = \frac{1}{2}\operatorname{Re}\left\{\boldsymbol{E}(\boldsymbol{k}_{\mathrm{t}}, z)\times \boldsymbol{H}^{*}(\boldsymbol{k}_{\mathrm{t}}, z)\right\}$$

Use the BAC-CAB rule and $\boldsymbol{k}^{\pm}\cdot\boldsymbol{\gamma}^{\pm}(\boldsymbol{k}_{\mathrm{t}}) = \boldsymbol{0}$ to obtain for the purely right-going waves

$$\langle \boldsymbol{S}(t)\rangle = \frac{1}{2\eta_0 k_0}\operatorname{Re}\boldsymbol{k}^{+}\left|\left(\mathbf{I}_2 - \frac{1}{k_z}\hat{\boldsymbol{z}}\boldsymbol{k}_{\mathrm{t}}\right)\cdot\boldsymbol{F}^{+}\right|^2$$

For the purely left-going waves, we have a similar expression:

$$\langle \boldsymbol{S}(t)\rangle = \frac{1}{2\eta_0 k_0}\operatorname{Re}\boldsymbol{k}^{-}\left|\left(\mathbf{I}_2 + \frac{1}{k_z}\hat{\boldsymbol{z}}\boldsymbol{k}_{\mathrm{t}}\right)\cdot\boldsymbol{F}^{-}\right|^2$$

In summary,

$$\begin{aligned}
\langle \boldsymbol{S}(t)\rangle &= \frac{1}{4\eta_0 k_0}\operatorname{Re}\boldsymbol{k}^{\pm}\left|\left(\mathbf{I}_2 \mp \frac{1}{k_z}\hat{\boldsymbol{z}}\boldsymbol{k}_{\mathrm{t}}\right)\cdot\boldsymbol{F}^{\pm}\right|^2 \\
&= \frac{1}{2\eta_0 k_0}\operatorname{Re}\boldsymbol{k}^{\pm}\left|\hat{\boldsymbol{e}}_{\parallel}F_{\parallel}^{\pm} + \hat{\boldsymbol{e}}_{\perp}F_{\perp}^{\pm} \mp \hat{\boldsymbol{z}}F_{\parallel}^{\pm}\frac{k_{\mathrm{t}}}{k_z}\right|^2 \\
&= \frac{\boldsymbol{k}_{\mathrm{t}} \pm \hat{\boldsymbol{z}}\operatorname{Re}k_z}{2\eta_0 k_0}\left(|F_{\parallel}^{\pm}|^2\left(1+\frac{k_{\mathrm{t}}^2}{|k_z|^2}\right) + |F_{\perp}^{\pm}|^2\right)
\end{aligned} \tag{10.25}$$

for purely right-going and left-going waves, respectively, where the projections $F_{\perp}^{\pm}$ and $F_{\parallel}^{\pm}$ are defined by $F_{\perp}^{\pm} = \hat{\boldsymbol{e}}_{\perp}\cdot\boldsymbol{F}^{\pm}$ and $F_{\parallel}^{\pm} = \hat{\boldsymbol{e}}_{\parallel}\cdot\boldsymbol{F}^{\pm}$. For a mixture of right- and left-going waves, interaction between the waves occurs and more terms have to be added to the power flux density. We also note that for a homogeneous (propagating) wave, the power flux density has a z component, but this component is lacking for a inhomogeneous (evanescent) wave.

10.3.2 Wave Splitting and Projection Dyadics

Wave splitting can also be viewed in the light of projection dyadics. Projection dyadics are introduced in Appendix F.2. Specifically, we have for the case where the eigenvalues are double and differ in sign, as in the vacuum case, see Section 10.6.1 and Example F.3 on page 668:

$$\mathrm{e}^{ik_0 z\mathbf{M}_0} = \frac{1}{2\lambda}\left\{\mathrm{e}^{ik_0 z\lambda}(\lambda\mathbf{I}_4 + \mathbf{M}_0) + \mathrm{e}^{-ik_0 z\lambda}(\lambda\mathbf{I}_4 - \mathbf{M}_0)\right\}$$

where $\lambda = k_z/k_0$ and the fundamental dyadic in vacuum, $\mathbf{M}_0$, is given by (10.14). Explicitly, we get

$$\mathrm{e}^{ik_0 z\mathbf{M}_0} = \mathbf{A}^{+}\mathrm{e}^{ik_z z} + \mathbf{A}^{-}\mathrm{e}^{-ik_z z} \tag{10.26}$$

where

$$\mathbf{A}^{+} = \frac{1}{2}\begin{pmatrix}\mathbf{I}_2 & -\mathbf{W}\\ -\mathbf{W}^{-1} & \mathbf{I}_2\end{pmatrix}, \qquad \mathbf{A}^{-} = \frac{1}{2}\begin{pmatrix}\mathbf{I}_2 & \mathbf{W}\\ \mathbf{W}^{-1} & \mathbf{I}_2\end{pmatrix}$$

and the dyadics $\mathbf{W}$ and $\mathbf{W}^{-1}$ are defined by (10.16) and (10.17) on page 527. Equation (10.26) is a decomposition of the propagation operator in two parts—one that gives propagation in the positive direction, $\mathbf{A}^+$, and one in the negative direction, $\mathbf{A}^-$. The action on the fields in vacuum is

$$\begin{pmatrix} \boldsymbol{E}_{xy}(z) \\ \boldsymbol{L}_{xy}(z) \end{pmatrix} = \mathrm{e}^{\mathrm{i}k_0 z \mathbf{M}_0} \cdot \begin{pmatrix} \boldsymbol{E}_{xy}(0) \\ \boldsymbol{L}_{xy}(0) \end{pmatrix}$$

or more explicitly

$$\begin{pmatrix} \boldsymbol{E}_{xy}(z) \\ \boldsymbol{L}_{xy}(z) \end{pmatrix} = \begin{pmatrix} \boldsymbol{F}^+(0) \\ -\mathbf{W}^{-1} \cdot \boldsymbol{F}^+(0) \end{pmatrix} \mathrm{e}^{\mathrm{i}k_z z} + \begin{pmatrix} \boldsymbol{F}^-(0) \\ \mathbf{W}^{-1} \cdot \boldsymbol{F}^-(0) \end{pmatrix} \mathrm{e}^{-\mathrm{i}k_z z}$$

where we have used the definition of the split fields, $\boldsymbol{F}^\pm(z)$, see also (10.18)

$$\boldsymbol{F}^\pm(z) = \frac{1}{2} \left\{ \boldsymbol{E}_{xy}(z) \mp \mathbf{W} \cdot \boldsymbol{L}_{xy}(z) \right\}$$

From these observations, we conclude that the decomposition (10.26) induces a wave splitting that is identical to (10.23). We have as above

$$\boldsymbol{F}^\pm(z) = \boldsymbol{F}^\pm(0) \mathrm{e}^{\pm \mathrm{i}k_z z}$$

10.4 Propagation of Fields—the Propagator Dyadic

In this section, the notion of propagator dyadic for a layered bianisotropic structure is introduced. We have already touched upon the topic in vacuum in the sections above, and we now treat the general problem.

The propagator dyadic maps the tangential electric and magnetic fields at two longitudinal z-positions to each other, e.g., the front surface fields of the structure to the fields at the rear surface of the structure.

Due to the linearity of the fundamental equation, the solution can be written in the form[2]

$$\begin{pmatrix} \boldsymbol{E}_{xy}(z) \\ \boldsymbol{L}_{xy}(z) \end{pmatrix} = \mathbf{P}(z, z_1) \cdot \begin{pmatrix} \boldsymbol{E}_{xy}(z_1) \\ \boldsymbol{L}_{xy}(z_1) \end{pmatrix} \tag{10.27}$$

The four-dimensional dyadic $\mathbf{P}(z, z_1)$ connects the fields at the two positions z and z_1 to each other. This dyadic is called the propagator dyadic, as it maps the field at z_1 to the fields at z. For later purposes, the propagator dyadic is decomposed into four two-dimensional dyadics, i.e.,

$$\mathbf{P}(z, z_1) = \begin{pmatrix} \mathbf{P}_{11}(z, z_1) & \mathbf{P}_{12}(z, z_1) \\ \mathbf{P}_{21}(z, z_1) & \mathbf{P}_{22}(z, z_1) \end{pmatrix}$$

[2] Formally, the fundamental equation, (10.11), can be solved by introducing the spatial ordering operator $\mathcal{S}$ as

$$\begin{pmatrix} \boldsymbol{E}_{xy}(z) \\ \boldsymbol{L}_{xy}(z) \end{pmatrix} = \mathcal{S} \exp\left\{ \mathrm{i}k_0 \int_{z_1}^{z} \mathbf{M}(z') \, \mathrm{d}z' \right\} \cdot \begin{pmatrix} \boldsymbol{E}_{xy}(z_1) \\ \boldsymbol{L}_{xy}(z_1) \end{pmatrix}$$

This operator is the space analog to the time ordering operator which appears in quantum mechanics [16].

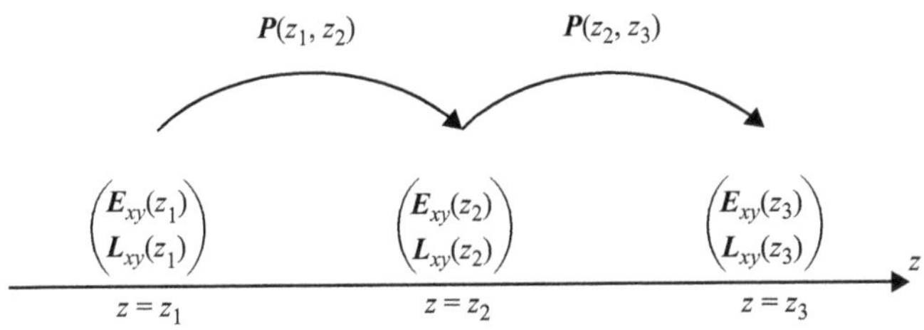

Figure 10.5 The propagator dyadic and its successive use.

The relation between the fields at z and z_1 then becomes

$$\begin{cases} \boldsymbol{E}_{xy}(z) = \mathbf{P}_{11}(z,z_1)\cdot\boldsymbol{E}_{xy}(z_1) + \mathbf{P}_{12}(z,z_1)\cdot\boldsymbol{L}_{xy}(z_1) \\ \boldsymbol{L}_{xy}(z) = \mathbf{P}_{21}(z,z_1)\cdot\boldsymbol{E}_{xy}(z_1) + \mathbf{P}_{22}(z,z_1)\cdot\boldsymbol{L}_{xy}(z_1) \end{cases}$$

Naturally, if $z = z_1$, then the propagator dyadic coincides with the identity operator, and therefore, it satisfies

$$\mathbf{P}(z_1,z_1) = \mathbf{I}_4$$

The fundamental equation (10.11) implies that $\mathbf{P}$ satisfies the same system of ODEs as the transverse fields. Explicitly, we have [229]

$$\begin{cases} \dfrac{\mathrm{d}}{\mathrm{d}z}\mathbf{P}(z,z_1) = \mathrm{i}k_0\mathbf{M}(z)\cdot\mathbf{P}(z,z_1) \\ \mathbf{P}(z_1,z_1) = \mathbf{I}_4 \end{cases}$$

where the dyadic $\mathbf{M}$ is the same as in (10.11). For a homogeneous material, an explicit solution of the propagator dyadic can often be obtained, and in Section 10.6 some of these cases are analyzed in detail.

The propagator dyadic can be used successively to propagate the fields from z_1 to z_3 via an intermediate position z_2, see Figure 10.5. We have

$$\mathbf{P}(z_3,z_1) = \mathbf{P}(z_3,z_2)\cdot\mathbf{P}(z_2,z_1)$$

We postpone the problem of calculating the propagator dyadics until later, and instead we concentrate on the applications of a propagator dyadic.

10.4.1 Reflection and Transmission Dyadics

In this section, the reflection and transmission dyadics for a general bianisotropic slab are computed. These dyadics are easy to obtain using the wave splitting and the propagator dyadic.

Recall that all the generating sources are located in the empty half-space, $z < 0$. Recall also that $\boldsymbol{F}^+(\boldsymbol{k}_t, 0, \omega)$ and $\boldsymbol{F}^-(\boldsymbol{k}_t, 0, \omega)$ are the incident and reflected tangential electric fields at $z = 0$, respectively, and that $\boldsymbol{F}^+(\boldsymbol{k}_t, d, \omega)$ is the transmitted tangential electric field at $z = d$. All these fields are associated with the transverse wave vector $\boldsymbol{k}_t$. Specifically, the total incident field $\boldsymbol{E}_\mathrm{i}(\boldsymbol{r},\omega)$ at $z<0$ (but to the right of the sources, i.e., $z_0<z<0$) in terms of $\boldsymbol{F}^+(\boldsymbol{k}_\mathrm{t},0,\omega)$ is ($z_0<z<0$)

$$\boldsymbol{E}_\mathrm{i}(\boldsymbol{r},\omega)=\frac{1}{4\pi^2}\iint\limits_{\mathbb{R}^2}\left(\mathbf{I}_2-\frac{1}{k_z}\hat{\boldsymbol{z}}\boldsymbol{k}_\mathrm{t}\right)\cdot\boldsymbol{F}^+(\boldsymbol{k}_\mathrm{t},0,\omega)\mathrm{e}^{\mathrm{i}\boldsymbol{k}^+\cdot\boldsymbol{r}}\,\mathrm{d}k_x\,\mathrm{d}k_y$$

by the use of (10.3), (10.20), and (10.23), and where, see (10.21)

$$\boldsymbol{k}^+=\boldsymbol{k}_\mathrm{t}+\hat{\boldsymbol{z}}k_z$$

Notice that all components of the field, not just the tangential ones, are evaluated.

Above, we have assumed the sources are located to the left of the slab. We now also assume there is an excitation from the right, $\boldsymbol{F}^-(\boldsymbol{k}_\mathrm{t},d,\omega)$, and we now try to find the relation between the excitation, $\boldsymbol{F}^+(0)$ and $\boldsymbol{F}^-(d)$, and the reaction $\boldsymbol{F}^-(0)$ and $\boldsymbol{F}^+(d)$. The relation between the fields at the end point $z=0$ and $z=d$ of the slab is obtained from (10.27) and (10.19)

$$\begin{pmatrix}\boldsymbol{E}_{xy}(d)\\ \boldsymbol{L}_{xy}(d)\end{pmatrix}=\mathbf{P}(d,0)\cdot\begin{pmatrix}\boldsymbol{E}_{xy}(0)\\ \boldsymbol{L}_{xy}(0)\end{pmatrix}=\mathbf{P}(d,0)\cdot\mathbf{S}^{-1}\cdot\begin{pmatrix}\boldsymbol{F}^+(0)\\ \boldsymbol{F}^-(0)\end{pmatrix}$$

Moreover, from (10.18)

$$\begin{aligned}\begin{pmatrix}\boldsymbol{F}^+(d)\\ \boldsymbol{F}^-(d)\end{pmatrix}&=\mathbf{S}\cdot\begin{pmatrix}\boldsymbol{E}_{xy}(d)\\ \boldsymbol{L}_{xy}(d)\end{pmatrix}=\mathbf{S}\cdot\mathbf{P}(d,0)\cdot\mathbf{S}^{-1}\cdot\begin{pmatrix}\boldsymbol{F}^+(0)\\ \boldsymbol{F}^-(0)\end{pmatrix}\\ &=\mathbf{G}(d,0)\cdot\mathbf{S}^{-1}\cdot\begin{pmatrix}\boldsymbol{F}^+(0)\\ \boldsymbol{F}^-(0)\end{pmatrix}=\mathbf{T}(d,0)\cdot\begin{pmatrix}\boldsymbol{F}^+(0)\\ \boldsymbol{F}^-(0)\end{pmatrix}\end{aligned}\tag{10.28}$$

where the two four-dimensional dyadics $\mathbf{G}$ and $\mathbf{T}$ are defined as

$$\mathbf{G}(d,0)=\mathbf{S}\cdot\mathbf{P}(d,0)\qquad\text{and}\qquad\mathbf{T}(d,0)=\mathbf{G}(d,0)\cdot\mathbf{S}^{-1}$$

where

$$\begin{cases}2\mathbf{G}_{11}=\mathbf{P}_{11}-\mathbf{W}\cdot\mathbf{P}_{21}\\ 2\mathbf{G}_{12}=\mathbf{P}_{12}-\mathbf{W}\cdot\mathbf{P}_{22}\\ 2\mathbf{G}_{21}=\mathbf{P}_{11}+\mathbf{W}\cdot\mathbf{P}_{21}\\ 2\mathbf{G}_{22}=\mathbf{P}_{12}+\mathbf{W}\cdot\mathbf{P}_{22}\end{cases}\quad\text{and}\quad\begin{cases}\mathbf{T}_{11}=\mathbf{G}_{11}-\mathbf{G}_{12}\cdot\mathbf{W}^{-1}\\ \mathbf{T}_{12}=\mathbf{G}_{11}+\mathbf{G}_{12}\cdot\mathbf{W}^{-1}\\ \mathbf{T}_{21}=\mathbf{G}_{21}-\mathbf{G}_{22}\cdot\mathbf{W}^{-1}\\ \mathbf{T}_{22}=\mathbf{G}_{21}+\mathbf{G}_{22}\cdot\mathbf{W}^{-1}\end{cases}\tag{10.29}$$

Especially, since $\mathbf{P}(0,0)=\mathbf{I}_4$

$$\mathbf{T}(0,0)=\mathbf{I}_4$$

Equation (10.28) gives the relation between the split fields at both ends of the slab, i.e., at $z=0$ and $z=d$. A more common representation (scattering representation) is to relate $\boldsymbol{F}^-(0)$ and $\boldsymbol{F}^+(d)$ (output fields) in terms of $\boldsymbol{F}^+(0)$ and $\boldsymbol{F}^-(d)$ (input fields). Since from (10.28)

$$\boldsymbol{F}^-(0)=\mathbf{T}_{22}^{-1}\cdot\boldsymbol{F}^-(d)-\mathbf{T}_{22}^{-1}\cdot\mathbf{T}_{21}\cdot\boldsymbol{F}^+(0)$$

we get

$$F^+(d) = \mathbf{T}_{11} \cdot F^+(0) + \mathbf{T}_{12} \cdot \left(\mathbf{T}_{22}^{-1} \cdot F^-(d) - \mathbf{T}_{22}^{-1} \cdot \mathbf{T}_{21} \cdot F^+(0)\right)$$

or in a matrix form

$$\begin{pmatrix} F^-(0) \\ F^+(d) \end{pmatrix} = \begin{pmatrix} -\mathbf{T}_{22}^{-1} \cdot \mathbf{T}_{21} & \mathbf{T}_{22}^{-1} \\ \mathbf{T}_{11} - \mathbf{T}_{12} \cdot \mathbf{T}_{22}^{-1} \cdot \mathbf{T}_{21} & \mathbf{T}_{12} \cdot \mathbf{T}_{22}^{-1} \end{pmatrix} \cdot \begin{pmatrix} F^+(0) \\ F^-(d) \end{pmatrix}$$

The arguments of all dyadics $\mathbf{T}_{ij}$ here are $\mathbf{T}_{ij} = \mathbf{T}_{ij}(d, 0)$, $i, j = 1, 2$.

If we again specialize to sources on the left-hand side of the slab, $z < 0$, then $F^-(d) = \mathbf{0}$ and the reflection and transmission dyadics are defined as

$$\begin{cases} F^-(k_t, 0, \omega) = \mathbf{r} \cdot F^+(k_t, 0, \omega) \\ F^+(k_t, d, \omega) = \mathbf{t} \cdot F^+(k_t, 0, \omega) \end{cases}$$

where $\mathbf{r}$ and $\mathbf{t}$ are the reflection and transmission dyadics for the tangential electric field, respectively. The reflection dyadic is

$$\mathbf{r} = -\mathbf{T}_{22}^{-1} \cdot \mathbf{T}_{21} \tag{10.30}$$

and similarly the transmission dyadic is

$$\mathbf{t} = \mathbf{T}_{11} - \mathbf{T}_{12} \cdot \mathbf{T}_{22}^{-1} \cdot \mathbf{T}_{21} = \mathbf{T}_{11} + \mathbf{T}_{12} \cdot \mathbf{r} \tag{10.31}$$

The reflected field $E_r(r, \omega)$ for $z \leq 0$ is given by

$$E_r(r, \omega) = \frac{1}{4\pi^2} \iint_{\mathbb{R}^2} \left(\mathbf{I}_2 + \frac{\hat{z}k_t}{k_z}\right) \cdot \mathbf{r} \cdot F^+(k_t, 0, \omega) e^{ik^- \cdot r} \, dk_x \, dk_y \tag{10.32}$$

and the transmitted field $E_t(r, \omega)$ for $z \geq d$ is

$$E_t(r, \omega) = \frac{1}{4\pi^2} \iint_{\mathbb{R}^2} \left(\mathbf{I}_2 - \frac{\hat{z}k_t}{k_z}\right) \cdot \mathbf{t} \cdot F^+(k_t, 0, \omega) e^{ik^+ \cdot r - ik_z d} \, dk_x \, dk_y \tag{10.33}$$

where $k^{\pm} = k_t \pm \hat{z}k_z$, see (10.21). Notice that these expressions contain all components of the field, not just the tangential ones.

10.4.2 Slab above Ground

For a slab that ends with a ground plane, see Figure 10.6, the boundary conditions imply, see (10.27) and (10.19)

$$\begin{pmatrix} \mathbf{0} \\ -\eta_0 J_S \end{pmatrix} = \mathbf{P}(d, 0) \cdot \begin{pmatrix} E_{xy}(0) \\ L_{xy}(0) \end{pmatrix} = \begin{pmatrix} \mathbf{P}_{11} - \mathbf{P}_{12} \cdot \mathbf{W}^{-1} & \mathbf{P}_{11} + \mathbf{P}_{12} \cdot \mathbf{W}^{-1} \\ \mathbf{P}_{21} - \mathbf{P}_{22} \cdot \mathbf{W}^{-1} & \mathbf{P}_{21} + \mathbf{P}_{22} \cdot \mathbf{W}^{-1} \end{pmatrix} \cdot \begin{pmatrix} F^+(0) \\ F^-(0) \end{pmatrix}$$

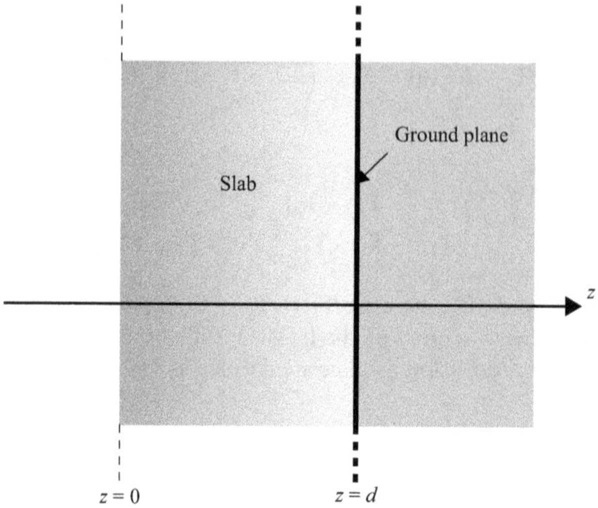

Figure 10.6 The geometry of a stratified structure backed with a ground plane. The region outside the slab to the left is vacuum.

since the tangential electric field at $z = d$ is $\boldsymbol{E}_{xy}(d) = \mathbf{0}$, and the surface current density on the ground plane is $\boldsymbol{J}_S = -\hat{z} \times \boldsymbol{H}_{xy}(d)$. We rewrite as

$$\mathbf{0} = \left(\mathbf{P}_{11} - \mathbf{P}_{12} \cdot \mathbf{W}^{-1}\right) \cdot \boldsymbol{F}^{+}(0) + \left(\mathbf{P}_{11} + \mathbf{P}_{12} \cdot \mathbf{W}^{-1}\right) \cdot \boldsymbol{F}^{-}(0)$$

and

$$-\eta_0 \boldsymbol{J}_S = \left(\mathbf{P}_{21} - \mathbf{P}_{22} \cdot \mathbf{W}^{-1}\right) \cdot \boldsymbol{F}^{+}(0) + \left(\mathbf{P}_{21} + \mathbf{P}_{22} \cdot \mathbf{W}^{-1}\right) \cdot \boldsymbol{F}^{-}(0)$$

We solve for the reflection dyadic of the grounded slab

$$\mathbf{r} = -\left(\mathbf{P}_{11}(d,0) + \mathbf{P}_{12}(d,0) \cdot \mathbf{W}^{-1}\right)^{-1} \cdot \left(\mathbf{P}_{11}(d,0) - \mathbf{P}_{12}(d,0) \cdot \mathbf{W}^{-1}\right) \tag{10.34}$$

and the surface current density at $z = d$ is

$$\boldsymbol{J}_S = \frac{1}{\eta_0}\left(\mathbf{P}_{22}(d,0) \cdot \mathbf{W}^{-1} \cdot (\mathbf{I}_2 - \mathbf{r}) - \mathbf{P}_{21}(d,0) \cdot (\mathbf{I}_2 + \mathbf{r})\right) \cdot \boldsymbol{F}^{+}(0) = \mathbf{C} \cdot \boldsymbol{F}^{+}(0)$$

where the dyadic $\mathbf{C}$ is defined as

$$\mathbf{C} = \frac{1}{\eta_0}\left(\mathbf{P}_{22}(d,0) \cdot \mathbf{W}^{-1} \cdot (\mathbf{I}_2 - \mathbf{r}) - \mathbf{P}_{21}(d,0) \cdot (\mathbf{I}_2 + \mathbf{r})\right) \tag{10.35}$$

Example 10.1

To see how this machinery works, we consider the very simple example of a perfectly conducting surface $z = z_1 = 0$ in vacuum. The fundamental dyadic in vacuum is

$$\mathbf{P}(z_1, z_1) = \mathbf{I}_4$$

i.e.,

$$\begin{cases} \mathbf{P}_{11}(z_1, z_1) = \mathbf{P}_{22}(z_1, z_1) = \mathbf{I}_2 \\ \mathbf{P}_{12}(z_1, z_1) = \mathbf{P}_{21}(z_1, z_1) = \mathbf{0} \end{cases}$$

From (10.34), we get

$$\mathbf{r} = -\left(\mathbf{P}_{11} + \mathbf{P}_{12} \cdot \mathbf{W}^{-1}\right)^{-1} \cdot \left(\mathbf{P}_{11} - \mathbf{P}_{12} \cdot \mathbf{W}^{-1}\right) = -\mathbf{I}_2$$

Similarly, from (10.35)

$$\mathbf{C} = \frac{1}{\eta_0}(\mathbf{P}_{22} \cdot \mathbf{W}^{-1} \cdot (\mathbf{I}_2 - \mathbf{r}) - \mathbf{P}_{21} \cdot (\mathbf{I}_2 + \mathbf{r})) = \frac{2}{\eta_0}\left(\hat{\boldsymbol{e}}_\parallel \hat{\boldsymbol{e}}_\parallel \frac{1}{\cos\theta_\mathrm{i}} + \hat{\boldsymbol{e}}_\perp \hat{\boldsymbol{e}}_\perp \cos\theta_\mathrm{i}\right)$$

■

10.4.3 Composition of Two Slabs

In this section, we derive an algorithm to compute the reflection and transmission dyadics of a slab, that is composed of two different subslabs, expressed in the individual reflection and transmission dyadics of the subslabs. This problem is closely related to the Redheffer[3] $*$-product [224] or a composition of transmission lines.

Let $\mathbf{r}_i^\pm$ and $\mathbf{t}_i^\pm$, $i = 1, 2$, denote the reflection and the transmission dyadics for two subslabs. Furthermore, let the split fields at the left boundary of the first slab be $\boldsymbol{F}_1^\pm$, on the intermediate boundary be $\boldsymbol{F}^\pm$, and on the right boundary be $\boldsymbol{F}_2^\pm$, see Figure 10.7. The relations between the outputs and inputs at each boundary are

$$\begin{cases} \boldsymbol{F}_1^- = \mathbf{r}_1^+ \cdot \boldsymbol{F}_1^+ + \mathbf{t}_1^- \cdot \boldsymbol{F}^- \\ \boldsymbol{F}^+ = \mathbf{t}_1^+ \cdot \boldsymbol{F}_1^+ + \mathbf{r}_1^- \cdot \boldsymbol{F}^- \end{cases}$$

and

$$\begin{cases} \boldsymbol{F}^- = \mathbf{r}_2^+ \cdot \boldsymbol{F}^+ + \mathbf{t}_2^- \cdot \boldsymbol{F}_2^- \\ \boldsymbol{F}_2^+ = \mathbf{t}_2^+ \cdot \boldsymbol{F}^+ + \mathbf{r}_2^- \cdot \boldsymbol{F}_2^- \end{cases}$$

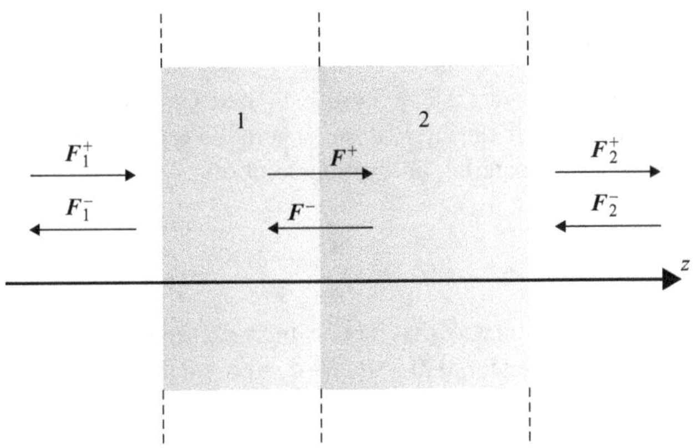

Figure 10.7 The composite slab and its two subslabs.

[3]Raymond Moos Redheffer (1921–2005), American mathematician.

Solve for the intermediate fields $F^{\pm}$, and we get

$$\begin{cases} F^+ = (\mathbf{I}_2 - \mathbf{r}_1^- \cdot \mathbf{r}_2^+)^{-1} \cdot (\mathbf{t}_1^+ \cdot F_1^+ + \mathbf{r}_1^- \cdot \mathbf{t}_2^- \cdot F_2^-) \\ F^- = \mathbf{r}_2^+ \cdot (\mathbf{I}_2 - \mathbf{r}_1^- \cdot \mathbf{r}_2^+)^{-1} \cdot (\mathbf{t}_1^+ \cdot F_1^+ + \mathbf{r}_1^- \cdot \mathbf{t}_2^- \cdot F_2^-) + \mathbf{t}_2^- \cdot F_2^- \end{cases}$$

The reflection and transmission dyadics of the composite slab, $\mathbf{r}^{\pm}$ and $\mathbf{t}^{\pm}$, are now extracted. We have

$$\begin{cases} F_1^- = \mathbf{r}^+ \cdot F_1^+ + \mathbf{t}^- \cdot F_2^- \\ F_2^+ = \mathbf{t}^+ \cdot F_1^+ + \mathbf{r}^- \cdot F_2^- \end{cases}$$

where

$$\begin{cases} \mathbf{r}^+ = \mathbf{r}_1^+ + \mathbf{t}_1^- \cdot \mathbf{r}_2^+ \cdot (\mathbf{I}_2 - \mathbf{r}_1^- \cdot \mathbf{r}_2^+)^{-1} \cdot \mathbf{t}_1^+ \\ \mathbf{t}^- = \mathbf{t}_1^- \cdot \mathbf{r}_2^+ \cdot (\mathbf{I}_2 - \mathbf{r}_1^- \cdot \mathbf{r}_2^+)^{-1} \cdot \mathbf{r}_1^- \cdot \mathbf{t}_2^- + \mathbf{t}_1^- \cdot \mathbf{t}_2^- \\ \mathbf{r}^- = \mathbf{r}_2^- + \mathbf{t}_2^+ \cdot (\mathbf{I}_2 - \mathbf{r}_1^- \cdot \mathbf{r}_2^+)^{-1} \cdot \mathbf{r}_1^- \cdot \mathbf{t}_2^- \\ \mathbf{t}^+ = \mathbf{t}_2^+ \cdot (\mathbf{I}_2 - \mathbf{r}_1^- \cdot \mathbf{r}_2^+)^{-1} \cdot \mathbf{t}_1^+ \end{cases}$$

These expressions give the total reflection and transmission dyadics in terms of the dyadics of the two subslabs. This procedure can be repeated to obtain the reflection and transmission dyadics of a slab composed of several subslabs.

10.5 Propagator Dyadics—Homogeneous Layers

The explicit expressions of the propagation dyadics are now addressed. We start with the simplest case and continue with more complex materials. First, we investigate the form of the propagator in a single homogeneous layer, then we apply this result to several layers.

10.5.1 Single Layer

The propagator of a single layer, (z_1, z), is investigated first. This amounts to solving the fundamental equation (10.11) in the interval (z_1, z) with the appropriate initial conditions when $z = z_1$. For a homogeneous material, an explicit solution of the propagator can be obtained. The result is

$$\mathbf{P}(z, z_1) = e^{ik_0(z-z_1)\mathbf{M}} \tag{10.36}$$

where $\mathbf{M}$ is the constant 4×4 matrix that models the material properties of the homogeneous material, see (10.12). The exponential of a square dyadic is defined in term of the Taylor series of the exponential, i.e.,

$$\exp \mathbf{A} = \sum_{n=0}^{\infty} \frac{1}{n!} \mathbf{A}^n$$

This series converges for all matrices since the exponential is an entire function.

Notice the very simple structure of the propagator in (10.36). The fundamental dyadic **M** contains all the wave propagation properties of the slab, and the exponential function propagates the field in the correct way from one position z_1 to another position z.

There are several ways to compute the propagator in the homogeneous case. We present two such methods in Sections 10.5.2 and 10.5.3, see also Comment 10.1. Notice also that the exponential of a square dyadic is a standard routine in e.g., MATLAB or Mathematica. However, caution should always be exercised when there is a large difference in the imaginary parts of the eigenvalues of **M** [298].

Comment 10.1

As an alternative to the two methods presented in this section to compute the propagator, we sketch an elegant algorithm to compute the exponential of a dyadic (matrix) **A**. This algorithm is named the Putzer (Eugene James Putzer (1929–2006), American mathematician) algorithm [220], and we give the algorithm without any derivation here in this comment. We refer to Putzer's original work for details [220].

The Putzer algorithm for a given n-dimensional dyadic **A** with n (distinct or not) eigenvalues $\lambda_1, \lambda_2, \ldots, \lambda_n$ reads:

$$e^{\mathbf{A}t} = \sum_{j=0}^{n-1} r_{j+1}(t)\mathbf{P}_j$$

where the dyadics $\mathbf{P}_j, j = 1, 2, \ldots, n$, are generated by a recursion procedure. The explicit recursion is

$$\begin{cases} \mathbf{P}_0 = \mathbf{I}_n \\ \mathbf{P}_j = \prod_{k=1}^{j} (\mathbf{A} - \lambda_k \mathbf{I}_n) = \mathbf{P}_{j-1}\left(\mathbf{A} - \lambda_j \mathbf{I}\right) \qquad j = 1, 2, \ldots, n-1 \end{cases}$$

and the scalar functions $r_j(t)$, $j = 1, 2, \ldots, n$, are generated by the following systems of ODE:

$$\begin{cases} \dfrac{\mathrm{d}r_1(t)}{\mathrm{d}t} = \lambda_1 r_1(t) \\ r_1(0) = 1 \end{cases}$$

and

$$\begin{cases} \dfrac{\mathrm{d}r_j(t)}{\mathrm{d}t} = \lambda_j r_j(t) + r_{j-1}(t) \\ r_j(0) = 0 \end{cases} \qquad j = 2, 3, \ldots, n$$

10.5.2 Homogeneous Layer—Distinct Eigenvalues (Projection Dyadics)

In this section, we present, in addition to the Putzer algorithm in Comment 10.1, an efficient method to compute the exponential of a matrix. The method employs the projection dyadics, which are summarized in Appendix F.2 on page 664, and the Cayley–Hamilton theorem in Appendix F.1.

We assume the eigenvalues, λ_k, $k = 1, 2, 3, 4$, of the fundamental dyadic $\mathbf{M}$ are all distinct. The projection dyadic associated with each eigenvalue is introduced in Appendix F.2. Define the projection dyadics $\mathbf{M}_k$, $k = 1, 2, 3, 4$:

$$\begin{cases} \mathbf{M}_1 = \dfrac{(\mathbf{M} - \lambda_2 \mathbf{I}_4) \cdot (\mathbf{M} - \lambda_3 \mathbf{I}_4) \cdot (\mathbf{M} - \lambda_4 \mathbf{I}_4)}{(\lambda_1 - \lambda_2)(\lambda_1 - \lambda_3)(\lambda_1 - \lambda_4)} \\ \mathbf{M}_2 = \dfrac{(\mathbf{M} - \lambda_1 \mathbf{I}_4) \cdot (\mathbf{M} - \lambda_3 \mathbf{I}_4) \cdot (\mathbf{M} - \lambda_4 \mathbf{I}_4)}{(\lambda_2 - \lambda_1)(\lambda_2 - \lambda_3)(\lambda_2 - \lambda_4)} \\ \mathbf{M}_3 = \dfrac{(\mathbf{M} - \lambda_1 \mathbf{I}_4) \cdot (\mathbf{M} - \lambda_2 \mathbf{I}_4) \cdot (\mathbf{M} - \lambda_4 \mathbf{I}_4)}{(\lambda_3 - \lambda_1)(\lambda_3 - \lambda_2)(\lambda_3 - \lambda_4)} \\ \mathbf{M}_4 = \dfrac{(\mathbf{M} - \lambda_1 \mathbf{I}_4) \cdot (\mathbf{M} - \lambda_2 \mathbf{I}_4) \cdot (\mathbf{M} - \lambda_3 \mathbf{I}_4)}{(\lambda_4 - \lambda_1)(\lambda_4 - \lambda_3)(\lambda_4 - \lambda_3)} \end{cases}$$

These projection dyadics satisfy the following orthogonality relation (details are presented in Appendix F.2):

$$\mathbf{M}_i \cdot \mathbf{M}_j = \delta_{ij} \mathbf{M}_i, \qquad i,j = 1, 2, 3, 4$$

and moreover, the exponential of a dyadic can be expressed in terms of these projection dyadics, viz.

$$e^{\mathbf{M}} = \sum_{k=1}^{4} e^{\lambda_k} \mathbf{M}_k$$

We apply this result to our propagation problem, (10.36), and find ($z - z_1$ is replaced by z)

$$e^{ik_0 z \mathbf{M}} = \sum_{k=1}^{4} e^{ik_0 z \lambda_k} \mathbf{M}_k$$

which reduces the problem of taking the exponential of a dyadic to finding the eigenvalues of the dyadic and to a few simple dyadic operations, which are easily implemented on a digital computer.

As a special case of the result above, the eigenvalues appear in two pairs, i.e., ($\lambda_+^2 \neq \lambda_-^2$)

$$\lambda_1 = -\lambda_4 = \lambda_+, \qquad \lambda_2 = -\lambda_3 = \lambda_- \tag{10.37}$$

We get, see Example F.2 on page 667

$$\begin{cases} \mathbf{M}_1 = \dfrac{(\mathbf{M}^2 - \lambda_-^2 \mathbf{I}_4) \cdot (\mathbf{M} + \lambda_+ \mathbf{I}_4)}{2\lambda_+(\lambda_+^2 - \lambda_-^2)} \\ \mathbf{M}_2 = -\dfrac{(\mathbf{M}^2 - \lambda_+^2 \mathbf{I}_4) \cdot (\mathbf{M} + \lambda_- \mathbf{I}_4)}{2\lambda_-(\lambda_+^2 - \lambda_-^2)} \\ \mathbf{M}_3 = \dfrac{(\mathbf{M}^2 - \lambda_+^2 \mathbf{I}_4) \cdot (\mathbf{M} - \lambda_- \mathbf{I}_4)}{2\lambda_-(\lambda_+^2 - \lambda_-^2)} \\ \mathbf{M}_4 = -\dfrac{(\mathbf{M}^2 - \lambda_-^2 \mathbf{I}_4) \cdot (\mathbf{M} - \lambda_+ \mathbf{I}_4)}{2\lambda_+(\lambda_+^2 - \lambda_-^2)} \end{cases}$$

and the exponential of **M** is

$$\begin{aligned}
e^{ik_0z\mathbf{M}} &= e^{ik_0z\lambda_+}\mathbf{M}_1 + e^{-ik_0z\lambda_+}\mathbf{M}_4 + e^{ik_0z\lambda_-}\mathbf{M}_2 + e^{-ik_0z\lambda_-}\mathbf{M}_3 \\
&= \frac{1}{\lambda_+^2 - \lambda_-^2}\left(\mathbf{M}^2 - \lambda_-^2\mathbf{I}_4\right)\cdot\left(\mathbf{I}_4\cos(k_0z\lambda_+) + \frac{i}{\lambda_+}\mathbf{M}\sin(k_0z\lambda_+)\right) \\
&\quad - \frac{1}{\lambda_+^2 - \lambda_-^2}\left(\mathbf{M}^2 - \lambda_+^2\mathbf{I}_4\right)\cdot\left(\mathbf{I}_4\cos(k_0z\lambda_-) + \frac{i}{\lambda_-}\mathbf{M}\sin(k_0z\lambda_-)\right)
\end{aligned} \tag{10.38}$$

For materials where (10.37) applies, λ_+^2 and λ_-^2 are eigenvalues of $\mathbf{M}^2$.

10.5.3 Homogeneous Layer—Distinct Eigenvalues

Cayley–Hamilton theorem, see Appendix F.1 on page 663, itself provides another useful way of computing the propagator of a single homogeneous layer, provided the eigenvalues of the fundamental dyadic, **M**, are distinct.

Since the exponential is an entire analytic function, the Cayley–Hamilton theorem gives, see Appendix F.1 and Theorem F.2 on page 663

$$e^{ik_0z\mathbf{M}} = q_0(k_0z)\mathbf{I}_4 + q_2(k_0z)\mathbf{M}\cdot\mathbf{M} + \left(q_1(k_0z)\mathbf{I}_4 + q_3(k_0z)\mathbf{M}\cdot\mathbf{M}\right)\cdot\mathbf{M}$$

The coefficients, $q_l(k_0z)$, $l = 1, 2, 3, 4$, are given by the system of linear equations

$$e^{ik_0z\lambda_l} = q_0(k_0z) + q_2(k_0z)\lambda_l^2 + \left(q_1(k_0z) + q_3(k_0z)\lambda_l^2\right)\lambda_l, \qquad l = 1, 2, 3, 4 \tag{10.39}$$

provided the eigenvalues, λ_l, $l = 1, 2, 3, 4$, of the fundamental dyadic, **M**, are distinct. For an isotropic material, where the eigenvalues are in pairs, the propagator can be obtained as a limit of the results obtained below, see Section 10.6 or Example F.4 on page 669.

Typically, for most materials, two eigenvalues, say λ_1 and λ_2, have positive real parts and two eigenvalues, λ_3 and λ_4, have negative real parts. These eigenvalues correspond to right-going and left-going waves, respectively, see also the wave splitting in Section 10.3.

In terms of the Vandermonde[4] matrix [80]

$$V = \begin{pmatrix} 1 & \lambda_1 & \lambda_1^2 & \lambda_1^3 \\ 1 & \lambda_2 & \lambda_2^2 & \lambda_2^3 \\ 1 & \lambda_3 & \lambda_3^2 & \lambda_3^3 \\ 1 & \lambda_4 & \lambda_4^2 & \lambda_4^3 \end{pmatrix} = \begin{pmatrix} v_{11} & v_{12} & v_{13} & v_{14} \\ v_{21} & v_{22} & v_{23} & v_{24} \\ v_{31} & v_{32} & v_{33} & v_{34} \\ v_{41} & v_{42} & v_{43} & v_{44} \end{pmatrix}$$

the system of equations (10.39) can be written as

$$e = V\cdot q, \qquad q = V^{-1}\cdot e$$

[4] Alexandre-Théophile Vandermonde (1735–1796), French musician, mathematician, and chemist.

where

$$e = \begin{pmatrix} e^{ik_0 z\lambda_1} \\ e^{ik_0 z\lambda_2} \\ e^{ik_0 z\lambda_3} \\ e^{ik_0 z\lambda_4} \end{pmatrix}, \qquad q = \begin{pmatrix} q_0(k_0 z) \\ q_1(k_0 z) \\ q_2(k_0 z) \\ q_3(k_0 z) \end{pmatrix}$$

The inverse of the matrix V is given by

$$V^{-1} = \frac{1}{\Delta}\begin{pmatrix} V_{11} & V_{21} & V_{31} & V_{41} \\ V_{12} & V_{22} & V_{32} & V_{42} \\ V_{13} & V_{23} & V_{33} & V_{43} \\ V_{14} & V_{24} & V_{34} & V_{44} \end{pmatrix}$$

where

$$\Delta = (\lambda_4 - \lambda_3)(\lambda_4 - \lambda_2)(\lambda_4 - \lambda_1)(\lambda_3 - \lambda_2)(\lambda_3 - \lambda_1)(\lambda_2 - \lambda_1)$$

is the Vandermonde determinant and $V_{ij} = (-1)^{i+j} D_{ij}$ is the algebraic complement of the matrix element v_{ij}. Here, the determinants D_{ij} are

$$D_{11} = \det\begin{pmatrix} \lambda_2 & \lambda_2^2 & \lambda_2^3 \\ \lambda_3 & \lambda_3^2 & \lambda_3^3 \\ \lambda_4 & \lambda_4^2 & \lambda_4^3 \end{pmatrix} \qquad D_{13} = \det\begin{pmatrix} 1 & \lambda_2 & \lambda_2^3 \\ 1 & \lambda_3 & \lambda_3^3 \\ 1 & \lambda_4 & \lambda_4^3 \end{pmatrix}$$

$$D_{12} = \det\begin{pmatrix} 1 & \lambda_2^2 & \lambda_2^3 \\ 1 & \lambda_3^2 & \lambda_3^3 \\ 1 & \lambda_4^2 & \lambda_4^3 \end{pmatrix} \qquad D_{14} = \det\begin{pmatrix} 1 & \lambda_2 & \lambda_2^2 \\ 1 & \lambda_3 & \lambda_3^2 \\ 1 & \lambda_4 & \lambda_4^2 \end{pmatrix}$$

$$D_{21} = \det\begin{pmatrix} \lambda_1 & \lambda_1^2 & \lambda_1^3 \\ \lambda_3 & \lambda_3^2 & \lambda_3^3 \\ \lambda_4 & \lambda_4^2 & \lambda_4^3 \end{pmatrix} \qquad D_{23} = \det\begin{pmatrix} 1 & \lambda_1 & \lambda_1^3 \\ 1 & \lambda_3 & \lambda_3^3 \\ 1 & \lambda_4 & \lambda_4^3 \end{pmatrix}$$

$$D_{22} = \det\begin{pmatrix} 1 & \lambda_1^2 & \lambda_1^3 \\ 1 & \lambda_3^2 & \lambda_3^3 \\ 1 & \lambda_4^2 & \lambda_4^3 \end{pmatrix} \qquad D_{24} = \det\begin{pmatrix} 1 & \lambda_1 & \lambda_1^2 \\ 1 & \lambda_3 & \lambda_3^2 \\ 1 & \lambda_4 & \lambda_4^2 \end{pmatrix}$$

$$D_{31} = \det\begin{pmatrix} \lambda_1 & \lambda_1^2 & \lambda_1^3 \\ \lambda_2 & \lambda_2^2 & \lambda_2^3 \\ \lambda_4 & \lambda_4^2 & \lambda_4^3 \end{pmatrix} \qquad D_{33} = \det\begin{pmatrix} 1 & \lambda_1 & \lambda_1^3 \\ 1 & \lambda_2 & \lambda_2^3 \\ 1 & \lambda_4 & \lambda_4^3 \end{pmatrix}$$

$$D_{32} = \det\begin{pmatrix} 1 & \lambda_1^2 & \lambda_1^3 \\ 1 & \lambda_2^2 & \lambda_2^3 \\ 1 & \lambda_4^2 & \lambda_4^3 \end{pmatrix} \qquad D_{34} = \det\begin{pmatrix} 1 & \lambda_1 & \lambda_1^2 \\ 1 & \lambda_2 & \lambda_2^2 \\ 1 & \lambda_4 & \lambda_4^2 \end{pmatrix}$$

$$D_{41} = \det \begin{pmatrix} \lambda_1 & \lambda_1^2 & \lambda_1^3 \\ \lambda_2 & \lambda_2^2 & \lambda_2^3 \\ \lambda_3 & \lambda_3^2 & \lambda_3^3 \end{pmatrix} \qquad D_{43} = \det \begin{pmatrix} 1 & \lambda_1 & \lambda_1^3 \\ 1 & \lambda_2 & \lambda_2^3 \\ 1 & \lambda_3 & \lambda_3^3 \end{pmatrix}$$

$$D_{42} = \det \begin{pmatrix} 1 & \lambda_1^2 & \lambda_1^3 \\ 1 & \lambda_2^2 & \lambda_2^3 \\ 1 & \lambda_3^2 & \lambda_3^3 \end{pmatrix} \qquad D_{44} = \det \begin{pmatrix} 1 & \lambda_1 & \lambda_1^2 \\ 1 & \lambda_2 & \lambda_2^2 \\ 1 & \lambda_3 & \lambda_3^2 \end{pmatrix}$$

For the important special case when ($\lambda_+^2 \neq \lambda_-^2$), see also (10.37)

$$\lambda_1 = -\lambda_4 = \lambda_+, \qquad \lambda_2 = -\lambda_3 = \lambda_-$$

one gets, see also (10.38)

$$\begin{aligned} e^{ik_0 z \mathbf{M}} &= \frac{1}{\lambda_+^2 - \lambda_-^2} \left(\mathbf{M}^2 - \lambda_-^2 \mathbf{I}_4\right) \cdot \left(\mathbf{I}_4 \cos(k_0 z \lambda_+) + \frac{i}{\lambda_+} \mathbf{M} \sin(k_0 z \lambda_+)\right) \\ &- \frac{1}{\lambda_+^2 - \lambda_-^2} \left(\mathbf{M}^2 - \lambda_+^2 \mathbf{I}_4\right) \cdot \left(\mathbf{I}_4 \cos(k_0 z \lambda_-) + \frac{i}{\lambda_-} \mathbf{M} \sin(k_0 z \lambda_-)\right) \end{aligned}$$

which is identical to the result in Section 10.5.2.

10.5.4 Several Layers

Let $z_j, j = 1, \dots, N-1$, be the location of $N-1$ parallel interfaces, see Figure 10.8, and let $\mathbf{M}_j, j = 1, \dots, N$, be the fundamental dyadics of the corresponding regions, respectively. It is assumed that all slabs are homogeneous and that regions $j = 1$ and $j = N$ are empty half-spaces,[5] $\mathbf{M}_1 = \mathbf{M}_N = \mathbf{M}_0$, see (10.14), i.e., $\boldsymbol{\epsilon} = \boldsymbol{\mu} = \mathbf{I}_3$ and $\boldsymbol{\xi} = \boldsymbol{\zeta} = \mathbf{0}$ in these half-spaces.

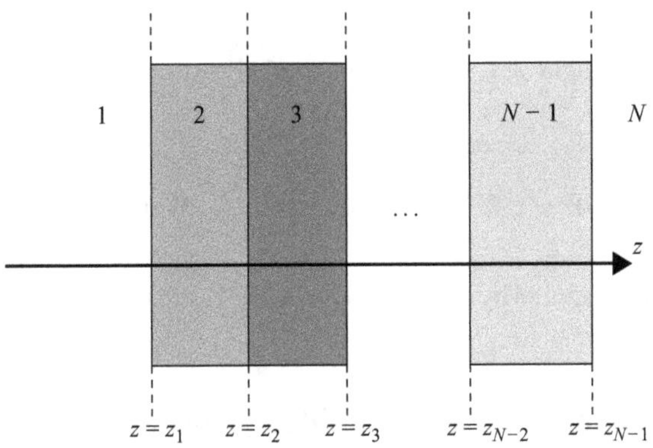

Figure 10.8 The geometry of a stratified structure of several layers. Regions 1 and N are empty half-spaces. Each layer is assumed to be homogeneous.

[5]A material half-space can be obtained by the limit process $z_{N-1} \to \infty$.

Since the tangential electric and magnetic fields are continuous at the boundaries, a cascade coupling technique can be applied. Using (10.36) repeatedly gives

$$\mathbf{P}(z_{N-1}, z_1) = \mathrm{e}^{\mathrm{i}k_0(z_{N-1}-z_{N-2})\mathbf{M}_{N-1}} \cdot \ldots \cdot \mathrm{e}^{\mathrm{i}k_0(z_3-z_2)\mathbf{M}_3} \cdot \mathrm{e}^{\mathrm{i}k_0(z_2-z_1)\mathbf{M}_2}$$

This is the exact propagator in terms of a product of propagator of the subslabs. We can replace the right-hand side with the exponential of a single matrix by the use of the Baker–Campbell–Hausdorff formula, see Appendix F.2.3 on page 669.

10.6 Examples

The fundamental dyadic for a general, stratified, bianisotropic material is given in Section 10.2. In this section, explicit expressions for the single-slab propagator are given for some important classes of linear and homogeneous materials.

10.6.1 Isotropic Media

The results for simple media are well known. For a homogeneous isotropic material ($\boldsymbol{\xi} = \boldsymbol{\zeta} = \mathbf{0}$, $\boldsymbol{\epsilon} = \epsilon \mathbf{I}_3$, $\boldsymbol{\mu} = \mu \mathbf{I}_3$), the fundamental dyadic is

$$\mathbf{M} = \begin{pmatrix} \mathbf{0} & -\mu \mathbf{I}_2 + \frac{1}{\epsilon k_0^2} \boldsymbol{k}_\mathrm{t}\boldsymbol{k}_\mathrm{t} \\ -\epsilon \mathbf{I}_2 - \frac{1}{\mu k_0^2} \mathbf{J} \cdot \boldsymbol{k}_\mathrm{t}\boldsymbol{k}_\mathrm{t} \cdot \mathbf{J} & \mathbf{0} \end{pmatrix}$$

The eigenvalues of the fundamental dyadic are

$$\lambda_+^2 = \lambda_-^2 = \lambda^2 = \epsilon\mu - k_\mathrm{t}^2 / k_0^2$$

The propagator of the single slab with degenerated eigenvalues is analyzed in Example F.3 on page 668. The result is

$$\mathbf{P} = \mathrm{e}^{\mathrm{i}k_0 z \mathbf{M}} = \mathbf{I}_4 \cos(k_0 z \lambda) + \frac{\mathrm{i}}{\lambda} \mathbf{M} \sin(k_0 z \lambda) \tag{10.40}$$

This result can also be obtained by a limit process of the results in Appendix F.2, see Example F.4 on page 669.

10.6.1.1 Single Isotropic Layer Embedded in Vacuum

The propagator for a single layer of thickness d is given by (10.40). Explicitly, in the $\{\hat{\boldsymbol{e}}_\|, \hat{\boldsymbol{e}}_\perp\}$-system, it is

$$\mathbf{P} = \mathrm{e}^{\mathrm{i}k_0 d \mathbf{M}} = \begin{pmatrix} \left(\hat{\boldsymbol{e}}_\|\hat{\boldsymbol{e}}_\| + \hat{\boldsymbol{e}}_\perp\hat{\boldsymbol{e}}_\perp\right)\cos(k_0 d\lambda) & -\mathrm{i}\left(\hat{\boldsymbol{e}}_\|\hat{\boldsymbol{e}}_\| \frac{\lambda}{\epsilon} + \hat{\boldsymbol{e}}_\perp\hat{\boldsymbol{e}}_\perp \frac{\mu}{\lambda}\right)\sin(k_0 d\lambda) \\ -\mathrm{i}\left(\hat{\boldsymbol{e}}_\|\hat{\boldsymbol{e}}_\| \frac{\epsilon}{\lambda} + \hat{\boldsymbol{e}}_\perp\hat{\boldsymbol{e}}_\perp \frac{\lambda}{\mu}\right)\sin(k_0 d\lambda) & \left(\hat{\boldsymbol{e}}_\|\hat{\boldsymbol{e}}_\| + \hat{\boldsymbol{e}}_\perp\hat{\boldsymbol{e}}_\perp\right)\cos(k_0 d\lambda) \end{pmatrix}$$

where $\lambda^2 = \epsilon\mu - k_t^2/k_0^2 = \epsilon\mu - \sin^2\theta_i$. Straightforward calculations show that principal blocks of the scattering dyadic, see (10.29), are

$$\begin{cases} 2\mathbf{T}_{11} = \hat{\boldsymbol{e}}_\parallel\hat{\boldsymbol{e}}_\parallel \left(2\cos(k_0 d\lambda) + \mathrm{i}\left(\dfrac{\lambda}{\epsilon\cos\theta_i} + \dfrac{\epsilon\cos\theta_i}{\lambda}\right)\sin(k_0 d\lambda)\right) \\ \qquad + \hat{\boldsymbol{e}}_\perp\hat{\boldsymbol{e}}_\perp \left(2\cos(k_0 d\lambda) + \mathrm{i}\left(\dfrac{\lambda}{\mu\cos\theta_i} + \dfrac{\mu\cos\theta_i}{\lambda}\right)\sin(k_0 d\lambda)\right) \\ 2\mathbf{T}_{22} = \hat{\boldsymbol{e}}_\parallel\hat{\boldsymbol{e}}_\parallel \left(2\cos(k_0 d\lambda) - \mathrm{i}\left(\dfrac{\lambda}{\epsilon\cos\theta_i} + \dfrac{\epsilon\cos\theta_i}{\lambda}\right)\sin(k_0 d\lambda)\right) \\ \qquad + \hat{\boldsymbol{e}}_\perp\hat{\boldsymbol{e}}_\perp \left(2\cos(k_0 d\lambda) - \mathrm{i}\left(\dfrac{\lambda}{\mu\cos\theta_i} + \dfrac{\mu\cos\theta_i}{\lambda}\right)\sin(k_0 d\lambda)\right) \end{cases}$$

and

$$\begin{cases} 2\mathbf{T}_{12} = \mathrm{i}\left(\hat{\boldsymbol{e}}_\parallel\hat{\boldsymbol{e}}_\parallel\left(-\dfrac{\lambda}{\epsilon\cos\theta_i} + \dfrac{\epsilon\cos\theta_i}{\lambda}\right) + \hat{\boldsymbol{e}}_\perp\hat{\boldsymbol{e}}_\perp\left(\dfrac{\lambda}{\mu\cos\theta_i} - \dfrac{\mu\cos\theta_i}{\lambda}\right)\right)\sin(k_0 d\lambda) \\ \mathbf{T}_{21} = -\mathbf{T}_{12} \end{cases}$$

where θ_i is the angle of incidence, see (10.4) on page 522. Consequently, the reflection and transmission dyadics, see (10.30) and (10.31), are

$$\begin{cases} \mathbf{r} = \hat{\boldsymbol{e}}_\parallel\hat{\boldsymbol{e}}_\parallel r_{\parallel\parallel} + \hat{\boldsymbol{e}}_\perp\hat{\boldsymbol{e}}_\perp r_{\perp\perp} \\ \mathbf{t} = \hat{\boldsymbol{e}}_\parallel\hat{\boldsymbol{e}}_\parallel t_{\parallel\parallel} + \hat{\boldsymbol{e}}_\perp\hat{\boldsymbol{e}}_\perp t_{\perp\perp} \end{cases} \tag{10.41}$$

where the components of the reflection dyadics are

$$r_{\parallel\parallel} = \frac{\mathrm{i}\left(-\frac{\lambda}{\epsilon\cos\theta_i} + \frac{\epsilon\cos\theta_i}{\lambda}\right)\sin(k_0 d\lambda)}{2\cos(k_0 d\lambda) - \mathrm{i}\left(\frac{\lambda}{\epsilon\cos\theta_i} + \frac{\epsilon\cos\theta_i}{\lambda}\right)\sin(k_0 d\lambda)} = r_{1\parallel}\frac{1 - \mathrm{e}^{2\mathrm{i}k_0\lambda d}}{1 - r_{1\parallel}^2\mathrm{e}^{2\mathrm{i}k_0\lambda d}}$$

and

$$r_{\perp\perp} = \frac{\mathrm{i}\left(\frac{\lambda}{\mu\cos\theta_i} - \frac{\mu\cos\theta_i}{\lambda}\right)\sin(k_0 d\lambda)}{2\cos(k_0 d\lambda) - \mathrm{i}\left(\frac{\lambda}{\mu\cos\theta_i} + \frac{\mu\cos\theta_i}{\lambda}\right)\sin(k_0 d\lambda)} = r_{1\perp}\frac{1 - \mathrm{e}^{2\mathrm{i}k_0\lambda d}}{1 - r_{1\perp}^2\mathrm{e}^{2\mathrm{i}k_0\lambda d}}$$

The components of the transmission dyadics are

$$t_{\parallel\parallel} = \frac{2}{2\cos(k_0 d\lambda) - \mathrm{i}\left(\frac{\lambda}{\epsilon\cos\theta_i} + \frac{\epsilon\cos\theta_i}{\lambda}\right)\sin(k_0 d\lambda)} = \frac{(1 - r_{1\parallel}^2)\mathrm{e}^{\mathrm{i}k_0\lambda d}}{1 - r_{1\parallel}^2\mathrm{e}^{2\mathrm{i}k_0\lambda d}}$$

and

$$t_{\perp\perp} = \frac{2}{2\cos(k_0 d\lambda) - \mathrm{i}\left(\frac{\lambda}{\mu\cos\theta_i} + \frac{\mu\cos\theta_i}{\lambda}\right)\sin(k_0 d\lambda)} = \frac{(1 - r_{1\perp}^2)\mathrm{e}^{\mathrm{i}k_0\lambda d}}{1 - r_{1\perp}^2\mathrm{e}^{2\mathrm{i}k_0\lambda d}}$$

Here, we have used

$$\begin{cases} r_{1\parallel} = \dfrac{\frac{1}{2}\left(\frac{\lambda}{\epsilon\cos\theta_{\mathrm{i}}} - \frac{\epsilon\cos\theta_{\mathrm{i}}}{\lambda}\right)}{1+\frac{1}{2}\left(\frac{\lambda}{\epsilon\cos\theta_{\mathrm{i}}} + \frac{\epsilon\cos\theta_{\mathrm{i}}}{\lambda}\right)} = \dfrac{\lambda - \epsilon\cos\theta_{\mathrm{i}}}{\lambda + \epsilon\cos\theta_{\mathrm{i}}} \\[2ex] r_{1\perp} = \dfrac{\frac{1}{2}\left(\frac{\mu\cos\theta_{\mathrm{i}}}{\lambda} - \frac{\lambda}{\mu\cos\theta_{\mathrm{i}}}\right)}{1+\frac{1}{2}\left(\frac{\mu\cos\theta_{\mathrm{i}}}{\lambda} + \frac{\lambda}{\mu\cos\theta_{\mathrm{i}}}\right)} = \dfrac{\mu\cos\theta_{\mathrm{i}} - \lambda}{\mu\cos\theta_{\mathrm{i}} + \lambda} \end{cases}$$

which are recognized as the Fresnel[6] reflection coefficients [101].

Similar to the definition of the angle of incidence, θ_{i}, we define the angle of transmission, θ_{t}, as (Snell's[7] law)

$$k\sin\theta_{\mathrm{t}} = k_0\sin\theta_{\mathrm{i}} \quad \text{or} \quad \sin\theta_{\mathrm{t}} = \frac{k_{\mathrm{t}}}{k}$$

where the wave number of the slab is $k = k_0\,(\epsilon\mu)^{1/2}$.

For a PEC-backed isotropic slab, (10.34) yields the appropriate reflection coefficients

$$\begin{cases} r_{\parallel\parallel} = -\dfrac{\cos(k_0 d\lambda) + \mathrm{i}\frac{\lambda}{\epsilon\cos\theta_{\mathrm{i}}}\sin(k_0 d\lambda)}{\cos(k_0 d\lambda) - \mathrm{i}\frac{\lambda}{\epsilon\cos\theta_{\mathrm{i}}}\sin(k_0 d\lambda)} = \dfrac{r_{1\parallel} - \mathrm{e}^{2\mathrm{i}k_0\lambda d}}{1 - r_{1\parallel}\mathrm{e}^{2\mathrm{i}k_0\lambda d}} \\[2ex] r_{\perp\perp} = -\dfrac{\cos(k_0 d\lambda) + \mathrm{i}\frac{\mu\cos\theta_{\mathrm{i}}}{\lambda}\sin(k_0 d\lambda)}{\cos(k_0 d\lambda) - \mathrm{i}\frac{\mu\cos\theta_{\mathrm{i}}}{\lambda}\sin(k_0 d\lambda)} = \dfrac{r_{1\perp} - \mathrm{e}^{2\mathrm{i}k_0\lambda d}}{1 - r_{1\perp}\mathrm{e}^{2\mathrm{i}k_0\lambda d}} \end{cases}$$

where $r_{1\parallel}$ and $r_{1\perp}$ are the Fresnel reflection coefficients.

Example 10.2

If the material in the slab is lossless, then the permittivity and the permeability, the eigenvalue λ, and also the Fresnel reflection coefficients are real numbers. The magnitudes of the reflection coefficients are

$$|r_{\parallel\parallel}|^2 = \frac{a_\parallel^2}{(1 - r_{1\parallel}^2)^2 + a_\parallel^2}, \qquad |r_{\perp\perp}|^2 = \frac{a_\perp^2}{(1 - r_{1\perp}^2)^2 + a_\perp^2}$$

where $a_\parallel = 2r_{1\parallel}\sin k_0\lambda d$ and $a_\perp = 2r_{1\perp}\sin k_0\lambda d$, respectively. Similarly, the magnitudes of the transmission coefficients are

$$|t_{\parallel\parallel}|^2 = \frac{(1 - r_{1\parallel}^2)^2}{(1 - r_{1\parallel}^2)^2 + a_\parallel^2}, \qquad |t_{\perp\perp}|^2 = \frac{(1 - r_{1\perp}^2)^2}{(1 - r_{1\perp}^2)^2 + a_\perp^2}$$

---

[6]Augustin-Jean Fresnel (1788–1827), French engineer and physicist.

[7]Willebrord Snellius (1580–1626), Dutch astronomer and mathematician.

Note that $\left|r_{\parallel\parallel}\right|^2 + \left|t_{\parallel\parallel}\right|^2 = 1$ and $|r_{\perp\perp}|^2 + |t_{\perp\perp}|^2 = 1$. This result is easily generalized to hold for the reflectivity R and the transmissivity T defined below in (10.44)

$$\begin{cases} R = \dfrac{|F_{\parallel}^{-}(0)|^2 + |F_{\perp}^{-}(0)|^2 \cos^2\theta_{\mathrm{i}}}{|F_{\parallel}^{+}(0)|^2 + |F_{\perp}^{+}(0)|^2 \cos^2\theta_{\mathrm{i}}} = \dfrac{\left|r_{\parallel\parallel}\right|^2 |F_{\parallel}^{+}(0)|^2 + |r_{\perp\perp}|^2 |F_{\perp}^{+}(0)|^2 \cos^2\theta_{\mathrm{i}}}{|F_{\parallel}^{+}(0)|^2 + |F_{\perp}^{+}(0)|^2 \cos^2\theta_{\mathrm{i}}} \\[2ex] T = \dfrac{|F_{\parallel}^{+}(d)|^2 + |F_{\perp}^{+}(d)|^2 \cos^2\theta_{\mathrm{i}}}{|F_{\parallel}^{+}(0)|^2 + |F_{\perp}^{+}(0)|^2 \cos^2\theta_{\mathrm{i}}} = \dfrac{\left|t_{\parallel\parallel}\right|^2 |F_{\parallel}^{+}(0)|^2 + |t_{\perp\perp}|^2 |F_{\perp}^{+}(0)|^2 \cos^2\theta_{\mathrm{i}}}{|F_{\parallel}^{+}(0)|^2 + |F_{\perp}^{+}(0)|^2 \cos^2\theta_{\mathrm{i}}} \end{cases}$$

The properties of the reflection and transmission coefficients above imply

$$R + T = 1$$

which is a consequence of energy conservation. ■

Example 10.3

The electromagnetic wave can be split into two parts with an optical device called a beam splitter. This is a component of vital importance in most interferometers. The generic beam splitter is in a form of a cube, made of two triangular glass prisms, see Figure 10.9, with the two glass prisms glued together using polyester, epoxy, adhesive or resin. The thickness of the resin is adjusted so that 50% (or some other ratio) of the intensity is reflected and transmitted.

Figure 10.9 Photo of a beam splitter and the glued area. Photo: G. Kristensson.

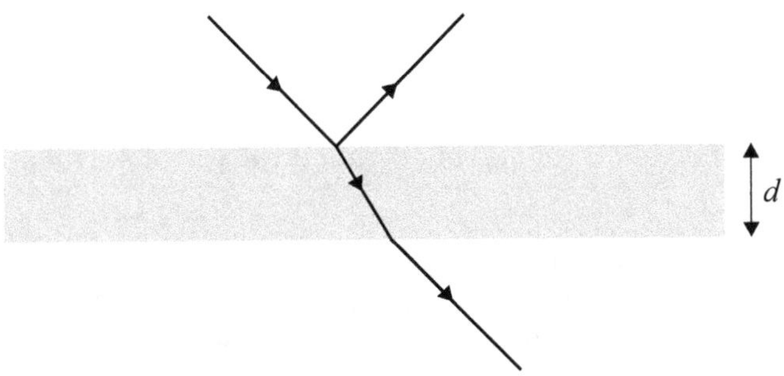

Figure 10.10 The thin film model for the beam splitter.

We model the beam splitter with a thin slab (thin film) with a wave incident at an angle of incidence of 45°, see Figure 10.10. This situation can be analyzed with the results above, if we interpret the permittivity ϵ and permeability μ as the values relative to the background material parameters. In this case, it is glass background.

We focus on one polarization of the incident wave, say the parallel polarization. The reflection and transmission coefficients for an incident angle of $\theta_{\mathrm{i}} = \pi/4$:

$$r_{\|\|} = r_{1\|} \frac{1 - \mathrm{e}^{2\mathrm{i}k\lambda d}}{1 - r_{1\|}^2 \mathrm{e}^{2\mathrm{i}k\lambda d}}$$

and

$$t_{\|\|} = \frac{(1 - r_{1\|}^2)\mathrm{e}^{\mathrm{i}k\lambda d}}{1 - r_{1\|}^2 \mathrm{e}^{2\mathrm{i}k\lambda d}}$$

where k is the wave number in glass, and where

$$r_{1\|} = \frac{\lambda - \epsilon/\sqrt{2}}{\lambda + \epsilon/\sqrt{2}}$$

and

$$\lambda^2 = \epsilon\mu - 1/2$$

The permittivity ϵ and permeability μ are the values relative to the glass background material.

Neglecting losses, the condition for a 50–50 split between reflected and transmitted intensity is $R = T$, i.e.,

$$\left|r_{\|\|}\right|^2 = \left|t_{\|\|}\right|^2$$

With the notation in Example 10.2, we get

$$\frac{a_{\|}^2}{(1 - r_{1\|}^2)^2 + a_{\|}^2} = \frac{(1 - r_{1\|}^2)^2}{(1 - r_{1\|}^2)^2 + a_{\|}^2}$$

where $a_{\|} = 2r_{1\|} \sin k\lambda d$. This condition is satisfied if

$$a_{\|}^2 = (1 - r_{1\|}^2)^2$$

or

$$2\sin^2 k\lambda d = \frac{1 + r_{1\|}^4}{2r_{1\|}^2} - 1$$

Note that the right-hand side is a decreasing function in the interval $r_{1\|}^2 \in [0, 1]$. For a given slab material, $r_{1\|}^2 \leq 1$, the right-hand side is a number in the interval $[0, \infty)$, and there is always a thickness d that solves the condition. The existence of a beam splitter for this polarization is therefore guaranteed. ∎

10.6.1.2 Half-Space Problem

The half-space problem is of special interest in many applications, and this section analyzes this geometry in detail.

The general incident and reflected fields in the empty half-space are

$$\begin{cases} \boldsymbol{E}_{\mathrm{i}}(\boldsymbol{r},\omega) = \dfrac{1}{4\pi^2} \displaystyle\iint\limits_{\mathbb{R}^2} \left(\mathbf{I}_2 - \dfrac{\hat{z}\boldsymbol{k}_{\mathrm{t}}}{k_z}\right) \cdot \boldsymbol{F}^{+}(\boldsymbol{k}_{\mathrm{t}}, 0, \omega) \mathrm{e}^{\mathrm{i}\boldsymbol{k}_{\mathrm{t}}\cdot\boldsymbol{r}_{\mathrm{c}} + \mathrm{i}k_z z}\, \mathrm{d}k_x\, \mathrm{d}k_y \\ \boldsymbol{E}_{\mathrm{r}}(\boldsymbol{r},\omega) = \dfrac{1}{4\pi^2} \displaystyle\iint\limits_{\mathbb{R}^2} \left(\mathbf{I}_2 + \dfrac{\hat{z}\boldsymbol{k}_{\mathrm{t}}}{k_z}\right) \cdot \mathbf{r} \cdot \boldsymbol{F}^{+}(\boldsymbol{k}_{\mathrm{t}}, 0, \omega) \mathrm{e}^{\mathrm{i}\boldsymbol{k}_{\mathrm{t}}\cdot\boldsymbol{r}_{\mathrm{c}} - \mathrm{i}k_z z}\, \mathrm{d}k_x\, \mathrm{d}k_y \end{cases}$$

The half-space reflection coefficients are

$$\begin{cases} r_{\parallel} = r_{\parallel\parallel} = \dfrac{\lambda - \epsilon\cos\theta_{\mathrm{i}}}{\lambda + \epsilon\cos\theta_{\mathrm{i}}} \\ r_{\perp} = r_{\perp\perp} = \dfrac{\mu\cos\theta_{\mathrm{i}} - \lambda}{\mu\cos\theta_{\mathrm{i}} + \lambda} \end{cases}$$

where $\lambda^2 = \epsilon\mu - \sin^2\theta_{\mathrm{i}}$. We identify these reflection coefficients as the Fresnel reflection coefficients.

Several examples illustrate the results in this section on reflection by an isotropic half-space.

Example 10.4

For a non-magnetic material, the Fresnel reflection coefficients simplify to

$$\begin{cases} r_{\parallel} = \dfrac{\left(\epsilon - \sin^2\theta_{\mathrm{i}}\right)^{1/2} - \epsilon\cos\theta_{\mathrm{i}}}{\left(\epsilon - \sin^2\theta_{\mathrm{i}}\right)^{1/2} + \epsilon\cos\theta_{\mathrm{i}}} \\ r_{\perp} = \dfrac{\cos\theta_{\mathrm{i}} - \left(\epsilon - \sin^2\theta_{\mathrm{i}}\right)^{1/2}}{\cos\theta_{\mathrm{i}} + \left(\epsilon - \sin^2\theta_{\mathrm{i}}\right)^{1/2}} \end{cases} \tag{10.42}$$

At normal incidence, the reflection coefficients coincide

$$r = r_{\parallel} = r_{\perp} = \frac{1 - \epsilon^{1/2}}{1 + \epsilon^{1/2}}$$

from which we can solve for ϵ. The result is

$$\epsilon = \left(\frac{1 - r}{1 + r}\right)^2$$

Provided we can measure both the modulus and the phase of r, the permittivity can be determined. The phase measurement of the reflection coefficient introduces some difficulties, especially at high frequencies, i.e., optical frequencies. If we can construct a measurement algorithm just based upon intensity measurements, this disadvantage is circumvented. Below in the next example, we develop such an algorithm. ■

Example 10.5
As in the previous example, we assume the half-space contains a homogeneous, non-magnetic material, and the Fresnel reflection coefficients are given by (10.42). We assume the incident field is linearly polarized along the direction $\hat{\boldsymbol{p}}$, i.e.,

$$\boldsymbol{E}_{\rm i} = E_0\hat{\boldsymbol{p}} = \left(\mathbf{I}_2 - \frac{\hat{z}\hat{\boldsymbol{e}}_\parallel \sin\theta_{\rm i}}{\cos\theta_{\rm i}}\right)\cdot \boldsymbol{F}^+(0) = \hat{\boldsymbol{e}}_\perp F_\perp^+(0) - \frac{\hat{z}\sin\theta_{\rm i} - \hat{\boldsymbol{e}}_\parallel\cos\theta_{\rm i}}{\cos\theta_{\rm i}} F_\parallel^+(0)$$

Moreover, we assume the incident wave is a propagating plane wave, i.e., no evanescent wave present and therefore $\cos\theta_{\rm i}$ is real. The reflected field is

$$\boldsymbol{E}_{\rm r} = \left(\mathbf{I}_2 + \frac{\hat{z}\hat{\boldsymbol{e}}_\parallel \sin\theta_{\rm i}}{\cos\theta_{\rm i}}\right)\cdot \mathbf{r}\cdot \boldsymbol{F}^+(0) = \hat{\boldsymbol{e}}_\perp r_\perp F_\perp^+(0) + \frac{\hat{z}\sin\theta_{\rm i} + \hat{\boldsymbol{e}}_\parallel\cos\theta_{\rm i}}{\cos\theta_{\rm i}} r_\parallel F_\parallel^+(0)$$

where we used (10.41)

$$\mathbf{r} = \hat{\boldsymbol{e}}_\parallel\hat{\boldsymbol{e}}_\parallel r_\parallel + \hat{\boldsymbol{e}}_\perp\hat{\boldsymbol{e}}_\perp r_\perp$$

We introduce unit vectors as above, see (10.21)

$$\begin{cases} \hat{\boldsymbol{k}}^\pm = \dfrac{\boldsymbol{k}_{\rm t} \pm \hat{z}k_z}{k} = \hat{\boldsymbol{e}}_\parallel \sin\theta_{\rm i} \pm \hat{z}\cos\theta_{\rm i} \\ \pm\hat{\boldsymbol{e}}_\perp \times \hat{\boldsymbol{k}}^\pm = \mp\hat{z}\sin\theta_{\rm i} + \hat{\boldsymbol{e}}_\parallel\cos\theta_{\rm i} \end{cases}$$

and we get

$$\begin{cases} \boldsymbol{E}_{\rm i} = F_\perp^+\hat{\boldsymbol{e}}_\perp + \dfrac{F_\parallel^+}{\cos\theta_{\rm i}}\hat{\boldsymbol{e}}_\perp \times \hat{\boldsymbol{k}}^+ \\ \boldsymbol{E}_{\rm r} = \hat{\boldsymbol{e}}_\perp r_\perp F_\perp^+ + \dfrac{r_\parallel F_\parallel^+}{\cos\theta_{\rm i}}\hat{\boldsymbol{e}}_\perp \times \hat{\boldsymbol{k}}^- \end{cases}$$

If the incident field is linearly polarized, the coefficients satisfy, see Section 1.3.7 on page 74

$$0 = {\rm i}\hat{\boldsymbol{k}}^+\cdot\left(\boldsymbol{E}_{\rm i}\times\boldsymbol{E}_{\rm i}^*\right) = \frac{2}{\cos\theta_{\rm i}}\,{\rm Im}\left(F_\perp^+F_\parallel^{+*}\right) \quad\Rightarrow\quad {\rm Im}\left(F_\perp^+F_\parallel^{+*}\right) = 0$$

which implies that $F_\parallel^+ = qF_\perp^+$, where q is a real number. The polarization state of the reflected field is determined by

$$\begin{aligned} {\rm i}\hat{\boldsymbol{k}}^-\cdot\left(\boldsymbol{E}_{\rm r}\times\boldsymbol{E}_{\rm r}^*\right) &= \frac{2}{\cos\theta_{\rm i}}\,{\rm Im}\left(r_\perp F_\perp^+ r_\parallel^* F_\parallel^{+*}\right) \\ &= \frac{2}{\cos\theta_{\rm i}}\left[{\rm Re}\left(r_\perp r_\parallel^*\right){\rm Im}\left(F_\perp^+F_\parallel^{+*}\right) + {\rm Im}\left(r_\perp r_\parallel^*\right){\rm Re}\left(F_\perp^+F_\parallel^{+*}\right)\right] \\ &= \frac{2}{\cos\theta_{\rm i}}\,{\rm Im}\left(r_\perp r_\parallel^*\right){\rm Re}\left(F_\perp^+F_\parallel^{+*}\right) = \frac{2q}{\cos\theta_{\rm i}}\,{\rm Im}\left(r_\perp r_\parallel^*\right){\rm Re}\left|F_\perp^+\right|^2 \end{aligned}$$

and the reflected field is in general elliptically polarized.

We form the quotient

$$r_\parallel / r_\perp = \rho e^{{\rm i}\alpha}$$

where ρ is the modulus of the quotient and α is its phase. Two extreme cases can easily be identified.

- Normal incidence $\theta_\mathrm{i} = 0$, for which $\rho = 1$ and $\alpha = 0$
- Grazing incidence $\theta_\mathrm{i} = \pi/2$, for which $\rho = 1$ and $\alpha = \pi$

Between these extreme values there is an angle $\theta_\mathrm{i} = \theta_\mathrm{c}$ where $\alpha = \pi/2$. This angle of incidence is of special interest below. For this angle of incidence, $r_\parallel = \mathrm{i}\rho r_\perp$, and the reflected field, following the notation in Example 1.21 on page 80, is, see (1.82)

$$\boldsymbol{E}_\mathrm{r} = A\mathrm{e}^{\mathrm{i}\alpha}\hat{\boldsymbol{e}}_1 + \underbrace{A\mathrm{e}^{\mathrm{i}\alpha}\frac{\mathrm{i}\rho q}{\cos\theta_\mathrm{c}}}_{B\mathrm{e}^{\mathrm{i}\beta}}\hat{\boldsymbol{e}}_2$$

where $A\mathrm{e}^{\mathrm{i}\alpha} = r_\perp F_\perp^+$, $\hat{\boldsymbol{e}}_1 = \hat{\boldsymbol{e}}_\perp$, and $\hat{\boldsymbol{e}}_2 = \hat{\boldsymbol{e}}_\perp \times \hat{\boldsymbol{k}}^-$. We conclude

$$\beta = \alpha + \pi/2 = \pi, \qquad B = A\frac{\rho q}{\cos\theta_\mathrm{c}}$$

From (1.83) on page 80, we also see that

$$\tan 2\vartheta = \frac{A^2\sin 2\alpha + B^2\sin 2\beta}{A^2\cos 2\alpha + B^2\cos 2\beta} = \frac{A^2\sin 2\alpha - B^2\sin 2\alpha}{A^2\cos 2\alpha - B^2\cos 2\alpha} = \tan 2\alpha \quad \Rightarrow \quad \vartheta = \alpha$$

The lengths of the half axes are, see (1.84) on page 80

$$\begin{cases} a = \sqrt{A^2\cos^2(\vartheta - \alpha) + B^2\cos^2(\vartheta - \beta)} = A \\ b = \sqrt{A^2\sin^2(\vartheta - \alpha) + B^2\sin^2(\vartheta - \beta)} = B \end{cases}$$

and the angle ψ between the $\hat{\boldsymbol{e}}_1$-axis and the half axes $\boldsymbol{a}$ is determined by, see (1.85) on page 81

$$\tan\psi = \frac{B\cos(\vartheta - \beta)}{A\cos(\vartheta - \alpha)} = 0$$

The polarization ellipse is horizontal and the quotient between its semi-axes (vertical to horizontal) is

$$\frac{b}{a} = \frac{B}{A} = \frac{\rho q}{\cos\theta_\mathrm{c}}$$

We proceed by calculating

$$\frac{1+\rho\mathrm{e}^{\mathrm{i}\alpha}}{1-\rho\mathrm{e}^{\mathrm{i}\alpha}} = \frac{r_\perp + r_\parallel}{r_\perp - r_\parallel} = \frac{(1-\epsilon)\cos\theta_\mathrm{c}\left(\epsilon - \sin^2\theta_\mathrm{c}\right)^{1/2}}{\epsilon\cos^2\theta_\mathrm{c} - \left(\epsilon - \sin^2\theta_\mathrm{c}\right)} = \frac{\cos\theta_\mathrm{c}\left(\epsilon - \sin^2\theta_\mathrm{c}\right)^{1/2}}{\sin^2\theta_\mathrm{c}}$$

Take the square on both sides, and we get

$$\left(\frac{1+\mathrm{i}\rho}{1-\mathrm{i}\rho}\right)^2 = \left(\frac{1+\rho\mathrm{e}^{\mathrm{i}\alpha}}{1-\rho\mathrm{e}^{\mathrm{i}\alpha}}\right)^2 = \frac{\cos^2\theta_\mathrm{c}\left(\epsilon - \sin^2\theta_\mathrm{c}\right)}{\sin^4\theta_\mathrm{c}}$$

from which we can extract the permittivity ϵ. We get

$$\epsilon = \sin^2\theta_c \left(1 + \left(\frac{1+i\rho}{1-i\rho}\right)^2 \tan^2\theta_c\right) = \sin^2\theta_c \left(1 + \left(\frac{1-\rho^2+2i\rho}{1+\rho^2}\right)^2 \tan^2\theta_c\right)$$

The real and imaginary parts are

$$\begin{cases} \operatorname{Re}\epsilon = \sin^2\theta_c \left(1 + \dfrac{\left(1-\rho^2\right)^2 - 4\rho^2}{\left(1+\rho^2\right)^2} \tan^2\theta_c\right) \\ \qquad = \sin^2\theta_c \dfrac{\left(1+\rho^4\right)\left(1+\tan^2\theta_c\right) + 2\rho^2\left(1-3\tan^2\theta_c\right)}{\left(1+\rho^2\right)^2} \\ \operatorname{Im}\epsilon = \sin^2\theta_c \dfrac{4\left(1-\rho^2\right)\rho}{\left(1+\rho^2\right)^2} \tan^2\theta_c \end{cases}$$

We observe that the real part is negative, provided

$$\left(1+\rho^4\right)\left(1+\tan^2\theta_c\right) + 2\rho^2\left(1-3\tan^2\theta_c\right) = \left(1+\rho^2\right)^2 + \tan^2\theta_c\left(1+\rho^4-6\rho^2\right) < 0$$

The reflected field is a completely polarized field, since the matrix of coherence, see (1.92) on page 88

$$[\mathbf{J}] = \begin{pmatrix} E_{01}E_{01}^* & E_{01}E_{02}^* \\ E_{02}E_{01}^* & E_{02}E_{02}^* \end{pmatrix} = A^2 \begin{pmatrix} 1 & \dfrac{-i\rho q}{\cos\theta_c} \\ \dfrac{i\rho q}{\cos\theta_c} & \dfrac{\rho^2 q^2}{\cos^2\theta_c} \end{pmatrix}$$

has zero determinant, and the Stokes parameters are, see (1.93) on page 90

$$\begin{cases} s_0 = J_{11} + J_{22} = A^2\left(1 + \dfrac{\rho^2 q^2}{\cos^2\theta_c}\right) \\ s_1 = J_{11} - J_{22} = A^2\left(1 - \dfrac{\rho^2 q^2}{\cos^2\theta_c}\right) \\ s_2 = J_{12} + J_{21} = 2\operatorname{Re} J_{12} = 0 \\ s_3 = i\left(J_{21} - J_{12}\right) = 2\operatorname{Im} J_{12} = -A^2 \dfrac{2\rho q}{\cos\theta_c} \end{cases}$$

■

Example 10.6
At normal incidence, the reflection coefficient is (the same for both polarizations)

$$r = \frac{\eta - 1}{\eta + 1} = \frac{1 - Z}{1 + Z}$$

provided the upper half-space is empty, and the lower half-space has a relative wave impedance $\eta = (\mu/\epsilon)^{1/2}$ or wave admittance $Z = 1/\eta$. The (complex) values of the wave admittance that give constant reflectivity R are circles in the complex Z-plane, since

$$\left|\frac{Z-1}{Z+1}\right|^2 = R \leq 1$$

is the equation for (harmonic) circles, see e.g., Reference [138, Appendix D]. The circles are centered at Z_c on the real axis, and radius Z_r, given by

$$\begin{cases} Z_c = \dfrac{1+R}{1-R} \in \mathbb{R} \\ Z_r = \dfrac{2\sqrt{R}}{1-R} \end{cases}$$

That is,

$$\begin{cases} \operatorname{Re} Z = Z_c + Z_r \cos\phi \\ \operatorname{Im} Z = Z_r \sin\phi \end{cases} \qquad \phi \in [0, 2\pi)$$

If the material is non-magnetic, $Z = \epsilon^{1/2}$, and

$$\begin{cases} \operatorname{Re}\epsilon = (\operatorname{Re} Z)^2 - (\operatorname{Im} Z)^2 = (Z_c + Z_r\cos\phi)^2 - Z_r^2 \sin^2\phi \\ \operatorname{Im}\epsilon = 2 \operatorname{Re} Z \operatorname{Im} Z = 2\,(Z_c + Z_r \cos\phi)\, Z_r \sin\phi \end{cases}$$

As ϕ varies, these equations trace a curve in the complex ϵ-plane, see Figure 10.11. This figure shows, without using a specific model, that highly reflecting, low loss materials, such as metals at optical frequencies, must have a large negative real part of the permittivity. That this is the case for the Drude model, we have already seen in Section 1.3.2.2, see (1.61) on page 53. ■

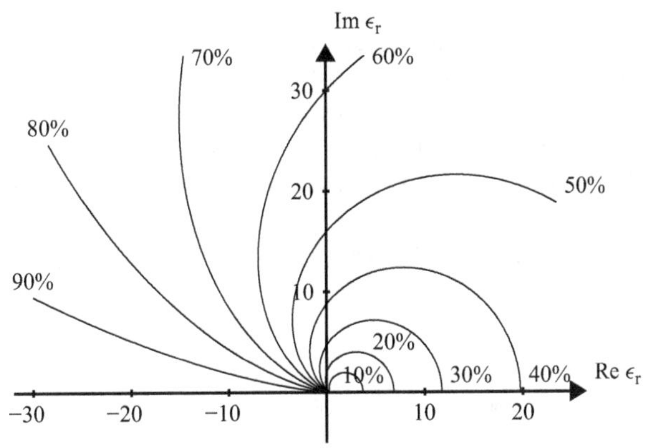

Figure 10.11 The possible values of the complex permittivity ϵ for different values of the reflectivity R.

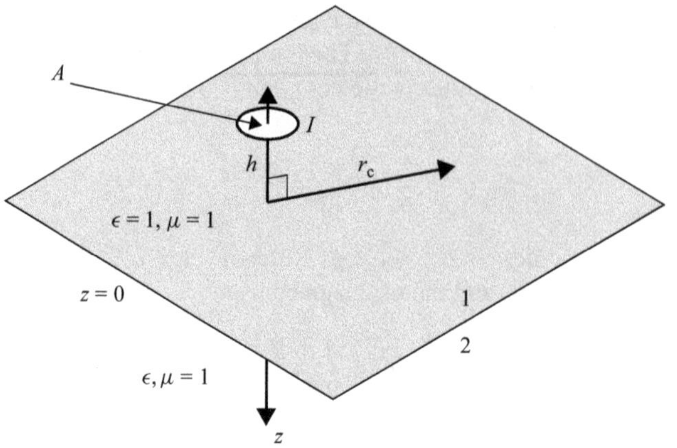

Figure 10.12 Vertical magnetic dipole above an interface separating vacuum from a homogeneous, isotropic, non-magnetic half-space with permittivity ϵ. The antenna, with magnetic dipole moment m, has the area A and carries the current I.

Example 10.7
In this example, we illustrate the half-space case with an example with many geophysical applications. Specifically, we compute the field radiated by a magnetic dipole above an isotropic half-space (region 2, $z > 0$), see Figure 10.12.[8] We assume the magnetic dipole is located in vacuum (region 1, $z < 0$), and moreover, that the material 2 below the interface is non-magnetic, $\mu = 1$ with a permittivity ϵ. This example is an important model for wave propagation of radio waves above ground or water.

The electric field from a magnetic dipole at the origin, with magnetic dipole moment $\boldsymbol{m} = -\hat{z}m$ (the dipole has strength m and is directed along the negative z-direction), in vacuum is [101, page 413]

$$\boldsymbol{E}_1(\boldsymbol{r}) = -\hat{\phi} k_0^2 \eta_0 m \frac{e^{ik_0 r}}{4\pi r}\left(1 - \frac{1}{ik_0 r}\right)\sin\theta = -i\omega\mu_0 m \nabla \times \left(\hat{z}\frac{e^{ik_0 r}}{4\pi r}\right)$$

The polar angle θ is the angle between $\hat{z}$ and the observation point $\boldsymbol{r}$. A magnetic dipole can be realized as a plane, closed loop that carries the current I. The dipole strength m, the current I, and the area A on the antenna are related by $m = IA$.

The Fourier transform w.r.t. the lateral x- and y-coordinates and identification of the x- and y-components of the electric field leads to (the details are left to the reader[9])

$$\boldsymbol{E}_1(\boldsymbol{r}) = \frac{1}{4\pi^2}\int_0^{2\pi}\int_0^{\infty} \boldsymbol{E}_{ixy}(\boldsymbol{k}_t) e^{ik_t r_c \cos(\psi - \phi)} e^{ik_z|z|}\, k_t \, dk_t \, d\psi$$

---

[8] In Examples 10.7 and 10.9, which both deal with geophysical examples, the positive z-axis is pointing vertically down.

[9] A representation of a spherical wave that is useful here is (the branch of the square root is chosen so that the imaginary part of the square root is non-negative)

$$\frac{e^{ik_0 r}}{4\pi r} = \frac{i}{4\pi}\int_0^{\infty} J_0(k_t r_c) e^{i(k_0^2 - k_t^2)^{1/2}|z|} \frac{k_t \, dk_t}{(k_0^2 - k_t^2)^{1/2}}$$

where

$$\boldsymbol{E}_{ixy}(\boldsymbol{k}_t) = -\frac{i\omega\mu_0 m k_t}{2k_z}\hat{\boldsymbol{e}}_\perp$$

and k_z is defined by (10.4), and where $\boldsymbol{k}_t = k_t\left(\hat{\boldsymbol{x}}\cos\psi + \hat{\boldsymbol{y}}\sin\psi\right)$.

Let the magnetic dipole be located on the height $h \geq 0$ above the interface $z = 0$ between the two isotropic materials, i.e., the antenna is located at $(0, 0, -h)$, see Figure 10.12. The incident electric field from the antenna at $(0, 0, -h)$, we get by replacing $z \rightarrow z + h$ in the expression of the dipole expression above, i.e.,

$$\boldsymbol{E}_{ixy}(\boldsymbol{r}) = \frac{1}{4\pi^2}\int_0^{2\pi}\int_0^\infty \boldsymbol{E}_{ixy}(\boldsymbol{k}_t)e^{ik_t r_c\cos(\psi-\phi)}e^{ik_z|z+h|}\,k_t\,dk_t\,d\psi$$

The incident field at the interface then is

$$\boldsymbol{E}_{ixy}(z = 0) = \frac{1}{4\pi^2}\int_0^{2\pi}\int_0^\infty \boldsymbol{E}_{ixy}(\boldsymbol{k}_t)e^{ik_t r_c\cos(\psi-\phi)}e^{ik_z h}\,k_t\,dk_t\,d\psi$$

Since the reflection dyadic is diagonal and the incident field only has a $\hat{\boldsymbol{e}}_\perp$-component, the total field in material 1 is

$$\begin{aligned}\boldsymbol{E}_{1xy}(\boldsymbol{r}) = \frac{1}{4\pi^2}\int_0^{2\pi}\int_0^\infty &\left(e^{ik_z|z+h|} + r_\perp(\boldsymbol{k}_t)e^{-ik_z(z-h)}\right) \qquad z \leq 0\\ &\times \boldsymbol{E}_{ixy}(\boldsymbol{k}_t)e^{ik_t r_c\cos(\psi-\phi)}\,k_t\,dk_t\,d\psi\end{aligned}$$

The special case when the antenna is located on the interface, $h = 0$, leads to ($z \leq 0$)

$$\begin{aligned}\boldsymbol{E}_{1xy}(\boldsymbol{r}) &= -\frac{i\omega\mu_0 m}{8\pi^2}\int_0^{2\pi}\int_0^\infty \left(1 + \frac{k_z - \tilde{k}_z}{k_z + \tilde{k}_z}\right)e^{ik_t r_c\cos(\psi-\phi)}e^{-ik_z z}\hat{\boldsymbol{e}}_\perp\frac{k_t^2}{k_z}\,dk_t\,d\psi\\ &= \hat{\boldsymbol{\phi}}\frac{\omega\mu_0 m}{2\pi}\int_0^\infty J_1(k_t r_c)e^{-ik_z z}\frac{k_t^2\,dk_t}{k_z + \tilde{k}_z} \qquad z \leq 0\end{aligned}$$

Here, we have introduced the longitudinal wave number in the material 2 defined as

$$\tilde{k}_z = \left(k_0^2\epsilon\mu - k_t^2\right)^{1/2} = \left(k^2 - k_t^2\right)^{1/2}$$

where the wave number in the half-space is $k = k_0(\epsilon\mu)^{1/2}$. Moreover, the integral representation of the Bessel function $J_1(z)$ has been used, see (B.3) in Appendix B

$$\frac{1}{2i\pi}\int_0^{2\pi} e^{iz\cos(\psi-\phi)}\begin{pmatrix}\cos\psi\\ \sin\psi\end{pmatrix}d\psi = J_1(z)\begin{pmatrix}\cos\phi\\ \sin\phi\end{pmatrix}$$

This is the exact expression of the electric field in material 1 when the antenna is located at the interface. The corresponding vertical component of the magnetic field, H_{1z}, is (the magnetic field has also a $\hat{\boldsymbol{r}}_c$-component, which is not given here)

$$H_{1z}(\boldsymbol{r}) = -\frac{im}{2\pi}\int_0^\infty J_0(k_t r_c)e^{-ik_z z}\frac{k_t^3\,dk_t}{k_z + \tilde{k}_z} \qquad z \leq 0$$

which is deduced by the use of (10.6) and the integral representation of $J_0(z)$

$$J_0(z) = \frac{1}{2\pi} \int_0^{2\pi} \mathrm{e}^{\mathrm{i}z\cos(\psi-\phi)}\,\mathrm{d}\psi$$

The remaining integral can be solved provided the observation point also lies on the surface, $z = 0$. This is the problem of van der Pol,[10] and the details in these computations are referred to Problem 10.4:

$$\begin{cases} E_{1\phi}(\boldsymbol{r}) = -\dfrac{2\mathrm{i}\omega\mu_0 m}{k_0^2 - k^2}\dfrac{\partial}{\partial r_\mathrm{c}}\left[\dfrac{1}{r_\mathrm{c}}\dfrac{\partial}{\partial r_\mathrm{c}}\left(\dfrac{\mathrm{e}^{\mathrm{i}k_0 r_\mathrm{c}}}{4\pi r_\mathrm{c}} - \dfrac{\mathrm{e}^{\mathrm{i}k r_\mathrm{c}}}{4\pi r_\mathrm{c}}\right)\right] \\ H_{1z}(\boldsymbol{r}) = -\dfrac{2m}{k_0^2 - k^2}\dfrac{1}{r_\mathrm{c}}\dfrac{\partial}{\partial r_\mathrm{c}}\left\{r_\mathrm{c}\dfrac{\partial}{\partial r_\mathrm{c}}\left[\dfrac{1}{r_\mathrm{c}}\dfrac{\partial}{\partial r_\mathrm{c}}\left(\dfrac{\mathrm{e}^{\mathrm{i}k_0 r_\mathrm{c}}}{4\pi r_\mathrm{c}} - \dfrac{\mathrm{e}^{\mathrm{i}k r_\mathrm{c}}}{4\pi r_\mathrm{c}}\right)\right]\right\} \end{cases} \quad z = 0 \qquad (10.43)$$

The dominating contribution to the field at the interface far away from the antenna, $r_\mathrm{c} \to \infty$, then is

$$\begin{cases} E_{1\phi}(\boldsymbol{r}) = \dfrac{\mathrm{i}\omega\mu_0 m}{2\pi\left(k_0^2 - k^2\right) r_\mathrm{c}^2}\left(k_0^2 \mathrm{e}^{\mathrm{i}k_0 r_\mathrm{c}} - k^2 \mathrm{e}^{\mathrm{i}k r_\mathrm{c}}\right) \\ H_{1z}(\boldsymbol{r}) = \dfrac{\mathrm{i}m}{2\pi\left(k_0^2 - k^2\right) r_\mathrm{c}^2}\left(k_0^3 \mathrm{e}^{\mathrm{i}k_0 r_\mathrm{c}} - k^3 \mathrm{e}^{\mathrm{i}k r_\mathrm{c}}\right) \end{cases} \quad z = 0$$

The terms inside the parentheses in these expressions are responsible for the interference effects that are present (for real k the period of the interference pattern is $2\pi/|k_0 - k|$). ■

Example 10.8

We investigate what conditions the transverse wave vector $\boldsymbol{k}_\mathrm{t}$ have to satisfy in order to give zero reflected field for some incident field. These $\boldsymbol{k}_\mathrm{t}$-values correspond to directions that quantify the Brewster[11] angles.

To obtain zero reflected field for some incident field, we must have

$$\mathbf{r}(\boldsymbol{k}_\mathrm{t}) \cdot \boldsymbol{E}_{\mathrm{i}xy}(\boldsymbol{k}_\mathrm{t}) = \boldsymbol{0}$$

or the incident field $\boldsymbol{E}_{\mathrm{i}xy}(\boldsymbol{k}_\mathrm{t})$ is an eigenvector to the reflection dyadic with zero eigenvalue. A necessary and sufficient condition for this to occur is

$$\det \mathbf{r}(\boldsymbol{k}_\mathrm{t}) = 0$$

This condition is general, and not only valid for an isotropic half-space, but also for a general bianisotropic material.

---

[10] Balthasar van der Pol (1889–1959), Dutch physicist.

[11] David Brewster (1781–1868), Scottish physicist, mathematician, and astronomer.

For an isotropic material, Fresnel's reflection coefficients imply that one of the following conditions have to be met:

$$\begin{cases} r_{\parallel} = \dfrac{\left(k_0^2\epsilon\mu - k_t^2\right)^{1/2} - \epsilon k_z}{\left(k_0^2\epsilon\mu - k_t^2\right)^{1/2} + \epsilon k_z} = 0 \\ r_{\perp} = \dfrac{\mu k_z - \left(k_0^2\epsilon\mu - k_t^2\right)^{1/2}}{\mu k_z + \left(k_0^2\epsilon\mu - k_t^2\right)^{1/2}} = 0 \end{cases}$$

or

$$\tilde{k}_z = \epsilon k_z \quad (\parallel\text{-case}) \qquad \text{and} \qquad \tilde{k}_z = \mu k_z \quad (\perp\text{-case})$$

where

$$\tilde{k}_z = \left(k_0^2\epsilon\mu - k_t^2\right)^{1/2} = \left(k^2 - k_t^2\right)^{1/2}$$

and the wave number in the half-space is $k = k_0(\epsilon\mu)^{1/2}$.

Take the squares of these condition, and we get

$$k^2 - k_t^2 = \epsilon^2\left(k_0^2 - k_t^2\right) \qquad \text{and} \qquad \mu^2\left(k_0^2 - k_t^2\right) = k^2 - k_t^2$$

Solve for k_t, which gives

$$k_t^2 = \frac{\omega^2}{c_0^2}\frac{\epsilon(\mu - \epsilon)}{1 - \epsilon^2} \qquad \text{and} \qquad k_t^2 = \frac{\omega^2}{c_0^2}\frac{\mu(\epsilon - \mu)}{1 - \mu^2}$$

If the material in the half-space is non-magnetic $\mu = 1$, then

$$k_t^2 = \frac{\omega^2}{c_0^2}\frac{\epsilon}{1 + \epsilon}$$

for the first equation, while the second equation has no solution unless $k = k_0$. An alternative way of expressing this result is to introduce the incident angle θ_i:

$$\tan^2\theta_i = \frac{k_0^2 - k_z^2}{k_z^2} = \frac{k_t^2}{k_0^2 - k_t^2} = \frac{\epsilon}{(1 + \epsilon) - \epsilon} = \epsilon$$

■

10.6.2 Biisotropic Media

Electromagnetic wave propagation in biisotropic media has a long history. Some important contributions are collected in Reference 165. For a homogeneous biisotropic slab ($\boldsymbol{\epsilon} = \epsilon\mathbf{I}_3$, $\boldsymbol{\mu} = \mu\mathbf{I}_3$, $\boldsymbol{\xi} = \xi\mathbf{I}_3$, $\boldsymbol{\zeta} = \zeta\mathbf{I}_3$), the fundamental dyadic is

$$\mathbf{M} = \begin{pmatrix} -\mathbf{J}\zeta + \frac{a\xi}{k_0^2}\boldsymbol{k}_t\boldsymbol{k}_t\cdot\mathbf{J} & -\mu\mathbf{I}_2 + \frac{a\mu}{k_0^2}\boldsymbol{k}_t\boldsymbol{k}_t \\ -\epsilon\mathbf{I}_2 - \frac{a\epsilon}{k_0^2}\mathbf{J}\cdot\boldsymbol{k}_t\boldsymbol{k}_t\cdot\mathbf{J} & \mathbf{J}\xi - \frac{a\zeta}{k_0^2}\mathbf{J}\cdot\boldsymbol{k}_t\boldsymbol{k}_t \end{pmatrix}$$

where $a^{-1} = \epsilon\mu - \xi\zeta$.

Two special cases are of interest, viz. reciprocal, biisotropic medium (isotropic chiral medium or Pasteur[12] medium), which is characterized by $\zeta = -\xi$, and non-reciprocal, achiral, biisotropic medium (Tellegen[13] medium), which is characterized by $\zeta = \xi$. The eigenvalues of **M** are all distinct unless the medium is Tellegen. Specifically, the eigenvalues are

$$\lambda_1 = -\lambda_4 = \lambda_+, \qquad \lambda_2 = -\lambda_3 = \lambda_-$$

where

$$\lambda_\pm^2 = n_\pm^2 - \frac{k_t^2}{k_0^2}, \qquad n_\pm = \sqrt{\mu\epsilon - \left(\frac{\xi+\zeta}{2}\right)^2} \pm \mathrm{i}\frac{\xi-\zeta}{2}$$

which correspond to right-going (left-going waves correspond to the associated negative values, λ_3 and λ_4) right- and left-hand circularly polarized plane waves in the medium, respectively.

To find the explicit form of the propagator, we need

$$\mathbf{M}\cdot\mathbf{M} = \begin{pmatrix} \mathbf{I}_2\left(\mu\epsilon - \zeta^2 - k_t^2/k_0^2\right) & \mathbf{J}\mu\left(\zeta-\xi\right) \\ \mathbf{J}\epsilon\left(\zeta-\xi\right) & \mathbf{I}_2\left(\mu\epsilon - \xi^2 - k_t^2/k_0^2\right) \end{pmatrix}$$

If the medium is not Tellegen, the single-slab propagator is given by (10.38). In particular, for Pasteur media ($\zeta = -\xi$), the propagator is

$$\mathbf{P} = \frac{1}{2}\begin{pmatrix} \mathbf{P}_+ & \mathrm{i}\sqrt{\frac{\mu}{\epsilon}}\mathbf{P}_+\cdot\mathbf{J} \\ \mathrm{i}\sqrt{\frac{\epsilon}{\mu}}\mathbf{J}\cdot\mathbf{P}_+ & -\mathbf{J}\cdot\mathbf{P}_+\cdot\mathbf{J} \end{pmatrix} + \frac{1}{2}\begin{pmatrix} \mathbf{P}_- & -\mathrm{i}\sqrt{\frac{\mu}{\epsilon}}\mathbf{P}_-\cdot\mathbf{J} \\ -\mathrm{i}\sqrt{\frac{\epsilon}{\mu}}\mathbf{J}\cdot\mathbf{P}_- & -\mathbf{J}\cdot\mathbf{P}_-\cdot\mathbf{J} \end{pmatrix}$$

where the dyadics ($n_\pm = \sqrt{\mu\epsilon} \pm \mathrm{i}\,\xi$)

$$\mathbf{P}_\pm = \mathbf{I}_2\cos(k_0 d\lambda_\pm) \pm \frac{1}{\lambda_\pm}\left(\mathbf{I}_2 n_\pm - \frac{1}{n_\pm k_0^2}\mathbf{k}_t\mathbf{k}_t\right)\cdot\mathbf{J}\sin(k_0 d\lambda_\pm)$$

For a Tellegen material ($\zeta = \xi$), (10.40) applies with $\lambda^2 = \mu\epsilon - \xi^2 - k_t^2/k_0^2$.

The technique presented in Section 10.4.1 can be used to obtain the reflection and transmission dyadics for a single biisotropic slab. For results, the reader is referred to Reference 164 and references given therein.

---

[12] Louis Pasteur (1822–1895), French chemist and microbiologist.

[13] Bernard D.H. Tellegen (1900–1990), Dutch electrical engineer.

10.6.3 Anisotropic Media

An anisotropic material is characterized by $\boldsymbol{\xi} = \boldsymbol{\zeta} = \mathbf{0}$, and the blocks of the fundamental dyadic are

$$\begin{cases} \mathbf{M}_{11} = -\dfrac{1}{\epsilon_{zz}k_0}\boldsymbol{k}_\mathrm{t}\boldsymbol{\epsilon}_z + \dfrac{1}{\mu_{zz}k_0}\mathbf{J}\cdot\boldsymbol{\mu}_\perp\boldsymbol{k}_\mathrm{t}\cdot\mathbf{J} \\ \mathbf{M}_{12} = \mathbf{J}\cdot\boldsymbol{\mu}_{\perp\perp}\cdot\mathbf{J} + \dfrac{1}{\epsilon_{zz}k_0^2}\boldsymbol{k}_\mathrm{t}\boldsymbol{k}_\mathrm{t} - \dfrac{1}{\mu_{zz}}\mathbf{J}\cdot\boldsymbol{\mu}_\perp\boldsymbol{\mu}_z\cdot\mathbf{J} \\ \mathbf{M}_{21} = -\boldsymbol{\epsilon}_{\perp\perp} + \dfrac{1}{\epsilon_{zz}}\boldsymbol{\epsilon}_\perp\boldsymbol{\epsilon}_z - \dfrac{1}{\mu_{zz}k_0^2}\mathbf{J}\cdot\boldsymbol{k}_\mathrm{t}\boldsymbol{k}_\mathrm{t}\cdot\mathbf{J} \\ \mathbf{M}_{22} = -\dfrac{1}{\epsilon_{zz}k_0}\boldsymbol{\epsilon}_\perp\boldsymbol{k}_\mathrm{t} + \dfrac{1}{\mu_{zz}k_0}\mathbf{J}\cdot\boldsymbol{k}_\mathrm{t}\boldsymbol{\mu}_z\cdot\mathbf{J} \end{cases}$$

If, in addition, the material is non-magnetic ($\boldsymbol{\mu} = \mathbf{I}_3$), further simplifications occur. We have

$$\begin{cases} \mathbf{M}_{11} = -\dfrac{1}{\epsilon_{zz}k_0}\boldsymbol{k}_\mathrm{t}\boldsymbol{\epsilon}_z \\ \mathbf{M}_{22} = -\dfrac{1}{\epsilon_{zz}k_0}\boldsymbol{\epsilon}_\perp\boldsymbol{k}_\mathrm{t} \end{cases} \qquad \begin{cases} \mathbf{M}_{12} = -\mathbf{I}_2 + \dfrac{1}{\epsilon_{zz}k_0^2}\boldsymbol{k}_\mathrm{t}\boldsymbol{k}_\mathrm{t} \\ \mathbf{M}_{21} = -\boldsymbol{\epsilon}_{\perp\perp} + \dfrac{1}{\epsilon_{zz}}\boldsymbol{\epsilon}_\perp\boldsymbol{\epsilon}_z - \dfrac{1}{k_0^2}\mathbf{J}\cdot\boldsymbol{k}_\mathrm{t}\boldsymbol{k}_\mathrm{t}\cdot\mathbf{J} \end{cases}$$

Finding explicit expressions for the single-slab propagator in the general anisotropic case is complicated. However, closed-form solutions can be obtained in some special cases, which can be valuable in many applications.

10.6.3.1 Non-Magnetic Uniaxial Media

A non-magnetic uniaxial medium has a general permittivity dyadic of the form

$$\boldsymbol{\epsilon} = \epsilon_1\left(\mathbf{I}_3 - \hat{\boldsymbol{u}}\hat{\boldsymbol{u}}\right) + \epsilon_2\hat{\boldsymbol{u}}\hat{\boldsymbol{u}}$$

where the unit vector $\hat{\boldsymbol{u}}$ defines an optical axis in the material. The permittivity ϵ_2 denotes the value of the permittivity along the optical axis, and ϵ_1 the permittivity in the plane perpendicular to the optical axis.

The unit vector $\hat{\boldsymbol{u}}$ has a general orientation in space, and it has a projection $\boldsymbol{u}_{xy} = \mathbf{I}_2\cdot\hat{\boldsymbol{u}}$ in the x–y plane, and a projection along the z-axis $u_z = \hat{\boldsymbol{u}}\cdot\hat{\boldsymbol{z}}$. The decomposition of the permittivity dyadic is, see Section 10.1.1

$$\begin{cases} \boldsymbol{\epsilon}_{\perp\perp} = \mathbf{I}_2\cdot\boldsymbol{\epsilon}\cdot\mathbf{I}_2 = \epsilon_1\mathbf{I}_2 + (\epsilon_2-\epsilon_1)\boldsymbol{u}_{xy}\boldsymbol{u}_{xy} \\ \boldsymbol{\epsilon}_z = \hat{\boldsymbol{z}}\cdot\boldsymbol{\epsilon}\cdot\mathbf{I}_2 = (\epsilon_2-\epsilon_1)u_z\boldsymbol{u}_{xy} \\ \boldsymbol{\epsilon}_\perp = \mathbf{I}_2\cdot\boldsymbol{\epsilon}\cdot\hat{\boldsymbol{z}} = (\epsilon_2-\epsilon_1)u_z\boldsymbol{u}_{xy} = \boldsymbol{\epsilon}_z \\ \epsilon_{zz} = \hat{\boldsymbol{z}}\cdot\boldsymbol{\epsilon}\cdot\hat{\boldsymbol{z}} = \epsilon_1 + (\epsilon_2-\epsilon_1)u_z^2 \end{cases}$$

The blocks of the fundamental dyadic are

$$\begin{cases} \mathbf{M}_{11} = -\dfrac{(\epsilon_2-\epsilon_1)u_z}{\epsilon_{zz}k_0}\boldsymbol{k}_{\mathrm{t}}\boldsymbol{u}_{xy} \\ \mathbf{M}_{12} = -\mathbf{I}_2 + \dfrac{1}{\epsilon_{zz}k_0^2}\boldsymbol{k}_{\mathrm{t}}\boldsymbol{k}_{\mathrm{t}} \\ \mathbf{M}_{21} = -\epsilon_1\mathbf{I}_2 - \dfrac{\epsilon_1(\epsilon_2-\epsilon_1)}{\epsilon_{zz}}\boldsymbol{u}_{xy}\boldsymbol{u}_{xy} - \dfrac{1}{k_0^2}\mathbf{J}\cdot\boldsymbol{k}_{\mathrm{t}}\boldsymbol{k}_{\mathrm{t}}\cdot\mathbf{J} \\ \mathbf{M}_{22} = -\dfrac{(\epsilon_2-\epsilon_1)u_z}{\epsilon_{zz}k_0}\boldsymbol{u}_{xy}\boldsymbol{k}_{\mathrm{t}} = \mathbf{M}_{11}^{\mathrm{t}} \end{cases}$$

where $\epsilon_{zz} = \epsilon_1 + (\epsilon_2 - \epsilon_1)u_z^2$. The matrix representation of the fundamental dyadic $\mathbf{M}$ in the $\{\hat{\boldsymbol{e}}_{\parallel}, \hat{\boldsymbol{e}}_{\perp}\}$-system is

$$\begin{pmatrix} -\frac{(\epsilon_2-\epsilon_1)}{\epsilon_{zz}}\frac{k_{\mathrm{t}}}{k_0}u_z u_{\parallel} & -\frac{(\epsilon_2-\epsilon_1)}{\epsilon_{zz}}\frac{k_{\mathrm{t}}}{k_0}u_z u_{\perp} & \frac{1}{\epsilon_{zz}}\frac{k_{\mathrm{t}}^2}{k_0^2} - 1 & 0 \\ 0 & 0 & 0 & -1 \\ -\epsilon_1 - \frac{\epsilon_1(\epsilon_2-\epsilon_1)}{\epsilon_{zz}}u_{\parallel}^2 & -\frac{\epsilon_1(\epsilon_2-\epsilon_1)}{\epsilon_{zz}}u_{\parallel}u_{\perp} & -\frac{(\epsilon_2-\epsilon_1)}{\epsilon_{zz}}\frac{k_{\mathrm{t}}}{k_0}u_z u_{\parallel} & 0 \\ -\frac{\epsilon_1(\epsilon_2-\epsilon_1)}{\epsilon_{zz}}u_{\parallel}u_{\perp} & -\epsilon_1 - \frac{\epsilon_1(\epsilon_2-\epsilon_1)}{\epsilon_{zz}}u_{\perp}^2 + \frac{k_{\mathrm{t}}^2}{k_0^2} & -\frac{(\epsilon_2-\epsilon_1)}{\epsilon_{zz}}\frac{k_{\mathrm{t}}}{k_0}u_z u_{\perp} & 0 \end{pmatrix}$$

where $u_{\parallel} = \hat{\boldsymbol{e}}_{\parallel}\cdot\boldsymbol{u}_{xy}$ and $u_{\perp} = \hat{\boldsymbol{e}}_{\perp}\cdot\boldsymbol{u}_{xy}$.

Normal Incidence: At normal incidence, $k_{\mathrm{t}} = 0$, the expression for the non-magnetic uniaxial material simplifies to

$$\mathbf{M} = \begin{pmatrix} \mathbf{0} & -\mathbf{I}_2 \\ -\epsilon_1\mathbf{I}_2 - \frac{\epsilon_1(\epsilon_2-\epsilon_1)}{\epsilon_{zz}}\boldsymbol{u}_{xy}\boldsymbol{u}_{xy} & \mathbf{0} \end{pmatrix}$$

and

$$\mathbf{M}\cdot\mathbf{M} = \begin{pmatrix} \epsilon_1\mathbf{I}_2 + \frac{\epsilon_1(\epsilon_2-\epsilon_1)}{\epsilon_{zz}}\boldsymbol{u}_{xy}\boldsymbol{u}_{xy} & \mathbf{0} \\ \mathbf{0} & \epsilon_1\mathbf{I}_2 + \frac{\epsilon_1(\epsilon_2-\epsilon_1)}{\epsilon_{zz}}\boldsymbol{u}_{xy}\boldsymbol{u}_{xy} \end{pmatrix}$$

The eigenvalues of $\mathbf{M}$ are all distinct. Specifically, (10.37) holds, where

$$\lambda_+^2 = \epsilon_1, \qquad \lambda_-^2 = \frac{\epsilon_1\epsilon_2}{\epsilon_{zz}} = \frac{\epsilon_1\epsilon_2}{\epsilon_1 + (\epsilon_2-\epsilon_1)u_z^2}$$

These eigenvalues correspond to right-going (left-going waves correspond to the associated negative values, λ_3 and λ_4) ordinary and extraordinary waves in the medium, respectively. The single-slab propagator is given by (10.38).

Optical Axis in the Normal Direction: If the optical axis is directed along the normal direction (the z-direction) in the non-magnetic uniaxial material, the projection of the unit vector $\hat{\boldsymbol{u}}$ is $\boldsymbol{u}_{xy} = \mathbf{0}$, and the material dyadic reads

$$\mathbf{M} = \begin{pmatrix} \mathbf{0} & -\mathbf{I}_2 + \frac{1}{\epsilon_2 k_0^2}\boldsymbol{k}_{\rm t}\boldsymbol{k}_{\rm t} \\ -\epsilon_1 \mathbf{I}_2 - \frac{1}{k_0^2}\mathbf{J}\cdot\boldsymbol{k}_{\rm t}\boldsymbol{k}_{\rm t}\cdot\mathbf{J} & \mathbf{0} \end{pmatrix}$$

and

$$\mathbf{M}\cdot\mathbf{M} = \begin{pmatrix} \epsilon_1\left(\mathbf{I}_2 - \frac{1}{\epsilon_2 k_0^2}\boldsymbol{k}_{\rm t}\boldsymbol{k}_{\rm t}\right) + \frac{1}{k_0^2}\mathbf{J}\cdot\boldsymbol{k}_{\rm t}\boldsymbol{k}_{\rm t}\cdot\mathbf{J} & \mathbf{0} \\ \mathbf{0} & \epsilon_1\left(\mathbf{I}_2 - \frac{1}{\epsilon_2 k_0^2}\boldsymbol{k}_{\rm t}\boldsymbol{k}_{\rm t}\right) + \frac{1}{k_0^2}\mathbf{J}\cdot\boldsymbol{k}_{\rm t}\boldsymbol{k}_{\rm t}\cdot\mathbf{J} \end{pmatrix}$$

Again, the eigenvalues of the material dyadics $\mathbf{M}$ are all distinct, and

$$\lambda_+^2 = \epsilon_1 - \frac{k_{\rm t}^2}{k_0^2}, \qquad \lambda_-^2 = \epsilon_1 - \frac{k_{\rm t}^2}{k_0^2}\frac{\epsilon_1}{\epsilon_2}$$

The single-slab propagator is given by (10.38)

$$\mathbf{P} = {\rm e}^{{\rm i}k_0 d\mathbf{M}} = \begin{pmatrix} \hat{\boldsymbol{e}}_\parallel\hat{\boldsymbol{e}}_\parallel \cos(k_0 d\lambda_-) & -{\rm i}\hat{\boldsymbol{e}}_\parallel\hat{\boldsymbol{e}}_\parallel \frac{\lambda_-}{\epsilon_1}\sin(k_0 d\lambda_-) \\ -{\rm i}\hat{\boldsymbol{e}}_\parallel\hat{\boldsymbol{e}}_\parallel \frac{\epsilon_1}{\lambda_-}\sin(k_0 d\lambda_-) & \hat{\boldsymbol{e}}_\parallel\hat{\boldsymbol{e}}_\parallel \cos(k_0 d\lambda_-) \end{pmatrix}$$
$$+ \begin{pmatrix} \hat{\boldsymbol{e}}_\perp\hat{\boldsymbol{e}}_\perp \cos(k_0 d\lambda_+) & -{\rm i}\hat{\boldsymbol{e}}_\perp\hat{\boldsymbol{e}}_\perp \frac{1}{\lambda_+}\sin(k_0 d\lambda_+) \\ -{\rm i}\hat{\boldsymbol{e}}_\perp\hat{\boldsymbol{e}}_\perp \lambda_+ \sin(k_0 d\lambda_+) & \hat{\boldsymbol{e}}_\perp\hat{\boldsymbol{e}}_\perp \cos(k_0 d\lambda_+) \end{pmatrix}$$

The derivation of the refection and transmission coefficients for a non-magnetic uniaxial material with optical axis in the normal direction is done in a similar way as the derivation in the isotropic case. We write in the same form as in (10.41), where now

$$\begin{cases} r_{\parallel\parallel} = r_{1\parallel}\dfrac{1 - {\rm e}^{2{\rm i}k_0\lambda_- d}}{1 - r_{1\parallel}^2 {\rm e}^{2{\rm i}k_0\lambda_- d}} \\[2ex] r_{\perp\perp} = r_{1\perp}\dfrac{1 - {\rm e}^{2{\rm i}k_0\lambda_+ d}}{1 - r_{1\perp}^2 {\rm e}^{2{\rm i}k_0\lambda_+ d}} \end{cases} \qquad \begin{cases} t_{\parallel\parallel} = \dfrac{(1 - r_{1\parallel}^2){\rm e}^{{\rm i}k_0\lambda_- d}}{1 - r_{1\parallel}^2 {\rm e}^{2{\rm i}k_0\lambda_- d}} \\[2ex] t_{\perp\perp} = \dfrac{(1 - r_{1\perp}^2){\rm e}^{{\rm i}k_0\lambda_+ d}}{1 - r_{1\perp}^2 {\rm e}^{2{\rm i}k_0\lambda_+ d}} \end{cases}$$

and

$$\begin{cases} r_{1\parallel} = \dfrac{\lambda_- - \epsilon_1\cos\theta_{\rm i}}{\lambda_- + \epsilon_1\cos\theta_{\rm i}} \\[2ex] r_{1\perp} = \dfrac{\cos\theta_{\rm i} - \lambda_+}{\cos\theta_{\rm i} + \lambda_+} \end{cases}$$

For a PEC-backed uniaxial slab, similar results hold. Explicitly, the result is

$$\begin{cases} r_{\parallel\parallel} = \dfrac{r_{1\parallel} - e^{2ik_0\lambda_- d}}{1 - r_{1\parallel} e^{2ik_0\lambda_- d}} \\[2ex] r_{\perp\perp} = \dfrac{r_{1\perp} - e^{2ik_0\lambda_+ d}}{1 - r_{1\perp} e^{2ik_0\lambda_+ d}} \end{cases}$$

10.7 Numerical Computations

We are now ready to illustrate the analysis presented in the previous sections in a series of numerical computations of increasing complexity. Since all computations rely on eigenvalues computations and simple matrix manipulations, the programming task is most easily done in a language that supports matrix manipulations. The numerical implementation is straightforward and causes no essential problem except for cases where strongly dissipative layers and evanescent waves are present. These cases deserve special attention [298].

10.7.1 Reflectivity and Transmissivity

The Poynting vectors associated with the right-going and the left-going plane waves were derived on page 531, see (10.25). The reflectivity R and the transmissivity T of the planar bianisotropic slab relate the reflected and transmitted power flux to the incident power flux. They are defined by (the end points of the slab are denoted z_1 and z_{N-1})

$$\begin{cases} R = \dfrac{|F_\parallel^-(z_1)|^2 + |F_\perp^-(z_1)|^2 \cos^2\theta_i}{|F_\parallel^+(z_1)|^2 + |F_\perp^+(z_1)|^2 \cos^2\theta_i} \\[2ex] T = \dfrac{|F_\parallel^+(z_{N-1})|^2 + |F_\perp^+(z_{N-1})|^2 \cos^2\theta_i}{|F_\parallel^+(z_1)|^2 + |F_\perp^+(z_1)|^2 \cos^2\theta_i} \end{cases} \tag{10.44}$$

respectively. Notice that these quantities depend on the angles θ_i and ϕ_i of the incident wave (or equivalently on the tangential wave vector $\boldsymbol{k}_t$). The reflectivity and the transmissivity can be expressed in terms of the components of the reflection and transmission dyadics for the electric field as

$$\begin{cases} \mathbf{r} = \hat{\boldsymbol{e}}_\parallel\hat{\boldsymbol{e}}_\parallel r_{\parallel\parallel} + \hat{\boldsymbol{e}}_\parallel\hat{\boldsymbol{e}}_\perp r_{\parallel\perp} + \hat{\boldsymbol{e}}_\perp\hat{\boldsymbol{e}}_\parallel r_{\perp\parallel} + \hat{\boldsymbol{e}}_\perp\hat{\boldsymbol{e}}_\perp r_{\perp\perp} \\ \mathbf{t} = \hat{\boldsymbol{e}}_\parallel\hat{\boldsymbol{e}}_\parallel t_{\parallel\parallel} + \hat{\boldsymbol{e}}_\parallel\hat{\boldsymbol{e}}_\perp t_{\parallel\perp} + \hat{\boldsymbol{e}}_\perp\hat{\boldsymbol{e}}_\parallel t_{\perp\parallel} + \hat{\boldsymbol{e}}_\perp\hat{\boldsymbol{e}}_\perp t_{\perp\perp} \end{cases}$$

where the angles θ_i and ϕ_i, and the polarization angle, χ, of the incident electric field at $z = z_1$ are defined by

$$\boldsymbol{E}_i(z_1) = E_0(z_1)\left(\hat{\boldsymbol{e}}_\perp \cos\chi + \hat{\boldsymbol{e}}_\perp \times \hat{\boldsymbol{k}}^+ \sin\chi\right)$$

The field $E_0(z_1)$ is a complex number that quantifies the amplitude and phase of the incident plane polarized wave at the front end, z_1, of the slab. The angle of polarization χ measures the angle between the electric field (linear polarization is assumed) and the plane of incidence, i.e., $\chi = 0$ gives the TE-wave and $\chi = \pi/2$ gives the TM-wave of incidence. The corresponding incident magnetic field is, see (10.20)

$$\eta_0 \boldsymbol{H}^{\mathrm{i}}(z_1) = E_0(z_1)\hat{\boldsymbol{k}}^+ \times \left(\hat{\boldsymbol{e}}_\perp \cos\chi + \hat{\boldsymbol{e}}_\perp \times \hat{\boldsymbol{k}}^+ \sin\chi\right)$$

where

$$\hat{\boldsymbol{k}}^+ = \frac{\boldsymbol{k}_{\mathrm{t}} + \hat{\boldsymbol{z}}k_z}{k_0}$$

Straightforward computations give the split components of the incident field, see (10.15)

$$\begin{cases} F_\parallel^+(z_1) = E_0(z_1)\cos\theta_{\mathrm{i}} \sin\chi \\ F_\perp^+(z_1) = E_0(z_1)\cos\chi \end{cases}$$

Here, the parallel components give the TM-wave and the perpendicular component the TE-wave. For a linearly polarized incident field, the reflectivity and the transmissivity are

$$\begin{cases} R = |r_{\parallel\parallel}\sin\chi + r_{\parallel\perp}\cos\chi/\cos\theta_{\mathrm{i}}|^2 + |r_{\perp\parallel}\cos\theta_{\mathrm{i}}\sin\chi + r_{\perp\perp}\cos\chi|^2 \\ T = |t_{\parallel\parallel}\sin\chi + t_{\parallel\perp}\cos\chi/\cos\theta_{\mathrm{i}}|^2 + |t_{\perp\parallel}\cos\theta_{\mathrm{i}}\sin\chi + t_{\perp\perp}\cos\chi|^2 \end{cases} \tag{10.45}$$

which in the absence of any cross-polarized terms, e.g., for isotropic materials, reduce to

$$\begin{cases} R = |r_{\parallel\parallel}|^2\sin^2\chi + |r_{\perp\perp}|^2\cos^2\chi \\ T = |t_{\parallel\parallel}|^2\sin^2\chi + |t_{\perp\perp}|^2\cos^2\chi \end{cases}$$

The perpendicular polarization (TE-wave) corresponds to $\chi = 0$ and the parallel polarization (TM-wave) to $\chi = \pi/2$, and for an angle $\chi \in (0, \pi/2)$, it corresponds to a mixture between the two polarizations.

10.7.2 Example—Dielectric Slab with Uniaxial Layers

This first example is relevant for radome applications in antenna technology. We illustrate the effect on the transmission properties (transmissivity T) when introducing one or several uniaxial layers in a dielectric slab. The specific parameters of the slab with only isotropic layers and with uniaxial reinforcements are given in Table 10.1. The uniaxial material parameters are obtained by a homogenization of the glass fiber reinforcements layers of the slab, which is a subject that is not covered in this textbook. More details on the homogenization procedure are found in the literature, see e.g., Reference 228.

The transmissivity for a series of incident angles and two different polarizations, TE and TM, is presented in Figure 10.13. The change of the permittivity in the lateral directions, i.e., ϵ_{xx} and ϵ_{yy}, alters the transmission properties even at normal incidence. We see that the effect of the uniaxial layers (glass fiber reinforcements) is not negligible. This is especially true at higher frequencies. At these higher frequencies, an extra transmission loss of several decibels is observed due to the glass fiber reinforcement.

Table 10.1 Material parameters for the dielectric slab in Section 10.7.2 which are presented in Figure 10.13.

Without reinforcing layer

	ϵ	d (mm)
Epoxy	3.65(1+i0.0320)	0.8
Rohacell	1.10(1+i0.0004)	6.4
Epoxy	3.65(1+i0.0320)	0.8

With reinforcing layer (E-glass, $\epsilon = 6.32(1+\mathrm{i}0.0037)$)

	$\epsilon_{xx} = \epsilon_{yy}$	ϵ_{zz}	d (mm)
Epoxy + E-glass (27%)	4.30(1+i0.0210)	4.20(1+i0.0253)	0.8
Rohacell	1.10(1+i0.0004)	1.10(1+i0.0004)	6.4
Epoxy + E-glass (27%)	4.30(1+i0.0210)	4.20(1+i0.0253)	0.8

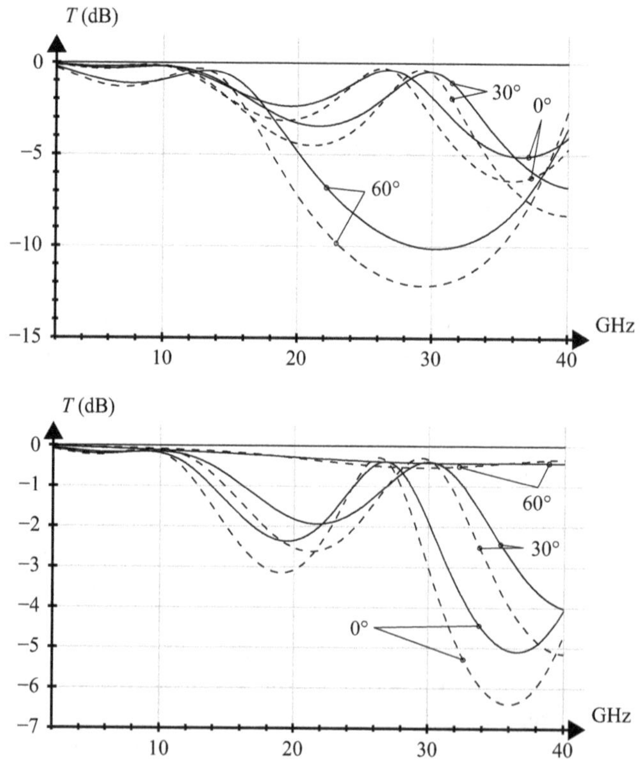

Figure 10.13 The transmissivity T for an isotropic slab (solid lines) and a uniaxial slab (broken lines) with data given in Table 10.1, as a function of frequency for different angles of incidence θ_i, and two generic polarizations, TE and TM, of the incident wave. Top: TE polarization, $\chi = 0$. Bottom: TM polarization, $\chi = \pi/2$.

10.7.3 Example—Bianisotropic Media

A more complex example, that illustrates the performance of the algorithm in the general bianisotropic case, is given next. An example of such a material is the Ω-material, which has been investigated thoroughly in the past, see e.g., Reference 165.

In Figures 10.14 and 10.15, we illustrate the reflectivity R and the transmissivity T, see (10.45), of a slab of Ω-material as a function of the two angles θ_i and ϕ_i (or equivalently $\boldsymbol{k}_\mathrm{t}$) for the two generic polarizations of the incident wave (TE- and TM-waves). The specific data for the material are given in Table 10.2. This material can be manufactured by putting small Ω-shaped elements in the x–y plane in a host medium [165].

From these computations, we clearly see that the transmissivity varies as a function of the incident direction ϕ_i, which shows that the slab lacks symmetry in the x–y plane. The difference between the two different planes of polarization is most striking at glancing incidence close to $\theta_\mathrm{i} = \pi/2$ and at $\phi_\mathrm{i} = \pi/2$.

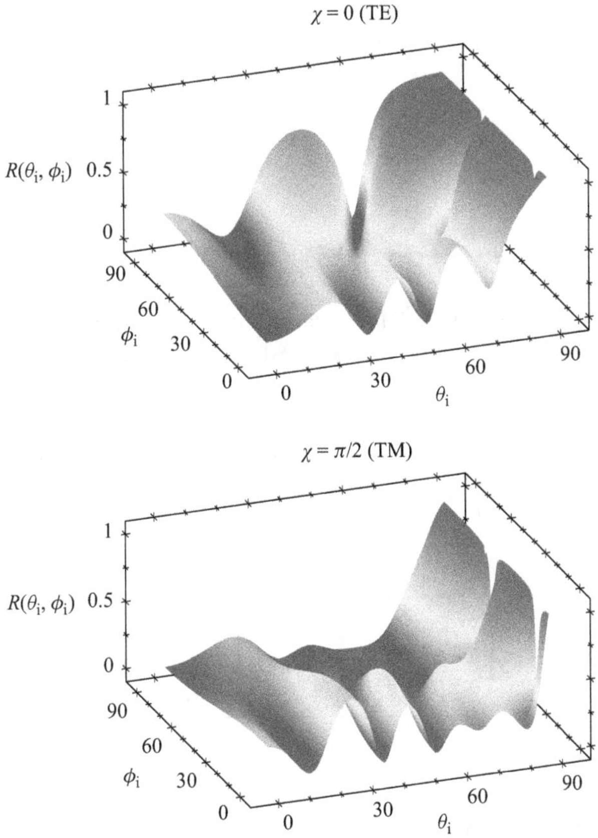

Figure 10.14 The reflectivity R for a bianisotropic slab given in Table 10.2. The two generic polarizations, TE and TM, of the incident wave are depicted.

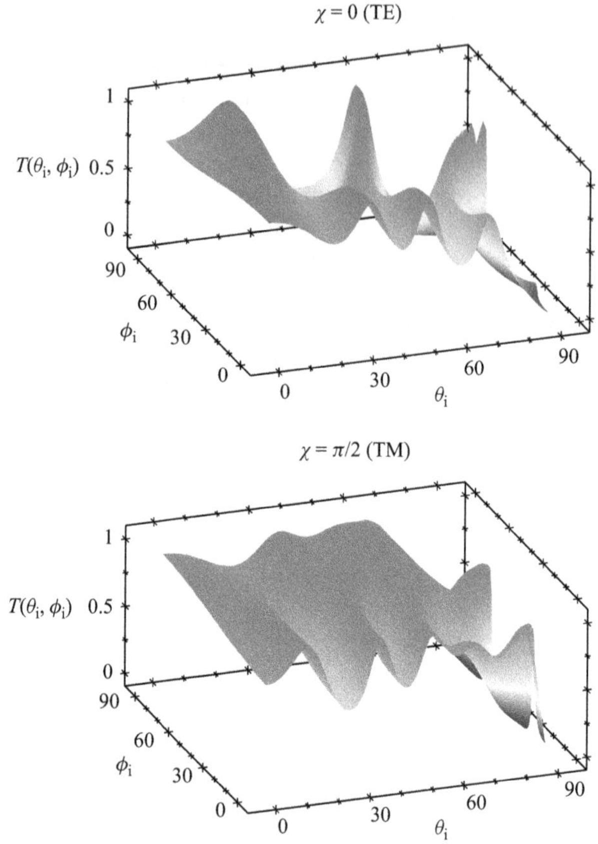

Figure 10.15 The transmissivity T for a bianisotropic slab given in Table 10.2. The two generic polarizations, TE and TM, of the incident wave are depicted.

Table 10.2 Material parameters for the bianisotropic material in Example 10.7.3 which is presented in Figures 10.14 and 10.15.

ϵ	μ	ξ	ζ	$k_0 d$
$\begin{pmatrix} 8 & 0 & 0 \\ 0 & 3 & 0 \\ 0 & 0 & 3 \end{pmatrix}$	$\begin{pmatrix} 1 & 0 & 0 \\ 0 & 1 & 0 \\ 0 & 0 & 1 \end{pmatrix}$	$\begin{pmatrix} 0 & 0 & \mathrm{i} \\ 0 & 0 & 0 \\ 0 & 0 & 0 \end{pmatrix}$	$\begin{pmatrix} 0 & 0 & 0 \\ 0 & 0 & 0 \\ -\mathrm{i} & 0 & 0 \end{pmatrix}$	8π

10.8 Asymptotic Analysis

As above, the two regions to the left and to the right of the slab ($z \in [0, d]$) are assumed empty, see Figure 10.4.

The general expression of the reflected field by a slab is, see (10.32)

$$\boldsymbol{E}_{\mathrm{r}}(\boldsymbol{r}) = \frac{1}{4\pi^2} \iint\limits_{\mathbb{R}^2} \left(\mathbf{I}_2 + \frac{\hat{\boldsymbol{z}}\boldsymbol{k}_{\mathrm{t}}}{k_z}\right) \cdot \mathbf{r}(\boldsymbol{k}_{\mathrm{t}}) \cdot \boldsymbol{F}^{+}(\boldsymbol{k}_{\mathrm{t}}, 0) \mathrm{e}^{\mathrm{i}\boldsymbol{k}^{-}\cdot\boldsymbol{r}} \, \mathrm{d}k_x \, \mathrm{d}k_y, \qquad z \le 0$$

where, see (10.21)

$$\boldsymbol{k}^{-} = \boldsymbol{k}_{\mathrm{t}} - \hat{\boldsymbol{z}} k_z$$

and where $\boldsymbol{F}^{+}(\boldsymbol{k}_{\mathrm{t}}, 0)$ is the Fourier transform (in the variables x and y) of the incident field $\boldsymbol{E}_{\mathrm{i}xy}$ evaluated on the surface $z = 0$, i.e.,

$$\boldsymbol{F}^{+}(\boldsymbol{k}_{\mathrm{t}}, 0) = \iint\limits_{\mathbb{R}^2} \boldsymbol{E}_{\mathrm{i}xy}(\boldsymbol{r})\big|_{z=0} \, \mathrm{e}^{-\mathrm{i}\boldsymbol{k}_{\mathrm{t}}\cdot\boldsymbol{r}_{\mathrm{c}}} \, \mathrm{d}x \, \mathrm{d}y$$

The longitudinal wave number is as usual

$$k_z = \left(k_0^2 - k_x^2 - k_y^2\right)^{1/2} = \left(k_0^2 - k_{\mathrm{t}}^2\right)^{1/2}$$

and the field point $\boldsymbol{r}$ in spherical coordinates is

$$\boldsymbol{r} = \boldsymbol{r}_{\mathrm{c}} + \hat{\boldsymbol{z}} z = r\left(\sin\theta \cos\phi \hat{\boldsymbol{x}} + \sin\theta \sin\phi \hat{\boldsymbol{y}} + \cos\theta \hat{\boldsymbol{z}}\right) = r\hat{\boldsymbol{r}}$$

Note that the polar angle $\theta \in (\pi/2, \pi]$.

In Example G.1 on page 685, we analyze this integral and its asymptotic behavior for a large argument in the exponential function. Far away from the origin, i.e., $k_0 r \to \infty$, we get to leading order, see (G.2) on page 686

$$\boldsymbol{E}_{\mathrm{r}}(\boldsymbol{r}) = -\frac{\mathrm{i}k_0 |\cos\theta|}{2\pi r} \left(\mathbf{I}_2 + \frac{\hat{\boldsymbol{z}}\boldsymbol{k}_{\mathrm{t}s}}{k_z}\right) \cdot \mathbf{r}(\boldsymbol{k}_{\mathrm{t}s}) \cdot \boldsymbol{F}^{+}(\boldsymbol{k}_{\mathrm{t}s}, 0) \mathrm{e}^{\mathrm{i}k_0 r}, \qquad \theta \in (\pi/2, \pi] \tag{10.46}$$

where the longitudinal wave number k_z in the reflection dyadic $\mathbf{r}(\boldsymbol{k}_{\mathrm{t}})$ is evaluated at the stationary point $\boldsymbol{k}_{\mathrm{t}s}$

$$\begin{cases} \boldsymbol{k}_{\mathrm{t}s} = k_0(\hat{\boldsymbol{x}} \sin\theta \cos\phi + \hat{\boldsymbol{y}} \sin\theta \sin\phi) \\ k_z = k_0 |\cos\theta| \end{cases}$$

The power transport at large distances from the origin then becomes

$$\begin{aligned} \langle \boldsymbol{S}(t) \rangle &= \frac{1}{2} \operatorname{Re}\left\{\boldsymbol{E}_{\mathrm{r}}(\boldsymbol{r}) \times \boldsymbol{H}_{\mathrm{r}}^{*}(\boldsymbol{r}, \omega)\right\} = \frac{1}{2k_0\eta_0} \operatorname{Re}\left\{\boldsymbol{E}_{\mathrm{r}}(\boldsymbol{r}) \times (\boldsymbol{k}^{-} \times \boldsymbol{E}_{\mathrm{r}}^{*}(\boldsymbol{r}))\right\} \\ &= \hat{\boldsymbol{r}} \frac{k_0^2}{8\pi^2 r^2 \eta_0} \left(\left|\mathbf{r}(\boldsymbol{k}_{\mathrm{t}s}) \cdot \boldsymbol{F}^{+}(\boldsymbol{k}_{\mathrm{t}s}, 0)\right|_{\parallel}^2 + \left|\mathbf{r}(\boldsymbol{k}_{\mathrm{t}s}) \cdot \boldsymbol{F}^{+}(\boldsymbol{k}_{\mathrm{t}s}, 0)\right|_{\perp}^2 \cos^2\theta \right) \end{aligned}$$

since $\boldsymbol{k}^{-} = \boldsymbol{k}_{\mathrm{t}s} - \hat{\boldsymbol{z}} k_z = k_0 \hat{\boldsymbol{r}}$.

Similarly, the transmitted field is, see (10.33)

$$\boldsymbol{E}_{\mathrm{t}}(\boldsymbol{r}) = \frac{1}{4\pi^2} \iint\limits_{\mathbb{R}^2} \left(\mathbf{I}_2 - \frac{\hat{\boldsymbol{z}}\boldsymbol{k}_{\mathrm{t}}}{k_z}\right) \cdot \mathbf{t} \cdot \boldsymbol{F}^{+}(\boldsymbol{k}_{\mathrm{t}}, 0) \mathrm{e}^{\mathrm{i}\boldsymbol{k}^{+}\cdot\boldsymbol{r} - \mathrm{i}k_z d} \, \mathrm{d}k_x \, \mathrm{d}k_y, \qquad z \ge d$$

The transmitted field evaluated far away from the surface $z = d$ is

$$E_t(r) = -\frac{ik_0 \cos\theta}{2\pi r} \cdot \left(\mathbf{I}_2 - \frac{\hat{z}k_{ts}}{k_z}\right) \cdot \mathbf{t}(k_{ts}) \cdot F^+(k_{ts}, 0)e^{-ik_0 d\cos\theta}e^{ik_0 r}, \qquad \theta \in [0, \pi/2) \quad (10.47)$$

where the stationary point for the transmitted field is

$$\begin{cases} k_{ts} = k_0(\hat{x}\sin\theta\cos\phi + \hat{y}\sin\theta\sin\phi) \\ k_z = k_0\cos\theta \end{cases}$$

The power transport at large distances from the origin is

$$\begin{aligned}\langle S(t)\rangle &= \frac{1}{2}\operatorname{Re}\left\{E_t(r) \times H_t^*(r,\omega)\right\} = \frac{1}{2k_0\eta_0}\operatorname{Re}\left\{E_t(r) \times (k^+ \times E_t^*(r))\right\} \\ &= \hat{r}\frac{k_0^2}{8\pi^2 r^2\eta_0}\left(\left|\mathbf{t}(k_{ts}) \cdot F^+(k_{ts},0)\right|_{\parallel}^2 + \left|\mathbf{t}(k_{ts}) \cdot F^+(k_{ts},0)\right|_{\perp}^2 \cos^2\theta\right)\end{aligned}$$

Example 10.9

A simple illustration of the result in this section is a vertical magnetic dipole above an isotropic half-space with permittivity ϵ and permeability μ. This is a configuration of great importance in many geophysical applications. The geometry is depicted in Figure 10.12 on page 554.

If we assume that the dipole is located at a height $h > 0$ above the surface $z = 0$, i.e., at $r = -h\hat{z}$, the incident field (the field has no z-component) is, see Example 10.7

$$E_i(r) = -\frac{1}{4\pi^2}\iint_{\mathbb{R}^2} \frac{i\omega\mu_0 m k_t}{2k_z}\hat{e}_\perp e^{ik_t\cdot r_c + ik_z|z+h|}\, dk_x\, dk_y, \qquad z \le 0$$

and the Fourier transform at the surface $z = 0$ is easily calculated

$$F^+(k_t, 0) = -i\hat{e}_\perp\frac{\omega\mu_0 m k_t}{2k_z}e^{ik_z h}$$

The dominant contribution to the incident field at large distance, $z \le -h$, is, see (G.2) on page 686

$$E_i(r) = -\hat{\phi}\frac{\omega\mu_0 m k_0 \sin\theta}{4\pi}\frac{e^{ik_0 r}}{r}e^{-ik_0 h|\cos\theta|}, \qquad \theta \in (\pi/2, \pi]$$

The $\perp$-components of the reflection and transmission dyadics are

$$r_\perp(k_t) = \frac{\mu k_z - \tilde{k}_z}{\mu k_z + \tilde{k}_z}, \qquad t_\perp(k_t) = 1 + r_\perp(k_t) = \frac{2\mu k_z}{\mu k_z + \tilde{k}_z}$$

where we used the notation

$$\tilde{k}_z = \left(k_0^2\epsilon\mu - k_t^2\right)^{1/2} = \left(k^2 - k_t^2\right)^{1/2}$$

and $k = k_0(\epsilon\mu)^{1/2}$.

We first deal with the reflected field $\boldsymbol{E}_{\mathrm{r}}(\boldsymbol{r})$ at $z \leq 0$. Evaluate $\boldsymbol{F}^{+}(\boldsymbol{k}_{\mathrm{t}}, 0)$ at the stationary point $\boldsymbol{k}_{\mathrm{t}s} = k_0(\hat{\boldsymbol{x}} \sin\theta \cos\phi + \hat{\boldsymbol{y}} \sin\theta \sin\phi)$. The result is ($|\boldsymbol{k}_{\mathrm{t}s}| = k_0 \sin\theta$)

$$\boldsymbol{F}^{+}(\boldsymbol{k}_{\mathrm{t}s}, 0) = -\mathrm{i}\hat{\boldsymbol{e}}_{\perp} \frac{\omega\mu_0 m \sin\theta}{2|\cos\theta|} \mathrm{e}^{\mathrm{i}k_0 h|\cos\theta|}, \qquad \theta \in (\pi/2, \pi]$$

Moreover, the reflection coefficient evaluated at the stationary point is

$$r_{\perp}(\boldsymbol{k}_{\mathrm{t}s}) = \frac{\mu|\cos\theta| - \sqrt{k^2/k_0^2 - \sin^2\theta}}{\mu|\cos\theta| + \sqrt{k^2/k_0^2 - \sin^2\theta}}, \qquad \theta \in (\pi/2, \pi]$$

and the dominant contribution to the reflected field in the region $z \leq -h$ far away from the interface is, see (10.46)

$$\boldsymbol{E}_{\mathrm{r}}(\boldsymbol{r}) = -\hat{\boldsymbol{\phi}} \frac{\mathrm{i}k_0|\cos\theta|}{2\pi r} r_{\perp}(\boldsymbol{k}_{\mathrm{t}s}) F_{\perp}^{+}(\boldsymbol{k}_{\mathrm{t}s}, 0) \mathrm{e}^{\mathrm{i}k_0 r}, \qquad \theta \in (\pi/2, \pi]$$

The asymptotic behavior of the total field therefore is

$$\boldsymbol{E}_1(\boldsymbol{r}) = -\frac{\mathrm{e}^{\mathrm{i}k_0 r}}{4\pi r} \hat{\boldsymbol{\phi}} \omega\mu_0 m k_0 \sin\theta \left(\mathrm{e}^{-\mathrm{i}k_0 h|\cos\theta|} + r_{\perp}(\boldsymbol{k}_{\mathrm{t}s}) \mathrm{e}^{\mathrm{i}k_0 h|\cos\theta|}\right), \qquad \theta \in (\pi/2, \pi]$$

The computation of the transmitted field below the interface follows the same line, see (10.47), but the wave number now is k instead of k_0. With appropriate alterations for this new wave number, we get

$$\boldsymbol{E}_{\mathrm{t}}(\boldsymbol{r}) = -\hat{\boldsymbol{\phi}} \frac{\mathrm{i}k\cos\theta}{2\pi r} t_{\perp}(\boldsymbol{k}_{\mathrm{t}s}) F_{\perp}^{+}(\boldsymbol{k}_{\mathrm{t}s}, 0) \mathrm{e}^{\mathrm{i}kr}, \qquad \theta \in [0, \pi/2)$$

where the stationary point now is $\boldsymbol{k}_{\mathrm{t}s} = k(\hat{\boldsymbol{x}} \sin\theta \cos\phi + \hat{\boldsymbol{y}} \sin\theta \sin\phi)$. Evaluate $\boldsymbol{F}^{+}$ at the stationary point. The result is ($|\boldsymbol{k}_{\mathrm{t}s}| = k \sin\theta$)

$$\boldsymbol{F}^{+}(\boldsymbol{k}_{\mathrm{t}s}, 0) = -\mathrm{i}\hat{\boldsymbol{\phi}} \frac{\omega\mu_0 m \sin\theta}{2\sqrt{k_0^2/k^2 - \sin^2\theta}} \mathrm{e}^{\mathrm{i}kh\sqrt{k_0^2/k^2 - \sin^2\theta}}, \qquad \theta \in [0, \pi/2)$$

where the transmission coefficient $t_{\perp}(\boldsymbol{k}_{\mathrm{t}s})$ evaluated at the stationary point is

$$t_{\perp}(\boldsymbol{k}_{\mathrm{t}s}) = \frac{2\mu\sqrt{k_0^2/k^2 - \sin^2\theta}}{\mu\sqrt{k_0^2/k^2 - \sin^2\theta} + \cos\theta}, \qquad \theta \in [0, \pi/2)$$

The explicit expression is

$$\boldsymbol{E}^{\mathrm{t}}(\boldsymbol{r}) = -\frac{\mathrm{e}^{\mathrm{i}kr}}{4\pi r} \hat{\boldsymbol{\phi}} \frac{\omega\mu_0 m \sin\theta k \cos\theta}{\sqrt{k_0^2/k^2 - \sin^2\theta}} t_{\perp}(\boldsymbol{k}_{\mathrm{t}s}) \mathrm{e}^{\mathrm{i}kh\sqrt{k_0^2/k^2 - \sin^2\theta}}, \qquad \theta \in [0, \pi/2)$$

In Figure 10.16, we illustrate the power flux density in two lossless, isotropic half-spaces. ■

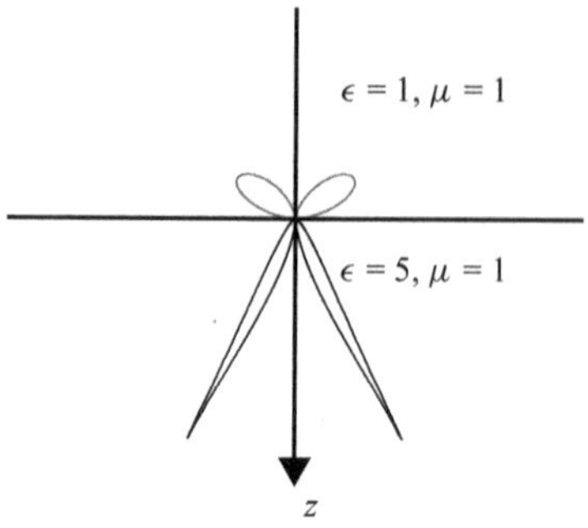

Figure 10.16 The far field radiation pattern of the total field in the half-spaces $z \leq 0$ and $z \geq 0$, respectively, from a vertical magnetic dipole. The dipole is located in medium 1 at $h = \lambda/4$ (λ is the wavelength in vacuum) above the surface $z = 0$. The parameters of the material below the interface are $\epsilon = 5$ and $\mu = 1$. Note that there is no transport of power along the surface, since the field along the surface attenuates as $1/r^2$ at large distances along the surface, and that most power is transmitted into the lower half-space. The scale is arbitrary.

10.9 The Green Dyadic

In the previous sections, the source region was located outside the slab. In this section, we generalize the analysis presented above to include a situation where the sources are located inside the slab. More explicitly, we aim at finding the Green dyadic for a plane-stratified medium.

In analogy to the scattering geometry described above, the positions of the interfaces are $z_j, j = -2, -1, 1, 2$. In total, there are four parallel interfaces, see Figure 10.17, and let $\mathbf{M}_j$, $j = -2, -1, 0, 1, 2$, be the fundamental dyadics of the corresponding five regions, respectively. It is assumed that all slabs are either homogeneous or have material parameters that vary with depth only, and moreover, that the regions $j = -2, 0, 2$ are empty, i.e., $\mathbf{M}_{-2} = \mathbf{M}_0 = \mathbf{M}_2$, see (10.14) on page 526, which means that $\boldsymbol{\epsilon} = \boldsymbol{\mu} = \mathbf{I}_3$ and $\boldsymbol{\xi} = \boldsymbol{\zeta} = \mathbf{0}$ in these regions. The other material dyadics $\mathbf{M}_{-1}$ and $\mathbf{M}_1$ model a general bianisotropic material. The intention is to imbed the sources in the empty region 0, and eventually, if necessary, let the thickness of this layer approach zero.

The Green dyadic for the electric field in an empty region ($k^2 = k_0^2$) satisfies (2.27) on page 113

$$\nabla \times \left(\nabla \times \mathbf{G}_{\mathrm{e}}(k_0, |\boldsymbol{r} - \boldsymbol{r}'|)\right) - k_0^2 \mathbf{G}_{\mathrm{e}}(k_0, |\boldsymbol{r} - \boldsymbol{r}'|) = \mathbf{I}_3 \delta(\boldsymbol{r} - \boldsymbol{r}') \tag{10.48}$$

where the source point is assumed to be located in region 0, i.e., $z_{-1} < z' < z_1$. We investigate the general solution to this equation. Any additional conditions on the solution, e.g., radiation conditions are applied later. The standard technique to solve the equation is to find a particular solution and then the solution to the homogeneous equation. We start with the particular solution.

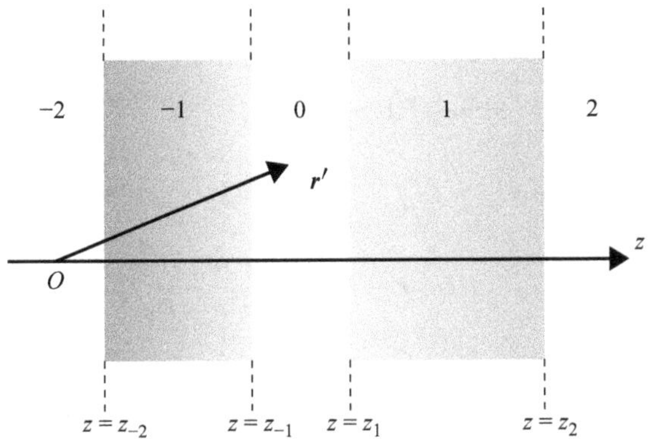

Figure 10.17 The geometry of a stratified structure of two layers. Regions −2, 0, and 2 are empty. The slabs −1 and 1 can be homogeneous or the material parameters vary w.r.t. depth z. The source point $\boldsymbol{r}'$ is located in the region 0 between the planes $z = z_{-1}$ and $z = z_1$.

10.9.1 Particular Solution or Free-Space Solution

A convenient way of finding a particular solution in free space to (10.48) is to use the explicit solution of the scalar Green function. From (2.28) on page 114, we have that one solution of (10.48) has the form

$$\mathbf{G}_{\mathrm{e}}(k_0, |\boldsymbol{r} - \boldsymbol{r}'|) = \left(\mathbf{I}_3 + \frac{1}{k_0^2}\nabla\nabla\right) g(k_0, |\boldsymbol{r} - \boldsymbol{r}'|)$$

where

$$\nabla^2 g(k_0, |\boldsymbol{r} - \boldsymbol{r}'|) + k_0^2 g(k_0, |\boldsymbol{r} - \boldsymbol{r}'|) = -\delta(\boldsymbol{r} - \boldsymbol{r}')$$

A Fourier transformation in the lateral variables, x and y, of this scalar equation gives an ODE in z. The Fourier transform of the Green function $g(k_0, |\boldsymbol{r} - \boldsymbol{r}'|)$ is in analogy with the notation above

$$g(\boldsymbol{k}_{\mathrm{t}}, z, \boldsymbol{r}') = \iint_{\mathbb{R}^2} g(k_0, |\boldsymbol{r} - \boldsymbol{r}'|) \mathrm{e}^{-\mathrm{i}\boldsymbol{k}_{\mathrm{t}}\cdot\boldsymbol{r}_{\mathrm{c}}} \, \mathrm{d}x \, \mathrm{d}y$$

The inverse Fourier transform is

$$g(k_0, |\boldsymbol{r} - \boldsymbol{r}'|) = \frac{1}{4\pi^2} \iint_{\mathbb{R}^2} g(\boldsymbol{k}_{\mathrm{t}}, z, \boldsymbol{r}') \mathrm{e}^{\mathrm{i}\boldsymbol{k}_{\mathrm{t}}\cdot\boldsymbol{r}_{\mathrm{c}}} \, \mathrm{d}k_x \, \mathrm{d}k_y$$

The solution, $g(\boldsymbol{k}_{\mathrm{t}}, z, \boldsymbol{r}')$, has to satisfy the following two requirements, see e.g., Reference 252:

$$\begin{cases} \lim\limits_{\epsilon\to 0^+} g(\boldsymbol{k}_{\mathrm{t}}, z' + \epsilon, \boldsymbol{r}') - g(\boldsymbol{k}_{\mathrm{t}}, z' - \epsilon, \boldsymbol{r}') = 0 \\ \lim\limits_{\epsilon\to 0^+} \dfrac{\mathrm{d}}{\mathrm{d}z} g(\boldsymbol{k}_{\mathrm{t}}, z' + \epsilon, \boldsymbol{r}') - \dfrac{\mathrm{d}}{\mathrm{d}z} g(\boldsymbol{k}_{\mathrm{t}}, z' - \epsilon, \boldsymbol{r}') = -\mathrm{e}^{-\mathrm{i}\boldsymbol{k}_{\mathrm{t}}\cdot\boldsymbol{r}'_{\mathrm{c}}} \end{cases}$$

A particular solution that also satisfies the standard out-going radiation conditions is

$$g(\boldsymbol{k}_t, z, \boldsymbol{r}') = \frac{i}{2k_z} e^{ik_z|z-z'|} e^{-i\boldsymbol{k}_t \cdot \boldsymbol{r}'_c}$$

Here, $z_{-1} < z' < z_1$ and $k_z = \left(k_0^2 - k_t^2\right)^{1/2}$, where $\operatorname{Im} k_z \geq 0$. This solution in physical space is

$$g(k_0, |\boldsymbol{r} - \boldsymbol{r}'|) = \frac{e^{ik_0|\boldsymbol{r}-\boldsymbol{r}'|}}{4\pi |\boldsymbol{r} - \boldsymbol{r}'|}$$

just as expected.

A particular solution $\mathbf{G}_e$ to (10.48) in region 0 in the Fourier domain is also easily obtained ($\nabla \longrightarrow i\boldsymbol{k}_t + \hat{z}\partial_z$)

$$\begin{aligned}
\mathbf{G}_e(\boldsymbol{k}_t, z, \boldsymbol{r}') &= \frac{1}{k_0^2}\left(\mathbf{I}_3 k_0^2 - \boldsymbol{k}_t\boldsymbol{k}_t + \hat{z}\hat{z}\frac{d^2}{dz^2} + i\left(\hat{z}\boldsymbol{k}_t + \boldsymbol{k}_t\hat{z}\right)\frac{d}{dz}\right) g(\boldsymbol{k}_t, z, \boldsymbol{r}') \\
&= \frac{i}{2k_0^2 k_z}\left(\mathbf{I}_3 k_0^2 - \boldsymbol{k}_t\boldsymbol{k}_t - k_z^2\hat{z}\hat{z}\right) e^{ik_z|z-z'|} e^{-i\boldsymbol{k}_t\cdot\boldsymbol{r}'_c} \\
&\quad - \operatorname{sign}(z - z')\frac{i}{2k_0^2}\left(\hat{z}\boldsymbol{k}_t + \boldsymbol{k}_t\hat{z}\right) e^{ik_z|z-z'|} e^{-i\boldsymbol{k}_t\cdot\boldsymbol{r}'_c} - \frac{1}{k_0^2}\hat{z}\hat{z}\delta(z - z') e^{-i\boldsymbol{k}_t\cdot\boldsymbol{r}'_c}
\end{aligned} \tag{10.49}$$

where the delta function $\delta(z - z')$ emerges from the differentiation w.r.t. z twice. This particular solution has the components, see Section 10.1.1

$$\mathbf{G}_e = \mathbf{G}_{e\perp\perp} + \hat{z}\boldsymbol{G}_{ez} + \boldsymbol{G}_{e\perp}\hat{z} + \hat{z}G_{ezz}\hat{z}$$

where

$$\begin{cases}
\mathbf{G}_{e\perp\perp} = \dfrac{i}{2k_z}\left(\mathbf{I}_2 - \dfrac{\boldsymbol{k}_t\boldsymbol{k}_t}{k_0^2}\right) e^{ik_z|z-z'|} e^{-i\boldsymbol{k}_t\cdot\boldsymbol{r}'_c} = \dfrac{i\mathbf{W}}{2k_0} e^{ik_z|z-z'|} e^{-i\boldsymbol{k}_t\cdot\boldsymbol{r}'_c} \\[2ex]
\boldsymbol{G}_{ez} = \boldsymbol{G}_{e\perp} = -\operatorname{sign}(z - z')\dfrac{i\boldsymbol{k}_t}{2k_0^2} e^{ik_z|z-z'|} e^{-i\boldsymbol{k}_t\cdot\boldsymbol{r}'_c} \\[2ex]
G_{ezz} = \dfrac{ik_t^2}{2k_0^2 k_z} e^{ik_z|z-z'|} e^{-i\boldsymbol{k}_t\cdot\boldsymbol{r}'_c} - \dfrac{1}{k_0^2}\delta(z - z') e^{-i\boldsymbol{k}_t\cdot\boldsymbol{r}'_c}
\end{cases}$$

Note that

$$\boldsymbol{k}_t\boldsymbol{k}_t + k_z^2\hat{z}\hat{z} + k_z \operatorname{sign}(z - z')\left(\hat{z}\boldsymbol{k}_t + \boldsymbol{k}_t\hat{z}\right) = \boldsymbol{k}^{\pm}\boldsymbol{k}^{\pm}, \qquad z \gtrless z'$$

where $\boldsymbol{k}^{\pm}$ are defined in (10.21)

$$\boldsymbol{k}^{\pm} = \boldsymbol{k}_t \pm \hat{z}k_z$$

10.9.1.1 Definition of Planar Vector Waves

The particular solution of (10.49) is the free-space solution of the electric Green dyadic expanded in plane waves. This solution is most conveniently expressed in terms of two sets of dimensionless, vector-valued plane waves, $\varphi_j^{\pm}(k_t; r)$ and $\varphi_j^{\pm\dagger}(k_t; r)$, $j = 1, 2$, which are defined as

$$\begin{cases} \varphi_1^{\pm}(\boldsymbol{k}_t; \boldsymbol{r}) = \dfrac{\hat{z} \times \boldsymbol{k}_t}{4\pi \mathrm{i} k_t} \mathrm{e}^{\mathrm{i}\boldsymbol{k}_t\cdot \boldsymbol{r}_c \pm \mathrm{i}k_z z} \\ \varphi_2^{\pm}(\boldsymbol{k}_t; \boldsymbol{r}) = \dfrac{\mp \boldsymbol{k}_t k_z + k_t^2 \hat{z}}{4\pi k_0 k_t} \mathrm{e}^{\mathrm{i}\boldsymbol{k}_t\cdot \boldsymbol{r}_c \pm \mathrm{i}k_z z} \end{cases}$$
$$\begin{cases} \varphi_1^{\pm\dagger}(\boldsymbol{k}_t; \boldsymbol{r}) = -\dfrac{\hat{z} \times \boldsymbol{k}_t}{4\pi \mathrm{i} k_t} \mathrm{e}^{-\mathrm{i}\boldsymbol{k}_t\cdot \boldsymbol{r}_c \pm \mathrm{i}k_z z} \\ \varphi_2^{\pm\dagger}(\boldsymbol{k}_t; \boldsymbol{r}) = \dfrac{\pm \boldsymbol{k}_t k_z + k_t^2 \hat{z}}{4\pi k_0 k_t} \mathrm{e}^{-\mathrm{i}\boldsymbol{k}_t\cdot \boldsymbol{r}_c \pm \mathrm{i}k_z z} \end{cases} \tag{10.50}$$

where, as above, $k_t = |\boldsymbol{k}_t|$, and k_z is defined by (10.4). Both solutions, $\varphi_j^{\pm}(\boldsymbol{k}_t; \boldsymbol{r})$ and $\varphi_j^{\pm\dagger}(\boldsymbol{k}_t; \boldsymbol{r})$, $j = 1, 2$, satisfy the electric field equation, i.e.,

$$\nabla \times \left(\nabla \times \varphi_j^{\pm}(\boldsymbol{k}_t; \boldsymbol{r})\right) - k_0^2 \varphi_j^{\pm}(\boldsymbol{k}_t; \boldsymbol{r}) = \boldsymbol{0}$$

The plus super index denotes an exponentially decreasing inhomogeneous (evanescent) wave as $z \to \infty$ and similar for the minus super index as $z \to -\infty$. The index $j = 1$ labels the TE-waves, and $j = 2$ labels the TM-waves. Notice that the symbol dagger (†) corresponds to the transformation $\boldsymbol{k}_t \to -\boldsymbol{k}_t$, i.e.,

$$\varphi_j^{\pm\dagger}(\boldsymbol{k}_t; \boldsymbol{r}) = \varphi_j^{\pm}(-\boldsymbol{k}_t; \boldsymbol{r})$$

and that $\boldsymbol{k}^{\pm} \cdot \varphi_j^{\pm}(\boldsymbol{k}_t; \boldsymbol{r}) = \boldsymbol{k}^{\mp} \cdot \varphi_j^{\pm\dagger}(\boldsymbol{k}_t; \boldsymbol{r}) = 0$, where $\boldsymbol{k}^{\pm} = \boldsymbol{k}_t \pm k_z \hat{z}$. Notice also that

$$\begin{cases} \nabla \times \varphi_1^{\pm}(\boldsymbol{k}_t; \boldsymbol{r}) = k_0 \varphi_2^{\pm}(\boldsymbol{k}_t; \boldsymbol{r}) \\ \nabla \times \varphi_2^{\pm}(\boldsymbol{k}_t; \boldsymbol{r}) = k_0 \varphi_1^{\pm}(\boldsymbol{k}_t; \boldsymbol{r}) \end{cases} \qquad \begin{cases} \nabla \times \varphi_1^{\pm\dagger}(\boldsymbol{k}_t; \boldsymbol{r}) = k_0 \varphi_2^{\pm\dagger}(\boldsymbol{k}_t; \boldsymbol{r}) \\ \nabla \times \varphi_2^{\pm\dagger}(\boldsymbol{k}_t; \boldsymbol{r}) = k_0 \varphi_1^{\pm\dagger}(\boldsymbol{k}_t; \boldsymbol{r}) \end{cases}$$

and that these planar vector waves satisfy the following orthogonality relations on S, which is defined as $z = z_0$:

$$\begin{cases} \displaystyle\iint_S \varphi_j^{\mp}(\boldsymbol{k}_t; \boldsymbol{r}) \cdot \varphi_{j'}^{\pm\dagger}(\boldsymbol{k}_t'; \boldsymbol{r})\, \mathrm{d}x\, \mathrm{d}y = \frac{1}{4} \delta_{jj'} \delta(\boldsymbol{k}_t - \boldsymbol{k}_t') \\ \displaystyle\iint_S \left(\hat{z} \times \varphi_j^{\mp}(\boldsymbol{k}_t; \boldsymbol{r})\right) \cdot \varphi_{j'}^{\pm\dagger}(\boldsymbol{k}_t'; \boldsymbol{r})\, \mathrm{d}x\, \mathrm{d}y = \mp \frac{k_z}{4\mathrm{i}k_0} \delta_{\bar{j}j'} \delta(\boldsymbol{k}_t - \boldsymbol{k}_t') \\ \displaystyle\iint_S \left(\hat{z} \times \varphi_j^{\pm}(\boldsymbol{k}_t; \boldsymbol{r})\right) \cdot \varphi_{j'}^{\pm\dagger}(\boldsymbol{k}_t'; \boldsymbol{r})\, \mathrm{d}x\, \mathrm{d}y = \pm \frac{(-1)^j k_z}{4\mathrm{i}k_0} \mathrm{e}^{\pm 2\mathrm{i}k_z z_0} \delta_{\bar{j}j'} \delta(\boldsymbol{k}_t - \boldsymbol{k}_t') \end{cases} \tag{10.51}$$

where the dual index $\bar{j}$ is defined $\bar{1} = 2$ and $\bar{2} = 1$.

The following dyadic identity is also useful:

$$\begin{aligned}
&\frac{\hat{z}\times\boldsymbol{k}_\mathrm{t}}{k_\mathrm{t}}\frac{\hat{z}\times\boldsymbol{k}_\mathrm{t}}{k_\mathrm{t}}+\frac{\mp\boldsymbol{k}_\mathrm{t}k_z+k_\mathrm{t}^2\hat{z}}{k_0k_\mathrm{t}}\frac{\mp\boldsymbol{k}_\mathrm{t}k_z+k_\mathrm{t}^2\hat{z}}{k_0k_\mathrm{t}}\\
&=\frac{\left(k_0^2-k_\mathrm{t}^2\right)\boldsymbol{k}_\mathrm{t}\boldsymbol{k}_\mathrm{t}+k_\mathrm{t}^2\left(k_0^2-k_z^2\right)\hat{z}\hat{z}\mp k_\mathrm{t}^2k_z\left(\hat{z}\boldsymbol{k}_\mathrm{t}+\boldsymbol{k}_\mathrm{t}\hat{z}\right)}{k_0^2k_\mathrm{t}^2}+\frac{\left(\hat{z}\times\boldsymbol{k}_\mathrm{t}\right)\left(\hat{z}\times\boldsymbol{k}_\mathrm{t}\right)}{k_\mathrm{t}^2}\\
&=\mathbf{I}_3-\frac{1}{k_0^2}\left\{\boldsymbol{k}_\mathrm{t}\boldsymbol{k}_\mathrm{t}+k_z^2\hat{z}\hat{z}\pm k_z\left(\hat{z}\boldsymbol{k}_\mathrm{t}+\boldsymbol{k}_\mathrm{t}\hat{z}\right)\right\}
\end{aligned}\tag{10.52}$$

since $\mathbf{I}_3=\boldsymbol{k}_\mathrm{t}\boldsymbol{k}_\mathrm{t}/k_\mathrm{t}^2+\left(\hat{z}\times\boldsymbol{k}_\mathrm{t}\right)\left(\hat{z}\times\boldsymbol{k}_\mathrm{t}\right)/k_\mathrm{t}^2+\hat{z}\hat{z}$ is a decomposition of the unit dyadic in cylindrical coordinates.

The electric Green dyadic in free space is

$$\mathbf{G}_\mathrm{e}(k_0,|\boldsymbol{r}-\boldsymbol{r}'|)=\frac{1}{4\pi^2}\iint_{\mathbb{R}^2}\mathbf{G}_\mathrm{e}(\boldsymbol{k}_\mathrm{t},z,\boldsymbol{r}')e^{\mathrm{i}\boldsymbol{k}_\mathrm{t}\cdot\boldsymbol{r}_\mathrm{c}}\,\mathrm{d}k_x\,\mathrm{d}k_y\tag{10.53}$$

For $z>z'$, we have, see (10.49) and (10.52)

$$\begin{aligned}
\mathbf{G}_\mathrm{e}(k_0,|\boldsymbol{r}-\boldsymbol{r}'|)&=\frac{1}{8\pi^2k_0^2}\iint_{\mathbb{R}^2}\Big\{\mathbf{I}_3k_0^2-\boldsymbol{k}_\mathrm{t}\boldsymbol{k}_\mathrm{t}-k_z^2\hat{z}\hat{z}\\
&\qquad-k_z\left(\hat{z}\boldsymbol{k}_\mathrm{t}+\boldsymbol{k}_\mathrm{t}\hat{z}\right)\Big\}\frac{\mathrm{i}}{k_z}e^{\mathrm{i}\boldsymbol{k}_\mathrm{t}\cdot(\boldsymbol{r}_\mathrm{c}-\boldsymbol{r}_\mathrm{c}')+\mathrm{i}k_z(z-z')}\,\mathrm{d}k_x\,\mathrm{d}k_y\\
&=2\mathrm{i}\sum_{j=1,2}\iint_{\mathbb{R}^2}\boldsymbol{\varphi}_j^+(\boldsymbol{k}_\mathrm{t};\boldsymbol{r})\boldsymbol{\varphi}_j^{-\dagger}(\boldsymbol{k}_\mathrm{t};\boldsymbol{r}')\frac{\mathrm{d}k_x\,\mathrm{d}k_y}{k_z},\qquad z>z'
\end{aligned}$$

and for $z<z'$, we have

$$\begin{aligned}
\mathbf{G}_\mathrm{e}(k_0,|\boldsymbol{r}-\boldsymbol{r}'|)&=\frac{1}{8\pi^2k_0^2}\iint_{\mathbb{R}^2}\Big\{\mathbf{I}_3k_0^2-\boldsymbol{k}_\mathrm{t}\boldsymbol{k}_\mathrm{t}-k_z^2\hat{z}\hat{z}\\
&\qquad+k_z\left(\hat{z}\boldsymbol{k}_\mathrm{t}+\boldsymbol{k}_\mathrm{t}\hat{z}\right)\Big\}\frac{\mathrm{i}}{k_z}e^{\mathrm{i}\boldsymbol{k}_\mathrm{t}\cdot(\boldsymbol{r}_\mathrm{c}-\boldsymbol{r}_\mathrm{c}')-\mathrm{i}k_z(z-z')}\,\mathrm{d}k_x\,\mathrm{d}k_y\\
&=2\mathrm{i}\sum_{j=1,2}\iint_{\mathbb{R}^2}\boldsymbol{\varphi}_j^-(\boldsymbol{k}_\mathrm{t};\boldsymbol{r})\boldsymbol{\varphi}_j^{+\dagger}(\boldsymbol{k}_\mathrm{t};\boldsymbol{r}')\frac{\mathrm{d}k_x\,\mathrm{d}k_y}{k_z},\qquad z<z'
\end{aligned}$$

which we combine to

$$\mathbf{G}_\mathrm{e}(k_0,|\boldsymbol{r}-\boldsymbol{r}'|)=2\mathrm{i}\sum_{j=1,2}\iint_{\mathbb{R}^2}\boldsymbol{\varphi}_j^\pm(\boldsymbol{k}_\mathrm{t};\boldsymbol{r})\boldsymbol{\varphi}_j^{\mp\dagger}(\boldsymbol{k}_\mathrm{t};\boldsymbol{r}')\frac{\mathrm{d}k_x\,\mathrm{d}k_y}{k_z}\tag{10.54}$$

where the upper (lower) is used if $z>z'$ $(z<z')$. This is the sought expansion in planar vector waves of the Green dyadic for the electric field in free space.

10.9.2 Homogeneous Solution in Free Space

The homogeneous solution to the Green dyadic for the electric field in an isotropic space $(k_1^2 = k_0^2)$ satisfies

$$\nabla \times \left(\nabla \times \mathbf{G}_\mathrm{e}(k_0, |\boldsymbol{r} - \boldsymbol{r}'|)\right) - k_0^2 \mathbf{G}_\mathrm{e}(k_0, |\boldsymbol{r} - \boldsymbol{r}'|) = \mathbf{0}$$

The Fourier transform of this equation and expansion of the cross products imply $(\nabla \longrightarrow \mathrm{i}\boldsymbol{k}_\mathrm{t} + \hat{\boldsymbol{z}}\partial_z$ and the BAC-CAB rule)

$$\left(\mathrm{i}\boldsymbol{k}_\mathrm{t} + \hat{\boldsymbol{z}}\frac{\mathrm{d}}{\mathrm{d}z}\right)\left(\left(\mathrm{i}\boldsymbol{k}_\mathrm{t} + \hat{\boldsymbol{z}}\frac{\mathrm{d}}{\mathrm{d}z}\right) \cdot \mathbf{G}_\mathrm{e}(\boldsymbol{k}_\mathrm{t}, z, \boldsymbol{r}')\right) - \left(k_z^2 + \frac{\mathrm{d}^2}{\mathrm{d}z^2}\right)\mathbf{G}_\mathrm{e}(\boldsymbol{k}_\mathrm{t}, z, \boldsymbol{r}') = \mathbf{0}$$

The Green dyadic is decomposed as described in Section 10.1.1, i.e.,

$$\mathbf{G}_\mathrm{e} = \mathbf{G}_{\mathrm{e}\perp\perp} + \hat{\boldsymbol{z}}\boldsymbol{G}_{\mathrm{e}z} + \boldsymbol{G}_{\mathrm{e}\perp}\hat{\boldsymbol{z}} + \hat{\boldsymbol{z}}G_{\mathrm{e}zz}\hat{\boldsymbol{z}}$$

This gives

$$\begin{aligned}&\left(\mathrm{i}\boldsymbol{k}_\mathrm{t} + \hat{\boldsymbol{z}}\frac{\mathrm{d}}{\mathrm{d}z}\right)\left(\mathrm{i}\boldsymbol{k}_\mathrm{t} \cdot \mathbf{G}_{\mathrm{e}\perp\perp} + \mathrm{i}\boldsymbol{k}_\mathrm{t} \cdot \boldsymbol{G}_{\mathrm{e}\perp}\hat{\boldsymbol{z}}\right) + \left(\mathrm{i}\boldsymbol{k}_\mathrm{t} + \hat{\boldsymbol{z}}\frac{\mathrm{d}}{\mathrm{d}z}\right)\left(\frac{\mathrm{d}}{\mathrm{d}z}\boldsymbol{G}_{\mathrm{e}z} + \frac{\mathrm{d}}{\mathrm{d}z}G_{\mathrm{e}zz}\hat{\boldsymbol{z}}\right)\\ &- \left(k_z^2 + \frac{\mathrm{d}^2}{\mathrm{d}z^2}\right)\left(\mathbf{G}_{\mathrm{e}\perp\perp} + \hat{\boldsymbol{z}}\boldsymbol{G}_{\mathrm{e}z} + \boldsymbol{G}_{\mathrm{e}\perp}\hat{\boldsymbol{z}} + \hat{\boldsymbol{z}}G_{\mathrm{e}zz}\hat{\boldsymbol{z}}\right) = \mathbf{0}\end{aligned}$$

This identity decomposes into four different parts or components. They are

$$\begin{cases} -\boldsymbol{k}_\mathrm{t}\boldsymbol{k}_\mathrm{t} \cdot \mathbf{G}_{\mathrm{e}\perp\perp} + \mathrm{i}\boldsymbol{k}_\mathrm{t}\dfrac{\mathrm{d}}{\mathrm{d}z}\boldsymbol{G}_{\mathrm{e}z} - \left(k_z^2 + \dfrac{\mathrm{d}^2}{\mathrm{d}z^2}\right)\mathbf{G}_{\mathrm{e}\perp\perp} = \mathbf{0} \\ \mathrm{i}\dfrac{\mathrm{d}}{\mathrm{d}z}\boldsymbol{k}_\mathrm{t} \cdot \mathbf{G}_{\mathrm{e}\perp\perp} - k_z^2\boldsymbol{G}_{\mathrm{e}z} = \mathbf{0} \\ -\boldsymbol{k}_\mathrm{t}\boldsymbol{k}_\mathrm{t} \cdot \boldsymbol{G}_{\mathrm{e}\perp} + \mathrm{i}\boldsymbol{k}_\mathrm{t}\dfrac{\mathrm{d}}{\mathrm{d}z}G_{\mathrm{e}zz} - \left(k_z^2 + \dfrac{\mathrm{d}^2}{\mathrm{d}z^2}\right)\boldsymbol{G}_{\mathrm{e}\perp} = \mathbf{0} \\ \mathrm{i}\dfrac{\mathrm{d}}{\mathrm{d}z}\boldsymbol{k}_\mathrm{t} \cdot \boldsymbol{G}_{\mathrm{e}\perp} - k_z^2 G_{\mathrm{e}zz} = 0 \end{cases}$$

Combining the first and the second equations gives

$$\begin{cases} \left(k_z^2 + \dfrac{\mathrm{d}^2}{\mathrm{d}z^2}\right)\mathbf{G}_{\mathrm{e}\perp\perp} = \mathbf{0} \\ \boldsymbol{G}_{\mathrm{e}z} = \dfrac{\mathrm{i}}{k_z^2}\dfrac{\mathrm{d}}{\mathrm{d}z}\boldsymbol{k}_\mathrm{t} \cdot \mathbf{G}_{\mathrm{e}\perp\perp} \end{cases}$$

Similarly, the third and the fourth equations imply

$$\begin{cases} \left(k_z^2 + \dfrac{\mathrm{d}^2}{\mathrm{d}z^2}\right)\boldsymbol{G}_{\mathrm{e}\perp} = \mathbf{0} \\ G_{\mathrm{e}zz} = \dfrac{\mathrm{i}}{k_z^2}\dfrac{\mathrm{d}}{\mathrm{d}z}\boldsymbol{k}_\mathrm{t} \cdot \boldsymbol{G}_{\mathrm{e}\perp} \end{cases}$$

The general solution to these systems is

$$\begin{cases} \mathbf{G}_{\mathrm{e}\perp\perp} = \mathbf{A}\mathrm{e}^{\mathrm{i}k_z z} + \mathbf{B}\mathrm{e}^{-\mathrm{i}k_z z} \\ \boldsymbol{G}_{\mathrm{e}z} = -\dfrac{1}{k_z}\boldsymbol{k}_\mathrm{t} \cdot \left(\mathbf{A}\mathrm{e}^{\mathrm{i}k_z z} - \mathbf{B}\mathrm{e}^{-\mathrm{i}k_z z}\right) \\ \boldsymbol{G}_{\mathrm{e}\perp} = \boldsymbol{a}\mathrm{e}^{\mathrm{i}k_z z} + \boldsymbol{b}\mathrm{e}^{-\mathrm{i}k_z z} \\ G_{\mathrm{e}zz} = -\dfrac{1}{k_z}\boldsymbol{k}_\mathrm{t} \cdot \left(\boldsymbol{a}\mathrm{e}^{\mathrm{i}k_z z} - \boldsymbol{b}\mathrm{e}^{-\mathrm{i}k_z z}\right) \end{cases} \tag{10.55}$$

where $\mathbf{A}$ and $\mathbf{B}$ are arbitrary two-dimensional dyadics in the x–y-plane (may depend on $\boldsymbol{r}'$ and other variables or parameters, but not z), and $\boldsymbol{a}$ and $\boldsymbol{b}$ are constant two-dimensional vectors in the x–y-plane.

10.9.3 General Solution

We are now ready to put the results from Sections 10.9.1 and 10.9.2 together and to find the solution that satisfies the boundary condition at the two interfaces $z = z_{-1}$ and $z = z_1$. The solution also has to satisfy the radiation conditions outside the slab.

The general solution to (10.48) is found by adding the general homogeneous solution, (10.55), and the particular solution, (10.49):

$$\begin{cases} \mathbf{G}_{\mathrm{e}\perp\perp} = \dfrac{\mathrm{i}}{2k_0}\left\{\mathbf{A}\mathrm{e}^{\mathrm{i}k_z(z-z')} + \mathbf{B}\mathrm{e}^{-\mathrm{i}k_z(z-z')} + \mathbf{W}\mathrm{e}^{\mathrm{i}k_z|z-z'|}\right\}\mathrm{e}^{-\mathrm{i}\boldsymbol{k}_\mathrm{t}\cdot\boldsymbol{r}'_\mathrm{c}} \\ \boldsymbol{G}_{\mathrm{e}z} = -\dfrac{\mathrm{i}}{2k_0}\left\{\dfrac{1}{k_z}\boldsymbol{k}_\mathrm{t}\cdot\left(\mathbf{A}\mathrm{e}^{\mathrm{i}k_z(z-z')} - \mathbf{B}\mathrm{e}^{-\mathrm{i}k_z(z-z')}\right) + \dfrac{\boldsymbol{k}_\mathrm{t}}{k_0}\operatorname{sign}(z-z')\mathrm{e}^{\mathrm{i}k_z|z-z'|}\right\}\mathrm{e}^{-\mathrm{i}\boldsymbol{k}_\mathrm{t}\cdot\boldsymbol{r}'_\mathrm{c}} \\ \boldsymbol{G}_{\mathrm{e}\perp} = \dfrac{\mathrm{i}}{2k_0}\left\{\boldsymbol{a}\mathrm{e}^{\mathrm{i}k_z(z-z')} + \boldsymbol{b}\mathrm{e}^{-\mathrm{i}k_z(z-z')} - \dfrac{\boldsymbol{k}_\mathrm{t}}{k_0}\operatorname{sign}(z-z')\mathrm{e}^{\mathrm{i}k_z|z-z'|}\right\}\mathrm{e}^{-\mathrm{i}\boldsymbol{k}_\mathrm{t}\cdot\boldsymbol{r}'_\mathrm{c}} \\ G_{\mathrm{e}zz} = -\dfrac{\mathrm{i}}{2k_0}\left\{\dfrac{1}{k_z}\boldsymbol{k}_\mathrm{t}\cdot\left(\boldsymbol{a}\mathrm{e}^{\mathrm{i}k_z(z-z')} - \boldsymbol{b}\mathrm{e}^{-\mathrm{i}k_z(z-z')}\right) - \dfrac{k_\mathrm{t}^2}{k_0k_z}\mathrm{e}^{\mathrm{i}k_z|z-z'|} - \dfrac{2\mathrm{i}}{k_0}\delta(z-z')\right\}\mathrm{e}^{-\mathrm{i}\boldsymbol{k}_\mathrm{t}\cdot\boldsymbol{r}'_\mathrm{c}} \end{cases}$$

where, for convenience, we have renormalized the dyadics $\mathbf{A}$ and $\mathbf{B}$ and the vectors $\boldsymbol{a}$ and $\boldsymbol{b}$ to obtain simple algebraic expressions in the calculations below (a factor of $\mathrm{i}\mathrm{e}^{-\mathrm{i}\boldsymbol{k}_\mathrm{t}\cdot\boldsymbol{r}_\mathrm{c}\pm k_z z'}/2k_0$ differs). We have also used the definition of the wave-splitting operator, see Section 10.3

$$\mathbf{I}_2 - \frac{\boldsymbol{k}_\mathrm{t}\boldsymbol{k}_\mathrm{t}}{k_0^2} = \frac{k_z}{k_0}\mathbf{W}(\boldsymbol{k}_\mathrm{t})$$

The two-dimensional dyadics $\mathbf{A}$ and $\mathbf{B}$ and the two-dimensional vectors $\boldsymbol{a}$ and $\boldsymbol{b}$ are determined by the radiation condition of the entire slab, i.e., in region -2 we have only left-going waves and in region 2 we have only right-going waves, see Section 10.3. In region 0, the correct relation between the right- and left-going parts of the solution is automatically fulfilled if we identify the right- and left-going waves of the Green dyadic and relate them with the reflection dyadic at the boundary of the slab, i.e., the reflection dyadic of the slab consisting of the region 1 for $z > z'$ and the region -1 for $z < z'$.

The action of $\mathbf{G}_e(\boldsymbol{k}_t, z, \boldsymbol{r}')$ on a vector field $\boldsymbol{J}(\boldsymbol{k}_t, z)$ is

$$\boldsymbol{F} = \mathbf{G}_e \cdot \boldsymbol{J} = \left(\mathbf{G}_{e\perp\perp} + \hat{z}\boldsymbol{G}_{ez} + \boldsymbol{G}_{e\perp}\hat{z} + \hat{z}G_{ezz}\hat{z}\right) \cdot \left(\boldsymbol{J}_{xy} + \hat{z}J_z\right)$$

We easily identify the transverse and normal components of the vector $\boldsymbol{F}$:

$$\begin{cases} \boldsymbol{F}_{xy} = \mathbf{G}_{e\perp\perp} \cdot \boldsymbol{J}_{xy} + \boldsymbol{G}_{e\perp} \cdot J_z \\ F_z = \boldsymbol{G}_{ez} \cdot \boldsymbol{J}_{xy} + G_{ezz}J_z \end{cases}$$

We proceed by decomposing the Green dyadic in right- and left-going waves. This decomposition is a generalization of the decomposition made in Section 10.3.

In analogy with (10.20), we have the decomposition

$$-\hat{z} \times \left(\hat{z} \times \mathbf{G}_e\right) = \mathbf{I}_2 \cdot \mathbf{G}_e = \mathbf{G}_e^+ + \mathbf{G}_e^-$$

where

$$\begin{cases} \mathbf{G}_e^+(z) = \dfrac{\mathrm{i}}{2k_0} \mathrm{e}^{-\mathrm{i}\boldsymbol{k}_t \cdot \boldsymbol{r}_c' + \mathrm{i}k_z(z-z')} \left(\mathbf{W} - \dfrac{\boldsymbol{k}_t\hat{z}}{k_0} + \mathbf{A} + \boldsymbol{a}\hat{z}\right) \\ \mathbf{G}_e^-(z) = \dfrac{\mathrm{i}}{2k_0} \mathrm{e}^{-\mathrm{i}\boldsymbol{k}_t \cdot \boldsymbol{r}_c' - \mathrm{i}k_z(z-z')} \left(\mathbf{B} + \boldsymbol{b}\hat{z}\right) \end{cases} \qquad z' < z < z_1$$

and

$$\begin{cases} \mathbf{G}_e^+(z) = \dfrac{\mathrm{i}}{2k_0} \mathrm{e}^{-\mathrm{i}\boldsymbol{k}_t \cdot \boldsymbol{r}_c' + \mathrm{i}k_z(z-z')} \left(\mathbf{A} + \boldsymbol{a}\hat{z}\right) \\ \mathbf{G}_e^-(z) = \dfrac{\mathrm{i}}{2k_0} \mathrm{e}^{-\mathrm{i}\boldsymbol{k}_t \cdot \boldsymbol{r}_c' - \mathrm{i}k_z(z-z')} \left(\mathbf{W} + \dfrac{\boldsymbol{k}_t\hat{z}}{k_0} + \mathbf{B} + \boldsymbol{b}\hat{z}\right) \end{cases} \qquad z_{-1} < z < z' \quad (10.56)$$

Note that $\mathbf{A}$ and $\mathbf{B}$ are two-dimensional dyadics in the x–y-plane. Similarly, the vectors $\boldsymbol{a}$ and $\boldsymbol{b}$ lie in the x–y-plane. We then have

$$\begin{aligned} \mathbf{G}_e(z) &= \left(\mathbf{I}_2 - \frac{\hat{z}\boldsymbol{k}_t}{k_z}\right) \cdot \mathbf{G}_e^+(z) + \left(\mathbf{I}_2 + \frac{\hat{z}\boldsymbol{k}_t}{k_z}\right) \cdot \mathbf{G}_e^-(z) \\ &= \boldsymbol{\gamma}^+(\boldsymbol{k}_t) \cdot \mathbf{G}_e^+(z) + \boldsymbol{\gamma}^-(\boldsymbol{k}_t) \cdot \mathbf{G}_e^-(z), \qquad z_{-1} < z < z_1, \quad z \neq z' \end{aligned}$$

where we used the projection dyadic $\gamma^{\pm}(\boldsymbol{k}_t)$ in (10.22), and

$$\frac{\boldsymbol{k}_t}{k_z} \cdot \mathbf{W} = \frac{\boldsymbol{k}_t}{k_0} \qquad (10.57)$$

The right- and left-going fields are related by the two reflection dyadics $\mathbf{r}_+$ and $\mathbf{r}_-$ for the two slabs $z > z_1$ and $z < z_{-1}$, respectively

$$\begin{cases} \mathbf{G}_e^-(\boldsymbol{k}_t, z_1, \boldsymbol{r}') = \mathbf{r}_+ \cdot \mathbf{G}_e^+(\boldsymbol{k}_t, z_1, \boldsymbol{r}') \\ \mathbf{G}_e^+(\boldsymbol{k}_t, z_{-1}, \boldsymbol{r}') = \mathbf{r}_- \cdot \mathbf{G}_e^-(\boldsymbol{k}_t, z_{-1}, \boldsymbol{r}') \end{cases}$$

or

$$\begin{cases} \mathbf{B} + b\hat{z} = \mathrm{e}^{2ik_z(z_1 - z')}\mathbf{r}_+ \cdot \left(\mathbf{W} - \dfrac{\boldsymbol{k}_t\hat{z}}{k_0} + \mathbf{A} + \boldsymbol{a}\hat{z}\right) \\ \mathbf{A} + \boldsymbol{a}\hat{z} = \mathrm{e}^{-2ik_z(z_{-1} - z')}\mathbf{r}_- \cdot \left(\mathbf{W} + \dfrac{\boldsymbol{k}_t\hat{z}}{k_0} + \mathbf{B} + b\hat{z}\right) \end{cases}$$

A decomposition in the transverse and normal parts gives the formal solution

$$\begin{cases} \mathbf{A} = \mathrm{e}^{-2ik_z(z_{-1} - z')}\mathbf{r}_- \cdot (\mathbf{W} + \mathbf{B}) \\ \mathbf{B} = \mathrm{e}^{2ik_z(z_1 - z')}\mathbf{r}_+ \cdot (\mathbf{W} + \mathbf{A}) \end{cases}$$

and

$$\begin{cases} \boldsymbol{a} = \mathrm{e}^{-2ik_z(z_{-1} - z')}\mathbf{r}_- \cdot \left(\dfrac{\boldsymbol{k}_t}{k_0} + \boldsymbol{b}\right) \\ \boldsymbol{b} = \mathrm{e}^{2ik_z(z_1 - z')}\mathbf{r}_+ \cdot \left(-\dfrac{\boldsymbol{k}_t}{k_0} + \boldsymbol{a}\right) \end{cases}$$

which we formally can solve for the unknown dyadics **A** and **B**, and the unknown vectors $\boldsymbol{a}$ and $\boldsymbol{b}$. The result is

$$\begin{cases} \mathbf{A} = \mathrm{e}^{-2ik_z(z_{-1} - z')}\left(\mathbf{I}_2 - \mathrm{e}^{2ik_z(z_1 - z_{-1})}\mathbf{r}_- \cdot \mathbf{r}_+\right)^{-1} \cdot \mathbf{r}_- \cdot \left(\mathbf{I}_2 + \mathrm{e}^{2ik_z(z_1 - z')}\mathbf{r}_+\right) \cdot \mathbf{W} \\ \mathbf{B} = \mathrm{e}^{2ik_z(z_1 - z')}\left(\mathbf{I}_2 - \mathrm{e}^{2ik_z(z_1 - z_{-1})}\mathbf{r}_+ \cdot \mathbf{r}_-\right)^{-1} \cdot \mathbf{r}_+ \cdot \left(\mathbf{I}_2 + \mathrm{e}^{-2ik_z(z_{-1} - z')}\mathbf{r}_-\right) \cdot \mathbf{W} \end{cases} \tag{10.58}$$

and

$$\begin{cases} \boldsymbol{a} = \mathrm{e}^{-2ik_z(z_{-1} - z')}\left(\mathbf{I}_2 - \mathrm{e}^{2ik_z(z_1 - z_{-1})}\mathbf{r}_- \cdot \mathbf{r}_+\right)^{-1} \cdot \mathbf{r}_- \cdot \left(\mathbf{I}_2 - \mathrm{e}^{2ik_z(z_1 - z')}\mathbf{r}_+\right) \cdot \dfrac{\boldsymbol{k}_t}{k_0} \\ \boldsymbol{b} = -\mathrm{e}^{2ik_z(z_1 - z')}\left(\mathbf{I}_2 - \mathrm{e}^{2ik_z(z_1 - z_{-1})}\mathbf{r}_+ \cdot \mathbf{r}_-\right)^{-1} \cdot \mathbf{r}_+ \cdot \left(\mathbf{I}_2 - \mathrm{e}^{-2ik_z(z_{-1} - z')}\mathbf{r}_-\right) \cdot \dfrac{\boldsymbol{k}_t}{k_0} \end{cases} \tag{10.59}$$

A special, very important, case is when there is no vacuum layer present, i.e., $z_{-1} = z_1$. (In fact, this situation is no loss of generality, since all other layers are general.) We have

$$\begin{cases} \mathbf{A} = (\mathbf{I}_2 - \mathbf{r}_- \cdot \mathbf{r}_+)^{-1} \cdot \mathbf{r}_- \cdot (\mathbf{I}_2 + \mathbf{r}_+) \cdot \mathbf{W} \\ \mathbf{B} = (\mathbf{I}_2 - \mathbf{r}_+ \cdot \mathbf{r}_-)^{-1} \cdot \mathbf{r}_+ \cdot (\mathbf{I}_2 + \mathbf{r}_-) \cdot \mathbf{W} \end{cases}$$

and

$$\begin{cases} \boldsymbol{a} = (\mathbf{I}_2 - \mathbf{r}_- \cdot \mathbf{r}_+)^{-1} \cdot \mathbf{r}_- \cdot (\mathbf{I}_2 - \mathbf{r}_+) \cdot \dfrac{\boldsymbol{k}_t}{k_0} \\ \boldsymbol{b} = -(\mathbf{I}_2 - \mathbf{r}_+ \cdot \mathbf{r}_-)^{-1} \cdot \mathbf{r}_+ \cdot (\mathbf{I}_2 - \mathbf{r}_-) \cdot \dfrac{\boldsymbol{k}_t}{k_0} \end{cases}$$

To summarize, the general solution of the Green dyadic in a bianisotropic slab is ($z_{-1} < z < z_1$)

$$
\begin{cases}
\mathbf{G}_{\mathrm{e}\perp\perp} = \dfrac{\mathrm{i}}{2k_0}\left\{\mathbf{A}\mathrm{e}^{\mathrm{i}k_z(z-z')} + \mathbf{B}\mathrm{e}^{-\mathrm{i}k_z(z-z')} + \mathbf{W}\mathrm{e}^{\mathrm{i}k_z|z-z'|}\right\}\mathrm{e}^{-\mathrm{i}\boldsymbol{k}_\mathrm{t}\cdot\boldsymbol{r}'_\mathrm{c}} \\
\boldsymbol{G}_{\mathrm{e}z} = -\dfrac{\mathrm{i}}{2k_0}\left\{\dfrac{1}{k_z}\boldsymbol{k}_\mathrm{t}\cdot\left(\mathbf{A}\mathrm{e}^{\mathrm{i}k_z(z-z')} - \mathbf{B}\mathrm{e}^{-\mathrm{i}k_z(z-z')}\right) + \dfrac{\boldsymbol{k}_\mathrm{t}}{k_0}\operatorname{sign}(z-z')\mathrm{e}^{\mathrm{i}k_z|z-z'|}\right\}\mathrm{e}^{-\mathrm{i}\boldsymbol{k}_\mathrm{t}\cdot\boldsymbol{r}'_\mathrm{c}} \\
\boldsymbol{G}_{\mathrm{e}\perp} = \dfrac{\mathrm{i}}{2k_0}\left\{\boldsymbol{a}\mathrm{e}^{\mathrm{i}k_z(z-z')} + \boldsymbol{b}\mathrm{e}^{-\mathrm{i}k_z(z-z')} - \dfrac{\boldsymbol{k}_\mathrm{t}}{k_0}\operatorname{sign}(z-z')\mathrm{e}^{\mathrm{i}k_z|z-z'|}\right\}\mathrm{e}^{-\mathrm{i}\boldsymbol{k}_\mathrm{t}\cdot\boldsymbol{r}'_\mathrm{c}} \\
G_{\mathrm{e}zz} = -\dfrac{\mathrm{i}}{2k_0}\left\{\dfrac{1}{k_z}\boldsymbol{k}_\mathrm{t}\cdot\left(\boldsymbol{a}\mathrm{e}^{\mathrm{i}k_z(z-z')} - \boldsymbol{b}\mathrm{e}^{-\mathrm{i}k_z(z-z')}\right) - \dfrac{k_\mathrm{t}^2}{k_0k_z}\mathrm{e}^{\mathrm{i}k_z|z-z'|} - \dfrac{2\mathrm{i}}{k_0}\delta(z-z')\right\}\mathrm{e}^{-\mathrm{i}\boldsymbol{k}_\mathrm{t}\cdot\boldsymbol{r}'_\mathrm{c}}
\end{cases}
$$

where $\mathbf{A}$, $\mathbf{B}$, $\boldsymbol{a}$, and $\boldsymbol{b}$ are explicitly given by (10.58) and (10.59). Finally, the Green dyadic $\mathbf{G}_\mathrm{e}(k_0, |\boldsymbol{r} - \boldsymbol{r}'|)$ for the bianisotropic slab is obtained by the inverse Fourier transform, see (10.53). A closed-form solution of this inverse Fourier transform is only known for a limit number of material configurations, and the solution to the general problem is not known. A few examples illustrate the theory.

Example 10.10
One important example is the case where the half-space $z < z_{-1}$ is vacuum, see the geometry in Figure 10.17. If the half-space is vacuum, we have $\mathbf{r}_- = \mathbf{0}$ and the dyadics $\mathbf{A}$ and $\mathbf{B}$ and the vectors $\boldsymbol{a}$ and $\boldsymbol{b}$ are simplified to

$$
\begin{cases}
\mathbf{A} = \mathbf{0} \\
\mathbf{B} = \mathrm{e}^{2\mathrm{i}k_z(z_1-z')}\mathbf{r}_+\cdot\mathbf{W}
\end{cases}
\qquad
\begin{cases}
\boldsymbol{a} = \mathbf{0} \\
\boldsymbol{b} = -\mathrm{e}^{2\mathrm{i}k_z(z_1-z')}\mathbf{r}_+\cdot\dfrac{\boldsymbol{k}_\mathrm{t}}{k_0}
\end{cases}
$$

and the Green dyadic is ($z < z_1$)

$$
\begin{cases}
\mathbf{G}_{\mathrm{e}\perp\perp} = \dfrac{\mathrm{i}}{2k_0}\left\{\mathbf{B}\mathrm{e}^{-\mathrm{i}k_z(z-z')} + \mathbf{W}\mathrm{e}^{\mathrm{i}k_z|z-z'|}\right\}\mathrm{e}^{-\mathrm{i}\boldsymbol{k}_\mathrm{t}\cdot\boldsymbol{r}'_\mathrm{c}} \\
\boldsymbol{G}_{\mathrm{e}z} = \dfrac{\mathrm{i}}{2k_0}\left\{\dfrac{\boldsymbol{k}_\mathrm{t}}{k_z}\cdot\mathbf{B}\mathrm{e}^{-\mathrm{i}k_z(z-z')} - \dfrac{\boldsymbol{k}_\mathrm{t}}{k_0}\operatorname{sign}(z-z')\mathrm{e}^{\mathrm{i}k_z|z-z'|}\right\}\mathrm{e}^{-\mathrm{i}\boldsymbol{k}_\mathrm{t}\cdot\boldsymbol{r}'_\mathrm{c}} \\
\boldsymbol{G}_{\mathrm{e}\perp} = \dfrac{\mathrm{i}}{2k_0}\left\{\boldsymbol{b}\mathrm{e}^{-\mathrm{i}k_z(z-z')} - \dfrac{\boldsymbol{k}_\mathrm{t}}{k_0}\operatorname{sign}(z-z')\mathrm{e}^{\mathrm{i}k_z|z-z'|}\right\}\mathrm{e}^{-\mathrm{i}\boldsymbol{k}_\mathrm{t}\cdot\boldsymbol{r}'_\mathrm{c}} \\
G_{\mathrm{e}zz} = \dfrac{\mathrm{i}}{2k_0}\left\{\dfrac{\boldsymbol{k}_\mathrm{t}}{k_z}\cdot\boldsymbol{b}\mathrm{e}^{-\mathrm{i}k_z(z-z')} + \dfrac{k_\mathrm{t}^2}{k_0k_z}\mathrm{e}^{\mathrm{i}k_z|z-z'|}\right\}\mathrm{e}^{-\mathrm{i}\boldsymbol{k}_\mathrm{t}\cdot\boldsymbol{r}'_\mathrm{c}} - \dfrac{1}{k_0^2}\delta(z-z')\mathrm{e}^{-\mathrm{i}\boldsymbol{k}_\mathrm{t}\cdot\boldsymbol{r}'_\mathrm{c}}
\end{cases}
$$

■

Example 10.11
In this example, we investigate the special form of the Green dyadic if the half-space $z < z_{-1}$ is vacuum, and the surface z_1 is perfectly conducting. This is a special case of Example 10.10,

with the reflection dyadic given by $\mathbf{r}_+ = -\mathbf{I}_2$, see Example 10.1. The Green dyadic then simplifies to $(z, z' < z_1)$

$$\begin{cases} \mathbf{G}_{\mathrm{e}\perp\perp} = \dfrac{\mathrm{i}}{2k_0}\mathbf{W}\left\{-\mathrm{e}^{\mathrm{i}k_z(2z_1-z-z')} + \mathrm{e}^{\mathrm{i}k_z|z-z'|}\right\}\mathrm{e}^{-\mathrm{i}\boldsymbol{k}_\mathrm{t}\cdot\boldsymbol{r}'_\mathrm{c}} \\ \boldsymbol{G}_{\mathrm{e}z} = \dfrac{\mathrm{i}\boldsymbol{k}_\mathrm{t}}{2k_0^2}\left\{-\mathrm{e}^{\mathrm{i}k_z(2z_1-z-z')} - \operatorname{sign}(z-z')\mathrm{e}^{\mathrm{i}k_z|z-z'|}\right\}\mathrm{e}^{-\mathrm{i}\boldsymbol{k}_\mathrm{t}\cdot\boldsymbol{r}'_\mathrm{c}} \\ \boldsymbol{G}_{\mathrm{e}\perp} = \dfrac{\mathrm{i}\boldsymbol{k}_\mathrm{t}}{2k_0^2}\left\{\mathrm{e}^{\mathrm{i}k_z(2z_1-z-z')} - \operatorname{sign}(z-z')\mathrm{e}^{\mathrm{i}k_z|z-z'|}\right\}\mathrm{e}^{-\mathrm{i}\boldsymbol{k}_\mathrm{t}\cdot\boldsymbol{r}'_\mathrm{c}} \\ G_{\mathrm{e}zz} = \dfrac{\mathrm{i}k_\mathrm{t}^2}{2k_zk_0^2}\left\{\mathrm{e}^{\mathrm{i}k_z(2z_1-z-z')} + \mathrm{e}^{\mathrm{i}k_z|z-z'|}\right\}\mathrm{e}^{-\mathrm{i}\boldsymbol{k}_\mathrm{t}\cdot\boldsymbol{r}'_\mathrm{c}} - \dfrac{1}{k_0^2}\delta(z-z')\mathrm{e}^{-\mathrm{i}\boldsymbol{k}_\mathrm{t}\cdot\boldsymbol{r}'_\mathrm{c}} \end{cases}$$

where we also used (10.57).

If we denote $\boldsymbol{r}'_0 = \boldsymbol{r}' + 2(z_1 - z')\hat{\boldsymbol{z}}$, which is the mirror point of the source point $\boldsymbol{r}'$ in the plane $z = z_1$. The explicit form of this Green dyadic then is, cf. (10.49) $(z, z' < z_1)$

$$\mathbf{G}_\mathrm{e}(k_0, \boldsymbol{r}, \boldsymbol{r}') = \left(\mathbf{I}_3 + \frac{1}{k_0^2}\nabla\nabla\right)\frac{\mathrm{e}^{\mathrm{i}k_0|\boldsymbol{r}-\boldsymbol{r}'|}}{4\pi\,|\boldsymbol{r}-\boldsymbol{r}'|} + \left(\mathbf{I}_3 + \frac{1}{k_0^2}\nabla\nabla\right)\frac{\mathrm{e}^{\mathrm{i}k_0|\boldsymbol{r}-\boldsymbol{r}'_0|}}{4\pi\,|\boldsymbol{r}-\boldsymbol{r}'_0|}\cdot(-\mathbf{R})$$

where the reflection dyadic $\mathbf{R} = \mathbf{I}_3 - 2\hat{\boldsymbol{z}}\hat{\boldsymbol{z}} = \hat{\boldsymbol{x}}\hat{\boldsymbol{x}} + \hat{\boldsymbol{y}}\hat{\boldsymbol{y}} - \hat{\boldsymbol{z}}\hat{\boldsymbol{z}}$, which reverses the z-component of the dyadic. Note that the dyadic $(-\mathbf{R})$ reverses the x–y-components of the currents and keep the direction of the z-component. ■

10.9.4 The Transmitted Field

In several applications, the field outside the slab is the most important quantity, i.e., the field at $z > z_2$ and $z < z_{-2}$. To obtain these fields, we operate with the appropriate transmission dyadic on the out-going part of the field. The out-going (right- and left-going) parts of the fields in the region $z_{-1} < z < z_1$ are, see (10.56)

$$\begin{cases} \mathbf{G}_\mathrm{e}^+(z) = \dfrac{\mathrm{i}}{2k_0}\mathrm{e}^{-\mathrm{i}\boldsymbol{k}_\mathrm{t}\cdot\boldsymbol{r}'_\mathrm{c}+\mathrm{i}k_z(z-z')}\left(\mathbf{W} - \dfrac{\boldsymbol{k}_\mathrm{t}\hat{\boldsymbol{z}}}{k_0} + \mathbf{A} + \boldsymbol{a}\hat{\boldsymbol{z}}\right), & z' < z < z_1 \\ \mathbf{G}_\mathrm{e}^-(z) = \dfrac{\mathrm{i}}{2k_0}\mathrm{e}^{-\mathrm{i}\boldsymbol{k}_\mathrm{t}\cdot\boldsymbol{r}'_\mathrm{c}-\mathrm{i}k_z(z-z')}\left(\mathbf{W} + \dfrac{\boldsymbol{k}_\mathrm{t}\hat{\boldsymbol{z}}}{k_0} + \mathbf{B} + \boldsymbol{b}\hat{\boldsymbol{z}}\right), & z_{-1} < z < z' \end{cases}$$

The fields outside the slab are obtained by an operation of the two transmission dyadics $\mathbf{t}_+$ and $\mathbf{t}_-$ for the two slabs $z > z'$ and $z < z'$, respectively. The transmitted Green dyadics (transverse parts) at $z = z_2$ and $z = z_{-2}$ are

$$\begin{cases} \mathbf{G}_\mathrm{e}^+(z_2) = \mathbf{t}^+\cdot\mathbf{G}_\mathrm{e}^+(z_1) \\ \mathbf{G}_\mathrm{e}^-(z_{-2}) = \mathbf{t}^-\cdot\mathbf{G}_\mathrm{e}^-(z_{-1}) \end{cases}$$

From these dyadics, all fields in the two half-spaces can be found, and

$$\begin{cases} \mathbf{G}_{\mathrm{e}}(z) = \mathrm{e}^{\mathrm{i}k_z(z-z_2)}\left(\mathbf{I}_2 - \dfrac{\hat{z}\boldsymbol{k}_{\mathrm{t}}}{k_z}\right)\cdot\mathbf{G}_{\mathrm{e}}^{+}(z_2), & z > z_2 \\ \mathbf{G}_{\mathrm{e}}(z) = \mathrm{e}^{-\mathrm{i}k_z(z-z_{-2})}\left(\mathbf{I}_2 + \dfrac{\hat{z}\boldsymbol{k}_{\mathrm{t}}}{k_z}\right)\cdot\mathbf{G}_{\mathrm{e}}^{-}(z_{-2}), & z < z_{-2} \end{cases}$$

or

$$\begin{cases} \mathbf{G}_{\mathrm{e}}(z) = \dfrac{\mathrm{i}}{2k_0}\mathrm{e}^{-\mathrm{i}\boldsymbol{k}_{\mathrm{t}}\cdot\boldsymbol{r}'_{\mathrm{c}}+\mathrm{i}k_z(z_1-z'+z-z_2)}\boldsymbol{\gamma}^{+}\cdot\mathbf{t}^{+}\cdot\left(\mathbf{W} - \dfrac{\boldsymbol{k}_{\mathrm{t}}\hat{z}}{k_0} + \mathbf{A} + \boldsymbol{a}\hat{z}\right), & z > z_2 \\ \mathbf{G}_{\mathrm{e}}(z) = \dfrac{\mathrm{i}}{2k_0}\mathrm{e}^{-\mathrm{i}\boldsymbol{k}_{\mathrm{t}}\cdot\boldsymbol{r}'_{\mathrm{c}}-\mathrm{i}k_z(z_{-1}-z'+z-z_{-2})}\boldsymbol{\gamma}^{-}\cdot\mathbf{t}^{-}\cdot\left(\mathbf{W} + \dfrac{\boldsymbol{k}_{\mathrm{t}}\hat{z}}{k_0} + \mathbf{B} + \boldsymbol{b}\hat{z}\right), & z < z_{-2} \end{cases}$$

where we used the projection dyadic $\gamma^{\pm}(\boldsymbol{k}_{\mathrm{t}})$ in (10.22). In each domain, the Green dyadic is obtained by an inverse Fourier transform, see (10.53). A closed-form solution of this inverse Fourier transform is only known for a limit number of material configurations, and the solution to the general problem is not known.

Problems for Chapter 10

Section 10.2 The Fundamental Equation

10.1 Verify the Hermitian adjoint of the material dyadic $\mathbf{M}$ in (10.13), i.e.,

$$\mathbf{M}_{11}^{\dagger} = \mathbf{M}_{22}, \qquad \mathbf{M}_{12}^{\dagger} = \mathbf{M}_{12}, \qquad \mathbf{M}_{21}^{\dagger} = \mathbf{M}_{21}, \qquad \mathbf{M}_{22}^{\dagger} = \mathbf{M}_{11}$$

Hint: Use $\left(\mathbf{J}\cdot\boldsymbol{\zeta}_{\perp\perp}\right)^{\dagger} = -\boldsymbol{\xi}_{\perp\perp}\cdot\mathbf{J}$, $\boldsymbol{\zeta}_{\perp}^{\dagger} = \boldsymbol{\xi}_z$, $\boldsymbol{\epsilon}_{\perp}^{\dagger} = \boldsymbol{\epsilon}_z$, etc. , and that $a^* = a$.

Section 10.3 Wave Splitting

10.2 Derive the expression of the total fields $\boldsymbol{E}$ and $\eta_0\eta\boldsymbol{H}$ in terms of the split fields $\boldsymbol{F}^{\pm}$ in (10.20) on page 528.

$$\begin{cases} \boldsymbol{E} = \left(\mathbf{I}_2 - \dfrac{1}{k_z}\hat{z}\boldsymbol{k}_{\mathrm{t}}\right)\cdot\boldsymbol{F}^{+} + \left(\mathbf{I}_2 + \dfrac{1}{k_z}\hat{z}\boldsymbol{k}_{\mathrm{t}}\right)\cdot\boldsymbol{F}^{-} \\ \eta_0\boldsymbol{H} = \dfrac{1}{k_0}\boldsymbol{k}^{+}\times\left(\mathbf{I}_2 - \dfrac{1}{k_z}\hat{z}\boldsymbol{k}_{\mathrm{t}}\right)\cdot\boldsymbol{F}^{+} + \dfrac{1}{k_0}\boldsymbol{k}^{-}\times\left(\mathbf{I}_2 + \dfrac{1}{k_z}\hat{z}\boldsymbol{k}_{\mathrm{t}}\right)\cdot\boldsymbol{F}^{-} \end{cases}$$

Section 10.4 Propagation of Fields—the Propagator Dyadic

10.3 Show that the reflection dyadic $\mathbf{r}$ in the form

$$\begin{aligned} \mathbf{r} = -\mathbf{T}_{22}^{-1}\cdot\mathbf{T}_{21} &= \left(\mathbf{P}_{11} + \mathbf{W}\cdot\mathbf{P}_{21} + (\mathbf{P}_{12} + \mathbf{W}\cdot\mathbf{P}_{22})\cdot\mathbf{W}^{-1}\right)^{-1} \\ &\cdot\left(\mathbf{P}_{11} + \mathbf{W}\cdot\mathbf{P}_{21} - (\mathbf{P}_{12} + \mathbf{W}\cdot\mathbf{P}_{22})\cdot\mathbf{W}^{-1}\right) \end{aligned}$$

also has the form

$$\mathbf{r} = \left(\mathbf{I}_2 + \mathbf{W} \cdot \left(\mathbf{P}_{21} - \mathbf{P}_{22} \cdot \mathbf{W}^{-1}\right) \cdot \left(\mathbf{P}_{11} - \mathbf{P}_{12} \cdot \mathbf{W}^{-1}\right)^{-1}\right)$$
$$\cdot \left(\mathbf{I}_2 - \mathbf{W} \cdot \left(\mathbf{P}_{21} - \mathbf{P}_{22} \cdot \mathbf{W}^{-1}\right) \cdot \left(\mathbf{P}_{11} - \mathbf{P}_{12} \cdot \mathbf{W}^{-1}\right)^{-1}\right)^{-1}$$

Section 10.6 Examples

*__10.4__ Show the result by van der Pol from 1931.

$$\int_0^\infty J_0(k_t r_c) \frac{k_t \, \mathrm{d}k_t}{\left(k_1^2 - k_t^2\right)^{1/2} + \left(k_2^2 - k_t^2\right)^{1/2}} = \frac{\mathrm{i}}{k_1^2 - k_2^2} \frac{1}{r_c} \frac{\partial}{\partial r_c} \left(\frac{\mathrm{e}^{\mathrm{i}k_1 r_c}}{r_c} - \frac{\mathrm{e}^{\mathrm{i}k_2 r_c}}{r_c} \right)$$

The branch of the complex square root in this expression is chosen so that the imaginary part of the square root is non-negative. Thereafter, show the result in (10.43) on page 556.

Hint: Use the Sommerfeld's result

$$\frac{\mathrm{e}^{\mathrm{i}kr}}{4\pi r} = \frac{\mathrm{i}}{4\pi} \int_0^\infty J_0(k_t r_c) \mathrm{e}^{\mathrm{i}\left(k^2 - k_t^2\right)^{1/2} |z|} \frac{k_t \, \mathrm{d}k_t}{\left(k^2 - k_t^2\right)^{1/2}}$$

Section 10.9 The Green Dyadic

*__10.5__ Solve the Green dyadic for the electric field in an isotropic space directly from the differential equation, i.e., solve

$$\nabla \times \left(\nabla \times \mathbf{G}_e(k_0, \boldsymbol{r}, \boldsymbol{r}')\right) - k_0^2 \mathbf{G}_e(k_0, \boldsymbol{r}, \boldsymbol{r}') = \mathbf{I}\delta(\boldsymbol{r} - \boldsymbol{r}')$$

by using the solution of the scalar equation

$$\frac{\mathrm{d}^2}{\mathrm{d}z^2} g(z) + k_z^2 g(z) = -\delta(z - z')$$

$$g(z) = \frac{\mathrm{i}}{2k_z} \mathrm{e}^{\mathrm{i}k_z |z - z'|}$$

APPENDIX A

Vectors and Linear Transformations

This appendix contains a short overview of the concept of vectors and linear transformations (dyadics) of vectors, and how these are represented in terms of their components. Transformations between different rotated coordinate systems are also reviewed as well as the corresponding transformation of the components of a dyadic. The appendix ends with a short overview of quaternions and their use to represent rotations.

A.1 Vectors

In this textbook, we denote vectors in italic boldface, e.g., $\boldsymbol{u}$. Vectors are in general functions of space and time coordinates, i.e., $\boldsymbol{u} = \boldsymbol{u}(\boldsymbol{r}, t)$, but in this appendix these variables are suppressed, since they are not essential for the analysis. A vector that depends on the space and time coordinates is called a vector field. The formal definition of a vector is made in Definition A.1 on page 590.

In a particular Cartesian coordinate system $(\hat{\boldsymbol{e}}_1, \hat{\boldsymbol{e}}_2, \hat{\boldsymbol{e}}_3)$, the representation of a vector $\boldsymbol{u}$ is[1]

$$\boldsymbol{u} = \hat{\boldsymbol{e}}_1 u_1 + \hat{\boldsymbol{e}}_2 u_2 + \hat{\boldsymbol{e}}_3 u_3$$

A "hat" or caret (ˆ) over a vector, e.g., $\hat{\boldsymbol{e}}_1$, denotes that the vector has unit length (unit vector). The length of a vector[2] is $|\boldsymbol{a}| = a = \sqrt{\boldsymbol{a}^* \cdot \boldsymbol{a}} = \sqrt{\boldsymbol{a} \cdot \boldsymbol{a}^*}$. The components of the vector, u_i, are obtained by the scalar product

$$u_i = \boldsymbol{u} \cdot \hat{\boldsymbol{e}}_i, \qquad i = 1, 2, 3$$

The components of the vector are often written as a column vector, e.g.,

$$[\boldsymbol{u}] = \begin{pmatrix} u_1 \\ u_2 \\ u_3 \end{pmatrix}$$

[1]We assume the unit vectors (basis vectors) $(\hat{\boldsymbol{e}}_1, \hat{\boldsymbol{e}}_2, \hat{\boldsymbol{e}}_3)$ are orthonormal and right-hand oriented.

[2]Note the difference between the scalar products $\boldsymbol{a} \cdot \boldsymbol{a}$ and $\boldsymbol{a} \cdot \boldsymbol{a}^*$. Complex-valued vectors have complex components, but the Cartesian coordinate system vectors are always real valued.

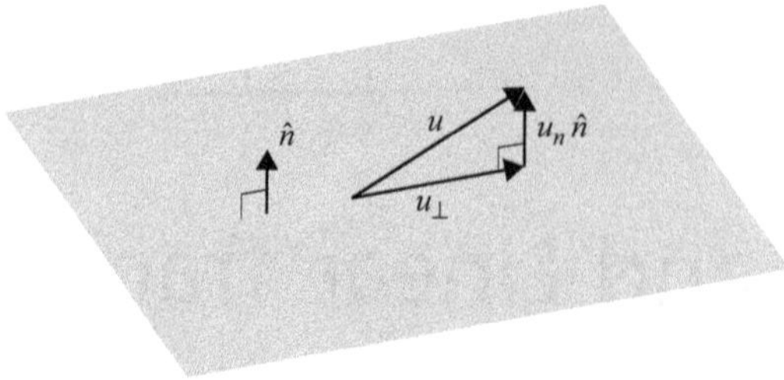

Figure A.1 Projection of a vector $\boldsymbol{u}$ in a component, $u_n\hat{\boldsymbol{n}}$, along the direction $\hat{\boldsymbol{n}}$ and a component, $\boldsymbol{u}_\perp$, perpendicular to this direction.

In this textbook, we use brackets around the vector, $[\boldsymbol{u}]$, to indicate that we refer to the components of the vector in a particular coordinate system and not the vector $\boldsymbol{u}$ itself. Notice that the vector $\boldsymbol{u}$ is geometric quantity, which is independent of coordinate system. The coordinate representation, $[\boldsymbol{u}]$, on the other hand, depends on the coordinate system we use to represent the vector in, i.e., the components of a vector are not the same in different coordinate systems, but the vector itself remains the same.

In wave propagation problems, the vectors are decomposed in its components along a given direction (usually the direction of propagation) $\hat{\boldsymbol{n}}$, and a component lying in the plane orthogonal to the direction $\hat{\boldsymbol{n}}$. We denote this decomposition or projection of a vector $\boldsymbol{u}$ as a vector sum of two parts $\boldsymbol{u}_\perp$ and $u_n\hat{\boldsymbol{n}}$, see Figure A.1

$$\boldsymbol{u} = \boldsymbol{u}_\perp + u_n\hat{\boldsymbol{n}}$$

where

$$\begin{cases} u_n = \boldsymbol{u} \cdot \hat{\boldsymbol{n}} \\ \boldsymbol{u}_\perp = \boldsymbol{u} - \hat{\boldsymbol{n}}(\boldsymbol{u} \cdot \hat{\boldsymbol{n}}) = -\hat{\boldsymbol{n}} \times (\hat{\boldsymbol{n}} \times \hat{\boldsymbol{u}}) \end{cases} \tag{A.1}$$

by the use of the BAC-CAB rule, $\boldsymbol{a} \times (\boldsymbol{b} \times \boldsymbol{c}) = \boldsymbol{b}(\boldsymbol{a} \cdot \boldsymbol{c}) - \boldsymbol{c}(\boldsymbol{a} \cdot \boldsymbol{b})$.

A.2 Linear Transformations, Matrices, and Dyadics

Often, we deal with linear mappings from one vector field to other vector field, i.e., mapping of a vector $\boldsymbol{u}$ to another vector $\boldsymbol{v}$—both in general functions of the space and time coordinates $\boldsymbol{r}$ and t. The most simple type of linear transformation is

$$\boldsymbol{v} = \boldsymbol{a}\underbrace{(\boldsymbol{b} \cdot \boldsymbol{u})}_{\text{scalar}}$$

The vector $\boldsymbol{u}$ is here mapped to a new vector $\boldsymbol{v}$ in a new direction along the vector $\boldsymbol{a}$. The scaling of the vector is made with the vector $\boldsymbol{b}$ by the scalar product $\boldsymbol{b} \cdot \boldsymbol{u}$. This mapping is a called a (simple) dyadic, and we use the symbol $\boldsymbol{ab}$ for this transformation—the two vectors $\boldsymbol{a}$ and $\boldsymbol{b}$ are juxtaposed without any sign between the vectors—together with the dyadic product (either with or without parenthesis around the transformation $\boldsymbol{ab}$) defined as

$$\boldsymbol{v} = (\boldsymbol{ab}) \cdot \boldsymbol{u} = \boldsymbol{ab} \cdot \boldsymbol{u} \stackrel{\text{def}}{=} \boldsymbol{a}(\boldsymbol{b} \cdot \boldsymbol{u})$$

Note that the vectors $\boldsymbol{a}$ and $\boldsymbol{b}$ in the transformation $\boldsymbol{ab}$ are forming a new quantity without any signs between the vectors. The components of the vector $\boldsymbol{v}$ are

$$v_i = a_i \sum_{j=1}^{3} b_j u_j, \qquad i = 1, 2, 3$$

Simple dyadics can be used as building blocks for general linear transformations of vector fields. The sum of two dyadics, e.g., $\mathbf{A} = \boldsymbol{a}_1\boldsymbol{b}_1 + \boldsymbol{a}_2\boldsymbol{b}_2$, is a new dyadic defined in the following natural way:

$$\boldsymbol{v} = \mathbf{A} \cdot \boldsymbol{u} = (\boldsymbol{a}_1\boldsymbol{b}_1 + \boldsymbol{a}_2\boldsymbol{b}_2) \cdot \boldsymbol{u} = \boldsymbol{a}_1\boldsymbol{b}_1 \cdot \boldsymbol{u} + \boldsymbol{a}_2\boldsymbol{b}_2 \cdot \boldsymbol{u}$$

Note that the linear transformation (dyadic) $\mathbf{A}$ is written in roman boldface to distinguish the quantity from the vector $\boldsymbol{A}$ (italic boldface).

In a specific coordinate system $(\hat{\boldsymbol{e}}_1, \hat{\boldsymbol{e}}_2, \hat{\boldsymbol{e}}_3)$, we represent the general linear transformation $\mathbf{A}$ from a vector field $\boldsymbol{u}$ to another vector field $\boldsymbol{v}$ by simple dyadics $\hat{\boldsymbol{e}}_i\hat{\boldsymbol{e}}_j$, $i, j = 1, 2, 3$.

$$\mathbf{A} = \sum_{i,j=1}^{3} A_{ij} \hat{\boldsymbol{e}}_i \hat{\boldsymbol{e}}_j \tag{A.2}$$

The action on a vector field $\boldsymbol{u}$ becomes

$$\boldsymbol{v} = \mathbf{A} \cdot \boldsymbol{u} \stackrel{\text{def}}{=} \sum_{i,j=1}^{3} \hat{\boldsymbol{e}}_i A_{ij} \left(\hat{\boldsymbol{e}}_j \cdot \boldsymbol{u}\right) = \sum_{i,j=1}^{3} \hat{\boldsymbol{e}}_i A_{ij} u_j$$

or in its components

$$v_i = \sum_{j=1}^{3} A_{ij} u_j, \qquad i = 1, 2, 3$$

The component representation of the linear transformation or dyadic $\mathbf{A}$ is represented by a matrix

$$[\mathbf{A}] = \begin{pmatrix} A_{11} & A_{12} & A_{13} \\ A_{21} & A_{22} & A_{23} \\ A_{31} & A_{32} & A_{33} \end{pmatrix}$$

We denote the matrix of the dyadic $\mathbf{A}$ by enclosing the dyadic by brackets, i.e., $[\mathbf{A}]$, to distinguish the coordinate representation of the dyadic from the dyadic $\mathbf{A}$ itself, which is independent of any coordinate representation. This convention is similar to the one we use for a vector $\boldsymbol{u}$ and its coordinate representation $[\boldsymbol{u}]$.

By the notation $\boldsymbol{u}_i = \sum_{j=1}^{3} A_{ij}\hat{\boldsymbol{e}}_j$, $i = 1, 2, 3$, we see that the general linear transformation $\mathbf{A}$ always can be written as

$$\mathbf{A} = \hat{\boldsymbol{e}}_1 \boldsymbol{u}_1 + \hat{\boldsymbol{e}}_2 \boldsymbol{u}_2 + \hat{\boldsymbol{e}}_3 \boldsymbol{u}_3$$

which shows that three dyadics are enough to represent a general linear transformation or dyadic.

So far, the action of vector, $\boldsymbol{u}$, on a dyadic, $\mathbf{A}$, has been from the right by a scalar multiplication. However, one advantage with the dyadic notation is seen if we let the action take place from the left. This action leads to the transpose operation. To this end, make a scalar multiplication from the left, that gives a new vector field $\boldsymbol{v}$ defined by, see (A.2)

$$\boldsymbol{v} = \boldsymbol{u} \cdot \mathbf{A} \stackrel{\text{def}}{=} \sum_{i,j=1}^{3} \left(\hat{\boldsymbol{e}}_i \cdot \boldsymbol{u}\right) A_{ij} \hat{\boldsymbol{e}}_j = \sum_{i,j=1}^{3} \hat{\boldsymbol{e}}_j A_{ij} u_i = \sum_{i=1}^{3} u_i \boldsymbol{u}_i$$

or in component form

$$v_i = \sum_{j=1}^{3} A_{ji} u_j, \qquad i = 1, 2, 3$$

We observe that a scalar multiplication from the left gives the transpose of the matrix $[\mathbf{A}]$. This operation defines the transposed dyadic, $\mathbf{A}^{\mathrm{t}}$.

$$\boldsymbol{v} = \mathbf{A}^{\mathrm{t}} \cdot \boldsymbol{u} \stackrel{\text{def}}{=} \boldsymbol{u} \cdot \mathbf{A}$$

and

$$[\mathbf{A}^{\mathrm{t}}] = [\mathbf{A}]^{\mathrm{t}}$$

In particular,

$$\boldsymbol{u}_1 \cdot (\mathbf{A} \cdot \boldsymbol{u}_2) = \left(\boldsymbol{u}_2 \cdot \mathbf{A}^{\mathrm{t}}\right) \cdot \boldsymbol{u}_1 = \boldsymbol{u}_2 \cdot \left(\mathbf{A}^{\mathrm{t}} \cdot \boldsymbol{u}_1\right)$$

for all complex-valued vectors $\boldsymbol{u}_1$ and $\boldsymbol{u}_2$.

The unit dyadic $\mathbf{I}_3$ maps all vectors onto itself, $\mathbf{I}_3 \cdot \boldsymbol{u} = \boldsymbol{u}$. The inverse dyadic, $\mathbf{A}^{-1}$, of a dyadic $\mathbf{A}$ is the dyadic that satisfies $\mathbf{A} \cdot \mathbf{A}^{-1} = \mathbf{A}^{-1} \cdot \mathbf{A} = \mathbf{I}_3$. The inverse $\mathbf{A}^{-1}$ has a matrix representation that is the inverse of the matrix representation of $\mathbf{A}$ in a specific coordinate system, i.e.,

$$[\mathbf{A}^{-1}] = [\mathbf{A}]^{-1} = \begin{pmatrix} A_{11} & A_{12} & A_{13} \\ A_{21} & A_{22} & A_{23} \\ A_{31} & A_{32} & A_{33} \end{pmatrix}^{-1}$$

if the coordinate representation of $\mathbf{A}$ is

$$[\mathbf{A}] = \begin{pmatrix} A_{11} & A_{12} & A_{13} \\ A_{21} & A_{22} & A_{23} \\ A_{31} & A_{32} & A_{33} \end{pmatrix}$$

Similarly, we define the (complex) conjugate of a dyadic, $\mathbf{A}^*$, and its Hermitian[3] adjoint,[4] $\mathbf{A}^\dagger$, of a dyadic $\mathbf{A}$ by the coordinate representations

$$[\mathbf{A}^*] = [\mathbf{A}]^* = \begin{pmatrix} A_{11}^* & A_{12}^* & A_{13}^* \\ A_{21}^* & A_{22}^* & A_{23}^* \\ A_{31}^* & A_{32}^* & A_{33}^* \end{pmatrix}, \qquad [\mathbf{A}^\dagger] = [\mathbf{A}]^\dagger = \begin{pmatrix} A_{11}^* & A_{21}^* & A_{31}^* \\ A_{12}^* & A_{22}^* & A_{32}^* \\ A_{13}^* & A_{23}^* & A_{33}^* \end{pmatrix}$$

We find that $\mathbf{A}^\dagger = \mathbf{A}^{*\mathrm{t}} = \mathbf{A}^{\mathrm{t}*}$. A dyadic is called symmetric if $\mathbf{A}^{\mathrm{t}} = \mathbf{A}$, and Hermitian if $\mathbf{A}^\dagger = \mathbf{A}$. If the dyadic satisfies $\mathbf{A}^{\mathrm{t}} = -\mathbf{A}$ ($\mathbf{A}^\dagger = -\mathbf{A}$), the dyadic is called anti-symmetric (anti-Hermitian). A general dyadic can be decomposed in a symmetric (Hermitian) and an anti-symmetric (anti-Hermitian) part as

$$\mathbf{A} = \mathbf{A}_{\mathrm{s}} + \mathbf{A}_{\mathrm{a}} \qquad (\mathbf{A} = \mathbf{A}_{\mathrm{H}} + \mathbf{A}_{\text{a-H}})$$

where

$$\begin{cases} \mathbf{A}_{\mathrm{s}} = \dfrac{1}{2}\left(\mathbf{A} + \mathbf{A}^{\mathrm{t}}\right) \\ \mathbf{A}_{\mathrm{a}} = \dfrac{1}{2}\left(\mathbf{A} - \mathbf{A}^{\mathrm{t}}\right) \end{cases} \qquad \left(\begin{cases} \mathbf{A}_{\mathrm{H}} = \dfrac{1}{2}\left(\mathbf{A} + \mathbf{A}^\dagger\right) \\ \mathbf{A}_{\text{a-H}} = \dfrac{1}{2}\left(\mathbf{A} - \mathbf{A}^\dagger\right) \end{cases}\right)$$

Similarly, a complex-valued dyadic can be decomposed in its real and imaginary parts as

$$\mathbf{A} = \mathbf{A}_{\mathrm{r}} + \mathrm{i}\mathbf{A}_{\mathrm{i}}$$

where

$$\begin{cases} \mathbf{A}_{\mathrm{r}} = \dfrac{1}{2}\left(\mathbf{A} + \mathbf{A}^\dagger\right) \\ \mathbf{A}_{\mathrm{i}} = \dfrac{1}{2\mathrm{i}}\left(\mathbf{A} - \mathbf{A}^\dagger\right) \end{cases}$$

Notice that both $\mathbf{A}_{\mathrm{r}}$ and $\mathbf{A}_{\mathrm{i}}$ are Hermitian dyadics, and that the off-diagonal elements of $\mathbf{A}_{\mathrm{r}}$ and $\mathbf{A}_{\mathrm{i}}$ in general are not real numbers (the diagonal elements are). For any $\boldsymbol{u} \in \mathbb{C}^3$

$$\begin{cases} \operatorname{Re}\left\{\boldsymbol{u}^* \cdot \mathbf{A} \cdot \boldsymbol{u}\right\} = \boldsymbol{u}^* \cdot \mathbf{A}_{\mathrm{r}} \cdot \boldsymbol{u} \\ \operatorname{Im}\left\{\boldsymbol{u}^* \cdot \mathbf{A} \cdot \boldsymbol{u}\right\} = \boldsymbol{u}^* \cdot \mathbf{A}_{\mathrm{i}} \cdot \boldsymbol{u} \end{cases}$$

Analogously, the vector product between a vector field $\boldsymbol{u}$ and a dyadic $\mathbf{A}$ is defined as

$$\mathbf{B} = \mathbf{A} \times \boldsymbol{u} \stackrel{\text{def}}{=} \sum_{i,j=1}^{3} \hat{\boldsymbol{e}}_i A_{ij} \left(\hat{\boldsymbol{e}}_j \times \boldsymbol{u}\right)$$

---

[3] Charles Hermite (1822–1901), French mathematician.

[4] Also called the Hermitian conjugate or the Hermitian transpose.

The vector product can also be applied from the left. We have

$$\mathbf{B} = \boldsymbol{u} \times \mathbf{A} \stackrel{\text{def}}{=} \sum_{i,j=1}^{3} \left(\boldsymbol{u} \times \hat{\boldsymbol{e}}_i\right) A_{ij} \hat{\boldsymbol{e}}_j$$

A.2.1 Projections

The projection of a vector $\boldsymbol{u}$ on a plane with unit normal vector $\hat{\boldsymbol{n}}$ is the linear mapping expressed as a dyadic, see (A.1)

$$\boldsymbol{u}_\perp = \boldsymbol{u} - \hat{\boldsymbol{n}} \left(\hat{\boldsymbol{n}} \cdot \boldsymbol{u}\right) = \mathbf{I}_\perp \cdot \boldsymbol{u}$$

where the projection dyadic $\mathbf{I}_\perp$ is

$$\mathbf{I}_\perp = \mathbf{I}_3 - \hat{\boldsymbol{n}}\hat{\boldsymbol{n}}$$

In the same manner, an arbitrary dyadic $\mathbf{A}$ can be decomposed into components parallel and perpendicular to a given fixed direction $\hat{\boldsymbol{n}}$. We obtain this decomposition by employing $\mathbf{I}_3 = \mathbf{I}_\perp + \hat{\boldsymbol{n}}\hat{\boldsymbol{n}}$

$$\boldsymbol{v} = \mathbf{A} \cdot \boldsymbol{u} = \left(\mathbf{I}_\perp + \hat{\boldsymbol{n}}\hat{\boldsymbol{n}}\right) \cdot \mathbf{A} \cdot \left(\mathbf{I}_\perp + \hat{\boldsymbol{n}}\hat{\boldsymbol{n}}\right) \cdot \boldsymbol{u}$$

This decomposition implies that every dyadic $\mathbf{A}$ can be written as:

$$\mathbf{A} = \mathbf{A}_{\perp\perp} + \hat{\boldsymbol{n}}\boldsymbol{A}_n + \boldsymbol{A}_\perp \hat{\boldsymbol{n}} + \hat{\boldsymbol{n}} A_{nn} \hat{\boldsymbol{n}}$$

where we introduced the notation

$$\begin{cases} \mathbf{A}_{\perp\perp} = \mathbf{I}_\perp \cdot \mathbf{A} \cdot \mathbf{I}_\perp \\ \boldsymbol{A}_n = \hat{\boldsymbol{n}} \cdot \mathbf{A} \cdot \mathbf{I}_\perp \\ \boldsymbol{A}_\perp = \mathbf{I}_\perp \cdot \mathbf{A} \cdot \hat{\boldsymbol{n}} \\ A_{nn} = \hat{\boldsymbol{n}} \cdot \mathbf{A} \cdot \hat{\boldsymbol{n}} \end{cases}$$

Note that the dyadic $\mathbf{A}_{\perp\perp}$ is a two-dimensional dyadic (mapping vectors in the plane with normal $\hat{\boldsymbol{n}}$ into the same plane). Similarly, the vectors $\boldsymbol{A}_n$ and $\boldsymbol{A}_\perp$ are two-dimensional vectors in the plane, and A_{nn} is a scalar. If we let the direction $\hat{\boldsymbol{n}}$ be the unit vector $\hat{\boldsymbol{e}}_3$, we obtain the coordinate representations of these quantities (we use the index 3 instead of n)

$$[\mathbf{A}_{\perp\perp}] = \begin{pmatrix} A_{11} & A_{12} & 0 \\ A_{21} & A_{22} & 0 \\ 0 & 0 & 0 \end{pmatrix}, \qquad [\boldsymbol{A}_\perp] = \begin{pmatrix} A_{13} \\ A_{23} \\ 0 \end{pmatrix}, \qquad [\boldsymbol{A}_3] = \begin{pmatrix} A_{31} \\ A_{32} \\ 0 \end{pmatrix}$$

Such a decomposition is proved useful in the analysis of wave propagation in planar structures.

A.3 Rotation of Coordinate System

Several textbooks deal with rotation of vectors and linear transformations in space, e.g., the excellent book by Kuipers [149], which also contains the concept of quaternions and its connections to rotations in $\mathbb{R}^3$. A primer on quaternions is found in Appendix A.3.2. For more details, we refer to Reference 149 and also the detailed paper by Diebel [55].

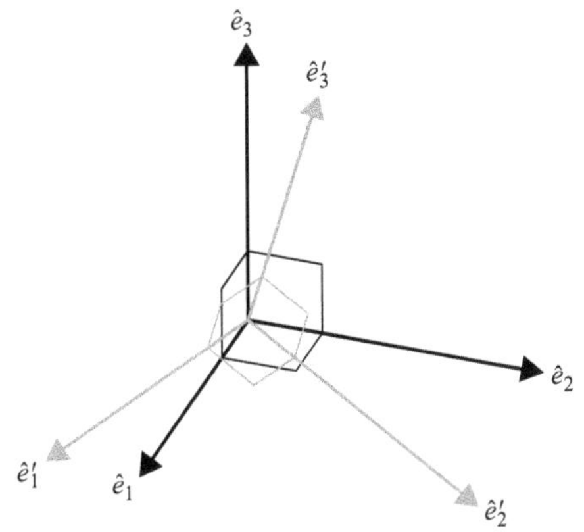

Figure A.2 Two rotated coordinate systems $(\hat{e}_1, \hat{e}_2, \hat{e}_3)$ and $(\hat{e}'_1, \hat{e}'_2, \hat{e}'_3)$ with a common origin. The original coordinate system is shown in black and the rotated coordinate system in gray.

We start this section with some more basic concepts on rotations of a coordinate systems. Two (Cartesian) coordinate systems $(\hat{e}_1, \hat{e}_2, \hat{e}_3)$ and $(\hat{e}'_1, \hat{e}'_2, \hat{e}'_3)$, both orthonormal and right-hand oriented, share the common origin. An example of two such systems is depicted in Figure A.2.

Since the unprimed set of unit vectors forms a basis, each of the primed unit vectors $(\hat{e}'_1, \hat{e}'_2, \hat{e}'_3)$ can be expressed as a linear combination of the unprimed unit vectors $(\hat{e}_1, \hat{e}_2, \hat{e}_3)$. We have

$$\begin{cases} \hat{e}'_1 = \hat{e}_1 a_{11} + \hat{e}_2 a_{12} + \hat{e}_3 a_{13} \\ \hat{e}'_2 = \hat{e}_1 a_{21} + \hat{e}_2 a_{22} + \hat{e}_3 a_{23} \\ \hat{e}'_3 = \hat{e}_1 a_{31} + \hat{e}_2 a_{32} + \hat{e}_3 a_{33} \end{cases}$$

or

$$\hat{e}'_i = \sum_{j=1}^{3} \hat{e}_j a_{ij}, \qquad i = 1, 2, 3$$

The components a_{ij}, $i, j = 1, 2, 3$, are the direction cosines[5] between the axes i and j:

$$a_{ij} = \hat{e}'_i \cdot \hat{e}_j, \qquad i, j = 1, 2, 3$$

In general, we have

$$\hat{e}'_i \cdot \hat{e}_j = a_{ij} \neq a_{ji} = \hat{e}'_j \cdot \hat{e}_i \qquad i, j = 1, 2, 3$$

[5]Also given as $a_{ij} = \cos(x'_i, x_j)$ where (x'_i, x_j) is the angle between $\hat{e}'_i$ and $\hat{e}_j$ directions.

Similarly, the unprimed unit vectors $\hat{e}_i$ can be written as a linear combination of the primed ones $(\hat{e}'_1, \hat{e}'_2, \hat{e}'_3)$. We get

$$\hat{e}_i = \hat{e}'_1 \left(\hat{e}'_1 \cdot \hat{e}_i\right) + \hat{e}'_2 \left(\hat{e}'_2 \cdot \hat{e}_i\right) + \hat{e}'_3 \left(\hat{e}'_3 \cdot \hat{e}_i\right) \quad i = 1, 2, 3$$

or expressed in the direction cosines a_{ij}

$$\begin{cases} \hat{e}_1 = \hat{e}'_1 a_{11} + \hat{e}'_2 a_{21} + \hat{e}'_3 a_{31} \\ \hat{e}_2 = \hat{e}'_1 a_{12} + \hat{e}'_2 a_{22} + \hat{e}'_3 a_{32} \\ \hat{e}_3 = \hat{e}'_1 a_{13} + \hat{e}'_2 a_{23} + \hat{e}'_3 a_{33} \end{cases}$$

In short

$$\hat{e}_i = \sum_{j=1}^{3} \hat{e}'_j a_{ji}, \qquad i = 1, 2, 3$$

From these expressions, we see that if the transformation $\hat{e}_i \to \hat{e}'_i$ is made with a_{ij} then the transformation $\hat{e}'_i \to \hat{e}_i$ is made with a_{ji}. The matrix $[\mathbf{A}]$ is therefore an orthogonal matrix, i.e., $[\mathbf{A}]^{-1} = [\mathbf{A}]^{\mathrm{t}}$, and its determinant satisfies $(\det[\mathbf{A}])^2 = 1$, which implies $\det[\mathbf{A}] = \pm 1$. Since we have assumed that the unit vectors $(\hat{e}'_1, \hat{e}'_2, \hat{e}'_3)$ and $(\hat{e}_1, \hat{e}_2, \hat{e}_3)$ both are right-hand oriented, the determinant is 1, i.e., $\det[\mathbf{A}] = 1$.

We are now ready to give the formal definition of a vector. A vector $\boldsymbol{u}$ is a geometric quantity with components (u_1, u_2, u_3) in the system $(\hat{e}_1, \hat{e}_2, \hat{e}_3)$, which are related to the components (u'_1, u'_2, u'_3) in the system $(\hat{e}'_1, \hat{e}'_2, \hat{e}'_3)$ by the direction cosines, a_{ij}, $i, j = 1, 2, 3$, in the following way:

$$u'_i = \sum_{j=1}^{3} a_{ij} u_j, \qquad i = 1, 2, 3 \tag{A.3}$$

or expressed as column vectors and standard matrix multiplication

$$\begin{pmatrix} u'_1 \\ u'_2 \\ u'_3 \end{pmatrix} = \begin{pmatrix} a_{11} & a_{12} & a_{13} \\ a_{21} & a_{22} & a_{23} \\ a_{31} & a_{32} & a_{33} \end{pmatrix} \begin{pmatrix} u_1 \\ u_2 \\ u_3 \end{pmatrix}$$

Definition A.1 *A physical quantity* $\boldsymbol{u}$ *which components* u_i *and* u'_i *in two rotated coordinate systems, respectively, are related to each other by* (A.3) *is called a* **vector**.

Comment A.1
A vector is more correctly called a polar vector, which distinguishes it from an axial vector which transforms by (A.3), where $\det[\mathbf{A}] = -1$. If the vector depends on the space coordinates, i.e., $\boldsymbol{u}$ is a vector field, then also the space coordinates are transformed as

$$u'_i(x'_1, x'_2, x'_3) = \sum_{j=1}^{3} a_{ij} u_j(x_1, x_2, x_3), \qquad i = 1, 2, 3$$

where x'_1, x'_2, x'_3 and x_1, x_2, x_3 are the components of the position vector in the primed and the unprimed systems, respectively.

This definition implies that for every vector $\boldsymbol{u}$ we have

$$\boldsymbol{u} = \hat{\boldsymbol{e}}_1' u_1' + \hat{\boldsymbol{e}}_2' u_2' + \hat{\boldsymbol{e}}_3' u_3' = \sum_{i=1}^{3} u_i' \hat{\boldsymbol{e}}_i' = \sum_{i,j=1}^{3} a_{ij} u_j \hat{\boldsymbol{e}}_i'$$

$$= \sum_{j=1}^{3} u_j \hat{\boldsymbol{e}}_j = \hat{\boldsymbol{e}}_1 u_1 + \hat{\boldsymbol{e}}_2 u_2 + \hat{\boldsymbol{e}}_3 u_3$$

By this definition, every vector becomes a quantity which is independent of the coordinate representation, just as we required in the beginning of this appendix.

In a similar way, we define a dyadic (or a tensor of the second kind).

Definition A.2 *A physical quantity* **D** *which components D_{ij} and D_{ij}' in two rotated coordinate systems, respectively, are related by*

$$D_{ij}' = \sum_{k,l=1}^{3} a_{ik} a_{jl} D_{kl}, \qquad i,j = 1,2,3$$

is called a **dyadic**. *This relation can also be expressed as similarity transformation*

$$[\mathbf{D}]' = [\mathbf{A}]\,[\mathbf{D}]\,[\mathbf{A}^{\mathrm{t}}]$$

A.3.1 Euler Angles

The two rotated coordinate systems, $(\hat{\boldsymbol{e}}_1, \hat{\boldsymbol{e}}_2, \hat{\boldsymbol{e}}_3)$ and $(\hat{\boldsymbol{e}}_1', \hat{\boldsymbol{e}}_2', \hat{\boldsymbol{e}}_3')$, are related to each other by the direction cosines. These relations can also be expressed in an alternative way by the three Euler angles α, β, and γ. These angles are defined by three consecutive rotations, see Figure A.3. The three rotations are explicitly given by:

1. A rotation with the angle α around the $\hat{\boldsymbol{e}}_3$ axis
2. A rotation with the angle β around the $\hat{\boldsymbol{e}}_2'$ axis
3. A rotation with the angle γ around the $\hat{\boldsymbol{e}}_3''$ axis

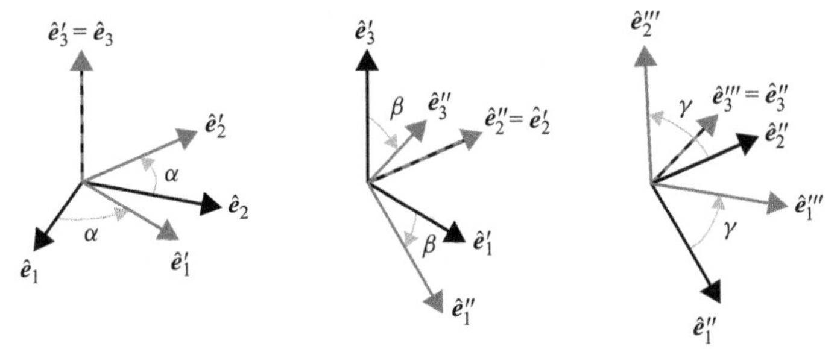

Figure A.3 Definition of the three Euler angles α, β and γ. The first rotation (left figure) gives rise to coordinate axes $(\hat{\boldsymbol{e}}_1', \hat{\boldsymbol{e}}_2', \hat{\boldsymbol{e}}_3')$, the second rotation (middle figure) gives rise to coordinate axis $(\hat{\boldsymbol{e}}_1'', \hat{\boldsymbol{e}}_2'', \hat{\boldsymbol{e}}_3'')$, and the final rotation (right figure) gives rise to coordinate axes $(\hat{\boldsymbol{e}}_1''', \hat{\boldsymbol{e}}_2''', \hat{\boldsymbol{e}}_3''')$. Each set of rotated axes is given in gray color.

The three different rotations are represented by the following matrices:

1. The first rotation is represented by

$$[\mathbf{R}_1] = \begin{pmatrix} \cos\alpha & \sin\alpha & 0 \\ -\sin\alpha & \cos\alpha & 0 \\ 0 & 0 & 1 \end{pmatrix}$$

2. The second rotation is represented by

$$[\mathbf{R}_2] = \begin{pmatrix} \cos\beta & 0 & -\sin\beta \\ 0 & 1 & 0 \\ \sin\beta & 0 & \cos\beta \end{pmatrix}$$

3. The third rotation is represented by

$$[\mathbf{R}_3] = \begin{pmatrix} \cos\gamma & \sin\gamma & 0 \\ -\sin\gamma & \cos\gamma & 0 \\ 0 & 0 & 1 \end{pmatrix}$$

In total, the rotation is made by

$$[\mathbf{A}] = [\mathbf{R}_3][\mathbf{R}_2][\mathbf{R}_1]$$

A.3.2 Quaternions

Quaternions are useful in representing rotations in $\mathbb{R}^3$, and they can be viewed as an extension of the complex numbers. An excellent introduction to the quaternions is found in Reference 149, and here we only give a short summary of their use.

Introduce the hyper-complex numbers of rank 4, which also is called the quaternions. These numbers were introduced 1843 by Hamilton[6] to extend the complex numbers.

We start by introducing an orthonormal basis **i**, **j**, **k**, in $\mathbb{R}^3$ together with the following multiplication algebra:

$$\begin{cases} \mathbf{i}^2 = \mathbf{j}^2 = \mathbf{k}^2 = \mathbf{ijk} = -1 \\ \mathbf{ij} = \mathbf{k} = -\mathbf{ji} \\ \mathbf{jk} = \mathbf{i} = -\mathbf{kj} \\ \mathbf{ki} = \mathbf{j} = -\mathbf{ik} \end{cases} \tag{A.4}$$

It is important not to mix this juxtaposed notation with the notion of dyadics above. This notation occurs only in this section. Notice the similarities with the multiplication rule for complex numbers $\mathrm{i}^2 = -1$.

A quaternion q is a four-tuple of real numbers, i.e.,

$$q = (q_0, q_1, q_2, q_3)$$

[6]Sir William Rowan Hamilton (1805–1865), Irish physicist, astronomer, and mathematician.

The first entry q_0 is called the scalar part of the quaternion and the remaining three are called the vector part of the quaternion. We adopt the notation

$$\boldsymbol{q} = \mathbf{i}q_1 + \mathbf{j}q_2 + \mathbf{k}q_3$$

and we write

$$q = q_0 + \boldsymbol{q} = q_0 + \mathbf{i}q_1 + \mathbf{j}q_2 + \mathbf{k}q_3$$

We see that it is possible to identify real vectors in $\mathbb{R}^3$ with quaternions with zero scalar parts. Such quaternions are called pure quaternions, i.e.,

$$\boldsymbol{q} \in \mathbb{R}^3 \quad \Leftrightarrow \quad q = 0 + \boldsymbol{q} \quad \text{(pure quaternion)}$$

Let q and p be two quaternions, i.e.,

$$q = q_0 + \boldsymbol{q}, \qquad p = p_0 + \boldsymbol{p}$$

The quaternions are equal provided $q_i = p_i$, $i = 0, 1, 2, 3$. We define addition of two quaternions as

$$q + p = p + q = q_0 + p_0 + \boldsymbol{p} + \boldsymbol{q}$$

Multiplication of two quaternions is more complex, and we use the product rules in (A.4).

$$\begin{aligned} pq &= (p_0 + \mathbf{i}p_1 + \mathbf{j}p_2 + \mathbf{k}p_3)(q_0 + \mathbf{i}q_1 + \mathbf{j}q_2 + \mathbf{k}q_3) \\ &= p_0(q_0 + \mathbf{i}q_1 + \mathbf{j}q_2 + \mathbf{k}q_3) + \mathbf{i}p_1(q_0 + \mathbf{i}q_1 + \mathbf{j}q_2 + \mathbf{k}q_3) \\ &\quad + \mathbf{j}p_2(q_0 + \mathbf{i}q_1 + \mathbf{j}q_2 + \mathbf{k}q_3) + \mathbf{k}p_3(q_0 + \mathbf{i}q_1 + \mathbf{j}q_2 + \mathbf{k}q_3) \end{aligned}$$

The scalar part of this product is

$$p_0q_0 - p_1q_1 - p_2q_2 - p_3q_3 = p_0q_0 - \boldsymbol{p} \cdot \boldsymbol{q}$$

The vector part of this product becomes

$$\begin{aligned} &p_0(\mathbf{i}q_1 + \mathbf{j}q_2 + \mathbf{k}q_3) + p_1(\mathbf{i}q_0 + \mathbf{k}q_2 - \mathbf{j}q_3) + p_2(\mathbf{j}q_0 - \mathbf{k}q_1 + \mathbf{i}q_3) + p_3(\mathbf{k}q_0 + \mathbf{j}q_1 - \mathbf{i}q_2) \\ &\quad = p_0\boldsymbol{q} + q_0\boldsymbol{p} + \mathbf{i}(p_2q_3 - p_3q_2) + \mathbf{j}(p_3q_1 - p_1q_3) + \mathbf{k}(p_1q_2 - p_2q_1) = p_0\boldsymbol{q} + q_0\boldsymbol{p} + \boldsymbol{p} \times \boldsymbol{q} \end{aligned}$$

The product pq therefore becomes

$$pq = p_0q_0 - \boldsymbol{p} \cdot \boldsymbol{q} + p_0\boldsymbol{q} + q_0\boldsymbol{p} + \boldsymbol{p} \times \boldsymbol{q} \tag{A.5}$$

In particular, the product of a quaternion q and a pure quaternion $\boldsymbol{v}$ (vector in $\mathbb{R}^3$) is

$$q\boldsymbol{v} = (q_0 + \boldsymbol{q})(0 + \boldsymbol{v}) = -\boldsymbol{q} \cdot \boldsymbol{v} + q_0\boldsymbol{v} + \boldsymbol{q} \times \boldsymbol{v} \tag{A.6}$$

Two more concepts are useful. The complex conjugate of a quaternion $q = q_0 + \boldsymbol{q} = q_0 + \mathbf{i}q_1 + \mathbf{j}q_2 + \mathbf{k}q_3$ is denoted q^* and defined as

$$q^* = q_0 - \boldsymbol{q} = q_0 - \mathbf{i}q_1 - \mathbf{j}q_2 - \mathbf{k}q_3$$

From (A.5) it is easy to prove that

$$(pq)^* = q^* p^*$$

Moreover, the norm of a quaternion $q = q_0 + \boldsymbol{q} = q_0 + \mathbf{i}q_1 + \mathbf{j}q_2 + \mathbf{k}q_3$ is denoted $|q|$ and defined as

$$|q| = \sqrt{q^* q}$$

This definition makes sense since, using the product rule (A.5), we have

$$q^* q = (q_0 - \boldsymbol{q})(q_0 + \boldsymbol{q}) = q_0^2 + \boldsymbol{q} \cdot \boldsymbol{q} + q_0 \boldsymbol{q} - q_0 \boldsymbol{q} - \boldsymbol{q} \times \boldsymbol{q} = q_0^2 + |\boldsymbol{q}|^2$$

which always is a non-negative real number. The reverse operation gives the same result

$$q q^* = q_0^2 + |\boldsymbol{q}|^2$$

The inverse of a quaternion q is denoted q^{-1} and satisfies

$$q q^{-1} = q^{-1} q = 1$$

Multiply from the left or right with q^*

$$\underbrace{q^* q}_{|q|^2} q^{-1} = q^*, \qquad q^{-1} \underbrace{q q^*}_{|q|^2} = q^*$$

This implies

$$q^{-1} = \frac{q^*}{|q|^2}$$

Notice the similarities with the reciprocal of a complex numbers $z^{-1} = z^*/|z|^2$.

To connect to a rotation of a vector $\boldsymbol{v}$ in $\mathbb{R}^3$, we evaluate

$$\begin{aligned} q\boldsymbol{v}q^* &= (-\boldsymbol{q} \cdot \boldsymbol{v} + q_0 \boldsymbol{v} + \boldsymbol{q} \times \boldsymbol{v})(q_0 - \boldsymbol{q}) = -q_0 \boldsymbol{q} \cdot \boldsymbol{v} + (q_0 \boldsymbol{v} + \boldsymbol{q} \times \boldsymbol{v}) \cdot \boldsymbol{q} + (\boldsymbol{q} \cdot \boldsymbol{v})\boldsymbol{q} \\ &+ q_0(q_0 \boldsymbol{v} + \boldsymbol{q} \times \boldsymbol{v}) - (q_0 \boldsymbol{v} + \boldsymbol{q} \times \boldsymbol{v}) \times \boldsymbol{q} = 2(\boldsymbol{q} \cdot \boldsymbol{v})\boldsymbol{q} + q_0(q_0 \boldsymbol{v} + 2\boldsymbol{q} \times \boldsymbol{v}) - |\boldsymbol{q}|^2 \boldsymbol{v} \end{aligned}$$

where we have used (A.6) and the product rule (A.5). Notice that $q\boldsymbol{v}q^*$ is a pure quaternion and therefore is associated with a vector $\boldsymbol{v}'$, i.e.,

$$\boldsymbol{v}' = (q_0^2 - |\boldsymbol{q}|^2)\boldsymbol{v} + 2(\boldsymbol{q} \cdot \boldsymbol{v})\boldsymbol{q} + 2q_0 \boldsymbol{q} \times \boldsymbol{v}$$

We conclude that $L_q(\boldsymbol{v}) = q\boldsymbol{v}q^*$ is a linear mapping on $\mathbb{R}^3$, and specialize to $q = \cos(\theta/2) + \hat{\boldsymbol{e}} \sin(\theta/2)$, where $\theta \in [-\pi, \pi]$ and $\hat{\boldsymbol{e}}$ is a real unit vector. The result is

$$\boldsymbol{v}' = \cos\theta \boldsymbol{v} + 2\hat{\boldsymbol{e}} \sin^2(\theta/2)(\hat{\boldsymbol{e}} \cdot \boldsymbol{v}) + \sin\theta \hat{\boldsymbol{e}} \times \boldsymbol{v}$$

The norm of this vector is

$$|v'|^2 = \cos^2\theta|v|^2 + 4\underbrace{(\cos\theta + \sin^2(\theta/2))}_{=\cos^2(\theta/2)}\sin^2(\theta/2)(\hat{e}\cdot v)^2 + \sin^2\theta\underbrace{|\hat{e}\times v|^2}_{|v|^2-(\hat{e}\cdot v)^2}$$
$$= |v|^2 + \sin^2\theta(\hat{e}\cdot v)^2 - \sin^2\theta(\hat{e}\cdot v)^2 = |v|^2$$

and if we let $v = \hat{e}$ we obtain

$$q\hat{e}q^* = 2\hat{e}\sin^2(\theta/2) + \underbrace{\cos\theta}_{\cos^2(\theta/2)-\sin^2(\theta/2)}\hat{e} = \hat{e}$$

We suspect that the mapping qvq^* with $q = \cos(\theta/2) + \hat{e}\sin(\theta/2)$ represents a rotation of the vector v to v' with rotation axis $\hat{e}$. We now prove that this is the case, and we also determine how large the angle of rotation is. To this end, decompose the vector $v = (\hat{e}\cdot v)\hat{e} + u$, where $u = v - (\hat{e}\cdot v)\hat{e}$ is orthogonal to $\hat{e}$, i.e., $u\cdot\hat{e} = 0$. We get

$$v' = L_q((\hat{e}\cdot v)\hat{e}) + L_q(u) = (\hat{e}\cdot v)\hat{e} + \underbrace{\cos\theta u + \sin\theta\hat{e}\times u}_{u'}$$

since $L_q(\hat{e}) = \hat{e}$. This proves that the rotation is θ in the positive $\hat{e}$-direction, see Figures A.4 and A.5. Note that both the quaternions $q = \cos(\theta/2)+\hat{e}\sin(\theta/2)$ and $q' = \cos((\theta+2\pi)/2)+\hat{e}\sin((\theta+2\pi)/2) = -q$ generate the same rotation and linear map qvq^*.

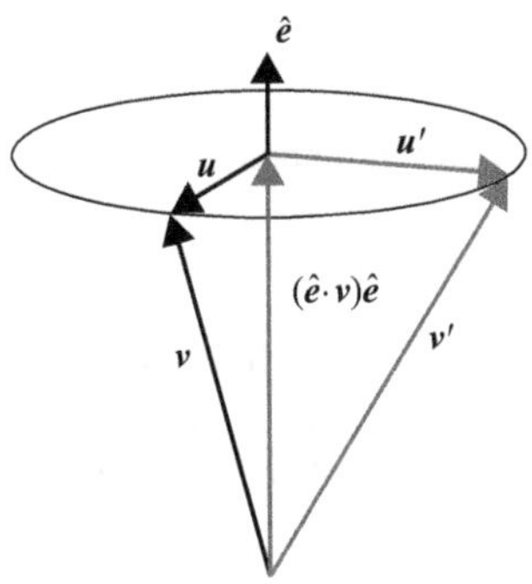

Figure A.4 The decomposition of the vector v.

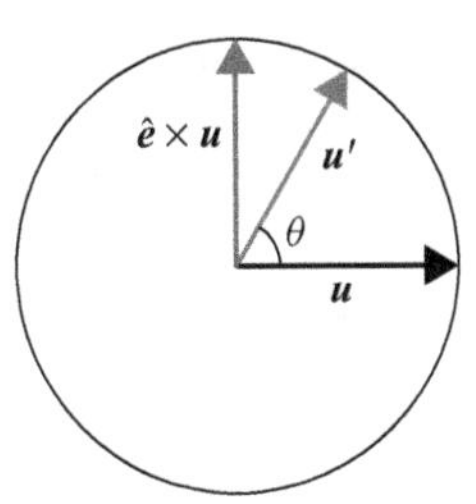

Figure A.5 The plane of rotation.

Example A.1
Let $\hat{\boldsymbol{e}} = (1,1,1)/\sqrt{3}$ be the axis of rotation and rotate the angle $\theta = 2\pi/3$. Then the appropriate quaternion is

$$q = \cos(\theta/2) + \hat{\boldsymbol{e}}\sin(\theta/2) = \frac{1}{2} + \frac{1}{2}\mathbf{i} + \frac{1}{2}\mathbf{j} + \frac{1}{2}\mathbf{k}$$

The rotation of the vector $\boldsymbol{v} = \mathbf{i}$ then is

$$L_q(\mathbf{i}) = q\mathbf{i}q^* = \frac{1}{4}(1+\mathbf{i}+\mathbf{j}+\mathbf{k})\mathbf{i}(1-\mathbf{i}-\mathbf{j}-\mathbf{k}) = \frac{1}{4}(1+\mathbf{i}+\mathbf{j}+\mathbf{k})(\mathbf{i}+1-\mathbf{k}+\mathbf{j}) = \mathbf{j}$$

■

Two consecutive rotations with angles α and β around a common axis $\hat{\boldsymbol{e}}$ combine to a single rotation around $\hat{\boldsymbol{e}}$ with angle $\alpha + \beta$. In fact, the appropriate quaternions p and q are

$$p = \cos(\alpha/2) + \hat{\boldsymbol{e}}\sin(\alpha/2), \qquad q = \cos(\beta/2) + \hat{\boldsymbol{e}}\sin(\beta/2)$$

and the product is

$$\begin{aligned} pq &= p_0q_0 - \boldsymbol{p}\cdot\boldsymbol{q} + p_0\boldsymbol{q} + q_0\boldsymbol{p} + \boldsymbol{p}\times\boldsymbol{q} = \cos(\alpha/2)\cos(\beta/2) - \sin(\alpha/2)\sin(\beta/2) \\ &\quad + \{\cos(\alpha/2)\sin(\beta/2) + \cos(\beta/2)\sin(\alpha/2)\}\,\hat{\boldsymbol{e}} = \cos((\alpha+\beta)/2) + \hat{\boldsymbol{e}}\sin((\alpha+\beta)/2) \end{aligned}$$

which proves the statement.

This observation generalizes to two arbitrary rotations as seen from the following analysis:

$$L_p(\boldsymbol{w}) = p\boldsymbol{w}p^*, \qquad L_q(\boldsymbol{v}) = q\boldsymbol{v}q^*$$

The product is then a single rotation quantified by the quaternion pq, i.e.,

$$L_p(L_q(\boldsymbol{v})) = pL_q(\boldsymbol{v})p^* = pq\boldsymbol{v}q^*p^* = pq\boldsymbol{v}(pq)^* = L_{pq}(\boldsymbol{v})$$

Now when we know that two consecutive rotations can be expressed as a single rotation, the appropriate question is: What is the common axis $\hat{\boldsymbol{e}}$ and angle of rotation θ for a product of two consecutive rotations—first a rotation α around the axis $\hat{\boldsymbol{a}}$ followed by a rotation β around the axis $\hat{\boldsymbol{b}}$? We have[7]

$$\begin{aligned} \cos(\theta/2) + \hat{\boldsymbol{e}}\sin(\theta/2) &= q_{\hat{e},\theta} = q_{\hat{b},\beta}q_{\hat{a},\alpha} = (\cos(\beta/2) + \hat{\boldsymbol{b}}\sin(\beta/2))(\cos(\alpha/2) + \hat{\boldsymbol{a}}\sin(\alpha/2)) \\ &= \cos(\beta/2)\cos(\alpha/2) - \hat{\boldsymbol{b}}\cdot\hat{\boldsymbol{a}}\sin(\beta/2)\sin(\alpha/2) \\ &\quad + \hat{\boldsymbol{a}}\cos(\beta/2)\sin(\alpha/2) + \hat{\boldsymbol{b}}\sin(\beta/2)\cos(\alpha/2) + \sin(\beta/2)\sin(\alpha/2)\hat{\boldsymbol{b}}\times\hat{\boldsymbol{a}} \end{aligned}$$

Identify the scalar and vector parts of this quaternion:

$$\theta = 2\arccos\Big(\cos(\beta/2)\cos(\alpha/2) - \hat{\boldsymbol{b}}\cdot\hat{\boldsymbol{a}}\sin(\beta/2)\sin(\alpha/2)\Big)$$

---

[7]To clarify the axis of rotation and angle of rotation, these data are added to the quaternion as a subscript.

and

$$\hat{\boldsymbol{e}} = \left\{\hat{\boldsymbol{a}} \cos(\beta/2)\sin(\alpha/2) + \hat{\boldsymbol{b}} \sin(\beta/2)\cos(\alpha/2) + \sin(\beta/2)\sin(\alpha/2)\hat{\boldsymbol{b}} \times \hat{\boldsymbol{a}}\right\} / \sin(\theta/2)$$

These two equations quantify the angle of rotation and the axis of rotation uniquely.

In Appendix A.3.1, we showed that rotations can be quantified as a 3×3 orthogonal matrix $[\mathbf{R}]$. We have in this section learned that the same rotation can be quantified by a quaternion q. In fact, any orthogonal matrix $[\mathbf{R}]$ can be identified with a quaternion q, and vice versa. The result is (for details and alternative formulations, we refer to Diebel [55])

$$[\mathbf{R}] = \begin{pmatrix} 2(q_0^2 + q_1^2) - 1 & 2(q_1 q_2 + q_0 q_3) & 2(q_1 q_3 - q_0 q_2) \\ 2(q_1 q_2 - q_0 q_3) & 2(q_0^2 + q_2^2) - 1 & 2(q_2 q_3 + q_0 q_1) \\ 2(q_1 q_3 + q_0 q_2) & 2(q_2 q_3 - q_0 q_1) & 2(q_0^2 + q_3^2) - 1 \end{pmatrix}$$

and for the inverse we have (with an overall sign ambiguity, since both q and $-q$ generate the same rotation, see above)

$$\begin{cases} q_0 = \pm\dfrac{\sqrt{\mathrm{Tr}(R) + 1}}{2} \\ q_1 = \dfrac{R_{23} - R_{32}}{4q_0} \\ q_2 = \dfrac{R_{31} - R_{13}}{4q_0} \\ q_3 = \dfrac{R_{12} - R_{21}}{4q_0} \end{cases}$$

APPENDIX B

Bessel Functions

In wave propagation problems, the Bessel differential equation often appears, especially in problems showing axial or spherical symmetries. This appendix collects some useful and important results for the solution of the Bessel differential equation. Moreover, the modified Bessel functions and the spherical Bessel and Hankel functions are presented. Additional results and some of the derivations are found in e.g., References 4, 134, 288.

B.1 Bessel and Hankel Functions

The Bessel[1] differential equation is

$$z^2 \frac{d^2}{dz^2} Z_n(z) + z \frac{d}{dz} Z_n(z) + (z^2 - n^2) Z_n(z) = 0 \tag{B.1}$$

where n is an integer.[2] There exist two linearly independent solutions to this differential equation. One is regular at the origin, $z = 0$, and this solution is the Bessel function $J_n(z)$ of order n. The argument z is a complex number. These solutions are often called cylindrical Bessel function of order n, which stresses the affinity to problems with axial symmetry. The Bessel functions $J_n(z)$ are defined real valued for a real argument z. The power series is (convergent everywhere in the complex z-plane)

$$J_n(z) = \sum_{k=0}^{\infty} \frac{(-1)^k}{k!(n+k)!} \left(\frac{z}{2}\right)^{n+2k} \tag{B.2}$$

We notice immediately that $J_n(z)$ is an even function for even n and odd for odd n, i.e.,

$$J_n(-z) = (-1)^n J_n(z)$$

A commonly used integral representation of the Bessel functions is

$$J_n(z) = \frac{1}{\pi} \int_0^{\pi} \cos(z \sin t - nt)\, dt = \frac{1}{2\pi} \int_0^{2\pi} e^{iz\cos t} e^{in(t - \frac{1}{2}\pi)}\, dt \tag{B.3}$$

[1] Friedrich Wilhelm Bessel (1784–1846), German mathematician and astronomer.

[2] A more general definition for e.g., complex-valued n is also possible, but the expressions and the results often differ.

Table B.1 Table of the roots ξ_{nj} to $J_n(z)$.

Order n	Root j			
	1	2	3	4
0	2.405	5.520	8.654	11.79
1	3.832	7.016	10.17	13.32
2	5.136	8.417	11.62	14.80
3	6.380	9.761	13.01	16.22
4	7.588	11.06	14.37	17.62

From this integral representation, we see that the Bessel functions for positive and negative *integer* orders, n, are related to each other

$$J_{-n}(z) = (-1)^n J_n(z)$$

The power series representation in (B.2) implies that for small arguments we have

$$J_n(z) = \frac{1}{n!}\left(\frac{z}{2}\right)^n + O(z^{n+2})$$

For large arguments, the dominant contribution is $(-\pi < \arg z < \pi)$

$$J_n(z) = \left(\frac{2}{\pi z}\right)^{1/2} \left\{P_n(z)\cos\left(z - \frac{n\pi}{2} - \frac{\pi}{4}\right) - Q_n(z)\sin\left(z - \frac{n\pi}{2} - \frac{\pi}{4}\right)\right\}$$

where the functions $P_n(z)$ and $Q_n(z)$ have the following asymptotic expansions ($\nu = 4n^2$):

$$\begin{cases} P_n(z) \sim 1 - \dfrac{(\nu-1)(\nu-9)}{2!(8z)^2} + \dfrac{(\nu-1)(\nu-9)(\nu-25)(\nu-49)}{4!(8z)^4} - \cdots \\[2ex] Q_n(z) \sim \dfrac{\nu-1}{8z} - \dfrac{(\nu-1)(\nu-9)(\nu-25)}{3!(8z)^3} + \cdots \end{cases} \tag{B.4}$$

The roots of the Bessel function $J_n(z)$ are all real, and the first roots, ξ_{nj}, are listed in Table B.1. The derivative of the Bessel function $J_n(z)$ has also only real roots, η_{nj}, and the first ones are listed in Table B.2. Larger roots (larger j values) are asymptotically given by

$$\xi_{nj} = j\pi + \left(n - \frac{1}{2}\right)\frac{\pi}{2}, \qquad \eta_{nj} = j\pi + \left(n - \frac{3}{2}\right)\frac{\pi}{2}$$

Another, linearly independent solution to the Bessel differential equation, which is real valued for real arguments, is the Neumann[3] function[4] $N_n(z)$. There are many alternative power

[3] Karl Gottfried Neumann (1832–1925), German mathematician.

[4] These solutions are also called Bessel functions of the second kind.

Table B.2 Table of the roots η_{nj} to $J_n'(z)$.

Order n	Root j			
	1	2	3	4
0	3.832	7.016	10.17	13.32
1	1.841	5.331	8.536	11.71
2	3.054	6.706	9.970	13.17
3	4.201	8.015	11.34	14.59
4	5.317	9.282	12.68	15.96

series expansions of the Neumann function. One is [134]

$$
\begin{aligned}
N_n(z) = \frac{2}{\pi}\left(\ln\left(\frac{z}{2}\right)+\gamma\right)J_n(z) &- \frac{1}{\pi}\sum_{k=0}^{\infty}(-1)^k\frac{\left(\frac{z}{2}\right)^{n+2k}}{k!(n+k)!}\left(\sum_{l=1}^{k}\frac{1}{l}+\sum_{l=1}^{k+n}\frac{1}{l}\right)\\
&- \frac{1}{\pi}\sum_{k=0}^{n-1}\frac{(n-k-1)!}{k!}\left(\frac{z}{2}\right)^{-n+2k}
\end{aligned}
$$

where the Euler constant $\gamma = 0.57721566\ldots$, and where the sum is defined as zero if the summation index exceeds the upper summation limit. This solution is singular at the origin $z = 0$. For small arguments, the dominant contribution is

$$
\begin{aligned}
N_0(z) &= \frac{2}{\pi}\left(\ln\left(\frac{z}{2}\right)+\gamma\right)\left(1+O(z^2)\right)\\
N_n(z) &= -\frac{(n-1)!}{\pi}\left(\frac{z}{2}\right)^{-n}\left(1+O(z^2)\right), \qquad n = 1, 2, 3, \ldots
\end{aligned}
$$

For large arguments, the Neumann functions have the following asymptotic expansion $(-\pi < \arg z < \pi)$:

$$
N_n(z) = \left(\frac{2}{\pi z}\right)^{1/2}\left\{P_n(z)\sin\left(z-\frac{n\pi}{2}-\frac{\pi}{4}\right)+Q_n(z)\cos\left(z-\frac{n\pi}{2}-\frac{\pi}{4}\right)\right\}
$$

where the functions $P_n(z)$ and $Q_n(z)$ are given by (B.4).

The solution of scattering problems often contains linear combinations of Bessel and Neumann functions, i.e., the Hankel[5] functions, $H_n^{(1)}(z)$ and $H_n^{(2)}(z)$ of the first and the second kind, respectively.[6] These are defined as

$$
\begin{aligned}
H_n^{(1)}(z) &= J_n(z) + \mathrm{i}N_n(z)\\
H_n^{(2)}(z) &= J_n(z) - \mathrm{i}N_n(z)
\end{aligned}
$$

[5] Hermann Hankel (1839–1873), German mathematician.

[6] These solutions are also called Bessel functions of the third kind.

The Hankel functions of the first and second kind have integral representations. A common set of representations is

$$H_n^{(1)}(z) = \frac{2}{\mathrm{i}\pi} \mathrm{e}^{-\mathrm{i}n\frac{\pi}{2}} \int_0^{\infty} \mathrm{e}^{\mathrm{i}z\cosh s} \cosh ns \, \mathrm{d}s, \qquad 0 < \arg z < \pi$$

$$H_n^{(2)}(z) = \frac{2\mathrm{i}}{\pi} \mathrm{e}^{\mathrm{i}n\frac{\pi}{2}} \int_0^{\infty} \mathrm{e}^{-\mathrm{i}z\cosh s} \cosh ns \, \mathrm{d}s, \qquad -\pi < \arg z < 0$$

From these integral representations, we see that the Hankel functions for positive and negative *integer* orders, n, are related to each other

$$H_{-n}^{(1)}(z) = (-1)^n H_n^{(1)}(z), \qquad H_{-n}^{(2)}(z) = (-1)^n H_n^{(2)}(z)$$

For large arguments, the Hankel functions have asymptotic expansions. These are

$$\begin{aligned} H_n^{(1)}(z) &= \left(\frac{2}{\pi z}\right)^{1/2} \mathrm{e}^{\mathrm{i}\left(z - \frac{n\pi}{2} - \frac{\pi}{4}\right)} \left(P_n(z) + \mathrm{i}Q_n(z)\right), \qquad -\pi < \arg z < 2\pi \\ H_n^{(2)}(z) &= \left(\frac{2}{\pi z}\right)^{1/2} \mathrm{e}^{-\mathrm{i}\left(z - \frac{n\pi}{2} - \frac{\pi}{4}\right)} \left(P_n(z) - \mathrm{i}Q_n(z)\right), \qquad -2\pi < \arg z < \pi \end{aligned} \tag{B.5}$$

where the functions $P_n(z)$ and $Q_n(z)$ are given by (B.4).

Solutions to the Bessel differential equation of different order are related to each other by recursion relations. Some of the more important ones are ($n = 0, 1, 2, \ldots; m = 0, 1, 2, \ldots$)

$$\begin{aligned} &Z_{n-1}(z) - Z_{n+1}(z) = 2Z_n'(z) \\ &Z_{n-1}(z) + Z_{n+1}(z) = \frac{2n}{z} Z_n(z) \\ &Z_{n+1}(z) = \frac{n}{z} Z_n(z) - Z_n'(z) \\ &Z_n'(z) = Z_{n-1}(z) - \frac{n}{z} Z_n(z) \\ &\left(\frac{\mathrm{d}}{z\,\mathrm{d}z}\right)^m \left[z^n Z_n(z)\right] = z^{n-m} Z_{n-m}(z) \\ &\left(\frac{\mathrm{d}}{z\,\mathrm{d}z}\right)^m \left[z^{-n} Z_n(z)\right] = (-1)^m z^{-n-m} Z_{n+m}(z) \end{aligned}$$

Here, $Z_n(z)$ is an arbitrary linear combination of $J_n(z)$, $N_n(z)$, $H_n^{(1)}(z)$, or $H_n^{(2)}(z)$. Specifically, we have

$$J_1(z) = -J_0'(z), \qquad H_1^{(1)}(z) = -H_0^{(1)\prime}(z)$$

which is frequently used in the analysis in this textbook.

Some useful indefinite integrals with solutions to the Bessel differential equation are $(n = 0, 1, 2, \ldots)$

$$\int x^{n+1} Z_n(x)\, \mathrm{d}x = x^{n+1} Z_{n+1}(x) = -x^{n+1}\left(Z_n'(x) - \frac{n}{x} Z_n(x)\right)$$

$$\int x^{-n+1} Z_n(x)\, \mathrm{d}x = -x^{-n+1} Z_{n-1}(x) = -x^{-n+1}\left(Z_n'(x) + \frac{n}{x} Z_n(x)\right)$$

$$\int x\,(Z_n(x))^2\, \mathrm{d}x = \frac{x^2}{2}\left[(Z_n(x))^2 - Z_{n-1}(x) Z_{n+1}(x)\right] = \frac{x^2}{2}\left(Z_n'(x)\right)^2 + \frac{1}{2}(x^2 - n^2)\,(Z_n(x))^2$$

As above, $Z_n(x)$ is an arbitrary linear combination of $J_n(x)$, $N_n(x)$, $H_n^{(1)}(x)$, or $H_n^{(2)}(x)$. Some additional, more complex, but useful integrals are $(n = 0, 1, 2, \ldots; m = 0, 1, 2, \ldots)$

$$\int \left[(\alpha^2 - \beta^2)x - \frac{m^2 - n^2}{x}\right] Z_m(\alpha x) Y_n(\beta x)\, \mathrm{d}x = \beta x Z_m(\alpha x) Y_{n-1}(\beta x) - \alpha x Z_{m-1}(\alpha x) Y_n(\beta x)$$
$$+ (m - n) Z_m(\alpha x) Y_m(\beta x)$$

$$\int x Z_m(\alpha x) Y_m(\beta x)\, \mathrm{d}x = \frac{\beta x Z_m(\alpha x) Y_{m-1}(\beta x) - \alpha x Z_{m-1}(\alpha x) Y_m(\beta x)}{\alpha^2 - \beta^2}$$

$$\int \frac{Z_m(\alpha x) Y_n(\alpha x)}{x}\, \mathrm{d}x = \alpha x \frac{Z_{m-1}(\alpha x) Y_n(\alpha x) - Z_m(\alpha x) Y_{n-1}(\alpha x)}{m^2 - n^2} - \frac{Z_m(\alpha x) Y_n(\alpha x)}{m + n}$$

Here, $Z_n(\alpha x)$ and $Y_n(\beta x)$ are arbitrary linear combinations of $J_n(x)$, $N_n(x)$, $H_n^{(1)}(x)$ or $H_n^{(2)}(x)$.

For the Bessel functions, $J_n(z)$, the Neumann functions, $N_n(z)$, and the Hankel functions, $H_n^{(1)}(z)$ or $H_n^{(2)}(z)$, we have for a complex argument z

$$\begin{cases} J_n(z^*) = (J_n(z))^* \\ N_n(z^*) = (N_n(z))^* \end{cases} \qquad \begin{cases} H_n^{(1)}(z^*) = \left(H_n^{(2)}(z)\right)^* \\ H_n^{(2)}(z^*) = \left(H_n^{(1)}(z)\right)^* \end{cases}$$

and the reflection formulas

$$\begin{cases} H_n^{(1)}(z\mathrm{e}^{\mathrm{i}\pi}) = -\mathrm{e}^{-\mathrm{i}n\pi} H_n^{(2)}(z) \\ H_n^{(2)}(z\mathrm{e}^{-\mathrm{i}\pi}) = -\mathrm{e}^{\mathrm{i}n\pi} H_n^{(1)}(z) \end{cases}$$

The Wronskian for $J_n(z)$ and $N_n(z)$ is

$$J_n(z) N_n'(z) - J_n'(z) N_n(z) = \frac{2}{\pi z}$$

and for $J_n(z)$ and $H_n^{(1)}(z)$ it is

$$J_n(z) H_n^{(1)\prime}(z) - J_n'(z) H_n^{(1)}(z) = \frac{2\mathrm{i}}{\pi z}$$

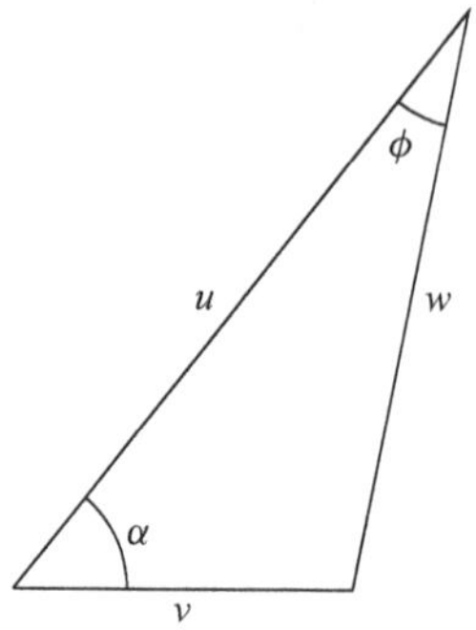

Figure B.1 The geometry of the quantities in the addition theorem.

The Wronskian for the Hankel functions is

$$H_n^{(2)}(z)H_n^{(1)\prime}(z) - H_n^{(2)\prime}(z)H_n^{(1)}(z) = \frac{4\mathrm{i}}{\pi z} \tag{B.6}$$

The Graf addition theorem[7] for Bessel functions is useful. Let $Z_n(x)$ be any linear combination of $J_n(x)$, $N_n(x)$, $H_n^{(1)}(x)$, and $H_n^{(2)}(x)$. The Graf addition theorem is

$$Z_n(w)\begin{pmatrix}\cos n\phi \\ \sin n\phi\end{pmatrix} = \sum_{k=-\infty}^{\infty} Z_{n+k}(u)J_k(v)\begin{pmatrix}\cos k\alpha \\ \sin k\alpha\end{pmatrix}, \qquad \left|v\mathrm{e}^{\pm\mathrm{i}\alpha}\right| < |u|$$

where w is

$$w = \sqrt{u^2 + v^2 - 2uv\cos\alpha}$$

The relations between the different quantities w, u, v, α, and ϕ are shown in Figure B.1.

B.1.1 Useful Integrals

Some integrals related to Bessel functions are derived in this subsection. We start with the integral representation for integer order n, (B.3)

$$\int_0^{2\pi} \mathrm{e}^{\mathrm{i}z\cos\phi}\mathrm{e}^{\mathrm{i}n\phi}\,\mathrm{d}\phi = 2\pi\mathrm{i}^n J_n(z)$$

From this, we easily conclude by a simple change of variables that

$$\int_0^{2\pi} \mathrm{e}^{\mathrm{i}z\cos(\phi-\alpha)}\mathrm{e}^{\mathrm{i}n\phi}\,\mathrm{d}\phi = \mathrm{e}^{\mathrm{i}n\alpha}\int_{-\alpha}^{2\pi-\alpha} \mathrm{e}^{\mathrm{i}z\cos\psi}\mathrm{e}^{\mathrm{i}n\psi}\,\mathrm{d}\psi = 2\pi\mathrm{i}^n J_n(z)\mathrm{e}^{\mathrm{i}n\alpha}$$

This integral is a function of the variables z and α.

---

[7] Johann Heinrich Graf (1852–1918), Swiss mathematician.

In scattering problems, the following integral is often encountered

$$I(\tau, r_c) = \int_0^{2\pi} \hat{\phi}\, e^{ir_c\cdot\tau} e^{in\phi}\, d\phi = \int_0^{2\pi} \hat{\phi}\, e^{ir_c\tau\cos(\phi-\alpha)} e^{in\phi}\, d\phi$$

where the position vector in the x–y-plane is denoted $\boldsymbol{r}_c = \hat{\boldsymbol{x}} r_c \cos\phi + \hat{\boldsymbol{y}} r_c \sin\phi$ and $\boldsymbol{\tau} = \hat{\boldsymbol{x}}\tau\cos\alpha + \hat{\boldsymbol{y}}\tau\sin\alpha$. This integral can be rewritten as

$$I(\tau, r_c) = \frac{1}{ir_c} \hat{z} \times \nabla_\tau \int_0^{2\pi} e^{ir_c\tau\cos(\phi-\alpha)} e^{in\phi}\, d\phi$$

where the nabla operator in the τ variable is denoted ∇_τ, and where we used

$$\hat{z} \times \nabla_\tau e^{ir_c\cdot\tau} = i\hat{z} \times r_c e^{ir_c\cdot\tau} = ir_c\hat{z} \times \hat{r}_c e^{ir_c\cdot\tau} = ir_c\hat{\phi} e^{ir_c\cdot\tau}$$

From this, we infer

$$I(\tau, r_c) = \frac{2\pi i^{n-1}}{r_c} \hat{z} \times \nabla_\tau J_n(r_c\tau) e^{in\alpha} = 2\pi i^{n-1} \hat{z} \times \left(\hat{\tau} J_n'(r_c\tau) e^{in\alpha} + in\hat{\alpha} \frac{J_n(r_c\tau)}{r_c\tau} e^{in\alpha} \right)$$

B.2 Modified Bessel Functions

The modified Bessel differential equation is

$$z^2 \frac{d^2}{dz^2} Z_n(z) + z \frac{d}{dz} Z_n(z) - (z^2 + n^2) Z_n(z) = 0$$

where n is assumed to be a non-negative integer. A comparison with the Bessel equation in Section B.1 shows that the solutions to the modified Bessel equation are Bessel functions with an argument iz. Two linearly independent solutions to the modified Bessel equations, commonly denoted by $I_n(z)$ and $K_n(z)$, are defined in terms of the Bessel functions as

$$\begin{cases} I_n(z) = i^{-n} J_n(iz) \\ K_n(z) = \dfrac{i^{n+1}\pi}{2} H_n^{(1)}(iz) \end{cases} \tag{B.7}$$

The terms in front of the Bessel and Hankel functions on the right-hand side make the functions $I_n(z)$ and $K_n(z)$ real valued for real arguments z. The function $I_n(z)$ is regular at $z = 0$, and $K_n(z)$ is singular as $z \to 0$.

From the properties of the Bessel functions, we see that $I_n(z)$ is an even function for even n and an odd function for odd n, i.e.,

$$I_n(-z) = (-1)^n I_n(z)$$

and that

$$I_{-n}(z) = (-1)^n I_n(z)$$

These results can also be obtained by the use of the power series representation of $I_n(z)$ at $z = 0$

$$I_n(z) = \sum_{k=0}^{\infty} \frac{1}{k!(n+k)!} \left(\frac{z}{2}\right)^{n+2k}$$

The corresponding power series representation of $K_n(z)$ at $z = 0$ is

$$K_n(z) = (-1)^{n+1} \left(\ln\left(\frac{z}{2}\right) + \gamma\right) I_n(z) + (-1)^n \frac{1}{2} \sum_{k=0}^{\infty} \frac{\left(\frac{z}{2}\right)^{n+2k}}{k!(n+k)!} \left(\sum_{l=1}^{k} \frac{1}{l} + \sum_{l=1}^{k+n} \frac{1}{l}\right)$$

$$+ \frac{1}{2} \sum_{k=0}^{n-1} (-1)^k \frac{(n-k-1)!}{k!} \left(\frac{z}{2}\right)^{-n+2k}$$

where the Euler constant $\gamma = 0.57721566\ldots$, and where all sums are defined zero if the integer in the sum is larger than the upper summation index. These power series representations give the dominating contribution as $z \to 0$

$$I_n(z) = \frac{1}{n!} \left(\frac{z}{2}\right)^n + O(z^{n+2})$$

and

$$\begin{cases} K_0(z) = -\left(\ln\left(\dfrac{z}{2}\right) + \gamma\right) + O(z^2) \\ K_n(z) = \dfrac{(n-1)!}{2} \left(\dfrac{z}{2}\right)^{-n} + \cdots, \qquad n \neq 0 \end{cases}$$

For large arguments, we have

$$I_n(z) = \frac{e^z}{(2\pi z)^{1/2}} P_n(z), \qquad -\frac{\pi}{2} < \arg z < \frac{\pi}{2}$$

and

$$K_n(z) = \left(\frac{\pi}{2z}\right)^{1/2} e^{-z} Q_n(z), \qquad -\frac{3\pi}{2} < \arg z < \frac{3\pi}{2}$$

where the functions $P_n(z)$ and $Q_n(z)$ have the following asymptotic expansions ($\nu = 4n^2$):

$$\begin{cases} P_n(z) \sim 1 - \dfrac{\nu-1}{8z} + \dfrac{(\nu-1)(\nu-9)}{2!(8z)^2} - \dfrac{(\nu-1)(\nu-9)(\nu-25)}{3!(8z)^3} + \cdots \\ Q_n(z) \sim 1 + \dfrac{\nu-1}{8z} + \dfrac{(\nu-1)(\nu-9)}{2!(8z)^2} + \dfrac{(\nu-1)(\nu-9)(\nu-25)}{3!(8z)^3} + \cdots \end{cases}$$

Useful integral representations of the modified Bessel functions are [206]

$$\begin{cases} I_n(z) = \dfrac{1}{\pi} \displaystyle\int_0^{\pi} e^{z \cos t} \cos nt \, dt \\ K_n(z) = \displaystyle\int_0^{\infty} e^{-z \cosh t} \cosh nt \, dt, \qquad |\arg z| < \pi/2 \end{cases} \tag{B.8}$$

Between the solutions to the modified Bessel equation of different order, there are recursion relations. For $I_n(z)$, $n = 0, 1, 2, \ldots$, the most important ones are $(m = 0, 1, 2, \ldots)$

$$I_{n-1}(z) + I_{n+1}(z) = 2I_n'(z)$$

$$I_{n-1}(z) - I_{n+1}(z) = \frac{2n}{z} I_n(z)$$

$$I_{n+1}(z) = I_n'(z) - \frac{n}{z} I_n(z)$$

$$I_n'(z) = I_{n-1}(z) - \frac{n}{z} I_n(z)$$

$$\left(\frac{\mathrm{d}}{z\,\mathrm{d}z}\right)^m \left[z^n I_n(z)\right] = z^{n-m} I_{n-m}(z)$$

$$\left(\frac{\mathrm{d}}{z\,\mathrm{d}z}\right)^m \left[z^{-n} I_n(z)\right] = z^{-n-m} I_{n+m}(z)$$

and for $K_n(z)$ the corresponding ones are $(n = 0, 1, 2, \ldots; m = 0, 1, 2, \ldots)$

$$K_{n-1}(z) + K_{n+1}(z) = -2K_n'(z)$$

$$K_{n+1}(z) - K_{n-1}(z) = \frac{2n}{z} K_n(z)$$

$$K_{n+1}(z) = \frac{n}{z} K_n(z) - K_n'(z)$$

$$K_n'(z) = -K_{n-1}(z) - \frac{n}{z} K_n(z)$$

$$\left(\frac{\mathrm{d}}{z\,\mathrm{d}z}\right)^m \left[z^n K_n(z)\right] = (-1)^m z^{n-m} K_{n-m}(z)$$

$$\left(\frac{\mathrm{d}}{z\,\mathrm{d}z}\right)^m \left[z^{-n} K_n(z)\right] = (-1)^m z^{-n-m} K_{n+m}(z)$$

B.3 Spherical Bessel and Hankel Functions

The spherical Bessel and Hankel functions show up in scattering problems, especially when we express a solution in the spherical coordinate system (r, θ, ϕ).

The spherical Bessel differential equation is

$$z^2 \frac{\mathrm{d}^2}{\mathrm{d}z^2} Z_n(z) + 2z \frac{\mathrm{d}}{\mathrm{d}z} Z_n(z) + (z^2 - n(n+1)) Z_n(z) = 0$$

where n is assumed to be a non-negative integer and the argument z a complex number.

A solution to this differential equation is the spherical Bessel functions $j_n(z)$, which are defined by the power series expansion[8]

$$j_n(z) = 2^n z^n \sum_{k=0}^{\infty} \frac{(-1)^k (k+n)!}{k!(2n+2k+1)!} z^{2k} = z^n \sum_{k=0}^{\infty} \frac{1}{k!(2n+2k+1)!!} \left(-\frac{z^2}{2}\right)^k \tag{B.9}$$

This solution is real valued for real arguments, it is regular at the origin, $z = 0$, and its power series is absolutely convergent in the entire complex plane. From the power series, we also conclude that $j_n(z)$ is an even function for even integers n and an odd function for odd n, i.e.,

$$j_n(-z) = (-1)^n j_n(z)$$

The other, linearly independent solution, which is real valued for real arguments, is the spherical Neumann function $y_n(z)$. Its power series is [1]

$$y_n(z) = -\frac{1}{z^{n+1}} \sum_{k=0}^{\infty} \frac{(2n-2k-1)!!}{k!} \left(\frac{z^2}{2}\right)^k$$

We see that this solution is singular at the origin $z = 0$.

In scattering problems, we often use linear combinations of the spherical Bessel and Neumann functions. These are the spherical Hankel functions, $h_n^{(1)}(z)$ and $h_n^{(2)}(z)$ of the first and second kind, respectively. These are defined as[9]

$$\begin{aligned} h_n^{(1)}(z) &= j_n(z) + \mathrm{i} y_n(z) \\ h_n^{(2)}(z) &= j_n(z) - \mathrm{i} y_n(z) \end{aligned}$$

The spherical Bessel, Neumann, and Hankel functions for $n = 0$ are

$$\begin{cases} j_0(z) = \dfrac{\sin z}{z} \\ y_0(z) = -\dfrac{\cos z}{z} \end{cases} \qquad \begin{cases} h_0^{(1)}(z) = -\dfrac{\mathrm{i}}{z}\mathrm{e}^{\mathrm{i}z} \\ h_0^{(2)}(z) = \dfrac{\mathrm{i}}{z}\mathrm{e}^{-\mathrm{i}z} \end{cases}$$

and

$$\frac{(z j_0(z))'}{z} = \frac{\cos z}{z}, \qquad \frac{\left(z h_0^{(1)}(z)\right)'}{z} = \frac{1}{z}\mathrm{e}^{\mathrm{i}z}$$

---

[8]Between the spherical Bessel function, $j_l(z)$, and the cylindrical Bessel function, $J_n(z)$, see Appendix B.1, there is a relation

$$j_l(z) = \left(\frac{\pi}{2z}\right)^{1/2} J_{l+1/2}(z)$$

[9]Between the spherical Hankel functions, $h_l^{(1,2)}(z)$, and the cylindrical Hankel function, $H_n^{(1,2)}(z)$, see Appendix B.1, there is a relation

$$h_l^{(1,2)}(z) = \left(\frac{\pi}{2z}\right)^{1/2} H_{l+1/2}^{(1,2)}(z)$$

For $n = 1$, they are

$$\begin{cases} j_1(z) = \dfrac{\sin z}{z^2} - \dfrac{\cos z}{z} \\ y_1(z) = -\dfrac{\cos z}{z^2} - \dfrac{\sin z}{z} \end{cases} \qquad \begin{cases} h_1^{(1)}(z) = e^{iz}\left(-\dfrac{1}{z} - \dfrac{i}{z^2}\right) \\ h_1^{(2)}(z) = e^{-iz}\left(-\dfrac{1}{z} + \dfrac{i}{z^2}\right) \end{cases}$$

and

$$\frac{(zj_1(z))'}{z} = -\frac{\sin z}{z^3} + \frac{\cos z}{z^2} + \frac{\sin z}{z}, \qquad \frac{(zh_1^{(1)}(z))'}{z} = e^{iz}\left(-\frac{i}{z} + \frac{1}{z^2} + \frac{i}{z^3}\right)$$

From the power series representation in (B.9), we obtain for small arguments

$$\begin{cases} j_n(z) = \dfrac{2^n n! z^n}{(2n+1)!} + O(z^{n+2}) \\ h_n^{(1)}(z) = -i\dfrac{(2n)!}{2^n n! z^{n+1}} + O(z^{-n+1}) \\ (zj_n(z))' = \dfrac{2^n (n+1)! z^n}{(2n+1)!} + O(z^{n+2}) \\ (zh_n^{(1)}(z))' = i\dfrac{(2n)!}{2^n (n-1)! z^{n+1}} + O(z^{-n+1}) \end{cases} \quad \text{as } z \to 0 \tag{B.10}$$

Specifically, for $n = 1$

$$\begin{cases} j_1(z) \approx \dfrac{z}{3}, \quad \text{as } z \to 0 \\ h_1^{(1)}(z) \approx \dfrac{1}{iz^2}, \quad \text{as } z \to 0 \end{cases} \qquad \begin{cases} \dfrac{(zj_1(z))'}{z} \approx \dfrac{2}{3}, \quad \text{as } z \to 0 \\ \dfrac{(zh_1^{(1)}(z))'}{z} \approx -\dfrac{1}{iz^3}, \quad \text{as } z \to 0 \end{cases}$$

Alternative expansions of the spherical Bessel and Neumann functions in finite trigonometric series are [1]

$$\begin{cases} j_n(z) = \dfrac{1}{z}\left\{p_n(z)\sin\left(z - \dfrac{n\pi}{2}\right) + q_n(z)\cos\left(z - \dfrac{n\pi}{2}\right)\right\} \\ y_n(z) = \dfrac{1}{z}\left\{q_n(z)\sin\left(z - \dfrac{n\pi}{2}\right) - p_n(z)\cos\left(z - \dfrac{n\pi}{2}\right)\right\} \end{cases} \tag{B.11}$$

The explicit expressions of the finite sums $p_n(z)$ and $q_n(z)$ are

$$\begin{cases} p_n(z) = \displaystyle\sum_{k=0}^{[n/2]} (-1)^k \frac{(n+2k)!}{(2k)!(n-2k)!} \frac{1}{(2z)^{2k}} = 1 - \frac{b_n}{z^2} + O\left(z^{-4}\right) \\ q_n(z) = \displaystyle\sum_{k=0}^{[(n-1)/2]} (-1)^k \frac{(n+2k+1)!}{(2k+1)!(n-2k-1)!} \frac{1}{(2z)^{2k+1}} = \frac{a_n}{z} + O\left(z^{-3}\right) \end{cases} \tag{B.12}$$

where $[\cdot]$ denotes the integer part of the argument, and where $a_n = n(n+1)/2$ and $b_n = (n+2)(n+1)n(n-1)/8$. Note that these series are finite and that the series, by definition, is zero if the upper limit of summation is a negative integer.

Similarly, for the spherical Hankel functions, we have

$$\begin{cases} h_n^{(1)}(z) = \dfrac{\mathrm{e}^{\mathrm{i}z-\mathrm{i}(n+1)\pi/2}}{z} \displaystyle\sum_{k=0}^{n} \dfrac{(n+k)!}{k!(n-k)!} \dfrac{1}{(-2\mathrm{i}z)^k} = \dfrac{\mathrm{e}^{\mathrm{i}z-\mathrm{i}(n+1)\pi/2}}{z} (p_n(z) + \mathrm{i}q_n(z)) \\ h_n^{(2)}(z) = \dfrac{\mathrm{e}^{-\mathrm{i}z+\mathrm{i}(n+1)\pi/2}}{z} \displaystyle\sum_{k=0}^{n} \dfrac{(n+k)!}{k!(n-k)!} \dfrac{1}{(2\mathrm{i}z)^k} = \dfrac{\mathrm{e}^{-\mathrm{i}z+\mathrm{i}(n+1)\pi/2}}{z} (p_n(z) - \mathrm{i}q_n(z)) \end{cases} \tag{B.13}$$

and the asymptotic behaviors at the origin and at infinity are

$$\begin{aligned} p_n(z) \pm \mathrm{i}q_n(z) &= \sum_{k=0}^{n} \frac{(n+k)!}{k!(n-k)!} \frac{1}{(\mp 2\mathrm{i}z)^k} \\ &= \begin{cases} \dfrac{(\pm\mathrm{i})^n}{2^n z^n} \dfrac{(2n)!}{n!} \left(1 \mp \mathrm{i}z + O\left(z^2\right)\right) \\ 1 \pm \mathrm{i}\dfrac{a_n}{z} + O\left(z^{-2}\right) \end{cases} \end{aligned} \tag{B.14}$$

where $a_n = n(n+1)/2$.

In scattering problems, another combination of spherical Bessel and Hankel functions occurs, see also Appendix B.3.3 and the definition of the Riccati–Bessel functions

$$\begin{cases} \dfrac{(zj_n(z))'}{z} = \dfrac{1}{z}\left\{P_n(z)\sin\left(z - \dfrac{n\pi}{2}\right) + Q_n(z)\cos\left(z - \dfrac{n\pi}{2}\right)\right\} \\ \dfrac{\left(zh_n^{(1)}(z)\right)'}{z} = \dfrac{\mathrm{e}^{\mathrm{i}z-\mathrm{i}(n+1)\pi/2}}{z} (P_n(z) + \mathrm{i}Q_n(z)) \\ \dfrac{\left(zh_n^{(2)}(z)\right)'}{z} = \dfrac{\mathrm{e}^{-\mathrm{i}z+\mathrm{i}(n+1)\pi/2}}{z} (P_n(z) - \mathrm{i}Q_n(z)) \end{cases} \tag{B.15}$$

where

$$\begin{cases} P_n(z) = p_n'(z) - q_n(z) = -\dfrac{a_n}{z} + O\left(z^{-3}\right) \\ Q_n(z) = q_n'(z) + p_n(z) = 1 - \dfrac{a_n + b_n}{z^2} + O\left(z^{-4}\right) \end{cases} \tag{B.16}$$

and the asymptotic behaviors at the origin and at infinity are

$$\begin{aligned} P_n(z) \pm \mathrm{i}Q_n(z) &= p_n'(z) \pm \mathrm{i}q_n'(z) \pm \mathrm{i}(p_n(z) \pm \mathrm{i}q_n(z)) \\ &= \begin{cases} -\dfrac{(\pm\mathrm{i})^n}{2^n z^{n+1}} \dfrac{(2n)!}{(n-1)!} \left(1 \mp \mathrm{i}z + O\left(z^2\right)\right) \\ \pm\mathrm{i}\left(1 \pm \mathrm{i}\dfrac{a_n}{z} + O\left(z^{-2}\right)\right) \end{cases} \end{aligned} \tag{B.17}$$

Between different orders of spherical Bessel functions, there are recursion relations. Some of the important are ($n = 0, 1, 2, \ldots; m = 0, 1, 2, \ldots$)

$$\begin{cases} f_{n-1}(z) + f_{n+1}(z) = \dfrac{2n+1}{z} f_n(z) \\ n f_{n-1}(z) - (n+1) f_{n+1}(z) = (2n+1) f_n'(z) \\ \dfrac{n+1}{z} f_n(z) + f_n'(z) = f_{n-1}(z) \\ n f_n(z) - z f_{n+1}(z) = z f_n'(z) \\ \left(\dfrac{\mathrm{d}}{z\,\mathrm{d}z}\right)^m \left[z^{n+1} f_n(z)\right] = z^{n-m+1} f_{n-m}(z) \\ \left(\dfrac{\mathrm{d}}{z\,\mathrm{d}z}\right)^m \left[z^{-n} f_n(z)\right] = (-1)^m z^{-n-m} f_{n+m}(z) \end{cases} \tag{B.18}$$

Here, $f_n(z)$ is any combination of a spherical Bessel function, $j_n(z)$, a spherical Neumann function, $y_n(z)$, or the Hankel functions, $h_n^{(1)}(z)$ or $h_n^{(2)}(z)$.

The Wronskian between the spherical Bessel function $j_n(z)$ and the spherical Neumann function $y_n(z)$ is

$$j_n(z) y_n'(z) - j_n'(z) y_n(z) = \frac{1}{z^2} \tag{B.19}$$

and between the spherical Bessel function $j_n(z)$ and the spherical Hankel function $h_n^{(1)}(z)$ the Wronskian is

$$j_n(z) h_n^{(1)\prime}(z) - j_n'(z) h_n^{(1)}(z) = \frac{\mathrm{i}}{z^2}$$

and between the spherical Hankel functions $h_n^{(1)}(z)$ and $h_n^{(2)}(z)$ the Wronskian is

$$h_n^{(2)}(z) h_n^{(1)\prime}(z) - h_n^{(2)\prime}(z) h_n^{(1)}(z) = \frac{2\mathrm{i}}{z^2}$$

A very useful lemma in the textbook is

Lemma B.1 *The functions $h_n^{(1)}(z)$ and $h_n^{(1)\prime}(z)$ have no real zeros.*

Proof: By inspection, we conclude that $x = 0$ is not a root to neither $h_n^{(1)}(z)$ nor $h_n^{(1)\prime}(z)$. A root $x \neq 0$ to $h_n^{(1)}(x) = 0$ implies that both the real and imaginary parts must be zero, i.e., $j_n(x) = y_n(x) = 0$, which is in conflict with (B.19). Similarly, a root $x \neq 0$ to $h_n^{(1)\prime}(x) = 0$ implies that $j_n'(x) = y_n'(x) = 0$, which again is in conflict with (B.19). □

This lemma shows that the roots of the spherical Hankel functions cannot lie on the real axis. In fact, they are all located symmetrically w.r.t. the imaginary axis in the lower complex half-plane. Specifically, the only root in the finite complex plane of $h_1^{(1)}(z)$ is $z = -\mathrm{i}$, and the roots in the finite complex plane of $(z h_1^{(1)}(z))'$ are $z = (\pm\sqrt{3} - \mathrm{i})/2$.

The roots of the spherical Hankel functions play an important role in the interpretation of scattering contributions. For spherical objects, the positions of the singularity expansion method (SEM) poles are related to the roots of combinations of the spherical Hankel and Bessel functions. For more details on this method, see Reference 113.

B.3.1 Integral Representations

The spherical Bessel and Hankel functions have the following useful integral representations:

$$j_n(z) = \frac{1}{2\mathrm{i}^n} \int_{-1}^{1} P_n(t) \mathrm{e}^{\mathrm{i}tz} \, \mathrm{d}t \tag{B.20}$$

and

$$\begin{cases} h_n^{(1)}(z) = -\dfrac{1}{\mathrm{i}^n} \displaystyle\int_{\Gamma_1} P_n(t) \mathrm{e}^{\mathrm{i}tz} \, \mathrm{d}t \\ h_n^{(2)}(z) = \dfrac{1}{\mathrm{i}^n} \displaystyle\int_{\Gamma_2} P_n(t) \mathrm{e}^{\mathrm{i}tz} \, \mathrm{d}t \end{cases} \tag{B.21}$$

where $P_n(t)$ are the Legendre polynomials, and the complex contours Γ_1 and Γ_2 are defined in Figure B.2, which for a real argument z can be specialized as in Figure B.3.

We prove these identities by the use of the differential equations of the spherical Bessel functions and the Legendre polynomials, i.e.,

$$z^2 f_n''(z) + 2z f_n'(z) + (z^2 - n(n+1)) f_n(z) = 0$$

and

$$(1 - t^2) P_n''(t) - 2t P_n'(t) + n(n+1) P_n(t) = 0 \tag{B.22}$$

respectively.

First prove that an integral

$$f_n(z) = C_n \int_{\Gamma} P_n(t) \mathrm{e}^{\mathrm{i}tz} \, \mathrm{d}t \tag{B.23}$$

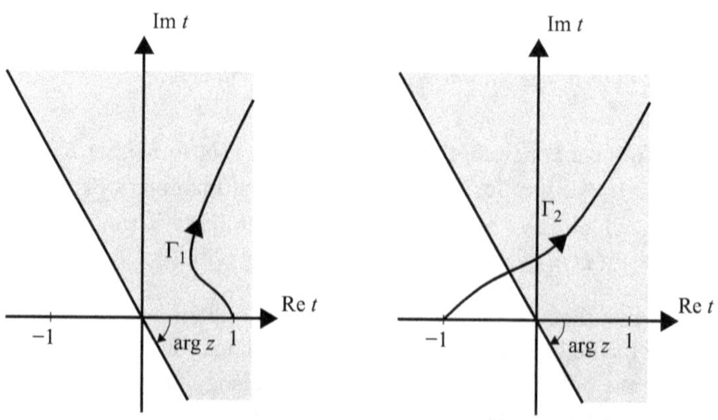

Figure B.2 The complex integration contours Γ_1 and Γ_2. The contours Γ_1 and Γ_2 can end at infinity anywhere in the shaded half-plane.

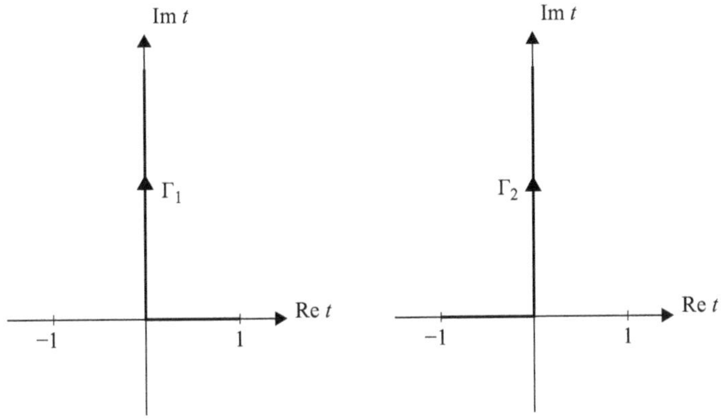

Figure B.3 The complex integration contours Γ_1 and Γ_2 for a real argument z.

satisfies the differential equation of the spherical Bessel functions under certain conditions. The contour Γ and the constant C_n are determined below. We get

$$\left[z^2\frac{\mathrm{d}^2}{\mathrm{d}z^2}+2z\frac{\mathrm{d}}{\mathrm{d}z}+z^2-n(n+1)\right]\int_\Gamma P_n(t)\mathrm{e}^{\mathrm{i}tz}\,\mathrm{d}t$$
$$=\int_\Gamma\left[-t^2z^2+2\mathrm{i}zt+z^2-n(n+1)\right]P_n(t)\mathrm{e}^{\mathrm{i}tz}\,\mathrm{d}t$$

The use of (B.22) and integration by parts give

$$\begin{aligned}&\left[z^2\frac{\mathrm{d}^2}{\mathrm{d}z^2}+2z\frac{\mathrm{d}}{\mathrm{d}z}+z^2-n(n+1)\right]\int_\Gamma P_n(t)\mathrm{e}^{\mathrm{i}tz}\,\mathrm{d}t\\
&=\int_\Gamma\left[-t^2z^2+2\mathrm{i}zt+z^2\right]P_n(t)\mathrm{e}^{\mathrm{i}tz}\,\mathrm{d}t+\int_\Gamma\left[(1-t^2)P_n''(t)-2tP_n'(t)\right]\mathrm{e}^{\mathrm{i}tz}\,\mathrm{d}t\\
&=\int_\Gamma\left[-t^2z^2+2\mathrm{i}zt+z^2+2\mathrm{i}tz+2\right]P_n(t)\mathrm{e}^{\mathrm{i}tz}\,\mathrm{d}t\\
&\quad+\left\{(1-t^2)P_n'(t)-2tP_n(t)\right\}\mathrm{e}^{\mathrm{i}tz}\Big|_{\partial\Gamma}-\int_\Gamma\left[\mathrm{i}z(1-t^2)-2t\right]P_n'(t)\mathrm{e}^{\mathrm{i}tz}\,\mathrm{d}t\\
&=(1-t^2)\left\{P_n'(t)-\mathrm{i}zP_n(t)\right\}\mathrm{e}^{\mathrm{i}tz}\Big|_{\partial\Gamma}\end{aligned}$$

We see that the integral in (B.23) is a solution to the differential equation of the spherical Bessel functions provided

$$(1-t^2)\left\{P_n'(t)-\mathrm{i}zP_n(t)\right\}\mathrm{e}^{\mathrm{i}tz}\Big|_{\partial\Gamma}=0$$

where $\partial\Gamma$ denotes the end points of the contour Γ.

We now choose the contour Γ such that the contributions at the end points vanish. This can be done in several ways. The first obvious way is to let the contour Γ be the linear segment

$[-1, 1]$. Then, at the end points, $z = \pm 1$, the contributions vanish. The solution in (B.23) then is regular at $z = 0$, and consequently, it must be a multiple of $j_n(z)$, i.e.,

$$j_n(z) = C_n \int_{-1}^{1} P_n(t)\mathrm{e}^{\mathrm{i}tz}\, \mathrm{d}t$$

To determine the constant C_n, we expand the exponential in the integrand and identify the lowest non-vanishing power of z on both sides in the identity. The spherical Bessel function has an expansion, see (B.10)

$$j_n(z) = \frac{2^n n!}{(2n+1)!} z^n + O(z^{n+2})$$

The integral has the expansion

$$\int_{-1}^{1} P_n(t)\mathrm{e}^{\mathrm{i}tz}\, \mathrm{d}t = \sum_{k=n}^{\infty} \frac{\mathrm{i}^k z^k}{k!} \int_{-1}^{1} P_n(t) t^k\, \mathrm{d}t$$

where we have used the orthogonality of the Legendre polynomials

$$\int_{-1}^{1} P_n(t) t^k\, \mathrm{d}t = 0, \qquad k = 0, 1, 2, \ldots, n-1$$

to show that the sum starts at $k = n$. The use of the series expansion of the Legendre polynomials in Appendix C.1 gives

$$t^n = \frac{2^n n! n!}{(2n)!} P_n(t) + \sum_{k=0}^{n-1} b_k(n) P_k(t)$$

and therefore, by orthogonality, the lowest non-vanishing power in z of the integral is

$$\int_{-1}^{1} P_n(t)\mathrm{e}^{\mathrm{i}tz}\, \mathrm{d}t = \frac{\mathrm{i}^n z^n}{n!} \frac{2^n n! n!}{(2n)!} \frac{2}{2n+1} + O(z^{n+2}) = \frac{\mathrm{i}^n 2^{n+1} n!}{(2n+1)!} z^n + O(z^{n+2})$$

From this, we can conclude that the undetermined coefficient C_n is

$$C_n 2\mathrm{i}^n = 1$$

and (B.20) is proved.

Another choice of the contour Γ leads to the spherical Hankel functions. Let Γ_1 and Γ_2 start at $t = \pm 1$, respectively, and end at infinity $t = \infty \mathrm{e}^{\mathrm{i}\phi}$, where $\phi \in (-\arg z, \pi - \arg z)$. Then again, the contributions at the end points vanish, since $\operatorname{Im}(tz) > 0$ at the end point at infinity. To determine what type of solution it is, we determine the behavior of these solutions as $|z| \to \infty$. The asymptotic behavior at large arguments is obtained by integration by parts. We get

$$\begin{cases} \displaystyle\int_{\Gamma_1} P_n(t)\mathrm{e}^{\mathrm{i}tz}\, \mathrm{d}t = \frac{1}{\mathrm{i}z} P_n(t)\mathrm{e}^{\mathrm{i}tz}\Big|_{\partial\Gamma_1} \left(1 + O\left(z^{-1}\right)\right) = -\frac{1}{\mathrm{i}z}\mathrm{e}^{\mathrm{i}z}\left(1 + O\left(z^{-1}\right)\right) \\ \displaystyle\int_{\Gamma_2} P_n(t)\mathrm{e}^{\mathrm{i}tz}\, \mathrm{d}t = \frac{1}{\mathrm{i}z} P_n(t)\mathrm{e}^{\mathrm{i}tz}\Big|_{\partial\Gamma_2} \left(1 + O\left(z^{-1}\right)\right) = -\frac{1}{\mathrm{i}z}(-1)^n\mathrm{e}^{-\mathrm{i}z}\left(1 + O\left(z^{-1}\right)\right) \end{cases}$$

A comparison with the leading behavior for the spherical Hankel functions in (B.13)

$$\begin{cases} h_n^{(1)}(z) = \dfrac{e^{iz-i(n+1)\pi/2}}{z}\left(1+O\left(z^{-1}\right)\right) \\ h_n^{(2)}(z) = \dfrac{e^{-iz+i(n+1)\pi/2}}{z}\left(1+O\left(z^{-1}\right)\right) \end{cases}$$

determines the specific constants, and completes the derivation of the integral representations in (B.21).

B.3.2 Modulus of a Spherical Hankel Function

The absolute value (modulus) of the spherical Hankel functions, $h_n^{(1)}(x)$ and $h_n^{(2)}(x)$, has special properties on the real axis, and in this section we show some of these properties. To this end, we specialize to a positive argument $x > 0$, and observe that on the positive real axis, we have

$$\left|h_n^{(1)}(x)\right|^2 = \left|h_n^{(2)}(x)\right|^2 = j_n^2(x) + y_n^2(x) = h_n^{(1)}(x)h_n^{(2)}(x)$$

We already know that $h_n^{(1)}(x)$ has no real roots, see Lemma B.1, but we also prove that its absolute value is a decreasing function on the positive real axis. In fact, we have

Lemma B.2 *For each fixed positive integer n, the functions*

$$\begin{cases} \left|h_n^{(i)}(x)\right|^2 \\ \left|xh_n^{(i)}(x)\right|^2 \end{cases} \qquad \begin{cases} \left|(xh_n^{(i)}(x))'\right|^2 + \left|xh_n^{(i)}(x)\right|^2 \\ \left|(xh_n^{(i)}(x))'\right|^2 + n(n+1)\left|h_n^{(i)}(x)\right|^2 \end{cases} \qquad i = 1, 2$$

are decreasing functions on the positive real axis. The limits as $x \to \infty$ are

$$\lim_{x\to\infty} \left|h_n^{(i)}(x)\right|^2 = 0, \qquad \lim_{x\to\infty} \left|xh_n^{(i)}(x)\right|^2 = 1, \qquad \lim_{x\to\infty} \left|(xh_n^{(i)}(x))'\right|^2 = 1$$

and

$$\begin{cases} \lim_{x\to\infty} \left|(xh_n^{(i)}(x))'\right|^2 + \left|xh_n^{(i)}(x)\right|^2 = 2 \\ \lim_{x\to\infty} \left|(xh_n^{(i)}(x))'\right|^2 + n(n+1)\left|h_n^{(i)}(x)\right|^2 = 1 \end{cases}$$

for $i = 1, 2$.

Proof: We now prove this lemma starting with Nicholson's integral[10,11] [288, (1) on page 444]

$$j_n^2(x) + y_n^2(x) = \frac{4}{\pi x}\int_0^\infty K_0(2x\sinh t)\cosh\left[(2n+1)t\right]\, dt, \qquad x > 0$$

[10] The identity is not restricted to integer values of n, but n can be an arbitrary complex number.

[11] John William Nicholson (1881–1955), English mathematician.

where $K_0(x)$ is the modified Bessel function, see (B.7). The integral representation in (B.8) shows that $K_0(x)$ is real valued and decreasing on the positive real axis. Thus, $|h_n^{(1,2)}(x)|^2$ is a decreasing function on the positive real axis, i.e.,

$$\frac{\mathrm{d}}{\mathrm{d}x}\left|h_n^{(1)}(x)\right|^2 < 0, \qquad x > 0$$

and the first part of the lemma is proved.

To prove that $|xh_n^{(1,2)}(x)|^2$ is a decreasing function, use Nicholson's integral and integrate by parts. We get

$$\begin{aligned}
\frac{\mathrm{d}}{\mathrm{d}x}\left(x^2 j_n^2(x) + x^2 y_n^2(x)\right) &= \frac{\mathrm{d}}{\mathrm{d}x}\left\{\frac{4x}{\pi}\int_0^\infty K_0(2x\sinh t)\cosh\left[(2n+1)t\right]\,\mathrm{d}t\right\} \\
&= \frac{4}{\pi}\int_0^\infty \left\{K_0(2x\sinh t) + 2x\sinh t K_0'(2x\sinh t)\right\}\cosh\left[(2n+1)t\right]\,\mathrm{d}t \\
&= \frac{4}{\pi}\left\{\tanh t\cosh\left[(2n+1)t\right]K_0(2x\sinh t)\right|_{t=0}^{t=\infty} \\
&\quad + \frac{4}{\pi}\int_0^\infty K_0(2x\sinh t)\left\{\cosh\left[(2n+1)t\right] - \frac{\mathrm{d}}{\mathrm{d}t}\left(\tanh t\cosh\left[(2n+1)t\right]\right)\right\}\mathrm{d}t\right\}
\end{aligned}$$

The first term on the right-hand side vanishes, since

$$\tanh t K_0(2x\sinh t) \approx -t\ln xt \to 0, \quad \text{as } t \to 0$$

and we obtain

$$\begin{aligned}
\frac{\mathrm{d}}{\mathrm{d}x}\left(x^2 j_n^2(x) + x^2 y_n^2(x)\right) = \frac{4}{\pi}\int_0^\infty & K_0(2x\sinh t)\cosh\left[(2n+1)t\right]\tanh t \\
& \times \{\tanh t - (2n+1)\tanh\left[(2n+1)t\right]\}\,\mathrm{d}t
\end{aligned}$$

The integrand is negative, since

$$(2n+1)\tanh\left[(2n+1)t\right] - \tanh t$$

is positive for $t \geq 0$ and n a positive integer, and we conclude that

$$\frac{\mathrm{d}}{\mathrm{d}x}\left|xh_n^{(1)}(x)\right|^2 = \frac{\mathrm{d}}{\mathrm{d}x}\left(x^2 j_n^2(x) + x^2 y_n^2(x)\right) < 0, \qquad x > 0$$

and the second part of the lemma is proved.

Using the differential equations, we obtain

$$\begin{aligned}
\frac{\mathrm{d}}{\mathrm{d}x}\left|(xh_n^{(1)}(x))'\right|^2 &= (xh_n^{(1)}(x))''(xh_n^{(2)}(x))' + (xh_n^{(1)}(x))'(xh_n^{(2)}(x))'' \\
&= \left(\frac{n(n+1)}{x^2} - 1\right)\left\{xh_n^{(1)}(x)(xh_n^{(2)}(x))' + (xh_n^{(1)}(x))'xh_n^{(2)}(x)\right\} \\
&= \left(\frac{n(n+1)}{x^2} - 1\right)\frac{\mathrm{d}}{\mathrm{d}x}\left|xh_n^{(1)}(x)\right|^2
\end{aligned}$$

From this identity, we see that $|(xh_n^{(1)}(x))'|^2$ is not monotonic,[12] but, on the other hand for $n = 1, 2, \ldots$

$$\frac{\mathrm{d}}{\mathrm{d}x}\left\{\left|(xh_n^{(1)}(x))'\right|^2 + \left|xh_n^{(1)}(x)\right|^2\right\} = \frac{n(n+1)}{x^2}\frac{\mathrm{d}}{\mathrm{d}x}\left|xh_n^{(1)}(x)\right|^2 < 0, \qquad x > 0$$

and

$$\begin{aligned}
&\frac{\mathrm{d}}{\mathrm{d}x}\left\{\left|(xh_n^{(1)}(x))'\right|^2 + n(n+1)\left|h_n^{(1)}(x)\right|^2\right\} \\
&\quad = \left(\frac{n(n+1)}{x^2} - 1\right)\frac{\mathrm{d}}{\mathrm{d}x}\left|xh_n^{(1)}(x)\right|^2 + n(n+1)\frac{\mathrm{d}}{\mathrm{d}x}\left|xh_n^{(1)}(x)\right|^2 - n(n+1)2x\left|h_n^{(1)}(x)\right|^2 \\
&\quad = \left(\frac{n(n+1)}{x^2} - 1 + n(n+1)\right)\frac{\mathrm{d}}{\mathrm{d}x}\left|xh_n^{(1)}(x)\right|^2 - n(n+1)2x\left|h_n^{(1)}(x)\right|^2 < 0, \qquad x > 0
\end{aligned}$$

and the remaining two parts of the lemma are proved. The limit values are easily proved by the properties of the spherical Hankel functions for large real arguments. □

We complete this section with a series expression of the absolute value of the spherical Hankel functions [1, 206]

$$\left|h_n^{(1)}(x)\right|^2 = \sum_{m=0}^{n} a_m(n)x^{-2m-2}$$

The coefficients, $a_m(n)$, all positive, are given by

$$a_m(n) = \frac{(2m)!(n+m)!}{2^{2m}(m!)^2(n-m)!} = \frac{(2m-1)!!(n+m)!}{(2m)!!(n-m)!}, \qquad m = 0, 1, 2, \ldots, n$$

B.3.3 Related Functions

Finally, we identify the differential equation for a spherical Bessel function times a power of z, i.e., $F_{\nu n}(z) = z^{\nu} f_n(z)$, where $f_n(z)$ is a spherical Bessel, Neumann, or Hankel function, and where $\nu = 0, \pm 1, \pm 2, \ldots$ is an integer. The function $F_{\nu n}(z)$ satisfies

$$z^2\frac{\mathrm{d}^2}{\mathrm{d}z^2}F_{\nu n}(z) + 2z(1-\nu)\frac{\mathrm{d}}{\mathrm{d}z}F_{\nu n}(z) + (z^2 - n(n+1) + \nu(\nu+1) - 2\nu)F_{\nu n}(z) = 0 \qquad \text{(B.24)}$$

The solution for $\nu = 1$ becomes particularly simple and important

$$z^2\frac{\mathrm{d}^2}{\mathrm{d}z^2}F_{1n}(z) + (z^2 - n(n+1))F_{1n}(z) = 0 \qquad \text{(B.25)}$$

[12] In fact, it has a minimum at $x = \sqrt{n(n+1)}$.

The functions F_{1n} are important in scattering problems and are called the Riccati–Bessel functions.[13] These are usually defined as [277]

$$\begin{cases} \psi_n(z) = z j_n(z) \\ \xi_n(z) = z h_n^{(1)}(z) \end{cases} \tag{B.26}$$

The Riccati–Bessel functions for $n = 0, 1$ are

$$\begin{cases} \psi_0(z) = \sin z \\ \xi_0(z) = -\mathrm{i}\mathrm{e}^{\mathrm{i}z} \end{cases} \qquad \begin{cases} \psi_1(z) = \dfrac{\sin z}{z} - \cos z \\ \xi_1(z) = -\mathrm{e}^{\mathrm{i}z}\left(1 + \dfrac{\mathrm{i}}{z}\right) \end{cases}$$

For small arguments, we have

$$\begin{cases} \psi_n(z) = \dfrac{2^n n! z^{n+1}}{(2n+1)!} + O(z^{n+3}) \\ \xi_n(z) = -\mathrm{i}\dfrac{(2n)!}{2^n n! z^n} + O(z^{-n+2}) \end{cases} \quad \text{as } z \to 0 \tag{B.27}$$

The Riccati–Bessel functions have a finite trigonometric series expansion. They are

$$\begin{cases} \psi_n(z) = p_n(z) \sin\left(z - \dfrac{n\pi}{2}\right) + q_n(z) \cos\left(z - \dfrac{n\pi}{2}\right) \\ \xi_n(z) = \mathrm{e}^{\mathrm{i}z - \mathrm{i}(n+1)\pi/2}\left(p_n(z) + \mathrm{i} q_n(z)\right) \end{cases} \tag{B.28}$$

The explicit expressions of the finite sums $p_n(z)$ and $q_n(z)$ are given in (B.12).

The derivatives of the Riccati–Bessel functions have the following asymptotic behavior for large arguments:

$$\begin{cases} \psi_n'(z) = P_n(z) \sin\left(z - \dfrac{n\pi}{2}\right) + Q_n(z) \cos\left(z - \dfrac{n\pi}{2}\right) \\ \xi_n'(z) = \mathrm{e}^{\mathrm{i}z - \mathrm{i}(n+1)\pi/2}\left(P_n(z) + \mathrm{i} Q_n(z)\right) \end{cases} \tag{B.29}$$

where $P_n(z)$ and $Q_n(z)$ are given in (B.16) and (B.17).

The Wronskian between $\psi_n(z)$ and $\xi_n(z)$ is

$$\psi_n(z)\xi_n'(z) - \psi_n'(z)\xi_n(z) = \mathrm{i} \tag{B.30}$$

Lemma B.3 *The functions $\xi_n(z)$ and $\xi_n'(z)$ have no real zeros.*

Proof: By inspection, we conclude that $x = 0$ is not a root to neither $\xi_n(z)$ nor $\xi_n'(z)$, see (B.28) and (B.29). The lack of a root $x \neq 0$ to $\xi_n(z) = z h_n^{(1)}(z)$ then follows directly from Lemma B.1 on page 611. Finally, the function $\xi_n'(x)$ has no real zeros, since a real root, $x \neq 0$, to $\xi_n(x)$ implies that $j_n(x) + x j_n'(x) = y_n(x) + x y_n'(x) = 0$, which is in conflict with (B.19). □

---

[13] Jacopo Francesco Riccati (1676–1754), Italian mathematician.

Many relations are derived from the properties of the spherical Bessel and Hankel functions above. For example, a set of useful recursion relations of the Riccati–Bessel functions is, see (B.18)

$$\begin{cases} g_{n+1}(z) + g_{n-1}(z) = \dfrac{2n+1}{z} g_n(z) \\ g_n'(z) = (n+1)\dfrac{g_n(z)}{z} - g_{n+1}(z) = g_{n-1}(z) - n\dfrac{g_n(z)}{z} \\ g_n'(z) = \dfrac{(n+1)g_{n-1}(z) - n g_{n+1}(z)}{2n+1} \end{cases} \qquad n = 0, 1, 2, \ldots$$

where $g_n(z)$ is any combination of the functions $\psi_n(z)$ and $\xi_n(z)$.

In connection to the transition matrix entries of a dielectric sphere in Section 8.2, we encountered the logarithmic derivative of the Riccati–Bessel functions, see (8.8) on page 420

$$D_n(z) = \frac{\psi_n'(z)}{\psi_n(z)}, \qquad G_n(z) = \frac{\xi_n'(z)}{\xi_n(z)}$$

The recursion relations above imply that we have the following recursion relation for these logarithmic derivatives:

$$D_{n-1}(z) = \frac{n}{z} - \frac{1}{D_n(z) + n/z}$$

Computation of $D_n(z)$ can conveniently be made by a downward recurrence scheme, which is proved numerically stable [119]. The start value is not critical, zero can be used. Computation of $G_n(x)$, with a real argument x, can be made by an upward recurrence relation [119]. The start value is $G_0(x) = \mathrm{i}$, and the upward recurrence scheme is

$$G_n(z) = -\frac{n}{z} - \frac{1}{n/z - G_{n-1}(z)}$$

More details on this topic are found in Reference 159.

APPENDIX C

Spherical Harmonics

In this appendix, we define the Legendre polynomials that are used in the book, and moreover, we list some of their most important properties. The derivations and more details are found in, e.g., Reference 4. In scattering theory, the spherical harmonics and the vector spherical harmonics play an important role, and after the definition of these functions, we also present some of their important properties. Specifically, the transformation properties between spherical and plane waves are derived.

C.1 Legendre Polynomials

We define the Legendre[1] polynomials, $P_n(x)$, from the generating function $F(x, t)$, i.e.,

$$F(x,t) = \frac{1}{\left(1 - 2xt + t^2\right)^{1/2}} = \sum_{n=0}^{\infty} P_n(x)t^n, \qquad |t| < 1$$

An alternative way of definition of the Legendre polynomials is by Rodrigues' formula[2]

$$P_n(x) = \frac{1}{2^n n!} \frac{\mathrm{d}^n}{\mathrm{d}x^n} \left(x^2 - 1\right)^n$$

This leads to a power series expansion

$$P_n(x) = \sum_{k=0}^{[n/2]} (-1)^k \frac{(2n-2k)!}{2^n k!(n-k)!(n-2k)!} x^{n-2k}$$

where $[x]$ is the integer part of the real number x.

---

[1]Adrien-Marie Legendre (1752–1833), French mathematician.

[2]Benjamin Olinde Rodrigues (1795–1851), French mathematician.

The first polynomials are

$$\begin{cases} P_0(x) = 1 \\ P_1(x) = x \\ P_2(x) = (3x^2 - 1)/2 \\ P_3(x) = (5x^3 - 3x)/2 \end{cases}$$

Between different orders of the Legendre polynomials, there are relations, viz. the recursion relations. The most important are

$$\begin{cases} (n+1)P_n(x) = P'_{n+1}(x) - xP'_n(x) \\ nP_n(x) = xP'_n(x) - P'_{n-1}(x) \\ (n+1)P_{n+1}(x) = (2n+1)xP_n(x) - nP_{n-1}(x) \\ (2n+1)P_n(x) = P'_{n+1}(x) - P'_{n-1}(x) \\ (1-x^2)P'_n(x) = (n+1)\left(xP_n(x) - P_{n+1}(x)\right) = n\left(P_{n-1}(x) - xP_n(x)\right) \end{cases}$$

With these recursion relations, we easily prove

$$P_n(1) = 1, \qquad P'_n(1) = \frac{n(n+1)}{2}$$

for all $n = 0, 1, 2, \ldots$. In general, the jth derivative evaluated at $x = 1$ is

$$P_n^{(j)}(1) = \frac{(n+j)!}{2^j j!(n-j)!}$$

The value at the origin is ($n = 0, 1, 2, 3, \ldots$)

$$P_{2n}(0) = (-1)^n \frac{(2n)!}{2^{2n}(n!)^2} = (-1)^n \frac{(2n-1)!!}{(2n)!!}, \qquad P_{2n+1}(0) = 0$$

Moreover, we have

$$P_n(-x) = (-1)^n P_n(x)$$

From the result above, we get the value at $x = -1$

$$P_n(-1) = (-1)^n, \qquad P'_n(-1) = (-1)^{n+1} \frac{n(n+1)}{2}$$

In general, the jth derivative evaluated at $x = -1$ is

$$P_n^{(j)}(-1) = (-1)^{n+j} \frac{(n+j)!}{2^j j!(n-j)!}$$

The Legendre polynomials are orthogonal on the interval $[-1, 1]$, i.e.,

$$\int_{-1}^{1} P_n(x)P_{n'}(x)\, dx = \frac{2\delta_{nn'}}{2n+1}$$

and they satisfy the differential equation

$$(1-x^2)P_n''(x) - 2xP_n'(x) + n(n+1)P_n(x) = 0$$

Moreover, the set of polynomials $\{P_n(x)\}_{n=0}^{\infty}$ is a complete orthogonal system on the interval $I = [-1, 1]$. This means that every square integrable function, defined in the interval $[-1, 1]$, has a convergent (generalized) Fourier series. The convergence is in general only in the $L^2(I)$-norm.

$$f(x) = \sum_{n=0}^{\infty} a_n P_n(x), \qquad x \in [-1, 1]$$

where the (Fourier) coefficients a_n are determined by the integrals

$$a_n = \frac{2n+1}{2} \int_{-1}^{1} f(x) P_n(x)\, \mathrm{d}x$$

Other useful integrals that involve Legendre polynomials—obtained by the recursion relations, orthogonality, and by iteration—are

$$\int_{-1}^{1} x P_{n'}(x) P_n(x)\, \mathrm{d}x = \frac{2(n+1)}{(2n+1)(2n+3)} \delta_{n',n+1} + \frac{2n}{(2n-1)(2n+1)} \delta_{n',n-1}$$

and

$$\int_{-1}^{1} P_n'(x) P_{n'}(x)\, \mathrm{d}x = \begin{cases} 1 - (-1)^{n+n'}, & n > n' \\ 0, & n \leq n' \end{cases} \tag{C.1}$$

The last integral implies

$$P_n'(x) = \sum_{n'=0}^{[(n-1)/2]} (2n - 4n' - 1) P_{n-2n'-1}(x) \tag{C.2}$$

C.1.1 Combinations of Legendre Polynomials

In the analysis of scattering by spherical objects, specific combinations of Legendre polynomials appear. We introduce the polynomials $\pi_n(x)$ and $\tau_n(x)$ defined as

$$\begin{cases} \pi_n(x) = P_n'(x) \\ \tau_n(x) = n(n+1)P_n(x) - xP_n'(x) = n^2 P_n(x) - \pi_{n-1}(x) \end{cases} \tag{C.3}$$

The first polynomials are:

$$\begin{cases} \pi_0(x) = 0 \\ \pi_1(x) = 1 \\ \pi_2(x) = 3x \\ \pi_3(x) = (15x^2 - 3)/2 \end{cases} \qquad \begin{cases} \tau_0(x) = 0 \\ \tau_1(x) = x \\ \tau_2(x) = 6x^2 - 3 \\ \tau_3(x) = 3x(15x^2 - 11)/2 \end{cases}$$

and special values are

$$\begin{cases} \pi_n(1) = \tau_n(1) = \dfrac{n(n+1)}{2} \\ \pi_n(-1) = (-1)^{n+1}\dfrac{n(n+1)}{2}, \quad \tau_n(-1) = (-1)^n \dfrac{n(n+1)}{2} \end{cases}$$

A useful integral is

$$\int_{-1}^{1} \pi_n(x)\pi_{n'}(x)(1-x^2)\,\mathrm{d}x = -\int_{-1}^{1} P_n(x) \underbrace{\frac{\mathrm{d}}{\mathrm{d}x}\left(P'_{n'}(x)(1-x^2)\right)}_{-n(n+1)P_n(x)} \mathrm{d}x = \frac{2n(n+1)\delta_{nn'}}{2n+1}$$

where we used the differential equation of the Legendre polynomials. Moreover, from (C.1) we get

$$\int_{-1}^{1} \pi_n(x)P_{n'}(x)\,\mathrm{d}x = \begin{cases} 2, & n > n' \text{ and } n - n' \text{ an odd integer} \\ 0, & \text{otherwise} \end{cases}$$

and

$$\int_{-1}^{1} \tau_n(x)P_{n'}(x)\,\mathrm{d}x = \frac{2n^2\delta_{nn'}}{2n+1} - \begin{cases} 2, & n > n'+1 \text{ and } n - n' \text{ an even integer} \\ 0, & \text{otherwise} \end{cases}$$

These last two integrals and (C.2) then give

$$\int_{-1}^{1} \pi_n(x)\pi_{n'}(x)\,\mathrm{d}x = \begin{cases} \min\{n,n'\}(\min\{n,n'\}+1), & n - n' \text{ an even integer} \\ 0, & \text{otherwise} \end{cases}$$

and

$$\int_{-1}^{1} \pi_n(x)\tau_{n'}(x)\,\mathrm{d}x = \begin{cases} n'(n'+1), & n > n' \text{ and } n - n' \text{ an odd integer} \\ -n(n+1), & n \le n'-1 \text{ and } n - n' \text{ an odd integer} \\ 0, & \text{otherwise} \end{cases}$$

and finally

$$\begin{aligned} \int_{-1}^{1} \tau_n(x)\tau_{n'}(x)\,\mathrm{d}x = &\ \frac{2n^2(n+1)^2\delta_{nn'}}{2n+1} \\ &- \begin{cases} \min\{n,n'\}(\min\{n,n'\}+1), & n - n' \text{ an even integer} \\ 0, & \text{otherwise} \end{cases} \end{aligned}$$

C.2 Associated Legendre Functions

The associated Legendre functions, defined for positive integers $l, m \geq 0$, are denoted $P_l^m(x)$ and defined by

$$P_l^m(x) = (1 - x^2)^{m/2} \frac{d^m}{dx^m} P_l(x), \qquad x \in [-1, 1]$$

where $P_l(x)$ is the Legendre polynomial defined in Appendix C.1. For a negative integer value of m, we use $(m = -1, -2, \ldots)$

$$P_l^m(x) = (-1)^{|m|} \frac{(l - |m|)!}{(l + |m|)!} P_l^{|m|}(x), \qquad x \in [-1, 1]$$

The integer l is a non-negative integer, i.e., $l = 0, 1, 2, 3, \ldots$, and the integer m is an integer in the interval $m = -l, -l + 1, \ldots, -1, 0, 1, \ldots, l$, since $|m| > l$ leads to $P_l^m(x) = 0$.

The functions are either even or odd, depending on the value of $l + m$.

$$P_l^m(-x) = (-1)^{l+m} P_l^m(x)$$

Between different associated Legendre function, there exist relations. These are either recursion relations in the index l or the azimuthal index m or a combination of these indices. Some of the useful recursion relations are:

$$\begin{cases} \dfrac{2mx}{\left(1 - x^2\right)^{1/2}} P_l^m(x) = P_l^{m+1}(x) + (l + m)(l - m + 1) P_l^{m-1}(x) \\ (2l + 1)(1 - x^2)^{1/2} P_l^m(x) = P_{l+1}^{m+1}(x) - P_{l-1}^{m+1}(x) \\ \qquad = (l + m)(l + m - 1) P_{l-1}^{m-1}(x) - (l - m + 1)(l - m + 2) P_{l+1}^{m-1}(x) \\ (1 - x^2)^{1/2} P_l^{m\prime}(x) = \dfrac{1}{2} P_l^{m+1}(x) - \dfrac{1}{2}(l + m)(l - m + 1) P_l^{m-1}(x) \end{cases}$$

Useful are also

$$\begin{cases} (1 - x^2) P_l^{m\prime}(x) = (l + m) P_{l-1}^m(x) - l x P_l^m(x) = (l + 1) x P_l^m(x) - (l - m + 1) P_{l+1}^m(x) \\ (2l + 1)(1 - x^2) P_l^{m\prime}(x) = (l + 1)(l + m) P_{l-1}^m(x) - l(l - m + 1) P_{l+1}^m(x) \\ (l - m + 1) P_{l+1}^m(x) = (2l + 1) x P_l^m(x) - (l + m) P_{l-1}^m(x) \end{cases}$$

We also note that

$$P_l^0(x) = P_l(x)$$

and

$$P_l^m(1) = \delta_{m0}$$

Other useful limits are

$$\lim_{x\to 1}\sqrt{1-x^2}P_l^{m\prime}(x) = -\delta_{m1}\frac{l(l+1)}{2}, \qquad \lim_{x\to 1}\frac{P_l^m(x)}{\sqrt{1-x^2}} = \delta_{m1}\frac{l(l+1)}{2}$$

The value at the origin is

$$P_l^m(0) = \begin{cases} (-1)^{(l-m)/2}\dfrac{(l+m)!}{2^l\left(\frac{l-m}{2}\right)!\left(\frac{l+m}{2}\right)!}, & l+m \text{ even} \\ 0, \quad l+m \text{ odd} \end{cases}$$

and the value of the derivative at the origin is

$$P_l^{m\prime}(0) = \begin{cases} 0, \quad l+m \text{ even} \\ (l+m)P_{l-1}^m(0) = -(l-m+1)P_{l+1}^m(0), \quad l+m \text{ odd} \end{cases}$$

The value of the second derivative at the origin can be expressed in the value of the function itself by the use of the differential equation below. We have

$$P_l^{m\prime\prime}(0) = \begin{cases} \left(m^2 - l(l+1)\right)P_l^m(0), & l+m \text{ even} \\ 0, \quad l+m \text{ odd} \end{cases}$$

The associated Legendre functions satisfy the following ODE

$$(1-x^2)\frac{\mathrm{d}^2}{\mathrm{d}x^2}P_l^m(x) - 2x\frac{\mathrm{d}}{\mathrm{d}x}P_l^m(x) + \left(l(l+1) - \frac{m^2}{1-x^2}\right)P_l^m(x) = 0$$

In many applications, it is convenient to introduce an angle $\theta \in [0,\pi]$ instead of the variable x. Define $x = \cos\theta$, and the definition of the associated Legendre functions become $(m \geq 0)$

$$P_l^m(\cos\theta) = \sin^m\theta\frac{\mathrm{d}^m}{\mathrm{d}\,(\cos\theta)^m}P_l(\cos\theta), \qquad \theta \in [0,\pi]$$

and the differential equation becomes

$$\frac{1}{\sin\theta}\frac{\mathrm{d}}{\mathrm{d}\theta}\left(\sin\theta\frac{\mathrm{d}P_l^m(\cos\theta)}{\mathrm{d}\theta}\right) + \left(l(l+1) - \frac{m^2}{\sin^2\theta}\right)P_l^m(\cos\theta) = 0$$

As with the Legendre polynomials, the associated Legendre functions are orthogonal over the interval $[-1,1]$.

$$\int_{-1}^{1}P_l^m(x)P_{l'}^m(x)\,\mathrm{d}x = \int_0^\pi P_l^m(\cos\theta)P_{l'}^m(\cos\theta)\sin\theta\,\mathrm{d}\theta = \frac{2\delta_{ll'}}{2l+1}\frac{(l+m)!}{(l-m)!}$$

and

$$\int_0^\pi\left(\frac{\mathrm{d}P_l^m(\cos\theta)}{\mathrm{d}\theta}\frac{\mathrm{d}P_{l'}^m(\cos\theta)}{\mathrm{d}\theta} + m^2\frac{P_l^m(\cos\theta)}{\sin\theta}\frac{P_{l'}^m(\cos\theta)}{\sin\theta}\right)\sin\theta\,\mathrm{d}\theta = \frac{2l(l+1)\delta_{ll'}}{2l+1}\frac{(l+m)!}{(l-m)!}$$

and

$$\int_0^{\pi} \left(\frac{dP_l^m(\cos\theta)}{d\theta} \frac{P_{l'}^m(\cos\theta)}{\sin\theta} + \frac{P_l^m(\cos\theta)}{\sin\theta} \frac{dP_{l'}^m(\cos\theta)}{d\theta} \right) \sin\theta \, d\theta = 0$$

Note that the m index is the same in both functions in the integrand. There is also another orthogonality relation for the associated Legendre functions, but this holds for an l index that is the same, $m \neq 0$, and with a weight function $(1 - x^2)^{-1}$.

$$\int_{-1}^{1} \frac{P_l^m(x) P_l^{m'}(x)}{1 - x^2} \, dx = \frac{\delta_{mm'}}{m} \frac{(l+m)!}{(l-m)!}$$

C.3 Spherical Harmonics

The spherical harmonics are denoted $Y_{\sigma ml}(\theta, \phi)$ and they are defined by

$$Y_{\sigma ml}(\theta, \phi) = \sqrt{\frac{\varepsilon_m}{2\pi}} \sqrt{\frac{2l+1}{2} \frac{(l-m)!}{(l+m)!}} P_l^m(\cos\theta) \begin{Bmatrix} \cos m\phi \\ \sin m\phi \end{Bmatrix} = C_{lm} P_l^m(\cos\theta) \begin{Bmatrix} \cos m\phi \\ \sin m\phi \end{Bmatrix}$$

where the indices σ, m, and l take the following values

$$\sigma = \begin{Bmatrix} \mathrm{e} \\ \mathrm{o} \end{Bmatrix}, \qquad m = 0, 1, 2, \ldots, l, \qquad l = 0, 1, \ldots$$

The normalization factor C_{lm} used in this textbook is

$$C_{lm} = \sqrt{\frac{\varepsilon_m}{2\pi}} \sqrt{\frac{2l+1}{2} \frac{(l-m)!}{(l+m)!}}$$

where the Neumann factor is defined as

$$\varepsilon_m = 2 - \delta_{m0}, \quad \text{i.e.,} \quad \begin{cases} \varepsilon_0 = 1 \\ \varepsilon_m = 2, \quad m > 0 \end{cases}$$

Our definition uses only non-negative integer of the azimuth index m. The values of $\sigma = \mathrm{e}$ (even) and $\sigma = \mathrm{o}$ (odd) correspond to even and odd trigonometric azimuth functions $\cos m\phi$ and $\sin m\phi$, respectively. In our definition, the spherical harmonics are real-valued function, which often is advantageous in numerical implementations, since the symmetries of the problem lead to simplifications. The definitions of the spherical harmonics differ between textbooks, and it is recommended that the explicit definition of these functions is checked carefully before use. The two main alternatives of definitions of the spherical harmonics are given in Comment C.1. The normalization coefficient often varies between different textbooks.

Comment C.1
In the literature, there are several ways of defining the spherical harmonics, each with its pros and cons. Commonly used are the complex combinations [4]

$$Y_{lm}(\theta, \phi) = C'_{lm} P_l^m(\cos\theta) \mathrm{e}^{\mathrm{i}m\phi}$$

where $m = -l, \ldots, -1, 0, 1, \ldots, l$. The real combinations of the azimuth functions, which are used in this book, are

$$Y_{\sigma ml}(\theta, \phi) = C_{lm} P_l^m(\cos\theta) \begin{Bmatrix} \cos m\phi \\ \sin m\phi \end{Bmatrix}$$

The transformations between the two sets are ($m \geq 0$)

$$\begin{cases} Y_{lm} = \dfrac{C'_{lm}}{C_{lm}} (Y_{\mathrm{e}ml} + \mathrm{i} Y_{\mathrm{o}ml}) \\ Y_{\sigma ml} = \dfrac{C_{lm}}{2C'_{lm}} \begin{Bmatrix} Y_{lm} + Y_{l-m} \\ -\mathrm{i}Y_{lm} + \mathrm{i}Y_{l-m} \end{Bmatrix} \end{cases}$$

Instead of writing the angles θ and ϕ in the argument, we often use the unit vector $\hat{\boldsymbol{r}} = \hat{\boldsymbol{x}} \sin\theta \cos\phi + \hat{\boldsymbol{y}} \sin\theta \sin\phi + \hat{\boldsymbol{z}} \cos\theta$, see Figure C.1, and the spherical harmonics are often denoted $Y_{\sigma ml}(\hat{\boldsymbol{r}})$.

Reflection of the unit position vector in the origin $\hat{\boldsymbol{r}} \to -\hat{\boldsymbol{r}}$, i.e., $\theta \to \pi - \theta, \phi \to \phi + \pi$, implies the parity conditions of the spherical harmonics

$$Y_{\sigma ml}(-\hat{\boldsymbol{r}}) = (-1)^l Y_{\sigma ml}(\hat{\boldsymbol{r}})$$

Special values, which are important in scattering problems, are

$$Y_{\sigma ml}(\hat{\boldsymbol{z}}) = \delta_{\sigma \mathrm{e}} \delta_{m0} \sqrt{\frac{2l+1}{4\pi}}$$

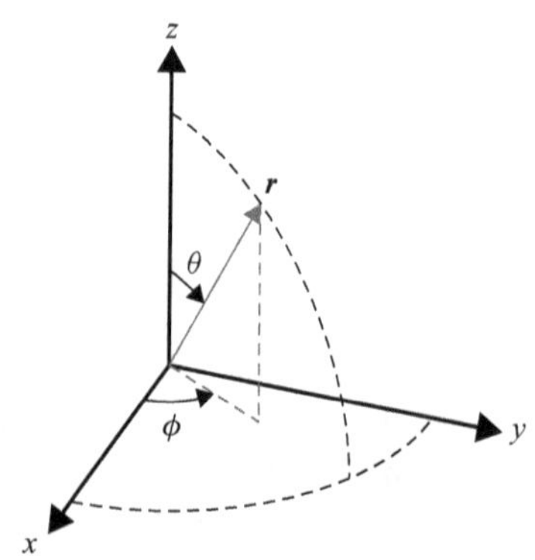

Figure C.1 Definition of the spherical angles θ and ϕ of the position vector $\boldsymbol{r}$.

Moreover, the lowest order spherical harmonics are

$$Y_{\sigma 00}(\hat{\boldsymbol{r}}) = \delta_{\sigma \mathrm{e}} \sqrt{\frac{1}{4\pi}}, \qquad \begin{cases} Y_{\sigma 01}(\hat{\boldsymbol{r}}) = \delta_{\sigma \mathrm{e}} \sqrt{\dfrac{3}{4\pi}} \cos\theta \\ Y_{\sigma 11}(\hat{\boldsymbol{r}}) = \sqrt{\dfrac{3}{4\pi}} \sin\theta \begin{Bmatrix} \cos\phi \\ \sin\phi \end{Bmatrix}, \quad \sigma = \begin{Bmatrix} \mathrm{e} \\ \mathrm{o} \end{Bmatrix} \end{cases}$$

The spherical harmonics are orthonormal over the unit sphere Ω, i.e.,

$$\iint_{\Omega} Y_{\sigma ml}(\hat{\boldsymbol{r}}) Y_{\sigma' m' l'}(\hat{\boldsymbol{r}})\, \mathrm{d}\Omega = \int_0^{2\pi} \mathrm{d}\phi \int_0^{\pi} \sin\theta\, \mathrm{d}\theta\, Y_{\sigma ml}(\theta,\phi) Y_{\sigma' m' l'}(\theta,\phi) = \delta_{\sigma\sigma'} \delta_{mm'} \delta_{ll'}$$

where $\mathrm{d}\Omega = \sin\theta\, \mathrm{d}\theta\, \mathrm{d}\phi$ is the surface element on the unit sphere. Moreover, the system $Y_{\sigma ml}(\hat{\boldsymbol{r}})$, $\sigma = \mathrm{e,o}$, $m = 0, 1, 2, \ldots, l$, $l = 0, 1, \ldots$, is a complete orthonormal system over the unit sphere, see Reference [38, page 24]. A square integrable function $f(\theta,\phi)$ defined on the unit sphere has a convergent expansion in the spherical harmonics $Y_{\sigma ml}(\hat{\boldsymbol{r}})$.

$$f(\theta,\phi) = \sum_{l=0}^{\infty} \sum_{m=0}^{l} \sum_{\sigma=\mathrm{e,o}} a_{\sigma ml} Y_{\sigma ml}(\theta,\phi), \qquad \theta \in [0,\pi], \quad \phi \in [0, 2\pi)$$

where the generalized Fourier coefficients $a_{\sigma ml}$ are determined by the integrals

$$a_{\sigma ml} = \int_0^{2\pi} \mathrm{d}\phi \int_0^{\pi} \sin\theta\, \mathrm{d}\theta\, f(\theta,\phi) Y_{\sigma ml}(\theta,\phi)$$

The convergence is in the square norm sense

$$\left\| f(\hat{\boldsymbol{r}}) - \sum_{l=0}^{N} \sum_{m=0}^{l} \sum_{\sigma=\mathrm{e,o}} a_{\sigma ml} Y_{\sigma ml}(\hat{\boldsymbol{r}}) \right\|^2 = \left\| f(\hat{\boldsymbol{r}}) \right\|^2 - \sum_{l=0}^{N} \sum_{m=0}^{l} \sum_{\sigma=\mathrm{e,o}} |a_{\sigma ml}|^2 \to 0, \qquad \text{as } N \to \infty$$

where the norm $\|\cdot\|$ is defined as

$$\left\| f(\hat{\boldsymbol{r}}) \right\|^2 = \int_0^{2\pi} \mathrm{d}\phi \int_0^{\pi} \sin\theta\, \mathrm{d}\theta\, f(\hat{\boldsymbol{r}}) f^*(\hat{\boldsymbol{r}}) = \iint_{\Omega} \left| f(\hat{\boldsymbol{r}}) \right|^2 \mathrm{d}\Omega$$

The completeness can also be stated as, see Appendix F.9.1 for more details

$$\int_0^{2\pi} \mathrm{d}\phi \int_0^{\pi} \sin\theta\, \mathrm{d}\theta\, f(\theta,\phi) Y_{\sigma ml}(\theta,\phi) = 0, \qquad \text{for all } \sigma, m, l \Rightarrow f(\theta,\phi) = 0$$

The spherical harmonics $Y_{\sigma ml}(\hat{\boldsymbol{r}})$ are eigenvectors to the Laplace operator ∇_{Ω}^2 on the unit sphere with eigenvalues $-l(l+1)$, i.e.,

$$\nabla_{\Omega}^2 Y_{\sigma ml}(\hat{\boldsymbol{r}}) = \frac{1}{\sin\theta} \frac{\partial}{\partial\theta} \sin\theta \frac{\partial}{\partial\theta} Y_{\sigma ml}(\hat{\boldsymbol{r}}) + \frac{1}{\sin^2\theta} \frac{\partial^2}{\partial\phi^2} Y_{\sigma ml}(\hat{\boldsymbol{r}}) = -l(l+1) Y_{\sigma ml}(\hat{\boldsymbol{r}}) \tag{C.4}$$

where ∇_Ω^2 denotes the Laplace operator on the unit sphere, i.e.,

$$\nabla_\Omega^2 = \frac{1}{\sin\theta}\frac{\partial}{\partial\theta}\sin\theta\frac{\partial}{\partial\theta} + \frac{1}{\sin^2\theta}\frac{\partial^2}{\partial\phi^2} \tag{C.5}$$

C.4 Vector Spherical Harmonics

A vector field $\boldsymbol{F}(\hat{\boldsymbol{r}})$ defined on the unit sphere has a series expansion in the spherical harmonics defined in Section C.3.

$$\boldsymbol{F}(\hat{\boldsymbol{r}}) = \sum_{l=0}^{\infty}\sum_{m=0}^{l}\sum_{\sigma=\mathrm{e,o}} \boldsymbol{a}_{\sigma ml} Y_{\sigma ml}(\hat{\boldsymbol{r}})$$

The drawback with this expansion is that the expansion coefficients $\boldsymbol{a}_{\sigma ml}$ are vector valued. However, if there is a suitable set of vector-valued basis functions, the expansion coefficients would be scalar. To accomplish this, let us introduce the vector spherical harmonics.

The vector spherical harmonics can be defined in several ways. In this textbook, we adopt the following definition:

$$\begin{cases} \boldsymbol{A}_{1\sigma ml}(\hat{\boldsymbol{r}}) = \dfrac{1}{\sqrt{l(l+1)}}\nabla\times\left(\boldsymbol{r}Y_{\sigma ml}(\hat{\boldsymbol{r}})\right) = \dfrac{1}{\sqrt{l(l+1)}}\nabla Y_{\sigma ml}(\hat{\boldsymbol{r}})\times\boldsymbol{r} \\ \boldsymbol{A}_{2\sigma ml}(\hat{\boldsymbol{r}}) = \dfrac{1}{\sqrt{l(l+1)}}r\nabla Y_{\sigma ml}(\hat{\boldsymbol{r}}) \\ \boldsymbol{A}_{3\sigma ml}(\hat{\boldsymbol{r}}) = \hat{\boldsymbol{r}}Y_{\sigma ml}(\hat{\boldsymbol{r}}) \end{cases} \tag{C.6}$$

The integer indices are $l = 0, 1, 2, 3, \ldots$, $m = 0, 1, \ldots, l$, and $\sigma = \mathrm{e,o}$. For $l = 0$, the first two expressions are not defined, since both the numerator and the denominator are zero. We define these vector spherical harmonics to be zero, i.e., $\boldsymbol{A}_{1\sigma 00}(\hat{\boldsymbol{r}}) = \boldsymbol{A}_{2\sigma 00}(\hat{\boldsymbol{r}}) = \boldsymbol{0}$. Alternatively, we can express the vector spherical harmonics in the spherical basis vectors $\{\hat{\boldsymbol{r}}, \hat{\boldsymbol{\theta}}, \hat{\boldsymbol{\phi}}\}$. The result is

$$\begin{cases} \boldsymbol{A}_{1\sigma ml}(\hat{\boldsymbol{r}}) = \dfrac{1}{\sqrt{l(l+1)}}\left(\hat{\boldsymbol{\theta}}\dfrac{1}{\sin\theta}\dfrac{\partial}{\partial\phi}Y_{\sigma ml}(\hat{\boldsymbol{r}}) - \hat{\boldsymbol{\phi}}\dfrac{\partial}{\partial\theta}Y_{\sigma ml}(\hat{\boldsymbol{r}})\right) \\ \boldsymbol{A}_{2\sigma ml}(\hat{\boldsymbol{r}}) = \dfrac{1}{\sqrt{l(l+1)}}\left(\hat{\boldsymbol{\theta}}\dfrac{\partial}{\partial\theta}Y_{\sigma ml}(\hat{\boldsymbol{r}}) + \hat{\boldsymbol{\phi}}\dfrac{1}{\sin\theta}\dfrac{\partial}{\partial\phi}Y_{\sigma ml}(\hat{\boldsymbol{r}})\right) \\ \boldsymbol{A}_{3\sigma ml}(\hat{\boldsymbol{r}}) = \hat{\boldsymbol{r}}Y_{\sigma ml}(\hat{\boldsymbol{r}}) \end{cases}$$

These vector spherical harmonics have the following properties:

$$\begin{cases} \hat{\boldsymbol{r}}\cdot\boldsymbol{A}_{\tau\sigma ml}(\hat{\boldsymbol{r}}) = 0, \quad \tau = 1, 2 \\ \hat{\boldsymbol{r}}\times\boldsymbol{A}_{3\sigma ml}(\hat{\boldsymbol{r}}) = \boldsymbol{0} \end{cases}$$

and

$$\begin{cases} \boldsymbol{A}_{1\sigma ml}(\hat{\boldsymbol{r}}) = \boldsymbol{A}_{2\sigma ml}(\hat{\boldsymbol{r}})\times\hat{\boldsymbol{r}} \\ \boldsymbol{A}_{2\sigma ml}(\hat{\boldsymbol{r}}) = \hat{\boldsymbol{r}}\times\boldsymbol{A}_{1\sigma ml}(\hat{\boldsymbol{r}}) \end{cases}$$

which we can summarize as

$$\hat{\boldsymbol{r}} \times \boldsymbol{A}_{\tau\sigma ml}(\hat{\boldsymbol{r}}) = (-1)^{\tau+1} \boldsymbol{A}_{\bar{\tau}\sigma ml}(\hat{\boldsymbol{r}}), \qquad \tau = 1, 2 \tag{C.7}$$

where we have introduced the dual index to τ, defined by $\bar{1} = 2$ and $\bar{2} = 1$.

Inversion of the argument, $\hat{\boldsymbol{r}} \to -\hat{\boldsymbol{r}}$, i.e., $\theta \to \pi - \theta$, $\phi \to \phi + \pi$, implies

$$\begin{cases} \boldsymbol{A}_{1\sigma ml}(-\hat{\boldsymbol{r}}) = (-1)^{l} \boldsymbol{A}_{1\sigma ml}(\hat{\boldsymbol{r}}) \\ \boldsymbol{A}_{2\sigma ml}(-\hat{\boldsymbol{r}}) = (-1)^{l+1} \boldsymbol{A}_{2\sigma ml}(\hat{\boldsymbol{r}}) \\ \boldsymbol{A}_{3\sigma ml}(-\hat{\boldsymbol{r}}) = (-1)^{l+1} \boldsymbol{A}_{3\sigma ml}(\hat{\boldsymbol{r}}) \end{cases} \tag{C.8}$$

With an argument $\hat{\boldsymbol{r}} = \hat{\boldsymbol{z}}$ ($\theta = 0$), the vector spherical harmonics have specific values.

$$\begin{cases} \boldsymbol{A}_{1\sigma ml}(\hat{\boldsymbol{z}}) = \delta_{m1} \sqrt{\dfrac{2l+1}{8\pi}} \begin{Bmatrix} -\hat{\boldsymbol{y}} \\ \hat{\boldsymbol{x}} \end{Bmatrix} = -\delta_{m1} \sqrt{\dfrac{2l+1}{8\pi}} \hat{\boldsymbol{z}} \times \begin{Bmatrix} \hat{\boldsymbol{x}} \\ \hat{\boldsymbol{y}} \end{Bmatrix} \\ \boldsymbol{A}_{2\sigma ml}(\hat{\boldsymbol{z}}) = \delta_{m1} \sqrt{\dfrac{2l+1}{8\pi}} \begin{Bmatrix} \hat{\boldsymbol{x}} \\ \hat{\boldsymbol{y}} \end{Bmatrix} \\ \boldsymbol{A}_{3\sigma ml}(\hat{\boldsymbol{z}}) = \delta_{\sigma \mathrm{e}} \delta_{m0} \sqrt{\dfrac{2l+1}{4\pi}} \hat{\boldsymbol{z}} \end{cases} \tag{C.9}$$

The lowest order vector spherical harmonics for $m = 0$ can be written explicitly as

$$\begin{cases} \boldsymbol{A}_{1\sigma 00}(\hat{\boldsymbol{r}}) = \boldsymbol{0} \\ \boldsymbol{A}_{2\sigma 00}(\hat{\boldsymbol{r}}) = \boldsymbol{0} \\ \boldsymbol{A}_{3\sigma 00}(\hat{\boldsymbol{r}}) = \delta_{\sigma \mathrm{e}} \sqrt{\dfrac{1}{4\pi}} \hat{\boldsymbol{r}} \end{cases}$$

for $l = 0$, and similarly for $l = 1$

$$\begin{cases} \boldsymbol{A}_{1\sigma 01}(\hat{\boldsymbol{r}}) = \delta_{\sigma \mathrm{e}} \sqrt{\dfrac{3}{8\pi}} \hat{\boldsymbol{\phi}} \sin\theta = -\delta_{\sigma \mathrm{e}} \sqrt{\dfrac{3}{8\pi}} \hat{\boldsymbol{r}} \times \hat{\boldsymbol{z}} \\ \boldsymbol{A}_{2\sigma 01}(\hat{\boldsymbol{r}}) = -\delta_{\sigma \mathrm{e}} \sqrt{\dfrac{3}{8\pi}} \hat{\boldsymbol{\theta}} \sin\theta = \delta_{\sigma \mathrm{e}} \sqrt{\dfrac{3}{8\pi}} \hat{\boldsymbol{r}} \times (\hat{\boldsymbol{z}} \times \hat{\boldsymbol{r}}) \\ \boldsymbol{A}_{3\sigma 01}(\hat{\boldsymbol{r}}) = \delta_{\sigma \mathrm{e}} \sqrt{\dfrac{3}{4\pi}} \hat{\boldsymbol{r}} \cos\theta = \delta_{\sigma \mathrm{e}} \sqrt{\dfrac{3}{4\pi}} \hat{\boldsymbol{r}} (\hat{\boldsymbol{z}} \cdot \hat{\boldsymbol{r}}) \end{cases} \tag{C.10}$$

Since $\hat{\boldsymbol{z}} = \hat{\boldsymbol{r}} \cos\theta - \hat{\boldsymbol{\theta}} \sin\theta$, this result implies that the unit vector $\hat{\boldsymbol{z}}$ can be expressed in the vector spherical harmonics for $l = 1$ and $m = 0$.

$$\hat{\boldsymbol{z}} = \sqrt{\frac{8\pi}{3}} \left(\boldsymbol{A}_{2\mathrm{e}01}(\hat{\boldsymbol{r}}) + \frac{1}{\sqrt{2}} \boldsymbol{A}_{3\mathrm{e}01}(\hat{\boldsymbol{r}}) \right)$$

The vector spherical harmonics for $m = 0$ for a general l index are

$$\begin{cases} \boldsymbol{A}_{1\sigma 0l}(\hat{\boldsymbol{r}}) = \delta_{\sigma \mathrm{e}} \dfrac{\sqrt{2l+1}}{\sqrt{4\pi l(l+1)}} \hat{\boldsymbol{\phi}} \sin\theta P_l'(\cos\theta) \\ \boldsymbol{A}_{2\sigma 0l}(\hat{\boldsymbol{r}}) = -\delta_{\sigma \mathrm{e}} \dfrac{\sqrt{2l+1}}{\sqrt{4\pi l(l+1)}} \hat{\boldsymbol{\theta}} \sin\theta P_l'(\cos\theta) \\ \boldsymbol{A}_{3\sigma 0l}(\hat{\boldsymbol{r}}) = \delta_{\sigma \mathrm{e}} \sqrt{\dfrac{2l+1}{4\pi}} \hat{\boldsymbol{r}} P_l(\cos\theta) \end{cases}$$

Moreover, the vector spherical harmonics for $l = 1$ and $m = 1$ are

$$\begin{cases} \boldsymbol{A}_{1\sigma 11}(\hat{\boldsymbol{r}}) = \sqrt{\dfrac{3}{8\pi}} \left(\hat{\boldsymbol{\theta}} \begin{Bmatrix} -\sin\phi \\ \cos\phi \end{Bmatrix} - \hat{\boldsymbol{\phi}} \cos\theta \begin{Bmatrix} \cos\phi \\ \sin\phi \end{Bmatrix} \right) = -\sqrt{\dfrac{3}{8\pi}} \hat{\boldsymbol{r}} \times \begin{Bmatrix} \hat{\boldsymbol{x}} \\ \hat{\boldsymbol{y}} \end{Bmatrix} \\ \boldsymbol{A}_{2\sigma 11}(\hat{\boldsymbol{r}}) = \hat{\boldsymbol{r}} \times \boldsymbol{A}_{1\sigma 11}(\hat{\boldsymbol{r}}) = \sqrt{\dfrac{3}{8\pi}} \hat{\boldsymbol{r}} \times \left(\begin{Bmatrix} \hat{\boldsymbol{x}} \\ \hat{\boldsymbol{y}} \end{Bmatrix} \times \hat{\boldsymbol{r}} \right) \\ \boldsymbol{A}_{3\sigma 11}(\hat{\boldsymbol{r}}) = \sqrt{\dfrac{3}{4\pi}} \hat{\boldsymbol{r}} \sin\theta \begin{Bmatrix} \cos\phi \\ \sin\phi \end{Bmatrix} = \sqrt{\dfrac{3}{4\pi}} \hat{\boldsymbol{r}} \left(\begin{Bmatrix} \hat{\boldsymbol{x}} \\ \hat{\boldsymbol{y}} \end{Bmatrix} \cdot \hat{\boldsymbol{r}} \right) \end{cases} \tag{C.11}$$

and for $l = 2$

$$\begin{cases} \boldsymbol{A}_{1\sigma 12}(\hat{\boldsymbol{r}}) = \sqrt{\dfrac{5}{8\pi}} \left(\hat{\boldsymbol{\theta}} \cos\theta \begin{Bmatrix} -\sin\phi \\ \cos\phi \end{Bmatrix} + \hat{\boldsymbol{\phi}} \left(1 - 2\cos^2\theta\right) \begin{Bmatrix} \cos\phi \\ \sin\phi \end{Bmatrix} \right) \\ \qquad = -\sqrt{\dfrac{5}{8\pi}} \left\{ \hat{\boldsymbol{r}} \times \hat{\boldsymbol{\phi}} \cos\theta \left(\begin{Bmatrix} \hat{\boldsymbol{x}} \\ \hat{\boldsymbol{y}} \end{Bmatrix} \cdot \hat{\boldsymbol{\phi}} \right) - \hat{\boldsymbol{r}} \times \hat{\boldsymbol{\theta}} \left(1 - 2\cos^2\theta\right) \left(\begin{Bmatrix} \hat{\boldsymbol{x}} \\ \hat{\boldsymbol{y}} \end{Bmatrix} \cdot \hat{\boldsymbol{r}}_{\mathrm{c}} \right) \right\} \\ \boldsymbol{A}_{2\sigma 12}(\hat{\boldsymbol{r}}) = \hat{\boldsymbol{r}} \times \boldsymbol{A}_{1\sigma 12}(\hat{\boldsymbol{r}}) = \\ \qquad = \sqrt{\dfrac{5}{8\pi}} \left(\hat{\boldsymbol{\phi}} \cos\theta \left(\begin{Bmatrix} \hat{\boldsymbol{x}} \\ \hat{\boldsymbol{y}} \end{Bmatrix} \cdot \hat{\boldsymbol{\phi}} \right) - \hat{\boldsymbol{\theta}} \left(1 - 2\cos^2\theta\right) \left(\begin{Bmatrix} \hat{\boldsymbol{x}} \\ \hat{\boldsymbol{y}} \end{Bmatrix} \cdot \hat{\boldsymbol{r}}_{\mathrm{c}} \right) \right) \\ \qquad = \sqrt{\dfrac{5}{8\pi}} \left(\cos\theta \begin{Bmatrix} \hat{\boldsymbol{x}} \\ \hat{\boldsymbol{y}} \end{Bmatrix} - \left(\hat{\boldsymbol{r}} \cos\theta + \hat{\boldsymbol{\theta}} \sin\theta \right) \left(\begin{Bmatrix} \hat{\boldsymbol{x}} \\ \hat{\boldsymbol{y}} \end{Bmatrix} \cdot \hat{\boldsymbol{r}} \right) \right) \\ \boldsymbol{A}_{3\sigma 12}(\hat{\boldsymbol{r}}) = \sqrt{\dfrac{15}{4\pi}} \hat{\boldsymbol{r}} \sin\theta\cos\theta \begin{Bmatrix} \cos\phi \\ \sin\phi \end{Bmatrix} = \sqrt{\dfrac{15}{4\pi}} \hat{\boldsymbol{r}} \cos\theta \left(\begin{Bmatrix} \hat{\boldsymbol{x}} \\ \hat{\boldsymbol{y}} \end{Bmatrix} \cdot \hat{\boldsymbol{r}} \right) \end{cases} \tag{C.12}$$

or more generally for all l index

$$\begin{cases} \boldsymbol{A}_{1\sigma 1l}(\hat{\boldsymbol{r}}) = \dfrac{\sqrt{2l+1}}{\sqrt{2\pi l(l+1)}} \left(\hat{\boldsymbol{\theta}} \pi_l(\cos\theta) \begin{Bmatrix} -\sin\phi \\ \cos\phi \end{Bmatrix} - \hat{\boldsymbol{\phi}} \tau_l(\cos\theta) \begin{Bmatrix} \cos\phi \\ \sin\phi \end{Bmatrix} \right) \\ \boldsymbol{A}_{2\sigma 1l}(\hat{\boldsymbol{r}}) = \dfrac{\sqrt{2l+1}}{\sqrt{2\pi l(l+1)}} \left(\hat{\boldsymbol{\phi}} \pi_l(\cos\theta) \begin{Bmatrix} -\sin\phi \\ \cos\phi \end{Bmatrix} + \hat{\boldsymbol{\theta}} \tau_l(\cos\theta) \begin{Bmatrix} \cos\phi \\ \sin\phi \end{Bmatrix} \right) \\ \boldsymbol{A}_{3\sigma 1l}(\hat{\boldsymbol{r}}) = \sqrt{\dfrac{2l+1}{2\pi l(l+1)}} \hat{\boldsymbol{r}} \sin\theta \pi_l(\cos\theta) \begin{Bmatrix} \cos\phi \\ \sin\phi \end{Bmatrix} \end{cases} \tag{C.13}$$

where the functions $\pi_l(x)$ and $\tau_l(x)$ are defined in (C.3).

In Chapter 7, we make use of a set of complex-valued vector spherical harmonics, $\boldsymbol{a}_n\left(\hat{\boldsymbol{r}}\right)$, defined as

$$\boldsymbol{a}_n(\hat{\boldsymbol{r}}) = \mathrm{i}^{l-\tau+2}\boldsymbol{A}_n(\hat{\boldsymbol{r}})$$

Special values for $l = 1$ are

$$\begin{aligned}\boldsymbol{a}_{\tau\sigma m1}\left(\hat{\boldsymbol{r}}\right) &= \delta_{\tau,1}\sqrt{\frac{3}{8\pi}}\hat{\boldsymbol{r}} \times \left(\hat{\boldsymbol{x}}\delta_{\sigma,\mathrm{e}}\delta_{m,1} + \hat{\boldsymbol{y}}\delta_{\sigma,\mathrm{o}}\delta_{m,1} + \hat{\boldsymbol{z}}\delta_{\sigma,\mathrm{e}}\delta_{m,0}\right) \\ &+ \mathrm{i}\delta_{\tau,2}\sqrt{\frac{3}{8\pi}}\hat{\boldsymbol{r}} \times \left(\left(\hat{\boldsymbol{x}}\delta_{\sigma,\mathrm{e}}\delta_{m,1} + \hat{\boldsymbol{y}}\delta_{\sigma,\mathrm{o}}\delta_{m,1} + \hat{\boldsymbol{z}}\delta_{\sigma,\mathrm{e}}\delta_{m,0}\right) \times \hat{\boldsymbol{r}}\right)\end{aligned}$$

The divergence and the curl of the vector spherical harmonics are obtained from the definitions of the vector spherical harmonics (C.6) and the use of the rules of the nabla operator. We have

$$\begin{cases} \nabla \cdot \boldsymbol{A}_{1\sigma ml}(\hat{\boldsymbol{r}}) = 0 \\ \nabla \cdot \boldsymbol{A}_{2\sigma ml}(\hat{\boldsymbol{r}}) = -\sqrt{l(l+1)}\dfrac{Y_{\sigma ml}(\hat{\boldsymbol{r}})}{r} \\ \nabla \cdot \boldsymbol{A}_{3\sigma ml}(\hat{\boldsymbol{r}}) = \dfrac{2Y_{\sigma ml}(\hat{\boldsymbol{r}})}{r} \end{cases} \tag{C.14}$$

and[3]

$$\begin{cases} \nabla \times \boldsymbol{A}_{1\sigma ml}(\hat{\boldsymbol{r}}) = \dfrac{\boldsymbol{A}_{2\sigma ml}(\hat{\boldsymbol{r}}) + \sqrt{l(l+1)}\boldsymbol{A}_{3\sigma ml}(\hat{\boldsymbol{r}})}{r} \\ \nabla \times \boldsymbol{A}_{2\sigma ml}(\hat{\boldsymbol{r}}) = -\dfrac{\boldsymbol{A}_{1\sigma ml}(\hat{\boldsymbol{r}})}{r} \\ \nabla \times \boldsymbol{A}_{3\sigma ml}(\hat{\boldsymbol{r}}) = \dfrac{\sqrt{l(l+1)}\boldsymbol{A}_{1\sigma ml}(\hat{\boldsymbol{r}})}{r} \end{cases} \tag{C.15}$$

Using these relations, we obtain the double curl of the vector spherical harmonics

$$\begin{cases} \nabla \times \left(\nabla \times \boldsymbol{A}_{1\sigma ml}(\hat{\boldsymbol{r}})\right) = \dfrac{l(l+1)\boldsymbol{A}_{1\sigma ml}(\hat{\boldsymbol{r}})}{r^2} \\ \nabla \times \left(\nabla \times \boldsymbol{A}_{2\sigma ml}(\hat{\boldsymbol{r}})\right) = -\dfrac{\sqrt{l(l+1)}\boldsymbol{A}_{3\sigma ml}(\hat{\boldsymbol{r}})}{r^2} \\ \nabla \times (\nabla \times \boldsymbol{A}_{3\sigma ml}(\hat{\boldsymbol{r}})) = \dfrac{l(l+1)\boldsymbol{A}_{3\sigma ml}(\hat{\boldsymbol{r}})}{r^2} \end{cases}$$

[3] We have here used $\nabla \times (\boldsymbol{a} \times \boldsymbol{b}) = \boldsymbol{a}(\nabla \cdot \boldsymbol{b}) - \boldsymbol{b}(\nabla \cdot \boldsymbol{a}) + (\boldsymbol{b} \cdot \nabla)\boldsymbol{a} - (\boldsymbol{a} \cdot \nabla)\boldsymbol{b}$ with $\boldsymbol{a} = \nabla Y_{\sigma ml}(\hat{\boldsymbol{r}})$ and $\boldsymbol{b} = \boldsymbol{r}$, and that $\nabla \cdot \boldsymbol{r} = 3$, $\nabla^2 Y_{\sigma ml}(\hat{\boldsymbol{r}}) = -r^{-2}l(l+1)Y_{\sigma ml}(\hat{\boldsymbol{r}})$, $\frac{\partial}{\partial r}\left(\nabla Y_{\sigma ml}(\hat{\boldsymbol{r}})\right) = -r^{-1}\nabla Y_{\sigma ml}(\hat{\boldsymbol{r}})$ and $(\boldsymbol{a} \cdot \nabla)\boldsymbol{r} = \boldsymbol{a}$ for an arbitrary vector $\boldsymbol{a}$.

and that

$$\begin{cases} \nabla_\Omega^2 \boldsymbol{A}_{1\sigma ml}(\hat{\boldsymbol{r}}) = -l(l+1)\boldsymbol{A}_{1\sigma ml}(\hat{\boldsymbol{r}}) \\ \nabla_\Omega^2 \boldsymbol{A}_{2\sigma ml}(\hat{\boldsymbol{r}}) = -l(l+1)\boldsymbol{A}_{2\sigma ml}(\hat{\boldsymbol{r}}) + 2\sqrt{l(l+1)}\boldsymbol{A}_{3\sigma ml}(\hat{\boldsymbol{r}}) \\ \nabla_\Omega^2 \boldsymbol{A}_{3\sigma ml}(\hat{\boldsymbol{r}}) = -(l(l+1)+2)\,\boldsymbol{A}_{3\sigma ml}(\hat{\boldsymbol{r}}) + 2\sqrt{l(l+1)}\boldsymbol{A}_{2\sigma ml}(\hat{\boldsymbol{r}}) \end{cases}$$

where ∇_Ω^2 denotes the Laplace operator on the unit sphere, see (C.5).

The vector spherical harmonics satisfy the orthogonal relation on the unit sphere.

$$\begin{aligned} &\iint_\Omega \boldsymbol{A}_{\tau\sigma ml}(\hat{\boldsymbol{r}}) \cdot \boldsymbol{A}_{\tau'\sigma'm'l'}(\hat{\boldsymbol{r}})\, \mathrm{d}\Omega \\ &\quad = \int_0^{2\pi} \mathrm{d}\phi \int_0^\pi \sin\theta\, \mathrm{d}\theta\, \boldsymbol{A}_{\tau\sigma ml}(\theta,\phi) \cdot \boldsymbol{A}_{\tau'\sigma'm'l'}(\theta,\phi) = \delta_{\tau\tau'}\delta_{\sigma\sigma'}\delta_{mm'}\delta_{ll'} \end{aligned}$$

The natural norm for vector-valued functions on the unit sphere is

$$\|\boldsymbol{F}(\hat{\boldsymbol{r}})\|^2 = \int_0^{2\pi} \mathrm{d}\phi \int_0^\pi \sin\theta\, \mathrm{d}\theta\, \boldsymbol{F}(\hat{\boldsymbol{r}}) \cdot \boldsymbol{F}^*(\hat{\boldsymbol{r}}) = \iint_\Omega |\boldsymbol{F}(\hat{\boldsymbol{r}})|^2\, \mathrm{d}\Omega$$

Every square integrable vector field defined on the unit sphere has a norm-convergent expansion in vector spherical harmonics. The completeness of these basis functions is found in, e.g., Reference [38, page 170]. The vector field $\boldsymbol{F}(\theta,\phi)$ defined on the unit sphere is expanded as

$$\boldsymbol{F}(\theta,\phi) = \sum_{l=0}^{\infty}\sum_{m=0}^{l}\sum_{\sigma=\mathrm{e,o}}\sum_{\tau=1}^{3} a_{\tau\sigma ml}\boldsymbol{A}_{\tau\sigma ml}(\theta,\phi), \qquad \theta\in[0,\pi],\ \ \phi\in[0,2\pi)$$

where the Fourier coefficients $a_{\tau\sigma ml}$ are determined by the integrals

$$a_{\tau\sigma ml} = \iint_\Omega \boldsymbol{F}(\hat{\boldsymbol{r}}) \cdot \boldsymbol{A}_{\tau\sigma ml}(\hat{\boldsymbol{r}})\, \mathrm{d}\Omega$$

Note that the expansion coefficients in this expansion are scalars. The convergence is in general only in the norm sense.

$$\begin{aligned} &\left\| \boldsymbol{F}(\hat{\boldsymbol{r}}) - \sum_{l=0}^{N}\sum_{m=0}^{l}\sum_{\sigma=\mathrm{e,o}}\sum_{\tau=1}^{3} a_{\tau\sigma ml}\boldsymbol{A}_{\tau\sigma ml}(\hat{\boldsymbol{r}}) \right\|^2 \\ &\quad = \|\boldsymbol{F}(\hat{\boldsymbol{r}})\|^2 - \sum_{l=0}^{N}\sum_{m=0}^{l}\sum_{\sigma=\mathrm{e,o}}\sum_{\tau=1}^{3} |a_{\tau\sigma ml}|^2 \to 0, \quad \text{as } N\to\infty \end{aligned}$$

C.5 Addition Theorem for the Legendre Polynomials

In this section, we prove the addition theorem for the Legendre polynomial $P_l(\cos\gamma)$.

$$P_l(\cos\gamma) = \frac{4\pi}{2l+1}\sum_{m=0}^{l}\sum_{\sigma=\mathrm{e,o}} Y_{\sigma ml}(\hat{\boldsymbol{r}})Y_{\sigma ml}(\hat{\boldsymbol{k}})$$

where

$$\cos\gamma = \hat{\boldsymbol{r}}\cdot\hat{\boldsymbol{k}} = \sin\theta\,\sin\alpha\,\cos(\phi-\beta) + \cos\theta\,\cos\alpha$$

and (θ,ϕ) and (α,β) are the spherical angles for the unit vectors $\hat{\boldsymbol{r}}$ and $\hat{\boldsymbol{k}}$, respectively, see Figure C.2.

To prove this formula, we start by expanding an arbitrary function $f_l(\hat{\boldsymbol{r}})$ that satisfies the equation

$$\nabla_\Omega^2 f_l(\hat{\boldsymbol{r}}) = -l(l+1)f_l(\hat{\boldsymbol{r}}) \tag{C.16}$$

in terms of spherical harmonics. We assume

$$f_l(\hat{\boldsymbol{r}}) = \sum_{\sigma'=\mathrm{e,o}}\sum_{l'=1}^{\infty}\sum_{m'=0}^{l'} a_{\sigma'm'l'}Y_{\sigma'm'l'}(\hat{\boldsymbol{r}})$$

where

$$a_{\sigma'm'l'} = \iint_\Omega f_l(\hat{\boldsymbol{r}})Y_{\sigma'm'l'}(\hat{\boldsymbol{r}})\,\mathrm{d}\Omega$$

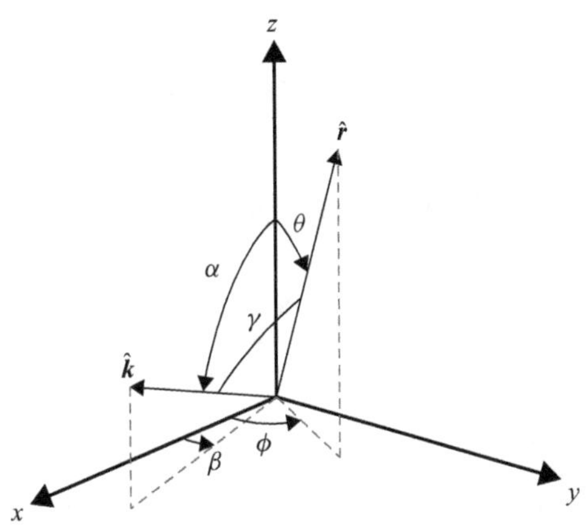

Figure C.2 The definition of the spherical angles (θ,ϕ), (α,β) and the angle γ. The unit vector $\hat{\boldsymbol{r}}$ has spherical angles (θ,ϕ), and the unit vector $\hat{\boldsymbol{k}}$ has spherical angles (α,β). The angle γ is the angle between the directions $\hat{\boldsymbol{r}}$ and $\hat{\boldsymbol{k}}$.

In this expansion only $l' = l$ contributes. To see this, form the quantity $\nabla^2_\Omega f_l(\hat{\boldsymbol{r}})$ in two ways. First by operating ∇^2_Ω on f_l and use the differential equation that f_l satisfies, and then second, by differentiating under the summation signs and use (C.4). By equating these two expressions, we get

$$-l(l+1) \sum_{\sigma'=\mathrm{e,o}} \sum_{l'=1}^{\infty} \sum_{m'=0}^{l'} a_{\sigma' m' l'} Y_{\sigma' m' l'}(\hat{\boldsymbol{r}}) = - \sum_{\sigma'=\mathrm{e,o}} \sum_{l'=1}^{\infty} \sum_{m'=0}^{l'} l'(l'+1) a_{\sigma' m' l'} Y_{\sigma' m' l'}(\hat{\boldsymbol{r}})$$

Since the expansion coefficients $a_{\sigma' m' l'}$ are uniquely defined, we conclude that they are zero unless $l' = l$, i.e.,

$$f_l(\hat{\boldsymbol{r}}) = \sum_{m=0}^{l} \sum_{\sigma=\mathrm{e,o}} a_{\sigma ml} Y_{\sigma ml}(\hat{\boldsymbol{r}})$$

In other words, the function $f_l(\hat{\boldsymbol{r}})$ is expressed in a series of spherical harmonics of order l. In particular, evaluate this expansion at $\theta = 0$ ($\hat{\boldsymbol{r}} = \hat{\boldsymbol{z}}$). We get, since $Y_{\sigma ml}(\theta = 0, \phi) = \sqrt{(2l+1)/4\pi}\, \delta_{m0} \delta_{\sigma \mathrm{e}}$

$$f_l(\hat{\boldsymbol{r}})\big|_{\theta=0} = \sum_{m=0}^{l} \sum_{\sigma=\mathrm{e,o}} a_{\sigma ml} Y_{\sigma ml}(\theta = 0, \phi) = \sqrt{\frac{2l+1}{4\pi}} a_{\mathrm{e}0l}$$

where

$$a_{\mathrm{e}0l} = \iint_\Omega f_l(\hat{\boldsymbol{r}}) Y_{\mathrm{e}0l}(\hat{\boldsymbol{r}})\, \mathrm{d}\Omega = \sqrt{\frac{2l+1}{4\pi}} \iint_\Omega f_l(\hat{\boldsymbol{r}}) P_l(\cos\theta)\, \mathrm{d}\Omega$$

Therefore,

$$\iint_\Omega f_l(\hat{\boldsymbol{r}}) P_l(\cos\theta)\, \mathrm{d}\Omega = \frac{4\pi}{2l+1} f_l(\hat{\boldsymbol{r}})\big|_{\theta=0} \tag{C.17}$$

We now apply this result to $P_l(\cos\gamma)$. Since $P_l(\cos\gamma)$ satisfies (C.16), we have an expansion

$$P_l(\cos\gamma) = \sum_{m=0}^{l} \sum_{\sigma=\mathrm{e,o}} a_{\sigma ml} Y_{\sigma ml}(\theta, \phi)$$

where

$$a_{\sigma ml} = \iint_\Omega P_l(\cos\gamma) Y_{\sigma ml}(\theta, \phi)\, \mathrm{d}\Omega(\theta, \phi) = \iint_\Omega Y_{\sigma ml}(\theta, \phi) P_l(\cos\gamma)\, \mathrm{d}\Omega(\gamma, \psi)$$

We have here used the fact that the surface measure, $\mathrm{d}\Omega(\theta, \phi) = \sin\theta\, \mathrm{d}\theta\, \mathrm{d}\phi$, on the unit sphere is invariant under rotations, and therefore the same as for the rotated coordinate system $d\Omega(\gamma, \psi) = \sin\gamma \; d\gamma \; d\psi$. From the result above, see (C.17), this integral can be evaluated as

$$a_{\sigma ml} = \frac{4\pi}{2l+1} Y_{\sigma ml}(\theta, \phi)|_{\gamma=0} = \frac{2l+1}{4\pi} Y_{\sigma ml}(\hat{\boldsymbol{k}})$$

and the final result is

$$P_l(\cos\gamma) = \frac{4\pi}{2l+1} \sum_{m=0}^{l} \sum_{\sigma=\mathrm{e,o}} Y_{\sigma ml}(\hat{\boldsymbol{r}}) Y_{\sigma ml}(\hat{\boldsymbol{k}}) \tag{C.18}$$

In particular, we get for $\hat{\boldsymbol{r}} = \hat{\boldsymbol{k}}$

$$\sum_{m=0}^{l} \sum_{\sigma=\mathrm{e,o}} Y_{\sigma ml}(\hat{\boldsymbol{r}}) Y_{\sigma ml}(\hat{\boldsymbol{r}}) = \frac{2l+1}{4\pi}$$

which is independent of the direction $\hat{\boldsymbol{r}}$.

C.6 Transformation Formulas

We can use (B.20), i.e.,

$$j_l(x) = \frac{1}{2\mathrm{i}^l} \int_{-1}^{1} P_l(t) \mathrm{e}^{\mathrm{i}tx} \, \mathrm{d}t$$

to establish the transformation between a plane wave with propagation direction $\hat{\boldsymbol{k}}$, i.e., $\mathrm{e}^{\mathrm{i}k\hat{\boldsymbol{k}}\cdot\boldsymbol{r}}$, and a regular spherical wave, $j_l(kr)Y_{\sigma ml}(\hat{\boldsymbol{r}})$. Also, the inverse transformation can be computed. We start by expanding $\exp\{\mathrm{i}kz\}$ in a Legendre series

$$\mathrm{e}^{\mathrm{i}kz} = \mathrm{e}^{\mathrm{i}kr\cos\theta} = \sum_{l=0}^{\infty} a_l P_l(\cos\theta)$$

The coefficients a_l in this expansion are found by

$$a_l = \frac{2l+1}{2} \int_0^{\pi} \mathrm{e}^{\mathrm{i}kr\cos\theta} P_l(\cos\theta) \sin\theta \; \mathrm{d}\theta = \mathrm{i}^l (2l+1) j_l(kr)$$

where we have used the identity (B.20). Thus, we have

$$\mathrm{e}^{\mathrm{i}kr\cos\theta} = \sum_{l=0}^{\infty} \mathrm{i}^l (2l+1) j_l(kr) P_l(\cos\theta)$$

Introduce the wave vector $\boldsymbol{k} = k\hat{\boldsymbol{k}}$ and the position vector $\boldsymbol{r} = r\hat{\boldsymbol{r}}$, and let $\cos\gamma = \hat{\boldsymbol{k}} \cdot \hat{\boldsymbol{r}}$. Then

$$\mathrm{e}^{\mathrm{i}\boldsymbol{k}\cdot\boldsymbol{r}} = \sum_{l=0}^{\infty} \mathrm{i}^l (2l+1) j_l(kr) P_l(\cos\gamma) = 4\pi \sum_{l=0}^{\infty} \sum_{m=0}^{l} \sum_{\sigma=\mathrm{e,o}} \mathrm{i}^l j_l(kr) Y_{\sigma ml}(\hat{\boldsymbol{r}}) Y_{\sigma ml}(\hat{\boldsymbol{k}})$$

where the addition theorem for the Legendre polynomial, (C.18), is used. This is the transformation between a regular spherical wave and a plane wave.

The inverse of this transformation is also obtained by multiplication by $Y_{\sigma ml}(\hat{\boldsymbol{k}})$, and integration over the unit sphere in the $\hat{\boldsymbol{k}}$-variables. Orthogonality gives

$$j_l(kr)Y_{\sigma ml}(\hat{\boldsymbol{r}}) = \frac{1}{4\pi \mathrm{i}^l} \iint\limits_{\Omega} \mathrm{e}^{\mathrm{i}\boldsymbol{k}\cdot\boldsymbol{r}} Y_{\sigma ml}(\hat{\boldsymbol{k}}) \, \mathrm{d}\Omega_{\hat{\boldsymbol{k}}}$$

We summarize,

$$\mathrm{e}^{\mathrm{i}\boldsymbol{k}\cdot\boldsymbol{r}} = 4\pi \sum_{l=0}^{\infty} \sum_{m=0}^{l} \sum_{\sigma=\mathrm{e,o}} \mathrm{i}^l j_l(kr) Y_{\sigma ml}(\hat{\boldsymbol{r}}) Y_{\sigma ml}(\hat{\boldsymbol{k}}) \qquad \text{plane} \longrightarrow \text{spherical}$$

$$j_l(kr)Y_{\sigma ml}(\hat{\boldsymbol{r}}) = \frac{1}{4\pi \mathrm{i}^l} \iint\limits_{\Omega} \mathrm{e}^{\mathrm{i}\boldsymbol{k}\cdot\boldsymbol{r}} Y_{\sigma ml}(\hat{\boldsymbol{k}}) \, \mathrm{d}\Omega_{\hat{\boldsymbol{k}}} \qquad \text{spherical} \longrightarrow \text{plane}$$

APPENDIX D

The Fourier and Other Transforms

A series of useful results related to Fourier transforms in one or several dimensions is collected in this appendix. This overview also contains the Poisson summation formula (Section D.1.2), the Hilbert transform (Section D.2), and Meĭman's theorem (Section D.3), that often are used in conjunction with Fourier transforms. The class of positive-definite functions and Herglotz functions is also reviewed in Sections D.4 and D.5, respectively. The appendix ends with two sections on the Watson transformation (Section D.6) and the location of zeros and poles (Section D.7).

D.1 The Fourier Transform

In this section, we summarize the most import properties of the Fourier transform in $\mathbb{R}^n$ used in this book. For mathematical details and proofs, we refer to the literature, see e.g., Reference 261. The definition of the Fourier transform of an integrable complex-valued function $f(\boldsymbol{x})$ is[1]

$$\hat{f}(\boldsymbol{\xi}) = \int_{\mathbb{R}^n} f(\boldsymbol{x}) \mathrm{e}^{-\mathrm{i}\boldsymbol{\xi}\cdot\boldsymbol{x}} \, \mathrm{d}x^n$$

where the volume element $\mathrm{d}x^n$ is a short-hand notation for $\mathrm{d}x^n = \mathrm{d}x_1 \ldots \mathrm{d}x_n$, and where $\boldsymbol{\xi} = \hat{\boldsymbol{e}}_1\xi_1 + \cdots + \hat{\boldsymbol{e}}_n\xi_n$.

The Fourier transform satisfies

$$\left|\hat{f}(\boldsymbol{\xi})\right| \leq \int_{\mathbb{R}^n} |f(\boldsymbol{x})| \, \mathrm{d}x^n$$

and it has an inverse

$$f(\boldsymbol{x}) = \frac{1}{(2\pi)^n} \int_{\mathbb{R}^n} \hat{f}(\boldsymbol{\xi}) \mathrm{e}^{\mathrm{i}\boldsymbol{\xi}\cdot\boldsymbol{x}} \, \mathrm{d}\xi^n$$

The Fourier transform of a derivative is

$$\widehat{\partial_i f}(\boldsymbol{\xi}) = \mathrm{i}\left(\boldsymbol{\xi} \cdot \hat{\boldsymbol{e}}_i\right)\hat{f}(\boldsymbol{\xi}), \qquad i = 1, 2, \ldots, n$$

[1] Note that this definition employs another sign convention than used for Fourier transforming the fields in the time domain in Chapter 1.

where $\hat{\boldsymbol{e}}_i$ is a unit vector in the x_i-direction. Parseval's formula[2] reads

$$\int_{\mathbb{R}^n} f(\boldsymbol{x})g^*(\boldsymbol{x})\, \mathrm{d}x^n = \frac{1}{(2\pi)^n} \int_{\mathbb{R}^n} \hat{f}(\boldsymbol{\xi})\hat{g}^*(\boldsymbol{\xi})\, \mathrm{d}\xi^n$$

Convolution transforms into products, i.e.,

$$\widehat{f * g}(\boldsymbol{\xi}) = \int_{\mathbb{R}^n}\int_{\mathbb{R}^n} f(\boldsymbol{x}-\boldsymbol{y})g(\boldsymbol{y})\, \mathrm{d}y^n \mathrm{e}^{-\mathrm{i}\boldsymbol{\xi}\cdot\boldsymbol{x}}\, \mathrm{d}x^n = \hat{f}(\boldsymbol{\xi})\hat{g}(\boldsymbol{\xi})$$

and inverse

$$\widehat{fg}(\boldsymbol{\xi}) = \frac{1}{(2\pi)^n} \int_{\mathbb{R}^n} \hat{f}(\boldsymbol{\xi}-\boldsymbol{\eta})\hat{g}(\boldsymbol{\eta})\, \mathrm{d}\eta^n = \frac{1}{(2\pi)^n} \hat{f} * \hat{g}(\boldsymbol{\xi})$$

D.1.1 Paley–Wiener Theorem

The Paley–Wiener[3] theorem guaranties that a causal function has a Fourier transform that is analytic in the upper complex half-plane [120].

Theorem D.1 (Paley–Wiener) *For $f(x) \in L^2(\mathbb{R})$, the following two statements are equivalent:*

1. *There exists a function $F(\xi)$, analytic in the upper half complex plane $\mathbb{C}_+ = \{z \in \mathbb{C} : \operatorname{Im} z > 0\}$, satisfying*

$$\int_{-\infty}^{\infty} |F(x+\mathrm{i}y)|^2\, \mathrm{d}x < \infty, \qquad y > 0$$

and

$$\lim_{y\to 0} \int_{-\infty}^{\infty} |F(x+\mathrm{i}y) - f(x)|^2\, \mathrm{d}x = 0$$

2. *The Fourier transform of $f(x)$ is causal, i.e.,*

$$\hat{f}(\xi) = 0, \qquad \xi < 0$$

D.1.2 The Poisson Summation Formula

In this section, we work in one dimension, $\mathbb{R}$. Start with the Fourier transform of the function $f(x)$, i.e.,

$$\hat{f}(\xi) = \int_{\mathbb{R}} f(x)\mathrm{e}^{-\mathrm{i}\xi x}\, \mathrm{d}x$$

---

[2] Marc-Antoine Parseval (1755–1836), French mathematician.

[3] Raymond Edward Alan Christopher Paley (1907–1933), English mathematician; Norbert Wiener (1894–1964), American mathematician and philosopher.

We also define, provided the sum converges

$$F(\xi) = \sum_{n=-\infty}^{\infty} \hat{f}(\xi + 2\pi n)$$

which is a 2π-periodic quantity, $F(\xi) = F(\xi + 2\pi)$, and therefore it has a Fourier series representation

$$F(\xi) = \sum_{n=-\infty}^{\infty} F_n e^{in\xi}$$

where the Fourier coefficients F_n are

$$\begin{aligned} F_n &= \frac{1}{2\pi}\int_0^{2\pi} F(\xi) e^{-in\xi}\, d\xi = \frac{1}{2\pi}\sum_{n'=-\infty}^{\infty}\int_0^{2\pi} \hat{f}(\xi + 2\pi n') e^{-in\xi}\, d\xi \\ &= \frac{1}{2\pi}\int_{\mathbb{R}} \hat{f}(\xi) e^{-in\xi}\, d\xi = f(n) \end{aligned}$$

by the inversion formula of the Fourier transform. We obtain

$$\sum_{n=-\infty}^{\infty} \hat{f}(\xi + 2\pi n) = \sum_{n=-\infty}^{\infty} f(n) e^{in\xi}$$

As a special case, let $\xi = 0$ and we get the Poisson[4] summation formula

$$\sum_{n=-\infty}^{\infty} \hat{f}(2\pi n) = \sum_{n=-\infty}^{\infty} f(n)$$

D.2 Hilbert Transform and Plemelj's Formulas

The Hilbert[5] transform plays an important role in this book. We have [27]

$$f(x) = \frac{1}{i\pi}\lim_{R\to\infty} P\int_{-R}^{R} \frac{f(x')}{x' - x}\, dx' = \frac{1}{i\pi} P\int_{-\infty}^{\infty} \frac{f(x')}{x' - x}\, dx', \qquad x \in \mathbb{R} \tag{D.1}$$

where $P\int\, dx'$ denotes Cauchy principal value defined by

$$P\int_{-\infty}^{\infty} \frac{f(x')}{x' - x}\, dx' = \lim_{\epsilon\to 0}\left(\int_{-\infty}^{x-\epsilon} \frac{f(x')}{x' - x}\, dx' + \int_{x+\epsilon}^{\infty} \frac{f(x')}{x' - x}\, dx'\right)$$

---

[4] Siméon-Denis Poisson (1781–1840), French mathematician.

[5] David Hilbert (1862–1943), German mathematician.

Sufficient conditions for a function $f(z)$ to satisfy the Hilbert transform are:

1. $f(z)$ is analytic in the upper half-plane, $\mathbb{C}_+ = \{z \in \mathbb{C} : \operatorname{Im} z > 0\}$
2. $|f(z)| \to 0$ as $|z| \to \infty$ in $\mathbb{C}_+ \cup \mathbb{R}$
3. $f(z)$ is continuous on the real axis

These conditions can be relaxed, but they suffice for the treatment in this book. Note that, by Liouville's[6] theorem [82, page 87] or [95, page 204], the only function that can satisfy these conditions in the entire complex plane is $f(z) = 0$.

Plemelj's formulas[7] are obtained by taking the real and imaginary parts of the Hilbert transform (D.1), i.e., $f(x) = f_r(x) + \mathrm{i}f_i(x)$, where the real and imaginary parts of the complex function $f(x)$ are denoted $f_r(x)$ and $f_i(x)$, respectively. We have

$$\begin{cases} f_r(x) = \dfrac{1}{\pi} P \displaystyle\int_{-\infty}^{\infty} \frac{f_i(x')}{x' - x} \, \mathrm{d}x' \\ f_i(x) = -\dfrac{1}{\pi} P \displaystyle\int_{-\infty}^{\infty} \frac{f_r(x')}{x' - x} \, \mathrm{d}x' \end{cases} \qquad x \in \mathbb{R} \tag{D.2}$$

We proceed and decompose $f_r(x)$ and $f_i(x)$ in their even and odd parts as

$$f_{r,i}(x) = f_{r,i}^{e}(x) + f_{r,i}^{o}(x), \qquad x \in \mathbb{R} \tag{D.3}$$

where

$$\begin{cases} f_{r,i}^{e}(x) = \dfrac{1}{2}\left(f_{r,i}(x) + f_{r,i}(-x)\right) \\ f_{r,i}^{o}(x) = \dfrac{1}{2}\left(f_{r,i}(x) - f_{r,i}(-x)\right) \end{cases} \qquad \begin{cases} f_{r,i}^{e}(-x) = f_{r,i}^{e}(x) \\ f_{r,i}^{o}(-x) = -f_{r,i}^{o}(x) \end{cases} \quad x \in \mathbb{R}$$

Using the symmetries of the even and the odd parts of $f_{r,i}(x)$, Plemelj's formulas then become

$$\begin{cases} f_r^{e}(x) = \dfrac{2}{\pi} P \displaystyle\int_{0}^{\infty} \frac{x' f_i^{o}(x')}{x'^2 - x^2} \, \mathrm{d}x' \\ f_r^{o}(x) = \dfrac{2x}{\pi} P \displaystyle\int_{0}^{\infty} \frac{f_i^{e}(x')}{x'^2 - x^2} \, \mathrm{d}x' \end{cases} \qquad \begin{cases} f_i^{e}(x) = -\dfrac{2}{\pi} P \displaystyle\int_{0}^{\infty} \frac{x' f_r^{o}(x')}{x'^2 - x^2} \, \mathrm{d}x' \\ f_i^{o}(x) = -\dfrac{2x}{\pi} P \displaystyle\int_{0}^{\infty} \frac{f_r^{e}(x')}{x'^2 - x^2} \, \mathrm{d}x' \end{cases} \tag{D.4}$$

since

$$\frac{1}{x' - x} + \frac{1}{x' + x} = \frac{2x'}{x'^2 - x^2}, \qquad \frac{1}{x' - x} - \frac{1}{x' + x} = \frac{2x}{x'^2 - x^2}$$

The requirements on a function to satisfy the Plemelj's formulas are conveniently summarized in the theorem by Titchmarsh[8] [261].

---

[6] Joseph Liouville (1809–1882), French mathematician.

[7] Josip Plemelj (1873–1967), Slovene mathematician.

[8] Edward Charles Titchmarsh (1899–1963), British mathematician.

Theorem D.2 (Titchmarsh) *If $f(x)$ is square integrable on the real axis, i.e., $f \in L^2(\mathbb{R})$, the following three conditions are equivalent:*

1. *The inverse Fourier transform of $f(x)$ vanishes for $x < 0$, i.e.,*

$$f(x) = \frac{1}{2\pi} \int_{-\infty}^{\infty} \hat{f}(\xi) \mathrm{e}^{-\mathrm{i}x\xi} \, \mathrm{d}\xi = 0, \qquad x < 0$$

2. *$f(x)$ is, for almost all x, the limit as $y \to 0^+$ of the function $f(z) = f(x + \mathrm{i}y)$, which is holomorphic in the upper half of the z-plane and satisfies*

$$\int_{-\infty}^{\infty} |f(x + \mathrm{i}y)|^2 \, \mathrm{d}x < \infty, \qquad y > 0$$

3. *The real and imaginary parts of $f(x) = f_\mathrm{r}(x) + \mathrm{i}f_\mathrm{i}(x)$ satisfy Plemelj's formulas*

$$\begin{cases} f_\mathrm{r}(x) = \dfrac{1}{\pi} P \displaystyle\int_{-\infty}^{\infty} \frac{f_\mathrm{i}(x')}{x' - x} \, \mathrm{d}x' \\ f_\mathrm{i}(x) = -\dfrac{1}{\pi} P \displaystyle\int_{-\infty}^{\infty} \frac{f_\mathrm{r}(x')}{x' - x} \, \mathrm{d}x' \end{cases}$$

D.2.1 Integral Identities

The following theorem is instrumental in the development of sum rules in this textbook, see Reference 13 for a more stringent proof. The theorem has some remote resemblance to Plemelj's formulas.

Theorem D.3 *Let $f(z)$ be an analytic function in $\mathbb{C}_+ = \{z \in \mathbb{C} : \operatorname{Im} z > 0\}$ satisfying the following series at the origin:*

$$f(z) = \sum_{n=-1}^{N-1} a_n z^n + O(z^N), \quad \text{as } |z| \to 0, \qquad 0 \le \arg z \le \pi$$

The sum is empty if the integer N is negative. Moreover, at infinity

$$f(z) = \sum_{m=1-M}^{1} b_m z^m + O(z^{-M}) = \sum_{m=-1}^{M-1} b_{-m} z^{-m} + O(z^{-M}), \quad \text{as } |z| \to \infty, \qquad 0 \le \arg z \le \pi$$

The sum is empty if the integer M is negative. The function $f(z)$ is assumed to have a well-defined limit as the argument approaches the real axis, and the coefficients a_n, $n = -1, 0, \ldots, N-2, N-1$, and b_m, $m = 1 - M, -M, \ldots, 0, 1$, are ***real*** *numbers.*

Then the following Cauchy principal value integral holds:

$$P \int_{-\infty}^{\infty} \frac{\operatorname{Im} f(x)}{x^p} \, \mathrm{d}x = \pi (a_{p-1} - b_{p-1}), \qquad p = 2 - M, 3 - M, \ldots, N - 1, N$$

It is assumed that the index interval $2 - M, 3 - M, \ldots, N - 1, N$ is not empty, i.e., $N + M \ge 2$.

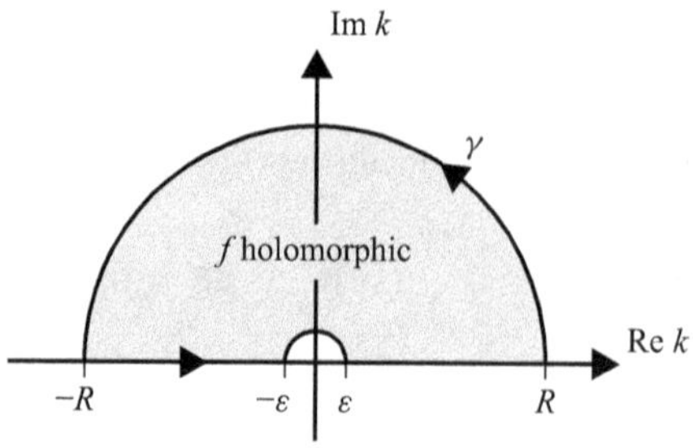

Figure D.1 The complex contour γ used in Theorem D.3.

Proof: We conclude that $zf(z)$ and $f(z)/z$ are bounded in a vicinity of the origin and infinity, respectively.

Integrate the function $f(z)/z^p$ w.r.t. to z along the closed complex contour γ in Figure D.1 for any integer p. Cauchy's theorem implies that the total integral over the contour γ is zero. This fact can be restated as

$$\int_{\varepsilon<|x|<R} \frac{f(x)}{x^p}\,\mathrm{d}x = \mathrm{i}\varepsilon^{1-p}\int_0^{\pi} f\left(\varepsilon \mathrm{e}^{\mathrm{i}\phi}\right)\mathrm{e}^{\mathrm{i}(1-p)\phi}\,\mathrm{d}\phi - \mathrm{i}R^{1-p}\int_0^{\pi} f\left(R\mathrm{e}^{\mathrm{i}\phi}\right)\mathrm{e}^{\mathrm{i}(1-p)\phi}\,\mathrm{d}\phi \quad \text{(D.5)}$$

Here, the small and the large semi-circles are parameterized by $z(\phi)=\varepsilon\mathrm{e}^{\mathrm{i}\phi}$ and $z(\phi)=R\mathrm{e}^{\mathrm{i}\phi}$, respectively, where $\phi\in[0,\pi]$. It is convenient to introduce the auxiliary integral I_p, defined as

$$I_p=\int_0^{\pi}\mathrm{e}^{\mathrm{i}p\phi}\,\mathrm{d}\phi=\begin{cases}\mathrm{i}\dfrac{1-(-1)^p}{p}, & p\neq 0\\ \pi, & p=0\end{cases}$$

With the asymptotic expansions in the theorem, the first and the second terms of the right-hand side of (D.5) then become

$$\mathrm{i}\varepsilon^{1-p}\int_0^{\pi} f\left(\varepsilon\mathrm{e}^{\mathrm{i}\phi}\right)\mathrm{e}^{\mathrm{i}(1-p)\phi}\,\mathrm{d}\phi=\mathrm{i}\sum_{n=-1}^{N-1}a_n\varepsilon^{1-p+n}I_{1-p+n}+O(\varepsilon^{1-p+N})$$

and

$$-\mathrm{i}R^{1-p}\int_0^{\pi} f\left(R\mathrm{e}^{\mathrm{i}\phi}\right)\mathrm{e}^{\mathrm{i}(1-p)\phi}\,\mathrm{d}\phi=-\mathrm{i}\sum_{m=1-M}^{1}b_mR^{1-p+m}I_{1-p+m}+O(R^{1-p-M})$$

respectively. Due to the assumptions made on the coefficients a_n and b_m, all terms in the sum on the right-hand side are real, except for the terms containing a factor I_0, which are purely imaginary. This situation occurs when $p=1+n$ and $p=1+m$, respectively.

When $\epsilon\to 0$ and $R\to\infty$, the rest terms $O(\varepsilon^{1-p+N})$ and $O(R^{1-p-M})$ vanish for $p=\ldots,N-1,N$ and $p=2-M,3-M,\ldots$, respectively. If $p=2-M,3-M,\ldots,N-1,N$, then both terms vanish. Therefore, the left-hand side of (D.5) has a well-behaved limit as

$\varepsilon \to 0$ and $R \to \infty$ if and only if $p = 2 - M, 3 - M, \ldots, N-1, N$, and the imaginary part of (D.5) reads

$$\lim_{\epsilon\to 0}\lim_{R\to\infty}\int_{\varepsilon<|x|<R}\frac{\operatorname{Im} f(x)}{x^p}\,\mathrm{d}x$$

$$= \lim_{\epsilon\to 0}\lim_{R\to\infty}\operatorname{Im}\left\{\mathrm{i}\varepsilon^{1-p}\int_0^\pi f\left(\varepsilon \mathrm{e}^{\mathrm{i}\phi}\right)\mathrm{e}^{\mathrm{i}(1-p)\phi}\,\mathrm{d}\phi - \mathrm{i}R^{1-p}\int_0^\pi f\left(R\mathrm{e}^{\mathrm{i}\phi}\right)\mathrm{e}^{\mathrm{i}(1-p)\phi}\,\mathrm{d}\phi\right\}$$

$$= \pi(a_{p-1} - b_{p-1}), \qquad p = 2 - M, 3 - M, \ldots, N-1, N$$

and we finally get

$$P\int_{-\infty}^{\infty}\frac{\operatorname{Im} f(x)}{x^p}\,\mathrm{d}x = \pi(a_{p-1} - b_{p-1}), \qquad p = 2 - M, 3 - M, \ldots, N-1, N$$

and the theorem is proved. □

D.3 Meĭman's Theorem

The location of roots of an analytic function $f(z)$ in the upper half-plane is of great importance in many applications. In particular, functions that are Fourier transforms of real-valued quantities, and thus satisfying the cross symmetry

$$f(z) = f^*(-z^*), \qquad \operatorname{Im} z \geq 0$$

are of interest. This cross symmetry implies that the functions $f(z)$ are real valued along the positive imaginary axis. Specifically, $f(0)$ is a real number. Meĭman's[9] theorem implies that the imaginary axis is the only place where the functions are real valued in the upper half-plane, provided the imaginary part of $f(z)$ is positive on the real axis.

Theorem D.4 (Meĭman) *Let the function $f(z)$ be analytic in the upper half-plane $\mathbb{C}_+$, and continuous on the real axis, satisfying*

$$|f(z)| \to 0 \text{ as } |z| \to \infty \text{ in } \mathbb{C}_+ \cup \mathbb{R}$$

and the cross symmetry

$$f(z) = f^*(-z^*) \text{ in } \mathbb{C}_+ \cup \mathbb{R}$$

Then, if $f(0) \neq 0$, and if $\operatorname{Im} f(x) > 0$ for all $x > 0$ (and by the cross symmetry $\operatorname{Im} f(x) < 0$ for all $x < 0$), $f(z)$ does not take any real values at any finite point in $\mathbb{C}_+$, except on the imaginary axis, where it decreases monotonically to zero as $z \to \mathrm{i}\infty$ along the imaginary axis.

A consequence of Meĭman's theorem is that the function $f(z)$ does not have any roots in the upper half-plane, including the real axis. For convenience, we give the proof.

[9] Naum Natanovich Meĭman (1912–2001), Soviet mathematician.

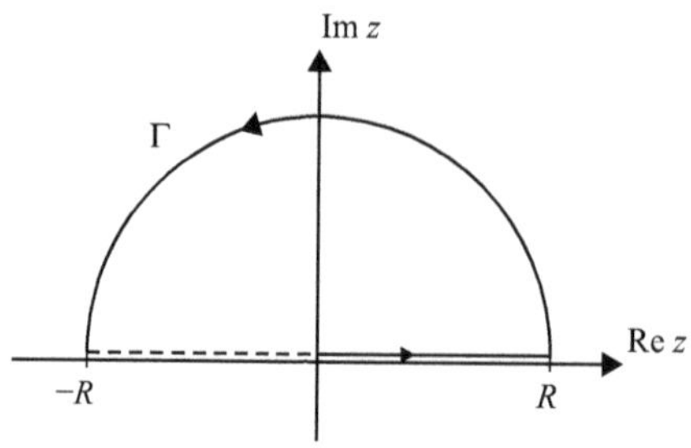

Figure D.2 The contour Γ in the complex z-plane. The dashed part of the contour denotes the negative real axis.

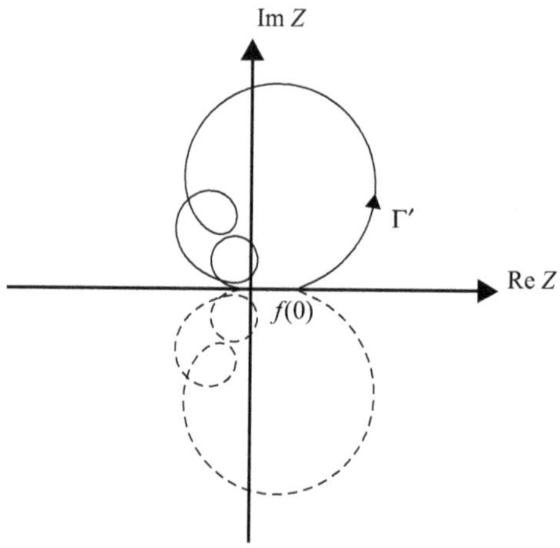

Figure D.3 The contour Γ′ in the complex $Z = f(z)$-plane, under the assumption that $f(0) > 0$. If $f(0) < 0$, the contour starts in the left half of the complex Z-plane. The solid part of the contour is the map of the positive real axis and the dashed part the map of the negative real axis.

Proof: For any real constant α, the function $f(z) - \alpha$ is analytic in $\mathbb{C}_+$ and the argument principle [82] implies that, see Comment D.1

$$\frac{1}{2\pi \mathrm{i}} \oint_{\Gamma} \frac{\mathrm{d}f(z)}{\mathrm{d}z} \frac{\mathrm{d}z}{f(z) - \alpha} = \frac{1}{2\pi \mathrm{i}} \oint_{\Gamma'} \frac{\mathrm{d}Z}{Z - \alpha}$$

is equal to the number of roots of the function $f(z) - \alpha$ in $\mathbb{C}_+$ inside the contour Γ, i.e., the number of points at which $f(z) = \alpha$ inside Γ. The appropriate contour Γ is shown in Figure D.2. The contour Γ′ is the map of the contour Γ under $f(z)$, i.e., $\Gamma' = f(\Gamma)$ and shown in Figure D.3, where we explicitly assume that $f(0) > 0$. The semi-circle in Figure D.2 is mapped to the origin in Figure D.3. The positive real axis, solid curve in Figure D.2, is mapped to a curve in the upper part of the complex Z-plane in Figure D.3, due to the assumption that $\operatorname{Im} f(z) > 0$ on the positive real axis (solid curve). The negative real axis is mapped to the dashed part of the curve in Figure D.3.

As a consequence, the contour Γ' does not intersect the real z-axis for any finite real value except at $f(0)$. From the argument principle, we then conclude that if $0<\alpha<f(0)$, so that α lies inside the contour Γ' (or $f(0)<\alpha<0$ if $f(0)<0$), there is only one root to $f(z)=\alpha$ in the upper half-plane, and if $0<f(0)<\alpha$ ($\alpha<f(0)<0$), there are no roots to $f(z)=\alpha$. By continuity, the root has to lie on the imaginary axis, since $f(z)$ varies from $f(0)>\alpha>0$ to zero as $z\in \mathrm{i}[0,\infty)$. The argument principle showed that this is the only root in the upper half-plane. Moreover, the variation along the imaginary axis must be monotonic, since a maximum or a minimum implies that there must be more than one root of $f(z)=\alpha$, which cannot happen by the arguments above. □

Comment D.1
The argument principle states that if $f(z)$ is a meromorphic function inside and on some closed contour γ, with no zeros or poles on γ, then the following formula holds:

$$N-P=\frac{1}{2\pi\mathrm{i}}\oint_{\gamma}\frac{f'(z)}{f(z)}\,\mathrm{d}z$$

where N and P denote the number of zeros and poles of $f(z)$ inside the contour γ, respectively. Each zero and pole is counted as many times as its multiplicity and order, respectively. This statement of the theorem assumes that the contour γ is simple, i.e., without self-intersections, and that it is oriented counterclockwise. If the curve is self-intersecting, the number of turns around the root or pole has to included.

D.3.1 Zeros in the Upper Complex Half-Plane

Related to Meĭman's theorem and the absence of zeros in the upper complex half-plane is the following theorem:

Theorem D.5 *Let the function $f(z)$ be a non-constant analytic function in the upper half-plane $\mathbb{C}_+$, and continuous in the closed domain $\mathbb{C}_+\cup\mathbb{R}$. Moreover, assume*

$$|f(z)|\to 0 \text{ uniformly as } |z|\to\infty \text{ in } \mathbb{C}_+\cup\mathbb{R}$$

and

$$\mathrm{Re}f(x)\geq 0,\qquad x\in\mathbb{R}$$

Then, $\mathrm{Re}f(z)>0$, for all $z\in\mathbb{C}_+$. In particular, there are no zeros of $f(z)$ in $\mathbb{C}_+$.

Proof: Denote the real part of f by u, i.e., $u(z)=\mathrm{Re}f(z)$. By the Cauchy–Riemann[10] equations, the real-valued function u is a harmonic function in the upper half-plane [82]. Let z_0 be an arbitrary point in $\mathbb{C}_+$, and let $\epsilon>0$ be an arbitrary real number. Since the function

[10] Bernhard Riemann (1826–1866), German mathematician.

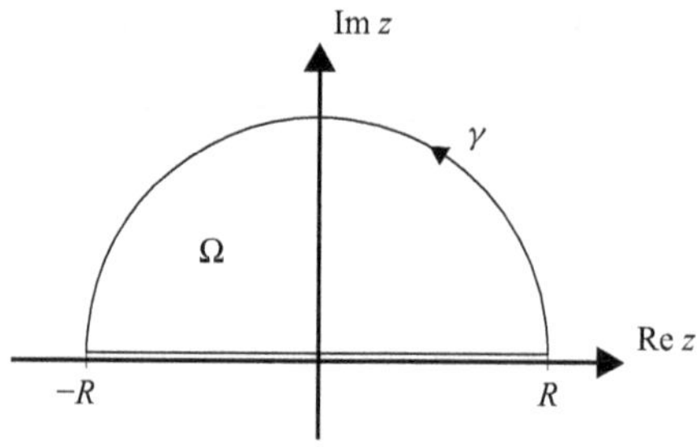

Figure D.4 The semi-circle γ in the upper complex half-plane used in the proof of Theorem D.5.

$|u(z)| \leq |f(z)| \to 0$ as $|z| \to \infty$ in $\mathbb{C}_+ \cup \mathbb{R}$, there is a $R_0 > |z_0|$ such that the values on the semi-circle γ of radius R satisfy, see Figure D.4

$$u(z) \geq -\epsilon, \qquad \text{for all } |z| = R \geq R_0$$

The minimum principle for harmonic functions implies that $u(z_0)$ is always larger (unless u is a constant) than its values on the boundary [251]. This shows that

$$u(z_0) > -\epsilon$$

Note that the condition on the real line $u(x) \geq 0$ is always satisfied. Since ϵ and z_0 are arbitrary, we conclude, together with the assumption on the real axis, that

$$u(z) \geq 0, \qquad \text{for all } z \in \mathbb{C}_+ \cup \mathbb{R}$$

We now repeat the arguments, but using the sharper bound, $u(z) \geq 0$, for all $z \in \mathbb{C}_+ \cup \mathbb{R}$. This holds in particular on the semi-circle γ, i.e.,

$$u(z) \geq 0, \qquad \text{for all } z \in \gamma$$

The minimum principle then shows that

$$u(z_0) > 0$$

which then proves the theorem, since z_0 is an arbitrary point in $\mathbb{C}_+$. □

The same conclusion can be made by focusing on the imaginary part, and we have the following corollary:

Corollary D.1 *Let the function $f(z)$ be a non-constant analytic function in the upper half-plane $\mathbb{C}_+$, and continuous in the closed domain $\mathbb{C}_+ \cup \mathbb{R}$. Moreover, assume*

$$|f(z)| \to 0 \text{ uniformly as } |z| \to \infty \text{ in } \mathbb{C}_+ \cup \mathbb{R}$$

and

$$\operatorname{Im} f(x) \geq 0, \qquad x \in \mathbb{R}$$

Then, $\operatorname{Im} f(z) > 0$, *for all $z \in \mathbb{C}_+$. In particular, there are no zeros of $f(z)$ in $\mathbb{C}_+$.*

D.4 Positive-Definite Functions

The definition and some properties of positive-definite functions are reviewed in this section. We start with the concept of a positive semi-definite matrix.

An $n \times n$ complex-valued matrix A is called a positive semi-definite matrix provided

$$\sum_{i=1}^{n} \sum_{j=1}^{n} A_{ij} z_i z_j^* \geq 0$$

for all $z_i \in \mathbb{C}$, $i = 1, 2, \ldots, n$ (the definition of a positive-definite matrix is given in Appendix F.3.1). If we let all but one of the numbers z_i be zero, say $z_i = 1$ and the rest zero, we see that all diagonal elements have to be real and non-negative, i.e., $A_{ii} \geq 0$. If we let two numbers, say z_i and z_j, $i \neq j$, be non-zero and the rest equal to zero, we obtain

$$A_{ii}|z_i|^2 + A_{jj}|z_j|^2 + A_{ij} z_i z_j^* + A_{ji} z_j z_i^* \geq 0 \tag{D.6}$$

The imaginary part of this expression has to be zero, which implies

$$(A_{ij} - A_{ji}^*) z_i z_j^* + (A_{ji} - A_{ij}^*) z_j z_i^* = 0$$

implying[11] that $A_{ij} = A_{ji}^*$, i.e., A_{ij} is Hermitian. We are now in a position to rewrite (D.6) with $z_j = 1$ as[12]

$$\left| \sqrt{A_{ii}} z_i + \frac{A_{ij}}{\sqrt{A_{ii}}} \right|^2 \geq \frac{|A_{ij}|^2}{A_{ii}} - A_{jj}$$

and with $z_i = -A_{ij}/A_{ii}$, we get

$$|A_{ij}|^2 \leq A_{ii} A_{jj}$$

which also holds if one of the diagonal elements is zero.

A positive-definite function is now defined. We restrict ourselves to continuous functions, but the definition can be generalized to hold for distributions as well.

Definition D.1 *A continuous, complex-valued function $f(x)$ is positive definite if for every positive integer n and every set of real numbers x_i, $i = 1, 2, \ldots, n$, the matrix A_{ij} defined as*

$$A_{ij} = f(x_i - x_j)$$

is a positive semi-definite matrix.

---

[11] Take e.g., $z_i = 1, i$ and $z_j = 1$.

[12] Use $|a + b|^2 = |a|^2 + |b|^2 + 2\,\mathrm{Re}\,(ab^*)$. If $A_{ii} = 0$, change the role of i and j. If both $A_{ii} = A_{jj} = 0$, then $A_{ij} = 0$.

From above, we get

$$\begin{cases} f(0) \geq 0 \\ f(-x) = f^*(x) \\ |f(x)| \leq f(0) \end{cases}$$

The definition of a positive-definite function can equally well be formulated as the following condition on the convolution integral:

$$\int_{-\infty}^{\infty} \left(\int_{-\infty}^{\infty} f(x-y)\phi^*(y)\, \mathrm{d}y \right) \phi(x)\, \mathrm{d}x \geq 0$$

for all continuous functions $\phi(x)$ with compact support, since the integral is a limit of Riemann sums

$$\sum_{i,j} f(x_i - x_j)\phi(x_i)\phi^*(x_j)\, \Delta x_i \Delta x_j$$

which is always positive if f is a positive-definite function.

A simple example of a positive-definite function is the exponential $\exp\{\mathrm{i}xk\}$, since

$$\sum_{i=1}^{n}\sum_{j=1}^{n} \mathrm{e}^{\mathrm{i}(x_i - x_j)k} z_i z_j^* = \sum_{i=1}^{n} \left| \mathrm{e}^{\mathrm{i}x_i k} z_i \right|^2 \geq 0$$

which can be generalized to the inverse Fourier transform of a positive function $F(k) \geq 0$. We have

$$\int_{-\infty}^{\infty}\int_{-\infty}^{\infty} \left(\int_{-\infty}^{\infty} F(k)\mathrm{e}^{\mathrm{i}k(x-y)}\, \mathrm{d}k \right) \phi(x)\phi^*(y)\, \mathrm{d}x\, \mathrm{d}y = \int_{-\infty}^{\infty} F(k) \left| \int_{-\infty}^{\infty} \phi(x)\mathrm{e}^{\mathrm{i}kx}\, \mathrm{d}x \right|^2 \mathrm{d}k \geq 0$$

The opposite is also true, i.e., a function with positive Fourier transform is a positive-definite function. This result is stated more precisely in a theorem by Bochner[13] [123, 124]. We have

Theorem D.6 (Bochner) *Positive-definite functions are exactly the functions of the form*

$$f(x) = \int_{-\infty}^{\infty} \mathrm{e}^{\mathrm{i}kx}\, \mathrm{d}\mu(k)$$

where μ is a finite positive Borel[14] *measure, i.e., $\mu(\mathbb{R}) < \infty$.*

In the main text, we often encounters causal functions $f(t)$, i.e., $f(t) = 0$ for $t < 0$. If we denote the Fourier transform of this function by

$$\hat{f}(k) = \int_0^{\infty} f(t)\mathrm{e}^{\mathrm{i}kt}\, \mathrm{d}t$$

---

[13] Salomon Bochner (1899–1982), American mathematician.

[14] Félix Édouard Justin Émile Borel (1871–1956), French mathematician.

If $f(t)$ is real valued, $\hat{f}(-k)=\hat{f}^*(k)$. Moreover, the Fourier transform of f can be extended uniquely into the upper complex plane of k, i.e., $\operatorname{Im} k \geq 0$, by Paley–Wiener theorem.

We extend the causal function $f(t)$ to an even function of t by defining $F(t)=f(|t|)$. The Fourier transform of $F(t)$ is related to $\hat{f}(k)$ as

$$\hat{F}(k) = \int_{-\infty}^{\infty} F(t)e^{ikt}\, dt = \int_{0}^{\infty} F(t)e^{ikt}\, dt + \int_{0}^{\infty} F(-t)e^{-ikt}\, dt = 2\operatorname{Re}\hat{f}(k)$$

Suppose the causal function $f(t)$ satisfies

$$\int_{-\infty}^{\infty} \left(\int_{-\infty}^{t} f(t-t')\phi(t')\, dt' \right) \phi(t)\, dt \geq 0$$

for all continuous, real-valued functions $\phi(x)$ with compact support. Then, $f(|t|)$ is a positive-definite function (with real-valued test functions $\phi(x)$), since

$$\int_{-\infty}^{\infty} \left(\int_{-\infty}^{\infty} F(t-t')\phi(t')\, dt' \right) \phi(t)\, dt = \int_{-\infty}^{\infty} \left(\int_{-\infty}^{t} f(t-t')\phi(t')\, dt' \right) \phi(t)\, dt + \int_{-\infty}^{\infty} \left(\int_{t}^{\infty} f(t'-t)\phi(t')\, dt' \right) \phi(t)\, dt$$

which is identical to

$$\int_{-\infty}^{\infty} \left(\int_{-\infty}^{\infty} F(t-t')\phi(t')\, dt' \right) \phi(t)\, dt = \int_{-\infty}^{\infty} \left(\int_{-\infty}^{t} f(t-t')\phi(t')\, dt' \right) \phi(t)\, dt + \int_{-\infty}^{\infty} \left(\int_{-\infty}^{t'} f(t'-t)\phi(t)\, dt \right) \phi(t')\, dt' \geq 0$$

As a consequence of Bochner's theorem, Theorem D.6, a causal function $f(t)$ that satisfies

$$\int_{-\infty}^{\infty} \left(\int_{-\infty}^{t} f(t-t')\phi(t')\, dt' \right) \phi(t)\, dt \geq 0$$

for all continuous, real-valued functions $\phi(x)$ with compact support, has a Fourier transform that has a positive real part, i.e.,

$$\operatorname{Re}\hat{f}(k) \geq 0$$

D.5 Herglotz Functions

Analytic functions play a central role in the theory of electromagnetics, and several functions encountered in this book have specific analytic properties. A special class of analytic functions is the Herglotz[15] functions, which are defined as

---

[15]Gustav Herglotz (1881–1953), German mathematician.

Definition D.2 *A function $f(z)$ is called a Herglotz function if*

1. *$f(z)$ is defined and analytic everywhere in the upper half-plane, $\mathbb{C}_+ = \{z \in \mathbb{C} : \operatorname{Im} z > 0\}$*
2. *$\operatorname{Im} f(z) \geq 0$ for all $z \in \mathbb{C}_+$*

i.e., the function maps the upper half-space into itself.

If $f(z)$ is a Herglotz function, then $-1/f(z)$ is a Herglotz function, and if $f(z)$ and $g(z)$ are Herglotz functions, the composition $f(g(z))$ is a new Herglotz function.

The properties of a Herglotz function are summarized in the following theorem [13, 203]:

Theorem D.7 *Let $f(z)$ be a Herglotz function. Then*

1. *$f(z)$ has a finite normal limit $f(x+\mathrm{i}0) = \lim_{\varepsilon \downarrow 0} f(x+\mathrm{i}\varepsilon)$ for a.e. $x \in \mathbb{R}$.*
2. *If $f(z)$ has a zero normal limit on a subset of $\mathbb{R}$ having positive Lebesque[16] measure, then $f \equiv 0$.*
3. *There exists a Borel measure μ on $\mathbb{R}$ satisfying*

$$\int_{\mathbb{R}} \frac{\mathrm{d}\mu(x)}{1+x^2} < \infty$$

such that the Nevanlinna[17] representation

$$f(z) = A + Bz + \int_{-\infty}^{\infty} \frac{1+zx}{(1+x^2)(x-z)} \, \mathrm{d}\mu(x), \qquad \operatorname{Im} z > 0$$

where

$$A = \operatorname{Re} f(\mathrm{i}) \qquad \text{and} \qquad B = \lim_{y \to \infty} \frac{f(\mathrm{i}y)}{\mathrm{i}y} \geq 0$$

holds.

4. *Let $(x_1, x_2) \subset \mathbb{R}$, then the Stieltjes[18] inversion formula for μ reads*

$$\frac{1}{2}\mu(\{x_1\}) + \frac{1}{2}\mu(\{x_2\}) + \mu((x_1, x_2)) = \frac{1}{\pi} \lim_{\varepsilon \downarrow 0} \int_{x_1}^{x_2} \operatorname{Im}(f(x+\mathrm{i}\varepsilon)) \, \mathrm{d}x$$

5. *The absolutely continuous part μ_{ac} of μ w.r.t. Lebesque measure $\mathrm{d}x$ on $\mathbb{R}$ is given by*

$$\mathrm{d}\mu_{ac}(x) = \frac{1}{\pi} \operatorname{Im}(f(x+\mathrm{i}0)) \, \mathrm{d}x$$

With enough regularity on the real axis, the Herglotz function has the integral representation

$$f(z) = A + Bz + \frac{1}{\pi} \int_{-\infty}^{\infty} \frac{(1+zx') \operatorname{Im} f(x')}{(1+x'^2)(x'-z)} \, \mathrm{d}x', \qquad \operatorname{Im} z > 0$$

---

[16]Henri Léon Lebesgue (1875–1941), French mathematician.

[17]Rolf Herman Nevanlinna (1895–1980), Finnish mathematician.

[18]Thomas Joannes Stieltjes (1856–1894), Dutch mathematician.

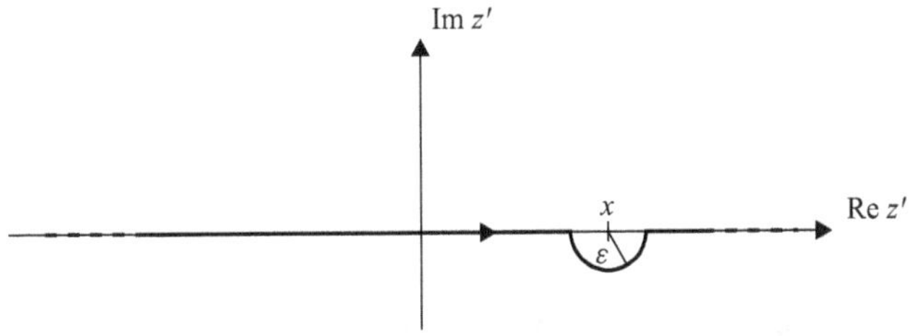

Figure D.5 The distortion of the integration path along the real axis.

where

$$A = \operatorname{Re} f(\mathrm{i}) \qquad \text{and} \qquad B = \lim_{y\to\infty} \frac{f(\mathrm{i}y)}{\mathrm{i}y} \geq 0$$

Again, if the function $f(z)$ has enough regularity in a neighborhood of the real axis, we can let $z = x + \mathrm{i}y$ approach the real axis, i.e., $y \to 0$. This process deserves some care, since the integrand is singular at the point, x, on the real axis. The contour of integration along the real axis is conveniently distorted by a semi-circle of radius ε and centered at x in the lower complex z' half-plane, see Figure D.5. In the limit $\varepsilon \to 0$, the straight parts of the integration are the Cauchy principal value of the integral

$$P\int_{-\infty}^{\infty} \frac{g(x')}{x'-x}\, \mathrm{d}x' = \lim_{\epsilon\to 0}\left(\int_{-\infty}^{x-\epsilon} \frac{g(x')}{x'-x}\, \mathrm{d}x' + \int_{x+\epsilon}^{\infty} \frac{g(x')}{x'-x}\, \mathrm{d}x'\right)$$

and the contributions from semi-circle are easily computed in this limit. We get

$$f(x) = A + Bx + \frac{1}{\pi} P\int_{-\infty}^{\infty} \frac{(1+xx')\operatorname{Im} f(x')}{(1+x'^2)(x'-x)}\, \mathrm{d}x' + \mathrm{i}\operatorname{Im} f(x)$$

or

$$f_{\mathrm{r}}(x) = A + Bx + \frac{1}{\pi} P\int_{-\infty}^{\infty} \frac{(1+xx') f_{\mathrm{i}}(x')}{(1+x'^2)(x'-x)}\, \mathrm{d}x'$$

where the real and imaginary parts of the complex function $f(x) = f_{\mathrm{r}}(x) + \mathrm{i} f_{\mathrm{i}}(x)$ are denoted $f_{\mathrm{r}}(x)$ and $f_{\mathrm{i}}(x)$, respectively.

We proceed by decomposing $f_{\mathrm{r}}(x)$ and $f_{\mathrm{i}}(x)$ in their even and odd parts as, see (D.3)

$$f_{\mathrm{r,i}}(x) = f_{\mathrm{r,i}}^{\mathrm{e}}(x) + f_{\mathrm{r,i}}^{\mathrm{o}}(x), \qquad x \in \mathbb{R}$$

Using the symmetries of the even and the odd parts of $f_{\mathrm{r,i}}(x)$, we obtain

$$f_{\mathrm{r}}^{\mathrm{o}}(x) = Bx + \frac{x}{\pi} P\int_{-\infty}^{\infty} \frac{f_{\mathrm{i}}^{\mathrm{e}}(x')}{x'^2 - x^2}\, \mathrm{d}x'$$

since

$$\frac{1+xx'}{x'-x} - \frac{1-xx'}{x'+x} = \frac{2x(1+x'^2)}{x'^2 - x^2}$$

This relation is related to the Plemelj's formulas.

Example D.1

A simple example illustrates the representation in this section. Assume the function $f(z)$ is the complex square root, i.e.,

$$f(z) = z^{1/2}$$

We fix the branch of the square root by assigning $f(x) = \sqrt{x}$ if $x > 0$ (branch cut along the positive real axis). This function is then a Herglotz function and the Nevanlinna representation is

$$f(z) = A + Bz + \int_{-\infty}^{\infty} \frac{1 + zx}{(1 + x^2)(x - z)} \, \mathrm{d}\mu(x), \qquad \operatorname{Im} z > 0$$

where

$$A = \operatorname{Re} f(\mathrm{i}) = \operatorname{Re} \frac{1 + \mathrm{i}}{\sqrt{2}} = \frac{1}{\sqrt{2}} \qquad \text{and} \qquad B = \lim_{y \to \infty} \frac{f(\mathrm{i}y)}{\mathrm{i}y} = \frac{1 + \mathrm{i}}{\mathrm{i}\sqrt{2}} \lim_{y \to \infty} \frac{\sqrt{y}}{y} = 0$$

and the measure

$$\frac{\mathrm{d}\mu(x)}{\mathrm{d}x} = \frac{1}{\pi} \lim_{\varepsilon \downarrow 0} \operatorname{Im}(f(x + \mathrm{i}\varepsilon)) = \frac{1}{\pi} \begin{cases} 0, & x > 0 \\ \sqrt{-x}, & x < 0 \end{cases}$$

The Nevanlinna representation becomes

$$z^{1/2} = \frac{1}{\sqrt{2}} + \frac{1}{\pi} \int_{-\infty}^{0} \frac{1 + zx}{(1 + x^2)(x - z)} \sqrt{-x} \, \mathrm{d}x, \qquad \operatorname{Im} z > 0$$

A simple change of variables, $x \to -t^2$, gives

$$z^{1/2} = \frac{1}{\sqrt{2}} + \frac{1}{\pi} \int_{-\infty}^{\infty} \frac{(zt^2 - 1)t^2}{(1 + t^4)(t^2 + z)} \, \mathrm{d}t, \qquad \operatorname{Im} z > 0$$

By the use of residue calculus (close the contour in the upper half-plane and use the residues at the poles at $t = (\pm 1 + \mathrm{i})/\sqrt{2}$ and $t = \mathrm{i}z^{1/2}$), we easily prove that this is an identity for $\operatorname{Im} z > 0$. ■

Other examples of Herglotz functions are

$$\alpha z, \quad \beta, \quad -\frac{\alpha}{z}, \quad \ln z, \quad \mathrm{i} \ln(1 - \mathrm{i}z)$$

where $\alpha > 0$ and $\operatorname{Im} \beta \geq 0$.

Herglotz functions have been generalized to dyadic-valued (matrix-valued) functions. We refer to the existing literature for more details on this topic [79].

D.6 The Watson Transformation

The Watson transformation is named after G.N. Watson.[19] Consider the integral

$$\int_C \frac{f(z)}{\sin \pi z}\, \mathrm{d}z$$

where the contour C is depicted in Figure D.6, and $f(z)$ is an analytic function inside and in a neighborhood of the contour C. The integral is assumed to be convergent.

The integrand has simple poles at the integers $n = 0, \pm 1, \pm 2, \ldots$ with residues

$$\operatorname{Res} \frac{f(z)}{\sin \pi z}\bigg|_{z=n} = (-1)^n \frac{f(n)}{\pi}$$

The residue theorem then implies that the integral can be written as

$$\int_C \frac{f(z)}{\sin \pi z}\, \mathrm{d}z = 2\mathrm{i} \sum_{n=-\infty}^{\infty} (-1)^n f(n)$$

Similarly, if the contour C', see Figure D.7, only encloses the non-negative integers $n = 0, 1, 2, \ldots$, we get

$$\int_{C'} \frac{f(z)}{\sin \pi z}\, \mathrm{d}z = 2\mathrm{i} \sum_{n=0}^{\infty} (-1)^n f(n)$$

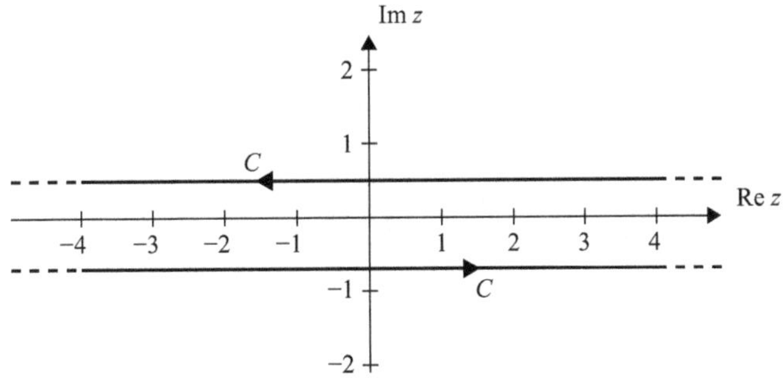

Figure D.6 The contour C.

[19] George Neville Watson (1886–1965), English mathematician.

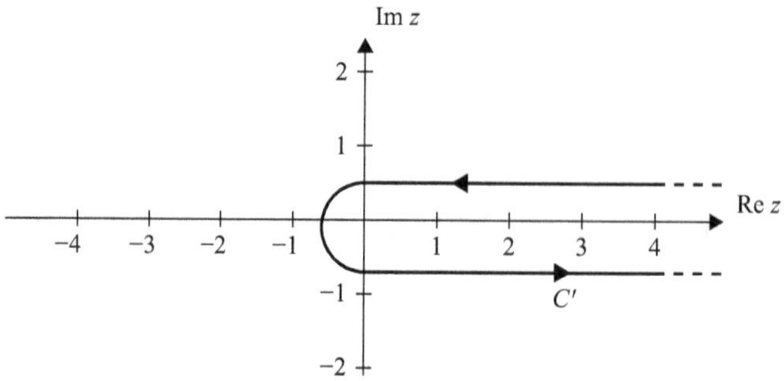

Figure D.7 The contour C'.

D.7 Zeros and Poles of an Analytic Function

The determination of the location of zeros of an analytic function is vital in the diagnostics of many scattering problems. The following theorem is useful for a numerical procedure to find the roots of an analytic function in a bounded complex domain:

Theorem D.8 *Let Ω be an open domain in the complex z-plane, and let $f(z)$ be an analytic function in Ω with a simple zero at z_0. Then*

$$z_0 = \frac{\displaystyle\oint_\gamma \frac{z\,\mathrm{d}z}{f(z)}}{\displaystyle\oint_\gamma \frac{\mathrm{d}z}{f(z)}}$$

where γ is any contour that lies inside Ω, and that encircles the zero z_0.

We give the proof of this theorem.

Proof: Since the zero is simple, the function $f(z)$ is

$$f(z) = g(z)(z - z_0), \qquad z \in \Omega$$

where $g(z)$ has no zeroes in Ω. The residue theorem then gives

$$\oint_\gamma \frac{z\,\mathrm{d}z}{f(z)} = 2\pi\mathrm{i}\,\mathrm{Res}\left.\frac{z}{f(z)}\right|_{z=z_0} = 2\pi\mathrm{i}\frac{z_0}{g(z_0)}$$

Similarly,

$$\oint_\gamma \frac{\mathrm{d}z}{f(z)} = 2\pi\mathrm{i}\frac{1}{g(z_0)}$$

and the theorem is proved. □

In a similar fashion, the location of the poles of an analytic function be located. The next theorem is analogous to Theorem D.8.

Theorem D.9 *Let Ω be an open domain in the complex z-plane, and let $f(z)$ be analytic in Ω, except for a simple pole at z_0. Then*

$$z_0 = \frac{\oint_\gamma zf(z)\,\mathrm{d}z}{\oint_\gamma f(z)}$$

where γ is any contour that lies inside Ω, and that encircles—but does not go through—the pole z_0.

The proof is very similar to Theorem D.8.

Proof: Since the pole is simple, the function $f(z)$ is

$$f(z) = \frac{g(z)}{z - z_0}, \qquad z \in \Omega$$

where $g(z)$ is analytic in Ω. The residue theorem then gives

$$\oint_\gamma zf(z)\,\mathrm{d}z = 2\pi\mathrm{i}\,\mathrm{Res}\; zf(z)|_{z=z_0} = 2\pi\mathrm{i}z_0 g(z_0)$$

Similarly,

$$\oint_\gamma f(z)\,\mathrm{d}z = 2\pi\mathrm{i}\,\mathrm{Res}\, f(z)|_{z=z_0} = 2\pi\mathrm{i}g(z_0)$$

and the theorem is proved. □

APPENDIX E

Relativity

If two observers are in motion relative to each other, they describe the same physical events with different space and time coordinates. The famous Lorentz transformation describes the relation between the two space–time coordinate systems. This section reviews some of the properties of this transformation that are relevant for electromagnetic waves.

E.1 Lorentz Transformation

Consider two reference frames K and K' in vacuum with a relative velocity $\boldsymbol{v}$ to each other. Then, the space–time coordinates in the two coordinate systems are related by the Lorentz transformation [101]

$$\begin{cases} c_0 t' = \gamma(c_0 t - \boldsymbol{\beta} \cdot \boldsymbol{r}) \\ \boldsymbol{r}' = \boldsymbol{r} + \dfrac{\gamma - 1}{\beta^2}(\boldsymbol{\beta} \cdot \boldsymbol{r})\boldsymbol{\beta} - \gamma \boldsymbol{\beta} c_0 t \end{cases} \tag{E.1}$$

where

$$\begin{cases} \boldsymbol{\beta} = \dfrac{\boldsymbol{v}}{c_0} \\ \beta = |\boldsymbol{\beta}| \end{cases} \qquad \gamma = \frac{1}{\sqrt{1 - \beta^2}}$$

The transformation in (E.1) relates an event (unprimed coordinates $(\boldsymbol{r}, t)$) in the reference frame K with an event (primed coordinates $(\boldsymbol{r}', t')$) in the reference frame K'.

E.2 Transformation of the Electromagnetic Fields

The electromagnetic field in vacuum in the two reference frames K and K', respectively, is related as (evaluated at a space–time point $(\boldsymbol{r}, t)$ and $(\boldsymbol{r}', t')$, respectively) [101]

$$\begin{cases} \boldsymbol{E}' = \gamma(\boldsymbol{E} + \boldsymbol{\beta} \times c_0 \boldsymbol{B}) - \dfrac{\gamma^2}{\gamma + 1}\boldsymbol{\beta}(\boldsymbol{\beta} \cdot \boldsymbol{E}) \\ \boldsymbol{B}' = \gamma(\boldsymbol{B} - \boldsymbol{\beta} \times \boldsymbol{E}/c_0) - \dfrac{\gamma^2}{\gamma + 1}\boldsymbol{\beta}(\boldsymbol{\beta} \cdot \boldsymbol{B}) \end{cases} \tag{E.2}$$

One immediate conclusion is that a pure electric (magnetic) field in one reference frame, transforms to both the electric and magnetic fields in the other reference frame.

E.3 Boundary Conditions at a Moving Interface

In Section 1.1.1, the boundary conditions of the electromagnetic fields at a stationary interface S were determined. In many technical applications, e.g., radar and space technology, we encounter situations where moving interfaces are important. In this section, we generalize the results in Section 1.1.1 to accommodate moving interfaces. The starting point for this generalization is the equations in (1.9) and (1.10) on page 5 i.e.,

$$\begin{cases} \iint\limits_S \hat{\nu} \times \boldsymbol{E} \, \mathrm{d}S = -\iiint\limits_V \frac{\partial \boldsymbol{B}}{\partial t} \, \mathrm{d}v \\ \iint\limits_S \hat{\nu} \times \boldsymbol{H} \, \mathrm{d}S = \iiint\limits_V \boldsymbol{J} \, \mathrm{d}v + \iiint\limits_V \frac{\partial \boldsymbol{D}}{\partial t} \, \mathrm{d}v \\ \iint\limits_S \boldsymbol{B} \cdot \hat{\nu} \, \mathrm{d}S = 0 \\ \iint\limits_S \boldsymbol{D} \cdot \hat{\nu} \, \mathrm{d}S = \iiint\limits_V \rho \, \mathrm{d}v \end{cases} \tag{E.3}$$

We are going to apply these integral identities to a pillbox V_h, which is fixed in the laboratory frame—the frame in which the surface locally moves with the velocity $\boldsymbol{v}$. The height of the pillbox is h, and the projected area on the surface S is denoted A. At time t, the interface S coincides with the pillbox (more precisely, the surface S between the two media intersects the pillbox V_h halfway) as shown in Figure E.1, which can be interpreted as a photograph of the geometry in the laboratory frame at time t. All (unprimed) field quantities are measured in the laboratory frame. Notice that we have not interchanged differentiation w.r.t. time and integration in these expressions yet.

As in Section 1.1.1, we intend to study the integrals in (E.3) in the limit $h \to 0$. The two scalar equations in (E.3) follow the same argument as in Section 1.1.1, *provided all quantities are interpreted in the laboratory frame*. We obtain

$$\begin{cases} \hat{\nu} \cdot (\boldsymbol{B}_1 - \boldsymbol{B}_2) = 0 \\ \hat{\nu} \cdot (\boldsymbol{D}_1 - \boldsymbol{D}_2) = \rho_S \end{cases} \tag{E.4}$$

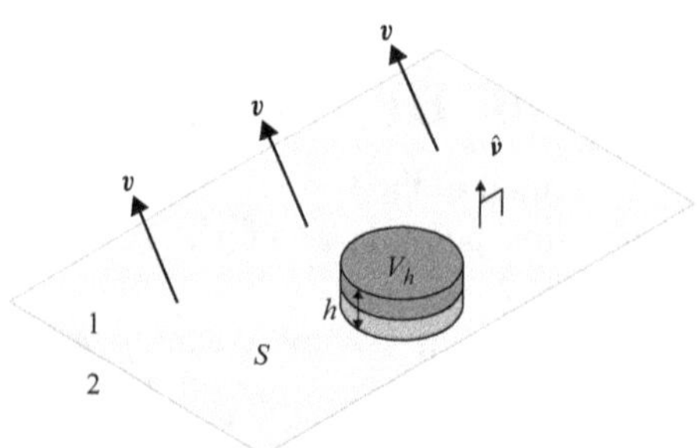

Figure E.1 Interface between two different media 1 and 2 moving at a relative velocity $\boldsymbol{v}$. The pillbox V_h is fixed in the laboratory frame and the surface S intersects the pillbox at time t as shown in the figure.

The two vector-valued equations in (E.3) have to be analyzed with some care. We expect the two integrals on the right-hand side

$$\iiint_V \frac{\partial \boldsymbol{B}}{\partial t}\, dv, \qquad \iiint_V \frac{\partial \boldsymbol{D}}{\partial t}\, dv$$

to have non-zero limits as the height of the pillbox h goes to zero. In fact, the integrand is expected to have variations that are greater than a simple jump discontinuity, which is the case for a stationary surface.

The coordinate representation of the surface is here denoted $\boldsymbol{r} = \boldsymbol{r}(t)$. The velocity of the surface is then

$$\boldsymbol{v} = \frac{d\boldsymbol{r}(t)}{dt}$$

To compute the integrals, we study the integral identity

$$\frac{d}{dt}\iiint_{V_h} \boldsymbol{B}(\boldsymbol{r}(t), t)\, dv = \iiint_{V_h} \left(\frac{\partial}{\partial t}\boldsymbol{B}(\boldsymbol{r}(t), t) + (\boldsymbol{v}\cdot\nabla)\,\boldsymbol{B}(\boldsymbol{r}(t), t)\right) dv \tag{E.5}$$

where we have used

$$\frac{d}{dt}\boldsymbol{B}(\boldsymbol{r}(t), t) = \lim_{\Delta t\to 0}\frac{\boldsymbol{B}(\boldsymbol{r}(t+\Delta t), t+\Delta t) - \boldsymbol{B}(\boldsymbol{r}(t), t)}{\Delta t} = \frac{\partial}{\partial t}\boldsymbol{B}(\boldsymbol{r}(t), t) + (\boldsymbol{v}\cdot\nabla)\,\boldsymbol{B}(\boldsymbol{r}(t), t)$$

The integral on the left-hand side in (E.5) has an integrand where the spatial variable is sliding with the velocity of the surface.[1] This integral does not show the rapid variation as the integral of the $\partial_t\boldsymbol{B}$ field. We expect this integral to have zero limit as $h \to 0$, i.e.,

$$\lim_{h\to 0}\iiint_{V_h} \left(\frac{\partial}{\partial t}\boldsymbol{B}(\boldsymbol{r}(t), t) + (\boldsymbol{v}\cdot\nabla)\,\boldsymbol{B}(\boldsymbol{r}(t), t)\right) dv = 0$$

Since, the vector identity (use (10) on the inside of the first cover)

$$\nabla \times (\boldsymbol{v}\times\boldsymbol{B}) = \boldsymbol{v}\,(\nabla\cdot\boldsymbol{B}) - (\boldsymbol{v}\cdot\nabla)\,\boldsymbol{B}$$

holds, provided $\boldsymbol{v}$ is constant in space, we get with the use of the analogous theorem of Gauss

$$\lim_{h\to 0}\iiint_{V_h} \left(\frac{\partial}{\partial t}\boldsymbol{B}(\boldsymbol{r}(t), t) + \boldsymbol{v}\,(\nabla\cdot\boldsymbol{B}(\boldsymbol{r}(t), t))\right) dv = \lim_{h\to 0}\iint_{S_h} \hat{\boldsymbol{v}}\times(\boldsymbol{v}\times\boldsymbol{B}(\boldsymbol{r}(t), t))\, dS$$

A similar expression holds for the $\boldsymbol{D}$ field.

[1] Observe that this integral is not identical to the integral over a pillbox in the rest frame, since no transformation of the field and the volume element are performed.

Apply to the flux densities $\boldsymbol{B}$ and $\boldsymbol{D}$, and use $\nabla \cdot \boldsymbol{B} = 0$ and $\nabla \cdot \boldsymbol{D} = \rho$. We get

$$\begin{cases} \lim\limits_{h\to 0} \iiint\limits_{V_h} \dfrac{\partial \boldsymbol{B}}{\partial t}\, \mathrm{d}v = \lim\limits_{h\to 0} \iint\limits_{S_h} \hat{\boldsymbol{\nu}} \times (\boldsymbol{v} \times \boldsymbol{B})\, \mathrm{d}S \\ \lim\limits_{h\to 0} \iiint\limits_{V_h} \dfrac{\partial \boldsymbol{D}}{\partial t}\, \mathrm{d}v = \lim\limits_{h\to 0} \iint\limits_{S_h} \left(\hat{\boldsymbol{\nu}} \times (\boldsymbol{v} \times \boldsymbol{D}) - \boldsymbol{v}\rho_S\right) \mathrm{d}S \end{cases}$$

Applied to (E.3), and we get (A is the projected area of the pillbox on the surface S)

$$\begin{cases} A\left[\hat{\boldsymbol{\nu}} \times (\boldsymbol{E}_1 - \boldsymbol{E}_2)\right] = -A\left[\hat{\boldsymbol{\nu}} \times (\boldsymbol{v} \times (\boldsymbol{B}_1 - \boldsymbol{B}_2))\right] \\ A\left[\hat{\boldsymbol{\nu}} \times (\boldsymbol{H}_1 - \boldsymbol{H}_2)\right] = A\boldsymbol{J}_S + A\left[\hat{\boldsymbol{\nu}} \times (\boldsymbol{v} \times (\boldsymbol{D}_1 - \boldsymbol{D}_2)) - \boldsymbol{v}\rho_S\right] \end{cases}$$

or

$$\begin{cases} \hat{\boldsymbol{\nu}} \times [\boldsymbol{E}_1 - \boldsymbol{E}_2 + \boldsymbol{v} \times (\boldsymbol{B}_1 - \boldsymbol{B}_2)] = \boldsymbol{0} \\ \hat{\boldsymbol{\nu}} \times [\boldsymbol{H}_1 - \boldsymbol{H}_2 - \boldsymbol{v} \times (\boldsymbol{D}_1 - \boldsymbol{D}_2)] = \boldsymbol{J}_S - \boldsymbol{v}\rho_S \end{cases}$$

These conditions can be simplified further by the use of the BAC-CAB rule and (E.4)

$$\begin{cases} \hat{\boldsymbol{\nu}} \times (\boldsymbol{E}_1 - \boldsymbol{E}_2) = \left(\boldsymbol{v} \cdot \hat{\boldsymbol{\nu}}\right)(\boldsymbol{B}_1 - \boldsymbol{B}_2) \\ \hat{\boldsymbol{\nu}} \times (\boldsymbol{H}_1 - \boldsymbol{H}_2) = \boldsymbol{J}_S - \left(\boldsymbol{v} \cdot \hat{\boldsymbol{\nu}}\right)(\boldsymbol{D}_1 - \boldsymbol{D}_2) \\ \hat{\boldsymbol{\nu}} \cdot (\boldsymbol{B}_1 - \boldsymbol{B}_2) = 0 \\ \hat{\boldsymbol{\nu}} \cdot (\boldsymbol{D}_1 - \boldsymbol{D}_2) = \rho_S \end{cases}$$

These equations generalize the result of Section 1.1.1, and they become identical to (1.12) when $\boldsymbol{v} = \boldsymbol{0}$.

The special case of a perfectly conducting material 2 implies

$$\begin{cases} \hat{\boldsymbol{\nu}} \times \boldsymbol{E}_1 = \left(\boldsymbol{v} \cdot \hat{\boldsymbol{\nu}}\right)\boldsymbol{B}_1 \\ \hat{\boldsymbol{\nu}} \times \boldsymbol{H}_1 = \boldsymbol{J}_S - \left(\boldsymbol{v} \cdot \hat{\boldsymbol{\nu}}\right)\boldsymbol{D}_1 \\ \hat{\boldsymbol{\nu}} \cdot \boldsymbol{B}_1 = 0 \\ \hat{\boldsymbol{\nu}} \cdot \boldsymbol{D}_1 = \rho_S \end{cases}$$

and, of course, when $\boldsymbol{v} = \boldsymbol{0}$ these equations become identical to (1.13).

Comment E.1

An alternative derivation of the boundary conditions can be derived from the boundary conditions in the stationary frame (see Section 1.1.1), the Lorentz transformation of the fields in (E.2), and the corresponding transformation of the surface charge density ρ_S and the surface current density $\boldsymbol{J}_S$. For details, the reader is encouraged to consult Reference 276.

APPENDIX F

Some Useful Mathematical Results

This appendix contains a series of useful mathematical results that are used in the book.

F.1 Cayley–Hamilton Theorem

The following theorems are important when computing the action of an entire function of a square dyadic [36]. The first theorem by Cayley[1] and Hamilton is fundamental.

Theorem F.1 (Cayley–Hamilton) *A quadratic, n-dimensional dyadic* $\mathbf{A}$ *satisfies its own characteristic equation* $p_{\mathbf{A}}(\lambda)$

$$\textit{If } p_{\mathbf{A}}(\lambda) = \det(\mathbf{A} - \lambda \mathbf{I}_n), \textit{ then } p_{\mathbf{A}}(\mathbf{A}) = \mathbf{0}$$

where $\mathbf{I}_n$ *is the unit dyadic in n dimensions.*

An explicit form of this theorem that is useful in many situations is

$$(\mathbf{A} - \lambda_1 \mathbf{I}_n)^{n_1} \cdot \ldots \cdot (\mathbf{A} - \lambda_k \mathbf{I}_n)^{n_k} \cdot \ldots \cdot (\mathbf{A} - \lambda_m \mathbf{I}_n)^{n_m} = \mathbf{0} \tag{F.1}$$

where n_k, $k = 1, \ldots, m$, is the multiplicity of the eigenvalue λ_k, $k = 1, \ldots, m$, and $\sum_{k=1}^{m} n_k = n$. The eigenvalues are numbered such that $\operatorname{Re} \lambda_1 \geq \operatorname{Re} \lambda_2 \geq \cdots \geq \operatorname{Re} \lambda_n$.

From this theorem, one can prove the following important theorem.

Theorem F.2 *Let* $\lambda_1, \ldots, \lambda_m$ *be the different eigenvalues of the n-dimensional dyadic* $\mathbf{A}$, *and* $n_1, \ldots, n_m$ *their multiplicity. If* $f(z)$ *is an entire function, then*

$$f(\mathbf{A}) = q(\mathbf{A})$$

where the uniquely defined polynomial q of degree $\leq n-1$ *is defined by the following conditions*

$$\frac{d^j q}{dz^j}(\lambda_k) = \frac{d^j f}{dz^j}(\lambda_k), \qquad j = 0, \ldots, n_k - 1, \;\; k = 1, \ldots, m$$

[1]Arthur Cayley (1821–1895), British mathematician.

F.2 Projection Dyadics

F.2.1 Distinct Eigenvalues

We start the review with the case when all eigenvalues are distinct. The following definition is central.

Definition F.1 *Let* $\mathbf{A}$ *be a quadratic, n-dimensional dyadic and, furthermore, that the dyadic has n distinct eigenvalues,* λ_k, $k = 1, 2, \ldots, n$. *The eigenvalues are numbered such that* $\operatorname{Re}\lambda_1 \geq \operatorname{Re}\lambda_2 \geq \cdots \geq \operatorname{Re}\lambda_n$. *Define the projection dyadics* $\mathbf{A}_k$, $k = 1, 2, \ldots, n$

$$\mathbf{A}_k = \frac{(\mathbf{A} - \lambda_1\mathbf{I}_n) \cdot \ldots \cdot (\mathbf{A} - \lambda_{k-1}\mathbf{I}_n) \cdot (\mathbf{A} - \lambda_{k+1}\mathbf{I}_n) \cdot \ldots \cdot (\mathbf{A} - \lambda_n\mathbf{I}_n)}{(\lambda_k - \lambda_1) \cdot \ldots \cdot (\lambda_k - \lambda_{k-1})(\lambda_k - \lambda_{k+1}) \cdot \ldots \cdot (\lambda_k - \lambda_n)}$$

From this definition, we immediately get (use $\det(\mathbf{AB}) = \det\mathbf{A}\det\mathbf{B}$)

$$\det\mathbf{A}_k = 0, \qquad k = 1, 2, \ldots, n$$

since $\det(\mathbf{A} - \lambda_k\mathbf{I}_n) = 0$, $k = 1, 2, \ldots, n$.

To justify that these dyadics are called projection dyadic, we prove

Theorem F.3 *The projection dyadics in Definition F.1 satisfy*

$$\mathbf{A}_i \cdot \mathbf{A}_j = \delta_{ij}\mathbf{A}_i, \qquad i,j = 1, 2, \ldots, n$$

Proof: The dyadic $\mathbf{A}$ satisfies Cayley–Hamilton theorem, see (F.1). We write this as

$$(\mathbf{A} - \lambda_1\mathbf{I}_n) \cdot \ldots \cdot (\mathbf{A} - \lambda_k\mathbf{I}_n) \cdot \ldots \cdot (\mathbf{A} - \lambda_n\mathbf{I}_n) = \mathbf{0}$$

Since the dyadics in this expression all commute, we rewrite this expression for every $k = 1, 2, \ldots, n$ as

$$\begin{aligned}\mathbf{A} \cdot (\mathbf{A} - \lambda_1\mathbf{I}_n) \cdot \ldots \cdot (\mathbf{A} - \lambda_{k-1}\mathbf{I}_n) \cdot (\mathbf{A} - \lambda_{k+1}\mathbf{I}_n) \cdot \ldots \cdot (\mathbf{A} - \lambda_n\mathbf{I}_n) \\ = \lambda_k(\mathbf{A} - \lambda_1\mathbf{I}_n) \cdot \ldots \cdot (\mathbf{A} - \lambda_{k-1}\mathbf{I}_n) \cdot (\mathbf{A} - \lambda_{k+1}\mathbf{I}_n) \cdot \ldots \cdot (\mathbf{A} - \lambda_n\mathbf{I}_n)\end{aligned}$$

which implies

$$\mathbf{A}_k \cdot \mathbf{A} = \mathbf{A} \cdot \mathbf{A}_k = \lambda_k\mathbf{A}_k, \qquad k = 1, 2, \ldots, n \tag{F.2}$$

In the product $\mathbf{A}_i \cdot \mathbf{A}_j$, $i \neq j$, there is always a sequence $(\mathbf{A} - \lambda_1\mathbf{I}_n) \cdot \ldots \cdot (\mathbf{A} - \lambda_k\mathbf{I}_n) \cdot \ldots \cdot (\mathbf{A} - \lambda_n\mathbf{I}_n)$, which implies that $\mathbf{A}_i \cdot \mathbf{A}_j = \mathbf{0}$, $i \neq j$. Moreover, we have with repeated use of (F.2)

$$\begin{aligned}(\mathbf{A} - \lambda_1\mathbf{I}_n) \cdot \ldots \cdot (\mathbf{A} - \lambda_{k-1}\mathbf{I}_n) \cdot (\mathbf{A} - \lambda_{k+1}\mathbf{I}_n) \cdot \ldots \cdot (\mathbf{A} - \lambda_n\mathbf{I}_n) \cdot \mathbf{A}_k \\ = (\lambda_k - \lambda_1) \cdot \ldots \cdot (\lambda_k - \lambda_{k-1})(\lambda_k - \lambda_{k+1}) \cdot \ldots \cdot (\lambda_k - \lambda_n)\mathbf{A}_k\end{aligned}$$

From this identity, we get $\mathbf{A}_k \cdot \mathbf{A}_k = \mathbf{A}_k$, $i = 1, 2, \ldots, n$, and the theorem is proved. □

The projection dyadics can be used to decompose the identity and the dyadic itself. We have

Theorem F.4 *With the assumptions specified in Definition F.1, we have*

$$\mathbf{I}_n = \sum_{i=1}^{n} \mathbf{A}_i, \qquad \mathbf{A} = \sum_{i=1}^{n} \lambda_i \mathbf{A}_i$$

Proof: Let $\hat{\boldsymbol{e}}_k$, $k = 1, 2, \dots, n$, be the distinct (not necessarily an orthonormal set) eigenvectors of the dyadic $\mathbf{A}$, i.e.,

$$\mathbf{A} \cdot \hat{\boldsymbol{e}}_k = \lambda_k \hat{\boldsymbol{e}}_k, \qquad k = 1, 2, \dots, n$$

and therefore

$$\begin{aligned} \mathbf{A}_i \cdot \hat{\boldsymbol{e}}_k &= \frac{(\mathbf{A} - \lambda_1 \mathbf{I}_n) \cdot \dots \cdot (\mathbf{A} - \lambda_{i-1} \mathbf{I}_n) \cdot (\mathbf{A} - \lambda_{i+1} \mathbf{I}_n) \cdot \dots \cdot (\mathbf{A} - \lambda_n \mathbf{I}_n)}{(\lambda_i - \lambda_1) \cdot \dots \cdot (\lambda_i - \lambda_{i-1})(\lambda_i - \lambda_{i+1}) \cdot \dots \cdot (\lambda_i - \lambda_n)} \cdot \hat{\boldsymbol{e}}_k \\ &= \frac{(\lambda_k - \lambda_1) \cdot \dots \cdot (\lambda_k - \lambda_{i-1})(\lambda_k - \lambda_{i+1}) \cdot \dots \cdot (\lambda_k - \lambda_n)}{(\lambda_i - \lambda_1) \cdot \dots \cdot (\lambda_i - \lambda_{i-1})(\lambda_i - \lambda_{i+1}) \cdot \dots \cdot (\lambda_i - \lambda_n)} \hat{\boldsymbol{e}}_k \\ &= \delta_{ik} \hat{\boldsymbol{e}}_k, \qquad i, k = 1, 2, \dots, n \end{aligned}$$

Summation gives

$$\sum_{i=1}^{n} \mathbf{A}_i \cdot \hat{\boldsymbol{e}}_k = \hat{\boldsymbol{e}}_k, \qquad k = 1, 2, \dots, n$$

Since the set of eigenvectors $\hat{\boldsymbol{e}}_k$, $k = 1, 2, \dots, n$, is a basis in $\mathbb{C}^n$, the action of an arbitrary element $\boldsymbol{x} \in \mathbb{C}^n$ is

$$\sum_{i=1}^{n} \mathbf{A}_i \cdot \boldsymbol{x} = \boldsymbol{x}$$

which proves that $\mathbf{I}_n = \sum_{i=1}^{n} \mathbf{A}_i$.

Similarly, we get by the use of (F.2)

$$\sum_{i=1}^{n} \lambda_i \mathbf{A}_i = \sum_{i=1}^{n} \mathbf{A} \cdot \mathbf{A}_i = \mathbf{A} \cdot \sum_{i=1}^{n} \mathbf{A}_i = \mathbf{A}$$

by the use of $\sum_{i=1}^{n} \mathbf{A}_i = \mathbf{I}_n$, and the theorem is proved. □

Theorem F.4 induces a decomposition of a general vector $\boldsymbol{u} \in \mathbb{C}^n$. We define the projections $\boldsymbol{u}_k = \mathbf{A}_k \cdot \boldsymbol{u}$, and we have from Theorem F.4

$$\boldsymbol{u} = \sum_{k=1}^{n} \boldsymbol{u}_k$$

The projections $\boldsymbol{u}_k$ satisfy, see Theorem F.3

$$\mathbf{A}_i \cdot \boldsymbol{u}_j = \delta_{ij}\boldsymbol{u}_j, \qquad i,j = 1, 2, \ldots, n$$

If the eigenvectors $\hat{\boldsymbol{e}}_k$, $k = 1, 2, \ldots, n$, form an orthonormal set, then an alternative representation of the identity dyadic $\mathbf{I}_n$ is

$$\mathbf{I}_n = \sum_{k=1}^{n} \hat{\boldsymbol{e}}_k \hat{\boldsymbol{e}}_k$$

A generalization of Theorem F.4 is

Theorem F.5 *With the assumptions specified in Definition F.1, we have*

$$\mathbf{A}^k = \sum_{i=1}^{n} \lambda_i^k \mathbf{A}_i, \qquad k = 1, 2, \ldots$$

Proof: The result in Theorem F.4 and repeated use of (F.2) give

$$\mathbf{A}^k = \mathbf{A}^{k-1} \sum_{i=1}^{n} \lambda_i \mathbf{A}_i = \mathbf{A}^{k-2} \sum_{i=1}^{n} \lambda_i \mathbf{A} \cdot \mathbf{A}_i = \mathbf{A}^{k-2} \sum_{i=1}^{n} \lambda_i^2 \mathbf{A}_i = \sum_{i=1}^{n} \lambda_i^k \mathbf{A}_i$$

which proves the theorem. □

Finally, we have

Theorem F.6 *Let $f(x)$ be an entire function in the complex plane $\mathbb{C}$. Then the dyadic-valued function $f(\mathbf{A})$, uniquely defined by its power series expansion, has the following expansion in the projection dyadics $\mathbf{A}_k$:*

$$f(\mathbf{A}) = \sum_{i=1}^{n} f(\lambda_i)\mathbf{A}_i$$

Note the following very important implication of this theorem: the left-hand side, which by definition contains arbitrary powers of the dyadic $\mathbf{A}$, can be represented by a polynomial of at most $n - 1$ powers of $\mathbf{A}$. This is a consequence of Cayley–Hamilton theorem, and has immediate implications on the numerical implementations of the dyadic-valued function $f(\mathbf{A})$.

Proof: The result follows immediately from the power series of $f(x)$ and Theorem F.5. □

Example F.1

An important example of this result is the exponential of a dyadic. We have from Theorem F.6

$$\mathrm{e}^{\mathbf{A}} = \sum_{i=1}^{n} \mathrm{e}^{\lambda_i} \mathbf{A}_i$$

■

Example F.2
An important special case is $n = 4$, and when the eigenvalues are arranged in pairs, i.e.,

$$\lambda_1 = -\lambda_4 = \lambda_+, \qquad \lambda_2 = -\lambda_3 = \lambda_-$$

where we assume that $\lambda_+^2 \neq \lambda_-^2$, and $\operatorname{Re}\lambda_1 \geq \operatorname{Re}\lambda_2 \geq 0$. In this case, λ_+^2 and λ_-^2 are eigenvalues of $\mathbf{A}^2$. We get

$$\begin{cases} \mathbf{A}_1 = \dfrac{(\mathbf{A}^2 - \lambda_-^2\mathbf{I}_4)\cdot(\mathbf{A} + \lambda_+\mathbf{I}_4)}{2\lambda_+(\lambda_+^2 - \lambda_-^2)} \\ \mathbf{A}_2 = -\dfrac{(\mathbf{A}^2 - \lambda_+^2\mathbf{I}_4)\cdot(\mathbf{A} + \lambda_-\mathbf{I}_4)}{2\lambda_-(\lambda_+^2 - \lambda_-^2)} \\ \mathbf{A}_3 = \dfrac{(\mathbf{A}^2 - \lambda_+^2\mathbf{I}_4)\cdot(\mathbf{A} - \lambda_-\mathbf{I}_4)}{2\lambda_-(\lambda_+^2 - \lambda_-^2)} \\ \mathbf{A}_4 = -\dfrac{(\mathbf{A}^2 - \lambda_-^2\mathbf{I}_4)\cdot(\mathbf{A} - \lambda_+\mathbf{I}_4)}{2\lambda_+(\lambda_+^2 - \lambda_-^2)} \end{cases}$$

and

$$e^{i\mathbf{A}} = e^{i\lambda_+}\mathbf{A}_1 + e^{-i\lambda_+}\mathbf{A}_4 + e^{i\lambda_-}\mathbf{A}_2 + e^{-i\lambda_-}\mathbf{A}_3 = \frac{\mathbf{A}^2 - \lambda_-^2\mathbf{I}_4}{\lambda_+^2 - \lambda_-^2}\cdot\left(\frac{i}{\lambda_+}\mathbf{A}\sin\lambda_+ + \mathbf{I}_4\cos\lambda_+\right)$$
$$- \frac{\mathbf{A}^2 - \lambda_+^2\mathbf{I}_4}{\lambda_+^2 - \lambda_-^2}\cdot\left(\frac{i}{\lambda_-}\mathbf{A}\sin\lambda_- + \mathbf{I}_4\cos\lambda_-\right)$$

The result in this example is useful in the solution of propagation problems in Chapter 10. ■

F.2.2 Diagonalizable Case

In the previous section, we assumed that all eigenvalues were distinct, and we applied the Cayley–Hamilton theorem in the analysis of projection dyadics. A formally similar analysis can be carried out for the case where some eigenvalues are multiple, provided the dyadic is not defective, i.e., the eigenvectors still span $\mathbb{C}^n$.

The quadratic, n-dimensional dyadic $\mathbf{A}$ is assumed to have m distinct eigenvalues, λ_k, $k = 1, \ldots, m$, where n_k denotes the multiplicity of the eigenvalues, and $\sum_{k=1}^m n_k = n$. Moreover, it is assumed that the corresponding eigenvectors span $\mathbb{C}^n$. In this case, the dyadic satisfies the minimal polynomial

$$(\mathbf{A} - \lambda_1\mathbf{I}_n)\cdot\ldots\cdot(\mathbf{A} - \lambda_k\mathbf{I}_n)\cdot\ldots\cdot(\mathbf{A} - \lambda_m\mathbf{I}_n) = \mathbf{0}$$

With this result, we define the projection dyadics $\mathbf{A}_k$, $k = 1, 2, \ldots, m$:

$$\mathbf{A}_k = \frac{(\mathbf{A} - \lambda_1\mathbf{I}_n)\cdot\ldots\cdot(\mathbf{A} - \lambda_{k-1}\mathbf{I}_n)\cdot(\mathbf{A} - \lambda_{k+1}\mathbf{I}_n)\cdot\ldots\cdot(\mathbf{A} - \lambda_m\mathbf{I}_n)}{(\lambda_k - \lambda_1)\cdot\ldots\cdot(\lambda_k - \lambda_{k-1})(\lambda_k - \lambda_{k+1})\cdot\ldots\cdot(\lambda_k - \lambda_m)}$$

Many of the results for the non-degenerated case are also true for the degenerated case, e.g.,

$$\det \mathbf{A}_k = 0, \qquad k = 1, 2, \ldots, m$$

and the projection dyadics satisfy

$$\mathbf{A}_i \cdot \mathbf{A}_j = \delta_{ij}\mathbf{A}_i, \qquad i,j = 1, 2, \ldots, m$$

and

$$\mathbf{A}^k = \sum_{i=1}^{m} \lambda_i^k \mathbf{A}_i, \qquad k = 0, 1, 2, \ldots$$

and for an analytic function $f(z)$ we have

$$f(\mathbf{A}) = \sum_{i=1}^{m} f(\lambda_i)\mathbf{A}_i$$

Example F.3
The four-dimensional case, $n = 4$, is very important in wave propagation problems. Let λ and $\tilde{\lambda}$ be the eigenvalues (double) and we assume the eigenvectors span $\mathbb{C}^4$. The minimal polynomial is $p(x) = (x - \lambda)(x - \tilde{\lambda})$. We therefore have

$$(\mathbf{A} - \lambda\mathbf{I}_4) \cdot (\mathbf{A} - \tilde{\lambda}\mathbf{I}_4) = \mathbf{A}^2 - (\lambda + \tilde{\lambda})\mathbf{A} + \lambda\tilde{\lambda}\mathbf{I}_4 = \mathbf{0}$$

and

$$\begin{cases} \mathbf{A}_1 = \dfrac{\mathbf{A} - \tilde{\lambda}\mathbf{I}_4}{\lambda - \tilde{\lambda}} \\ \mathbf{A}_2 = \dfrac{\mathbf{A} - \lambda\mathbf{I}_4}{\tilde{\lambda} - \lambda} \end{cases}$$

We also have

$$f(\mathbf{A}) = f(\lambda)\mathbf{A}_1 + f(\tilde{\lambda})\mathbf{A}_2$$

In particular, $\tilde{\lambda} = -\lambda$ and $f(\lambda) = \exp \mathrm{i}\lambda$

$$\mathrm{e}^{\mathrm{i}\mathbf{A}} = \frac{1}{2\lambda}\left\{\mathrm{e}^{\mathrm{i}\lambda}(\mathbf{A} + \lambda\mathbf{I}_4) - \mathrm{e}^{-\mathrm{i}\lambda}(\mathbf{A} - \lambda\mathbf{I}_4)\right\} = \frac{\mathrm{i}}{\lambda}\mathbf{A}\sin\lambda + \mathbf{I}_4\cos\lambda$$

■

Example F.4

The result in Example F.3 can also be obtained from Example F.2 as a limiting process. Let $\lambda_+ = \lambda$, $\lambda_- = \lambda + h$, and let $h \to 0$. We get

$$\begin{aligned} e^{i\mathbf{A}} &= -\frac{\mathbf{A}^2 - (\lambda^2 + 2h\lambda + h^2)\mathbf{I}_4}{2h\lambda + h^2} \cdot \left(\frac{i}{\lambda}\mathbf{A}\sin\lambda + \mathbf{I}_4\cos\lambda\right) \\ &\quad + \frac{\mathbf{A}^2 - \lambda^2\mathbf{I}_4}{2h\lambda + h^2} \cdot \left(\frac{i}{\lambda + h}\mathbf{A}\sin(\lambda + h) + \mathbf{I}_4\cos(\lambda + h)\right) \\ &\to \frac{i}{\lambda}\mathbf{A}\sin\lambda + \mathbf{I}_4\cos\lambda + \frac{\mathbf{A}^2 - \lambda^2\mathbf{I}_4}{2\lambda} \cdot \left(\frac{i}{\lambda}\mathbf{A}\cos\lambda - \frac{i}{\lambda^2}\mathbf{A}\sin\lambda - \mathbf{I}_4\sin\lambda\right) \\ &= \frac{i}{\lambda}\mathbf{A}\sin\lambda + \mathbf{I}_4\cos\lambda \end{aligned}$$

since $\mathbf{A}^2 - \lambda^2\mathbf{I}_4 = \mathbf{0}$, from above. ■

F.2.3 Baker–Campbell–Hausdorff Formula

Let $\mathbf{A}_i$, $i = 1, 2$, be two dyadics, and $[\mathbf{A}_1, \mathbf{A}_2] = \mathbf{A}_1 \cdot \mathbf{A}_2 - \mathbf{A}_2 \cdot \mathbf{A}_1$ be the commutator of the dyadics. Then we construct a dyadic $\mathbf{A}$, such that

$$e^{\mathbf{A}} = e^{\mathbf{A}_1} \cdot e^{\mathbf{A}_2}$$

by

$$\mathbf{A} = \mathbf{A}_1 + \mathbf{A}_2 + \frac{1}{2}[\mathbf{A}_1, \mathbf{A}_2] + \frac{1}{3!}\{[[\mathbf{A}_1, \mathbf{A}_2], \mathbf{A}_2] + [[\mathbf{A}_2, \mathbf{A}_1], \mathbf{A}_1]\} + \cdots$$

This is the Baker–Campbell–Hausdorff[2] formula.

In the long-wave limit, the approximation $\mathbf{A} = \mathbf{A}_1 + \mathbf{A}_2$ can be used to homogenize a two-component structure of bianisotropic layers.

F.3 Hermitian Forms

Let $\mathbf{A}$ be a two-dimensional,[3] complex-valued dyadic, i.e., it represent a linear mapping from $\mathbb{C}^2$ into itself. Every dyadic of this kind has a decomposition in two Hermitian dyadics $\mathbf{X}$ and $\mathbf{Y}$, such that $\mathbf{A} = \mathbf{X} + i\mathbf{Y}$. The explicit expressions of the Hermitian dyadics $\mathbf{X}$ and $\mathbf{Y}$ are

$$\begin{cases} \mathbf{X} = \frac{1}{2}\left(\mathbf{A} + \mathbf{A}^\dagger\right) \\ \mathbf{Y} = \frac{1}{2i}\left(\mathbf{A} - \mathbf{A}^\dagger\right) \end{cases}$$

---

[2] Henry Frederick Baker (1866–1956), British mathematician; John Edward Campbell (1862–1924), British mathematician; and Felix Hausdorff (1868–1942), Polish mathematician.

[3] Several of the results in this section can be generalized to n dimensions.

We easily see that $\mathbf{X} = \mathbf{X}^\dagger$ and $\mathbf{Y} = \mathbf{Y}^\dagger$, where † denotes Hermitian adjoint. We denote $\mathbf{X}$ and $\mathbf{Y}$ the real and the imaginary parts of the dyadic $\mathbf{A}$, respectively.

The dyadic $\mathbf{A}$ generates a sesquilinear form by

$$q(\boldsymbol{x},\boldsymbol{y}) = \boldsymbol{x}^* \cdot \mathbf{A} \cdot \boldsymbol{y}, \qquad \boldsymbol{x},\boldsymbol{y} \in \mathbb{C}^2$$

The sesquilinear form $q(\boldsymbol{x},\boldsymbol{y})$ is a complex-valued function defined on $\mathbb{C}^2 \times \mathbb{C}^2$. A quadratic form $Q(\boldsymbol{x}) = q(\boldsymbol{x},\boldsymbol{x})$ is also generated by the dyadic $\mathbf{A}$. The sesquilinear form $q(\boldsymbol{x},\boldsymbol{y})$ can always be retrieved from the corresponding quadratic form $Q(\boldsymbol{x})$ by the following identity

$$q(\boldsymbol{x},\boldsymbol{y}) = Q((\boldsymbol{x}+\boldsymbol{y})/2) - Q((\boldsymbol{x}-\boldsymbol{y})/2) - \mathrm{i}Q((\boldsymbol{x}+\mathrm{i}\boldsymbol{y})/2) + \mathrm{i}Q((\boldsymbol{x}-\mathrm{i}\boldsymbol{y})/2)$$

The dyadic $\mathbf{A}$ is Hermitian, if and only if the quadratic form $Q(\boldsymbol{x})$ is real valued.

The Hermitian dyadic $\mathbf{X}$ has two real eigenvalues, $\lambda_{1,2}$, determined by

$$\lambda^2 - \lambda \operatorname{tr}\mathbf{X} + \det\mathbf{X} = 0$$

with solutions

$$\lambda_{1,2} = \frac{\operatorname{tr}\mathbf{X} \pm \sqrt{(\operatorname{tr}\mathbf{X})^2 - 4\det\mathbf{X}}}{2}$$

The eigenvalues are of different sign if and only if $\lambda_1\lambda_2 = \det\mathbf{X} < 0$. In terms of components of the dyadic $\mathbf{X}$ in the orthonormal basis $\hat{e}_{1,2}$, the determinant is

$$\det\mathbf{X} = \left(\hat{e}_1^* \cdot \mathbf{X} \cdot \hat{e}_1\right)\left(\hat{e}_2^* \cdot \mathbf{X} \cdot \hat{e}_2\right) - \left|\hat{e}_1^* \cdot \mathbf{X} \cdot \hat{e}_2\right|^2$$

F.3.1 Positive-Definite Dyadics and Positive-Definite Matrices

An n-dimensional complex-valued dyadic $\mathbf{A}$ is said to be a positive dyadic if

$$\boldsymbol{x}^* \cdot \mathbf{A} \cdot \boldsymbol{x} > 0, \qquad \text{for all } \mathbf{0} \neq \boldsymbol{x} \in \mathbb{C}^n$$

An $n \times n$ complex-valued matrix A is called a positive matrix provided

$$\sum_{i=1}^{n}\sum_{j=1}^{n} A_{ij} z_i z_j^* > 0$$

for all $z_i \in \mathbb{C}$, $i = 1, 2, \ldots, n$, not all z_i zero. The negative definite, positive semi-definite, and negative semi-definite dyadics and matrices are defined in the same way, except that the inequalities above are required to be negative, non-negative, and non-positive, respectively.

F.4 Möbius Transform

The Möbius transform, or fractional linear transform, is defined as the mapping [82]

$$t(z) = \frac{az+b}{cz+d}, \qquad ad - bc \neq 0$$

This transformation maps the extended complex plane, $\mathbb{C} \cup \{\infty\}$, to itself, i.e., $t : \mathbb{C} \cup \{\infty\} \mapsto \mathbb{C} \cup \{\infty\}$. Suppose $t_1(z)$ and $t_2(z)$ are two Möbius transforms. Then the composite transformations $t_1(t_2(z))$ and $t_2(t_1(z))$ are Möbius transforms. More explicitly, if

$$\begin{cases} t_1(z) = \dfrac{az+b}{cz+d}, & ad - bc \neq 0 \\ t_2(z) = \dfrac{Az+B}{Cz+D}, & AD - BC \neq 0 \end{cases}$$

then

$$(t_2 \circ t_1)(z) = t_2(t_1(z)) = \frac{A(az+b)/(cz+d)+B}{C(az+b)/(cz+d)+D} = \frac{\alpha z + \beta}{\gamma z + \delta}$$

where

$$\begin{cases} \alpha = Aa + Bc \\ \beta = Ab + Bd \\ \gamma = Ca + Dc \\ \delta = Cb + Dd \end{cases} \tag{F.3}$$

Notice that

$$\alpha\delta - \gamma\beta = (ad - bc)(AD - BC) \neq 0$$

Every Möbius transform can be identified by a complex two-dimensional non-singular matrix

$$t_1 \leftrightarrow \begin{pmatrix} a & b \\ c & d \end{pmatrix}, \qquad t_2 \leftrightarrow \begin{pmatrix} A & B \\ C & D \end{pmatrix}$$

Then by (F.3), we have by matrix multiplication

$$t_2 \circ t_1 \leftrightarrow \begin{pmatrix} \alpha & \beta \\ \gamma & \delta \end{pmatrix} = \begin{pmatrix} A & B \\ C & D \end{pmatrix} \begin{pmatrix} a & b \\ c & d \end{pmatrix}$$

Therefore, the coefficients in the composite Möbius transformation are obtained by simple 2×2 matrix algebra.

From this, we immediately see that the inverse of a Möbius transform is identified by the inverse of the matrix, i.e.,

$$t \leftrightarrow \begin{pmatrix} a & b \\ c & d \end{pmatrix}, \qquad t^{-1} \leftrightarrow \begin{pmatrix} a & b \\ c & d \end{pmatrix}^{-1} = \frac{1}{ad - bc} \begin{pmatrix} d & -b \\ -c & a \end{pmatrix}$$

which leads to the inverse transformation

$$t^{-1}(z) = \frac{-dz + b}{cz - a}$$

F.5 Solid Angle

In several occasions in this book, we use the concept of solid angle. The formal definition is

Definition F.2 *The solid angle $\Omega(\mathbf{r})$ that a surface S subtends at a point $\mathbf{r} \notin S$ is the projection of the surface on the unit sphere centered at $\mathbf{r}$, i.e.,*

$$\Omega(\boldsymbol{r}) = \iint_S \frac{\hat{\boldsymbol{\nu}}(\boldsymbol{r}') \cdot (\boldsymbol{r}' - \boldsymbol{r})}{|\boldsymbol{r}' - \boldsymbol{r}|^3} \, \mathrm{d}S'$$

The angle is positive for that part of the surface that is pointing away (in the direction of $\hat{\boldsymbol{\nu}}$) from the point $\mathbf{r}$, and negative for that part that is pointing toward $\mathbf{r}$.

F.6 Helmholtz' Theorem

Helmholtz' theorem is a decomposition of a vector field as a sum of two unique parts—one longitudinal or irrotational part and one transverse or solenoidal part. If the vector field is $\boldsymbol{F}(\boldsymbol{r})$, which we assume is defined and twice differentiable in a volume V, the decomposition reads

$$\boldsymbol{F}(\boldsymbol{r}) = \boldsymbol{F}_\mathrm{l}(\boldsymbol{r}) + \boldsymbol{F}_\mathrm{t}(\boldsymbol{r})$$

where the fields $\boldsymbol{F}_\mathrm{l}(\boldsymbol{r})$ and $\boldsymbol{F}_\mathrm{t}(\boldsymbol{r})$ satisfy

$$\nabla \times \boldsymbol{F}_\mathrm{l}(\boldsymbol{r}) = \boldsymbol{0}, \qquad \nabla \cdot \boldsymbol{F}_\mathrm{t}(\boldsymbol{r}) = 0$$

One way of deriving this decomposition is to start with the relation

$$\boldsymbol{F}(\boldsymbol{r}) = \iint_V \boldsymbol{F}(\boldsymbol{r}')\delta(\boldsymbol{r} - \boldsymbol{r}') \, \mathrm{d}v', \qquad \boldsymbol{r} \in V$$

Use, see Section 2.1.2 on page 103 (let $k \to 0$ in (2.8) on page 103)

$$\nabla^2 \frac{1}{4\pi \, |\boldsymbol{r} - \boldsymbol{r}'|} = -\delta(\boldsymbol{r} - \boldsymbol{r}')$$

and (2.6) on page 102, we get

$$\begin{aligned}
\boldsymbol{F}(\boldsymbol{r}) &= -\iiint_V \nabla^2 \frac{\boldsymbol{F}(\boldsymbol{r}')}{4\pi \, |\boldsymbol{r} - \boldsymbol{r}'|} \, \mathrm{d}v' = -\nabla^2 \iiint_V \frac{\boldsymbol{F}(\boldsymbol{r}')}{4\pi \, |\boldsymbol{r} - \boldsymbol{r}'|} \, \mathrm{d}v' \\
&= \nabla \times \left(\nabla \times \iiint_V \frac{\boldsymbol{F}(\boldsymbol{r}')}{4\pi \, |\boldsymbol{r} - \boldsymbol{r}'|} \, \mathrm{d}v' \right) - \nabla \left(\nabla \cdot \iiint_V \frac{\boldsymbol{F}(\boldsymbol{r}')}{4\pi \, |\boldsymbol{r} - \boldsymbol{r}'|} \, \mathrm{d}v' \right), \qquad \boldsymbol{r} \in V
\end{aligned}$$

Since $\boldsymbol{F}(\boldsymbol{r}')$ only depends on the primed coordinates, we get

$$\begin{aligned}
\nabla \times \frac{\boldsymbol{F}(\boldsymbol{r}')}{4\pi \, |\boldsymbol{r} - \boldsymbol{r}'|} &= \left(\nabla \frac{1}{4\pi \, |\boldsymbol{r} - \boldsymbol{r}'|} \right) \times \boldsymbol{F}(\boldsymbol{r}') = -\left(\nabla' \frac{1}{4\pi \, |\boldsymbol{r} - \boldsymbol{r}'|} \right) \times \boldsymbol{F}(\boldsymbol{r}') \\
&= -\nabla' \times \frac{\boldsymbol{F}(\boldsymbol{r}')}{4\pi \, |\boldsymbol{r} - \boldsymbol{r}'|} + \frac{\nabla' \times \boldsymbol{F}(\boldsymbol{r}')}{4\pi \, |\boldsymbol{r} - \boldsymbol{r}'|}
\end{aligned}$$

and

$$\nabla \cdot \frac{\boldsymbol{F}(\boldsymbol{r}')}{4\pi\,|\boldsymbol{r}-\boldsymbol{r}'|} = \left(\nabla \frac{1}{4\pi\,|\boldsymbol{r}-\boldsymbol{r}'|}\right) \cdot \boldsymbol{F}(\boldsymbol{r}') = -\left(\nabla' \frac{1}{4\pi\,|\boldsymbol{r}-\boldsymbol{r}'|}\right) \cdot \boldsymbol{F}(\boldsymbol{r}')$$
$$= -\nabla' \cdot \frac{\boldsymbol{F}(\boldsymbol{r}')}{4\pi\,|\boldsymbol{r}-\boldsymbol{r}'|} + \frac{\nabla' \cdot \boldsymbol{F}(\boldsymbol{r}')}{4\pi\,|\boldsymbol{r}-\boldsymbol{r}'|}$$

Apply these results and use Gauss' theorem and an analogous theorem to rewrite the expression of $\boldsymbol{F}(\boldsymbol{r})$ above. Denoting the bounding surface of V by S and $\hat{\boldsymbol{\nu}}$ its outward unit normal vector, we get

$$\boldsymbol{F}(\boldsymbol{r}) = \nabla \times \iiint_V \frac{\nabla' \times \boldsymbol{F}(\boldsymbol{r}')}{4\pi\,|\boldsymbol{r}-\boldsymbol{r}'|}\,\mathrm{d}v' - \nabla \iiint_V \frac{\nabla' \cdot \boldsymbol{F}(\boldsymbol{r}')}{4\pi\,|\boldsymbol{r}-\boldsymbol{r}'|}\,\mathrm{d}v'$$
$$- \nabla \times \iint_S \frac{\hat{\boldsymbol{\nu}}(\boldsymbol{r}') \times \boldsymbol{F}(\boldsymbol{r}')}{4\pi\,|\boldsymbol{r}-\boldsymbol{r}'|}\,\mathrm{d}S' + \nabla \iint_S \frac{\hat{\boldsymbol{\nu}}(\boldsymbol{r}') \cdot \boldsymbol{F}(\boldsymbol{r}')}{4\pi\,|\boldsymbol{r}-\boldsymbol{r}'|}\,\mathrm{d}S', \qquad \boldsymbol{r} \in V$$

We can now identify

$$\boldsymbol{F}_{\mathrm{l}}(\boldsymbol{r}) = -\nabla \iiint_V \frac{\nabla' \cdot \boldsymbol{F}(\boldsymbol{r}')}{4\pi\,|\boldsymbol{r}-\boldsymbol{r}'|}\,\mathrm{d}v' + \nabla \iint_S \frac{\hat{\boldsymbol{\nu}}(\boldsymbol{r}') \cdot \boldsymbol{F}(\boldsymbol{r}')}{4\pi\,|\boldsymbol{r}-\boldsymbol{r}'|}\,\mathrm{d}S'$$

and

$$\boldsymbol{F}_{\mathrm{t}}(\boldsymbol{r}) = \nabla \times \iiint_V \frac{\nabla' \times \boldsymbol{F}(\boldsymbol{r}')}{4\pi\,|\boldsymbol{r}-\boldsymbol{r}'|}\,\mathrm{d}v' - \nabla \times \iint_S \frac{\hat{\boldsymbol{\nu}}(\boldsymbol{r}') \times \boldsymbol{F}(\boldsymbol{r}')}{4\pi\,|\boldsymbol{r}-\boldsymbol{r}'|}\,\mathrm{d}S'$$

These fields satisfy

$$\nabla \times \boldsymbol{F}_{\mathrm{l}}(\boldsymbol{r}) = \boldsymbol{0}, \qquad \nabla \cdot \boldsymbol{F}_{\mathrm{t}}(\boldsymbol{r}) = 0$$

If the surface recedes to infinity, the volume V becomes $\mathbb{R}^3$, and if the field $\boldsymbol{F}(\boldsymbol{r})$ is regular—more precisely $\lim_{r\to\infty} \nabla \times \boldsymbol{F}(\boldsymbol{r}) = \boldsymbol{0}$ and $\lim_{r\to\infty} \nabla \cdot \boldsymbol{F}(\boldsymbol{r}) = 0$—the surface integrals vanish and we obtain

$$\boldsymbol{F}_{\mathrm{l}}(\boldsymbol{r}) = -\nabla \iiint_{\mathbb{R}^3} \frac{\nabla' \cdot \boldsymbol{F}(\boldsymbol{r}')}{4\pi\,|\boldsymbol{r}-\boldsymbol{r}'|}\,\mathrm{d}v'$$

and

$$\boldsymbol{F}_{\mathrm{t}}(\boldsymbol{r}) = \nabla \times \iiint_{\mathbb{R}^3} \frac{\nabla' \times \boldsymbol{F}(\boldsymbol{r}')}{4\pi\,|\boldsymbol{r}-\boldsymbol{r}'|}\,\mathrm{d}v'$$

We can also write the vector $\boldsymbol{F}(\boldsymbol{r})$ as a sum of two potentials—one scalar and one vector—as

$$\boldsymbol{F}(\boldsymbol{r}) = -\nabla\phi(\boldsymbol{r}) + \nabla \times \boldsymbol{A}(\boldsymbol{r})$$

where

$$\phi(\boldsymbol{r}) = \iiint\limits_{\mathbb{R}^3} \frac{\nabla' \cdot \boldsymbol{F}(\boldsymbol{r}')}{4\pi\,|\boldsymbol{r}-\boldsymbol{r}'|}\,\mathrm{d}v'$$

and

$$\boldsymbol{A}(\boldsymbol{r}) = \iiint\limits_{\mathbb{R}^3} \frac{\nabla' \times \boldsymbol{F}(\boldsymbol{r}')}{4\pi\,|\boldsymbol{r}-\boldsymbol{r}'|}\,\mathrm{d}v'$$

F.6.1 Uniqueness of the Decomposition

We now prove that the decomposition above is unique.

Let $\phi(\boldsymbol{r})$ and $\boldsymbol{a}(\boldsymbol{r})$ be a given scalar and vector fields, respectively. Both fields are defined and twice continuously differentiable in the volume V, which is assumed to be simply connected, and with a bounding surface S and outward pointing unit normal $\hat{\boldsymbol{\nu}}$.

The field $\boldsymbol{F}(\boldsymbol{r})$ satisfying

$$\nabla \cdot \boldsymbol{F}(\boldsymbol{r}) = \phi(\boldsymbol{r}), \quad \nabla \times \boldsymbol{F}(\boldsymbol{r}) = \boldsymbol{a}(\boldsymbol{r}), \qquad \boldsymbol{r} \in V \tag{F.4}$$

with a given boundary value, i.e., $\boldsymbol{F}(\boldsymbol{r}) = \boldsymbol{b}(\boldsymbol{r})$, $\boldsymbol{r} \in S$, where $\boldsymbol{b}(\boldsymbol{r})$ is a given field defined on the surface S, is then uniquely determined.

To prove the uniqueness of $\boldsymbol{F}(\boldsymbol{r})$, we assume there are two different fields, $\boldsymbol{F}_1(\boldsymbol{r})$ and $\boldsymbol{F}_2(\boldsymbol{r})$, that satisfy the relations in (F.4). The difference $\boldsymbol{D}(\boldsymbol{r}) = \boldsymbol{F}_1(\boldsymbol{r}) - \boldsymbol{F}_2(\boldsymbol{r})$ then satisfies

$$\nabla \cdot \boldsymbol{D}(\boldsymbol{r}) = 0, \quad \nabla \times \boldsymbol{D}(\boldsymbol{r}) = \boldsymbol{0}, \qquad \boldsymbol{r} \in V$$

and $\boldsymbol{D}(\boldsymbol{r}) = \boldsymbol{0}$, $\boldsymbol{r} \in S$. The vanishing curl of $\boldsymbol{D}(\boldsymbol{r})$ implies that there exists a potential $\psi(\boldsymbol{r})$, defined in V, such that $\boldsymbol{D}(\boldsymbol{r}) = -\nabla\psi(\boldsymbol{r})$, and moreover, due to vanishing divergence of $\boldsymbol{D}(\boldsymbol{r})$, the potential satisfies Laplace equation $\nabla^2\psi(\boldsymbol{r}) = 0$ in V. The divergence theorem

$$\iint\limits_S \hat{\boldsymbol{\nu}} \cdot (\chi \nabla\psi)\,\mathrm{d}S = \iiint\limits_V \nabla \cdot (\chi\nabla\psi)\,\mathrm{d}V = \iiint\limits_V \left(\chi\nabla^2\psi + \nabla\chi \cdot \nabla\psi\right)\mathrm{d}v$$

is now applied to $\chi(\boldsymbol{r}) = \psi(\boldsymbol{r})$. We get

$$\iiint\limits_V |\boldsymbol{D}|^2\,\mathrm{d}v = \iiint\limits_V |\nabla\psi|^2\,\mathrm{d}v = \iint\limits_S \hat{\boldsymbol{\nu}} \cdot (\psi\nabla\psi)\,\mathrm{d}S = -\iint\limits_S \hat{\boldsymbol{\nu}} \cdot (\psi\boldsymbol{D})\,\mathrm{d}S = 0$$

and we conclude that $\boldsymbol{D}(\boldsymbol{r}) = \boldsymbol{0}$ in V, and thus $\boldsymbol{F}_1(\boldsymbol{r}) = \boldsymbol{F}_2(\boldsymbol{r})$ in V.

This uniqueness result also proves that a given vector field $\boldsymbol{F}(\boldsymbol{r})$ in V has a unique decomposition in a transverse and a longitudinal part given by Helmholtz' theorem. This is an immediate consequence of the uniqueness theorem just proved, since the transverse and longitudinal parts satisfy

$$\nabla \cdot \boldsymbol{F}_{\mathrm{t}}(\boldsymbol{r}) = 0, \quad \nabla \times \boldsymbol{F}_{\mathrm{t}}(\boldsymbol{r}) = \nabla \times \boldsymbol{F}(\boldsymbol{r}), \qquad \boldsymbol{r} \in V$$

and

$$\nabla \cdot \boldsymbol{F}_1(\boldsymbol{r}) = \nabla \cdot \boldsymbol{F}(\boldsymbol{r}), \quad \nabla \times \boldsymbol{F}_1(\boldsymbol{r}) = \boldsymbol{0}, \qquad \boldsymbol{r} \in V$$

and both these problems are uniquely solvable.

F.7 The Translation Matrices

Let $\boldsymbol{r}' = \boldsymbol{r} + \boldsymbol{d}$, see Figure 7.2 on page 367. The spherical vector wave with origin at O and O', respectively, is related by

$$\begin{cases} \boldsymbol{v}_n(k\boldsymbol{r}') = \sum_{n'} \mathcal{R}_{nn'}(k\boldsymbol{d})\boldsymbol{v}_{n'}(k\boldsymbol{r}), & \text{for all } \boldsymbol{d} \\ \boldsymbol{u}_n(k\boldsymbol{r}') = \sum_{n'} \mathcal{R}_{nn'}(k\boldsymbol{d})\boldsymbol{u}_{n'}(k\boldsymbol{r}), & r > d \\ \boldsymbol{u}_n(k\boldsymbol{r}') = \sum_{n'} \mathcal{P}_{nn'}(k\boldsymbol{d})\boldsymbol{v}_{n'}(k\boldsymbol{r}), & r < d \end{cases}$$

Denote the spherical coordinates of $\boldsymbol{r}$, $\boldsymbol{r}'$, and $\boldsymbol{d}$ by (r, θ, ϕ), (r', θ', ϕ'), and (d, η, ψ), respectively. Here $r, r', d \in [0, \infty)$, $\theta, \theta', \eta \in [0, \pi]$, and $\phi, \phi', \psi \in [0, 2\pi)$. The translation matrices for a translation $\boldsymbol{d}$ ($d \geq 0$) are [23]

$$\begin{aligned}
\mathcal{P}_{1\sigma ml,1\sigma m'l'}(k\boldsymbol{d}) &= (-1)^{m'} C_{ml,m'l'}(d,\eta)\cos(m-m')\psi \\
&\quad + (-1)^{\sigma} C_{ml,-m'l'}(d,\eta)\cos(m+m')\psi \\
\mathcal{P}_{1\sigma ml,1\sigma' m'l'}(k\boldsymbol{d}) &= (-1)^{m'+\sigma'} C_{ml,m'l'}(d,\eta)\sin(m-m')\psi \\
&\quad + C_{ml,-m'l'}(d,\eta)\sin(m+m')\psi, \qquad \sigma \neq \sigma' \\
\mathcal{P}_{1\sigma ml,2\sigma' m'l'}(k\boldsymbol{d}) &= (-1)^{m'+\sigma} D_{ml,m'l'}(d,\eta)\cos(m-m')\psi \\
&\quad - D_{ml,-m'l'}(d,\eta)\cos(m+m')\psi, \qquad \sigma \neq \sigma' \\
\mathcal{P}_{1\sigma ml,2\sigma m'l'}(k\boldsymbol{d}) &= (-1)^{m'} D_{ml,m'l'}(d,\eta)\sin(m-m')\psi \\
&\quad + (-1)^{\sigma} D_{ml,-m'l'}(d,\eta)\sin(m+m')\psi \\
\mathcal{P}_{2\sigma ml,\tau\sigma' m'l'}(k\boldsymbol{d}) &= \mathcal{P}_{1\sigma ml,\bar{\tau}\sigma' m'l'}(k\boldsymbol{d}), \qquad \tau = 1, 2
\end{aligned}$$

where

$$\begin{aligned}
C_{ml,m'l'}(d,\eta) &= \frac{(-1)^{m+m'}}{2}\sqrt{\frac{\varepsilon_m \varepsilon_{m'}}{4}} \\
&\quad \times \sum_{\lambda=|l-l'|}^{l+l'} (-1)^{(l'-l+\lambda)/2}(2\lambda+1)\sqrt{\frac{(2l+1)(2l'+1)(\lambda-(m-m'))!}{l(l+1)l'(l'+1)(\lambda+(m-m'))!}} \\
&\quad \times \begin{pmatrix} l & l' & \lambda \\ 0 & 0 & 0 \end{pmatrix} \begin{pmatrix} l & l' & \lambda \\ m & -m' & m'-m \end{pmatrix} \left[l(l+1) + l'(l'+1) - \lambda(\lambda+1)\right] \\
&\quad \times h_\lambda^{(1)}(kd) P_\lambda^{m-m'}(\cos\eta)
\end{aligned}$$

$$
\begin{aligned}
D_{ml,m'l'}(d,\eta) = {} & \frac{(-1)^{m+m'}}{2}\sqrt{\frac{\varepsilon_m\varepsilon_{m'}}{4}} \\
& \times \sum_{\lambda=|l-l'|+1}^{l+l'} (-1)^{(l'-l+\lambda+1)/2}(2\lambda+1)\sqrt{\frac{(2l+1)(2l'+1)(\lambda-(m-m'))!}{l(l+1)l'(l'+1)(\lambda+(m-m'))!}} \\
& \times \begin{pmatrix} l & l' & \lambda-1 \\ 0 & 0 & 0 \end{pmatrix}\begin{pmatrix} l & l' & \lambda \\ m & -m' & m'-m \end{pmatrix}\sqrt{\lambda^2-(l-l')^2} \\
& \times \sqrt{(l+l'+1)^2-\lambda^2}\, h_\lambda^{(1)}(kd)P_\lambda^{m-m'}(\cos\eta)
\end{aligned}
$$

and where the Neumann factor is defined as above as $\varepsilon_m = 2 - \delta_{m0}$, and where $\begin{pmatrix} \cdot & \cdot & \cdot \\ \cdot & \cdot & \cdot \end{pmatrix}$ denotes Wigner[4] 3-j symbol [59], see subsection below, and

$$
(-1)^\sigma = \begin{cases} 1, & \sigma = \mathrm{e} \\ -1, & \sigma = \mathrm{o} \end{cases}
$$

The translation matrix $\mathcal{R}_{nn'}(k\boldsymbol{r})$ is identical to $\mathcal{P}_{nn'}(k\boldsymbol{r})$ but with $h_\lambda^{(1)}(kd)$ replaced with $j_\lambda(kd)$.

Translation in the opposite direction is identical to the transpose of the translation matrices [213], i.e.,[5]

$$
\mathcal{R}^{\mathrm{t}}(k\boldsymbol{d}) = \mathcal{R}(-k\boldsymbol{d}), \qquad \mathcal{P}^{\mathrm{t}}(k\boldsymbol{d}) = \mathcal{P}(-k\boldsymbol{d})
$$

Additional useful identities are [213]

$$
\begin{cases} \mathcal{P}(k(\boldsymbol{d}_1+\boldsymbol{d}_2)) = \mathcal{P}(k\boldsymbol{d}_1)\mathcal{R}(k\boldsymbol{d}_2) \\ \mathcal{R}(k\boldsymbol{d}_2)\mathcal{P}(k\boldsymbol{d}_1) = \mathcal{P}(k\boldsymbol{d}_1)\mathcal{R}(k\boldsymbol{d}_2) \end{cases} \quad d_2 < d_1
$$

F.7.1 Wigner 3-j Symbol

In problems where translations of spherical vector waves occur, the Wigner 3-j symbol is of paramount importance. They are denoted as

$$
\begin{pmatrix} j_1 & j_2 & j_3 \\ m_1 & m_2 & m_3 \end{pmatrix}
$$

---

[4]Eugene Paul Wigner (1902–1995), Hungarian-American theoretical physicist and mathematician.

[5]Multiple use of the translation properties and the completeness of the regular spherical vector waves imply

$$
\begin{aligned}
\boldsymbol{v}_n(k\boldsymbol{r}) = \boldsymbol{v}_n(k(\boldsymbol{r}-\boldsymbol{d}+\boldsymbol{d})) &= \sum_{n'} \mathcal{R}_{nn'}(-k\boldsymbol{d})\boldsymbol{v}_{n'}(k(\boldsymbol{r}+\boldsymbol{d})) \\
&= \sum_{n'n''} \mathcal{R}_{nn'}(-k\boldsymbol{d})\mathcal{R}_{n'n''}(k\boldsymbol{d})\boldsymbol{v}_{n''}(k\boldsymbol{r})
\end{aligned}
$$

and consequently, $\mathcal{R}^{-1}(k\boldsymbol{d}) = \mathcal{R}(-k\boldsymbol{d})$. Note that this relation holds for the infinite matrix, and for every finite-dimensional realization of the property, the relation is approximative.

Here, m_1, m_2, and m_3 are integers and j_1, j_2, and j_3 non-negative integers (in fact, also half integers are permitted, but not used in electromagnetic problems). Some of its properties are

1.

$$\begin{pmatrix} j_1 & j_2 & j_3 \\ m_1 & m_2 & m_3 \end{pmatrix} = 0 \quad \text{if not} \begin{cases} m_1 + m_2 + m_3 = 0 \\ \text{and} \\ |j_1 - j_2| \le j_3 \le j_1 + j_2 \\ \text{(triangle inequality)} \end{cases}$$

2. Cyclic permutations

$$\begin{pmatrix} j_1 & j_2 & j_3 \\ m_1 & m_2 & m_3 \end{pmatrix} = \begin{pmatrix} j_3 & j_1 & j_2 \\ m_3 & m_1 & m_2 \end{pmatrix} = \begin{pmatrix} j_2 & j_3 & j_1 \\ m_2 & m_3 & m_1 \end{pmatrix}$$

$$\begin{pmatrix} j_1 & j_2 & j_3 \\ m_1 & m_2 & m_3 \end{pmatrix} = (-1)^{j_1+j_2+j_3} \begin{pmatrix} j_2 & j_1 & j_3 \\ m_2 & m_1 & m_3 \end{pmatrix} = (-1)^{j_1+j_2+j_3} \begin{pmatrix} j_1 & j_2 & j_3 \\ -m_1 & -m_2 & -m_3 \end{pmatrix}$$

3.

$$\begin{pmatrix} j_1 & j_2 & j_3 \\ 0 & 0 & 0 \end{pmatrix} = 0 \quad \text{if } j_1 + j_2 + j_3 \text{ odd integer}$$

or explicitly,

$$\begin{pmatrix} j_1 & j_2 & j_3 \\ 0 & 0 & 0 \end{pmatrix} = \begin{cases} (-1)^g \sqrt{\frac{(2g-2j_1)!(2g-2j_2)!(2g-2j_3)!}{(2g+1)!}} \frac{g!}{(g-j_1)!(g-j_2)!(g-j_3)!}, & J = 2g \\ 0, & J = 2g+1 \end{cases}$$

where $J = j_1 + j_2 + j_3$.

4. If $j_1 + j_2 + j_3$ is even

$$\begin{pmatrix} j_1 & j_2 & j_3 \\ 1 & -1 & 0 \end{pmatrix} = \begin{pmatrix} j_1 & j_2 & j_3 \\ 0 & 0 & 0 \end{pmatrix} \frac{j_3(j_3+1) - j_1(j_1+1) - j_2(j_2+1)}{2\sqrt{j_1(j_1+1)j_2(j_2+1)}}$$

If $j_1 + j_2 + j_3$ is odd

$$\begin{pmatrix} j_1 & j_2 & j_3 \\ 1 & -1 & 0 \end{pmatrix} = \begin{pmatrix} j_1 & j_2 & j_3 - 1 \\ 0 & 0 & 0 \end{pmatrix} \frac{\sqrt{j_3^2 - (j_1 - j_2)^2}\sqrt{(j_1 + j_2 + 1)^2 - j_3^2}}{2\sqrt{j_1(j_1+1)j_2(j_2+1)}}$$

Evaluation of the Wigner 3-j symbol

$$\begin{pmatrix} j_1 & j_2 & j_3 \\ m_1 & -m_1 & 0 \end{pmatrix}$$

is most conveniently done by recursion. We present two useful methods [291].

Recursion 1 A three-term recursion relation for the Wigner 3-j symbol is given by [240, 266] $(m_1 + m_2 + m_3 = 0)$

$$j_1 A(j_1+1)\begin{pmatrix} j_1+1 & j_2 & j_3 \\ m_1 & m_2 & m_3 \end{pmatrix} + B(j_1)\begin{pmatrix} j_1 & j_2 & j_3 \\ m_1 & m_2 & m_3 \end{pmatrix} + (j_1+1)A(j_1)\begin{pmatrix} j_1-1 & j_2 & j_3 \\ m_1 & m_2 & m_3 \end{pmatrix} = 0$$

where

$$\begin{cases} A(j_1) = \sqrt{j_1^2 - (j_2 - j_3)^2}\sqrt{(j_2 + j_3 + 1)^2 - j_1^2}\sqrt{j_1^2 - m_1^2} \\ B(j_1) = -(2j_1 + 1)\left[j_2(j_2+1)m_1 - j_3(j_3+1)m_1 - j_1(j_1+1)(m_3 - m_2)\right] \end{cases}$$

If we denote

$$f(j_1) = \begin{pmatrix} j_1 & j_2 & j_3 \\ m_1 & m_2 & m_3 \end{pmatrix}$$

the recursion relation is

$$j_1 A(j_1+1) f(j_1+1) + B(j_1) f(j_1) + (j_1+1) A(j_1) f(j_1 - 1) = 0$$

which is a stable recursion formula for increasing values of j_1. The range of j_1 is finite, and the smallest and largest values are

$$\begin{cases} j_{1\,\mathrm{min}} = \max\{|j_2 - j_3|, |m_1|\} \\ j_{1\,\mathrm{max}} = j_2 + j_3 \end{cases}$$

We observe that

$$A(j_{1\,\mathrm{min}}) = A(j_{1\,\mathrm{max}+1}) = 0$$

so the start and the end of the recursion relation is a two-term recursion

$$j_{1\,\mathrm{min}} A(j_{1\,\mathrm{min}} + 1) f(j_{1\,\mathrm{min}} + 1) + B(j_{1\,\mathrm{min}}) f(j_{1\,\mathrm{min}}) = 0$$

and

$$B(j_{1\,\mathrm{max}}) f(j_{1\,\mathrm{max}}) + (j_{1\,\mathrm{max}} + 1) A(j_{1\,\mathrm{max}}) f(j_{1\,\mathrm{max}} - 1) = 0$$

The recursion relation satisfies the normalization condition

$$\sum_{j_1} (2j_1 + 1) f(j_1)^2 = 1$$

Recursion 2 A second three-term recursion relation for the Wigner 3-j symbol is [240] $(m_1 + m_2 + m_3 = 0)$

$$C(m_2+1)\begin{pmatrix} j_1 & j_2 & j_3 \\ m_1 & m_2+1 & m_3-1 \end{pmatrix} + D(m_2)\begin{pmatrix} j_1 & j_2 & j_3 \\ m_1 & m_2 & m_3 \end{pmatrix}$$
$$+ C(m_2)\begin{pmatrix} j_1 & j_2 & j_3 \\ m_1 & m_2-1 & m_3+1 \end{pmatrix} = 0$$

where

$$\begin{cases} C(m_2) = \sqrt{j_2 - m_2 + 1}\sqrt{j_2 + m_2}\sqrt{j_3 + m_3 + 1}\sqrt{j_3 - m_3} \\ D(m_2) = j_2(j_2+1) + j_3(j_3+1) - j_1(j_1+1) + 2m_2m_3 \end{cases}$$

If we denote

$$g(m_2) = \begin{pmatrix} j_1 & j_2 & j_3 \\ m_1 & m_2 & -m_1 - m_2 \end{pmatrix}$$

we have

$$C(m_2+1)g(m_2+1) + D(m_2)g(m_2) + C(m_2)g(m_2-1) = 0$$

The range of m_2 is finite, and the smallest and largest values are

$$\begin{cases} m_{2\min} = \max\{-j_2, -j_3 - m_1\} \\ m_{2\max} = \min\{j_2, j_3 - m_1\} \end{cases}$$

The start and the end of the recursion relation is a two-term recursion.

$$C(m_{2\min}+1)g(m_{2\min}+1) + D(m_{2\min})g(m_{2\min}) = 0$$

and

$$D(m_{2\max})g(m_{2\max}) + C(m_{2\max})g(m_{2\max}-1) = 0$$

The recursion relation satisfies the normalization condition

$$\sum_{m_2} (2j_1+1)g(m_2)^2 = 1$$

F.8 Volterra Equations

Let $K(x,x')$ and $g(x)$ be continuous functions on $[a,b] \times [a,b]$ and $[a,b]$, respectively. The Volterra[6] equation then is [96, 167]

$$f(x) = g(x) + \int_a^x K(x,x')f(x')\,\mathrm{d}x', \qquad x \in [a,b]$$

[6]Vito Volterra (1860–1940), Italian mathematician and physicist.

The function $K(x,x')$ in the integral is called the integral kernel of the integral equation. The Volterra integral equation has a unique solution $f(x)$ [96, 167]. Moreover, the solution is continuous on $[a,b]$.

Useful in this textbook is also the dyadic version of the Volterra integral equation. Let $\mathbf{K}(x,x')$ be a continuous, dyadic-valued function on $[a,b]\times[a,b]$, and let $\mathbf{g}(x)$ be a continuous, dyadic-valued function on $[a,b]$. The Volterra dyadic-valued equation

$$\mathbf{f}(x) = \mathbf{g}(x) + \int_a^x \mathbf{K}(x,x')\cdot\mathbf{f}(x')\,\mathrm{d}x', \qquad x\in[a,b]$$

has a unique continuous dyadic-valued solution $\mathbf{f}(x)$ on $[a,b]$ [132].

F.9 Vectors and Linear Operators in Hilbert Spaces

This section contains the basic definitions of vectors and linear operators in Hilbert spaces. These results are useful in the analysis presented in this textbook. No proofs are given here, since there are many excellent textbooks on the subject, e.g., References 56, 132, 195, 250.

Definition F.3 *Let X be a complex (or real) linear space. Then a scalar product on X is a map $(\cdot,\cdot)_X : X\times X\to\mathbb{C}$ such that*

$$\begin{aligned}
&(u,u)_X \geq 0 && \textit{(positivity)}\\
&(u,u)_X = 0 \ \textit{if and only if}\ u=0 && \textit{(definiteness)}\\
&(u,v)_X = (v,u)_X^* && \textit{(conjugate symmetry)}\\
&(\alpha u+\beta v,w)_X = \alpha(u,w)_X+\beta(v,w)_X && \textit{(linearity)}
\end{aligned}$$

for all $u,v,w\in X$ and $\alpha,\beta\in\mathbb{C}$ (or $\mathbb{R}$). This map is linear in the first argument and anti-linear (conjugate-linear) in the second.

The norm associated with X is

$$\|u\|_X = \sqrt{(u,u)_X}$$

The scalar product satisfy

Theorem F.7 (Cauchy–Schwarz)

$$|(u,v)_X| \leq \sqrt{(u,u)_X(v,v)_X} = \|u\|_X\,\|v\|_X$$

for all $u,v\in X$ and equality if and only if u and v are linearly dependent, i.e., $u=\alpha v$ for some complex constant α.

F.9.1 Function Spaces

In terms of the scalar product

$$(u, v)_{L^2(\Omega)} = \iiint_{\Omega} u(\boldsymbol{r}) v^*(\boldsymbol{r}) \, \mathrm{d}v$$

we define the function space $L^2(\Omega)$, consisting of all complex-valued function defined in an open region $\Omega \in \mathbb{R}^3$ such that

$$\|u\|^2_{L^2(\Omega)} = (u, u)_{L^2(\Omega)} = \iiint_{\Omega} |u(\boldsymbol{r})|^2 \, \mathrm{d}v < \infty$$

or

$$L^2(\Omega) = \left\{u : \|u\|_{L^2(\Omega)} < \infty\right\}$$

Similarly, we define the function space $\boldsymbol{L}^2(\Omega) = L^2(\Omega; \mathbb{C}^3)$ as all complex vector-valued function with finite L^2 norm, i.e.,

$$\boldsymbol{L}^2(\Omega) = \left\{\boldsymbol{u} : \|\boldsymbol{u}\|_{L^2(\Omega)} < \infty\right\}$$

with the scalar product

$$(\boldsymbol{u}, \boldsymbol{v})_{L^2(\Omega)} = \iiint_{\Omega} \boldsymbol{u}(\boldsymbol{r}) \cdot \boldsymbol{v}^*(\boldsymbol{r}) \, \mathrm{d}v$$

The function space $L^2(\partial\Omega)$ consists of all complex-valued function defined on the surface $\partial\Omega$ with finite L^2 norm, i.e.,

$$L^2(\partial\Omega) = \left\{u : \|u\|_{L^2(\partial\Omega)} < \infty\right\}$$

with the scalar product

$$(u, v)_{L^2(\partial\Omega)} = \iint_{\partial\Omega} u(\boldsymbol{r}) v^*(\boldsymbol{r}) \, \mathrm{d}S$$

The space of tangential vectors on the surface $\partial\Omega$ is also relevant. The function space $\boldsymbol{L}^2_{\mathrm{t}}(\partial\Omega)$ consists of all tangential, complex vector-valued function defined on the surface $\partial\Omega$ with finite $\boldsymbol{L}^2$ norm, i.e.,

$$\boldsymbol{L}^2_{\mathrm{t}}(\partial\Omega) = \left\{\boldsymbol{u} : \|\boldsymbol{u}\|_{L^2_{\mathrm{t}}(\partial\Omega)} < \infty; \boldsymbol{u} \cdot \hat{\boldsymbol{\nu}} = 0\right\}$$

where $\hat{\boldsymbol{\nu}}$ is the outward pointing unit vector of the boundary $\partial\Omega$, and where the scalar product is

$$(\boldsymbol{u}, \boldsymbol{v})_{L^2_{\mathrm{t}}(\partial\Omega)} = \iint_{\partial\Omega} \boldsymbol{u}(\boldsymbol{r}) \cdot \boldsymbol{v}^*(\boldsymbol{r}) \, \mathrm{d}S$$

F.9.1.1 Basis

Let H be a Hilbert space with scalar product $(u, v)_H$. A system $\{\phi_j\}$ is a Schauder[7] basis of H if and only if every $f \in H$ can be uniquely represented as

$$f = \sum_{j=1}^{\infty} c_j \phi_j$$

where the series converges in the norm of H.

The system $\{\phi_j\}$ is a Riesz[8] basis of H if and only if $\phi_j = T\psi_j$ where $\{\psi_j\}$ is an orthonormal basis of H and T is a linear, bounded and boundedly invertible operator, i.e., T^{-1} is bounded and defined on all of H.

F.9.1.2 Completeness and Closedness

Let H be a Hilbert space with scalar product $(u, v)_H$. A system $\{\phi_j\}, j \in I$, where I is some countable index set, is closed in H if and only if for every $f \in H$

$$(f, \phi_j)_H = 0, \quad \text{for all } j \in I \quad \Rightarrow \quad f = 0$$

A system $\{\phi_j\}$ is complete in H if and only if for every $f \in H$ and every $\epsilon > 0$ there are constants $c_j(\epsilon, f)$ and an integer $N(\epsilon, f)$ such that

$$f_\epsilon = \sum_{j=1}^{N(\epsilon, f)} c_j(\epsilon, f)\phi_j \quad \Rightarrow \quad \|f - f_\epsilon\|_H < \epsilon$$

The two concepts—closedness and completeness—are equivalent in a normed linear space [50, Theorem 11.1.7, page 263] or [263, pages 90–95].

F.9.1.3 Linear Independence

A countable, infinite set of functions is linearly independent if every finite subset is linearly independent. More precisely, the system $\{\phi_j\}$ is linearly independent if for every non-empty subset of $\{\phi_j\}$, say $\{\phi_1, \phi_2, \ldots, \phi_N\}$, we have

$$\sum_{j=1}^{N} c_j \phi_j = 0 \quad \Rightarrow \quad c_j = 0$$

---

[7] Juliusz Paweł Schauder (1899–1943), Polish mathematician.

[8] Frigyes Riesz (1880–1956), Hungarian mathematician.

APPENDIX G

Asymptotic Evaluation of Integrals

G.1 One-Dimensional Case

In this section, we compute the dominant contribution to the integral

$$I(\xi) = \int_a^b g(x)e^{i\xi f(x)}\, dx \tag{G.1}$$

for large values of the real parameter ξ. The limits a and b are assumed to be real and can be $\pm\infty$. The function $f(x)$ is assumed to be real valued in the interval $[a, b]$. The function $g(x)$ can be complex valued.

An important special case of this integral occurs when $f(x) = x$. The behavior of the integral in this special case, for large values on ξ, is given by the Riemann–Lebesque's lemma

$$\int_a^b g(x)e^{i\xi x}\, dx \to 0, \qquad \xi \to \infty$$

provided the function $g(x)$ is a Lebesque integrable function on the interval $[a, b]$. The integral vanishes for large values on ξ due to the exponential function that causes the integrand to oscillate rapidly, and the positive and the negative contributions cancel. The Riemann–Lebesque's lemma can be generalized to the integral in (G.1) provided $g(x)$ is a Lebesque integrable function on the interval $[a, b]$, and if $f(x)$ is continuously differentiable on $[a, b]$, and not constant on any subinterval of $[a, b]$. We then have

$$I(\xi) = \int_a^b g(x)e^{i\xi f(x)}\, dx \to 0, \qquad \xi \to \infty$$

The function $f(x)$ is critical for how fast $I(\xi)$ vanishes in the limit as $\xi \to \infty$. If $f'(x) \neq 0$ in $[a, b]$, then we can integrate by parts in (G.1). We get

$$I(\xi) = \left.\frac{g(x)}{i\xi f'(x)}e^{i\xi f(x)}\right|_{x=a}^{x=b} - \frac{1}{i\xi}\int_a^b \frac{d}{dx}\frac{g(x)}{f'(x)}e^{i\xi f(x)}\, dx$$

1

Provided the derivative $\frac{\mathrm{d}}{\mathrm{d}x}\frac{g(x)}{f'(x)}$ is a Lebesque integrable function, the Riemann–Lebesque's lemma shows that

$$I(\xi) = \frac{1}{\mathrm{i}\xi}\left(\frac{g(b)}{f'(b)}\mathrm{e}^{\mathrm{i}\xi f(b)} - \frac{g(a)}{f'(a)}\mathrm{e}^{\mathrm{i}\xi f(a)}\right)(1 + o(1)), \qquad \xi \to \infty$$

and we conclude that $I(\xi)$ vanishes as $1/\xi$ for large values on ξ.

If there are one (or several points) $x = x_s$ where $f'(x_s) = 0$, then the function $f(x)$ oscillates slower in a neighborhood of such a point. The points $x = x_s$ for which $f'(x_s) = 0$ are called stationary points to the integral. The Riemann–Lebesque's lemma shows that $I(\xi)$ vanishes also in this case, but the integral goes to zero slower than $1/\xi$ for large values on ξ.

To investigate how fast $I(\xi)$ vanishes if there is a stationary point x_s in the interval $[a, b]$, we expand the function $f(x)$ in a Taylor series around the stationary point x_s. The expansion up to second order is ($f'(x_s) = 0, f''(x_s) \neq 0$)

$$f(x) = f(x_s) + \frac{1}{2}f''(x_s)(x - x_s)^2 + \cdots$$

The pertinent integral we must solve is

$$I(\xi) = \int_{-\infty}^{\infty} g(x_s)\mathrm{e}^{\mathrm{i}\xi\left(f(x_s)+\frac{1}{2}f''(x_s)(x-x_s)^2\right)}\,\mathrm{d}x + O(1/\xi)$$

where we have neglected higher order terms in the Taylor expansion, evaluated the function $g(x)$ in the stationary point, and extended the limits of integration to infinity. The error we then make in these approximations are of order $1/\xi$, which is negligible in comparison with the term we are about to compute in $I(\xi)$

$$I(\xi) = g(x_s)\mathrm{e}^{\mathrm{i}\xi f(x_s)}\int_{-\infty}^{\infty} \mathrm{e}^{\mathrm{i}\xi\frac{1}{2}f''(x_s)t^2}\,\mathrm{d}t + O(1/\xi)$$

The remaining integral can be calculated exactly, see (G.3) on page 688. The result is

$$I(\xi) = g(x_s)\mathrm{e}^{\mathrm{i}\xi f(x_s)}\mathrm{e}^{\pm\mathrm{i}\pi/4}\left|\frac{2\pi}{\xi f''(x_s)}\right|^{1/2} + O(1/\xi)$$

where the plus (minus) sign holds for the case when $f''(x_s) > 0$ ($f''(x_s) < 0$).

G.2 Multi-Dimensional Case

The two-dimensional case uses results from the one-dimensional case in Section G.1. Here, we only state the results without proofs. An excellent treatment of the subject is found in Reference 68, see also Reference 114.

In two dimensions, the following integral is of interest

$$I(\xi) = \iint_{\mathbb{R}^2} g(u, v)\mathrm{e}^{\mathrm{i}\xi f(u,v)}\,\mathrm{d}u\,\mathrm{d}v$$

Assume $f(u, v)$ has a simple stationary point at (u_s, v_s), i.e.,

$$\frac{\partial f}{\partial u}(u_s, v_s) = \frac{\partial f}{\partial v}(u_s, v_s) = 0$$

and define a matrix Q_{ij}

$$Q_{ij} = \begin{pmatrix} \frac{\partial^2 f}{\partial u^2}(u_s, v_s) & \frac{\partial^2 f}{\partial u \partial v}(u_s, v_s) \\ \frac{\partial^2 f}{\partial v \partial u}(u_s, v_s) & \frac{\partial^2 f}{\partial v^2}(u_s, v_s) \end{pmatrix}$$

Then we have

$$I(\xi) \sim g(u_s, v_s) e^{i\xi f(u_s, v_s)} \frac{2\pi}{\xi} \frac{e^{i\pi\sigma/4}}{\sqrt{Q}}, \qquad \xi \to \infty$$

where

$$Q = \det Q_{ij}$$

is the determinant of the matrix Q_{ij} and

$$\sigma = \operatorname{sign} \lambda_1 + \operatorname{sign} \lambda_2$$

where λ_i, $i = 1, 2$, are the eigenvalues of Q_{ij}. If $Q > 0$ at the stationary point, then the surface is elliptic at that point, i.e., all points on S lie on the same side of the tangent plane.

Example G.1

An important example of the results above is (the wave number k is assumed to be a positive real number)

$$\boldsymbol{F}(\boldsymbol{r}) = \frac{1}{4\pi^2} \iint_{\mathbb{R}^2} \boldsymbol{F}(\boldsymbol{k}_t) e^{i\boldsymbol{k}_t \cdot \boldsymbol{r}_c + ik_z(\boldsymbol{k}_t)z} \, dk_x \, dk_y$$

evaluated for large distances from the origin, i.e., $\xi = kr = k|\boldsymbol{r}| \to \infty$ and $z \geq 0$. Here, the field vector $\boldsymbol{r}$ is

$$\boldsymbol{r} = \boldsymbol{r}_c + \hat{\boldsymbol{z}}z = r(\hat{\boldsymbol{x}} \sin\theta \cos\phi + \hat{\boldsymbol{y}} \sin\theta \sin\phi + \hat{\boldsymbol{z}} \cos\theta) = r\hat{\boldsymbol{r}}$$

and the Fourier variable (transverse wave vector) $\boldsymbol{k}_t$ is

$$\boldsymbol{k}_t = k_x \hat{\boldsymbol{x}} + k_y \hat{\boldsymbol{y}}$$

We assume that $\theta \in [0, \pi/2]$. The longitudinal wave number k_z is defined as

$$k_z(\boldsymbol{k}_t) = \left(k^2 - k_t^2\right)^{1/2} = \left(k^2 - k_x^2 - k_y^2\right)^{1/2}, \qquad k_t = |\boldsymbol{k}_t|$$

and the branch of the square root is chosen such that $\operatorname{Im} k_z \geq 0$.

In this example, the (phase) function $f(u, v)$ is ($(u, v) = (k_x, k_y)/k$)

$$f(u, v) = u \sin\theta \cos\phi + v \sin\theta \sin\phi + \sqrt{1 - u^2 - v^2} \cos\theta$$

The stationary point (u_s, v_s) is determined by

$$(u_s, v_s) = (\sin\theta\cos\phi, \sin\theta\sin\phi)$$

and evaluated at the stationary point f is

$$f(u_s, v_s) = 1$$

The matrix Q_{ij} evaluated at the stationary point (u_s, v_s) is

$$\begin{cases} Q_{11} = -1 - \dfrac{\sin^2\theta\cos^2\phi}{\cos^2\theta} \\ Q_{12} = Q_{21} = -\dfrac{\sin^2\theta\cos\phi\sin\phi}{\cos^2\theta} \\ Q_{22} = -1 - \dfrac{\sin^2\theta\sin^2\phi}{\cos^2\theta} \end{cases}$$

and its determinant Q is

$$Q = \frac{1}{\cos^2\theta}$$

The eigenvalues are

$$\begin{cases} \lambda_1 = -1 \\ \lambda_2 = -\dfrac{1}{\cos^2\theta} \end{cases}$$

Therefore, at large distances from the origin, the leading contribution is

$$\boldsymbol{F}(\boldsymbol{r}) = \frac{-ik\cos\theta}{2\pi r}\boldsymbol{F}(\boldsymbol{k}_{ts})e^{ikr}, \qquad kr \to \infty \tag{G.2}$$

where

$$\boldsymbol{k}_{ts} = k(\hat{\boldsymbol{x}}\sin\theta\cos\phi + \hat{\boldsymbol{y}}\sin\theta\sin\phi)$$

Notice the resemblance with the far field amplitude in the main text of the book. ∎

Example G.2
We also illustrate the general theory above by the evaluation of the following integral

$$I(\xi) = \iint_D G(x,y)e^{i\xi(1-r/z_0)}\,dx\,dy, \qquad \text{as } \xi \to \infty$$

where D is a finite domain centered at the origin in the x–y plane, and z_0 is a fixed distance. Here, $f(x,y) = 1 - \sqrt{x^2 + y^2 + z_0^2}/z_0$, with a stationary point at $x_s = y_s = 0$. To leading order, the domain of integration D can be extended to the entire x–y plane, without changing the form of the dominant contribution. Moreover, the matrix

$$Q_{ij} = \begin{pmatrix} \frac{\partial^2 f}{\partial x^2}(x_s, y_s) & \frac{\partial^2 f}{\partial x \partial y}(x_s, y_s) \\ \frac{\partial^2 f}{\partial y \partial x}(x_s, y_s) & \frac{\partial^2 f}{\partial y^2}(x_s, y_s) \end{pmatrix} = \begin{pmatrix} -1/z_0^2 & 0 \\ 0 & -1/z_0^2 \end{pmatrix}$$

with double eigenvalue $\lambda = -1/z_0^2$. We obtain

$$\iint_D G(x,y) e^{i\xi(1-r/z_0)} \, dx \, dy = \frac{2\pi z_0^2}{i\xi} G(0,0) + o(1/\xi), \qquad \text{as } \xi \to \infty$$

We use the result of this example in Chapters 4 and 10. ■

G.3 Computation of an Integral

The integral

$$I = \int_{-\infty}^{\infty} e^{iax} x^{\alpha-1} \, dx, \qquad 0 < \alpha < 1, \ a > 0$$

is evaluated in this section. The restriction in α is needed for convergence of the integral. The integral I is a generalized integral and we write it as a limit, i.e.,

$$I = \lim_{\substack{r \to 0 \\ R \to \infty}} \int_r^R e^{iax} x^{\alpha-1} \, dx = \lim_{\substack{r \to 0 \\ R \to \infty}} \int_{C_1} e^{iaz} z^{\alpha-1} \, dz$$

where the contour C_1 is defined in Figure G.1. Cauchy's theorem gives

$$I = \lim_{\substack{r \to 0 \\ R \to \infty}} \int_{C_2+C_3+C_4} e^{iaz} z^{\alpha-1} \, dz$$

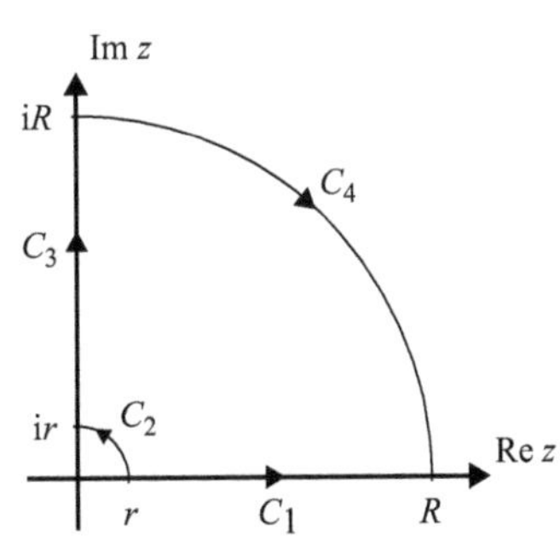

Figure G.1 Integration contours.

The contours C_i, $i = 2, 3, 4$, are also defined in Figure G.1. Jordan's[1] lemma shows that the integration along the contour C_4 does not contribute in the limit $R \to \infty$, provided $\alpha < 1$. Similarly, integration along C_2 does not contribute in the limit $r \to 0$, provided $\alpha > 0$. We get

$$I = \lim_{\substack{r\to 0\\ R\to\infty}} \int_{C_3} \mathrm{e}^{\mathrm{i}az} z^{\alpha-1}\,\mathrm{d}z = \lim_{\substack{r\to 0\\ R\to\infty}} \int_r^R \mathrm{e}^{-at}(\mathrm{i}t)^{\alpha-1}\,\mathrm{i}\,\mathrm{d}t$$

$$= \mathrm{e}^{\mathrm{i}\alpha\pi/2} \int_0^\infty \mathrm{e}^{-at} t^{\alpha-1}\,\mathrm{d}t = \mathrm{e}^{\mathrm{i}\alpha\pi/2} a^{-\alpha} \int_0^\infty \mathrm{e}^{-u} u^{\alpha-1}\,\mathrm{d}u$$

A comparison with the definition of the Gamma function $\Gamma(z)$ [1]

$$\Gamma(z) = \int_0^\infty \mathrm{e}^{-u} u^{z-1}\,\mathrm{d}u$$

gives

$$I = \mathrm{e}^{\mathrm{i}\alpha\pi/2} a^{-\alpha}\Gamma(\alpha)$$

This integral can be used to compute the integral ($a > 0$, $\alpha = 1/2$)

$$\int_{-\infty}^{\infty} \mathrm{e}^{\pm \mathrm{i}az^2}\,\mathrm{d}z = 2\int_0^\infty \mathrm{e}^{\pm \mathrm{i}az^2}\,\mathrm{d}z = \int_0^\infty \mathrm{e}^{\pm \mathrm{i}at} \frac{\mathrm{d}t}{\sqrt{t}} = \mathrm{e}^{\pm \mathrm{i}\frac{\pi}{4}} \frac{1}{\sqrt{a}}\Gamma(1/2) = \mathrm{e}^{\pm \mathrm{i}\frac{\pi}{4}} \sqrt{\frac{\pi}{a}} \tag{G.3}$$

[1] Marie Ennemond Camille Jordan (1838–1922), French mathematician.

APPENDIX H

The Nabla Operator in Curvilinear Coordinate Systems

In this appendix, some important expressions with the nabla operator in the Cartesian, cylindrical, and spherical are collected.

H.1 Cartesian Coordinate System

The Cartesian coordinates (x, y, z) is the most basic coordinate system. The gradient and the Laplace operator of a scalar field $\psi(x, y, z)$ in this coordinate system are

$$\nabla\psi = \hat{\boldsymbol{x}}\frac{\partial\psi}{\partial x} + \hat{\boldsymbol{y}}\frac{\partial\psi}{\partial y} + \hat{\boldsymbol{z}}\frac{\partial\psi}{\partial z}$$

$$\nabla^2\psi = \frac{\partial^2\psi}{\partial x^2} + \frac{\partial^2\psi}{\partial y^2} + \frac{\partial^2\psi}{\partial z^2}$$

The divergence, the curl, and the Laplace operator of a vector field $\boldsymbol{A}(x, y, z) = \hat{\boldsymbol{x}}A_x(x, y, z) + \hat{\boldsymbol{y}}A_y(x, y, z) + \hat{\boldsymbol{z}}A_z(x, y, z)$ are

$$\nabla\cdot\boldsymbol{A} = \frac{\partial A_x}{\partial x} + \frac{\partial A_y}{\partial y} + \frac{\partial A_z}{\partial z}$$

$$\nabla\times\boldsymbol{A} = \hat{\boldsymbol{x}}\left(\frac{\partial A_z}{\partial y} - \frac{\partial A_y}{\partial z}\right) + \hat{\boldsymbol{y}}\left(\frac{\partial A_x}{\partial z} - \frac{\partial A_z}{\partial x}\right) + \hat{\boldsymbol{z}}\left(\frac{\partial A_y}{\partial x} - \frac{\partial A_x}{\partial y}\right)$$

$$\nabla^2\boldsymbol{A} = \hat{\boldsymbol{x}}\nabla^2 A_x + \hat{\boldsymbol{y}}\nabla^2 A_y + \hat{\boldsymbol{z}}\nabla^2 A_z$$

H.2 Circular Cylindrical (Polar) Coordinate System

We now treat two curvilinear coordinate systems, and we start with the circular cylindrical coordinate system (r_c, ϕ, z) defined by

$$\begin{cases} r_c = \sqrt{x^2 + y^2} \\ \phi = \begin{cases} \arccos\frac{x}{\sqrt{x^2+y^2}} & y \geq 0 \\ 2\pi - \arccos\frac{x}{\sqrt{x^2+y^2}} & y < 0 \end{cases} \\ z = z \end{cases}$$

The gradient and the Laplace operator of a scalar field $\psi(r_c, \phi, z)$ in this coordinate system are

$$\nabla\psi = \hat{\boldsymbol{r}}_c \frac{\partial\psi}{\partial r_c} + \hat{\boldsymbol{\phi}} \frac{1}{r_c}\frac{\partial\psi}{\partial\phi} + \hat{\boldsymbol{z}}\frac{\partial\psi}{\partial z}$$

$$\nabla^2\psi = \frac{1}{r_c}\frac{\partial}{\partial r_c}\left(r_c\frac{\partial\psi}{\partial r_c}\right) + \frac{1}{r_c^2}\frac{\partial^2\psi}{\partial\phi^2} + \frac{\partial^2\psi}{\partial z^2}$$

The divergence, the curl, and the Laplace operator of a vector field decomposed in cylindrical components, i.e., $\boldsymbol{A}(r_c, \phi, z) = \hat{\boldsymbol{r}}_c A_{r_c}(r_c, \phi, z) + \hat{\boldsymbol{\phi}} A_\phi(r_c, \phi, z) + \hat{\boldsymbol{z}} A_z(r_c, \phi, z)$ are

$$\nabla\cdot\boldsymbol{A} = \frac{1}{r_c}\frac{\partial}{\partial r_c}\left(r_c A_{r_c}\right) + \frac{1}{r_c}\frac{\partial A_\phi}{\partial\phi} + \frac{\partial A_z}{\partial z}$$

$$\nabla\times\boldsymbol{A} = \hat{\boldsymbol{r}}_c\left(\frac{1}{r_c}\frac{\partial A_z}{\partial\phi} - \frac{\partial A_\phi}{\partial z}\right) + \hat{\boldsymbol{\phi}}\left(\frac{\partial A_{r_c}}{\partial z} - \frac{\partial A_z}{\partial r_c}\right) + \hat{\boldsymbol{z}}\frac{1}{r_c}\left(\frac{\partial}{\partial r_c}\left(r_c A_\phi\right) - \frac{\partial A_{r_c}}{\partial\phi}\right)$$

$$\nabla^2\boldsymbol{A} = \hat{\boldsymbol{r}}_c\left(\nabla^2 A_{r_c} - \frac{A_{r_c}}{r_c^2} - \frac{2}{r_c^2}\frac{\partial A_\phi}{\partial\phi}\right) + \hat{\boldsymbol{\phi}}\left(\nabla^2 A_\phi - \frac{A_\phi}{r_c^2} + \frac{2}{r_c^2}\frac{\partial A_{r_c}}{\partial\phi}\right) + \hat{\boldsymbol{z}}\nabla^2 A_z$$

H.3 Spherical Coordinate System

The spherical coordinate system (r, θ, ϕ) (polar angle θ and the azimuth angle ϕ) is defined by

$$\begin{cases} r = \sqrt{x^2+y^2+z^2} \\ \theta = \arccos\frac{z}{\sqrt{x^2+y^2+z^2}} \\ \phi = \begin{cases} \arccos\frac{x}{\sqrt{x^2+y^2}} & y \geq 0 \\ 2\pi - \arccos\frac{x}{\sqrt{x^2+y^2}} & y < 0 \end{cases} \end{cases}$$

The gradient and the Laplace operator of a scalar field $\psi(r, \theta, \phi)$ in this coordinate system are

$$\nabla\psi = \hat{\boldsymbol{r}}\frac{\partial\psi}{\partial r} + \hat{\boldsymbol{\theta}}\frac{1}{r}\frac{\partial\psi}{\partial\theta} + \hat{\boldsymbol{\phi}}\frac{1}{r\sin\theta}\frac{\partial\psi}{\partial\phi}$$

$$\begin{aligned}\nabla^2\psi &= \frac{1}{r^2}\frac{\partial}{\partial r}\left(r^2\frac{\partial\psi}{\partial r}\right) + \frac{1}{r^2\sin\theta}\frac{\partial}{\partial\theta}\left(\sin\theta\frac{\partial\psi}{\partial\theta}\right) + \frac{1}{r^2\sin^2\theta}\frac{\partial^2\psi}{\partial\phi^2} \\ &= \frac{1}{r}\frac{\partial^2}{\partial r^2}(r\psi) + \frac{1}{r^2\sin\theta}\frac{\partial}{\partial\theta}\left(\sin\theta\frac{\partial\psi}{\partial\theta}\right) + \frac{1}{r^2\sin^2\theta}\frac{\partial^2\psi}{\partial\phi^2}\end{aligned}$$

and the divergence, the curl, and the Laplace operator of a vector field $\boldsymbol{A}(r,\theta,\phi) = \hat{\boldsymbol{r}}A_r(r,\theta,\phi) + \hat{\boldsymbol{\theta}}A_\theta(r,\theta,\phi) + \hat{\boldsymbol{\phi}}A_\phi(r,\theta,\phi)$ are

$$\nabla \cdot \boldsymbol{A} = \frac{1}{r^2}\frac{\partial}{\partial r}\left(r^2 A_r\right) + \frac{1}{r\sin\theta}\frac{\partial}{\partial \theta}\left(\sin\theta A_\theta\right) + \frac{1}{r\sin\theta}\frac{\partial A_\phi}{\partial \phi}$$

$$\begin{aligned}\nabla \times \boldsymbol{A} = {} & \hat{\boldsymbol{r}}\frac{1}{r\sin\theta}\left(\frac{\partial}{\partial\theta}\left(\sin\theta A_\phi\right) - \frac{\partial A_\theta}{\partial\phi}\right) + \hat{\boldsymbol{\theta}}\frac{1}{r}\left(\frac{1}{\sin\theta}\frac{\partial A_r}{\partial\phi} - \frac{\partial}{\partial r}\left(rA_\phi\right)\right) \\ & + \hat{\boldsymbol{\phi}}\frac{1}{r}\left(\frac{\partial}{\partial r}\left(rA_\theta\right) - \frac{\partial A_r}{\partial\theta}\right)\end{aligned}$$

$$\begin{aligned}\nabla^2 \boldsymbol{A} = {} & \hat{\boldsymbol{r}}\left(\nabla^2 A_r - \frac{2A_r}{r^2} - \frac{2}{r^2}\frac{\partial A_\theta}{\partial\theta} - \frac{2\cot\theta}{r^2}A_\theta - \frac{2}{r^2\sin\theta}\frac{\partial A_\phi}{\partial\phi}\right) \\ & + \hat{\boldsymbol{\theta}}\left(\nabla^2 A_\theta - \frac{A_\theta}{r^2\sin^2\theta} + \frac{2}{r^2}\frac{\partial A_r}{\partial\theta} - \frac{2\cos\theta}{r^2\sin^2\theta}\frac{\partial A_\phi}{\partial\phi}\right) \\ & + \hat{\boldsymbol{\phi}}\left(\nabla^2 A_\phi - \frac{A_\phi}{r^2\sin^2\theta} + \frac{2}{r^2\sin\theta}\frac{\partial A_r}{\partial\phi} + \frac{2\cos\theta}{r^2\sin^2\theta}\frac{\partial A_\theta}{\partial\phi}\right)\end{aligned}$$

APPENDIX I

Notation

Appropriate notation leads to a more easily understood, systematic, and structured text, and in the same token, good notation implies a tendency of making less errors and slips in the analysis. Most of the notation is explained at the place in the text where it is introduced, but some more general notation that is often used is collected in this appendix.

I.1 Sets

- The set of real numbers is denoted $\mathbb{R}$, and the set of positive real numbers $\mathbb{R}_+$.
- The set of complex numbers is denoted $\mathbb{C}$, and the set of complex numbers in the upper half-plane $\mathbb{C}_+ = \{z \in \mathbb{C} : \operatorname{Im} z > 0\}$.

I.2 Volumes and Surfaces

- The general symbol for a volume in $\mathbb{R}^3$ is V. It is in general an open set. The closed set of V is denoted $\overline{V}$. We use the notion volume, region, and domain alternately in the text.
- The bounding surface of V is generally denoted S, and the outward pointing unit normal is denoted $\hat{\boldsymbol{\nu}}$. The surface S is assumed smooth, unless stated otherwise. More precisely, the different parts of the surface have a continuously differentiable parametrization

$$\boldsymbol{r} = \boldsymbol{r}(u, v), \quad (u, v) \in \mathcal{D} \subseteq \mathbb{R}^2$$

  and moreover,

$$\frac{\partial \boldsymbol{r}}{\partial u} \times \frac{\partial \boldsymbol{r}}{\partial v} \neq \boldsymbol{0}$$

  in the domain $\mathcal{D}$.
- The difference between sets is denoted by ($\setminus$), e.g., $A \setminus B$ or $\mathbb{R}^3 \setminus \overline{V}$, i.e., the open set of all points outside V.

I.3 Vectors and Transformations

In addition to the notation presented below, the reader is also encouraged to consult Appendix A on vectors and transformations.

- Vector-valued quantities (mostly in $\mathbb{R}^3$) are denoted in italic boldface, e.g., $\boldsymbol{a}$ and $\boldsymbol{b}$, and vectors of unit length have a “hat” or caret (ˆ) over a symbol, e.g., $\hat{\boldsymbol{x}}$ and $\hat{\boldsymbol{r}}_{\mathrm{c}}$.

- The position vector in $\mathbb{R}^3$ is denoted $\boldsymbol{r} = x\hat{\boldsymbol{x}} + y\hat{\boldsymbol{y}} + z\hat{\boldsymbol{z}}$, and the position vector in the x–y-plane is denoted $\boldsymbol{r}_c = x\hat{\boldsymbol{x}} + y\hat{\boldsymbol{y}}$.
- The (Euclidean) scalar product between two vectors, $\boldsymbol{a}$ and $\boldsymbol{b}$, is denoted in the usual standard way by a dot $(\cdot)$, i.e., $\boldsymbol{a} \cdot \boldsymbol{b}$. If the vectors are complex valued (complex components and real-valued unit vectors), the appropriate scalar product is $\boldsymbol{a}^* \cdot \boldsymbol{b} = \boldsymbol{b} \cdot \boldsymbol{a}^*$, where the star $*$ denotes the complex conjugate of the vector. For complex-valued functions ($L_2(\Omega)$ functions defined on a domain Ω), the scalar product is defined in the standard way

$$(f, g) = \iiint\limits_{\Omega} f(\boldsymbol{r}) g^*(\boldsymbol{r})\,\mathrm{d}v$$

- We make a distinction between a vector $\boldsymbol{a}$ and its representation in components in a specific coordinate system, and denote the components as a column vector or with brackets around the vector, i.e.,

$$[\boldsymbol{a}] = \begin{pmatrix} a_x \\ a_y \\ a_z \end{pmatrix}$$

where the vector is

$$\boldsymbol{a} = \hat{\boldsymbol{x}} a_x + \hat{\boldsymbol{y}} a_y + \hat{\boldsymbol{z}} a_z$$

- Linear vector-valued transformations (dyadics) are denoted in roman boldface, e.g., $\mathbf{A}$. A linear transformation $\mathbf{A}$ acting on a vector field $\boldsymbol{a}$ gives a new vector field $\boldsymbol{b}$ and we use the notation

$$\boldsymbol{b} = \mathbf{A} \cdot \boldsymbol{a}$$

In a specific coordinate system, the linear transformation $\mathbf{A}$ is represented by a 3×3 matrix $[\mathbf{A}]$, where we again use brackets around $\mathbf{A}$ to emphasize that we refer to its components. The components of the vector $\boldsymbol{b}$ are then

$$[\boldsymbol{b}] = [\mathbf{A}] \cdot [\boldsymbol{a}]$$

or

$$\begin{pmatrix} b_x \\ b_y \\ b_z \end{pmatrix} = \begin{pmatrix} A_{xx} & A_{xy} & A_{xz} \\ A_{yx} & A_{yy} & A_{yz} \\ A_{zx} & A_{zy} & A_{zz} \end{pmatrix} \begin{pmatrix} a_x \\ a_y \\ a_z \end{pmatrix}$$

- The identity and the null dyadics in n dimensions are denoted $\mathbf{I}_n$ and $\mathbf{0}_n$, respectively, and the corresponding matrix representations are denoted $[\mathbf{I}_n]$ and $[\mathbf{0}_n]$, respectively. In three dimensions, we have

$$[\mathbf{I}_3] = \begin{pmatrix} 1 & 0 & 0 \\ 0 & 1 & 0 \\ 0 & 0 & 1 \end{pmatrix} \qquad [\mathbf{0}_3] = \begin{pmatrix} 0 & 0 & 0 \\ 0 & 0 & 0 \\ 0 & 0 & 0 \end{pmatrix}$$

and in two dimensions the corresponding notation is

$$[\mathbf{I}_2] = \begin{pmatrix} 1 & 0 \\ 0 & 1 \end{pmatrix} \qquad [\mathbf{0}_2] = \begin{pmatrix} 0 & 0 \\ 0 & 0 \end{pmatrix}$$

- The dyadic $\mathbf{J} = \hat{z} \times \mathbf{I}_2$ (matrix $[\mathbf{J}]$) performs a projection of a vector onto the x–y-plane followed by a rotation of the vector of $\pi/2$ in the x–y-plane,

$$[\mathbf{J}] = \begin{pmatrix} 0 & -1 \\ 1 & 0 \end{pmatrix}$$

- The transpose of a matrix is denoted by a superscript "t" (^{t}) and the Hermitian adjoint of a matrix with the superscript dagger (†), i.e.,

$$A^{\mathrm{t}}_{ij} = A_{ji}$$
$$A^{\dagger}_{ij} = A^{*}_{ji}$$

- Six-dimensional vectors and dyadics are written in sans serif boldface to distinguish them from the ordinary three-dimensional vectors, i.e., e for a six-dimensional vector, and D for a six-dimensional dyadics.

I.4 Symbols and Functions

- In Appendix F.9.1 on page 681, we give the definitions on some common function spaces.
- We use the symbols o and O (small and large Ordo, respectively) defined by

$$f(x) = o\left(g(x)\right), \quad x \to a \qquad \Leftrightarrow \qquad \lim_{x \to a} \frac{f(x)}{g(x)} = 0$$

and

$$f(x) = O\left(g(x)\right), \quad x \to a$$
$$\Leftrightarrow$$
$$\frac{f(x)}{g(x)} \quad \text{bounded in a neighborhood of } a$$

- The right limit value of a function $f(x)$ at a point a is denoted $f(a^+)$, i.e.,

$$f(a^+) = \lim_{\substack{x \to a \\ x > a}} f(x)$$

Similarly, the left limit is denoted $f(a^-)$ with definition

$$f(a^-) = \lim_{\substack{x \to a \\ x < a}} f(x)$$

- The symbol $[x]$ denotes the integer part of the real number x.
- The symbol ■ denotes the end of an example.

- The Heaviside step function, $H(t)$, is defined in the usual way as

$$H(t) = \begin{cases} 0, & t < 0 \\ 1, & t \geq 0 \end{cases}$$

- The Kronecker[1] delta (function) symbol δ_{ij}, is defined as

$$\delta_{ij} = \begin{cases} 1, & i = j \\ 0, & i \neq j \end{cases}$$

- The delta function (distribution) $\delta(x - x_0)$ is defined as

$$\int_{-\infty}^{\infty} \delta(x - x_0) f(x) \, \mathrm{d}x = f(x_0)$$

I.5 Real and Imaginary Parts of Numbers and Dyadics

- The real and imaginary parts of a complex number $z = x + \mathrm{i}y$ are denoted $\mathrm{Re}\, z$ and $\mathrm{Im}\, z$, respectively, i.e.,

$$\mathrm{Re}\, z = x = \frac{1}{2}(z + z^*)$$
$$\mathrm{Im}\, z = y = \frac{1}{2\mathrm{i}}(z - z^*)$$

  A star (*) is used to denote the complex conjugate of a complex number, i.e., $z^* = x - \mathrm{i}y$.
- The real and imaginary parts of a dyadic can also be defined in terms of their Hermitian adjoint. We adopt

$$\mathrm{Re}\, \mathbf{A} = \frac{1}{2}(\mathbf{A} + \mathbf{A}^\dagger)$$
$$\mathrm{Im}\, \mathbf{A} = \frac{1}{2\mathrm{i}}(\mathbf{A} - \mathbf{A}^\dagger)$$

  where the dagger (†) denotes the Hermitian adjoint. We see that the real and imaginary parts satisfy

$$(\mathrm{Re}\, \mathbf{A})^\dagger = \mathrm{Re}\, \mathbf{A}$$
$$(\mathrm{Im}\, \mathbf{A})^\dagger = \mathrm{Im}\, \mathbf{A}$$

  i.e., $\mathrm{Re}\, \mathbf{A}$ and $\mathrm{Im}\, \mathbf{A}$ are both Hermitian dyadics and that

$$\mathbf{A} = \mathrm{Re}\, \mathbf{A} + \mathrm{i}\, \mathrm{Im}\, \mathbf{A}$$

---

[1] Leopold Kronecker (1823–1891), German mathematician.

I.6 Curvilinear Coordinates

We also refer to Appendix H.

- The cylindrical coordinate system (r_c, ϕ, z) is defined by

$$\begin{cases} r_c = \sqrt{x^2 + y^2} \\ \phi = \begin{cases} \arccos \frac{x}{\sqrt{x^2+y^2}} & y \geq 0 \\ 2\pi - \arccos \frac{x}{\sqrt{x^2+y^2}} & y < 0 \end{cases} \\ z = z \end{cases}$$

The domains of the coordinates (r_c, ϕ, z) are $r_c \in [0, \infty)$, $\phi \in [0, 2\pi)$, and $z \in (-\infty, \infty)$.

- The spherical coordinate system (r, θ, ϕ) is defined as

$$\begin{cases} r = \sqrt{x^2 + y^2 + z^2} \\ \theta = \arccos \frac{z}{\sqrt{x^2 + y^2 + z^2}} \\ \phi = \begin{cases} \arccos \frac{x}{\sqrt{x^2+y^2}} & y \geq 0 \\ 2\pi - \arccos \frac{x}{\sqrt{x^2+y^2}} & y < 0 \end{cases} \end{cases}$$

The domains of the coordinates (r, θ, ϕ) are $r \in [0, \infty)$, $\theta \in [0, \pi]$, and $\phi \in [0, 2\pi)$.

APPENDIX J

Units and Constants

The explicit form of the equations in electromagnetism varies depending on the system of units that we use. The SI system is the one that is used in most literature nowadays, and this textbook is no exception. The relevant constants in the SI system that are used in the text are collected in this appendix.

The speed of light in vacuum c_0 has the value (exact value)

$$c_0 = 299\,792\,458 \text{ m/s}$$

The constants μ_0 and ϵ_0 denote the permeability and the permittivity of vacuum, respectively. Their exact values are

$$\mu_0 = 4\pi \times 10^{-7} \text{ N/A}^2$$
$$\epsilon_0 = \frac{1}{c_0^2 \mu_0} \text{ F/m}$$

Approximative values of these constants are

$$\mu_0 \approx 12.566370614\ldots \times 10^{-7} \text{ N/A}^2$$
$$\epsilon_0 \approx 8.854187817\ldots \times 10^{-12} \text{ F/m}$$

The wave impedance of vacuum is denoted

$$\eta_0 = \sqrt{\frac{\mu_0}{\epsilon_0}} = c_0\mu_0 = 299\,792\,458 \cdot 4\pi \times 10^{-7}\ \Omega \approx 376.730313\ \Omega$$

The charge of the electron, $-e$, and its mass, m, have the values

$$e \approx 1.6021766208(98) \times 10^{-19} \text{ C}$$
$$m \approx 9.10938291(40) \times 10^{-31} \text{ kg}$$
$$e/m \approx 1.75881963 \times 10^{11} \text{ C/kg}$$

Frequencies can be expressed in units of eV (electron volt), 1 eV $= 2.418 \times 10^{14}$ Hz or vacuum wavelength 1.23984193 μm.

The Planck constant[1] is $h = 6.626070040(81) \times 10^{-34}$ J s.

[1]Max Karl Ernst Ludwig Planck (1858–1947), German theoretical physicist.

Bibliography

[1] Abramowitz, M. and Stegun, I.A. (editors). *Handbook of Mathematical Functions*. Applied Mathematics Series No. 55. National Bureau of Standards, Washington, DC, 1970.

[2] Adam, J.A. The mathematical physics of rainbows and glories. *Phys. Rep.*, **356**, 229–365, 2002.

[3] Aden, A.L. and Kerker, M. Scattering of electromagnetic waves from two concentric spheres. *J. Appl. Phys.*, **22**, 1242–1246, 1951.

[4] Arfken, G. *Mathematical Methods for Physicists*, 3rd ed. Academic Press, Orlando, FL, 1985.

[5] Ashcroft, N.W. and Mermin, N.D. *Solid State Physics*. Holt, Rinehart and Winston, New York, NY, 1976.

[6] Asvestas, J. and Kleinman, R. Electromagnetic scattering by indented screens. *IEEE Trans. Antennas Propag.*, **42**, 22–30, 1994.

[7] Athanasiadis, C., Martin, P., and Stratis, I.G. Electromagnetic scattering by a homogeneous chiral obstacle: boundary integral equations and low-chirality approximations. *SIAM J. Appl. Math.*, **59**, 1745–1762, 1999.

[8] Aydin, K. and Hizal, A. On the completeness of the spherical vector wave functions. *J. Math. Anal. Appl.*, **117**, 428–440, 1986.

[9] Barber, P.W. and Hill, S.C. *Light Scattering by Particles: Computational Methods*. World Scientific Publisher, Singapore, 1990.

[10] Berg, M., Sorensen, C., and Chakrabarti, A. A new explanation of the extinction paradox. *J. Quant. Spectrosc. Radiat. Transfer*, **112**, 1170–1181, 2011.

[11] Berg, M.J., Subedi, N.R., Anderson, P.A., and Fowler, N.B. Using holography to measure extinction. *Opt. Lett.*, **39**, 3993–3996, 2014.

[12] Bernland, A. *Integral Identities for Passive Systems and Spherical Waves in Scattering and Antenna Problems*. Ph.D. thesis, Lund University, Lund, Sweden, 2012.

[13] Bernland, A., Luger, A., and Gustafsson, M. Sum rules and constraints on passive systems. *J. Phys. A: Math. Theor.*, **44**, 145205, 2011.

[14] Bhattacharyya, A.K. Control of radar cross-section and cross-polarisation characteristics of an isotropic chiral sphere. *Electron. Lett.*, **26**, 1066–1067, 1990.

[15] Björkberg, J. and Kristensson, G. Electromagnetic scattering by a perfectly conducting elliptic disk. *Can. J. Phys.*, **65**, 723–734, 1987.

[16] Bjorken, J.D. and Drell, S.D. *Relativistic Quantum Fields*. McGraw-Hill, New York, NY, 1965.

[17] Bohren, C.F. Light scattering by an optically active sphere. *Chem. Phys. Lett.*, **29**, 458–462, 1974.

[18] Bohren, C.F. and Huffman, D.R. *Absorption and Scattering of Light by Small Particles*. John Wiley & Sons, New York, NY, 1983.

[19] Bokut', B.V. and Fedorov, F.I. On the theory of optical activity in crystals. III. General equation of normals. *Opt. Spectrosc.*, **6**, 342, 1959.

[20] Borghese, F., Denti, P., Saija, R., Toscano, G., and Sindoni, O. Multiple electromagnetic scattering from a cluster of spheres. I. Theory. *Aerosol Sci. Technol.*, **3**, 227–235, 1984.

[21] Born, M. and Wolf, E. *Principles of Optics*, 7th ed. Cambridge University Press, Cambridge, UK, 1999.

[22] Bossavit, A. *Électromagnetisme, en vue de la Modélisation*, vol. 14. Springer-Verlag, Paris, 2004.

[23] Boström, A., Kristensson, G., and Ström, S. Transformation properties of plane, spherical and cylindrical scalar and vector wave functions. In V.V. Varadan, A. Lakhtakia, and V.K. Varadan (editors), *Field Representations and Introduction to Scattering*, volume 1 of *Acoustic, Electromagnetic and Elastic Wave Scattering*, Chapter 4, pages 165–210. Elsevier Science Publishers, Amsterdam, 1991.

[24] Brillouin, L. The scattering cross section of spheres for electromagnetic waves. *J. Appl. Phys.*, **20**, 1110–1125, 1949.

[25] Bringi, V.N., Varadan, V.V., and Varadan, V.K. Coherent wave attenuation by a random distribution of particles. *Radio Sci.*, **17**, 946–952, 1982.

[26] Bruno, O.P. and Reitich, F. High-order methods for high-frequency scattering applications. In *Modeling and Computations in Electromagnetics*, pages 129–163. Springer, Berlin, 2008.

[27] Byron, F.W. and Fuller, R.W. *Mathematics of Classical and Quantum Physics*, vol. 2. Addison-Wesley, Reading, MA, 1970.

[28] Caloz, C. and Itoh, T. *Electromagnetic Metamaterials: Transmission Line Theory and Microwave Applications*. John Wiley & Sons, Hoboken, NJ, 2005.

[29] Čerenkov, P.A. Visible radiation produced by electrons moving in a medium with velocities exceeding that of light. *Phys. Rev.*, **52**, 378–379, 1937.

[30] Cessenat, M. *Mathematical Methods in Electromagnetism*. Series on *Advances in Mathematics for Applied Sciences*, vol. 41. World Scientific Publisher, Singapore, 1996.

[31] Chapman, S.J., Lawry, J.M.H., Ockendon, J.R., and Tew, R.H. On the theory of complex rays. *SIAM Rev.*, **41**, 417–509, 1999.

[32] Cheng, D.K. *Field and Wave Electromagnetics*, vol. 2. Addison-Wesley, Reading, MA, 1989.

[33] Chew, W.C. *Waves and Fields in Inhomogeneous Media*. IEEE Press, Piscataway, NJ, 1995.

[34] Chew, W.C., Tong, M.S., and Hu, B. *Integral Equation Methods for Electromagnetic and Elastic Waves*, vol. 12. Morgan & Claypool, San Rafael, CA, 2008.

[35] Cock, H.F. A comparison of the dielectric behaviour of pure water and human blood at microwave frequencies. *Br. J. Appl. Phys.*, **3**, 249–255, 1952.

[36] Coddington, E.A. and Carlson, R. *Linear Ordinary Differential Equations*. SIAM, Philadelphia, PA, 1997.

[37] Collin, R. Dyadic Green's function expansions in spherical coordinates. *Electromagnetics*, **6**, 183–207, 1986.

[38] Colton, D. and Kress, R. *Integral Equation Methods in Scattering Theory*. John Wiley & Sons, New York, NY, 1983.

[39] Colton, D. and Kress, R. The unique solvability of the null field equations of acoustics. *Quart. J. Mech. Appl. Math.*, **36**, 87–95, 1983.

[40] Colton, D. and Kress, R. *Inverse Acoustic and Electromagnetic Scattering Theory*, 3rd ed. Springer-Verlag, Berlin, 2013.

[41] Condon, E.U. Theories of optical rotatory power. *Rev. Mod. Phys.*, **9**, 432–457, 1937.

[42] Condon, E.U., Altar, W., and Eyring, H. One-electron rotatory power. *J. Chem. Phys.*, **5**, 753–775, 1937.

[43] Costabel, M. and Dauge, M. Singularities of electromagnetic fields in polyhedral domains. *Arch. Rational Mech. Anal.*, **151**, 221–276, 2000.

[44] Cullen, J.A. Surface currents induced by short-wavelength radiation. *Phys. Rev.*, **109**, 1863–1867, 1958.

[45] Dallas, A.G. Basis properties of traces and normal derivatives of spherical-separable solutions of the Helmholtz equation. Technical Report No. 2000-6:1-42, Department of Mathematical Sciences, University of Delaware, 2000.

[46] Dallas, A.G. On the convergence and numerical stability of the second Waterman scheme for approximation of the acoustic field scattered by a hard object. Technical Report No. 2000-7:1-35, Department of Mathematical Sciences, University of Delaware, 2000.

[47] Dassios, G. and Kleinman, R. *Low Frequency Scattering*. Oxford University Press, Oxford, 2000.

[48] Dautray, R. and Lions, J.L. *Mathematical Analysis and Numerical Methods for Science and Technology. Volume 1: Physical Origins and Classical Methods*. Springer-Verlag, Berlin, 2012.

[49] Dautray, R. and Lions, J.L. *Mathematical Analysis and Numerical Methods for Science and Technology. Volume 3: Spectral Theory and Applications*. Springer-Verlag, Berlin, 2012.

[50] Davis, P.J. *Interpolation and Approximation*. Dover Publications, New York, NY, 1975.

[51] de Hoop, A.T. A time-domain energy theorem for scattering of plane electromagnetic waves. *Radio Sci.*, **19**, 1179–1184, 1984.

[52] de Hoop, A.T. A time-domain energy theorem for scattering of plane acoustic waves in fluids. *J. Acoust. Soc. Am.*, **77**, 11–14, 1985.

[53] de Hoop, A.T. *Handbook of Radiation and Scattering of Waves*. Academic Press, San Diego, CA, 1995.

[54] de Rooij, W.A. and van der Stap, C.C.A.H. Expansion of Mie scattering matrices in generalized spherical functions. *Astron. Astrophys.*, **131**, 237–248, 1984.

[55] Diebel, J. Representing attitude: Euler angles, unit quaternions, and rotation vectors. *Matrix*, **58**, 1–35, 2006.

[56] Doicu, A., Eremin, Y., and Wriedt, T. *Acoustic and electromagnetic scattering analysis using discrete sources*. Academic Press, New York, 2000.

[57] Doicu, A., Wriedt, T., and Eremin, Y.A. *Light Scattering by Systems of Particles*. Springer-Verlag, Berlin, 2006.

[58] Du, H. Mie-scattering calculation. *Appl. Opt.*, **43**, 1951–1956, 2004.

[59] Edmonds, A.R. *Angular Momentum in Quantum Mechanics*, 3rd ed. Princeton University Press, Princeton, NJ, 1974.

[60] Egeland, A. and Larsen, T.R. Measurements of the Earth-ionosphere cavity resonance frequencies at high latitude. *Phys. Norv.*, **2**, 85–97, 1967.

[61] Elliott, R.S. *Electromagnetics: History, Theory, and Applications*. IEEE Press, New York, NY, 1993.

[62] Engheta, N. and Zablocky, P.G. A step towards determining transient response of chiral materials: Kramers–Kronig relations for chiral parameters. *Electron. Lett.*, **26**, 2132–2133, 1990.

[63] Englman, R. and Ruppin, R. Optical lattice vibrations in finite ionic crystals: III. *J. Phys. C: Solid State Phys.*, **1**, 1515–1531, 1968.

[64] Espenscheid, W.F., Willis, E., Matijevic, E., and Kerker, M. Aerosol studies by light scattering. IV. Preparation and particle size distribution of aerosols consisting of concentric spheres. *J. Colloid Sci.*, **20**, 501–521, 1965.

[65] Evans, L.C. *Partial Differential Equations*. American Mathematical Society, Providence, RI, 1998.

[66] Fedorov, F.I. On the theory of optical activity in crystals. I. The law of conservation of energy and the optical activity tensors. *Opt. Spectrosc.*, **6**, 49–53, 1959.

[67] Fedorov, F.I. On the theory of optical activity in crystals. II. Crystals of cubic symmetry and plane classes of central symmetry. *Opt. Spectrosc.*, **6**, 237–240, 1959.

[68] Felsen, L.B. and Marcuvitz, N. *Radiation and Scattering of Waves*. IEEE Press, Piscataway, NJ, 1994.

[69] Fenn, R.W. and Oser, H. Scattering properties of concentric soot—water spheres for visible and infrared light. *Appl. Opt.*, **4**, 1504–1509, 1965.

[70] Fikioris, J. and Waterman, P. Multiple scattering of waves. III. The electromagnetic case. *J. Quant. Spectrosc. Radiat. Transfer*, **123**, 8–16, 2013.

[71] Foldy, L.L. The multiple scattering of waves. I. General theory of isotropic scattering by randomly distributed scatterers. *Phys. Rev.*, **67**, 107–119, 1945.

[72] Ford, G.W. and Werner, S.A. Scattering and absorption of electromagnetic waves by a gyrotropic sphere. *Phys. Rev. B*, **18**, 6752–6769, 1978.

[73] Frank, I.M. and Tamm, I.Y. Coherent visible radiation from fast electrons passing through matter. *Dokl. Akad. Nauk SSSR*, **14**, 109–114, 1937.

[74] Fuller, K.A. Scattering of light by coated spheres. *Opt. Lett.*, **18**, 257–259, 1993.

[75] Gai, H., Wang, J., and Tian, Q. Modified Debye model parameters of metals applicable for broadband calculations. *Appl. Opt.*, **46**, 2229–2233, 2007.

[76] Ganesh, M. and Hawkins, S.C. A far-field based T-matrix method for two dimensional obstacle scattering. *ANZIAM J.*, **51**, C215–C230, 2010.

[77] Ganesh, M. and Hawkins, S.C. Three-dimensional electromagnetic scattering T-matrix computations. *J. Comp. Appl. Math.*, **234**, 1702–1709, 2010.

[78] Ganesh, M., Hawkins, S.C., and Hiptmair, R. Convergence analysis with parameter estimates for a reduced basis acoustic scattering T-matrix method. *IMA J. Numer. Anal.*, **32**, 1348–1374, 2012.

[79] Gesztesy, F. and Tsekanovskii, E. On matrix-valued Herglotz functions. *Math. Nachr.*, **218**, 61–138, 2000.

[80] Golub, G.H. and van Loan, C.F. *Matrix Computations*. The Johns Hopkins University Press, Baltimore, MD, 1983.

[81] Grant, E.H., Buchanan, T.J., and Cook, H.F. Dielectric behavior of water at microwave frequencies. *J. Chem. Phys.*, **26**, 156–161, 1957.

[82] Greene, R.E. and Krantz, S.G. *Function Theory of One Complex Variable*, 3rd ed. American Mathematical Society, Providence, RI, 2006.

[83] Greenleaf, A., Kurylev, Y., Lassas, M., and Uhlmann, G. Cloaking devices, electromagnetic wormholes, and transformation optics. *SIAM Rev.*, **51**, 3–33, 2009.

[84] Greenleaf, A., Kurylev, Y., Lassas, M., and Uhlmann, G. Invisibility and inverse problems. *Bull. Am. Math. Soc.*, **46**, 55–97, 2009.

[85] Gustafsson, M. Sum rule for the transmission cross section of apertures in thin opaque screens. *Opt. Lett.*, **34**, 2003–2005, 2009.

[86] Gustafsson, M. Time-domain approach to the forward scattering sum rule. *Proc. R. Soc. Lond. A*, **466**, 3579–3592, 2010.

[87] Gustafsson, M. and Sjöberg, D. Sum rules and physical bounds on passive metamaterials. *New J. Phys.*, **12**, 043046, 2010.

[88] Gustafsson, M. and Sjöberg, D. Physical bounds and sum rules for high-impedance surfaces. *IEEE Trans. Antennas Propag.*, **59**, 2196–2204, 2011.

[89] Gustafsson, M., Sohl, C., Karlsson, A., and Kristensson, G. A time-domain approach to the extinction paradox for scattering of electromagnetic waves. In *Proceedings of the XXIXth URSI General Assembly, Chicago, USA*, International Union of Radio Science, 2008. Chicago, IL, USA, August 7–16.

[90] Gustafsson, M., Vakili, I., Bayer Keskin, S.E., Sjöberg, D., and Larsson, C. Optical theorem and forward scattering sum rule for periodic structures. *IEEE Trans. Antennas Propag.*, **60**, 3818–3826, 2012.

[91] Hale, G.M. and Querry, M.R. Optical constants of water in the 200-nm to 200-μm wavelength region. *Appl. Opt.*, **12**, 555–563, 1973.

[92] Hansen, J.E. (editor). *Spherical Near-Field Antenna Measurements*. Number 26 in IEE electromagnetic waves series. Peter Peregrinus Ltd., Stevenage, UK, 1988.

[93] Hanson, G.W. and Yakovlev, A.B. *Operator Theory for Electromagnetics: An Introduction*. Springer-Verlag, New York, NY, 2002.

[94] Hergert, W. and Wriedt, T. *The Mie Theory: Basics and Applications*, vol. 169. Springer, Berlin, 2012.

[95] Hille, E. *Analytic Function Theory*, 2nd ed., vol. 1. Chelsea Publishing Company, New York, NY, 1982.

[96] Hochstadt, H. *Integral Equations*. John Wiley & Sons, New York, NY, 1973.

[97] Hovenier, J., van de Hulst, H.C., and Van der Mee, C. Conditions for the elements of the scattering matrix. *Astron. Astrophys.*, **157**, 301–310, 1986.

[98] Irvine, W.M. and Pollack, J.B. Infrared optical properties of water and ice spheres. *Icarus*, **8**, 324–360, 1968.

[99] Ishimaru, A. *Electromagnetic Wave Propagation, Radiation, and Scattering*. Prentice-Hall, Inc., Englewood Cliffs, New Jersey, 1991.

[100] Ishimaru, A. *Wave Propagation and Scattering in Random Media*. IEEE Press, New York, NY, 1997.

[101] Jackson, J.D. *Classical Electrodynamics*, 3rd ed. John Wiley & Sons, New York, NY, 1999.

[102] James, G.L. *Geometrical Theory of Diffraction for Electromagnetic Waves*, 3rd ed. Peter Peregrinus Ltd., Stevenage, UK, 1986.

[103] Jefimenko, O.D. *Electricity and Magnetism: An Introduction to the Theory of Electric and Magnetic Fields*, 2nd ed. Electret Scientific Company, Waynesburg, PA, 1989.

[104] Jin, J.M. *Theory and Computation of Electromagnetic Fields*. John Wiley & Sons, Hoboken, NJ, 2011.

[105] John, F. *Partial Differential Equations*, 4th ed. Springer-Verlag, New York, NY, 1982.

[106] Johnson, B.R. Light scattering by a multilayer sphere. *Appl. Opt.*, **35**, 3286–3296, 1996.

[107] Jones, D.S. Removal of an inconsistency in the theory of diffraction. *Proc. Camb. Phil. Soc.*, **48**, 733–741, 1952.

[108] Jones, D.S. Approximate methods in high-frequency scattering. *Proc. R. Soc. Lond. A*, **239**, 338–348, 1957.

[109] Jones, D.S. High-frequency scattering of electromagnetic waves. *Proc. R. Soc. Lond. A*, **240**, 206–213, 1957.

[110] Jones, D.S. *The Theory of Electromagnetism*. Pergamon, Oxford, 1964.

[111] Jones, D.S. Diffraction of a high-frequency plane electromagnetic wave by a perfectly conducting circular disc. *Proc. Camb. Phil. Soc.*, **61**, 247–270, 1965.

[112] Jones, D.S. Scattering by inhomogeneous dielectric particles. *Quart. J. Mech. Appl. Math.*, **38**, 135–155, 1985.

[113] Jones, D.S. *Methods in Electromagnetic Wave Propagation*, 2nd ed. IEEE Press, Piscataway, NJ, 1994.

[114] Jones, D.S. and Kline, M. Asymptotic expansion of multiple integrals and the method of stationary phase. *J. Math. and Phys.*, **37**, 1–28, 1958.

[115] Karlsson, A. On the time domain version of the optical theorem. *Am. J. Phys.*, **68**, 344–349, 2000.

[116] Karlsson, A. Some results extracted from the time domain version of the optical theorem. *Radio Sci.*, **37**, VIC 9–1–VIC 9–10, 2002.

[117] Karlsson, A. and Kristensson, G. Electromagnetic scattering from subterranean obstacles in a stratified ground. *Radio Sci.*, **18**, 345–356, 1983.

[118] Karlsson, A. and Kristensson, G. Constitutive relations, dissipation and reciprocity for the Maxwell equations in the time domain. *J. Electromagnet. Waves Appl.*, **6**, 537–551, 1992.

[119] Kattawar, G.W. and Plass, G.N. Electromagnetic scattering from absorbing spheres. *Appl. Opt.*, **6**, 1377–1382, 1967.

[120] Katznelson, Y. *An Introduction to Harmonic Analysis*. Cambridge University Press, Cambridge, UK, 2004.

[121] Keller, J.B. Diffraction of a convex cylinder. *IRE Trans. Antennas Propag.*, **4**, 312–321, 1956.

[122] Keller, J.B. and Lewis, R.M. Asymptotic methods for partial differential equations: the reduced wave equation and Maxwell's equations. In *Surveys in Applied Mathematics*, pages 1–82. Springer, Berlin, 1995.

[123] King, F.W. *Hilbert Transforms*, vol. 1. Cambridge University Press, Cambridge, UK, 2009.

[124] King, F.W. *Hilbert Transforms*, vol. 2. Cambridge University Press, Cambridge, UK, 2009.

[125] Kitamura, R., Pilon, L., and Jonasz, M. Optical constants of silica glass from extreme ultraviolet to far infrared at near room temperature. *Appl. Opt.*, **46**, 8118–8133, 2007.

[126] Kleinman, R.E. and Senior, T.B.A. Rayleigh Scattering. In V.K. Varadan and V.V. Varadan (editors), *Low and High Frequency Asymptotics*, volume 2 of *Acoustic, Electromagnetic and Elastic Wave Scattering*, chapter 1, pages 1–70. North-Holland, Amsterdam, 1986.

[127] Kong, J.A. *Electromagnetic Wave Theory*. John Wiley & Sons, New York, NY, 1986.

[128] Kouyoumjian, R.G. Asymptotic high-frequency methods. *Proc. IEEE*, **53**, 864–876, 1965.

[129] Kouyoumjian, R.G. and Pathak, P.H. A uniform geometrical theory of diffraction for an edge in a perfectly conducting surface. *Proc. IEEE*, **62**, 1448–1461, 1974.

[130] Kraus, J.D. *Electromagnetics*, 4th ed. McGraw-Hill, New York, NY, 1992.

[131] Kreiss, H.O. and Lorenz, J. *Initial-Boundary Value Problems and the Navier–Stokes Equations*. Academic Press, San Diego, CA, 1989.

[132] Kress, R. *Linear Integral Equations*, 2nd ed. Springer-Verlag, Heidelberg, 1999.

[133] Kress, R. Chapter 1.4.1 – Specific Theoretical Tools. In R. Pike and P. Sabatier (editors), *Scattering*, pages 175–190, Academic Press, London, 2002.

[134] Kreyszig, E. *Advanced Engineering Mathematics*, 10th ed. John Wiley & Sons, New York, NY, 1993.

[135] Kristensson, G. A uniqueness theorem for Helmholtz' equation: penetrable media with an infinite interface. *SIAM J. Math. Anal.*, **11**, 1104–1117, 1980.

[136] Kristensson, G. Electromagnetic scattering from buried inhomogeneities—a general three-dimensional formalism. *J. Appl. Phys.*, **51**, 3486–3500, 1980.

[137] Kristensson, G. On the generation of surface waves in frequency selective structures. Technical Report LUTEDX/(TEAT-7118)/1–38/(2003), Lund University, Department of Electrical and Information Technology, P.O. Box 118, S-221 00 Lund, Sweden, 2003. http://www.eit.lth.se.

[138] Kristensson, G. *Second Order Differential Equations—Special Functions and Their Classification*. Springer-Verlag, London, 2010.

[139] Kristensson, G. The polarizability and the capacitance change of a bounded object in a parallel plate capacitor. *Phys. Scr.*, **86**, 035405, 2012.

[140] Kristensson, G. Electromagnetic scattering by a bounded obstacle in a parallel plate waveguide. *J. Quant. Spectrosc. Radiat. Transfer*, **123**, 92–102, 2013.

[141] Kristensson, G. Coherent scattering by a collection of randomly located obstacles—an alternative integral equation formulation. *J. Quant. Spectrosc. Radiat. Transfer*, **164**, 97–108, 2015.

[142] Kristensson, G., Åkerberg, M., and Poulsen, S. Scattering from a frequency selective surface supported by a bianisotropic substrate. *Prog. Electromagn. Res.*, **35**, 83–114, 2002.

[143] Kristensson, G., Larsson, C., Sohl, C., and Gustafsson, M. Bounds on metamaterials—theoretical and experimental results. In S. Zouhdi, A. Sihvola, and A.P. Vinogradov (editors), *Metamaterials and Plasmonics: Fundamentals, Modelling, Applications*, NATO Science for Peace and Security Series, pages 21–33. Springer, Berlin, 2009.

[144] Kristensson, G., Poulsen, S., and Rikte, S. Propagators and scattering of electromagnetic waves in planar bianisotropic slabs—an application to frequency selective structures. *Prog. Electromagn. Res.*, **48**, 1–25, 2004.

[145] Kristensson, G., Ramm, A.G., and Ström, S. Convergence of the T-matrix approach in scattering theory. II. *J. Math. Phys.*, **24**, 2619–2631, 1983.

[146] Kristensson, G., Waller, P., and Derneryd, A. Radiation efficiency and surface waves for patch antennas on inhomogeneous substrates. *IEE Proc. Microwaves Antennas Propag.*, **150**, 477–483, 2003.

[147] Kristensson, G. and Waterman, P.C. The T-matrix for acoustic and electromagnetic scattering by circular disks. *J. Acoust. Soc. Am.*, **72**, 1612–1625, 1982.

[148] Kristensson, G. and Wellander, N. Some a priori estimates of solutions to the maxwell equations. *Math. Methods Appl. Sci.*, **38**, 2194–2215, 2015.

[149] Kuipers, J. *Quaternions and Rotation Sequences: A Primer with Applications to Orbits, Aerospace, and Virtual Reality*. John Wiley & Sons, Hoboken, NJ, 2002.

[150] Lai, H.M., Wong, W.Y., and Wong, W.H. Extinction paradox and actual power scattered in light beam scattering: a two-dimensional study. *J. Opt. Soc. Am. A*, **21**, 2324–2333, 2004.

[151] Lakhtakia, A., Varadan, V.K., and Varadan, V.V. *Time-Harmonic Electromagnetic Fields in Chiral Media*, volume 335 of *Lecture Notes in Physics*. Springer-Verlag, New York, NY, 1989.

[152] Landau, L.D., Lifshitz, E.M., and Pitaevskiĭ, L.P. *Electrodynamics of Continuous Media*, 2nd ed. Pergamon Press, Oxford, 1984.

[153] Lane, J.A. and Saxton, J.A. Dielectric dispersion in pure polar liquids at very high radio-frequencies. I. Measurements on water, methyl and ethyl alcohols. *Proc. R. Soc. Lond. A*, **A213**, 400–408, 1952.

[154] Larsson, C., Sohl, C., Gustafsson, M., and Kristensson, G. Measuring the extinction cross section. In *Third European Conference on Antennas and Propagation*, Berlin, Germany, pages 3633–3636, 2009.

[155] Lax, M. Multiple scattering of waves. *Rev. Mod. Phys.*, **23**, 287–310, 1951.

[156] Lax, M. Multiple scattering of waves. II. The effective field in dense systems. *Phys. Rev.*, **85**, 621–629, 1952.

[157] Lax, P.D. and Phillips, R.S. *Scattering Theory*. Academic Press, New York, NY, 1967.

[158] Leis, R. *Initial Boundary Value Problems in Mathematical Physics*. Teubner & Wiley, Stuttgart, 1986.

[159] Lentz, W.J. Generating Bessel functions in Mie scattering calculations using continued fractions. *Appl. Opt.*, **15**, 668–671, 1976.

[160] Li, J.L.W., Ong, W.L., and Zheng, K.H.R. Anisotropic scattering effects of a gyrotropic sphere characterized using the T-matrix method. *Phys. Rev. E*, **85**, 036601, 2012.

[161] Lindell, I.V. *Methods for Electromagnetic Field Analysis*. IEEE Press and Oxford University Press, Oxford, 1995.

[162] Lindell, I.V. and Olyslager, F. Analytic Green dyadic for a class of nonreciprocal anisotropic media. *IEEE Trans. Antennas Propag.*, **45**, 1563–1565, 1997.

[163] Lindell, I.V. and Olyslager, F. Green dyadics for self-dual bi-anisotropic media. *J. Electromagnet. Waves Appl.*, **14**, 153–163, 2000.

[164] Lindell, I.V. and Sihvola, A.H. Quasi-static analysis of scattering from a chiral sphere. *J. Electromagnet. Waves Appl.*, **4**, 1223–1231, 1990.

[165] Lindell, I.V., Sihvola, A.H., Tretyakov, S.A., and Viitanen, A.J. *Electromagnetic Waves in Chiral and Bi-Isotropic Media*. Artech House, Boston, MA, 1994.

[166] Lindell, I.V. and Weiglhofer, W.S. Green dyadic for a uniaxial bianisotropic medium. *IEEE Trans. Antennas Propag.*, **42**, 1013–1016, 1994.

[167] Logan, J.D. *Applied Mathematics*, 3rd ed. John Wiley & Sons, New York, NY, 2006.

[168] Ludwig, D. Uniform asymptotic expansion of the field scattered by a convex object at high frequencies. *Comm. Pure Appl. Math.*, **20**, 103–138, 1967.

[169] Mackowski, D.W. Analysis of radiative scattering for multiple sphere configurations. *Proc. R. Soc. London, Ser. A*, **433**, 599–614, 1991.

[170] Mackowski, D.W., Altenkirch, R.A., and Menguc, M.P. Internal absorption cross sections in a stratified sphere. *Appl. Opt.*, **29**, 1551–1559, 1990.

[171] Markoš, P. and Soukoulis, C.M. *Wave Propagation: From Electrons to Photonic Crystals and Left-Handed Materials*. Princeton University Press, Princeton, NJ, 2008.

[172] Martin, P.A. On the null-field equations for the exterior problems of acoustics. *Quart. J. Mech. Appl. Math.*, **33**, 385–396, 1980.

[173] Martin, P.A. Acoustic scattering and radiation problems, and the null-field method. *Wave Motion*, **4**, 391–408, 1982.

[174] Martin, P.A. *Multiple Scattering: Interaction of Time-Harmonic Waves with N Obstacles*, volume 107 of *Encyclopedia of Mathematics and Its Applications*. Cambridge University Press, Cambridge, UK, 2006.

[175] Martin, T. and Wallin, M. Validation method for numerical simulations of large objects using the T-matrix for a collection of scatterers. In *International Conference on Electromagnetics in Advanced Applications (ICEAA)*, Turin, Italy, pages 1127–1130, 2015.

[176] Mathur, N.C. and Yeh, K.C. Multiple scattering of electromagnetic waves by random scatterers of finite size. *J. Math. Phys.*, **5**, 1619–1628, 1964.

[177] Maxwell, J.C. A dynamical theory of the electromagnetic field. *Philos. Trans. R. Soc. Lond.*, **155**, 459–512, 1865.

[178] Maxwell, J.C. *A Treatise on Electricity and Magnetism*, vol. 1. Dover Publications, New York, NY, 1954.

[179] Maxwell, J.C. *A Treatise on Electricity and Magnetism*, vol. 2. Dover Publications, New York, NY, 1954.

[180] McNamara, D.A., Pistorius, C.W.I., and Malherbe, J.A.G. *Introduction to the Uniform Geometrical Theory of Diffraction*. Artech House, Boston, MA, 1990.

[181] Michaeli, A. High-frequency electromagnetic fields near a smooth convex surface in the shadow region. *J. Electromagnet. Waves Appl.*, **7**, 533–546, 1993.

[182] Mishchenko, M.I., Travis, L.D., and Lacis, A.A. *Scattering, Absorption, and Emission of Light by Small Particles*. Cambridge University Press, Cambridge, 2002.

[183] Mishchenko, M.I. Multiple scattering, radiative transfer, and weak localization in discrete random media: unified microphysical approach. *Rev. Geophys.*, **46**, RG2003, 2008.

[184] Mishchenko, M.I. Directional radiometry and radiative transfer: The convoluted path from centuries-old phenomenology to physical optics. *J. Quant. Spectrosc. Radiat. Transfer*, **146**, 4–33, 2014.

[185] Mishchenko, M.I. *Electromagnetic Scattering by Particles and Particle Groups*. Cambridge University Press, Cambridge, UK, 2014.

[186] Mishchenko, M.I., Hovenier, J.W., and Travis, L.D. *Light Scattering by Nonspherical Particles: Theory, Measurements, and Applications*. Academic Press, San Diego, CA, 2000.

[187] Mishchenko, M.I., Travis, L.D., and Lacis, A.A. *Multiple Scattering of Light by Particles: Radiative Transfer and Coherent Backscattering*. Cambridge University Press, Cambridge, UK, 2006.

[188] Mishchenko, M.I., Travis, L.D., and Mackowski, D.W. T-matrix computations of light scattering by nonspherical particles: a review. *J. Quant. Spectrosc. Radiat. Transfer*, **55**, 535–575, 1996.

[189] Mishchenko, M.I., Videen, G., Babenko, V.A., Khlebtsov, N.G., and Wriedt, T. T-matrix theory of electromagnetic scattering by particles and its applications: a comprehensive reference database. *J. Quant. Spectrosc. Radiat. Transfer*, **88**, 357–406, 2004.

[190] Mishchenko, M.I., Videen, G., Babenko, V.A., Khlebtsov, N.G., and Wriedt, T. Comprehensive T-matrix reference database: a 2004–06 update. *J. Quant. Spectrosc. Radiat. Transfer*, **106**, 304–324, 2007.

[191] Mishchenko, M.I., Videen, G., Khlebtsov, N.G., and Wriedt, T. Comprehensive T-matrix reference database: a 2012–2013 update. *J. Quant. Spectrosc. Radiat. Transfer*, **123**, 145–152, 2013.

[192] Mishchenko, M.I., Videen, G., Khlebtsov, N.G., Wriedt, T., and Zakharova, N.T. Comprehensive T-matrix reference database: a 2006–07 update. *J. Quant. Spectrosc. Radiat. Transfer*, **109**, 1447–1460, 2008.

[193] Mishchenko, M.I., Zakharova, N.T., Khlebtsov, N.G., Wriedt, T., and Videen, G. Comprehensive thematic T-matrix reference database: a 2013–2014 update. *J. Quant. Spectrosc. Radiat. Transfer*, **146**, 349–354, 2014.

[194] Mishchenko, M.I., Zakharova, N.T., Videen, G., Khlebtsov, N.G., and Wriedt, T. Comprehensive T-matrix reference database: a 2007–2009 update. *J. Quant. Spectrosc. Radiat. Transfer*, **111**, 650–658, 2010.

[195] Monk, P. *Finite Element Methods for Maxwell's Equations*. Oxford University Press, Oxford, 2003.

[196] Morse, P.M. and Feshbach, H. *Methods of Theoretical Physics*, vol. 2. McGraw-Hill, New York, NY, 1953.

[197] Müller, C. *Foundations of the Mathematical Theory of Electromagnetic Waves*. Springer-Verlag, Berlin, 1969.

[198] Müller, C. Boundary values and diffraction problems. In *Inst. Naz. Alta Math. Sympos. Math.*, **18**, 353–367, 1976.

[199] Nédélec, J.C. *Acoustic and Electromagnetic Equations: Integral Representations for Harmonic Problems*, vol. 144. Springer-Verlag, Berlin, 2001.

[200] Newton, R.G. Optical theorem and beyond. *Am. J. Phys*, **44**, 639–642, 1976.

[201] Newton, R.G. *Scattering Theory of Waves and Particles*, 2nd ed. Dover Publications, New York, NY, 2002.

[202] Nussenzveig, H.M. High-frequency scattering by an impenetrable sphere. *Ann. Phys.*, **34**, 23–95, 1965.

[203] Nussenzveig, H.M. *Causality and Dispersion Relations*. Academic Press, London, 1972.

[204] Nussenzveig, H.M. Uniform approximation in scattering by spheres. *J. Phys. A: Math. Gen.*, **21**, 81, 1988.

[205] Ogawa, T., Tanaka, Y., Miura, T., and Yasuhara, M. Observations of natural ELF and VLF electromagnetic noises by using ball antennas. *J. Geomagn. Geoelectr.*, **18**, 443–454, 1966.

[206] Olver, F.W.J., Lozier, D.W., Boisvert, R.F., and Clark, C.W. *NIST Handbook of Mathematical Functions*. Cambridge University Press, Cambridge, UK, 2010.

[207] Olyslager, F. Time-harmonic two and three-dimensional Green's dyadics for general uniaxial bianisotropic media. *IEEE Trans. Antennas Propag.*, **43**, 430–434, 1995.

[208] Olyslager, F. Time-harmonic two- and three-dimensional closed form Green's dyadics for gyrotropic, bianisotropic, and anisotropic media. *Electromagnetics*, **17**, 369–386, 1997.

[209] Olyslager, F. and Jakoby, B. Time-harmonic two- and three-dimensional Green dyadics for a special class of gyrotropic bianisotropic media. *IEE Proc. Microwaves Antennas Propag.*, **143**, 413–416, 1996.

[210] Orfanidis, S.J. Electromagnetic Waves and Antennas, 2014. Available from http://www.ece.rutgers.edu/~orfanidi/ewa/.

[211] Pena, O. and Pal, U. Scattering of electromagnetic radiation by a multilayered sphere. *Comput. Phys. Commun.*, **180**, 2348–2354, 2009.

[212] Pendry, J., Holden, A., Stewart, W., and Youngs, I. Extremely low frequency plasmons in metallic mesostructures. *Phys. Rev. Lett.*, **76**, 4773, 1996.

[213] Peterson, B. and Ström, S. T-matrix for electromagnetic scattering from an arbitrary number of scatterers and representations of E(3). *Phys. Rev. D*, **8**, 3661–3678, 1973.

[214] Peterson, B. and Ström, S. T-matrix formulation of electromagnetic scattering from multilayered scatterers. *Phys. Rev. D*, **10**, 2670–2684, 1974.

[215] Poulsen, S. Scattering from frequency selective surfaces: a continuity condition for entire domain basis functions and an improved set of basis functions for crossed dipole. *IEE Proc.-H Microwaves Antennas Propag.*, **146**, 234–240, 1999.

[216] Poulsen, S. Scattering from frequency selective surfaces: an efficient set of V-dipole basis functions. *IEEE Trans. Antennas Propag.*, **51**, 540–548, 2003.

[217] Poulsen, S. *Stealth Radomes*. Ph.D. thesis, Department of Electroscience, Lund University, Lund, Sweden, 2006.

[218] Price, C. and Melnikov, A. Diurnal, seasonal and inter-annual variations in the Schumann resonance parameters. *J. Atmos. Sol.-Terr. Phys.*, **66**, 1179–1185, 2004. SPECIAL—Space Processes and Electrical Changes in Atmospheric Layers.

[219] Price, C., Pechony, O., and Greenberg, E. Schumann resonances in lightning research. *J. Lightning Res.*, **1**, 1–15, 2007.

[220] Putzer, E.J. Avoiding the Jordan canonical form in the discussion of linear systems with constant coefficients. *Am. Math. Month.*, **73**, 2–7, 1966.

[221] Rakić, A.D., Djurišić, A.B., Elazar, J.M., and Majewski, M.L. Optical properties of metallic films for vertical-cavity optoelectronic devices. *Appl. Opt.*, **22**, 5271–5283, 1998.

[222] Ramm, A.G. Convergence of the T-matrix approach to scattering theory. *J. Math. Phys.*, **23**, 1123–1125, 1982.

[223] Ramm, A.G. *Scattering by an Obstacle*. D. Reidel, Dordrecht, 1986.

[224] Redheffer, R. On the relation of transmission-line theory on scattering and transfer. *J. Math. and Phys.*, **41**, 1–41, 1962.

[225] Reif, F. *Fundamentals of Statistical and Thermal Physics*. McGraw-Hill, New York, NY, 1965.

[226] Reitz, J.R., Milford, F.J., and Christy, R.W. *Foundations of Electromagnetic Theory*, 4th ed. Addison-Wesley Publishing Company, Boston, MA, 2009.

[227] Rheinstein, J. Scattering of electromagnetic waves from dielectric coated conducting spheres. *IEEE Trans. Antennas Propag.*, **12**, 334–340, 1964.

[228] Rikte, S., Andersson, M., and Kristensson, G. Homogenization of woven materials. *Arch. Elektron. Übertragungstech.*, **53**, 261–271, 1999.

[229] Rikte, S., Kristensson, G., and Andersson, M. Propagation in bianisotropic media—reflection and transmission. *IEE Proc. Microwaves Antennas Propag.*, **148**, 29–36, 2001.

[230] Roach, G.F., Stratis, I.G., and Yannacopoulos, A.N. *Mathematical Analysis of Deterministic and Stochastic Problems in Complex Media Electromagnetics*. Princeton University Press, Princeton, NJ, 2012.

[231] Rother, T. and Kahnert, M. *Electromagnetic Wave Scattering on Nonspherical Particles: Basic Methodology and Simulations*, 2nd ed., vol. 145. Springer-Verlag, Berlin, 2014.

[232] Rother, T., Kahnert, M., Doicu, A., and Wauer, J. Surface Green's function of the Helmholtz equation in spherical coordinates. *Prog. Electromagn. Res.*, **38**, 47–95, 2002.

[233] Rother, T. and Wauer, J. Case study about the accuracy behavior of three different T-matrix methods. *Appl. Opt.*, **49**, 5746–5756, 2010.

[234] Ruppin, R. Extinction properties of a sphere with negative permittivity and permeability. *Solid State Commun.*, **116**, 411–415, 2000.

[235] Ruppin, R. Surface polaritons of a left-handed medium. *Phys. Lett. A*, **277**, 61–64, 2000.

[236] Ruppin, R. Surface polaritons of a left-handed material slab. *J. Phys.: Condens. Matter*, **13**, 1811–1818, 2001.

[237] Ruppin, R. and Englman, R. Optical lattice vibrations in finite ionic crystals: II. *J. Phys. C: Solid State Phys.*, **1**, 630–643, 1968.

[238] Russakoff, G. A derivation of the macroscopic Maxwell equations. *Am. J. Phys*, **38**, 1188–1195, 1970.

[239] Saxon, D. Tensor scattering matrix for the electromagnetic field. *Phys. Rev.*, **100**, 1771, 1955.

[240] Schulten, K. and Gordon, R.G. Exact recursive evaluation of $3j$- and $6j$-coefficients for quantum-mechanical coupling of angular momenta. *J. Math. Phys.*, **16**, 1961–1970, 1975.

[241] Sha, W.E.I. and Chew, W.C. Frequency-independent scattering for the large sphere. In *The 2009 Asia Pacific Microwave Conference (APMC 2009)*, Singapore, pages 854–857, 2009.

[242] Sha, W.E.I. and Chew, W.C. High frequency scattering by an impenetrable sphere. *Progress In Electromagnetics Research*, **97**, 291–325, 2009.

[243] Sihvola, A.H., Ylä-Oijala, P., Järvenpää, S., and Avelin, J. Polarizabilities of platonic solids. *IEEE Trans. Antennas Propag.*, **52**, 2226–2233, 2004.

[244] Silver, S. *Microwave Antenna Theory and Design*, volume 12 of *Radiation Laboratory Series*. McGraw-Hill, New York, NY, 1949.

[245] Sleator, F.B. The disc. In J.J. Bowman, T.B.A. Senior, and P.L.E. Uslenghi (editors), *Electromagnetic and Acoustic Scattering by Simple Shapes*, Chapter 14, pages 528–588. North-Holland, Amsterdam, 1987.

[246] Sohl, C., Gustafsson, M., and Bernland, A. Some paradoxes associated with a recent summation rule in scattering theory. Technical Report LUTEDX/(TEAT-7163)/1–6/(2008), Department of Electrical and Information Technology, Lund University, Lund, Sweden, 2008. Available from http://www.eit.lth.se.

[247] Sohl, C., Gustafsson, M., and Kristensson, G. The integrated extinction for broadband scattering of acoustic waves. *J. Acoust. Soc. Am.*, **122**, 3206–3210, 2007.

[248] Sohl, C., Gustafsson, M., and Kristensson, G. Physical limitations on broadband scattering by heterogeneous obstacles. *J. Phys. A: Math. Theor.*, **40**, 11165–11182, 2007.

[249] Sohl, C., Gustafsson, M., and Kristensson, G. Physical limitations on metamaterials: restrictions on scattering and absorption over a frequency interval. *J. Phys. D: Appl. Phys.*, **40**, 7146–7151, 2007.

[250] Stakgold, I. *Boundary Value Problems of Mathematical Physics*, vol. 1. Macmillan, London, 1967.

[251] Stakgold, I. *Boundary Value Problems of Mathematical Physics*, vol. 2. Macmillan, New York, NY, 1968.

[252] Stakgold, I. *Green's Functions and Boundary Value Problems*. John Wiley & Sons, New York, NY, 1998.

[253] Stratton, J.A. *Electromagnetic Theory*. McGraw-Hill, New York, NY, 1941.

[254] Stratton, J.A. and Chu, L.J. Diffraction theory of electromagnetic waves. *Phys. Rev.*, **56**, 99–107, 1939.

[255] Ström, S. and Zheng, W. Basic features of the null field method for dielectric scatterers. *Radio Sci.*, **22**, 1273–1281, 1987.

[256] Ström, S. and Zheng, W. The null field approach to electromagnetic scattering from composite objects. *IEEE Trans. Antennas Propag.*, **36**, 376–383, 1988.

[257] Ström, S. and Zheng, W. Null-field computations of radar cross sections of composite objects. *Proc. IEEE*, **77**, 761–769, 1989.

[258] Taylor, J.R. *Scattering Theory: The Quantum Theory of NonRelativistic Collisions*. Robert E. Krieger Publishing Company, Malabar, FL, 1983.

[259] Tishkovets, V.P., Petrova, E., and Mishchenko, M. Scattering of electromagnetic waves by ensembles of particles and discrete random media. *J. Quant. Spectrosc. Radiat. Transfer*, **112**, 2095–2127, 2011.

[260] Tishkovets, V.P. Incoherent and coherent backscattering of light by a layer of densely packed random medium. *J. Quant. Spectrosc. Radiat. Transfer*, **108**, 454–463, 2007.

[261] Titchmarsh, E.C. *Introduction to the Theory of Fourier Integrals*, 2nd ed. Oxford University Press, Oxford, 1948.

[262] Toon, O.B. and Ackerman, T.P. Algorithms for the calculation of scattering by stratified spheres. *Appl. Opt.*, **20**, 3657–3660, 1981.

[263] Tricomi, F.G. *Integral Equations*. Dover Publications, New York, NY, 1957.

[264] Tsang, L. and Kong, J.A. *Scattering of Electromagnetic Waves: Advanced Topics*. John Wiley & Sons, New York, NY, 2001.

[265] Tsang, L., Mandt, C.E., and Ding, K.H. Monte Carlo simulations of the extinction rate of dense media with randomly distributed dielectric spheres based on solution of Maxwell's equations. *Opt. Lett.*, **17**, 314–316, 1992.

[266] Tuzun, R.E., Burkhardt, P., and Secrest, D. Accurate computation of individual and tables of 3-j and 6-j symbols. *Comput. Phys. Commun.*, **112**, 112–148, 1998.

[267] Twersky, V. On multiple scattering of waves. *J. Res. Nat. Bur. Stand. D*, **64**, 715–730, 1960.

[268] Twersky, V. On propagation in random media of discrete scatterers. *Proc. Symp. Appl. Math*, **16**, 84–116, 1964.

[269] Twersky, V. Coherent scalar field in pair-correlated random distributions of aligned scatterers. *J. Math. Phys.*, **18**, 2468–2486, 1977.

[270] Ung, B. and Sheng, Y. Interference of surface waves in a metallic nanoslit. *Opt. Express*, **15**, 1182–1190, 2007.

[271] Vakili, I., Gustafsson, M., Sjöberg, D., Seviour, R., Nilsson, M., and Nordebo, S. Sum rules for parallel-plate waveguides: experimental results and theory. *IEEE Trans. Microwave Theory Tech.*, **62**, 2574–2582, 2014.

[272] Valagiannopoulos, C.A. An overview of the Watson transformation presented through a simple example. *Prog. Electromagn. Res.*, **75**, 137–152, 2007.

[273] van Bladel, J. Some remarks on Green's dyadic for infinite space. *IRE Trans. Antennas Propag.*, **6**, 563–566, 1961.

[274] van Bladel, J. Lorenz or Lorentz. *IEEE Antennas Propag. Mag.*, **33**, 69, 1991.

[275] van Bladel, J. *Singular Electromagnetic Fields and Sources*. IEEE Press and Oxford University Press, Oxford, 1991.

[276] van Bladel, J. *Electromagnetic Fields*, 2nd ed. IEEE Press, Piscataway, NJ, 2007.

[277] van de Hulst, H.C. *Light Scattering by Small Particles*. John Wiley & Sons, New York, NY, 1957.

[278] Varadan, V.V., Bringi, V.N., and Varadan, V.K. Frequency dependent dielectric constants of discrete random media. In R. Burridge, S. Childress, and G. Papanicolaou (editors), *Macroscopic Properties of Disordered Media*, volume 154 of *Lecture Notes in Physics*, pages 272–284. Springer-Verlag, Heidelberg, 1982.

[279] Varadan, V.V., Ma, Y., and Varadan, V.K. Propagator model including multipole fields for discrete random media. *J. Opt. Soc. Am. A*, **2**, 2195–2201, 1985.

[280] Varadan, V.K., Bringi, V.N., and Varadan, V.V. Coherent electromagnetic wave propagation through randomly distributed dielectric scatterers. *Phys. Rev. D*, **19**, 2480–2489, 1979.

[281] Wang, R.T. and van de Hulst, H.C. Rainbows: Mie computations and the airy approximation. *Appl. Opt.*, **30**, 106–117, 1991.

[282] Waterman, P.C. Matrix formulation of electromagnetic scattering. *Proc. IEEE*, **53**, 805–812, 1965.

[283] Waterman, P.C. Scattering by dielectric obstacles (integral equation method for electromagnetic scattering by perfectly conducting obstacles modified for volume scattering regions

involving disparities in dielectric constant, permeability or conductivity). *Alta Freq.*, **38**, 348–352, 1968.

[284] Waterman, P.C. New formulation of acoustic scattering. *J. Acoust. Soc. Am.*, **45**, 1417–1429, 1969.

[285] Waterman, P.C. Symmetry, unitarity, and geometry in electromagnetic scattering. *Phys. Rev. D*, **3**, 825–839, 1971.

[286] Waterman, P.C. The T-matrix revisited. *J. Opt. Soc. Am. A*, **24**, 2257–2267, 2007.

[287] Waterman, P.C. and Truell, R. Multiple scattering of waves. *J. Math. Phys.*, **2**, 512–537, 1961.

[288] Watson, G.N. *A Treatise on the Theory of Bessel Functions*, 2nd ed. Cambridge University Press, Cambridge, UK, 1966.

[289] Weder, R. *Spectral and Scattering Theory for Wave Propagation in Perturbed Stratified Media.* Springer-Verlag, New York, 1991.

[290] Wellander, N. and Kristensson, G. Estimates of scattered electromagnetic fields. *Math. Method Appl. Sci.*, **37**, 167–172, 2014.

[291] Weniger, E.J. and Steinborn, E.O. Programs for the coupling of spherical harmonics. *Comput. Phys. Commun.*, **25**, 149–157, 1982.

[292] Wertheim, M.S. Exact solution of the Percus–Yevick integral equation for hard spheres. *Phys. Rev. Lett.*, **10**, 321–323, 1963.

[293] West, R., Gibbs, D., Tsang, L., and Fung, A. Comparison of optical scattering experiments and the quasi-crystalline approximation for dense media. *J. Opt. Soc. Am. A*, **11**, 1854–1858, 1994.

[294] Wiscombe, W.J. Improved Mie scattering algorithms. *Appl. Opt.*, **19**, 1505–1509, 1980.

[295] Yafaev, D.R. *Mathematical Scattering Theory*. American Mathematical Society, Providence, RI, 1992.

[296] Yaghjian, A.D. Electric dyadic Green's functions in the source region. *Proc. IEEE*, **68**, 248–263, 1980.

[297] Yaghjian, A.D. Equivalence of surface current and aperture field integrations for reflector antennas. *IEEE Trans. Antennas Propag.*, **32**, 1355–1358, 1984.

[298] Yang, H.Y.D. A spectral recursive transformation method for electromagnetic waves in generalized anisotropic layered media. *IEEE Trans. Antennas Propag.*, **45**, 520–526, 1997.

[299] Yang, W. Improved recursive algorithm for light scattering by a multilayered sphere. *Appl. Opt.*, **42**, 1710–1720, 2003.

[300] Young, A.T. Rayleigh scattering. *Phys. Today*, **35**, 42–48, 1982.

[301] Yung, E.K.N. and Hu, B.J. Scattering by conducting sphere coated with chiral media. *Microwave Opt. Techn. Lett.*, **35**, 288–293, 2002.

[302] Zachmanoglou, E.C. and Thoe, D.W. *Introduction to Partial Differential Equations with Applications*. Dover Publications, Mineola, NY, 1986.

[303] Zakharova, N.T., Videen, G., and Khlebtsov, N.G. Comprehensive T-matrix reference database: a 2009–2011 update. *J. Quant. Spectrosc. Radiat. Transfer*, **113**, 1844–1852, 2012.

[304] Zemanian, A.H. *Distribution Theory and Transform Analysis: An Introduction to Generalized Functions, with Applications*. Dover Publications, New York, NY, 1987.

[305] Zemanian, A.H. *Realizability Theory for Continuous Linear Systems*. Dover Publications, New York, NY, 1995.

[306] Zheng, W. and Ström, S. The null field approach to electromagnetic scattering from composite objects: the case of concavo-convex constituents. *IEEE Trans. Antennas Propag.*, **37**, 373–383, 1989.

[307] Zheng, W. and Ström, S. The null-field approach to electromagnetic resonance of composite objects. *Comput. Phys. Commun.*, **68**, 157–174, 1991.

Answers to Problems

1.1 Apply the theorem of divergence (Gauss' theorem) to the vector field $\boldsymbol{B} = \boldsymbol{A} \times \boldsymbol{a}$, where $\boldsymbol{a}$ is an arbitrary constant vector.

1.2 On the surface of the conductor, we have $\boldsymbol{S} = -\hat{\boldsymbol{r}}_{\mathrm{c}} \frac{1}{2} a\sigma E^2$ where the electric field on the surface of the conductor is related to the current by $I = \pi a^2 \sigma E$. The terms in Poynting's theorem are

$$\iint_S \boldsymbol{S} \cdot \hat{\boldsymbol{\nu}} \, \mathrm{d}S = -\pi a^2 l \sigma E^2$$

$$\iiint_V \boldsymbol{E} \cdot \boldsymbol{J} \, \mathrm{d}v = \pi a^2 l \sigma E^2$$

1.3

$$\boldsymbol{P}(t) = \begin{cases} \boldsymbol{0} & t \leq 0 \\ \epsilon_0 \alpha \tau \boldsymbol{E}_0 \left[1 - \mathrm{e}^{-t/\tau}\right] & 0 < t < T \\ \epsilon_0 \alpha \tau \boldsymbol{E}_0 \mathrm{e}^{-t/\tau} \left[\mathrm{e}^{T/\tau} - 1\right] & t \geq T \end{cases}$$

1.4

$$[\boldsymbol{\Sigma}](t) = \omega_{\mathrm{p}}^2 \mathrm{e}^{-\nu t} \begin{pmatrix} \cos \omega_{\mathrm{g}} t & \sin \omega_{\mathrm{g}} t & 0 \\ -\sin \omega_{\mathrm{g}} t & \cos \omega_{\mathrm{g}} t & 0 \\ 0 & 0 & 1 \end{pmatrix}$$

1.5

$$\begin{cases} \chi_{xx}(t) = \chi_{yy}(t) = \dfrac{\omega_{\mathrm{p}}^2}{\nu^2 + \omega_{\mathrm{g}}^2} \left(\omega_{\mathrm{g}} \mathrm{e}^{-\nu t} \sin \omega_{\mathrm{g}} t + \nu \left(1 - \mathrm{e}^{-\nu t} \cos \omega_{\mathrm{g}} t\right)\right) \\ \chi_{xy}(t) = -\chi_{yx}(t) = \dfrac{\omega_{\mathrm{p}}^2}{\nu^2 + \omega_{\mathrm{g}}^2} \left(\omega_{\mathrm{g}} \left(1 - \mathrm{e}^{-\nu t} \cos \omega_{\mathrm{g}} t\right) - \nu \mathrm{e}^{-\nu t} \sin \omega_{\mathrm{g}} t\right) \\ \chi_{xz}(t) = \chi_{yz}(t) = \chi_{zx}(t) = \chi_{zy}(t) = 0 \\ \chi_{zz}(t) = \dfrac{\omega_{\mathrm{p}}^2}{\nu} \left(1 - \mathrm{e}^{-\nu t}\right) \end{cases}$$

1.6

$$\begin{cases} \boldsymbol{A} = \hat{\boldsymbol{x}} + \mathrm{i}\hat{\boldsymbol{y}} \\ \boldsymbol{B} = (\hat{\boldsymbol{x}} + \xi\hat{\boldsymbol{y}}) + \mathrm{i}(-\xi\hat{\boldsymbol{x}} + \hat{\boldsymbol{y}}) \end{cases}$$

where ξ is an arbitrary real number.

1.8

$$2\alpha\beta + \alpha^2 + \omega^2 \geq 0 \text{ which implies } \quad \beta \geq -\frac{\alpha}{2}$$

1.9

$$\alpha^2 - \beta^2 + \omega^2 \geq 0 \text{ which implies } \quad |\beta| \leq \alpha$$

1.10

$$[\boldsymbol{\epsilon}] = \begin{pmatrix} \epsilon & \mathrm{i}\epsilon_{\mathrm{g}} & 0 \\ -\mathrm{i}\epsilon_{\mathrm{g}} & \epsilon & 0 \\ 0 & 0 & \epsilon_z \end{pmatrix}$$

where

$$\begin{cases} \epsilon = 1 - \dfrac{\omega_{\mathrm{p}}^2(\omega + \mathrm{i}\nu)}{\omega\left(\omega^2 - \omega_{\mathrm{g}}^2 - \nu^2 + 2\mathrm{i}\nu\omega\right)} \\ \epsilon_{\mathrm{g}} = -\dfrac{\omega_{\mathrm{p}}^2\omega_{\mathrm{g}}}{\omega\left(\omega^2 - \omega_{\mathrm{g}}^2 - \nu^2 + 2\mathrm{i}\nu\omega\right)} \\ \epsilon_z = 1 - \dfrac{\omega_{\mathrm{p}}^2}{\omega(\omega + \mathrm{i}\nu)} \end{cases}$$

1.11

$$\begin{cases} \boldsymbol{\mu} = \begin{pmatrix} \mu & \mathrm{i}\mu_g & 0 \\ -\mathrm{i}\mu_g & \mu & 0 \\ 0 & 0 & 1 \end{pmatrix} \\ [\boldsymbol{\epsilon}] = \epsilon\,[\mathbf{I}]_3 \end{cases}$$

where

$$\begin{cases} \mu = 1 - \dfrac{\omega_0\omega_m}{\omega^2 - \omega_0^2} \\ \mu_g = \dfrac{\omega\omega_m}{\omega^2 - \omega_0^2} \end{cases}$$

The two frequencies, ω_0 (the gyromagnetic frequency) and ω_m (the saturation frequency), are explicitly given by

$$\begin{cases} \omega_0 = -g\mu_0 H_0 \\ \omega_m = -g\mu_0 M_0 \end{cases}$$

1.12 (a) Linearly polarized field.
(b) Right circular polarized field.
(c) Left circular polarized field.
(d) An ellipse with half axes a and b along the $\hat{e}_1$- and the $\hat{e}_2$-axes, respectively. The field is right-handed elliptic polarized and $\boldsymbol{E}(t = 0) = \hat{e}_1 a$.
(e) An ellipse with half axes $\sqrt{2}a \cos \pi/8$ and $\sqrt{2}a \sin \pi/8$, respectively. The latter axis is tilted 45° against the positive $\hat{e}_1$-axis. The field is right-handed elliptic polarized and $\boldsymbol{E}(t = 0) = a\left(\hat{e}_1 + \hat{e}_2/\sqrt{2}\right)$.

1.13 (a)

$$\boldsymbol{E}_0 = \hat{e}_1 E_1 + \hat{e}_2 E_2 \qquad E_1, E_2 \text{ complex number}$$

can be written as

$$\boldsymbol{E}_0 = \underbrace{\frac{E_1 - \mathrm{i}E_2}{2}(\hat{e}_1 + \mathrm{i}\hat{e}_2)}_{\text{RCP}} + \underbrace{\frac{E_1 + \mathrm{i}E_2}{2}(\hat{e}_1 - \mathrm{i}\hat{e}_2)}_{\text{LCP}}$$

(b)

$$|a| = |b|$$

i.e.,

$$\boldsymbol{E}_0 = a(\boldsymbol{E}_+ + \mathrm{e}^{\mathrm{i}\alpha}\boldsymbol{E}_-)$$

2.1 $\boldsymbol{E}(r, \omega)$ and $\boldsymbol{H}(r, \omega)$ are zero for $r > a$.

2.2

$$\begin{aligned}
\boldsymbol{E} &= \frac{pk^2}{\epsilon_0}\frac{\mathrm{e}^{\mathrm{i}kr}}{4\pi r}\left\{\left[3\hat{\boldsymbol{r}}(\hat{z}\cdot\hat{\boldsymbol{r}}) - \hat{z}\right]\left(\frac{1}{k^2r^2} - \frac{\mathrm{i}}{kr}\right) + \hat{\boldsymbol{r}} \times (\hat{z} \times \hat{\boldsymbol{r}})\right\} \\
\boldsymbol{H} &= -\mathrm{i}k\omega p\frac{\mathrm{e}^{\mathrm{i}kr}}{4\pi r}\left(\mathrm{i} - \frac{1}{kr}\right)\hat{\boldsymbol{r}} \times \hat{z} \\
\langle \boldsymbol{S}(t)\rangle &= \frac{|p|^2k^3\omega}{32\pi^2r^2\epsilon_0}\hat{\boldsymbol{r}}\left(1 - (\hat{z}\cdot\hat{\boldsymbol{r}})^2\right) \\
P &= \frac{|p|^2k^3\omega}{12\pi\epsilon_0}
\end{aligned}$$

2.3 Use the spherical symmetry of the problem.

2.4 Use the addition theorem for the Green function in spherical waves.

2.5

$$v > c_0\sqrt{\frac{\omega^2 - \omega_0^2}{\omega^2 - 2\omega_0^2}} \approx 0.66c_0$$

2.10

$$A_1 = \begin{pmatrix} 0 & 0 & 0 & 0 & 0 & 0 \\ 0 & 0 & 0 & 0 & 0 & -1 \\ 0 & 0 & 0 & 0 & 1 & 0 \\ 0 & 0 & 0 & 0 & 0 & 0 \\ 0 & 0 & 1 & 0 & 0 & 0 \\ 0 & -1 & 0 & 0 & 0 & 0 \end{pmatrix}$$

$$A_2 = \begin{pmatrix} 0 & 0 & 0 & 0 & 0 & 1 \\ 0 & 0 & 0 & 0 & 0 & 0 \\ 0 & 0 & 0 & -1 & 0 & 0 \\ 0 & 0 & -1 & 0 & 0 & 0 \\ 0 & 0 & 0 & 0 & 0 & 0 \\ 1 & 0 & 0 & 0 & 0 & 0 \end{pmatrix}$$

and

$$A_3 = \begin{pmatrix} 0 & 0 & 0 & 0 & -1 & 0 \\ 0 & 0 & 0 & 1 & 0 & 0 \\ 0 & 0 & 0 & 0 & 0 & 0 \\ 0 & 1 & 0 & 0 & 0 & 0 \\ -1 & 0 & 0 & 0 & 0 & 0 \\ 0 & 0 & 0 & 0 & 0 & 0 \end{pmatrix}$$

4.1

$$\boldsymbol{F}(\hat{\boldsymbol{r}}) = \hat{\boldsymbol{r}} \times \left(\boldsymbol{K}(\hat{\boldsymbol{r}}) \times \hat{\boldsymbol{r}}\right)$$

where

$$\boldsymbol{K}(\hat{\boldsymbol{r}}) = \hat{z}\frac{\mathrm{i}I_0 k \eta_0}{2\pi \sin^2\theta}\left[\cos\left(\frac{kl}{2}\cos\theta\right) - \cos\left(\frac{kl}{2}\right)\right]$$

θ is the polar angle of $\hat{\boldsymbol{r}}$ in spherical coordinates.

4.2

$$\boldsymbol{F}(\hat{\boldsymbol{r}}) = \hat{\boldsymbol{r}} \times \left(\boldsymbol{K}(\hat{\boldsymbol{r}}) \times \hat{\boldsymbol{r}}\right)$$

where

$$\boldsymbol{K}(\hat{\boldsymbol{r}}) = \hat{\phi}\frac{\mathrm{i}ka^2\eta_0}{2a}J_1(ka\sin\theta)$$

with long wavelength approximation

$$\boldsymbol{K}(\hat{\boldsymbol{r}}) = \hat{\phi}\frac{mk^2\eta_0\sin\theta}{4\pi}, \qquad m = I\pi a^2$$

4.5

$$\begin{cases} M_{11} = \frac{1}{2}\left\{|S_{\parallel\parallel}|^2 + |S_{\parallel\perp}|^2 + |S_{\perp\parallel}|^2 + |S_{\perp\perp}|^2\right\} \\ M_{12} = \frac{1}{2}\left\{|S_{\parallel\parallel}|^2 - |S_{\parallel\perp}|^2 + |S_{\perp\parallel}|^2 - |S_{\perp\perp}|^2\right\} \\ M_{13} = \operatorname{Re}\left\{S_{\parallel\parallel}S^*_{\parallel\perp} + S_{\perp\parallel}S^*_{\perp\perp}\right\} \\ M_{14} = -\operatorname{Im}\left\{S_{\parallel\parallel}S^*_{\parallel\perp} + S_{\perp\parallel}S^*_{\perp\perp}\right\} \\ M_{21} = \frac{1}{2}\left\{|S_{\parallel\parallel}|^2 + |S_{\parallel\perp}|^2 - |S_{\perp\parallel}|^2 - |S_{\perp\perp}|^2\right\} \\ M_{22} = \frac{1}{2}\left\{|S_{\parallel\parallel}|^2 - |S_{\parallel\perp}|^2 - |S_{\perp\parallel}|^2 + |S_{\perp\perp}|^2\right\} \\ M_{23} = \operatorname{Re}\left\{S_{\parallel\parallel}S^*_{\parallel\perp} - S_{\perp\parallel}S^*_{\perp\perp}\right\} \\ M_{24} = -\operatorname{Im}\left\{S_{\parallel\parallel}S^*_{\parallel\perp} - S_{\perp\parallel}S^*_{\perp\perp}\right\} \\ M_{31} = \operatorname{Re}\left\{S_{\parallel\parallel}S^*_{\perp\parallel} + S_{\parallel\perp}S^*_{\perp\perp}\right\} \\ M_{32} = \operatorname{Re}\left\{S_{\parallel\parallel}S^*_{\perp\parallel} - S_{\parallel\perp}S^*_{\perp\perp}\right\} \\ M_{33} = \operatorname{Re}\left\{S_{\parallel\parallel}S^*_{\perp\perp} + S_{\parallel\perp}S^*_{\perp\parallel}\right\} \\ M_{34} = -\operatorname{Im}\left\{S_{\parallel\parallel}S^*_{\perp\perp} - S_{\parallel\perp}S^*_{\perp\parallel}\right\} \\ M_{41} = \operatorname{Im}\left\{S_{\parallel\parallel}S^*_{\perp\parallel} + S_{\parallel\perp}S^*_{\perp\perp}\right\} \\ M_{42} = \operatorname{Im}\left\{S_{\parallel\parallel}S^*_{\perp\parallel} - S_{\parallel\perp}S^*_{\perp\perp}\right\} \\ M_{43} = \operatorname{Im}\left\{S_{\parallel\parallel}S^*_{\perp\perp} + S_{\parallel\perp}S^*_{\perp\parallel}\right\} \\ M_{44} = \operatorname{Re}\left\{S_{\parallel\parallel}S^*_{\perp\perp} - S_{\parallel\perp}S^*_{\perp\parallel}\right\} \end{cases}$$

5.3

$$\boldsymbol{F}(\hat{\boldsymbol{r}}, \tau) = \frac{\mu_0\mu}{4\pi}\hat{\boldsymbol{r}} \times \left(\hat{\boldsymbol{r}} \times \frac{\partial}{\partial\tau}\iiint_{V_s} \boldsymbol{J}_s(\boldsymbol{r}', \tau + \hat{\boldsymbol{r}}\cdot\boldsymbol{r}'/c)\,\mathrm{d}v'\right)$$

6.1

$$\begin{cases} S_{\parallel\parallel} = k_0^2 a^3 \dfrac{\epsilon - 1}{\epsilon + 2}\cos\theta \\ S_{\parallel\perp} = S_{\perp\parallel} = 0 \\ S_{\perp\perp} = k_0^2 a^3 \dfrac{\epsilon - 1}{\epsilon + 2} \end{cases} \qquad \begin{cases} \sigma_s|_{\text{unpol}} = \dfrac{8\pi k_0^4 a^6}{3}\left|\dfrac{\epsilon - 1}{\epsilon + 2}\right|^2 \\ P|_{\text{unpol}} = \dfrac{\sin^2\theta}{1 + \cos^2\theta} \end{cases}$$

θ is the angle between $\hat{\boldsymbol{k}}_i$ and $\hat{\boldsymbol{r}}$.

6.2

$$\begin{cases} S_{\parallel\parallel} = k_0^2 a^3 \left(\cos\theta - \dfrac{1}{2}\right) \\ S_{\parallel\perp} = S_{\perp\parallel} = 0 \\ S_{\perp\perp} = k_0^2 a^3 \left(1 - \dfrac{1}{2}\cos\theta\right) \end{cases} \qquad \begin{cases} \sigma_s|_{\text{unpol}} = \dfrac{10\pi k_0^4 a^6}{3} \\ P|_{\text{unpol}} = \dfrac{3\sin^2\theta}{5\cos^2\theta + 5 - 8\cos\theta} \end{cases}$$

θ is the angle between $\hat{\boldsymbol{k}}_{\text{i}}$ and $\hat{\boldsymbol{r}}$.

6.3

$$\boldsymbol{F}_{\text{PO}}(\hat{\boldsymbol{r}}) = \hat{\boldsymbol{r}} \times \left(\boldsymbol{K}(\hat{\boldsymbol{r}}) \times \hat{\boldsymbol{r}}\right)$$
$$\sigma_{\text{ext}} = \sigma_{\text{s}} = 2ab$$

where

$$\boldsymbol{K}(\hat{\boldsymbol{r}}) = \hat{\boldsymbol{x}}\frac{\text{i}E_0 kab}{2\pi}\frac{\sin\left(\frac{ka}{2}\sin\theta\cos\phi\right)}{\frac{ka}{2}\sin\theta\cos\phi}\frac{\sin\left(\frac{kb}{2}\sin\theta\sin\phi\right)}{\frac{kb}{2}\sin\theta\sin\phi}$$

θ is the polar angle and ϕ the azimuthal angle of $\hat{\boldsymbol{r}}$ in spherical coordinates.

6.4

$$\boldsymbol{F}_{\text{PO}}(\hat{\boldsymbol{r}}) = \hat{\boldsymbol{r}} \times \left(\boldsymbol{K}(\hat{\boldsymbol{r}}) \times \hat{\boldsymbol{r}}\right)$$
$$\sigma_{\text{ext}} = \sigma_{\text{s}} = 2\pi a^2$$

where

$$\boldsymbol{K}(\hat{\boldsymbol{r}}) = \text{i}ka^2\boldsymbol{E}_0\frac{J_1(ka\sin\theta)}{ka\sin\theta}$$

θ is the polar angle of $\hat{\boldsymbol{r}}$ in spherical coordinates.

6.5

$$\boldsymbol{F}_{\text{GO}}(\hat{\boldsymbol{z}}) = \text{i}\frac{ka^2}{2}\boldsymbol{E}_0$$

6.6

$$\boldsymbol{F}_{\text{PO}}(\hat{\boldsymbol{r}}) = \text{i}\frac{E_0 khb}{2\pi}\text{e}^{\text{i}kF(1+\cos\theta)}(\hat{\boldsymbol{\theta}}\cos\theta\sin\phi + \hat{\boldsymbol{\phi}}\cos\phi)\frac{\sin\left(\frac{kh}{2}\sin\theta\sin\phi\right)}{\frac{kh}{2}\sin\theta\sin\phi}$$
$$\cdot\sqrt{\frac{32kF}{\text{i}\pi(k^2b^2 + 16k^2F^2)}}\text{e}^{\text{i}kb\frac{b}{8F}\sin^2\theta/2}\frac{1}{1 - \frac{b^2\tan^2\theta/2}{16F^2\cos^2\phi}}$$
$$\cdot\left\{\frac{\sin\left(\frac{kb}{2}\sin\theta\cos\phi\right)}{\frac{kb}{2}\sin\theta\cos\phi} - \frac{\cos\left(\frac{kb}{2}\sin\theta\cos\phi\right)}{4\text{i}kF\cos^2\theta/2\cos^2\phi}\right\}$$

6.7

$$\boldsymbol{F}_{\mathrm{GO}}(\hat{\boldsymbol{r}}) = -\mathrm{i}E_0\hat{\boldsymbol{r}} \times \left[(\hat{z}+\hat{\boldsymbol{r}}) \times \hat{\boldsymbol{y}}\right] \frac{khb}{4\pi} \mathrm{e}^{\mathrm{i}k\xi(1-\cos\theta)+2\mathrm{i}kF} \sqrt{\frac{32kF}{\mathrm{i}\pi(k^2b^2+16k^2F^2)}}$$
$$\cdot \frac{\sin\left(\frac{kh}{2}\sin\theta\sin\phi\right)}{\frac{kh}{2}\sin\theta\sin\phi} \frac{\sin\left(\frac{kb}{2}\sin\theta\cos\phi\right)}{\frac{kb}{2}\sin\theta\cos\phi}$$

6.8

$$\boldsymbol{F}_{\mathrm{PO}}(\hat{z}) = \mathrm{i}\frac{khb}{2\pi}\mathrm{e}^{-2\mathrm{i}kl}\hat{z} \times \left\{\hat{z} \times \left(\boldsymbol{E}_0 - 2\hat{\boldsymbol{x}}E_{0x}\right)\right\}$$

6.9

$$\boldsymbol{F}_{\mathrm{PO}}(-\hat{\boldsymbol{k}}_{\mathrm{i}}) = -\frac{a}{2}\mathrm{e}^{-2\mathrm{i}ka}\boldsymbol{E}_0\left(1 - \frac{\mathrm{i}}{2ka}\left(1 - \mathrm{e}^{2\mathrm{i}ka}\right)\right)$$

6.11

$$\boldsymbol{F}_{\mathrm{PO}}(\hat{z}) = \mathrm{i}\hat{\boldsymbol{y}}kF\mathrm{e}^{2\mathrm{i}kF}\frac{k^2p}{2\pi\epsilon_0}\frac{a^2}{a^2+4F^2}$$

6.12

$$\boldsymbol{F}_{\mathrm{GO}}(\hat{z}) = \mathrm{i}\hat{\boldsymbol{y}}kF\mathrm{e}^{2\mathrm{i}kF}\frac{k^2p}{2\pi\epsilon_0}\frac{a^2}{a^2+4F^2}$$

6.13

$$\boldsymbol{F}_{\mathrm{PO}}\left(\hat{\boldsymbol{r}}\right) = \mathrm{i}\frac{E_0h}{2\pi}\sqrt{\frac{2ka}{\mathrm{i}\pi}}\mathrm{e}^{\mathrm{i}ka}(\hat{\boldsymbol{\theta}}\cos\theta\sin\phi + \hat{\boldsymbol{\phi}}\cos\phi)\frac{\sin\left(\frac{kh}{2}\sin\theta\sin\phi\right)}{\frac{kh}{2}\sin\theta\sin\phi}$$
$$\cdot \int_{-kb/2}^{kb/2}\frac{1}{\sqrt{k^2a^2-t^2}}\mathrm{e}^{-\mathrm{i}\left(t\sin\theta\cos\phi - \sqrt{k^2a^2-t^2}\cos\theta\right)}\,\mathrm{d}t$$

For small values of kb this expression becomes

$$\boldsymbol{F}_{\mathrm{PO}}\left(\hat{\boldsymbol{r}}\right) = \mathrm{i}\frac{E_0h}{2\pi}\sqrt{\frac{2ka}{\mathrm{i}\pi}}\mathrm{e}^{\mathrm{i}ka(1+\cos\theta)}(\hat{\boldsymbol{\theta}}\cos\theta\sin\phi + \hat{\boldsymbol{\phi}}\cos\phi)\frac{\sin\left(\frac{kh}{2}\sin\theta\sin\phi\right)}{\frac{kh}{2}\sin\theta\sin\phi}$$
$$\cdot \int_{-kb/2}^{kb/2}\frac{1}{ka}\mathrm{e}^{-\mathrm{i}t\sin\theta\cos\phi}\,\mathrm{d}t$$

which is identical to the corresponding result for the cylindrical parabolic reflector provided $F = a$.

7.3

$$E_{\mathrm{i}}(\boldsymbol{r}) = -\frac{\eta_0\eta}{\mathrm{i}k}\nabla\times\left\{\nabla\times\iint_{S_{\mathrm{i}}} g(k,|\boldsymbol{r}-\boldsymbol{r}'|)(\hat{\boldsymbol{\nu}}(\boldsymbol{r}')\times \boldsymbol{H}_{\mathrm{i}}(\boldsymbol{r}'))\,\mathrm{d}S'\right\}$$
$$+\nabla\times\iint_{S_{\mathrm{i}}} g(k,|\boldsymbol{r}-\boldsymbol{r}'|)(\hat{\boldsymbol{\nu}}(\boldsymbol{r}')\times \boldsymbol{E}_{\mathrm{i}}(\boldsymbol{r}'))\,\mathrm{d}S', \qquad r<R$$

$$a_{\tau n} = -\eta_0\eta k^2\iint_{S_{\mathrm{i}}} \boldsymbol{u}_{\tau n}(k\boldsymbol{r}')\cdot(\hat{\boldsymbol{\nu}}(\boldsymbol{r}')\times \boldsymbol{H}_{\mathrm{i}}(\boldsymbol{r}'))\,\mathrm{d}S'$$
$$+\mathrm{i}k^2\iint_{S_{\mathrm{i}}} \boldsymbol{u}_{\overline{\tau} n}(k\boldsymbol{r}')\cdot(\hat{\boldsymbol{\nu}}(\boldsymbol{r}')\times \boldsymbol{E}_{\mathrm{i}}(\boldsymbol{r}'))\,\mathrm{d}S'$$

7.6 The electric field is proportional to $\boldsymbol{u}_{2\sigma 11}\,(k(\boldsymbol{r}-\boldsymbol{r}_{\mathrm{s}}))$, where $\sigma=\mathrm{e}$ holds for the electric dipole oriented along $\hat{\boldsymbol{x}}$, and $\sigma=\mathrm{o}$ holds for the electric dipole oriented along $\hat{\boldsymbol{y}}$.

$$a_n = \mathcal{P}_{n,2\mathrm{e}11}(k\boldsymbol{r}_{\mathrm{s}}), \quad \text{dipole along } \hat{\boldsymbol{x}}, \quad a_n = \mathcal{P}_{n,2\mathrm{o}11}(k\boldsymbol{r}_{\mathrm{s}}), \quad \text{dipole along } \hat{\boldsymbol{y}}$$

The corresponding magnetic dipole expansion coefficients a_n are obtained by letting $\tau=1$.

7.8

$$\omega_{l''} = \frac{4\pi}{k^2}(2l''+1)\sum_{l,l'=0}^{\infty}\begin{pmatrix} l & l' & l'' \\ 0 & 0 & 0\end{pmatrix}^2\left\{A_lA_{l'}^*|\hat{\boldsymbol{r}}_{\mathrm{c}}\cdot\hat{\boldsymbol{p}}_{\mathrm{e}}|^2 + B_lB_{l'}^*|\hat{\boldsymbol{\phi}}\cdot\hat{\boldsymbol{p}}_{\mathrm{e}}|^2\right\}$$

in terms of Wigner 3-j symbol [59], and where

$$A_{l'} = \sum_{l=\max\{l',1\}}^{\infty}\frac{2l+1}{l(l+1)}(a_{ll'}t_{1l}+b_{ll'}t_{2l})$$
$$B_{l'} = \sum_{l=\max\{l',1\}}^{\infty}\frac{2l+1}{l(l+1)}(a_{ll'}t_{2l}+b_{ll'}t_{1l})$$

and where

$$a_{ll'} = \begin{cases} 2l'+1, & l'<l \text{ and } l-l' \text{ odd integer} \\ 0, & l'\geq l \text{ or } l-l' \text{ even integer}\end{cases}$$

and

$$b_{ll'} = l^2\delta_{ll'} - \begin{cases} 2l'+1, & l'<l-1 \text{ and } l-l' \text{ even integer} \\ 0, & l'\geq l-1 \text{ or } l-l' \text{ odd integer}\end{cases}$$

8.3

$$\mathbf{S}(\hat{z}, \hat{z}) = \frac{\mathbf{I}_2}{2k}\left(x^3 + \frac{113x^5}{90} + O(x^6)\right)$$

8.4

$$\begin{cases} t_{1l} = \dfrac{f_{1l}}{a_{1l}} = -\dfrac{\frac{Z}{\mathrm{i}\eta\eta_0}\psi_l'(ka) + \psi_l(ka)}{\frac{Z}{\mathrm{i}\eta\eta_0}\xi_l'(ka) + \xi_l(ka)} \\ t_{2l} = \dfrac{f_{2l}}{a_{2l}} = -\dfrac{\psi_l'(ka) - \frac{Z}{\mathrm{i}\eta\eta_0}\psi_l(ka)}{\xi_l'(ka) - \frac{Z}{\mathrm{i}\eta\eta_0}\xi_l(ka)} \end{cases}$$

8.6

$$\frac{\mathrm{d}\sigma}{\mathrm{d}\Omega}\left(\hat{\boldsymbol{r}} = -\hat{\boldsymbol{k}}_{\mathrm{i}}\right) = \frac{\pi}{k^2}\left|\sum_{l=1}^{\infty}(2l+1)(-1)^l\left(t_{1l} - t_{2l}\right)\right|^2$$

8.7 To lowest order in ka:

$$\boldsymbol{E}_1(\boldsymbol{r}, \omega) = \frac{3\epsilon}{\epsilon_1 + 2\epsilon}\boldsymbol{E}_0$$

8.8

$$t_{\tau l} = -\frac{m_\tau\left(\psi_l(k_1 b) + t_{\tau l}^{\mathrm{C}}\xi_l(k_1 b)\right)\psi_l'(kb) - \left(\psi_l'(k_1 b) + t_{\tau l}^{\mathrm{C}}\xi_l'(k_1 b)\right)\psi_l(kb)}{m_\tau\left(\psi_l(k_1 b) + t_{\tau l}^{\mathrm{C}}\xi_l(k_1 b)\right)\xi_l'(kb) - \left(\psi_l'(k_1 b) + t_{\tau l}^{\mathrm{C}}\xi_l'(k_1 b)\right)\xi_l(kb)}$$

where the transition matrix of the core is

$$t_{\tau l}^{\mathrm{C}} = -\frac{m_\tau\psi_l(k_1 a)\psi_l'(ka) - \psi_l'(k_1 a)\psi_l(ka)}{m_\tau\xi_l(k_1 a)\psi_l'(ka) - \xi_l'(k_1 a)\psi_l(ka)}$$

where

$$m_\tau = \delta_{\tau 1}\frac{\eta_1}{\eta} + \delta_{\tau 2}\frac{\eta}{\eta_1}$$

To first order in $\varepsilon = k(b-a)$

$$t_{\tau l} = \mathrm{i}\varepsilon\left\{\left(m_\tau\frac{k_1}{k} - 1\right)\left(\psi_l'(kb)\right)^2 + \left(\frac{l(l+1)}{(kb)^2}\left(1 - \frac{k}{m_\tau k_1}\right) + \frac{k_1}{m_\tau k} - 1\right)\left(\psi_l(kb)\right)\right\}^2 + O(\varepsilon^2)$$

and

$$\begin{cases} \gamma_{\mathrm{m}} = \dfrac{4\pi b^3}{3}\left(1 - \dfrac{a}{b}\right)\dfrac{(\mu_1 - \mu)(2\mu_1 + \mu)}{\mu\mu_1} \\ \gamma_{\mathrm{e}} = \dfrac{4\pi b^3}{3}\left(1 - \dfrac{a}{b}\right)\dfrac{(\epsilon_1 - \epsilon)(2\epsilon_1 + \epsilon)}{\epsilon\epsilon_1} \end{cases}$$

8.9

$$\left(\delta_{\tau 1}\frac{\eta}{\eta_2}+\delta_{\tau 2}\frac{\eta_2}{\eta}\right)\left(\psi_l(kb)+t^{\mathrm{C}}_{\tau l}\xi_l(kb)\right)\xi_l'(k_2 b)=\left(\psi_l'(kb)+t^{\mathrm{C}}_{\tau l}\xi_l'(kb)\right)\xi_l(k_2 b)$$

where the core transition matrix is

$$t^{\mathrm{C}}_{\tau l}=-\frac{\psi_l(ka)\psi_l'(k_1 a)-m_\tau\psi_l'(ka)\psi_l(k_1 a)}{\xi_l(ka)\psi_l'(k_1 a)-m_\tau\xi_l'(ka)\psi_l(k_1 a)}$$

where

$$m_\tau=\delta_{\tau 1}\frac{\eta_1}{\eta}+\delta_{\tau 2}\frac{\eta}{\eta_1}$$

Lowest order frequencies are: $f=8.55-1.38\mathrm{i}$, $15.6-2.00\mathrm{i}$, and $22.6-2.50\mathrm{i}$.

8.11

$$\sigma_{\mathrm{s}}=\frac{8\pi k^4a^6}{3}\left\{\frac{\left|(\epsilon_1+2\epsilon)(\mu_1-\mu)-\chi^2-\kappa^2\right|^2}{\left|(\epsilon_1+2\epsilon)(\mu_1+2\mu)-\chi^2-\kappa^2\right|^2}+\frac{\left|(\epsilon_1-\epsilon)(\mu_1+2\mu)-\chi^2-\kappa^2\right|^2+3\Pi\left(|\chi|^2-|\kappa|^2\right)}{\left|(\epsilon_1+2\epsilon)(\mu_1+2\mu)-\chi^2-\kappa^2\right|^2}\right\}$$

$$\sigma_{\mathrm{ext}}=4\pi ka^3\,\mathrm{Im}\,\frac{(\epsilon_1+2\epsilon)(\mu_1-\mu)+(\epsilon_1-\epsilon)(\mu_1+2\mu)-2\chi^2-2\kappa^2+6\Pi\chi}{(\epsilon_1+2\epsilon)(\mu_1+2\mu)-\chi^2-\kappa^2}$$

10.5

$$\begin{cases}\mathbf{G}_{\mathrm{e}\perp\perp}=\dfrac{\mathrm{i}}{2k_z}\left(\mathbf{I}_2-\dfrac{\boldsymbol{k}_{\mathrm{t}}\boldsymbol{k}_{\mathrm{t}}}{k_0^2}\right)\mathrm{e}^{\mathrm{i}k_z|z-z'|}\mathrm{e}^{\mathrm{i}\boldsymbol{k}_{\mathrm{t}}\cdot\boldsymbol{r}_{\mathrm{c}}'}\\[2ex] \boldsymbol{G}_{\mathrm{e}z}=-\dfrac{\mathrm{i}\boldsymbol{k}_{\mathrm{t}}}{2k_0^2}\,\mathrm{sign}(z-z')\mathrm{e}^{\mathrm{i}k_z|z-z'|}\mathrm{e}^{\mathrm{i}\boldsymbol{k}_{\mathrm{t}}\cdot\boldsymbol{r}_{\mathrm{c}}'}\\[2ex] \boldsymbol{G}_{\mathrm{e}\perp}=-\dfrac{\mathrm{i}\boldsymbol{k}_{\mathrm{t}}}{2k_0^2}\,\mathrm{sign}(z-z')\mathrm{e}^{\mathrm{i}k_z|z-z'|}\mathrm{e}^{\mathrm{i}\boldsymbol{k}_{\mathrm{t}}\cdot\boldsymbol{r}_{\mathrm{c}}'}\\[2ex] G_{\mathrm{e}zz}=-\dfrac{1}{k_z^2}\delta(z-z')\mathrm{e}^{\mathrm{i}\boldsymbol{k}_{\mathrm{t}}\cdot\boldsymbol{r}_{\mathrm{c}}'}+\dfrac{\mathrm{i}k_{\mathrm{t}}^2}{2k_0^2k_z}\mathrm{e}^{\mathrm{i}k_z|z-z'|}\mathrm{e}^{\mathrm{i}\boldsymbol{k}_{\mathrm{t}}\cdot\boldsymbol{r}_{\mathrm{c}}'}\end{cases}$$

Index

Transformation of Unit Vectors

Cylindrical coordinates (r_c, ϕ, z)

$$\begin{cases} r_c = \sqrt{x^2+y^2} \\ \phi = \begin{cases} \arccos \frac{x}{\sqrt{x^2+y^2}} & y > 0 \\ 2\pi - \arccos \frac{x}{\sqrt{x^2+y^2}} & y < 0 \end{cases} \\ z = z \end{cases}$$

Spherical coordinates (r, θ, ϕ)

$$\begin{cases} r = \sqrt{x^2+y^2+z^2} \\ \theta = \arccos \frac{z}{\sqrt{x^2+y^2+z^2}} \\ \phi = \begin{cases} \arccos \frac{x}{\sqrt{x^2+y^2}} & y > 0 \\ 2\pi - \arccos \frac{x}{\sqrt{x^2+y^2}} & y < 0 \end{cases} \end{cases}$$

$(r, \theta, \phi) \longrightarrow (x, y, z)$

$$\begin{cases} \hat{\boldsymbol{r}} = \hat{\boldsymbol{x}} \sin\theta \cos\phi + \hat{\boldsymbol{y}} \sin\theta \sin\phi + \hat{\boldsymbol{z}} \cos\theta \\ \hat{\boldsymbol{\theta}} = \hat{\boldsymbol{x}} \cos\theta \cos\phi + \hat{\boldsymbol{y}} \cos\theta \sin\phi - \hat{\boldsymbol{z}} \sin\theta \\ \hat{\phi} = -\hat{\boldsymbol{x}} \sin\phi + \hat{\boldsymbol{y}} \cos\phi \end{cases}$$

$(x, y, z) \longrightarrow (r, \theta, \phi)$

$$\begin{cases} \hat{\boldsymbol{x}} = \hat{\boldsymbol{r}} \sin\theta \cos\phi + \hat{\boldsymbol{\theta}} \cos\theta \cos\phi - \hat{\phi} \sin\phi \\ \hat{\boldsymbol{y}} = \hat{\boldsymbol{r}} \sin\theta \sin\phi + \hat{\boldsymbol{\theta}} \cos\theta \sin\phi + \hat{\phi} \cos\phi \\ \hat{\boldsymbol{z}} = \hat{\boldsymbol{r}} \cos\theta - \hat{\boldsymbol{\theta}} \sin\theta \end{cases}$$

$(r_c, \phi, z) \longrightarrow (x, y, z)$

$$\begin{cases} \hat{\boldsymbol{r}}_c = \hat{\boldsymbol{x}} \cos\phi + \hat{\boldsymbol{y}} \sin\phi = (\hat{\boldsymbol{x}}x + \hat{\boldsymbol{y}}y)/\sqrt{x^2+y^2} \\ \hat{\phi} = -\hat{\boldsymbol{x}} \sin\phi + \hat{\boldsymbol{y}} \cos\phi = (-\hat{\boldsymbol{x}}y + \hat{\boldsymbol{y}}x)/\sqrt{x^2+y^2} \\ \hat{\boldsymbol{z}} = \hat{\boldsymbol{z}} \end{cases}$$

$(x, y, z) \longrightarrow (r_c, \phi, z)$

$$\begin{cases} \hat{\boldsymbol{x}} = \hat{\boldsymbol{r}}_c \cos\phi - \hat{\phi} \sin\phi \\ \hat{\boldsymbol{y}} = \hat{\boldsymbol{r}}_c \sin\phi + \hat{\phi} \cos\phi \\ \hat{\boldsymbol{z}} = \hat{\boldsymbol{z}} \end{cases}$$

$(r, \theta, \phi) \longrightarrow (r_c, \phi, z)$

$$\begin{cases} \hat{\boldsymbol{r}} = \hat{\boldsymbol{r}}_c \sin\theta + \hat{\boldsymbol{z}} \cos\theta \\ \hat{\boldsymbol{\theta}} = \hat{\boldsymbol{r}}_c \cos\theta - \hat{\boldsymbol{z}} \sin\theta \\ \hat{\phi} = \hat{\phi} \end{cases}$$

$(r_c, \phi, z) \longrightarrow (r, \theta, \phi)$

$$\begin{cases} \hat{\boldsymbol{r}}_c = \hat{\boldsymbol{r}} \sin\theta + \hat{\boldsymbol{\theta}} \cos\theta \\ \hat{\phi} = \hat{\phi} \\ \hat{\boldsymbol{z}} = \hat{\boldsymbol{r}} \cos\theta - \hat{\boldsymbol{\theta}} \sin\theta \end{cases}$$

Important Vector Identities

(1) $(\boldsymbol{a} \times \boldsymbol{c}) \times (\boldsymbol{b} \times \boldsymbol{c}) = \boldsymbol{c}\,((\boldsymbol{a} \times \boldsymbol{b}) \cdot \boldsymbol{c})$

(2) $(\boldsymbol{a} \times \boldsymbol{b}) \cdot (\boldsymbol{c} \times \boldsymbol{d}) = (\boldsymbol{a} \cdot \boldsymbol{c})(\boldsymbol{b} \cdot \boldsymbol{d}) - (\boldsymbol{a} \cdot \boldsymbol{d})(\boldsymbol{b} \cdot \boldsymbol{c})$

(3) $\boldsymbol{a} \times (\boldsymbol{b} \times \boldsymbol{c}) = \boldsymbol{b}(\boldsymbol{a} \cdot \boldsymbol{c}) - \boldsymbol{c}(\boldsymbol{a} \cdot \boldsymbol{b})$ (BAC-CAB rule)

(4) $\boldsymbol{a} \cdot (\boldsymbol{b} \times \boldsymbol{c}) = \boldsymbol{b} \cdot (\boldsymbol{c} \times \boldsymbol{a}) = \boldsymbol{c} \cdot (\boldsymbol{a} \times \boldsymbol{b})$

Integration Formulas

Stokes' Theorem and Analogous Theorems

(1) $$\iint_S (\nabla \times \boldsymbol{A}) \cdot \hat{\boldsymbol{\nu}}\, \mathrm{d}S = \int_C \boldsymbol{A} \cdot d\boldsymbol{r}$$ (Stokes' theorem)

(2) $$\iint_S \hat{\boldsymbol{\nu}} \times \nabla\varphi\, \mathrm{d}S = \int_C \varphi\, d\boldsymbol{r}$$

(3) $$\iint_S (\hat{\boldsymbol{\nu}} \times \nabla) \times \boldsymbol{A}\, \mathrm{d}S = \int_C d\boldsymbol{r} \times \boldsymbol{A}$$

Gauss' Theorem and Analogous Theorems

(1) $$\iiint_V \nabla \cdot \boldsymbol{A}\, \mathrm{d}v = \iint_S \boldsymbol{A} \cdot \hat{\boldsymbol{\nu}}\, \mathrm{d}S$$ (Gauss' theorem)

(2) $$\iiint_V \nabla\varphi\, \mathrm{d}v = \iint_S \varphi\hat{\boldsymbol{\nu}}\, \mathrm{d}S$$

(3) $$\iiint_V \nabla \times \boldsymbol{A}\, \mathrm{d}v = \iint_S \hat{\boldsymbol{\nu}} \times \boldsymbol{A}\, \mathrm{d}S$$

Green's Theorems

(1) $$\iiint_V \left(\psi\nabla^2\varphi - \varphi\nabla^2\psi\right) \mathrm{d}v = \iint_S (\psi\nabla\varphi - \varphi\nabla\psi) \cdot \hat{\boldsymbol{\nu}}\, \mathrm{d}S$$

(2) $$\iiint_V \left(\psi\nabla^2\boldsymbol{A} - \boldsymbol{A}\nabla^2\psi\right) \mathrm{d}v$$

$$= \iint_S \left(\nabla\psi \times (\hat{\boldsymbol{\nu}} \times \boldsymbol{A}) - \nabla\psi\,(\hat{\boldsymbol{\nu}} \cdot \boldsymbol{A}) - \psi\,(\hat{\boldsymbol{\nu}} \times (\nabla \times \boldsymbol{A})) + \hat{\boldsymbol{\nu}}\psi(\nabla \cdot \boldsymbol{A})\right) \mathrm{d}S$$

Vector Operations with ∇

$$(1)\quad \nabla(\varphi+\psi)=\nabla\varphi+\nabla\psi$$
$$(2)\quad \nabla(\varphi\psi)=\psi\nabla\varphi+\varphi\nabla\psi$$
$$(3)\quad \nabla(\boldsymbol{a}\cdot\boldsymbol{b})=(\boldsymbol{a}\cdot\nabla)\,\boldsymbol{b}+(\boldsymbol{b}\cdot\nabla)\,\boldsymbol{a}+\boldsymbol{a}\times(\nabla\times\boldsymbol{b})+\boldsymbol{b}\times(\nabla\times\boldsymbol{a})$$
$$(4)\quad \nabla(\boldsymbol{a}\cdot\boldsymbol{b})=-\nabla\times(\boldsymbol{a}\times\boldsymbol{b})+2\,(\boldsymbol{b}\cdot\nabla)\,\boldsymbol{a}+\boldsymbol{a}\times(\nabla\times\boldsymbol{b})+\boldsymbol{b}\times(\nabla\times\boldsymbol{a})+\boldsymbol{a}\,(\nabla\cdot\boldsymbol{b})-\boldsymbol{b}\,(\nabla\cdot\boldsymbol{a})$$

$$(5)\quad \nabla\cdot(\boldsymbol{a}+\boldsymbol{b})=\nabla\cdot\boldsymbol{a}+\nabla\cdot\boldsymbol{b}$$
$$(6)\quad \nabla\cdot(\varphi\boldsymbol{a})=\varphi\,(\nabla\cdot\boldsymbol{a})+(\nabla\varphi)\cdot\boldsymbol{a}$$
$$(7)\quad \nabla\cdot(\boldsymbol{a}\times\boldsymbol{b})=\boldsymbol{b}\cdot(\nabla\times\boldsymbol{a})-\boldsymbol{a}\cdot(\nabla\times\boldsymbol{b})$$

$$(8)\quad \nabla\times(\boldsymbol{a}+\boldsymbol{b})=\nabla\times\boldsymbol{a}+\nabla\times\boldsymbol{b}$$
$$(9)\quad \nabla\times(\varphi\boldsymbol{a})=\varphi(\nabla\times\boldsymbol{a})+(\nabla\varphi)\times\boldsymbol{a}$$
$$(10)\quad \nabla\times(\boldsymbol{a}\times\boldsymbol{b})=\boldsymbol{a}\,(\nabla\cdot\boldsymbol{b})-\boldsymbol{b}\,(\nabla\cdot\boldsymbol{a})+(\boldsymbol{b}\cdot\nabla)\,\boldsymbol{a}-(\boldsymbol{a}\cdot\nabla)\,\boldsymbol{b}$$
$$(11)\quad \nabla\times(\boldsymbol{a}\times\boldsymbol{b})=-\nabla(\boldsymbol{a}\cdot\boldsymbol{b})+2\,(\boldsymbol{b}\cdot\nabla)\,\boldsymbol{a}+\boldsymbol{a}\times(\nabla\times\boldsymbol{b})+\boldsymbol{b}\times(\nabla\times\boldsymbol{a})+\boldsymbol{a}\,(\nabla\cdot\boldsymbol{b})-\boldsymbol{b}\,(\nabla\cdot\boldsymbol{a})$$

$$(12)\quad \nabla\cdot\nabla\varphi=\nabla^2\varphi=\Delta\varphi$$
$$(13)\quad \nabla\times(\nabla\times\boldsymbol{a})=\nabla(\nabla\cdot\boldsymbol{a})-\nabla^2\boldsymbol{a}$$
$$(14)\quad \nabla\times(\nabla\varphi)=\boldsymbol{0}$$
$$(15)\quad \nabla\cdot(\nabla\times\boldsymbol{a})=0$$
$$(16)\quad \nabla^2(\varphi\psi)=\varphi\nabla^2\psi+\psi\nabla^2\varphi+2\nabla\varphi\cdot\nabla\psi$$

$$(17)\quad \nabla r=\hat{\boldsymbol{r}}$$
$$(18)\quad \nabla\times\boldsymbol{r}=\boldsymbol{0}$$
$$(19)\quad \nabla\times\hat{\boldsymbol{r}}=\boldsymbol{0}$$
$$(20)\quad \nabla\cdot\boldsymbol{r}=3$$
$$(21)\quad \nabla\cdot\hat{\boldsymbol{r}}=\frac{2}{r}$$
$$(22)\quad \nabla\times(\boldsymbol{a}\times\boldsymbol{r})=2\boldsymbol{a},\qquad \boldsymbol{a}\text{ constant vector}$$
$$(23)\quad \nabla(\boldsymbol{a}\cdot\boldsymbol{r})=\boldsymbol{a},\qquad \boldsymbol{a}\text{ constant vector}$$
$$(24)\quad (\boldsymbol{a}\cdot\nabla)\,\boldsymbol{r}=\boldsymbol{a}$$
$$(25)\quad (\boldsymbol{a}\cdot\nabla)\,\hat{\boldsymbol{r}}=\frac{1}{r}\big(\boldsymbol{a}-\hat{\boldsymbol{r}}(\boldsymbol{a}\cdot\hat{\boldsymbol{r}})\big)=\frac{\boldsymbol{a}_\perp}{r}$$
$$(26)\quad \nabla^2(\boldsymbol{r}\cdot\boldsymbol{a})=2\nabla\cdot\boldsymbol{a}+\boldsymbol{r}\cdot\left(\nabla^2\boldsymbol{a}\right)$$

$$(27)\quad \nabla u(f)=(\nabla f)\frac{\mathrm{d}u}{\mathrm{d}f}$$
$$(28)\quad \nabla\cdot\boldsymbol{F}(f)=(\nabla f)\cdot\frac{\mathrm{d}\boldsymbol{F}}{\mathrm{d}f}$$
$$(29)\quad \nabla\times\boldsymbol{F}(f)=(\nabla f)\times\frac{\mathrm{d}\boldsymbol{F}}{\mathrm{d}f}$$

www.ingramcontent.com/pod-product-compliance
Lightning Source LLC
LaVergne TN
LVHW080807220926
848485LV00011B/331